LEDA

A platform for Combinatorial and Geometric Computing

The LEDA software library is a comprehensive software platform for combinatorial and geometric computing that is in use at more than 1500 institutions worldwide. LEDA transforms the theoretical insights and ideas of the researchers working in the area of algorithms into working software that can be used by computer scientists, engineers, and applied mathematicians for practical application to such topics as optimization, scheduling, CAD, computer graphics, and compiler construction.

The present book, written by the authors of the LEDA system, gives a detailed description of its architecture and its implementation. There are also chapters on how to use LEDA, each one accompanied by a demo program which can be obtained through the Internet or Web.

Like the system itself, this authoritative and unique account will be essential for all current LEDA users as the definitive guide to its further use, but it is also self contained enough to attract newcomers to what is already one of the most useful software resouces for algorithm designers and developers.

KURT MEHLHORN is Director of the Max-Planck-Institut für Informatik, Saarbrücken, Germany.

STEFAN NÄHER is Professor of Computer Science at the Martin-Luther Universität, Halle, Germany.

To
Ena and Ulli
Uli, Steffi, Tim, and Tim

LEDA

A Platform for Combinatorial
and Geometric Computing

KURT MEHLHORN
STEFAN NÄHER

CAMBRIDGE
UNIVERSITY PRESS

PUBLISHED BY THE PRESS SYNDICATE OF THE UNIVERSITY OF CAMBRIDGE
The Pitt Building, Trumpington Street, Cambridge, United Kingdom

CAMBRIDGE UNIVERSITY PRESS
The Edinburgh Building, Cambridge CB2 2RU, UK http://www.cup.cam.ac.uk
40 West 20th Street, New York, NY 10011-4211, USA http://www.cup.org
10 Stamford Road, Oakleigh, Melbourne 3166, Australia
Ruiz de Alarcón 13, 28014 Madrid, Spain

First published 1999

Printed in the United Kingdom at the University Press, Cambridge

Typeset by the author [CRC]

A catalogue record of this book is available from the British Library

Library of Congress cataloguing in publication data applied for

Mehlhorn, Kurt, 1949–
Leda: a platform for combinatorial and geometric computing /
Kurt Mehlhorn, Stefan Näher.
p. cm.
Includes bibliographical references (p.
ISBN 0 521 56329 1 (hb)
1. C++ (Computer program language). 2. LEDA (Computer file)
I. Näher, Stefan. II. Title.
QA76.73.C153M44 2000
005.13′3–dc21 99-24952 CIP

ISBN 0 521 563291 hardback

Disclaimer of Warranty

The programs listed in this book are provided 'as is' without warranty of any kind. We make no warranties, express or implied, that the programs are free of error, or are consistent with any particular standard of merchantability, or that they will meet your requirements for any particular application. They should not be relied upon for solving a problem whose incorrect solution could result in injury or loss of property. If you do use the programs or procedures in such a manner, it is at your own risk. The authors and publisher disclaim all liability for direct, incidental or consequential damages resulting from your use of the programs in this book.

To the best of the authors' and publisher's knowledge, none of the programs included in this book in itself contains any date-sensitive elements that will cause Year 2000-related problems. However, this statement implies no warranty in this matter on the part of the authors or publisher.

Contents

Preface

LEDA (Library of Efficient Data Types and Algorithms) is a C++ library of combinatorial and geometric data types and algorithms. It offers

Data Types, such as random sources, stacks, queues, maps, lists, sets, partitions, dictionaries, sorted sequences, point sets, interval sets, ...,

Number Types, such as integers, rationals, bigfloats, algebraic numbers, and linear algebra.

Graphs and Supporting Data Structures, such as node- and edge-arrays, node- and edge-maps, node priority queues and node partitions, iteration statements for nodes and edges, ...,

Graph Algorithms, such as shortest paths, spanning trees, flows, matchings, components, planarity, planar embedding, ...,

Geometric Objects, such as points, lines, segments, rays, planes, circles, polygons, ...,

Geometric Algorithms, such as convex hulls, triangulations, Delaunay diagrams, Voronoi diagrams, segment intersection, ..., and

Graphical Input and Output.

The modules just mentioned cover a considerable part of combinatorial and geometric computing as treated in courses and textbooks on data structures and algorithms [AHU83, dBKOS97, BY98, CLR90, Kin90, Kle97, NH93, Meh84b, O'R94, OW96, PS85, Sed91, Tar83a, van88, Woo93].

From a user's point of view, LEDA is a platform for combinatorial and geometric computing. It provides *algorithmic intelligence* for a wide range of applications. It eases a programmer's life by providing powerful and easy-to-use data types and algorithms which can be used as building blocks in larger programs. It has been used in such diverse areas as code optimization, VLSI design, robot motion planning, traffic scheduling, machine learning and computational biology. The LEDA system is installed at more than 1500 sites.

We started the LEDA project in the fall of 1988. The project grew out of several considerations.

- We had always felt that a significant fraction of the research done in the algorithms area was eminently practical. However, only a small part of it was actually used. We frequently heard from our former students that the intellectual and programming effort needed to implement an advanced data structure or algorithm is too large to be cost-effective. We concluded that *algorithms research must include implementation if the field wants to have maximum impact*.
- We surveyed the amount of code reuse in our own small and tightly connected research group. We found several implementations of the same balanced tree data structure. Thus there was constant reinvention of the wheel even within our own small group.
- Many of our students had implemented algorithms for their master's thesis. Work invested by these students was usually lost after the students graduated. We had no depository for implementations.
- The specifications of advanced data types which we gave in class and which we found in text books, including the one written by one of the authors, were incomplete and not sufficiently abstract to allow to combine implementations easily. They contained phrases of the form: "Given a pointer to a node in the heap its priority can be decreased in constant amortized time". Phrases of this kind imply that a user of a data structure has to know its implementation. As a consequence combining implementations is a non-trivial task. We performed the following experiment. We asked two groups of students to read the chapters on priority queues and shortest path algorithms in a standard text book, respectively, and to implement the part they had read. The two parts would not fit, because the specifications were incomplete and not sufficiently abstract.

We started the LEDA project to overcome these shortcomings by creating a platform for combinatorial and geometric computing. *LEDA should contain the major findings of the algorithms community in a form that makes them directly accessible to non-experts having only limited knowledge of the area.* In this way we hoped to reduce the gap between research and application.

The LEDA system is available from the LEDA web-site.

`http://www.mpi-sb.mpg.de/LEDA/leda.html`

A commercial version of LEDA is available from Algorithmic Solutions Software GmbH.

`http://www.algorithmic-solutions.de`

LEDA can be used with almost any C++ compiler and is available for UNIX and WINDOWS systems. The LEDA mailing list (see the LEDA web page) facilitates the exchange of information between LEDA users.

This book provides a comprehensive treatment of the LEDA system and its use. We treat the architecture of the system, we discuss the functionality of the data types and algorithms available in the system, we discuss the implementation of many modules of the system, and we give many examples for the use of LEDA. We believe that the book is useful to five types of readers: readers with a general interest in combinatorial and geometric computing, casual users of LEDA, intensive users of LEDA, library designers and software engineers, and students taking an algorithms course.

The book is structured into fourteen chapters.

Chapter 1, Introduction, introduces the reader to the use of LEDA and gives an overview of the system and our design goals.

Chapter 2, Foundations, discusses the basic concepts of the LEDA system. It defines key concepts, such as type, object, variable, value, item, copy, linear order, and running time, and it relates these concepts to C++. We recommend that you read this chapter quickly and come back to it as needed. The detailed knowledge of this chapter is a prerequisite for the intensive use of LEDA. The casual user should be able to satisfy his needs by simply modifying example programs given in the book. The chapter draws upon several sources: object-oriented programming, abstract data types, and efficient algorithms. It lays out many of our major design decisions which we call LEDA axioms.

Chapters 3 to 12 form the bulk of the book. They constitute a guided tour of LEDA. We discuss numbers, basic data types, advanced data types, graphs, graph algorithms, embedded graphs, geometry kernels, geometry algorithms, windows, and graphwins. In each chapter we introduce the functionality of the available data types and algorithms, illustrate their use, and give the implementation of some of them.

Chapter 13, Implementation, discusses the core part of LEDA, e.g., the implementation of parameterized data types, implementation parameters, memory management, and iteration.

Chapter 14, Documentation, discusses the principles underlying the documentation of LEDA and the tools supporting it.

The book can be read without having the LEDA system installed. However, access to the LEDA system will greatly increase the *joy of reading*. The demo directory of the LEDA system contains numerous programs that allow the reader to exercise the algorithms discussed in the book. The demos give a feeling for the functionality and the efficiency of the algorithms, and in a few cases even animate them.

The book can be read from cover to cover, but we expect few readers to do it. We wrote the book such that, although the chapters depend on each other as shown in Figure A, most chapters can be read independently of each other. We sometimes even repeat material in order to allow for independent reading.

All readers should start with the chapters Introduction and Foundations. In these chapters we give an overview of LEDA and introduce the basic concepts of LEDA. We suggest that you read the chapter on foundations quickly and come back to it as needed.

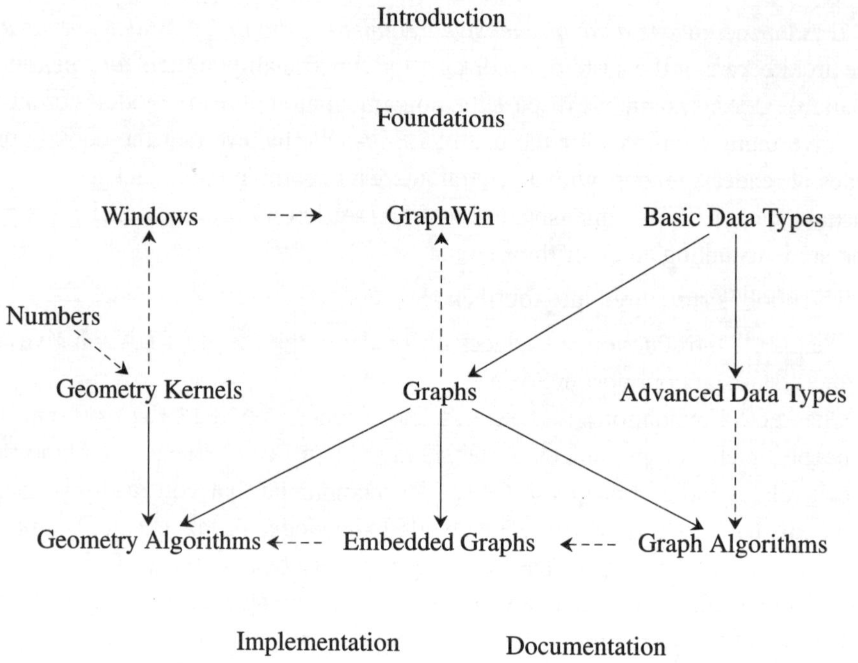

Figure A The dependency graph between the chapters. A dashed arrow means that partial knowledge is required and a solid arrow means that extensive knowledge is required. Introduction and Foundations should be read before all other chapters and Implementation and Documentation can be read independently from the other chapters.

The chapter on basic data types (list, stacks, queues, array, random number generators, and strings) should also be read by every reader. The basic data types are ubiquitous in the book.

Having read the chapters Introduction, Foundations and Basic Data Types, the reader may take different paths depending on interest.

Casual users of LEDA should read the chapters treating their domain of interest, and *intensive users of LEDA* should also read the chapter on implementation.

Readers interested in Data Structures should read the chapters on advanced data types, on implementation, and some of the sections of the chapter on geometric algorithms. The chapter on advanced data types treats dictionaries, search trees and hashing, priority queues, partitions, and sorted sequences, and the chapter on implementation discusses, among other things, the realization of parameterized data types. The different sections in the chapter on advanced data types can be read independently. In the chapter on geometric algorithms we recommend the section on dynamic Delaunay triangulations; some knowledge of graphs and computational geometry is required to read it.

Readers interested in Graphs and Graph Algorithms should continue with the chapter on graphs. From there one can proceed to either the chapter on graph algorithms or the chapter on embedded graphs. Within the chapter on graph algorithms the sections can be read independently. However, the chapter on embedded graphs must be read from front to rear. Some knowledge of priority queues and partitions is required for some of the sections on graph algorithms.

Readers interested in Computational Geometry can continue with either the chapter on graphs or the chapter on geometry kernels. Both chapter are a prerequisite for the chapter on geometric algorithms. The chapter on geometry kernels requires partial knowledge of the chapter on numbers. The chapter on geometric algorithms splits into two parts that can be read independently. The first part is on convex hulls, Delaunay triangulations, and Voronoi diagrams, and the second part is on line segment intersection and polygons.

Geometric algorithms are dull without graphical input and output. The required knowledge is provided by the chapter on windows. The section on the Voronoi demo in the chapter on geometric algorithms gives a comprehensive example for the interplay between geometric data types and algorithms and the window class.

Readers interested in Algorithm Animation should read the chapter on windows and graphwin, the section on animating strongly connected components in the chapter on graph algorithms, the section on the Voronoi demo in the geometric algorithms chapter, and study the many programs in the xlman subdirectory of the demo directory.

Readers interested in Software Libraries should read the chapters on foundations, on implementation, and on documentation. They should also study some other chapters at their own choice.

Readers interested in developing a LEDA Extension Package should read the chapters on implementation and documentation in addition to the chapters related to their domain of algorithmic interest.

For all the algorithms discussed in the book, we also derive the required theory and give the proof of correctness. However, sometimes our theoretical treatment is quite compact and tailored to our specific needs. We refer the reader to the textbooks [AHU83, Meh84b, Tar83a, CLR90, O'R94, Woo93, Sed91, Kin90, van88, NH93, PS85, BY98, dBKOS97] for a more comprehensive view.

LEDA is implemented in C++ and we expect our readers to have some knowledge of it. We are quite conservative in our use of C++ and hence a basic knowledge of the language suffices for most parts of the book. The required concepts include classes, objects, templates, member functions, and non-member functions and are typically introduced in the first fifty pages of a C++ book [LL98, Mur93, Str91]. Only the chapter on implementation requires the reader to know more advanced concepts like inheritance and virtual functions.

The book contains many tables showing *running times*. All running times were determined on an ULTRA-SPARC with 300 MHz CPU and 256 MByte main memory. LEDA and all programs were compiled with CC (optimization flags -DLEDA_CHECKING_OFF and -O).

We welcome *feedback* from our readers. A book of this length is certain to contain errors. If you find any errors or have other constructive suggestions, we would appreciate hearing from you. Please send any comments concerning the book to `ledabook@mpi-sb.mpg.de`

For comments concerning the system use

`ledares@mpi-sb.mpg.de`

or sign up for the LEDA discussion group. We will maintain a list of corrections on the web.

We received financial support from a number of sources. Of course, our home institutions deserve to be mentioned first. We started LEDA at the Universität des Saarlandes in Saarbrücken, in the winter 1990/1991 we both moved to the Max-Planck-Institut für Informatik, also in Saarbrücken, and in the fall of 1994 Stefan Näher moved to the Martin-Luther Universität in Halle. Our work was also supported by the Deutsche Forschungsgemeinschaft (Sonderforschungsbereich SFB 124 VLSI-Entwurf und Parallelität und Schwerpunktprogramm Effiziente Algorithmen und ihre Anwendungen), by the Bundesministerium für Forschung und Technologie (project SOFTI), and by the European Community (projects ALCOM, ALCOM II, ALCOM-IT, and CGAL).

Discussions with many colleagues, bug reports, experience reports (positive and negative), suggestions for changes and extensions, and code contributions helped to shape the project. Of course, we could not have built LEDA without the help of many other persons. We want to thank David Alberts, Ulrike Bartuschka, Christoph Burnikel, Ulrich Finkler, Stefan Funke, Evelyn Haak, Jochen Könemann, Ulrich Lauther, Andreas Luleich, Mathias Metzler, Michael Müller, Michael Muth, Markus Neukirch, Markus Paul, Thomas Papanikolaou, Stefan Schirra, Christian Schwarz, Michael Seel, Jack Snoeyink, Ken Thornton, Christian Uhrig, Michael Wenzel, Joachim Ziegler, Thomas Ziegler, and many others for their contributions.

Special thanks go to Christian Uhrig, the chief officer of Algorithmic Solutions GmbH, to Michael Seel, who is head of the LEDA-group at the MPI, and to Ulrich Lauther from Siemens AG, our first industrial user.

Evelyn Haak typeset the book. Actually, she did a lot more. She made numerous suggestions concerning the layout, she commented on the content, and she suggested changes. Holger Blaar, Stefan Funke, Gunnar Klau, Volker Priebe, Michael Seel, René Weißkircher, Mark Ziegelmann, and Joachim Ziegler proof-read parts of the book. We want to thank them for their many constructive comments. Of course, all the remaining errors are ours.

Finally, we want to thank David Tranah from Cambridge University Press for his support and patience.

We hope that you enjoy reading this book and that LEDA eases your life as a programmer.

Stefan Näher
Halle, Germany
April, 1999

Kurt Mehlhorn
Saarbrücken, Germany
April, 1999

1

Introduction

In this chapter we introduce the reader to LEDA by showing several short, but powerful, programs, we give an overview of the structure of the LEDA system, we discuss our design goals and the approach that we took to reach them, and we give a short account of the history of LEDA.

1.1 Some Programs

We show several programs to give the reader a first impression of LEDA. In each case we will first state the algorithm and then show the program. It is not essential to understand the algorithms in full detail; our goal is to show:

- how easily the algorithms are transferred into programs and
- how natural and elegant the programs are.

In other words,

Algorithm + LEDA = Program.

The directory LEDAROOT/demo/book/Intro (see Section 1.2) contains all programs discussed in this section.

1.1.1 *Word Count*

We start with a very simple program. Our task is to read a sequence of strings from standard input, to count the number of occurrences of each string in the input, and to print a list of all occurring strings together with their frequencies on standard output. The input is ended by the string "end".

In our solution we use the LEDA types *string* and dictionary arrays (*d_arrays*). The parametrized data type dictionary array (*d_array*<I, E>) realizes arrays with index type I and element type E. We use it with index type *string* and element type *int*.

⟨*word_count.c*⟩≡

```
#include <LEDA/d_array.h>
#include <LEDA/string.h>
main()
{ d_array<string,int> N(0);
  string s;
  while ( true )
  { cin >> s;
    if ( s == "end" ) break;
    N[s]++;
  }
  forall_defined(s,N) cout << "\n" << s << " " << N[s];
}
```

We give some more explanations. The program starts with the include statement for dictionary arrays and strings. In the first line of the main program we define a dictionary array N with index type *string* and element type *int* and initialize all entries of the array to zero. Conceptually, this creates an infinite array with one entry for each conceivable string and sets all entries to zero; the implementation of d_arrays stores the non-zero entries in a balanced search tree with key type string. In the second line we define a string s. The while-loop does most of the work. We read a string s; if the string is equal to "end", we break from the loop. Otherwise, we increment the entry $N[s]$ of the array N by one. The iteration *forall_defined*(s, N) in the last line successively assigns all strings to s for which the corresponding entry of N was touched during execution. For each such string, the string and its frequency are printed on the standard output. A new line is used for each pair. On input

```
stefan
stefan
kurt
end
```

the program will print

```
kurt 1
stefan 2
```

1.1.2 *Shortest Paths*

Dijkstra's shortest path algorithm [Dij59] takes a directed graph $G = (V, E)$, a node $s \in V$, called the source, and a non-negative cost function on the edges $cost : E \to R_{\geq 0}$. It computes for each node $v \in V$ the distance from s, see Figure 1.1. A typical text book

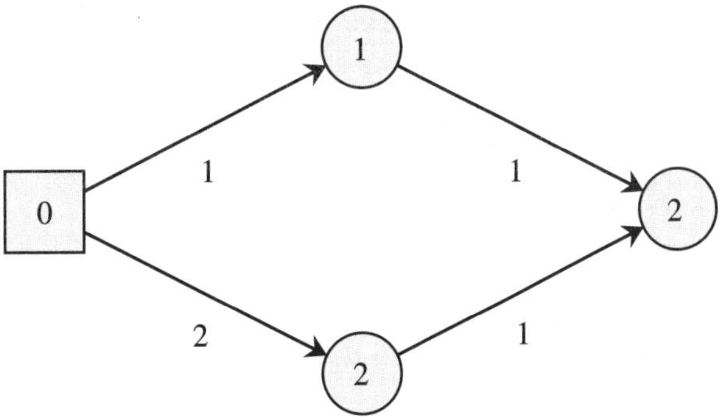

Figure 1.1 A shortest path in a graph. Each edge has a non-negative cost. The cost of a path is the sum of the cost of its edges. The source node s is indicated as a square. For each node the length of the shortest path from s is shown.

presentation of the algorithm is as follows (we will prove the correctness of the algorithm in Section 6.6):

```
set dist(s) to 0.
set dist(v) to infinity for v different from s.
declare all nodes unreached.
while there is an unreached node
{ let u be an unreached node with minimal dist-value.          (*)
  declare u reached.
  forall edges e = (u,v) out of u
     set dist(v) = min( dist(v), dist(u) + cost(e) )
}
```

The text book presentation will then continue to discuss the implementation of line (*). It will state that the pairs $\{(v, dist(v));\ v \text{ unreached}\}$ should be stored in a priority queue, e.g., a Fibonacci heap, because this will allow the selection of an unreached node with minimal distance value in logarithmic time. It will probably refer to some other chapter of the book for a discussion of priority queues.

We next give the corresponding LEDA program; it is very similar to the pseudo-code above. In fact, after some experience with LEDA you should be able to turn the pseudo-code into code within a few minutes.

⟨*DIJKSTRA.c*⟩≡

```
#include <LEDA/graph.h>
#include <LEDA/node_pq.h>
void DIJKSTRA(const graph &G, node s,
              const edge_array<double>& cost,
              node_array<double>& dist)
{ node_pq<double> PQ(G);
  node v; edge e;
  forall_nodes(v,G)
```

```
  { if (v == s) dist[v] = 0; else dist[v] = MAXDOUBLE;
    PQ.insert(v,dist[v]);
  }
  while ( !PQ.empty() )
  { node u = PQ.del_min();
    forall_out_edges(e,u)
      { v = target(e);
        double c = dist[u] + cost[e];
        if ( c < dist[v] )
        {  PQ.decrease_p(v,c);  dist[v] = c;  }
      }
  }
}
```

We give some more explanations. We start by including the graph and the node priority queue data types. The function *DIJKSTRA* takes a graph *G*, a node *s*, an *edge_array cost*, and a *node_array dist*. Edge arrays and node arrays are arrays indexed by edges and nodes, respectively. We declare a priority queue *PQ* for the nodes of graph *G*. It stores pairs $(v, dist[v])$ and is initially empty. The *forall_nodes*-loop initializes *dist* and *PQ*. In the main loop we repeatedly select a pair $(u, dist[u])$ with minimal distance value and then iterate over all out-going edges to update distance values of neighboring vertices.

We next incorporate the shortest path program into a small demo. We generate a random graph with *n* nodes and *m* edges and choose the edge costs as random number in the range [0 .. 100]. We call the function above and report the running time.

⟨*dijkstra_time.c*⟩≡

```
⟨DIJKSTRA.c⟩
main()
{ int n = read_int("number of nodes = ");
  int m = read_int("number of edges = ");
  graph G;
  random_graph(G,n,m);
  edge_array<double> cost(G);
  node_array<double> dist(G);
  edge e; forall_edges(e,G) cost[e] = ((double) rand_int(0,100));
  float T = used_time();
  DIJKSTRA(G,G.first_node(),cost,dist);
  cout << "\n\nThe shortest path computation took " <<
          used_time(T) << " seconds.\n\n";
}
```

On a graph with 10000 nodes and 100000 edges the computation takes less than a second.

1.1.3 *Curve Reconstruction*

The reconstruction of a curve from a set of sample points is an important problem in computer vision. Amenta, Bern, and Eppstein [ABE98] introduced a reconstruction algorithm which they called CRUST. Figure 1.2 shows a point set and the curves reconstructed by

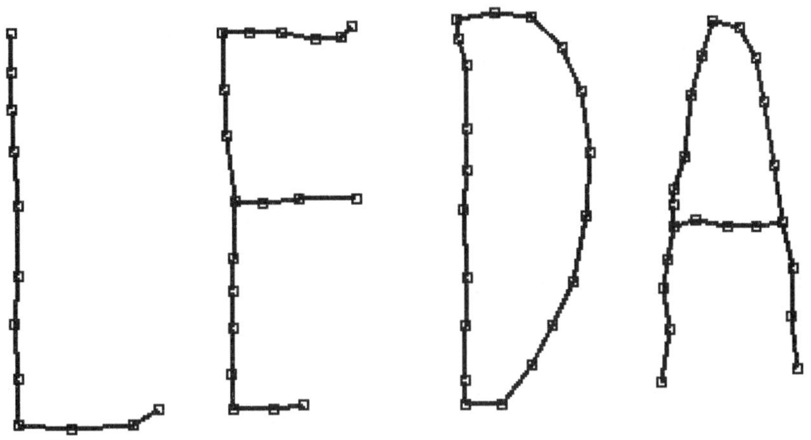

Figure 1.2 A set of points in the plane and the curve reconstructed by CRUST. The figure was generated by the program presented in Section 1.1.3.

their algorithm. The algorithm *CRUST* takes a list S of points and returns a graph G. CRUST makes use of Delaunay diagrams and Voronoi diagrams (which we will discuss in Sections 10.4 and 10.5) and proceeds in three steps:

- It first constructs the Voronoi diagram *VD* of the points in S.
- It then constructs a set $L = S \cup V$, where V is the set of vertices of *VD*.
- Finally, it constructs the Delaunay triangulation *DT* of L and makes G the graph of all edges of *DT* that connect points in S.

The algorithm is very simple to implement[1].

⟨*crust.c*⟩≡

```
#include <LEDA/graph.h>
#include <LEDA/map.h>
#include <LEDA/float_kernel.h>
#include <LEDA/geo_alg.h>
void CRUST(const list<point>& S, GRAPH<point,int>& G)
{
  list<point> L = S;
  GRAPH<circle,point> VD;
  VORONOI(L,VD);
```

[1] In 97 the authors attended a conference, where Nina Amenta presented the algorithm. We were supposed to give a presentation of LEDA later in the day. We started the presentation with a demo of algorithm CRUST.

```
  // add Voronoi vertices and mark them
  map<point,bool> voronoi_vertex(false);
  node v;
  forall_nodes(v,VD)
  { if (VD.outdeg(v) < 2) continue;
    point p = VD[v].center();
    voronoi_vertex[p] = true;
    L.append(p);
  }
  DELAUNAY_TRIANG(L,G);
  forall_nodes(v,G)
    if (voronoi_vertex[G[v]]) G.del_node(v);
}
```

We give some explanations. We start by including graphs, maps, the floating point geometry kernel, and the geometry algorithms. In CRUST we first make a copy of S in L. Next we compute the Voronoi diagram *VD* of the points in L. In LEDA we represent Voronoi diagrams by graphs whose nodes are labeled with circles. A node v is labeled by a circle passing through the defining sites of the vertex. In particular, *VD*[v].*center*( ) is the position of the node v in the plane. Having computed *VD* we iterate over all nodes of *VD* and add all finite vertices (a Voronoi diagram also has nodes at infinity, they have degree one in our graph representation of Voronoi diagrams) to L. We also mark all added points as vertices of the Voronoi diagram. Next we compute the Delaunay triangulation of the extended point set in G. Having computed the Delaunay triangulation, we collect all nodes of G that correspond to vertices of the Voronoi diagram in a list *vlist* and delete all nodes in *vlist* from G. The resulting graph is the result of the reconstruction.

We next incorporate CRUST into a small demo which illustrates its speed. We generate n random points in the plane and construct their crust. We are aware that it does really make sense to apply CRUST to a random set of points, but the goal of the demo is to illustrate the running time.

⟨*crust_time.c*⟩≡

```
⟨crust.c⟩
main()
{ int n = read_int("number of points = ");
  list<point> S;
  random_points_in_unit_square(n,S);
  GRAPH<point,int> G;
  float T = used_time();
  CRUST(S,G);
  cout << "\n\nThe crust computation took " <<
          used_time(T) << " seconds.\n\n";
}
```

For 3000 points the computation takes less than a second.

1.1.4 *A Curve Reconstruction Demo*

We use the program of the preceding section for a small interactive demo.

⟨*crust_demo.c*⟩≡

```
#include <LEDA/window.h>
⟨crust.c⟩
main()
{ window W; W.display();
  W.set_node_width(2); W.set_line_width(2);
  point p;
  list<point> S;
  GRAPH<point,int> G;
  while ( W >> p )
  { S.append(p);
    CRUST(S,G);
    node v; edge e;
    W.clear();
    forall_nodes(v,G) W.draw_node(G[v]);
    forall_edges(e,G) W.draw_segment(G[source(e)], G[target(e)]);
  }
}
```

We give some more explanations. We start by including the window type. In the main program we define a window and open its display. A window will pop up. We state that we want nodes and edges to be drawn with width two. We define the list S and the graph G required for CRUST. In each iteration of the while-loop we read a point in W (each click of the left mouse button enters a point), append it to S and compute the crust of S in G. We then draw G by drawing its vertices and its edges. Each edge is drawn as a line segment connecting its endpoints. Figure 1.2 was generated with the program above.

1.1.5 *Discussion*

We hope that you are impressed by the programs which we have just shown you. In each case only a few lines of code were necessary to achieve complex functionality and, moreover, the code is elegant and readable. We conclude that LEDA is ideally suited for rapid prototyping as summarized in the equation

$$\text{Algorithm} + \text{LEDA} = \text{Program}.$$

The data structures and algorithms in LEDA are efficient. For example, the computation of shortest paths in a graph with 10000 nodes and 100000 edges and the computation of the crust of 3000 points took less than a second each. Thus

$$\text{Algorithm} + \text{LEDA} = \text{Efficient Program}.$$

`Acknowledgements`	`acknowledgements`
`README`	`information about LEDA`
`INSTALL`	`this file`
`CHANGES`	`most recent changes`
`FIXES`	`bug fixes since last release`
`LEPS/`	`LEDA extension packages`
`Manual/`	`user manual`
`Makefile`	`make script`
`confdir/`	`configuration directory`
`lconfig`	`configuration command`
`cmd/`	`commands`
`incl/`	`include directory`
`src/`	`source files`
`demo/`	`demo programs`
`test/`	`test programs`
`data/`	`data files`

Figure 1.3 The top level of the LEDA root directory. Depending on the version of LEDA that is installed at your system, some of the files may be missing or empty.

1.2 The LEDA System

The LEDA system can be downloaded from the LEDA web-site.

`http://www.mpi-sb.mpg.de/LEDA/leda.html`

A commercial version of LEDA is available from Algorithmic Solutions Software GmbH.

`http://www.algorithmic-solutions.de`

At both places you will also find an installation guide.

Figure 1.3 shows the top level of the LEDA directory; some files may be missing or empty depending on the version of LEDA that is installed at your system. We use LEDAROOT to denote the path name of the LEDA directory. In this section we will discuss essential parts of the LEDA directory tree.

README and INSTALL tell you how to install the system. In the remainder of this section all path-names will be relative to the LEDA root directory.

1.2.1 *The Include Directory*

The include directory `incl/LEDA` contains:

- all header files of the LEDA system,
- subdirectory `templates` for the template versions of network algorithms,
- subdirectory `generic` for the kernel independent versions of geometric algorithms,
- subdirectory `impl` for header files of different implementations of dictionaries and priority queues,

- subdirectory `thread` for the classes needed to make LEDA thread-safe,
- subdirectory `sys` for the classes that adapt LEDA to different compilers and systems, and
- subdirectories `bitmaps` and `pixmaps` for bitmaps and pixel maps.

1.2.2 *The Source Code Directory*

The source code directory `src` contains the source code of LEDA. If you have downloaded an object code package, as you probably have, this directory will be empty. Otherwise, it has one subdirectory for each of the major parts of LEDA: basic data types, numbers, dictionaries, priority queues, graphs, graph algorithms, geometry kernels, geometry algorithms, windows, ...

1.2.3 *The LEDA Manual*

The directory `Manual` contains the LaTeX-sources of the LEDA manual. You may make the manual by typing "make" in this directory. This requires that certain additional tools are installed at your system. Alternatively, and we recommend the alternative, you may download the LEDA manual from our web-site. There are two versions of the LEDA manual available on our web-site:

- A paper version in the form of either a ps-file or a dvi-file.
- An HTML-version.

1.2.4 *The Demo Directory*

The directory `demo` contains demos. All demos mentioned in this book are contained in either the subdirectory `xlman` or the subdirectory `book`. We call the demos in the former directory xlman-demos.

All xlman-demos have a graphical user interface and can be accessed through the xlman-utility, see Section 1.2.5. Of course, one can also call them directly in directory xlman. You will find many screenshots in this book; many of them are screenshots of xlman-demos.

The demos in the `book`-directory typically have an ASCII-interface and demonstrate running times. The book-directory is structured according to the chapters of this book.

1.2.5 *Xlman*

Xlman gives you on-line access to the xlman-demos and the LEDA manual (if xdvi is installed at your system). Figure 1.4 shows a screenshot of xlman.

1.2.6 *LEDA Extension Packages*

LEDA extension packages (LEPs) extend LEDA into particular application domains and areas of algorithmics not covered by the core system. Anybody may contribute a LEDA extension package. At the time of writing this there are LEDA extension packages for:

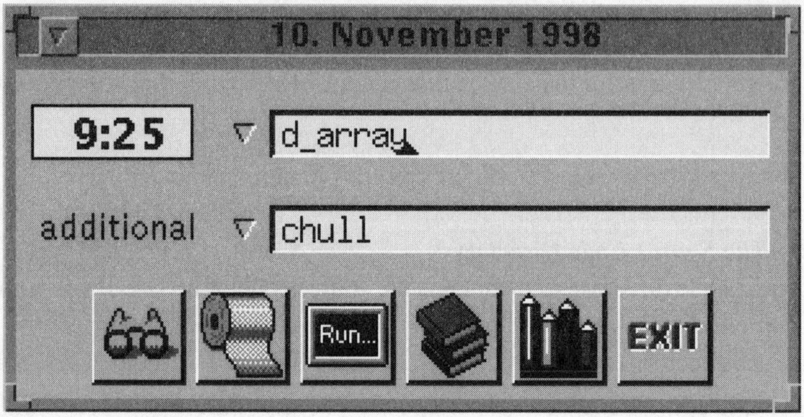

Figure 1.4 A screen shot of xlman. The upper text line shows the name of a LEDA manual page, and the lower text line shows the name of an LEP manual page (this line may be missing in your installation). The six buttons at the bottom have the following functionality: on-line access to manual pages, printing manual pages, running LEDA demos, access to LEDA documents, xlman configuration, and exit. Some of the functionality relies on other tools, e.g., xdvi, and may be missing on your system.

- abstract Voronoi diagrams (by Michael Seel),
- higher-dimensional geometry (by Kurt Mehlhorn, Michael Müller, Stefan Näher, Stefan Schirra, Michael Seel, and Christian Uhrig),
- dynamic graph algorithms (by David Alberts, Umberto Nanni, Guilio Pasqualone, Christos Zaroliagis, Pippo Cattaneo, and Guiseppe F. Italiano),
- graph iterators (by Marco Nissen and Karsten Weihe),
- external memory computations (by Andreas Crauser),
- PQ-trees (by Sebastian Leipert), and
- SD-trees (by Peter Hilpert).

LEDA extension packages must satisfy a set of basic requirements which guarantee compatibility with the LEDA philosophy; the requirements are defined on our web-page.

1.3 The LEDA Web-Site

The LEDA web-site (`http://www.mpi-sb.mpg.de/LEDA/leda.html`) is an important source of information about LEDA. We mentioned already that it allows you to download the most recent version of the system and the manual. It also gives you information about the people behind LEDA and latest news, and it contains pointers to other systems which are either built on top of LEDA or which we have used successfully together with LEDA. We will discuss some of these systems in the next section.

1.4 Systems that Go Well with LEDA

Although LEDA covers many aspects of combinatorial and geometric computing, it cannot cover all of them. In our own work we therefore also use other systems.

In the realm of exact solution of NP-complete problems we use LEDA together with ABACUS, a branch-and-cut framework for polyhedral optimization, and with CPLEX and SoPLEX, two solvers for linear programs. ABACUS was developed by Michael Jünger, Stefan Reinelt, and Stefan Thienel.

For graph drawing we use AGD, a library of automatic graph drawing, and GDToolkit , a toolkit for graph drawing. AGD is a joint effort of Petra Mutzel's group at the MPI, Stefan Näher's group in Halle, and Michael Jünger's group in Cologne. GDToolkit was developed by Guiseppe Di Battista's group in Rome. We say a bit more about AGD and GDToolkit in Section 8.1.

For computational geometry we also use CGAL, a computational geometry algorithms library. CGAL is a joint effort of ETH Zürich Freie Universität Berlin, INRIA Sophia Antipolis, Martin-Luther Universität Halle-Wittenberg, Max-Planck-Institut für Informatik and Universität des Saarlandes, RISC Linz Tel-Aviv University, and Universiteit Utrecht. We will say more about CGAL in Section 9.11.

1.5 Design Goals and Approach

We had four major goals for the design of LEDA:

- Ease of use.
- Extensibility.
- Correctness.
- Efficiency.

We next discuss our four goals and how we tried to reach them.

We wanted the library to reduce the gap between the algorithms community and the "rest of the world" and therefore *ease of use* was a major concern. We wanted the library to be useable without intimate knowledge of our field of research; a basic course in data structures and algorithms should suffice. We also wanted the data types and algorithms of LEDA to be useable without any knowledge of their implementation.

We invented the item concept, see Section 2.2, as an abstraction of the concept of "pointer to a container in a data type" and used it for the specification of all container-based data types. We formulated rules (see Chapter 2) that capture key concepts, such as copy constructor, assignment, and compare functions, uniformly for all data types. We introduced a powerful graph type, see Chapter 6, which supports the natural and elegant formulation of graph and network algorithms and is also the basis for many of the geometric algorithms.

Ease of use also means easy access to information. The LEDA manual, see Chapter 14, gives precise and readable specifications for the LEDA data types and algorithms. The specifications are short (typically not more than a page), general (so as to allow several implementations) and abstract (so as to hide all details of the implementation). All specifications follow a common format, see Section 2.1. We developed tools that support the production of manual pages and documentations. Finally, we wrote this book that gives a comprehensive view of LEDA.

Combinatorial and geometric computing is a diverse area and hence it is impossible for a library to provide ready-made solutions for all application problems. For this reason it is important that LEDA is easily extensible and can be used as a platform for further software development. LEDA itself is a good example for the *extensibility* of LEDA. The advanced data types and algorithms discussed in Chapters 5, 7, 8, and 10 are built on top of the basic data types introduced in Chapters 3, 4, 6, and 9. The basic data types in turn rest on a conceptual framework described in Chapter 2 and the implementation principles discussed in Chapter 13.

Incorrect software is hard to use at best and dangerous at worst. We underestimated the difficulties of achieving *correctness*. After all, any publication in our area proves the correctness of the described algorithms and going from a correct algorithm to a correct program is tedious and time-consuming, but hardly an intellectual challenge. So we thought, when we started the project. We now think differently.

Many of the algorithms in LEDA are quite intricate and therefore difficult to implement correctly. Programmers make mistakes and we are no exception. How do we guard against errors? Many of our implementations are carefully documented (this book contains many examples), we test extensively, as does our large user community, and we have recently adopted the philosophy that programs should give sufficient justification for their answers to allow checking, see Section 2.14. We have developed program checkers for many of our programs.

The correct implementation of geometric programs was particularly difficult, as the theoretical underpinning was insufficient. Geometric algorithms are typically derived under two simplifying assumptions: (1) the underlying machine model is the *real RAM* which can compute with real numbers in the sense of mathematics and (2) inputs are in general position. However, the number types *int* and *double* offered by programming languages are only crude approximations of real numbers and practical inputs are frequently degenerate. Our approach is to formulate geometric algorithms such that they work for all inputs, see Chapter 10, and to realize the real RAM (as far as it is needed for computational geometry) by exact number types, see Chapter 4, and an exact and yet efficient geometry kernel, see Chapter 9.

Efficiency was our fourth design goal. It may surprise some readers that we list it last. However, efficiency without correctness is meaningless and efficiency without ease of use is a questionable blessing. We achieve efficiency by the use of efficient algorithms and their careful implementation.

Our implementations are usually based on the asymptotically most efficient algorithms known for a particular problem. In many cases we even implemented different algorithmic approaches. For example, there are several shortest path, matching, and flow algorithms, there are several convex hull, line segment intersection, and Delaunay diagram algorithms, and there are several realizations of dictionaries and priority queues. In the case of data types, the implementation parameter mechanism allows the convenient selection of an implementation. For example, the declarations

```
dictionary<string,int> D1;
dictionary<string,int,skip_list> D2;
```

declare *D1* as a dictionary from *string* to *int* with the default implementation and select the skip list implementation for *D2*.

The description of many algorithms leaves considerable freedom for the implementor, i.e., a description typically defines a family of algorithms all with the same asymptotic worst case running time and leaves decisions that do not affect worst case running time to the implementor. The decisions may, however, dramatically affect the running time on inputs that are not worst case. We have carefully explored the available opportunities; Sections 7.6 and 7.10 give particularly striking examples. We found it useful to concentrate on the best and average case after getting the worst case "right".

LEDA has its own memory manager. It provides efficient implementations of the *new* and *delete* operators.

How efficient are the programs in LEDA? We give many tables of running times in this book which show that LEDA programs are able to solve large problem instances. We made comparisons, see Tables 1.1 and 1.2, and other people did, see for example [Ski98, Ski]. The comparisons show that our running times are competitive, despite the fact that LEDA is more like a decathlon athlete than a specialist for a particular discipline.

1.6 History

We started the project in the fall of 1988. We spent the first six months on specifications and on selecting our implementation language. Our test cases were priority queues, dictionaries, partitions, and algorithms for shortest paths and minimum spanning trees. We came up with the item concept as an abstraction of the notion "pointer into a data structure". It worked successfully for the three data types mentioned above and we are now using it for most data types in LEDA. Concurrently with searching for the correct specifications

```
Number of list entries: 100000

LIST<INT>          LEDA       STL
build list         0.020 sec  0.040 sec
pop and push       0.030 sec  0.030 sec
reversing          0.020 sec  0.030 sec
copy constr        0.050 sec  0.050 sec
assignment         0.020 sec  0.040 sec
clearing           0.000 sec  0.020 sec
sorting            0.130 sec  0.400 sec
sorting again      0.140 sec  0.330 sec
merging            0.030 sec  0.080 sec
unique             0.080 sec  0.080 sec
unique again       0.000 sec  0.010 sec
iteration          0.000 sec  0.000 sec
---------------------------------------
total              0.520 sec  1.110 sec


LIST<CLASS>        LEDA       STL
build list         0.090 sec  0.030 sec
pop and push       0.100 sec  0.030 sec
reversing          0.070 sec  0.030 sec
copy constr        0.140 sec  0.060 sec
assignment         0.120 sec  0.030 sec
clearing           0.080 sec  0.020 sec
sorting            0.770 sec  0.510 sec
sorting again      0.900 sec  0.380 sec
merging            0.200 sec  0.090 sec
unique             0.250 sec  0.100 sec
unique again       0.010 sec  0.000 sec
iteration          0.010 sec  0.000 sec
---------------------------------------
total              2.740 sec  1.280 sec
```

Table 1.1 A comparison of the list data type in LEDA and in the implementation of the Standard Template Library [MS96] that comes with the GNU C++ compiler. The upper part compares *list<int>* and the lower part compares *list<class>*, where the objects of type *class* require several words of storage. LEDA lists are faster for small objects and slower for large objects. This table was generated by the program stl_vs_leda in LEDAROOT/demo/stl. You can perform your own comparisons if your C++ compiler comes with an implementation of the STL. All running times are in seconds.

we investigated several languages for their suitability as our implementation platform. We looked at Smalltalk, Modula, Ada, Eiffel, and C++. We wanted a language that supported abstract data types and type parameters (polymorphism) and that was widely available. We wrote sample programs in each language. Based on our experiences we selected C++ because of its flexibility, expressive power, and availability.

A first publication about LEDA appeared in the conferences MFCS 1989 and ICALP 1990 [MN89, NM90]. Stefan Näher became the head of the LEDA project and he is the

Type of network	LEDA	CG
random (4000 nodes 28000 edges)	0.31	0.11
CG1 (8002 nodes 12003 edges)	9.26	4.20
CG2 (8002 nodes 12001 edges)	0.11	0.73
AMO (4001 nodes 7998 edges)	0.05	1.74

Table 1.2 A comparison of the maxflow implementation in LEDA and the one by Cherkassky and Goldberg [CG97, CG]. The latter implementation is generally considered the best code available. We used four different kinds of graphs: random graphs and graphs generated by three different generators. The generators CG1 and CG2 were suggested by Cherkassky and Goldberg and the generator AMO was suggested by Ahuja, Magnanti, and Orlin. The generators are discussed in detail in Section 7.10. All running times are in seconds. You may perform your own experiments by running the program flow_test in LEDAROOT/demo/book/Intro and following the instructions given in MAXFLOW_README in the same directory.

main designer and implementer of LEDA. Progress reports appeared in [MN92, Näh93, MN94b, MN95, BKM+95, MNU97, MN98b].

In the second half of 1989 and during 1990 Stefan Näher implemented a first version of the combinatorial part (= data structures and graph algorithms) of LEDA (Version 1.0). Then there were releases 2.0, 2.1, 2.2, 3.0, 3.1, 3.2, 3.3, 3.4, 3.5, 3.6, and 3.7. With the appearance of this book we will release version 4.0. Each new release offered new functionality, increased efficiency, and removed bugs.

LEDA runs on many different platforms (Unix, Windows, OS/2) and with many different compilers.

In early 1995 LEDA Software GmbH was founded to market a commercial version of LEDA. Christian Uhrig became the Chief Executive Officer. The company was renamed to Algorithmic Solutions Software GmbH in late 1997 to reflect the fact that it not only markets LEDA, but also other systems like, for example, AGD and CGAL, and that it also develops algorithmic solutions for specific needs.

The research version is used at more than 1500 academic and research sites. Try the website `http://www.mpi-sb.mpg.de/LEDA/DOWNLOADSTAT` to find out whether the system is already in use at your site.

2

Foundations

We discuss the foundations of the LEDA system. We introduce some key concepts, such as type, object, variable, value, item, and linear order, we relate these concepts to our implementation base C++, and we put forth our major design decisions. A superficial knowledge of this chapter suffices for a first use of LEDA. We recommend that you read it quickly and come back to it as needed.

The chapter is structured as follows. We first discuss the specification of data types. Then we treat the concept "copy of an object" and its relation to assignment and parameter passing by value. The other kinds of parameter passing come next and sections on iteration statements follow. We then tie data types to the class mechanism of C++. Type parameters, linear orders, equality, hashed types, and implementation parameters are the topics of the next sections. Finally, we discuss some helpful small functions, management, error handling, header and implementation files, compilation flags, and program checking.

2.1 Data Types

The most important concept is that of a *data type* or simply *type*. A type T consists of a set of *values*, which we denote $val(T)$, a set of *objects*, which we denote $obj(T)$, and a set of functions that can be applied to the objects of the type. An object may or may not have a *name*. A named object is also called a *variable* and an object without a name is called an *anonymous object*. An object is a region of storage that can hold a value of the corresponding type.

The set of objects of a type varies during execution of a program. It is initially empty, it grows as new objects are created (either by variable definitions or by applications of the *new* operator), and it shrinks as objects are destroyed.

The values of a type form a set that exists independently of any program execution. We define it using standard mathematical concepts and notation. When we refer to the values of a type without reference to an object, we also use *element* or *instance*, e.g., we say that the number 5 is a value, an element, or an instance of type *int*.

An object always holds a value of the appropriate type. The object is initialized when it is created and the value may be modified by functions operating on the object. For an object x we use x also to denote the value of x. This is a misuse of notation to which every programmer is accustomed to.

In LEDA the specification (also called definition) of a data type consists of four parts: a definition of the instances of the type, a description of how to create an object of the type, the definition of the operations available on the objects of the type, and information about the implementation. In the LEDA manual the four parts appear under the headers *Definition*, *Creation*, *Operations*, and *Implementation*, respectively. Sometimes, there is also a fifth section illustrating the use of the data type by an example. As an example we give the complete specification of the parameterized data type *stack*<E> in Figure 2.1.

2.1.1 *Definition*

The first section of a specification defines the instances of the data type using standard mathematical concepts and notation. It also introduces notation that is used in later sections of the specification. We give some examples:

- An instance of type *string* is a finite sequence of characters. The length of the sequence is called the *length* of the string.
- An instance of type *stack*<E> is a sequence of elements of type E. One end of the sequence is designated as its *top* and all insertions into and deletions from a stack take place at its top end. The length of the sequence is called the *size* of the stack. A stack of size zero is called *empty*.
- An instance of type *array*<E> is an injective mapping from an interval $I = [a\,..\,b]$ of integers into the set of variables of type E. We call I the index set and E the element type of the array. For an array A we use $A(i)$ to denote the variable indexed by i, $a \le i \le b$.
- An instance of type *set*<E> is a set of elements of type E. We call E the element type of the set; E must be linearly ordered. The number of elements in the set is called the *size* of the set and a set of size zero is called *empty*.
- An instance of type *list*<E> is a sequence of list items (predefined item type *list_item*). Each item contains an element of type E. We use $\langle x \rangle$ to denote an item with content x.

Most data types in LEDA are *parameterized*, e.g., stacks, arrays, lists, and sets can be used for an arbitrary element type E and we will later see that dictionaries are defined in terms of a key type and an information type. A concrete type is obtained from a parameterized

Stacks (stack)

1. Definition

An instance S of the parameterized data type *stack<E>* is a sequence of elements of data type E, called the element type of S. Insertions or deletions of elements take place only at one end of the sequence, called the top of S. The size of S is the length of the sequence, a stack of size zero is called the empty stack.

2. Creation

stack<E> S;	declares a variable S of type *stack<E>*. S is initialized with the empty stack.

3. Operations

E	S.top( )	returns the top element of S. *Precondition*: S is not empty.
void	S.push(E x)	adds x as new top element to S.
E	S.pop( )	deletes and returns the top element of S. *Precondition*: S is not empty.
int	S.size( )	returns the size of S.
bool	S.empty( )	returns true if S is empty, false otherwise.
void	S.clear( )	makes S the empty stack.

4. Implementation

Stacks are implemented by singly linked linear lists. All operations take time $O(1)$, except clear which takes time $O(n)$, where n is the size of the stack.

Figure 2.1 The specification of the type *stack<E>*.

type by substituting concrete types for the type parameter(s); this process is called *instantiation of the parameterized type*. So *array<string>* are arrays of strings, *set<int>* are sets of integers, and *stack<set<int> * >* are stacks of pointers to sets of integers. Frequently, the actual type parameters have to fulfill certain conditions, e.g., the element type of sets must be linearly ordered. We discuss type parameters in detail in Section 2.8.

2.1.2 *Creation*

We discuss how objects are created and how their initial value is defined. We will see that an object either has a name or is anonymous. We will also learn how the lifetime of an object is determined.

A *named object* (also called variable) is introduced by a C++ *variable definition*. We give some examples.

```
string s;
```

introduces a variable s of type *string* and initializes it to the empty string.

```
stack<E> S;
```

introduces a variable S of type *stack<E>* and initializes it to the empty stack.

```
b_stack<E> S(int n);
```

introduces a variable S of type *b_stack<E>* and initializes it to the empty stack. The stack can hold a maximum of n elements.

```
set<E> S;
```

introduces a variable S of type *set<E>* and initializes it to the empty set.

```
array<E> A(int l,int u);
```

introduces a variable A of type *array<E>* and initializes it to an injective function $a : [l\,..\,u] \longrightarrow obj(E)$. Each object in the array is initialized by the default initialization of type E; this concept is defined below.

```
list<E> L;
```

introduces a variable L of type *list<E>* and initializes it to the empty list.

```
int i;
```

introduces a variable of type *int* and initializes it to some value of type *int*.

We always give variable definitions in their generic form, i.e., we use formal type names for the type parameters (E in the definitions above) and formal arguments for the arguments of the definition (*int a*, *int b*, and *int n* in the definitions above). Let us also see some concrete forms.

```
string s("abc");          // initialized to "abc"
set<int> S;               // initialized to empty set of integers
array<string> A(2,5);     // array with index set [2..5],
                          // each entry is set to the empty string
b_stack<int> S(100);      // a stack capable of holding up to 100
                          // ints; initialized to the empty stack
```

The most general form of a variable definition in C++ is

```
T<T1,...,Tk> y(x1,...,xl).
```

It introduces a variable with name y of type *T<T1, ..., Tk>* and uses arguments *x1*, ..., *xl* to determine the initial value of y. Here T is a parameterized type with k type parameters and *T1*, ..., *Tk* are concrete types. If any of the parameter lists is empty the corresponding pair of brackets is to be omitted.

Two kinds of variable definitions are of particular importance: the definition with default initialization and the definition with initialization by copying. A *definition with default initialization* takes no argument and initializes the variable with the *default value* of the type. The default value is typically the "simplest" value of the type, e.g., the empty string, the empty set, the empty dictionary, … . We define the default value of a type in the section with header Creation. Examples are:

```
string s;         // initialized to the empty string
stack<int> S;     // initialized to the empty stack
array<string> A; // initialized to the array with empty index set
```

The built-in types such as *char*, *int*, *float*, *double*, and all pointer types are somewhat an exception as they have no default value, e.g., the definition of an integer variable initializes it with some integer value. This value may depend on the execution history. Some compilers will initialize i to zero (more generally, 0 casted to the built-in type in question), but one should not rely on this[1].

We can now also explain the definition of an array. Each variable of the array is initialized by the default initialization of the element type. If the element type has a default value (as is true for all LEDA types), this value is taken and if it has no default value (as is true for all built-in types), some value is taken. For example, *array<list<E> >* $A(1, 2)$ defines A as an array of lists of element type E. Each entry of the array is initialized with the empty list.

A *definition with initialization by copying* takes a single argument of the same type and initializes the variable with a copy of the argument. The syntactic form is

```
T<T1,...,Tk> y(x)
```

where x refers to a value of type *T<T1, ..., Tk>*, i.e., x is either a variable name or more generally an expression of type *T<T1, ..., Tk>*. An alternative syntactic format is

```
T<T1,...,Tk> y = x.
```

We give some examples.

```
stack<int> P(S);     // initialized to a copy of S
set<string> U(V);    // initialized to a copy of V
string s = t;        // initialized to a copy of t
int i = j;           // initialized to a copy of j
int h = 5;           // initialized to a copy of 5
```

We have to postpone the general definition of what constitutes a copy to Section 2.3 and give only some examples here. A copy of an integer is the integer itself and a copy of a string is the string itself. A copy of an array is an array with the same index set but new variables. The initial values of the new variables are copies of the values of the corresponding old variables.

LEDA Rule 1 *Definition with initialization by copying is available for every LEDA type. It initializes the defined variable with a copy of the argument of the definition.*

[1] The C++ standard defines that variables specified static are automatically zero-initialized and that variables specified automatic or register are not guaranteed to be initialized to a specified value.

How long does a variable live? The *lifetime* of a named variable is either tied to the block containing its definition (this is the default rule) or is the execution of the entire program (if the variable is explicitly defined to be static). The first kind of variable is called *automatic* in C++ and the second kind is called *static*. Automatic variables are created and initialized each time the flow of control reaches their definition and destroyed on exit from their block. Static variables are created and initialized when the program execution starts and destroyed when the program execution ends.

We turn to *anonymous objects* next. They are created by the operator *new*; the operator returns a pointer to the newly created object. The general syntactic format is

```
new T<T1,...,Tk> (x1,...,xl);
```

where T is a parameterized type, $T1$, ..., Tk are concrete types, and $x1$, ..., xl are the arguments for the initialization. Again, if any of the argument lists is empty then the corresponding pair of brackets is omitted. The expression returns a pointer to a new object of type $T\langle T1, ..., Tk\rangle$. The object is initialized as determined by the arguments $x1, \ldots, xl$. We give an example.

```
stack<int> *sp = new stack<int>;
```

defines a pointer variable *sp* and creates an anonymous object of type *stack<int>*. The stack is initialized to the empty stack and *sp* is initialized to a pointer to this stack.

The lifetime of an object created by *new* is not restricted to the scope in which it is created. It extends till the end of the execution of the program unless the object is explicitly destroyed by the *delete* operator; *delete* can only be applied to pointers returned by *new* and if it is applied to such a pointer, it destroys the object pointed to. We say more about the destruction of objects in Section 2.3.

2.1.3 *Operations*

Every type comes with a set of operations that can be applied to the objects of the type. The definition of an operation consists of two parts: the definition of its interface (= syntax) and the definition of its effect (= semantics).

We specify the *interface of an operation* essentially by means of the C++ function declaration syntax. In this syntax the result type of the operation is followed by the operation name which in turn is followed by the argument list specifying the type of each argument. The result type of an operation returning no result is *void*. We extend this syntax by prefixing the operation name by the name of an object to which the operation is being applied. This facilitates the definition of the semantics. For example

```
void S.insert(E x);
```

defines the interface of the insert operation for type *set<E>*; *insert* takes an argument x of type E and returns no result. The operation is applied to the set (with name) S.

```
E& A[int i];
```

defines the interface of the access operation for type *array<E>*. Access takes an argument i of type *int* and returns a variable of type E. The operation is applied to array A.

```
E S.pop();
```

defines the interface of the pop operation for type *stack<E>*. It takes no argument and returns an element of type E. The operation is applied to stack S.

```
int s.pos(string s1);
```

defines the interface of the *pos* operation for type *string*. It takes an argument *s1* of type *string* and returns an integer. The operation is applied to string s.

The *semantics of an operation* is defined using standard mathematical concepts and notation. The complete definitions of our four example operations are:

void	*S.insert*(*E x*)	adds x to S.
E&	*A*[*int i*]	returns the variable $A(i)$. *Precondition*: $a \le i \le b$.
E	*S.pop*( )	removes and returns the top element of S. *Precondition*: S is not empty.
int	*s.pos*(*string s1*)	returns -1 if *s1* is not a substring of s and returns the minimal i, $0 \le i \le$ *s.length*( )-1, such that *s1* occurs as a substring of s starting at position i, otherwise.

In the definition of the semantics we make use of the notation introduced in sections Definition and Creation. For example, in the case of arrays the section Definition introduces $A(i)$ as the notation for the variable indexed by i and introduces a and b as the array bounds.

Frequently, an operation is only defined for a subset of all possible arguments, e.g., the *pop* operation on stacks can only be applied to a non-empty stack. The *precondition* of an operation defines which conditions the arguments of an operation must satisfy. If the precondition of an operation is violated then the effect of the operation is undefined. This means that *everything can happen*. The operation may terminate with an error message or with an arbitrary result, it may not terminate at all, or it may result in abnormal termination of the program. Does LEDA check preconditions? Sometimes it does and sometimes it does not. For example, we check whether an array index is out of bounds or whether a pop from an empty stack is attempted, but we do not check whether item *it* belongs to dictionary D in *D.inf*(*it*). Checking the latter condition would increase the running time of the operation form constant to logarithmic and is therefore not done. More generally, we do not check preconditions that would change the order of the running time of an operation. All checks can be turned off by the compile-time flag `-DLEDA_CHECKING_OFF`.

All types offer the assignment operator. For type T this is the operator

```
T& operator=(const T&).
```

The assignment operator is not listed under the operations of a type since all types have it and since its semantics is defined in a uniform way as we will see in Section 2.3.

Our implementation base C++ allows overloading of operation and function names and it allows optional arguments. We use both mechanisms. An *overloaded function name* denotes different functions depending on the types of the arguments. For example, we have two translate operations for points:

```
point p.translate(vector v);
point p.translate(double alpha,double dist);
```

The first operation translates p by vector v and the second operation translates p in direction *alpha* by distance *dist*.

An *optional argument* of an operation is given a default value in the specification of the operation. C++ allows only trailing arguments to be optional, i.e., if an operation has k arguments, $k \geq 1$, then the last l, $l \geq 0$, may be specified to be optional. An example is the insert operation into lists. If L is a *list*$<E>$ then

```
list_item L.insert(E x,list_item it, int dir = after)
```

inserts x before ($dir ==$ *before*) or after ($dir ==$ *after*) item *it* into L. The default value of *dir* is *after*, i.e., *L.insert*(x, *it*) is equivalent to *L.insert*(x, *it*, *after*).

2.1.4 ***Implementation***

Under this header we give information about the implementation of the data type. We name the data structure used, give a reference, list the *running time* of the operations, and state the *space requirement*. Here is an example.

The data type list is realized by doubly linked linear lists. All operations take constant time except for the following operations: *search* and *rank* take linear time $O(n)$, *item*(i) takes time $O(i)$, *bucket_sort* takes time $O(n + j - i)$ and *sort* takes time $O(n \cdot c \cdot \log n)$ where c is the time complexity of the compare function. n is always the current length of the list. The space requirement is $16 + 12n$ bytes.

It should be noted that the time bounds do not include the time needed for parameter passing. The cost of passing a reference parameter is bounded by a constant and the cost of passing a value parameter is the cost of copying the argument. We follow the custom to account for parameter passing at the place of call.

Similarly, the space bound does not include the extra space needed for the elements contained in the set, it only accounts for the space required by the data structure that realizes the set. The extra space needed for an element is zero if the element fits into one machine word and is the space requirement of the element otherwise. This reflects how parameterized data types are implemented in LEDA. Values that fit exactly into one word are stored directly in the data structure and values that do not fit exactly are stored indirectly through a pointer. The details are given in Section 13.1.

The information about the space complexity allows us to compute the exact space requirement of a list of size n. We give some examples. A set of type *list<int>* and *list<list<int>* $*$ >

requires $16 + 12n$ bytes since integers and pointers fit exactly into a word. A list of type *list<list<int> >* where the i-th list has n_i elements, $1 \leq i \leq n$, requires $16 + 12n + \sum_{1 \leq i \leq n}(16 + 12n_i)$ bytes.

The information about time complexity is less specific than that for space. We only give *asymptotic bounds*, i.e., bounds of the form $O(f(n))$ where f is a function of n. A bound of this form means that there are constants c_1 and c_2 (independent of n) such that the running time on an instance of size n is bounded by $c_1 + c_2 \cdot f(n)$. The constants c_1 and c_2 are not explicitly given. An asymptotic bound does not let us predict the actual running time on a particular input (as c_1 and c_2 are not available); it does, however, give a feeling for the behavior of an algorithm as n grows. In particular, if the running time is $O(n)$ then an input of twice the size requires at most twice the computing time, if the running time is $O(n^2)$ then the computing time at most quadruples, and if it is $O(\log n)$ then the computing time grows only by an additive constant as n doubles. Thus asymptotic bounds allow us to extrapolate running times from smaller to larger problem instances.

Why do we not give explicit values for the constants c_1 and c_2? The answer is simple, we do not know them. They depend on the machine and compiler which you use (which we do not know) and even for a fixed machine and compiler it is very difficult to determine them, as machines and compilers are complex objects with complex behavior, e.g., machines have pipelines, multilevel memories, and compilers use sophisticated optimization strategies. It is conceivable that program analysis combined with a set of simple experiments allows one to determine good approximations of the constants, see [FM97] for a first step in this direction.

Our usual notion of running time is worst-case running time, i.e., if an operation is said to have running time $O(f(n))$ then it is guaranteed that the running time is bounded by $c_1 + c_2 \cdot f(n)$ for every input of size n and some constants c_1 and c_2. Sometimes, running times are classified as being expected (also called average) or amortized. We give some examples.

The expected access time for maps is constant. This assumes that a random set is stored in the map.

The expected time to construct the convex hull of n points in 3-dimensional space is $O(n \log n)$. The algorithm is randomized.

The amortized running time of *insert* and *decrease_prio* in priority queues is constant and the amortized running time of *delete_min* is $O(\log n)$.

In the remainder of this section we explain the terms expected and amortized. An *amortized* time bound is valid for a sequence of operations but not for an individual operation. More precisely, assume that we execute a sequence $op_1, op_2, \ldots, op_m$ of operations on an object D, where op_1 constructs D. Let n_i be the size of D before the i-th operation and assume that the i-th operation has amortized cost $O(T_i(n_i))$. Then the total running time for the sequence $op_1, op_2, \ldots, op_m$ is

$$O(m + \sum_{1 \leq i \leq m} T_i(n_i)),$$

i.e., a summation of the amortized time bounds for the individual operations yields a bound for the sequence of the operations. Note that this does not preclude that the i-th operation takes much longer than $T_i(n_i)$ for some i, it only states that the entire sequence runs in the bound stated. However, if the i-th operation takes longer than $T_i(n_i)$ then the preceding operations took less than their allowed time.

We give an example: in priority queues (with the Fibonacci heap implementation) the amortized running time of *insert* and *decrease_prio* is constant and the amortized cost of *delete_min* is $O(\log n)$. Thus an arbitrary sequence of n *insert*, n *delete_min*, and m *decrease_prio* operations takes time $O(m + n \log n)$.

We turn to *expected* running times next. There are two ways to compute expected running times. Either one postulates a probability distribution on the inputs or the algorithm is randomized, i.e., uses random choices internally.

Assume first that we have a probability distribution on the inputs, i.e., if x is any conceivable input of size n then $prob(x)$ is the probability that x actually occurs as an input. The expected running time $\bar{T}(n)$ is computed as a weighted sum $\bar{T}(n) = \sum_x prob(x) \cdot T(x)$, where x ranges over all inputs of size n and $T(x)$ denotes the running time on input x. We refer the reader to any of the textbooks [AHU83, CLR90, Meh84b] for a more detailed treatment. We usually assume the *uniform distribution*, i.e., if x and y are two inputs of the same size then $prob(x) = prob(y)$. It is time for an example.

The expected access time for maps is constant. A *map<I, E>* realizes a partial function m from some type I to some other type E; the index type I must be either the type *int* or a pointer or item type. Let D be the domain of m, i.e., the set of arguments for which m is defined. The uniform distribution assumption is then that all subsets D of I of size n are equally likely. The average running time is computed with respect to this distribution.

Two words of caution are in order at this point. Small average running time does not preclude the possibility of outliers, i.e., inputs for which the actual running time exceeds the average running time by a large amount. Also, average running time is stated with respect to a particular probability distribution on the inputs. This distribution is probably not the distribution from which your inputs are drawn. So be careful.

A *randomized* algorithm uses random choices to control its execution. For example, one of our convex hull algorithms takes as input a set of points in the plane, permutes the points randomly, and then computes the hull in an incremental fashion. The running time and maybe also the output of a randomized algorithm depends on the random choices made. Averaging over the random choices yields the expected running time of the algorithm. Note that we are only averaging with respect to the random choices made by the algorithm, and do not average with respect to inputs. In fact, time bound of randomized algorithms are worst-case with respect to inputs. As of this writing all randomized algorithms in LEDA are of the so-called *Las Vegas* style, i.e., their output is independent of the random choices made. For example, the convex hull algorithm always computes the convex hull. If the output of a randomized algorithm depends on the random choices then the algorithm is called *Monte Carlo* style. An example of a Monte Carlo style randomized algorithm is the primality tests of Solovay and Strassen [SS77] and Rabin [Rab80]. They take two integers n

and s and test the primality of n. If the algorithms declare n non-prime then n is non-prime. If they declare n prime then this answer is correct with probability at least $1-2^{-s}$, i.e., there is chance that the answer is incorrect. However, this chance is miniscule (less than 2^{-100} for $s = 100$). The expected running time is $O(s \log^3 n)$.

2.2 Item Types

Item types are ubiquitous in LEDA. We have dic_items (= items in dictionaries), pq_items (= items in priority queues), nodes and edges (= items in graphs), points, segments, and lines (= basic geometric items), and many others. What is an item?

Items are simply addresses of containers and item variables are variables that can store items. In other words, item types are essentially C++ pointer types. We say essentially, because some item types are not implemented as pointer types. We come back to this point below.

A (value of type) *dic_item* is the address of a dic_container and a (value of type) *point* is the address of a point_container. A dic_container has a key and an information field and additional fields that are needed for the data structure underlying the dictionary and a point_container has fields for the x- and y-coordinate and additional fields for internal use. In C++ notation we have as a first approximation (the details are different):

```
class dic_container
{ K key;
  I inf;
  // additional fields required for the underlying data structure
}
typedef dic_container* dic_item;
class point_container
{ double x, y;
  // additional fields required for internal use
}
typedef point_container* point;
   // Warning: this is NOT the actual definition of point
```

We distinguish between *dependent* and *independent* item types. The containers corresponding to a dependent item type can only live as part of a collection of containers, e.g., a dictionary-container can only exist as part of a dictionary, a priority-queue-container can only exist as part of a priority queue, and a node-container can only exist as part of a graph. A container of an independent item type is self-sufficient and needs no "parent type" to justify its existence. Points, segments, and lines are examples of independent item types. We discuss the common properties of all item types now and treat the special properties of dependent and independent item types afterwards. We call an item of an independent or dependent item type an independent or dependent item, respectively.

An item is the address of a container. We refer to the values stored in the container as *attributes* of the item, e.g., a point has an x- and a y-coordinate and a dic_item has a key and an information. We have functions that allow us to read the attributes of an item. For a point p, *p.xcoord*( ) returns the x-coordinate of the point, for a segment s, *s.start*( ) returns the start point of the segment, and for a dic_item *it* which is part of a dictionary D, *D.key*(*it*) returns the key of the item. Note the syntactic difference: for dependent items the parent object is the main argument of the access function and for independent items the item itself is the main argument.

We will systematically blur the distinction between items and containers. The previous paragraph was the first step. We write "a point has an x-coordinate" instead of the more verbose "a point refers to a container which stores an x-coordinate" and "a dic_item has a key" instead of the more verbose "a dic_item refers to a container that stores a key". We also say "a dic_item which is part of a dictionary D" instead of the more verbose "a dic_item that refers to a container that is part of a dictionary D". We will see more examples below. For example, we say that an insert *D.insert*(k, i) into a dictionary "adds an item with key k and information i to the dictionary and returns it" instead of the more verbose "adds a container with key k and information i to the dictionary and returns the address of the container". Our shorthand makes many statements shorter and easier to read but can sometimes cause confusion. Going back to the longhand should always resolve the confusion.

We said above that item types are essentially C++ pointer types. The actual implementation may be different and frequently is. In the current implementation of LEDA all dependent item types are realized directly as pointer types, e.g., the type *dic_item* is defined as *dic_container*∗, and all independent item types are realized as classes whose only data member is a pointer to the corresponding container class.

The reason for the distinction is storage management which is harder for containers associated with independent item types. For example, a dictionary-container can be returned to free store precisely if it is either deleted from the dictionary containing it or if the lifetime of the dictionary containing it ends. Both situations are easily recognized. On the other hand, a point-container can be returned to free store if no point points to it anymore. In order to recognize this situation we make every point-container know how many points point to it. This is called a reference count. The count is updated by the operations on points, e.g., an assignment $p = q$ increases the count of the container pointed to by q and decreases the count of the container pointed to by p. When the count of a container reaches zero it can be returned to free store. In order to make all of this transparent to the user of type *point* it is necessary to encapsulate the pointer in a class and to redefine the pointer operations assignment and access. This technique is known under the name *handle types* and is discussed in detail in Section 13.7.

All item types offer the assignment operator and the equality predicate. Assume that T is an item type and that *it1* and *it2* are variables of type T. The assignment

```
it1 = it2;
```

assigns the value of *it2* to *it1* and returns a reference to *it1*. This is simply the assignment between pointers. In the case of handle types the assignment has the side effect of updating the reference counters of the objects pointed to by *it1* and *it2*.

The equality predicate (operator *bool operator* == (*const T&*, *const T&*)) is more subtle. For dependent item types it is the equality between values (i.e., pointers) but for independent item types it is usually defined differently. For example, two points in the Euclidean plane are equal if they agree in their Euclidean coordinates.

```
point p(2.0,3.0);  // a point with coordinates 2.0 and 3.0
point q(2.0,3.0);  // another point with the same coordinates
p == q;            // evaluates to true
```

Note that p and q are not equal as pointers. They point to distinct point-containers. However, they agree in their Euclidean coordinates and therefore the two points are said to be equal. For independent item types we also have the *identity* predicate (realized by function *bool identical*(*const T&*, *const T&*)). It tests for equality of values (i.e., pointers). Thus *identical*(p, q) evaluates to false. We summarize in:

LEDA Rule 2

(a) *For independent item types the identity predicate is equality between values. The equality predicate is defined individually for each item type. It is usually equality between attributes.*

(b) *For dependent item types the equality predicate is equality between values.*

2.2.1 *Dependent Item Types*

Many advanced data types in LEDA are defined as collections of items, e.g., a dictionary is a collection of dic_items and a graph is defined in terms of nodes and edges. This collection usually has some combinatorial structure imposed on it, e.g., it may be arranged in the form of a sequence, or in the form of a tree, or in the form of a general graph. We give some examples.

An instance of type *dictionary*$<K, I>$ is a collection of dic_items, each of which has an associated key of type K and an associated information of type I. The keys of distinct items are distinct. We use $\langle k, i\rangle$ to denote an item with key k and information i.

An instance of type *list*$<E>$ is a sequence of list_items, each of which has an associated information of type E. We use $\langle e\rangle$ to denote an item with information e.

An instance of type *sortseq*$<K, I>$ is a sequence of seq_items, each of which has an associated key of type K and an associated information of type I. The key type K must be linearly ordered and the keys of the items in the sequence increase monotonically from front to rear. We use $\langle k, i\rangle$ to denote an item with key k and information i.

An instance of type *graph* is a list of nodes and a list of edges. Each edge has a source node and a target node. We use (v, w) to denote an edge with source v and target w.

An instance of type *partition* is a collection of partition_items and a partition of these items into so-called *blocks*.

In all examples above an instance of the complex data type is a collection of items. This collection has some combinatorial structure: lists and sorted sequences are sequences of items, the items of a partition are arranged into disjoint blocks, and the nodes and edges of a graph form a graph. The items have zero or more attributes: dic_items and seq_items have a key and an information, an edge has a source and a target node, whereas a partition_item has no attribute. An attribute either helps to define the combinatorial structure, as in the case of graphs, or associates additional information with an item, as in the case of dictionaries, lists, and sorted sequences. The combinatorial structure is either defined by referring to standard mathematical concepts, such as set, sequence, or tree, or by using attributes, e.g., an edge has a source and a target. The values of the attributes belong to certain types; these types are usually type parameters. The type parameters and the attribute values may have to fulfill certain constraints, e.g., sorted sequences require their key type to be linearly ordered, dictionaries require the keys of distinct items to be distinct, and the keys of the items in a sorted sequence must be monotonically increasing from front to rear.

Many operations on dictionaries (and similarly, for the other complex data types of LEDA) have items in their interface, e.g., an *insert* into a dictionary returns an item, and a *change_inf* takes an item and a new value for its associated information. Why have we chosen this design which deviates from the specifications usually made in data structure text books? The main reason is efficiency.

Consider the following popular alternative. It defines a dictionary as a partial function from some type K to some other type I, or alternatively, as a set of pairs from $K \times I$, i.e., as the graph of the function. In an implementation each pair (k, i) in the dictionary is stored in some location of memory. It is frequently useful that the pair (k, i) cannot only be accessed through the key k but also through the location where it is stored, e.g., we may want to lookup the information i associated with key k (this involves a search in the data structure), then compute with the value i a new value i', and finally associate the new value with k. This either involves another search in the data structure or, if the lookup returned the location where the pair (k, i) is stored, it can be done by direct access. Of course, the second solution is more efficient and we therefore wanted to support it in LEDA.

We provide direct access through dic_items. A dic_item is the address of a dictionary container and can be stored in a dic_item variable. The key and information stored in a dictionary container can be accessed directly through a dic_item variable.

Doesn't this introduce all the dangers of pointers, e.g., the potential to change information which is essential to the correct functioning of the underlying data structure? The answer is no, because the access to dictionary containers through dictionary items is restricted, e.g., the access to a key of a dictionary container is read-only. In this way, items give the efficiency of pointers but exclude most of their misuse, e.g., given a dic_item its associated key and information can be accessed in constant time, i.e., we have the efficiency of pointer access, but the key of a dic_item cannot be changed (as this would probably corrupt the underlying data structure), i.e., one of the dangers of pointers is avoided. The wish to have the efficiency of pointer access without its dangers was our main motivation for introducing items into the signatures of operations on complex data types.

Let us next see some operations involving items. We use dictionaries as a typical example. The operations

```
dic_item D.lookup(K k);
I        D.inf(dic_item it);
void     D.change_inf(dic_item it,I j);
```

have the following semantics: *D.lookup*(k) returns the item[2], say *it*, with key k in dictionary D, *D.inf*(*it*) extracts the information from *it*, and a new information j is associated with *it* by *D.change_inf*(*it*, j). Note that only the first operation involves a search in the data structure realizing D and that the other two operations access the item directly.

Let us have a look at the insert operation for dictionaries next:

```
dic_item D.insert(K k,I i);
```

There are two cases to consider. If D contains an item *it* whose key is equal to k then the information associated with *it* is changed to i and *it* is returned. If D contains no such item, then a *new* container, i.e., a container which is not part in any dictionary, is added to D, this container is made to contain (k, i), and its address is returned. In the specification of dictionaries all of this is abbreviated to

dic_item *D.insert*(*K k*, *I i*)	associates the information i with the key k. If there is an item $\langle k, j\rangle$ in D then j is replaced by i, else a new item $\langle k, i\rangle$ is added to D. In both cases the item is returned.

For any dependent item type the set of values of the type contains the special value *nil*[3]. This value never belongs to any collection and no attributes are ever defined for it. We use it frequently as the return value for function calls that fail in some sense. For example *D.lookup*(k) returns *nil* if there is no item with key k in D.

Containers corresponding to dependent item types cannot exist outside collections. Assume, for example, that the container referred to by dic_item *it* belongs to some dictionary D and is deleted from D by *D.del_item*(*it*). This removes the container from D and destroys it. It is now illegal[4] to access the fields of this container.

LEDA Rule 3 *It is illegal to access the attributes of an item which refers to a container that has been destroyed or to access the attributes of the item nil.*

In the definition of operations involving items this axiom frequently appears in the form of a precondition.

I *D.inf*(*dic_item it*)	returns the information of item *it*. *Precondition*: *it* must belong to dictionary D.

[2] The operation returns *nil* if there is no item with key k in D.
[3] Recall that all dependent item types are pointer types internally.
[4] Of course, as in ordinary life, illegal actions can be performed anyway. The outcome of an illegal action is hard to predict. You may be lucky and read the values that existed before the container was destroyed, or you may be unlucky and read some random value, or you might get caught and generate a segmentation fault.

2.2.2 *Independent Item Types*

We now come to independent item types. Points, lines, segments, integers, rationals, and reals are examples of independent item types. We discuss points.

A point is an item with two attributes of type double, called the x- and y-coordinate of the point, respectively[5]. We use (a, b) to denote a point with x-coordinate a and y-coordinate b.

Note that we are not saying that a point is a pair of doubles. We say: a point is an item and this item has two double attributes, namely the coordinates of the point. In other words, a point is logically a pointer to a container that contains two doubles (and additional fields for internal use). This design has several desirable implications:

- Assignment between points takes constant time. This is particularly important for types where the attributes are large, e.g., arbitrary precision integers.
- Points can be tested for identity (= same pointer value) and for equality (= same attribute values). The identity test is cheap.
- The storage management for points and all other independent item types is transparent to the LEDA user.

We have functions to query the attributes of a point: *p.xcoord*( ) returns the x-coordinate and *p.ycoord*( ) returns the y-coordinate. We also have operations to construct new points from already constructed points, e.g.,

```
point p.translate(double a,double b);
```

returns a new point $(p.xcoord(\,) + a,\ p.ycoord(\,) + b)$, i.e., it returns an item with attributes $p.xcoord(\,) + a$ and $p.ycoord(\,) + b$. It is important to note that *translate* does not change the point p. In fact, there is no operation on points that changes the attributes of an already existing point. This is true for all independent item types.

LEDA Rule 4 *Independent item types offer no operations that allow to change attributes; the attributes are immutable.*

We were led to this rule by programs of the following kind (which is not a LEDA program):

```
q = p;
p.change_x(a); // change x-coordinate of p to a
```

After the assignment q and p point to the same point-container and hence changing p's x-coordinate also changes q's x-coordinate, a dangerous side-effect that can lead to errors that are very hard to find[6]. We therefore wanted to exclude this possibility of error. We explored two alternatives. The first alternative redefines the semantics of the assignment statement to mean component-wise assignment and the second alternative forbids operations that change

[5] There are also points with rational coordinates and points in higher dimensional space.

[6] Both authors spent many hours finding errors of this kind.

attributes. We explored both alternatives in a number of example programs, adopted the second alternative[7], and casted it into the rule above.

A definition of an independent item always initializes all attributes of the item. For example,

```
point p(2.0,3.0);
point q;            // q has coordinates but it is not known which.
```

defines a point p with coordinates (2.0, 3.0) and a point q. The coordinates of point q are defined but their exact value is undetermined. This is the same convention as for built-in types.

LEDA Rule 5 *The attributes of an independent item are always defined. In particular, definition with default initialization initializes all attributes. A type may specify the initial values but it does not have to.*

We explored alternatives to this rule. For example, we considered the rule that the initial value of an attribute is always the default value of the corresponding type. This rule sounds elegant but we did not adopt it because of the following example. We mentioned already that the default value of type *double* is undefined and that the default value of type *rational* is zero. Thus a point with rational coordinates (type *rat_point*) would be initialized to the origin and a point with floating point coordinates (type *point*) would be initialized to some unspecified point. This would be confusing and a source of error. The rule above helps to avoid this error by encouraging the practice that objects of an independent item type are to be initialized explicitly.

2.3 Copy, Assignment, and Value Parameters

We now come to a central concept of C++ and hence LEDA, the notion of a *copy*. Its importance stems from the fact that several other key concepts are defined in terms of it, namely assignment, creation with initialization by copying, parameter passing by value, and function value return. We give these definitions first and only afterwards define what it means to copy a value. At the end of the section we also establish a relation between destruction and copying.

We distinguish between primitive types and non-primitive types. All built-in types, all pointer types, and all item types are primitive. For primitive types the definition of a copy is trivial, for non-primitive types the definition is somewhat involved. Fortunately, most LEDA users will never feel the need to copy a non-primitive object and hence can skip the non-trivial parts of this section.

We start by defining assignment and creation with initialization by copying in terms of copying. This will also reveal a close connection between assignment and creation with

[7] This does not preclude the possibility that other examples would have led us to a different conclusion.

initialization. The designers of C++ decided that definition with initialization is defined in terms of copy and we decided that assignment should also be defined in terms of copy. Observe that C++ allows one to implement the assignment operator for a class in an arbitrary way. We decided that the assignment operator should have a uniform semantics for all LEDA types.

LEDA Rule 6 *An assignment* `x = A` *assigns a copy of the value of expression A to the variable x.*

C++ Axiom 1 *A definition* `T x = A` *creates a new variable x of type T and initializes it with a copy of the value of A. An alternative syntactic form is* `T x(A)`. *The statement* `new T(A)` *returns a pointer to a newly created anonymous object of type T. The object is initialized with a copy of the value of A.*

The axioms above imply that the code fragments `T x; x = A` and `T x = A` are equivalent, i.e., creation with default initialization followed by an assignment is equivalent to creation with initialization by copying[8]. The next axiom ties parameter passing by value and value return to definition with initialization and hence to copying.

C++ Axiom 2
a) A value parameter of type T and name x is specified as `T x`. *Let A be an actual parameter, i.e., A is an expression of type T. Parameter passing is equivalent to the definition* `T x = A`.
b) Let f be a function with return type T and let `return A` *be a return statement in the body of f; A is an expression of type T. Function value return is equivalent to the definition* `T x = A` *where x is a name invented by the compiler. x is called a temporary variable.*

Now that we have seen so many references to the notion of copy of a value, it is time to define it. A copy of a natural number is simply the number itself. More generally, this is true for all so-called *primitive* types.

LEDA Rule 7

(a) *All built-in types, all pointer types, and all item types are primitive.*
(b) *A copy of a value of a primitive type is the value itself.*

We conclude, that the primitive types behave exactly like the built-in types and hence if you understand what copy, assignment, parameter passing by value, and function value return mean for the built-in types, you also understand them for all primitive types. For non-primitive types the definition of a copy is more complex and making a copy is usually a non-constant time operation. Fortunately, the copy operation for non-primitive types is rarely needed. We give the following advice.

Advice: Avoid assignment, initialization by copying, parameter passing by value, and

[8] This assumes that both kinds of creations are defined for the type T.

function value return for non-primitive types. Also exercise care when using a non-primitive type as an actual type parameter.

```
// read on, if you plan to use any of the statements below
L1 = L2;                            // L1 and L2 are lists
int f(list<int> A);                 // non-primitive value parameter
list<int> f();                      // non-primitive return value
dictionary<string,list<int> > D; // non-primitive type parameter
```

The values of non-primitive types exhibit structure, e.g., a value of type *stack*<*E*> is a sequence of elements of type E, a value of type *array*<*E*> is a set of variables of type E indexed by an interval of integers, and a value of type *list*<*E*> is a sequence of list items each with an associated element of type E. Therefore, non-primitive types are also called *structured.* A copy of a value of a structured type is similar but not identical to the original in the same sense as the Xerox-copy of a piece of paper is similar but not identical to the original; it has the same content but is on a different piece of paper.

We distinguish two kinds of structured types, *item-based* and *non-item-based.* A structured type is called item-based if its values are defined as collections of items. Dictionaries, sorted sequences, and lists are examples of item-based structured types, and arrays and sets are examples of non-item-based structured types. We also say *simple-structured* type instead of non-item-based structured type.

LEDA Rule 8

(a) A value x of a simple-structured type is a set or sequence of elements or variables of some type E. A copy of x is a component-wise copy.

(b) A copy of a variable is a new variable of the same type, initialized with a copy of the value of the original.

We give some examples. Copying the stack (1, 4, 2) produces the stack (1, 4, 2), copying an *array*<*int*> with index set [1 .. 3] means creating three new integer variables indexed by the integers one to three and initializing the variables with copies of the values of the corresponding variable in the original, and copying a *stack*<*dictionary*<K, I> $*$ > produces a stack with the same length and the same pointer values. The following code fragment shows that a copy of a value of a structured type is distinct from the original.

```
array<int> A(0,2);
array<int> B = A;
int* p = A[0];
int* q = B[0];
p == q;           // evaluates to false
```

We next turn to item-based structured types.

LEDA Rule 9 *A value of an item-based structured type is a structured collection of items each of which has zero or more attributes. A copy of such a value is a collection of new items, one for each item in the original. The combinatorial structure imposed on the new items is isomorphic to the structure of the original. Every attribute of a new item which* *does not encode combinatorial structure is set to a copy of the corresponding attribute of the corresponding item in the original.*

Again we give some examples. Copying a *list<E>* of length 5 means creating five new list items, arranging these items in the form of a list, and setting the contents of the i-th new item, $1 \leq i \leq 5$, to a copy of the contents of the i-th item in the original. To copy a graph (type *graph*) with n nodes and m edges means creating n new nodes and m new edges and creating the isomorphic graph structure on them. To copy a *GRAPH<E1, E2>*[9] means copying the underlying graph and associating with each new node or edge a copy of the variable associated with the corresponding original node or edge. According to LEDA Rule 8 this means creating a new variable and initializing it with a copy of the value of the old variable.

The programming language literature sometimes uses the notions of *shallow* and *deep copy*. We want to relate these notions to the LEDA concept of a copy. Consider a structure *node_container* consisting of a pointer to a node container and a pointer to some other type.

```
class node_container
{ node_container* succ;
  E*              content;
}
```

Such a structure may, for example, arise in the implementation of a singly linked list; one pointer is used for the successor node and the other pointer is used for the the content, i.e., the list has type *list<E *>* for some type E. A shallow copy of a node is a new node whose two fields are initialized by component-wise assignment. A deep copy of a node is a copy of the entire region of storage reachable from the node, i.e., both kinds of pointers are followed when making a deep copy. In other words, a shallow copy follows no pointer, a deep copy follows all pointers. Our notion of copying is more semantically oriented. Copying a *list<E *>* of n items means creating n new items (this involves following the successor pointers), establishing a list structure on them, and setting the content attribute of each item to a copy of the contents of the corresponding item in the original. Since the type $E*$ is primitive (recall that all pointer types are primitive) this is tantamount to setting the contents of any new item to the contents of the corresponding old item. In particular, no copying of values of type E takes place. In other words, when making a copy of a *list<E *>* we follow successor pointers as if making a deep copy, but we do not follow the $E*$ pointers as if making a shallow copy.

Parameter passing by value involves copying. Since most arguments to operations on complex data types have value parameters, this has to be taken into account when reading the specifications of operations on data types. Consider, for example, the operation *D.insert*(k, i) for dictionaries. It takes a key k and an information i, adds a new item $\langle k, i\rangle$ to D and returns the new item[10]. Actually, this is not quite true. The truth is that the new

[9] A *GRAPH<E1, E2>* is a graph where each node and edge has an associated variable of type *E1* and *E2*, respectively.

[10] We assume for simplicity, that D contains no item with key k.

item contains a copy of k and a copy of i. For primitive types a value and a copy of it are identical and hence the sentence specifying the semantics of *insert* can be taken literally. For non-primitive types copies and originals are distinct and hence the sentence specifying the semantics of *insert* is misleading. We should say "adds a new item ⟨copy of k, copy of i⟩ to D" instead of "adds a new item $\langle k, i\rangle$ to D". We have decided to suppress the words "copy of" for the sake of brevity[11]. The following example shows the effect of copying.

```
dictionary<string,dictionary<int,int> > M;
dictionary<int,int> D;
dic_item it = D.insert(1,1);

M.insert("Ulli",D);
M.lookup("Ulli").inf(it);    // illegal
D.change_inf(it,2);
M.lookup("Ulli").access(1); // returns 1
D.insert(2,2);
M.lookup("Ulli").lookup(2); // returns nil
```

The insertion of D into M stores a copy of D in M. The item *it* belongs to D but not to the copy of D. Thus querying its *inf*-attribute in the copy of D returned by *M.lookup*("*Ulli*") is illegal. The operation *D.change_inf*(*it*, 2) changes the *inf*-attribute of *it* to 2; this has no effect on the copy of D stored in M and hence the access operation in the next line returns 1. Similarly, the second insertion into D has no effect on the copy and hence the lookup in the last line returns *nil*.

When the lifetime of an object ends it is *destructed*. The lifetime of a named object ends either at the end of the block where it was defined (this is the default rule) or when the program terminates (if declared static). The life of an anonymous object is ended by a call of *delete*. We need to say what it means to destruct an object. For LEDA-objects there is a simple rule.

LEDA Rule 10 *When a LEDA-object is destructed the space allocated for the object is freed. This is exactly the space that would be copied when a copy of the object were made.*

2.4 More on Argument Passing and Function Value Return

C++ knows two kinds of parameter passing, by value and by reference. Similarly, a function may return its result by value or by reference. We have already discussed value arguments and value results. We now review reference arguments and reference results and at the end of the section discuss functions as arguments. This section contains no material that is

[11] In the early versions of LEDA only primitive types were allowed as type parameters and hence there was no need for the words "copy of". When we allowed non-primitive types as type parameters we decided to leave the specification of *insert* and many other operations unchanged and to only make one global remark.

specific for LEDA; it is just a short review of reference parameters, reference results, and function arguments in C++.

The specification of a formal parameter has one of the three forms:

```
T x            (value parameter of type T),
T& x           (reference parameter of type T),
const T& x     (constant reference parameter of type T).
```

The qualifier `const` in the last form specifies that it is illegal to modify the value of the parameter in the body of the procedure. The compiler attempts to verify that this is indeed the case. Let A be the actual parameter corresponding to formal parameter x. Parameter passing is tantamount to the definition `T x = A` in the case of a value parameter and to the definition `T& x = A` in the case of a reference parameter. We already know the semantics of `T x = A`: a new variable x of type T is created and initialized with a copy of the value of expression A. The definition `T& x = A` does not define a new variable. Rather it introduces x as an additional name for the object denoted by A. Note that the argument A must denote an object in the case of a reference parameter. In either case the lifetime of x ends when the function call terminates.

Argument passing by reference must be used for parameters whose value is to be changed by the function. For arguments that are not to be changed by the function one may use either a value parameter[12] or a constant reference parameter. Note, however, that passing by value makes a copy of the argument and that copying a "large" value, e.g., a graph, list, or array, is expensive. Moreover, we usually want the function to work on the original of a value and not on a copy. We therefore advise to specify arguments of non-primitive types either as reference parameters or as constant reference parameters and to use value parameters only for primitive types. In our own code we very rarely pass objects of non-primitive type by value. If we do then we usually add the comment: "Yes, we actually want to work on a copy".

An example for the use of a constant reference parameter is

```
void DIJKSTRA(const graph& G, node s, const edge_array<int>& cost,
              node_array<int>& dist, node_array<edge>& pred)
```

This function[13] takes a graph G, a node s of G, a non-negative cost function on the edges of G, and computes the distance of each vertex from the source (in *dist*). Also for each vertex $v \neq s$, $pred[v]$ is the last edge on a shortest path from s to v. The constant qualifiers ensure that *DIJKSTRA* does not change G and *cost* (although they are reference parameters). What would happen if we changed G to a value parameter? Well, we would pass a copy of G instead of G itself. Since a copy of a graph has new nodes and edges, s is not a node of the copy and *cost* is not defined for the edges of a copy. The function would fail if G was passed by value. Thus, it is essential that G is passed by reference.

Parameter passing moves information into a function and function value return moves

[12] It is legal to assign to a variable that is defined as a value parameter. Such an assignment does not affect the value of the actual parameter.

[13] See Section 6.6 for a detailed discussion of this function.

information out of a function. Consider the call of a function f with return type T or $T\&$ for some type T and assume that the call terminates with the return statement `return A`. The call is equivalent to the definition of a temporary t which is initialized with A, i.e., `return A` amounts to either `T t = A` or `T& t = A`. The temporary replaces the function call.

Let us go through an example. Let T be any type. We define four functions with the four combinations of return value and parameter specification.

```
T f1(T x)   {  return x; }
T f2(T& x)  {  return x; }
T& f3(T& x) {  return x; }
T& f4(T x)  {  return x; }
        // illegal, since a reference to a local variable is returned
```

Let y and z be objects of type T. The statement

```
z = f1(y);
```

copies y three times, first from y to the formal parameter x (value argument), then from x to a temporary t (value return), and finally from t to z (assignment). In

```
z = f2(y);
```

y is copied only twice, first from y to a temporary (value return) and then from the temporary into z (assignment).

```
z = f3(y);
```

copies y once, namely from y into z (assignment). Since *f3* returns a reference to an object of type T it can also be used on the left-hand side of an assignment. So

```
f3(y) = z;
```

assigns z to y.

Some operations take *functions as arguments*. A function argument f with result type T and argument types *T1*, ..., *Tk* is specified as

```
T(*f)(T1,T2,...,Tk)
```

The $*$ reflects the fact that a pointer to the function is passed. As a concrete example let us look at the bucket sort operation on lists with element type E:

```
void L.bucket_sort(int i,int j,int(*f)(E&));
```

requires a function f with a reference parameter of type E that maps each element of L into $[i\,..\,j]$. It sorts the items of L into increasing order according to f, i.e., item $\langle x\rangle$ is before $\langle y\rangle$ after the call if either $f(x) < f(y)$ or $f(x) = f(y)$ and $\langle x\rangle$ precedes $\langle y\rangle$ before the call.

2.5 Iteration

For many data types, LEDA offers *iteration macros* that allow to iterate over the elements of a collection. These macros are similar to the C++ *for*-statement. We give some examples. For all item-based types we have

```
forall_items(it,D)
  { /* the items in D are successively assigned to it */ }
```

This iteration successively assigns all items in *D* to *it* and executes the loop body for each one of them. For lists and sets we also have iteration statements that iterate over elements.

```
// L is a list<point>
point p;
forall(p,L)
  { /* the elements of L are successively assigned to p */ }.
```

For graphs we have statements to iterate over all nodes, all edges, all edges adjacent to a given node, ..., for example:

```
forall_nodes(v,G)
  { /* the nodes of G are successively assigned to v*/ }
forall_edges(e,G)
  { /* the edges of G are successively assigned to e*/ }
forall_adj_edges(e,v)
  { /* all edges adjacent to v are successively assigned to e */ }
```

It is dangerous to modify a collection while iterating over it. We have

LEDA Rule 11 *An iteration over the items in a collection C must not add new items to C. It may delete the item under the iterator, but no other item. The attributes of the items in C can be changed without restriction.*

We give some examples:

```
// L is a list<int>
// delete all occurrences of 5
forall(it,L)
  if ( L[it] == 5 ) L.del(it);
forall(it,L)
  if ( L[it] == 5 ) L.del(L.succ(it));  // illegal
// add 1 to the elements following a 5
forall(it,L)
  if ( L[it] == 5 ) L[L.succ(it)]++;
forall(it,L)
  L.append(1);  // infinite loop
// G is a graph;
//add a new node s and edges (s,v) for all nodes of G
node s = G.new_node();
node v;
forall_nodes(v,G) if (v != s) G.new_edge(s,v);
```

The iterations statements in LEDA are realized by macro expansion. This will be discussed in detail in Section 13.9. We give only one example here to motivate the rule above and the rules to follow. The *forall_items* loop for lists

```
forall_items(it,L) { <<body>> }
```

expands into a C++ for-statement. The expansion process introduces a new variable *loop_it* of type *list_item* and initializes it with the first item of *L*; a distinct variable is generated for every loop by the expansion process. In each iteration of the loop, *loop_it* is assigned to *it*, *loop_it* is advanced, and the loop body is executed. The loop terminates when *it* has the value *nil*.

```
for (list_item loop_it = (L).first_item();
     it = loop_it, loop_it = (L).next_item(loop_it), it;  )
{ <<body>> }
```

The fact that we use macro expansion to reduce the forall-loop to a C++ for-loop has two consequences.

LEDA Rule 12 *Break and continue statements can be used in forall-loops.*

We give an example.

```
list_item it;
forall_items(it,L) if ( L[it] == 5 ) break;
if ( it ) // there is an occurrence of 5 in L
else      // there is no occurrence of 5 in L
```

There is second consequence which is less pleasing. Consider

```
edge e;
forall(e,G.all_edges()) { <<body>> }
```

where the function *G.all_edges*( ) returns a list of all edges of *G*. The expansion process will generate

```
for (list_item loop_it = (G.all_edges()).first_item();
     it=loop_it,loop_it=(G.all_edges()).next_item(loop_it),it;)
{ <<body>> }
```

and hence the function *G.all_edges*( ) is called in every iteration of the loop. This is certainly not what is intended.

LEDA Rule 13 *The data type argument in an iteration statement must not be a function call that produces an object of the data type but an object of the data type itself.*

The correct way to write the loop above is

```
list<edge> E = G.all_edges();
edge e;
forall(e,E) { <<body>> }
```

or even simpler

```
forall_edges(e,G) { <<body>> }
```

2.6 STL Style Iterators

STL (Standard Template Library [MS96]) is a library of basic data types and algorithms that is part of the C++ standard. STL has a concept called *iterators* that is related to, but different from LEDA's item concept. In STL the forall-items loop for a *list<int>* is written as

```
for (list<int>::iterator it = L.begin(); it != L.end(); it++)
{ <<body>> }
```

In the loop body the content of the iterator can be accessed by $*it$; in LEDA one writes $L[it]$ to access the content of *it*.

Many LEDA data structures offer also STL style iterators. This feature is still experimental and we refer the user to the manual for details.

2.7 Data Types and C++

LEDA's implementation base is C++. We show in this section how abstract data types can be realized by the *class mechanism* of C++. We do so by giving a complete implementation of the data type stack which we specified at the beginning of this chapter. We also give the reader a first impression of LEDA's structure and we introduce the reader to Lweb and noweb.

A C++ class consists of *data members* and *function members*. The data members define how the values of the class are represented and the function members define the operations available on the class. Classes may be parameterized. We now define a parameterized class *stack<E>* that realizes the LEDA data type with the same name.

⟨*stack.c*⟩≡

```
template <class E> // E is the type parameter of stack
class stack
{ private:
    ⟨data members⟩
  public:
    ⟨function members⟩
};
```

Figure 2.2 Lweb: lweave transforms a file source.lw into a file source.tex; notangle extracts program files. Lweb is a dialect of noweb [Ram94].

The definition of a class consists of a private part and a public part; the private part is only visible within the class and the public part is also visible outside the class. We declare the data members private to the class and hence invisible outside the class. This emphasizes the fact that we are defining an abstract data type and hence it is irrelevant outside the class how a value is represented in the machine and how the operations are implemented. To further emphasize this fact we give an implementation of stacks in this section that is different from the one actually used in LEDA. The function members are the interface of the class and hence public.

It is time to give more information about Lweb. *Lweb* is the literate programming tool which we use to produce manual pages, implementation reports, and which we used to produce this book. It is dialect of *noweb* [Ram94]. It allows us to write a program and its documentation into a single file (usually with extension .lw) and offers two utilities to produce two views of this file, one for a human reader and one for the C++ compiler: *lweave* typesets program and documentation and creates a file with extension .tex which can then be further processed using TeX and LaTeXand *notangle* extracts the program and puts it into a file (usually with extension .c or .cc or .h). Figure 2.2 visualizes the process.

We postpone the discussion of lweave to Chapter 14 and only discuss notangle here. A noweb-file[14] consists of documentation chunks and code chunks. A documentation chunk starts with `@` followed by a blank or by a carriage return in column one of a line and a code chunk starts with ⟨*name of chunk*⟩= in column one of a line. Code chunks are given names. If several chunks are given the same name they are concatenated. Code chunks are referred to by ⟨*name of chunk*⟩.

In this section we have already defined a chunk ⟨*stack.c*⟩. It refers to chunks ⟨*data members*⟩ and ⟨*function members*⟩ which will be defined below. The command

```
notangle -Rstack.c Foundations.lw > stack.c
```

will extract the chunk `stack.c` (the "R" stands for root) from the file Foundations.lw (the name of the file containing this chapter) and write it into stack.c.

We come back to stacks. We represent a *stack<E>* by a C++ array A of type E and two integers sz and n with $n < sz$. The array A has size sz and the stack consists of elements

14 As far as notangle is concerned there is no difference between a noweb-file (usually with extension .nw) and a Lweb-file.

$A[0]$, $A[1]$, ..., $A[n]$ with $A[n]$ being the top element of the stack. The stack is empty if $n = -1$.

⟨*data members*⟩≡

```
E* A;
int sz;
int n;
```

The function members correspond to the operations available on stacks. We start with the constructors. There are two ways to create a stack: *stack*<*E*> *S* creates an empty stack and *stack*<*E*> $S(X)$ creates a stack whose initial value is a copy of X. The corresponding function members are the so-called *default constructor* and so-called *copy constructor*, respectively. In C++ a constructor has the same name as the class itself, i.e., the constructors of class T have name T. The default constructor has no argument and the copy constructor has a constant reference argument of type T.

⟨*function members*⟩≡

```
stack()                    // default constructor
{ /* we start with an array of ten elements */
  A = new E[10];
  sz = 10;
  n = -1;
}
stack(const stack<E>& X) // copy constructor
{ sz = X.sz;
  A = new E[sz];
  n = X.n;
  for(int i = 0; i <= n; i++) A[i] = X.A[i];
}
```

We give some more functions: *empty* returns *true* if the stack is empty, *top* returns the top element of a non-empty stack, *push* adds an element to a stack, *pop* deletes an element from a non-empty stack and returns it, and = performs assignment. We let *top* check its precondition and call an error-handler when it is violated. However, *pop* does not check its precondition. Recall that LEDA does not promise to check all preconditions.

⟨*function members*⟩+≡

```
int empty()  { return (n == -1); }
E top()
{ if ( n == -1) error_handler(1,"stack::top: stack is empty");
  return A[n];
}
E pop()  { return A[n--]; }
```

A *push* first checks whether there is still room in the array. If not, it doubles the size of A. In either case it increases n and assigns x to the new top element of the stack.

⟨*function members*⟩+≡

```
void push(const E& x)
{ if (n + 1 == sz)
  { sz = 2 * sz;
    E* B = A;
    A = new E[sz];
    for (int i = 0; i <= n; i++) A[i] = B[i];
    delete[] B;
  }
  A[++n] = x;
}
```

An assignment first checks for the trivial assignment $S = S$, then destroys the old value of the left-hand side, copies the right-hand side into the left-hand side, and finally returns a reference to the left-hand side.

⟨*function members*⟩+≡

```
stack<E>& operator=(const stack<E>& X)
{ if (this != &X)
  { delete[] A;
    sz = X.sz;
    A = new E[sz];
    n = X.n;
    for (int i=0; i<=n; i++) A[i] = X.A[i];
  }
  return (*this);
}
```

When the lifetime of a stack ends the array A needs to be deleted.

⟨*function members*⟩+≡

```
~stack()  { delete[] A; }
```

This completes the definition of class *stack<E>*. The class essentially realizes the data type *stack<E>* as defined on page 18; we invite the reader to complete the implementation by writing the code for *clear*.

Our implementation of the stack data type wastes space. Imagine that we perform 1000 pushes followed by 1000 pops. The pushes will increase the size of A to at least 1000 but A does not shrink again during the pops. The LEDA implementation of stacks uses space in a more thrifty way; its space requirement is proportional to the number of elements in the stack.

In this section we gave the reader a first impression of how the data types of LEDA are implemented in C++. Chapter 13 gives the details.

2.8 Type Parameters

Most data types in LEDA are parameterized. We have lists over an arbitrary element type E and dictionaries over any linearly ordered key type K and any information type I. Any class that provides a certain small set of functions can be used as an actual type argument: one must be able to create a variable of the type and initialize it either with the default value (default constructor) or with a copy of an already existing value (copy constructor). One must be able to perform assignment (operator =), to read a value of the type from an input stream (function *Read*), and to print a value onto an output stream (function *Print*). Finally, when the lifetime of an object ends one must be able to destruct it (destructor). Sometimes, type arguments need to have additional abilities. Linearly ordered types have to support comparisons between their elements, hashed types have to support hashing, and numerical types have to support arithmetic.

LEDA Rule 14 *Any actual type argument must provide the following six functions:*

a default constructor	`T::T()`
a copy constructor	`T::T(const T&)`
an assignment operator	`T& T::operator=(const T&)`
a read function	`void Read(T&,istream&)`
a print function	`void Print(const T&,ostream&)`
a destructor	`T::~T().`

A linearly ordered type must in addition provide

a compare function	`int compare(const T&,const T&).`

A hashed type must in addition provide

a hash function	`int Hash(const T&)`
an equality operator	`bool operator ==(const T&,const T&).`

A numerical type must in addition have the basic arithmetic functions addition, subtraction, and multiplication, and the standard comparison operators.

We have already discussed the default constructor, the copy constructor, the destructor, and the assignment operator. The functions *Read* and *Print* read an object of type T from an input stream and print it to an output stream, respectively. Equality and the functions *compare*, *Hash* are discussed in the next section and number types are discussed in Chapter 4. We next give the complete definition of a linearly ordered class *pair*.

⟨*definition of class pair*⟩≡

```
class pair
{ double x, y;
public:
  pair() {  x = y = 0; }
  pair(const pair& p)  {  x = p.x; y = p.y; }
  friend void Read(pair& p,istream& is) { is >> p.x >> p.y; }
  friend void Print(const pair& p,ostream& os)
                                        { os << p.x << " " << p.y; }
  friend int compare(const pair&,const pair&);
};
```

```
int compare(const pair& p,const pair& q)
{ if (p.x < q.x) return - 1;
  if (p.x > q.x) return  1;
  if (p.y < q.y) return - 1;
  if (p.y > q.y) return  1;
  return 0;
}
```

We need to make two remarks about the definition of the class *pair*. (1) The functions *Read*, *Print*, and *compare* are not member functions of the class, but global functions. They are declared as friends of *pair* so that they can access the private data of the class. (2) We did not define two of the required functions, namely the assignment operator and the destructor ~*pair*. The reason is that C++ will generate them automatically. More precisely, if no copy constructor, assignment operator, or destructor is defined then the default version is used. The default version copies component-wise, assigns component-wise, and destructs component-wise, respectively. Thus the definition of the copy constructor could also be omitted from class *pair*.

The type *pair* can be used as the key type in a dictionary, i.e., we may define

```
dictionary<pair,int> D;
```

What happens if one uses a class *T* as an actual type parameter without defining one of the required functions (that are not generated automatically)? The C++ compiler will produce an error message that it cannot match certain functions. For example, the compiler used by the first author produces

```
LEDA/dictionary.h:52: no match for
'_IO_ostream_withassign & << const pair & '
```

when given the following program

⟨*parameterized_data_type_test.c*⟩≡

```
#include <LEDA/dictionary.h>
class pair
{ double x;
  double y;
public:
  pair() {  x = y = 0; }
  pair(const pair& p)  {  x = p.x; y = p.y; }
};
main(){
  dictionary<pair,int> D;
}
```

2.9 Memory Management

LEDA provides an efficient *memory management system* that is used for all node, edge, and item types and that can easily be customized for user-defined classes by means of the `LEDA_MEMORY` macro. One simply has to add the macro call `LEDA_MEMORY(T)` to the definition of class T. This call creates *new* and *delete* operators for the class T that rely on LEDA's memory manager. The main advantages over the built-in *new* and *delete* operators are:

- Memory is allocated in big chunks and thus frequent and costly calls to the memory allocator are avoided.
- Memory returned by the *delete* operator is reused by later calls of the *new* operator, i.e., the manager provides garbage collection.

The implementation of LEDA's memory manager is discussed in Section 13.8. The definition of our class *pair* now reads as follows. We advise the reader to follow this scheme in the definition of his classes.

⟨*refined definition of class pair*⟩≡
```
class pair
{ private:
    double x, y;
  public:
    pair() {  x = y = 0; }
    /* pair uses the default versions of copy constructor,
       assignment operator, and destructor */
    friend void Read(pair& p,istream& is) { is >> p.x >> p.y; }
    friend void Print(const pair& p,ostream& os)
      { os << p.x << " " << p.y; }
    friend int compare(const pair&,const pair&);
  LEDA_MEMORY(pair);
};
```

2.10 Linearly Ordered Types, Equality and Hashed Types

Algorithms frequently need to compare objects: a geometric algorithm may have to determine whether one line is above another line at a certain x-value, a sorting algorithm needs to compare the objects it is supposed to sort, and a shortest path algorithm needs to compare the lengths of two paths. Also, many data types such as dictionaries, priority queues, and sorted sequences need to compare the objects of their key type. The appropriate mathematical concept is a linear order.

A binary relation $\leq$ (less than or equal) on a set S is called a *linear order* if the following three conditions hold for all x, y, $z \in S$:

- $x \le x$ (reflexivity).
- $x \le y$ and $y \le z$ implies $x \le z$ (transitivity).
- $x \le y$ or $y \le x$ (anti-symmetry).

Note that the "or" in the third condition is not exclusive. We may have $x \le y$ and $y \le x$ even if x and y are distinct. Here is an example. For non-vertical lines g and h, define $g \le h$ if the intersection of g with the y-axis is below or equal to the intersection of h with the y-axis. Then $g \le h$ and $h \le g$ iff g and h intersect the y-axis in the same point.

We call x and y *equivalent* if $x \le y$ and $y \le x$ and we say that x is *strictly less than* y and write $x < y$ or $y > x$ if $x \le y$ and x and y are not equivalent. Note that for any two elements x and y exactly one of the following three relations holds: x is strictly less than y, x is equivalent to y, or y is strictly less than x.

In LEDA, a function *int cmp(const T&, const T&)* is said to realize a linear order on the type T if there is a linear order $\le$ on T such that for all x and y in T

$$cmp(x, y) \begin{cases} < 0, & \text{if } x < y \\ = 0, & \text{if } x \text{ is equivalent to } y \\ > 0, & \text{if } x > y \end{cases}$$

LEDA Rule 15 *A type T is called* linearly ordered *if the function*

```
int compare(const T&,const T&)
```

is defined for the type T and realizes a linear order on T. If compare(x, y) returns zero for two objects x and y then they are called compare-equivalent *or simply* equivalent.

Note that we have adopted the syntactic convention that the function with the name *compare* defines the order on T. This is in line with similar conventions already used in C++, e.g., that constructors have the same name as the type.

For many primitive data types a function *compare* is predefined and defines the so-called *default ordering* of the type. The default ordering is the usual "less than or equal" for the numerical types, the lexicographic ordering for strings, and the lexicographic ordering of the Cartesian coordinates for points. For all other types T there is no default ordering, and the user has to define the function *compare* if a linear order on T is required. We already gave an example in the preceding section.

A weaker concept than linear orders is equivalence relation. A binary relation R defines an *equivalence relation* on a set S if the following three conditions hold for all $x, y, z \in S$:

- xRx (reflexivity).
- xRy and yRz implies xRz (transitivity).
- xRy implies yRx (symmetry).

We have already seen an equivalence relation, namely compare-equivalence. The relation R defined by $x R y$ if $compare(x, y) == 0$ defines an equivalence relation. We also require

LEDA Rule 16 *If the equality operator*

```
bool operator==(const T&,const T&)
```

is defined for a class T then it defines an equivalence relation on T. We call x and y equal if $x == y$ evaluates to true.

We require *no* relationship between equality and compare-equivalence, i.e., two objects may be equal but not compare-equivalent or compare-equivalent but not equal. However, for all LEDA types with predefined *compare* and == the two notions agree. On the other hand, there are applications where it is natural to distinguish between the two concepts. For example, a plane sweep algorithm for line segment intersection (cf. Section 10.7.2) compares segments by the y-coordinate of their intersection with a vertical sweep line and thus two segments can be compare-equivalent without being equal.

We next turn to hashed types. A hashed type T must provide the equality operator and the function *int Hash(const T&)*. Of course, the hash function should not tell objects apart that are equal.

LEDA Rule 17 *For any hashed type and any objects x and y of type T: if $x == y$ then $Hash(x) == Hash(y)$.*

There is one further point that we have to make. Recall that, for example, a dictionary stores copies of keys (and informations) and that for structured types a copy of a value is distinct from the original. It is possible to write compare functions and equality operators that distinguish between a value and a copy of the value. This would lead to a disaster, e.g., a lookup in a dictionary would fail to find a stored key. We therefore have

LEDA Rule 18 *A value and a copy of a value must be compare-equivalent and equal.*

For primitive types, this axiom is trivially fulfilled since a copy is identical to the original.

In some situations it is useful to have more than one linear order for a type T. For example, we might want to have two dictionaries *D1* and *D2* with key type *pair*. In *D1* the pairs are to be ordered by the lexicographic ordering of their Cartesian coordinates and in *D2* by the lexicographic ordering of their polar coordinates. The dictionary *D1* is easy to define. We simply write

```
dictionary<pair,int> D1,
```

but how can we define the second dictionary? After all, we have the syntactic convention that the function with the name *compare* defines the order on a type. There are two solutions, one old and one added recently.

The first solution is to define an equivalent type with the alternative ordering. The code sequence

```
int pol_cmp(const point& x,const point& y)
{ /* compute lexicographic ordering by polar coordinates */ }
DEFINE_LINEAR_ORDER(point,pol_cmp,pol_point);
dictionary<pol_point,int> D2;
```

first defines the ordering by polar coordinates and then defines a type *pol_point* by a call to the DEFINE_LINEAR_ORDER macro. The type *pol_point* is equivalent to the type *point*, in particular, a pol_point can be assigned to a point and vice versa. However, the ordering on the type *pol_point* is given by the function *pol_cmp*. The last line defines the desired dictionary *D2*.

The second solution makes the linear order an additional argument of any data type that requires a linearly ordered type, e.g.,

```
dictionary<point,int> D(pol_cmp);
```

declares a dictionary *D* that uses the function *pol_cmp* for comparing points.

Instead of passing a function to the dictionary, one can also pass a class which has a function operator and is derived from the class *leda_cmp_base*. This variant is helpful when the compare function depends on a global parameter. We give an example. More examples can be found in Sections 10.7.2 and 10.3. Assume that we want to compare edges of a graph *GRAPH<point, int>* (in this type every node has an associated point in the plane; the point associated with a node v is accessed as $G[v]$) according to the distance of their endpoints. We write

⟨*compare_example*⟩≡

```
class cmp_edges_by_length: public leda_cmp_base<edge> {
  const GRAPH<point,int>& G;
public:
  cmp_edges_by_length(const GRAPH<point,int>& g): G(g){}
  int operator()(const edge& e, const edge& f) const
  { point pe = G[G.source(e)]; point qe = G[G.target(e)];
    point pf = G[G.source(f)]; point qf = G[G.target(f)];
    return compare(pe.sqr_dist(qe),pf.sqr_dist(qf));
  }
};
main(){
  GRAPH<point,int> G;
  cmp_edges_by_length cmp(G);
  list<edge> E = G.all_edges();
  E.sort(cmp);
}
```

The class *cmp_edges_by_length* has a function operator that takes two edges e and f of a graph G and compares them according to their length. The graph G is a parameter of the constructor. In the main program we define *cmp(G)* as an instance of *cmp_edges_by_length* and then pass *cmp* as the compare object to the sort function of *list<edge>*. In the implementation of the sort function a comparison between two edges is made by writing *cmp*(e, f), i.e., for the body of the sort function there is no difference whether a function or a compare object is passed to it.

The example above illustrates a nice feature of literate programming. We gave a named program chunk that illustrates a concept of LEDA. Of course, we want to make sure that the program fragment is correct and hence we want to execute it. To this effect we enclose it into a larger program chunk which we can extract and compile. We usually do not show the enclosing program chunk, i.e., we enclose it into a LaTeX command `\ignore` that makes it invisible to LaTeXby expanding to the empty string. We show the construction once:

```
\ignore{
<<compare_test.c>>=

#include <LEDA/graph.h>
#include <LEDA/point.h>

<<compare_example>>

@ }%end ignore
```

2.11 Implementation Parameters

Some data types in LEDA, e.g., dictionary, priority queue, d_array, and sorted sequence, come with several implementations. A user of such a data type can choose a particular implementation by giving the name of the implementation as an additional parameter, e.g., *_d_array*<I, E, *skiplist*> selects the skiplist implementation of dictionary arrays. Note that the type name now starts with an underscore. This is necessary since C++ does not allow us to overload templates. The following program uses the skiplist implementation of dictionary arrays to count word occurrences in the input stream.

```
#include <LEDA/d_array.h>
#include <LEDA/impl/skiplist.h>
main()
{ _d_array<string,int,skiplist> N(0);
  // d_array<string,int> N(0) selects default implementation
  string s;
  while (cin >> s) N[s]++;
  forall_defined(s,N) cout << s << " " << N[s] << endl;
}
```

The types with and without implementation parameter are closely related.

Any type *_T*<*T1*, ..., *Tk*, *xyz_impl*> is derived (in the C++ sense of the word) from the corresponding "normal" parameterized type T<*T1*, ..., *Tk*>. This allows us, for example, to pass an instance of type *_T*<*T1*, ..., *Tk*, *xyz_impl*> as an argument to a function with a formal parameter of type T<*T1*, ..., *Tk*>&, a feature that allows us to execute even pre-compiled algorithms with different implementations of data types. We give an example. We define a procedure *word_count* that has a parameter of type *d_array<string, int>*.

```
void word_count(d_array<string,int>& N)
{ string s;
  while (cin >> s) N[s]++;
  forall_defined(s,N) cout << s << " " << N[s] << endl;
}
```

Any implementation of d_arrays can be passed to *word_count*.

```
d_array<string,int> N1(0);
word_count(N1);
_d_array<string,int,skiplist> N2(0);
word_count(N2);
```

The section “Implementation Parameters” of the LEDA manual surveys the implementation parameters currently available. Section 13.6 discusses the realization of implementation parameters. The latter section also describes how a LEDA user may add his own implementation of a data type to the system.

2.12 Helpful Small Functions

There are a number of small, but helpful, functions. We mention some of them here and refer the reader to the section “Miscellaneous Functions” of the LEDA manual for the full list.

```
int i = read_int("i = ");
```

prints “`i = `” (more generally, its string argument) on standard output and then reads an integer from standard input. Similar functions exist to read strings, character, and doubles.

The function *used_time* is very helpful for running time experiments. For example, the chunk

```
float T = used_time();  // sets T to the current cpu time
// an experiment
cout << used_time(T);
   // sets T to the current cpu time and returns the difference
   // to the previous value of T
// another  experiment
cout << used_time(T);
```

will print the cpu time used in each of two experiments.

The function

```
void print_statistics();
```

prints a summary of the currently used memory. For example, the program

⟨*memory_statistic*⟩≡

```
list<point> L;
{ for (int i = 0; i < 100000; i++) L.append(point());
  list<point> L1 = L;
}
print_statistics();
```

produces

```
STD_MEMORY_MGR (memory status)
     +--------------------------------------------------+
     |   size     used     free    blocks     bytes     |
     +--------------------------------------------------+
     |    12    100000   100214      294    2402568     |
     |    20        27      381        1       8160     |
     |    40    100002       77      493    4003160     |
     +--------------------------------------------------+
     |   time:  0.53 sec              space: 6300.92 kb |
     +--------------------------------------------------+
```

The statistics tell us that space for a total of 100000 + 100214 records of size 12 bytes (= list nodes), for a total of 27 + 381 records of size 20, and for a total of 100002 + 77 records of size 40 (= points) was allocated. It also gives information on which of these records are currently used and which are free. In our example, the records for the nodes of *L* and the points in *L* are still allocated and the records for the nodes of *L1* have already been freed. Observe that the program allocates space for 200000 list nodes, but only for 100000 structures to contain representations of points; read Section 2.2.2 to understand why. Space is allocated in blocks of 8160 bytes. The next to last column shows the number of allocated blocks for the structures of the different sizes and the last column shows the space consumption in bytes. Our program required about 6.3 megabytes. It ran for 0.53 seconds.

The functions

```
T    leda_min(const T& a, const T& b);
T    leda_max(const T& a, const T& b);
void leda_swap(T& a, T& b);
```

return the minimum, the maximum, and swap the values of their arguments, respectively. They can be used for any type T.

Finally, the function

```
double truncate(double x, int k = 10);
```

returns a double whose mantissa is truncated after $k - 1$ bits after the binary point, i.e., if $x \neq 0$ then the binary representation of the mantissa of the result has the form d.ddddddd, where the number of d's is equal to k.

2.13 Error Handling

The error handler

```
error_handler(int i, char* s);
```

writes s to the diagnostic output (cerr) and terminates the program abnormally if $i \neq 0$. The function

```
leda_assert(bool b, int i, char* s);
```

calls *error_handler*(i, s) if b is *false* and has no effect otherwise. Users can provide their own error handling function *handler* by calling

```
set_error_handler(handler);
```

After this function call *handler* is used instead of the default error handler. *handler* must be a function of type *void handler*(*int*, *char*∗). The parameters are replaced by the error number and the error message, respectively.

2.14 Program Checking

Programming is an error-prone task. How do we make sure that the programs in LEDA are correct? We take the following measures:

- We start from correct algorithms as described in the large literature on data structures and algorithms.
- We try to document our programs carefully. This book contains many examples of carefully documented programs. We try to document so carefully that we can show our programs around and give them to colleagues to read. Don Knuth coined the name "literate programming" for this style of programming.
- We test extensively and our large user community tests.
- We use program checking [SM90, BK89, BLR90, MNS+96].

In this section we concentrate on the last item. Consider a program P that computes a function f. We call P *checkable* if for any input x it returns y, the alleged value of $f(x)$, and maybe additional information I that makes it easy to verify that indeed $y = f(x)$. By easy to verify we mean two things. Firstly, there must be a simple program C (a checking program) that, given x, y, and I, checks whether indeed $y = f(x)$. The program C should be so simple that its correctness is "obvious". Secondly, the running time of C on inputs x, y, and I should be no larger than the running time of P on x. This guarantees that the checking program C can be used without severe penalty in running time.

We give some examples.

Consider a program that takes an $m \times n$ matrix A and an m vector b and is supposed to check whether the linear system $A \cdot x = b$ has a solution. As stated, the program is supposed to return a boolean value indicating whether the system is solvable or not. This program is not checkable. In order to make it checkable, we extend the interface.

On input A and b the program returns either:

- "the system is solvable" and a vector x such that $A \cdot x = b$ or
- "the system is unsolvable" and a vector c such that $c^T \cdot A = 0$ and $c^T \cdot b \neq 0$.

The extended program is easy to check. If it answers "the system is solvable", we check that $A \cdot x = b$ and if it answers "the system is unsolvable", we check that $c^T \cdot A = 0$ and $c^T \cdot b \neq 0$. Thus the check amounts to a matrix-vector and a vector-vector product which are fast and also easy to program. We leave it as an exercise to prove that the vector c exists, when the system is solvable, and only remark that Gaussian elimination will produce it.

The second example is planarity testing. The task is to decide whether a graph is planar. A witness of planarity is a planar embedding and a witness of non-planarity is a Kuratowski subgraph. The details can be found in Section 8.7. The planarity test played an important role in the development of LEDA. A first implementation of it was added to LEDA in 1991. The implementation had been tested on a small number of graphs. In 1993 we were sent a graph together with a planar drawing of it. However, our program declared the graph non-planar. It took us some days to discover the bug. More importantly, we realized that a complex question of the form "is this graph planar" deserves more than a yes-no answer. We adopted the thesis that

a program should justify (prove) its answers in a way
that is easily checked by the user of the program.

By now many functions in LEDA justify their answers and come with checkers, see Sections 5.5.3, 10.3, 10.4.3, 10.4.6, 10.5.3, and all sections in Chapter 7.

What do we gain by program checking?

First, the answer of a program can be verified for any single problem instance. This is much less than program verification which gives a guarantee for all problem instances, but it is assuring.

Second, a user of a program can develop trust in the program with little intellectual investment. A user of a linear systems solver does not need to understand the intricacies of Gaussian elimination. For any program run, she can convince herself of the correctness of the computation by a simple matrix-vector and vector-vector product. The program for the latter two tasks is so simple, that it is even conceivable to verify them formally. See [BSM97] for a first example of a verified checker.

Third, a developer of a program can give compelling evidence of its correctness without revealing any details of the implementation. It suffices to publish the interface of the functions, to define what constitutes a witness, and to publish the checking program.

Fourth, program checking allows us to use a potentially incorrect program as if it were correct. If a program operates correctly on a particular instance, fine, and if it operates incorrectly, it is caught by the checker. Thus, if all subroutines of a function f are checked, no checker of a subroutine fires, and an error occurs during the execution of f, the error must be in f. This feature of program checking is extremely useful during the debugging phase of program development.

Fifth, program checking supports testing. Traditionally testing is restricted to problem instances for which the solution is known by other means. Program checking allows one to test on *any* instance. For example, we use the following program (among others) to check our algorithm to compute maximal matching in graphs (see Section 7.7).

```
for (int n = 0; n < 100; n++)
  for (int m = 0; m < 100; m++)
  { random_graph(G,n,m); // random graph with n nodes and m edges
    list<edge> M = MAX_CARD_MATCHING(G,OSC);
    CHECK_MAX_CARD_MATCHING(G,M,OSC);
  }
```

Sixth, a checker can only be written if the problem at hand is rigorously defined. We noticed that some of our specifications contained hidden assumptions which were revealed during the design of the checker. For example, an early version of our biconnected components algorithm assumed that the graph contains no isolated nodes.

The papers [SM90, BS94, SM91, BSM97, BS95, BSM95, SWM95, BK89, BLR90, BW96, WB97, AL94, MNS+96, DLPT97] contain further material on program checking.

2.15 Header Files, Implementation Files, and Libraries

The specifications of all LEDA types and algorithms are contained in the header files in directory LEDAROOT/incl/LEDA. In order to use a particular LEDA type or algorithms one must include the appropriate header file.

```
#include <LEDA/list.h>        // to use lists
#include <LEDA/dictionary.h> // to use dictionaries
#include <LEDA/point.h>       // to use points
#include <LEDA/graph_alg.h>  // to use the graph algorithms
#include <LEDA/geo_alg.h>    // to use the geometric algorithms
```

The implementations of all LEDA data types and algorithms are contained in the .c-files collected in the various subdirectories of LEDAROOT/src. They are pre-compiled into four libraries (libL.a, libG.a, libP.a, libWx.a) which can be linked with C++ application programs. The section "Using LEDA" of the LEDA manual describes how this is done.

2.16 Compilation Flags

The compilation flag -DLEDA_CHECKING_OFF turns off all checking of preconditions.

3

Basic Data Types

The basic data types *stack*, *queue*, *list*, *array*, *random number*, *tuple*, and *string* are ubiquitous in computing. Most readers are probably thoroughly familiar with them already. All sections of this chapter can be read independently.

3.1 Stacks and Queues

A *stack* is a last-in-first-out store for the elements of some type E and a queue is a first-in-first-out store. Both data types store sequences of elements of type E; they differ in the set of operations that can be performed on the sequence. In a stack one end of the sequence is designated as the *top* of the stack and all queries and updates on a stack operate on the top end of the sequence. In a *queue* all insertions occur at one end, the *rear* of the queue, and all deletions occur at the other end, the *front* of the queue. The definitions

```
stack<E> S;
queue<E> Q;
```

define a stack S and a queue Q for the element type E, respectively. Both structures are initially empty. The following operations are available on stacks. If x is an object of type E then the insertion *S.push*(x) adds x as the new top element. We can inspect the contents of a stack: *S.top*( ) returns the top element and *S.pop*( ) deletes and returns the top element. Of course, both operations are illegal if S is empty. The call *S.empty*( ) returns *true* if the stack is empty and *false* otherwise and *S.size*( ) returns the number of elements in the stack. So *S.empty*( ) is equivalent to *S.size*( ) $== 0$. All elements of a stack can be removed by *S.clear*( ).

We illustrate stacks by a program to evaluate a simple class of expressions. The character 1 is an expression and if E_1 and E_2 are expressions then $(E_1 + E_2)$ and $(E_1 * E_2)$ are expressions. Thus, $(1 + 1)$ and $((1 + 1) * (1 + (1 + 1)))$ are expressions, but $1 + 1$ and $(1+2)$ are not. The former is not an expression since it is not completely bracketed and the latter is not an expression since we only allow the constant 1 as an operand. We will ask you in the exercises to evaluate more complex expressions. There is a simple algorithm to evaluate expressions. It uses two stacks, a *stack<int> S* to hold intermediate results and a *stack<char> Op* to hold operator symbols. Initially, both stacks are empty. The expression is scanned from left to right. Let c be the current character scanned. If c is an open bracket, we do nothing, if c is a 1, we push it onto S, if c is a $+$ or $*$, we push it onto Op, and if c is a closing bracket, we remove the two top elements from S, say x and y, and the top element from Op, say op, and push the value $x\ op\ y$ onto S. When an expression is completely scanned, its value is the top element of S, in fact, it is the only element in S. The following program assumes that a well-formed expression followed by a dot is given on standard input. It prints the value of the expression onto standard output.

⟨*stack_demo.c*⟩≡

```
#include<LEDA/stack.h>
main()
{ char c;
  stack<int> S;  stack<char> Op;
  while ( (c = read_char("next symbol = ")) != '.' )
  { switch(c)
     {  case '(' : break;
        case '1' : { S.push(1); break; }
        case '+' : { Op.push(c); break; }
        case '*' : { Op.push(c); break; }
        case ')' : { int x = S.pop();  int y = S.pop();
                     char op = Op.pop();
                     if ( op = '+' ) S.push(x+y); else S.push(x*y);
                     break;
                   }
     }
  }
  cout << "\n\nvalue = " << S.pop() << "\n\n";
}
```

On input $((1+1)*(1+(1+1)))$ this program prints 6, on input $(1+(1+1))$ it prints 3, and on input () it crashes because it attempts to pop from an empty stack. This is bad software engineering practice and we will ask you in the exercises to remedy this shortcoming.

We turn to queues. The two ends of a queue are called the *front* and the *rear* of the queue, respectively. An insertion *Q.append*(x) appends x at the rear, *Q.top*( ) returns the front element, and *Q.pop*( ) deletes and returns the front element. Of course, the latter two calls require Q to be non-empty. The function *Q.empty*( ) checks for emptiness and *Q.size*( ) returns the number of elements in the queue. *Q.clear*( ) removes all elements from the queue.

Queues and stacks are implemented as singly linked lists. All operations take constant time except *clear*, which takes linear time. The space requirement is linear. LEDA also offers bounded queues and stacks, for example,

```
b_stack<E> S(n);
```

defines a stack S that can hold up to n elements. Bounded stacks and queues are implemented by arrays and hence always use the same amount of space independently of the actual number of elements stored in them. They are preferable to unbounded queues and stacks when the maximal size is known beforehand and the number of elements stored in the data structure is always close to the maximal size.

In the remainder of this section we show how to implement a queue by two stacks. This is to demonstrate the versatility of stacks, to illustrate that the same abstract data type can be implemented in many ways, to give an example of an amortized analysis of a data structure, and to amuse the user; it is not the implementation of queues used in LEDA. We use two stacks *Sfront* and *Srear* and split the queue into two parts: If $a_1, \ldots, a_m$ is the current content of *Sfront* and $b_1, \ldots, b_n$ is the current contents of *Srear* with a_m and b_n being the top elements, respectively, then $a_m, \ldots, a_1, b_1, \ldots, b_n$ is the current contents of the queue. Appending an element to the queue is realized by pushing it onto *Srear*. Popping an element from the queue is realized by popping an element from *Sfront*. If *Sfront* is empty, we first move all elements from *Srear* to *Sfront* (by popping from *Srear* and pushing onto *Sfront*). Note that this will reverse the sequence as it should be.

⟨*strange_queue.h*⟩≡

```
#include <LEDA/stack.h>
template<class E>
class queue {
  stack<E> Sfront, Srear;
public:
  queue<E>(){ } // initialization to empty queue
  void append(const E& x){ Srear.push(x); }
  E pop()
  { if ( Sfront.empty() )
       { while ( !Srear.empty() ) Sfront.push(Srear.pop()); }
    if ( Sfront.empty() ) error_handler(1,"queue: pop from empty queue");
    return Sfront.pop();
  }
  bool empty() { return Sfront.empty() && Srear.empty(); }
  int size()   { return Sfront.size() + Srear.size(); }
};
```

It is interesting to analyze the time complexity of this queue implementation. We claim that a sequence of n queue operations takes total time $O(n)$. To see this we note first that the constructor and the operations *append*, *empty*, and *size* run in constant time. A *pop* operation may take an arbitrary amount of time. More precisely, it takes constant time

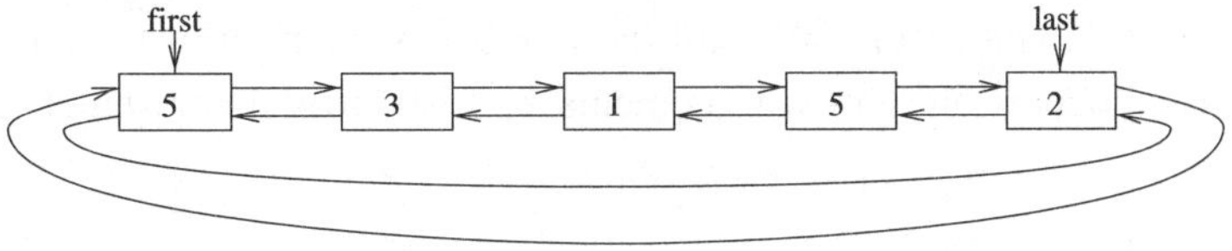

Figure 3.1 A list of five integers.

plus time proportional to the number of elements moved from *Srear* to *Sfront*. Since each element is moved at most once from *Srear* to *Sfront*, we incur a constant cost per element for moving elements from *Srear* to *Sfront*. We conclude that the time spent in all *pop* operations is linear.

Exercises for 3.1

1 Implement the type *stack*.

2 Implement the type *queue*.

3 Extend the expression evaluator such that it complains about illegal inputs.

4 Extend the expression evaluator such that it can handle arbitrary integers as operands.

5 Extend the expression evaluator such that it can handle expressions that are not completely bracketed. The usual precedence rules should be applied, i.e., $a + b * c$ is interpreted as $(a + (b * c))$. More specifically, the evaluator should be able to handle all expressions that are generated by the following four rules:

A factor is either an integer or a bracketed expression.

A term is either a factor or a factor times a term.

An expression is either a term or a term plus an expression.

That's all.

3.2 Lists

Lists are a simple, yet powerful, data type. It is difficult to implement a combinatorial or geometric algorithm without using lists. Moreover, the implementation of several LEDA data types, e.g., stacks, queues, and graphs, is based on lists. In this section we discuss lists for unordered and ordered element types, we sketch the implementation of lists, and in the final subsection we treat singly linked lists.

3.2.1 *Basics*

```
list<E> L;
```

declares a list L for elements of type E and initializes it to the empty list. Generally, a list L over element type E (type *list*<E>) is a sequence of items (of predefined type *list_item*), each holding an element of type E. Figure 3.1 shows a list of integers. It consists of five items shown as rectangular boxes. The contents of the first item is 5, the contents of the second item is 3, and so on. We call the number of items in a list the length of the list and use $\langle x \rangle$ to denote an item with contents x. Lists offer an extremely rich repertoire of operations.

```
L.empty();
```

checks L for emptiness. Let's assume that L is non-empty. Then

```
E         x = L.head();
list_item it = L.first();
```

assign the contents of the first item of L to x and the first item to *it*. Please pause for a moment to grasp the difference. *L.first*( ) returns the first item and *L.head*( ) returns the contents of the first item. Thus, if L is the list of Figure 3.1, the value of x is now 5 and the value of *it* is the first box. The content of the item (box) *it* can be accessed by *L.contents*(*it*) or $L[it]$. So

```
x == L.contents(it)
```

evaluates to *true* and so do

```
        3 == L.contents(L.succ(L.first());
L.last() != L.first();
      nil == L.pred(L.first());
L.tail() == L[L.cyclic_pred(L.first())];
L.last() == L.cyclic_pred(L.first()).
```

We need to explain these expressions a bit further. For a list L, *L.head*( ) and *L.tail*( ) return the contents of the first and last item of L, respectively (5 and 2 in our example) and *L.first*( ) and *L.last*( ) return the first and last item of L, respectively (the first and the fifth box in our example). The items in a list can be viewed as either arranged linearly or arranged cyclically. The operations *succ* and *pred* support the linear view of a list and the operations *cyclic_succ* and *cyclic_pred* support the cyclic view. Thus, if *it* is an item of a list L different from the last item then *L.succ*(*it*) returns the successor item of *it* and *L.succ*(*L.last*( )) returns *nil* and if *it* is different from the first item then *L.pred*(*it*) returns the predecessor item of *it* and *L.pred*(*L.first*( )) returns *nil*. *L.cyclic_pred*(*it*) and *L.cyclic_succ*(*it*) return the cyclic predecessor and successor, respectively, where the cyclic predecessor of the first item is the last item. So in the next to last expression above both sides evaluate to the contents of the last item of L and in the last expression both sides evaluate to the last item of L.

We further illustrate the use of items by the member function *print*. It takes two arguments, an output stream O and a character *space* and prints the elements of a list separated by *space* onto O. The default value of *space* is the space character. It requires that the type E offers a function *Print*(x, O) that prints an object x of type E onto O, see Section 2.8 for a discussion of the *Print*-function for type parameters.

```
template<class E>
void list<E>::print(ostream& O, char space = " ")
  { list_item it = first();
```

```
    while ( it != nil )
    { Print(contents(it),O);
      if ( it != last_item() ) O << space;
      it = succ(it);
    }
  }
```

Note how *it* steps through the items of the list. It starts at the first item. In the general step, we first print the contents of *it* and then advance *it* to its successor item. We do so until *it* falls off the list.

Iterating over the items or elements of a list is a very frequently occurring task and therefore LEDA offers corresponding iteration macros. The iteration statements

```
forall(x,L) << body >>
```

and

```
forall_items(it,L) << body >>
```

step through the elements and items of L, respectively, and execute *body* for each one of them. Thus,

```
list_item it;
forall_items(it,L) Print(L[it],cout);
E x;
forall(x,L) Print(x,cout);
```

prints the elements of L twice. The *forall_items* loop is a macro that expands into

```
for (list_item loop_it = L.first();
      it = loop_it, loop_it = L.next_item(loop_it), it; )
{ << body >> }
```

and the *forall* loop is a macro that essentially expands into

```
for ( list_item it = L.first(); it; it = L.succ(it) )
{ x = L[it];
  << body >>;
}
```

As one can see from the expansions both iteration statements work in time proportional to the length of the list. However, since the assignment $x = L[it]$ may be a costly operation (if E is a complicated type) it is usually more efficient to use the *forall_items* loop. The fact that the iteration statements for lists (and any other LEDA data type, for that matter) are realized as macros is a possible source for programming errors; we advise *to never write forall_items(it, f( )), where f is a function that produces a list*, see Sections 2.5 and 13.9 for details.

Next, we turn to update operations on lists.

```
L[it] = x;
```

changes the contents of the item *it* and

```
L.append(x);
```

adds a new item $\langle x \rangle$ after the last item of L and returns the item. We may store the item for later use:

```
list_item it = L.append(x);
```

The operations

```
L.del_item(it);
L.pop();
L.Pop();
```

remove the item *it*, the first item, and the last item of L, respectively. Each operation returns the contents of the item removed. So we may write $x = L.pop(\,)$. The program fragment

```
list<int> L;
L.append(5);
L.append(3);
list_item it = L.append(1);
L.append(5);
L.append(2);
```

builds the list of Figure 3.1 and assigns the third item of L to *it*. So $L[it]$ evaluates to 1 and *L.del_item(it)* removes the third item from L, i.e., L consists of four items with contents 5, 3, 5, and 2, respectively, after the call.

Two lists L and *L1* of the same type can be combined by

```
L.conc(L1,dir);
```

where *dir* determines whether *L1* is appended to the rear end (*dir* = *LEDA::after*) or front end (*dir* = *LEDA::before*) of L; *before* and *after* are predefined constants. As a side effect, *conc* clears the list *L1*. The lists L and *L1* must be distinct list objects. A list L can be split into two parts. If *it* is an item of L then

```
L.split(it,L1,L2,dir);
```

splits L before (*dir* = *LEDA::before*) or after (*dir* = *LEDA::after*) item *it* into lists *L1* and *L2*. The lists *L1* and *L2* must be distinct list objects. It is allowed, however, that one of them is equal to L. If L is distinct from *L1* and *L2* then L is empty after the split. *Split* and *conc* take constant time. Given *split* and *conc*, it is easy to write a function *splice*[1] that inserts a list *L1* after item *it* into a list L. If *it* = *nil*, *L1* is added to the front of L.

```
if ( it == nil )
   L.conc(L1,LEDA::before);
else
 { list<E> L2;
   L.split(it,L,L2,LEDA::after);
   L.conc(L1,LEDA::after);
   L.conc(L2,LEDA::after);
 }
```

[1] *splice* is a member function of *lists* and so there is no need to define it at the user level. We give its implementation in order to illustrate *split* and *conc*.

The *apply* operator applies a function to all elements of a list, i.e., if f is a function defined for objects of type E then

```
L.apply(f);
```

performs the call $f(x)$ for all items $\langle x \rangle$ of L. The element x is passed by reference. For example, if L is a list of integers then

```
void incr(int& i) { i++; }
L.apply(incr);
```

increases all elements of L by one. *apply* takes linear time plus the time for the function calls.

LEDA provides many ways to reorder the elements of a list.

```
L.reverse_items();
```

reverses the items in L and

```
L.permute();
```

randomly permutes the items of L. Both functions take linear time and both functions are good examples to illustrate the difference between items and their contents. The call *L.reverse_items*( ) does not change the set of items comprising the list L and it does not change the contents of any item, it changes the order in which the items are arranged in the list. The last item becomes the first, the next to last item becomes the second, and so on. Thus,

```
list_item it = L.first();
L.reverse_items();
bool b = ( it == L.last() );
```

assigns *true* to b.

For contrast, we give a piece of code that reverses the contents of the items but leaves the order of the items unchanged. It makes use of a function *leda_swap* that swaps the contents of two variables of the same type. We use two items *it0* and *it1* which we position initially at the first and last item of L. We interchange their contents and advance both of them. We do so as long as the items are distinct and *it0* is not the successor of *it1*. The former test guarantees termination for a list of odd length and the latter test guarantees termination for a list of even length. If the list is empty the first and the last item are *nil* and the former test guarantees that the loop body is not entered.

```
/* this is not the implementation of reverse_items */
list_item it0 = L.first();
list_item it1 = L.last();
while ( it0 != it1 && it0 != L.succ(it1) )
{ leda_swap(L[it0],L[it1]);
  it0 = L.succ(it0);
  it1 = L.pred(it1);
}
```

The above code implements

```
L.reverse().
```

We turn to sorting. We will discuss general sorting methods in the next section and discuss bucket sorting now. If f is an integer-valued function on E then

```
L.bucket_sort(f);
```

sorts L into increasing order as prescribed by f. More precisely, *bucket_sort* rearranges the items of L such that the f-values are non-decreasing after the sort and such that the relative order of two items with the same f-value is unchanged by the sort. Such a sort is called *stable*. For an example, assume that we apply *bucket_sort* to the list L of Figure 3.1 with f the identity function. This will make the third item the first item, the fifth item the second item, the second item the third item, the first item the fourth item, and the fourth item the fifth item. *bucket_sort* takes time $O(n + r - l)$, where n is the length of the list and l and r are the minimum and maximum value of $f(e)$ as e ranges over the elements of the list.

We give an application of bucket sort. Assume that L is a list of edges of a graph G (type *list<edge>*) and that *dfs_num* is a numbering of the nodes of G (type *node_array<int>*). Our goal is to reorder L such that the edges are ordered according to the number of the source of the edge, i.e., all edges out of the node with smallest number come first, then all edges out of the node with second smallest number, and so on. For an edge e of a graph G, *G.source*(e) returns the source node of the edge and hence *dfs_num*[*G.source*(e)] is the number of the source of the edge. We define a function *ord* that, given an edge e, returns *dfs_num*[*G.source*(e)] and then call *bucket_sort* with this function.

```
int ord(edge e){ return dfs_num[G.source(e)]; };
L.bucket_sort(ord);
```

3.2.2 *Lists for Ordered Sets*

Recall that a type E is linearly ordered if the function *int compare*(*const E*&, *const E*&) is defined and establishes a linear order on E, cf. Section 2.10. For lists over linearly ordered element types additional operations are available.

```
list_item L.search(E x);
```

searches for an occurrence of x in L. It uses *compare* to compare x with the elements of L. If x occurs in L, the leftmost occurrence is returned and if x does not occur in L, *nil* is returned. The running time of *search* is proportional to the distance of the leftmost occurrence of x from the front of the list. We next show how to use *search* in a primitive but highly effective implementation of the *set* data type, the so-called *self-organizing list implementation*. We realize a set over type E (type *so_set<E>*) as a list over E and use *search* to realize the *member* operation; the prefix "so" stands for self-organizing. We will make the member operation more effective by rearranging the list after each successful access. We use the operation *move_to_front*(*it*) that takes an item *it* of a list, removes it from its current position, and makes it the first element of the list. The effect of moving each accessed item to the front of the list is to collect the frequently accessed items near the front of the list. Since the access time in a list is linear in the distance from the front, this strategy keeps the expected access time small. We refer the reader to [Meh84a, III.6.1.1] for the theory of self-organizing lists and turn to the implementation. We derive *so_set<E>* from *list<E>* and accordingly define a *so_set_item* as a new name for a *list_item*. We realize the membership test by *search* followed by *move_to_front* (if the search was successful), we realize *insert* by a membership test followed by append (if the membership test returned false). The other member functions are self-explanatory.

⟨*so_set.h*⟩≡

```
#include <LEDA/list.h>
typedef list_item so_set_item;
template <class E>
class so_set: private list<E>{
public:
  bool member(const E& e)
  { list_item it = search(e);
    if (it) { move_to_front(it); }
    return ( it != nil );
  }
  void insert(const E& e)                { if (!member(e)) append(e); }
  so_set_item first() const              { return list<E>::first(); }
  so_set_item succ(so_set_item it) const { return list<E>::succ(it); }
  E contents(so_set_item it) const    { return list<E>::contents(it); }
};
```

We give an application of our new data type. We read the file containing the source of this chapter, insert all its words into a *so_set*, and finally print the first thirty words in the set.

⟨*so_set_demo*⟩≡

```
main(){
so_set<string> S;
file_istream I("datatype.lw");
string s;
float T = used_time();
while ( I >> s ) S.insert(s);
cout << "time required = " << used_time(T);
so_set_item it = S.first();
for (int i = 0; i < 30; i++)
  { cout << (i % 5 == 0 ? "\n" : " ") << S.contents(it);
    it = S.succ(it);
  }
}
```

The output of this program is:

```
time required = 13.58
} \end{exercises} respectively. and $s$
of length the are $m$
$n$ where $O(n+m)$ time in
runs program that Show substring
a is $p$ if only
success this @ else p.length())
```

As expected, we see frequent English words, because the move-to-front-heuristic tends to keep them near the front of the list, and words that occurred near the end of the text, because they were accessed last.

We turn to merging and sorting. If *cmp* defines a linear order on the element type of *L* then

```
L.sort(cmp); L1.sort(cmp);
L.merge(L1,cmp)
```

sorts *L* and *L1* according to the linear order and then merges the two sorted lists. If we call the functions without the *cmp*-argument

```
L.sort(); L1.sort();
L.merge(L1);
```

the default order on the element type is used. Merging two lists of length n and m, respectively, takes time $O(n+m)$ and sorting a list of n elements takes expected time $O(n \log n)$. Let us verify this fact experimentally. We start with n equal to 128000 and repeatedly double n. For each value of n we generate a list of length n, make two copies of the list and merge them, and we permute the items of the list and then sort the list. For each value of n we output n, the measured running time for the merge and the sort, respectively, and the running time divided by n and $n \log n$, respectively.

⟨*sort_merge_times*⟩≡

```
main()
{ int min, max;
  ⟨sort merge times: read max⟩
  for (int n = min; n <= max; n = 2*n)
  { list<int> L;
    for (int j = 0; j < n; j++) L.append(j);
    list<int> L1 = L;
    list<int> L2 = L;
    float T1 = used_time();
    L1.merge(L2);
    T1 = used_time(T1);
    L.permute();
    float T2 = used_time();
    L.sort();
```

	Merging		Sorting	
n	time	normalized	time	normalized
128000	0.07	0.547	0.64	0.425
256000	0.15	0.586	1.35	0.423
512000	0.3	0.586	3.15	0.468
1024000	0.58	0.566	6.31	0.445

Table 3.1 The table produced by the experiment. All running times are in seconds. The normalized time is the $10^6 T/n$ in the case of merging and $10^6 T/(n \log n)$ in the case of sorting. The normalized time of sorting grows slowly. This is due to the increased memory access time for larger inputs. You can produce your own table by running sort_merge_times.

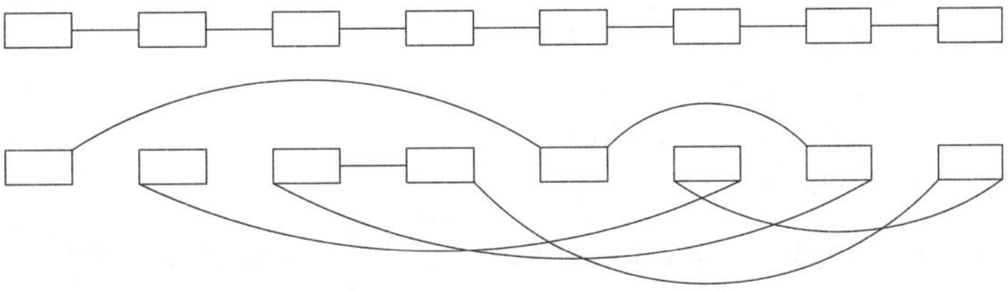

Figure 3.2 The list L before and after the call of *permute*.

```
    T2 = used_time(T2);
    ⟨sort merge times: produce table⟩
  }
}
```

Table 3.1 shows the outcome of the experiment. Does it confirm our statement that the running time of merge is $\Theta(n)$ and that the running time of sort is $\Theta(n \log n)$? In the case of merging one may say yes, since the numbers in the third column of our table are essentially constant, however, in the case of sorting the answer is a definite no, since the numbers in the last column of our table certainly grow. Why is this so? The explanation lies in the influence of cache memory on the running time of algorithms.

The internal memory of modern computers is organized hierarchically. There are at least two levels of internal memory, a small and very fast first-level memory (usually called cache) and a larger and slower second-level memory (usually called main memory). On many machines the hierarchy consists of more than two levels, see [HP90] for an excellent account of computer architecture. In the example above we first allocate a list of n items: this puts the items consecutively into storage. Then we change the order of the items in the list randomly. This leaves the items where they are and changes the links, i.e., after *permute* the links jump around widely in memory, see Figure 3.2. The job of *sort* is to untangle this

Build	Traverse	Permute	Traverse
0.59	0.16	2.77	0.44

Table 3.2 Illustration of cache effects on running time: We built a list of 1000000 items, traversed it, permuted it, and traversed it again. You may perform your own experiments with the cache_effects demo.

mess. In doing so, it frequently has to access items that are not in the fastest memory. This explains the last column of our table, at least qualitatively.

Next, we attempt a quantitative explanation. Consider the following program:

```
list<int> L;
for (int i = 0; i < 1000000; i++) L.append(i);
// L.permute();
float T = used_time();
list_item it = L.first();
while (it != nil) it = L.succ(it);
cout << used_time(T);
```

We make the following assumptions (see [HP90] for a justification): It takes ten machine instructions to execute one iteration of the while-loop. Memory is organized in two levels and the first level can hold 10000 items. An access to an item that is in first level is serviced immediately and an access to an item that is not in the first level costs an additional twenty machine cycles. An access to an item in second level moves this item and the seven items following it in second-level memory from second-level memory to first-level memory. An access to an item that is not in first-level memory in called a *cache miss*.

What behavior will we see? First assume that the list is permuted. Since the first level memory can hold only 10000 items it is unlikely that the successor of the current item is also in memory. We should therefore expect that each iteration of the loop takes thirty machine cycles, ten for the instructions executed in the loop and twenty for the transport of an item into fast memory. Next assume that the list is not permuted. Now we will incur the access time for slow memory only once in eight iterations and hence eight iterations will take a total of 100 machine cycles. In contrast, the eight iterations will take a total of 240 machine cycles on the permuted list. Thus, permuting the list will make the program about 2.4 times slower for large n. For $n = 10000$ we will see no slowdown yet, as the entire list fits in fast memory. For very large n we will see a slowdown of 2.4 and for intermediate n we will see a slowdown less than 2.4.

Table 3.2 shows actual measurements.

3.2.3 *The Implementation of Lists*

Lists are implemented as doubly linked lists. Each item corresponds to a structure (type *dlink*) with three fields, one for the contents of the item and one each for the predecessor and the successor item, and the list itself is realized by a structure (type *dlist*) containing pointers to the first and last item of the list and additional bookkeeping information. The space requirement of a list of n items is $16 + 12n$ bytes plus the space needed for the elements of the list. The contents of an item is either stored directly in the item (if it fits into four bytes) or is stored through a pointer, i.e., the e-field of a *dlink* either contains the contents of the item or a pointer to the contents of the item. In the former case there is no extra space needed for the elements of the list and in the latter case additional space for n objects of type E is needed (here E denotes the type of the objects stored in the list). All of this is discussed in detail in the chapter on implementation.

⟨*storage layout for lists*⟩≡

```
typedef dlink* list_item;
class dlink {
  dlink* succ;
  dlink* pred;
  GenPtr e;              // for the contents of the item
  // space: 3 words = 12 bytes
};

class dlist {
   dlink* h;                          // head
   dlink* t;                          // tail
   link* iterator                     // iterator, historical
   int count;                         // length of list
   // space: four words = 16 bytes
   ⟨member functions of class dlist⟩
};
```

There is no space to show the implementations of all member functions. We show only the implementation of bucket sort. The implementation is very low-level and therefore hard to understand. *Bucket_sort* assumes that a function *ord* and integers i and j are given such that *ord* maps the elements of the list into the range $[i .. j]$. It uses an array *bucket* of linear lists; *bucket*$[i]$ points to the end of the i-th bucket list as shown in Figure 3.3. Initially, all bucket lists are empty. The algorithm runs through the items of the list to be sorted, computes for each item x the index $k = ord(x \rightarrow e)$ of the bucket into which the item (recall that $x \rightarrow e$ contains the object stored in item x) belongs, and appends the item to the appropriate bucket. Afterwards, it joins all bucket lists into a single list. This is done from right to left.

⟨*list: bucket sort*⟩≡

```
void dlist::bucket_sort(int i, int j)
{
  if (h == nil) return; // empty list
  int n = j-i+1;
  register list_item* bucket = new list_item[n+1];
  register list_item* stop = bucket + n;
```

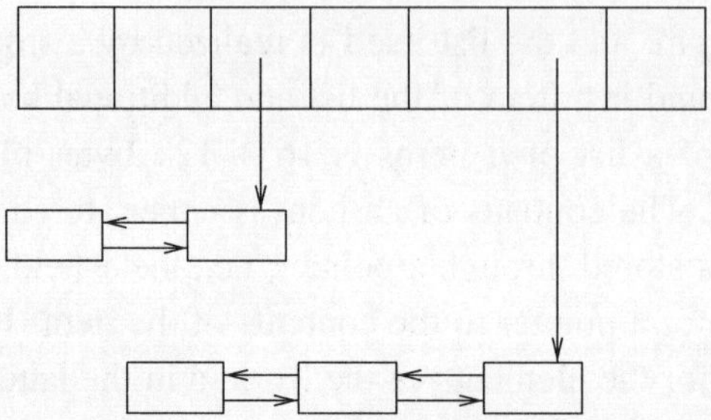

Figure 3.3 Illustration of bucket sort. We have two non-empty buckets. The list items are shown as rectangular boxes, successor pointers go from left to right, and predecessor pointers go from right to left. The pointers from the bucket array to the rears of the bucket lists are shown vertically.

```
register list_item* p;
register list_item q;
register list_item x;
for(p = bucket; p <= stop; p++)  *p = 0;
while (h)
{ x = h;
  h = h->succ;
  int k = ord(x->e);
  if ( k >= i && k <= j )
   { // add x at end of k-th bucket
     p = bucket + k - i;
     x->pred = *p;
     if (*p) (*p)->succ = x;
     *p = x;
   }
  else
     error_handler(1,"bucket_sort: value out of range") ;
}
for(p = stop; *p == 0; p--);
// now p points to the end of the rightmost non-empty bucket
// make it the new tail of the list.
t = *p;
t->succ = nil;
for(q = *p; q->pred; q = q->pred);
         // now q points to the start of this bucket
// link buckets together from right to left:
// q points to the start of the last bucket
// p points to end of the next bucket
while( --p >= bucket )
  if (*p)
  { (*p)->succ = q;
    q->pred = *p;
    for(q = *p; q->pred; q = q->pred);
  }
```

```
  h = q;   // head = start of leftmost non-empty bucket
  delete[] bucket;
}
```

Aren't you glad that one of us wrote this program?

3.2.4 *Singly Linked Lists*

LEDA also offers singly linked lists (type *slist*) in which each item only knows its successor. They require space $16 + 8n$ bytes but offer a smaller repertoire of operations. Singly linked lists are used to implement stacks and queues.

Exercises for 3.2

1 Implement queues by singly linked lists.

2 Implement more operations on lists, e.g., *conc* or *merge*.

3 Write a procedure that reverses the order of the items in a list.

4 Extend the data type *so_set* to a dictionary. Realize a dictionary from K to I as a list of pointer to pairs (*list<two_tuple<K, I> * >*). Then proceed in analogy to the text.

5 (Topological sorting) Let L be a list of pairs of integers in the range from 1 to n. Compute an ordering of the integers 1 to n such that if (x, y) is any pair in the list then x precedes y in the ordering, or decide that there is no such ordering. So if n is 4 and L is (2, 1), (1, 4), (3, 4) then 2, 3, 1, 4 is a possible ordering. Hint: 2 can go first because it does not appear as the second component of any pair.

6 Redo the calculation for the slowdown due to cache misses for the case that an iteration of the loop takes 100 clock cycles instead of ten.

7 Find out what a cache miss costs on the machine that you are using.

3.3 Arrays

Arrays are what they are supposed to be: collections of variables of a certain type E that are indexed by either an interval or a two-dimensional box of integers. The declarations

```
array<string> A(3,5);
array<string> B(10);
array2<int>   C(1,2,4,6);
```

define two one-dimensional arrays and one two-dimensional arrays: A is a one-dimensional array of strings with index set $[3..5]$, B is a one-dimensional array of strings with index set $[0..9]$, and C is a two-dimensional array of integers with index set $[1..2] \times [4..6]$, respectively. Each entry is initialized with the appropriate default value. So each entry of A and B is initialized to the empty string and each entry of C is initialized to some integer.

We use the standard C++ subscript operator for the selection of variables in one-dimensional arrays. So $A[4]$ evaluates to the variable with index 4 in A. For two-dimensional arrays we need to use round brackets since C++ does not allow the use of angular brackets with two arguments. So $C(1, 5)$ evaluates to the variable with index $(1, 5)$ in C. Arrays check whether their indices are legal (this can be turned off by the compiler flag `-DLEDA_CHECKING_OFF`) and hence we get an error in the following assignment:

```
A[6] = "Kurt" //  "ERROR array:: index out of range"
```

An array knows its index set. The calls *A.low*( ) and *A.high*( ) return the lower and upper index bound of A, respectively. For two-dimensional arrays we have the corresponding functions *low1*, *high1*, *low2*, and *high2*.

We illustrate arrays by two sorting functions: straight insertion sort and merge sort. Both operate on an *array*<*E*> *A* and assume that the element type E is linearly ordered by the function *compare*, see Section 2.10. We use $[l \,..\, h]$ to denote the index range of A. Straight insertion sort follows a very simple strategy; it sorts increasingly larger initial segments of A. Assume that we have already sorted an initial segment $A[l], \ldots, A[i-1]$ of A for some i. Initially, $i = l + 1$. In the incremental step we add $A[i]$ to the sorted initial segment by inserting it at the proper position. We determine j with $A[j] \le A[i] < A[j+1]$, move $A[j+1], \ldots, A[i-1]$ one position to the right, and put $A[i]$ into $A[j+1]$, see Figure 3.4. Straight insertion sort is a stable sorting method. Its running time is quadratic.

⟨*straight_insertion_sort*⟩≡

```
template<class E>
void straight_insertion_sort(array<E>& A)
{ int l = A.low();
  int h = A.high();
  for (int i = l + 1; i <= h; i++)
     { E x = A[i];
       int j = i - 1;
       while ( j >= l && compare(x,A[j]) < 0 )
       { A[j+1] = A[j];
         j--;
       }
       A[j+1] = x;
     }
}
```

We turn to merge sort. It is much more efficient than straight insertion sort and runs in time $O(n \log n)$ on an array of size n. The underlying strategy is also simple. Merge sort operates in phases. At the beginning of the k-th phase, $k \ge 0$, the array is partitioned into sorted blocks of size 2^k. These blocks are paired and any pair is merged into a single sorted block. In the program below we use K to denote 2^k and we use an auxiliary array B with the same index set as A. In even phases the merge step reads from A and writes into B, and in odd phases the roles of A and B are interchanged. In this way the data moves back and forth between A and B. If it ends up in B at the end of *merge_sort*, we need to copy it back to A. We use a boolean variable *even_phase* that is true iff the next phase is even. The actual merging is done by the function *merge*. A call *merge*(X, Y, i, K, h) takes the blocks of X

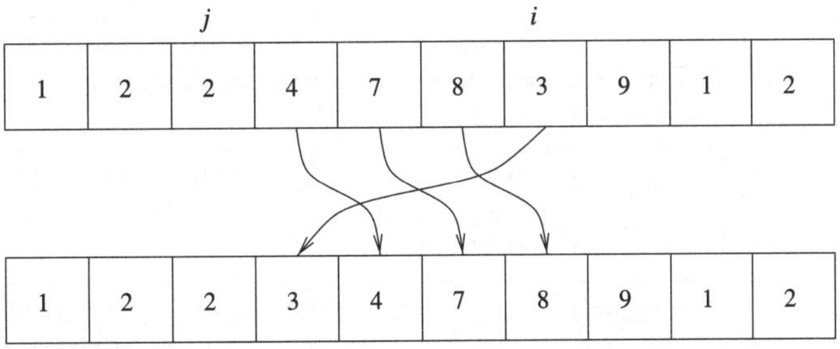

Figure 3.4 We insert $A[i]$ into the already sorted initial segment by inserting it into the proper position, say position $j+1$, and moving elements $A[j+1], \ldots, A[i-1]$ one element to the right.

starting at positions i and $i+K$, respectively, and merges them into the block of Y starting at position i and having the combined size of the two blocks to be merged. The last element of the two blocks to be merged is to be found at position h; this information is important if the size of A is not a power of two.

⟨*merge_sort*⟩≡

```
  ⟨merge routine⟩
  template<class E>
  void merge_sort(array<E>& A)
  { int l = A.low();  int h = A.high(); int n = h - l + 1;
    array<E> B(l,h);
    bool even_phase = true;
    for (int K = 1; K < n; K = 2*K)
    { for (int i = l; i <= h; i = i + 2*K)
      { if ( even_phase ) merge(A,B,i,K,h);
        else merge(B,A,i,K,h);
      }
      even_phase = !even_phase;
    }
    if ( !even_phase )
      { for (int i = l; i <= h; i++) A[i] = B[i]; }
  }
```

It remains to define *merge*(X, Y, i, K, h). Our goal is to fill the block of Y starting at position i and extending to position m where $m = \min(i + 2K - 1, h)$ from the two blocks of X starting at positions i and $i + K$, respectively. The two blocks in X extend to $ml = \min(i + K - 1, h)$ and m, respectively. We maintain one index in each of the three blocks to control the merging process: The index j indicates the position in Y that is to be filled next and the indices il and ih point to the smallest remaining elements in the two blocks of X. We always move the smaller of $X[il]$ and $X[ih]$ to $Y[j]$. We break ties in favor of $X[il]$. This makes merge sort a stable sorting method.

n	2	4	8	16	32	64	128	256
insertion sort	0.83	3.27	13.5					
merge sort	0.03	0.06	0.14	0.29	0.66	1.28	2.85	5.99
insertion sort	0.47	1.74	7.07					
merge sort	0.02	0.03	0.09	0.17	0.38	0.78	1.66	3.49
member function	0.01	0.01	0.03	0.07	0.15	0.3	0.6	1.33

Table 3.3 Running times of our sorting routines and the member function *sort*. All running times are in seconds and for an array of $1000n$ integers. Insertion sort and merge sort have been compiled without and with the flag `-DLEDA_CHECKING_OFF`. You may produce you own table by calling array_sort_times.

⟨*merge routine*⟩≡

```
#include <LEDA/misc.h>  // to include Min
template<class E>
void merge(array<E>& X, array<E>& Y, int i, int K, int h)
{ int il = i; int ih = i + K;
  int ml = Min(i + K - 1,h); int m = Min(i + 2*K - 1,h);
  for (int j = i; j <= m ; j++)
    { if ( ih <= m && ( il > ml || compare(X[ih],X[il]) < 0 ) )
        { Y[j] = X[ih]; ih++; }
      else
        { Y[j] = X[il]; il++; }
    }
}
```

Table 3.3 shows the running times of our two sorting procedures in comparison to the member function *sort* for the task of sorting an array of n *ints*. Observe how the running time of insertion sort explodes. Since its running time grows proportional to n^2, it quadruples whenever n is doubled. In contrast, the running time of the two other methods is $O(n \log n)$ and hence basically doubles whenever n is doubled. The member function *sort* beats our implementation of merge sort because it exploits the fact that the objects to be sorted are *ints*, see Section 13.5.

Arrays are implemented by C++ arrays. There are important differences, however:

- The index sets may be arbitrary intervals of integers and arrays check whether their indices are legal. The index check can be turned off by the compiler flag `-DLEDA_CHECKING_OFF`.
- The entries of an array are initialized to the default value of the entry type.
- An assignment $A = B$ assigns a copy of B to A, i.e., A is made to have the same

number of variables as B and these variables are initialized with copies of the values of the corresponding variables in A. Thus, it is perfectly legal to assign an array of size 100 to an array of size 5.

- One-dimensional arrays offer some additional higher level functions which we discuss next.

We can reorder the elements of a one-dimensional array according to a linear order on the type E. The linear order may either be the default order of the type or be given by a compare function. Thus,

```
A.sort();
```

sorts the entries of our array A according to the lexicographic ordering on strings. On a sorted array we may use binary search.

```
A.binary_search("Stefan");
```

returns the index $i \in [3..5]$ containing `"Stefan"` if there is such an index and returns $A.low() - 1$ if there is no such index. We can permute the entries of an array by

```
A.permute();
```

The space required for an array of n elements is n times the space required for an object of type E. All access operations on arrays take constant time, *sort* takes time $O(n \log n)$ and *binary_search* takes time $O(\log n)$.

In many applications one needs arrays with large but sparsely used index sets, e.g., we may have 10^4 indices in the range from 0 to 10^9. In this situation it would be a complete waste of space and time to allocate an array of 10^9 elements and therefore a different data structure is called for. The data types *map* and *h_array* are appropriate. They will be discussed in Section 5.1.

Exercises for 3.3

1 Implement other sorting routines for arrays. Candidates are bubble sort, shell sort, heap sort, quick sort, and others.

2 Implement the type *array* by C++ arrays.

3 (Sparse arrays) Use lists to realize arrays whose index ranges are the integers from 0 to 2^{20}. Call the type *sparse_array*<E>. The constructor for the class should have an argument of type E. All elements of the array are initialized with this value. The time efficiency of your method is not important. However, the space requirement should be proportional to the number of indices for which the subscript operator was executed.

3.4 Compressed Boolean Arrays (Type int_set)

Boolean arrays are often used to represent sets. In this situation one also wants to perform the set operations *union*, *intersection*, and *complement* besides the usual operations on arrays (read the value of an entry or set the value of an entry). The data type *int_set* provides these operations in a space- and time-efficient way. It stores boolean arrays as bit-vectors, i.e., λ entries are stored in a single word on a machine with word size λ, and it uses the parallelism available at the word level to perform the set operations for λ entries in a single machine instruction. A speed-up of about λ is thus obtained for the set operations. On the other hand, reading or setting a single entry takes slightly longer than for an array.

```
int_set S(n),T(n),R(n);
```

defines S, T, and R as subsets of $[0\,..\,n-1]$ and initializes them to the empty set; the alternative definition *int_set* $S(a, b)$ defines S as a subset of $[a\,..\,b]$. If x is an integer with $0 \le x \le n-1$ then

```
S.insert(x);
S.del(x);
S.member(x);
```

inserts x into S, deletes x from S, and tests for membership of x, respectively. *S.clear*( ) makes S the empty set. The set operations union, intersection, and complement are denoted by the corresponding logical operator. So

```
S = T | R;
S = T & R;
S = ~T;
```

assigns the union of T and R, the intersection of T and R, and the complement of T to S, respectively. We also have the shorthands $S \mathrel{|}= R$ for $S = S \mid R$ and $S \mathrel{\&}= R$ for $S = S \,\&\, R$. Note that the shorthands are more efficient than the verbose versions since the verbose versions first construct a temporary object and then copy that object into the left-hand side (except if your compiler is clever). The space requirement of int_sets is $O(n/\lambda)$; *insert*, *del*, and *member* take time $O(1)$, and the other operations take time $O(n/\lambda)$.

As an application of compressed boolean arrays we give an algorithm for the multiplication of boolean matrices that runs in time $O(n^3/\lambda)$. Let A and B be boolean matrices with index sets $[0\,..\,n-1] \times [0\,..\,n-1]$, and let C be their product, i.e.,

$$C(i, k) = \bigvee_{j=0}^{n-1} A(i, j) \wedge B(j, k)$$

for all i and k. The obvious method to obtain C from A and B takes time $O(n^3)$. We can obtain a faster algorithm by observing that for each i, $0 \le i < n$, the i-th row of C is the bit-wise *or* of certain rows of B, namely those that are selected by the i-th row of A. If we represent the rows of B and C as compressed arrays we obtain each row of C in time $O(n^2/\lambda)$ and hence can multiply two matrices in time $O(n^3/\lambda)$.

We give the details. First we compute a compressed version of B.

```
array<int_set*> B_compressed(0,n-1);
int i;
for (i = 0; i < n; i++)
```

```
  { B_compressed[i] = new int_set(0,n-1);
    for (int j = 0; j < n; j++)
       if ( B(i,j) ) B_compressed[i]->insert(j);
  }
```

Next we perform the multiplication. We compute each row first in compressed form and then expand it into C.

```
int_set compressed_row(0,n-1);
for (i = 0; i < n; i++)
  { for (int j = 0; j < n; j++)
       if (A(i,j)) compressed_row |= *B_compressed[j];
    for (j = 0; j < n; j++)
       C(i,j) = compressed_row.member[j];
    compressed_row.clear();
  }
```

Exercise for 3.4

1 Compare the method described above with the following variant of the traditional method.

```
for (int i = 0; i < n; i++)
  for (int k = 0; k < n; k++)
    { C(i,k) = false;
      for (int j = 0; j < n; j++)
        if ( A(i,j) && B(j,k) )
          { C(i,k) = true;
            break;
          }
    }
```

How do the two algorithms perform when A and B contain only zeros and ones, respectively? Is there a way to combine the advantages of both methods?

3.5 Random Sources

We frequently need random values in our programs. A *random source* provides an unbounded stream of integers in some range [*low* .. *high*], where *high* and *low* are *ints* with $low \le high$ and $high - low < 2^{31}$. The size restriction comes from the fact that the implementation of random sources uses *long*s. The definition

```
random_source S(7,319);
```

defines a random source S and sets its range to [7 .. 319]. Ranges of the form $[0 .. 2^p]$ are particularly useful. Therefore we have also the definition

```
random_source S(p);
```

that sets the range to $[0 \mathinner{..} 2^p - 1]$ ($1 \le p \le 31$ is required) and the definition *random_source S* that sets the range to $[0 \mathinner{..} 2^{31} - 1]$. The random source *rand_int* is already defined in the header file `random.h`; it has range $[0 \mathinner{..} 2^{31} - 1]$. A random value is extracted from a source by the operator $\gg$. So

```
S >> x >> y;
```

extracts two integers in the range $[low \mathinner{..} high]$ and assigns them to x and y; this assumes that x and y are defined as ints. Note that we are using the C++ input stream syntax for random sources, i.e., $S \gg x$ assigns to x and returns a reference to S.

We may also extract characters, unsigned integers, bools, and doubles from a random source. For the first three types this works as follows: first an integer from the range $[low \mathinner{..} high]$ is extracted and then this integer is converted to the appropriate type. Thus, if b is a boolean variable then $S \gg b$ extracts a truth value. Note that the value of b is not uniformly distributed if $high - low + 1$ is an odd number. In particular, if $low = 0$ and $high = 2$ then we should expect the value *false* about twice as often as the value *true* (as 0 and 2 are converted to *false* and only 1 is converted to *true*). We *recommend* to extract characters and boolean values only from sources whose range spans a power of two. If a source S is asked for a double d by $S \gg d$ then a random integer $u \in [0 \mathinner{..} 2^{31} - 1]$ is extracted and $u/(2^{31} - 1)$ is assigned to d, i.e., the value assigned to d lies in the unit interval.

The range of a random source can be changed either permanently or for a single operation: The operations *S.set_range*(*low*, *high*) and *S.set_range*(p) change the range of S to $[low \mathinner{..} high]$ and $[0 \mathinner{..} 2^p - 1]$, respectively, and $S(low, high)$ and $S(p)$ change the range for a single operation and return an integer in $[low \mathinner{..} high]$ and $[0 \mathinner{..} 2^p - 1]$, respectively.

Of course, the stream of integers generated by a random source is only pseudo-random. It is generated from a *seed* that can either be supplied by the user (by *S.set_seed*(s)) or is generated automatically from the internal clock. If a seed is supplied then the source behaves deterministically; this is particularly useful during debugging. If no seed is supplied the sequence produced depends on the time of the day.

In the remainder of this section we describe several uses and the implementation of random sources.

A Chance Experiment: We use random sources for a chance experiment that is relevant to the analysis of merge sort for secondary memory; see [Moo61] and [Knu81, section 5.4.1]. Assume that we have to sort a set S that is too large to fit into main memory. Merge sort for external memory approaches this problem in two phases. In the first phase it partitions S and sorts each subset and in the second phase it merges the sorted subsets (usually called *runs*). Of course, it is desirable that the number of runs produced in the first phase is kept small, or in other words, that the runs produced in the first phase are long. Assume that M elements of S can be kept in main memory. Then runs of length M can be produced by reading M elements into main memory and sorting them. Longer runs can be produced by a method called *replacement selection*. This method partitions its internal memory into a priority queue Q and a reservoir R that can together store M elements. The production of runs starts by reading M elements into the priority queue. A run is generated by repeated selection of the minimum element *Q_min* from Q. This element is added to the current run (and written to secondary memory) and the spot freed in main memory is filled by the next element x from S. If x is smaller than *Q_min* then x is added to R and it is added to Q otherwise. We continue until Q becomes empty. When this is the case, the elements in R are moved to Q and the production of the next run starts. Each run produced by replacement selection has length at least M. The two extreme situations arise when S is sorted: if S is sorted in descending order then each run has exactly length M and if S is sorted in ascending order then a single run will be produced.

The program below simulates the behavior of replacement selection for a set S of random *doubles*. We maintain a priority queue Q and a stack S. We initialize Q with M random doubles and R to the empty stack. Then we start the production of runs. In each iteration we remove the smallest element *Q_min* from Q and then produce a new random double x. If $x <$ *Q_min* we add x to Q, and we add it to R otherwise. When Q is empty we move all elements from R to Q and start the production of the next run. For each run we record the quotient of the length of the run and M.

⟨*runlength*⟩≡
```
main(){
int M, n;
⟨read M and n⟩
p_queue<double,int> Q;   // second type parameter is not used
stack<double> R;
random_source S;
double x;
int i;
for (i = 0; i < M; i++) { S >> x; Q.insert(x,0); }
array<double> RL(1,n);   // RL[i] = length of i-th run
for (i = 1; i <= n; i++)
{ // production of i-th run
  int runlength = 0;
  while ( !Q.empty() )
  { double Q_min = Q.del_min(); runlength++ ;
    S >> x;
    if (x < Q_min) R.push(x);
    else           Q.insert(x,0);
  }
  RL[i] = (double)runlength / M;
  while ( !R.empty() ) Q.insert(R.pop(),0);
}
⟨produce table runlength⟩
}
```

Table 3.4 shows the output of a sample run; we used $M = 10^5$. The length of the i-th run

Round	Length	Round	Length	Round	Length	Round	Length
1	1.717	6	1.998	11	2	16	2.001
2	1.95	7	1.998	12	2	17	2.002
3	1.998	8	2.002	13	1.999	18	1.992
4	2.002	9	1.997	14	2.002	19	2
5	1.996	10	2	15	2	20	2.003

Table 3.4 Run formation by replacement selection, we used $M = 10^5$ and $n = 20$. You may perform your own experiments by calling program runlength.

seems to converge to $2M$ as n grows. We refer the reader to [Moo61] and [Knu81, section 5.4.1] for a proof of this fact.

We give a second interpretation of the chance experiment above. Consider a circular track on which a snow plow is operating. When the snow plow starts to operate there are M snow flakes on the track (at random locations). In every time unit the snow plow removes one snow flake and one new flake falls (at a random location). We compute how many snow flakes the snow plow removes in its i-th circulation of the track.

Random Permutations and Graphs: We show how to generate more complex random objects, namely random permutations and random graphs.

Let A be an array. We want to permute the elements of A randomly. Let $a_0, \ldots, a_{n-1}$ be the elements of A. We can generate a random permutation of these elements by selecting a random element and putting it into the last position of the permutation, selecting a random element from the remaining elements and putting it into the next to last position of the permutation, and so on. In the program below we realize this process in-place. We keep an index j into A, initially $j = n - 1$. We maintain the invariant that the elements in position 0 to j have not been selected for the permutation yet and that positions $j + 1$ to $n - 1$ contain the part of the permutation that has been produced so far. In order to fill the next position of the permutation we choose a random integer i in $[0 .. j]$ and interchange $A[i]$ and $A[j]$. We obtain

```
random_source S;
for (int j = n - 1; j >= 1; j--)  leda_swap(A[j],A[S(0,j)]);
```

where *leda_swap* interchanges its arguments. The method just described is used in operation *permute*( ) of types *array* and *list*.

Our next task is to generate a random graph with n nodes and m edges. This is very easy. We start with an empty graph G, then add n nodes to G, and finally choose m pairs of random nodes and create an edge for each one of them. A node can be added to a graph G by *G.new_node*( ). This call also returns the newly added node. We store the nodes in an *array<node>* V. In order to add a random edge we choose two random integers, say l and k, in $[0..n-1]$ and then add the edge from $V[l]$ to $V[k]$ to G.

```
random_source S;
graph G;              //empty graph
array<node> V(0,n-1);
  for (int i = 0; i < n; i++) V[i] = G.new_node();
  for (int i = 0; i < m; i++)
        G.new_edge( V[S(0,n-1)] , V[S(0,n-1)] );
```

The program above realizes the function *random_graph*(G, n, m). LEDA also offers functions to generate other types of random graphs, e.g., random planar graphs. We discuss these generators in later chapters.

Non-Uniform Distributions: We show how to generate integers according to an arbitrary discrete probability distribution. The method that we are going to describe is called the *alias-method* and has been invented by Walker [Wal77]. Let $w[0..n-1]$ be an array of positive integers. For all i, $0 \le i < n$, we interpret $w[i]$ as the weight of i. Our goal is to generate i with probability $w[i]/W$, where $W = w[0] + \ldots + w[n-1]$. We start with the simplifying assumption that n divides W and let $K = W/n$. We will remove this restriction later. We view W as an n by K arrangement of squares, n columns of K squares each and label $w[i]$ squares by i for all i, $0 \le i \le n$, see Figure 3.5. In order to generate an integer we select a random square and return its label. This makes the generation of a random integer a constant time process. The drawback of this method is that it requires space W. The space requirement can be improved to $O(n)$ by observing that there is always a labeling of the squares such that at most two different labels are used in any column. This can be seen as follows. Call a weight *small* if it is less than or equal to K and call it *large* otherwise. Clearly, there is at least one small weight. Let $w[i]$ be an arbitrary small weight. If $w[i]$ is equal to K then we assign an entire column to i and if $w[i]$ is less than K then we take an arbitrary large weight (there must be one!), say $w[j]$, and assign $w[i]$ squares to i and $K - w[i]$ squares to j. We also reduce $w[j]$ by $K - w[i]$. In either case, we have reduced the number of weights by one and are left with $n-1$ weights whose sum is $K(n-1)$. Proceeding in this way we label each column by at most two numbers.

We still need to remove the assumption that n divides W. We redefine K as $K = \lceil W/n \rceil$ and add an additional weight $w[n] = K(n+1) - W$. This yields $n+1$ weights whose sum is equal to $K(n+1)$. We can now construct a labeling as described above. We also need to modify the generation process slightly, because it is now possible that the number n is generated. When this happens we declare the generation attempt a failure and repeat. The probability of success is $W/(K(n+1))$ and hence the expected number of iterations required is $K(n+1)/W$. We need to bound this quantity. We have $W \ge n$ since each weight $w[i]$ it at least one and we have $Kn < W + n$ and hence $W > (K-1)n$ by the definition of K. Thus if $K = 1$ then $K(n+1)/W \le (n+1)/n \le 2$ and if $K \ge 2$ then $K(n+1)/W \le K(n+1)/((K-1)n) \le 4$. In either case we conclude that the expected number of iterations required is bounded by 4.

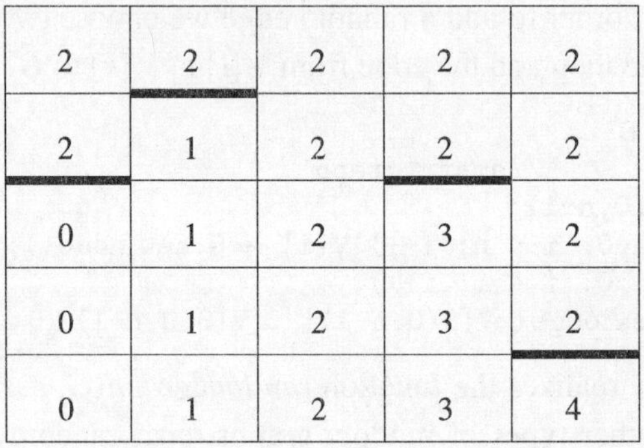

Figure 3.5 Illustration of alias-method. We have $n = 5$, $w = (3, 4, 14, 3, 1)$, and $K = 5$. The labeling shown is succinctly encoded by the vectors $T = (3, 4, 5, 3, 1)$, $L = (0, 1, 2, 3, 4)$, and $U = (2, 2, \_, 2, 2)$: for each column j the lowest T_j squares are labeled L_j and the highest $K - T_j$ squares are labeled U_j.

We turn to an implementation. We define a class *random_variate*. Its constructor takes an *array<int>* w of non-negative integers and index range $[l\,..\,h]$ and sets up the vectors T, L, and U and the integer K defined above. Its member function *generate* generates any integer $i \in [l\,..\,h]$ with probability $w[i]/W$ where $W = \sum_i w[i]$.

⟨*definition of class random_variate*⟩≡

```
class random_variate{
  array<int> T, L, U;
  int l, h, n, K;
public:
  random_variate(const array<int>& w) { ⟨random variate: constructor⟩ }
  int generate()                      { ⟨random variate: generate⟩ }
};
```

The constructor operates in two phases. In the first phase we compute the total weight W, the number n of non-zero weights, the integer K, and an array *array<int>* $u(l, h+1)$ with the additional weight $u[h+1] = K(n+1) - W$.

⟨*random variate: constructor*⟩≡

```
l =  w.low(); h = w.high();
int W = 0;
array<int> u(l,h+1);
n = 0;  // number of non-zero weights
int i;
for (i = l; i <= h; i++)
{ W += u[i] = w[i];
  if ( u[i] < 0 )
     error_handler(1,"random variate: negative weight");
  if ( u[i] > 0 ) n++;
}
```

```
if ( n == 0 ) error_handler(1,"random_variate: no non-zero weight");
K = W/n + (W % n == 0? 0 : 1);
u[h + 1] = K*(n+1) - W; n++;
```

In the second phase we set up the arrays T, L, and U. We use two stacks *Small* and *Large*: In *Small* we store all all i such that $u[i]$ is small and in *Large* we store all i such that $u[i]$ is large. We store the labeling in three arrays T, L, and U such that for every column c, $0 \leq c \leq n-1$, squares 1 to $T[c]$ are labeled $L[c]$ and squares $T[c]+1$ to K are labeled $U[c]$.

⟨*random variate: constructor*⟩+≡

```
stack<int> Small,Large;
for (i = 1; i <= h + 1; i++)
{ if ( u[i] == 0 )  continue;
  if ( u[i] <= K )  Small.push(i);
  else              Large.push(i);
}
U = T = L = array<int>(n);
for (int c = 0; c < n; c++)
{ int i = Small.pop();
  T[c] = u[i];
  L[c] = i;
  if ( u[i] < K )
  { int j = Large.pop();
    U[c] = j;
    u[j] -= (K-u[i]);
    if ( u[j] <= K ) Small.push(j); else Large.push(j);
  }
}
```

The generator chooses a random *row* and a random *column* and looks up the table entry defined by this row and column. If the table entry is different from $h+1$, it is returned. Otherwise the process is repeated.

⟨*random variate: generate*⟩≡

```
int r;
do { int row = rand_int(1,K);
     int column = rand_int(0,n-1);
     r = (row <= T[column] ? L[column] : U[column]);
   }
while (r == h + 1);
return r;
```

Random Walks in Graphs (Simulating Markov Chains): We give an application of class *random_variate*. We perform a random walk on a graph. Let $G=(V,E)$ be a directed graph and for each edge e let $w[e]$ be a non-negative weight. We start our walk in an arbitrary node of G and move according to the following rule: Suppose that we are currently in node v and let $e_0, \ldots, e_{d-1}$ be the edges out of v. We follow edge e_i with probability proportional to $w[e_i]$ for all i, $0 \leq i < d$. If there is no edge out of v the walk terminates. We define a class *markov_chain* that allows us to simulate such a process.

⟨*definition of class markov_chain*⟩≡

```
class markov_chain {
  graph& G;
  int    N;
  node_array<int> visits;
  node   vcur;
  node_array<array<node> > neighbors;
  node_array<random_variate*> variate;
public:
  markov_chain(const graph& g, const edge_array<int>& w,
                          node s = nil): G(g)
    { ⟨markov chain: constructor⟩ }
  void step(int T = 1) { ⟨markov chain: step⟩ }
  int  number_of_visits(node v) { return visits[v]; }
  ⟨markov chain: further member functions⟩
};
```

The constructor takes a graph G, an edge array of weights, and a start vertex. If no start vertex is specified the first node of G is taken as the start vertex. The function *step*(T) performs T steps of the random walk and the function *number_of_visits*(v) returns the number of visits to node v. We give the details below.

The constructor sets up the required data structures. We build two data structures for each node v: an *array<node> neighbors*$[v]$ that stores for each i, $0 \leq i < outdeg(v)$, the target of the i-th edge out of v and a random variate *variate*$[v]$ that produces i with probability proportional to the weight of the i-th edge out of v. We set up both data structures by scanning through the edges out of v, collecting the target of the edges out of v in *neighbors*$[v]$ and their weights in a temporary array *weights*. Then we use the latter array to construct the random variate for v.

⟨*markov chain: constructor*⟩≡

```
N = 0;
visits = node_array<int>(G,0);
vcur = s; if ( s == nil) vcur = G.first_node();
neighbors = node_array<array<node> >(G);
variate   = node_array<random_variate*>(G);

node v; edge e;
forall_nodes(v,G)
{ if (G.outdeg(v) == 0) continue;
  neighbors[v] = array<node>(G.outdeg(v));
  array<int> weights(G.outdeg(v));
  int i = 0;
```

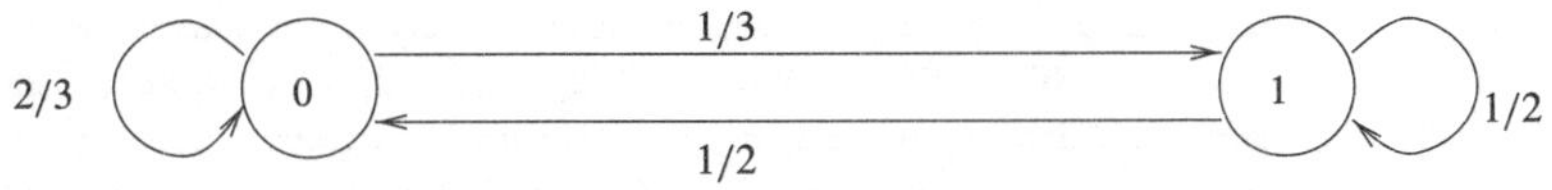

Figure 3.6 A graph with two nodes. The edge probabilities are shown next to each edge.

```
    forall_adj_edges(e,v)
    { neighbors[v][i] = G.target(e);
      weights[i] = w[e];
      i++;
    }
    variate[v] = new random_variate(weights);
  }
```

Given these data structures it is easy to perform T steps of the walk. If the outdegree of the current node is zero we stay put. Otherwise, we generate a neighbor at random and move to the neighbor.

⟨*markov chain: step*⟩≡

```
if (T <= 0 ) return;
for (int i = 0; i < T; i++)
{ if ( G.outdeg(vcur) == 0) return;
  vcur = neighbors[vcur][variate[vcur] -> generate()];
  visits[vcur]++;
  N++;
}
```

Let us perform a random walk on the graph shown in Figure 3.6.

⟨*random_walk_example*⟩≡

```
main(){
graph G;
node v0 = G.new_node();
node v1 = G.new_node();
edge e00 = G.new_edge(v0,v0); edge e01 = G.new_edge(v0,v1);
edge e10 = G.new_edge(v1,v0); edge e11 = G.new_edge(v1,v1);
edge_array<int> weight(G);
weight[e00] = 2; weight[e01] = 1;
weight[e10] = 1; weight[e11] = 1;
while( true )
{ int N = read_int("number of steps = ");
  markov_chain M(G,weight);
  M.step(N);
  cout << "# of visits of v0 = " << M.number_of_visits(v0) <<"\n";
  cout << "# of visits of v1 = " << M.number_of_visits(v1) <<"\n";
}
}
```

n	1	10	100	1000	10000	100000	1000000	10000000
v_0	0	3	63	570	6058	60180	600704	6003568
v_1	1	7	37	430	3942	39820	399296	3996432

Table 3.5 The statistics of a random walk on the graph of Figure 3.6. Each column gives the number of visits to both nodes in the first n steps of the walk. You may perform your own experiments by calling random_walk.

Table 3.5 shows a sample output of this program. There is a simple analytical explanation for the output based on the theory of Markov chains, see [KSK76] for an introduction to Markov chains. Let $p_{i,n}$ be the relative frequency of node i during the first n steps of the random walk. It is known that the $p_{i,n}$ converge to so-called stationary probabilities π_i and that the stationary probabilities satisfy a system of linear equations directly related to the transition graph. For each node j there is an equation expressing π_j as a sum over all edges directed into j. The contribution to this sum of an edge (i, j) is $q_{ij} \cdot \pi_i$, where q_{ij} is the transition probability of the edge. In our example we obtain:

$$\begin{aligned} \pi_0 &= 2/3 \cdot \pi_0 + 1/2 \cdot \pi_1 \\ \pi_1 &= 1/3 \cdot \pi_0 + 1/2 \cdot \pi_1. \end{aligned}$$

This system has solution $\pi_0 = 6/10$ and $\pi_1 = 4/10$. In Table 3.5 we see the convergence of the visit frequencies to the stationary probabilities.

Dynamic Random Variates: We generalize the class *random_variate* to a class called *dynamic_random_variate* which offers an additional operation *set_weight* that allows the user to change weights dynamically. More precisely, if R is a dynamic random variate with weight vector w and i is in the index range of w then *set_weight*(i, g) changes $w[i]$ to g; g is an arbitrary non-negative integer. The generation process of dynamic random variates is less efficient than the one for (static) random variates; it takes time $O(\log n)$, where n is the size of the index range of w.

The implementation is fairly simple. We put the weights into the leaves of a balanced binary tree with n leaves and $n - 1$ internal nodes. In each node we store the sum of the weights of the leaves in its subtree. In particular, $W = \sum_i w[i]$ is stored in the root of the tree. A weight change amounts to updating the weights along one leaf to root path. In order to generate a random variate we choose a random integer s in $[0 \,..\, W - 1]$. If s is less than the total weight of the left subtree, we proceed recursively to the left subtree and if s is larger or equal to the total weight of the left subtree, we subtract the weight of the left subtree and proceed recursively to the right subtree. In this way, changing a weight and generating a random variate takes time proportional to the height of the tree. If a balanced tree is used the height is $O(\log n)$.

A particularly simple implementation results when the nodes of the tree are numbered

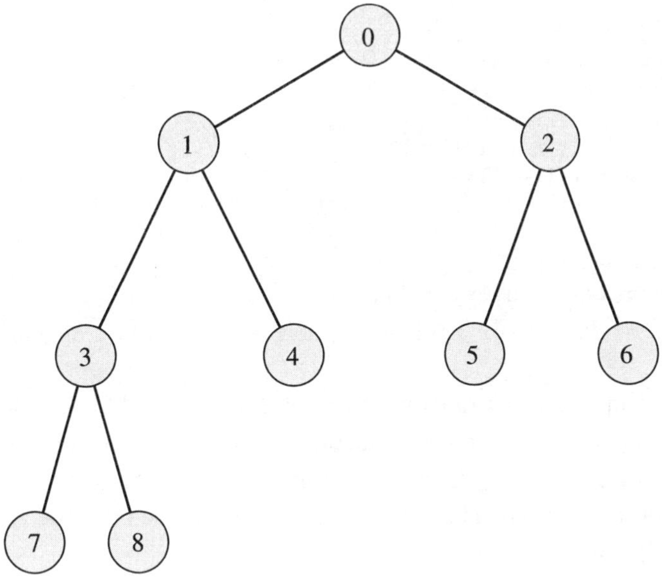

Figure 3.7 A tree with five leaves and a total of nine nodes. The number of each node is shown. The children of node i have numbers $2i$ and $2i+1$.

with the integers 1 to $2n-1$ in preorder, i.e., the root is given the number 1, the children of the node with number i, $1 \le i < n$ have numbers $2i$ and $2i+1$, and the leaves are numbered n to $2n-1$. See Figure 3.7 for an example. The parent of node i, $2 \le i \le 2n-1$ has number $\lfloor i/2 \rfloor$.

In the implementation we use an *array<int>* u with index range $[1 \,..\, 2n-1]$ to store the tree.

⟨definition of class dynamic_random_variate⟩≡

```
class dynamic_random_variate{
private:
  array<int> u;
  int n, h, l;
public:
  dynamic_random_variate(const array<int>& w)
  { ⟨dynamic random variate: constructor⟩ }
  int generate()              { ⟨dynamic random variate: generate⟩ }
  int set_weight(int i, int g) { ⟨dynamic random variate: set weight⟩ }
};
```

The constructor stores the weight vector w in the entries n to $2n-1$ of u and then fills each entry u_i, $n-1 \ge i \ge 1$ as the sum of the entries of its children.

⟨*dynamic random variate: constructor*⟩≡

```
l = w.low(); h = w.high(); n = h - l + 1;
u = array<int>(1,2*n - 1);
int i;
for (i = 0; i < n ; i++)
{ u[n + i] = w[l + i];
  if ( u[n + i] < 0 ) error_handler(1,"dynamic variate: negative weight");
}
for (i = n - 1; i > 0; i--)
  u[i] = u[2*i] + u[2*i + 1];
if (u[1] == 0 ) error_handler(1,"dynamic variate: no non-zero weight");
```

The generator chooses a random integer s in $[0\,..\,W-1]$ and then walks down a path in the tree. When the walk reaches node i, s is a random integer in $[0\,..\,u[i]-1]$. If i is a leaf we return $l+(i-n)$ since the leaf numbered i corresponds to entry $l+(i-n)$ of weight vector w. If i is not a leaf and $s < u[2i]$, we proceed to child $2i$ and if $s \geq u[2i]$, we subtract $u[2i]$ from s and proceed to child $2i+1$.

⟨*dynamic random variate: generate*⟩≡

```
int s = rand_int(0,u[1] - 1);
int i = 1;
while ( i < n )
{ int j = 2*i;
  if ( s < u[j] )
    i = j;
  else
    { i = j + 1;
      s -= u[j];
    }
}
return l + i - n;
```

In order to change weight i to g we walk the path from leaf $n+(i-l)$ to the root and change all entries of u along the path by $delta = g - u[i]$. The old value of $u[i]$ is returned.

⟨*dynamic random variate: set weight*⟩≡

```
int ui = u[i];
i = n + (i - l);
int delta = g - u[i];
if ( g < 0 ) error_handler(1,"dynamic variate: negative weight");
while (i > 1)
{ u[i] += delta;
  i = i/2;
}
u[1] += delta;
if ( u[1] == 0 ) error_handler(1,"dynamic variate: no positive weight");
return ui;
```

n	Static	Dynamic
100	32.02	52.7
10000	41.07	90.34

Table 3.6 Running time of random variate generation: We set up a weight vector with n entries and then generated 10^7 random variates according to it. We used classes *random_variate* and *dynamic_random_variate*.
You can make your own experiments using the random_variate_demo.

Table 3.6 illustrates the speed of our two methods for generating random variates. Surprisingly, the $O(\log n)$ method is faster than the constant time method.

Dynamic Markov Chains: The use of dynamic random variates instead of static random variates in Markov chain data type yields a dynamic Markov chain data type which also supports the change of edge weights.

Simulating a Supermarket Check-Out: We use dynamic random variates to simulate a supermarket check-out. We consider a supermarket with n check-out stations. We assume that there is a queue (maybe empty) in front of every check-out station and use $q[i]$ to denote the queue length in front of the i-th check-out station. Servicing a customer at a check-out station takes either 1 (probability $2/3$) or 2 (probability $1/3$) time units. Thus the average servicing time is $4/3$ time units.

We assume that $3n/4$ customers arrive at every time unit. Customers tend to choose check-out stations with short queues. We assume that a customer chooses queue i with probability proportional to $1/(1+q[i])$.

In the program we define random variates R and S; S is a static random variate which models the distribution of service times and R is a dynamic random variate which yields check-out stations. In R we use $\lfloor M/(1+q[i])\rfloor$ as the weight of i, where M is a large constant. In each time step we first generate $3n/4$ customers. For each customer we choose the service length by calling *S.generate*( ) and the service station by calling *R.generate*( ). We update the queue lengths after each generation of a customer.

We collect all customers requiring short service in a list *short_service* and all customers requiring long service in a list *long_service*. After having generated the new customers we service all customers in *short_service*, update queue lengths appropriately, and move all customers in *long_service* to *short_service*.

⟨*supermarket check-out*⟩≡

```
array<int> q(n);
array<int> w(n);
int M = 10000;
for (int i = 0; i < n; i++) { q[i] = 0; w[i] = M; }
```

```
dynamic_random_variate R(w);
array<int> w1(1,2); w1[1] = 2; w1[2] = 1;
random_variate S(w1);
list<int> short_service, long_service;
for (int t = 0; t < T; t++)
{ for (int k = 0; k < 3*n/4; k++)
  { int i = R.generate(); q[i]++; R.set_weight(i,M/(1 + q[i]));
    if ( S.generate() == 1)
      short_service.append(i);
    else
      long_service.append(i);
  }
  int i;
  forall(i,short_service)
  { q[i]--; R.set_weight(i,M/(1 + q[i])); }
  short_service.clear();
  short_service.conc(long_service);
  ⟨report queue lengths⟩
}
```

Implementation of random_source: Our implementation of random sources follows the description in [Knu81, Vol2, section 3.2.2]. We first give the mathematics and then the program. Internally, we always generate a sequence of integers in the range $[0 \,..\, 2^{31} - 1]$. We define 32 unsigned longs $X_0, X_1, \ldots, X_{31}$ by

$$X_0 = seed$$

and

$$X_i = (1103515245 \cdot X_{i-1} + 12345) \bmod m$$

for $1 \leq i \leq 31$. Here $m = 2^{32}$. We extend this sequence by

$$X_i = (X_{i-3} + X_{i-32}) \bmod m$$

for $i \geq 32$. In this way an infinite sequence $X_0, X_1, \ldots$ of unsigned longs is obtained. Following [Knu81, Vol2, section 3.2.2], we discard the first 320 elements of this sequence (they are considered as a warm-up phase of the generator) and we also drop the right-most bit of each number (since it is the least random). Thus, the i-th number output by the internal generator is

```
(X[i + 320] >> 1) & 0x7fffffff.
```

We next show how to generate a number uniformly at random in $[low \,..\, high]$. Let X be a number produced by the internal generator. Then $low + X \bmod (high - low + 1)$ is a number in the range $[low \,..\, high]$. However, this number is not uniformly distributed (consider the case where $low = 0$ and $high = 2^{31} - 2$ and observe that in this case the number 0 is generated with probability twice as large as any other number). We therefore proceed differently. Our approach is based on the observation that if X is a random number in $[0 .. 2^{31}-1]$ and p is an integer less than 32 then $X \bmod 2^p$ is a random number in $[0 .. 2^p-1]$. Let $\mathit{diff} = \mathit{high} - \mathit{low}$ and let p be such that $2^{p-1} \le \mathit{diff} < 2^p$. We generate random numbers X using the internal source until $X \bmod 2^p \le \mathit{diff}$ and then output $\mathit{low} + X \bmod 2^p$. Since $2^{p-1} \le \mathit{diff}$ at most two X's have to be tried on average. The complete program follows.

⟨*generation of a random number in [low..high]*⟩≡

```
int diff = high - low;
/* compute pat = 2^p - 1 with 2^{p-1} <= diff < 2^p */
unsigned long pat = 1;
while (pat <= diff) pat <<= 1;
pat--;
/* pat = 0...01...1 with exactly p ones.
   Now, generate random x in [0 .. pat]
   until x <= diff and return low + x    */
unsigned long x = internal_source() & pat;
while ( x > diff) x = internal_source() & pat;
return (int)(low + x);
```

Exercises for 3.5

1 Add an operator ≫ to the type *random_source* that allows you to extract a random point in the two-dimensional unit square.

2 Consider the following program.

```
int i,j,x;
array<int> A(0,n-1);
for (i = 0; i < n; i++)
  { while (true)
      { x = rand_int(0,n-1);
        for (j = 0; j < i && x != A[j]; j++) ; // empty body
        if (j == i) break;
      }
    A[i] = x;
  }
```

a) Does it generate a random permutation of the integers 0 to $n-1$?
b) What is the expected running time of the program?

3 Change the random graph generator such that it generates graphs without self-loops, i.e., no edges (v, v), and without parallel edges, i.e., no two edges with the same source and target.

4 Let $d_0, \ldots, d_{n-1}$ be non-negative integers whose sum is even. Generate a random undirected graph where node i has degree d_i for all i, $0 \le i < n$. Hint: Create an array A of length $2m = \sum_i d_i$, write the integer i into d_i entries of A for all i, permute A, and then generate the edge $(A[2j], A[2j+1])$ for all j, $0 \le j < m$.

5 Balls and bins: Throw n balls randomly into m bins, i.e, choose n random integers in the range $[0 .. m-1]$ and tabulate how often each number is chosen. Perform the experiment

with $n = 10^6$ and $m = 100$, $m = 1000, \ldots, m = 10^6$. If you want to understand the outcome of the experiment analytically consult [MR95].

6 Use classes *random_variate* and *so_set* to perform the following experiment. Let w be any vector of n non-negative integers with $w_0 \geq w_1 \geq \ldots \geq w_{n-1}$. Store the integers 0 to $n-1$ in a *so_set* and perform N access operations. For each i, $0 \leq i < n$ access i with probability proportional to w_i. Determine the total cost of all accesses where the cost of an access is the distance of the accessed item from the front of the list (you need to modify *so_set::member* slightly in order to get this information) and compare it to $C = N \sum_i w_i(i+1)/W$ where $W = \sum_i w_i$. Note that C is the expected cost of the accesses if the list were arranged in order of decreasing weight.

3.6 Pairs, Triples, and such

A tuple is an aggregation of variables of arbitrary types. LEDA offers two-tuples, three-tuples, and four-tuples. We use two-tuples as our running example in this section. For any types A and B and objects a and b belonging to these types the declarations

```
two_tuple<A,B> p;
two_tuple<A,B> q(a,b);
```

define a two-tuple p and a two-tuple q, respectively. The components of p are initialized to the default values of A and B, respectively, and the components of q are initialized with copies of a and b, respectively. The operations *first* and *second* return the two variables contained in a two-tuple. So we may write

```
a = p.first();
p.second() = b;
```

The operators $==$, $\ll$, $\gg$ and the functions *compare* and *Hash* are defined for two-tuples. They assume that the corresponding functions are defined for the component types. The operators $\ll$ and $\gg$ read and write a two-tuple, respectively, the operator $==$ realizes component-wise equality, *compare* amounts to the lexicographic ordering of two-tuples and *Hash* returns the bitwise exclusive or of the hash values of the components. All of these functions and operators are defined as template functions. For example,

```
template <class A, class B>
int compare(const two_tuple<A,B>& p, const two_tuple<A,B>& q)
{ int s = compare(p.first(),q.first());
  if (s != 0) return s;
  return compare(p.second(),q.second());
}
```

If one uses two-tuples in a situation that requires the compare function for two-tuples, e.g., if one defines a *list<two_tuple<int, int> > L* and then calls *L.sort( )*, it is wise to give the compiler a hint that it should make the compare function for *two_tuple<int, int>*. In the following program this is done by defining a variable p of type *two_tuple<int, int>* and calling *compare*(p, p).

⟨*two_tuple_test*⟩≡

```
main()
{ list< two_tuple<int,int> > L;
  two_tuple<int,int> p;
  compare(p,p); // dummy compare
  L.sort();
}
```

3.7 Strings

A *string* is a sequence of characters, where a character is an element of the C++ type *char*. The number of characters in a string is called the length of the string and the characters in a string are numbered starting at zero. So u is the character at position one in *Kurt*. The string of length zero is called the empty string; it is the default value of the type. Strings are related to the *char$*$* type of C++. There are, however, two significant differences:

- The value of a variable of type string is a sequence of characters, it is not a pointer. In particular, assignment and parameter passing by value work properly for strings. Strings are a primitive type, see Section 2.3.
- Strings offer a large number of additional operations, e.g., pattern matching, substring replacement, and comparison according to the lexicographic ordering. We have to admit, however, that some programming languages, e.g., PERL and AWK, offer much more elaborate string classes.

Let us see strings at work.

```
string s("Stefan");
```

defines a string variable s and initializes it with the value `"Stefan"`.

```
string t = s + s;
```

defines another string variable t and initializes it to `"StefanStefan"`; the operator $+$ is the concatenation operation on strings. The expression $t(2, 5)$ returns the substring of t starting at position 2 and ending at position 5. Since we start counting at 0 this is the string `"efan"`. We can also search for the occurrence of one string in another string: If a and b are strings then *a.pos*(b) searches for an occurrence of b in a. If b does not occur in a then *pos* returns -1 and if b does occur then it returns the first position in a at which b occurs. Thus

```
t.pos("efa");
```

returns 2, i.e., the first position in t at which an occurrence of `"efa"` starts, and *t.pos("Kurt")* returns -1. Another useful operation on strings is substring replacement. It comes in several forms: $a.replace(i, j, b)$ returns $a(0, i-1) + b + a(j+1, a.length() - 1)$, i.e., b is substituted for the substring $a(i, j)$, and $a.replace(b1, b2, n)$ replaces the n-th occurrence of $b1$ in a by $b2$, and finally $a.replace\_all(b1, b2)$ replaces all occurrences of $b1$ in a by $b2$. It is important to notice that all three versions do not change the string a. Rather, they return a new string. So after

```
string u = t.replace(2,5,"Kurt");
string v = t.replace(s,"Kurt",2);
```

we have a string u with value `"StKurtStefan"`, i.e., the substring of t starting at position 2 and ending at position 5 is replaced by `"Kurt"`, and a string v with value `"StefanKurt"`, i.e., the second call of *replace* returns a string in which the second occurrence of s in t is replaced by `"Kurt"`.

The operator $<$ realizes the lexicographic ordering of strings. So

```
(t < (s + s + s ));
```

evaluates to true since `"StefanKurt"` precedes `"StefanStefanStefan"` in the lexicographic ordering of strings. Many other operations on strings can be found in the manual.

Strings are implemented by C++ character vectors. All operations on strings that do not involve pattern matching take linear time. Pattern matching takes quadratic time. More precisely, it takes time $O(nm)$ in the worst case to search for a string of length m in a string of length n. There are $O(n+m)$ pattern matching algorithms, see for example [CLR90].

3.8 Making Simple Demos and Tables

This book contains many tables. For many of these tables there is also a corresponding demo which allows the reader to perform experiments on his or her own. We wanted to have a single program that handles both cases. In this section we describe the IO-interface used in these programs.

The program below serves as the random_variate_demo and also produces Table 3.6. It makes all its input and output through *IO_interface I*. The program can be executed in two modes: in book-mode it produces a table[2] and in demo-mode it realizes the random_variate_demo. The demo-mode is the default and the book-mode is selected at compile-time by compiling with the flag `-DBOOK`[3].

[2] This book is typeset using LaTeX and hence the program generates a sequence of LaTeX-commands that produce a table.

[3] An alternative design would be to use an integer variable to distinguish between the cases and set the variable through a command line argument.

⟨*random_variate_demo.c*⟩≡

```
#include <LEDA/random_variate.h>
#include <LEDA/IO_interface.h>

main()
{ IO_interface I("Random Variates");

  I.write_demo("This demo illustrates the speed of classes \
random variate and dynamic random variate. \nYou will be asked \
to input integers n and N. We set up the weight vector w with \
w[2] = 2, w[3] = 3, ..., w[n+1] = n + 1 and generate N random \
variates according to this weight vector.");

  int n, N;
  n = I.read_int("n = ",100);
  N = I.read_int("N = ",100000);
  if ( n < 1 ) error_handler(1,"n must be at least one");

#ifdef BOOK
N = 10000000;
for (n = 100; n <= 10000; n = n*n)
{ I.write_table("\n ", n);
#endif
  array<int> w(2, 1 + n);
  array<double> Rfreq(2,n+1), Qfreq(2,n+1);
  int W = 0; int i;
  for (i = 2; i < n + 2; i++) { W += w[i] = i; Qfreq[i] = Rfreq[i] = 0; }

  dynamic_random_variate R(w);
  random_variate Q(w);

  float T = used_time(); float UT;
  for (i = 0; i < N; i++) Qfreq[Q.generate()]++;

  UT = used_time(T);
  I.write_demo("static random variate, time = ",UT);
  I.write_table(" & ",UT);

  for (i = 0; i < N; i++) Rfreq[R.generate()]++;

  UT = used_time(T);
  I.write_demo("dynamic random variate, time = ",UT);
  I.write_table(" & ",UT, " \\\\ \\hline");

  I.write_demo("We report some frequencies.");

  for (i = n + 1; i >= Max(2,n - 3); i--)
  { I.write_demo("relative frequency, i = ",i);
    I.write_demo(0,", w[i]/W = ",((double)w[i])/W);
    I.write_demo(1,"generated freq, static variate = ",  Qfreq[i]/N);
    I.write_demo(1,"generated freq, dynamic variate = ", Rfreq[i]/N);
  }
#ifdef BOOK
}
#endif
}
```

The output statements come in two kinds: *write_table* and *write_demo*. The output statement *write_xxx* produces output when executed in *xxx*-mode and produces no output otherwise. Thus, the introductory text that explains the demo is output in demo-mode, but is suppressed in book-mode. The output statements come in different forms:

```
I.write_xxx(string mes);
I.write_xxx(string mes, double T, string mes2 = "");
I.write_xxx(string mes, int T,    string mes2 = "");
I.write_xxx(int k, string mes);
I.write_xxx(int k, string mes, double T, string mes2 = "");
I.write_xxx(int k, string mes, int T,    string mes2 = "");
```

The first form outputs the string *mes* and the second and the third form output the string *mes*, followed by the number T, followed by the optional string *mes2*. The output is preceded by an empty line. The last three forms allow a finer control over the positioning of the output; the output is preceded by k line feeds, i.e., with $k = 0$ the output is printed on the same line as the previous output, with $k = 1$ the output is printed on a new line, and with $k = 2$ the output is preceded by an empty line.

The input statement

```
int I.read_int(string mes, int n = 0);
```

returns n in book-mode and asks for an integer input with prompt *mes* in demo-mode.

The precision of the output of double-values is controlled by a precision parameter p. It is set to 4 by default and can be changed by

```
I.set_precision(int prec);
```

We come to the implementation. It is quite simple.

⟨*IO_interface*⟩≡

```
⟨definition of IO_interface_book⟩
⟨definition of IO_interface_demo⟩
#ifdef BOOK
#define IO_interface IO_interface_book
#else
#define IO_interface IO_interface_demo
#endif
```

We define classes *IO_interface_book* and *IO_interface_demo* in the obvious way (see LEDAROOT/incl/LEDA/IO_interface.h for the details) and define *IO_interface* as one of them depending on the compile-time flag.

4

Numbers and Matrices

Numbers are at the origin of computing. We all learn about integers, rationals, and real numbers during our education. Unfortunately, the number types *int*, *float*, and *double* provided by C++ are only crude approximations of their mathematical counterparts: there are only finitely many numbers of each type and for floats and doubles the arithmetic incurs rounding errors. LEDA offers the additional number types *integer*, *rational*, *bigfloat*, and *real*. The first two are the exact realization of the corresponding mathematical types and the latter two are better approximations of the real numbers. Vectors and matrices are one- and two-dimensional arrays of numbers, respectively. They provide the basic operations of linear algebra.

4.1 Integers

C++ provides the integral types *short*, *int*, and *long*. All three types come in signed and unsigned form. Let w be the word size of the machine and let $m = 2^w$. Most current workstations have $w = 32$ or $w = 64$. Unsigned ints and signed ints use w bits, shorts use at most that many bits, and longs use at least that many bits.

The unsigned integers consist of the integers between 0 and $m - 1$ (both inclusive) and arithmetic is modulo m.

The signed integers form an interval [`MININT`,`MAXINT`], where `MININT` and `MAXINT` are predefined constants; under UNIX they are available in the systems file `limits.h`. On most machines signed integers are represented in two's complement. Then `MININT` $= -2^{w-1}$ and `MAXINT` $= 2^{w-1} - 1$. The conversion from signed ints to unsigned ints adds a suitable multiple of m so as to bring the number into the interval $[0 .. m - 1]$. If numbers are represented in two's complement this conversion does not change the bit pattern. The conversion from unsigned int to signed int is machine dependent.

An arithmetic operation on signed integers may produce a result outside the range of representable numbers; one says that the operation underflows or overflows. The treatment of overflow and underflow is implementation dependent, in particular, it is not guaranteed that they lead to a runtime error, in fact they usually do not. On the author's workstations the summation MAXINT + MAXINT has result -2, since adding $011\ldots1$ to itself yields $11\ldots10$, which is the representation of -2 in two's complement. We give an example of the disastrous effect that an undetected overflow might have.

Some network algorithms are easier to state if the integers are augmented by the value ∞. For example, in a shortest path algorithm it is convenient to initialize the distance labels to ∞. In an implementation it is tempting to use MAXINT as the implementation of ∞ and to forget that it does not quite have the properties of ∞. In particular, MAXINT + 1 = MININT on the author's workstations which is drastically different from mathematics' $\infty+1=\infty$. This difference led to the following error in one of the first author's programs[1]. He implemented Dijkstra's shortest path algorithm (its working is discussed in Section 6.6) as follows:

```
void DIJKSTRA(const graph& G, node s, const edge_array<int>& cost,
              node_array<int>& dist)
{ node_pq PQ(G);
  node v; edge e;
  forall_nodes(v,G) dist[v] = MAXINT;
  dist[s] = 0;
  forall_nodes(v,G) PQ.insert(v,dist[v]);
  while (!PQ.empty())
  { node v = PQ.delete_min();
    forall_adj_edges(e,v)
    { node w = G.target(e);
      if (dist[v] + cost[e] < dist[w])
      { dist[w] = dist[v] + cost[e];
        PQ.decrease_p(w,dist[w]);
      }
    }
  }
}
```

This program works fine when all nodes are reachable from s and all edge costs are in [0 .. MAXINT/n], where n is the number of nodes of G. However, consider the execution on the graph shown in Figure 4.1. When node v is removed from the queue, we have $dist[v] = dist[w] =$ MAXINT. We compute MAXINT + 1 which is MININT and hence decrease w's distance to MININT, a serious error. A correct implementation inserts only s into the queue initially and replaces the innermost block by

[1] The second author insists that he has never made this particular mistake.

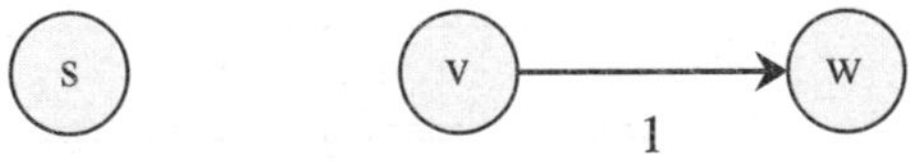

Figure 4.1 An example, where the naive use of MAXINT as a substitute for ∞ in Dijkstra's algorithm has a disastrous effect.

```
int c = dist[v] + cost[e];
if (dist[w] == MAXINT) PQ.insert(w,c);
else PQ.decrease_p(w,c);
dist[w] = c;
```

We come to the LEDA type *integer*. It realizes the mathematical type integer. The arithmetic operations $+$, $-$, $*$, $/$, $+=$, $-=$, $*=$, $/=$, $-$ (unary), $++$, $--$, the modulus operation (%, %=), bitwise AND (&&, &&=), bitwise OR (||, ||=), the complement operator ($\sim$), the shift operators ($\ll$, $\gg$), the comparison operators $<$, $\leq$, $>$, $\geq$, $==$, $!=$, and the stream operators are available. These operations never overflow and always yield the exact result. Of course, they may run out of memory. The following program computes the product of the first n integers.

```
integer factorial(int n) // computes 1 * 2 * ... * n
{ integer fac = 1; //automatic conversion from int
  for (int i = 2; i <= n; i++)  fac = fac*i;
  return fac;
}
```

Integers also provide some useful mathematical functions, e.g., *sqrt*(a) returns $\lfloor\sqrt{a}\rfloor$, *log*(a) returns $\lfloor\log a\rfloor$, and *gcd*(a, b) returns the greatest common divisor of a and b. We refer the reader to the manual pages for a complete listing.

Integers are essentially implemented by a vector of unsigned longs. The sign and the size are stored in extra variables. The implementation of integers is very efficient and compares well with other implementations. This is particularly true on SPARC machines since we have implemented several time critical functions not only in C++ but also in SPARC assembler code. When integers are used on SPARC machines the faster assembler code is executed. The running time of addition is linear and the running time of multiplication is $O(L^{\log 3})$, where L is the length of the operands. The following program verifies the latter fact experimentally. It repeatedly squares an integer n and measures the time needed for each squaring operation. In each iteration it prints the current length of n (= number of binary digits), the time needed for the iteration and the quotient of the running time of this and the previous iteration.

⟨*multiplication_times*⟩≡

```
main()
{ integer n;
  ⟨multiplication times: read n⟩
  int i;
```

n	Running time	T/T_prev
167587	0.63	3
335173	1.88	2.984
670346	5.74	3.053
1340691	17.43	3.037

Table 4.1 The time required to multiply two n bit integers. The multiplication times demo allows you to perform your own experiments.

```
  for (i = 0; i < 11; i++) n = n * n;
  float T_prev = 0;
  for (i = 0; i <= 5; i++)
  { float T = used_time();
    n = n * n;
    T = used_time(T);
    ⟨multiplication times: report times⟩
    T_prev = T;
  }
}
```

Table 4.1 shows a sample output of this program. Since n is squared in each iteration, its length L essentially doubles in each iteration. Thus, if the running time of an iteration is $c \cdot L^\alpha$ for some constants c and α then the running time of the next iteration is $c \cdot (2L)^\alpha$ and hence the quotient is $c \cdot (2L)^\alpha/(c \cdot L^\alpha) = 2^\alpha$. The measured quotient is about 3. Thus, $\alpha \approx \log 3$.

Integers are used a lot in LEDA's geometric algorithms. We briefly hint at the use now and treat it in detail in Chapter 9. Consider three points p, q, and r in the plane and let l denote the line through p and q and oriented from p to q. For any point s use s_1 and s_2 to denote its Euclidean coordinates. The test of whether r lies to the right of l, on l, or to the left of l is tantamount to determining the sign of the determinant

$$\begin{vmatrix} 1 & 1 & 1 \\ p_1 & q_1 & r_1 \\ p_2 & q_2 & r_2 \end{vmatrix}.$$

If the coordinates of our points are floating point numbers and the determinant[2] is evaluated

[2] Note that the determinant is zero if and only if the third column is a linear combination of the first two columns, i.e., if there are reals λ and μ such that $\lambda + \mu = 1$ and $r = \lambda p + \mu q$. In other words, if $r = p + \mu(q - p)$ for some μ. This shows that the determinant is zero if and only if r lies on the line through p and q. We still need to argue that the sign distinguishes the two half-planes defined by the line. Consider two points r and r' and the line segment from r to r'. The value of the determinant changes continuously as one moves from r to r' and hence assumes value 0 if r and r' lie on different sides of the line and does not assume value 0 if r and r' lie on the same side of the line. Since the determinant is a linear function we conclude that the two sides of the line are distinguished by the sign of the determinant.

with floating point arithmetic we may incur rounding error and determine the sign of the determinant incorrectly. This is a frequent source of error in the implementation of geometric algorithms, as we will see in Sections 4.4 and 9.6 . If the coordinates are integers then the determinant can be evaluated exactly and the correct sign can be determined. This feature facilitates the correct implementation of geometric algorithms enormously.

Exercises for 4.1

1 Write a procedure *random_integer(int L)* that returns a random integer of length L.

2 The greatest common divisor of two numbers x and y with $x \geq y \geq 0$ can be computed by the recursion $\gcd(x, y) = x$ if $y = 0$ and $\gcd(x, y) = \gcd(y, x \bmod y)$ if $y > 0$. Implement this algorithm, run it on integers of various lengths, and count the number of recursive calls. Relate the number of recursive calls to the length of y. Prove that the number of recursive calls is at most proportional to the length of y. Hint: Assume $x > y$ and let $x_0 = x$ and $x_1 = y$. For $i > 1$ and $x_{i-1} \neq 0$ let $x_i = x_{i-2} \bmod x_{i-1}$. Let $x_k = 0$ be the last element in the sequence just defined. Relate this sequence to the gcd-algorithm. Show that $x_{k-1} > 0$ and $x_{i-2} \geq x_{i-1} + x_i$ for $i < k$. Conclude that x_{k-j} is at least as large as the j-th Fibonacci number.

3 The standard algorithm for multiplying two L-bit integers has running time $O(L^2)$. LEDA uses the so-called Karatsuba-method ([KO63]) that runs in time $O(L^{\log 3})$. In order to multiply two numbers x and y it writes $x = x_1 \cdot 2^{L/2} + x_2$ and $y = y_1 \cdot 2^{L/2} + y_2$, where x_1, x_2, y_1, and y_2 have $L/2$ bits. Then it computes $z = (x_1 + x_2) \cdot (y_1 + y_2)$ and observes that $x \cdot y = x_1 \cdot y_1 \cdot 2^L + (z - x_1 y_1 - x_2 y_2) \cdot 2^{L/2} + x_2 y_2$. In this way only three multiplications of $L/2$-bit integers are needed to multiply two L-bit integers. The standard algorithm requires four. Implement Karatsuba's algorithm and time it as described in the text. Compare to the member function *operator$*$*.

4 In program ⟨*multiplication_times.c*⟩ let *n_old* be the value of n before the assignment $n = n * n$. Extend the program such that it also computes n/n_old and *sqrt*(n) in each iteration. Measure the execution times and compute quotients of successive execution times. Try to explain your findings.

5 Develop algorithms for integer division and integer square root based on Newton's iteration.

4.2 Rational Numbers

A rational number is the quotient of two integers. Well, that is the mathematical definition and it is also the definition in LEDA. The arithmetic operations $+$, $-$, $*$, $/$, $+=$, $-=$, $*=$, $/=$, $-$ (unary), $++$, $--$ are available on rationals. In addition, there are functions to extract the numerator and denominator, to cancel out the greatest common divisor of numerator and denominator, to compute squares and powers, to round rationals to integers, and many others.

LEDA's rational numbers are not necessarily normalized, i.e., numerator and denominator of a rational number may have a common factor. A call *p.normalize*( ) normalizes p.

This involves a gcd-computation to find the common factor in numerator and denominator and two divisions to remove them. Since normalization is a fairly costly process we do not do it automatically. It is, however, advisable to do it once in a while in a computation involving rational numbers.

Exercises for 4.2

1 Write a program to solve linear systems of equations using Gaussian elimination. Use rational numbers as the underlying number type. Make two versions of the program: in one version you keep all intermediate results in reduced form by calling *x.normalize*( ) for each intermediate result and in the other version you make no attempt to keep the numbers normalized. Run examples and determine the lengths of the numerators and denominators in the solution vector.

2 Investigate the question raised in the first item theoretically. Assume that all coefficients of the linear system are integers of length at most L and let n be the number of equations in the system. Show that the entries of the solution vector can be expressed as rational numbers in which the lengths of the numerator and the denominator are bounded by a polynomial in n and L. (Hint: Show first that the value of an n by n determinant of a matrix with integer entries of absolute value at most 2^L is bounded by $n!2^{nL}$. Then use Cramer's rule to express the entries of the solution vector as quotients of determinants.) Extend the result to all intermediate results occurring in Gaussian elimination. Conclude that Gaussian elimination has running time polynomial in n and L if all intermediate values are normalized. Why does this not imply that Gaussian elimination runs in polynomial time without normalization of intermediate results?

3 Implement Gaussian elimination with floating point arithmetic. Find examples where the result of the floating point computation deviates widely from the exact result. Use the program of the first item to compute the exact result.

4.3 Floating Point Numbers

Floating point numbers are the computer science version of mathematics' real numbers. C++ offers single (type *float*) and double (type *double*) precision floating point numbers and LEDA offers in addition arbitrary precision floating point numbers (type *bigfloat*). Floating point arithmetic on most workstations adheres to the so-called IEEE floating point standard [IEE87], which we review briefly.

A floating point number consists of a sign s, a mantissa m, and an exponent e. In double format s has one bit, m consists of 52 bits $m_1, \ldots, m_{52}$, and e consists of the remaining 11 bits of a double word. The number represented by the triple (s, m, e) is defined as follows:

- e is interpreted as an integer in $[0 \mathinner{.\,.} 2^{11} - 1] = [0 \mathinner{.\,.} 2047]$.
- If $m_1 = \ldots = m_{52} = 0$ and $e = 0$ then the number is $+0$ or -0 depending on s.
- If $1 \le e \le 2046$ then the number is $s \cdot (1 + \sum_{1 \le i \le 52} m_i 2^{-i}) \cdot 2^{e-1023}$.

- If some m_i is non-zero and $e = 0$ then the number is $s \cdot \sum_{1 \le i \le 52} m_i 2^{-i} 2^{-1023}$. This is a so-called denormalized number.
- If all m_i are zero and $e = 2047$ then the number is $+\infty$ or $-\infty$ depending on s.
- In all other cases the triple represents NaN (= not a number).

The largest positive double (except for ∞) is MAXDOUBLE $= (2 - 2^{-52}) \cdot 2^{1023}$ and the smallest positive double is MINDOUBLE $= 2^{-52} \cdot 2^{-1023}$. Both constants are predefined in the systems file `value.h`. Arithmetic on floating point numbers is only approximate. For example,

```
float x = 123456789;
cout << (x + 1) - x;
```

will output 0 and not 1, the reason being that a nine-digit decimal number does not fit into a single precision floating point number. Thus, *cout* $\ll x$ will not reproduce 123456789. Although floating point arithmetic is inherently inexact, the IEEE standard guarantees that the result of any arithmetic operation is close to the exact result, usually as close as possible. Consider, for example, an addition $x + y$. If one of the arguments is NaN or the addition has no defined result, e.g., $-\infty + \infty$, then the result is NaN. Otherwise let z be the exact result. If $|z| >$ MAXDOUBLE, as for example, in $\infty + (-5)$ or in MAXDOUBLE $+ 1$, the result is $\pm\infty$, if $z <$ MINDOUBLE then the result is zero, and if MINDOUBLE $\le z \le$ MAXDOUBLE then the result is a floating point number $\tilde{z}$ which is closest to z. In particular,

$$|z - \tilde{z}| \le 2^{-53} |\tilde{z}|$$

since the error is at most 1 in the 53rd position after the binary point. The number $eps = 2^{-53}$ is frequently called the *precision* of double precision floating point arithmetic.

There is a rich body of literature on floating point arithmetic, see, for example, [DH91, Gol90, Gol91]. We do not pursue the properties of floats and doubles any further and turn to bigfloats instead.

The LEDA type *bigfloat* extends the built-in floating point types. The mantissa m and the exponent e of a bigfloat are arbitrary integers (type *integer*) and the number represented by a pair (m, e) is $m \cdot 2^e$. In addition, there are the special values ± 0 , $\pm\infty$, and NaN (= not a number). Arithmetic on bigfloats is governed by two parameters: the *mantissa length* and the *rounding mode*. Both parameters can either be set globally or for a single operation.

```
bigfloat::set_global_prec(212);
bigfloat::set_rounding_mode(TO_ZERO);
```

sets the mantissa length to 212 and the rounding mode to TO_ZERO. The arithmetic on bigfloats is defined as follows: let z be the exact result of an arithmetic operation. The mantissa of the result is obtained by rounding z to the prescribed number of binary places as dictated by the rounding mode. The available rounding modes are TO_NEAREST (round to the nearest representable number), TO_P_INF (round towards positive infinity), TO_N_INF (round towards negative infinity), TO_ZERO (round towards zero), TO_INF (round away from zero), and EXACT. For example[3], if the mantissa length is 3 and $z = 54371$ then the rounded value of z is

54400 if the rounding mode is TO_NEAREST or TO_P_INF or TO_INF, is
54300 if the rounding mode is TO_N_INF or TO_ZERO, and is
54371 if the rounding mode is EXACT.

The rounding mode EXACT applies only to addition, subtraction, and multiplication. In this mode the precision parameter is ignored and no rounding takes place. Since the exponents of bigfloats are arbitrary integers, arithmetic operations never underflow or overflow. However, exceptions may occur, e.g., division by zero or taking the square root of a negative number. They are handled according to the IEEE floating point standard, e.g., $5/0$ evaluates to ∞, $-5/0$ evaluates to $-\infty$, $\infty + 5$ evaluates to ∞, and $0/0$ evaluates to NaN.

The following inequality captures the essence of bigfloat arithmetic. If z is the exact result of an arithmetic operation and $\tilde{z}$ is the computed value then

$$|z - \tilde{z}| \le 2^{-prec}|\tilde{z}|,$$

where *prec* is the mantissa length in use. With rounding mode TO_NEAREST the error bound is $2^{-prec-1}|\tilde{z}|$.

We illustrate bigfloats by a program that computes an approximation of Euler's number $e \approx 2.71$. Let m be an integer. Our goal is to compute a bigfloat z such that $|z - e| \le 2^{-m}$. Euler's number is defined as the value of the infinite series $\sum_{n\ge 0} 1/n!$. The simplest strategy to approximate e is to sum a sufficiently large initial fragment of this sum with a sufficiently long mantissa, so as to keep the total effect of the rounding errors under control. Assume that we compute the sum of the first n_0 terms with a mantissa length of *prec* bits for still to be determined values of n_0 and *prec*, i.e., we execute the following program.

```
bigfloat::set_rounding_mode(TO_ZERO);
bigfloat::set_precision(prec);
bigfloat z = 2;
integer fac = 2;
int n = 2;
while (n < n0)
  { // fac = n! and z approximates 1/0! + ... + 1/(n-1)!
    z = z + 1/bigfloat(fac);
    n++; fac = fac * n;
  }
```

Let z_0 be the final value of z. Then z_0 is the value of $\sum_{n<n_0} 1/n!$ computed with bigfloat arithmetic with a mantissa length of *prec* binary places. We have incurred two kinds of errors in this computation: a truncation error since we summed only an initial segment of an infinite series and a rounding error since we used floating point arithmetic to sum the initial segment. Thus,

$$|e - z_0| \le |e - \textstyle\sum_{n<n_0} 1/n!| + |\sum_{n<n_0} 1/n! - z_0|$$

3 We use decimal notation instead of binary notation for this example.

$$= \sum_{n \geq n_0} 1/n! + |\sum_{n<n_0} 1/n! - z_0|$$

The first term is certainly bounded by $2/n_0!$ since, for all $n \geq n_0$, $n! = n_0! \cdot (n_0+1) \cdot \ldots \cdot n \geq n_0! \cdot 2^{n-n_0}$ and hence $\sum_{n \geq n_0} 1/n! \leq 1/n_0! \cdot (1 + 1/2 + 1/4 + \ldots) \leq 2/n_0!$. What can we say about the total rounding error? We observe that we use one floating point division and one floating point addition per iteration and that there are $n_0 - 2$ iterations. Also, since we set the rounding mode to `TO_ZERO` the value of z always stays below e and hence stays bounded by 3. Thus, the results of all bigfloat operations are bounded by 3 and hence each bigfloat operation incurs a rounding error of at most $3 \cdot 2^{-prec}$. Thus

$$|e - z_0| \leq 2/n_0! + 2n_0 \cdot 3 \cdot 2^{-prec}.$$

We want the right-hand side to be less than 2^{-m-1}; it will become clear in a short while why we want the error to be bounded by 2^{-m-1} and not just 2^{-m}. This can be achieved by making both terms less than 2^{-m-2}. For the first term this amounts to $2/n_0! \leq 2^{-m-2}$. We choose n_0 minimal with this property and observe that if we use the expression *fac.length*( ) $< m + 3$ as the condition of our while loop then this n_0 will be the final value of n; *fac.length*( ) returns the number of bits in the binary representation of *fac*. From $n_0! \geq 2^{n_0}$ and the fact that n_0 is minimal with $2/n_0! \leq 2^{-m-2}$ we conclude $n_0 \leq m + 3$ and hence $6n_0 2^{-prec} \leq 6(m+3) \cdot 2^{-prec} \leq 2^{-m-2}$ if $prec \geq 2m$; actually, $prec \geq m + \log(m+3) + 5$ suffices. The following program implements this strategy and computes z_0 with $|e - z_0| \leq 2^{-m-1}$.

We could output z_0, but z_0 is a number with $2m$ binary places and hence suggests a quality of approximation which we are not guaranteeing. Therefore, we round z_0 to the nearest number with a mantissa length of $m + 3$ bits. Since $z_0 \leq 3$ this will introduce an additional error of at most $3 \cdot 2^{-m-3} \leq 2^{-m-1}$. We conclude that the program below computes the desired approximation of Euler's number. This program is available as Euler_demo.

⟨*Euler_demo*⟩≡

```
main(){
int m;
⟨Euler: read m⟩
bigfloat::set_precision(2*m);
bigfloat::set_rounding_mode(TO_ZERO);
bigfloat z = 2;
integer fac = 2;
int n = 2;
while ( fac.length() < m + 3 )
  { // fac = n! and z approximates 1/0! + 1/1! + ... + 1/(n-1)!
    z = z + 1/bigfloat(fac);
    n++; fac = fac * n;
  }
// |z - e| <= 2^{m-1} at this point
z = round(z,m+3,TO_NEAREST);
⟨Euler: output z⟩
}
```

Exercises for 4.3

1 Compute π with an error less than 2^{-200}.

2 Assume that for i, $1 \le i \le 8$, x_i is an integer with $|x_i| \le 2^{20}$. Evaluate the expression $((x_1+x_2)\cdot(x_3+x_4))\cdot x_5+(x_6+x_7)\cdot x_8$ with double precision floating point arithmetic. Derive a bound for the maximal difference between the exact result and the computed result.

4.4 Algebraic Numbers

The data type *real* is LEDA's best approximation to mathematics' real numbers. It supports exact computation with k-th roots for arbitrary natural number k, the rational operators $+$, $-$, $*$, and $/$, and the comparison operators $==$, $!=$, $<$, $\le$, $\ge$, and $>$. Let us see a small example.

```
real x = (sqrt(17) - sqrt(12)) * (sqrt(17) + sqrt(12)) - 5;
cout << sign(x);
```

Note that the exact value of the expression defining x is 0. The distinctive feature of reals is that *sign*(x) actually evaluates to zero. More generally, if E is any expression with integer operands and operators $+$, $-$, $*$, $/$, and function calls *sqrt*(x) and *root*(x, k) where x is a *real* and k is a positive integer then the data type real is able to determine the sign of E. We want to *stress that reals compute the sign of an expression in the mathematical sense* and not the sign of an approximation of an expression. This is in sharp contrast to the evaluation of an expression with floating point arithmetic. Floating point arithmetic incurs rounding error and hence, in general, cannot compute the sign of an expression correctly.

Why are we so concerned about the sign of expressions? The reason is that many programs contain conditional statements that branch on the sign of an expression and that such programs may go astray if the wrong decision about the sign is made. We give two examples, both arising in computational geometry. Further examples can be found in Section 9.6.

In the first example we consider the lines

$$l_1:\ y = 9833\cdot x/9454 \qquad \text{and} \qquad l_2:\ y = 9366\cdot x/9005.$$

Both lines pass through the origin and the slope of l_1 is slightly larger than the slope of l_2, see Figure 4.2. At $x = 9454\cdot 9005$ we have $y_1 = 9833\cdot 9005 = 9366\cdot 9454+1 = y_2+1$.

The following program runs through multiples of 0.001 between 0 and 1 and computes the corresponding y-values y_1 and y_2. It compares the two y-values and, if the outcome of the comparison is different than in the previous iteration, prints x together with the current outcome.

```
int last_comp = -1;
float a = 9833; float b = 9454;
float c = 9366; float d = 9005;
```

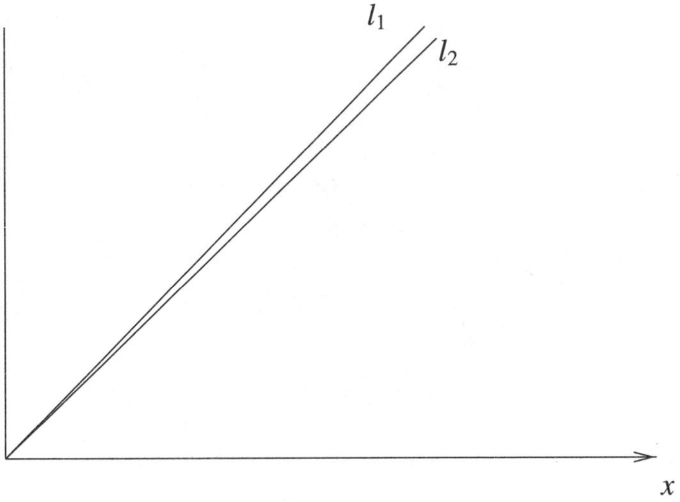

Figure 4.2 The lines l_1 and l_2: Both lines pass through the origin and l_1 is slightly steeper than l_2.

```
for (float x = 0; x < 1; x = x + 0.001)
{ float y1 = a*x/b;    float y2 = c*x/d; // l1 is steeper
  int comp = (y1 < y2? -1 : (y1 == y2? 0 : +1));
  if (comp != last_comp)
  { cout <<"\nAt " << x << ": ";
    if (comp == -1) cout << "l1 is below l2";
    if (comp ==  0) cout << "l1 intersects l2";
    if (comp == +1) cout << "l1 is above l2";
  }
  last_comp = comp;
}
```

Clearly, we should expect the program to print:

```
At 0.000: l1 intersects l2
At 0.001: l1 is above l2
```

Well, the actual output on the first author's workstation contains the following lines[4]:

```
At 0.000: l1 intersects l2
At 0.003: l1 is above l2
At 0.004: l1 intersects l2
At 0.005: l1 is above l2
At 0.008: l1 intersects l2
At 0.009: l1 is below l2
...
At 0.993: l1 intersects l2
At 1.000: l1 is below l2
```

[4] If the program is run on the same author's notebook, it produces the correct result. The explanation for this behavior is that on the notebook double precision arithmetic is used to implement floats. According to the C++ standard floats must not offer more precision than doubles; they are not required to provide less. You may use the braided lines demo to find out how the program behaves on your machine.

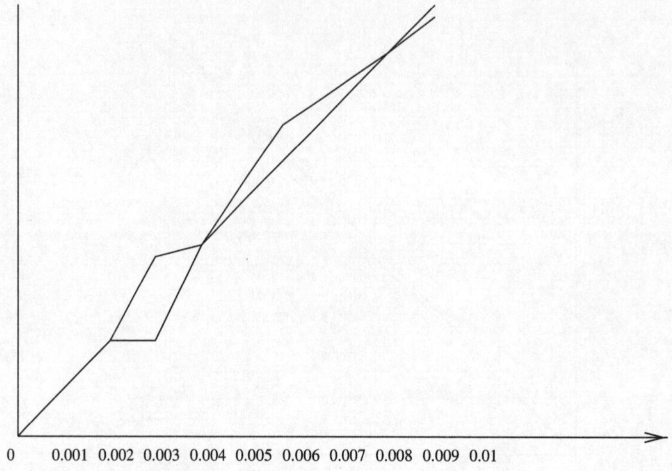

Figure 4.3 y_1 is equal to y_2 for $x = 0.001 \cdot i$ and i equal to 0, 1, and 2, is larger for i equal to 3, is equal for i equal to 4, is larger for i equal to 5, 6, and 7, is equal for i equal to 8, and is smaller for i equal to 9.

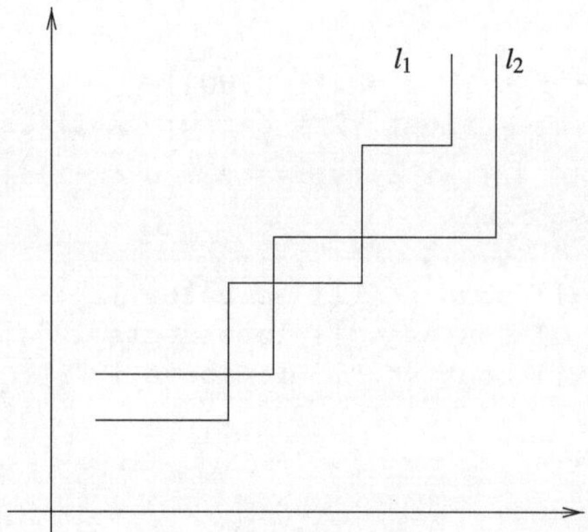

Figure 4.4 Lines as step functions and their multiple intersections.

We conclude that the lines intersect many times and are interlaced as shown in Figure 4.3. Observe that floating point arithmetic gives the wrong relationship between y_1 and y_2 not only for x close to zero but even for fairly large values of x. Thus the lines behave very differently from mathematical lines. Lyle Ramshaw coined the name *verzopfte Geraden (braided lines)* for the effect. Figure 4.4 explains the effect. The type *float* consists of only a finite number of values and hence a line is really a step function as shown in the figure. The width of the steps of our two lines l_1 and l_2 are distinct and hence the lines intersect.

The problem of braided lines is easily removed by the use of an exact number type; e.g., if *float* is replaced by *rational* in the program above, the output becomes what it should be:

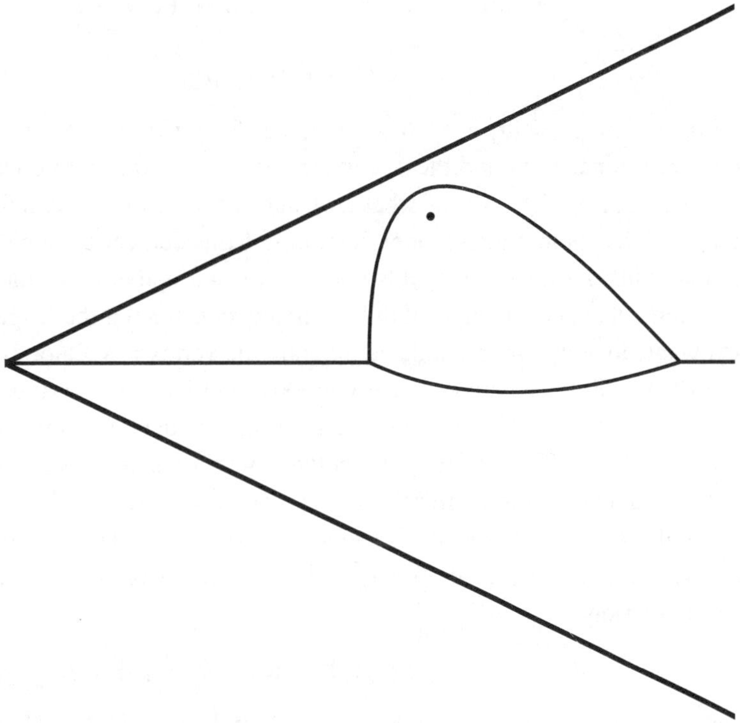

Figure 4.5 The Voronoi diagram of two lines l_1 and l_2 and a point p. The Voronoi diagram consists of parts of the angular bisector of l_1 and l_2 and of parts of the parabolas defined by p and l_1 and l_2, respectively. The Voronoi vertices are centers of circles passing through the point and touching the lines.

```
0/1: l1 intersects l2
1/1000: l1 is above l2
```

The second example goes beyond rational arithmetic and arises in the computation of Voronoi diagrams of line segments and points. Voronoi diagrams of line segments will be discussed in Section 10.5.5, and we assume for this paragraph that the reader has an intuitive understanding of Voronoi diagrams. For i, $1 \le i \le 2$, let $l_i : a_i x + b_i y + c_i = 0$ be a line in two-dimensional space and let $p = (0, 0)$ be the origin, cf. Figure 4.5. There are two circles passing through p and touching l_1 and l_2. These circles have centers $v_{1,2} = (x_{v_{1,2}}/z_v, y_{v_{1,2}}/z_v)$ where[5]

$$x_{v_{1,2}} = a_1 c_2 + a_2 c_1 \pm \sqrt{2 c_1 c_2 (\sqrt{N} + D)}$$

[5] The reader may compute these coordinates by solving the following equations for x_v/z_v and y_v/z_v.

$$\begin{aligned}(x_v/z_v)^2 + (y_v/z_v)^2 &= (a_1 x_v/z_v + b_1 y_v/z_v + c_1)^2/(a_1^2 + b_1^2)\\ (x_v/z_v)^2 + (y_v/z_v)^2 &= (a_2 x_v/z_v + b_2 y_v/z_v + c_2)^2/(a_2^2 + b_2^2)\end{aligned}$$

$$y_{v_{1,2}} = b_1c_2 + b_2c_1 \pm sign(S)\sqrt{2c_1c_2(\sqrt{N} - D)}$$

$$z_v = \sqrt{N} - a_1a_2 - b_1b_2$$

and $S = a_1b_2 + a_2b_1$, $N = (a_1^2 + b_1^2)(a_2^2 + b_2^2)$, and $D = a_1a_2 - b_1b_2$; in these expressions the $+$ in $\pm$ corresponds to v_1 and the $-$ corresponds to v_2. Consider now a third line l and let v be one of v_1 or v_2. The test of whether l intersects the circle centered at v is crucial for most algorithms computing Voronoi diagrams. Consider, for example, an incremental algorithm that adds the lines and points one by one and updates the diagram after every addition. Assume that such an algorithm has already constructed the diagram for p, l_1 and l_2 and next wants to add l. In the updated diagram the vertex v will not exist if l intersects the interior of the circle centered at v, v will exist and have degree four if l touches the circle centered at v, and v will exist and have the same incident edges if l does not intersect the circle centered at v. The question of whether l intersects, touches, or misses the circle centered at v is tantamount to comparing $dist(v, p)$ with $dist(v, l)$. We may also compare the squares of these numbers instead. The square of $dist(v, p)$ is $(x_v^2 + y_v^2)/z_v^2$ and the square of $dist(v, l)$ is $(ax_v/z_v + by_v/z_v + c)^2/(a^2 + b^2)$. In other words, we need to compute the sign of the expression

$$R = (ax_v + by_v + cz_v)^2 - (a^2 + b^2)(x_v^2 + y_v^2).$$

The following procedure takes inputs $a_1, b_1, \ldots, c$ and $pm \in \{-1, +1\}$ and performs this comparison; pm is used to select one of v_1 and v_2.

```
int INCIRCLE(integer a1, integer b1, integer c1, integer a2,
integer b2, integer c2, integer a, integer b, integer c, int pm)
{ real RN = sqrt((a1 * a1 + b1 * b1) * (a2 * a2 + b2 * b2));
  real A  = a1 * c2 + a2 * c1;
  real B  = b1 * c2 + b2 * c1;
  real C  = 2 * c1 * c2;
  real D  = a1 * a2 - b1 * b2;
  real S  = a1 * b2 + a2 * b1;
  real xv = A + pm * sqrt(C * (RN + D));
  real yv = B + pm * sign(S) * sqrt(C * (RN - D));
  real zv = RN - (a1 * a2 + b1 * b2);
  real P  = a * xv + b * yv + c * zv;
  real R  = P * P - (a * a + b * b) * (xv * xv + yv * yv);
  return sign(R);
}
```

How do reals work? The sign computation is based on the concept of a *separation bound*. A separation bound for an expression E is an *easily computable* number $sep(E)$ such that

$$val(E) \neq 0 \text{ implies } |val(E)| \geq sep(E),$$

where $val(E)$ denotes the value of E. Thus $|val(E)| < sep(E)$ implies $val(E) = 0$. Given a separation bound there is a simple strategy to determine the sign of $val(E)$:

- Compute an approximation A of $val(E)$ with $|A - val(E)| < sep(E)/2$ by evaluating E with *bigfloat* arithmetic with sufficient mantissa length. The required mantissa length can be determined by an error analysis in the same way, as we determined the mantissa length required for the computation of Euler's number with an error less than 2^{-m} in the preceding section. We stress that this error analysis is automated in the data type *real* and is invisible to the user.
- If $|A| \geq sep(E)/2$ then return the sign of A and if $|A| < sep(E)/2$ then return zero.

The correctness of this approach can be seen as follows:

If $|A| \geq sep(E)/2$ then $|A - val(E)| < sep(E)/2$ implies that $val(E)$ and A have the same sign.

If $|A| < sep(E)/2$ then $|A - val(E)| < sep(E)/2$ implies $|val(E)| < sep(E)$. Thus, $val(E) = 0$ by the definition of a separation bound.

Next, we give the separation bound that is used in LEDA. First, we need to define precisely what we mean by an expression. For simplicity, we deal only with expressions without divisions, although *reals* also handle divisions. An expression E is an acyclic directed graph (dag) in which each node has indegree at most two, in which each node of indegree 0 is labeled by a non-negative integer, each node of indegree 1 is labeled either by $-$ (unary minus) or by $root_k$ for some natural number k, and each node of indegree 2 is labeled by either a $+$ or a $*$. Figure 4.6 shows an expression. We define the *degree* $deg(E)$ of E as the product of the k's over all nodes labeled by root operations. The expression of Figure 4.6 has degree 4. We define the *bound* $b(E)$ of E as the value of the expression $\hat{E}$ which is obtained from E by removing all nodes labeled with a unary minus and connecting their input node directly to their outputs. In our example, we have $b(E) = (\sqrt{17} + \sqrt{12})(\sqrt{17} + \sqrt{12}) + 5$.

Theorem 1 ([BFMS97]) *Let E be an expression. Then $val(E) \leq b(E)$ and either*

$$val(E) = 0 \quad or \quad |val(E)| \geq b(E)^{1-deg(E)}.$$

We give a proof of a special case. Assume that A, B, and C are natural numbers. How close to zero can $A\sqrt{B} - C$ be, if non-zero? We have

$$\begin{aligned} |A\sqrt{B} - C| &= |A\sqrt{B} - C| \cdot (A\sqrt{B} + C)/(A\sqrt{B} + C) \\ &= |A^2B - C^2|/(A\sqrt{B} + C) \\ &\geq 1/(A\sqrt{B} + C), \end{aligned}$$

where the last inequality follows from the assumption that the value of our expression is different from zero and from the fact that $A^2B - C^2$ is an integer. The expression above has degree 2 and its b-value is equal to $A\sqrt{B} + C$. Thus, the derived bound corresponds precisely to the statement of the theorem.

It is worthwhile to restate the theorem in terms of the binary representation of $val(E)$. Let $L = \log b(E)$. Then $|val(E)| \leq 2^L$ and, if $val(E) \neq 0$, $|val(E)| \geq 2^{L\cdot(1-deg(E))}$. Thus, if $val(E) \neq 0$, then the binary representation of $val(E)$ either contains a non-zero digit in

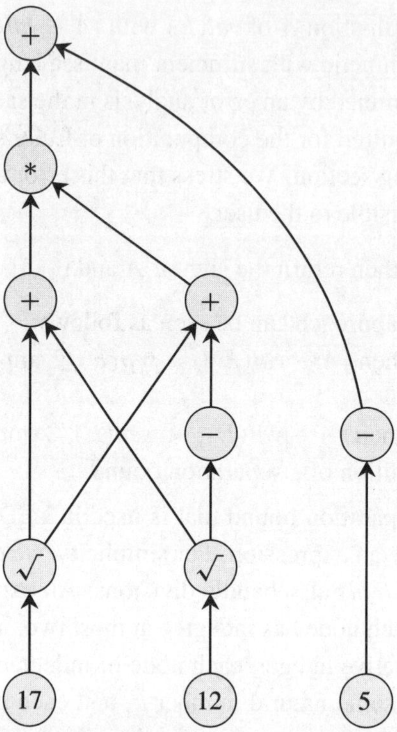

Figure 4.6 An expression dag E. The expression has degree 4 and computes $(\sqrt{17}+\sqrt{12})\cdot(\sqrt{17}-\sqrt{12})-5$.

the L digits before the binary point or a non-zero digit in the first $(deg(E)-1)\cdot L$ digits after the binary point. Conversely, if all of these digits are zero then $val(E)$ is zero. In the sequel we will rephrase this statement as: It suffices to inspect the first $deg(E)\cdot L$ bits of the binary representation of $val(E)$.

We give two applications of the theorem above. They are illustrated by the two real number demos, respectively.

First, let x be an arbitrary integer and consider the expression

$$E_1 = (\sqrt{x+5}+\sqrt{x})(\sqrt{x+5}-\sqrt{x})-5.$$

Then $deg(E_1)=4$ and $b(E_1)<4(x+5)+5$. Let $L_1=\log(4(x+5)+5)$. By the theorem above it therefore suffices to inspect the first $4L_1$ bits of the binary representation of $val(E_1)$ in order to determine its sign. So if x has 100 binary digits it certainly suffices to inspect 412 digits of $val(E_1)$. This is illustrated by the program below. It asks for an integer L and then constructs a random integer x with L decimal digits. It then computes the signs of E_1+5 and E_1.

L	80	160	320	640	1280	2560	5120
A	0.01	0.01	0.03	0.05	0.14	0.41	1.47
B	0.04	0.07	0.21	0.66	2.24	8.03	29.73

Table 4.2 The running times for computing the signs of $A = (\sqrt{x+5}+\sqrt{x})(\sqrt{x+5}-\sqrt{x})$ and $B = A - 5$ for x being a random integer with L decimal digits. Note that the time for computing the sign of A is much smaller than the time for computing the sign of B. This reflects the fact that a crude approximation of A allows us to conclude that A is positive and that about $4L$ digits of B need to be computed in order to allow the conclusion that B is zero. You may perform your own experiments by calling the first real number demo.

```
⟨real_demo1⟩≡
  ⟨real demo1: read L⟩
  integer x = 0;
  while (L > 0)
  { x = x*10 + rand_int(0,9);
    L--;
  }
  float T = used_time();
  real X = x;
  real SX = sqrt(X);
  real SXP = sqrt(X+5);
  real A = (SXP + SX) * (SXP - SX);
  real B = A - 5;
  int A_sign = A.sign(); float TA = used_time(T);
  int B_sign = B.sign(); float TB = used_time(T);
  ⟨real demo1: output signs and report running times⟩
```

Table 4.2 shows the running times of this program for $L = 80$, 160, 320, and so on.

Next, consider the expression

$$E_2 = (2^{2^k}+1)^{2^{-k}} - 2,$$

i.e., the number 2 is squared k times, 1 is added , square roots are taken k times, and finally 2 is subtracted. This yields a number slightly above 0. In fact[6],

$$\begin{aligned} val(E_2) &= (2^{2^k}+1)^{2^{-k}} - 2 = 2((1+2^{-2^k})^{2^{-k}} - 1) \\ &= 2(\exp(2^{-k}\ln(1+2^{-2^k})) - 1) \approx 2(\exp(2^{-k}2^{-2^k}) - 1) \\ &\approx 2(1+2^{-k}2^{-2^k} - 1) = 2^{1-k-2^k}, \end{aligned}$$

i.e., the first non-zero bit in the binary expansion of $val(E_2)$ is about $k+2^k$ positions after the binary point. What does the theorem above say? We have $deg(E_2) = 2^k$ and $b(E_2) \le 5$ and hence by the theorem it suffices to inspect the first $2^k \log 5$ bits of the binary representation

[6] We use the estimates $\ln(1+x) \approx x$ and $e^x \approx 1+x$ for x close to zero.

of $val(E_2)$. That's an overestimate by about a factor of two. The following program chunk illustrates this example. It asks for an integer k and then computes the sign of the expression E_2. It also shows the *bigfloat* approximation of E_2 that is computed in the sign computation.

⟨*real_demo2*⟩≡
```
int k = I.read_int("k = ");
float T = used_time();
real E = 2;
int i;
for (i = 0; i < k; i++)  { E = E*E; }
E = E + 1;
for (i = 0; i < k; i++) { E = sqrt(E); }
E = E - 2;
I.write_demo("The sign of E is ",E.sign(),".");
I.write_demo("This took ",used_time(T)," seconds.");
I.write_demo("An approximation of E: " + to_string(E.to_bigfloat()));
```

We close this section with a brief discussion of the implementation of reals. The data type *real* stores objects of type real by their expression dags, i.e., every operation on reals adds a node to the expression dag and records the arithmetic operation to be performed at the node and the inputs to the node. Thus the dag of Figure 4.6 is built for the expression $(\sqrt{17}+\sqrt{12})(\sqrt{17}-\sqrt{12})-5$. Whenever the sign of a real number has to be determined, a separation bound is computed as described in Theorem 1 and then a bigfloat computation is performed to determine the sign.

We sketch how the bigfloat computation is performed; for details we refer the reader to [BMS96].We set a parameter l to some small integer and compute an approximation A of $val(E)$ with $|A - val(E)| < 2^{-l}$. In order to compute such an approximation an error analysis along the lines of the preceding section is performed (this is fully automated) and then a bigfloat computation with the appropriate mantissa length is performed. If $|A| \geq 2 \cdot 2^{-l}$ then $val(E)$ and A have the same sign and we may return the sign of A. If $|A| < 2 \cdot 2^{-l}$ we double l and repeat. We continue in this fashion until $2^{-l} \leq sep(E)/2$, where $sep(E)$ is the separation bound. Table 4.2 illustrates the effect of this optimization: For the expression A a crude approximation allows us to decide the sign and hence $sign(A)$ is computed quickly, however, for expression B one has to go all the way to the quality of approximation prescribed by the separation bound.

We close with a warning. Reals are not a panacea. Although they allow in principle to compute the sign of any expression involving addition, subtraction, multiplication, division, and arbitrary roots, you may have to wait a long time for the answer when the expression is complex. The paper [BFMS99] discusses the use of *reals* in geometric computations.

Exercises for 4.4

1 Compute the sign of $E = (2^{2^k}+1)^{2^{-k}} - 2$ for different values of k. You may use program real_demo2 for this purpose. Don't be too ambitious. Try to predict the growth rate of the running time before performing the experiment.

2 Let $E = E_1 + E_2$ and assume that you want an approximation A of $val(E)$ such that $|val(E) - A| \leq \varepsilon$. Determine ε_1 and ε_2 and a precision *prec* such that computation of bigfloat approximations A_i of $val(E_i)$ with an error $|A_i - val(E_i)| \leq \varepsilon_i$ and summation of A_1 and A_2 with precision *prec* yields the desired approximation A of $val(E)$.

3 As above for $E = E_1 \cdot E_2$, $E = E_1/E_2$, and $E = \sqrt{E_1}$. Solutions to exercises 2. and 3. can be found in [BMS96].

4 Let p_1 and p_2 be two points in the plane, let l be a line, and consider the circle passing through p_1 and p_2 and touching l. Write a procedure that determines the position of a third point p_3 with respect to this circle.

4.5 Vectors and Matrices

Vectors and matrices are one- and two-dimensional arrays of numbers, respectively. Let n and m be integers. An n-dimensional vector v is a one-dimensional arrangement of n variables of some number type N; the variables are indexed from 0 to $n - 1$ and $v[i]$ denotes the variable with index i. An $n \times m$ matrix M is a two-dimensional arrangement of $n \cdot m$ variables of some number type N; the variables are indexed by pairs (i, j) with $0 \leq i \leq n - 1$ and $0 \leq j \leq m - 1$. We use $M(i, j)$ to denote the variable indexed by i and j and call n and m the number of rows and columns of M, respectively. Observe that as for two-dimensional arrays we use round brackets for the subscript operator in matrices. We have currently vectors and matrices with entries of type *double* (types *vector* and *matrix*) and type *integer* (types *integer_vector* and *integer_matrix*). Vectors and matrices over an arbitrary number type are part of the LEP for higher-dimensional geometry. We use the latter types in all our examples. The definitions

```
integer_vector v(m);
integer_matrix M(n,m);
```

define an m-vector v and an $n \times m$-matrix M, respectively. All entries of v and M are initialized to zero. The following procedure multiplies a matrix M by a vector v.

```
integer_vector integer_matrix::operator*(const integer_matrix& M,
                                         const integer_vector& v)
{ int n = M.dim1(); // # of rows of M
  int m = M.dim2(); // # of columns of M
  if (m != v.dim()) error_handler(1, "incompatible dimensions");
  integer_vector result(n);
  for (int i = 0; i < n; i++)
    for (int j = 0; j < m; j++) result[i] += M(i,j) * v[j];
  return result;
}
```

In the context of

```
integer_vector v(5);
integer_vector r;          // a 0 - dimensional vector
integer_matrix M(3, 5);
```

we may now write

```
r = M * v.
```

Note that we defined r as an empty vector. The assignment $r = M * v$ assigns the result of the multiplication $M * v$ to r. This involves allocation of memory (for three variables of type *integer*) and component-wise assignment. Vectors are internally represented as a pair consisting of an int *dim*, containing the dimension of the vector, and a pointer v to a C++ array containing the components of the vector. The code for the assignment operator is as follows:

```
integer_vector& integer_vector::operator=(const integer_vector& vec)
{ if (dim != vec.dim())
  { /* this vector does not yet have the right dimension */
    delete v;
    dim = vec.dim();
    v = new integer[dim];
  }
  for (int i = 0; i < dim; i++) v[i] = vec[i];
  return *this;
}
```

Vectors and matrices are similar to one- and two-dimensional C++ arrays of numbers, respectively. The main differences are as follows:

- Vectors and matrices know their dimension(s). Assignment is component-wise assignment. It allocates space automatically.
- Vectors and matrices check whether indices are legal. The checks can be turned off.
- Vectors and matrices are somewhat slower than their C++ counterparts.
- Vectors and matrices offer a large number of operations of linear algebra.

The basic operations of linear algebra are vector and matrix addition and multiplication, and multiplication by a scalar. For example, $M+N$ denotes the component-wise addition of two matrices M and N, $M * N$ denotes matrix multiplication, $M * v$ denotes matrix-vector product, $v * w$ is the scalar product of two vectors, and $v * 5$ multiplies each entry of v by the scalar 5.

We turn to the more advanced functions of linear algebra. Let M be an $n \times m$ integer matrix and let b be an n integer vector. Let x be an integer vector and let D be an integer variable. The call

```
linear_solver(M,b,x,D);
```

returns true if the linear system $M \cdot z = b$ has a solution and returns false otherwise. If the system is solvable then the vector $(1/D) \cdot x$ is a solution of the system. Why do we return the solution in this strange format? The solution vector of the system $M \cdot z = b$ has rational entries. We provide a common denominator in D and the numerator in x. For example, the system

$$\begin{aligned} 3z_0 + z_1 &= 5 \\ z_0 + z_1 &= 2 \end{aligned}$$

has the solution $z = (3/2, 1/2)$. We return this solution as $x = (3, 1)$ and $D = 2$.

The main use of linear algebra within LEDA is the exact implementation of geometric primitives; e.g., we solve a linear system to determine the equation of a hyperplane through a set of points and we compute a determinant to determine the orientation of a sequence of points. We use matrices and vectors over integers for that purpose. We hardly use vectors and matrices over doubles within LEDA and therefore have not optimized the robustness of our linear system solver. We do not recommend to use our procedures for serious numerical analysis. Much better codes are available in the numerical analysis literature. A good source of codes is the book [FPTV88].

A linear system $M \cdot z = b$ may have more than one solution, may have exactly one solution, or no solution at all. The call *linear_solver*(M, b, x, D, s_vecs, c) gives complete information about the solution space of the system $M \cdot z = b$:

- If the system is unsolvable then c is an n-vector such that $c^T \cdot M = 0$ and $c^T \cdot b \neq 0$, i.e., c specifies a linear combination of the equations such that the left-hand side of the resulting equation is identically zero and the right-hand side is non-zero. For example, for the system

$$\begin{aligned} z_0 + z_1 &= 5 \\ 2z_0 + 2z_1 &= 4 \end{aligned}$$

  the vector $c = (-2, 1)^T$ *proves* that the system is unsolvable.

- If the system is solvable then $(1/D) \cdot x$ is a solution and *s_vecs* is an $m \times d$ matrix for some d whose columns span the solution space of the corresponding homogeneous system $M \cdot z = 0$. Let col_j denote the j-th column of *s_vecs*. Then any solution to $M \cdot z = b$ can be written as

$$(1/D) \cdot x + \sum_{o \leq j < d} \lambda_j \cdot col_j$$

  for some reals λ_j, $0 \leq j < d$. You may extract the j-th column of *s_vecs* by *s_vecs.col*(j).

The rank of a matrix is the maximal number of linearly independent rows (or columns). The call

```
rank(M);
```

returns the rank of M.

From now on we assume that M is a square matrix, i.e., an $n \times n$ matrix for some n. A square matrix is called *invertible* or *non-singular* if there is a matrix N such that $M \cdot N = N \cdot M = I$, where I is the $n \times n$ identity matrix; the matrix N is called the inverse of M and is usually denoted by M^{-1}. A matrix without an inverse is called *singular*. A matrix is singular if and only if its determinant is equal to zero. The call

```
integer D = determinant(M);
```

returns the determinant of M. The inverse of an integer matrix has, in general, rational entries.

```
integer_matrix N = inverse(M,D);
```

assigns a common denominator of the entries of the inverse to D and returns the matrix of numerators in N, i.e, $(1/D) \cdot N$ is the inverse of M. The function *inverse* requires that M is non-singular and hence should only be used if M is known to be non-singular. The call

```
inverse(M,N,D,c);
```

returns true if M has an inverse and false otherwise. In the former case $(1/D) \cdot N$ is the inverse of M and in the latter case c is a non-zero vector with $c^T \cdot M = 0$. Note that such a vector proves that M is singular.

The LU-decomposition of a matrix is the decomposition as a product of a lower and an upper diagonal matrix.

```
LU_decomposition(M,L,U,q);
```

computes a lower diagonal matrix L, an upper diagonal matrix U, and a permutation q of $[0 .. n-1]$ (represented as an *array<int>*) such that for all i, $0 \le i < n$, the $q[i]$-th column of $L \cdot M$ is equal to the i-th column of U.

Exercises for 4.5

1 Write a procedure that determines whether a homogeneous linear system has a non-trivial solution.

2 Write a function that computes the equation of a hyperplane passing through a given set of d points in d-dimensional Euclidean space.

5

Advanced Data Types

We discuss some of the advanced data types of LEDA: dictionary arrays, hashing arrays, maps, priority queues, partitions, and sorted sequences. For each type we give its functionality, discuss its performance and implementation, and describe applications.

5.1 Sparse Arrays: Dictionary Arrays, Hashing Arrays, and Maps

Sparse arrays are arrays with an infinite or at least very large index set of which only a "sparse" subset is in actual use. We discuss the sparse array types of LEDA and the many implementations available for them. We start with the functionality and then discuss the performance guarantees given by the different types and implementations. We also give an experimental comparison. We advise on how to choose an implementation satisfying the needs of a particular application and discuss the implementation of *maps* in detail.

5.1.1 *Functionality*

Dictionary arrays (type *d_array*<I, E>), hashing arrays (type *h_array*<I, E>), and maps (type *map*<I, E>) realize arrays with large or even unbounded index set I and arbitrary entry type E. Examples are arrays indexed by points, strings, or arbitrary integers. We refer to d_arrays, h_arrays, and maps as *sparse array types*; another common name is *associative arrays*. The sparse array types have different requirements for the index type: dictionary arrays work only for linearly ordered types (see Section 2.10), hashing arrays work only for hashed types (see Section 2.8), and maps work only for pointer and item types and the type int. They also differ in their performance guarantees and functionality. Figure 5.1 shows the manual page of maps and Table 5.1 summarizes the properties of our sparse array types. Before we discuss them we illustrate the sparse array types by small examples.

	d_arrays	h_arrays	Maps
index type	linearly ordered	hashed	int or pointer or item type
access time	$O(\log n)$ worst case	$O(1)$ expected	$O(1)$ expected
forall_defined loop	sorted	unsorted	unsorted
persistence of variables	yes	no	no
undefine operation	available	available	not available

Table 5.1 Properties of d_arrays, h_arrays, and maps. The meaning of the various rows is explained in the text.

In the first example we use a d_array to build a small English–German dictionary and to print all word pairs in the dictionary.

```
d_array<string,string> dic;
dic["hello"] = "hallo";
dic["world"] = "Welt";
dic["book"]  = "Buch";

string s;
forall_defined(s,dic) cout << s << " " << dic[s] << "\n";
```

The *forall_defined* loop iterates over all indices of the array that were used as a subscript prior to the loop. The iteration is according to the order defined by the *compare* function of the index type; recall that dictionary arrays work only for linearly ordered types. In the case of strings the default *compare* function defines the lexicographic ordering and hence the program outputs:

```
book Buch
hello hallo
world Welt
```

In the second example we use a h_array to read a sequence of strings from standard input, to count the multiplicity of each string in the input, and to output the strings together with their multiplicities. H_arrays work only for hashed types and hence we need to define a hash function for strings. We define a very primitive hash function that maps the empty string to zero and any non-empty string to its leading character (for a string x, $x[0]$ returns the leading character of x).

```
int Hash(const string& x) { return (x.length() > 0) ? x[0] : 0; }
h_array<string,int> N(0); // default value 0
while (cin >> s) N[s]++;
forall_defined(s,N) cout << s << " " << N[s] << "\n";
```

1. Definition

An instance M of the parameterized data type *map<I, E>* is an injective mapping from the data type I, called the index type of M, to the set of variables of data type E, called the element type of M. I must be a pointer, item, or handle type or the type int. We use $M(i)$ to denote the variable indexed by i. All variables are initialized to *xdef*, an element of E that is specified in the definition of M. A subset of I is designated as the domain of M. Elements are added to *dom*(M) by the subscript operator.

Related data types are *d_arrays*, *h_arrays*, and *dictionaries*.

2. Creation

map<I, E> M;	creates an injective function m from I to the set of unused variables of type E, sets *xdef* to the default value of type E (if E has no default value then *xdef* is set to an unspecified element of E), and initializes M with m.
map<I, E> M(E x);	creates an injective function m from I to the set of unused variables of type E, sets *xdef* to x, and initializes M with m.

3. Operations

$E\&$	$M[I\ i]$	returns the variable $M(i)$ and adds i to *dom*(M). If M is a const-object then $M(i)$ is read-only and i is not added to *dom*(M).
bool	M.defined(I i)	returns true if $i \in$ *dom*(M).
void	M.clear( )	makes M empty.
void	M.clear(E x)	makes M empty and sets *xdef* to x.

Iteration

forall_defined(i, M) { "the indices i with $i \in$ *dom*(M) are successively assigned to i" }

forall(x, M) { "the entries $M[i]$ with $i \in$ *dom*(M) are successively assigned to x" }

4. Implementation

Maps are implemented by hashing with chaining and table doubling. Access operations $M[i]$ take expected time $O(1)$.

Figure 5.1 The manual page of data type *map*.

There are two further remarks required about this code fragment. First, in the definition of N we defined a default value for all entries of N: all entries of N are initialized to this default value. Second, hashed types have no particular order defined on their elements and hence the *forall_defined* loop for h_arrays steps through the defined indices of the array in no particular order.

In the third example we assume that we are given a list of segments in *seglist* and that we want to associate a random bit with each segment. A *map<segment, bool>* serves well for this purpose.

```
map&lt;segment,bool&gt; color;
segment seg;
forall(seg,seglist) color[seg] = rand_int(0,1);
```

After these introductory examples we turn to the detailed discussion of our sparse array types. An object A of a sparse array type is characterized by three quantities:

- An injective mapping from the index type into the variables of type E. For an index i we use $A(i)$ to denote the variable selected by i.

- An element *xdef* of type E, the default value of all variables in the array. It is determined in one of three ways. If the definition of the array has an argument, as, for example, in

  ```
  h_array<int,int> N(0);
  ```

  then this argument is *xdef*. If the definition of the array has no argument but the entry type of the array has a default value[1], as, for example, in

  ```
  d_array<string,string> D;
  ```

  then this default value is *xdef*. If the definition of the array has no argument and the entry type of the array has no default value, as, for example, in

  ```
  map<point,int> color;
  ```

  then *xdef* is some arbitrary value of E. This value may depend on the execution history.

- A subset $dom(A)$ of the index set, the so-called *domain* of A. All variables outside the domain have value *xdef*. Indices are added to the domain by the subscript operation and are deleted from the domain by the *undefine* operation. Maps have no *undefine* operation and put some indices in the domain even if they were not accessed[2]. D_arrays and h_arrays start with an empty domain and indices are added to the domain only by the subscript operation.

We come to the operations defined on sparse arrays. We assume that A belongs to one of our sparse array types and that I is a legal index type for this sparse array type as defined in the first row of Table 5.1. The subscript operator *operator*[] comes in two kinds:

```
const E& operator[](const I& i) const
E&       operator[](const I& i)
```

[1] This is the case for all but the built-in types of C++.

[2] These indices are used as sentinels in the implementation and allow us to make maps faster than the other sparse array types. We refer the reader to Section 5.2 for details.

The first version applies to const-objects and the second version applies to non-const-objects. Both versions return the variable $A(i)$. The first version allows only read access to the variable and the second version also allows us to modify the value of the variable. The second version adds i to the domain of A and the first version does not. How is the selection between the two versions made? Recall that in C++ every member function of a class X has an implicit argument referring to an instance of the object. This implicit argument has type `const X<I,E>*` for the first version of the subscript operator and has type `X<I,E>*` for the second version of the access operator; here X stands for one of the sparse array types. Thus depending on whether the subscript operator is applied to a constant sparse array or a modifiable sparse array either the first or the second version of the subscript operator is selected. Consider the following examples.

```
const map<int,int> M1;
      map<int,int> M2;
int x;
x = M1[5];                           // first  version
x = M2[5];                           // second version
x = ((const map<int,int>) M2)[7];    // first  version
```

Observe that the first version of the subscript operator is used in the first and the last call since *M1* is a constant map and since *M2* is cast to a constant map in the last line. The second version of the subscript operator is used in the second access. It is tempting but wrong to say (Kurt has made this error many times) that the use of the variable $A(i)$ dictates the selection: an access on the left-hand side of an assignment uses the second version (since the type `E&` is needed) and an access on the right-hand side of an assignment uses the second version (since the type `const E&` suffices). We emphasize, *the rule just stated is wrong*. In C++ the return type of a function plays no role in the selection of a version of an overloaded function; the selection is made solely on the basis of the argument types. We continue the example above.

```
    x = M2[5];                       // second version
M2[5] = x;                           // second version
    x = M1[5];                       // first version
M1[5] = x;                           // first version, illegal
```

The last assignment is illegal, since the first version of the access operator is selected for the constant map *M1*. It returns a constant reference to the variable *M1*(5), to which no assignment is possible.

```
bool A.defined(I i)
```

returns true if $i \in dom(A)$ and returns false otherwise. Finally, the operation

```
void A.undefine(I i)
```

removes i from $dom(A)$ and sets $A(i)$ to $xdef$. This operation is not available for maps.

Sparse arrays offer an iteration statement

```
forall_defined(i,A)
{ the elements of dom(A) are successively assigned to i }
```

which iterates over the indices in *dom*(A). In the case of d_arrays the indices are scanned in increasing order (recall that the index type of a d_array must be linearly ordered), in the case of h_arrays and maps the order is unspecified. The iteration statement

```
forall(x,A)
{ A[i] for i in dom(A) is successively assigned to x }
```

iterates over the values of the entries in *dom*(A).

5.1.2 *Performance Guarantees and Implementation Parameters*

Sparse arrays are one of the most widely studied data type and many different realizations with different performance guarantees have been proposed for them. We have included several into the LEDA system and give the user the possibility to choose an implementation through the implementation parameter mechanism.

```
_d_array<string,int,rs_tree> D1(0);
_d_array<string,int,rb_tree> D2(0);
_d_array<int,    int,dp_hashing> H;
```

defines three sparse arrays realized by randomized search trees, red-black trees, and dynamic perfect hashing, respectively. We now survey the available implementations; see also Tables 5.2 and 5.3. The implementations fall into two classes, those requiring a linearly ordered index type and those requiring a hashed index type. We use n to denote the size of the domain of the sparse array.

Implementations requiring a Linearly Ordered Index Type: This class of implementations contains deterministic and randomized implementations. The deterministic implementations are (a, b)-trees [Meh84a], *AVL*-trees [AVL62], $BB[\alpha]$-trees [NR73, BM80, Meh84a], red-black-trees [GS78, Meh84a], and unbalanced trees. The corresponding implementation parameters are *ab_tree*, *avl_tree*, *bb_tree*, *rb_tree*, and *bin_tree*, respectively. Except for unbalanced trees, all deterministic implementations guarantee $O(\log n)$ insertion, lookup, and deletion time. The actual running times of all deterministic implementations (except for unbalanced trees) are within a factor of two to three of one another. The unbalanced tree implementation can deteriorate to linear search and guarantees only linear insertion, lookup, and deletion time, as is clearly visible from the right part of Table 5.2. It should not be used.

The randomized implementations are skiplists [Pug90b] (*skiplist*) and randomized search trees [AS89] (*rs_tree*). Both implementations guarantee an expected insertion, deletion, and lookup time of $O(\log n)$. The expectations are taken with respect to the internal coin flips of the data structures.

Among the implementations requiring a linearly ordered index type ab-trees and skiplists

	Random integers				Sorted integers			
	insert	lookup	delete	total	insert	lookup	delete	total
ch_hash	0.23	0.09	0.18	0.5	0.2	0.05	0.12	0.37
dp_hash	1.48	0.21	1.08	2.77	1.37	0.21	1.02	2.6
map	0.15	0.04	—	0.19	0.15	0.05	—	0.2
skiplist	0.78	0.54	0.54	1.86	0.43	0.16	0.14	0.73
rs_tree	1.04	0.71	0.76	2.51	0.42	0.19	0.2	0.81
bin_tree	0.83	0.59	0.62	2.04	2704	1354	0.1501	4058
rb_tree	0.9199	0.54	0.74	2.2	0.6499	0.1802	0.3	1.13
avl_tree	0.8599	0.55	0.7	2.11	0.45	0.2	0.2402	0.8901
bb_tree	1.23	0.52	1	2.75	0.6399	0.2	0.3301	1.17
ab_tree	0.5898	0.25	0.4502	1.29	0.22	0.1399	0.2	0.5598
array					0.01001	0.01001	—	0.02002

Table 5.2 The performance of various implementations of sparse arrays. Hashing with chaining (*ch_hash*) and dynamic perfect hashing (*dp_hash*) are implementations of h_arrays, *map* is the implementation of map, and skiplists (*skiplist*), randomized search trees (*rs_tree*), unbalanced binary trees (*bin_tree*), red-black-trees (*rb_tree*), AVL-trees (*avl_tree*), BB[α]-trees (*bb_tree*), and 2-4-trees (*ab_trees*) are implementations of d_arrays. Running times are in seconds. We performed 10^5 insertions followed by 10^5 lookups followed by 10^5 deletions. We used random keys of type *int* in $[0 .. 10^7]$ for the left half of the table and we used the keys $0, 1, 2, \ldots$ for the right half of the table. Maps are the fastest implementation followed by hashing with chaining. Among the implementations of *d_arrays* ab-trees and skiplists are currently the most efficient. Observe the miserable performance of the *bin_tree* implementation for the sorted insertion order. For comparison we also included arrays for the second test.

are currently the most efficient. We give the details of the skiplist implementation in Section 5.7.

All implementations use linear space, e.g., the skiplist implementation requires $76n/3 + O(1) = 25.333n + O(1)$ bytes.

Implementations requiring a Hashed Index Type: There are two implementations: Hashing with chaining and dynamic perfect hashing.

Hashing with chaining is a deterministic data structure. Figure 5.2 illustrates it. It consists of a table and a singly linked list for each table entry. The table size T is a power of two such that $T = 1024$ if $n < 1024$ and $T/2 \le n \le 2T$ if $n \ge 1024$. The i-th list contains all x in the domain of the sparse array such that $i = Hash(x) \bmod T$. Let l_i be the number of

Random doubles				
	insert	lookup	delete	total
skiplist	3.09	2.36	1.95	7.4
rs_tree	3.81	2.69	2.48	8.98
bin_tree	2.85	1.94	2.15	6.94
rb_tree	2.75	1.82	2.28	6.85
avl_tree	2.82	1.89	2.24	6.95
bb_tree	4.06	1.88	3.81	9.75
ab_tree	2.09	1.51	1.61	5.21

Table 5.3 The performance of various implementations of sparse arrays. Running times are in seconds. We performed 10^5 insertions followed by 10^5 lookups followed by 10^5 deletions. We used random keys of type *double* in $[0\,..\,2^{31}]$.

elements in the i-th list and let k be the number of empty lists. The space requirement for hashing with chaining is $12(n+k)$ bytes.

We justify this formula. An item in a singly linked list requires twelve bytes; four bytes for the pointer to the successor and four bytes each for the key and the information (if a key or information does not fit into four bytes the space for the key or information needs to be added, see Section 13.4). There are T list items in the table and $l_i - 1$ extra items in the i-th list, if $l_i \geq 1$. Next observe that

$$\sum_{i;l_i \geq 1} (l_i - 1) = \sum_i (l_i - 1) + k = n - T + k.$$

The space required is therefore $12(T + n - T + k) = 12(n + k)$ bytes.

If the hash function behaves like a random function, i.e., its value is a random number in $[0\,..\,T-1]$, the probability that the i-th list is empty is equal to $(1 - 1/T)^n$ and hence the expected value of k is equal to $T(1-1/T)^n = T(1-1/T)^{T(n/T)} \approx Te^{-n/T}$; here, we used the approximation $(1 - 1/T)^T \approx e^{-1}$. The expected space requirement of hashing with chaining is therefore equal to $12(n + Te^{-n/T})$ bytes. The time to search for an element x, to insert it, or to delete it is $O(1)$ plus the time to search in the linear list to which x is hashed. The latter time is linear in the worst case. For random indices the expected length of each list is n/T and hence all operations take constant expected time for random indices.

After an insertion or deletion it is possible that the invariant relating T and n is violated. In this situation a so-called *rehash* is performed, i.e., the table size is doubled or halved and all elements are moved to the new table.

Dynamic perfect hashing [FKS84, DKM$^+$94] uses randomization. It is the implementation with the theoretically best performance. The operation *defined* takes constant worst

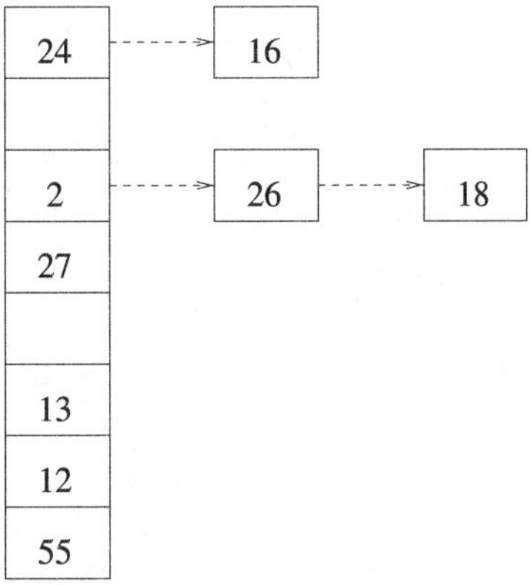

Figure 5.2 Hashing with chaining: The table size is 8 and the domain of the sparse array is $\{2, 12, 13, 16, 18, 24, 26, 27, 55\}$. The hash function $H(x)$ is the identity function $H(x) = x$ and hence any number x is stored in the list with index $x \bmod 8$.

case time and the operation $A[i]$ takes constant expected amortized time or constant worst case time depending on whether it is the first access with index i or not. This requires some explanation. Dynamic perfect hashing uses a two-level hashing scheme. A first-level hash function hashes the domain to some number T of buckets. T is chosen as in the case of hashing with chaining. As above, let l_i be the number of elements in the domain that are hashed to the i-th bucket. In the second level a separate table of size l_i^2 is allocated to the i-th bucket and a perfect hash function is used to map the elements in the i-th bucket to their private table, see Figure 5.3. In [FKS84, DKM+94] it is shown that suitable hash functions exist and can be found by random selection from a suitable class of hash functions. It is also shown in these papers that the space requirement of the scheme is linear, although with a considerably larger constant factor than for hashing with chaining. An access operation requires the evaluation of two hash functions and hence takes constant time in the worst case. An insertion (= first access to $A[i]$ for some index i) may require a rehash on either the second level or the first level of the data structure. Rehashes are costly but rare and hence the expected amortized time for an insert or delete is constant.

Experiments show that hashing with chaining is usually superior to dynamic perfect hashing and hence we have chosen hashing with chaining as the default implementation of *h_array<I, E>*.

Maps: Maps are implemented by hashing with chaining. Since the index type of a map must be an item or pointer type or the type int and since maps do not support the *undefine* operation, three optimizations are possible with respect to hashing with chaining as described above. First, items and pointers are interpreted as integers and the identity function is used as the hash function, i.e., an integer x is hashed to $x \bmod T$ where T is the table size. Since T is chosen as a power of two, evaluation of this hash function is very fast. Second, the list elements are not allocated in free store but are all stored in an array. This allows

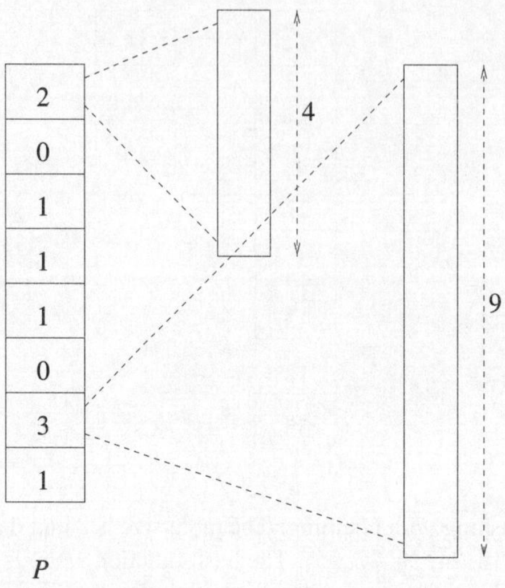

Figure 5.3 Dynamic perfect hashing: The first-level table P has size 8. For each entry of this table the number of elements hashed to this entry are indicated. If l, $l > 1$, elements are hashed to an entry then a second-level table of size l^2 is used to resolve the collisions. The sizes of the two second-level tables that are required in our example are also indicated.

for a faster realization of the rehash operation. Third, since the keys are integers a particularly efficient implementation of the access operation is possible. Section 5.2 contains the complete implementation of maps.

An Experimental Comparison: We give an experimental comparison of all sparse array types. We perform three kinds of experiments. In the first one, we use random integer keys in the range $[0 .. 10^7]$, in the second one, we use the keys $0, 1, \ldots$, and in the third one, we use random double keys. In each case we perform 10^5 insertions, followed by 10^5 lookups, followed by 10^5 deletions. Tables 5.2 and 5.3 summarize the results.

The following program performs the first two experiments and generates Table 5.2. In the main program we first define sparse arrays, one for each implementation, and two arrays A and B of size 10^5. We fill A with random integers and we fill B with the integers $0, 1, \ldots$. Then we call the function *dic_test* for each sparse array; *dic_test* first inserts $A[0]$, $A[1]$, $\ldots$, then looks up $A[0]$, $A[1]$, $\ldots$, and finally deletes $A[0]$, $A[1]$, $\ldots$. It then performs the same sequence of operations with B instead of A. For each sparse array type it produces a row of Table 5.2. The chunks ⟨*map test*⟩ and ⟨*array test*⟩ perform the same tests for maps[3] and arrays, respectively. We leave their details to the reader.

⟨*dic_performance.c*⟩≡

```
#include <LEDA/_d_array.h>
#include <LEDA/map.h>
#include <LEDA/array.h>
```

[3] Since maps do not support delete operations, we need two maps *M1* and *M2*, one for the experiment with A and one for the experiment with B.

```
#include <LEDA/IO_interface.h>
#include <LEDA/impl/ch_hash.h>
#include <LEDA/impl/dp_hash.h>
#include <LEDA/impl/avl_tree.h>
#include <LEDA/impl/bin_tree.h>
#include <LEDA/impl/rs_tree.h>
#include <LEDA/impl/rb_tree.h>
#include <LEDA/impl/skiplist.h>
#include <LEDA/impl/ab_tree.h>
#include <LEDA/impl/bb_tree.h>
int N;
int* A; int* B;
IO_interface I;
void dic_test(d_array<int,int>& D, string name)
{
  I.write_table("\n " + name);
  float T; float T0 = T = used_time();
  int i;
  for(i = 0; i < N; i++)  D[A[i]] = 0;
  I.write_table(" & ",used_time(T));
  for(i = 0; i < N; i++)  int* ptr = &D[A[i]];
  I.write_table(" & ",used_time(T));
  for(i = 0; i < N; i++)  D.undefine(A[i]);
  I.write_table(" & ",used_time(T));
  I.write_table(" & ",used_time(T0));
  ⟨same for B⟩
}
⟨map test⟩
int main()
{
  _d_array<int,int,ch_hash> CHH_DIC;
  _d_array<int,int,dp_hash> DPH_DIC;
  map<int,int> M1, M2;
  _d_array<int,int,avl_tree> AVL_DIC;
  _d_array<int,int,bin_tree> BIN_DIC;
  _d_array<int,int,rb_tree>  RB_DIC;
  _d_array<int,int,rs_tree>  RS_DIC;
  _d_array<int,int,skiplist> SK_DIC;
  _d_array<int,int,bb_tree>  BB_DIC;
  _d_array<int,int,ab_tree>  AB_DIC;
  N  = 100000;
  A = new int[N]; B = new int[N];
  int i;
  for(i = 0; i < N; i++) { A[i] = rand_int(0,10000000); B[i] = i; }
  dic_test(CHH_DIC,"ch\\_hash");
  dic_test(DPH_DIC,"dp\\_hash"); I.write_table(" \\hline");
  map_test(M1,M2,  "map");       I.write_table(" \\hline");
  dic_test(SK_DIC, "skiplist");
```

```
    dic_test(RS_DIC ,"rs\\_tree");
    dic_test(BIN_DIC,"bin\\_tree");
    dic_test(RB_DIC ,"rb\\_tree");
    dic_test(AVL_DIC,"avl\\_tree");
    dic_test(BB_DIC ,"bb\\_tree");
    dic_test(AB_DIC ,"ab\\_tree"); I.write_table(" \\hline");
    ⟨array test⟩
}
```

5.1.3 *Persistence of Variables*

We stated above that an access operation

```
E& A[I i]
```

returns the variable $A(i)$. Thus, one can write

```
E& x = A[5];
<some statements not touching A[5]>;
A[5] = y;
if ( x == y ) { .... }
```

and expect that the test $x == y$ returns true. This is not necessarily the case for h_arrays and maps as these types do not guarantee that different accesses to $A[5]$ return the same variable and *we therefore recommend never to establish a pointer or a reference to a variable contained in a map or h_array.* Given the efficiency of h_arrays and maps there is really no need to do so. The fact that the identity of variables is not preserved is best explained by recalling the implementation of h_arrays and maps. They use an array of linked lists where the size of the array is about the size of the domain of the sparse array. Whenever the invariant linking the size of the table and the size of the domain is violated the content of the sparse array is rehashed. In the process of rehashing new variables are allocated for some of the entries of the sparse array. Of course, the values of the entries are moved to the new variables. Thus, the content of $A(i)$ is preserved but not the variable $A(i)$.

D_arrays behave differently. Variables in d_arrays are persistent, i.e, the equality test in the code sequence above is guaranteed to return true.

5.1.4 *Choosing an Implementation*

LEDA gives you the choice between many implementations of sparse arrays. Which is best in a particular situation?

Tables 5.2 and 5.3 show that in certain situations maps are faster than h_arrays which in turn are faster than d_arrays. On the other hand the slower data types offer an increased functionality. This suggests using the type whose functionality just suffices in a particular application.

There are, however, other considerations to be taken into account. Maps and h_arrays perform well only for random inputs, they can perform miserably for non-random inputs. For maps a bad example is easily constructed. Use the indices $1024i$ for $i = 0, 1, \ldots$.

Since maps use the hash function $x \longrightarrow x \bmod T$ where T is the table size, and T is always a power of two these keys will not be distributed evenly by the hash function and hence the performance of maps will be much worse than for random inputs. In the case of h_arrays the situation is not quite as bad since you may overwrite the default hash function. For example, you may want to use

```
int Hash(int x){ return x/1024; }
```

if you know that the indices are multiples of 1024.

Which implementations are we using ourselves? We usually use maps to associate information with item types such as points and segments, we use d_arrays or dictionaries when the order on the indices is important for the application, and we use h_arrays when we know a hash function suitable for the application.

If you are not happy with any of the implementations provided in LEDA you may provide your own. Section 13.6 explains how this is done.

5.2 The Implementation of the Data Type Map

We give the complete implementation of the data type *map*. This section is for readers who want to understand the internals of LEDA. Readers that "only" want to use LEDA may skip this section without any harm.

We follow the usual trichotomy in the definition of LEDA's parameterized data types as explained in Section 13.4. Familiarity with this section is required for some of the fine points of this section. We define two classes, namely the abstract data type class *map*<*I*, *E*> and the implementation class *ch_map*, in three files, namely map.h, ch_map.h, and _ch_map.c. The abstract data type class has template parameters I and E and the implementation class stores *GenPtrs* (= *void*∗). In map.h we define the abstract data type class and implement it in terms of the implementation class. This implementation is fairly direct; its main purpose is to translate between the untyped view of the implementation class and the typed view of the abstract data type class. In ch_map.h and _ch_map.c, respectively, we define and implement the implementation class.

We first give the global structure of LEDAROOT/incl/LEDA/map.h.

⟨*map.h*⟩+≡

```
template<class I, class E>
class map : private ch_map {

 E xdef;

 void copy_inf(GenPtr& x)   const { LEDA_COPY(E,x);  }
 void clear_inf(GenPtr& x)  const { LEDA_CLEAR(E,x); }
 void init_inf(GenPtr& x)   const { x = leda_copy((E&)xdef); }

 public:
```

```
  typedef ch_map::item item;
  ⟨member functions of map⟩
};
```

We give some explanations. We derive the abstract data type class *map* from the implementation class *ch_map* and give it an additional data member *xdef*, which stores the default value of the variables of the map. Therefore, an instance of *map* consists of an instance of *ch_map* and a variable *xdef* of type E. The private function members *copy_inf*, *clear_inf*, and *init_inf* correspond to virtual functions of the implementation class and redefine them. The first two are required by the LEDA method for the implementation of parameterized data types and are discussed in Section 13.4. The third function is used to initialize an entry to a copy of *xdef*.

The public member functions will be discussed below. They define the user interface of maps as given in Table 5.1.

We come to our implementation class *ch_map*. It is based on the data structure hashing with chaining. Hashing with chaining uses an array of singly linked lists and therefore we introduce a container for list elements, which we call *ch_map_elem*. A *ch_map_elem* stores an unsigned long k, a generic pointer i, and a pointer to the successor container. We refer to k as the key-field and to i as the inf-field of the container. This nomenclature is inspired by dictionaries. Keys correspond to indices (type I) in the abstract data type class and infs correspond to elements (type E) in the abstract data type class.

A pointer to a *ch_map_elem* is called a *ch_map_item*.

The flag `__exportC` is used during a precompilation step. On UNIX-systems it is simply deleted and on Windows-systems it is replaced by appropriate key words that are needed for the generation of dynamic libraries.

⟨*ch_map_elem*⟩≡

```
class __exportC ch_map_elem
{
  friend class __exportC ch_map;
  unsigned long    k;
  GenPtr           i;
  ch_map_elem*  succ;
};
typedef ch_map_elem*  ch_map_item;
```

Next we discuss the data members of the implementation class.

⟨*data members of ch_map*⟩≡

```
ch_map_elem STOP;
ch_map_elem* table;
ch_map_elem* table_end;
ch_map_elem* free;
int table_size;
int table_size_1;
```

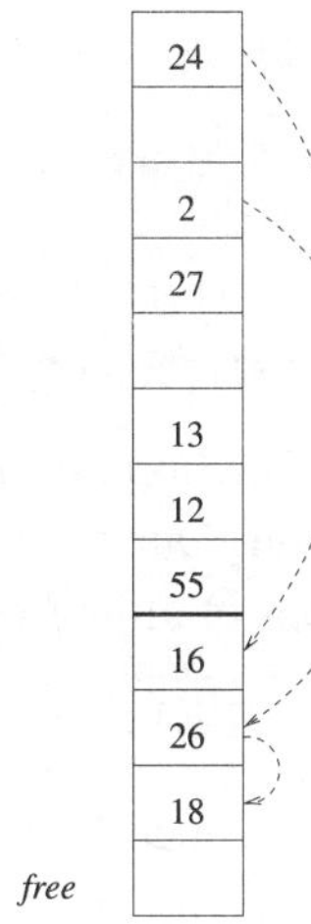

Figure 5.4 A hash table of size 12. The last four locations are used as an overflow area and the first eight locations correspond to eight linear lists. The set stored is $\{2, 12, 13, 16, 18, 24, 26, 27, 55\}$ and any number x is stored in the list with index $x \bmod 8$. If the i-th list contains more than one element then the first element is stored in the i-th table position and all other elements are stored in the overflow area. In the example, three elements are hashed to the second list and hence two of them are stored in the overflow area. The variable *free* points to the first free position in the overflow area.

We use a *table* of map elements of size $f \cdot T$ where T is a power of two and f is a number larger than one, see Figure 5.4. We use $f = 1.5$ in our implementation. The first T elements of the table correspond to the headers of T linear lists and the remaining $(f-1)T$ elements of the table are used as an overflow area to store further list elements. The variable *free* always points to the first unused map element in the overflow area. When the overflow area is full we move to a table twice the size. We use *table_size* to store T and *table_size_1* to store $T - 1$.

The main use of maps is to associate information with objects. Thus the most important operation for maps is the access operation with keys that are already in the table (the data structure literature calls such accesses *successful searches*) and we designed maps so that successful searches are particularly fast. An access for a key x involves the evaluation of a hash function plus the search through a linear list. Our hash function simply extracts the last log *table_size* bits from the binary representation of x.

⟨*HASH function*⟩≡

```
ch_map_elem*  HASH(unsigned long x)  const
{ return table + (x & table_size_1);  }
```

Why do we dare to take such a simple hash function? Let U be the set of unsigned longs. We assume, as is customary in the analysis of hashing, that a random subset $S \subseteq U$ of size n is stored in the hash table. Let $m =$ *table_size* denote the size of the hash table and for all

$i, 0 \leq i < m$, let s_i be the number of elements in S that are hashed to position i. Then

$$s_0 + s_1 + \ldots + s_{m-1} = n$$

and hence

$$\mathrm{E}[s_0] + \mathrm{E}[s_1] + \ldots + \mathrm{E}[s_{m-1}] = n$$

by linearity of expectations. A hash function is called *fair* if the same number of elements of U are hashed to every table position. Our hash function is fair. For a fair hash function symmetry implies that the expectations of all the s_i's are the same. Hence

$$\mathrm{E}[s_i] = n/m$$

for all i. No hash function can do better since $\sum_i \mathrm{E}[s_i] = n$. We conclude that any fair hash function yields the optimal expectations for the $\mathrm{E}[s_i]$. For the sake of speed the simplest fair hash function should be used. This is exactly what we do.

We mentioned already that our main goal was to make access operations as fast as possible. We will argue in the next three paragraphs that most successful accesses are accesses to elements which are stored in the first position of the list containing them. Let k denote the number of empty lists. Then $T - k$ lists are non-empty and hence there are $T - k$ elements which are first in their list. If n denotes the number of elements stored in the table the fraction of elements that are first in their list is $(T - k)/n$. We want to estimate this fraction for random keys and immediately before and after a rehash. We move to a new table when the overflow area is full. At this time, there are $(f - 1)T$ elements stored in the overflow area and $T - k$ elements in the first T positions of the table. Thus $n = fT - k$ at the time of a rehash.

For random keys the expected number of empty lists is $k = T \cdot (1 - 1/T)^n \approx Te^{-n/T}$. For random keys we will therefore move to a new table when $n \approx T \cdot (f - e^{-n/T})$ or $n/T + e^{-n/T} \approx f$. For $f = 1.5$ we get $n \approx 1.2T$, i.e., when about $1.2T$ elements are stored in the table we expect to move to a new table.

When $n \approx 1.2T$ about $0.7T$ elements are stored in the first T slots of the table and about $0.5T$ elements are stored in the overflow area of the table. Thus about $0.7/1.2 \approx 58\%$ of the successful searches go to the first element in a list. Immediately after a rehash we have $n \approx 0.6T$ (since $n \approx 1.2T$ before the rehash and a rehash doubles the table size) and the expected number of empty lists is $Te^{-0.6} \approx 0.55T$. Thus $0.45/0.6 \approx 75\%$ of the successful searches go to the first element in a list. In either case a significant fraction of the successful searches goes to the first element in a list.

How can we make accesses to first elements fast? A key problem is the encoding of empty lists. We explored two possibilities. In both solutions we use a special list element *STOP* as a sentinel. In the first solution we maintain the invariant that the i-th list is empty if the successor field of *table*[i] is nil and that the last entry of a non-empty list points to *STOP*. This leads to the following code for an access operation:

```
inline GenPtr& ch_map::access(unsigned long x)
{ ch_map_item p = HASH(x);
  if ( p->succ == nil)
    { p->k = x;
      init_inf(p->i);    // initializes p->i to xdef
      p->succ = &STOP;
      return p->i;
    }
   else
    { if ( p->k == x ) return p->i;
    }
    return access(p,x);
}
```

In this code, $access(p, x)$ handles the case that the list for x is non-empty and that the first element does not contain x. This code has two weaknesses. First, it tests each list for emptiness although successful searches always go to non-empty lists and, second, it needs to change the successor pointer of $table[i]$ to *&STOP* after the first insert into the i-th list.

In the second solution we encode the fact that the i-th list is empty in the key field of $table[i]$. Let NULLKEY and NONNULLKEY be keys that are hashed to zero and some non-zero value, respectively. In our implementation we use 0 for NULLKEY and 1 for NONNULLKEY. We use the special keys NULLKEY and NONNULLKEY to encode empty lists. More specifically, we maintain:

- $table[0].k =$ NONNULLKEY, i.e., the first entry of the zero-th list is unused. The information field of this entry is arbitrary.
- $table[i].k =$ NULLKEY iff the i-th list is empty for all i, $i > 0$, and
- the last entry of a non-empty list points to *STOP* and if the i-th list is empty then $table[i]$ points to *STOP*.

Observe that the zero-th list is treated somewhat unfairly. We leave its first position unused and thus make it artificially non-empty. Figure 5.5 illustrates the items above.

Consider a search for x and let p be the hash-value of x. If x is stored in the first element of the p-th list we have a successful search, and the p-th list is empty iff the key of the first element of the p-th list is equal to NULLKEY. Observe that this is true even for p equal to zero, because the first item guarantees that NULLKEY is not stored in the first element of list 0. We obtain the following code for the access operation:

⟨*inline functions*⟩≡

```
inline GenPtr& ch_map::access(unsigned long x)
{ ch_map_item p = HASH(x);
  if ( p->k == x ) return p->i;
  else
  { if ( p->k == NULLKEY )
    { p->k = x;
      init_inf(p->i);  // initializes p->i to xdef
```

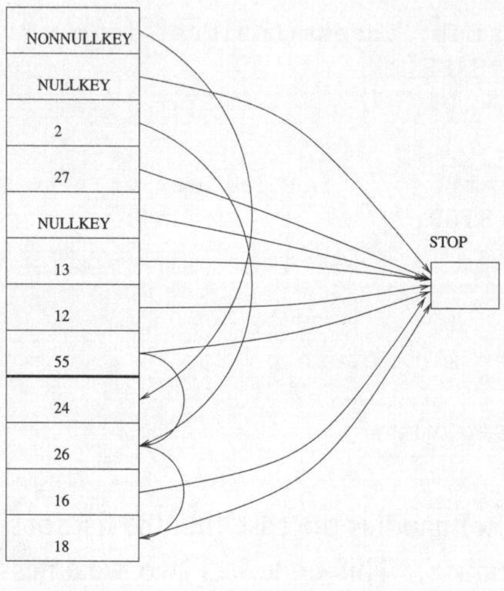

Figure 5.5 The realization of the hash table of Figure 5.4 in *ch_map*. The first entry of the zero-th list containts `NONNULLKEY` (whether the zero-th list is empty or not), empty lists other than the zero-th list contain `NULLKEY` in their first element, and each list points to `STOP`.

```
      return p->i;
    }
    else
      return access(p,x);
  }
}
```

Note that a successful search for a key x that is stored in the first position of its list is very fast. It evaluates the hash function, makes one equality test between keys, and returns the information associated with the key. If x is not stored in the first position of its table, we need to distinguish cases: if the list is empty we store $(x, xdef)$ in the first element of the list (note that the call $init\_inf(p \to i)$ sets the inf-field of p to *xdef*), and if the list is non-empty we call $access(p, x)$ to search for x in the remainder of the list. We will discuss this function below.

Our experiments show that the second design is about 10% faster than the first and we therefore adopted it for maps. In the implementation of h_arrays by hashing with chaining we use the first solution. Since h_arrays use non-trivial hash functions that may require substantial time for their evaluation, the second solution looses its edge over the first in the case of h_arrays.

We can now give an overview over LEDAROOT/incl/LEDA/impl/ch_map.h.

⟨*ch_map.h*⟩≡

```
#ifndef LEDA_CH_MAP_H
#define LEDA_CH_MAP_H
#include <LEDA/basic.h>
⟨ch_map_elem⟩
```

```
class __exportC ch_map
{
   const unsigned long NULLKEY;
   const unsigned long NONNULLKEY;
   ⟨data members of ch_map⟩
   virtual void clear_inf(GenPtr&)  const { }
   virtual void copy_inf(GenPtr&)   const { }
   virtual void init_inf(GenPtr&)   const { }
   ⟨HASH function⟩
   ⟨private member functions of ch_map⟩
   protected:
   typedef ch_map_item item;
   ⟨protected member functions of ch_map⟩
};
⟨inline functions⟩
#endif
```

We have already explained the data members. The virtual function members *clear_inf*, *copy_inf*, and *init_inf* are required by the LEDA method for the implementation of parameterized data types. We saw already how they are redefined in the definition of *map*.

The protected and private member functions will be discussed below. The protected member functions are basically in one-to-one correspondence to the public member functions of the abstract data type class and the private member functions define some basic functionality that is needed for the protected member functions, e.g., rehashing to move to a larger table.

We come to the file LEDAROOT/src/dic/_ch_map.c. There is little to say about it at this point except that is contains the implementation of class *ch_map*.

⟨*_ch_map.c*⟩≡

```
#include <LEDA/impl/ch_map.h>
⟨implementation of ch_map⟩
```

Having defined all data members and the global structure of all files we can start to implement functions. We start with the private members of *ch_map*.

⟨*private member functions of ch_map*⟩≡

```
void init_table(int T);
```

initializes a table of size T (T is assumed to be a power of two) and makes all lists (including list zero) empty. This is trivial to achieve. We allocate a new table of size fT and set all data members accordingly. We also initialize *table*[0].*k* to `NONNULLKEY`, *table*[i].*k* to `NULLKEY` for all i, $1 \le i < table\_size$, and let *table*[i].*succ* point to `STOP` for all i, $0 \le i < table\_size$. This initializes all lists to empty lists.

⟨*implementation of ch_map*⟩≡

```
void ch_map::init_table(int T)
{
  table_size = T;
  table_size_1 = T-1;
  table = new ch_map_elem[T + T/2];
  free = table + T;
  table_end = table + T + T/2;
  for (ch_map_item p = table; p < free; p++)
  { p->succ = &STOP;
    p->k = NULLKEY;
  }
  table->k = NONNULLKEY;
}
```

⟨*private member functions of ch_map*⟩+≡

```
void rehash();
```

moves to a table twice the current size. We do so by first moving all elements stored in the first T elements of the table and then all elements in the overflow area. Note that this strategy has two advantages over moving the elements list after list: First, we do not have to care about collisions when moving the elements in the first T table positions (because the element in position i is moved to either position i or $T + i$ in the new table depending on the additional bit that the new hash function takes into account), and second, locality of reference is better (since we move all elements by scanning the old table once).

When moving the elements from the overflow area we make use of the member function *insert*. We define it inline. It takes a pair (x, y) and moves it to the list for key x. If the first element of the list is empty, we move (x, y) there, and if the first element is non-empty, we move (x, y) to position *free*, insert it after the first element of the list, and increment *free*.

⟨*private member functions of ch_map*⟩+≡

```
inline void insert(unsigned long x, GenPtr y);
```

⟨*implementation of ch_map*⟩+≡

```
inline void ch_map::insert(unsigned long x, GenPtr y)
{ ch_map_item q = HASH(x);
  if ( q->k == NULLKEY )
    { q->k = x;
      q->i = y;
    }
  else
   { free->k = x;
     free->i = y;
     free->succ = q->succ;
     q->succ = free++;
   }
}
```

In *rehash* we first initialize the new table (this puts NONNULLKEY into the first entry of the zero-th list) and then move elements. We first move the elements in the main part of the table (*table*[0] is unused and hence the loop for moving elements starts at *table* + 1) and then the elements in the overflow area.

⟨*implementation of ch_map*⟩+≡

```
void ch_map::rehash()
{
  ch_map_item old_table = table;
  ch_map_item old_table_mid = table + table_size;
  ch_map_item old_table_end = table_end;

  init_table(2*table_size);

  ch_map_item p;

  for(p = old_table + 1; p < old_table_mid; p++)
  { unsigned long x = p->k;
    if ( x != NULLKEY ) // list p is non-empty
    { ch_map_item q = HASH(x);
      q->k = x;
      q->i = p->i;
    }
  }

  while (p < old_table_end)
  { unsigned long x = p->k;
    insert(x,p->i);
    p++;
  }

  delete[] old_table;
}
```

⟨*private member functions of ch_map*⟩+≡

```
GenPtr& access(ch_map_item p, unsigned long x);
```

searches for x in the list starting at p. The function operates under the precondition that the list is non-empty and x is not stored in p. The function is called by the inline function $access(x)$.

We search down the list starting at p. If the search reaches STOP, we have to insert x. If the table is non-full, we insert x at position *free*, and if the table is full, we rehash and recompute the hash value of x. If x now hashes to an empty list, we put it into the first entry of the list, and otherwise, we put it at *free*.

⟨*implementation of ch_map*⟩+≡

```
GenPtr& ch_map::access(ch_map_item p, unsigned long x)
{
  STOP.k = x;
  ch_map_item q = p->succ;
  while (q->k != x) q = q->succ;
  if (q != &STOP) return q->i;

  // index x not present, insert it
```

```
    if (free == table_end)    // table full: rehash
    { rehash();
      p = HASH(x);
    }
    if (p->k == NULLKEY)
    { p->k = x;
      init_inf(p->i);  // initializes p->i to xdef
      return p->i;
    }
    q = free++;
    q->k = x;
    init_inf(q->i);     // initializes q->i to xdef
    q->succ = p->succ;
    p->succ = q;
    return q->i;
}
```

We come to the protected member functions of *ch_map*. We start with some trivial stuff.

⟨*protected member functions of ch_map*⟩≡

```
unsigned long key(ch_map_item it) const { return it->k; }
GenPtr&        inf(ch_map_item it) const { return it->i; }
```

Constructors and Assignment: We start with the implementation class.

⟨*protected member functions of ch_map*⟩+≡

```
ch_map(int n = 1);
ch_map(const ch_map& D);
ch_map& operator=(const ch_map& D);
```

The default constructor initializes a data structure of size $\min(512, 2^{\lceil \log n \rceil})$. The copy constructor initializes a table of the same size as D and then copies all elements from D to the new table. Elements from the first part of the table are moved if their key is different from NULLKEY and elements from the second part of the table are always moved. The assignment operator works in the same way but clears and destroys the old table first.

⟨*implementation of ch_map*⟩+≡

```
ch_map::ch_map(int n) : NULLKEY(0), NONNULLKEY(1)
{
  if (n < 512)
     init_table(512);
  else
   { int ts = 1;
     while (ts < n) ts <<= 1;
     init_table(ts);
    }
}

ch_map::ch_map(const ch_map& D) : NULLKEY(0), NONNULLKEY(1)
```

```
{
  init_table(D.table_size);
  for(ch_map_item p = D.table + 1; p < D.free; p++)
  { if (p->k != NULLKEY || p >= D.table + D.table_size)
    { insert(p->k,p->i);
      D.copy_inf(p->i);  // see chapter Implementation
    }
  }
}
ch_map& ch_map::operator=(const ch_map& D)
{
  clear_entries();
  delete[] table;
  init_table(D.table_size);
  for(ch_map_item p = D.table + 1; p < D.free; p++)
  { if (p->k != NULLKEY || p >= D.table + D.table_size)
    { insert(p->k,p->i);
      copy_inf(p->i);    // see chapter Implementation
    }
  }
  return *this;
}
```

The constructors of the abstract data type class simply call the appropriate constructor of the implementation class.

⟨*member functions of map*⟩≡

```
map()  { }
map(E x,int table_sz) : ch_map(table_sz), xdef(x) { }
map(E x) : xdef(x) { }
map<I,E>& operator=(const map<I,E>& M)
{ ch_map::operator=((ch_map&)M);
  xdef = M.xdef;
  return *this;
}
map(const map<I,E>& M): ch_map((ch_map&)M), xdef(M.xdef) { }
```

Destruction: We follow our canonical design for constructors, see Section 13.4.3. On the level of the implementation class, we define a function *clear_entries* that clears the information field of all used entries, a function *clear* that first clears the entries of the table and destroys the table and then reinitializes the table to its default size (*clear* is not used but we define it for the sake of uniformity), and the destructor that simply deletes *table*. Note that our canonical design ensures that *clear_entries* is called before any call of the destructor and hence only *table* must be destroyed by the destructor. Following standard practice (see [ES90, page278]) we declare the destructor virtual.

⟨*protected member functions of ch_map*⟩+≡

```
void clear_entries();
void clear();
virtual ~ch_map() { delete[] table; }
```

⟨*implementation of ch_map*⟩+≡

```
void ch_map::clear_entries()
{ for(ch_map_item p = table + 1; p < free; p++)
    if (p->k != NULLKEY || p >= table + table_size)
    clear_inf(p->i);  // see chapter Implementation
 }

void ch_map::clear()
{ clear_entries();
  delete[] table;
  init_table(512);
}
```

The destructor of the abstract data type class first calls *clear_entries* and then the destructor of the implementation class.

⟨*member functions of map*⟩+≡

```
~map() { clear_entries(); }
```

Access Operations: We have already defined the operation *access*(x) that searches for x and, if unsuccessful, inserts x into the table. *Lookup* only searches; it returns the item corresponding to a key x, if there is one, and *nil* otherwise.

⟨*protected member functions of ch_map*⟩+≡

```
GenPtr& access(unsigned long x);
ch_map_item lookup(unsigned long x) const;
```

⟨*implementation of ch_map*⟩+≡

```
ch_map_item ch_map::lookup(unsigned long x) const
{ ch_map_item p = HASH(x);
  ((unsigned long &)STOP.k) = x;  // cast away const
  while (p->k != x) p = p->succ;
  return (p == &STOP) ? nil : p;
}
```

The abstract data type class uses these functions in the obvious way.

⟨*member functions of map*⟩+≡

```
const E& operator[](const I& i) const
{ ch_map_item p = lookup(ID_Number(i));
  return (p) ? LEDA_CONST_ACCESS(E,ch_map::inf(p)) : xdef;
}
```

```
E& operator[](const I& i)
{ return LEDA_ACCESS(E,access(ID_Number(i))); }

bool defined(const I& i) const { return lookup(ID_Number(i)) != nil; }
```

In the above, *LEDA_ACCESS*(E, i) returns the value of i converted to type E, see Section 13.4.5, and *ID_number*(i) returns the ID-number of i.

⟨*member functions of map*⟩+≡

```
void clear()      { ch_map::clear(); }
void clear(E x) { ch_map::clear(); xdef = x; }
```

Iteration: The implementation of the iteration statements follows the general strategy described in Section 13.9. The implementation class provides two functions that return the first used item and the used item following a used item, respectively. Both functions are simple. The first item in the hash table is always unused and hence *first_item* returns *next_item*(*table*). We come to *next_item*(*it*). Let *it* be any item. If *it* is *nil*, we return *nil*. So assume otherwise. To find the next used item we advance *it* one or more times until we are either in the overflow area or have reached an item whose key is not equal to NULLKEY. If the resulting value of *it* is less than *free* we return it and otherwise we return *nil*.

⟨*protected member functions of ch_map*⟩+≡

```
ch_map_item first_item() const;
ch_map_item next_item(ch_map_item it) const;
```

⟨*implementation of ch_map*⟩+≡

```
ch_map_item ch_map::first_item() const
{ return next_item(table); }

ch_map_item ch_map::next_item(ch_map_item it) const
{ if ( it == nil ) return nil;
  do { it++; }
  while ( it < table + table_size && it->k == NULLKEY);
  return ( it < free ? it : nil);
}
```

The abstract data type class must provide the functions *first_item*, *next_item*, *inf*, *key*. All four functions reduce to the corresponding function in the implementation class.

⟨*member functions of map*⟩+≡

```
item first_item() const          { return ch_map::first_item(); }
item next_item(item it) const { return ch_map::next_item(it); }

E inf(item it) const
{ return LEDA_CONST_ACCESS(E,ch_map::inf(it)); }

I key(item it) const
{ return LEDA_CONST_ACCESS(I,(GenPtr)ch_map::key(it)); }
```

Exercises for 5.2

1 The unbalanced tree implementation of sparse arrays deteriorates to linear lists in the case of a sorted insertion order. In particular, if the keys 1, 2, ..., n are inserted in this order then each insertion appends the key to be inserted at the end of the list. Try to explain the row for *bin_trees* in the lower half of Table 5.2 in view of this sentence.

2 Use maps and the indices $1024i$ for $i = 0, 1, \ldots$.

3 Use h_arrays and the indices $1024i$ for $i = 0, 1, \ldots$. Define your own hash function.

4 Design a hash function for strings. The function should depend on all characters of a string.

5 Extend the implementation of h_arrays such that variables become persistent. (Hint: do not store the array variables directly in the hash table but access them indirectly through a pointer). What price do you pay in terms of access and insert time?

6 Provide a new implementation of d_arrays or h_arrays and perform the experiments of Table 5.2.

5.3 Dictionaries and Sets

Dictionaries and sets are essentially another interface to d_arrays and therefore we can keep this section short.

A *dictionary* is a collection of items (type *dic_item*) each holding a key of some linearly ordered type K and an information from some type I. Note that we now use I for the information type and no longer for the index type. We illustrate dictionaries by a program that reads a sequence of strings from standard input, counts the number of occurrences of each string, and prints all strings together with their multiplicities.

```
dictionary<string,int> D;
string s;
dic_item it;
while (cin >> s)
  { it = D.lookup(s);
    if (it == nil) D.insert(s,1);
    else D.change_inf(it, D.inf(it) + 1);
  }
forall_dic_items(it, D)
    cout << D.key(it) << " " << D.inf(it) << "\n";
```

In the while-loop we first search for s in the dictionary. The lookup returns *nil* if s is not part of the dictionary and returns the unique item with key s otherwise. In the first case we insert the item $\langle s, 1\rangle$ into the dictionary. In the second case we increment the information associated with s.

Dictionaries are frequently used to realize sets. In this situation the information associated with an element in the dictionary is irrelevant, the only thing that counts is whether a key belongs to the dictionary or not. The data type *set* is appropriate in this situation. A set S of integers is declared by *set<int>* S. The number 5 is added by $S.insert(5)$, the number 8 is tested for membership by *S.member*(8), and the number 3 is deleted by *S.delete*(3). The operation *S.choose*( ) returns some element of the set. Of course, *choose* requires the set to be non-empty.

We will discuss an extension of dictionaries in a later section: *Sorted sequences*. Sorted sequences extend dictionaries by more fully exploiting the linear order defined on the key type. They offer queries to find the next larger element in a sequence and also operations to merge and split sequences.

LEDA also contains extensions of dictionaries to geometric objects such as points and parallel line segments. We discuss a dictionary type for points in Section 10.6. For more dictionary types for geometric objects we refer the reader to the manual.

Exercises for 5.3

1 Implement dictionaries in terms of d_arrays. Are you encountering any difficulties?

2 Implement d_arrays in terms of dictionaries. Are you encountering any difficulties?

5.4 Priority Queues

Priority queues are an indispensable ingredient for many network and geometric algorithms. Examples are Dijkstra's algorithm for the single-source shortest-path problem (cf. Section 6.6), and the plane sweep algorithm for line segment intersection (cf. Section 10.7.2). We start with the basic properties of priority queues, and then discuss the many implementations of priority queues in LEDA. We give recommendations about which priority queue to choose in a particular situation.

5.4.1 *Functionality*

A priority queue Q over a priority type P and an information type I is a collection of items (type *pq_item*), each containing a priority from type P and an information from type I. The type P must be linearly ordered. A priority queue organizes its items such that an item with minimum priority can be accessed efficiently.

```
p_queue<P,I> Q;
```

defines a priority queue Q with priority type P and information type I and initializes Q to the empty queue. A new item $\langle p, i \rangle$ is added by

```
Q.insert(p,i);
```

and

```
pq_item it = Q.find_min();
```

returns an item of minimal priority and assigns it to *it* (*find_min* returns *nil* if Q is empty). Frequently, we do not only want to access an item with minimal information but also want to delete it.

```
P p = Q.del_min();
```

deletes an item with minimum priority from Q and assigns its priority to p (Q must be non-empty, of course). An arbitrary item *it* can be deleted by

```
Q.del_item(it);
```

The fields of an item are accessed by *Q.prio*(*it*) and *Q.inf*(*it*), respectively. The operation *Q.insert*(p, i) adds a new item $\langle p, i\rangle$ and returns the item; so we may store it for later use:

```
pq_item it = Q.insert(p,i);
```

There are two ways to change the content of an item. The information can be changed arbitrarily:

```
Q.change_inf(it,i1);
```

makes *i1* the new information of item *it*. The priority of an item can only be decreased:

```
Q.decrease_p(it,p1);
```

makes *p1* the new priority of item *it*. The operation raises an error if *p1* is larger than the current priority of *it*. There is no way to increase the priority of an item[4]. Finally, there are the operations

```
Q.empty();
Q.size();
Q.clear();
```

that test for emptiness, return the number of items, and clear a queue, respectively.

Let us see priority queues at work. We read a sequence of doubles from standard input and store them in a priority queue. We then repeatedly extract the minimum element from the queue until the queue is empty. The net effect is to sort the input sequence into increasing order.

```
p_queue<double,int> Q; //the information type is irrelevant
double x;
while (cin >> x) Q.insert(x,0);
while (! Q.empty()) cout << Q.del_min << "\n";
```

A more sophisticated use of priority queues is *discrete event simulation*. We have a set of events associated with points in time. An event associated with time t is to be executed at time t. The execution of an event may create new events that are to be executed at later moments of time. Priority queues support discrete event simulation in a very natural way; one only has to store all still to be executed events together with their scheduled time in a priority queue (with time playing the role of the priority) and to always extract and execute the event with the minimal scheduled time.

[4] The fact that priorities can be decreased but not increased is dictated by the implementations. There are implementations that support very efficient decrease of priorities but there are no implementations that support efficient decrease and increase.

Name	Prio	Args	Running times *insert*	*delete_min*	*decrease_p*	*create, destruct*
f_heap	general	—	$O(\log n)$	$O(\log n)$	$O(1)$	$O(1)$
p_heap	general	—	$O(\log n)$	$O(\log n)$	$O(1)$	$O(1)$
k_heap	general	$N, k = 2$	$O(\log_k n)$	$O(k \log_k n)$	$O(\log_k n)$	$O(N)$
bin_heap	general	—	$O(\log n)$	$O(\log n)$	$O(\log n)$	$O(1)$
list_pq	general	—	$O(1)$	$O(n)$	$O(1)$	
b_heap	int, $[l\,..\,h]$	l, h	$O(1)$	$O(h - l)$	$O(h - l)$	$O(h - l)$
r_heap	int	C	$O(\log C)$	$O(\log C)$	$O(1)$	$O(\log C)$
m_heap	int	C	$O(1)$	$O(min - p\_min)$	$O(1)$	$O(C)$

Table 5.4 Properties of different priority queue implementations: the second column indicates whether the priorities can come from an arbitrary linearly ordered type (general) or must be integers, the third column indicates the arguments of the constructor, and the remaining columns indicate the running times of the various priority queue operations. *B_heaps* can only handle integer priorities from a fixed range $[l\,..\,h]$ and *r_heaps* and *m_heap* maintain a variable *p_min* and priorities must be integers in the range $[p\_min\,..\,p\_min + C - 1]$. *B_heaps* also support a *delete_max* operation. More detailed explanations are given in the text.

5.4.2 *Performance Guarantees and Implementation Parameters*

LEDA provides many implementations of priority queues. The implementations include Fibonacci heaps [FT87], pairing heaps [SV87], k-ary heaps and binary heaps [Meh84a, III.5.3.1], lists[5], buckets[6], redistributive heaps [AMOT90], and monotone heaps [Meh84c, IV.7.2]. Fibonacci heaps are the default implementation and other implementations can be selected using the implementation parameter mechanism. The implementation parameters are *f_heap*, *p_heap*, *k_heap*, *bin_heap*, *list_pq*, *b_heap*, *r_heap*, and *m_heap*, respectively. Fibonacci heaps support *insert*, *del_item* and *del_min* in time $O(\log n)$, *find_min*, *decrease_p*, *change_inf*, *inf*, *size*, and *empty* in time $O(1)$, and *clear* in time $O(n)$, where n denotes the current size of the queue. The time bounds are amortized. The space requirement of Fibonacci heaps is linear in the size of the queue. We give their implementation in Section 13.10.

Table 5.4 surveys the properties of the other implementations. Some implementations allow any linearly ordered type for the priority type (this is indicated by the word general) and some work only for a prespecified range of integer priorities. The constructors take zero or more arguments. For all priority queues that work only for a subset of the integers the set of admissible priorities is defined by constructor arguments. k-ary heaps require that

[5] In the list implementation the items of the queue are stored as an unordered list. This makes *delete_min* and *find_min* linear time processes (linear search through the entire list) and trivializes all other operations.

[6] In the bucket implementation we have an array of linear lists; the list with index i contains all items whose priority is equal to i. This scheme requires the priorities to be integers from a prespecified range.

an upper bound N for the maximal size of the queue and the parameter k is specified in the constructor; the default value of k is 2.

Redistributive heaps and monotone heaps do only support monotone use of the priority queue. The use of a priority queue is *monotone* if the priority argument in any *insert* or *decrease_p* operation is at least as large as the priority returned by the last *delete_min* or *find_min* operation. Dijkstra's shortest-path algorithm uses its priority queue in a monotone way. *R_heaps* and *m_heaps* maintain a variable *p_min* that is initialized to the priority of the first insertion and that is updated to the priority returned by any *delete_min* or *find_min* operation. Only priorities in the range $[\mathit{p\_min} .. \mathit{p\_min} + C - 1]$ can be inserted into the queue, where C is specified in the constructor. In *m_heaps* the cost of a *delete_min* is the difference between the result of this *delete_min* operation and the preceding one[7].

The *b_heap* implementation allows one to ask for the maximum priority and not only for the minimum priority. This is sometimes called a *double-sided* priority queue. For integer priorities there are realizations known that have an even better performance than *r_heaps*. The papers [AMOT90] and [CGS97] describe realizations where *insert* and *delete_min* take time $O(\sqrt{\log C})$ and $O((\log C)^{1/3+\varepsilon})$ for arbitrary $\varepsilon > 0$, respectively.

In order to select an implementation different from the default implementation, a declaration

```
_p_queue<K,int,prio_impl> Q(parameters);
```

has to be used, where *parameters* denotes the list of parameters required by the implementation, e.g.,

```
_p_queue<int,int,r_heap> Q(100000);
```

selects the *r_heap* implementation and sets C to 100000.

A priority queue with a particular implementation is, of course, still a priority queue and can hence be used wherever a priority queue can be used. We give an example. We write a procedure *dijkstra* that takes a graph G, a node s, an *edge_array<int> cost* of edge weights, and a *p_queue<int, node> PQ*, and solves the single-source shortest-path problem for the specified source node. The distances are returned in a *node_array<int> dist*. The edge costs must be non-negative.

⟨*dijkstra*⟩≡

```
void dijkstra(graph& G, node s, const edge_array<int>& cost,
              node_array<int>& dist, p_queue<int,node>& PQ)
{ node_array<pq_item> I(G);
  node v;
  forall_nodes(v,G)
    dist[v] = MAXINT;
  dist[s] = 0;
```

7 The *m_heap* implementation uses an array of size C of linear lists and a variable *p_min* which is initialized to the priority of the first insertion. An item with priority i is stored in the list with index $i \bmod C$. Since priorities are allowed only from the range $[\mathit{p\_min} .. \mathit{p\_min} + C - 1]$ this implies that each list contains only items with the same priority. A *delete_min* or *find_min* operation advances *p_min* cyclically until a non-empty list is found.

```
    I[s] = PQ.insert(0,s);
    while (! PQ.empty())
    { pq_item it = PQ.find_min();
      node u = PQ.inf(it);
      int du = dist[u];
      edge e;
      forall_adj_edges(e,u)
      { v = G.target(e);
        int c = du + cost[e];
        if (c < dist[v])
        { if (dist[v] == MAXINT)
            I[v] = PQ.insert(c,v);
          else
            PQ.decrease_p(I[v],c);
          dist[v] = c;
        }
      }
      PQ.del_item(it);
    }
  }
```

We give some explanations; the correctness of the algorithm is shown in Section 6.6. Dijkstra's algorithm keeps a tentative distance value for each node and a set of active nodes. For a node v its tentative distance value is stored in $dist[v]$ and the set of pairs $(dist[v], v)$, where v is an active node, is stored in the priority queue PQ. Each active node v knows the *pq_item* containing the pair $(dist[v], v)$; it is stored in entry $I[v]$ of the *node_array<pq_item>* I. Initially, only the source node s is active and its distance from s is zero. In each iteration of the loop the pair with minimum distance value is deleted from PQ, say the pair (du, u) and all edges e leaving u are scanned. An edge $e = (u, v)$ allows us to reach node v through a path of cost $c = du + cost[e]$. If c is smaller than the cost of the best path known to v so far, this change is recorded in $dist[v]$ and the priority queue is informed about the change. More precisely, if no path to v was known so far, i.e., $dist[v]$ is still equal to *MAXINT*, a new pair (c, v) is inserted into the priority queue and the item returned is stored in $I[v]$ and if some path was already known then the priority of node v in the queue is updated. Note that in the latter case $I[v]$ contains the item for v in PQ.

We turn to the analysis of the running time. It can be shown (see Section 7.5.3) that each node is inserted and deleted from the priority queue at most once; of course, nodes that cannot be reached from s are never inserted into the queue. The algorithm therefore performs at most n *insert*, *empty*, *find_min*, and *delete_min* operations and at most m *decrease_p* operations. Here n and m denote the number of nodes and edges of G, respectively. The time spent outside the calls to the priority queue is $O(n + m)$ since array accesses take constant time and since the time to scan through all edges leaving a node u is proportional to the outdegree of the node. It is fair also to include the time for the construction and the destruction of the queue (although this happens outside procedure *dijkstra*). The total running time is therefore bounded by $O(n+m+n\cdot(T_{insert}+T_{empty}+T_{find\_min}+T_{delete\_min})+m\cdot T_{decrease\_p}+$

	Worst case running time	Expected running time
f_heap	$O(m + n\log n)$	$O(m + n\log n)$
p_heap	$O(m + n\log n)$	$O(m + n\log n)$
k_heap	$O(m\log_k n + nk\log_k n)$	$O(m + n(\log(2m/n) + k)\log_k n)$
bin_heap	$O(m\log n + n\log n)$	$O(m + n\log(m/n)\log n)$
list_pq	$O(m + n^2)$	$O(m + n^2)$
b_heap	$O((m + n)nM)$	$O((m + n)nM)$
r_heap	$O(m + n\log M)$	$O(m + n\log M)$
m_heap	$O(m + max\_dist + M)$	$O(m + max\_dist + M)$

Table 5.5 Asymptotic running times of Dijkstra's algorithm with different priority queue implementations. In order to keep the formulae simple we assumed $n \le m$. For the last three rows the edge weights must be integral and from the range $[0\,..\,M-1]$. The rows for *b_heaps* and *m_heaps* require some explanation. Note that the maximal priority ever removed from the queue is bounded by $(n-1)M$ since a shortest path consists of at most $n-1$ edges. Thus one can use *b_heaps* with $l = 0$ and $h = nM$. For *r_heaps* and *m_heaps* we observe that the fact that edge costs are bounded by M guarantees that all priorities in the queue come from the range $[p\_min\,..\,p\_min + M - 1]$ and hence we can use these implementations with $C = M$. In *m_heaps* the cost of a *delete_min* is $O(min - p\_min)$, where *min* and *p_min* are the results of the current and the previous *delete_min* operations. The sum of the differences $min - p\_min$ over all *delete_min* operations is bounded by the maximal distance of any node from the source.

$T_{create} + T_{destruct})$ where T_X is the time bound for operation X. Note that the expression above is an upper bound on the running time. The actual number of *decrease_p* operations may be smaller than m. In fact, it can be shown that for random graphs and random edge weights the expected number of *decrease_p* operations is $O(\min(m, n\log(2m/n)))$, see [Nos85]. We can now use Table 5.4 to estimate the asymptotic running time of *dijkstra* with different implementations of the priority queue.

The result is shown in Table 5.5. The first five lines contain the implementations that work for arbitrary non-negative real edge weights. The best worst case and average case time is $O(m + n\log n)$; they are achieved by *f_heaps* and *p_heaps*. For dense graphs with $m = n^{1+\varepsilon}$ for some positive ε, *k_heaps* with $k = n^{1/\varepsilon}$ achieve a worst case time[8] of $O((1/\varepsilon)m)$ which is competitive with the above for ε bounded away from zero. The expected running time[9] of *bin_heaps* is competitive for $m = \Omega(n\log(m/n)\log n)$. The last three lines of the table contain implementations that work only for integral edge weights. In these lines we use M to denote 1 plus the maximal weight of any edge. The best worst case and

[8] The worst case running time of *k_heaps* is $O(nk\log_k n + m\log_k n)$. For $k = n^{1/\varepsilon}$ we have $\log_k n = \log n/\log k = 1/\varepsilon$ and $nk = n^{1+1/\varepsilon} = m$.

[9] In *bin_heaps* the cost of a *decrease_key* is $O(\log n)$. The expected number of *decrease_key* operations is $n\log(2m/n)$. Thus, if $m \ge n\log(2m/n)\log n$ the running time is $O(m)$.

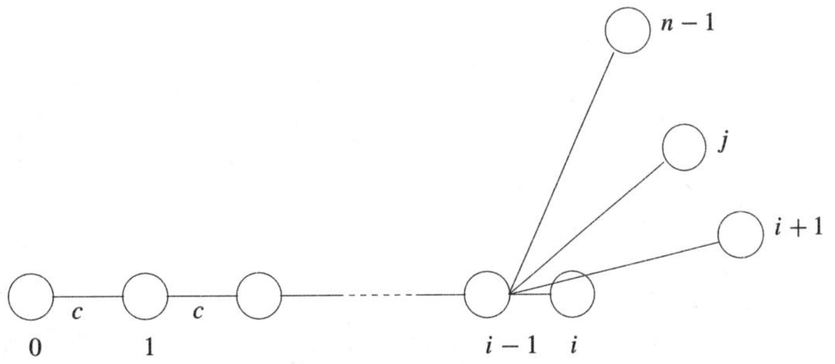

Figure 5.6 A worst case graph for Dijkstra's algorithm. All edges $(i, i+1)$ have cost c and an edge (i, j) with $i+1 < j$ has cost $c_{i,j}$. The $c_{i,j}$ are chosen such that the shortest path tree with root 0 is the path $0, 1, \ldots, n-1$ and such that the shortest path tree that is known after removing node $i-1$ from the queue is as shown. Among the edges out of node $i-1$ the edge $(i-1, i)$ is the shortest, the edge $(i-1, n-1)$ is the second shortest, and the edge $(i-1, i+1)$ is the longest.

average case time is $O(m + n \log M)$ achieved by *r_heaps*. For $M = O(1)$ the *m_heap* implementation is competitive. The heap implementations described in [AMOT90] and [CGS97] yield a running time of $O(m + n\sqrt{\log M})$ and $O(m + n(\log M)^{1/3+\varepsilon})$ for arbitrary $\varepsilon > 0$, respectively.

How do the different implementations compare experimentally? We will perform experiments with random graphs and with worst case graphs. Before reporting running times we construct a graph with n nodes and m edges that forces Dijkstra's algorithm into $m - n + 1$ *decrease_p* operations; observe that this number is the maximal possible since the distance of s is never decreased and since for any node v different from s the first edge into v that is scanned leads to an *insert* but not to a *decrease_p* operation. The construction works for all m and n with $m \le n(n-1)/2$. Let c be any non-negative integer. The graph consists of:

- the nodes $0, 1, \ldots, n-1$,
- the $n-1$ edges $(i, i+1)$, $0 \le i < n-1$, each having cost c, and
- the first $m' = m - (n-1)$ edges in the sequence $(0, 2), (0, 3), \ldots, (0, n-1), (1, 3), (1, 4), \ldots, (1, n-1), (2, 4), \ldots$. The edge (i, j) in this sequence is given cost $c_{i,j}$ to be defined below.

We will define the $c_{i,j}$ such that the shortest path tree with respect to node 0 is the path $[0, 1, \ldots, n-1]$, such that the nodes are removed from the queue in the order of their node number, and such that the shortest path tree that is known after removing node i from the queue is as shown in Figure 5.6. The shortest path from 0 to i has cost ic and the path $[0, 1, \ldots, i-1, i, j]$ has cost $ic + c_{i,j}$, see Figure 5.6.

When node 0 is removed from the queue all other nodes are put into the queue. The priority of node 1 is equal to c and the priority of node j, $j > 1$, is equal to $c_{0,j}$. Generally, just prior to the removal of node i the queue contains nodes i to $n-1$: Node i has priority

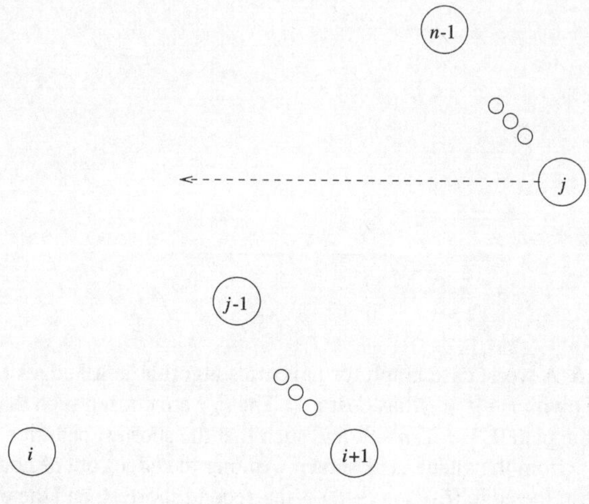

Figure 5.7 The effect of scanning the edges out of node i. When the scanning starts the nodes $i+1$ to $n-1$ are in the queue and we have $dist[n-1] < \ldots < dist[i+1]$. Just prior to the scanning of edge (i, j) we have the situation shown; in this figure distance values are indicated as x-coordinates. Scanning (i, j) will make $dist[j]$ the smallest priority in the queue. The edges out of i are scanned in the order $(i, i+2), \ldots, (i, n-1), (i, i+1)$.

ic and node j, $j > i$, has priority $(i-1)c + c_{i-1,j}$. We now remove node i from the queue and scan through the edges out of i. We postulate that we look at the edges in the order $(i, i+2), (i, i+3)$, $(i, n-1), (i, i+1)$.

Under what conditions will each edge (i, j) cause a *decrease_p* operation and, moreover, will the new priority given to node j by this edge be the smallest priority in the queue? This will be the case if the $c_{i,j}$ are chosen such that

$$\begin{aligned} ic + c_{i,i+2} &< (i-1)c + c_{i-1,n-1}, \\ c_{i,j} &< c_{i,j-1} \qquad \text{for all } j, i+2 < j \le n-1, \\ \text{and} \qquad c = c_{i,i+1} &< c_{i,n-1}. \end{aligned}$$

Note that the first inequality implies that the edge $(i, i+2)$ causes a *decrease_p* operation, that the second inequality implies that the edge (i, j) causes a *decrease_p* operation for all $j, i+2 < j \le n-1$, and that the third inequality implies that the edge $(i, i+1)$ causes a *decrease_p* operation. Also note that this choice of edge costs implies that before the scan of the edges out of i we have $dist[n-1] < \ldots < dist[i+1]$ and that consideration of edge (i, j) will make $dist[j]$ the smallest value in the queue, i.e., before (i, j) is considered we have $dist[j-1] < \ldots < dist[i+2] < dist[n-1] < \ldots < dist[j] < dist[i+1]$ and after (i, j) is considered we have $dist[j] < dist[j-1] < \ldots < dist[i+2] < dist[n-1] < \ldots < dist[j+1] < dist[i+1]$, see Figure 5.7. In this way each edge scan causes a major change in the priority queue.

How can we choose $c_{i,j}$'s satisfying these inequalities? We suggest the following strategy. We first determine the m' additional edges to be used and then assign the edge costs to the additional edges in reverse order. Note that the last edge can be given cost $c+1$, and that $c_{i,j}$ can be put to $c_{i,j+1}+1$ if $j < n-1$ and can be put to $c_{i+1,i+3}+c+1$ if $j = n-1$. The following program realizes this strategy and returns the largest cost assigned to any edge.

⟨*worst case generator*⟩≡

```
int DIJKSTRA_GEN(GRAPH<int,int>& G, int n, int m, int c = 0)
{ G.clear();
  array<node> V(n);
  int i;
  for (i = 0; i < n; i++) V[i] = G.new_node(i);
  stack<edge> S;
  int m1 = m - (n - 1);
  i = 0;
  int j = i + 2;
  while (m1 > 0)
  { if (j < n )
    { S.push(G.new_edge(V[i],V[j])); m1--; j++; }
    else
    { i++; j = i + 2;
      if (j == n)
      error_handler(1,"DIJKSTRA_GEN: m can be at most n*(n-1)/2");
    }
  }
  edge e = S.pop();
  int last_c = G[e] = c + 1;
  while (!S.empty())
  { e = S.pop();
    int j = G[G.target(e)];
    if (j == n-1)
      last_c = G[e] = last_c + c + 1;
    else
      last_c = G[e] = last_c + 1;
  }
 for (i = 0; i < n-1; i++) G.new_edge(V[i], V[i+1], c);
 return last_c;
}
```

A further remark about this program is required. The *new_edge* operation appends the new edge to the adjacency list of the source node and hence the adjacency list of any node i will be ordered $(i, i+2), \ldots, (i, n-1), (i, i+1)$, as desired.

We come to the experimental comparison of our different priority queue implementations. We refer the reader to [CGS97] for more experimental results. It is easy to time *dijkstra* with a particular implementation, e.g.,

⟨*generate a section of table: Dijkstra timings*⟩≡

```
{ p_queue<int,node> fheap;                        K = "fheap";
  dijkstra(G,s,cost,dist,fheap);
}
```

⟨*report time for heap of kind K*⟩

```
{ _p_queue<int,node,p_heap> pheap;                K = "pheap";
  dijkstra(G,s,cost,dist,pheap);
}
⟨report time for heap of kind K⟩
{ int d = m/n;                          // degree for k_heap
  if ( d < 2 ) d = 2;
  _p_queue<int,node,k_heap> kheap(n,d);           K = "kpeap";
  dijkstra(G,s,cost,dist,kheap);
}
⟨report time for heap of kind K⟩
{ _p_queue<int,node,bin_heap> binheap(n);         K = "binheap";
  dijkstra(G,s,cost,dist,binheap);
}
⟨report time for heap of kind K⟩
if (i != 2)  // listheaps are too slow for section 2 of table
{
  { _p_queue<int,node,list_pq> listheap;          K = "listheap";
    dijkstra(G,s,cost,dist,listheap);
  }
  ⟨report time for heap of kind K⟩
}
else cout << "& - " ; cout.flush();
{ _p_queue<int,node,r_heap> rheap(C);             K = "rheap";
  dijkstra(G,s,cost,dist,rheap);
}
⟨report time for heap of kind K⟩
{ _p_queue<int,node,m_heap> mheap(C);             K = "mheap";
  dijkstra(G,s,cost,dist,mheap);
}
⟨report time for heap of kind K⟩
```

generates one section of Table 5.6. We have enclosed the experiment in a block such that the time for the destruction of the queue is also measured. Table 5.6 shows the results of our experiments. You can perform your own experiments with the priority queue demo.

We see that *p_heaps* are consistently better than *f_heaps* and that *r_heaps* are in many situations even better. The exception is when the ratio m/n is very small, the maximal edge weight is large, and we use the worst case graph. In the latter situation, the $n \log M$ term in the running time dominates. For random graphs *bin_heaps* are competitive. *K_heaps* are worse than *bin_heaps* on random graphs (because our choice of k is bad for random graphs) and are competitive for worst case graphs. *List_pq* cannot be run for large values of n because of the n^2-term in the running time. *M_heaps* do surprisingly well even for large edge weights. This is due to the fact that the M-term in the running time does not really harm *m_heaps* in our experiments because of the large value of m.

5.4.3 *Choosing an Implementation*

LEDA gives you the choice between many implementations of priority queues. Which is best in a particular situation?

Instance	*f_heap*	*p_heap*	*k_heap*	*bin_heap*	*list_pq*	*r_heap*	*m_heap*
s,r,S	0.36	0.34	0.35	0.34	0.51	0.33	0.35
s,r,L	0.38	0.36	0.37	0.34	0.54	0.35	0.54
s,w,S	1.86	1.09	3.77	1.38	1	0.76	2.68
s,w,L	1.87	1.1	3.68	1.34	1	0.77	8.49
m,r,S	1.24	0.94	1.14	0.94	31.6	0.83	0.94
m,r,L	1.39	1.13	1.28	1.02	23	0.93	1.22
m,w,S	2.36	1.44	4.94	1.77	22.7	0.99	2.78
m,w,L	2.36	1.45	4.84	1.74	21.7	1.03	3.29
l,r,S	4.96	3.19	5.2	3.36	-	2.52	2.52
l,r,L	6.61	4.81	6.4	4.49	-	3.76	3.38
l,w,S	3.32	2.56	9.17	3.79	-	1.63	3.11
l,w,L	2.91	1.92	7.65	3.22	-	2.57	2.55

Table 5.6 Running times of Dijkstra's algorithm with different priority queue implementations. We used graphs with $m = 500000$ edges and either $n = 2000$, $n = 20000$, or $n = 200000$ nodes. The three cases are distinguished by the labels s, m, and l, respectively. For each combination of n and m we generated four graphs. Two random graphs (r) with random edge weights in $[0 .. M - 1]$, where $M = 100$ or $M = 100000$, and two worst case graphs (w) with $c = 0$ or $c = 10000$. The two cases for M and c are distinguished by the labels S and L, respectively. So s,r,L indicates that we used 2000 nodes, a random graph, and M equal to 100000. In the *k_heap* implementation we set $k = \max(2, m/n)$, as this minimizes the worst case running time.

Tables 5.4 and 5.6 suggest to use either *p_heaps*, *bin_heaps*, or *r_heaps*. *R_heaps* are the data structure of choice if the use of the queue is monotone and the parameter C is such that $\log C$ is not much larger that $\log n$. If the keys are not integers or $\log C$ is much larger than $\log n$, one should use either *bin_heaps* or *p_heaps*. The former are to be preferred when the number of *decrease_p* operations is not too large and the latter is to be preferred otherwise.

If you are not happy with any of the implementations provided in LEDA, you may provide your own. Section 13.6 explains how this is done.

Exercises for 5.4

1 Consider a graph with two nodes v and w and one edge (v, w) of cost M. What is the running time of the different versions of *dijkstra* on this graph as a function of M. Verify your result experimentally.

2 Implement hot queues as described in [CGS97].

3 Time Dijkstra's algorithm with *k_heaps* for different values of k. Do so for random graphs and also for worst case graphs. Which value of k works best?

4 Use priority queues to sort a set of n random integers or random doubles. Compare the different queue implementations. In the case of *k_heaps* try different values of k. Compare your findings for *k_heaps* with the experiments in [LL97].

5.5 Partition

We discuss the data type partition: its functionality, its implementation, and a non-trivial application in the realm of program checking.

5.5.1 *Functionality*

A partition P consists of a finite set of items of type *partition_item* and a decomposition of this set into disjoint sets called blocks. Figure 5.8 visualizes a partition. The declaration

```
partition P;
```

declares a partition P and initializes it to the empty partition, i.e., there are no items in P yet.

```
P.make_block();
```

adds a new item to P, makes this item a block by itself, and returns the item; see Figure 5.9. We may store the returned item for later use.

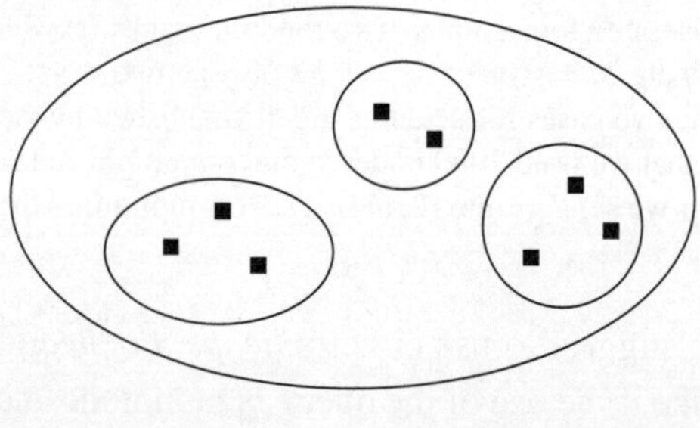

Figure 5.8 A partition P of eight items into three blocks. Partition items are indicated as solid squares and blocks are indicated as ellipses enclosing the items constituting the block.

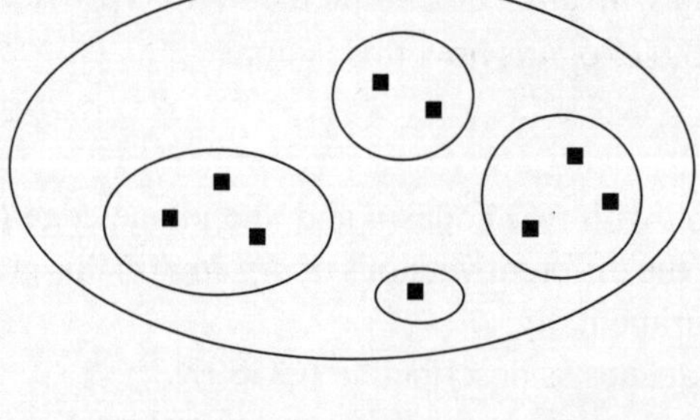

Figure 5.9 The partition of Figure 5.8 after a *make_block* operation.

```
partition_item it = P.make_block();
```

There are several ways to query a partition and to modify it.

```
P.same_block(it1,it2);
```

returns *true* if the partition items *it1* and *it2* belong to the same block of P and *false* otherwise.

```
P.union_blocks(it1,it2);
```

combines the blocks containing items *it1* and *it2*, respectively.

For each block one of its elements is designated as the "canonical" item of the block.

```
P.find(it);
```

returns the "canonical" element of the block containing *it*. Note that *it* and *P.find*(*it*) belong to the same block of P and that if *it1* and *it2* belong to the same block then *P.find*(*it1*) and *P.find*(*it2*) return the same item. Thus

```
P.same_block(it1,it2) == (P.find(it1) == P.find(it2))
```

is a fancy way to write the constant *true*.

If L is a list of partition items then

```
P.split(L);
```

splits all blocks consisting of items in L into singleton blocks. L must be a union of blocks of P.

We give a small example program to see partitions at work. We maintain a partition P of n items. We start with the partition into singleton blocks and then repeat the following step until the largest block has reached size $9n/10$. We choose two items at random and merge the blocks containing them (this has no effect if the two items belong already to the same block). During the experiment we keep track of the block sizes. Whenever the size of the maximal block reaches $in/100$ for some i, $i \geq 1$, we report the number of steps and the size of the two largest components.

In order to facilitate the selection of two random items we store all items of the partition in an *array<partition_item> Item*. This reduces the selection of a random partition item to the selection of a random integer.

We keep track of the block sizes in a *sortseq<int, int> freq*; see Section 5.6. We store for each block size s the number k of blocks having size s in *freq*. Initially, all blocks have size 1 and there are n blocks of size 1.

⟨*giant_component_demo*⟩≡

```
main(){
   ⟨giant component demo: read n⟩
   partition P;
   array<partition_item> Item(n);
   sortseq<int,int> freq;
```

```
for (int i = 0; i < n; i++) Item[i] = P.make_block();
int iteration = 0; int step = 1; int max_size = 1;
freq.insert(1,n);
while ( max_size < n/2 )
{ int v = rand_int(0,n-1);
  int w = rand_int(0,n-1);
  iteration++;
  if ( P.same_block(Item[v],Item[w]) ) continue;
  seq_item it = freq.lookup(P.size(Item[v]));
  freq[it]--;
  if ( freq[it] == 0 ) freq.del_item(it);
  it = freq.lookup(P.size(Item[w]));
  freq[it]--;
  if ( freq[it] == 0 ) freq.del_item(it);
  P.union_blocks(Item[v],Item[w]);
  int size = P.size(Item[v]);
  it = freq.lookup(size);
  if (it) freq[it]++; else freq.insert(size,1);
  it = freq.max();
  max_size = freq.key(it);
  int second_size = freq.key(freq.pred(it));
  while (max_size >= step*n/100 )
  { ⟨giant component demo: report step⟩
    step++;
  }
}
}
```

Part of the output of a sample run of the program above with $n = 10^6$ is as follows:

The maximal block size jumped above 0.16n after 542386 iterations. The maximal size of a block is 160055 and the second largest size of a block is 715.

The maximal block size jumped above 0.17n after 545700 iterations. The maximal size of a block is 170030 and the second largest size of a block is 722.

The maximal block size jumped above 0.18n after 548573 iterations. The maximal size of a block is 180081 and the second largest size of a block is 330.

The maximal block size jumped above 0.19n after 552784 iterations. The maximal size of a block is 190008 and the second largest size of a block is 336.

The maximal block size jumped above 0.20n after 556436 iterations. The maximal size of a block is 200003 and the second largest size of a block is 380.

Observe that it took more than 500 000 iterations until the largest block reached size $0.16n$, and only 4 000 additional iterations until the largest block reached size $0.17n, \ldots$. Moreover, the size of the largest block is much larger than the size of the second largest block. In fact, the second largest block is tiny compared to the largest block. This phenomenon is called *the evolution of the giant component* in the literature on random graphs, see [ASE92] for an analytical treatment of the phenomenon. You may perform your own experiments with the giant component demo. Qualitatively, the phenomenon of the giant component is easy to explain. At any time during the execution of the algorithm the probability to merge two blocks of size k_1 and k_2, respectively, is proportional to k_1k_2 since k_1k_2 is the number of pairs that can be formed by choosing one item in each block. Thus the two blocks most likely to be merged are the largest and the second largest block. Merging them makes the largest block larger and the second largest block smaller (as the third largest block becomes the second largest). Although we knew about the phenomenon before we wrote the demo we were surprised to see how dominating the largest block is.

There are two variants of the partition data type: *Partition* and *node_partition*. A node partition is a partition of the nodes of a particular graph. It is very useful for graph algorithms and we will discuss it in Section 6.6. A *Partition<I>* is a partition where one can associate an information of type I with every item of the partition. The operation

```
partition_item it = P.make_block(i);
```

creates an item with associated information i and makes the item a new block of P, the operation

```
P.inf(it);
```

returns the information of item *it* and

```
P.change_inf(it, i1);
```

changes the information of *it* to *i1*. The type *Partition* is appropriate whenever one wants to associate information with either the items or the blocks of a partition. In the latter case one simply associates the information with the canonical item of the block. We give one such application in Section 5.5.3.

5.5.2 *The Implementation*

Partitions are implemented by the so-called *union-find data structure with weighted union and path compression*. This data structure is a collection of *partition_nodes* which are arranged into a set of trees, see Figure 5.10 for an example. Each block of the partition corresponds to a tree. A *partition_item* is a pointer to a *partition_node*. Each partition node contains a pointer to its parent and each root node knows the size of the tree rooted at it. This is called the *size* of the root. A partition node also contains a field *next* that is used to link all nodes of a partition into a singly linked list. The definition of class *partition_node* is as follows:

⟨*partition_node*⟩≡

```
class partition_node {
  friend class partition;
  partition_node* parent;
  partition_node* next;
  int size;
```

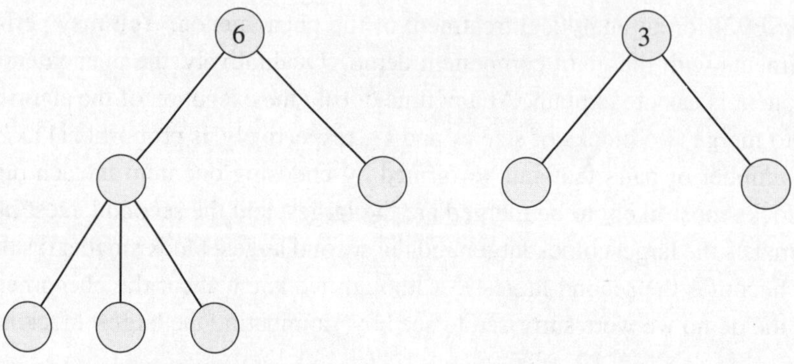

Figure 5.10 The representation of a partition with two blocks of six and three items, respectively. All edges are directed upwards. The size of root nodes is indicated inside the node. All nodes are also linked into a singly linked list. This list is not shown.

```
public:
  partition_node(partition_node* n)  { parent = 0; size = 1; next = n; }
  LEDA_MEMORY(partition_node)
};
typedef partition_node* partition_item;
```

The constructor constructs a node with no parent and size one. We will see its use below, where the use of the field *next* and the argument *n* will also become clear.

We come to class partition. It has only one data member *used_items* that points to the first item in the linear list of all items comprising the partition.

⟨*partition.h*⟩≡

```
#include <LEDA/basic.h>
⟨partition_node⟩
class partition {
  partition_item used_items;   // list of used partition items
public:
  ⟨member functions of partition⟩
};
```

In order to create an empty partition we set *used_items* to *nil* and in order to destroy a partition we go through the list of items comprising the partition and delete all of them.

⟨*member functions of partition*⟩≡

```
partition() { used_items = nil; }
~partition()
{ while (used_items)
  { partition_item p = used_items;
    used_items = used_items->next;
```

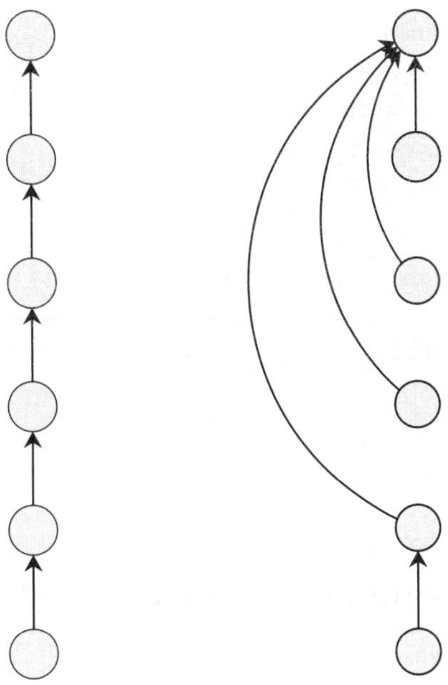

Figure 5.11 Path compression: All edges are directed upwards and the path compression was initiated by an operation *find*(p). After the path compression all ancestors of p including p point directly to the root of the tree containing p.

```
    delete p;
  }
}
```

In order to make a new block we allocate a new *partition_node*, append it to the front of the list of items comprising the partition, and return a pointer to the new node. Observe that we defined the constructor of class *partition_node* such that this works nicely.

⟨*member functions of partition*⟩+≡

```
partition_item make_block()
{ used_items = new partition_node(used_items);
  return used_items;
}
```

We come to function *find*(*partition_item* p). It returns the root of the tree representing the block containing p. This root is easy to find, we only have to follow the chain of parent pointers starting at p. We do slightly more. Once we have determined the *root* of the tree containing p we traverse the path starting at p a second time and change the parent pointer of all nodes on the path to *root*, see Figure 5.11. This is called path compression; it makes the current find operation a bit more expensive but saves all later find operations from traversing the path from p to *root*.

⟨*member functions of partition*⟩+≡

```
partition_item find(partition_item p)
{ // find with path compression
  partition_item x = p->parent;
  if (x == 0) return p;
  partition_item root = p;
  while (root->parent) root = root->parent;
  while (x != root)  // x is equal to p->parent
  { p->parent = root;
    p = x;
    x = p->parent;
  }
  return root;
}
```

The function *same_block*(p, q) returns *find*$(p) ==$ *find*(q).

⟨*member functions of partition*⟩+≡

```
bool  same_block(partition_item p, partition_item q)
{ return find(p) == find(q); }
```

In order to unite the blocks containing items p and q we first determine the roots of the trees containing these items. If the roots are the same then there is nothing to do. If the roots are different, we make one of them the child of the other. We follow the so-called weighted union rule and make the lighter root the child of the heavier root. This rule tends to keep trees shallow[10].

⟨*member functions of partition*⟩+≡

```
void  union_blocks(partition_item p, partition_item q)
{ // weighted union
  p = find(p);
  q = find(q);
  if ( p == q ) return;
  if (p->size > q->size)
      { q->parent = p;
        p->size += q->size; }
  else { p->parent = q;
        q->size += p->size; }
}
```

Despite its simplicity the implementation of *partition* given above is highly effective. A sequence of n *make_block* and m other operations takes time $O((m+n)\alpha(m+n,n))$

[10] We show that the depth of all trees is logarithmically bounded in their size. For any non-negative integer d let s_d be the minimal size of a root whose tree has depth d. Then $s_0 = 1$. A tree of depth d arises by making the root of a tree of depth $d-1$ the child of another root. The former root has size at least s_{d-1} and the latter root has at least this size by the weighted union rule. Thus $s_d \geq 2s_{d-1}$ and hence $s_d \geq 2^d$.

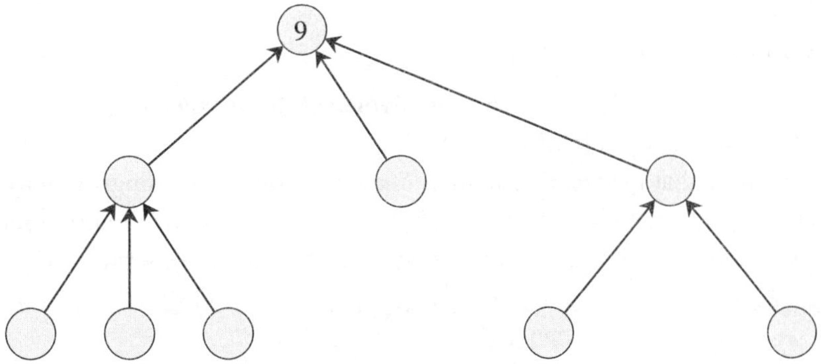

Figure 5.12 The weighted union rule: When the trees of Figure 5.10 are united the root of size 3 is made a child of the root of size 6.

[Tar75]. Here α is the so-called inverse Ackermann function; this function is extremely slowly growing and has value less than 5 even for $n = m = 10^{100}$, see [CLR90, Chapter 22] or [Meh84a, III.8.3].

5.5.3 *An Application of Partitions: Checking Priority Queues*

This section is joint work with Uli Finkler.

We will describe a checker for priority queues; this section assumes knowledge of the data type *p_queue*, see Section 5.4. We define a class *checked_p_queue*<P, I> that can be wrapped around any priority queue *PQ* to check its behavior, see Figure 5.13. The resulting object behaves like *PQ*, albeit a bit slower, if *PQ* operates correctly. However, if *PQ* works incorrectly then this fact will be revealed ultimately. In other words the layer of software that we are going to design behaves like a watch-dog. It monitors the behavior of *PQ* and is silent if *PQ* works correctly. However, if *PQ* behaves incorrectly, the watch-dog barks.

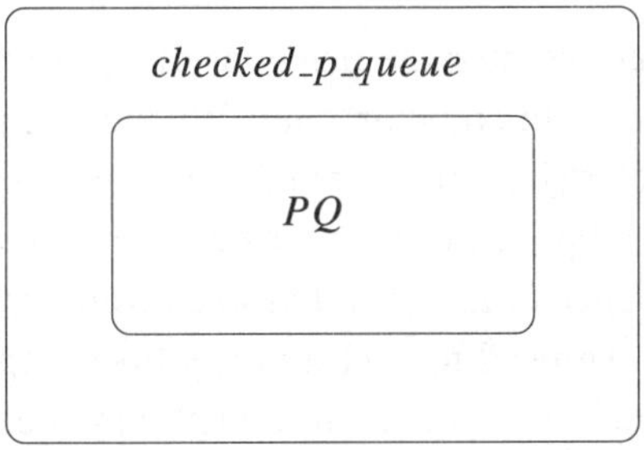

Figure 5.13 The class *checked_p_queue* wraps around a priority queue *PQ* and monitors its behavior. It offers the functionality of a priority queue.

How can the class *checked_p_queue* be used? Suppose we have designed a class *new_impl* which is a new implementation of priority queues. Using the implementation parameter mechanism we can write

```
_p_queue<P,I,new_impl> PQ;
```

to declare a *p_queue*<*P*, *I*> which is implemented by *new_impl*. We may use *PQ* in any application using a *p_queue*<*P*, *I*>.

Assume now that *new_impl* is faulty. Then an application using *PQ* may go astray and we will have to locate the bug. Is it in *PQ* or is it in the application program? The use of *checked_p_queues* facilitates the debugging process greatly. We write

```
_p_queue<P,I,list_item,new_impl> PQ;
checked_p_queue<P,I> CPQ(PQ);
```

and use *CPQ* in the application program. If *PQ* works incorrectly, *CPQ* will tell us. There is *no* change required in the application program since *checked_p_queue* is publicly derived from *p_queue* and hence can be used wherever a *p_queue* can be used, for example,

```
void f(p_queue<P,I>&) { ...}
p_queue<P,I> PQ;                          f(PQ);
_p_queue<P,I,new_impl> PQI;               f(PQI);
_p_queue<P,list_item,new_impl>PQI1;
checked_p_queue<P,I> CPQ(PQI1);           f(CPQ);
```

Observe that the information type of *PQI1* is *list_item* instead of *I*, i.e., we are checking a *p_queue*<*P*, *list_item*> instead of a *p_queue*<*P*, *I*>. This is a slight weakness of our solution. We believe that it is only a slight weakness because the information type *I* plays a minor role in the implementation of priority queues. Moreover, it can be overcome, see the exercises.

In the remainder of this section we give the implementation of the class *checked_p_queue*. The implementation is involved and reading this section certainly requires some stamina. We decided to put this section into the book because we strongly believe that the work on checkers is highly important for software libraries. Section 2.14 contains a general discussion on program checking.

The Idea: How can one monitor the behavior of a priority queue? Without concern for efficiency a solution is easy to come up with. Whenever a *delete_min* or *find_min* operation is performed all items of *PQ* are inspected and it is confirmed that the reported priority is indeed the minimum of all priorities in the queue. This solution does the job but defeats the purpose as it makes *delete_min* and *find_min* linear time operations. Our goal is a solution that adds only a small overhead to each priority queue operation. Our solution performs the checking of the items in the queue in a lazy way, i.e., when a *delete_min* or *find_min* operation is performed it is only recorded that all items currently in the queue must have a priority at least as large as the priority reported. The actual checking is done later. Note that this design implies that an error will not be detected immediately anymore but only ultimately.

Consider Figure 5.14. The top part of this figure shows the items in a priority queue from left to right in the order of their time. The *time* of a *pq_item it* is the time of the last *decrease_p* operation on *it* or, if there was none, the time of the addition of *it* to *PQ*. The

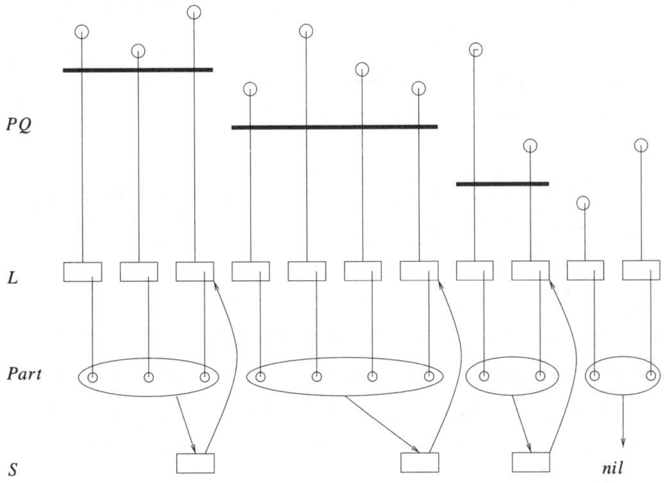

Figure 5.14 In the top part of the figure the items in a priority queue are shown as circles in the xy-plane. The x-coordinate corresponds to the time of an item and the y-coordinate corresponds to the priority of an item. The lower bounds for the priorities are indicated as heavy horizontal lines. The lower bound for the last two items is $-\infty$. The lower part of the figure illustrates our design of class *checked_p_queue*. The list L has one item for each item in *PQ*, the list S has one item for each step of L except for the step with lower bound $-\infty$ and the partition *Part* has one item for each item in L and one block for each step. The blocks of *Part* are indicated as ellipses. The information of the canonical item of a block of *Part* is the *S_item* associated with the block (*nil* for the block with lower bound $-\infty$). Each *S_item* knows the last *L_item* in its step.

vertical coordinate indicates the priority. With each item of the priority queue we have an associated lower bound. The *lower bound* for an item *it* is the maximal priority reported by any *delete_min* or *find_min* operation that took place after the time of *it*. We observe that *PQ* operates correctly if the priority of all *pq_items* is at least as large as their lower bound. We can therefore check *PQ* by comparing the priority of an item with its lower bound whenever an item is deleted from *PQ* or the time of an item is changed through a *decrease_p* operation.

How can we efficiently maintain the lower bounds of the items in the queue? We observe that lower bounds are monotonically decreasing from left to right, i.e., if the time of *it* is smaller than the time of *it'* then the lower bound for *it* is at least as large as the lower bound for *it'*. This observation follows immediately from the definition of the lower bounds and leads to the staircase-like form of the lower bounds shown in Figure 5.14. We call a maximal segment of items with the same lower bound a *step*.

How does the system of lower bounds evolve over time? When a new item is added to the queue its associated lower bound is $-\infty$ and when a *find_min* or *delete_min* operation reports a priority of value p all lower bounds smaller than p are increased to p, i.e., all steps of value at most p are removed and replaced by a single step of value p. Since the staircase of lower bounds is falling from left to right this amounts to replacing a certain number of steps at the end of the staircase by a single step, see Figure 5.15.

How can we represent a staircase of lower bounds such that it can be updated efficiently

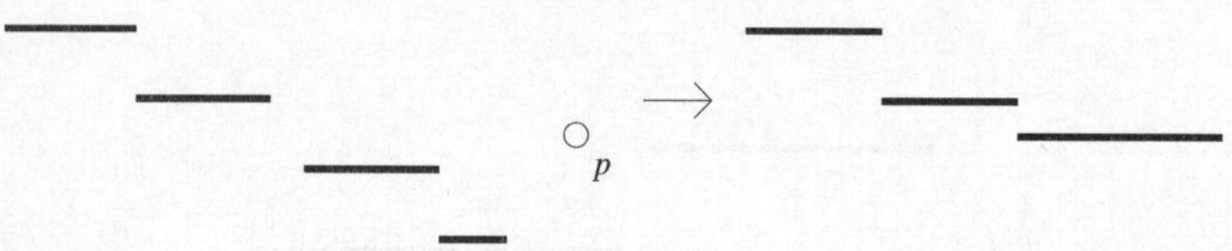

Figure 5.15 Updating the staircase of lower bounds after reporting a priority of p. All steps whose associated lower bound is at most p are replaced by a single step whose associated lower bound is p.

and such that lower bounds can be looked up efficiently? We keep a list L of *check_objects* and a list S of *step_objects*. We have one *check_object* in L for each item in *PQ* and order L according to the times of the corresponding item in *PQ*. We have one *step_object* in S for each step of our staircase of lower bounds except for the step whose associated lower bound is $-\infty$, see Figure 5.14.

A *check_object* is a quadruple consisting of a priority p, an information i, a *pq_item* and a *partition_item*. We explain the use of the partition item below. We mentioned already that check objects are in one-to-one correspondence to the items in *PQ* (if *PQ* operates correctly). The check object o corresponding to a *pq_item p_it* with associated priority p and associated information i contains p, i, and *p_it* as its first three components. We store p and i in the check object to guarantee *data integrity*, i.e., the checking layer stores its own copies of the pairs stored in the priority queue and hence can check whether *PQ* tampers with this data. In fact, we will not store the information i in *PQ* at all. We will rather use the information field of the item *p_it* of *PQ* to store the item of L containing o. In other words the queue to be checked will be of type *p_queue*<*P*, *L_item*> where *L_item* is a synonym for *list_item* that we reserve for the items in L. We use *Lit* as the canonical name of an *L_item*.

A *step_object* is a pair consisting of a priority and an *L_item*. The priority is the lower bound associated with the step and the *L_item* is the last item in L that belongs to the step. The list S will play a crucial role when we update our set of lower bounds after a *del_min* or *find_min* operation. When a priority p is reported by a *del_min* or *find_min* all steps whose *step_object* has a priority of at most p are merged into a single step. These steps constitute a final segment of S. We use *S_item* as the name of the items in S and use *s_it* as the canonical name of an *S_item*.

For the efficient lookup of lower bounds we use a *Partition*<*S_item*> *Part* with one item for each item in L and one block for each step of L. The information associated with the canonical element of a step is *nil*, if the step's lower bound is $-\infty$, and is the *S_item* corresponding to the step otherwise. The fourth component of each check object is the partition item corresponding to the check object.

Let us summarize. A checked priority queue consists of a priority queue, the lists L and S, a partition *Part*, and two integer counters *phase_length* and *op_count* (their use will be explained below). The items of L are in one-to-one correspondence to the items of *PQ* (if *PQ* operates correctly). All operations on *PQ* go through the checking layer, e.g., an operation *insert*(p, i) causes the checking layer to update its internal data structures, in particular, to add an item to L, and to forward the insert request to PQ. The new *L_item* will be returned by the insert operation.

The Class checked_p_queue: We fix the definitions of the data structures of the checking layer in the following layout for the class *checked_p_queue*$<P, I>$.

⟨*checked_p_queue.h*⟩≡

```
#ifndef LEDA_CHECKED_P_QUEUE_H
#define LEDA_CHECKED_P_QUEUE_H
#include <LEDA/p_queue.h>
#include <LEDA/list.h>
#include <LEDA/partition.h>
#include <assert.h>
#include <LEDA/tuple.h>
template <class P, class I>
class checked_p_queue : public p_queue<P,I>
{
  typedef  four_tuple<P,I,pq_item,partition_item> check_object;
  list<check_object> L;
  typedef list_item L_item;
  typedef two_tuple<P,L_item> step_object;
  list<step_object> S;
  typedef list_item S_item;
  Partition<S_item> Part;
  int phase_length, op_count;
  p_queue<P,L_item>* PQ;
  ⟨private member functions of class checked_p_queue⟩
  /* the default copy constructor and assignment operator work
     incorrectly, we make them unaccessible by
     declaring them private                                      */
  checked_p_queue(const checked_p_queue<P,I>& Q);
  checked_p_queue<P,I>& operator=(const checked_p_queue<P,I>& Q);
public:
  checked_p_queue(p_queue<P,L_item>& PQ_ext) // constructor
  { PQ = &PQ_ext;
    assert(PQ->empty());
    phase_length = 4; op_count = 0;
  }
  ⟨member functions of class checked_p_queue⟩
};
#endif
```

Observe that *checked_p_queue*$<P, I>$ is publicly derived from *p_queue*$<P, I>$ and hence will offer the same functions as *p_queues*. The private data members are a pointer to the *p_queue*$<P, L\_item>$ to be checked, the lists L and S, the partition *Part*, and two integers *phase_length* and *op_count*; we will explain the latter two data members below.

The constructor of *checked_p_queue* gets a reference to the queue to be checked and stores it in *PQ*. It also initializes *phase_length* to four and *op_count* to zero. The queue to be checked must be empty (but, of course, there is no guarantee that the emptiness test does not lie). All other data members are initialized by their default constructor.

The member functions of *checked_p_queue* split into private and public member functions. The public member functions are exactly the public member functions of the base class *p_queue* except for the copy constructor and the assignment operator. We were too lazy to implement them. Since C++ provides default implementations of both functions and since the default implementations are incorrect we declared both functions private to make them unaccessible.

The private member functions are used in the implementation of the public member functions. In order to motivate their definitions we give an overview of the implementations of the public member functions *insert* and *delete_min*. In this overview we concentrate on the interplay between the checking layer and *PQ* and do not give any details on how the staircase of lower bounds is manipulated.

An $insert(p, i)$ is realized as follows. The checking layer creates a check object o containing the pair (p, i) and appends o to L. Let *L_it* be new *L_item*. It then inserts the pair (p, L_it) into *PQ*. *PQ* returns an item *p_it* which the checking layer records in o. The checking layer also creates a new partition item corresponding to o. The new item either forms a block of its own (if the step with lower bound $-\infty$ is empty) or is joined into the step with lower bound $-\infty$. The checking layer then returns *L_it* as the result of the *insert*.

A *del_min* is realized as follows. The checking layer forwards the request to *PQ* and *PQ* returns a pair (p, L_it). Let $o = (p', i, p\_it, part\_it)$ be the checking object stored in *L_it*. The checker verifies that $p = p'$ and that p satisfies the lower bound associated with o, it updates the staircase of lower bounds, and it finally returns p.

We want to stress that the checking layer is responsible for the communication with the environment and that the checking layer stores all the pairs (p, i) that are in the priority queue. It forwards all requests from the environment to *PQ*. In a *del_min* operation it uses *PQ* as an *oracle*. The checking layer has no own means to answer minimum queries. It therefore asks *PQ* to point out the correct item. It maintains the system of lower bounds in order to find out whether *PQ* ever lied to it. The checker discovers lies by checking the lower bounds of items whenever an item is deleted or the priority of an item is changed.

We want to bound the delay between a lie and its discovery. For this purpose the checker has a private member function *periodic_check*. This operation goes through all elements of L and checks the lower bound of every element. *Periodic_check* is called after the 2^l-th operation performed on the priority queue for all $l \geq 2$. It is also called after the last operation performed on the priority queue. The integers *phase_length* and *op_count* are used to control the periodic checks. We divide the execution into phases. We use *phase_length* for the length of the current phase and use *op_count* to count the number of operations in the current phase. When *op_count* reaches *phase_length* we check all lower bounds, double *phase_length*, and reset *op_count* to zero.

The discussion above implies that we need to make two assumptions about the behavior of *PQ*:

- All calls to member functions of *PQ* must terminate. They may give wrong answers but they must terminate. It is beyond our current implementation to guarantee termination. A solution would require non-trivial but standard modification of the implementation of *PQ*. One can guard against run-time errors (e.g., invalid addresses) by compiling *PQ* with the debugging option and one can guard against infinite loops by specifying an upper bound on the execution time of each member function of *PQ*. The latter requires a worst case analysis of the running time of *PQ*'s member functions.
- All calls of $PQ \rightarrow inf$ must return valid *L_items*. One may guard against invalid *L_items* by compiling *checked_p_queues* with the debugging option. An alternative solution is described at the end of this section.

Private Member Functions: We are now ready for the definition of the private member functions. The first group provides natural access to the components of *check_objects* and *step_objects*.

⟨*private member functions of class checked_p_queue*⟩≡

```
P& prio(L_item l_it)                                 { return L[l_it].first(); }
const P&  prio(L_item l_it) const                    { return L[l_it].first(); }
I& inf(L_item l_it)                                  { return L[l_it].second(); }
const I&  inf(L_item l_it) const                     { return L[l_it].second(); }
pq_item& pq_it(L_item l_it)                          { return L[l_it].third(); }
pq_item  pq_it(L_item l_it) const                    { return L[l_it].third(); }
partition_item& part_it(L_item l_it)                 { return L[l_it].fourth(); }
partition_item  part_it(L_item l_it) const { return L[l_it].fourth(); }

P& prio_of_S_item(S_item s_it)                       { return S[s_it].first(); }
P  prio_of_S_item(S_item s_it) const                 { return S[s_it].first(); }
L_item& L_it(S_item s_it)                            { return S[s_it].second(); }
L_item  L_it(S_item s_it)  const                     { return S[s_it].second(); }
```

The second group supports the navigation in the data structures of the checker.

The canonical partition item corresponding to an *L_item Lit* is obtained by performing *Part.find* on the associated partition item.

The information associated with the canonical item is obtained by applying *Part.inf* to the canonical item.

The item *Lit* belongs to the step with lower bound $-\infty$ if the canonical information is equal to *nil* and belongs to a step with a defined lower bound otherwise.

The last item in the step containing *Lit* is either the last item of *L* (if *Lit* is unrestricted) or is the *L_item* stored in the *S_item* given by *canonical_inf*(*Lit*).

An item is the only item in its step if it is the last item in its step and is either the first item of *L* or its predecessor item in *L* is also the last item in its step.

All functions above are *const*-functions. They use operations *find* and *inf* of class *Partition* which are not *const*-functions. We therefore write ((*Partition<S_item>∗*) &*Part*) → instead of *Part.* to cast away *const* when calling one of these functions[11].

⟨*private member functions of class checked_p_queue*⟩+≡

```
partition_item canonical_part_it(L_item l_it) const
{ return ((Partition<S_item>*) &Part)->find(part_it(l_it)); }
S_item canonical_inf(L_item l_it) const
{ return ((Partition<S_item>*) &Part)->inf(canonical_part_it(l_it)); }
bool is_unrestricted(L_item l_it) const
{ return canonical_inf(l_it) == nil; }
bool is_restricted(L_item l_it) const
{ return ! is_unrestricted(l_it); }
L_item  last_item_in_step(L_item l_it) const
{ if ( is_restricted(l_it) )
      return L_it(canonical_inf(l_it));
  return L.last();
}
bool is_last_item_in_step(L_item l_it) const
{ return ( last_item_in_step(l_it) == l_it) ; }
bool is_only_item_in_step(L_item l_it) const
{ return (is_last_item_in_step(l_it) &&
    ( L.pred(l_it) == nil || is_last_item_in_step(L.pred(l_it))));
}
```

We put the functions above to their first use by writing a function that tests the validity of the data structures of the checking layer. This function is for debugging purposes only. The data structures must satisfy the following conditions:

- The sizes of *L* and *PQ* must be equal.
- Each item *Lit* in *L* points to an item in *PQ* which points back to *Lit*.
- The items in *L* can be partitioned into segments such that in each segment the value of *canonical_inf* is constant. Except for maybe the last segment, the *canonical_inf* is equal to an item in *S* and this item points back to the last *L_item* in the segment. In the last segment the *canonical_inf* is *nil*. The last segment may be empty and all other segments are non-empty.

⟨*private member functions of class checked_p_queue*⟩+≡

```
void validate_data_structure() const
{
#ifdef VALIDATE_DATA_STRUCTURE
  assert( PQ->size() == L.size() );
  L_item l_it;
  forall_items(l_it,L)
     { assert( pq_it(l_it) != nil ) ;
```

[11] It is tempting to write the cast as ((*Partition<S_item>*) *Part*). but this would amount to a call of the copy constructor of *Partition* and hence be a disaster.

```
      assert( PQ->inf(pq_it(l_it)) == l_it );
    }
  l_it = L.first();
  S_item s_it = S.first();
  while (s_it)
    { assert(canonical_inf(l_it) == s_it);
      while (l_it != L_it(s_it) )
        { l_it = L.succ(l_it);
          assert(l_it != nil);
          assert(canonical_inf(l_it) == s_it);
        }
      s_it = S.succ(s_it);
      l_it = L.succ(l_it);
    }
  while (l_it)
    { assert(canonical_inf(l_it) == nil);
      l_it = L.succ(l_it);
    }
#endif
}
```

The final group of private member functions checks lower bounds and update the staircase of lower bounds.

An *Litem Lit* satisfies its lower bound if either *Lit* is unrestricted or the priority of the step containing *Lit* is no larger than the priority of *p_it*.

⟨*private member functions of class checked_p_queue*⟩+≡

```
void check_lower_bound(L_item l_it) const
{ assert(is_unrestricted(l_it) ||
         compare(prio_of_S_item(canonical_inf(l_it)), prio(l_it)) <= 0 );
}
```

The function *periodic_check* is called at the end of every public member function. It increases *op_count* and when *op_count* has reached *phase_length* checks all lower bounds, doubles *phase_length*, and resets *op_count* to zero.

⟨*private member functions of class checked_p_queue*⟩+≡

```
void periodic_check()
{ if ( ++op_count == phase_length )
    { L_item l_it;
      forall_items(l_it,L) check_lower_bound(l_it);
      phase_length = 2*phase_length;
      op_count = 0;
    }
}
```

Finally, we show how to update lower bounds, see Figure 5.14. Let p be a priority. We move all lower bounds that are smaller than p up to p. This amounts to removing all items in S whose associated lower bound is less than or equal to p and adding a new item with priority p to S. We give more details.

If L is empty or the last step in our staircase of lower bounds extends to the end of the list and has a priority at least as large as p then there is nothing to do.

So assume otherwise. We scan S form its right end (= rear end) and remove items as long as their priority is at most p. Whenever we remove an item *s_it* from S we join the step corresponding to *s_it* with the step after it (it it exists). Finally, we add an item to S representing a step with priority p and ending at *L.last*( ) and make the item the canonical information of all items in the last step.

⟨*private member functions of class checked_p_queue*⟩+≡

```
void update_lower_bounds(P p)
{ if ( L.empty() ||
      ( !S.empty() && compare(prio_of_S_item(S.last()),p) >= 0
        && L_it(S.last()) == L.last()))  return;
  S_item s_it;
  while ( !S.empty()  &&
          compare(prio_of_S_item(s_it = S.last()),p) <= 0 )
    { L_item l_it = L_it(s_it);
      if ( L.succ(l_it) )
        Part.union_blocks(part_it(l_it),part_it(L.succ(l_it)));
      S.pop_back();
    }
  Part.change_inf(canonical_part_it(L.last()),
        S.append(step_object(p,L.last())));
}
```

After all this preparatory work we come to the public member functions.

The Insert Operation: To insert a new item $\langle p, i\rangle$ we append to L a new check object $(p, i, p\_it, part\_it)$; *p_it* is a new item in *PQ* created by the insertion of $(p,\ -)$ and *part_it* is a new item in *Part*. The lower bound of the new item is $-\infty$ and hence the information associated with *part_it* is *nil*. Let *l_it* be the new item in L. We store *l_it* as the information of *p_it*.

If there was already a step with lower bound $-\infty$, we add the new item to this block. Finally, we call *periodic_check* and return *l_it* (after casting it to *pq_item*)[12].

⟨*member functions of class checked_p_queue*⟩≡

```
pq_item insert(const P& p, const I& i)
{ pq_item p_it = PQ->insert(p,(L_item) 0);
  L_item last_l_it = L.last(); // last item in old list

  partition_item pa_it = Part.make_block((S_item) 0);
  list_item l_it = L.append(check_object(p,i,p_it,pa_it));
  PQ->change_inf(p_it,l_it);
  if (last_l_it && is_unrestricted(last_l_it) )
```

[12] The cast from *L_item* to *pq_item* is necessary since early in the design of LEDA we made the decision that the global type *pq_item* is the return type of *insert*. It would be more elegant to have *pq_item* as a type local to *p_queue*.

```
        Part.union_blocks(part_it(l_it),part_it(last_l_it));
    periodic_check();
    validate_data_structure();
    return (pq_item) l_it;
  }
```

The Find_min Operation: In order to perform a *find_min* operation we perform a *find_min* operation on *PQ* and extract an item *l_it* in *L* from the answer. Having received this advice from *PQ* we check the lower bound for *l_it* and update the system of lower bounds using the priority of *l_it*.

Since *checked_p_queue* is derived from *p_queue*, since *find_min* is a *const*-function of *p_queue*, and since *update_lower_bounds* and *perodic_check* are not, we need to cast away the *const*.

⟨*member functions of class checked_p_queue*⟩+≡

```
  pq_item find_min() const
  { L_item l_it = PQ->inf(PQ->find_min());
    check_lower_bound(l_it);
    ((checked_p_queue<P,I>*)this)->update_lower_bounds(prio(l_it));
    ((checked_p_queue<P,I>*)this)->periodic_check();
    validate_data_structure();
    return (pq_item) l_it;
  }
```

The Delete Operation: To delete an item *p_it* we check its lower bound, we delete it from *PQ*, and we delete the corresponding *L_item l_it* from *L*. If *l_it* is restricted and is the only item in its step, we delete the item in *S* representing the step and if *l_it* is the last item in its step but not the only item in its step, we change the *L_it*-field of *canonical_inf*(*l_it*) to the predecessor of *l_it* in *L*. We should also delete the item corresponding to *p_it* from *Part*. Unfortunately, *partition* does not offer a delete operation. We comment on this point at the end of the section.

⟨*member functions of class checked_p_queue*⟩+≡

```
  void del_item(pq_item p_it)
  { L_item l_it = (L_item) p_it;
    check_lower_bound(l_it);
    if ( is_restricted(l_it) )
    { if ( is_only_item_in_step(l_it) )
        S.del_item(canonical_inf(l_it));
      else if (is_last_item_in_step(l_it) )
              L_it(canonical_inf(l_it)) = L.pred(l_it);
    }
    PQ->del_item(pq_it(l_it));
    L.del_item(l_it);
```

```
    periodic_check();
    validate_data_structure();
  }
```

To perform a *del_min* operation we perform a *find_min* on *PQ* and then a *del_min* on the item returned. Finally, we update the lower bound according to the priority of the item returned.

⟨*member functions of class checked_p_queue*⟩+≡

```
  P del_min()
  { L_item l_it = PQ->inf(PQ->find_min());
    P p = prio(l_it);
    del_item((pq_item)l_it);
    update_lower_bounds(p);
    periodic_check();
    validate_data_structure();
    return p;
  }
```

Miscellaneous Functions: The functions *prio*, *inf*, *change_inf*, *size* and *empty* reduce to appropriate functions of the checking layer.

⟨*member functions of class checked_p_queue*⟩+≡

```
  const P& prio(pq_item it) const
  { ((checked_p_queue<P,I>*)this) -> periodic_check();
    return prio((L_item) it);
  }
  const I& inf(pq_item it) const
  { ((checked_p_queue<P,I>*)this) -> periodic_check();
    return inf((L_item) it);
  }
  void change_inf(pq_item it, const I& i)
  { periodic_check();
    inf((L_item) it) = i ;
  }
  int  size() const
  { ((checked_p_queue<P,I>*)this) -> periodic_check();
    return L.size();
  }
  bool empty() const
  { ((checked_p_queue<P,I>*)this) -> periodic_check();
    return L.empty();
  }
```

The Decrease_p Operation: In order to perform a *decrease_p* on item *Lit* we check whether the current priority satisfies its lower bound and we check whether the *decrease_p* operation actually decreases the priority of *Lit*. If so, we change the priority of *Lit* and forward the change to *PQ*.

The new lower bound for the item *Lit* is $-\infty$. If the old lower bound was also $-\infty$ then no action is required. Otherwise we must move *Lit* from its current position in L to the last position in L. This affects the step that contained *Lit*. If *Lit* was the only item in the step, we remove the step altogether and if *Lit* was the last, but not the only, item in its step, we record that *Lit*'s predecessor is the new last element in the step.

In order to move *Lit* to the last position of L we split L into three pieces (the items before *Lit*, *Lit*, and the items after *Lit*) and then reassemble the pieces. We allocate a new partition item for *Lit* and set its information to *nil* (since the new lower bound for *Lit* is $-\infty$). If the step with lower bound $-\infty$ was non-empty, we add *Lit* to this step.

⟨*member functions of class checked_p_queue*⟩+≡

```
void decrease_p(pq_item p_it, const P& p)
{ L_item l_it = (L_item) p_it;
  check_lower_bound(l_it);
  assert( compare(p,prio(l_it)) <= 0 );
  prio(l_it) = p;
  PQ->decrease_p(pq_it(l_it),p);
  if ( is_restricted(l_it) )
  { if ( is_only_item_in_step(l_it) ) S.del_item(canonical_inf(l_it));
    else if (is_last_item_in_step(l_it) )
            L_it(canonical_inf(l_it)) = L.pred(l_it);
    list<check_object> L1, L_it;
    L.split(l_it,L,L1,LEDA::before);
    L1.split(l_it,L_it,L1,LEDA::after);
    L.conc(L1);
    list_item last_it = L.last();
    L.conc(L_it);
    part_it(l_it) = Part.make_block((S_item) 0);
    if (last_it && is_unrestricted(last_it) )
      Part.union_blocks(part_it(l_it),part_it(last_it));
  }
  periodic_check();
  validate_data_structure();
}
```

The Clear Operation and the Destructor: Finally, to clear our data structure we check the lower bounds of all items and then clear for *PQ*, *L*, *S*, and *Part*. The destructor calls *clear*.

⟨*member functions of class checked_p_queue*⟩+≡

```
void clear()
{ L_item l_it;
  forall_items(l_it,L) check_lower_bound(l_it);
```

```
    PQ->clear(); L.clear(); S.clear(); Part.clear();
  }
  ~checked_p_queue() { clear(); }
```

Efficiency: We have now completed the definition of our checker for priority queues. How much overhead does it add? The body of any function of class *checked_p_queue* consists of a call of the same function of *PQ* plus a constant number of calls to functions of *L*, *S*, and *Part*, a call to *periodic_check* plus (maybe) a call of *update_lower_bounds*.

Update_lower_bounds adds at most one element to S (and no other function does) and removes zero or more entries from S. We conclude that the total number of elements added to S and hence removed from S is bounded by the number of operations on PQ. A call of *update_lower_bounds* that removes k elements from S has cost $O(1+k)$ plus the cost for $O(1+k)$ operations on a partition. We conclude that all calls of *update_lower_bounds* contribute a linear number of operations on *Part*. Therefore each call to *update_lower_bounds* contributes a constant number of operations on *Part* in the amortized sense.

The cost of a call to *periodic_check* is also amortized constant. This follows from the fact that the number of elements in the queue is at most twice *phase_length*, that the cost of a call is either $O(1)$ or $O(phase\_length)$, and that the latter alternative occurs only in every *phase_length*-th call to *perodic_check*.

We conclude that the *amortized overhead for each operation on PQ is a constant number of operations on lists and partitions.* Operations on lists require constant time and operations on partitions requires $\alpha(n)$ time.

An Experiment: The following program compares unchecked and checked priority queues experimentally. We generate an array of n random doubles and then use a binary heap to sort them. We first use the binary heap directly and then wrap it into a *checked_p_queue*. The running time of the checked version is about two times the running time of the unchecked version, e.g., it takes about 6.1 seconds to sort 100000 doubles with the unchecked version and slightly more than 12 seconds with the checked version.

```
⟨checked_p_queue_demo.c⟩≡
  ⟨checked_p_queue demo: includes⟩
  main(){
  ⟨checked_p_queue demo: read n⟩
  array<double> A(n);
  random_source S;
  for (int i = 0; i < n; i++) S >> A[i];
  float T = used_time();
  { _p_queue<double,int,bin_heap> PQ(n);
    for (int i = 0; i < n; i++) PQ.insert(A[i],0);
    while ( !PQ.empty() ) PQ.del_min();
  }
  float T1 = used_time(T);
```

```
{ _p_queue<double,list_item,bin_heap> PQ(n);
  checked_p_queue<double,int> CPQ(PQ);
  for (int i = 0; i < n; i++) CPQ.insert(A[i],0);
  while ( !CPQ.empty() ) CPQ.del_min();
}
float T2 = used_time(T);
⟨checked_p_queue demo: report times⟩
}
```

We made a similar test with the priority queue in Dijkstra's algorithm and observed a slowdown by a factor of about 2.5.

Final Remarks: We close this section with a discussion of some alternatives and improvements to our design.

The overhead introduced by our design is a constant number of operations on lists and partitions for each priority queue operation. Since operations on partitions take slightly super-linear time this invalidates the $O(1)$ upper bound for the *decrease_p* operation in the *f_heap* and *p_heap* implementation of priority queues. This can be remedied as follows. The class *checked_p_queue* uses the type *Partition* in a very special way. The blocks of L partition L into contiguous segments and all unions are between adjacent segments. For this special situation there is a realization of partitions that supports all operations in constant time, see [GT85].

Partitions do not offer an operation that deletes items and hence the *del_item* operation of *checked_p_queue* can only delete the items in *PQ*, L, and S, but cannot delete the item in *Part*. This shortcoming can be remedied by giving partitions a *del_item* operation. We briefly sketch the implementation. We perform deletions in a lazy way. When an item is to be deleted it is marked for deletion. We also keep track of the total number of items in the partition and the number of items that are marked for deletion. When more than three-quarters of the items are marked for deletion the partition data structure is cleaned. We go through all items (recall that they are linked into a singly linked list) and perform a find operation for each item. This makes all trees depth one. Then we delete all marked items except those that are the root of a non-trivial tree.

In our realization the checker puts some trust into *PQ*, namely that $PQ \rightarrow inf$ always returns a valid *L_item*. This shortcoming can be overcome by introducing a level of indirection into the data structure. We add an *array<L_item> A*. When the queue has size n precisely the first n entries of this array are used. When an item *p_it* of *PQ* stores a list item *Lit* in the current design it stores some integer $i \in [0 .. n-1]$ in the new design and $A[i]$ contains *Lit*. In this way the index-out-of-bounds check for arrays allows us to check for an invalid pointer. When an item is deleted from the queue and this item corresponds to position i of A, this position is first swapped with position $n - 1$ and then the last entry is removed. We leave the details to the reader. This solution is inspired by [AHU74, exercise 2.12].

The class *checked_p_queue* catches errors of the underlying priority queue eventually (at the latest at the next call of *periodic_check*) but not immediately. Is there a solution which guarantees immediate error detection? Yes and No. **Yes**, because we could simply put a correct priority queue implementation into the checker, and **no**, because it can be shown that no data structure whose running time has a smaller order of magnitude than the running time of priority queues can guarantee immediate error detection.

Exercises for 5.5

1 Modify the program checked_p_queue_demo so that you can experiment with different implementations of priority queues and not only with the binary heap implementation.

2 Implement the copy constructor and the assignment operator of our class *checked_p_queue*.

3 Modify the implementation of class *checked_p_queue*, so as to remove the assumption that $PQ \rightarrow inf$ always returns a valid *L_item*.

4 Modify the implementation of class *checked_p_queue* so that the queue to be checked has type *p_queue*<P, I>. Hint: Use a map to make the correspondence between *pq_items* and the items of L.

5 Use checked priority queues instead of priority queues in Dijkstra's algorithm as discussed in Section 5.4.

6 Add an operation *del_item* to the types *partition* and *Partition*<E>. Follow the sketch at the end of Section 5.5.3.

7 In the extract minimum problem we are given a permutation of the integers 1 to n interspersed with the letter E, e.g., 6,E,1,4,3,E,E,5,2,E,E,E is a possible input sequence. The E's are processed from left to right. Each E extracts the smallest number to its left which has not been extracted by a previous E. The output in our example would therefore be 6,1,3,2,4,5. Solve the problem using a priority queue. In the off-line version of this problem the input sequence is completely known before the first output needs to be produced. Solve the problem with the partition data type (Hint: Determine first which E outputs the number 1, then which E outputs 2, ...).

8 Implement the data structure of [GT85]. Make it available as a LEDA extension package.

5.6 Sorted Sequences

Sorted sequences are a versatile data type. We discuss their functionality in this section, give their implementation by means of skiplists in the next section, and apply them to Jordan sorting in the last section of this chapter.

A *sorted sequence* is a sequence of items in which each item has an associated key from a linearly ordered type K and an associated information from an arbitrary type I. We call K the key type and I the information type of the sorted sequence and use $\langle k, i \rangle$ to denote an item with associated key k and information i. The keys of the items of a sorted sequence must be in strictly increasing order, i.e., if $\langle k, i \rangle$ is before $\langle k', i' \rangle$ in the sequence then k is before k' in the linear order on K. Here comes a sorted sequence of type *sortseq*<*string*, *int*>:

$\langle Ena, 7 \rangle$ $\langle Kurt, 4 \rangle$ $\langle Stefan, 2 \rangle$ $\langle Ulli, 8 \rangle$

Sorted sequences offer a wide range of operations. They can do almost everything lists, dictionaries, and priority queues can do and they can do many other things. They even do all these things with the same asymptotic efficiency. Of course, there is a price to pay: Sorted sequences require more space (about $23.33n$ bytes for a sequence of n items) and the constant factors in the time bounds are larger. So please use sorted sequences only if you need their power.

We discuss the functionality of sorted sequences in several steps. In each step we introduce some operations and then give a small program using these operations. We start with the operations that we know already from dictionaries and priority queues, then turn to so-called finger searches, and finally discuss operations for splitting and merging sorted sequences.

Basic Functionality: Sorted sequences come in two kinds. The definitions

```
 sortseq<K,I> S;
_sortseq<K,I,ab_tree> T;
```

define S and T as sorted sequences with key type K and information type I. For T the ab_tree implementation of sorted sequences is chosen and for S the default implementations of types *sortseq* is chosen. The type *_sortseq*<*K*, *I*, *IMPL*> offers only a subset of the operations of *sortseq*<*K*, *I*>; in particular it does not offer any of the finger search operations. The items in a sorted sequence have type *seq_item*. The following implementations of *sortseqs* are currently available: skiplists [Pug90b], randomized search trees [AS89], $BB(\alpha)$-trees [NR73], *ab*-trees [AHU74, HM82], and red-black-trees [GS78]. They are selected by the implementation parameters skiplist, rs_tree, bb_tree, ab_tree, and rb_tree, respectively. Skiplists are the default implementation. We have mentioned already that sorted sequences extend dictionaries, lists, and priority queues, in particular we have the following operations:

K	S.key(*seq_item it*)	returns the key of item *it*. *Precondition*: *it* is an item in S.
I	S.inf(*seq_item it*)	returns the information of item *it*. *Precondition*: *it* is an item in S.
seq_item	S.lookup(K k)	returns the item with key k (*nil* if no such item exists in S).
seq_item	S.locate(K k)	returns the item $\langle k', i\rangle$ in S such that k' is minimal with $k' \geq k$ (*nil* if no such item exists).
seq_item	S.locate_succ(K k)	equivalent to $S.locate(k)$.
seq_item	S.succ(K k)	equivalent to $S.locate(k)$.

seq_item	S.locate_pred(*K k*)	returns the item $\langle k', i\rangle$ in S such that k' is maximal with $k' \le k$ (*nil* if no such item exists).
seq_item	S.pred(*K k*)	equivalent to *S.locate_pred*(k).
seq_item	S.min_item()	returns the item with minimal key (*nil* if S is empty).
seq_item	S.max_item()	returns the item with maximal key (*nil* if S is empty).
seq_item	S.succ(*seq_item it*)	returns the successor item of *it* in the sequence containing *it* (*nil* if there is no such item).
seq_item	S.pred(*seq_item x*)	returns the predecessor item of *it* in the sequence containing *it* (*nil* if there is no such item).
seq_item	S.insert(*K k, I i*)	associates information i with key k: If there is an item $\langle k, j\rangle$ in S then j is replaced by i, otherwise a new item $\langle k, i\rangle$ is added to S. In both cases the item is returned.
int	S.size()	returns the size of S.
bool	S.empty()	returns true if S is empty, false otherwise.
void	S.clear()	makes S the empty sorted sequence.
void	S.del(*K k*)	removes the item with key k from S (null operation if no such item exists).
void	S.del_item(*seq_item it*)	removes the item *it* from the sequence containing *it*.
void	S.change_inf(*seq_item it, I i*)	makes i the information of item *it*.

The operations *key*, *inf*, *succ*, *pred*, *max*, *min*, *del_item*, *change_inf*, *size*, and *empty* take constant time, *lookup*, *locate*, *locate_pred*, and *del* take logarithmic time, and *clear* takes linear time.

We come to our first program. We read a sequence of strings (terminated by "stop") and build a sorted sequence of type *sortseq<string, int>* for them[13]. Then we read a pair (*s1*, *s2*) of strings and output all input strings larger than or equal to *s1* and smaller than or equal to *s2*. This is done as follows. If *s2* is smaller than *s1* then there are no such strings. Assume otherwise and let item *last* contain the largest string less than or equal to *s2* and let *first* contain the smallest string larger or equal to *s1*. If either *first* or *last* does not exist or *last* is

[13] Observe that a sorted sequence needs an information type; we do not need informations in this application and have chosen the information type *int*; any other type would work equally well.

the predecessor of *first* then the answer is empty. Otherwise it consists of all strings that are stored in the items starting at *first* and ending at *last*.

⟨*sortseq_demo1.c*⟩≡

```
#include <LEDA/sortseq.h>
main()
{ sortseq<string,int> S;
  string s1,s2;
  cout << "Input a sequence of strings terminated by stop.\n";
  while (cin >> s1 && s1 != "stop") S.insert(s1, 0);
  while ( true )
  { cout << "\nInput a pair of strings.\n\n";
    cin >> s1 >> s2;
    cout << "All strings s with " <<
                                    s1 <<" <= s <= " << s2 <<":\n";
    if ( s2 < s1 ) continue;
    seq_item last  = S.locate_pred(s2);
    seq_item first = S.locate(s1);
    if ( !first || !last || first == S.succ(last) ) continue;
    seq_item it = first;
    while ( true )
    { cout << "\n" << S.key(it);
      if ( it == last ) break;
      it = S.succ(it);
    }
  }
}
```

The running time of this program is $O(n \log n + m \log n + L)$, where n denotes the number of strings put into the sorted sequence, m denotes the number of queries, and L is the total number of strings in all answers. In this time bound we have assumed for simplicity that a comparison between strings takes constant time and that a string can be printed in constant time. Both assumptions require that the strings have bounded length.

Finger Search: All search operations discussed so far take logarithmic time. *Finger search* opens the possibility for sub-logarithmic search time. It requires that the position of the key k to be searched for is approximately known. Let *it* be an item of the sorted sequence S; in the context of finger search we call *it* a *finger* into S. The operations

```
S.finger_locate(k);
S.finger_locate_from_front(k);
S.finger_locate_from_rear(k);
S.finger_locate(it, k);
```

have exactly the same functionality as the operation *locate*, i.e., all of them return the leftmost item *it′* in S having a key at least at large as k. They differ in their running time. If *it′* is the d-th item in a list of n items then the first three operations run in time $O(\log \min(d, n-d))$, $O(\log d)$, and $O(\log(n-d))$, respectively[14]. In other words,

[14] For the remainder of this section we assume $\log x$ to mean $\max(0, \log x)$.

finger_locate_from_front is particularly efficient for searches near the beginning of the sequence, *finger_locate_from_end* is particularly efficient for searches near the end of the sequence, and *finger_locate* is particularly efficient for searches near either end of the sequence (however, with a larger constant of proportionality); it runs the two former functions in parallel and stops as soon as one of them stops. The operation *S.finger_locate*(it, k) runs in time $O(\log \min(d, n-d))$ where d is the number of items in S between it and it'. For example, if it is the 5th item of S and it' is the 17th item then $d = 17 - 5 = 12$.

After a fast search we also want to insert fast. That's the purpose of the operation *insert_at*. Assume that *it* is an item of S and k is a key and that *it* is either the rightmost item in S with $key(it) < k$ or the leftmost item with $key(it) > k$. Then

```
S.insert_at(it, k, i)
```

adds $\langle k, i\rangle$ to S in time $O(1)$. If k's relation to the key of *it* is known then it is more efficient to use

```
S.insert_at(it, k, i, dir)
```

with *dir* equal to *LEDA::before* or *LEDA::after*.

We give an application of finger searching to sorting. More precisely, we give a sorting algorithm which runs fast on inputs that are nearly sorted. Let n and f be integers with $0 \le f \ll n$ and consider the sequence

$$n-1, n-2, \ldots, n-f, 0, 1, 2, \ldots, n-f-1.$$

We store this sequence in a list L and sort it in five different ways: four versions of insertion sort and, for comparison, the built-in sorting routine for lists. The easiest way to build a sorted sequence S from L is to call *S.insert* for each element of L. As before, we must give our sorted sequence an information type; we use the type *int* and hence insert the pair $(k, 0)$ for each element k of L.

⟨*repeated insertion sort*⟩≡

```
forall(k,L) S.insert(k, 0);
```

The running time of repeated insertion sort is $O(n \log n)$.

Let us take a closer look where the insertions are taking place for our input sequence. In the first f insertions the new element is always inserted at the beginning of the sequence and in the remaining $n - f$ insertions the new element is always inserted before the f-th element from the end of the sequence. Since $f \ll n$ it should be more efficient to search for the place of insertion from the rear end of the sequence.

⟨*finger search from rear end*⟩≡

```
forall(k, L)
{ if (S.empty()) it = S.insert(k, 0);
  else
  { seq_item it = S.finger_locate_from_rear(k);
    if (it) S.insert_at(it,k,0,LEDA::before);
```

```
    else S.insert_at(S.max_item(),k,0,LEDA::after);
  }
}
```

With finger search from the rear end each search takes time $O(\log f)$ and hence the total running time becomes $O(n \log f)$. The same running time results if we use the version of finger search that does not need to be told from which end of the sequence it should search.

⟨*finger search from both ends*⟩≡

```
forall(k, L)
{ if (S.empty()) it = S.insert(k, 0);
  else
  { seq_item it = S.finger_locate(k);
    if (it) S.insert_at(it,k,0,LEDA::before);
    else S.insert_at(S.max_item(),k,0,LEDA::after);
  }
}
```

We can do even better by observing that each insertion takes place next to the previous insertion. Hence it is wise to remember the position of the last insertion and to start the finger search from there.

⟨*finger search from last insertion*⟩≡

```
forall(k, L)
{ if (S.empty()) it = S.insert(k, 0);
  else
  { it = S.finger_locate(it,k);
    it = ( it ? S.insert_at(it,k,0,LEDA::before) :
                S.insert_at(S.max_item(),k,0,LEDA::after) );
  }
}
```

With this version of finger search each search takes constant time and hence a total running time of $O(n)$ results.

Table 5.7 shows the running times of our four versions of insertion sort in comparison to the built-in sorting routine for lists (*L.sort*( )) for $n = 500000$ and $f = 50$. We made the comparison for the key types *int*, *double*, and *four_tuple<int, int, int, int>* to study the influence of the cost of comparing two keys. The table shows that insertion sort with finger search is superior to repeated insertion sort for nearly sorted input sequences and that the advantage becomes larger (as is to be expected from the asymptotic analysis) as comparisons become more expensive. The table also shows that in the case of very expensive comparisons insertion sort with finger search can even compete with quicksort (which is the algorithm used in the sorting routine for lists).

It is worthwhile to take a more abstract view of the programs above. The less mathematically inclined reader may skip the next two paragraphs. Let $k_1, \ldots, k_n$ be a sequence of distinct keys from a linearly ordered type K. An *inversion* is a pair of keys that is not

	Repeated insertion	Finger search from rear	Finger search from both ends	Finger search from last insertion	List sort
int	5.45	4.78	4.7	2.98	2.22
double	6.28	5.1	7.12	3.28	2.53
quads	22.1	13.9	16.8	6.3	14.8

Table 5.7 Running times of the four versions of insertion sort and of the sorting routine *L.sort( )* for lists for $n = 500000$ and $f = 50$. The sorting routine for lists uses quicksort with the middle element of the list as the splitting element. It runs in time $O(n \log n)$. Three different key types were used: *int*, *double*, and the type *four_tuple<int, int, int, int>* where an integer i was represented as the quadruple $(0, 0, 0, i)$. This ensures that comparisons between quadruples are expensive. You may perform your own experiments with the sortseq sort demo.

in ascending order, i.e., a pair (i, j) of indices with $1 \le i < j \le n$ and $k_i > k_j$. We use F to denote the total number of inversions and use f_j to denote the number of inversions involving j as their second component, i.e.,

$$f_j = |\{i\,;\ i < j \text{ and } k_i > k_j\}|$$

If F is zero then the sequence is already sorted. The maximal value of F is $n(n-1)/2$. We show that insertion sort with finger search from the rear runs in time $O(n(1 + \log(F/n)))$ on a sequence with F inversions. So the worst case is $O(n \log n)$, the best case is $O(n)$, and the running time degrades smoothly as F increases. A sequence with a "small" value of F is sometimes called *nearly sorted*. Thus, insertion sort with finger search is fast on nearly sorted sequences.

Assume that we have already sorted $k_1, \ldots, k_{j-1}$ and next want to insert k_j. As in our programs above we use S to denote the resulting sorted sequence. Each key in $k_1, \ldots, k_{j-1}$ which is larger than k_j causes an inversion and hence the number of keys in $k_1, \ldots, k_{j-1}$ larger than k_j is equal to f_j. Thus, k_j needs to be inserted at the f_j-th position from the rear end of S. A finger search from the rear end of S determines this position in time $O(\log f_j)$. We conclude that the total running time of insertion sort with finger search from the rear end is

$$O(\sum_{1 \le j \le n} 1 + \log f_j) = O(n + \log \prod_{1 \le j \le n} f_j).$$

Subject to the constraint $\sum_{1 \le j \le n} f_j = F$, the product $\prod_{1 \le j \le n} f_j$ is maximized if all f_j's are equal and hence are equal to F/n. The claimed time bound of $O(n \cdot (1 + \log(F/n)))$ follows.

Split: There are several operations to combine and split sequences. If S is a sorted sequence and *it* is an item of S then

```
S.split(it, T, U, dir)
```

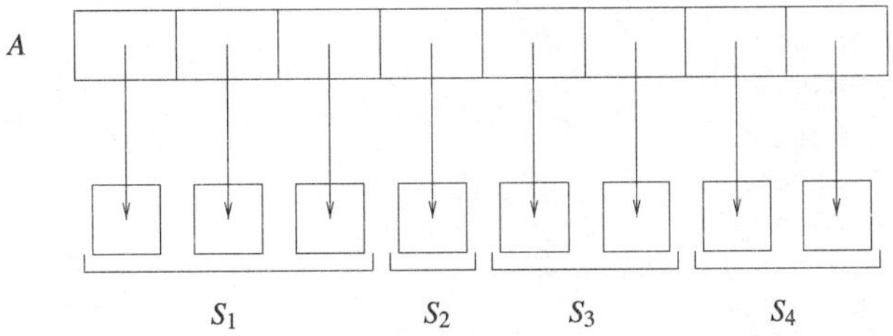

Figure 5.16 A sequence S of eight items that has been split into a sequence of length three, a sequence of length one, and two sequences of length two. The entry $A[i]$ of the array A contains a pointer to the i-th container of S. The sequences S_1, S_3, and S_4 need to be split further. In the sortseq_split program there will be a task in the task stack for each one of them. The task for S_3 has the form (pointer to S_3, 4, 5).

splits S after (if $dir = LEDA::after$) or before (if $dir = LEDA::before$) *it* and returns the two fragments in T and U. More precisely, if S is equal to

$$x_1, \ldots, x_{k-1}, it, x_{k+1}, \ldots, x_n$$

and *dir* is *LEDA::after* then $T = x_1, \ldots, x_{k-1}, it$ and $U = x_{k+1}, \ldots, x_n$ after the split. If *dir* is before then U starts with *it* after the split. The two sequences T and U must name distinct objects, but S may be one of T or U. If S is distinct from T and U then S is empty after the split. The running time of *split* is $O(\log n)$ for *_sortseqs* and is $O(1 + \log\min(k, n-k))$ for *sortseqs*.

We sketch an application of splitting in order to show the difference between the two time bounds. Assume that S is a sorted sequence of length n and consider the following process to split S into n sequences of length 1 each. We start with S and as long as we have a sequence of length larger than 2 we split this sequence at an arbitrary item.

In the following program we construct a sorted sequence S of n items and store its items in an array A. We also maintain a stack of "tasks". A task is a triple consisting of a pointer to a subsequence of S plus the indices of the first and the last item in the subsequence, see Figure 5.16. Initially there is only one task, namely, the triple $(S, 1, n)$. In each iteration of the loop we take the topmost task from the stack. If the sequence has less than two elements and hence requires no further split, we simply delete it. Otherwise, we split it at a random element and create tasks for the two parts. We continue until there are no tasks left.

⟨*sortseq_split*⟩≡
```
main(){
⟨sortseq split: read n⟩
typedef sortseq<int,int> int_seq;
array<seq_item> A(n);
int_seq* S = new int_seq();;
for (int i = 0; i < n; i++) A[i] = S->insert(i,0);
typedef three_tuple<int_seq*,int,int> task;
```

```
  stack<task> TS;
  TS.push(task(S,0,n-1));
  float UT = used_time();
  while ( !TS.empty() )
  { task t = TS.pop();
    int_seq* S = t.first();
    int l = t.second();
    int r = t.third();
    if ( r - l + 1 < 2 ) { delete S; continue; }
    int_seq* T = new int_seq();
    int_seq* U = new int_seq();
    int m = rand_int(l,r-1);
    S->split(A[m],*T,*U,LEDA::after);
    delete S;
    TS.push(task(T,l,m));
    TS.push(task(U,m+1,r));
  }
  ⟨sortseq split: report time⟩
  }
```

We show that the running time of this program is linear in n. We do so for arbitrary choice of the splitting index m and not only for random choice of m. The less mathematically inclined reader may skip the analysis. We use $T(n)$ to denote the maximal running time of the program on a sequence of n items. Then $T(1) = c$ and

$$T(n) \le \max_{1 \le m < n} T(m) + T(n-m) + c(1 + \log \min(m, n-m))$$

for $n > 1$ and a suitable constant c. The recurrence relation reflects the fact that it takes time $c(1+\log \min(m, n-m))$ to split a sequence of length n into sequences of length m and $n-m$ and additional time $T(m)$ and $T(n-m)$ to split these sequences further into sequences of length 1. We need to take the maximum with respect to m since we are interested in the worst case time. We show $T(n) \le c(5n - 2 - 2\log(n+1))$ for all n by induction on n. This is certainly true for n equal to 1. So assume $n > 1$ and let m maximize the right-hand side in the recurrence relation above. Because of the symmetry of the right-hand side in m and $n - m$ we may assume $m \le n/2$. Then

$$\begin{aligned}
T(n) &\le T(m) + T(n-m) + c(1 + \log \min(m, n-m)) \\
&\le c(5m - 2 - 2\log(m+1) + 5(n-m) - 2 - 2\log(n-m+1) + 1 + \log m) \\
&< c(5n - 2 - \log(m+1) - 2\log(n-m+1) - 1) \\
&\le c(5n - 2 - 2\log(n+1)),
\end{aligned}$$

where the first inequality is our recurrence relation, the second inequality follows from the induction hypothesis, the third inequality is simple arithmetic, and the last inequality follows from the fact that $1 + \log(m+1) + 2\log(n-m+1) \ge 2\log(n+1)$ for all m with $1 \le m \le n/2$. To see this, observe first that the second derivative of $f(m) = 1 + \log(m + 1) + 2\log(n-m+1)$ is negative and hence $\min_{1 \le m \le n/2} f(m) = \min(f(1), f(n/2))$.

Observe next that $f(1) \geq 2\log(n+1)$ and $f(n/2) \geq 2\log(n+1)$. This completes the induction.

Concatenation and Merging: We turn to concatenation and merging of sequences.

```
S.conc(T,dir)
```

appends T to the rear (if $dir = LEDA::after$) or front (if $dir = LEDA::before$) of S and makes T empty. Of course, we may apply *conc* with $dir = LEDA::after$ only if the key of the last item in S is smaller than the key of the first item in T and with $dir = LEDA::before$ only if the key of the last item in T is smaller than the key of the first item in S. The running time of *conc* is $O(\log(n+m))$ for *_sortseqs* and is $O(1+\log\min(n,m))$ for *sortseqs* where n and m are the lengths of the sequences to be concatenated. *Merge* generalizes *conc*.

```
S.merge(T)
```

merges the list T into the list S and makes T empty. For example, if $S = \langle 5,.\rangle\ \langle 7,.\rangle\ \langle 8,.\rangle$ and $T = \langle 6,.\rangle\ \langle 9,.\rangle$ are sequences with key type *int* then $S = \langle 5,.\rangle\ \langle 6,.\rangle\ \langle 7,.\rangle\ \langle 8,.\rangle\ \langle 9.,\rangle$ after the merge. Of course, S and T can only be merged if the keys of all items are distinct. The time to merge two sequences of lengths n and m, respectively, is $O(\log\binom{n+m}{n})$; *merge* is only supported by *sortseqs*.

We sketch how *merge* is implemented, we compare *merge* with two less sophisticated approaches to merging, and we show how to use *merge* in a robust version of merge sort. We start with a sketch of the implementation. Assume that the sequences S and T are to be merged and that the number of elements in T is at most the number of elements in S. We insert the elements of T one by one into S, starting with the first element of T. In order to locate the position of an element of T in S we use a finger search starting from the position of the last insertion (starting from the first element of S instead of the first element of T).

```
sortseq_item finger = S.min_item();
sortseq_item     it = T.min_item();

while ( it )
{ finger = S.finger_locate(finger,T.key(it));
  S.insert_at_item(finger,T.key(it),T.inf(it));
  it = T.succ(it);
}
```

The running time of this program is easy to analyze. We use m to denote the number of elements in T and n to denote the number of elements in S. Assume that the i-th element of T is to be inserted after the f_i-th element of S for all i with $1 \leq i \leq m$. Set $f_0 = 0$. The finger search that determines the position of the i-th element of T in S takes time $O(\log d_i)$ where $d_i = f_i - f_{i-1}$ is the number of elements of S that are between the position of insertion for the i-th and the $(i-1)$-th element. Clearly, $\sum_i d_i \leq n$. The total time for merging T into S is

$$\sum_i O(1+\log d_i) = O(m + \log \prod_i d_i).$$

Subject to the constraint $\sum_i d_i \leq n$, the product $\prod_i d_i$ is maximal if all d_i are equal to n/m. The running time is therefore $O(m + mlog(n/m)) = O(\log \binom{n+m}{n})$. To see the last equality observe first that

$$1 + \log(n/m) = 1 + \log(n+m)/m \leq 2\log((n+m)/m)$$

since $n + m \geq 2m$ and observe next that $m\log((n+m)/m) = \log((n+m)/m)^m$ and $((n+m)/m)^m \leq \binom{n+m}{m}$.

We next compare *merge* to two less sophisticated merge routines. Let T and U be sorted sequences of length n and m, respectively. There are two ways to merge U into T that come to mind immediately. The first method inserts the elements of U one by one into T. This takes time $O(m\log(n+m))$. The second method scans both files simultaneously from front to rear and inserts the elements of U as they are encountered during the scan. This takes time $O(n+m)$. In the following programs we assume that T and U are of type *sortseq<K, int>*.

⟨*three merging routines*⟩≡

```
template < class K >
void merging_by_repeated_insertion(sortseq<K,int>& T, sortseq<K,int>& U)
{ seq_item it = U.min_item();
  while ( it )
  { T.insert(U.key(it),U.inf(it));
    it = U.succ(it);
  }
}

template < class K >
void merging_by_scanning(sortseq<K,int>& T, sortseq<K,int>& U)
{ seq_item it1 = T.min_item();
  seq_item it2 = U.min_item();

  while ( it2 && compare(U.key(it2),T.key(it1)) < 0 )
  { T.insert_at(it1,U.key(it2),U.inf(it2),LEDA::before);
    it2 = U.succ(it2);
  }
  seq_item succ1 = T.succ(it1);
  while ( it2 )
  { K k2 = U.key(it2);
    while ( succ1 && compare(T.key(succ1),k2) < 0 )
    { it1 = succ1;
      succ1 = T.succ(succ1);
    }
    it1 = T.insert_at(it1,k2,U.inf(it2),LEDA::after);
    it2 = U.succ(it2);
  }
}

template < class K >
void merging_by_finger_search(sortseq<K,int>& T, sortseq<K,int>& U)
{ T.merge(U); }
```

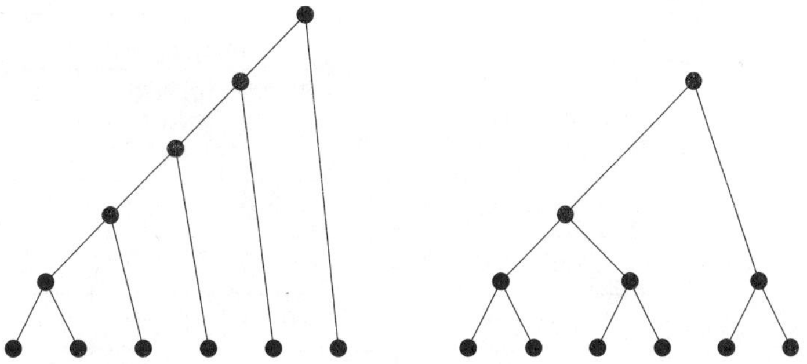

Figure 5.17 Two patterns for merging six sequences of length one. The merge pattern on the left is unbalanced: it first merges two sequences of length one, then merges the resulting sequence of length two with a sequence of length one, then merges the resulting sequence of length three with a sequence of length one, ... The second merge pattern is balanced: it first forms three sequences of length two, then merges two of them to a sequence of length four, and finally merges the sequence of length four with the remaining sequence of length two.

How do the three routines compare theoretically and experimentally? Let us consider three cases: $m = 1$, $m = n$, and $m = n/\log n$. Merging by repeated insertion takes time $O(\log n)$, $O(n \log n)$, and $O(n)$, respectively, merging by scanning takes $O(n)$ in all three cases, and merging based on finger search takes time $O(\log n)$, $O(n)$, and $O(m \log(n/m)) = O(n \log\log n / \log n)$, respectively. We see that merging based on finger search is never worse than the two other methods (it has a larger constant of proportionality, though) and that it is superior to both methods in two of the cases. Table 5.8 shows an experimental comparison of the three methods.

Robust Merge Sort: We use our three merging routines in a version of merge sort. In order to sort a set of n elements, merge sort starts with n sequences of length 1 (which are trivially sorted) and then uses merging to combine them into a single sorted sequence of length n. The *merge pattern*, i.e., the way in which the n sequences are combined into a single sequence can be visualized by a binary tree with n leaves and $n - 1$ internal nodes. The n leaves correspond to the n initial sequences and each internal node corresponds to a merging operation. In this way we associate with every internal node the sorted sequence that results from merging the two sequences associated with its children. Figure 5.17 shows two merging patterns.

How do our three merging routines behave? In the balanced merging pattern we perform about $n/2^k$ merges between sequences having length 2^k each and hence obtain a total running time of

$$O\Big(\sum_{0 \le k < \log n} (n/2^k) M(2^k, 2^k)\Big),$$

where $M(x, y)$ is the time to merge two sequences of length x and y. For merging by

		Merging by		
		repeated insertion	scanning	finger search
int	$m = 1$	0	0.75	0
	$m = 10$	0	0.767	0
	$m = 100$	0	0.7	0
	$m = 1000$	0.0333	0.767	0.05
	$m = 10000$	0.4	0.883	0.267
	$m = 100000$	3.65	1.78	1.75
double	$m = 1$	0	0.817	0
	$m = 10$	0	0.8	0
	$m = 100$	0.0167	0.817	0.0167
	$m = 1000$	0.05	0.833	0.0333
	$m = 10000$	0.433	0.95	0.317
	$m = 100000$	4.2	2.02	2.02
quadruple	$m = 1$	0	2.58	0
	$m = 10$	0	2.6	0
	$m = 100$	0.0167	2.67	0.0333
	$m = 1000$	0.183	2.63	0.15
	$m = 10000$	1.65	2.82	1.03
	$m = 100000$	15.8	4.38	6.6

Table 5.8 Running times of the three versions of merging for $n = 500000$ and different values of m. The sequence T consisted of the first n even integers and the sequence U consisted of the integers $2(n/m)i + 1$ for $i = 1, \ldots, m$. Three different key types were used: *int*, *double*, and the type *four_tuple<int, int, int, int>* where an integer i was represented as the quadruple $(0, 0, 0, i)$. This ensures that comparisons between quadruples are expensive. You may perform your own experiments with the sortseq merge demo.

repeated insertion we have $M(x, x) = O(x \log x)$ and hence obtain a total running time of

$$O(\sum_{0 \le k < \log n} (n/2^k)2^k k) = O(n \sum_{0 \le k < \log n} k) = O(n \log^2 n).$$

For merging by scanning and merging by finger search we have $M(x, x) = O(x)$ and hence

	Unbalanced merge tree	Balanced merge tree
merging by repeated insertion	5.07	11.5
merging by scanning	3.44e+03	9.82
merging by finger search	5.9	8.73

Table 5.9 This table was generated by program sortseq_merge_sort. You can perform your own experiments with the sortseq merge demo. Merging by finger search comes in shortly after the winner for both merge patterns.

obtain a total running time of

$$O(\sum_{0 \leq k < \log n} (n/2^k)2^k) = O(n \log n).$$

We conclude that the latter two merging methods perform optimally in the case of a balanced merging pattern but that merging by repeated insertion does not.

Let us turn to the unbalanced merging pattern. It builds a sequence of length i by merging a sequence of length $i-1$ and a sequence of length 1 for all i, $2 \leq i \leq n$. We obtain a total running time of

$$O(\sum_{2 \leq i \leq n} M(i, 1)).$$

For merging by repeated insertion and merging by finger search we have $M(x, 1) = O(\log x)$ and hence obtain a total running time of

$$O(\sum_{2 \leq i \leq n} \log i) = O(n \log n).$$

For merging by scanning we have $M(x, 1) = O(x)$ and hence obtain a total running time of

$$O(\sum_{2 \leq i \leq n} i) = O(n^2).$$

We conclude that the two former merging methods perform optimally in the case of an unbalanced merging pattern but that merging by scanning does not. *Only merging by finger searching performs optimally for both merge patterns.*

Table 5.9 shows an experimental comparison. You may perform your own experiments by calling the sortseq merge demo. This program generates n sorted sequences of length one and puts pointers to them into an array A (*intseq* is an abbreviation for *sortseq<int, int>*.). It permutes A to make sorting non-trivial.

⟨*fill A*⟩≡

```
for (i = 0; i < n; i++)
{ A[i] = new int_seq;
  A[i]->insert(i,0);
}
A.permute();
```

It then uses either the unbalanced merge pattern or the balanced merge pattern to merge the n sequences into a single sequence (*merge* is any one of our three merging routines).

⟨*unbalanced merge pattern*⟩≡

```
for (i = 1; i < n; i++)
{ merge(*A[0],*A[i]);
  delete A[i];
}
```

⟨*balanced merge pattern*⟩≡

```
while (n > 1)
{ int k = n/2;
  for (i = 0; i < k; i++)
  { merge(*A[i],*A[k + i]);
    delete A[k+i];
  }
  if ( 2 * k < n ) // n is odd
    { A[k] = A[n - 1]; n = k + 1; }
  else
    { n = k; }
}
```

We close our discussion of merging by showing that merge sort with merging by finger search has running time $O(n \log n)$ for every merge pattern. Recall that a merge pattern is a binary tree T with n leaves and that every internal node of T corresponds to a merge operation. For an internal node let $s(v)$ be the length of the sorted sequence that is the result of the merge operation at node v and for a leaf v let $s(v)$ be equal to one. With this notation the cost of the merge at a node v with children x and y is

$$O(\log \binom{s(v)}{s(x)}) = O(\log(s(v)!/(s(x)!s(y)!))) = O(\log s(v)! - \log s(x)! - \log s(y)!)$$

and the total running time of merge sort is obtained by summing this expression over all nodes v of T. In this sum every node z except for the root and the leaves contributes twice: it contributes $\log s(z)!$ when z is considered as a parent and it contributes $-\log s(z)!$ when z is considered as a child. The two contributions cancel. Therefore everything that remains is the contribution of the root (which is $\log n!$) and the contribution of the leaves (which is $-n \log 1$). We conclude that the total running time is $O(n \log n)$ independent of the merge pattern T.

Operations on Subsequences: We want to mention two further operations. Let a and b be two items in a sorted sequence S with a being equal to or before b. Then

```
S.reverse_items(a,b)
```

reverses the subsequence of items in S starting at a and ending at b, i.e., if

$$S = it_1, it_2, \ldots, it_{i-1}, it_i, it_{i+1}, \ldots, it_{j-1}, it_j, it_{j+1}, \ldots, it_n$$

before the operation and $a = it_i$ and $b = it_j$ then

$$S = it_1, it_2, \ldots, it_{i-1}, it_j, it_{j-1}, \ldots, it_{i+1}, it_i, it_{j+1}, \ldots, it_n$$

after the operation. We will see an application of *reverse_items* in a plane sweep algorithm for segment intersection in Section 10.7.2. *Reverse_items* runs in time proportional to the number of items that are reversed. *Reverse_items* is also available under the name *flip_items*.

The operation

```
S.delete_subsequence(a,b,T)
```

removes the subsequence starting at a and ending at b from S and assigns it to T. The running time is $O(\log \min(m, n - m))$ where n is the number of items in *IT* and m is the number of items that are removed. We will see an application of *delete_subsequence* in Section 5.8 on Jordan sorting.

Sequences and Items: Many of the operations on *sortseqs* take items as arguments, e.g.,

```
S.finger_locate(finger,x)
```

locates x in S by searching from the item *finger*. What happens if *finger* is not an item in S but in some other *sortseq IT*?

The complete specification of *finger_locate* is as follows (and this is, of course, the specification that is given in the manual). Let *IT* be the sorted sequence containing *finger*. Then

```
S.finger_locate(finger,x)
```

is equivalent to

```
IT.finger_locate(finger,x)
```

provided that *IT* has the same type as S. If *IT* and S have different types the semantics of *S.finger_locate(finger*, x) is undefined.

A similar statement holds for all other operations having items as arguments. So

```
S.reverse_items(a,b)
```

is applied to the sequence containing the items a and b (of course, a and b must belong to the same sequence).

If the items determine the sequence to which the operation is applied, why does one have to specify a sequence at all? We explored the alternative to make *finger_locate* a static member function of *sortseq*<K, I> and to write

```
sortseq<K,I>::finger_locate(finger,x);
```

We decided against it because in most applications of sorted sequences there is no problem in providing the sequence as an argument and in these situations it is clearer if the sequence is provided as an argument. The price to pay is that in the rare situation where the sequence is not known (the program in Section 5.8 is the only program we have ever written where this happens) one has to "invent" S, i.e., to declare a dummy sequence S and to apply *finger_locate* to it.

Exercise for 5.6

1 A *run* in a sequence of keys is a sorted subsequence. Let $L = k_1, \ldots, k_n$ be any sequence and let k be the number of runs in L, i.e., k is one larger than the number of i with $k_i > k_{i+1}$. Show that insertion sort with finger search from the position of the last insertion sorts a sequence consisting of k runs in time $O(n(1 + \log k))$.

5.7 The Implementation of Sorted Sequences by Skiplists

We first describe the skiplist data structure. Skiplists were invented by W. Pugh [Pug90a, Pug90b] and our implementation is based on his papers. We go beyond his papers by also providing implementations for finger searches, merging, and deletion of subsequences. We start with an overview of the data structure and then outline the content of the files skiplist.h, _skiplist.c, and sortseq.h. In the bulk of the section we give the implementations of the different operations on skiplists.

5.7.1 *The Skiplist Data Structure*

A skiplist is a sequence of *skiplist_nodes*, see Figure 5.18. We also say *tower* instead of *skiplist_node*. In a skiplist for a sequence of n elements we have $n + 2$ towers, n towers corresponding to the elements of the sequence and two towers called *header* and *STOP* that serve as sentinels. We refer to the former towers as *proper* and to the latter as *improper*.

A tower contains the following information:

— a *key*,
— an *inf*ormation,
— an integer *height*,
— an array *forward* of *height* + 1 pointers to towers,
— a *backward* pointer, and
— a *pred*ecessor pointer.

The keys of the proper towers in a skiplist are strictly increasing from front to rear of the sequence. The sentinels *header* and *STOP* have no keys stored in them although, logically, their keys are $-\infty$ and ∞, respectively. It would make life somewhat easier if the key type K provided the elements $-\infty$ and ∞. Because not all key types do, we have decided to

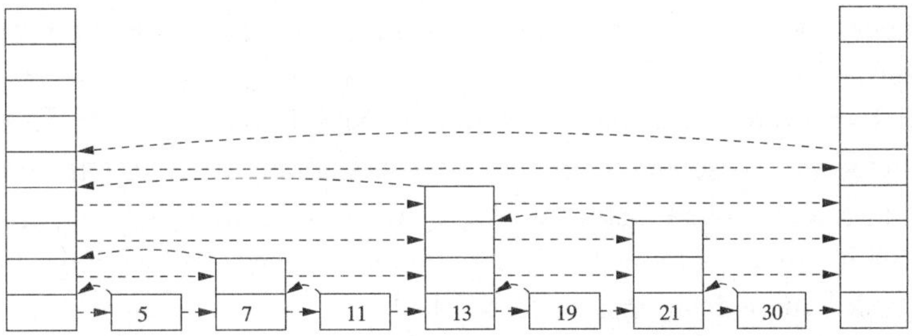

Figure 5.18 A skiplist: The sequence of keys stored in the sequence is 5, 7, 11, 13, 19, 21, 30. The proper towers have height 0, 1, 0, 3, 0, 2, and 0, respectively. Their keys are shown at the bottom of the towers. The two improper towers *header* and *STOP* are the first and last tower, respectively. They have no keys. The forward pointers point horizontally to the right. The backward pointers are shown as curved arcs and the predecessor pointers are not shown. All forward pointers that have no proper tower to point to, point to *STOP*. An object of type skiplist contains pointers to *header* and *STOP*. The header points back to the skiplist object.
A search for 19 proceeds as follows. We start in the header and consider the forward pointer at height 3 (= maximal height of a proper tower) out of the header. It ends in a tower with key 13. Since $19 > 13$ we move forward to the tower with key 13 and consider its forward pointer at height 3. It ends in *STOP* (which has key ∞) and so we drop down to height two. The forward pointer at height 2 out the tower with key 13 ends in the tower with key 21. Since $19 < 21$ we drop down to the height one, ...

store no keys in the sentinels. When formulating invariants we will however assume that the keys of *header* and *STOP* are $-\infty$ and ∞, respectively.

Skiplists represent the sequence stored at different levels of granularity. The tower of height at least zero represent the entire sequence, the towers of height at least one represent a subsequence, the towers of height at least two represent a subsequence of the subsequence, ... The operations on skiplists gain their efficiency by exploiting the different levels of granularity; Figure 5.18 sketches a search for key 19 in our example skiplist. Observe that the search first locates 19 with respect to the list represented by the towers of height at least 3, i.e., the list $(-\infty, 13, +\infty)$, then with respect to the list represented by the towers of height at least 2, i.e., the list $(-\infty, 7, 13, 21, +\infty)$, ...

The height of a proper tower is chosen probabilistically when the tower is created. We will explain this in more detail below. The height of a proper tower is always non-negative. The height of *STOP* is -1 and the height of *header* is equal to *MaxHeight*. We set *MaxHeight* to 32 in our implementation. When we choose the heights of proper towers we will make sure that their height is smaller than *MaxHeight*. The sentinels *header* and *STOP* can therefore be recognized by their height. *Headers* are the only items with height equal to *MaxHeight* and *STOP* nodes are the only items with negative height.

A *header* stores information in addition to the ones listed above: the data member *true_height* is one plus the maximal height of any proper tower (it is zero if there are no proper towers) and the member *myseq* stores a pointer to the skiplist to which *header* belongs. The *header* has type *header_node*, where a *header_node* is an *skiplist_node* with the two additional fields just mentioned.

A pointer to a *skiplist_node* is called an *sl_item* and a pointer to a *header_node* is called a *large_item*.

In the definitions below the flag *SMEM* (= simple memory management) allows us to choose between two schemes for memory allocation. If *SMEM* is defined, the obvious memory allocation scheme is used and *forward* is realized as an array of *sl_items* and if *SMEM* is not defined, a refined and more efficient memory allocation scheme is used. This is explained in more detail in Section 5.7.4.

The flag `__exportC` is used for preprocessing purposes. On UNIX-systems it is simply deleted and on Windows-systems it is replaced by flags which are needed to generate dynamic libraries.

⟨*definition of classes skiplist_node and header_node*⟩≡

```
class __exportC header_node;
class __exportC skiplist_node;
typedef skiplist_node* sl_item;
typedef header_node*     large_item;
const int MaxHeight = 32;
class __exportC skiplist_node
{ friend class __exportC skiplist;
  static leda_mutex mutex_id_count;
  static unsigned long id_count;
  GenPtr key;
  GenPtr inf;
  int    height;
  unsigned long id;             // id number
  sl_item pred;
  sl_item backward;
#ifdef SMEM
  sl_item* forward;  // array of forward pointers
#else
  sl_item forward[1];
#endif
  friend unsigned long ID_Number(skiplist_node* p){return p->id;}
};
class __exportC header_node : public skiplist_node
{ friend class __exportC skiplist;
#ifndef SMEM
  sl_item more_forward_pointers[MaxHeight];
#endif
  int true_height;
  skiplist* myseq;
};
```

A header node can be viewed as a *skiplist_node* and as a *header_node*. If v is an *sl_item* which is known to be a *large_item* (because $v \rightarrow height = MaxHeight$) then we can cast v to a large item by (*large_item*)v and access the skiplist containing v by ((*large_item*)v) $\rightarrow$ *myseq*.

We can now complete the definition of the skiplist data structure by defining the values

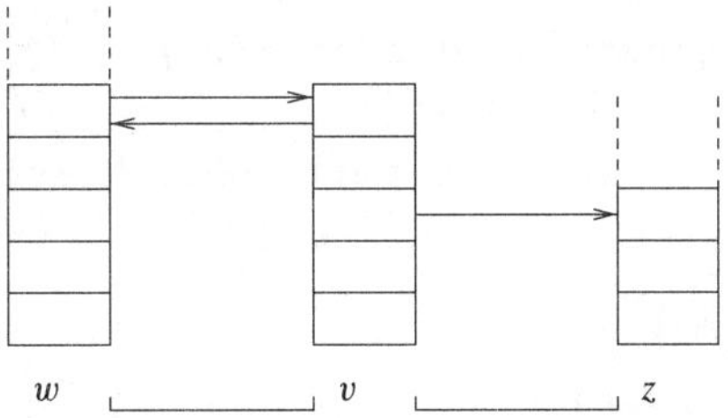

Figure 5.19 Forward and backward pointers: $v \rightarrow$ *forward*[2] points to the closest successor tower of height at least 2 and $v \rightarrow$ *backward* points to the closest predecessor tower of height at least 4.

of the various pointers stored in a tower, see Figure 5.19. Let v be any tower and let h be the height of v (view *header* as a tower of height *true_height* for this paragraph). Then:

- for all i, $0 \le i \le h$, the i-th forward pointer of v points to closest successor tower of height at least i (to *STOP* if there is no such tower),
- the backward pointer points to the node w with the highest forward pointer into v, i.e., the h-th forward pointer of w points to v,
- and the predecessor pointer of v points to the tower immediately preceding v.

The procedure *validate_data_structure* checks the invariants in time $O(\mathit{true\_height} \cdot n)$.

⟨*miscellaneous*⟩≡

```
void skiplist::validate_data_structure()
{ assert(header == header->myseq->header);
  assert (header->height == MaxHeight);
  assert(STOP->height == -1);

  int max_proper_height = -1;
  sl_item p = (sl_item) header;

  while (p != STOP)
  { assert(p->height >= 0);
    if (p != header && p->height > max_proper_height)
                                max_proper_height = p->height;
    p = p->forward[0];
  }
  assert(header->true_height == max_proper_height + 1);

  p = (sl_item) header;
  while (p != STOP)
  { sl_item q = p->forward[0];
    assert(p == q->pred);                     //condition three

    if (p != header && q != STOP)             //check order
                               assert(cmp(p->key,q->key) < 0);

    for(int h=0; h<=Min(p->height,header->true_height);h++)
    { sl_item r = p->forward[0];
```

```
      while (r->height < h && r != STOP) r = r->forward[0];
      assert ( r == p->forward[h]);          //condition one
      if ( h == r->height ) assert(r->backward == p);
    }                                        //condition two
    p = q;
  }
  assert(STOP->backward == (sl_item) header);
}
```

As a preview for later sections we describe briefly how one can search for a key x in a skiplist. We keep a node v and a height h such that $v \rightarrow key < x$ and $x \leq v \rightarrow forward[h] \rightarrow key$. Initially, $v = header$ and $h = true\_height$. In the basic search step we find a node v with the same property and h one less. This is easy to achieve. We only have to start a walk at node v taking forward pointers at height $h - 1$.

```
h--;
w = v->forward[h];
while (key > w->key)
{ v = w;
  w = v->forward[h];
}
```

The while-loop re-establishes the invariant $v \rightarrow key < x \leq v \rightarrow forward[h] \rightarrow key$. Continuing in this way down to $h = 0$ locates x among the items in the skiplist. The complete program for a search in a skiplist is therefore as follows:

```
sl_item v = header;
int h = header->true_height;
while ( h > 0 )
{ h--;
  w = v->forward[h];
  while (w != STOP && key > w->key)
  { v = w;
    w = v->forward[h];
  }
}
```

The search in skiplists is efficient because skiplists represent the underlying sequence at different levels of granularity. The forward pointers at level 0 represent the entire sequence, the forward pointers at level 1 represent the subsequence formed by the towers of height at least 1, the forward pointers at level 2 represent the subsequence formed by the towers of height at least 2, ... In a search we locate x with respect to the subsequence of towers of height at least h for decreasing values of h. This is trivial at the highest level and requires only little additional work for each smaller value of h.

The height of a proper tower is chosen probabilistically when the tower is created. It is set to h with probability $p^h(1 - p)$ where p with $0 < p < 1$ is a parameter that is fixed when the skiplist is created. In our implementation we use 1/4 as the default value for p. We draw three easy consequences from this probabilistic definition of height.

The probability that a proper tower has height h or more is $\sum_{k\geq h} p^k(1-p) = p^h$ and therefore the expected value of *height* can be computed as[15]:

$$\mathrm{E}[height] = \sum_{h\geq 1} p^h = p\sum_{h\geq 0} p^h = p/(1-p).$$

Since the space requirement for a tower of height h is $(6+h)\cdot 4$ bytes plus the space for the key and the information we conclude that the expected space requirement for a skiplist of n items is about $(6+p/(1-p))4n$ bytes plus the space for the keys and informations. For $p = 1/4$ we have $\mathrm{E}[h] = 1/3$ and hence the expected space requirement for a skiplist of n items is about $76/3n = 25.333n$ bytes. The refined memory allocation scheme needs a bit more, see Section 5.7.4.

The fact that p^h is the probability that a proper tower has height h or more implies that the probability that some tower in a collection of n proper towers has height h or more is at most $\min(1, np^h)$. This is one for $h \leq \log_{1/p} n$ and is at most p^l for $h = \lceil \log_{1/p} n\rceil + l$. Since *true_height* is one plus the maximal height of any proper tower, we can compute the expected value of *true_height* as:

$$\begin{aligned}
\mathrm{E}[true_height] &= \sum_{h\geq 1}\mathrm{prob}(true_height \geq h) = \sum_{h\geq 1}\mathrm{prob}(\text{maximal height} \geq h-1)\\
&\leq \sum_{h\geq 0}\min(1, p^h) \leq \sum_{0\leq h<\lceil \log_{1/p} n\rceil} 1 + \sum_{h\geq \lceil \log_{1/p} n\rceil} p^h\\
&\leq 1 + \log_{1/p} n + \sum_{l\geq 0} p^l = 1 + \log_{1/p} n + 1/(1-p).
\end{aligned}$$

Finally, if v is any tower then the probability that $v \rightarrow backward$ has height larger than v is p. Observe that $v \rightarrow backward$ has at least the height of v and that the conditional probability that a tower has height $h+1$ or more given that it has height h or more is $p^{h+1}/p^h = p$. Thus, the probability that $v \rightarrow backward$ has height larger than v is p.

We use this observation to bound the cost of a search. Consider a search for a key x and let $v_0, v_1, \ldots, v_k$ be the path traced by the variable v in the program above. Then $v_0 = header$ and $v_i = v_{i+1}\rightarrow backward$. By the above, the probability that the height of v_i is larger than the height of v_{i+1} is p and hence the expected number of nodes traversed at any particular height is $1/p$. We start at height zero and end at height *true_height*. The expected length of the path is therefore bounded by

$$1/p\cdot(1+\log_{1/p} n + 1/(1-p)).$$

This concludes our discussion of skiplist nodes.

We turn to the class representing skiplists. In an *skiplist* we store the items *header* and *STOP* and some quantities related to the random process: *prob* contains the parameter p in use, and *randomBits* contains an integer whose last *randomsLeft* bits are random. We use

[15] If X is a random variable which assumes non-negative integer values and $q_h = \mathrm{prob}(X \geq h)$ and $p_h = \mathrm{prob}(X = h)$ for all $h \geq 0$ then $\mathrm{E}[X] = \sum_{h\geq 0} p_h \cdot h = \sum_{h\geq 1} p_h \cdot h = \sum_{h\geq 1}(q_h - q_{h+1})\cdot h = \sum_{h\geq 1} q_h$.

randomBits as the random source in the construction of skiplist nodes. Whenever all bits in *randomBits* are used up we refill it using the LEDA random number generator.

⟨*data members of class skiplist*⟩≡

```
large_item  header;
sl_item STOP;
float prob;
int randomBits;
int randomsLeft;
```

⟨*private member functions of class skiplist*⟩≡

```
void fill_random_source()
{ randomBits = rand_int(0,MAXINT-1);
  randomsLeft = 31;
}
```

5.7.2 *The Files sortseq.h, skiplist.h, and _skiplist.c*

The definition of type *sortseq<K, I>* follows the strategy laid out in Section 13.4. We define two classes: an abstract data type class *sortseq<K, I>* and an implementation class *skiplist*. The class *sortseq<K, I>* is a parameterized class with type parameters K and I. The keys and infs in the implementation class are generic pointers.

The implementation class is defined in incl/LEDA/impl/skiplist.h and src/dict/_skiplist.c. We have already seen the chunks ⟨*definition of classes skiplist_node and header_node*⟩ and ⟨*data members of class skiplist*⟩. In the other chunks of skiplist.h we define a set of virtual functions that are later redefined in the abstract data type class and we define the functions that realize all operations on sorted sequences. The virtual functions are discussed in Section 5.7.3 and the other functions are discussed starting in Sections 5.7.5. In _skiplist.c we assemble the implementations of all member functions (except for the trivial ones which are given directly in the header file).

The compile-time constant SMEM is explained in Section 5.7.4.

⟨*skiplist.h*⟩≡

```
#ifndef SKIPLIST_H
#define SKIPLIST_H
// #define SMEM  remove comment for use of simple memory scheme
#include <LEDA/basic.h>
#include <assert.h>
⟨definition of classes skiplist_node and header_node⟩
class __exportC skiplist
{ ⟨data members of class skiplist⟩
  ⟨virtual functions of class skiplist⟩
  ⟨private member functions of class skiplist⟩
public:
  ⟨public member functions of class skiplist⟩
};
```

```
⟨implementation of inline functions⟩
#endif
```

⟨_skiplist.c⟩≡

```
#include <LEDA/impl/skiplist.h>
⟨memory management⟩
⟨constructors and related functions⟩;
⟨search functions⟩;
⟨insert and delete functions⟩;
⟨concatenate and related functions⟩;
⟨miscellaneous⟩;
```

The abstract data type class is derived from the implementation class (which we rename as IMPL to save ink) and an *seq_item* is nothing but an *sl_item*. The definition of *sortseq*<*K*, *I*> has two large sections: in ⟨*redefinition of virtual functions*⟩ all virtual functions of the implementation class are redefined (see Section 5.7.3) and in ⟨*public member functions of sortseq*⟩ all operations on sorted sequences are defined by calling the corresponding function of the implementation class (see Section 5.7.10).

⟨sortseq.h⟩≡

```
#ifndef SORTSEQ_H
#define SORTSEQ_H
#if !defined(LEDA_ROOT_INCL_ID)
#define LEDA_ROOT_INCL_ID 360010
#include <LEDA/REDEFINE_NAMES.h>
#endif
#include <LEDA/basic.h>
#include <LEDA/impl/skiplist.h>
#define IMPL skiplist
typedef sl_item seq_item;
template<class K, class I>
class sortseq : public virtual IMPL {
  ⟨redefinition of virtual functions⟩
public:
  ⟨public member functions of sortseq⟩
};
#if LEDA_ROOT_INCL_ID == 360010
#undef LEDA_ROOT_INCL_ID
#include <LEDA/UNDEFINE_NAMES.h>
#endif
#endif
```

5.7.3 *Virtual Functions and their Redefinition*

The class *skiplist* has virtual functions *cmp*, *clear_key*, *clear_inf*, *copy_key*, *copy_inf*, *print_key*, *print_inf* and *key_type_id*. All of them are redefined in *sortseq*<*K*, *I*>.

⟨*virtual functions of class skiplist*⟩≡

```
virtual int cmp(GenPtr x, GenPtr y) const
{ error_handler(1,"cmp should never be called"); return 0; }
virtual void copy_key(GenPtr&)  const  { }
virtual void copy_inf(GenPtr&)  const  { }
virtual void clear_key(GenPtr&) const
{ error_handler(1,"clear_key should never be called"); }
virtual void clear_inf(GenPtr&) const
{ error_handler(1,"clear_inf should never be called"); }
virtual void print_key(GenPtr)  const
{ error_handler(1,"print_key should never be called"); }
virtual void print_inf(GenPtr)  const
{ error_handler(1,"print_inf should never be called"); }
virtual int key_type_id() const
{ error_handler(1,"key_type_id should never be called");
  return 0;
}
```

⟨*redefinition of virtual functions*⟩≡

```
leda_cmp_base<K> cmp_def;
const leda_cmp_base<K> *cmp_ptr;
int cmp (GenPtr x, GenPtr y) const
{ return (*cmp_ptr) (LEDA_CONST_ACCESS(K,x), LEDA_CONST_ACCESS(K,y)); }
int ktype_id;
int key_type_id () const { return ktype_id; }
void clear_key(GenPtr& x) const   { LEDA_CLEAR(K,x); }
void clear_inf(GenPtr& x) const   { LEDA_CLEAR(I,x); }
void copy_key(GenPtr& x)  const   { LEDA_COPY(K,x);  }
void copy_inf(GenPtr& x)  const   { LEDA_COPY(I,x);  }
void print_key(GenPtr x)  const   { LEDA_PRINT(K,x,cout);  }
void print_inf(GenPtr x)  const   { LEDA_PRINT(I,x,cout);  }
```

What are these virtual functions good for? The implementation class uses them to manipulate keys and information fields. It calls *cmp* to compare two keys, it calls *copy_key*, *clear_key*, or *print_key* to copy, destroy or print a key (and analogously an inf), respectively, and it calls *key_type_id* to determine the kind of the key type (integer, double, or otherwise). The latter function allows us to optimize the treatment of integer and double keys. Keys and informations are stored as generic pointers in the implementation class and only the abstract class knows K and I. All virtual functions are redefined in the abstract class. For example, $cmp(x, y)$ is redefined as $LEDA\_COMPARE(K, x, y)$ which in turn amounts to converting x and y to type K and then calling the compare function of type K. Similar statements hold for the other virtual functions, see Section 13.4.

Except for *copy_key* and *copy_inf* the virtual functions are only called in their redefined form. In order to double-check we have included appropriate asserts into the bodies of the virtual functions. *Copy_key* and *clear_key* are also called by the copy-constructor of *skiplist*

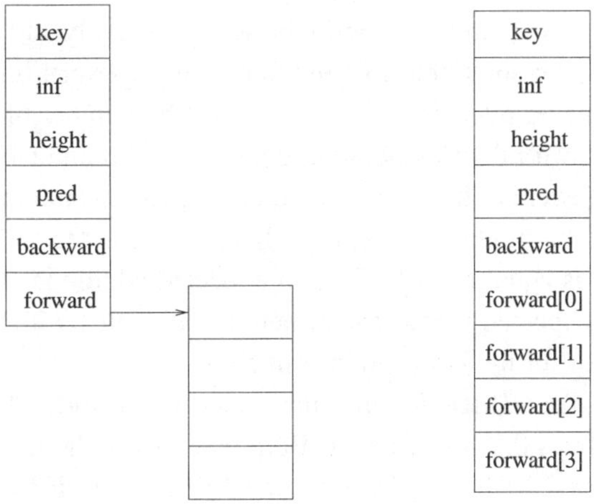

Figure 5.20 A skiplist node with four forward pointers. The left part shows the simple memory management scheme and the right part shows the refined memory management scheme.

and their original versions are used there. For this reason the original versions of *copy_key* and *copy_inf* are defined as functions with no effect.

5.7.4 *Memory Management*

We implemented two schemes for memory management: a simple scheme and a refined scheme. The refined scheme increases the speed of our implementation by almost a factor of two (if insertions and deletions have about the same frequency as lookups). The simple scheme can be selected by defining the constant `SMEM` in skiplist.h. Both schemes are illustrated by Figure 5.20.

In the simple scheme we construct an array of $h + 1$ forward pointers by

```
forward = new sl_item[h+1];
```

This calls the built-in new function and does not use LEDA's memory manager. An access to a forward pointer goes through a level of indirection as shown in Figure 5.20. The refined scheme avoids this level of indirection.

In the refined scheme we observe that the space required for a tower of height h is the size of an *skiplist_node* plus h times the size of a pointer. Recall that a node has already room for one forward pointer and that a tower of height h has $h + 1$ forward pointers. This suggests using the LEDA memory manager to allocate

$$int(sizeof(skiplist\_node)) + (h) * int(sizeof(skiplist\_node*))$$

bytes for a node of height h. Since C++ does not check array bounds and *forward* is the last field in *skiplist_node* this is equivalent to allocating space for the data member of an *skiplist_node* and an array *forward* of $h + 1$ pointers.

The scheme just described has the disadvantage that it leads to *true_height* different node sizes. The life of the LEDA memory manager becomes simpler if the number of different node sizes is small. We therefore modify the scheme slightly and round h to the next power of two if $h > 2$. We show that this modification uses very little additional space. The modified scheme never allocates more than twice the number of forward pointers that are actually needed and it allocates no additional forward pointer if $h \leq 2$. Since p^h is the probability that a tower has height h or more, the additional number of forward pointers per tower required by the modified scheme is therefore bounded by $\sum_{h \geq 3} p^h = p^3/(1-p)$. For $p = 1/4$ this is equal to $1/48$, i.e., an expected additional $1/12$ bytes per tower. We conclude that the expected space requirement for a skiplist with n items is about $25.42 \cdot n$ bytes plus the space for the keys and informations.

The macro *NEW_NODE*(v, h) allocates space for a node of height h and the macro *FREE_NODE*(v) frees that space again. Both macros use the LEDA memory management scheme. The macros *NEW_HEADER*(v) and *FREE_HEADER*(v) do the same for header nodes. Recall that a header always contains *MaxHeight* + 1 forward pointers.

⟨*memory management*⟩≡

```
inline int NODE_SIZE(int l)
{ int l1 = 0;
  if ( l > 0 )  // compute smallest power of two >= l
  { l1 = 1;
    while (l1 < l) l1 <<= 1;
  }
  return int(sizeof(skiplist_node))+
               (l1)*int(sizeof(skiplist_node*));
}

#define NEW_NODE(v,l)                                      \
v = (sl_item)std_memory.allocate_bytes(NODE_SIZE(l));      \
v->height = l;

#define FREE_NODE(v)                                       \
std_memory.deallocate_bytes(v,NODE_SIZE(v->height))

inline int HEADER_SIZE()
{ int l1 = 1;
  while (l1 < MaxHeight) l1 <<= 1;
  return int(sizeof(header_node))+
             (l1)*int(sizeof(skiplist_node*));
}

#define NEW_HEADER(v)                                      \
v = (large_item)std_memory.allocate_bytes(HEADER_SIZE());\
v->height = MaxHeight;

#define FREE_HEADER(v)                                     \
std_memory.deallocate_bytes(v,HEADER_SIZE())
```

5.7.5 *Construction, Assignment and Destruction*

The class *skiplist* has two constructors. The first constructor constructs an empty skiplist and the second constructor copies its argument.

Let us look more closely at the first constructor. We allocate a tower of height *MaxHeight* for *header* and a tower of height -1 for *STOP*. The *true_height* of the *header* is 0 and hence only the level 0 forward pointer of *header* is initialized.

The copy constructor first constructs an empty skiplist and then copies its argument L element by element. Since the constructor of class *skiplist* uses the trivial versions of the virtual functions *copy_key* and *copy_inf*, the calls of *copy_key* and *copy_inf* in *insert_at_item* have no effect, and we therefore have to use L's version of these functions to do the copying. This is a problem which arises in the implementation of all copy constructors; see Section 13.1 for a general discussion. *insert_at_item* is defined in Section 5.7.8.

The default constructor takes constant time and the copy constructor takes linear expected time plus the time to copy n keys and informations.

⟨*constructors and related functions*⟩≡
```
skiplist::skiplist(float p)
{ prob = p;
  randomsLeft = 0;
#ifdef SMEM
  header = new header_node;
  header->forward = new sl_item[MaxHeight+1];
  header->height = MaxHeight;
  STOP = new skiplist_node;
  STOP->height = -1;
#else
  NEW_HEADER(header);
  NEW_NODE(STOP,-1);
#endif
  header->true_height = 0;
  header->myseq = this;
  STOP->backward= (sl_item) header;
  STOP->pred= (sl_item) header;
  header->forward[0] = STOP;
}
skiplist::skiplist(const skiplist& L)
{ prob = L.prob;
  randomsLeft = 0;
#ifdef SMEM
  header = new header_node;
  header->forward = new sl_item[MaxHeight+1];
  header->height = MaxHeight;
  STOP = new skiplist_node;
  STOP->height = -1;
#else
  NEW_HEADER(header);
  NEW_NODE(STOP,-1);
#endif
  header->true_height = 0;
  header->myseq = this;
  STOP->backward= (sl_item) header;
  STOP->pred= (sl_item) header;
```

```
  header->forward[0] = STOP;
  sl_item p = L.STOP->pred;
  while (p!= L.header)
  { insert_at_item(header,p->key,p->inf);
    L.copy_key(p->key);
    L.copy_inf(p->inf);
    p = p->pred;
  }
}
```

We come to the assignment operator, the function *clear*, and the destructor. The assignment operator first clears the skiplist and then copies its argument. The *clear* function deletes all nodes of a skiplist and the destructor first calls *clear* and then deletes the two non-proper towers.

It would not do to copy the body of *clear* into the destructor since *~skiplist* uses the trivial versions of the virtual functions *clear_key* and *clear_inf* and hence does not know how to destroy a key or inf. This is a problem which arises in the implementation of all destructors; see Section 13.4.3 for a general discussion.

All three functions take linear expected time plus the time to copy or clear n keys and informations.

⟨*constructors and related functions*⟩+≡

```
skiplist& skiplist::operator=(const skiplist& L)
{ clear();
  sl_item p = L.STOP->pred;
  while (p!= L.header)
  { insert_at_item(header,p->key,p->inf,after);
    p = p->pred;
   }
  return *this;
 }
void skiplist::clear()
{ register sl_item p,q;
  p = header->forward[0];
  while(p!=STOP)
  { q = p->forward[0];
    clear_key(p->key);
    clear_inf(p->inf);
#ifdef SMEM
    delete p->forward;
    delete p;
#else
    FREE_NODE(p);
#endif
    p = q;
   }
 header->true_height = 0;
 header->forward[0] = STOP;
 STOP->pred= (sl_item) header;
```

```
}

skiplist::~skiplist()
{ clear();
#ifdef SMEM
  delete header->forward;
  delete header;
  delete STOP;
#else
  FREE_HEADER(header);
  FREE_NODE(STOP);
#endif
}
```

5.7.6 *Search Operations*

Skiplists offer a wide variety of search operations. We first give a fairly general search function called *search* and then derive the other search functions from it. *Search* takes a *key*, an item v and an integer h and returns a node q and an integer l. The node v has height at least h and *key* is known to lie between $v \to key$ (exclusive) and $v \to forward[h] \to key$ (inclusive). In the formulation of this precondition we used our simplifying assumption that the keys of *header* and *STOP* are $-\infty$ and ∞, respectively. *Search* finds the unique node q such that *key* lies between $q \to pred \to key$ (exclusive) and $q \to key$ (inclusive). If *key* is equal to $q \to key$ then $l \geq 0$; otherwise, $l < 0$.

The principle underlying *search* is simple. It maintains items p and q and a height k, $k \geq -1$, such that p's height is at least $k + 1$, q is the level $k + 1$ successor of p and $p \to key < key \leq q \to key$. Initially $k = h - 1$. If k is -1 then q is returned. If $k \geq 0$ then we search through level k starting at $p \to forward[k]$ to determine the new p and q.

```
q = p->forward[k];
while (key > q->key)    {  p = q; q = p->forward[k]; }
```

The basic strategy can be slightly optimized as follows. Before making a comparison between keys we check whether the current q has height k (otherwise, it is already known that $key \leq q \to key$). This optimization is worthwhile when a comparison between keys is considerably more expensive than a comparison between integers. This is the case when the comparison is made by calling *cmp* and it is not the case when the comparison is made by the operator $<$ for *ints* or *doubles*.

The expected running time of *search* is $O(1 + h)$ since $h + 1$ levels are visited and since the expected time spent on each level is constant. The easiest way to see the latter fact is to traverse the search path backwards and to recall that after following a constant expected number of backward pointers a higher tower is reached.

We give three versions of *search*, one called *gen_search* and working for arbitrary key type K, one called *double_search* and working only for keys of type *double*, and one called *int_search* and only working for keys of type *int*. *Search* selects the appropriate version by switching on the value of *key_type_id*. A general discussion of this optimization strategy can be found in Section 13.5.

⟨*search functions*⟩≡

```
sl_item skiplist::search(sl_item v, int h, GenPtr key, int& l) const
{ switch (key_type_id()) {
  case INT_TYPE_ID:    return int_search(v,h,key,l);
  case DOUBLE_TYPE_ID: return double_search(v,h,key,l);
  default:             return gen_search(v,h,key,l);
  }
}

sl_item skiplist::gen_search(sl_item v, int h, GenPtr key, int& l) const
{ register sl_item p = v;
  register sl_item q = p->forward[h];
  l = 0;
#ifdef CHECK_INVARIANTS
  assert(p->height == MaxHeight || cmp(key,p->key) > 0);
  assert(q->height < 0 || cmp(key,q->key) <= 0);
#endif
  if (q->height >= 0 && cmp(key,q->key) == 0)  return q;
  int k = h - 1;
  int c = -1;
  while (k >=0)
  { /* p->key < key < p->forward[k+1]->key and c = -1 */
    q = p->forward[k];
    while (k == q->height && (c = cmp(key,q->key)) > 0)
    { p = q;
      q = p->forward[k];
    }
    if (c == 0) break;
    k--;
  }
  l = k;
#ifdef CHECK_INVARIANTS
  p = q->pred;
  assert(p->height == MaxHeight || cmp(key,p->key) > 0);
  assert(q->height <  0 || cmp(key, q->key) <= 0);
  assert(l >= 0 && cmp(key,q->key) == 0 ||
  ( l < 0 && (q->height < 0 || cmp(key,q->key) < 0)));
#endif
  return q;
}
```

In the versions of *search* for integer and double keys we perform the following optimizations: we avoid the call of *cmp* and call the comparison operators $<$, $\leq$, $=$, ... instead. Moreover, we drop the comparison `k == q->height`, as it does not pay for integer keys.

⟨*search functions*⟩+≡

```
sl_item skiplist::int_search(sl_item v, int h, GenPtr key, int& l) const
{ sl_item p = v;
  sl_item q = p->forward[h];
  l = 0;
  int ki = LEDA_ACCESS(int,key);
  if ( q->height >= 0 && ki == LEDA_ACCESS(int,q->key) ) return q;
  int k = h - 1;
  STOP->key = key;
  while (k >= 0)
  { /* p->key < key <= p->forward[k+1]->key */
    q = p->forward[k];
    while ( ki > LEDA_ACCESS(int,q->key) )
    { p = q;
      q = p->forward[k];
    }
    if ( ki == LEDA_ACCESS(int,q->key) && q != STOP ) break;
    k--;
  }
  l = k;

#ifdef CHECK_INVARIANTS
  p = q->pred;
  assert(p->height==MaxHeight || ki>LEDA_ACCESS(int,p->key));
  assert(q->height <  0 || ki <= LEDA_ACCESS(int,q->key));
  assert(l >= 0 && ki == LEDA_ACCESS(int,q->key) ||
  ( l < 0 && (q->height<0 || ki<LEDA_ACCESS(int,q->key))));
#endif
  return q;
}
```

We refrain from showing the version for double keys. For all other search functions we will only show the generic version.

It is easy to derive the other search functions from the basic routine *search*. The call *locate_succ*(k) returns the item $\langle k1, i \rangle$ with $k \le k1$ and $k1$ minimal (*nil* if there is no such item), *locate_pred* is symmetric to *locate_succ*, *locate* is synonymous to *locate_succ* and *lookup*(k) returns the item $\langle k, i \rangle$ (*nil* if there is no such item). All operations in this section take logarithmic time.

⟨*search functions*⟩+≡

```
sl_item skiplist::locate_succ(GenPtr key) const
{ int l;
  sl_item q = search(header,header->true_height,key,l);
  return (q == STOP) ? 0 : q;
}
sl_item skiplist::locate(GenPtr key) const { return locate_succ(key); }
sl_item skiplist::locate_pred(GenPtr key) const
{ int l;
  sl_item q = search(header,header->true_height,key,l);
```

```
    if (l < 0) q = q->pred;
    return (q == header) ? 0 : q;
  }

  sl_item skiplist::lookup(GenPtr key) const
  { int k;
    sl_item q = search(header,header->true_height,key,k);
    return (k < 0) ? 0 : q;
  }
```

5.7.7 *Finger Searches*

We describe four versions of finger search.

The first three versions take a *key* and locate an item q and an integer l such that $q \rightarrow pred \rightarrow key < key \leq q \rightarrow key$ and $l \geq 0$ iff $key = q \rightarrow key$ and run in time $O(\log d)$, $O(\log(n-d))$, and $O(\log \min(d, n-d))$, respectively, if q is the d-th item in a list of n items. We first show how to obtain the time bounds $O(\log d)$ and $O(\log(n-d))$, respectively.

To achieve the first bound we compare *key* with the key of $header \rightarrow forward[k]$ for k equal to 0, 1, ... until a key at least as large as *key* is found. When this is the case we start a standard search at level k from the header.

```
k = 0;
while ( k < true_height )
{ if ( key <= header->forward[k]->key ) break;
  k++;
}
search(header,k,key,l);
```

Since the expected maximal height among the first d towers is $O(\log d)$ the expected maximal value of k is $O(\log d)$ and the time bound follows.

In order to achieve the second bound we compare *key* with the key of the rightmost tower q_k of height at least k for k equal to 0, 1, ... until a key smaller than *key* is found. When this is the case we start a standard search at level k from q_k. We can find q_k from q_{k-1} by following an expected constant number of backward pointers.

```
k = 0;
q = STOP->pred;
while ( k < true_height )
{ if ( key > q->key ) break;
  k++;
  while ( q->height < k ) q = q->backward;
}
search(q,k,key,l);
```

Since the expected maximal height among the last $n-d$ towers is $O(\log(n-d))$ the expected maximal value of k is $O(\log(n-d))$ and the time bound follows.

In order to obtain the minimum of both time bounds we perform the two searches simultaneously (also called dove-tailed), i.e., we merge the two loop bodies into one, and stop as soon as one of the two searches tells us to stop.

As in the case of standard searches we provide optimizations for keys of type *int* or *double*.

⟨*search functions*⟩+≡

```
sl_item skiplist::finger_search_from_front(GenPtr key, int& l) const
{ switch (key_type_id()) {
  case INT_TYPE_ID:    return int_finger_search_from_front(key,l);
  case DOUBLE_TYPE_ID: return double_finger_search_from_front(key,l);
  default:             return gen_finger_search_from_front(key,l);
  }
}
sl_item skiplist::gen_finger_search_from_front(GenPtr key, int& l) const
{ sl_item q = STOP->pred;
  int th = header->true_height;
  if (th == -1) return STOP;
  l = 0;
  int k = 0;
  int c1;
  while ( k < th )
  { if ( cmp(key,header->forward[k]->key) <= 0 ) break;
    k++;
  }
  return search(header,k,key,l);
}
```

and similarly

⟨*search functions*⟩+≡

```
sl_item skiplist::gen_finger_search_from_rear(GenPtr key, int& l) const
{ sl_item q = STOP->pred;
  int th = header->true_height;
  if (th == -1) return STOP;
  l = 0;
  int k = 0;
  while ( k < th )
  { if ( cmp(key, q->key) > 0 ) break;
    k++;
    while (k > q->height)  q = q->backward;
  }
  return search(q,k,key,l);
}
```

and

⟨*search functions*⟩+≡

```
sl_item skiplist::gen_finger_search(GenPtr key, int& l) const
{ sl_item q = STOP->pred;
  int th = header->true_height;
  if (th == -1) return STOP;
  l = 0;
```

```
    int k = 0;
    int c1,c2;
    while ( k < th )
    { c1 = cmp(key,header->forward[k]->key);
      c2 = cmp(key, q->key);
      if ( c1 <= 0 || c2 > 0 ) break;
      k++;
      while (k > q->height)  q = q->backward;
    }
    if (c1 <= 0)
      return search(header,k,key,l);
    else
      return search(q,k,key,l);
  }
```

The fourth version of finger search takes an item v and a *key* and returns an integer l and an item q such that $q \to pred \to key < key \leq q \to key$ and $l \geq 0$ iff $key = q \to key$. It runs in time $O(\log \min(d, n-d))$ where d is the number of items between v and q. The search is performed in the skiplist containing v and not in the skiplist which is given by *this*; recall the discussion in the paragraph preceding Section 5.7. This implies that we must not use the variables *header*, *STOP*, nor *true_height* in the program below. However, once we have determined the STOP node or the header node of the skiplist containing v (recall that STOP nodes are the only towers with negative height and that header nodes are the only towers with height *MaxHeight*) we can find the skiplist containing v as follows: if p is the header node of the skiplist containing v then $((large\_item)\ p) \to myseq$ is the skiplist containing v and if p is the STOP node of the skiplist containing v then $p \to backward$ is the corresponding header node and we are back to the situation where we know the header node.

The strategy used by *finger_search* is simple. If v is either the header or the STOP node of the skiplist containing v then we simply call the first version of finger search. So assume otherwise.

Assume first that *key* is larger than the key of v. For $k \geq 0$ let p_k be the rightmost tower to the left of or equal to v that has height k or more. We find the minimal k such that either p_k is a header node or $p_k \to forward[k]$ is a STOP node or *key* lies between the key of p_k and $p_k \to forward[k]$. In the first case we finish the search by calling the first version of finger search and in the last two cases (note that the second case is really a special case of the third case under the convention that the key of STOP is ∞) we start a standard search from p_k at level k. If *key* is smaller than the key of v, we use the symmetric strategy.

The running time of *finger_search* is readily determined. Assume for simplicity that q is to the right of v (the other case being symmetric) and that v is the n_1-th item in the sequence. Then v and q split the list into three parts of length n_1, $n_2 = d$, and $n_3 = n - n_1 - n_2$, respectively. Use h_i to denote the maximal height of a tower in the i-th part. Then $E[h_i] = \log n_i + O(1)$. The maximal value assumed by the variable k is equal to $h_0 = \min(h_1, h_2) = \log \min(d, n-d) + O(1)$. If the backward walk reaches the header then $h_1 \leq h_2$ and the second part of the search is the dove-tailed search of the preceding section that takes time $\min(\max(h_1, h_2), h_3)) = \min(h_2, h_3) = \log \min(d, n-d) + O(1)$. If the backward walk does not reach the header then the second part of the search is a standard search that takes time $O(h_0)$ as well.

As before we have three versions of *finger_search*, one for general keys, one for keys of type *int*, and one for keys of type *double*.

⟨*search functions*⟩+≡

```
sl_item skiplist::gen_finger_search(sl_item v, GenPtr key, int& l) const
{ l = 0;
  sl_item p = v;
  if ( p->height < 0 ) p = p->backward;
  // if p was a STOP node then it is a header now
  if ( p->height == MaxHeight )
    return ((large_item) p)->myseq->finger_search(key,l);
  int dir = cmp(key, v->key);
  if  ( dir == 0 ) return v;
  int k = 0;
  int c ;
  if (dir > 0)
  { while ( p->height < MaxHeight && p->forward[k]->height  >= 0 &&
              (c = cmp(key,p->forward[k]->key )) >= 0 )
    { if ( c == 0 ) return p->forward[k];
      k++;
      while ( k > p->height ) p = p->backward;
    }
    if ( p->height == MaxHeight )
      return ((large_item)p)->myseq->finger_search(key,l);
  }
  else
  { while ( p->height < MaxHeight && p->forward[k]->height >= 0 &&
              (c = cmp(key, p->key)) <= 0 )
    { if ( c == 0 )  return p;
      k = p->height;
      p = p->backward;
    }
    if (p->forward[k]->height  < 0 )
    { p = p->forward[k]->backward;
      return ((large_item)p)->myseq->finger_search(key,l);
    }
  }
#ifdef CHECK_INVARIANTS
assert(p->height == MaxHeight || cmp(key, p->key) > 0);
assert(p->forward[k]->height < 0 ||
                cmp(key, p->forward[k]->key) < 0);
#endif
  return search(p,k,key,l);
}
```

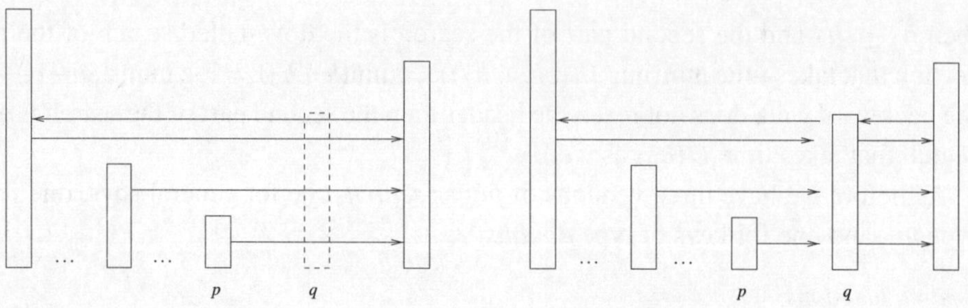

Figure 5.21 Insertion of a tower q after a tower p. All pointers that are "intersected" by the new tower are redirected.

5.7.8 *Insertions and Deletions*

We discuss the various procedures to insert into and to delete from a skiplist.

The procedure *insert_item_at_item*(q, p, *dir*) inserts the item q after and before p, respectively, as prescribed by *dir*. This requires to redirect pointers as shown in Figure 5.21. The *true_height* of the header is also adjusted to the maximum of the old height and 1 plus the height of the new item.

The running time of *insert_item_at_item* is proportional to the height of the new item. The expected height of the new item is constant.

⟨*insert and delete functions*⟩≡

```
void skiplist::insert_item_at_item(sl_item q, sl_item p, int dir)
{ if (dir == before) p = p->pred;
  /* insert item q immediately after item p */
  sl_item x;
  q->pred = p;
  p->forward[0]->pred = q;
  for (int k = 0; k <= q->height; k++ )
  { while (k > p->height) p = p->backward;
    x = p->forward[k];
    if (p->height == MaxHeight && x->height < 0 )
    {/* we have reached header and STOP and need to
           increase true_height */
     ((large_item) p)->true_height = k + 1;
     p->forward[k+1] = x;
    }
    q->forward[k] = x;
    p->forward[k] = q;
    if ( x->height == k ) x->backward = q;
   }
  q->backward = p;
}
```

The function *insert_at_item*(p, *key*, *inf*) modifies the skiplist in the vicinity of item p. If p's key is equal to *key* then its information is changed to *inf*. Otherwise a new item is created and inserted before or after *p* as dictated by *key*. The height of the new node is chosen randomly by a call *randomLevel*( ). The expected running time is constant.

⟨*insert and delete functions*⟩+≡

```
sl_item skiplist::insert_at_item(sl_item p, GenPtr key, GenPtr inf)
{ sl_item q;
  if (p->height < 0) p = p->pred;
  else
  { if ( p->height < MaxHeight )
    { int c = cmp(key,p->key);
      if (c == 0)
      { clear_inf(p->inf);
        copy_inf(inf);
        p->inf = inf;
        return p;
      }
      if ( c<0 ) p = p->pred;
    }
  }
  int k = randomLevel();
  if ( k >= MaxHeight ) k = MaxHeight - 1;
#ifdef SMEM
  q = new skiplist_node;
  q->forward = new sl_item[k+1];
  q->height = k;
#else
  NEW_NODE(q,k);
#endif
  copy_key(key);
  copy_inf(inf);
  q->key = key;
  q->inf = inf;
  insert_item_at_item(q,p,after);
  return q;
}
int skiplist::randomLevel()
{ int height = 0;
  int b = 0;
  if ( prob == 0.25 )
  { while ( b == 0 )
    { b = randomBits&3;      // read next two random bits
      randomBits >>= 2;
      randomsLeft -= 2;
      if ( b == 0 ) height++;
                          // increase height with prob 0.25
      if (randomsLeft < 2) fill_random_source();
    }
  }
  else                // user defined prob.
  { double p;
```

```
    rand_int >> p;
    while ( p < prob )
    { height++;
      rand_int >> p;
    }
  }
  return height;
}
```

There is also a version of *insert_at_item* which inserts before or after p as directed by *dir*. The expected running time is again constant.

⟨*insert and delete functions*⟩+≡

```
sl_item skiplist::insert_at_item(sl_item p,
                                 GenPtr key, GenPtr inf, int dir)
{ sl_item q;
  int k = randomLevel();
#ifdef SMEM
  q = new skiplist_node;
  q->forward = new sl_item[k+1];
  q->height = k;
#else
  NEW_NODE(q,k);
#endif
  copy_key(key);
  copy_inf(inf);
  q->key = key;
  q->inf = inf;
  insert_item_at_item(q,p,dir);
  return q;
}
```

This completes the discussion of the insertion procedures which insert at a given item.

Insert(k, i) inserts a new item $\langle k, i\rangle$ or changes the information of the item with key k (if there is such an item) and *del*(k) removes the item with key k.

⟨*insert and delete functions*⟩+≡

```
sl_item skiplist::insert(GenPtr key, GenPtr inf)
{ int k;
  sl_item p = search(header,header->true_height,key,k);
  if ( k >= 0 )
  { clear_inf(p->inf);
    copy_inf(inf);
    p->inf  = inf;
    return p;
  }
  p = insert_at_item(p,key,inf,before);
  return p;
}
```

Remove_item removes an item and *del_item* removes an item, frees its storage and also adjusts the height of the skiplist if required. The first function is used in the second and in *reverse_items*. A call *reverse_items*(p, q) with p equal or left of q reverses the subsequence with endpoints p and q.

Reverse_item has expected running time $O(d)$, where d is the length of the subsequence to be reversed. The other functions run in constant expected time.

⟨*insert and delete functions*⟩+≡
```
void skiplist::remove_item(sl_item q)
{
  if (q->height == MaxHeight || q->height < 0)
    error_handler(1,"cannot remove improper item");
  sl_item p = q->backward;
  sl_item x;
  for(int k = q->height; k >= 0; k--)
  { while ( p->forward[k] != q ) p = p->forward[k];
    x = q->forward[k];
    p->forward[k] = x;
    if ( x->height == k ) x->backward = p;
   }
  x->pred = p;
}

void skiplist::del_item(sl_item q)
{
  if (q->height == MaxHeight || q->height < 0)
    error_handler(1,"cannot delete improper item");
  remove_item(q);
  clear_key(q->key);
  clear_inf(q->inf);
  sl_item p = q->forward[q->height];
#ifdef SMEM
  delete q->forward;
  delete q;
#else
  FREE_NODE(q);
#endif
  if ( p->height < 0 )
  { large_item r = (large_item) p->backward;
    int& h = r->true_height;
    while( h > 0 && r->forward[h - 1] == p) h--;
  }
}

void skiplist::del(GenPtr key)
{ int k;
  sl_item q = search(header,header->true_height,key,k);
  if ( k>=0 ) del_item(q);
}

void skiplist::reverse_items(sl_item p, sl_item q)
{ sl_item r;
```

```
  while ( p != q )
  { r = p;
    p = p->forward[0];
    remove_item(r);
    insert_item_at_item(r,q,after);
  }
}
```

5.7.9 *Concatenate, Split, Merge and Delete_Subsequence*

We discuss concatenation, splitting, merging, and the deletion of subsequences.

Concatenation: We describe how to concatenate two skiplists of size n_1 and n_2, respectively, in time

$$O(\log \min(n_1, n_2)).$$

Assume that the two lists to be concatenated are given by *this* and *S1*. We first make sure that *this* is the higher list (by swapping *header* and *STOP* of *this* and *S1*, if necessary) and then append *S1* to either the front or the rear of *this*. Assume that we need to append *S1* to the rear of *this*, the other case being symmetric.

There are two strategies for performing the concatenation. The first strategy places the skiplists next to each other and then removes the STOP node of the left list and the header of the second list. The work required is proportional to the height of the higher list.

The second strategy places *S1* between the last element of *this* and the STOP node of *this* and then the header node and the STOP node of *S1*. The work required is proportional to the smaller height.

We use the second strategy. The details are as follows. For any k less than the height of *S1* the k-th forward pointer out of the rightmost tower in *this* of height at least k is redirected to the first item in *S1* of height at least k and the k-th forward pointer out of the rightmost tower in *S1* of height at least k is redirected to the STOP node of *this*, see Figure 5.22.

The running time of *conc* is proportional to the smaller of the two heights and is therefore $O(\log \min(n_1, n_2))$.

⟨*concatenate and related functions*⟩≡

```
void skiplist::conc(skiplist& S1, int dir)
{ if (header->true_height < S1.header->true_height)
  { leda_swap(header->myseq,S1.header->myseq);
    leda_swap(header,S1.header);
    leda_swap(STOP,S1.STOP);
    dir = ((dir == after) ? before : after);
  }
  if (S1.STOP->pred == S1.header)  return;
  /* S1 is non-empty and since height >= S1.height this is
     also non-empty */
  if (dir == after)
```

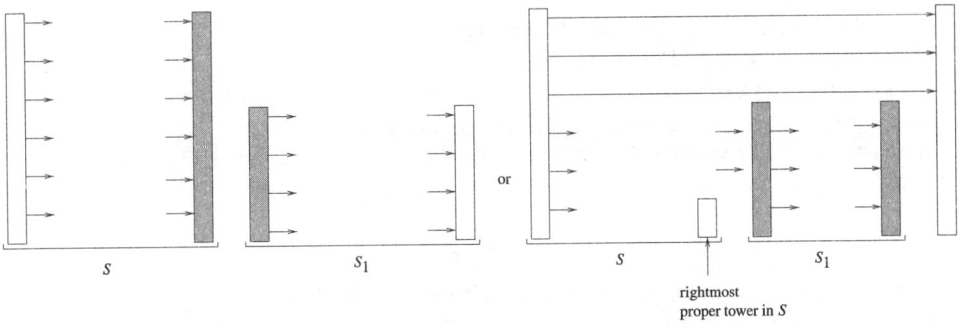

Figure 5.22 Concatenation of two skiplists S and S_1. S_1 is assumed to have smaller height and is appended at the rear of S. Only the header and STOP nodes of the lists are shown; their *true_height* is indicated as the height of the corresponding rectangles. The left part illustrates the first strategy and the right part illustrates the second strategy. The shaded towers are removed.

```
{ sl_item p = STOP->pred;
  sl_item q = S1.STOP->pred;
  assert(cmp(p->key, S1.header->forward[0]->key) < 0);
  STOP->pred = q;
  S1.header->forward[0]->pred = p;
  for (int k = 0; k < S1.header->true_height; k++)
  { /* p and q are the rightmost items of height at
       least k in this and S1, respectively */
    sl_item r = S1.header->forward[k];
    p->forward[k] = r;
    if ( r->height == k ) r->backward = p;
       q->forward[k] = STOP;
    while (p->height == k) p = p->backward;
    while (q->height == k) q = q->backward;
  }
}
else
{ sl_item q = S1.STOP->pred;
  assert(cmp(q->key, header->forward[0]->key) < 0);
  S1.header->forward[0]->pred= (sl_item) header;
  header->forward[0]->pred = q;
  for (int k = 0; k < S1.header->true_height; k++)
  { // q is the rightmost item of height at least k in S1
    sl_item r = header->forward[k];
    q->forward[k] = r;
    if (r->height == k) r->backward = q;
    r = S1.header->forward[k];
    header->forward[k] = r;
    if (r->height == k) r->backward= (sl_item) header;
    while (q->height == k) q = q->backward;
  }
}
S1.header->true_height = 0;
```

```
  S1.STOP->pred = (sl_item) S1.header;
  S1.header->forward[0] = S1.STOP;
#ifdef CHECK_INVARIANTS
  this->check_data_structure("this in conc");
  check_data_structure(S1,"S1 in conc");
#endif
}
```

We need to explain the last call of *check_data_structure*. In *delete_subsequence* we call *conc* with an argument *S1* that is locally defined within *delete_subsequence*. This *S1* is a skiplist but not a *sortseq* and hence its virtual functions have never been redefined. We therefore use *this* as the implicit argument of the last call of *check_data_structure* and in this way give it access to the redefined versions of the virtual functions.

Split: *S.split_at_item*$(p, S1, S2, dir)$ splits the skiplist containing p before or after item p into lists *S1* and *S2* as directed by *dir* in time proportional to the logarithm of the shorter result. We use P to denote the skiplist containing p. Clearly, *S1* and *S2* must be distinct, but one of them may be equal to P . If both of them are different from P then P is empty after the split. The primary argument S may be any skiplist. It must have the same type as P, *S1*, and *S2*.

A method whose running time is proportional to the logarithm of the size of P is easy to describe. We simply erect two new improper towers before or after p.

In order to obtain a running time that is proportional to the height of the smaller result list, we have to reuse *header* and *STOP* of P for the larger output list. We proceed as follows. We first determine the lower of the two outputs by simultaneously walking from p and its successor (this assumes *dir* == *after*) to *header* and *STOP* until one of the two walks reaches its destination. Let *max_lev* be the maximal level reached, i.e., both sublists contain a tower of height *max_lev* and for one of the sublists this is the maximal height.

Assume first *max_lev* is the maximal height of a tower in *S1*, i.e., in the left sublist. Then $1 + max\_lev$ is the height of *S1* and *height* is the height of *S2* after the split. We want to reuse *header* and *STOP* of P for *S2*. We interchange *header* and *STOP* of *S2* and P (this makes *S2* the input list and, if P and *S2* are distinct, makes P empty) and then remove *S1* from *S2*.

To remove *S1* from *S2* we do the following for each k, $0 \le k \le max\_lev$. Let *p_k* be the rightmost item in *S1* of height at least k. The k-th forward pointer out of *S1.header* is redirected to the destination of the k-th forward pointer out of *S2.header*, the k-th forward pointer out of *S2.header* is redirected to the destination of the k-th forward pointer out of *p_k* and the k-th forward pointer out of *p_k* is redirected to *S1.STOP*.

Assume next that *max_lev* is the maximal height of a tower in *S2*. Then the height of *S2* is $1 + max\_lev$ after the split and *height* is the height of *S1* after the split. We interchange *header* and *STOP* of *S1* and P and then remove *S2* from *S1* in a way similar to the one described above.

The running time is $O(max\_lev)$ and, if n_1 and n_2 denote the sizes of the two parts, respectively, then the expected value of *max_lev* is

$$O(\log \min(n_1, n_2)).$$

⟨*concatenate and related functions*⟩+≡

```
void skiplist::split_at_item(sl_item p,skiplist& S1,
                                       skiplist& S2,int dir)
{ if (dir == before) p = p->pred;
  sl_item p1 = p;
  sl_item p2 = p->forward[0];
  int max_lev = -1;
  while ( p1->height < MaxHeight && p2->height >= 0 )
  { /* p1 and p2 are proper towers of height
        larger than max_lev    */
    max_lev++;
    while (p1->height == max_lev) p1 = p1->backward;
    while (p2->height == max_lev) p2 = p2->forward[max_lev];
  }
  /* we have seen proper towers of height max_lev on both
     sides of the split and either p1 or p2 is a sentinel */
  large_item pheader;
  if (p1->height == MaxHeight)
    pheader = (large_item) p1;
  else
    pheader = (large_item) p2->backward;
  skiplist* Pp = pheader->myseq;
  if (Pp != &S1)  S1.clear();
  if (Pp != &S2)  S2.clear();
  if (p1->height == MaxHeight)
  { /* we reuse pheader and pSTOP for S2 */
    if (Pp != &S2)
    { leda_swap(Pp->header->myseq, S2.header->myseq);
      leda_swap(Pp->header,S2.header);
      leda_swap(Pp->STOP,S2.STOP);
    }
    S1.header->true_height = 1+max_lev;
    p1 = p;
    for (int k =0; k <= max_lev; k++)
    { // p1 is the rightmost item in S1 of height at least k
      sl_item q = S2.header->forward[k];
      S1.header->forward[k] = q;
      if (q->height == k) q->backward = (sl_item) S1.header;
      S2.header->forward[k] = p1->forward[k];
      if (p1->forward[k]->height == k)
         p1->forward[k]->backward = (sl_item) S2.header;
      p1->forward[k] = S1.STOP;
      while (k == p1->height) p1 = p1->backward;
    }
    S1.header->forward[max_lev + 1] = S1.STOP;
```

```
    /* the next line sets the predecessor of S1.STOP
       correctly if S1 is non-empty; if it is empty
       the last line corrects the mistake */
    S1.STOP->pred = p;
    S2.header->forward[0]->pred = (sl_item) S2.header;
    S1.header->forward[0]->pred = (sl_item) S1.header;
  }
  else
  { /* we want to reuse pheader and pSTOP for S1 */
    if (Pp != &S1)
    { leda_swap(Pp->header->myseq,S1.header->myseq);
      leda_swap(Pp->header,S1.header);
      leda_swap(Pp->STOP,S1.STOP);
    }
    S2.header->true_height = 1 + max_lev;
    p1 = p;
    p2 = S1.STOP->pred;
    for (int k =0; k <= max_lev; k++)
    { /* p1 and p2 are the rightmost items in S1 and S2
         of height at least k, respectively */
      sl_item q = p1->forward[k];
      S2.header->forward[k] = q;
      if (q->height == k) q->backward = (sl_item) S2.header;
      p1->forward[k] = S1.STOP;
      p2->forward[k] = S2.STOP;
      while (k == p1->height) p1 = p1->backward;
      while (k == p2->height) p2 = p2->backward;
    }
    S2.header->forward[max_lev + 1] = S2.STOP;
    /* the next line sets the predecessor of S2.STOP
       correctly if S2 is non-empty; if it is empty then
       the next line corrects the mistake */
    S2.STOP->pred = S1.STOP->pred;
    S2.header->forward[0]->pred = (sl_item) S2.header;
    S1.STOP->pred = p;
    S1.header->forward[0]->pred = (sl_item) S1.header;
  }
  if (Pp != &S1 && Pp != &S2)
  { /* P is empty if distinct from S1 and S2 */
    Pp->header->forward[0] = Pp->STOP;
    Pp->STOP->pred = Pp->STOP->backward =
                               (sl_item) Pp->header;
    Pp->header->true_height = 0;
  }
#ifdef CHECK_INVARIANTS
  this->check_data_structure("this in split");
  Pp->check_data_structure("P in split");
  check_data_structure(S1,"S1 in split");
  check_data_structure(S2,"S2 in split");
#endif
}
```

Merge: We describe how to merge two skiplists of length $n1$ and $n2$, respectively, in time $O(\binom{n1+n2}{n1})$.

Assume that the lists to be merged as given by *this* and by *S1*. We first determine the shorter list (by stepping through both lists in lock-step fashion and stopping as soon as the end of the shorter list is reached) and make sure that *this* is the larger list (by interchanging *header* and *STOP* of *this* and *S1* otherwise). We then erect a finger at the first item of *this* and consider the items of *S1* one by one. We locate the item by a finger search, insert the item into *this* and advance the finger to the point of insertion.

For the running time analysis we assume without loss of generality that $n_1 \le n_2$. For i, $1 \le i \le n_1$, let d_i be the stride of the finger search when inserting the i-th item of *S1* into *this*. Then $n_2 = \sum_i d_i$ and the total running time is $O(n_1 + \sum_i \log d_i)$. This sum is maximal if all the d_i's are equal to n_2/n_1 and is hence bounded by

$$O(n_1(1 + \log(n_2/n_1))) = O(\binom{n1+n2}{n1}).$$

⟨*concatenate and related functions*⟩+≡

```
void skiplist::merge(skiplist& S1)
{ sl_item p= (sl_item) header;
  sl_item q = S1.header;
  while ( p->height  >= 0 && q->height >= 0 )
  { p = p->forward[0];
    q = q->forward[0];
  }
  if (q->height >= 0)
  { /* swap if this is shorter than S1 */
    leda_swap(header->myseq,S1.header->myseq);
    leda_swap(header,S1.header);
    leda_swap(STOP,S1.STOP);
  }
  /* now S1 is at most as long as this */
  sl_item finger= (sl_item) header;
  p = S1.header->forward[0];
  while (p->height >= 0)
  { sl_item q = p->forward[0];
    int l;
    finger = finger_search(finger,p->key,l);
    if (l >= 0) error_handler(1,"equal keys in merge");
    insert_item_at_item(p,finger,before);
    finger = p; // put finger at newly inserted item
    p = q;
  }

  S1.header->true_height = 0;
  S1.STOP->pred = (sl_item) S1.header;
  S1.header->forward[0] = S1.STOP;
#ifdef CHECK_INVARIANTS
  check_data_structure("this in merge");
```

```
  S1.check_data_structure("S1 in merge");
#endif
}
```

Deletion of Subsequences: We describe how to delete a subsequence from a skiplist. More precisely, if a and b are items in a list P with a left of or equal to b then the call *S.delete_subsequence*(a, b, *S1*) deletes the subsequence starting at a and ending at b from P and assigns it to *S1*. The running time is $O(\log \min(n_1, n - n_1))$ where n and n_1 are the length of P and *S1* respectively. *S* only provides the type.

The items a and b split P into three parts. We first determine the lowest of the parts by simultaneously walking from $a \to pred$ and from b to the left and from $b \to forward[0]$ to the right until we reach *header*, a tower left of a, or *STOP*, respectively.

If either the first or the last subsequence is lowest then the operation can be reduced to two splits and one conc. If what is to become *S1* is lowest we directly insert *S1*'s *header* and *STOP* before a and after b, respectively.

Let h_i be the height of the i-th part. Then

$$\mathrm{E}[h_2] = O(\log n_1),\ \mathrm{E}[h_1] = O(\log(n - n_1)),\ \text{and}\ \mathrm{E}[h_3] = O(\log(n - n_1)).$$

The time to determine the lowest part is $\min(h_1, h_2, h_3)$. If h_2 is smallest then the running time of actually deleting the subsequence is $O(h_2)$. If h_2 is not the smallest then the times for the two splits and one conc are $\min(\max(h_1, h_2), h_3))$, $\min(h_1, h_2)$, and $\min(h_1, h_3)$, respectively. All three quantities are bounded by $\min(h_2, \max(h_1, h_3))$. The expected running time is therefore

$$O(\log \min(n_1, n - n_1))$$

in both cases.

⟨*concatenate and related functions*⟩+≡

```
void skiplist::delete_subsequence(sl_item a,
  sl_item b,skiplist& S1)
{ S1.clear();
  sl_item p1 = a->pred;
  sl_item p2 = b;
  sl_item p3 = b->forward[0];
  int k = -1;
  while ( p1->height < MaxHeight && p3->height >= 0 &&
          p2->height < MaxHeight && cmp(p2->key,a->key) >= 0 )
  { k++;
    while ( p1->height == k)  p1 = p1->backward;
    while ( p2->height == k)  p2 = p2->backward;
    while ( p3->height == k)  p3 = p3->forward[k];
  }
  if (p1->height == MaxHeight || p3->height  < 0)
  { if (p1->height < MaxHeight) p1 = p3->backward;
    skiplist* Pp = ((large_item) p1)->myseq;
```

```
    skiplist S2,S3;
    split_at_item(b,S2,S3,after);
    split_at_item(a,*Pp,S1,before);
    Pp->conc(S3,after);
    return;
  }
  // the middle list is the lowest and we have to do some work
  p1 = a->pred;
  p2 = b;
  /* correct predecessor pointers */
  a->pred = (sl_item) S1.header;
  S1.STOP->pred = b;
  b->forward[0]->pred = p1;
  /* height of S1 */
  S1.header->true_height = 1 + k;
  S1.header->forward[1+k] = S1.STOP;
  for (int i = 0; i <= k; i++)
  { /* p1 and p2 are the rightmost items of height at least
       i in the first and second part, respectively */
    sl_item q = p1->forward[i];
    S1.header->forward[i] = q;
    if (q->height == i) q->backward = S1.header;
    q = p2->forward[i];
    p1->forward[i] = q;
    if (q->height == i) q->backward = p1;
    p2->forward[i] = S1.STOP;
    while (i == p1->height)  p1 = p1->backward;
    while (i == p2->height)  p2 = p2->backward;
  }
}
```

It takes a lot of trivial stuff to complete the implementation of *skiplist*. We do not include it here to save space.

5.7.10 *Member Functions of Class* sortseq

The purpose of the file LEDAROOT/incl/LEDA/sortseq.h is to define the abstract data type class *sortseq* and to implement the abstract functions in terms of the concrete functions. We follow the general technique discussed in Section 13.4. Every abstract function (e.g. *lookup*) calls the concrete function with the same name after converting any arguments of type K or I to a generic pointer (by means of function *leda_cast*) and after converting any argument of type *sortseq*$\langle K, I\rangle$ to a *skiplist* (by a cast). Similarly, any result of type K or I is converted back from generic pointer (by means of the *LEDA_ACCESS* macro). Two examples should suffice to show the principle.

```
K key(seq_item it)    const { return LEDA_ACCESS(K,IMPL::key(it)); }
seq_item lookup(K k) const { return IMPL::lookup(leda_cast(k)); }
```

5.7.11 *A Final Word*

We have given the implementation of the data type *sortseq*. We glossed over some of the trivial stuff. The complete source code can be found in the LEDA source code directory.

Exercises for 5.7

1. Implement operations *union*, *intersection*, *setminus*, and *setdifference* for sorted sequences. Start from the implementation of *merge*.
2. Add the implementation parameter mechanism to the type *sortseq*<*K*, *I*>. Follow the construction of the type *_sortseq*<*K*, *I*, *IMPL*>.
3. Add the finger search operations to the *ab*-tree implementation or the randomized search tree implementation of sorted sequences. Inspect [Meh84a] and [AS89] for the relevant theory.

5.8 An Application of Sorted Sequences: Jordan Sorting

Let C be a Jordan curve in the plane[16] that is nowhere tangent to the x-axis. Let $x_1, x_2, \ldots, x_n$ be the abscissas of the intersection points of C with the x-axis, listed in the order the points occur on C (see Figure 5.23). Call a sequence $x_1, x_2, \ldots, x_n$ of real numbers obtainable in this way a *Jordan sequence*. The reader should convince himself at this point that the sequence 1, 3, 4, 2 is not a Jordan sequence. We describe a linear time algorithm to recognize and sort Jordan sequences due to Hoffmann et al. [HMRT85]. The Jordan demo allows you to exercise the algorithm.

As a sorting algorithm, *Jordan_sort* is not competitive with general purpose sorting algorithms, like quicksort and mergesort, despite its linear running time. We include the *Jordan_sort* program in the book as an example of how much LEDA simplifies the implementation of complex algorithms.

The Jordan sorting problem arises in the following context. Suppose we are given a simple polygon (as a sequence of edges) and a line and are asked to compute the points of intersection in the order they occur on the line. A traversal of the polygon produces the intersections in the order they occur on the polygon. Sorting the sequence of intersections produces the order on the line.

A Jordan sequence together with its intersections with the x-axis gives rise to two nested sets of parentheses, simply cut the plane at the x-axis into two half-planes (see Figure 5.24). We call a matching pair of parentheses a *bracket*. A nested set of brackets gives rise to an ordered forest in a natural way. Each bracket corresponds to a node of the tree and the children of a node correspond to the brackets directly nested within a bracket. The ordering of the children of a bracket corresponds to the left to right ordering of the subbrackets. We can turn the ordered forest into a tree by adding a fictitious bracket $(-\infty, +\infty)$. Figure 5.25

[16] A Jordan curve is a curve without self-intersections, i.e., a continuous injective mapping from the unit interval into the plane.

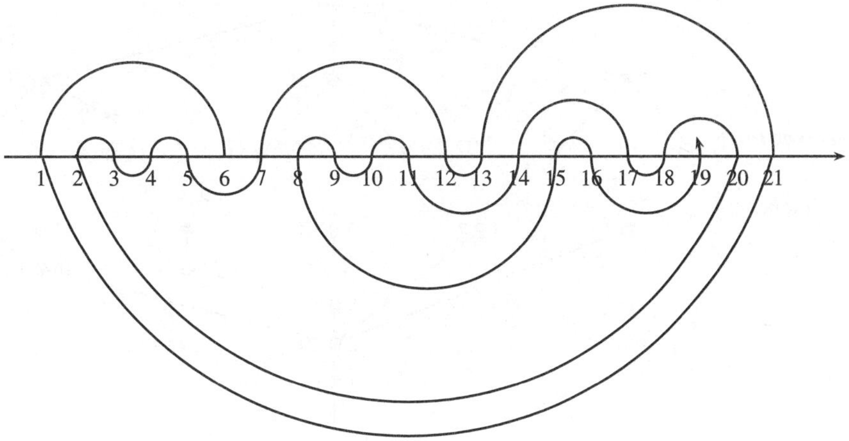

Figure 5.23 A Jordan curve and its intersections with the x-axis: The curve intersects the x-axis 21 times. We assumed for the drawing that the abscissas of the intersections are the integers 1 to 21. As the curve is traversed starting at 6 the sequence 6, 1, 21, 13, 12, 7, 5, 4, 3, 2, 20, 18, 17, 14, 11, 10, 9, 8, 15, 16, 19 is obtained.

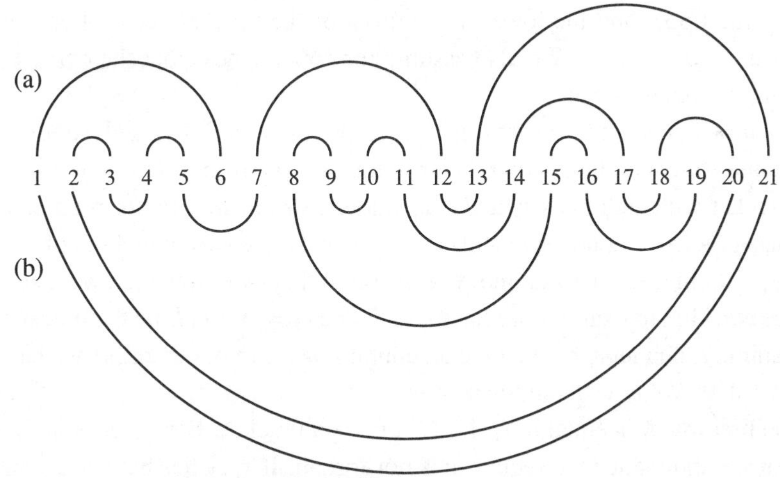

Figure 5.24 The nested parentheses corresponding to the Jordan curve of Figure 5.23; each pair of parentheses is drawn as a half-circle: (a) The parentheses corresponding to the upper half-plane; (b) The parentheses corresponding to the lower half-plane.

shows the ordered trees corresponding to the brackets of Figure 5.24. We call these trees the *lower* and the *upper tree*, respectively.

To sort a Jordan sequence $x_1, x_2, \ldots, x_n$ we process the numbers x_i in increasing order

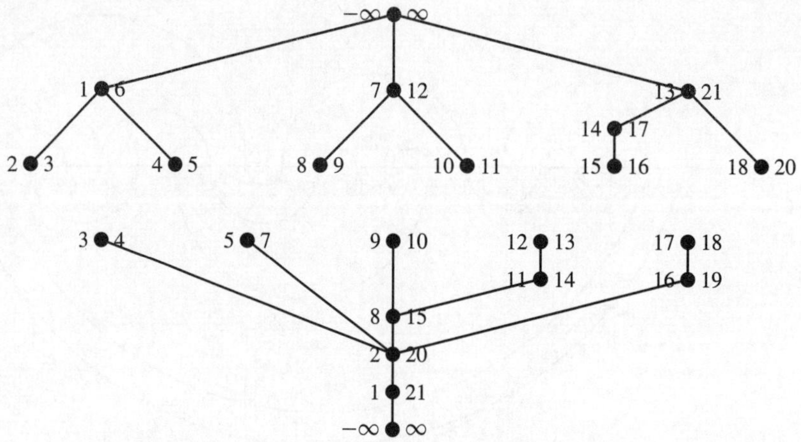

Figure 5.25 The upper and lower tree for the Jordan curve of Figure 5.23. The smaller and larger element of each bracket is on either side of the corresponding tree node.

on i, constructing three objects simultaneously: the sorted list of the numbers so far processed, and the upper and lower tree of the brackets corresponding to the numbers so far processed. Figure 5.26 shows the state of the algorithm after having processed number 8 in our example.

Initially, the upper and the lower tree consist of the bracket $(-\infty, +\infty)$ and the initial sorted list is $-\infty, x_1, +\infty$. We also assume for concreteness that the curve C crosses the x-axis from bottom to top at x_1.

Assume now that we have processed $x_1, \ldots, x_i$ for some $i \geq 1$ and want to process x_{i+1} next. Assume for concreteness that the crossing at x_i is from top to bottom. So we have to insert a bracket with endpoints x_i and x_{i+1} into the lower tree. In our running example this is the bracket (8,15). Let l_i and r_i with $l_i < x_i < r_i$ be the two neighbors of x_i in the sorted list; if one of them is equal to x_1 and we insert into the lower tree then we take the neighbor in distance two. In our example we have $l_i = 7$ and $r_i = 9$. Let l_i be the bracket in the lower tree containing l_i and let r_i be the bracket containing r_i. In our example we have $l_i = (5, 7)$ and $r_i = (9, 10)$. We now distinguish cases.

Assume first that l_i is equal to r_i, i.e., (l_i, r_i) is a bracket. If x_{i+1} does not lie between l_i and r_i then we abort since the sequence is not Jordan. If x_{i+1} lies between l_i and r_i then we make the bracket $(\min(x_i, x_{i+1}), \max(x_i, x_{i+1}))$ the single child of (l_i, r_i) and insert x_{i+1} at the appropriate position into the sorted list.

Assume next that l_i is not equal to r_i. Then one of the two brackets, call it T_i, does not contain x_i. We locate x_{i+1} in the ordered sequence of siblings of T_i. Two cases can occur: either x_{i+1} is contained in one of the siblings of T_i or it is not. If x_{i+1} is contained in a sibling of T_i then we abort since the sequence is not Jordan. If x_{i+1} is not contained in a sibling of T_i then we change the lower tree as follows. We create a new node corresponding to bracket $(\min(x_i, x_{i+1}), \max(x_i, x_{i+1}))$, make all siblings of T_i that are enclosed in the

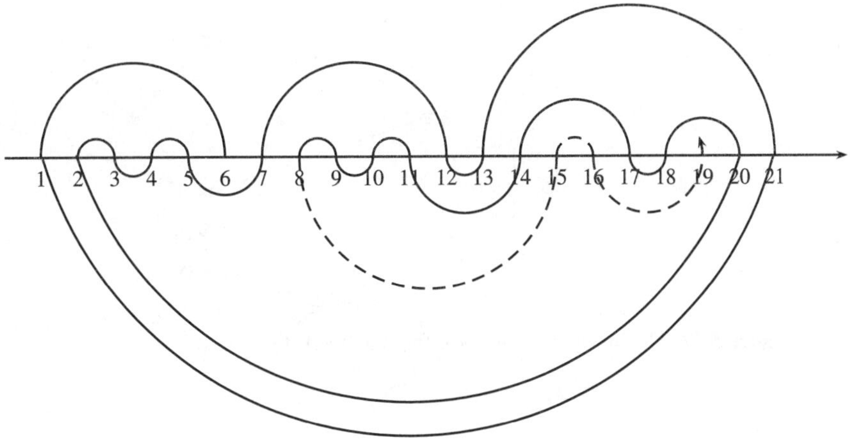

(a)

(b) 1 2 3 4 5 6 7 8 9 10 11 12 13 14 17 18 20 21

(c)

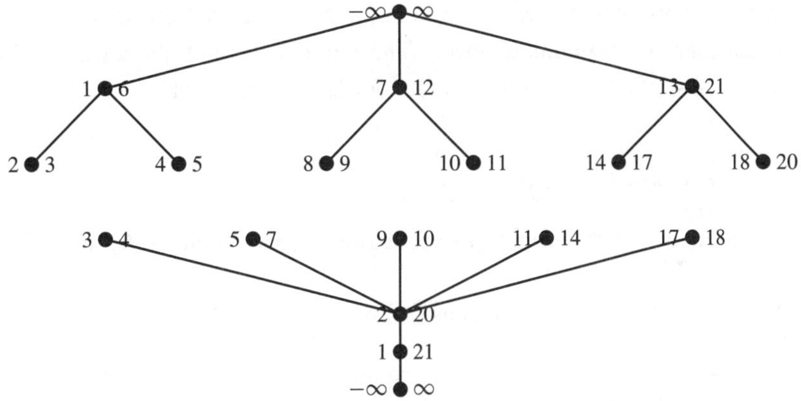

Figure 5.26 (a) The Jordan curve after reaching point 8. (b) The sorted sequence of points processed so far. (c) The lower and upper tree.

new bracket children of the new bracket, and add the new bracket to the list of siblings of T_i. We also insert x_{i+1} at the appropriate position into the sorted list of numbers processed so far.

In our example neither l_i nor r_i contains x_i and so either one of them can be T_i. The ordered list of siblings of T_i is (3, 4), (5, 7), (9, 10), (11, 14), (17, 18) and number 15 lies between brackets (11, 14) and (17, 18). So we make (9, 10) and (11, 14) children of the new bracket (8, 15) and let (8, 15) take their place in the list of children of bracket (2, 20). We also insert 15 between 14 and 17 into the sorted list of numbers processed so far. Figure 5.27 shows the lower tree after inserting the bracket (8, 15).

We proceed to describe the implementation of a procedure

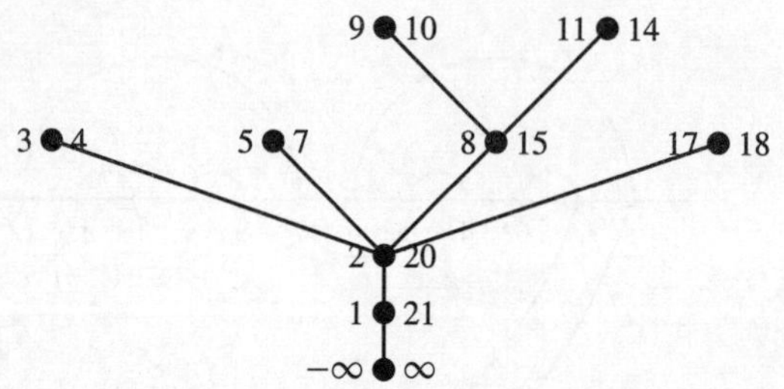

Figure 5.27 The lower tree after inserting the bracket (8,15).

```
bool Jordan_sort(const list<double>& In, list<double>& Out,
                                        window* Window = 0);
```

It takes a sequence *In* of doubles and tests whether the sequence is Jordan. If so, it returns the sorted output sequence in *Out*. If the third argument is non-nil then the execution of the algorithm is animated in *Window*. We define three files: the file Jordan.h contains the declaration of procedure *Jordan_sort*, the file Jordan.c contains its implementation, and the file Jordan_demo.c contains a demo. The latter file is not shown in the book, but can be found in LEDAROOT/demo. It includes Jordan.c as a subfile.

⟨*Jordan.h*⟩≡

```
#include <LEDA/list.h>
class window;
bool Jordan_sort(const list<double>&, list<double>&, window* Window = 0);
```

The global structure of Jordan.c is as follows:

⟨*Jordan.c*⟩≡

```
#include <LEDA/list.h>
#include <LEDA/window.h>
#include <LEDA/sortseq.h>
⟨Jordan.h⟩
⟨global variables⟩
⟨data structure⟩
⟨global functions⟩
⟨procedure Jordan sort⟩
```

As outlined above, we construct three data structures simultaneously: the sorted list L of the intersections processed so far and the upper and lower tree of brackets. We define appropriate classes. While reading these class definitions the reader may want to inspect Figure 5.28; it shows how the subtree of the upper tree rooted at the bracket (7, 12) is represented.

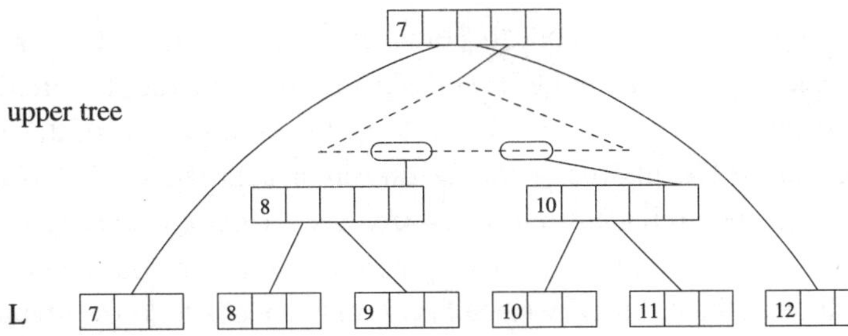

Figure 5.28 The representation of the subtree of the upper tree rooted at the bracket (7, 12). This bracket contains subbrackets (8, 9) and (10, 11). The items (= *Intersections*) of the list L are shown as rectangular boxes with three fields and brackets are shown as rectangular boxes with five fields. Solid lines correspond to pointers. Each intersection points to the bracket containing it which in turn points back to the intersection. Each bracket contains a *children_seq*. The *children_seq* contained in the bracket (7, 12) is shown as a dotted triangle. It has two items corresponding to the two subbrackets (8, 9) and (9, 10). The key of each item is a pointer to the subbracket and each subbracket stores in *pos_among_sibs* the item representing it in the *children_seq* of its parent. This allows, for example, the bracket (8, 9) to find the bracket (10, 11), namely if b is a pointer to the former bracket then $b \rightarrow$ *pos_among_sibs* is an item in the *children_seq* of bracket (7, 12), and the successor of this item is the item corresponding to (10, 11).

⟨*data structure*⟩≡

```
class intersection;
typedef intersection* Intersection;

class bracket;
typedef bracket*      Bracket;

list<Intersection> L;
```

We defined the list L as a list of pointers to intersections rather than a list of intersections as this will avoid frequent copying of intersections. Each intersection needs to know the bracket containing it in either tree. Therefore, an intersection contains its abscissa (a *double*) and pointers to the brackets in the two trees containing it. The constructor constructs an intersection with a particular x-coordinate.

⟨*data structure*⟩+≡

```
class intersection{
  public:
  double x;
  Bracket containing_bracket_in[2];

  intersection(double xcoord)
  { x = xcoord;
    containing_bracket_in[upper] = nil;
    containing_bracket_in[lower] = nil;
  }
};
```

A node of either tree corresponds to a bracket. A bracket needs to know its two endpoints (as items in L), its position among its siblings (a *seq_item*), and its sorted sequence of sub-brackets (a *sortseq<Bracket, int>*). We also store the x-coordinate of the left endpoint of the bracket. In order to save ink we use *children_seq* as an abbreviation for *sortseq<Bracket, int>*. The information type *int* in *children_seq* is irrelevant. *Children_seqs* need to be able to compare brackets. Brackets are compared by comparing the x-coordinates of their left endpoints. Because of the circularity (the class *bracket* needs to know about *children_seq* and *children_seq* needs a function *compare* for *Brackets*) we declare *compare* at the beginning of the next program chunk and define it at its end.

Bracket has two constructors. The first constructor takes two items a and b in L and the indicator *side* and constructs a bracket with endpoints a and b. The left endpoint is the endpoint with the smaller x-coordinate. Its x-coordinate is stored in *left_x*. The list item corresponding to the left endpoint is stored in *endpt*[*left*] and the appropriate reverse pointer is stored in the list item. The same holds true for the right endpoint. The *children* and the *pos_among_sibs* fields will be filled later.

The second constructor initializes only *left_x*. It is used to convert an x-coordinate into a bracket so that we can search[17] for the x-coordinate in a *children_seq*.

A bracket contains a number x, if x lies between the abscissa of the endpoints of the bracket.

⟨*data structure*⟩+≡

```
int compare(const Bracket&,const Bracket&);
typedef sortseq<Bracket,int> children_seq;
class bracket{
  public:
  double left_x;
  list_item endpt[2];
  children_seq children;
  seq_item pos_among_sibs;
  bracket(list_item a, list_item b, SIDE side)
  { if (L[a]->x > L[b]->x) leda_swap(a,b);
    left_x = L[a]->x;
    endpt[left] = a;
    L[a]->containing_bracket_in[side] = this;
    endpt[right] = b;
    L[b]->containing_bracket_in[side] = this;
  }
  bracket(double x){ left_x = x; }
  bool contains(double x)
  { return ( L[endpt[left]]->x < x && x < L[endpt[right]]->x ); }
```

[17] The key type of *children_seq* is *Bracket* and hence we can only search for a *Bracket* in a *children_seq*. We will have to search for a *double* and can do so only by converting the *double* into a *bracket*. This slight inconvenience would not arise if the search functions in a *sortseq*<K, I> would use a comparison function *compare*(*const* K&, *const* $K1$&) where the second argument type is allowed to be different from the first and in this way allowed to search for any object that can be compared with the keys of the sorted sequence.

```
};
int compare(const Bracket & b1,const Bracket & b2)
{ return compare(b1->left_x,b2->left_x); }
```

We complete the definition of the data structure with the definition of some global variables. We need a representation of ∞ for the bracket $(-\infty, \infty)$, we need a variable *side* that tells us in which tree we are working in, and we need the first abscissa *x1* and the corresponding item *x1_item* in L. We also define enumeration types {*upper*, *lower*} and {*left*, *right*} that are used to distinguish the upper and lower tree and the left and right endpoint of a bracket.

⟨*global variables*⟩≡

```
const double infty = MAXDOUBLE;
double x1;
list_item x1_item;
enum SIDE {upper,lower};
SIDE side;
enum {left,right};
```

We can now give the global structure of the Jordan sorting procedure. It takes a list *In* of *doubles* and decides whether it is Jordan. If so, it also produces a sorted version *Out* of *In*.

If the input list has length at most one then sorting is trivial. If it has length at least two then we first initialize L with $-\infty$, x_1, ∞ and build trivial upper and lower trees. Then we insert the elements of *In* one by one alternately into the lower or upper tree; the variable *side* keeps track of where we are. At the end we produce the sorted output list *Out*.

⟨*procedure Jordan sort*⟩≡

```
bool Jordan_sort(const list<double>& In, list<double>& Out,
                                                  window* Window)
{ if ( In.length() <= 1 ) {  Out = In; return true; }
  ⟨initialize L with x1 and construct trivial lower and upper trees⟩;
  /* we now process x_2 up to x_n */
  list_item it = In.succ(In.first()); // the second item
  side = upper;
  while (it)
  { ⟨process next input⟩;
    it = In.succ(it);
    side = ((side == upper)? lower : upper); // change sides
  }
  ⟨produce the output by copying L to Out⟩;
  return true;
}
```

We now discuss the three phases of *Jordan_sort*: initialization, processing an input, and producing the output list.

We initialize the list L with $-\infty$, x_1, ∞, and the upper and lower trees with the brackets

$(-\infty, \infty)$. We also store x_1 in *x1* and the corresponding item of L in *x1_item*. We set *xi_item* to *x1_item*; generally, *xi_item* corresponds to the last number inserted into L. This was called x_i in the discussion above.

⟨*initialize L with x1 and construct trivial lower and upper trees*⟩≡

```
x1 = In.head();
L.clear();
list_item minus_infty_item = L.append(new intersection(-infty));
list_item xi_item = x1_item = L.append(new intersection(x1));
list_item plus_infty_item = L.append(new intersection(infty));
bracket upper_root(minus_infty_item,plus_infty_item,upper);
bracket lower_root(minus_infty_item,plus_infty_item,lower);
```

We turn to the insertion part. The number to be inserted is $x = In[it]$. This was called x_{i+1} in the discussion above. Recall that *xi_item* is the item of list L holding *xi*. So the new bracket has endpoints x and *xi*. The new bracket needs to be inserted into the *side* tree.

We first determine the items *l_item* and *r_item* to the left and to the right of the current item and their corresponding intersections l and r; if one of them is equal to *x1* and *side* == *lower*, we skip it, since there is no bracket in the lower tree containing *x1*. We also retrieve the brackets *lB* and *rB* containing l and r. Then we distinguish cases according to whether the brackets *lB* and *rB* are identical or not and branch to the two sub-cases. Both sub-cases modify the list L and the *side* tree and set *x_item* to the item of L containing the new intersection.

After returning from the two sub-cases we update *xi_item*.

If *Window* is non-nil we also draw an appropriate half-circular arc into it. We divide the plane in half at $y = 50$ and draw red arcs in the upper half and black arcs in the lower half. The operation *draw_arc*(*x1*, *y1*, *x2*, *y2*, *r*, *c*) of class *window* draws a counterclockwise oriented circular arc starting in (*x1*, *y1*), ending in (*x2*, *y2*), and having radius *r* and color *c*.

⟨*process next input*⟩≡

```
double x = In[it];
double xi = L[xi_item]->x;
if (x == xi || x == x1) return false;
list_item l_item = L.pred(xi_item);
if (l_item == x1_item && side == lower) l_item = L.pred(l_item);
list_item r_item = L.succ(xi_item);
if (r_item == x1_item && side == lower) r_item = L.succ(r_item);
Intersection l = L[l_item];
Intersection r = L[r_item];
Bracket lB = l->containing_bracket_in[side];
Bracket rB = r->containing_bracket_in[side];
list_item    x_item;
if (Window != nil)
{ double r = (xi - x)/2; if (r < 0) r = -r;
 if ( side == upper)
   Window->draw_arc(point(xi,50),point((xi+x)/2,50+r), point(x,50),red);
```

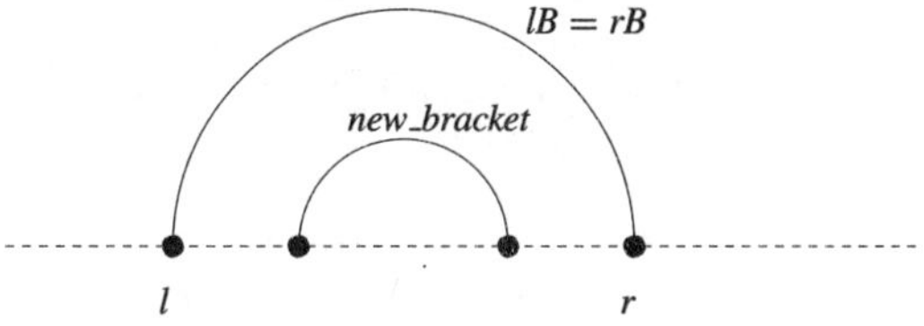

Figure 5.29 The two neighboring brackets *lB* and *rB* are identical and hence the new bracket becomes their child.

```
  else
    Window->draw_arc(point(xi,50),point((xi+x)/2,50-r), point(x,50),black);
    }
int dir = ( x > xi ? LEDA::after : LEDA::before);
if (lB == rB)
  { ⟨lB and rB are identical⟩ }
else
  { ⟨lB and rB are distinct⟩ }
xi_item = x_item;
```

If the brackets *lB* and *rB* are identical then we only need to check whether the bracket contains the new abscissa x. If not, we abort because the input sequence is not Jordan. Otherwise we insert x next to *xi* into list L, create a new bracket, and make it the only child of *lB*, see Figure 5.29.

⟨*lB and rB are identical*⟩≡

```
if (!(lB->contains(x))) return false;
x_item = L_insert(x,xi_item,dir);
Bracket new_bracket = new bracket(x_item,xi_item,side);
new_bracket->pos_among_sibs = lB->children.insert(new_bracket,0);
```

The procedure *L_insert* is essentially identical to *L.insert*. A small difference arises from the fact that *x1_item* is not part of a bracket on the lower side and hence if the new intersection is to be inserted next to *x1* then its position with respect to *x1* is not yet known.

⟨*global functions*⟩≡

```
list_item L_insert(double x, list_item it, int dir)
{ if ( side == lower &&
      (dir == LEDA::before && L.pred(it) == x1_item && x < x1) ||
      (dir == LEDA::after &&  L.succ(it) == x1_item && x > x1) )
      it = x1_item;
  return L.insert(new intersection(x),it,dir);
}
```

We come to the case in which the brackets *lB* and *rB* are not identical. We distinguish two cases.

If x lies between l and r then the new bracket (xi, x) does not enclose any brackets. We

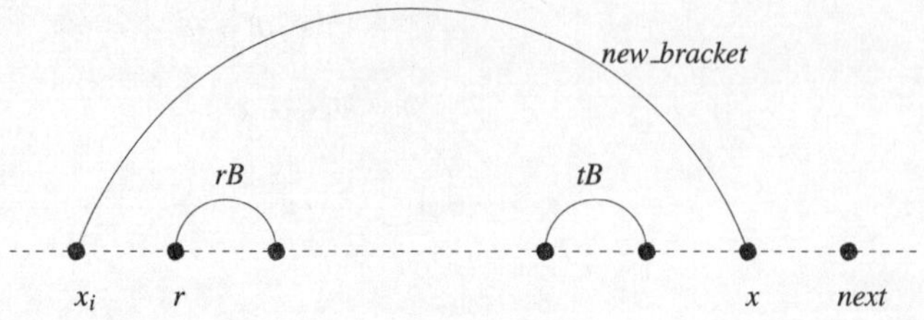

Figure 5.30 The new bracket extends to the right and x is larger than r; *tB* is the rightmost sibling of *rB* whose left endpoint is less than x. x must not be contained in *tB* and it must be contained in the parent bracket of *tB*. The children of the new bracket start at *rB* and end at *tB*.

therefore only have to insert x either before or after *xi* into L and make the new bracket either the left sibling of *rB* (if *lB* contains x) or the right sibling of *lB* (otherwise).

If x does not lie between l and r, we have to work harder.

⟨*lB and rB are distinct*⟩≡

```
children_seq S; // just for the type
if ( l->x < x && x < r->x )
{ x_item = L_insert(x,xi_item,dir);
  Bracket new_bracket = new bracket(xi_item,x_item,side);
  new_bracket->pos_among_sibs =
     ( lB->contains(x) ?
       S.insert_at(rB->pos_among_sibs,new_bracket,0,LEDA::before) :
       S.insert_at(lB->pos_among_sibs,new_bracket,0,LEDA::after) );
}
else
  if ( dir == LEDA::after )
  { ⟨new bracket has subbrackets and extends to the right⟩ }
  else
  { ⟨new bracket has subbrackets and extends to the left⟩ }
```

We come to the case that the new bracket extends to the right and that x is at least as large as r. Let *tB* be the rightmost sibling of *rB* whose left endpoint is less than or equal to x, cf. Figure 5.30. We determine *tB* by a finger search starting at *rB*. The right endpoint of *tB* must be smaller than x and x must be contained in the parent bracket of *tB* (which is also the parent bracket of *rB*); otherwise the sequence is not Jordan. The latter is guaranteed if x is smaller than the x-coordinate of the successor item of the right endpoint of *tB* (we skip *xl* if *side* == *lower*, as *xl* is not an endpoint of a bracket in the lower tree). Assume that both conditions hold. We add x after the right endpoint of *tB* to L, insert the new bracket (*xi*, x) before *rB*, delete the subsequence starting at *rB* and ending at *tB*, and make the subsequence the children sequence of the new bracket.

⟨*new bracket has subbrackets and extends to the right*⟩≡

```
Bracket query_bracket = new bracket(x);
seq_item x_pos = S.finger_locate_pred(rB->pos_among_sibs, query_bracket);
Bracket tB = S.key(x_pos);
```

```
list_item next = L.succ(tB->endpt[right]);
if ( next == x1_item && side == lower ) next = L.succ(next);
if ( x <= L[tB->endpt[right]]->x || x >= L[next]->x ) return false;
x_item = L_insert(x,tB->endpt[right],LEDA::after);
Bracket new_bracket = new bracket(xi_item,x_item,side);
new_bracket->pos_among_sibs =
  S.insert_at(rB->pos_among_sibs,new_bracket,0,LEDA::before);
S.delete_subsequence(rB->pos_among_sibs, x_pos, new_bracket->children);
```

If the new bracket has subbrackets and extends to the left we proceed symmetrically to the case above, i.e, we replace *pred* by *succ* and vice-versa, less than by greater than, ...

⟨*new bracket has subbrackets and extends to the left*⟩≡

```
Bracket query_bracket = new bracket(x);
seq_item x_pos = S.finger_locate_succ(lB->pos_among_sibs, query_bracket);
Bracket tB = S.key(x_pos);
list_item next = L.pred(tB->endpt[left]);
if ( next == x1_item && side == lower ) next = L.pred(next);
if ( x >= L[tB->endpt[left]]->x || x <= L[next]->x ) return false;
x_item = L_insert(x,tB->endpt[left],LEDA::before);
Bracket new_bracket = new bracket(x_item,xi_item,side);
new_bracket->pos_among_sibs =
  S.insert_at(tB->pos_among_sibs,new_bracket,0,LEDA::before);
S.delete_subsequence(x_pos,lB->pos_among_sibs, new_bracket->children);
```

Preparing the output is easy. After deleting the sentinels $-\infty$ and ∞ the output is available in L. We copy it to *Out*.

⟨*produce the output by copying L to Out*⟩≡

```
Out.clear();
L.pop(); L.Pop();
forall_items(it,L) Out.append(L[it]->x);
```

We described an algorithm to recognize and to sort Jordan sequences. The algorithm runs in linear time, see [HMRT85] for a proof[18]. As a sorting algorithm, *Jordan_sort* is not competitive with general purpose sorting algorithms, like quicksort and mergesort, despite its linear running time. We included the *Jordan_sort* program in the book as an example of how much LEDA simplifies the implementation of complex algorithms.

[18] The idea underlying the proof is as follows: in each iteration of Jordan sort a new bracket is constructed. This takes time $O(\log \min(k, m-k))$ where k is the number of subbrackets and $m-k$ is the number of siblings of the new bracket. One then proceeds as in the analysis of repeated splits on page 188.

6

Graphs and their Data Structures

The graph data type is one of the central data types in LEDA. In the first two sections we give a gentle introduction to it. Each of the remaining sections is devoted to a particular aspect of the graph data type: node and edge arrays, node and edge maps, node lists, node priority queues, node partitions, undirected graphs, graph generators, input and output, iteration statements, basic graph properties, parameterized graphs, and time and space complexity.

6.1 Getting Started

A *directed graph* $G = (V, E)$ consists of a set V of nodes or vertices and a set E of edges. Figure 6.1 shows a directed graph. Every edge e has a *source node source*(e) and a *target node target*(e). In our figures we draw an edge e as an arrow starting at *source*(e) and ending at *target*(e). We refer to the source and the target of an edge as the *endpoints* of the edge. An edge is said to be *incident* to its endpoints. We also say that an edge e is an edge *out of source*(e) and *into target*(e). The edges out of v are also called the edges *adjacent* to v. For an edge e with source node v and target node w we will write (v, w).

The declarations

```
graph G;
node  v, w;
edge  e, f;
```

declare variables G, v, w, e and f of type *graph*, *node*, and *edge*, respectively. The values of these variables are graphs, nodes, and edges, respectively; G is initialized to the empty graph, i.e., a graph with no node and no edge, and the initial values of v, w, e, and f are unspecified (since nodes and edges are pointer types). The special value *nil* is not a node or

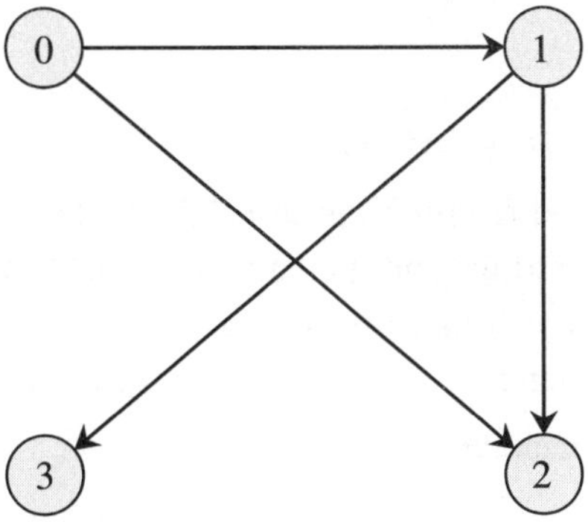

Figure 6.1 A directed graph.

edge of any graph and can be used to initialize nodes and edges with a definite value, as, for example, in

```
node v = nil;
```

Graph algorithms frequently need to iterate over the nodes and edges of a graph and the edges incident to a particular node. The iteration statement

```
forall_nodes(v,G){ }
```

iterates over all nodes of a graph, i.e., the nodes of G are successively assigned to v and the body of the loop is executed once for each value of v. Similarly,

```
forall_edges(e,G){ }
```

iterates over all edges e of G. There are three ways to iterate over the edges incident to a node v. The iteration statements

```
forall_out_edges(e,v){ }
forall_adj_edges(e,v){ }
```

iterate over all edges e out of v, i.e, all edges whose source node is equal to v,

```
forall_in_edges(e,v) { }
```

iterates over all edges e into v, i.e., over all edges whose target node is equal to v, and

```
forall_inout_edges(e,v){ }
```

iterates over all edges e into and out of v. So

```
int s = 0;
forall_edges(e,G) s++;
```

computes the number of edges of G. This number is also available as *G.number_of_edges*( ).

In many situations it is useful to associate additional information with the nodes and edges of a graph. LEDA offers several ways to do so. We briefly discuss *node arrays*, *edge arrays*, and *parameterized graphs*. We will give more details and also discuss node and edge maps later.

The declarations

```
node_array<string> name(G);
edge_array<int>    length(G,1);
```

introduce arrays *name* and *length* indexed by the nodes and edges of the G, respectively. The entries of *name* are strings and the entries of *length* are integers. All entries of *name* are initialized to the empty string (= the default value of *string*) and all entries of *length* are initialized to 1. If v is a vertex of G and e is an edge of G we may now write

```
name[v]   = "Saarbruecken";
length[e] = 5;
```

The following piece of code numbers the nodes of a graph with the integers 0 to $n - 1$, where n is the number of nodes of G. As is customary in the literature on graph algorithms we will usually write n for the number of nodes and m for the number of edges.

```
node_array<int> number(G);
int count = 0;
forall_nodes(v,G) number[v] = count++;
```

A second method to associate information with nodes and edges is to use so-called *parameterized graphs*. The declaration

```
GRAPH<string,int> H;
```

declares H as a parameterized graph where a string variable is associated with every vertex of H and an integer is associated with every edge of H. We may now write

```
H[v] = "Saarbruecken";
H[e] = 5;
```

to associate the string "*Saarbruecken*" with v and the integer 5 with e. Of course, both operations are only legal if v and e actually denote a vertex and edge of H, respectively.

There is an important difference between the two methods of associating information with nodes and edges. Node and edge arrays work only for static graphs, i.e., when a new node or edge is added to a graph it will not have a corresponding entry in the node and edge arrays of the graph (in Section 6.3 this condition will be relaxed somewhat). Parameterized graphs, on the contrary, are fully dynamic. Information can be associated with new edges and nodes without any restriction. In this sense, parameterized graphs are more flexible. Also, the access to the information stored in the nodes and edges of a parameterized graph is somewhat more efficient than the access to the information stored in a node or edge array. On the other hand, the great strength of node and edge arrays is that an arbitrary number of them can be defined for a graph.

It's time to learn how to build non-trivial graphs. A graph can be altered by adding and deleting nodes and edges. For example,

```
graph G;
G.new_node();
G.new_node();
node v;
forall_nodes(v,G) cout << G.outdeg(v);
```

makes G a graph with two nodes and no edge and then outputs the outdegree[1] of all nodes, i.e., outputs the number 0 twice. In order to add an edge we need to specify its source and its target. For example,

```
node w = G.first_node();
G.new_edge(w,G.succ_node(w));
```

will add an edge whose source and target are the first and second node of G respectively; note that LEDA internally orders the nodes of a graph in the order in which they were added to G. *G.first_node*( ) returns the first node in this ordering and *G.succ_node*(w) returns the node added immediately after w. There is a more interesting way to add edges. The operation *G.new_node*( ) does not only add a new node to the graph G but also returns the new node. We can remember the new node in a variable of type *node*. So

```
graph G;
node v0 = G.new_node();
node v1 = G.new_node();
node v2 = G.new_node();
node v3 = G.new_node();
G.new_edge(v0,v1); G.new_edge(v0,v2);
G.new_edge(v1,v2); G.new_edge(v1,v3);
```

creates the graph of Figure 6.1.

Let us do something more ambitious next. Suppose that we created a graph G and that we want to make an isomorphic copy H of it. Moreover, we want every node and edge of H to know its original in G. Here is an elegant way to do this. We use parameterized graphs, node arrays and edge arrays.

```
void CopyGraph(GRAPH<node,edge>& H, const graph& G)
{ H.clear();  // reset H to the empty graph
  node_array<node> copy_in_H(G);
  node v;
  forall_nodes(v,G) copy_in_H[v] = H.new_node(v);
  edge e;
  forall_edges(e,G)
    H.new_edge(copy_in_H[source(e)],copy_in_H[target(e)],e);
}
```

We define H as a parameterized graph where a node can be associated with each node and an edge can be associated with each edge. We also define a node array *copy_in_H* for G that allows us to associate a node with every node of G. We then iterate over the nodes of G. For every node v of G the operation *H.new_node*(v) adds a new node to H and associates v with the new node. Note that the *new_node* operation for a parameterized graph has an argument, namely the information that is to be associated with the new node. The operation *H.new_node*(v) also returns the new node. We remember it in *copy_in_H*[v]. The overall effect of the *forall_nodes*-loop is to give H as many nodes as G and to establish bidirectional

[1] The outdegree of a vertex v is the number of edges e with *source*(e) $= v$.

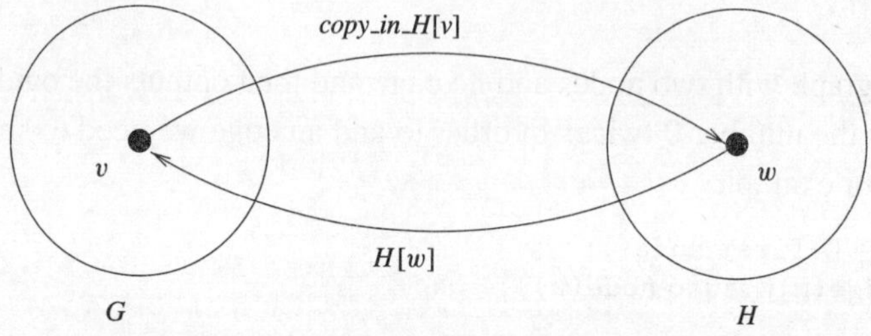

Figure 6.2 A graph G and an isomorphic copy H of it. Each node v of G knows its partner in H through *copy_in_H*$[v]$ and each node w of H knows its partner in G through $H[w]$.

links between the nodes of G and H: in particular, we have $H[copy\_in\_H[v]] = v$ for all nodes v of G and $copy\_in\_H[H[w]] = w$ for all nodes w of H, see Figure 6.2. It is now easy to add the edges. We iterate over the edges of G. For every edge e we add an edge to H that runs from $copy\_in\_H[source(e)]$ to $copy\_in\_H[target(e)]$ and also make e the information associated with the new edge. Observe that $H.new\_edge(x, y, inf)$ adds an edge from node x to node y and associates the information *inf* with it.

Exercise for 6.1

1 Write a program that makes a copy of a graph G with all edges reversed, i.e., for every edge $e = (v, w)$ in G there should be an edge from the copy of w to the copy of v in H.

6.2 A First Example of a Graph Algorithm: Topological Ordering

A graph is called *acyclic* if it contains no cycle. A cycle is a path that closes on itself, i.e., a sequence $e_0, e_1, \ldots, e_k$ of edges such that $target(e_i) = source(e_{i+1 \bmod k+1})$ for all i, $0 \le i \le k$. The graph in Figure 6.1 is acyclic. The nodes of an acyclic graph can be numbered such that all edges run from smaller to higher numbered nodes. The function

```
bool TOPSORT(const graph& G, node_array<int>& ord);
```

returns true if G is acyclic and false if G contains a cycle. In the former case it also returns a topological ordering of the nodes of G in *ord*.

The procedure works by repeatedly removing nodes of indegree zero and numbering the nodes in the order of their removal.

In the example of Figure 6.1 we first number node 0. Removing node 0 makes the indegree of node 1 zero and hence this node is numbered next. Removal of node 1 makes the indegree of node 2 zero, ...

For reasons of efficiency we keep track of the current indegree of all nodes and also maintain the list of nodes whose current indegree is zero.

```
#include <LEDA/graph.h>
#include <LEDA/queue.h>
bool procedure TOPSORT(const graph& G,node_array<int>& ord)
{ ⟨initialization⟩
  ⟨removing nodes of indegree zero⟩
}
```

In the initialization phase we determine the indegree of all nodes and initialize a queue of nodes of indegree zero.

⟨*initialization*⟩≡

```
node_array<int> INDEG(G);
queue<node>     ZEROINDEG;
node v,w;
forall_nodes(v,G)
  if ( (INDEG[v] = G.indeg(v)) == 0 ) ZEROINDEG.append(v);
```

In the main phase of the algorithm we consider the nodes of indegree zero in turn. When a vertex v is considered we number it and we decrease the indegrees of all adjacent nodes by one. Nodes whose indegree becomes zero are added to the rear of *ZEROINDEG*.

⟨*removing nodes of indegree zero*⟩≡

```
int count = 0;
node_array<int> node_ord(G);
while (!ZEROINDEG.empty())
{
  v = ZEROINDEG.pop();
  node_ord[v] = ++count;
  forall_out_edges(e,v)
  { node w = G.target(e);
    if ( --INDEG[w] == 0 ) ZEROINDEG.append(w);
  }
}
return (count == G.number_of_nodes());
```

TOPSORT considers every edge of G only once and hence has running time $O(n + m)$. In the section on depth-first search (see Section 7.3) we will see an alternative program for topological sorting.

6.3 Node and Edge Arrays and Matrices

Node and edge arrays and matrices are the main means of associating information with the nodes and edges of a graph. The declarations

```
node_array<E> A(G);
node_array<E> B(G,E x);
```

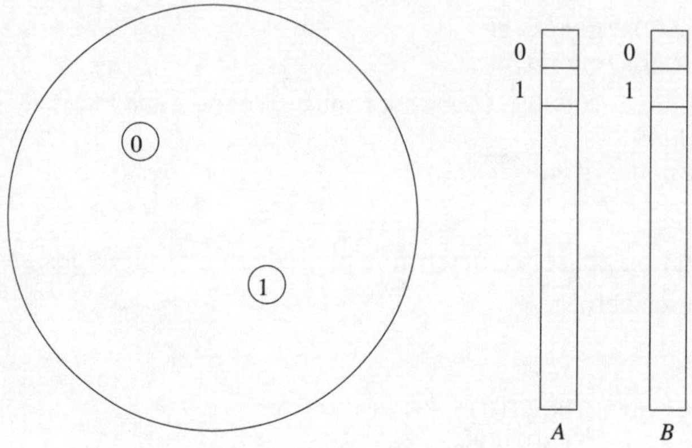

Figure 6.3 The realization of node arrays: Node arrays A and B are realized by regular arrays. The nodes of a graph are numbered and the node numbers are used as the indices into the arrays.

declare node arrays A and B for the nodes of G, respectively. The elements of A are initialized with the default value of E and the elements of B are initialized to x. Edge arrays are declared in a similar way. So

```
node_array<bool> visited(G,false);
```

declares a node array *visited* and initializes all its entries to false. The cost of declaring a node array for G is proportional to the number of nodes of G and the cost of declaring an edge array is proportional to the number of edges.

Node and edge arrays are a very flexible way of associating information with the nodes and edges of a graph: any number of node or edge arrays can be defined for a graph and they can be defined at any time during execution.

Node and edge arrays are implemented as follows. The nodes and edges of a graph are numbered in the order of their construction, starting at zero. We call the number of a node or edge its *index*. The index of a node v or edge e is available as $index(v)$ and $index(e)$, respectively. Node and edge arrays are realized by standard arrays. The node and edge indices are used to index into the arrays, see Figure 6.3.

The access to an entry $A[v]$ of a node array A (similarly, edge arrays) requires two accesses to memory, first the structure representing the node v is accessed to determine $index(v)$ and second the entry $A[index(v)]$ is accessed.

When the number of node and edge arrays that are needed for a graph is known, the following alternative is possible. Assume that *n_slots* node arrays and *e_slots* edge arrays are needed. The constructor

```
graph G(int n_slots, int e_slots);
```

constructs a graph where the structures representing nodes have room for the entries of *n_slots* node arrays and the structures representing edges have room for the entries of *e_slots* edge arrays. In order to use one of the slots for a particular array, one writes:

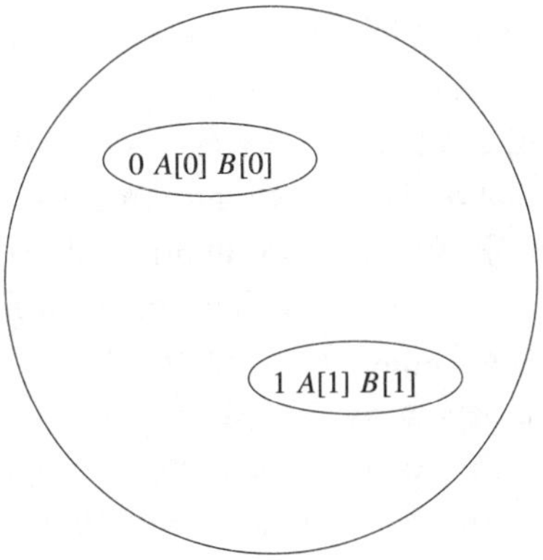

Figure 6.4 The alternative realization of node arrays: In a graph G constructed by *graph* $G(2, 0)$, every node has room for the entries of two node arrays.

```
node_array<E> A;
A.use_node_data(G, E x);
```

This will reserve one of the slots in the node structures for A and initialize all entries of the array to x. If no slot is available, the node array is realized by a standard array. Figure 6.4 illustrates the alternative. The alternative realization of node and edge arrays is frequently, but not always, faster (see the next section), as only one access to memory is needed to access an entry of a node or edge array, but it is also less convenient, as the number of node and edge arrays that can use the alternative is fixed at the time of the construction of the graph.

We recommend that you experiment with the alternative design during the optimization phase of program development.

Node and edge arrays, as discussed so far, are primarily useful for static graphs.

```
node_array<int> dist(G);
node v = G.new_node();
dist[v] = 5;
```

is illegal and produces the error message "*node_array*[v] not defined for v". We next discuss node and edge arrays for dynamic graphs. We have to admit, though, that we hardly use node and edge arrays for dynamic graphs ourselves. We prefer node and edge maps and parameterized graphs.

```
node_array<E> A(graph G, int n, E x);
```

declares A as a node array of size n for the nodes of G and initializes all entries of A to x; x must be specified even if it is the default value of E.

The constructor requires that $n \geq |V|$. The array A has room for $n - |V|$ additional nodes, i.e., for the nodes created by the next $n - |V|$ calls of *G.new_node*( ). In this way one can have the convenience and efficiency of node arrays also for dynamic graphs. Deletion of nodes is no problem for node arrays.

The following doubling and halving strategy is useful for node and edge arrays on dynamic graphs. Suppose that n_0 is the current number of nodes of G and that we want to create a node array A for G. We make A an array of size $2n_0$ and initialize two counters *ins_count* and *del_count* to zero. We increment *ins_count* for every *G.new_node*( ) operation and *del_count* for every *G.del_node*( ) operation. When *ins_count* reaches n_0 or *del_node* reaches $n_0/2$ we allocate a new array B of size $2(n_0 + \mathit{ins\_count} - \mathit{del\_count})$ and move the contents of A to B. This scheme ensures that node arrays are always at least 25% utilized and that the overhead for moving information around increases the running time by only a constant factor (since the cost of moving is $O(n_0)$ and since there are $\Omega(n_0)$ *new_node* and *del_node* operations between reorganizations of the node array).

We next turn to node matrices. The definition

```
node_matrix<int> M(G,0);
```

defines M as a two-dimensional matrix indexed by pairs of nodes of G and initializes all entries of G to zero. This takes time $O(n^2)$, where n is the number of nodes of G. The space requirement for a node matrix is quadratic in the number of nodes. So they should only be used for small graphs.

```
M(v,w) = 1;
```

sets the entry for pair (v, w) to one.

A node matrix can also be viewed as a node array of node arrays, i.e., the type *node_matrix<E>* is equivalent to the type *node_array<node_array<E> >*. This view is reflected in the operation

```
M[v];
```

which returns a node array.

We give an example of the use of node matrices. The following three-liner checks whether a graph is bidirected (also called symmetric), i.e., whether for every edge $e = (v, w)$ the reversed edge (w, v) is also present.

```
node_matrix<bool> M(G,false);
forall_edges(e,G) M(G.source(e),G.target(e)) = true;
forall_edges(e,G)
{ if ( !M(G.target(e),G.source(e)) ) error_handler(1,"not bidirected"); }
```

The program above has running time $\Theta(n^2 + m)$, $\Theta(n^2)$ for initializing M and $\Theta(m)$ for iterating over all edges twice. As we will see later there is also an $O(m)$ algorithm for the same task. It is available as

```
bool Is_Bidirected(G);
```

6.4 Node and Edge Maps

Nodes and edge maps are an alternative to node arrays. The declarations

```
node_map<E> A(G);
node_map<E> B(G,E x);
```

declare node maps A and B for the nodes of G, respectively. The elements of A are initialized with the default value of E and the elements of B are initialized to x. Edge maps are declared in a similar way. So

```
node_map<bool> visited(G,false);
```

declares a node map *visited* and initializes all its entries to false.

What is the difference between node and edge arrays and node and edge maps? Node and edge maps use hashing (see Section 5.1.2). The declaration of a node or edge map has constant cost (compare this to the linear cost for node and edge arrays) and the access to an entry of a node or edge map has constant expected cost.

Table 6.1 compares three ways of associating information with the nodes of a graph, the standard version of node arrays, the version of node arrays that makes use of a data slot in the node, and node maps. The table was produced by the program below. We give the complete program because the numbers in the table are somewhat surprising. We create a graph with n nodes and no edge and iterate R times over the nodes of the graph. In each iteration we access the information associated with the node. We iterate over the nodes once in their natural order and once in random order.

⟨*node_arrays_versus_node_maps*⟩≡

```
main(){
⟨node arrays versus node maps: read n and R⟩
graph G; graph G1(1,0); node v; int j;
random_graph(G,n,0); random_graph(G1,n,0);
float T = used_time();
float TA, TB, TM, TAP, TBP, TMP;
{ node_array<int> A(G,0);
  for ( j = 0; j < R; j++ )
    forall_nodes(v,G) A[v]++;
  TA = used_time(T);
}

{ node_array<int> A;
  A.use_node_data(G1,0);
  for ( j = 0; j < R; j++ )
    forall_nodes(v,G1) A[v]++;
  TB = used_time(T);
}
{ node_map<int> A(G,0);
  for ( j = 0; j < R; j++ )
    forall_nodes(v,G) A[v]++;
  TM = used_time(T);
```

Linear scan			Random scan		
array	node data	map	array	node data	map
3.25	4.39	3.48	8.96	5.9	9.56

Table 6.1 Node arrays versus node maps: The table shows the output of the program node_arrays_versus_node_maps.c. We used a node array (columns one and four), a node data slot (columns two and five), and a node map (columns three and six). We used a graph with one million nodes and $R = 10$. The nodes were scanned in linear order and in random order. The node_array_versus_node_maps demo allows you to perform your own experiments.

```
}
array<node> perm(n); array<node> perm1(n);
int i = 0;
forall_nodes(v,G) perm[i++] = v;
i = 0;
forall_nodes(v,G1) perm1[i++] = v;
perm.permute(); perm1.permute();
used_time(T);
{ node_array<int> A(G,0);
  for ( j = 0; j < R; j++ )
    for(i = 0; i < n; i++) A[perm[i]]++;
  TAP = used_time(T);
}
{ node_array<int> A;
  A.use_node_data(G1,0);
  for ( j = 0; j < R; j++ )
    for(i = 0; i < n; i++) A[perm1[i]]++;
  TBP = used_time(T);
}
{ node_map<int> A(G,0);
  for ( j = 0; j < R; j++ )
    for(i = 0; i < n; i++) A[perm[i]]++;
  TMP = used_time(T);
}
⟨node arrays versus node maps: report running times⟩
}
```

In the random scan over the nodes, node data slots outperform node arrays which in turn outperform node maps. This was to be expected, since node data slots avoid one level of indirection, and since maps have the overhead of hashing. Maps are only slightly slower than arrays due to our very efficient realization of maps, see Section 5.1.2. In the linear scan the situation is different. Node data slots are the slowest and maps are even closer to arrays. We believe that this is due to caching. We compare node arrays and node data slots. When node data slots are used, the node structures are larger, and hence fewer of them fit into a cache line. Node arrays use the cache more effectively in the linear scan because they can use one cache line for node structures and one cache line for the array itself and only the cache lines for the array itself are written. Thus the number of write-faults reduces. A similar explanation applies to node maps. Since it requires knowledge of the implementation of maps, we do not give it here.

We recommend to use node and edge maps in situations where a sparse map on nodes or edges, respectively, has to be maintained. If more than half of the entries are actually used, it is better to use node arrays.

We next turn to two-dimensional node maps. The definition

```
node_map2<int> M(G,0);
```

defines M as a two-dimensional map indexed by pairs of nodes of G and initializes all entries of G to zero. This takes constant time.

```
M(v,w) = 1;
```

sets the entry for pair (v, w) to one. The space requirement for a two-dimensional node map is proportional to the number of entries used.

We give an example for the use of two-dimensional node maps. The following three-liner checks whether a graph is bidirected (also called symmetric), i.e., whether for every edge $e = (v, w)$ the reversed edge (w, v) is also present.

```
node_map2<bool> M(G,false);
forall_edges(e,G) M(G.source(e),G.target(e)) = true;
forall_edges(e,G)
{ if ( !M(G.target(e),G.source(e)) ) error_handler(1,"not symmetric"); }
```

The program above has running time $O(m)$, $O(1)$ for initializing M and $O(m)$ for iterating over all edges twice. The space requirement is $O(m)$. Observe, that this is much better than what we obtained with node arrays in the preceding section if $m \ll n^2$.

Exercises for 6.4

1 Write a program that checks whether a graph is symmetric and, if so, computes an edge array *reversal* that stores for each edge a reversal of the edge. The source of *reversal*[e] must be equal to the target of e and vice versa.

2 Extend the program of the previous item so that it can also handle parallel edges. We want *reversal*[*reversal*[e]] $= e$ for all edges e.

3 Extend the program of the previous item so that it can also handle self-loops. We want *reversal*[e] $\neq e$ for all e.

6.5 Node Lists

A node list is a combination of a doubly linked list of nodes and a node map which gives, for each node, its position in the list, see Figure 6.5. *A node can be contained in a node list*

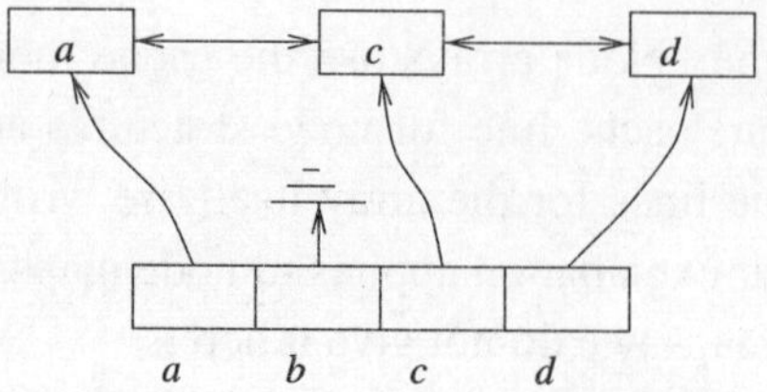

Figure 6.5 Node lists: A node list for a graph with four nodes a, b, c, and d. The node list contains the nodes a, c, and d in this order. The top part of the figure shows a doubly linked list and the lower part of the figure indicates a node map. The node map maps each node contained in the node list to the list item containing the node.
In a *snode_list* a singly linked list is used instead of a doubly linked list.

at most once. It can be contained in several node lists, but in each particular node list it can appear only once.

```
node_list L(G);
```

creates a node list for the graph G and initializes it with the empty list. Node lists offer all the usual list operations, e.g., *append*, *push*, *pop*, *insert*, *head*, *tail*, *pred*, *succ*, *cyclic_pred*, *cyclic_succ*, *empty*, and the possibility to iterate over the nodes in the list. In addition, node lists offer constant time member ship test.

The related data type *snode_list* is the combination of a singly linked list and a node map. It offers all the operations of singly linked lists plus constant time member ship test.

A prime example for the use of node lists is breadth-first search. The goal is to explore the nodes of a graph starting from some source node s in order of increasing distance from s. The distance of a node v from s is the smallest number of edges in a path from s to v.

The following program realizes breadth-first search. We collect the nodes of G in a *snode_list* Q in the order in which they are reached. We always explore the edges out of the first unprocessed node in Q. Whenever a node is encountered that has not been reached before (= is not in Q) we add it to the rear of Q.

```
snode_list Q;
Q.append(s);
node v = Q.head();
while ( v != nil )
{ edge e;
  forall_adj_edges(e,v)
  { node w = G.target(e);
    if ( !Q.member(w) ) Q.append(w);
  }
  v = Q.succ(v);
}
```

We will discuss breadth-first search in more detail in the chapter on graph algorithms.

Exercises for 6.5

1 Give an implementation of *snode_list* that uses a *node_map<node> succ_node*, two nodes *first_node* and *last_node*, and an integer *size*.

2 Give an implementation of *node_list* that uses two maps from nodes to nodes, namely, *succ_node* and *pred_node*, two nodes *first_node* and *last_node*, and an integer *size*.

6.6 Node Priority Queues and Shortest Paths

The declaration

```
node_pq<P> Q(G);
```

declares a *node priority queue* Q with priority type P for G and initializes it to the empty queue. A node priority queue with priority type P is a partial function from the nodes of G to the set P. The set P must be linearly ordered. If $Q(v)$ is defined we call it the priority of node v. We use dom Q to denote the set of nodes for which $Q(v)$ is defined, the *domain* of Q. Node priority queues allow us to manipulate the function Q by insertion, deletion, and (restricted) modification of values, and they allow us to select a node with smallest priority.

We next discuss some of the operations available on node priority queues in more detail, then show how to use them in an implementation of Dijkstra's algorithm for the single-source shortest-path problem, and finally show how node priority queues are implemented in terms of node arrays and general priority queues.

We come to the operations available on node priority queues:

```
node Q.find_min();
```

returns a node $v \in$ dom Q with minimal associated priority (*nil* if Q is empty),

```
bool Q.member(node v);
```

checks whether node v is contained in the queue Q, i.e., if $v \in$ dom Q,

```
void Q.insert(node v, P p);
```

adds the node v with associated priority p to the queue Q (the effect of this operation is unspecified if v is already contained in Q) and

```
void Q.decrease_p(node v, P p);
```

makes p the new priority of node v (the effect of this operation is unspecified if v is not contained in Q or p is larger than the old priority associated with v).

The implementation of node priority queues is based on priority queues and node arrays. The operations *find_min* and *decrease_p* take constant time, all other operations take time $O(\log s)$ where s is the current size of Q. The space requirement is proportional to the number of nodes of G. We give the details of the implementation at the end of the section.

We illustrate the use of node priority queues on Dijkstra's single-source shortest-path algorithm. Let G be a graph, let *edge_array<NT> cost* be a non-negative cost function[2] on the edges of G, and let s be a node of G. For any node v of G let $\mu(v)$ be the cost of a shortest path from s to v, where the cost (or length) of a path is the sum of the costs of its edges; if there is no path from s to v then $\mu(v) = \infty$. We use $cost(p)$ to denote the cost of a path p.

The task is to compute μ in a *node_array<NT> dist* and a *node_array<edge> pred* which contains for each node $v \neq s$ the last edge of a shortest path from s to v. We need to be more precise. Observe that not every number type has a representation for ∞, and hence the previous sentence does not specify how the algorithm should report the fact that $\mu(v) = \infty$ for a node v. We refine the specification to the following:

- If v is reachable from s then $dist[v] = \mu(v)$.
- $pred[s] = nil$.
- If $v \neq s$ and v is reachable from s then $pred[v]$ is the last edge of a shortest path from s to v.
- If $v \neq s$ and v is not reachable from s then $pred[v] = nil$.

Dijkstra's algorithm [Dij59] "simulates" the following physical process. Imagine the graph as a network of uni-directional wires, imagine that current is injected into the network at node s and time zero, and imagine that current spreads with unit speed. Thus current requires $cost[e]$ time units to spread across an edge e. In this model, the current will reach every node v at time $\mu(v)$.

In order to carry out the simulation, we turn the nodes of the network into active components. As soon as current reaches a node u, say at time $t = \mu(u)$, the node sends a message to each node v with $e = (u, v) \in E$ with the content:

You will receive current through edge e at time $t + cost[e]$.

Every node v keeps track of all the messages sent to it. More precisely, a node keeps track of the earliest time at which current will reach it, i.e., whenever a node v receives a message, it checks whether the message promises it an earlier delivery time and, if so, the node updates its time estimate. In our implementation we keep the current time estimate of node v in $dist[v]$ and we keep the edge through which the node will receive current at time $dist[v]$ in $pred[v]$. If v has received no message yet we have $pred[v] = nil$.

The simulation is driven by a global clock which we call wall time. At any time t there will be a set S of nodes which have already been reached by the current and which have accordingly sent messages to their neighbors, and there will be the set $V \setminus S$ of the remaining nodes which have not been reached yet by the current wave. Each node in $V \setminus S$ has received zero or more messages and keeps track of its earliest delivery time. Clearly, the node which

[2] *NT* denotes an arbitrary number type.

is reached next by the current is the node $u \in V \setminus S$ with the smallest delivery time, i.e., the smallest value *dist*$[u]$. It is the next node to send out messages.

In an implementation the crucial question is how to find the node v with minimal *dist*-value among the nodes in $V \setminus S$. The data type node priority queue is ideally suited for that purpose. Simply have a *node_pq*<*NT*> *P* with

$$\text{dom}\, P = \{v\ ;\ v \in V \setminus S \text{ and } pred[v] \neq nil\}$$

and $P(v) = dist[v]$ for any $v \in \text{dom}\, P$, i.e., P contains all nodes outside S which have received at least one message and records, for each such node, the earliest delivery time to the node. Then *P.del_min*( ) returns the desired node and deletes it from P. The complete program follows.

⟨*dijkstra.t*⟩+≡

```
template <class NT>
void DIJKSTRA_T(const graph& G, node s, const edge_array<NT>& cost,
                node_array<NT>& dist, node_array<edge>& pred)
{
  node_pq<NT>  PQ(G);
  node v; edge e;
  dist[s] = 0;
  PQ.insert(s,0);
  forall_nodes(v,G) pred[v] = nil;
  while (!PQ.empty())
  { node u = PQ.del_min(); // add u to S
    NT du = dist[u];
    forall_adj_edges(e,u)
    { v = G.opposite(u,e); // makes it work for ugraphs
      NT c = du + cost[e];
      if (pred[v] == nil && v != s )
        PQ.insert(v,c); // first message to v
      else if (c < dist[v]) PQ.decrease_p(v,c); // better path
           else continue;
      dist[v] = c;
      pred[v] = e;
    }
  }
}
```

The program runs in time $O(m + n \log n)$ since every node is deleted from the queue at most once and *del_min* has cost $O(\log n)$ and since every other operation is executed at most $O(n + m)$ times and has constant amortized cost.

In the remainder of this section we show how to implement node priority queues in terms of node arrays and priority queues. The construction is very simple. We realize a *node_prio*<*P*> *NPQ* for a graph *G* by a *p_queue*<*P*, *node*> *PQ* and a *node_array*<*pq_item*> *item_of* such that:

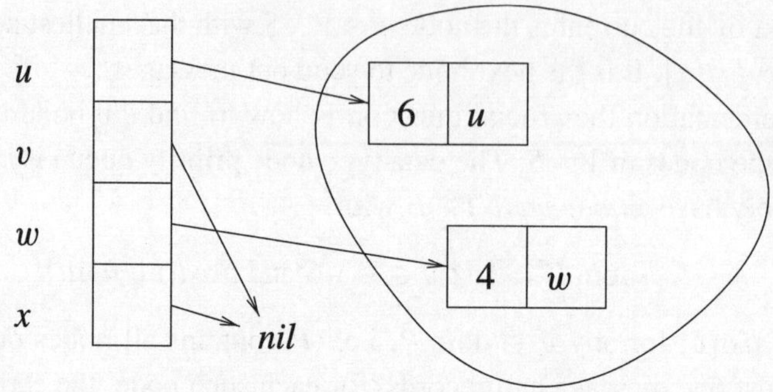

Figure 6.6 A node priority queue for a graph with four nodes u, v, w, and x. The priority of u is 6, the priority of w is 4, and v and x have no entry in the queue.

- if a node v is stored in *NPQ* with priority p then there is an item $pit = \langle p, v\rangle$ in *PQ* and $item\_of[v] = pit$.
- if a node v is not contained in *NPQ* then $item\_of[v] = nil$.

Figure 6.6 illustrates these invariants and node_pq.c shows the complete implementation.

⟨*node_pq.c*⟩≡

```
#include <LEDA/graph.h>
#include <LEDA/p_queue.h>

template <class P> class node_pq {
private:
  p_queue<P,node> PQ;
  node_array<pq_item> item_of;
public:
  node_pq(const graph& G): item_of(G,nil) { }
  ~node_pq() { }
  void insert(node v, P p)      { item_of[v]= PQ.insert(p,v); }
  P prio(node v)                { return PQ.prio(item_of[v]); }
  void decrease_p(node v, P p) { PQ.decrease_p(item_of[v],p); }
  void del(node v)
  { PQ.del_item(item_of[v]);
    item_of[v] = nil;
  }
  node find_min()               { return PQ.inf(PQ.find_min()); }
  node del_min()
  { node v= PQ.inf(PQ.find_min());
    PQ.del_min();
    item_of[v] = nil;
    return v;
  }
  ⟨node_pq::other operations⟩
};
```

Only a few words are required to explain this code. We construct a *node_pq*<*P*> for a graph G by constructing an empty priority queue PQ and a node array *item_of* for G and by initializing all entries of *item_of* to *nil*. The former is done by the default constructor of priority queues and requires no code and the latter is achieved by the constructor call *item_of*(G, nil). In order to insert a pair (v, p) we insert the pair (p, v) into PQ and store the item that is returned in *item_of*$[v]$. In order to look up the priority of a node v we return *PQ.prio*(*item_of*$[v]$), ...

Exercises for 6.6

1 Modify Dijkstra's algorithm such that it does not start with a single source node s but with a set L of sources. It is supposed to compute $\mu(L, v)$ for all nodes v where $\mu(L, v)$ is the minimum distance from a node in L to v.

2 (Single sink shortest path). Let s and t be distinct nodes in a directed graph with non-negative edge weights. The goal is to compute a shortest path from s to t. Assume that there is heuristic information available which gives for any node v a *lower bound* $lb(v)$ for the length of a shortest path from v to t. Modify Dijkstra's algorithm such that $dist(v) + lb(v)$ is used as the priority of node v.

3 Use the algorithm of the previous item to compute shortest paths in graphs embedded into the plane, e.g., Delaunay diagrams (see Section 10.4). Define the cost of an edge as the Euclidean distance between its endpoints and let $lb(v)$ for any node v be the Euclidean distance between v and t. Which improvement in running time results from the use of heuristic information?

4 Implement operations *member*, *clear*, *size*, and *empty* of *node_pq*.

6.7 Undirected Graphs

In an *undirected* graph the edges have no direction. Mathematically speaking, an edge in an undirected graph is an unordered pair $\{v, w\}$ of nodes and an edge in a directed graph is an ordered pair (v, w) of nodes. As for directed graphs, we call v and w the endpoints of the edge. The endpoints of an edge in an undirected graph must be distinct (since an edge is a set of vertices of cardinality two).

6.7.1 *Viewing Directed Graphs as Undirected Graphs*

Every directed graph without self-loops can be viewed as an undirected graph.

For an edge e and an endpoint v of e

```
G.opposite(v,e)
```

returns the other endpoint of e, i.e., returns $target(e)$ if $v = source(e)$ and returns $source(e)$ otherwise.

The iteration statement

```
forall_inout_edges(e,v){ }
```

iterates over all edges e having v as one of their endpoints. It iterates first over all edges out of v and then over all edges into v.

The iteration *forall_inout_edges* and the function *opposite* can also be applied to graphs with self-loops. Observe, however, that the iteration statement will consider a self-loop $e = (v, v)$ twice, once as an edge, whose source is equal to v, and once as an edge, whose target is equal to v.

It is our experience that the two statements above suffice to deal with undirected graphs. We can foresee one situation where they do not suffice: if one wants to iterate over the edges incident to v in some mixed order, first some edges out of v, then some edges into v, then again some edges out of v, ... We will see in Section 6.11 that the order of the out-edges and the order of the in-edges can be modified. Nevertheless, out-edges always come before in-edges in the *forall_inout_edges* iteration. If a more flexible scanning order is required, the following operation is useful:

```
G.make_undirected();
```

appends for every node v the list of in-edges of v to the list of out-edges of v and removes all self-loops. All edges incident to any node are now in a single list and hence can be rearranged freely using the operations to be described in Section 6.11.

```
G.make_directed();
```

partially reverses the operation above. It moves, for every node v, all edges e with $target(e) = v$ from the list of out-edges of v to the list of in-edges of v. Note that the operation does not reinsert self-loops.

6.7.2 *The Data Type ugraph*

We also have a data type *ugraph*. We use it very rarely. Ugraphs offer the same operations as graphs but the *new_edge* operation is interpreted differently. For example,

```
ugraph G;
node v = G.new_node(); node w = G.new_node();
edge e = G.new_edge(v,w);
```

creates an undirected graph with two nodes and one edge. The edge e is inserted into the out-lists of v and w (which in this context is better called the list of adjacent edges). Thus

```
e == G.first_adj_edge(v) && e == G.first_adj_edge(w)
```

evaluates to true. As for directed graphs the functions *source*( ) and *target*( ) yield the two endpoints of an edge, so *G.source*(e) returns v and *G.target*(e) returns w. Note that the role of the two nodes v and w in the definition of the edge e is not symmetric: v is made the source of e because it is mentioned first, and w is made the target of e because it is mentioned second.

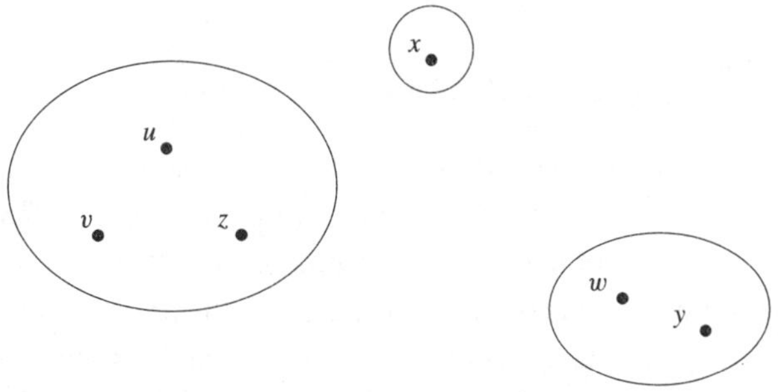

Figure 6.7 A node partition for a graph with six nodes u, v, w, x, y, and z: u, v, and z are in the same block, w and y are in the same block and x is a block of its own.

6.8 Node Partitions and Minimum Spanning Trees

We discuss node partitions. We first discuss their functionality and then illustrate their use in Kruskal's minimum spanning tree algorithm.

A *node partition* is a partition of the nodes of a graph G, i.e., a family of pairwise disjoint sets (called *blocks*) whose union is the set of nodes of G, see Figure 6.7 for an example.

```
node_partition P(G);
```

declares P as a node partition for G and initializes it to the finest partition of G, i.e., every node v of G forms its own block $\{v\}$. Node partitions offer the following operations:

```
bool P.same_block(node v,node w);
```

returns *true* iff v and w belong to the same block of P,

```
void P.union_blocks(node v,node w);
```

combines the blocks containing v and w. Each block has a *canonical representative*. The canonical representative of a block is some element in the block; it is not specified which. The operations

```
node P.find(node v);
node P(node v);
```

return the canonical representative of the block containing v. So, in the example of Figure 6.7, $P.find(x)$ returns x (for the singleton block there is no choice of canonical element) and $P.find(u)$ and $P.find(v)$ return the same element of block $\{u, v, z\}$ (it is not specified which). When the functional notation $P(v)$ is used for the find operation it is convenient to name the partition after the name for the canonical element; for example, in the matching algorithm of Section 7.7 we will call the node partition *base*. After a union operation the data structure chooses the canonical representative of the block formed (among the elements of the block). We can make v the canonical representative of the block containing v by

```
void P.make_rep(node v);
```

The operation

```
void P.split(list<node> T);
```

splits the blocks containing the nodes in T into singleton blocks. The operation requires that T is a union of blocks of P. So, in the example of Figure 6.7 we can apply split with the argument $\{u, v, z, x\}$ but not with the argument $\{u, y\}$ and not with the argument $\{u, v\}$.

The implementation of node partitions is based on the data types *partition* and *node_array*. A sequence of m operations (except for split) on a node partition of n nodes takes time $O((n+m)\alpha(n))$ where α is the functional inverse of the Ackermann function. The function α is extremely slowly growing, in particular $\alpha(n) \leq 5$ for $n \leq 10^{100}$. The running time of node partitions is therefore linear for all practical purposes. A split takes time proportional to the size of T.

We turn to Kruskal's minimum spanning tree algorithm.

Let G be a graph whose edges have an associated cost of some number type and let *cmp* be a function that compares edges according to their cost, i.e., *cmp(e1, e2)* returns -1, 0, and $+1$, respectively, if the cost of *e1* is smaller than, equal to, or larger than the cost of *e2*. A subset T of the edges of G is called a *spanning forest* of G if any two nodes that are connected in G are also connected using only edges in T and if the subgraph (V, T) is acyclic. A spanning forest of a connected graph is a tree. The cost of a spanning forest is the sum of the costs of its edges. A *minimum spanning forest* is a spanning forest of minimal cost, see Figure 6.8 for an example. Kruskal [Kru56] discovered a very simple method for computing minimum cost spanning forests; it is customary to refer to his algorithm as a spanning tree algorithm although it will not compute a tree on a graph consisting of more than one connected component.

Kruskal's algorithm starts with an empty set T of edges and considers the edges of G in order of increasing cost. When considering an edge $e = \{u, v\}$ it checks whether addition of e to T would close a cycle. If it does not close a cycle then e is added to T and if it closes a cycle then e is discarded. In this way, T gradually evolves into a minimum spanning forest.

We give a proof. Less mathematically inclined readers may skip the proof. For the following argument let $e_1, e_2, \ldots, e_m$ be the sequence of edges of G ordered in order of increasing cost and let F_0 be the lexicographically smallest minimum spanning forest[3]. We show that $T \cap \{e_1, \ldots, e_i\} = F_0 \cap \{e_1, \ldots, e_i\}$ for all i, $0 \leq i \leq m$, by induction on i. This is clearly true for $i = 0$. Consider $i > 0$. If e_i closes a cycle with respect to $T \cap \{e_1, \ldots, e_i\}$ then it closes a cycle with respect to F_0 and hence e_i belongs to neither of the two sets. If e_i does not close a cycle with respect to $T \cap \{e_1, \ldots, e_i\}$ then it is added to T. We need to show $e_i \in F_0$. Since F_0 is a spanning forest there must be a path p in F_0 connecting the endpoints of e_i and since the endpoints of e_i are not connected by the edges in $T \cap \{e_1, \ldots, e_{i-1}\} = F_0 \cap \{e_1, \ldots, e_{i-1}\}$ there must be an edge e_j with $j \geq i$ in

[3] We may view a spanning forest as a string over $\{0, 1\}$ of length m where a 1 in the i-th position indicates that e_i belongs to the spanning forest and a 0 indicates that it does not. The lexicographic ordering on these strings defines an ordering on spanning forests.

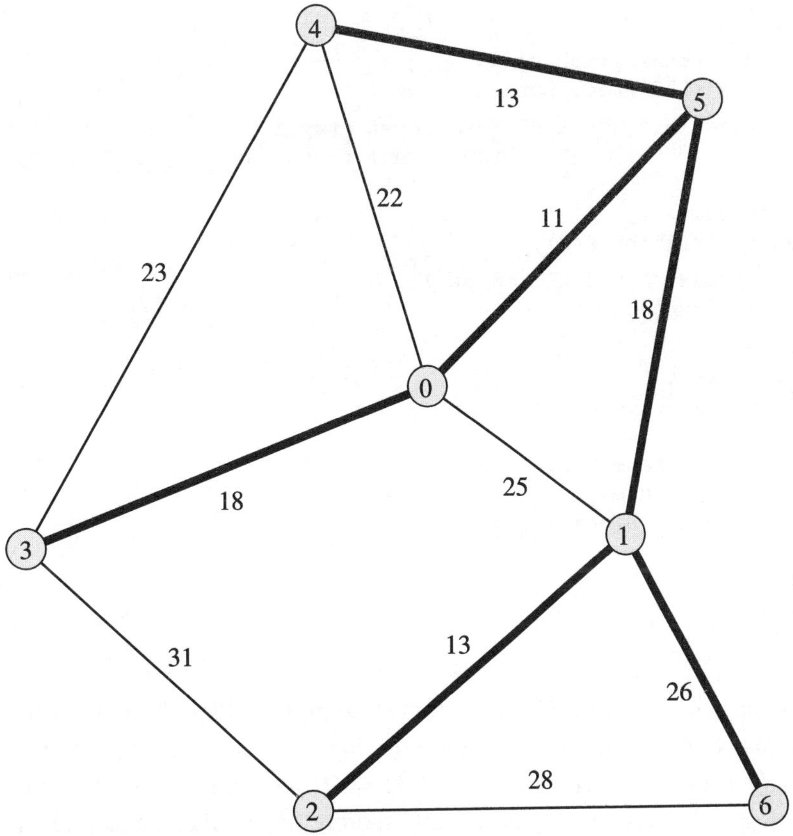

Figure 6.8 A minimum spanning forest in a graph G. The edges in the minimum are indicated in bold. The cost of each edge is indicated. This figure was generated with the spanning tree demo in xlman.

p. If $j = i$ we are done. So assume $j > i$ and consider $F' = F_0 \setminus e_j \cup e_i$. The cost of F' is at most the cost of F_0, F' is a spanning forest (since the removal of e_j splits one component of F_0 into two components each containing one of the endpoints of e_i and hence the addition of e_i glues them together again), and F' is lexicographically smaller than F_0, a contradiction. Thus $j = i$.

In an implementation the crucial question is how to check whether an edge e should be added to T. The data type *node_partition* is ideally suited for that purpose. We maintain the connected components of T as a node partition P, i.e., two nodes of G belong to the same block of P iff they are connected by a path of edges of T. Then an edge $e = \{u, v\}$ closes a cycle with respect to T iff u and v belong to the same block of P, i.e., if $P.same\_block(u, v)$. If e does not close a cycle we add e to T and update P by uniting the blocks containing u and v ($P.union\_blocks(u, v)$). We obtain the following algorithm:

⟨*Kruskal.c*⟩≡

```
#include <LEDA/graph.h>
#include <LEDA/node_partition.h>
list<edge> MIN_SPANNING_TREE(const graph& G,
                             int (*cmp)(const edge&, const edge&))
{
  list<edge> T;
  node_partition P(G);
  list<edge> L = G.all_edges();
  L.sort(cmp);
  edge e;
  forall(e,L)
  { node u = source(e);
    node v = target(e);
    if (! P.same_block(u,v))
    { T.append(e);
      P.union_blocks(u,v);
    }
  }
  return T;
}
```

The running time of Kruskal's algorithm is $O((n+m)\log(n+m))$, where m is the number of edges of G, since it takes time $O(m \log m)$ to sort the edges by cost and since the *forall_edges*-loop has cost $O((n+m)\alpha(n)) = O((n+m)\log(n+m))$. Kruskal's algorithm is efficient, but there are asymptotically more efficient algorithms known. In particular, there is a randomized algorithm with linear running time [KKT95].

The algorithm in LEDA combines Kruskal's algorithm with a heuristic and works in three phases. In the first phase it selects the $3n$ cheapest edges and runs Kruskal's algorithm on them. This yields a forest T. In the second phase it goes through the remaining edges and discards all edges that do not connect distinct components of T; this amounts to a *same_block* operation for each edge. In the third phase the still remaining edges are sorted by cost and are considered for inclusion in T in order of increasing cost. The hope underlying this heuristic is that the $3n$ edges selected in the first phase will already form a large part of the spanning tree and hence most remaining edges are discarded in the second phase. A saving results since the edges discarded in the second phase do not have to be sorted. In particular, if the third phase is empty the running time is $O((n+m)\alpha(n))$.

Table 6.2 shows some running times of the minimum spanning tree algorithm.

Exercises for 6.8

1 Experiment with the following modification of Kruskal's algorithm. First select the cn edges of smallest cost for some small constant c, say $c = 3$. Run Kruskal's algorithm on them. Then scan through the remaining edges and discard all edges that close a cycle. Sort the remaining edges in order of increasing cost and proceed with Kruskal's algorithm.

2 Implement Prim's minimum spanning tree algorithm. Let G be a connected graph and

n	m	Time
25000	250000	2.843
50000	500000	6.414
100000	1000000	13.83

Table 6.2 Running time of minimum spanning tree algorithm: For each n and m we generated 10 random graphs with n and m edges and random edge weights in [0 .. 100000] and ran MIN_SPANNING_TREE on them. You may perform your own experiments by running the minspantree_time demo.

let s be an arbitrary node of G. Prim's algorithm grows a minimum spanning tree from s. It maintains a subset S of the nodes of G and a set T of edges that comprise a minimum spanning tree of S. Initially, $S = \{s\}$ and $T = \emptyset$. For each node $v \notin S$ let $dist(v)$ be the smallest cost of an edge connecting v to a node in S. In each iteration Prim's algorithm selects the node $v \notin S$ with the smallest *dist*-value and adds it to S. What is an appropriate data structure for the *dist*-values and how can the *dist*-values be updated upon the addition of a node to S?

3 Implement *node_partitions*.

6.9 Graph Generators

Constructing graphs by a sequence of *new_node* and *new_edge* operations is a boring process, at least for humans. LEDA offers some *graph generators*.

```
complete_graph(graph& G, int n);
```

makes G the complete graph on n nodes. A graph G is *complete* if for every pair (v, w) of distinct nodes there is an edge e with $source(e) = v$ and $target(e) = w$. A complete graph on n nodes has $n(n-1)$ edges.

```
random_graph(graph& G, int n, int m, bool no_anti_parallel_edges,
             bool loopfree, bool no_parallel_edges);
```

makes G a random graph with n nodes and m edges in the so-called $G_{n,m}$-model of random graphs. A graph in this model consists of n nodes and m random edges. A random edge is generated by selecting a random element from a candidate set C defined as follows:

- C is initialized to the set of all n^2 pairs (v, w) of nodes, if *loopfree* is false, and to the set of all $n(n-1)$ pairs of distinct nodes, if *loopfree* is true.
- Upon selection of a pair (v, w) from C the pair is removed from C, when *no_parallel_edges* is true, and the reversed pair (w, v) is removed from C, when *no_anti_parallel_edges* is true.

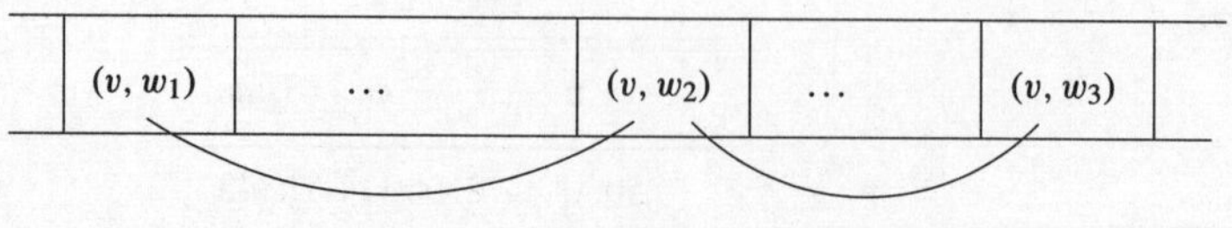

Figure 6.9 The storage layout of a graph generated by *random_graph_noncompact*. Memory is indicated as a horizontal band with low addresses at the left and high addresses at the right. Observe that the edges contained in any adjacency list spread over a large area of memory.

Several special cases of *random_graph* are available. The following pairs of calls are equivalent:

```
random_graph(G,n,m);
random_graph(G,n,m,false,false,false);
random_simple_graph(G,n,m);
random_graph(G,n,m,false,false,true);
random_simple_loopfree_graph(G,n,m);
random_graph(G,n,m,false,true,true);
random_simple_undirected_graph(G,n,m);
random_graph(G,n,m,true,true,true);
```

We give two implementations of *random_graph*. The first implementation works only for the case that all flags are set to false. The second implementation is to be preferred and we give the first implementation mainly for didactic reasons. The first implementation makes n calls of *new_node* and then m calls of *new_edge*(v, w) for random nodes v and w.

⟨*random_graph.c*⟩+≡

```
void random_graph_noncompact(graph& G, int n, int m)
{
  node*  V = new node[n];
  int i;
  G.clear();
  for(i=0; i<n; i++) V[i] = G.new_node();
  for(i = 0; i < m; i++)
    G.new_edge(V[rand_int(0,n-1)],V[rand_int(0,n-1)]);
  delete[] V;
}
```

Figure 6.9 indicates the storage layout generated by *random_graph_noncompact*. The edges are stored in the order in which they are generated. This implies that the edges belonging to any particular adjacency list are spread over a large area of memory and hence makes the layout not well suited for the most frequent iteration statement in graph algorithms: the iteration over the edges out of a node. A compact layout, which stores for each node all edges out of the node consecutively, is much better. A quantitative comparison will be given later in the section.

We turn to the function *random_graph_compact* that generates a representation where all edges out of any node are stored consecutively. It also supports the flags *no_anti_parallel_edges*,

...	(v, w_1)	(v, w_2)	(v, w_3)	...

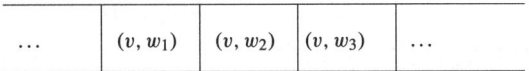

Figure 6.10 The storage layout of a graph generated by *random_graph_compact*. Memory is indicated as a horizontal band with low addresses at the left and high addresses at the right. Observe that the edges contained in any adjacency list are stored next to each other.

loopfree, and *no_parallel_edges*. In the generation process we distinguish cases according to whether the candidate set C is modified during the generation process or not.

We first deal with the simple case that the candidate set C is not modified by the process. We choose the edges in two phases. In the first phase we choose the source node of each edge and hence determine the out-degree of each node. In the second phase we iterate over the nodes of the graph and generate for each node the required number of outgoing edges. In this way all edges out of a node are generated consecutively. The running time is $O(n+m)$.

⟨*random_graph.c*⟩+≡

```
void random_graph_compact(graph& G, int n, int m,
                          bool no_anti_parallel_edges,
                          bool loopfree, bool no_parallel_edges)
{ if ( n == 0 && m > 0 )
    error_handler(1,"random graph: m to big");
  if ( n == 1 && m > 0 && loopfree )
    error_handler(1,"random graph: m to big");
  node*  V = new node[n];
  int* deg = new int[n];
  int i;
  G.clear();
  for (i = 0; i < n; i++)
  { V[i] = G.new_node();
    deg[i] = 0;
  }
  if ( !no_anti_parallel_edges && !no_parallel_edges )
  {
    for (i = 0; i < m; i++) deg[rand_int(0,n-1)]++;
    for (i = 0; i < n; i++)
    { node v = V[i];
      int  d = deg[i];
      while ( d > 0 )
      { int j = rand_int(0,n-1);
        if ( loopfree && j == i ) continue;
        G.new_edge(v,V[j]);
        d--;
      }
    }
  }
  else
  { ⟨random graph: difficult case⟩ }
```

```
  delete[] V;
  delete[] deg;
}
```

We come to the case where the candidate set C is modified during the generation process. In this situation we have to work harder.

We first check whether m is too large. If only parallel edges are forbidden then m can be at most n^2, if parallel edges and self-loops are forbidden then m can be at most $n(n-1)$, if parallel and anti-parallel edges are forbidden then m can be at most $n + n(n-1)/2$, and if parallel edges and anti-parallel edges and self-loops are forbidden then m can be at most $n(n-1)/2$.

For the generation process we maintain a *node_map2<bool>* C with the following properties:

- If *loopfree* is false then $C(v, w) =$ *true* iff $(v, w) \in C$.
- If *loopfree* is true then for all v and w with $v \neq w$: $C(v, w) =$ *true* iff $(v, w) \in C$, i.e., the map C is equal to the set C except on the diagonal. This relaxed "equality" removes the obligation to set $C(v, v)$ to false for all v.

We build the graph as follows. We generate a random pair (v, w) of nodes. If it does not belong to the candidate set, we discard it, and if it belongs to the candidate set, we add it to the graph and update the candidate set accordingly. We build the graph temporarily as an array E of lists of nodes. Once we have constructed all edges of the graph in E we actually construct G.

⟨*random graph: difficult case*⟩≡

```
⟨random graph: check whether m is too big⟩
node_map2<bool> C(G,true);
array<list<node> > E(n);
int i = m;
while ( i > 0 )
{ int vi = rand_int(0,n-1);
  node v = V[vi];
  node w = V[rand_int(0,n-1)];
  if ( (v == w && loopfree) || !C(v,w) ) continue;
  E[vi].append(w);
  if ( no_parallel_edges ) C(v,w) = false;
  if ( no_anti_parallel_edges ) C(w,v) = false;
  i--;
}
for (i = 0; i < n; i++)
{ node v = V[i];
  node w;
  forall(w,E[i]) G.new_edge(v,w);
}
```

Random	Simple	Simple loopfree	Simple undirected
10.97	21.05	20.98	24.24

Table 6.3 Running time of random graph generation: We generated a random graph with $n = 10^5$ nodes and $m = 10^6$ edges. The first column shows the running time with all flags set to false, and the other columns show the time to generate a simple graph, a simple loopfree graph, and a simple undirected graph, respectively. You may perform you own experiments using the random graph demo.

What is the running time of the generation process? The less mathematically inclined reader may skip the remainder of this section. We do the analysis for the case that no parallel edges are allowed and leave the other cases to the reader. In this situation the maximal number of edges is $M = n^2$ and each edge generated decreases the number of candidate edges by one. Thus there are $M - j$ candidate edges when j edges have already been generated, and hence an expected number of $M/(M - j)$ iterations are needed to generate a candidate. We conclude that the expected total number of iterations required to generate m edges is

$$\sum_{0 \le j < m} M/(M - j).$$

If $m > M/2$ this sum is less than (we use the estimate $\sum_{1 \le j \le k} 1/j \approx \ln k$)

$$2m \sum_{M-m+1 \le j \le M} 1/j = O(m(\ln M - \ln(M - m))) = O(m \ln(M/(M - m)))$$

and if $m < M/2$ this sum is $O(m)$. In either case the running time is $O(m(1 + \ln(M/(M - m))))$.

We still need to implement the check of whether m is too big. This check is non-trivial to implement due to the danger of overflow. Note that n^2 may be a number which does not fit into an *int*. We therefore cannot simply compute the upper bound for the number of edges in a variable of type *int*. We use a variable of type *double* instead. This will work as long as $n \le 2^{26}$, which is safe for some time to come. We only show one case of the check.

⟨*random graph: check whether m is too big*⟩≡
```
double md = m; double nd = n;
if ( no_parallel_edges && !loopfree &&
     !no_anti_parallel_edges && md > nd*nd)
  error_handler(1,"random graph: m too big");
```
⟨*random graph: more checks whether m is too big*⟩

Table 6.3 shows the running time of our random graph generators.

The storage representation of a graph can have significant impact on the running time of graph algorithms. We give an example. We generate a random graph with either one of the

n	m	Compact	Non-compact
100000	1000000	0.34	0.85

Table 6.4 Influence of representation on running time: We generated a random graph with n nodes and m edges with our two random graph generators and then ran ⟨*determine number of edges*⟩ on both of them. Observe that the running time is more than double for the non-compact representation. You may perform your own experiments by running the compact_versus_noncompact_representation demo.

two generators above and then count the number of edges in the graph by iterating over all the edges out of all nodes.

⟨*determine the number of edges*⟩≡
```
count = 0;
forall_nodes(v,G)
  forall_adj_edges(e,v) count++;
```

Table 6.4 shows the running times for the compact and the non-compact representation. The difference is huge. The running time for the non-compact representation is more than double the running time for the compact representation. Similar but not as striking differences can be obtained for other graph algorithms. The effect is less pronounced for other graph algorithms because they usually do more than incrementing a counter in the *forall_adj_edges*-loop.

The difference in speed is due to the influence of cache memory. It makes access to consecutive locations faster than access to random locations. We discuss the influence of cache memory on running time in some detail in Section 3.2.2.

In earlier versions of LEDA we used *random_graph_noncompact* as our random graph generator. When we moved to *random_graph_compact* the running time of all our graph algorithms improved significantly.

```
random_graph(graph& G, int n, double p);
```

makes G a random graph with n nodes and an expected number of $p \cdot n \cdot (n-1)$ edges. The graph is generated by the following experiment. First n nodes are created and then for any pair (v, w) of distinct nodes the edge (v, w) is added to G with probability p. In the graph literature this model of random graphs is called the $G_{n,p}$-model. The running time is $O(n^2)$. Graphs generated according to the $G_{n,p}$-model behave similar to graphs generated according to the $G_{n,pn(n-1)}$-model.

```
random_bigraph(graph& G, int a, int b, int m,
               list<node>& A, list<node>& B);
```

makes G a random bipartite graph with a nodes on the one side, b nodes on the other side, and m edges directed from the A-side to the B-side. The nodes on the two sides are returned in A and B.

The generators for planar graphs are treated in the chapter on embedded graphs, see Section 8.9.

Exercises for 6.9

1 Compare the compact and the non-compact representation of graphs for other graph algorithms.

2 Let $o = (o_0, \ldots, o_{n-1})$ and $i = (i_0, \ldots, i_{n-1})$ be vectors of non-negative integers with $\sum_{0 \le j < n} o_j = \sum_{0 \le j < n} i_j$. Show that there is a graph with n nodes and o_j edges out of node j and i_j edges into node j for all j, $0 \le j \le n-1$. Generate a random graph of this kind. Hint: Use the class *dynamic_random_variate* of Section 3.5. Set up random variates S and T according to the weight vectors o and i, respectively. Use S to choose sources and T to choose targets. After every generation of an edge decrement the weight of its source and its target.

3 Use *random_graph*(G, n, m) to generate a random graph and test the graph for simplicity (using *Is_Simple*(G)). Try to find the value of m (in relation to n) where about 50% of the generated graphs are simple. If you want to understand the experiment, read up on the so-called Birthday paradox, see for example [Fel68] or [MR95].

4 Write a $O(n+m)$ generator for random graphs in the $G_{n,p}$-model. Hint: Reduce the problem to generating a graph in the $G_{n,m}$-model. Let p_m be the probability that a random graph in the $G_{n,p}$-model has m edges. Show that the probability is maximal for $m \approx pn(n-1)$ by considering the quotient p_m/p_{m+1}. Also show that the probability falls off quickly as one goes away from $m \approx pn(n-1)$. The idea is now to generate m according to the distribution given by the p_m's and to call *random_graph*(G, n, m) afterwards. The problem with this approach is that the p_m's are numbers with long representations. A possible way around this problem is to write each p_m as a sum $p_{m,1} + p_{m,2} + \ldots$ where for each m the $p_{m,i}$ decrease exponentially in i. Consider the collection $\{p_{m,i}\,; 0 \le m \le n(n-1),\ i \ge 0\}$ and order it approximately by size. Generate m according to this distribution and then call *random_graph*(G, n, m). Provide your solution as an LEP.

6.10 **Input and Output**

We discuss how to write graphs to a file (or standard output) and how to read graphs from a file. We support two formats, the format shown in Figure 6.11 (henceforth called the standard representation) and the GML-format [Him97]. We will not formally define either format.

```
G.write();
```

writes the standard representation of G on standard output.

```
G.write(string s);
```

writes G onto the file with name s and

```
LEDA.GRAPH
void
void
4
|{}|
|{}|
|{}|
|{}|
5
1 2 0 |{}|
1 3 0 |{}|
2 3 0 |{}|
2 4 0 |{}|
3 4 0 |{}|
```

Figure 6.11 The standard representation of the graph of Figure 6.1. In the case of a parameterized graph the node and edge labels are enclosed in the angular brackets.

```
G.write_gml(string s,...);
```

writes G in gml-format. The additional arguments of *write_gml* can be used to fine-tune the way nodes and edges are output.

```
G.read(string s);
G.read_gml(string s, ...);
```

read a graph G from the file with name s. Either the standard representation or the GML-representation is expected.

The following piece of code is useful during the debugging phase of a graph algorithm.

```
while (true)
{ generate G;
  G.write("graph.gw");
  run graph algorithm on G;
  check result and abort if incorrect;
}
```

If the program aborts, a witness that falsifies the algorithm can be found in the file with name graph.gw.

There are several ways to inspect the witness graph:

- One can visually inspect the file to which the graph was written. This is tedious even for very small graphs.
- One can load the graph into a graph window. This is the most convenient method and we give more details below.
- One can send it through a graph drawing algorithm, see Section 8.1, and display the result.

We give more details on how to load a graph into a graph window, see Chapter 12 for more information about the graphwin type. The following piece of code assumes that the graph written has an integer node label and an integer edge label and that a parameterized graph was used. We define a graph *GRAPH<int, int> G* and read it from the file. We then define a *GraphWin gw* for *G*. We tell *gw* that we want the so-called data labels of the nodes and edges displayed, we open the display and put *gw* into edit mode[4]. When this program is executed, a window will pop up in which the graph *G* is displayed. The nodes of *G* will appear at random positions. The layout can be modified by dragging nodes around.

⟨*simple_visualization.c*⟩≡

```
#include <LEDA/graphwin.h>
main()
{
  GRAPH<int,int> G;
  G.read("graph.gw");
  GraphWin gw(G);
  node v; edge e;
  gw.set_node_label_type(data_label);
  gw.set_edge_label_type(data_label);
  gw.display();
  gw.edit();
}
```

Actually, there is no need even to write the program above. Call any of the programs starting with "gw" in xlman and use the file menu to load the graph.

6.11 Iteration Statements

Iterating over the nodes and edges of a graph or all the edges incident to a particular node is an essential component of any graph algorithm. Accordingly, we have seen iteration statements already many times in this chapter. In this section we treat them in detail. We first give a precise definition of the semantics, then discuss the possibility of hiding and unhiding edges and the possibilities of changing the order of iteration, and finally discuss which modifications of a graph are legal during iteration.

6.11.1 *Basics*

In order to understand the iteration statements we need to learn a bit about the representation of graphs in LEDA. A graph is a collection of nodes and edges which are arranged into several lists:

- The nodes are arranged into a list of nodes.

[4] If the statement *gw.edit*( ) is omitted, the program will briefly flash the graph and then terminate.

- The edges are arranged into a list of edges.
- In directed graphs two lists of edges are associated with every node v:

  $$adj\_edges(v) = \{e \in E\ ;\ v = source(e)\}\,,$$

  i.e., the list of edges starting in v, and

  $$in\_edges(v) = \{e \in E\ ;\ v = target(e)\}\,,$$

  i.e., the list of edges ending in v. The list $adj\_edges(v)$ is called the adjacency list of node v. For directed graphs we often use $out\_edges(v)$ as a synonym for $adj\_edges(v)$.
- In undirected graphs only the list $adj\_edges(v)$ is defined for every node v. Here it contains all edges incident to v, i.e.,

  $$adj\_edges(v) = \{e \in E\ ;\ v \in \{source(e), target(e)\}\}\,.$$

  An undirected graph must not contain self-loops, i.e., it must not contain an edge whose source is equal to its target.

The semantics of the iteration statements for graphs now reduces to the semantics of the iteration statements for lists.

```
forall_nodes(v,G)         { }
forall_rev_nodes(v,G)     { }
```

iterate over the list of nodes in either forward or backward direction,

```
forall_edges(e,G)         { }
forall_rev_edges(e,G)     { }
```

iterate over the list of edges in either forward or backward direction,

```
forall_adj_edges(e,v)     { }
forall_out_edges(e,v)     { }
forall_in_edges(e,v)      { }
forall_inout_edges(e,v) { }
```

iterate over the lists $adj\_edges(v)$, $out\_edges(v)$, $in\_edges(v)$, and $out\_edges(v)$ followed by $in\_edges(v)$, respectively, and

```
forall_adj_nodes(u,v)     { }
```

iterates over the other endpoint, i.e., $G.opposite(v, e)$, of all edges e in $adj\_edges(v)$.

6.11.2 *Modification during Iteration*

The rules are simple:

- It is unsafe to modify an object while iterating over it.
- However, the item under the iterator can be removed from the object.

In our experience the exception covers most of the situations where one wants to perform modifications during an iteration.

The following piece of code iterates over the edges of a graph and deletes all edges whose cost is negative.

```
forall_edges(e,G) if ( cost[e] < 0 ) G.del_edge(e);
```

The following piece of code is an infinite loop as new edges are appended to the list of edges during iteration.

```
forall_edges(e, G) G.new_edge(G.target(e), G.source(e));
```

A safe way to add the reversal of every edge to G is to write:

```
list<edge> L = G.all_edges();
forall(e, L) G.new_edge(G.target(e), G.source(e));
```

6.11.3 *Hiding and Restoring Edges*

Sometimes it is convenient to remove edges only temporarily from a graph. For this purpose we have the concept of a hidden edge.

```
G.hide_edge(e);
```

removes e temporarily from G until restored by

```
G.restore_edge(e);
```

The implementation is simple. *Hide_edge*(e) deletes e from G and stores it in a list of hidden edges and *restore_edge*(e) removes e from the list of hidden edges and puts it back into the list of real edges. The list of all hidden edges is available as *G.hidden_edges*( ), one can ask whether an edge e is hidden (*G.is_hidden*(e)), ...

The following lines of code hides all edges with negative cost, then runs some graph algorithm on the resulting graph, and finally restores all edges.

```
forall_edges(e,G) if ( cost[e] < 0 ) G.hide_edge(e);
// some graph algorithm
G.restore_all_edges();
```

The operations *hide_edge* and *restore_edge* change the order of the adjacency lists and hence should be used with *extreme care on embedded graphs*.

6.11.4 *Rearranging Nodes and Edges*

The lists of nodes and edges may be arranged by sorting. There are many different ways to sort. We go through the possibilities for nodes and remark that a similar set of sorting routines exists for edges.

```
G.sort_nodes(int (*cmp)(const node&, const node&));
```

sorts the nodes according to the compare function *cmp* and

```
G.sort_nodes(const node_array<NT>& A);
```

sorts the nodes according to the values in the node array A (the type *NT* must be a number type). The running time of both functions is $O(n \log n)$.

```
G.sort_nodes(const list<node>& vl);
```

assumes that *vl* is a permutation of the nodes of G. This permutation is taken as the new node ordering. The running time is linear.

```
G.bucket_sort_nodes(int (*ord)(const node&));
```

uses bucket sort to sort the nodes according to the values of the function $ord(v)$. The running time is $O(n + (b - a + 1))$ where a and b are the minimal and maximal values of *ord*, respectively.

```
void bucket_sort_nodes(const node_array<int>& A);
```

uses bucket sort with the ordering function $ord(v) = A[v]$.

Sorting the set of nodes rearranges the list of nodes. Subsequent *forall_nodes* loops iterate over the nodes in the modified order.

Sorting the set of edges rearranges the list of edges and the adjacency lists of all nodes. Subsequent *forall_edges*, *forall_adj_edges* and *forall_out_edges* loops iterate over the nodes in the modified order.

For example, if *cost* is an edge array that assigns an integer or double valued cost to every edge, then

```
G.sort_edges(cost);
```

rearranges the list of all edges and also the adjacency lists of all nodes in order of increasing cost.

6.12 Basic Graph Properties and their Algorithms

We define some basic graph properties and give the algorithms that decide them. For some of the algorithms we give the implementation. Many of the functions discussed in this section are illustrated by Figure 6.12 and by the submenu "test" of menu "graph" of any xlman-demo starting with the characters "gw".

6.12.1 *Functionality*

The function

```
void CopyGraph(GRAPH<node,edge>& H, const graph& G);
```

constructs an isomorphic copy H of G. For each node v of H the corresponding node in G is stored in $H[v]$ and for each edge e of H the corresponding edge of G is stored in $H[e]$. The mapping $v \longrightarrow H[v]$ is a bijection from the nodes of H to the nodes of G and for each

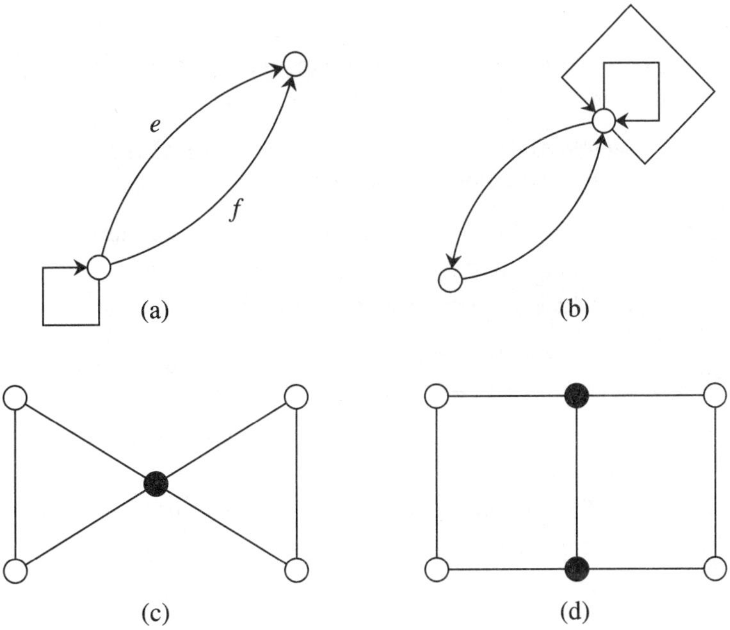

Figure 6.12 Illustration of basic graph properties: The graph (a) is not simple (the edges e and f are parallel) and has a self-loop h. The graph (b) is simple and bidirected. The graph (c) is connected but not biconnected (the full node is an articulation point). The graph (d) is biconnected but not triconnected (the full nodes form a split pair).

edge $e = (v, w)$ of H we have $source(H[e]) = H[v]$ and $target(H[e]) = H[w]$. We have already seen the implementation of *CopyGraph* in Section 6.1.

A graph is called *simple* iff is has no parallel edges, i.e., no two distinct edges e and f with the same source and sink, and a graph is called *loopfree* if it has no self-loop, i.e., no edge whose source is equal to its sink.

```
bool Is_Simple(const graph& G);
```

returns true if G is simple and returns false otherwise.

A directed graph $G = (V, E)$ is called *bidirected* if for every edge e the reversed edge $(target(e), source(e))$ also belongs to G, more precisely, if there is a bijection $rev : E \longrightarrow E$ such that:

- $source(e) = target(rev(e))$ and $target(e) = source(rev(e))$ for every $e \in E$ and
- $rev(e) \neq e$ for every $e \in E$.

The condition $rev(e) \neq e$ ensures that a self-loop cannot be its own reversal. A bidirected graph has an even number of edges. The main use of bidirected graphs is in the representation of embedded graphs, the topic of Chapter 8.

The calls

```
bool Is_Bidirected(const graph& G);
bool Is_Bidirected(const graph& G, edge_array<edge>& rev);
```

check whether G is bidirected. The second version also computes an appropriate bijection between the edges of G (if it exists).

```
void         Make_Bidirected(graph& G, list<edge>& R)
list<edge> Make_Bidirected(graph& G)
```

adds edges to G to make it bidirected. The added edges are returned in R or as the result of the function. An alternative to *Make_Bidirected* are the member functions *G.make_bidirected* and *G.make_map*( ) which are discussed in Section 8.2.

```
bool Is_Acyclic(const graph& G);
bool Is_Acyclic(const graph& G, list<edge>& L);
```

return true if the G is acyclic and return false otherwise. The second version also returns a list of edges whose removal makes G acyclic. We have already seen an implementation of the first version of *Is_Acyclic* in Section 6.2. The second version performs a depth-first search on G (see Section 7.3) and returns the list of back edges.

A *path in a directed graph* is a sequence

$$[v_0, e_1, v_1, e_2, v_2, \ldots, v_{k-1}, e_k, v_k]$$

of nodes and edges such that $source(e_i) = v_{i-1}$ and $target(e_i) = v_i$ for all i, $1 \le i \le k$. We call v_0 the source of the path and v_k the target of the path. The number of edges in the path is called the cardinality or length of the path. We will frequently abuse notation and write

$$[e_1, e_2, \ldots, e_k]$$

or

$$[v_0, v_1, v_2, \ldots, v_{k-1}, v_k]$$

instead of the more verbose notation above. A path is *simple* if all nodes (except maybe for the source and the target of the path) are pairwise distinct. A *cycle* is a path whose source is equal to its target.

A *path in an undirected graph* is a sequence

$$[v_0, e_1, v_1, e_2, v_2, \ldots, v_{k-1}, e_k, v_k]$$

of nodes and edges such that $\{source(e_i), target(e_i)\} = \{v_{i-1}, v_i\}$ for all i, $1 \le i \le k$ and $e_{i-1} \ne e_i$ for all i, $1 < i \le k$. We call v_0 and v_k the endpoints of the path. The number of edges in the path is called the cardinality or length of the path. We will frequently abuse notation and write

$$[v_0, v_1, v_2, \ldots, v_{k-1}, v_k]$$

instead of the more verbose notation above.

If G is a graph and e is an edge of G then $G \setminus e$ is the graph that results from removing e from G. If v is a node of G then $G \setminus v$ denotes the graph that results from removing v and all edges incident to v from G.

An undirected graph G is *connected* if for any two nodes v and w of G there is a path from v to w in G. An *articulation point* of an undirected graph G is any node of G such that $G \setminus v$ is not connected. An undirected graph is called *biconnected* if is has no articulation point. A *split pair* of an undirected graph is a pair $\{s_1, s_2\}$ of nodes such that $G \setminus \{s_1, s_2\}$ is not connected.

```
bool Is_Connected(const graph& G);
```

returns true if G (viewed as an undirected graph) is connected and returns false otherwise.

```
void        Make_Connected(graph& G,list<edge>& L);
list<edge> Make_Connected(graph& G);
```

make G connected by adding edges and return the list of inserted edges. The number of edges added is minimal.

```
void        Make_Biconnected(graph& G,list<edge>& L);
list<edge> Make_Biconnected(graph& G);
```

make G biconnected by adding edges and return the list of inserted edges.

```
bool Is_Biconnected(const graph& G);
bool Is_Biconnected(const graph& G, node& s);
```

test whether G is biconnected. The second version returns an articulation point in s if the graph is not biconnected.

A (directed or undirected) graph is *bipartite* if the nodes of the graph can be colored with two colors such that every edge of G connects nodes with different colors.

```
bool Is_Bipartite(const graph& G);
bool Is_Bipartite(const graph& G, list<node>& A, list<node>& B);
```

return true if G is bipartite and return false otherwise. The second version also returns a bipartition of the nodes of G in A and B (if the graph is bipartite).

A graph is *planar* if it can be drawn into the plane such that all nodes are placed at distinct points in the plane and such that no two edges cross.

```
bool Is_Planar(const graph& G);
```

returns true if G is planar and returns false otherwise. We will see a lot more of planar graphs in Chapter 8.

All functions above have linear running time $O(n + m)$.

```
bool Is_Triconnected(const graph& G);
bool Is_Triconnected(const graph& G, node& s1, node& s2);
```

returns true if G (viewed as an undirected graph) is triconnected and returns false otherwise. The second version returns a split pair in *s1* and *s2* if the graph is not triconnected. The running time is $O(n(n + m))$.

Table 6.5 reports some running times of the basic graph algorithms.

n	G	L	C	B	S	D	A	N	T
1000	0.07	0.01	0.01	0.03	0.04	0.1	0.01	0	17.9
10000	1.08	0.03	0.29	0.63	0.48	1.85	0.28	0.01	3342

Table 6.5 Speed of basic graph algorithms: We generated a random graph with n nodes and $m = 10n$ edges and then ran various graph algorithms on it:
G = generation of random graph,
L = time for removing self-loops,
C = time for testing connectedness,
B = time for testing biconnectedness,
S = time for testing simplicity,
D = time for testing bidirectedness,
A = time for testing acyclicity,
N = time for testing bipartiteness,
T = time for testing triconnectivity.
The time for testing bipartiteness is so small because a violation to bipartiteness is found very quickly in a random graph. For bipartite graphs the running time will be about the time to test connectedness. You may perform your own experiments by running the speed of basic graph algorithms demo.

6.12.2 *Implementations*

We give the implementation of the function *Is_Bidirected*.

We make two copies of the edges of G in lists *EST* and *ETS* and sort both lists.

In the sorted version of *EST* the edges are sorted by their source node, and edges with equal source node are sorted by their target node, i.e., all edges out of the first node come first, then all out of the second node, ... Within each group of edges the ordering is by target node.

In the sorted version of *ETS* the edges are sorted by their target node, and edges with equal target node are sorted by their source node, i.e., all edges into the first node come first, then all into the second node, ...

We use bucket sort for both sorts. This will play a role below.

Figure 6.13 shows an example. After having sorted the two lists the i-th edge of *EST* is the reversal of the i-th edge of *ETS* for all i (if G is bidirected).

Self-loops cause a small problem. As described so far, a self-loop can be matched with itself. There is a simple remedy. We use the fact that bucket sort is stable, i.e., the relative order of parallel edges is not changed.

Suppose now that we reverse *ETS* before the sorting step. Consider all self-loops incident to a particular node v, say $e_1, e_2, \ldots, e_k$. In *EST* they will appear exactly in the same order as in the original list of edges and in *ETS* they will appear in the reversed order. We match the i-th edge of one sequence with the i-th edge of the other sequence. When k is even we obtain a legal matching and when k is odd we will attempt to match one of the edges with itself. This leads to the following program.

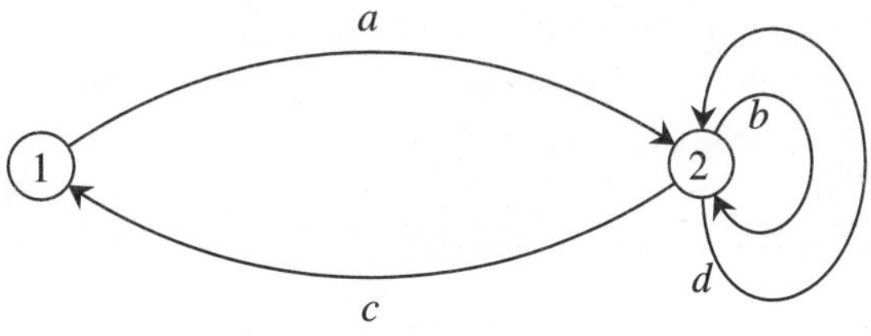

Figure 6.13 The lists *EST* and *ETS* in the implementation of Is_Bidirected: It is assumed that the original edge list of G is $E = (a, b, c, d)$. Observe that the edges b and d are parallel. In *EST* the edges are sorted by source, and edges with equal source are sorted by target. Parallel edges appear in the same order as in E. Thus *EST* $= (a, c, b, d)$. In *ETS* the edges are sorted by target, and edges with equal source are sorted by source. Parallel edges appear in the reverse order as in E. Thus *EST* $= (c, a, d, b)$.

The program uses the fact that the nodes of a graph are internally numbered and that $index(v)$ returns the number of a node v.

```
static int edge_ord1(const edge& e) { return index(source(e)); }
static int edge_ord2(const edge& e) { return index(target(e)); }

bool Is_Bidirected(const graph& G, edge_array<edge>& reversal)
{
  int n = G.max_node_index();

  edge e,r;

  list<edge> EST = G.all_edges();
  EST.bucket_sort(0,n,&edge_ord2);
  EST.bucket_sort(0,n,&edge_ord1);

  list<edge> ETS = G.all_edges();
  ETS.reverse();                                      //crucial
  ETS.bucket_sort(0,n,&edge_ord1);
  ETS.bucket_sort(0,n,&edge_ord2);

  // merge EST and ETS to find corresponding edges

  while (! EST.empty() && ! ETS.empty())
  { e = EST.pop();
    r = ETS.pop();

    if ( target(r) == source(e) && source(r) == target(e)
         && e != r )
       reversal[e] = r;
    else return false;
  }

  return true;
}
```

Exercises for 6.12

1 Give an implementation of the function *Is_Simple*. Use a *node_map2*.

2 Implement a function that tests whether a graph has a self-loop.
3 Implement the function *Make_Acyclic*. Read Section 7.3 first.
4 As above, but for function *Is_Connected*.
5 As above, but for function *Make_Connected*.
6 Provide a better implementation of the triconnectedness test. A linear time algorithm is described in [HT73]. Provide it as an LEP.

6.13 Parameterized Graphs

Parameterized graphs are another convenient way to associate information with the nodes and edges of a graph.

```
GRAPH<vtype,etype> G;
```

declares G as a parameterized graph and initializes G to the empty graph. With every node of G a variable of type *vtype* is associated and with every edge of G a variable of type *etype* is associated. The variables associated with nodes or edges can be accessed using array notation, i.e., $G[v]$ and $G[e]$ return the variables associated with node v and edge e, respectively. We have illustrated the use of parameterized types already in Section 6.1. We will see extensive use of parameterized graphs in the chapters on embedded graphs and on geometry. Here we want to discuss the relationship between parameterized graphs and graphs.

All operations defined on instances of the data type *graph* are also defined on instances of any parameterized graph type *GRAPH<vtype, etype>*, i.e., instances of a parameterized graph type can be used wherever an instance of the data type *graph* can be used, in particular, as arguments to functions with formal parameters of type *graph&*. If a function f(*graph& G*) is called with an argument Q of type *GRAPH<vtype, etype>* then inside f only the basic graph structure of Q (the adjacency lists) can be accessed. The node and edge entries are hidden.

The operations

```
node_array<vtype>& G.node_data()
edge_array<etype>& G.edge_data()
```

make the information associated with the nodes (edges) of G available as a node array (edge array) of type *node_array<vtype>* (*edge_array<etype>*). These operations are extremely useful when one wants to run a graph algorithm that requires a node or edge array as a parameter on a parameterized graph where one has stored the appropriate information in the nodes and edges, respectively. For example,

```
GRAPH<int,int> G;
node_array<edge> pred(G);
DIJKSTRA(G,G.first_node(), G.edge_data(), G.node_data(), pred);
```

runs Dijkstra's algorithm on G taking the edge data of G as the edge costs and storing the node distances in the nodes of G.

We have four different ways to associate information with the nodes, and similarly with the edges, of a graph in this section: node arrays, node data slots, node maps, and parameterized graphs. We use all four of them in our own work. We use parameterized graphs when the node information is an essential part of the graph. For example, we use the type *GRAPH<point, ...>* for graphs embedded into the plane; the position of any node v is given as $G[v]$. If the information is only temporarily associated with the node, as, for example, in a graph algorithm, we use node arrays and node maps. We use node maps for sparse arrays, where only a fraction of the nodes need an entry, and we use node arrays for dense arrays. We use node data slots, if speed is of utmost importance and node information is accessed many times and in random order, and we use standard node arrays otherwise. Standard node arrays are the most convenient and most widely used way to associate information with nodes.

6.14 Space and Time Complexity

Graphs are represented in their adjacency lists representation and hence the space requirement is $O(n + m)$, where n and m are the number of nodes and edges of the graph, respectively. Most operations on graphs take constant time except, of course, those which change or inspect the entire graph. The iterators take time proportional to the number of objects they iterate over, so *forall_edges*(e, G) takes time $O(m)$. We give some more information about the constant factors involved.

The space requirement of a *graph* or *GRAPH* with n nodes and m edges is $O(1) + 44m + 52n$ bytes, i.e., a graph with 10^4 nodes and 10^5 edges needs about 5 megabytes. For *GRAPH<T1, T2>* where an object of type *T1* or *T2* needs more than one word of storage one also has to account for the information associated with the nodes and edges. For example, a point requires 8 bytes and hence a *GRAPH<point, int>* requires an additional $8n$ bytes.

There is a trade-off between the space requirement of graphs and the functionality offered by them. We give some examples. Our graphs are fully dynamic, i.e., nodes and edges can be added and deleted at any time, and hence the adjacency information of every node is stored in a doubly linked list. For static graphs the adjacency information could be stored in an array. Our graphs support the dynamic addition of additional node and edge labels (in the form of node and edge arrays and maps) and hence every node or edge needs to have an integer index. This index could be saved if all node and edge labels have to be declared at the time of the construction of the graph.

We turn to running time. There is a large number of tables with running times of graph algorithms in this book. The tables prove that it is possible to solve problems on fairly large graphs using our algorithms. Moreover, the time bounds achieved by (most of) our algorithms are competitive with what other researchers report.

Exercises for 6.14

1 Implement a version of directed graphs where each node only knows about its outgoing edges but not about its incoming edges and where the adjacency lists are stored as singly linked lists and hence can only be traversed from front to rear. Make the graph class compatible with LEDA's graphs and provide it as an LEP.

2 Implement static directed graphs where all edges are stored in a single array, all edges in a adjacency list are stored consecutively, and each node has two pointers into the array, one to the first edge of its adjacency list, and one to the edge after the last edge of its adjacency list.

7

Graph Algorithms

LEDA offers a wide variety of graph algorithms. Starting in the third section of this chapter we discuss depth-first and breadth-first search, algorithms to compute graph decompositions, and algorithms for shortest paths, matchings in bipartite and general graphs, maximum flows, and minimum cuts. For each class of algorithms we first discuss their functionality and then discuss implementations. In many cases we also derive a checker of correctness.

The first two sections of this chapter are orthogonal to the other sections of the chapter. They deal with general considerations for algorithms on weighted graphs. In Section 7.1 we discuss the use of template functions for such algorithms and in Section 7.2 we discuss the requirements on the underlying arithmetic. Both sections can be skipped on first reading.

7.1 Templates for Network Algorithms

Many graph algorithms operate on graphs whose nodes or edges have an associated weight from some number type. For example, the single-source shortest-path algorithm operates on an edge-weighted graph and computes for each node its distance from the source. The algorithm works for any linearly ordered number type. It is natural to formulate it as a template function.

```
template <class NT>
bool DIJKSTRA_T(const graph& G, node s, const edge_array<NT>& c,
                node_array<NT>& dist, node_array<edge>& pred);
```

The template parameter *NT* can be instantiated with any number type. The number type must, of course, satisfy certain syntactic and semantic requirements, e.g., there must be a linear ordering defined on it and addition must be monotone. The most frequent instantiations are with the built-in number types *int* and *double* and the LEDA number types *integer* and *real*. It is desirable that:

- the most frequent instantiations are pre-compiled, as this reduces the compilation time of application programs and allows us to distribute object code instead of source code to all those users, who do not need instantiations with other number types, and that
- the pre-instantiated versions can be used side by side with the template version.

We describe our mechanism to achieve these goals. We use the shortest-path algorithm as our running example. We write three files: dijkstra.h, dijkstra.t, and _dijkstra.c, which are contained in the directories LEDAROOT/incl/LEDA, LEDAROOT/incl/LEDA/templates, and LEDAROOT/src, respectively.

The file dijkstra.h contains the prototypes of all functions. We distinguish the template version and the pre-instantiated versions of a function by the suffix _T in the function name. Thus

⟨*dijkstra.h*⟩≡

```
#ifndef DIJKSTRA_H
#define DIJKSTRA_H
#include <LEDA/graph.h>
template <class NT>
void DIJKSTRA_T(const graph& G, node s, const edge_array<NT>& c,
                node_array<NT>& dist, node_array<edge>& pred);
/* next come the pre-instantiated versions */
void DIJKSTRA(const graph& G, node s, const edge_array<int>& c,
              node_array<int>& dist, node_array<edge>& pred);
// and, similarly, for double, ...
#endif
```

The file dijkstra.t contains the definition of the template function.

⟨*dijkstra.t*⟩≡

```
#include <LEDA/dijkstra.h>
template <class NT>
void DIJKSTRA_T(const graph& G, node s, const edge_array<NT>& c,
                node_array<NT>& dist, node_array<edge>& pred)
{
  /* implementation of DIJKSTRA_T */
}
```

The file _dijkstra.c contains the implementations of the instantiations in terms of the template function.

⟨*_dijkstra.c*⟩≡

```
#include <LEDA/templates/dijkstra.t>
void DIJKSTRA(const graph& G, node s, const edge_array<int>& c,
              node_array<int>& dist, node_array<edge>& pred)
{
  DIJKSTRA_T(G,s,c,dist,pred);
}
// and, similarly, for double ...
```

Observe the include statement. As mentioned already, all files containing definitions of template functions are collected in the subdirectory *templates* of the LEDA include directory.

The file _dijkstra.c is pre-compiled into the object file _dijkstra.o, which is included in one of the object libraries of the LEDA system.

We next discuss how to use the pre-instantiated and the template versions of the shortest-path algorithm.

In order to use one of the pre-instantiated versions, one includes dijkstra.h into the application program, for example,

⟨*foo.c*⟩≡

```
#include <LEDA/dijkstra.h>
// define G, s, c, dist, pred with number type int
DIJKSTRA(G,s,c,dist,pred);
```

In order to use the template version, one includes templates/dijkstra.t into the application program, as, for example, in

⟨*foo.c*⟩+≡

```
#include <LEDA/templates/dijkstra.t>
// define G, s, c, dist, pred for any number type NT
DIJKSTRA_T(G,s,c,dist,pred);
// define G, s, c, dist, pred for number type int
// and use template version
DIJKSTRA_T(G,s,c,dist,pred);
// use pre-instantiated version
DIJKSTRA(G,s,c,dist,pred);
```

Observe that there is no problem to use one of the pre-instantiated versions and the template version side by side in an application program such as foo.c.

We nevertheless recommend a different strategy. We suggest that the t-files are not included directly into application programs, as t-files may contain the definitions of auxiliary functions which might clobber the name space of the application program. We rather recommend to define intermediate files as shown next.

In order to instantiate DIJKSTRA_T for a particular number type, say the LEDA number type *real*, we recommend defining files

⟨*real_dijkstra.h*⟩≡

```
#include <LEDA/real.h>

void DIJKSTRA(const graph& G, node s, const edge_array<real>& c,
              node_array<real>& dist, node_array<edge>& pred)
```

and

⟨*real_dijkstra.c*⟩≡

```
#include "real_dijkstra.h"
#include <LEDA/templates/dijkstra.t>

void DIJKSTRA(const graph& G, node s, const edge_array<real>& c,
              node_array<real>& dist, node_array<edge>& pred)
{
  DIJKSTRA_T(G,s,c,dist,pred);
}
```

to include the former in application programs, to pre-compile the latter, and to add the object file real_dijkstra.o to the set of objects for the linker. The alternative strategy has the advantage of introducing no extraneous names into application programs.

We summarize: functions whose name ends with _T are function templates. In order to use them one must include a file LEDA/templates/X.t. The pre-instantiated functions have the same name except for the _T. In order to use them one needs to include a file LEDA/X.h.

7.2 Algorithms on Weighted Graphs and Arithmetic Demand

Many algorithms of this chapter operate on weighted graphs and work for any number type *NT*. The algorithms use additions, subtractions, comparisons, and in rare cases multiplication and division. The correctness proofs of the algorithms rely on the laws of arithmetic and hence the algorithms are only correct if the implementation of the number type obeys the laws of arithmetic.

The two most commonly used number types are *int* and *double*. Unfortunately, both types do not guarantee that the basic arithmetic operations obey their mathematical laws. For example, *int*-arithmetic may overflow and wrap around[1] and *double*-arithmetic incurs rounding error, see Chapter 4. It is therefore not at all obvious that an instantiation of a network algorithm with types *int* or *double* will work correctly. Sections 4.1 and 7.10.5 contain examples of what can go wrong.

We use the following two-step approach to guarantee correctness.

[1] Execute `cout << MAXINT + MAXINT;`

Step 1: We analyze the arithmetic demand of our algorithms. We state clearly which operations must be supported by the number type (that's easy to do, since a simple inspection of the code suffices) and we prove theorems of the following form: if all input weights are integers whose absolute value is bounded by B, then all numbers handled by the algorithm are integers whose absolute value is bounded by $f \cdot B$. We call such an algorithm *f-bounded*. For example, we will show that the maximum weight bipartite matching algorithm is 3-bounded and that the maximum weight assignment algorithm is $4n$-bounded, where n is the number of nodes of the bipartite graph.

Step 2 for type int: In the instantiation of a network algorithm for type *int*, we check that all input weights w satisfy $f \cdot w \leq MAXINT$. If not, we write an appropriate message to diagnostic output. If yes, step 1 guarantees correctness of the computation.

We give an example. We mentioned already that the maximum weight bipartite matching algorithm is 3-bounded. The instantiation is therefore as follows:

⟨*instantiation for ints*⟩≡

```
list<edge> MAX_WEIGHT_BIPARTITE_MATCHING(graph& G,
                         const edge_array<int>& c, node_array<int>& pot)
{ int W = MAXINT/3;
  check_weights(G,c,-W,W,"MWBM<int>");
  return MAX_WEIGHT_BIPARTITE_MATCHING_T(G,c,pot);
}
```

where

⟨*scale_weights.h*⟩+≡

```
inline bool check_weights(const graph& G, const edge_array<int>& c,
                          int lb, int ub, string inf)
{ edge e;
  bool all_edges_ok = true;
  forall_edges(e,G)
    if ( c[e] < lb || c[e] > ub ) all_edges_ok = false;
  if ( !all_edges_ok ) cerr << inf << ": danger of overflow.\n";
  return all_edges_ok;
}
```

There is a similar function for node arrays.

Step 2 for type double: The problem with *double*-arithmetic is round-off error. Round-off errors invalidate the correctness and termination proof and hence a "naive" instantiation of a network algorithm with the number type *double* may run forever, terminate with a run-time error, terminate with an incorrect result, or terminate with the correct result.

It would be nice if we could guarantee that no rounding occurs during a computation, as this will guarantee termination and the absence of run-time errors. It does not guarantee by itself that the result produced has any relationship to the correct result. We come back to this point below.

We can avoid rounding by scaling the input weights appropriately. We replace any input weight w by $sign(w) \cdot \lfloor |w| \cdot S \rfloor / S$, where the *scaling parameter* $S = 2^s$ is a suitable power of two. We use the *same* scaling parameter for all input weights. This has the effect that, after scaling, all input weights are of the form $w' \cdot 2^{-s}$, where w' is an integer. Hence floating point arithmetic will incur no rounding error as long as all intermediate results are of the form $z \cdot 2^{-s}$, where z is an integer that fits into the mantissa of a floating point number. It remains to choose s.

Let C be the maximum absolute value of any input weight. Since the division by 2^s effects only the exponent of a floating point number, we may as well assume that every input weight w is replaced by $sign(w) \lfloor |w| \cdot S \rfloor$. This will turn all inputs into integers and hence step 1 guarantees that the absolute value of all intermediate results is bounded by $f \cdot \lfloor C \cdot S \rfloor$ in the case of an f-bounded algorithm. If we choose s such that all intermediate results can be represented exactly as a double precision floating point number then the computation will incur no rounding error. This is the case if

$$f \cdot \lfloor C \cdot S \rfloor < 2^{53},$$

since double precision floating point arithmetic can represent all integers in the range $[-(2^{53} - 1) .. 2^{53} - 1]$. Observe that double precision floating point arithmetic uses a 52-bit mantissa and that a floating point number with mantissa $m_1 m_2 \ldots m_{52}$ and exponent 52 represents the integer

$$(1 + \sum_{1 \le i \le 52} m_i 2^{-i}) \cdot 2^{52}.$$

The inequality $f \cdot \lfloor C \cdot S \rfloor < 2^{53}$ is certainly satisfied if

$$f \cdot C \cdot S < 2^{53}$$

or

$$s < 53 - \log(f \cdot C).$$

We summarize:

Lemma 1 *Consider an f-bounded algorithm, let C be the maximum absolute value of any input weight, and let S be a power of two such that $f \cdot C \cdot S < 2^{53}$. If every input weight w is replaced by $sign(w) \lfloor |w| \cdot S \rfloor$, then the algorithm will incur no rounding error in a computation with doubles and hence computes the correct result for the scaled inputs weights.*

What is the relationship between the result for the scaled input weights and the result for the original input weights? We can make no general claim. However, there are many situations where one can claim that the result for the scaled inputs is a good approximation for the result on the unscaled inputs. For all but one network problem considered in this chapter, namely min-cost flow, the objective value is a sum of input weights; for example, the cost of a shortest path is a sum of edge weights, the cost of a matching is a sum of edge weights, and the maximum flow in a network is the minimum capacity of a cut and hence a sum of edge weights. Assume that the objective value is the sum of at most L weights. For any set of at most L weights the sum of the scaled weights and the sum of the unscaled weights differs by at most L/S, since for any individual weight the difference is at most $1/S$. If S is chosen as the largest power of two such that $S < 2^{53}/(f \cdot C)$, then $S \geq 2^{52}/(f \cdot C)$ and hence the maximum absolute error in the objective function is at most $L \cdot f \cdot C \cdot 2^{-52}$. We summarize in:

Lemma 2 *Under the hypothesis of the preceding lemma and the additional assumption that the algorithm computes an objective value, which is the sum of at most L input weights, the maximum absolute error in the objective function is at most* $L \cdot f \cdot C \cdot 2^{-52}$.

Let us give an example. Consider the maximum weighted matching algorithm for bipartite graphs. This algorithm is 3-bounded and the value of a matching is the sum of at most n edges, where n is the number of nodes of the graph. The maximum absolute error is therefore at most $3 \cdot C \cdot 2^{-52}$.

Observe that Lemma 2 bounds the absolute error in the objective function, but not the relative error. We can make no general claims about the relative error. It must be studied individually for each algorithm.

In order to compute s and to scale the input weights, we use the functions *frexp*, *ldexp*, and *floor* from the math-library. Let $x = f \cdot C$.

```
double frexp(double x, int* exp);
```

returns a double y such that y is a double with magnitude in the interval $[1/2, 1)$ or 0, and x equals y times 2 raised to the power *exp* (more precisely, **exp*). If x is 0, both parts of the result are 0.

Thus, if x is non-zero, then $\log|x| = exp - \epsilon$ where $0 < \epsilon \leq 1$ and hence

$$53 - \log(f \cdot C) = 53 - exp + \epsilon.$$

We therefore choose s as

$$s = 53 - exp.$$

If $C = 0$ and hence $x = 0$, the choice of s is arbitrary. We will set s to 53 in this case. The following procedures implement the computation of s and S. We also compute $1/S$, as it will be convenient to have it around.

⟨*scale_weights.h*⟩+≡

```
#include <math.h>
inline int compute_s(double f, double C)
{
  int exp;
  double x = frexp(f*C,&exp);
  return 53 - exp;
}
inline double compute_S(double f, double C, double& one_over_S)
```

```
{
  int exp;
  double x = frexp(f*C,&exp);
  one_over_S = ldexp(1,exp - 53);
  return ldexp(1,53 - exp);
}
```

where

```
double ldexp(double x, int exp);
```

computes the quantity $x \cdot 2^{exp}$.

How can we compute $w' = sign(w) \cdot \lfloor |w| \cdot S \rfloor / S$? We use

```
double floor(double x);
```

which computes the largest integral value not greater than x.

⟨*scale_weights.h*⟩+≡

```
inline double scale_weight(double w, double S, double one_over_S)
{
  if ( w == 0 ) return 0;
  int sign_w = +1;
  if ( w < 0 ) { sign_w = -1; w = -w; }
  return sign_w * floor(w * S) * one_over_S;
}
```

Let us see scaling at work. We use again the weighted matching algorithm for bipartite graphs. The instantiation for number type *double* is as follows.

⟨*instantiation for double*⟩≡

```
list<edge> MAX_WEIGHT_BIPARTITE_MATCHING(graph& G,
                    const edge_array<double>& c, node_array<double>& pot)
{ edge_array<double> c1(G);
  scale_weights(G,c,c1,3.0,"MWBM<double>");
  return MAX_WEIGHT_BIPARTITE_MATCHING_T(G,c1,pot);
}
```

where

⟨*scale_weights.h*⟩+≡

```
inline bool scale_weights(const graph& G, const edge_array<double>& c,
                          edge_array<double>& c1, double f)
{ edge e;
  double C = 0;
  forall_edges(e,G) C = leda_max(C,fabs(c[e]));
  double one_over_S;
  double S = compute_S(f,C,one_over_S);
  bool no_scaling = true;
```

```
    forall_edges(e,G)
    { c1[e] = scale_weight(c[e],S,one_over_S);
      if ( c[e] != c1[e] ) no_scaling = false;
    }
    return no_scaling;
  }
  inline bool scale_weights(const graph& G, const edge_array<double>& c,
                            edge_array<double>& c1, double f, string inf)
  { bool no_scaling = scale_weights(G,c,c1,f);
    if ( no_scaling == false ) cerr << inf << ": scaling was required";
    return no_scaling;
  }
```

We also offer a function that replaces a weight vector by its scaled version.

⟨*scale_weights.h*⟩+≡

```
inline bool scale_weights(const graph& G, edge_array<double>& c,
                          double f)
{ edge_array<double> c0 = c;
  return scale_weights(G,c0,c,f);
}
```

There are also analogous functions for node arrays.

How does scaling interact with program checking? We showed in Lemmas 1 and 2 that a computation with doubles computes the exact result for the scaled weights and that the result for the scaled weights is frequently a good approximation of the result for the unscaled weights. We should not expect them to be equal. It is therefore nonsense to check whether a double computation produced the correct result for the unscaled weights if scaling took place.

For example, in the program

```
list<edge> M = MAX_WEIGHT_BIPARTITE_MATCHING(G,c,pot);
CHECK_MWBM(G,c,M,pot);
```

the call of CHECK_MWBM may fail. Indeed, it is very likely to fail if scaling took place in the computation of the maximum weight matching.

We recommend the following strategy of using program checking together with a computation with doubles. *The scaling should be done on the level of the user program.* To this end, each network algorithm comes with a function that replaces all input weights by their scaled versions.

For example, in <*mwb_matching.h*> we also define a function

```
bool MWBM_SCALE_WEIGHTS(const graph& G, edge_array<double>& c)
{
  return scale_weights(G,c,3.0);
}
```

that replaces the cost vector c by a scaled version. One may then write

```
MWBM_SCALE_WEIGHTS(G,c);
list<edge> M = MAX_WEIGHT_BIPARTITE_MATCHING(G,c,pot);
CHECK_MWBM(G,c,M,pot);
```

and checking will work.

The remainder of this section may be skipped. It is worthwhile to study in more detail what it means to replace w by $w' = sign(w) \cdot \lfloor |w| \cdot S \rfloor / S$. Clearly, if $w = 0$ then $w' = 0$. So assume $w \neq 0$. By symmetry, it suffices to study the case $w > 0$.

Lemma 3 *Let $0 < w = x \cdot 2^e$ with $1/2 \le |x| < 1$, e integral, and let $w_1 w_2 \ldots w_{52}$ be the mantissa of the floating point representation of w. Let s be an integer, let $S = 2^s$, and let $w' = \lfloor w \cdot S \rfloor / S$. If $e + s \le 0$ then $w' = 0$. If $e + s > 0$ then w' is obtained from w by replacing the mantissa by $w_1 \ldots w_{e+s-1} 0 \ldots 0$.*

Proof We have $w = x \cdot 2^e$ with $1/2 \le |x| < 1$. If $e + s \le 0$ then $w' = 0$. So assume $e + s > 0$. We have $2 \cdot x = 1 + \sum_{1 \le i \le 52} w_i 2^{-i}$ and hence

$$\begin{aligned}
\lfloor w \cdot S \rfloor &= \lfloor x \cdot 2^{e+s} \rfloor = \lfloor 2 \cdot x \cdot 2^{e+s-1} \rfloor \\
&= \lfloor (1 + \sum_{1 \le i \le 52} w_i 2^{-i}) \cdot 2^{e+s-1} \rfloor \le (1 + \sum_{1 \le i \le e+s-1} w_i 2^{-i}) \cdot 2^{e+s-1} \\
&= (1 + \sum_{1 \le i \le e+s-1} w_i 2^{-i}) / (2 \cdot 2^e \cdot 2^s)
\end{aligned}$$

and hence

$$w' = \lfloor w \cdot S \rfloor / S = (1 + \sum_{1 \le i \le e+s-1} w_i 2^{-i}) / (2 \cdot 2^e),$$

i.e., w' has the same exponent as w and mantissa $w_1 \ldots w_{e+s-1} 0 \ldots 0$. □

Let us consider two special cases.

If all input weights are integers, then the scaling will not change any input as long as $f \cdot C < 2^{53}$. This is as for *ints*, but with *MAXINT* replaced by $2^{53} - 1$.

For the second case we assume that all input weights are less than one. We may assume w.l.o.g. that $1/2 \le C < 1$. Then $s = 53 - k$ where $k = \lfloor \log f \rfloor + 1$ or $k = \lfloor \log f \rfloor$. If w is any input weight and w has binary representation

$$0.w w_1 w_2 \ldots$$

then w' has binary representation

$$0.w_1 w_2 \ldots w_{53-k} 000 \ldots,$$

i.e., the binary representation is truncated after the $(53 - k)$-th bit. In this way the scaled weights leave k bits of the mantissa unused. The unused bits can be used to compute intermediate results without rounding error.

7.3 Depth-First Search and Breadth-First Search

Depth-first search and breadth-first search are two powerful methods to explore a graph in a systematic way. Both methods start at some node v of a directed graph G and visit all nodes that can be reached from v. They differ in the order in which they visit the nodes.

Depth-first search always explores edges out of the node most recently reached by the search. When it has exhausted all edges out of a node it backtracks to the node from which the node was reached.

Depth-first search is most easily formulated as a recursive procedure *dfs* that takes a node v as an argument (and additional arguments depending on the application of depth-first search). A call *dfs*$(v, \ldots)$ first labels v as reached and then makes recursive calls for all nodes w such that (v, w) is an edge out of v and node w is not yet reached. A depth-first search on a graph G induces two numberings of the vertices of G, one in the order in which the nodes are reached by the search and one in the order in which the calls to *dfs* are completed. The two numbers associated with a node are usually called its *depth-first search number* and its *completion number*. Depth-first search can also be used to partition the edges of G into so-called *tree*, *forward*, *backward*, and *cross* edges.

In the program below we use node arrays *dfsnum* and *compnum* to record the two numberings and we use a list T to collect tree edges. The sets of forward, backward, and cross edges are determined implicitly, as we will discuss later. We define two procedures, a recursive procedure *dfs*$(v$, *dfsnum*, *compnum*, $T)$ and a master *DFS_NUM*$(G$, *dfsnum*, *compnum*$)$. A call *dfs*$(v, \ldots)$ visits and numbers all vertices reachable from v that were not reached previously. We maintain the invariant that *dfsnum*$[v] = -1$ iff v was not visited yet. The master procedure *DFS_NUM* initializes the variables and then iterates over all nodes. For every node v that was not reached yet it calls *dfs*$(v, \ldots)$. The call *dfs*$(v, \ldots)$ sets *dfsnum*$[v]$ to the current value of *dfsnum_counter*, and then iterates over all edges out of v. Each edge (v, w) to an unreached node w is added to T and leads to a recursive call *dfs*$(w, \ldots)$. When the edges out of v are exhausted *compnum*$[v]$ is set to the current value of *compnum_counter*.

⟨*dfs*⟩+≡

```
static int dfsnum_counter;
static int compnum_counter;

static void dfs(node v, node_array<int>& dfsnum, node_array<int>& compnum,
                list<edge>& T )
{ dfsnum[v] = ++dfsnum_counter;
  edge e;
  forall_adj_edges(e,v)
   { node w = target(e);
     if (dfsnum[w] == -1)
      { T.append(e);
        dfs(w,dfsnum,compnum,T);
      }
   }
  compnum[v] = ++compnum_counter;
}
```

```
list<edge> DFS_NUM(const graph& G, node_array<int>& dfsnum,
                                   node_array<int>& compnum)
{
  list<edge> T;
  dfsnum_counter = compnum_counter = 0;
  dfsnum.init(G,-1);  // declares all nodes unreached
  node v;
  forall_nodes(v,G)
    if (dfsnum[v] == -1) dfs(v,dfsnum,compnum,T);
  return T;
}
```

Figure 7.1 shows the result of a run of *DFS_NUM*. A call *DFS_NUM*$(G, \ldots)$ partitions the edges of G into four classes in a natural way; the four classes are also shown in Figure 7.1. An edge $e = (v, w)$ is called a *tree edge* if *dfs*$(w, \ldots)$ is called when the edge e is scanned in *dfs*$(v, \ldots)$; we use T to denote the set of tree edges. The tree T is the call tree of procedure *dfs*. An edge $e = (v, w)$ is called a *forward edge* if it is parallel to a path of tree edges, but is not a tree edge, i.e., $v \xrightarrow[T]{+} w$ and $e \notin T$; it is called a *backward edge* (or back edge) if it is anti-parallel to a path of tree edges, i.e., $w \xrightarrow[T]{*} v$; and it is called a *cross edge* in all other cases. The two numberings of the vertices can be used to classify the edges[2]. An edge (v, w) is a:

- tree or forward edge iff $dfsnum[v] < dfsnum[w]$ and $compnum[v] > compnum[w]$,
- backward edge iff $dfsnum[v] \geq dfsnum[w]$ and $compnum[v] \leq compnum[w]$,
- cross edge iff $dfsnum[v] > dfsnum[w]$ and $compnum[v] > compnum[w]$.

Let us see why this is true. We only give an intuitive argument and refer the reader to [Meh84c, IV.5] and [CLR90, chapter 23] for more detailed discussions.

If two calls C and D of *dfs* are nested within one another, say D is nested within C, then C starts before D and ends after D, i.e., the dfs-number of the node corresponding to C is smaller than the dfs-number of the node corresponding to D and the completion-number of the node corresponding to C is larger than the completion number of the node corresponding to D. This explains the characterization of tree, forward, and backward edges.

If two calls C and D are not nested within one another and, say C starts after D, then C starts after the completion of D and hence the dfs-number of the node corresponding to C is larger than the dfs-number of the node corresponding to D and the same holds for completion-numbers. This fact together with the observation that a cross edge always runs from a node reached later to a node reached earlier explains the characterization of cross edges.

Depth-first search considers every edge of the graph G exactly once and hence runs in linear time $O(n + m)$, where $n = |V|$ and $m = |E|$.

[2] There is no standard convention concerning self-loops. We classify self-loops as back edges.

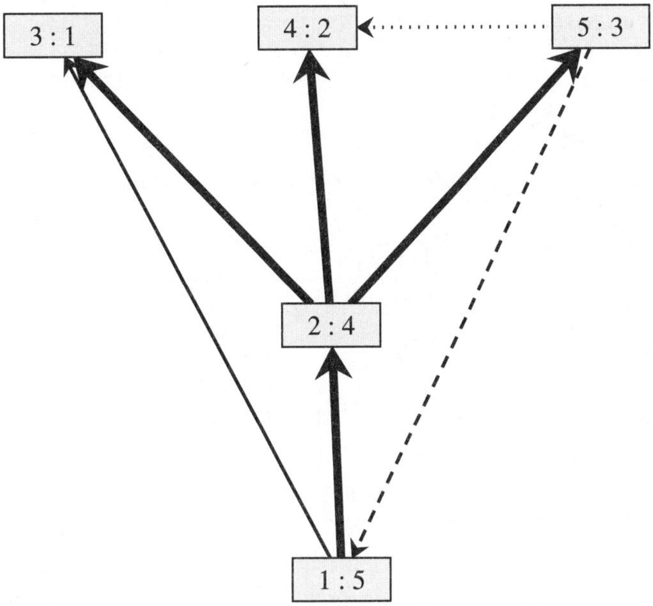

Figure 7.1 Depth-first search: The search started at the bottom-most node. For each node the dfs- and the completion-number are shown inside the node. Tree edges are shown as thick solid edges, forward edges are shown as thin solid edges, backward edges are shown as dashed edges, and cross edges are shown a dotted edges. It is customary to draw dfs-trees such that tree edges are directed upwards and cross edges are directed from right to left. Observe how dfs-numbers increase along every tree path and how completion-numbers decrease. Also observe that cross edges go from nodes with higher dfs- and completion-number to nodes with lower dfs- and completion-number. You may generate your own figures by calling the xlman-demo gw_dfs.

Why should one be interested in the classification of the edges into tree, forward, backward, and cross edges? Here is one reason. A depth-first search on an acyclic graph does not find any backward edges. Thus $compnum[v] > compnum[w]$ for any edge (v, w), i.e., all edges go from higher to lower completion numbers. In other words, *compnum* is a topological numbering of the graph.

We turn to breadth-first search. It explores the edges in the order in which their source vertex is reached. It uses a queue Q to store the vertices in the order in which they are reached and always explores edges out of the first node of the queue. When all edges out of the first node are scanned, the first node is popped from the queue and exploration from the new first node is started. BFS can be used to label the vertices with their distance from a particular node s, i.e., to compute a *node_array<int> dist* such that $dist[w] = d$ iff there is a path from s to w of length d and d is the smallest integer with this property.

⟨*bfs*⟩≡

```
void BFS(const graph& G, node s, node_array<int>& dist)
{ queue<node> Q;
  node v,w;
```

```
    forall_nodes(w,G) dist[w] = -1;
    dist[s] = 0;
    Q.append(s);
    while (!Q.empty())
    { v = Q.pop();
      forall_adj_nodes(w,v)
        if (dist[w] < 0)
        { Q.append(w);
          dist[w] = dist[v] + 1;
        }
    }
}
```

The correctness of BFS is easy to establish. Clearly, if $dist[w] = d$ then there is a path of length d from s to w. On the other hand, if $s = v_0, v_1, \ldots, v_l = w$ is a path from s to w of length l then $dist[v_i] \le i$ for all i, $1 \le i \le l$.

Exercises for 7.3

1 Why can there be no edge (v, w) in a depth-first search with $dfsnum[v] < dfsnum[w]$ and $compnum[v] < compnum[w]$?

2 Write a procedure based on depth-first search that tests a graph for acyclicity. If the graph is acyclic it should also compute a so-called topological numbering of the vertices of G, i.e., a labeling of the nodes of G such that for all edges of G the label of the source node is smaller than the label of the target node.

3 Use the program LEDAROOT/demo/xlman/gw_dfs.c as the basis of a program that illustrates BFS.

7.4 Reachability and Components

We start with an overview of the algorithms that compute reachability information and simple structural information of directed and undirected graphs: transitive closure, connected and biconnected components, and strongly connected components. Then we discuss the details of the strongly connected components algorithm, and finally we describe an animation of this algorithm.

7.4.1 *Functionality*

We deal with basic problems concerning reachability in directed and undirected graphs. We first consider directed graphs and later turn to undirected graphs.

Let $G = (V, E)$ be a directed graph and let v and w be two vertices of G. Recall that w is *reachable* from v if there is a path in G from v to w, i.e., if either $v = w$ or there is a sequence $e_1, \ldots, e_k$ of edges of G with $k \ge 1$, $v = source(e_1)$, $w = target(e_k)$, and $target(e_i) = source(e_{i+1})$ for all i, $1 \le i < k$.

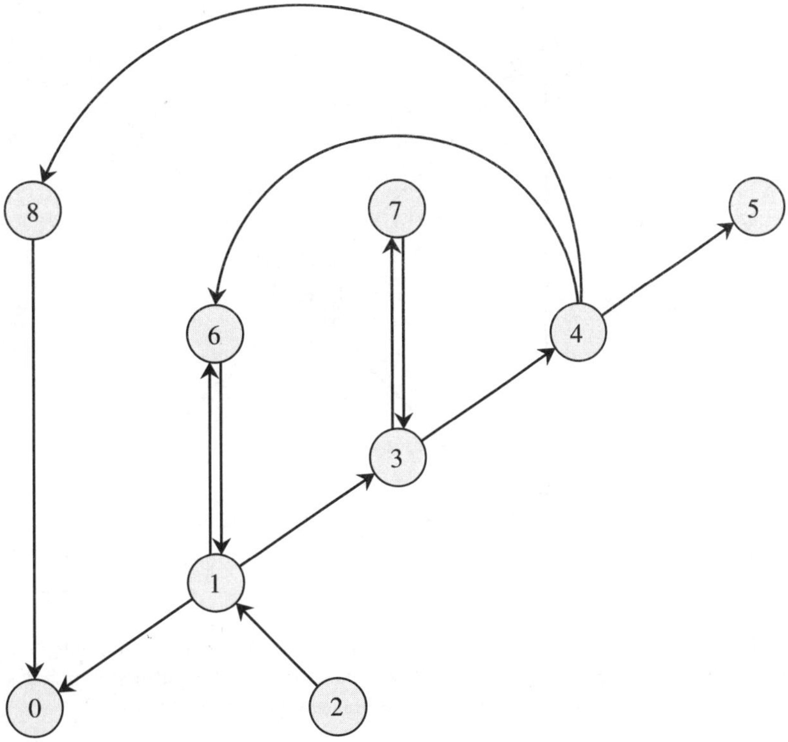

Figure 7.2 A graph with five strongly connected components. The five components are induced by the node sets $C_0 = \{8\}$, $C_1 = \{5\}$, $C_2 = \{1, 3, 4, 6, 7\}$, $C_3 = \{0\}$, and $C_4 = \{1\}$. The xlman-demo gw_scc_anim illustrates strongly connected components.

The graph $G^* = (V, E^*)$ where $E^* = \{(v, w);\ w \text{ is reachable from } v\}$ is called the *reflexive transitive closure* of G. The procedure

```
graph TRANSITIVE_CLOSURE(const graph& G);
```

computes G^* from G in time $O(n^2 + m_{red} \cdot n)$ where $n = |V|$ and m_{red} is the number of edges in a transitive reduction of G. A *transitive reduction* of G is a minimal (with respect to set inclusion of edges) subgraph of G with the same transitive closure as G. In an acyclic graph, m_{red} is the number of edges (v, w) of G such that there is no path of length two or more from v to w in G. For random graphs in the $G_{n,p}$-model and arbitrary value of p, $E(m_{red}) = O(n)$ and hence the expected running time of the transitive closure algorithm is $O(n^2)$, see [Meh84c, IV.3].

A directed graph G is called *strongly connected* if from any node of G there is a path to any other node of G. A *strongly connected component* (scc) of a graph G is a maximal strongly connected subgraph. Figure 7.2 shows a graph with five strongly connected components. Shrinking the strongly connected components of a graph to single nodes gives rise to an acyclic graph $G_s = (V_s, E_s)$ with

$$V_s = \{C;\ C \text{ is an scc of } G\}$$

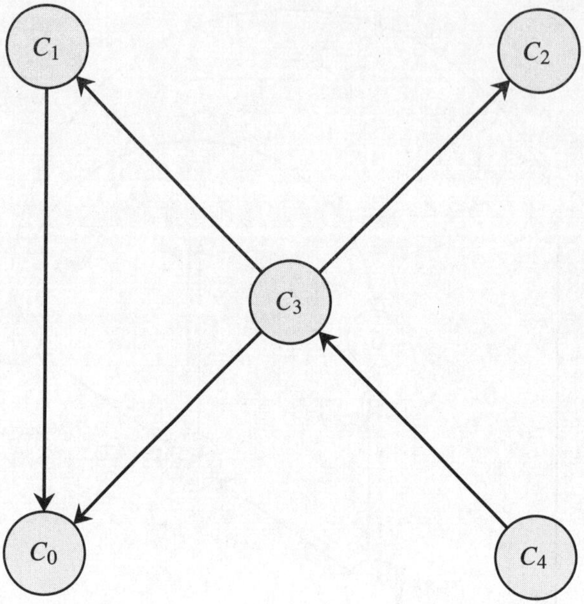

Figure 7.3 The graph obtained by shrinking the sccs of the graph in Figure 7.2 to single nodes. The given numbering of the sccs will be obtained if a first depth-first search is started in node 0 (it will only reach 0) and a second depth-first search is started in node 2.

and

$$E_s = \{(C, D);\, C, D \in V_s \text{ and there exists } (v, w) \in E \text{ with } v \in C \text{ and } w \in D\}$$

Figure 7.3 shows the shrunken graph obtained from the graph of Figure 7.2.

The procedure

```
int STRONG_COMPONENTS(const graph& G, node_array<int>& comp_num)
```

returns the number of strongly connected components of G and computes a *node_array<int>* *comp_num* with the following properties[3]:

- For all nodes v of G: $0 \le comp\_num[v] <$ number of sccs of G.
- $comp\_num[v] = comp\_num[w]$ iff the vertices v and w belong to the same strongly connected component.
- If (v, w) is an edge of G then $comp\_num[v] \ge comp\_num[w]$.

In other words, the array *comp_num* encodes the strongly connected components of G and moreover induces a topological ordering of the shrunken graph. The scc demo illustrates the strongly connected components algorithm. The demo allows one to construct a graph interactively. After every edit step the strongly connected components are recomputed and highlighted by a color and numbering code. Procedure *STRONG_COMPONENTS* runs in

[3] Observe that *comp_num* stands for component number and *compnum* stands for completion number.

linear time $O(n+m)$, where $n = |V|$ and $m = |E|$; its implementation is given in the next section.

The transitive closure algorithm uses the strongly connected components algorithm as a subroutine: it first computes the sccs, then the shrunken graph, then the transitive closure of the shrunken graph, and finally the transitive closure of the full graph. We give the simple procedure for computing the shrunken graph *SG* corresponding to a graph G. We first call the strong components algorithm for G and give *SG* one vertex for each scc of G. We then iterate over the edges of G and add an edge to *SG* for each edge (v, w) of G where v and w belong to distinct sccs. Finally, we remove parallel edges by calling *Make_Simple(SG)*.

⟨*shrunken_graph*⟩≡

```
graph SHRUNKEN_GRAPH(const graph& G)
{ node_array<int> comp_num(G);
  int N = STRONG_COMPONENTS(G, comp_num);
  graph SG;
  array<node> V(N);
  for (int i = 0; i < N; i++) V[i] = SG.new_node();
  edge e;
  forall_edges(e,G)
  { node v = G.source(e); node w = G.target(e);
    if (comp_num[v] > comp_num[w] )
      SG.new_edge(V[comp_num[v]],V[comp_num[w]]);
  }
  Make_Simple(SG);
  return SG;
}
```

We turn to undirected graphs. The data type *ugraph* represents undirected graphs. Alternatively, directed graphs may be interpreted as undirected graphs, see Section 6.7. In the early versions of LEDA we used *ugraphs* as the argument of all graph algorithms that operate on undirected graphs. We now prefer to use *graphs* and to let the algorithms interpret them as undirected graphs. In the discussion of the algorithms we talk about undirected graphs, of course.

Let $G = (V, E)$ be an undirected graph. It is called *connected* if for any two vertices v and w there is a path from v to w in G, i.e., either $v = w$ or there is a sequence $v_1, \ldots, v_k$ of vertices such that $v = v_1$, $w = v_k$, and $\{v_i, v_{i+1}\}$ is an edge of G for all i, $1 \le i < k$. A component of G is a maximal connected subgraph of G. The procedure

```
int COMPONENTS(const graph& G, node_array<int>& comp_num)
```

computes the number of connected components, say N, of G and an array *comp_num* such that $0 \le comp\_num[v] < N$ for all vertices v and $comp\_num[v] = comp\_num[w]$ iff the vertices v and w belong to the same connected component of G. It runs in linear time $O(n+m)$.

A connected undirected graph $G = (V, E)$ is called *biconnected* if $G - v$ is connected

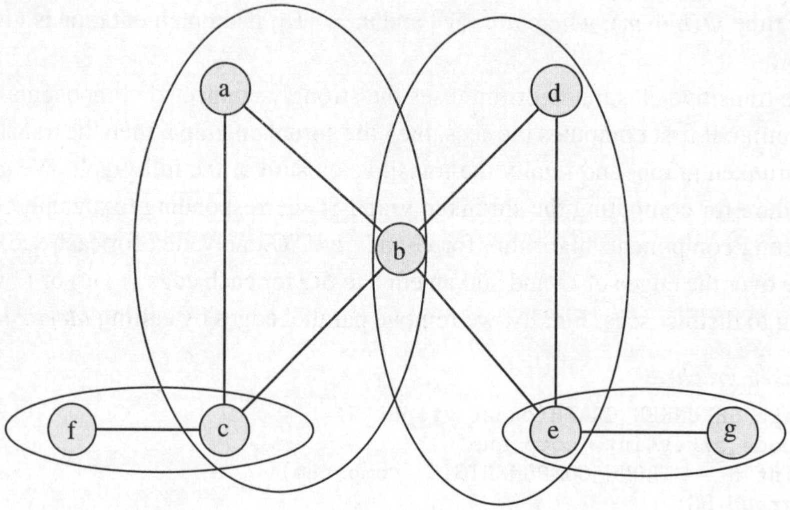

Figure 7.4 A graph with four bccs. The bccs are indicated by ovals. They have edge sets $\{\{f, c\}\}$, $\{\{e, g\}\}$,$\{\{a, b\}, \{b, c\}, \{a, c\}\}$, and $\{\{b, d\}, \{b, e\}, \{e, d\}\}$, respectively. The articulation points are the nodes b, c, and e.

for every $v \in V$. Here

$$G - v = (V - v, \{e; e \in E \text{ and } v \notin e\})$$

is the graph obtained by removing the vertex v and all edges incident to v from G. For graphs with at least three nodes the following alternative definition is useful: G is biconnected if for any distinct vertices v and w there are two vertex-disjoint paths connecting v and w. A *biconnected component* (bcc) is a maximal biconnected subgraph. A vertex a is called an *articulation point* of G if $G - a$ is not connected. Figure 7.4 shows a graph with four biconnected components.

Let G be an undirected graph and let $G_1 = (V_1, E_1), \ldots, G_m = (V_m, E_m)$ be the biconnected components of G. We claim that $E = E_1 \cup \ldots \cup E_m$ and $|V_i \cap V_j| \leq 1$ and $E_i \cap E_j = \emptyset$ for $i \neq j$. To see this, note first that for each edge $\{v, w\} \in E$ the graph consisting of vertices v and w and the single edge $\{v, w\}$ is biconnected, and hence contained in one of the biconnected components of G. It remains to show that any two distinct bccs share at most one vertex (this also implies that they can share no edge). Assume otherwise, i.e., we have distinct bccs G_i and G_j and a pair $\{v, w\}$ of nodes belonging to both. Since G_i and G_j are maximal biconnected subgraphs, the subgraph $G' = (V_i \cup V_j, E_i \cup E_j)$ is not biconnected and hence has an articulation point, say a. Let x and y be vertices in different components of $G' - a$. Since a is neither an articulation point in G_i nor in G_j, the graphs $G_i - a$ and $G_j - a$ are connected and hence x and y cannot both be vertices in the same graph G_i or G_j. We may assume w.l.o.g. that $x \in V_i$ and $y \in V_j$. Since a cannot be equal to both v and w we may assume $v \neq a$. Since $G_i - a$ and $G_j - a$ are connected, a path exists from x to v in $G_i - a$ and from y to v in $G_j - a$. Hence a path exists from x to y in

$G' - a$ and we have reached a contradiction. We conclude that the bccs of a graph partition the edges.

The procedure

```
int BICONNECTED_COMPONENTS(const graph& G, edge_array<int>& comp_num)
```

returns the number of bccs of the undirected version of G and computes an edge array *comp_num* such that $comp\_num[e] = comp\_num[f]$ iff the edges e and f belong to the same biconnected component of G. The running time is $O(n + m)$.

We give more details. Let c be the number of biconnected components and let c' be the number of biconnected components containing at least one edge; $c - c'$ is the number of isolated nodes in G, i.e., the number of nodes v that are not connected to a node different from v. The function returns c and labels each edge of G (which is not a self-loop) by an integer in $[0 .. c' - 1]$. Two edges receive the same label iff they belong to the same biconnected component. The edge labels are returned in *comp_num*. Be aware that self-loops receive no label since self-loops are ignored when interpreting a graph as an undirected graph.

The nodes of a biconnected graph can be numbered in a special way which is useful for many algorithms on biconnected graphs. Imagine the following physical experiment. G is a biconnected graph and s and t are any two nodes of G that are connected by an edge. We replace all edges of G by rubber bands and then pull s and t apart. Since G has no articulation point, this will exert force on every node of G and order the nodes of G along the line from s to t. We number the nodes from 1 to n starting with s and proceeding towards t. Every node v of G, except for s and t, will have a smaller numbered and a higher numbered neighbor. Such a numbering is called an *st-numbering* of G. The function

```
void ST_NUMBERING(graph& G, node_array<int>& stnum, list<node>& stlist)
```

numbers the nodes of G with the integers 1 to n (the number of any node v is returned in $stnum[v]$ and the ordered list of nodes is returned in *stlist*) such that every node v with $1 < stnum[v] < n$ is connected to a node with smaller number and to a node with higher number, and such that the nodes with numbers 1 and n are connected by an edge. The running time is $O(n + m)$. We will see an application of st-numbering in Section 8.7.

7.4.2 *Strongly Connected Components: An Implementation*

We give a program to compute the strongly connected components of a directed graph. An animation of this program is available as the xlman-demo gw_scc_anim. The algorithm is an extension of depth-first search and was first described in [CM96]; alternative algorithms are described in [Tar72] and [Sha81].

Consider a depth-first search on G and use $G_c = (V_c, E_c)$ to denote the subgraph already explored, i.e., V_c is the set of nodes v for which $dfs(v, \ldots)$ has been called and E_c consists of all edges e which have been explored in one of the calls of *dfs*. The algorithm maintains the strongly connected components of G_c. In order to derive the algorithm we first introduce some notation and then state some properties of G_c.

We call a vertex $v \in V$ *completed* if the call $dfs(v, \ldots)$ has been completed, *unreached* if

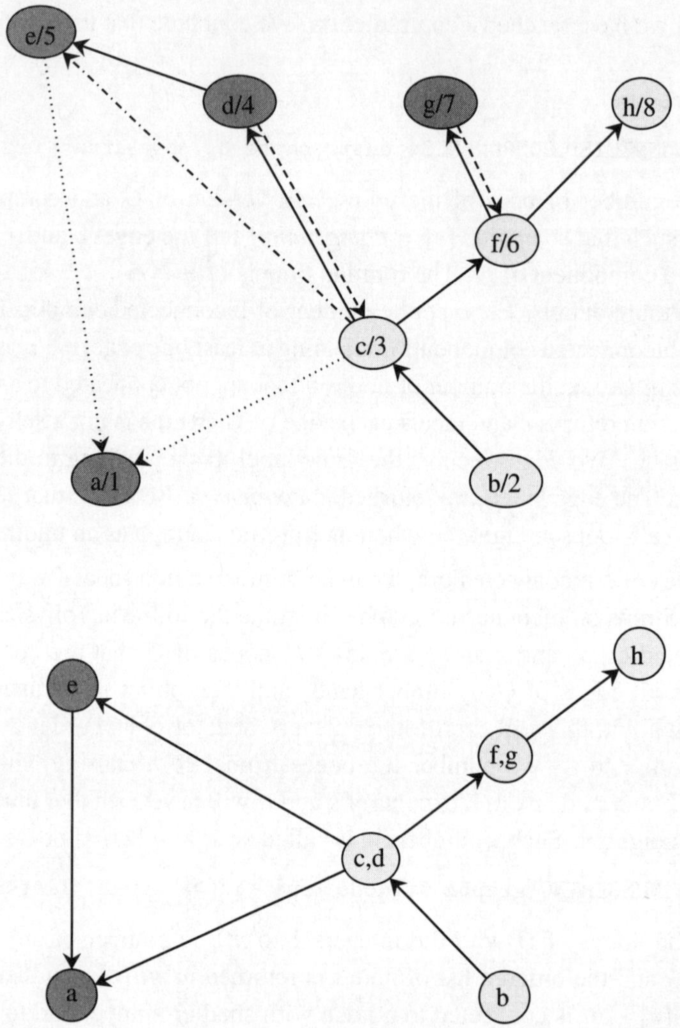

Figure 7.5 A snapshot of depth-first search on the graph of Figure 7.2 and the shrunken graph corresponding to it.
A first dfs was started at node a and a second dfs was started at node b. The upper part shows the snapshot of dfs; it is assumed that the search has just reached node h and is starting to explore the edges out of h. The edge (h, i) and the node i have not been seen yet and the depth-first search numbers of the nodes are indicated. The node h is the current node. Completed nodes are shown shaded.
The shrunken graph is shown in the lower part of the figure. The components $\{a\}$ and $\{e\}$ are permanent and all other components are tentative. The permanent components are shown shaded. The tentative components form a path P in the shrunken graph and h belongs to the last component of P. The roots of the tentative components are the vertices b, c, f, and h. They lie on a common tree path of the depth-first search tree of G.

the call $dfs(v, \ldots)$ has not been started yet, and *active* otherwise, i.e., if the call has already been started but not yet completed. All active nodes lie on a single path in G and this path corresponds to the recursion stack of depth-first search. We call the last node of this path the *current node*. We call an scc of G_c *permanent* if all its vertices are completed and we call it *tentative* if this is not the case. The *root* of an scc is the node in the scc with the smallest depth-first search number. Figure 7.5 illustrates these concepts. In this example the shrunken graph of G_c exhibits considerable structure:

(1) There is no edge $(v, w) \in E$ with v belonging to a permanent scc and w not belonging to a permanent scc. In particular, all vertices reachable from a vertex in a permanent scc are completed.

(2) The tentative sccs form a path P in the shrunken graph and the current node is contained in the last scc of this path.

(3) If C and C' are distinct tentative sccs with C preceding C' on P then all vertices in C have smaller dfs-number than all vertices in C'.

(4) Let C be a tentative scc of G_c and let r be its root. Then all vertices in C and all nodes in all successors of C on P are tree descendants of r in the depth-first search tree, i.e., the name root is justified.

We will show below that all four properties hold true generally and not only for our running example. The four properties will be invariants of the algorithm to be developed. The first invariant implies that the permanent sccs of G_c are actually sccs of G, i.e., it is justified to call them permanent. This observation is so important that it deserves to be stated as a lemma.

Lemma 4 *A permanent scc of G_c is an scc of G.*

Proof Let v be a vertex in a permanent scc of G_c and let w be a node of G such that v and w belong to the same scc of G. Thus there is a cycle C in G passing through v and w. If v and w do not belong to the same scc of G_c, one of the edges of C does not belong to G_c. The source node of this edge cannot be completed and hence does not lie in a permanent component. Since v lies in a permanent component, there must be an edge (x, y) on C such that x lies in a permanent component, but y does not. This is a contradiction to our first invariant. □

Invariants (2) to (4) suggest a simple method to represent the tentative sccs of G_c. We simply keep a sequence *unfinished* of all vertices in tentative sccs in increasing order of dfs-number and a sequence *roots* of all roots of tentative sccs. In our example *unfinished* is b, c, d, f, g, h, and *roots* is b, c, f, h. For both sequences the data type *stack<node>* is appropriate.

We can now start to write code. As already mentioned the program is an extension of depth-first search and has the same global structure. As in Section 7.3 we define two procedures: *STRONG_COMPONENTS* is the main procedure and *SCC_DFS* is an auxiliary procedure. Both procedures make use of the stacks *unfinished* and *roots* and the node arrays *dfsnum* and *comp_num*: *dfsnum*[v] is the dfs-number of v for all reached nodes and is -1 for all unreached nodes; *comp_num*[v] is the number of the scc containing v for all nodes belonging to permanent sccs and is -1 for all other nodes. The variables *dfscount* and *comp_count* keep track of the used dfs-numbers and component numbers, respectively.

STRONG_COMPONENTS defines and initializes all variables and then iterates over all nodes of G. It calls *SCC_DFS*($v, \ldots$) for each unreached node v. A call *SCC_DFS*($v, \ldots$) assigns the next dfs-number to v and makes v a tentative scc of its own. It then explores all edges out of v. Finally, it returns from the call.

⟨*SCC*⟩≡

```
void SCC_DFS(node v, const graph& G, node_array<int>& dfsnum,
             node_array<int>& comp_num, stack<node>& unfinished,
             stack<node>& roots, int& dfscount, int& comp_count)
{ dfsnum[v] = dfscount++;
  ⟨make v a tentative scc of its own⟩
  node w;
  forall_adj_nodes(w,v){ ⟨explore edge (v,w)⟩ }
  ⟨return from the call for node v⟩
}
int STRONG_COMPONENTS(const graph& G, node_array<int>& comp_num)
{ stack<node> unfinished;
  stack<node> roots;
  node_array<int> dfsnum(G, - 1);
  node v;
  forall_nodes(v,G) comp_num[v] = - 1;
  int dfscount = 0;
  int comp_count = 0;
  forall_nodes(v,G)
    if (dfsnum[v] == -1)
      SCC_DFS(v,G,dfsnum,comp_num,unfinished,roots,dfscount,comp_count);
  return comp_count;
}
```

A call *SCC_DFS*($v, \ldots$) makes v a tentative scc of its own since G_c contains no edges out of v yet. This amounts to adding v to the top of *unfinished* and *roots*. Thus

⟨*make v a tentative scc of its own*⟩≡

```
unfinished.push(v);
roots.push(v);
```

It is easy to check that all invariants are maintained.

We come to the exploration of an edge $e = (v, w)$. If e is a tree edge (this is the case iff *dfsnum*[w] $= -1$) we simply initiate a recursive call. If e is a non-tree edge and w belongs to a permanent scc (this will be the case if *dfsnum*[w] ≥ 0 and *comp_num*[w] ≥ 0), then, by Lemma 4, no action is required to maintain the invariants. If e is a non-tree edge and w belongs to a tentative scc (this will be the case if *dfsnum*[w] ≥ 0 and *comp_num*[w] $= -1$)

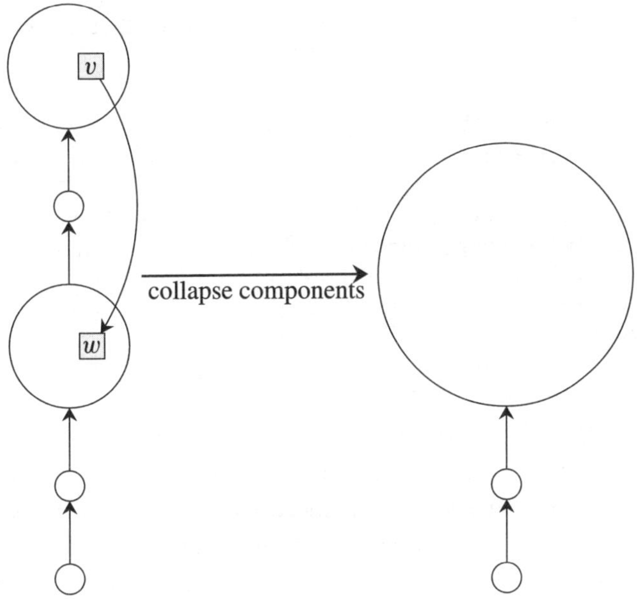

Figure 7.6 The path of tentative sccs and the effect of exploring an edge (v, w), where w belongs to a tentative scc. All tentative sccs on the path from the tentative scc containing w to the tentative scc containing v are collapsed into a single scc.

then some final segment of the path of tentative sccs collapses to a single scc (cf. Figure 7.6). Thus

⟨*explore edge (v,w)*⟩≡

```
if (dfsnum[w] == - 1)
   SCC_DFS(w,G,dfsnum,comp_num,unfinished,roots,dfscount,comp_count);
else if (comp_num[w] == - 1) { ⟨merge sccs⟩ }
```

We give the details of merging sccs. Assume that w belongs to a tentative scc with root r. Then r is the topmost root in *roots* with $dfsnum[r] \le dfsnum[w]$ (by invariant (3)). Any root r' above r ceases to be a root since $v \longrightarrow w \stackrel{*}{\longrightarrow} r \stackrel{*}{\longrightarrow} r' \stackrel{*}{\longrightarrow} v$. Note that $w \stackrel{*}{\longrightarrow} r$ since w and r belong to the same scc, and $r \stackrel{*}{\longrightarrow} r' \stackrel{*}{\longrightarrow} v$ since the shrunken graph of tentative sccs is a path. Thus

⟨*merge sccs*⟩≡

```
while (dfsnum[roots.top()] > dfsnum[w]) roots.pop();
```

What do we have to do when we return from a call, say for node v? The completion of v completes an scc if v is a root (by invariant (4)) and v is a root iff v = *roots.top*( ) (since the call for the topmost root is completed before the call of any other root contained in *roots*, again by invariant (4)). If v is a root the scc of v consists of all nodes in *unfinished* whose *dfsnum* is at least as large as v's *dfsnum* (by invariant (3)). We simply pop these nodes from *unfinished* and define their *comp_num*. Lemma 4 tells us that this scc is also an scc of the final graph.

⟨*return from the call for node v*⟩≡

```
if (v == roots.top())
{ do
  { w = unfinished.pop();
    comp_num[w] = comp_count;
  } while ( w != v);
  comp_count++;
  roots.pop();
}
```

Invariants (2), (3), and (4) are clearly maintained. For invariant (1) this can be seen as follows. Let C be the scc with root v. Then C is the last scc of the path P of tentative sccs and hence all other tentative sccs are predecessors of C on P. Thus there can be no edge in E_c from a vertex in C to a vertex in any other tentative scc. Since all nodes in C are completed, all edges $(x, y) \in E$ with $x \in C$ are also edges in E_c and invariant (1) holds.

7.4.3 *Strongly Connected Components: An Animation*

We describe an animation of the algorithm of the preceding section. The animation is available as the xlman-demo gw_scc_anim . The animation consists of two parts. In the first part the user can interactively construct a directed graph G; after every edit operation of the user the strongly connected components of G are recomputed and shown in number and color code, i.e., nodes belonging to the same scc are shown in the same color and with the same integer label. In the second part the execution of our scc-algorithm on the graph constructed in the first section is animated. Figure 7.7 shows a screen-shot. The overall structure of the program is as follows:

⟨*gw_scc_anim.c*⟩≡

```
#include <LEDA/graph_alg.h>
#include <LEDA/graphwin.h>
⟨display functions for part one⟩
⟨display functions for part two⟩
⟨help panels⟩
int main(){
GraphWin gw("SCC Animation Demo");
gw.display();            // open display
gw.set_directed(true);
int h_menu = gw.get_menu("Help");
gw_add_simple_call(gw,about_scc_anim1, "About SCC: phase 1",h_menu);
gw_add_simple_call(gw,about_scc_anim2, "About SCC: phase 2",h_menu);
gw_add_simple_call(gw,about_scc_anim_basics, "About SCC: basics",h_menu);
gw_add_simple_call(gw,about_scc_anim_data_structures,
                                "About SCC: data structures",h_menu);
⟨part one of demo⟩
```

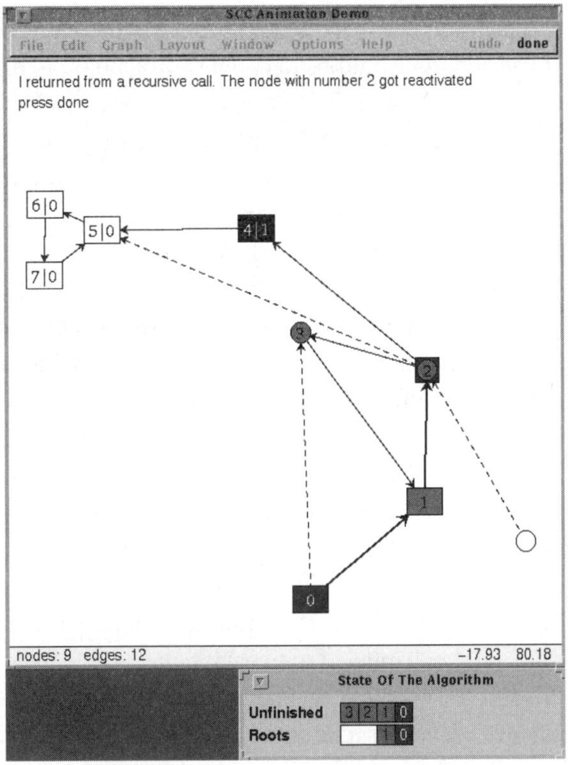

Figure 7.7 A screen-shot of the second part of gw_scc_anim. Explored nodes are labeled with their depth-first search number and nodes in permanent sccs are labeled with their depth-first search number and the number of the scc containing them. Explored edges are drawn solid and unexplored edges are drawn dashed. The nodes in permanent sccs are shown in the left half of the window and the other nodes are shown in the right half of the screen. The node with depth-first number 2 is the currently active node and there are two tentative components, one consisting of node 0 and the other one consisting of nodes 1, 2, and 3. There is one unreached node. The stacks *unfinished* and *roots* are indicated at the bottom of the screen-shot. The text at the top of the window explains the actions of the algorithm.

```
⟨part two of demo⟩
return 0;
}
```

The animation is based on the data type *GraphWin*; this data type is a combination of graphs and windows and is discussed in Chapter 12. Most of the current section can be appreciated without knowledge of GraphWins, as we explain the used features of GraphWin as we go along. However, the explanations of GraphWin will be kept short and hence readers without knowledge of GraphWin will miss some of the fine points. We hope that all readers will enjoy the demo so much that they will also study GraphWin.

In *main* we first define a *GraphWin gw* and then inform *gw* that we are dealing with directed graphs. We then set up the help menu. *GraphWin* has already a predefined help menu. We get its number and add three buttons to it. The corresponding functions *about_scc_anim1*, ... are defined in the program chunk ⟨*help panels*⟩. We do not show it as the help buttons of gw_scc_anim should give sufficient information. Having set up the help buttons we start part one of the demo. It makes use of the display functions defined in the corresponding chunk. The same holds true for the second part of the demo.

We come to the first part of the demo. Thanks to the powerful *GraphWin* data type it is extremely simple to write. A *GraphWin* always has an associated graph and moreover it maintains information about how to display the constituents of this graph: for example, for a node it maintains the position of the node, the color of the node, and the shape of the node (circle, square, rectangle, or ellipse), and for an edge it maintains the style of the edge (solid or dashed or dotted) and the color and the width of the edge. The display information can be modified.

In *display_scc* we first get the current graph G from *gw* and then compute the strongly connected components of G. We then set for each node v of G the color of v to the component number of v modulo 16 (as we rely only on the availability of 16 different colors) and we set the so-called user label[4] of v to the component number of v. We also inform *gw* that we want the user label to be displayed with each node.

We want *display_scc* to be called whenever the graph associated with *gw* is modified. This is easy to achieve. It is possible to associate functions with a *GraphWin* (so-called handlers) that are called whenever a node or edge is added or deleted. For example, *gw.set_del_edge_handler*(*display_scc*) informs *gw* that the function *display_scc* is to be called whenever an edge is deleted. The handlers for the addition of a node or edge are syntactically required to have a second argument which is a node or edge, respectively. We therefore need to wrap *display_scc* accordingly before defining the new edge and the new node handler.

After having set the handlers we open the display, show the help information for phase one, and put *gw* into edit mode. The call *gw.edit*( ) is terminated by a click on the done-button of *gw*.

⟨*display functions for part one*⟩≡

```
void display_scc(GraphWin& gw)
{ graph& G = gw.get_graph();
  node_array<int> comp_num(G);
  int N = STRONG_COMPONENTS(G,comp_num);
  node v;
  forall_nodes(v,G)
  { gw.set_color(v,comp_num[v]%16);
    gw.set_user_label(v,string("%d",comp_num[v]));
    gw.set_label_type(v,user_label);
  }
}

void new_edge_handler(GraphWin& gw, edge)  { display_scc(gw); }
```

4 In *GraphWin* each node has a number of predefined labels; one of them is called the user label.

```
void new_node_handler(GraphWin& gw, node)  { display_scc(gw); }
```

⟨*part one of demo*⟩≡

```
gw.set_init_graph_handler(display_scc);
gw.set_del_edge_handler(display_scc);
gw.set_del_node_handler(display_scc);
gw.set_new_node_handler(new_node_handler);
gw.set_new_edge_handler(new_edge_handler);
about_scc_anim1(gw);  // inform user about phase 1
gw.message("\\blue Construct or load a graph and press done.");
wait(1.75);
gw.message("");
gw.edit();            // enter edit mode
```

We come to part two of the demo. The goal of part two is to animate the strongly connected components algorithm of the preceding section. The idea behind the animation is as follows. We use a split design for the main window. The right half of the window shows all tentative components of G_c and all unexplored nodes and the left half of the screen shows all permanent components. Also, unreached nodes are shown as white empty circles and unexplored edges are shown dashed. The code below sets up the initial configuration of this design and also displays some textual information for the user (which we do not show here to save space).

We first get the coordinates of the window boundaries and then move the contents of *gw* to the right half of the screen. We then create the initial drawing of the demo. For each node v we set the color to white, state that the node is to be drawn as a circle of radius *small_width* (*small_width* is defined in program chunk ⟨*display functions for part two*⟩), state that the displayed information is the user label, set the user label to the empty string, and compute the position to which v is moved once it belongs to a permanent component. We also set the style of all edges to dashed. We then call the *STRONG_COMPONENTS* function of the preceding section; of course, this function needs to be augmented by display actions and therefore needs additional arguments, namely, *gw* and *perm_pos*.

⟨*part two of demo*⟩≡

```
gw.disable_calls();     // disable buttons
about_scc_anim2(gw);
graph&  G = gw.get_graph();
window& W = gw.get_window();
node_array<point> perm_pos(G);
double xmin = gw.get_xmin() + W.pix_to_real(20);
          // coordinate of left boundary plus 20 pixels
double xmax = gw.get_xmax() - W.pix_to_real(20);
double ymin = gw.get_ymin() + W.pix_to_real(30);
double ymax = gw.get_ymax() - W.pix_to_real(20);
double dx = xmax - xmin;
double dy = ymax - ymin;
```

```
gw.place_into_box(xmin+dx/2,ymin,xmax,ymax-dy/5);
         // move everything to right half of screen
gw.set_flush(false);  // changes are accumulated
node v;
forall_nodes(v,G)
{ gw.set_color(v,white);
  gw.set_label(v,user_label); gw.set_user_label(v,"");
  gw.set_shape(v,circle_node);
  gw.set_node_width(small_width);
  double xcoord = gw.get_position(v).xcoord();
  double ycoord = gw.get_position(v).ycoord();
  perm_pos[v] = point(xcoord-dx/2,ycoord);
}
edge e;
forall_edges(e,G) gw.set_style(e,dashed_edge);
gw.redraw();          // all changes are performed now
gw.set_flush(true);
⟨more information about part two⟩
node_array<int> comp_num(G);
STRONG_COMPONENTS(G,comp_num,gw,perm_pos);
gw.message("\\bf Wasn\'t this a nice demo ?");
wait(1);
gw.message("");
gw.fill_window();
gw.enable_calls();             // enable buttons
gw.edit();
```

We come to the display functions used for part two. We display nodes in two sizes: roots and nodes in permanent components are shown as large rectangles and all other nodes are shown as small circles. All nodes in the same strongly connected components are colored with the same color. For permanent components we use the color corresponding to the component number and for tentative components we use the height of the root of the component in the *roots*-stack. In order to keep the colors for permanent and tentative components separate (or at least approximately so) we add an integer *color_shift* to all colors of tentative components.

The demo can be run in either of two modes. In step mode the next action is triggered by a click on the done-button and in continuous mode the animation is run to completion without user interaction. The choice of mode is controlled by the variable *step* and the procedure *message* which we use to write messages *msg* into *gw*. If *step* is true, *msg* is displayed until the done-button is pressed. If *step* is true and the exit button is pressed, *step* is set to false and the demo runs to completion (since *message* has no effect when *step* is false).

We define a window *state_win* (in addition to the window associated with *gw*) and use it to display state information. The state information is generated by the function *state_info*.

It draws the stacks[5] *unfinished* and *roots* as sequences of rectangles into *state_win*. Each rectangle is labeled with the dfs-number of the node it represents. The stacks *unfinished* and *roots* are displayed in a way that equal elements are aligned (recall that *roots* is a subsequence of *unfinished*).

⟨*display functions for part two*⟩≡

```
static int  small_width = 20;
static int  large_width = 36;
static int  color_shift = 5;

static bool step = true;

void message(GraphWin& gw, string msg)
{ msg += "\\5 \\blue press done \\black";
  if (step && !gw.wait(msg)) step = false;
}

static window state_win(320,60,"State Of The Algorithm");

static void state_redraw(window* wp) { wp->flush_buffer(); }

static color text_color(color col)
{ if (col==black || col==red || col==blue || col==violet ||
      col==brown || col==pink || col==blue2 || col==grey3)
      return white;
  else
      return black;
}

void state_info(GraphWin& gw, const list<node>& unfinished,
                              const list<node>& roots,
                              const node_array<int>& dfsnum,
                              node cur_v)
{
  if (!state_win.is_open())
  { state_win.set_bg_color(grey1);
    state_win.set_redraw(state_redraw);
    state_win.display(-gw.get_window().xpos()+8,0);
    state_win.init(0,320,0);
    state_win.start_buffering();
   }
  state_win.clear();

  double th = state_win.text_height("H");

  double x0 = state_win.text_width("Unfinished") + 2*th;
  double y1 = state_win.ymax() - 1.75*th;
  double y2 = state_win.ymax() - 3.20*th;

  double d = 18;

  state_win.draw_text(5,y1+(d+th)/2,"Unfinished");
  state_win.draw_text(5,y2+(d+th)/2,"Roots");

  list_item r_it = roots.first();

  double x = x0;
```

[5] In contrast to the preceding section we realize both stacks as lists, the reason being that we need to iterate over all elements in both stacks and that stacks do not support iteration over their elements (they probably should).

```
    list_item u_it;
    forall_items(u_it, unfinished)
    { node v = unfinished[u_it];
      color col = gw.get_color(v);
      int   dn  = dfsnum[v];
      state_win.draw_box(x,y1,x+d,y1+d,col);
      state_win.draw_rectangle(x,y1-1,x+d,y1+d,black);
      state_win.draw_ctext(x+d/2,y1+d/2,string("%d",dn),text_color(col));
      if ( v == roots[r_it] )
      { state_win.draw_box(x,y2,x+d,y2+d,col);
        state_win.draw_rectangle(x,y2-1,x+d,y2+d,black);
        state_win.draw_ctext(x+d/2,y2+d/2,string("%d",dn),
                                                    text_color(col));
        r_it = roots.succ(r_it);
       }
      else
        state_win.draw_box(x+1,y2,x+d,y2+d,white);
      x += d;
     }
    state_win.draw_rectangle(x0,y1-1,x,y1+d,black);
    state_win.draw_rectangle(x0,y2-1,x,y2+d,black);
    state_win.flush_buffer();
}
```

The functions *STRONG_COMPONENTS* and *SCC_DFS* have the same overall structure as in the preceding section, but are augmented by display actions. At the beginning of a call *SCC_DFS*$(v,\ldots)$ we call *gw.select*(v) to highlight v and at the end of the call we call *gw.deselect*(v) to unhighlight v. In the *forall_adj_edges_loop* we color the edge explored red and make it solid.

⟨*display functions for part two*⟩+≡

```
void SCC_DFS(node v, const graph& G, node_array<int>& dfsnum,
             node_array<int>& comp_num, list<node>& unfinished,
             list<node>& roots, int& dfscount, int& comp_count,
             GraphWin& gw, const node_array<point>& perm_pos)
{ gw.select(v);
  ⟨new node v was reached⟩
  node w; edge e;
  forall_adj_edges(e,v)
  { w = G.target(e);
    gw.set_style(e,solid_edge);
    gw.set_color(e,red);
    string msg = "I am exploring the red edge.\\3 ";
    if (dfsnum[w] == - 1) { ⟨tree edge and recursive call⟩ }
    else if (comp_num[w] == - 1)
           { ⟨non-tree edge into tentative component⟩ }
         else
           { ⟨non-tree edge into permanent component⟩ }
  }
```

```
    if (v == roots.head()) { ⟨v is a root⟩ }
    gw.deselect(v);
  }
```

In *STRONG_COMPONENTS* we inform the user about every new call of *SCC_DFS* except for the first.

⟨*display functions for part two*⟩+≡

```
int STRONG_COMPONENTS(const graph& G, node_array<int>& comp_num,
                GraphWin& gw, const node_array<point>& perm_pos)
{ list<node> unfinished;
  list<node> roots;
  node_array<int> dfsnum(G,-1);
  node v;
  forall_nodes(v,G) comp_num[v] = -1;
  int dfscount = 0;
  int comp_count = 0;
  forall_nodes(v,G)
    if (dfsnum[v] == -1)
    { SCC_DFS(v,G,dfsnum,comp_num,unfinished,roots,dfscount,
                                   comp_count,gw,perm_pos);
      message(gw,"This was a return from an outermost call\\3
                I am looking for an unreached node and \\n\
                (if successful) start a new search from it.");
    }
  return comp_count;
}
```

When a new node is reached it is given a dfs-number and is pushed on *unfinished* and *roots*. The new node forms a tentative strongly connected component of its own. We set the color of v to the size of the *roots*-stack (shifted by *color_shift* so as to avoid too much overlap with the colors used for permanent components), we set the user label of v to its dfs-number, and we set the shape and width of v to a large rectangular shape (so as to indicate that v is a root). We build up a string to explain our actions, hand it to *message* to display it, and call *state_info* to update the state information.

⟨*new node v was reached*⟩≡

```
dfsnum[v] = dfscount++;
unfinished.push(v);
roots.push(v);
gw.set_color(v,(color_shift + roots.size())%16);
gw.set_user_label(v,string("%d",dfsnum[v]));
gw.set_shape(v,rectangle_node);
gw.set_width(v,large_width);
string msg;
msg += "A new node has been reached.\\3 ";
msg += "It got the dfs-number ";
msg += string("%d ",dfsnum[v]);
msg += "and it is the new current node.\\3 ";
```

```
msg += "It is the root of a new tentative component.";
state_info(gw,unfinished,roots,dfsnum,v);
message(gw,msg);
```

A tree edge $e = (v, w)$ leads to a recursive call. We inform the reader about this fact by textual output, we unhighlight v as it ceases to be a current node, and we emphasize the edge e (by increasing its width and setting its color to blue); in this way the tree path to the current node is always shown as a path of thick blue edges. Then we make the recursive call. After the return from the recursive call, we de-emphasize e and highlight v (again), and we inform the reader that we just returned from a recursive call and that v became active again.

⟨*tree edge and recursive call*⟩≡

```
msg += "It's a tree edge and I am making a recursive call.";
message(gw,msg);
state_info(gw,unfinished,roots,dfsnum,v);
gw.deselect(v);
gw.set_color(e,blue);
gw.set_width(e,2);
SCC_DFS(w,G,dfsnum,comp_num,unfinished,roots,dfscount,
                                   comp_count,gw, perm_pos);
gw.set_width(e,1);
gw.set_color(e,black);
gw.select(v);
state_info(gw,unfinished,roots,dfsnum,0);
message(gw,"I returned from a recursive call. The node with \
number " + string("%d ",dfsnum[v]) + " got reactivated");
```

A non-tree edge $e = (v, w)$ into a tentative component may close a cycle involving several tentative components. These components are merged into one. More precisely, all components whose root has a dfs-number larger than *dfsnum*$[w]$ cease to exist. We inform the user about this fact by textual output and then start popping *roots*. Whenever a node is popped from *roots* its shape and width are changed to a small circle. We put a *wait*(0.25) statement into the loop that pops from *roots* so that different roots are visibly popped one after the other. Once all roots are popped we recolor the nodes in the newly formed scc and give state information. Finally, we change the color of *e* back to black.

⟨*non-tree edge into tentative component*⟩≡

```
msg += "It's a non-tree edge into a tentative component. This edge may \
       merge several components into one.\\n More precisely: all \
       components whose root is larger than " + string("%d ",dfsnum[w]);
msg += "cease to exist and are merged into the component \
       containing the node with dfs-number " + string("%d. ",dfsnum[w]);
msg += "Algorithmically, this amounts to removing all roots \
       larger than " + string("%d ",dfsnum[w]);
msg += "from the stack of roots. I do so one by one. Removal of a node \
```

```
          from the stack of roots turns its shape from rectangular \
          to circular.";
  message(gw,msg);
  state_info(gw,unfinished,roots,dfsnum,v);
  while (dfsnum[roots.head()] > dfsnum[w])
  { node z = roots.pop();
    gw.set_shape(z,circle_node);
    gw.set_width(z,small_width);
    state_info(gw,unfinished,roots,dfsnum,v);
    wait(0.25);
  }
  node u;
  forall(u,unfinished)
    if (dfsnum[u] >= dfsnum[roots.head()] )
      gw.set_color(u,(color_shift + roots.size())%16);
  state_info(gw,unfinished,roots,dfsnum,0);
  message(gw,string("Now all roots are removed and the newly formed \
                    component has been recolored. The current \
                    node is still: %d.", dfsnum[v]));
  gw.set_color(e,black);
```

A non-tree edge *e* into a permanent component requires no action. We inform the user and change the color of *e* back to black.

⟨*non-tree edge into permanent component*⟩≡

```
  msg += "It's a non-tree edge into a permanent component. I do nothing.";
  message(gw,msg);
  state_info(gw,unfinished,roots,dfsnum,v);
  gw.set_color(e,black);
```

When a call *SCC_DFS*($v, \ldots$) for a root v is completed a permanent component has been found. We inform the reader accordingly. All nodes in the permanent component are moved to the left half of the window (by setting their position as given by *perm_pos*, the shape and width is changed to a large rectangular shape, the user label is set to a pair consisting of dfs-number and component number, and the color is set to the color corresponding to the component number.

⟨*v is a root*⟩≡

```
  string msg = "Node " + string("%d",dfsnum[v]) + " has been \
         completed. It is a root and hence we have identified \
         a permanent component. \\3 \
         The permanent component consists of all nodes  in \
         unfinished whose dfs-number is at least as large as "
         + string("%d",dfsnum[v]) + ". \\3 \
         I move all nodes in the component to the left and \
         indicate their dfs-number and their component number.";
  state_info(gw,unfinished,roots,dfsnum,0);
```

```
message(gw,msg);
do { w = unfinished.pop();
     if (v == w) roots.pop();
     comp_num[w] = comp_count;
     gw.set_shape(w,rectangle_node);
     gw.set_width(w,large_width);
     gw.set_color(w,comp_count%16);
     gw.set_user_label(w,string("%d | %d", dfsnum[w],comp_num[w]));
     state_info(gw,unfinished,roots,dfsnum,0);
     gw.set_position(w,perm_pos[w]);
} while ( w != v);
comp_count++;
```

Enjoy the animation.

Exercises for 7.4

1 Modify the algorithm for the computation of strongly connected components to compute biconnected components of undirected graphs. Hint: Define the root of a biconnected component as the node in the component with the second largest dfs-number. Then proceed as for strongly connected components.

2 Part one of the animation of strongly connected components is unsatisfactory as color changes are not "local". It would be desirable to have the following behavior: after the addition or deletion of a node or edge only the colors of those nodes change whose containing strongly connected component has changed. Modify the animation to achieve this behavior.

3 Animate the biconnected components algorithm of the first item.

4 Extend the first part of the animation of strongly connected components so that the shrunken graph is also visualized. A reasonable approach seems to represent each vertex of the shrunken graph by the convex hull of the vertices of the corresponding strongly connected component.

5 Define the shrunken graph of an undirected graph with respect to its biconnected components as follows. There is a vertex for each biconnected component and for each articulation point. A vertex standing for a component is connected to a vertex representing an articulation point if the articulation point is contained in the component. Show that the shrunken graph is a tree and give a program that computes it.

7.5 Shortest Paths

We introduce the shortest-path problem and describe the functionality of our various shortest-path programs. We discuss a checker for the single-source shortest-path problem and derive a generic shortest-path algorithm. We give algorithms and their implementations for acyclic networks, for the single-source problem with arbitrary edge costs, for the single-source single-sink problem, for the all-pairs problem, and for the minimum cost to profit

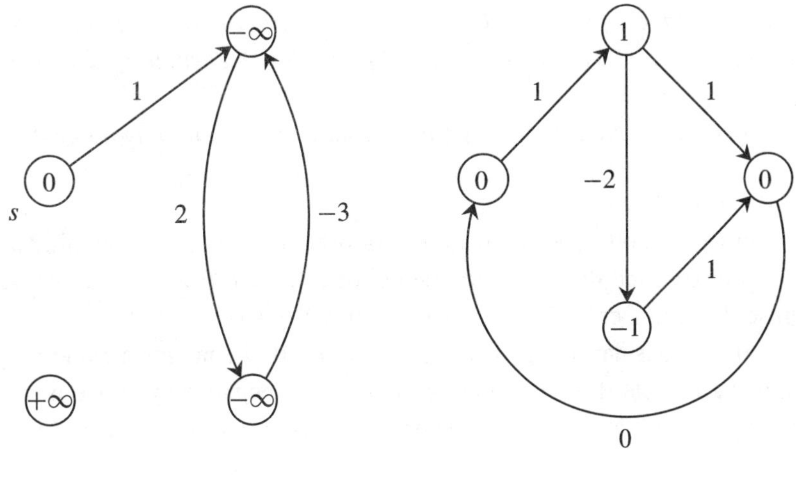

Figure 7.8 The node labels indicate $\mu(s, .)$. The graph on the left contains a negative cycle and also a node that is not reachable from s. Therefore there are node labels equal to $\pm\infty$. The graph on the right contains no negative cycle and all nodes are reachable from s. Therefore all node labels are finite.

ratio cycle problem; an algorithm for the single-source problem with non-negative edge costs was already given in Section 6.6. We also give experimental results about the running times of the various implementations.

7.5.1 *Functionality*

Let $G = (V, E)$ be a directed graph and let $c : E \longrightarrow I\!R$ be a *cost* function on the edges of G. We will also say *length* instead of cost. We extend the cost function to *paths* in the natural way: the cost (or length) of a path is the sum of the costs of its constituent edges, i.e., if $p = [e_1, e_2, \ldots, e_k]$ is a path then $c(p) = \sum_{1 \le i \le k} c(e_i)$. We will abuse notation and write $c(u, v)$ instead of $c(e)$ for $e = (u, v)$. For every vertex $v \in G$ the trivial path consisting of no edge is a path from v to v; its cost is zero. A *cycle* is a non-trivial path from v to v for some node v. A *negative cycle* is a cycle whose cost is negative.

For two vertices v and w we use $\mu(v, w)$ to denote the minimal cost of a path from v to w, i.e.,

$$\mu(v, w) = \inf\{c(p)\ ;\ p \text{ is a path from } v \text{ to } w\}\,.$$

The infimum of the empty set is defined as $+\infty$, i.e., $\mu(v, w) = +\infty$ if w is not reachable from v. Figure 7.8 illustrates this definition. The set of paths from v to w is in general an infinite set and hence it is not clear whether $\mu(v, w)$ is actually achieved by a path from v to w. The following lemma gives information about the existence of shortest paths.

Lemma 5

(a) *If w is not reachable from v then $\mu(v, w) = +\infty$.*

(b) *If there is a path from v to w containing a negative cycle then $\mu(v, w) = -\infty$.*

(c) If w is reachable from v and there is no path from v to w passing through a negative cycle then $-\infty < \mu(v, w) < +\infty$ and $\mu(v, w)$ is the length of a simple path from v to w.

(d) If $\mu(v, w) = -\infty$ then there is a path from v to w containing a negative cycle.

Proof Part (a) is true by definition.

For part (b), we observe that if there is a path from v to w containing a negative cycle then by going around the cycle sufficiently often a path from v to w whose cost is below any prescribed number is obtained. Thus $\mu(v, w) = -\infty$.

For part (c) consider any path p from v to w. If p contains a cycle let p' be obtained by removing a cycle from p. Since p contains no negative cycle we have $c(p') \le c(p)$. Continuing in this way we obtain a simple path from v to w whose cost is at most the cost of p. Thus

$$\mu(v, w) = \inf\{c(p)\ ;\ p \text{ is a simple path from } v \text{ to } w\}\,.$$

The number of simple paths from v to w is finite and hence $\mu(v, w) = c(p)$ for some simple path p.

We turn to part (d). If $\mu(v, w) = -\infty$ then w is reachable from v. If there is no path from v to w containing a negative cycle then $\mu(v, w) > -\infty$ by part (c). □

We distinguish between the *single-source single-sink shortest-path problem*, the *single-source shortest-path problem*, and the *all-pairs shortest-path problem*. The first problem asks for the computation of $\mu(s, t)$ for two specified nodes s and t and will be discussed in Section 7.5.6. The second problem asks to compute $\mu(s, v)$ for a specified node s and all v and the third problem asks to compute $\mu(s, v)$ for all nodes s and v. The single-source problem is the basis for the solutions to the other two problems and hence we discuss it first.

In our discussion of the single-source problem we use s to denote the source and we write $\mu(v)$ instead of $\mu(s, v)$. The following characterization of the function μ is extremely useful for the correctness proofs of shortest-path algorithms[6].

Lemma 6

(a) We have

$$\mu(s) = \min(0, \min\{\mu(u) + c(e)\ ;\ e = (u, s) \in E\})$$

and

$$\mu(v) = \min\{\mu(u) + c(e)\ ;\ e = (u, v) \in E\}$$

for $v \neq s$.

(b) If d is a function from V to $\mathbb{R} \cup \{-\infty, +\infty\}$ with

- *$d(v) \ge \mu(v)$ for all $v \in V$,*

[6] In this characterization and for the remainder of the section we use the following definitions for the arithmetic and order on $\mathbb{R} \cup \{-\infty, +\infty\}$: $-\infty < x < +\infty$, $+\infty + x = +\infty$, and $-\infty + x = -\infty$ for all $x \in \mathbb{R}$.

- *$d(s) \le 0$, and*
- *$d(v) \le d(u) + c(u, v)$ for all $e = (u, v) \in E$*

then $d(v) = \mu(v)$ for all $v \in V$.

Proof For part (a) we consider only the case $v \neq s$ and leave the case $v = s$ to the reader. Any path p from s to v consists of a path from s to some node u plus an edge from u to v. Thus

$$\begin{aligned} \mu(v) &= \inf\{c(p) \; ; \; p \text{ is a path from } s \text{ to } v\} \\ &= \min_u \inf\left\{c(p') + c(e) \; ; \; p' \text{ is a path from } s \text{ to } u \text{ and } e = (u, v) \in E\right\} \\ &= \min\{\mu(u) + c(e) \; ; \; e = (u, v) \in E\}. \end{aligned}$$

For part (b) we assume for the sake of a contradiction that $d(v) > \mu(v)$ for some v. Then $\mu(v) < +\infty$. We distinguish cases.

If $\mu(v) > -\infty$, let $[s = v_0, v_1, \ldots, v_k = v]$ be a shortest path from s to v. We have $\mu(s) = 0 = d(s)$, $\mu(v_i) = \mu(v_{i-1}) + c(v_{i-1}, v_i)$ for $i > 0$, and $\mu(v) < d(v)$. Thus, there is a least $i > 0$ with $\mu(v_i) < d(v_i)$ and hence

$$d(v_i) > \mu(v_i) = \mu(v_{i-1}) + c(v_i, v_{i-1}) = d(v_{i-1}) + c(v_i, v_{i-1}),$$

a contradiction.

If $\mu(v) = -\infty$, let $[s = v_0, v_1, \ldots, v_i, \ldots, v_j, \ldots, v_k = v]$ be a path from s to v containing a negative cycle. Such a path exists by Lemma 5. Assume that the subpath from v_i to v_j is a negative cycle. If $d(v) > \mu(v)$ then $d(v) > -\infty$ and hence $d(v_l) > -\infty$ for all l, $0 \le l \le k$. Thus,

$$\begin{aligned} d(v_i) &= d(v_j) && \text{since } v_i = v_j \\ &\le d(v_{j-1}) + c(v_{j-1}, v_j) \\ &\le d(v_{j-2}) + c(v_{j-2}, v_{j-1}) + c(v_{j-1}, v_j) \\ &\vdots \\ &\le d(v_i) + \textstyle\sum_{l=i}^{j-1} c(v_l, v_{l+1}), \end{aligned}$$

and hence $\sum_{l=i}^{j-1} c(v_l, v_{l+1}) \ge 0$, a contradiction to the fact that the subpath from v_i to v_j is a negative cycle. □

We split the set of vertices of G into three sets:

$$\begin{aligned} V^- &= \{v \in V \; ; \; \mu(v) = -\infty\}, \\ V^f &= \{v \in V \; ; \; -\infty < \mu(v) < +\infty\}, \text{ and} \\ V^+ &= \{v \in V \; ; \; \mu(v) = +\infty\}. \end{aligned}$$

The vertex s belongs to V^f if there is no negative cycle passing through s and it belongs to V^- otherwise; in the latter case V^f is empty. The set V^+ consists of all vertices that are not

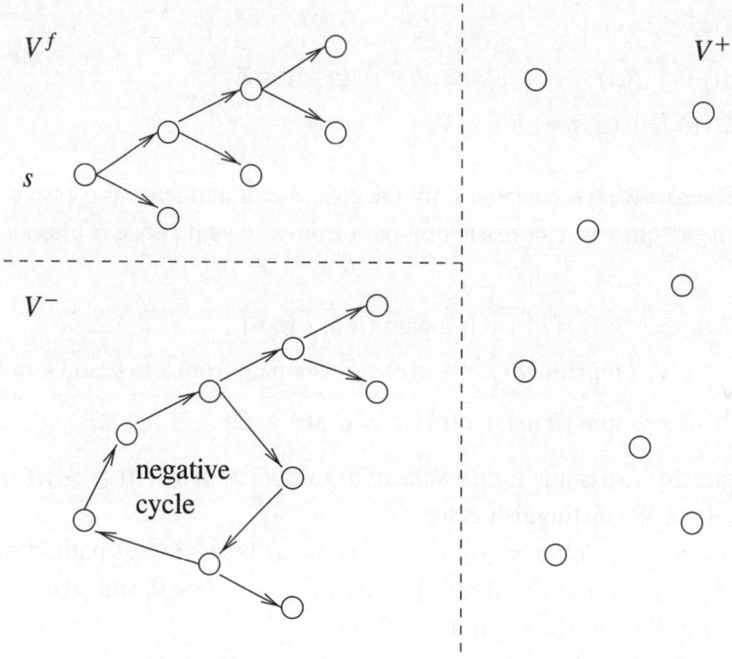

Figure 7.9 A solution to a single-source problem: It consists of a shortest-path tree on V^f, a collection of negative cycles plus trees emanating from them on V^-, a set V^+ of isolated nodes, and the values $\mu(v)$ for $v \in V^f$.

reachable from s. A *shortest-path tree* with respect to s is a tree defined on V^f such that for any $v \in V^f$ the tree path from s to v is a shortest path from s to v.

We next define the output convention for the single-source shortest-path problem. What do we want to know? Certainly, $\mu(v)$ for all nodes v. However, knowing $\mu(v)$ is usually not enough. If $v \in V^f$, it is useful to know a shortest path from s to v and if $v \in V^-$, it is useful to know the negative cycle that "puts" v into V^-. Our algorithms therefore also produce a shortest-path tree on V^f and a collection of negative cycles plus trees emanating from them on V^-, see Figure 7.9. The exact definition is as follows[7]:

The solution to a single-source shortest-path problem (G, s, c) is a pair (*dist*, *pred*), where *dist* is a *node_array*<*NT*> and *pred* is a *node_array*<*edge*>. Let

$$P = \{pred[v]\ ;\ v \in V \text{ and } pred[v] \neq nil\}\,.$$

The pair must have the following properties:

- $s \in V^f$ iff $pred[s] = nil$ and $s \in V^-$ iff $pred[s] \neq nil$.
- For $v \neq s$: $v \in V^+$ iff $pred[v] = nil$ and $v \in V^f \cup V^-$ iff $pred[v] \neq nil$.

[7] We further comment on our output convention after its definition.

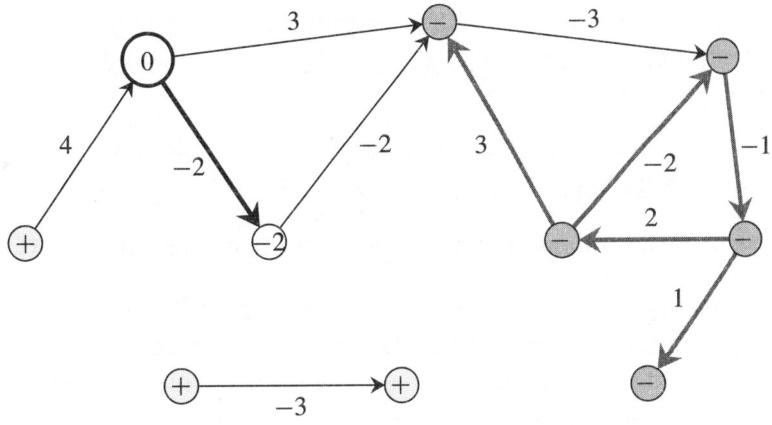

Figure 7.10 The output of a single-source shortest-path problem. The source node s is shown bigger than all other nodes. Its *dist*-label is zero. Edge costs are indicated. For every node v with $pred[v] \neq nil$ the edge $pred[v]$ is shown in bold. For the nodes in V^f the *dist*-value is shown inside the node. For nodes $v \in V^+ \cup V^-$ the set containing v is indicated by a + or −. V^+ consists of all nodes $v \neq s$ with $pred[v] = nil$, V^f consists of all nodes that are reachable from s by a P-path, and V^- consists of all nodes that lie on a P-cycle or are reachable from a P-cycle by the P-path. All P-cycles have negative cost. You may generate your own figures with the xlman-demo gw_shortest_path.

- $v \in V^f$ if v is reachable from s by a P-path[8] and $s \in V^f$. P restricted to V^f forms a shortest-path tree and $dist[v] = \mu(v)$ for $v \in V^f$.
- All P-cycles have negative cost and $v \in V^-$ iff v lies on a P-cycle or is reachable from a P-cycle by a P-path.

Figure 7.10 shows an example. Observe that our output convention leaves the value of $dist[v]$ unspecified for $v \in V^+ \cup V^-$. We have made this choice because most number types have no representation for $+\infty$ and $-\infty$. In the *absence of negative cycles* our output convention simplifies to the following:

- For $v \neq s$: $v \in V^f$ iff $pred[v] \neq nil$ and $v \in V^+$ otherwise.
- $pred[s] = nil$.
- P is a shortest-path tree on V^f and $dist[v] = \mu(v)$ for $v \in V^f$.

Our output convention for the single-source shortest-path problem is non-standard. Most papers on the shortest-path problem do not define precisely how negative cycles are reported and this was also true for early versions of LEDA. We have defined our output convention such that:

- the return value of a single-source algorithm consists of a pair (*dist*, *pred*), as is customary for single-source algorithms. We played with the idea to add an output

[8] A P-path is a path all of whose edges belong to P and a P-cycle is a cycle all of whose edges belong to P.

parameter, which indicates for every node its membership to the sets V^+, V^f, and V^-. We decided against it, because we wanted to stick with the traditional interface of shortest-path algorithms,

- it can be checked in linear time whether a pair (*dist*, *pred*) is a solution to the shortest-path problem (G, s, c), see Section 7.5.2,
- shortest-path algorithms can satisfy it with little additional effort.

We turn to algorithms. All algorithms are function templates that work for an arbitrary number type *NT*. We use the convention that names of function templates for graph algorithms end with _*T*. In order to use the templates one must include LEDA/templates/shortest_path.t. LEDA also contains pre-compiled instantiations for the number types *int* and *double*. The function names for the instantiated versions are *without* the suffix _*T*. In order to use the instantiated versions one must include LEDA/graph_alg.h. Section 7.1 discusses the relationship between templates and instantiated versions in more detail.

Acyclic Graphs:

```
void ACYCLIC_SHORTEST_PATH_T(const graph& G, node s,
                             const edge_array<NT>& c,
                             node_array<NT>& dist,
                             node_array<edge>& pred)
```

solves the problem in time $O(n + m)$ for acyclic graphs, see Section 7.5.4. As always, we use n to denote the number of nodes of G and m to denote the number of edges of G.

Non-Negative Edge Costs:

```
void DIJKSTRA_T(const graph& G, node s, const edge_array<NT>& c,
                node_array<NT>& dist, node_array<edge>& pred)
```

solves the problem in time $O(m + n \log n)$ if all edge costs are non-negative. We have discussed this function already in Section 6.6. If all edge costs are equal to one then breadth-first search, see Section 7.3, solves the problem in linear time.

General Edge Costs:

```
bool BELLMAN_FORD_T(const graph& G, node s, const edge_array<NT>& c,
                    node_array<NT>& dist, node_array<edge>& pred)
```

solves the problem in time $O(n \cdot m)$ for arbitrary edge costs. It returns false if $\mu(v) = -\infty$ for some vertex v. Otherwise, it returns *true*.

We also have a procedure

```
bool SHORTEST_PATH_T(const graph& G, node s, const edge_array<NT>& c,
                     node_array<NT>& dist, node_array<edge>& pred)
```

that tests whether one of the two special cases applies and, if so, applies the efficient procedure applicable to the special case. If none of the special cases applies, *BELLMAN_FORD_T* is called. The implementation of *SHORTEST_PATH_T* is simple.

⟨*SP.t*⟩≡

```
template <class NT>
bool SHORTEST_PATH_T(const graph& G, node s, const edge_array<NT>& c,
                     node_array<NT>& dist, node_array<edge>& pred )
{ if ( Is_Acyclic(G) )
  { ACYCLIC_SHORTEST_PATH_T(G,s,c,dist,pred);
    return true;
  }
  bool non_negative = true;
  edge e;
  forall_edges(e,G) if (c[e] < 0) non_negative = false;
  if (non_negative) { DIJKSTRA_T(G,s,c,dist,pred);
                      return true;
                    }
  return BELLMAN_FORD_T(G,s,c,dist,pred);
}
```

The Single-Sink Problem: The single-source single-sink shortest-path problem asks for the computation of a shortest path from a specified node s, the source, to a specified node t, the sink.

```
NT DIJKSTRA_T(const graph& G, node s, node t,
              const edge_array<NT>& c, node_array<edge>& pred)
```

computes a shortest path from s to t and returns its length. The cost of all edges must be non-negative. The return value is unspecified if there is no path from s to t. The array *pred* allows one to trace a shortest path from s to t in reverse order, i.e., *pred*$[t]$ is the last edge on the path. If there is no path from s to t or if $s = t$ then *pred*$[t] =$ *nil*. The worst case running time is $O(m + n \log n)$, but frequently much better. The implementation is discussed in Section 7.5.6.

The All-Pairs Problem: The all-pairs shortest-path problem asks for the computation of the complete distance function μ.

```
bool ALL_PAIRS_SHORTEST_PATHS_T(graph& G, edge_array<NT> c,
                                node_matrix<NT> DIST)
```

returns *true* if G has no negative cycle and returns *false* otherwise. In the latter case all values returned in *DIST* are unspecified. In the former case we have for all v and w: if $\mu(v, w) < \infty$ then $DIST(v, w) = \mu(v, w)$ and if $\mu(v, w) = \infty$, the value of $DIST(v, w)$ is unspecified. The procedure runs in time $O(nm + n^2 \log n)$.

Our output convention for the all-pairs problem is somewhat unsatisfactory. It is dictated by the fact that many number types have no representation of $+\infty$. An alternative solution is to also return a node matrix *PRED* of edges in analogy to the single-source problem.

7.5.2 A Checker for Single-Source Shortest-Path Algorithms

We develop a program *CHECK_SP_T*$(G, s, c, dist, pred)$ that checks whether $(dist, pred)$ is a correct solution to the shortest-path problem (G, s, c). If not, the program aborts (with the error message "assertion failed") and if so, the program returns a *node_array<int> label* with $label[v] < 0$ if $v \in V^-$, $label[v] = 0$ if $v \in V^f$, and $label[v] > 0$ if $v \in V^+$.

Let $P = \{pred[v]\,;\, pred[v] \neq nil\}$ be the set of edges defined by the *pred*-array and define

$$\begin{aligned} U^+ &= \{v\,;\, v \neq s \text{ and } pred[v] = nil\}, \\ U^f &= \emptyset, \text{ if } pred[s] \neq nil, \\ U^f &= \{v\,;\, v \text{ is reachable from } s \text{ by a } P\text{-path}\}, \text{ if } pred[s] = nil, \\ U^- &= \{v\,;\, v \text{ lies on a } P\text{-cycle or is reachable from a } P\text{-cycle by a } P\text{-path}\}. \end{aligned}$$

We perform the following checks:

(1) $v \in U^+$ iff v is not reachable from s in G.
(2) All P-cycles have negative cost.
(3) There is no edge $(v, w) \in E$ with $v \in U^-$ and $w \in U^f$.
(4) For all $e = (v, w) \in E$: if $v \in U^f$ and $w \in U^f$ then $dist[v] + c[e] \geq dist[w]$.
(5) For all $v \in U^f$: if $v = s$ then $dist[v] = 0$ and if $v \neq s$ then $dist[v] = dist[u] + c[pred[v]]$ where u is the source of $pred[v]$.

Lemma 7 *If $(dist, pred)$ satisfies the five conditions above then it is a solution to the shortest-path problem (G, s, c).*

Proof Observe first that $v \in V^+$ iff v is not reachable from s. Thus (1) implies that $U^+ = V^+$ and hence $U^f \cup U^- = V^f \cup V^-$. We next show that $U^- \subseteq V^-$. Consider any $v \in U^-$. By definition of U^- there is a P-cycle, call it C, from which v is reachable. Moreover, the cost of C is negative by (2) and there is a node on C that is reachable from s by (1). Thus $\mu(s, v) = -\infty$ and hence $v \in V^-$. Thus $U^- \subseteq V^-$ and therefore $U^f \supseteq V^f$. Assume for the sake of a contradiction that the latter inclusion is proper and let $v \in U^f \setminus V^f$ be arbitrary. Then $v \in V^-$ and hence there is a path p from s to v containing a negative cycle, say C. By (3) there is no edge (x, y) with $x \in U^-$ and $y \in U^f$. We conclude that all vertices on p belong to U^f. This implies that (4) holds for all edges of C. Let $e_0, e_1, \ldots, e_{k-1}$ with $e_i = (v_i, v_{i+1})$ be the edges of C. Then $v_0 = v_k$. We have

$$dist[v_0] + c(C) = dist[v_0] + \sum_{0 \leq i < k} c[e_i]$$

$$\begin{aligned} &\geq \; dist[v_1] + \sum_{1 \leq i < k} c[e_i] \;\geq\; \dots \;\geq\; dist[v_k] \\ &= \; dist[v_0] \end{aligned}$$

by repeated application of (4). Thus $c(C) \geq 0$, a contradiction. We have now shown that $U^+ = V^+$, $U^f = V^f$, and $U^- = V^-$. We still need to show that P restricted to V^f is a shortest-path tree. Consider any $v \in V^f$. Condition (5) implies that $dist[v]$ is the length of the P-path from s to v and (4) implies that the length of any path from s to v is at least $dist[v]$. Thus P is a shortest-path tree. □

We come to the implementation. We start with condition (1). We use depth-first search to determine all nodes reachable from s and we check whether for all nodes v different from s: $pred[v] = nil$ iff v is not reachable from s. We give all nodes that are not reachable from s the label *PLUS*; *PLUS* is an element of an enumeration type that we use to classify nodes. All nodes start with the label *UNKNOWN*. The other members of the enumeration type will be explained below.

⟨*condition one*⟩≡

```
enum{ NEG_CYCLE = -2,  ATT_TO_CYCLE = -1, FINITE = 0, PLUS = 1,
      CYCLE = 2, ON_CUR_PATH = 3, UNKNOWN = 4  };
node_array<int> label(G,UNKNOWN);
node_array<bool> reachable(G,false);
DFS(G,s,reachable);
node v;
forall_nodes(v,G)
{ if (v != s)
  { assert( (pred[v] == nil) == (reachable[v] == false));
    if (reachable[v] == false) label[v] = PLUS;
  }
}
```

Next we compute the sets U^f and U^-. Consider any node $v \notin U^+$. Tracing the path $[v, source(pred[v]), source(pred[source(pred[v])]), \ldots]$ until either a node is encountered twice or until the path cannot be extended further (it must end in s in the latter case because s is the only node outside U^+ which may have no incoming P-edge) allows us to classify all nodes on the path. In the former case v and all nodes on the path belong to U^- and in the latter case all of them belong to U^f. For the sequel it is useful to have a finer classification of the nodes in U^- into nodes lying on a P-cycle (label *CYCLE*) and nodes attached to a cycle by a P-path (label *ATT_TO_CYCLE*) and so we will compute the finer classification.

Of course, we do not want to trace the same path several times. We therefore stop tracing a path once a node is reached whose label is known (more precisely, is different from *UNKNOWN*). As we trace a path all nodes on the path are put onto a stack S and are given the label *ON_CUR_PATH*.

We initialize the classification step by giving s the label *FINITE* if its *pred*-value is *nil*.

⟨*classification of nodes*⟩≡

```
if (pred[s] == nil) label[s] = FINITE;
forall_nodes(v,G)
{ if ( label[v] == UNKNOWN )
  { stack<node> S;
    node w = v;
    while ( label[w] == UNKNOWN )
    { label[w] = ON_CUR_PATH;
      S.push(w);
      w = G.source(pred[w]);
    }
    ⟨label all nodes on current path⟩
  }
}
```

When a node w is encountered whose label is different from *UNKNOWN* we distinguish cases: if w is labeled *FINITE*, i.e., $v \in U^f$, then all nodes on the path belong to U^f, and if w is labeled *CYCLE* or *ATT_TO_CYCLE*, i.e., $v \in U^-$, then all nodes on the path (except for w) are attached to a cycle but do not lie on a cycle themselves, and if w belongs to the current path then the situation is as shown in Figure 7.11. This leads to the following code.

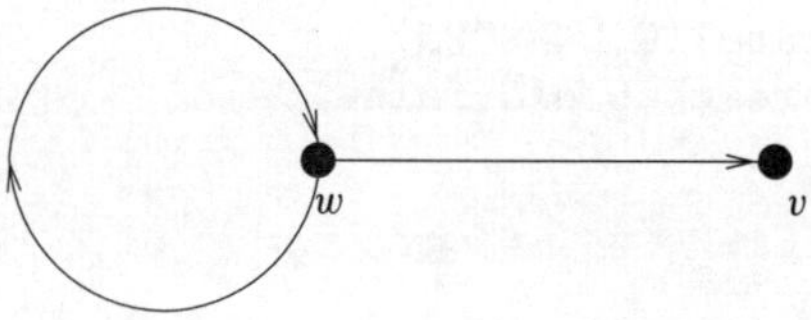

Figure 7.11 A cycle and a path emanating from it. The search started in v and w is the first node encountered twice.

⟨*label all nodes on current path*⟩≡

```
int t = label[w];
if ( t == ON_CUR_PATH )
{ node z;
  do { z = S.pop();
       label[z] = CYCLE;
     }
  while ( z != w );
  while ( !S.empty() ) label[S.pop()] = ATT_TO_CYCLE;
}
else // t is CYCLE, ATT_TO_CYCLE, or FINITE
{ if ( t == CYCLE ) t = ATT_TO_CYCLE;
  while ( !S.empty() ) label[S.pop()] = t;
}
```

We next check that all P-cycles have negative cost. Given our classification of nodes this is fairly simple. For every cycle node we trace the cycle containing it and compute its cost. We assert that the cost is negative. If so, we promote all nodes on the cycle to label *NEG_CYCLE*; this guarantees that every cycle is traced only once.

⟨*condition two*⟩≡

```
forall_nodes(v,G)
{ if ( label[v] == CYCLE )
  { node w = v;
    NT cycle_length = 0;
    do
    { cycle_length += c[pred[w]];
      label[w] = NEG_CYCLE;
      w = G.source(pred[w]);
    } while (w != v);
    assert(cycle_length < 0);
  }
}
```

Conditions (3), (4), and (5) are trivial to check.

⟨*conditions three, four, and five*⟩≡

```
if ( label[s] == FINITE ) assert(dist[s] == 0);

edge e;
forall_edges(e,G)
{ node v = G.source(e);
  node w = G.target(e);
  if ( label[w] == FINITE )
  { assert( label[v] == FINITE || label[v] == PLUS);
    if ( label[v] == FINITE )
    { assert( dist[v] + c[e] >= dist[w] );
      if ( e == pred[w] ) assert( dist[v] + c[e] == dist[w] );
    }
  }
}
```

Putting it all together we obtain:

⟨*check_sp.t*⟩+≡

```
template <class NT>
node_array<int> CHECK_SP_T(const graph& G, node s,
                           const edge_array<NT>& c,
                           const node_array<NT>& dist,
                           const node_array<edge>& pred)
{ ⟨condition one⟩
  ⟨classification of nodes⟩
  ⟨condition two⟩
  ⟨conditions three, four, and five⟩
  return label;
}
```

7.5.3 *A Generic Single-Source Shortest-Path Algorithm*

We derive a generic shortest-path algorithm. All our implementations for the single-source problem will be instances of the generic algorithm and the correctness proofs and running time claims of our implementations will be consequences of the lemmas derived in this section.

In Lemma 6 we gave a characterization of shortest-path distances.

Let $d : V \longrightarrow I\!R \cup \{-\infty, \infty\}$ be a function with

$$\begin{array}{llll} (1) & d(v) & \geq & \mu(v) \quad\quad \text{for all } v \in V \\ (2) & d(s) & \leq & 0 \\ (3) & d(v) & \leq & d(u) + c(u, v) \quad \text{for all } e = (u, v) \in E \end{array}$$

Then $d(v) = \mu(v)$ for all $v \in V$.

The generic algorithm maintains a function d satisfying (1) and (2) and aims at establishing (3). We call $d(v)$ the *tentative distance label* of v.

$d(s) = 0$; $d(v) = \infty$ for $v \neq s$;
$\pi(v) = nil$ for all $v \in V$;
while there is an edge $e = (u, v) \in E$ with $d(v) > d(u) + c(e)$
{ $d(v) = d(u) + c(e)$;
  $\pi(v) = e$;
}

We will refer to the body of the while-loop as *relaxing*[9] *edge* e. Besides the tentative distance labels the generic algorithm maintains for each node v the edge $\pi(v)$ that defined $d(v)$.

It is easy to see that (1) and (2) are invariants of the algorithm. We only have to observe that $d(v)$ never increases (and hence $d(s) \leq 0$ always) and that $d(v) < +\infty$ implies that $d(v)$ is the length of some path from s to v (and hence $d(v) \geq \mu(v)$ always). When the algorithm terminates we also have (3). Thus, $d(v) = \mu(v)$ for all $v \in V$ when the algorithm terminates. A lot more can be said about the generic algorithm.

Lemma 8 *The following is true at any time during the execution of the generic algorithm*[10]. *Let*

$$P = \{e\,; e = \pi(v) \in E \text{ for some } v \in V\}.$$

(a) *$d(s) = 0$ iff $\pi(s) = nil$ and $d(v) < \infty$ iff $\pi(v) \neq nil$ for $v \neq s$.*
(b) *If $\pi(v) = e = (u, v)$ then $d(v) \geq d(u) + c(e)$.*
(c) *If $\pi(v) \neq nil$ then v either lies on a P-cycle, or is reachable from a P-cycle by a P-path, or is reachable from s by a P-path. If $\pi(s) \neq nil$ then s lies on a P-cycle.*

[9] Think of $e = (u, v)$ as a rubber band that wants to keep v within distance $c(e)$ of u. If $d(v) > d(u) + c(e)$ the rubber band is under tension. Setting $d(v)$ to $d(u) + c(e)$ relaxes it.

[10] Observe the similarity of items (a), (d), (e), (f), and (g) with the four bullets in the definition of our output convention.

(d) *P-cycles have negative cost.*
(e) *If v lies on a P-cycle or is reachable from a P-cycle then $\mu(v) = -\infty$.*
(f) *If $v \in V^f$ and $d(v) = \mu(v)$ then there is a P-path from s to v and this path has cost $\mu(v)$.*
(g) *If $d(v) = \mu(v)$ for all $v \in V^f$ then P defines a shortest-path tree on V^f.*

Proof (a) We start with $d(s) = 0$, $d(v) = \infty$ for all v with $v \neq s$, and $\pi(v) = nil$ for all v. When $d(v)$ is decreased, $\pi(v)$ is set, and when $\pi(v)$ is set, $d(v)$ is decreased.

(b) Consider the moment of time when $\pi(v)$ was set most recently. At this moment we had $d(v) = d(u) + c(u, v)$, $d(v)$ has not changed since then, and $d(u)$ can only have decreased.

(c) Consider any node u, $u \neq s$, with $(u, v) \in P$ for some v. Then $\pi(v) = (u, v)$ and hence $d(v) < \infty$ by part (a). Then $d(u) + c(u, v) \leq d(v)$ by part (b) and hence $d(u) < \infty$. Thus, $\pi(u) \neq nil$ by part (a). We conclude that any node u, $u \neq s$, with an outgoing P-edge has also an incoming P-edge. Thus s is the only node which may have outgoing P-edges but no incoming P-edge.

(d) Let $[e_0, \ldots, e_{k-1}]$ with $e_i = (v_i, v_{i+1})$ and $v_0 = v_k$ be a P-cycle. We may assume w.l.o.g. that $\pi(v_k) = e_{k-1}$ is the edge in the cycle that was added to P last. Just prior to the addition of e_{k-1} we have

$$d(v_{i+1}) \geq d(v_i) + c(e_i) \text{ for all } i, 0 \leq i \leq k-2$$

by part (b) and

$$d(v_k) > d(v_{k-1}) + c(e_{k-1}).$$

Summation yields

$$\sum_{0 \leq i < k} d(v_{i+1}) > \sum_{0 \leq i < k} (d(v_i) + c(e_i))$$

and hence (since $v_k = v_0$ and thus $d(v_k) = d(v_0)$)

$$\sum_{0 \leq i < k} c(e_i) < 0.$$

(e) Any node v with $\pi(v) \neq nil$ has $d(v) < \infty$ and is hence reachable from s in G. Any P-cycle has negative cost. Thus $\mu(v) = -\infty$ for any node v lying on a P-cycle or being reachable from a P-cycle.

(f) Assume $V^f \neq \emptyset$ and consider any node $v \in V^f$ with $d(v) = \mu(v)$. For $v = s$ there is nothing to show. For $v \neq s$, $d(v) = \mu(v) < \infty$ implies $\pi(v) \neq nil$. From (c) and (e) we conclude that v is reachable from s by a P-path $p = [e_0, \ldots, e_{k-1}]$ with $e_i = (v_i, v_{i+1})$, $v_0 = s$, and $v_k = v$. From (b) we conclude

$$d(v_{i+1}) \geq d(v_i) + c(e_i) \text{ for } i, 0 \leq i < k$$

and hence

$$d(v_k) \geq d(v_0) + c(p) = c(p),$$

where the last equality follows from $v_0 = s \in V^f$ and hence $d(s) = \mu(s) = 0$ by (1) and (2). Thus, $c(p) \le d(v) = \mu(v)$ and we must have equality since no path from s to v can be shorter than $\mu(v)$.

(g) This follows immediately from part (f). □

There are two major problems with the generic algorithm:

- In the presence of negative cycles it will never terminate (since the d-values are always the length of some path and hence cannot reach $-\infty$).
- Even in the absence of negative cycles the running time can be exponential, see [Meh84c, page40] for an example.

We address the second problem in the remainder of this section and deal with the first problem in Section 7.5.7. When we decrease the distance label $d(v)$ of a node v in the generic algorithm this may create additional violations of (3), namely for the edges out of v. This suggests maintaining a set U of nodes with

$$U \supseteq \{u\,;\, d(u) < \infty \text{ and } \exists (u, v) \in E \text{ with } d(u) + c(u, v) < d(v)\}$$

and to rewrite the algorithm[11] as:

$d(s) = 0$; $d(v) = \infty$ for $v \ne s$;
$U = \{s\}$;
while $U \ne \emptyset$
{ select $u \in U$ and remove it;
  forall edges $e = (u, v)$
  { **if** $d(u) + c(e) < d(v)$
    { add v to U;
      $d(v) = d(u) + c(e)$;
      $\pi(v) = e$;
    }
  }
}

We are left with the decision of which node u to select from U. There is always an optimal choice.

Lemma 9 *(Existence of optimal choice)*

(a) *As long as $d(v) > \mu(v)$ for some $v \in V^f$: for any $v \in V^f$ with $d(v) > \mu(v)$ there is a $u \in U$ with $d(u) = \mu(u)$ and lying on a shortest path from s to v.*

(b) *When a node u is removed from U with $d(u) = \mu(u)$ then it is never added to U again.*

[11] We reuse the name generic shortest-path algorithm for the modified version of the algorithm.

Proof (a) Let $[s = v_0, v_1, \ldots, v_k = v]$ be a shortest path from s to v. Then $\mu(s) = 0 = d(s)$ and $d(v_k) > \mu(v_k)$. Let i be minimal such that $d(v_i) > \mu(v_i)$. Then $i > 0$, $d(v_{i-1}) = \mu(v_{i-1})$ and

$$d(v_i) > \mu(v_i) = \mu(v_{i-1}) + c(v_{i-1}, v_i) = d(v_{i-1}) + c(v_{i-1}, v_i).$$

Thus, $v_{i-1} \in U$.

(b) We have $d(u) \geq \mu(u)$ always. Also, when u is added to U then $d(u)$ is decreased. Thus, if a node u is removed from U with $d(u) = \mu(u)$ it will never be added to U at a later time. □

There are two important special cases of the single-source problem where the existence claim of an optimal choice can be made algorithmic. Both cases deal with graphs where the structure of the graphs excludes negative cycles: graphs with non-negative edge costs and acyclic graphs.

Lemma 10 *(Algorithmic optimal choice)*

(a) *If $c(e) \geq 0$ for all $e \in E$ then $d(u) = \mu(u)$ for the node $u \in U$ with minimal $d(u)$.*

(b) *If G is acyclic and $u_0, u_1, \ldots, u_{n-1}$ is a topological order of the nodes of G, i.e., if $(u_i, u_j) \in E$ then $i < j$, then $d(u) = \mu(u)$ for the node $u = u_i \in U$ with i minimal.*

Proof Assume $d(u) > \mu(u)$ for the node chosen in either part (a) or (b). By the preceding lemma there is a node $z \in U$ lying on a shortest path from s to u with $d(z) = \mu(z)$. We now distinguish cases.

In part (a) we have $\mu(z) \leq \mu(u)$ since all edge costs are assumed to be non-negative. Thus, $d(z) < d(u)$, contradicting the choice of u.

In part (b) we have $z = u_j$ for some $j < i$, contradicting the choice of u. □

Part (a) of the lemma above is the basis of Dijkstra's algorithm, see Section 6.6, and part (b) is the basis of a linear time algorithm for acyclic graphs, which we will discuss in the next section.

In our shortest-path programs we use a *node_array<NT> dist* to represent the function d and a *node_array<edge> pred* for the function π. Since most number types have no representation of $+\infty$ we will not be able to maintain equality between d and *dist*. We exploit the fact that the equivalence

$$d(v) = +\infty \text{ iff } v \neq s \text{ and } \pi(v) = nil$$

holds in the generic algorithm and use it for the representation of $+\infty$. We maintain the following relationship between (d, π) and (*dist*, *pred*): for all nodes v:

- *pred*$[v] = \pi(v)$ and
- *dist*$[v] = d(v)$, if $d(v) < \infty$, and *dist*$[v]$ arbitrary, if $d(v) = +\infty$.

With this representation a comparison $d < d(v)$ with $d \in \mathbb{R}$ can be realized as:

```
(pred[v] == nil && v != s) || d < dist[v]
```

We remark that the alternative

```
(v != s && pred[v] == nil) || d < dist[v]
```

is less efficient. All but one node v is different from s and hence the test `v != s` evaluates to true most of the time; thus the test `pred[v] == nil` will also be performed most of the time in the second line. In the first line, the test `pred[v] == nil` evaluates to true only when the first edge into v is considered (since $d(v) < +\infty$ afterwards) and hence evaluates to false in the majority of the cases (at least if the average indegree is larger than two). Thus the test `v != s` will not be made in the majority of the cases.

The general rule is that in a conjunction of tests one should start with the test that evaluates to false most often and that in a disjunction of tests one should start with the test that evaluates to true most often. Please, do not use this rule blindly since interchanging the order of tests may change the semantics (since C++ evaluates a test from left to right and terminates the evaluation once the value of the test is known). In the example above, it would be unwise (why?) to change the expression into

```
d < dist[v] || (pred[v] == nil && v != s)
```

7.5.4 *Acyclic Graphs*

We show how topological sorting can be used to solve the single-source shortest-path problem in acyclic graphs in linear time $O(n + m)$. Let G be an acyclic graph and assume that $v_1, v_2, \ldots, v_n$ is an ordering of the nodes such that $(v_i, v_j) \in E$ implies $i \leq j$. Such an ordering is easy to compute.

⟨*acyclic graphs: establish topological order*⟩≡

```
node_array<int> top_ord(G);
TOPSORT(G,top_ord);  // top_ord is now a topological ordering of G
int n = G.number_of_nodes();
array<node> v(1,n);
node w;
forall_nodes(w,G) v[top_ord[w]] = w;  // top_ord[v[i]] == i for all i
```

The call TOPSORT(...) numbers the nodes of G with the integers 1 to n such that all edges go from lower numbered to higher numbered nodes. In the forall_node-loop we store the node with number i in $v[i]$.

It is now easy to implement the generic single-source algorithm. Let $k = top\_ord[s]$. Nodes v_i with $i < k$ are not reachable from s. We step through the nodes in the order $v_k, v_{k+1}, \ldots$ and maintain the set U implicitly. Assume we have reached node i. Then U consists of all nodes v_j with $j \geq i$ and $dist(v_j) < +\infty$. For $j > k$ the latter condition is equivalent to $pred[v_j] \neq nil$. If $v[i]$ is equal to s or has a defined predecessor edge we propagate $dist[v[i]]$ over all edges out of $v[i]$ and proceed to the next node.

⟨*acyclic_sp.t*⟩+≡

```
template <class NT>
void ACYCLIC_SHORTEST_PATH_T(const graph& G, node s,
                             const edge_array<NT>& c,
                             node_array<NT>& dist,
                             node_array<edge>& pred)
{
  ⟨acyclic graphs: establish topological order⟩
  forall_nodes(w,G) pred[w] = nil;
  dist[s] = 0;
  for(int i = top_ord[s]; i <= n; i++)
  { node u = v[i];
    if ( pred[u] == nil && u != s ) continue;
    edge e;
    NT du = dist[u];
    forall_adj_edges(e,u)
    { node w = G.target(e);
      if ( pred[w] == nil || du + c[e] < dist[w])
      { pred[w] = e;
        dist[w] = du + c[e];
      }
    }
  }
}
```

The correctness follows immediately from the remarks preceding the program and Lemma 10. The running time is $O(n + m)$ since each node and each edge is considered at most once.

7.5.5 *Non-Negative Edge Costs*

Dijkstra's algorithm for the shortest-path problem with non-negative edge costs was already treated in Section 6.6.

7.5.6 *The Single-Source Single-Sink Problem*

The single-source single-sink shortest-path problem is probably the most natural shortest-path problem. The goal is to find a shortest path from a given source node s to a given sink node t.

We describe the so-called *bidirectional search algorithm* (an alternative approach is discussed in the exercises). The algorithm assumes that edge costs are non-negative. The worst case running time of the algorithm is $O(m + m \log n)$; the observed running time is frequently much better.

The bidirectional search algorithm runs two instances of Dijkstra's algorithm (see Section 6.6) concurrently, one to find shortest-path distances from s and one to find shortest-path distances to t. The first instance is simply Dijkstra's algorithm and the second instance

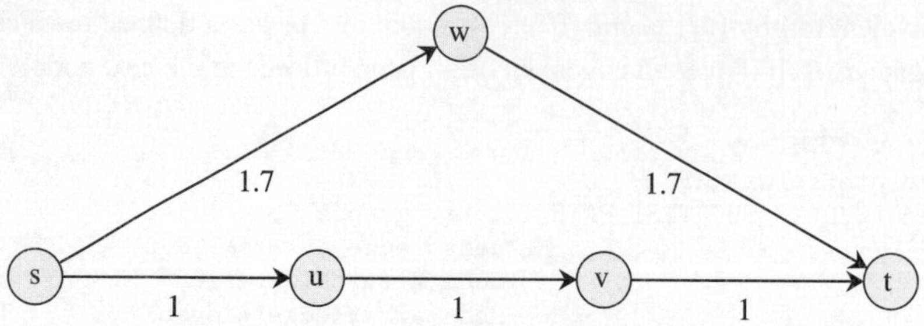

Figure 7.12 Termination of the bidirectional shortest-path algorithm: In our implementation we alternately add nodes to K_s and K_t. In the example we add s to K_s, t to K_t, u to K_s, v to K_t, w to K_s, w to K_t, and v to K_s. The algorithm terminates when v is added to K_s. It does not terminate when w is added to K_t, although $w \in K_s \cap K_t$ at this point of time. Observe that $d_t(v) = 1$ after adding t to K_t and $d_s(v) = 2$ after adding u to K_s. Thus $D = 3$ after adding u to K_s and hence w does not realize D when it is added to $K_s \cap K_t$.

is a symmetric version of Dijkstra's algorithm, where the search starts at t and shortest-path distances are propagated across the edges *into* a node instead of the edges out of a node.

We use $d_s(v)$ to denote the tentative distance from s to v and $d_t(v)$ to denote the tentative distance from v to t. Initially, $d_s(s) = d_t(t) = 0$, $d_s(v) = \infty$ for $v \neq s$, and $d_t(v) = \infty$ for $v \neq t$. The algorithm maintains

$$D = \min_v (d_s(v) + d_t(v))$$

which is the shortest known length of a path from s to t.

Let K_s and K_t be the set of nodes that were removed from the priority queue in the shortest-path calculations from s and t, respectively. We know from Section 7.5.3 that

$$\begin{aligned} d_s(v) &= \mu(s, v) \text{ for } v \in K_s \\ d_s(v) &= \min\{\mu(s, u) + c(u, v); u \in K_s\} \text{ for } v \notin K_s \\ d_t(v) &= \mu(v, t) \text{ for } v \in K_t \\ d_t(v) &= \min\{c(v, u) + \mu(u, t); u \in K_t\} \text{ for } v \notin K_t \end{aligned}$$

The bidirectional algorithm terminates when D is realized by a node in $K_s \cap K_t$ or when both queues become empty. In the former case D is the shortest-path distance from s to t, and in the latter case there is no path from s to t. Figure 7.12 illustrates the termination condition.

Theorem 2 *The bidirectional search algorithm is correct.*

Proof If there is no path from s to t then there is never a node in $K_s \cap K_t$ and hence the algorithm terminates when both queues become empty. Thus the algorithm is correct if there is no path from s to t.

So let us assume that there is a path from s to t. Let $p = [s = v_0, v_1, \ldots, v_{k-1}, v_k = t]$ be a shortest path from s to t.

We argue first that the event that D is realized by a node in $K_s \cap K_t$ will occur. This is easy to see. Observe first that all nodes on p are reachable from s as well as t. When a node v_h on p is added to $K_s \cap K_t$, we have

$$\mu(s,t) \le D \le d_s(v_h) + d_t(v_h) = \mu(s, v_h) + \mu(v_h, t) = \mu(s,t),$$

and hence the event that D is realized by a node in $K_s \cap K_t$ will occur at the latest when a node on p is added to $K_s \cap K_t$.

It remains to show that $D = c(p)$ when the event actually occurs. Assume otherwise, i.e., $c(p) < D$ when the algorithm terminates. Then there is no node of p in $K_s \cap K_t$ at the time of termination.

Consider the time of termination, let $w \in K_s \cap K_t$ be the node with $D = d_s(w) + d_t(w)$, let i be minimal with $v_{i+1} \notin K_s$, and let j be maximal with $v_{j-1} \notin K_t$. Both indices exist since $v_0 = s$ is the first node to be added to K_s and $v_k = t$ is the first node to be added to K_t. We have $i < j$ by our assumption that no node of p is added to $K_s \cap K_t$ and hence $d_s(w) \le d_s(v_{i+1})$, since $w \in K_s$ and $v_{j-1} \notin K_s$, and $d_t(w) \le d_t(v_{j-1})$, since $w \in K_t$ and $v_{j-1} \notin K_t$. If $i + 1 \le j - 1$, we have

$$c(p) \ge \mu(s, v_{i+1}) + \mu(v_{j-1}, t) = d_s(v_{i+1}) + d_t(v_{j-1}) \ge d_s(w) + d_t(w) = D,$$

and if $i + 1 = j$, we have with $v = v_{i+1} = v_j$

$$c(p) = \mu(s, v) + \mu(v, t) = d_s(v) + d_t(v) \ge D.$$

□

We turn to the implementation. We distinguish the two versions of Dijkstra's algorithm by indices 0 and 1 and provide two copies of the required data structures in arrays with index set $\{0, 1\}$.

⟨*single sink: data structures*⟩≡

```
array<node> terminal(2);
terminal[0] = s; terminal[1] = t;
array<node_pq<NT>* >  PQ(2);
PQ[0] = new node_pq<NT>(G);
PQ[1] = new node_pq<NT>(G);
PQ[0]->insert(terminal[0],0);
PQ[1]->insert(terminal[1],0);
array<node_array<NT> > dist(2);
dist[0] = dist[1] = node_array<NT>(G);
dist[0][s] = dist[1][t] = 0;
array<node_array<edge> > Pred(2);
Pred[0] = Pred[1] = node_array<edge>(G,nil);
bool D_equals_infinity = (s != t? true : false);
NT D = 0;
```

We store the tentative distances $d_s(v)$ and $d_t(v)$ in *dist*[0][v] and *dist*[1][v], respectively, we use *PQ*[0] and *PQ*[1] as the priority queue in the search from s and t, respectively, we use *Pred*[0][v] to record the edge into v that defines $d_s(v)$, and we use *Pred*[1][v] to record the edge out of v that defines $d_t(v)$.

We initialize D to infinity if $s \neq t$, and to zero otherwise. Since we cannot assume that the number type *NT* provides the value $+\infty$ we use a boolean flag to indicate this special value.

A remark is in order about the declarations above. We declared *dist* as an array of node arrays and *PQ* as an array of pointers to priority queues. Why did we make this distinction? In order to declare an *array*<*T*> for some type T, T must provide a default constructor, a copy constructor, and some other operations, e.g., the input and output operators $\ll$ and $\gg$, see Section 2.8. Node arrays provide all required functions except for the input and output operators and those are easily defined in the current file, since the missing functions are non-member functions of node arrays. The situation is different for node priority queues; they define only a few of the required functions and, in particular, a member function is missing. We cannot add the member function in this file. Moreover, in the case of *dist* it is more important to have an array of node arrays instead of an array of pointers to node arrays, since having an array of pointers to node arrays would force us to write either `(*dist[i])[v]` or `dist[i]->operator[](v)` instead of `dist[i][v]`.

⟨*dijkstra_single_sink.t*⟩≡

```
template <class T>
ostream& operator<<(ostream& o,const node_array<T>&) { return o; }
template <class T>
istream& operator>>(istream& i,node_array<T>&)       { return i; }
```

The structure of the single-source single-sink program is as described above. We run both instances of Dijkstra's algorithm concurrently, and terminate when either both queues become empty or when we encounter a node $u \in K_s \cap K_t$ with $D = d_s(u) + d_t(u)$. In the former case there is no path from s to t. According to our output convention for the single-source single-sink problem this fact is recorded by having *pred*[t] $=$ *nil* in the return values.

⟨*dijkstra_single_sink.t*⟩+≡

```
template<class NT>
NT DIJKSTRA_T(const graph& G, node s, node t,
              const edge_array<NT>& cost, node_array<edge>& pred)
{
  ⟨single sink: data structures⟩
  while ( !PQ[0]->empty() || !PQ[1]->empty() )
  { for (int i = 0; i < 2; i++)
    { if ( PQ[i]->empty() ) continue;
      node u = PQ[i]->del_min();
      ⟨return if u is in Ks and Kt and D = d_s(u) + d_t(u)⟩
      ⟨relax edges out of u, if i = 0, or into u, if i = 1⟩
```

```
    }
  }
  pred[t] = nil; // no path from s to t
  return D;
}
```

The relaxation of edges is copied from Section 6.6 with two small modifications.

In the search for shortest paths from s we iterate over the edges out of u, and in the search for shortest paths to t we iterate over the edges into u.

Whenever the dist-value of a node is improved we check whether this leads to an improvement of D.

⟨*relax edges out of u, if i = 0, or into u, if i = 1*⟩≡

```
for ( edge e = (i == 0? G.first_adj_edge(u): G.first_in_edge(u));
      e != nil;
      e = (i == 0? G.adj_succ(e): G.in_succ(e)) )
{ node v = (i == 0? G.target(e) : G.source(e) );
  NT c = dist[i][u] + cost[e];
  if ( Pred[i][v] == nil && v != terminal[i] )
     PQ[i]->insert(v,c); // first path to v
  else if (c < dist[i][v]) PQ[i]->decrease_p(v,c); // better path
       else continue;
  dist[i][v] = c;
  Pred[i][v] = e;
  if ( ( v == terminal[1-i] || Pred[1-i][v] != nil )
       // dist[1-i][v] is defined iff true
     && ( D_equals_infinity || dist[0][v] + dist[1][v] < D ))
  { D_equals_infinity = false;
    D = dist[0][v] + dist[1][v];
  }
}
```

How can we check whether $u \in K_s \cap K_t$? Assume w.l.o.g. that $i = 0$. Then $u \in K_s$ since we have just removed it from $PQ[0]$. Also, we have $u \in K_t$ if u has been in $PQ[1]$, but is not there anymore. u has been or still is in $PQ[1]$ if either $u = t$ or $Pred[1][u]$ is defined, and u is not in $PQ[1]$ if $PQ[1] \rightarrow member(u)$ returns false.

If $u \in K_s \cap K_t$ and $D = d_s(u) + d_t(u)$ we terminate the computation, record the path in the predecessor array, and return D as the length of the shortest path from s to t. In order to record the path in the *pred*-array we trace the two "half paths" from u to s and from u to t, respectively. When tracing the latter path we observe that $Pred[1]$ stores out-edges and not in-edges.

⟨*return if u is in Ks and Kt and D = d_s(u) + d_t(u)*⟩≡

```
if ( (u == terminal[1-i] || Pred[1-i][u] != nil) &&
     !PQ[1-i]->member(u) && dist[0][u] + dist[1][u] == D )
{ // have found shortest path from s to t.
  // trace path from u to s
  node z = u;
```

n	m	Single-sink	Dijkstra
10000	500000	0.118	0.736

Table 7.1 A comparison of the running time of the single-sink algorithm with the running time of the standard version of Dijkstra's algorithm. The standard version computes the distance from the source to all other vertices and then extracts the distance value of the sink.

```
  while ( z != s ) z = G.source(pred[z] = Pred[0][z]);
  // trace path from u to t
  z = u;
  edge e;
  while ( (e = Pred[1][z] ) != nil) { pred[z = G.target(e)] = e; }
  return D;
}
```

Table 7.1 compares the running times of the single-source single-sink algorithm presented in this section and the standard version of Dijkstra's algorithm.

7.5.7 *General Networks: The Bellman–Ford Algorithm*

We derive and implement a single-source shortest-path algorithm for arbitrary edge costs. The algorithm is due to Bellman [Bel58] and Ford. We will refer to the algorithm as the basic Bellman–Ford algorithm[12]. In Section 7.5.3 we studied a generic shortest-path algorithm. Let us recall what we know:

- The algorithm maintains a set U containing all nodes u for which there is an edge (u, v) with $d(u) + c(u, v) < d(v)$. U may also contain other nodes.
- In each iteration the algorithm selects some node in U and relaxes all edges out of it.
- As long as $d(v) > \mu(v)$ for some $v \in V^f$, there is a node $u \in U$ with $d(u) = \mu(u)$ (Lemma 9). We use the phrase that not all finite distance values are determined to mean that $d(v) > \mu(v)$ for some $v \in V^f$.
- When a node u is removed from U with $d(u) = \mu(u)$ it will never be added to U again at a later stage.
- Let $P = \{e\,;\, e = \pi(v) \in E$ for some $v \in V\}$. All P-cycles are negative and if $d(v) = \mu(v)$ for all $v \in V^f$ then P defines a shortest-path tree on V^f.

What is a good strategy for selecting from U? We know that U contains a perfect choice (at least as long as not all finite distance values are determined), but we do not know which node in U is the perfect choice. In order to play it safe we should therefore not discriminate against any node in U. A way to achieve fairness is to organize the computation in phases.

[12] We will study a refined version in the next section.

Let U_i be the set U at the beginning of phase i, $i \geq 0$; U_0 is equal to $\{s\}$. In phase i we remove all vertices in U_i from U. Newly added vertices are inserted into U_{i+1}. In this way we guarantee that at least one finite distance value is determined in each phase (if there is one that is still to be determined) and hence all finite distance values are determined after at most n phases.

In the program below we realize the set U by a queue Q. During phase i all nodes in U_i are at the front of the queue and all nodes in U_{i+1} are at the rear of the queue. We separate U_i and U_{i+1} by the marker *nil*. We count the number of phases in *phase_count*. Whenever the marker appears at the front of Q we increment *phase_count*. In order to avoid putting nodes several times into Q we keep a *node_array<bool> in_Q* with *in_Q*$[v]$ = *true* iff $v \in Q$.

We terminate the algorithm when Q becomes empty or when phase n is reached. In the former case we have $d(v) = \mu(v)$ for all v and in the latter case we have $d(v) = \mu(v)$ for all $v \in V^+ \cup V^f$. We will deal with the nodes in V^- in a postprocessing step.

⟨*bellman_ford_basic.t*⟩≡

```
#include <LEDA/graph_alg.h>
#include <LEDA/b_queue.h>
⟨BF: helper⟩
template <class NT>
bool BELLMAN_FORD_B_T(const graph& G, node s, const edge_array<NT>& c,
                      node_array<NT>& dist, node_array<edge>& pred )
{ int n = G.number_of_nodes();
  int phase_count = 0;
  b_queue<node> Q(n+1);
  node_array<bool> in_Q(G,false);
  node u,v;
  edge e;
  forall_nodes(v,G) pred[v] = nil;
  dist[s] = 0;
  Q.append(s); in_Q[s] = true;
  Q.append((node) nil); // end marker
  while( phase_count < n )
  { u = Q.pop();
    if ( u == nil)
    { phase_count++;
      if ( Q.empty() ) return true;
      Q.append((node) nil);
      continue;
    }
    else in_Q[u] = false;
    NT du = dist[u];
    forall_adj_edges(e,u)
    { v = G.opposite(u,e);  // makes it also work for ugraphs
      NT d = du + c[e];
      if ( (pred[v] == nil && v != s) || d < dist[v] )
      { dist[v] = d; pred[v] = e;
        if ( !in_Q[v] ) { Q.append(v); in_Q[v] = true; }
```

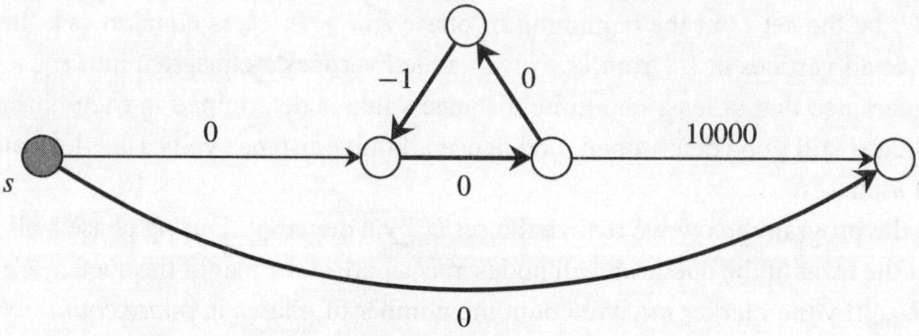

Figure 7.13 The situation at the beginning of phase $n = |V| = 5$. Note that v belongs to V^-, that P contains a negative cycle from which v is reachable in G, but that v is not reachable from this cycle by a P-path. Running the algorithm for another 10000 phases will establish the output convention; the quantity 10000 reflects the fact that traversing the cycle 10000 times creates a path of cost −10000.

```
      }
    }
  }
  ⟨BF: postprocessing⟩
  return false;
}
```

We turn to the postprocessing step required when U is non-empty after n phases. Figure 7.13 shows that our output convention is not automatically satisfied. As the figure shows there may be nodes in V^- that are not reachable yet from a P-cycle by a P-path.

How can we establish our output convention that all nodes in V^- are reachable from a P-cycle by a P-path? We could run the algorithm for more phases until a path containing a negative cycle has been discovered for all nodes in V^-. This may take very long as Figure 7.13 shows. We need a better method. In the following lemma we show that Figure 7.14 describes the situation at the beginning of phase n. The argument is with respect to the generic algorithm with the selection rule of the Bellman–Ford algorithm.

For an integer k, $k \geq 0$, let

$$\mu_k(v) = \min\{c(p)\ ;\ p \text{ is a path from } s \text{ to } v \text{ consisting of at most } k \text{ edges}\}.$$

Lemma 11 *After n phases:*

(a) *$d(v) \leq \mu_n(v)$ and if $v \in U$ then $d(v) < \mu_{n-1}(v)$.*
(b) *$s \in V^f$ iff $\pi(s) = nil$.*
(c) *Every $u \in U$ lies either on a P-cycle or on a P-path emanating from a P-cycle.*
(d) *Every $v \in V^-$ is reachable in G from a $u \in U$.*
(e) *If $\pi(s) \neq nil$ then the output convention is already satisfied.*

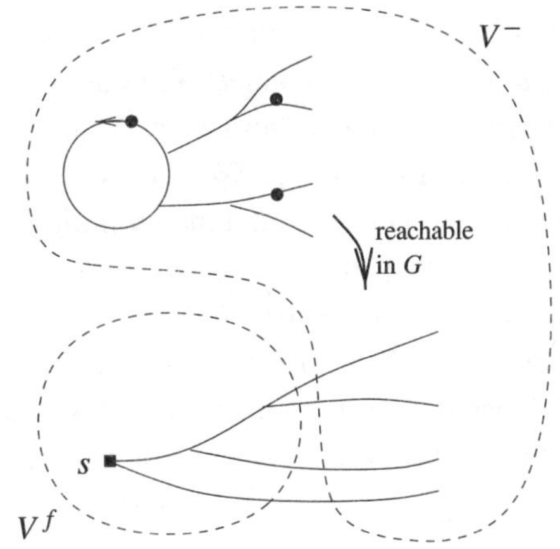

Figure 7.14 The situation at the beginning of phase n: Some nodes in V^- are still reachable from s by a P-path and some are already contained in a P-cycle or lie on a P-path emanating from a P-cycle. *All* nodes in U (nodes in U are shown as solid circles) belong to the latter category by part (d) of Lemma 11. All nodes in V^- are reachable in G from a node in U by part(e) of Lemma 11.

Proof (a) Let $p = [s = v_0, v_1, \ldots, v_k = v]$ be any path starting in s. Then $d(v_i) \le \sum_{0<j\le i} c(v_{j-1}, v_j)$ at the beginning of phase i and hence $d(v) \le \mu_{n-1}(v)$ at the beginning of phase $n-1$ and $d(v) \le \mu_n(v)$ at the beginning of phase n. If v is added to U in phase $n-1$, $d(v)$ is decreased and hence $d(v) < \mu_{n-1}(v)$ at the beginning of phase n.

(b) If $s \in V^f$ then $d(s) = 0$ and hence $\pi(s) = nil$. If $s \notin V^f$ then there is a negative cycle passing through s and hence $\mu_n(s) < 0$. Thus, $d(s) < 0$ and hence $\pi(s) \ne nil$.

(c) If $\pi(s) \ne nil$ part (c) follows from Lemma 8, part (c). So assume $s \in V^f$ and assume that there is a $u \in U$ that is reachable from s by a P-path, say p. Then $d(u) \ge c(p) \ge \mu_{n-1}(u)$, a contradiction to part (a).

(d) Let $v \in V^-$ be arbitrary. Since $\mu(v) = -\infty$ there must be a path p from s to v with $c(p) < d(v)$. Let p_i be the path consisting of the first i edges of p and let v_i be the target node of p_i. Let k be minimal such that $c(p_k) < d(v_k)$. Then $k > 0$ since $c(p_0) = 0$ and $d(v_0) = d(s) \le 0$ and hence $c(p_{k-1}) \ge d(v_{k-1})$. Thus,

$$d(v_k) > c(p_k) = c(p_{k-1}) + c(v_{k-1}, v_k) \ge d(v_{k-1}) + c(v_{k-1}, v_k)$$

and hence $v_{k-1} \in U$.

(e) If $\pi(s) \ne nil$ then part (c) of Lemma 8 tells us that every node reachable from s lies either on a P-cycle or a P-path emanating from a P-cycle. □

Parts (a), (d), and (e) of the lemma above are the key for the postprocessing step. If $\pi(s) \ne nil$ we are done. So assume $\pi(s) = nil$, i.e., $V^f \ne \emptyset$, and let R be the set of nodes that are reachable from s by a P-path. Then $R \supseteq V^f$ but this inclusion may be proper, see Figure 7.14. All nodes in R that are reachable from a node $u \in U$ belong to V^- and hence their π-values have to be changed. We can do so by performing a depth-first search from each node $u \in U$. Whenever a node in R is reached we change its π-value to the edge which led to the node. In this way we connect all nodes in $R \cap V^-$ to the nodes in U and hence, by part (c), make them reachable from P-cycles by P-paths.

How can we determine the nodes in R? We simply perform a depth-first search from s on the subgraph defined by P. This can be done by hiding all edges not in P, performing a depth-first search, and restoring (= unhiding) all edges in P. In the program below the nodes in R are labeled true in the node array *in_R*.

In the program chunk below the cast ((*graph**) &G) turns G from a const-object to a non-const-object. The cast is required since *hide_edge* and *restore_all_edges* modify the graph and the cast is safe since *restore_all_edges* restores the original situation.

⟨*BF: postprocessing*⟩≡

```
if (pred[s] != nil) return false;
node_array<bool> in_R(G,false);
forall_edges(e,G)
  if (e != pred[G.target(e)]) ((graph*) &G)->hide_edge(e);
DFS(G,s,in_R); // sets in_R[v] = true for v in R
((graph*) &G)->restore_all_edges();
node_array<bool> reached_from_node_in_U(G,false);
forall_nodes(v,G)
  if (in_Q[v] && !reached_from_node_in_U[v])
     Update_pred(G,v,in_R,reached_from_node_in_U,pred);
```

where

⟨*BF: helper*⟩≡

```
inline void Update_pred(const graph& G, node v,
                        const node_array<bool>& in_R,
                        node_array<bool>& reached_from_node_in_U,
                        node_array<edge>& pred)
{ reached_from_node_in_U[v] = true;
  edge e;
  forall_adj_edges(e,v)
    { node w = G.target(e);
      if ( !reached_from_node_in_U[w] )
        { if ( in_R[w] ) pred[w] = e;
          Update_pred(G,w,in_R,reached_from_node_in_U,pred);
        }
    }
}
```

The running time of the Bellman–Ford algorithm is $O(nm)$. This can be seen as follows. There are at most n phases and the running time of each phase is proportional to the sum of the outdegrees of the nodes removed from Q in the phase. This implies that the cost of any one phase is $O(m)$ and the bound follows.

A somewhat tighter analysis is as follows. Let D be the maximal number of edges on any shortest path. We have $D < n$ if V^- is empty and $D = \infty$ otherwise. Then Q is empty after phase D and hence the running time is $O(\min(D, n) \cdot m)$.

For many graphs D is much smaller than n. Examples are complete graphs with edge costs chosen uniformly at random from $[0 .. 1]$. In this case $D = O(\log^2 n)$ with high probability [CFMP97]; the expected running time is therefore $O(n^2 \log^2 n)$ for complete graphs with random edge costs. More generally, it is an experimental fact that the Bellman–Ford algorithm is efficient for almost any kind of random graph.

However, there are also graphs where the worst case running time is actually achieved. We give one example in the next section and one now.

A first example are graphs with negative cycles. If V^- is non-empty then the algorithm always uses n phases and a high running time results. We will show in the next but one section how negative cycles can frequently be recognized earlier.

7.5.8 *A Difficult Graph*

The goal of this section is to construct a graph with non-negative edge costs that forces the algorithm of the preceding section into its worst case running time.

The running time analysis given above tells us that a running time of $\Omega(nm)$ results if a fixed fraction of the nodes is removed and added to the queue in each iteration. The Bellman–Ford algorithm uses a breadth-first scanning strategy, i.e., essentially explores paths in the order of their number of edges. Thus if we ensure that paths consisting of more edges have smaller cost we will ensure that every node is added to the queue many times.

We will define the graph in two steps. In the first step we will allow edges of negative cost and in the second step we will remove them. Figure 7.15 shows our worst case example. The graph has nodes $0, \ldots, L-1, L, \ldots, L+K-2$ where $L = 2^l$ is a power of two. We will fix K and L later.

On nodes $L-1$ to $L+K-2$ we have the complete graph in which all edge costs are zero. This makes $(K-1)^2$ edges. On the first L nodes we have the edge $(0, L-1)$, the edges $(L - L/2^j, L - L/2^{j+1})$ and $(L - L/2^{j+1}, L-1)$ for all $j, 0 \leq j < l-1$, and the L edges $(i, i+1)$ for all $i, 0 \leq i < L-1$. This makes for no more than $2L$ edges.

We claim that for any r, $1 \leq r < L$, there is exactly one path from node 0 to node $L-1$ consisting of r edges. This is certainly true for $r = 1$. So assume that $r > 1$. We construct the path as follows. If $r > L/2$ we use $L/2$ edges to go from 0 to $L/2$ and if $r \leq L/2$ we use one edge. In either case we are left with the task of constructing a path from $L/2$ to $L-1$ consisting of r' edges, where $1 \leq r' < L/2$. This path is constructed by applying the argument recursively.

How do we assign edge costs to the edges (i, j) with $0 \leq i < j < L$? We want an assignment which favors paths with more edges. This suggests assigning cost -1 to every edge as this makes sure that the cost of a path consisting of k edges is equal to $-k$. Thus paths with more edges are shorter than paths with fewer edges. We said at the beginning that we will construct a graph with non-negative edge costs and now we have set the cost

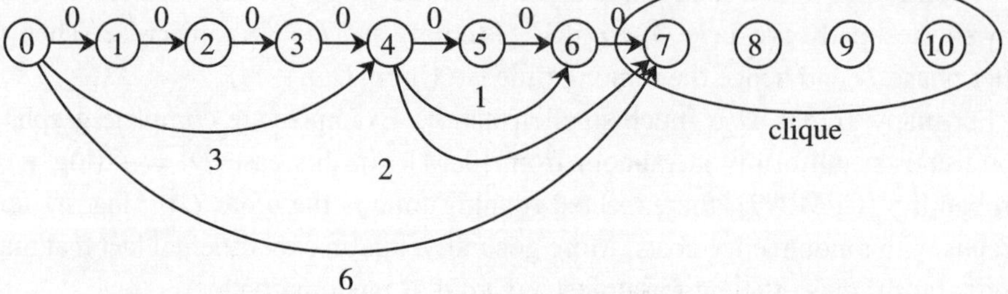

Figure 7.15 The graph generated by *BF_GEN* for $L = 8$ and $K = 4$. The K nodes labeled $L-1$ to $L+K-2$ form a complete directed graph in which all edge costs are zero. The edges in this clique are not shown.

of some edges to -1. This is easily corrected. We set the cost of edge (i, j) to $j - i - 1$. Then all edges have non-negative cost and the cost of a path from 0 to $L-1$ consisting of k edges has cost $L - 1 - k$. Thus we are again favoring paths with more edges over paths with fewer edges.

The total number of edges in our graph is certainly less than $2L + K^2$ and the number of nodes is $L + K - 1$. With $K = \lfloor\sqrt{m/2}\rfloor$ and L the largest power of two no larger than $n/2$, we get a graph with at most $n + m/2$ edges and $n/2 + \sqrt{m/2}$ nodes. This is less than m and n, respectively, if $m \geq 2n$ and $m \leq n^2/2$.

The following procedure *BF_GEN* realizes the construction just outlined. For the edge costs there is the choice between non-negative and arbitrary edge costs. If $m \geq 2n$ and $m \leq n^2/2$ then the constructed graph has at most n nodes and at most m edges.

⟨*_BF_GEN.c*⟩≡

```
#include <LEDA/array.h>
#include <LEDA/graph_alg.h>
void BF_GEN(GRAPH<int,int>& G, int n, int m,
                           bool non_negative)
{ G.clear();
  int K = 1; while ( (K+1)*(K+1) <= m/2 ) K++;
  int l = 0; int L = 1;
  while ( 2*L <= n/2 ) {l++; L = 2*L; }
  array<node> V(n);
  int i, j;
  for (i = 0; i < n; i++) V[i] = G.new_node(i);
  for (i = L - 1; i < L - 1 + K; i++)
    for (j = L - 1; j < L - 1 + K; j++)
      if ( j != i ) G.new_edge(V[i], V[j], 0);
  for (i = 0; i < L - 1; i++) G.new_edge(V[i], V[i+1], 0);
  G.new_edge(V[0],V[L-1],(non_negative? L-1-1  :-1));
  int powj = 1;
  for (j = 0; j < l-1; j++)
  { int x = L - L/powj;
```

```
        int y = L - L/(2*powj);
        G.new_edge(V[x],V[y],  (non_negative? y-x-1 :   -1));
        G.new_edge(V[y],V[L-1],(non_negative? L-1-y-1 : -1));
        powj *= 2;
      }
    }
```

How does our algorithm of the previous section do on the graphs generated by *BF_GEN*? There will be L phases and in each phase the K nodes $L-1, \ldots, L+K-2$ will be removed from the queue and hence K^2 edges will be scanned in each phase. Since $L \geq n/4$ and $K^2 \geq m/4$ the running time is $\Omega(nm)$.

Table 7.2 shows the running times of the basic and the refined version of the Bellman–Ford algorithm (the refined version is the subject of the next section), the time for checking the output, and, if applicable, the running time of Dijkstra's algorithm. We observe that the basic version beats the refined version for random inputs and that both of them are almost competitive with Dijkstra's algorithm for random inputs with non-negative edge costs. The situation changes completely for graphs with negative cycles and graphs generated by *BF_GEN*.

For random graphs with negative cycles the running time of the basic version explodes because it always executes n phases on such graphs. The refined version behaves much better.

For graphs generated by *BF_GEN* the basic version shows the claimed $\Omega(nm)$ behavior. Doubling n (more than) quadruples the running time; the fact that the running time more than quadruples is due to cache effects. Again, the refined version behaves much better. Its running time seems to less than triple if n is doubled. We will explain this effect at the end of Section 7.5.9. Dijkstra's algorithm performs much better than either version of the Bellman–Ford algorithm.

In all cases the time needed to verify the computation is no larger than the time required to compute the result.

There are more shortest-path algorithms than the ones treated in this book, see [AMO93], and some of them have an edge over the algorithms in LEDA in certain situations. The papers [CG96, CGR94, MCN91] contain extensive experimental comparisons of various shortest-path algorithms. The algorithms that we have selected for LEDA are the asymptotically most efficient and also exhibit excellent actual running times.

7.5.9 *A Refined Bellman–Ford Algorithm*

We describe a variant of the Bellman–Ford algorithm due to Tarjan [Tar81]. The worst case running time of the variant is also $O(nm)$. However, the algorithm is frequently much faster than the basic Bellman–Ford algorithm, as Table 7.2 shows, and the algorithm is never much slower. It is available as *BELLMAN_FORD_T*[13].

[13] This is clearly a misnomer. However, we want to keep the name *BELLMAN_FORD_T* for our currently best implementation for the single-source problem with arbitrary edge costs.

Instance	*BF_Basic*	*Bellman_Ford*	*Dijkstra*	*Checking*
n, $n = 10000$	0.3	0.57	0.22	0.31
n, $n = 20000$	0.69	1.36	0.57	0.69
n, $n = 40000$	1.98	3.59	1.47	1.69
c, $n = 10000$	0.3	0.63	—	0.3
c, $n = 20000$	0.81	1.63	—	0.7
c, $n = 40000$	2.02	3.72	—	1.68
r, $n = 2000$	20.2	0.08	—	0.03
r, $n = 4000$	73.15	0.17	—	0.08
r, $n = 8000$	462.5	0.54	—	0.18
g, $n = 4000$	7.52	0.42	0.01001	0.04999
g, $n = 8000$	30.66	1.17	0.04004	0.07996
g, $n = 16000$	131.5	3.24	0.07001	0.19

Table 7.2 Running times of different shortest-path algorithms. We used four different kinds of graphs. Random graphs (generated by *random_graph*(G, n, m)) with random non-negative edge costs in [0 .. 1000], random graphs with arbitrary edge costs but no negative cycles (we chose for each node v a random node potential $pot[v] \in [0 .. 1000]$ and for each edge $e = (v, w)$ a random cost $c[e] \in [0 .. 1000]$ and then set the cost of e to $pot[v] + c[e] - pot[w]$; this generates arbitrary edge costs but no negative cycles as the potentials cancel along any cycle, see Section 7.5.10.), random graphs with random edge costs in [−100 .. 1000], and graphs generated by *BF_GEN*. In the table above the four types of graphs are indicated by the labels n, c, r, and g, respectively. For each type we generated graphs with three different values of n and $m = 8n$. Observe that the graphs in the top half of the table are much larger than the graphs in the lower half of the table. The column *BF_Basic* stands for the basic version of the Bellman–Ford algorithm. You may generate your own version of this table by calling shortest_path_time in the demo-directory.

The variant maintains the shortest-path tree[14] not only implicitly in the form of the *pred*-array but also explicitly. We use T to denote the shortest-path tree. The algorithm uses T to overcome two weaknesses of the basic Bellman–Ford algorithm. Consider the scanning of an edge $e = (v, w)$ and assume that it reduces $dist[w]$ to $dist[v] + c[e]$. In the basic algorithm the only action is to add w to Q (if it is not already there). In the variant we do more:

- The fact that a shorter path to w has been discovered implies that shorter paths exist for all nodes in T_w (= the subtree of T rooted at w). Thus there is no need to propagate the current distance labels of these nodes any further (as smaller distance labels will be

[14] We ignore the possibility of a negative cycle for the moment.

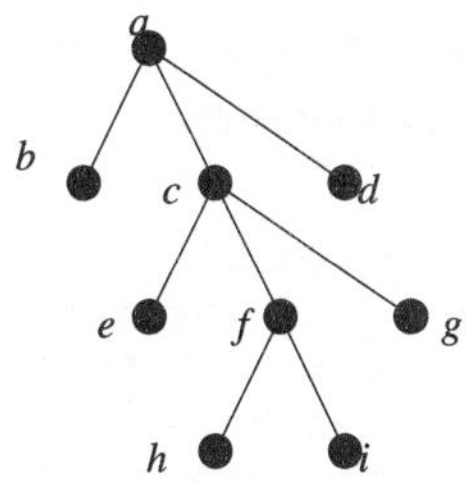

Figure 7.16 The pre-order traversal of the tree shown yields the sequence $a, b, c, e, f, h, i, g, d$.

propagated sometime in the future) and hence all nodes in T_w can be removed from Q and T. Upon removal of T_w from Q and T, w is added to Q and made a child of v in T. This modification introduces a distance related component into the otherwise purely breadth-first scanning strategy of the Bellman–Ford algorithm.

- If w is an ancestor of v or, equivalently, v is a descendant of w then a negative cycle has been detected and all nodes reachable from v can be added to V^-. This modification replaces the indirect way of recognizing negative cycles used in the basic algorithm ("more than n phases") by a direct method.

We come to the details. We use T to denote the current shortest-path tree. It is rooted at s and if w is a child of v in T then $pred[w] = (v, w)$. Conversely, if $pred[w] \neq nil$ then w was already added to T at least once; it may or may not belong to T currently. The tree T is represented by its list of vertices in pre-order traversal, see Figure 7.16, i.e., a single node tree is represented by that node and a tree with root r and subtrees $T_1, \ldots, T_k$ is represented by r, followed by the list for $T_1, \ldots,$ followed by the list for T_k. We use a *list<node>* T to represent the shortest-path tree, a *node_array<int> t_degree* to store the degree of each node, and a *node_array<list_item> pos_in_T* to store the position of each node in the list T. For nodes $v \notin T$ we have $t\_degree[v] = 0$ and $pos\_in\_T[v] = nil$ and for nodes $v \in T$ we have $T[pos\_in\_T[v]] = v$.

The queue Q is also realized as a list of nodes. Every node knows its position in Q . We use a *node_array<list_item> pos_in_Q* for that purpose. If a node v belongs to Q then $pos\_in\_Q[v]$ is its position in Q and if a node v does not belong to Q then $pos\_in\_Q[v] = nil$.

We use $w\_item = pos\_in\_T[w]$ to denote the item corresponding to node w in T. We define a procedure *delete_subtree*$(w\_item, \ldots)$ that deletes all nodes in the subtree T_w from T and Q and returns the item following T_w in T. In Figure 7.16 a call *delete_subtree*$(f\_item, \ldots)$ would delete the subtree T_f and return the item corresponding to g.

If w has no children ($t\_degree[w] = 0$), we simply delete w from T and maybe also from Q. If w has children, the idea is to remove the subtrees of the children by recursive calls. The first child is easy to find; it is the node immediately after w in the list T. The second child (if the degree of w is more than one) is the first node after the sublist representing the first subtree of T_w. This is precisely the node returned by the first recursive call of *delete_subtree* and hence a simple loop removes all subtrees of T_w.

The procedure *delete_subtree* uses Q, T, *pos_in_Q*, *pos_in_T* and *t_degree*. We make them parameters. We will initialize them below.

⟨*BF: auxiliary functions*⟩≡

```
inline list_item BF_delete_subtree(list_item w_item, list<node>& Q,
          list<node>& T, node_array<int>& t_degree,
          node_array<list_item>& pos_in_Q,
          node_array<list_item>& pos_in_T)
{ list_item child = T.succ(w_item);
  node w = T[w_item];
  while (t_degree[w] > 0)
  { t_degree[w]--;
    child = BF_delete_subtree(child,Q,T,t_degree,pos_in_Q,pos_in_T);
  }
  pos_in_T[w] = nil;
  T.del_item(w_item);
  if ( pos_in_Q[w])
  { Q.del_item(pos_in_Q[w]);
    pos_in_Q[w] = nil;
  }
  return child;
}
```

As in the basic algorithm we operate in phases. For the zeroth phase we initialize Q and T with s.

⟨*BF: initialize T, Q, dist, and pred*⟩≡

```
node_array<list_item> pos_in_Q(G,nil);
node_array<int>       t_degree(G,0);
node_array<list_item> pos_in_T(G,nil);
node v;
forall_nodes(v,G) pred[v] = nil;
dist[s] = 0;
list<node> Q;  pos_in_Q[s] = Q.append(s);
list<node> T;  pos_in_T[s] = T.append(s);
```

During the k-th phase, $k \geq 0$, we maintain the following invariants. They refine the invariants of the basic algorithm. We use $\mu_k(v)$ to denote the length of a shortest path from s to v consisting of at most k edges.

(1) For every node v, *dist*[v] is the cost of some path from s to v, and if v belongs to T then *dist*[v] is the cost of the tree path from s to v and *pred*[v] is the tree edge ending in v.

(2) If v has been in T at least once, but is not in T now, then $\mu(v) <$ *dist*[v], i.e, its current distance label is not its true distance label.

(3) Only leaves of T belong to Q, and these leaves have depth k or $k+1$ in T. The nodes of depth k precede the nodes of depth $k+1$ in Q.

(4) The algorithm maintains a *node_array<bool> in_Vm* such that the following items hold for every node v:

(a) *in_Vm*$[v] =$ *true* implies $v \in V^-$.
(b) If every path defining $\mu_k(v)$ contains a negative cycle then *in_Vm*$[v] =$ *true*.
(c) If *in_Vm*$[v] =$ *true* and w is reachable from v in G then *in_Vm*$[w] =$ *true*.

(5) If v is a node in $T \setminus Q$ then $dist[v] + c[e] \geq dist[w]$ for all edges $e = (v, w)$ with *in_Vm*$[w] =$ *false*, i.e., if v is in T but not in Q then its outgoing edges are relaxed. Observe that *in_Vm*$[w] =$ *true* implies $\mu(w) = -\infty$ and hence may be interpreted as "$dist[w] = -\infty$".

(6) For every node v with *in_Vm*$[v] =$ *false*, $dist(v) \leq \mu_k(v)$.

Phase k ends when Q contains no node of depth k anymore[15] and the algorithm terminates when Q is empty.

Let v be the first node in Q and let k be its depth in T. The goal is to remove v from Q without violating the invariants. We explain the required actions first and then give the code. We suggest that the code is read in parallel to the explanation.

We scan all edges $e = (v, w)$ out of v. If *in_Vm*$[w] =$ *true* then there is nothing to do (by invariants (5) and (6)). So assume otherwise. We compare $dist[v] + c[e]$ and $dist[w]$. There are two cases to consider.

If $dist[w] \leq dist[v] + c[e]$ then there is nothing to do, i.e, all invariants hold already. This is obvious if we have inequality or $w \in T$. So assume that we have equality and w does not belong to T. Don't we have to add w to T? No! Observe that $dist[w] = dist[v] + c[e]$ implies $dist[w] < \infty$. Thus w has been in T at least once, and hence (2) implies $\mu(w) < dist[w]$. Thus the invariants also hold in this case.

If $dist[v] + c[e] < dist[w]$ then $\mu(z) < dist[z]$ for all nodes z in T_w. Thus, we may remove w and all its descendants from T and Q, set $dist[w]$ to $dist[v] + c[e]$ and $pred[w]$ to e.

If v was not in T_w and hence v is still in T at this point we make w a child of v and add w to Q. This maintains all invariants. In order to make w a child of v, we simply insert it immediately after v into the list T and increment the degree of v.

If v belonged to T_w then we discovered a negative cycle consisting of the tree path from w to v followed by the edge e. We move all nodes reachable from v in G to V^-.

⟨*bellman_ford.t*⟩≡

```
#include <assert.h>
⟨BF: auxiliary functions⟩
template <class NT>
bool BELLMAN_FORD_T(const graph& G, node s,
                           const edge_array<NT> & c,
                           node_array<NT> & dist,
                           node_array<edge>& pred)
{ ⟨BF: initialize T, Q, dist, and pred⟩
  node_array<bool> in_Vm(G,false); // for V_minus
  bool no_negative_cycle = true;
```

[15] The algorithm does not keep track of node depths and phase numbers; we only use them in the correctness proof.

```
  while (!Q.empty())
  { // select a node v from Q
    node v = Q.pop(); pos_in_Q[v] = nil;
    edge e;
    forall_adj_edges(e,v)
    { node w = G.target(e);
      if ( in_Vm[w] ) continue;
      NT d = dist[v] + c[e];
      if ( ( pred[w] == nil && w != s ) || d < dist[w])
      { dist[w] = d;
        // remove the subtree rooted at w from T and Q
        // if w has a parent, decrease its degree
        if (pos_in_T[w])
        { BF_delete_subtree(pos_in_T[w],Q,T,t_degree,
                                         pos_in_Q,pos_in_T);
          if (pred[w] != nil) t_degree[G.source(pred[w])]--;
        }
        pred[w] = e;
        if (pos_in_T[v] == nil) // v belonged to T_w
        { no_negative_cycle = false;
          ⟨move v and all nodes reachable from it to Vm⟩
        }
        else
        { // make w a child of v and add w to Q
          pos_in_T[w] = T.insert(w,pos_in_T[v],after);
          t_degree[v]++;
          pos_in_Q[w] = Q.append(w);
        }
      }
    }
  }
#ifndef LEDA_CHECKING_OFF
CHECK_SP_T(G,s,c,dist,pred);
#endif
  return no_negative_cycle;
}
```

We still need to complete the case that a negative cycle is detected. When v belonged to T_w we discovered a negative cycle. After setting $pred[w] = e = (v, w)$ this negative cycle is already recorded in the *pred*-array. What remains is to add all nodes that are reachable from v to V^- and to set their *pred*-values accordingly. We want to do so without destroying the negative cycle just found.

This is readily achieved. We first set *in_Vm* to *true* for all nodes on the cycle and then in a second pass over the cycle call $add\_to\_Vm(G, z, \ldots)$ for all nodes z of the cycle. In $add\_to\_Vm(G, z, \ldots)$ we scan all edges out of z. For each edge $e = (z, w)$, where w does not belong to V^- yet, we remove all nodes in T_w from T and Q, we add w to V^-, set $pred[w]$ to e, and make a recursive call.

⟨*move v and all nodes reachable from it to Vm*⟩≡

```
node z = v;
do
{ in_Vm[z] = true;
  z = G.source(pred[z]);
} while (z != v);
do
{ BF_add_to_Vm(G,z,in_Vm,pred,Q,T,t_degree,pos_in_Q,pos_in_T);
  z = G.source(pred[z]);
} while (z != v);
```

where

⟨*BF: auxiliary functions*⟩+≡

```
inline void BF_add_to_Vm(const graph& G, node z,
               node_array<bool>& in_Vm,
               node_array<edge>& pred,
               list<node>& Q, list<node>& T,
               node_array<int>& t_degree,
               node_array<list_item>& pos_in_Q,
               node_array<list_item>& pos_in_T)
{ edge e;
  forall_adj_edges(e,z)
    { node w = G.target(e);
      if ( !in_Vm[w] )
        { if (pos_in_T[w])
            { BF_delete_subtree(pos_in_T[w],Q,T,t_degree,
                                          pos_in_Q,pos_in_T);
              if (pred[w] != nil) t_degree[G.source(pred[w])]--;
            }
            pred[w] = e;
            in_Vm[w] = true;
            BF_add_to_Vm(G,w,in_Vm,pred,
                          Q,T,t_degree,pos_in_Q,pos_in_T);
        }
    }
}
```

This completes the description of the algorithm. We still have to complete the correctness proof and establish the $O(nm)$ running time.

Lemma 12 *The refined Bellman–Ford algorithm solves the single-source shortest-path problem in time $O(nm)$.*

Proof The nodes in V^+ are never reached and hence are treated correctly.

Next consider the nodes in V^-. Invariant (4) tells us that *in_Vm* is set to true only for nodes in V^-. We need to show that *in_Vm* is set to true for all nodes in V^- at some point during the execution. Let $v \in V^-$ be arbitrary. If every path defining $\mu_n(v)$ contains a negative cycle or if v is reachable from such a node then *in_V*$[v]$ is set to true by invariant

(4). We need to show that this is indeed the case. If $v \in V^-$ then there must be an integer $N > n$ and a path $[v_0 = s, v_1, \ldots, v_N = v]$ from s to v such that this path is shorter than any path from s to v with less than N edges. The prefix consisting of the first i edges of this path is a path to v_i that is shorter than any path to v_i with less than i edges. In particular, $\mu_n(v_n) < \mu_{n-1}(v_n)$ and hence any path to v_n defining $\mu_n(v_n)$ contains a negative cycle.

Finally, consider a node in V^f and assume that $\mu(v) = \mu_k(v)$. Then $dist[v] = \mu(v)$ after phase k, the tree path from s to v has cost $\mu(v)$, and the tree path is recorded in the *pred*-array by invariants (1), (2), and (6).

The two preceding paragraphs establish that there are at most $n + 1$ phases. Since each node is removed from Q at most once in each phase the running time is $O(nm)$. □

Table 7.2 shows the running times of the refined Bellman–Ford algorithm on the graphs generated by *BF_GEN*. The running time seems to triple if n is doubled. This can be explained as follows. At the beginning of each phase the nodes L to $L - K - 2$ are children of node $L - 1$ in the shortest-path tree and the nodes $L - 1$ to $L - K - 2$ (and some nodes smaller than $L - 1$) are in Q. In the basic algorithm all nodes $L - 1$ to $L - K - 2$ are removed from the queue and their outgoing edges are scanned. This results in $\Omega(m)$ edge scans per phase. In the refined algorithm the discovery of a better path to node $L - 1$ causes the nodes L to $L - K - 2$ to be removed from Q and T without(!!) scanning their edges. When the edges out of node $L - 1$ are scanned they are again added to Q and T. In this way only the edges out of node $L - 1$ are scanned in each phase. Thus only $\Theta(K) = \Theta(\sqrt{m})$ edges are scanned in each phase and the total running time is therefore $\Theta(n\sqrt{m})$. In particular, for $m = 8n$ as in Table 7.2, the running time grows like $n^{3/2}$ and hence about triples when n is doubled[16].

7.5.10 *The All-Pairs Problem*

The all-pairs shortest-path problem is the task to compute $\mu(v, w)$ for all pairs of nodes v and w. This could be solved by solving the single-source problem with respect to each v. We describe a better method based on so-called *node potentials*; the improved method applies whenever G has no negative cycles. We will see further uses of the node potential method in the section on matchings.

A node potential assigns a number $pot(v)$ to each vertex v. The *transformed* or *reduced* edge costs $\bar{c}$ with respect to a potential function *pot* are defined by

$$\bar{c}(e) = pot(v) + c(e) - pot(w)$$

for each edge $e = (v, w) \in E$. Consider a path $p = [e_0, \ldots, e_{k-1}]$ and let $e_i = (v_i, v_{i+1})$. Then

$$\bar{c}(p) \quad = \sum_{0 \le i < k} \bar{c}(e_i) \quad = \sum_{0 \le i < k} (pot(v_i) + c(e_i) - pot(v_{i+1}))$$

16 The authors initially assumed that the running time of the refined algorithm would also grow like nm on the *BF_GEN*-examples and were surprised to learn from the experiments that this is not the case. It took us some time to understand why not.

$$= \quad pot(v_0) + \sum_{0 \le i < k} c(e_i) - pot(v_k) \quad = \quad pot(v_0) + c(p) - pot(v_k),$$

i.e., the cost of p with respect to $\bar{c}$ is the cost of p with respect to c plus the potential difference between the source and the target of the path. This difference is independent(!!) of the particular path p and only depends on the endpoints of the path. Thus for any two paths p and q with the same source and the same target, $\bar{c}(p) \le \bar{c}(q)$ iff $c(p) \le c(q)$, i.e., the relative order of path costs is not changed by the transformation.

Assume now that G has no negative cycles and that all nodes of G are reachable from some node s. We claim that $pot(v) = \mu(s, v)$ is a node potential such that all reduced costs with respect to it are non-negative. This is easily seen. Observe first that $\mu(s, v)$ is finite for all v if G has no negative cycles and all nodes are reachable from s. The reduced costs are therefore well defined. Observe next that for any edge $e = (v, w)$ we have $\mu(s, v) + c(e) \ge \mu(s, w)$ and hence

$$\bar{c}(e) = \mu(s, v) + c(e) - \mu(s, w) \ge 0.$$

The observations above suggest the following strategy to solve the all-pairs problem. We first solve the single-source problem with respect to some node s from which all nodes of G are reachable. If G has a negative cycle, we stop. Otherwise we use the distances from s to transform the edge costs into non-negative ones and solve the single-source problem for each node v of G. Finally, we translate the computed distances back to the original edge costs, i.e., for each pair (v, w) we set

$$dist(v, w) = dist1(v, w) + pot(w) - pot(v),$$

where *dist* and *dist1* denote the distances with respect to the original and the transformed distance function.

How do we choose s? We add a new vertex s to G and add edges (s, v) of length 0 for all vertices of G. Observe that this does not create any additional cycles; in particular, it does not create any negative cycles. We use the distances $\mu(s, v)$ as our potential function.

⟨*all_pairs.t*⟩≡

```
#include <LEDA/graph_alg.h>

template <class NT>
bool ALL_PAIRS_SHORTEST_PATHS_T(graph&G, const edge_array<NT>& c,
                                node_matrix<NT>& DIST)
{ edge e;
  node v,w;

  node s = G.new_node();
  forall_nodes(v,G) if ( v != s ) G.new_edge(s,v);

  edge_array<NT>    c1(G);
  forall_edges(e,G) c1[e] = (G.source(e) == s? 0 : c[e]);
  node_array<NT>    dist1(G);
  node_array<edge>  pred(G);

  if (!BELLMAN_FORD_T(G,s,c1,dist1,pred)) return false;

  G.del_node(s);
```

```
  forall_edges(e,G)
    c1[e] = dist1[G.source(e)] + c[e] - dist1[G.target(e)];
  // (G,c1) is a non-negative network; for every node v
  // compute row DIST[v] of the distance matrix DIST
  // by a call of DIJKSTRA_T(G,v,c1,DIST[v])
  forall_nodes(v,G) DIJKSTRA_T(G,v,c1,DIST[v],pred);
  // correct the entries of DIST
  forall_nodes(v,G)
  { NT dv = dist1[v];
    forall_nodes(w,G) DIST(v,w) += (dist1[w] - dv);
  }
 return true;
}
```

7.5.11 *Minimum Cost to Profit Ratio Cycles*

We consider a graph G with two weight functions defined on its edges: a function p that assigns a profit to each edge and a function c that assigns a cost to each edge. For a cycle C we use

$$p(C) = \sum_{e \in C} p(e), \qquad c(C) = \sum_{e \in C} c(e), \qquad \lambda(C) = c(C)/p(C)$$

to denote the profit, the cost, and cost to profit ratio of the cycle, respectively. Our goal is to find a cycle that minimizes the cost to profit ratio[17]. We use λ^* and C^* to denote the minimum ratio and a cycle realizing it, respectively, i.e.,

$$\lambda^* = \lambda(C^*) = \min\{\lambda(C)\ ;\ C \text{ is a cycle}\}.$$

Figure 7.17 shows an example. We will define a function

```
rational MINIMUM_RATIO_CYCLE(graph& G,
                             const edge_array<int>& c,
                             const edge_array<int>& p,
                             list<edge>& C_opt);
```

that returns the ratio and the list of edges (in *C_opt*) of a minimum cost to profit ratio cycle. The program returns zero if there is no cycle in G; also the empty list is returned in *C_opt* in this case. The procedure runs in time $O(nm \log(n \cdot C \cdot P))$ where C and P are the maximum cost and profit of any edge, respectively. Observe that edge costs and profits are assumed to be integral. We assume that there are no cycles of cost zero or less with respect to either c or p.

Lawler [Law66] has shown that λ^* and C^* can be found by binary search and repeated shortest-path calculations.

Let λ be a real parameter and consider the cost function c_λ defined by

$$c_\lambda(e) = c(e) - \lambda \cdot p(e)$$

[17] For some readers it may seem more natural to maximize the ratio $p(C)/c(C)$. However, maximizing $p(C)/c(C)$ is the same as minimizing $c(C)/p(C)$ if the cost and profit of all cycles are positive.

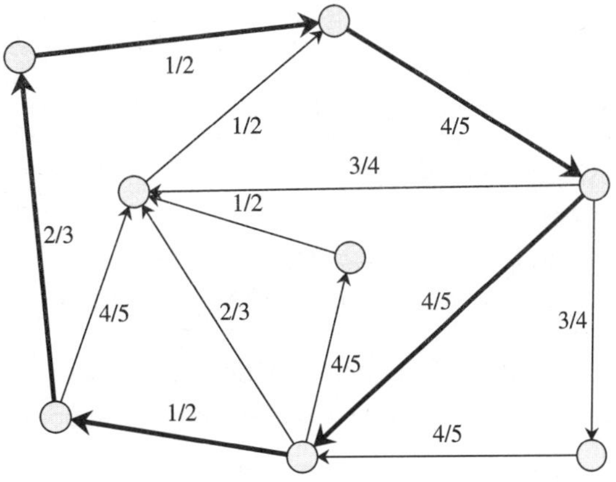

Figure 7.17 An example of a minimum cost to profit ratio cycle. Edge labels are of the form "cost/profit". The optimal cycle is shown in bold. It has cost 12 and profit 17. This figure was generated with the xlman-demo gw_minimum_ratio_cycle. The program minimum_ratio_cycle in LEDAROOT/demo/book/Graph illustrates the execution of *MINIMUM_RATIO_CYCLE*.

for all edges e. We can compare λ with the unknown λ^* by solving a shortest-path problem with cost function c_λ.

If $\lambda > \lambda^*$ then

$$c_\lambda(C^*) = c(C^*) - \lambda \cdot p(C^*) = (\lambda(C^*) - \lambda) \cdot p(C^*) < 0,$$

i.e., there is a negative cycle.

If $\lambda \le \lambda^*$ and C is any cycle then

$$c_\lambda(C) = c(C) - \lambda \cdot p(C) = (\lambda(C) - \lambda) \cdot p(C) \ge (\lambda^* - \lambda) \cdot p(C) \ge 0,$$

i.e., there is no negative cycle.

We capture this argument in the following procedure. It takes a rational *lambda* and returns true if *lambda* is greater than λ^*. The implementation is simple. It assumes that s is a node from which all other nodes of G are reachable. We set up the cost function c_λ and then test for a negative cycle. It is important that all nodes are reachable from s (otherwise, a negative cycle could hide in a part of the graph that is unreachable from s).

We have performed one optimization. The costs $c_\lambda(e)$ for $e \in E$ are rational numbers, all with the same denominator. We therefore multiply all costs with their common denominator and work in integers.

⟨*minimum ratio cycle: compare*⟩≡

```
bool greater_than_lambda_star(const graph& G, node s,
                              const edge_array<int>& c,
                              const edge_array<int>& p,
                              rational lambda)
```

```
{ edge_array<integer> cost(G);
  edge e;
  integer num = lambda.numerator();
  integer denom = lambda.denominator();
  forall_edges(e,G) cost[e] = denom*c[e] - num*p[e];
  node_array<integer> dist(G);
  node_array<edge> pred(G);
  return !BELLMAN_FORD_T(G,s,cost,dist,pred);
}
```

We next show how to use the compare function above in a binary search for λ^*. Let P_{max} and C_{max} be the maximum profit and cost of any edge, respectively. Then

$$p(C) \in [1 \mathinner{..} n \cdot P_{max}] \text{ and } c(C) \in [1 \mathinner{..} n \cdot C_{max}].$$

Thus $\lambda(C)$ is a rational number whose denominator is in the former range and whose numerator is in the latter range. If C_1 and C_2 are cycles with $\lambda(C_1) = a/b \neq c/d = \lambda(C_2)$ then

$$|\lambda(C_1) - \lambda(C_2)| = |a/b - c/d| = |ad - cb|/(bd) \geq 1/(bd) \geq 1/(n \cdot P_{max})^2.$$

Let $\delta = 1/(n \cdot P_{max})^2$. We now have all the ingredients for a binary search. We start with the half-open interval $[\lambda_{min} \mathinner{..} \lambda_{max}) = [0 \mathinner{..} 1 + n \cdot C_{max})$ (it is convenient to maintain the invariant $\lambda_{min} \leq \lambda^* < \lambda_{max}$) and then repeatedly compare $\lambda = (\lambda_{min} + \lambda_{max})/2$ with λ^*. If $\lambda > \lambda^*$ we set λ_{max} to λ and if $\lambda \leq \lambda^*$ we set λ_{min} to λ. In this way we maintain the invariant $\lambda_{min} \leq \lambda^* < \lambda_{max}$. We continue until $\lambda_{max} - \lambda_{min} \leq \delta$. Then $\lambda_{min} \leq \lambda^* < \lambda_{min} + \delta$ and hence there is no cycle C with $\lambda^* < \lambda(C) < \lambda_{max}$. We will use this observation below to extract C^* and λ^*.

The following procedure summarizes the discussion. We first add a new node s and edges (s, v) for all $v \in V$ to our graph (this makes all nodes reachable from s) and then perform the binary search. Whenever a midpoint is computed in the binary search we normalize its representation, i.e., cancel out common factors of numerator and denominator. This is important to keep the representations of the rationals small.

⟨*_minimum_ratio_cycle.c*⟩≡

```
#include <LEDA/templates/shortest_path.t>
#include <LEDA/rational.h>
⟨minimum ratio cycle: compare⟩
rational MINIMUM_RATIO_CYCLE(graph& G,
                             const edge_array<int>& c,
                             const edge_array<int>& p,
                             list<edge>& C_opt)
{ node v; edge e;
  ⟨additional variables for demos⟩ // for minimum ratio cycle demo
  C_opt.clear();
  if ( Is_Acyclic(G) ) return rational(0);
  node s = G.new_node();
```

```
  forall_nodes(v,G) if (v != s) G.new_edge(s,v);
  edge_array<int> c1(G);
  edge_array<int> p1(G);
  int Cmax = 0; int Pmax = 0;
  forall_edges(e,G)
  { if (G.source(e) == s) { c1[e] = p1[e] = 0; }
    else
    { c1[e] = c[e]; p1[e] = p[e];
      Cmax = Max(Cmax,c[e]);
      Pmax = Max(Pmax,p[e]);
    }
  }
  int n = G.number_of_nodes();
  ⟨minimum ratio cycle: check precondition⟩
  integer int_n(n);
  integer int_Pmax(Pmax);
  rational lambda_min(integer(0));
  rational lambda_max(int_n * integer(Cmax) + integer(1));
  rational delta(1,int_n * int_n * int_Pmax * int_Pmax);
  while (lambda_max - lambda_min > delta)
  { rational lambda = (lambda_max + lambda_min)/2;
    lambda.normalize();   // important
    ⟨report progress in demos⟩
    if ( greater_than_lambda_star(G,s,c1,p1,lambda) )
     lambda_max = lambda;
    else
     lambda_min = lambda;
  }
  rational lambda_opt;
  { ⟨minimum ratio cycle: determine lambda_opt and C_opt⟩ }
  G.del_node(s);
  return lambda_opt;
}
```

When the binary search terminates we have

$$\lambda_{max} - \lambda_{min} \leq \delta \text{ and } \lambda_{min} \leq \lambda^* < \lambda_{max}$$

and hence there can be no cycle C with $\lambda^* < \lambda(C) < \lambda_{max}$. Let $\lambda = \lambda_{max}$. Since $\lambda^* < \lambda_{max}$, there is a negative cycle with respect to c_λ. Let C be any negative cycle with respect to c_λ. Then $\lambda(C) < \lambda = \lambda_{max}$ and hence $\lambda(C) = \lambda^*$. We conclude that any negative cycle with respect to c_λ is an optimal cycle.

A negative cycle with respect to c_λ is easy to find. We set up the cost function c_λ and run BELLMAN_FORD_T. We then run CHECK_SP_T on the output. It labels all nodes lying on a negative cycle by -2. We pick any such node and trace the cycle containing it.

⟨*minimum ratio cycle: determine lambda_opt and C_opt*⟩≡

```
edge_array<integer> cost(G);
node v; edge e;
integer num = lambda_max.numerator();
integer denom = lambda_max.denominator();
forall_edges(e,G) cost[e] = denom*c1[e] - num*p1[e];
node_array<integer> dist(G);
node_array<edge> pred(G);
BELLMAN_FORD_T(G,s,cost,dist,pred);
node_array<int> label = CHECK_SP_T(G,s,cost,dist,pred);
forall_nodes(v,G) if (label[v] == -2) break;
int P = 0; int C = 0;
node z = v;
do { P += p[pred[z]]; C += c[pred[z]];
     C_opt.append(pred[z]);
     z = G.source(pred[z]);
   } while ( z != v);
lambda_opt = rational(C)/rational(P);
```

We still need to show how to check the precondition $p(C) > 0$ and $c(C) > 0$ for all cycles C. We discuss the latter condition. Consider the cost function c_λ defined by $c'(e) = c(e) - 1/n$ for all edges e. Clearly, if there is no negative cycle with respect to c' then there is no cycle of length zero or less with respect to c. Conversely, if $c(C) > 0$ and hence $c(C) \geq 1$ for all C then $c'(C) = c(C) - |C|/n \geq 1 - n/n \geq 0$ and there is no negative cycle with respect to c'.

We can therefore misuse our comparison function to check the precondition.

⟨*minimum ratio cycle: check precondition*⟩≡

```
edge_array<int> unit_cost(G,1);
rational one_over_n(integer(1),integer(n));
if (greater_than_lambda_star(G,s,c1,unit_cost,one_over_n))
   error_handler(1,"cycle of cost zero or less wrt c");
if (greater_than_lambda_star(G,s,p1,unit_cost,one_over_n))
   error_handler(1,"cycle of cost zero or less wrt p");
```

The running time of the algorithm is $O(nm \log(n \cdot P_{max} \cdot C_{max}))$. This can be seen as follows. The binary search starts with an interval of length $nC_{max} + 1$ and ends with an interval of length $1/(n \cdot P_{max})^2$. The length of the interval is halved in each iteration and hence the number of iterations is $O(\log(n \cdot P_{max} \cdot C_{max}))$. Each iteration takes time $O(nm)$.

The technique used in our program for the minimum ratio cycle problem is called *parametric search*. Parametric search is applicable in the following situation:

- One searches for the threshold value λ^* of a monotone predicate $P(\lambda)$ of one real argument λ. A predicate P is monotone if

$$\lambda_1 < \lambda_2 \text{ and } P(\lambda_1) \text{ imply } P(\lambda_2),$$

and the threshold value of P is

$$\lambda^* = \inf\{\lambda \,;\, P(\lambda)\}\,.$$

In the problem of this section $P(\lambda)$ holds if there is a negative cycle with respect to the cost function c_λ.

- There is a decision procedure for $P(\lambda)$.
- There is a master procedure that drives the search for λ^*. We used binary search as the master procedure in this section.

We refer the reader to [Meg83] and [AST94] for further applications of parametric search.

Parametric search has high demands on the underlying arithmetic. You can get an impression of the arithmetic demand of the minimum ratio cycle procedure by calling the program minimum_ratio_cycle in LEDAROOT/demo/book/Graph. The paper [SSS97] discusses an application of the number class *real* to parametric search.

Exercises for 7.5

1 (Single-pair shortest-path problem) Let s and t be distinct nodes in a directed graph with non-negative edge costs. The goal is to compute a shortest path from s to t. Assume that there is heuristic information available which gives, for any node v, a *lower bound* $lb(v)$ for the length of a shortest path from v to t. Modify Dijkstra's algorithm such that $dist(v) + lb(v)$ is used as the priority of node v.

2 Show that the condition $d(v) \geq \mu(v)$ for all v in part (b) of Lemma 6 is essential, i.e., the claim does not hold without it.

3 Investigate the following shortest-path algorithm. Split the input graph G into G^- consisting of all edges of negative cost and $G^{\geq 0}$ consisting of all edges of non-negative cost. What can you say when G^- is not acyclic? If G^- is acyclic then run alternately the acyclic shortest-path algorithm on G^- and Dijkstra's algorithm on $G^{\geq 0}$. In each case the distance labels output by the preceding run must be taken as the initial distance labels for the next run. Modify the programs accordingly.

4 Consider the following version of the Bellman–Ford algorithm. It iterates over all edges on the graph n times. Whenever an edge $e = (v, w)$ is considered, $d(w)$ is set to the minimum of $d(w)$ and $d(v) + c(e)$.

```
dist[s] = 0;
forall_nodes(v,G) pred[v] = nil;
for(int i = 0; i < n; i++)
  forall_edges(e,G)
  { node v = G.source(e);
    node w = G.target(e);
    if ( v != s && pred[v] == nil) continue;
    // dist[v] is finite
    d = dist[v] + cost[e];
    if ( pred[w] == nil && v != s || d < dist[w] )
    { dist[w] = d; pred[w] = e; }
  }
```

Show that the algorithm computes all finite distances correctly (Hint: show that $d(v)$ is bounded above by the length of a shortest path consisting of at most k edges after the k-th iteration.). Modify the algorithm so that it satisfies our output convention. Implement the algorithm and compare its running time to the implementations of the Bellman–Ford algorithm given in the text. What is best case running time of the algorithm?

5 In all our algorithms we implemented the test $c < d(w)$ in a somewhat clumsy way due to the fact that $d(w)$ may be $+\infty$ and that most number types have no representation for $+\infty$. Show that nC where C is the largest cost of any edge can be taken as an approximation of $+\infty$. Modify the algorithms accordingly and time them in comparison to the algorithms in the text.

6 Our algorithm for determining minimum ratio cycles uses binary search. It starts with an interval of length $nC_{max} + 1$ and stops as soon as the length of the interval becomes $1/(nP_{max})^2$ or less. Thus there are $\log(n^3 P_{max}^2 C_{max})$ iterations and hence the algorithm handles rational numbers with denominator as large as $n^3 P_{max}^2 C_{max}$. This is unnecessarily large since the $\lambda(C)$ are rational numbers whose denominator is bounded by nP_{max}. Explore the possibility that the values of λ are restricted to rational numbers whose denominator is bounded by nP_{max}. This requires us to write a function that "rounds" a rational number to the closest rational number whose denominator is bounded by some prescribed integer. Inspect the function *small_rational_near* of class *rational* to see how such a function can be realized.

7 Define a number class *NT_star*. The definition is with respect to a fixed graph G with integral weight functions c and p. Let λ^* be the minimum cost to profit ratio of a cycle in G. Each number of this class is represented by a pair of integers. Addition is componentwise and there is no multiplication. Zero has both its components equal to zero. A pair (a, b) is less than (equal to, larger than) a pair (c, d) if $a + \lambda^* b < (=, >) c + \lambda^* d$. Implement the compare function as follows. Let $\lambda = (c - a)/(b - d)$ and use the comparison between λ and λ^* (realized by a shortest-path computation as in the text). The number type maintains an interval $[\lambda_{min} .. \lambda_{max}]$ containing λ^*. Whenever a comparison is performed this interval is updated. Use the number type in a shortest-path computation on the graph G. What will the final interval be?

7.6 Bipartite Cardinality Matching

We start with the problem definition and the functionality of the bipartite matching algorithms. We describe a checker and then lay the foundations of matching algorithms. In the bulk of the section we discuss the implementations of several matching algorithms and derive some general implementation principles. We close with an experimental comparison of our implementations.

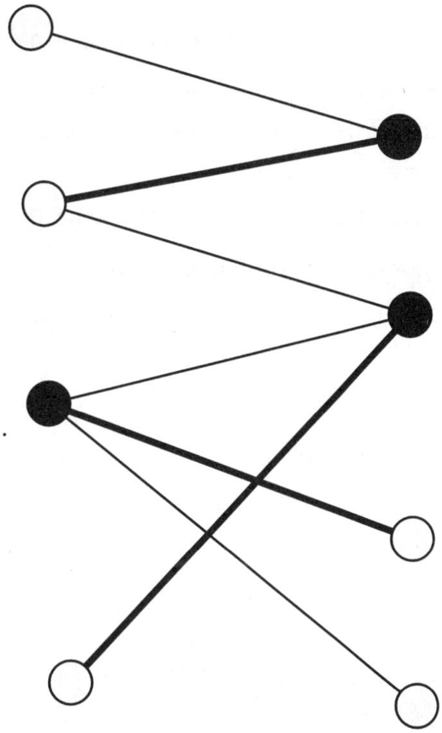

Figure 7.18 A graph and a maximum matching: The bold edges form a matching of cardinality three. The filled nodes form a node cover of cardinality three; a node cover is a set of nodes containing at least one endpoint of every edge. The node cover proves the optimality of the matching. This figure was generated with the xlman-demo gw_mcb_matching.

7.6.1 *Concepts and Functionality*

Let $G = (V, E)$ be a graph. A *matching* M is a subset of the edges no two of which share an endpoint, see Figure 7.18. The cardinality $|M|$ of a matching M is the number of edges in M.

A node v is called *matched* with respect to a matching M if there is an edge in M incident to v and it is called *free* or *unmatched* otherwise. An edge $e \in M$ is called a *matching* edge. A matching is called *perfect* if all nodes of G are matched. For a matched node v the unique node w connected to v by a matching edge is called the *mate* of v.

In this section we assume that G is *bipartite*, i.e., that there is a partition $V = A \dot{\cup} B$ of the nodes of G such that every edge of G has one endpoint in A and one endpoint in B. Matchings in general graphs are the topic of Section 7.7. The procedure

```
bool Is_Bipartite(const graph& G, list<node>& A, list<node>& B)
```

tests whether G is bipartite and if so computes an appropriate partition of the nodes in lists A and B. It runs in time $O(n + m)$.

The procedure

```
list<edge> MAX_CARD_BIPARTITE_MATCHING(graph& G);
```

returns a maximum cardinality matching; the graph G must be bipartite. The worst case and average case running time of the algorithm are $O(\sqrt{n} \cdot m)$ and $O(m \log n)$, respectively. The variant

```
list<edge> MAX_CARD_BIPARTITE_MATCHING(graph& G, node_array<bool>& NC);
```

returns in addition a proof of optimality in the form of a node cover *NC*.

A *node cover* is a set U of nodes such that for every edge (v, w) of G at least one of the endpoints is in U.

Lemma 13 *Let M be a matching and let U be a node cover. Then* $|M| \leq |U|$.
If $|M| = |U|$ *then M is a maximum cardinality matching and U is a minimum cardinality node cover.*

Proof Since U is a node cover, each edge $e \in M$ has at least one endpoint in U. We assign an endpoint in U to each edge in M; for an edge in M having both endpoints in U the choice of the endpoint is arbitrary. Each node is assigned at most once since every node v has at most one edge in M incident to it. Hence, $|M| \leq |U|$.

If $|M| = |U|$ then M is a maximum cardinality matching, since no matching can have cardinality larger than $|U|$, and U is a minimum cardinality node cover, since no node cover can have cardinality smaller than $|M|$. □

We will later show that in bipartite graphs there is always a node cover and a matching of the same cardinality. Lemma 13 is the basis for a checker for maximum cardinality matchings in bipartite graphs. The checker takes a set M of edges and a set *NC* of nodes, and checks that M is a matching, *NC* is a node cover, and that the cardinality of M is equal to the cardinality of *NC*.

⟨*_mcb_matching*⟩≡

```
static bool False(string s)
{ cerr << "CHECK_MCB: " + s +"\n"; return false; }
bool CHECK_MCB(const graph& G,const list<edge>& M,
               const node_array<bool>& NC)
{ node v; edge e;
  // check that M is a matching
  node_array<int> deg_in_M(G,0);
  forall(e,M)
  { deg_in_M[G.source(e)]++;
    deg_in_M[G.target(e)]++;
  }
  forall_nodes(v,G)
   if ( deg_in_M[v] > 1 ) return False("M is not a matching");
  // check size(M) = size(NC)
  int K = 0;
  forall_nodes(v,G) if (NC[v]) K++;
```

```
    if ( K != M.size() ) return False("M is smaller than node cover");
    // check that NC is a node cover
    forall_edges(e,G)
      if ( ! (NC[G.source(e)] || NC[G.target(e)]) )
        return False("NC is not a node cover");
    return true;
}
```

7.6.2 *Concepts for Maximum Matching Algorithms*

We introduce the concepts of alternating and augmenting paths that are crucial for all matching algorithms. A large part of the section applies not only to bipartite graphs but to all graphs. We will clearly state when we restrict attention to bipartite graphs.

A simple path $p = [e_0, e_1, \ldots, e_{k-1}]$ from v to w in G is called an *alternating* path with respect to a matching M if:

- the edges in p are alternately in M and not in M,
- exactly one of e_0 and e_{k-1} is a matching edge if $v = w$,
- either e_0 is a matching edge or v is free and either e_{k-1} is a matching edge or w is free if $v \neq w$.

Figure 7.19 shows examples. The importance of alternating paths stems from:

Lemma 14 *If p is an alternating path with respect to M then $M' = M \oplus p = (M \setminus p) \cup (p \setminus M)$ is also a matching.*

Proof Consider any node z. We need to show that at most one edge of M' is incident to z. This is obvious if z does not lie on p or if z is not an endpoint of p or if p is a cycle. So assume that z is an endpoint of p and p is not a cycle, say $z = v \neq w$. Since p is simple, it contains only one edge incident to v, namely e_0. Moreover, if $e_0 \notin M$ then v is free with respect to M. Thus at most one edge of M' is incident to v. □

If p is alternating with respect to M then $M \oplus p$ has cardinality one larger than M if both endpoints of p are free, has the same cardinality as M if exactly one endpoint is free, and has cardinality one smaller than M if no endpoint is free.

An alternating path p is called *augmenting* if both endpoints of p are free. For an augmenting path the cardinality of the matching $M \oplus p$ is one larger than the cardinality of M. If M does not have maximum cardinality then there is always an augmenting path, as the next lemma shows; if M is "far" from optimality there are many augmenting paths (even short ones).

Lemma 15 *Let M and M' be matchings in a graph G. We have the following:*

- *$M \oplus M'$ consists of alternating paths and alternating cycles.*

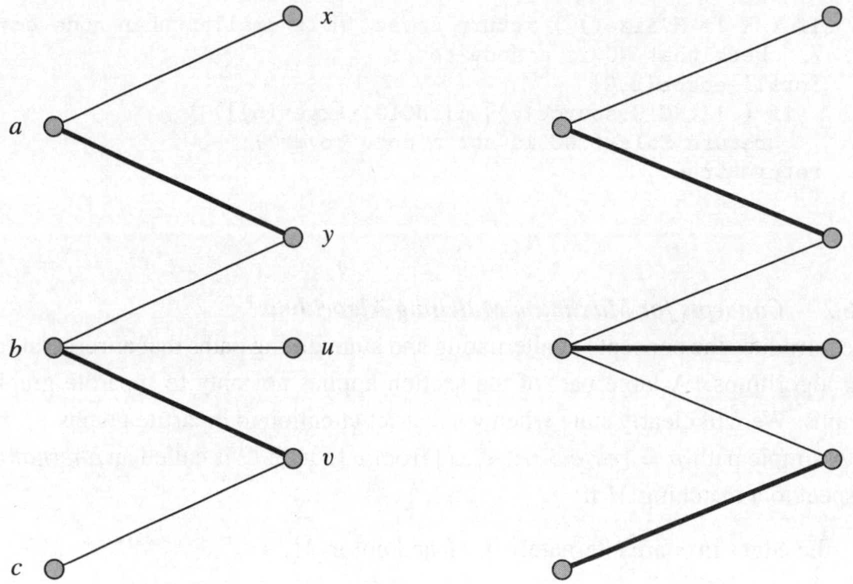

Figure 7.19 Alternating paths: The edges of a matching M are shown in bold. The paths $p_1 = [a, y, b, v]$, $p_2 = [u, b, v]$, and $p_3 = [u, b, v, c]$ are alternating with respect to M, but the path $p_4 = [a, y, b]$ is not. Augmenting M by p_1 decreases the size of the matching (as both endpoints of p_1 are matched), augmenting by p_2 leaves the size of the matching unchanged (as exactly one of the endpoints of p_2 is matched), and augmentation by p_3 increases the size of the matching by one (as both endpoints of p_3 are free). The right half of the figure shows the matching obtained by augmenting by p_3.

- *If $|M| < |M'|$ then there is at least one augmenting path in G with respect to M.*
- *Let $d = |M'| - |M|$. Then there is at least one augmenting path of length at most n/d and there are at least $d/2$ augmenting paths of length at most $2n/d$.*

Proof Consider the graph with edge set $M \oplus M'$. In this graph each node has degree zero, one, or two, and hence the graph consists of paths, cycles, and isolated nodes. Since M and M' are matchings, the edges of M and M' alternate on every path and cycle.

An alternating cycle contains the same number of edges of M and M'. Thus, if $|M| < |M'|$, then there must be at least one path in $M \oplus M'$ which contains more edges of M' than of M. Such a path contains one more edge of M' than of M and hence the first and the last edge of the path belong to M'. Thus the path is augmenting with respect to M.

The argument in the previous paragraph actually shows that there must be d paths in $M \oplus M'$ which contain more edges of M' than of M. Thus there are d augmenting paths with respect to M. The paths are node-disjoint and hence contain at most n edges in total. Thus their average length is at most n/d and there are at least $d/2$ paths whose length is at most $2n/d$. □

It is worthwhile looking at a numerical example. Assume that M is empty and that G

allows for a perfect matching. Taking M' as a perfect matching we have $d = n/2$ and hence there are at least $n/4$ augmenting paths of length at most 2.

Corollary 3 *Let M be a matching in a graph G. M is a maximum cardinality matching in G iff there is no augmenting path in G with respect to M.*

Proof Clearly, if there is an augmenting path p with respect to M then M is not a maximum cardinality matching.

Assume conversely, that M is not a maximum cardinality matching. Then there is a matching M' such that $|M| < |M'|$. Lemma 15 implies the existence of an augmenting path with respect to M. □

Corollary 3 immediately suggests an algorithm for finding maximum matchings.

```
M = some matching;
while there is an augmenting path p with respect to M
{ augment M by p; }
```

In the remainder of this section we concentrate on bipartite graphs. In a bipartite graph $G = (A \dot{\cup} B, E)$ there is a particularly simple method for finding augmenting paths. We direct all free edges from A to B and all matching edges from B to A. The existence of an augmenting path is then tantamount to the existence of a path from a free node in A to a free node in B. Also, augmentation by a path p is trivial. One simply reverses the direction of all edges on the path. Observe that this correctly records that the endpoints of p are now matched and that M was replaced by $M \oplus p$, see Figure 7.20. *We will use this "directed" view in all our implementations of bipartite matching algorithms.*

Before we turn to implementations we make the observation that it suffices to search for augmenting paths only from vertices in A and from each vertex only once, i.e., the algorithm above can be modified to:

```
M = some matching;
forall nodes v in A
{ if there is an augmenting path p with respect to M starting in v
    { augment M by p; }
}
```

We prove that the modified algorithm is correct. We observe first that the set of nodes in A that are matched in $M \oplus p$ are exactly the nodes that are matched in M plus the source node of p.

Let M_0 be the initial matching, let A_0 be the nodes in A that are matched in M_0 and let $v_1, v_2, \ldots, v_k$ be the vertices in $A \setminus A_0$ in the order in which they are considered. For all i, $i \geq 1$, let M_i be equal to M_{i-1} if there is no augmenting path p_i with respect to M_{i-1}

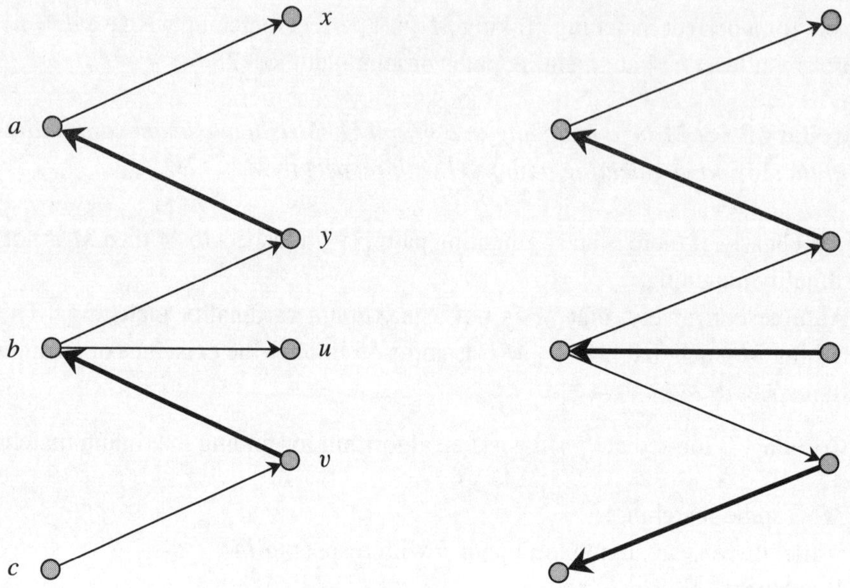

Figure 7.20 The edges of a matching M are shown in bold. Matching edges are directed from right to left and non-matching edges are directed from left to right. The path $p = [c, v, b, u]$ is an augmenting path with respect to M. Augmenting M by p yields the matching $M \oplus p$ shown in the right half of the figure.

starting in v_i and let it be $M_{i-1} \oplus p_i$ otherwise. Let $A_i = A_0 \cup \{v_1, \ldots, v_i\}$ and let G_i be the subgraph spanned by $V_i = A_i \,\dot{\cup}\, B$.

Lemma 16 *For all i: M_i is a maximum cardinality matching in G_i.*

Proof The claim is certainly true for $i = 0$ as all nodes in A_0 are matched. So consider $i \geq 1$ and assume that the claim is true for M_{i-1}. Let k_i be the maximum cardinality of a matching in the subgraph spanned by V_i. If $k_i = k_{i-1}$ then the claim clearly holds for i. So assume that $k_i > k_{i-1}$ and let M^* be an optimal matching in G_i. Then v_i must be matched in M^* (otherwise there would be a matching of cardinality k_i in G_{i-1}, a contradiction to the optimality of M_{i-1}) and hence $M_{i-1} \oplus M^*$ contains a path p starting in v_i. The path starts with an edge in M^* and is alternating with respect to M_{i-1}; we consider the maximal length path of this form. If p also ends with an edge in M^* then p is augmenting with respect to M_{i-1} and hence the cardinality of M_i is one larger than the cardinality of M_{i-1}. Thus M_i is optimal. If p ends with an edge in M_{i-1} then $M^* \oplus p$ has the same cardinality as M^* and does not match v_i. Thus there is a matching of cardinality k_i in G_{i-1}, a contradiction to the optimality of M_{i-1}. □

7.6.3 *Translating between the Directed and the Undirected View*

We stated in the previous section that augmenting paths in bipartite graphs are particularly easy to find if one adopts a directed view: all matching edges are directed from B to A, all non-matching edges are directed from A to B, and augmentation by a path p means to reverse all its edges. We take this directed view in all our implementations of bipartite matching algorithms. However, we do not want to impose this directed view on the users of matching algorithm. For them an "undirected" view is more appropriate. In this section we discuss how to translate between the two views.

We postulate the following common interface for all our implementations:

- The node set is partitioned into disjoint sets A and B (given as lists of nodes).
- All edges are directed from A to B.
- The implementations are allowed to modify the graph in two ways: they may reorder adjacency lists and they may change the orientation of edges during execution. At termination, all edges must again[18] be directed from A to B. However, the ordering of the adjacency lists may be arbitrary.

In this section we show how to prepare this input format and how to restore the original graph.

We determine a bipartition $V = A \dot{\cup} B$ of V by calling *Is_Bipartite*(G, A, B). This call will return true iff G is bipartite and compute A and B if G is bipartite. We then orient all edges from A to B. Having oriented all edges from A to B we compute a maximum matching by calling one of our matching algorithms. After returning from the matching algorithm we restore the original orientation of all edges and the original order of all adjacency lists.

We give more details. We deal with the edge orientations first. We collect all edges out of nodes in B in a list *edges_out_of_B* and reverse the orientation of all of them (operation *rev_edge*). After return from the matching algorithm we again reverse all edges in *edges_out_of_B* and thus restore their original orientation.

We come to the orderings of the adjacency lists. We number all edges according to their original order and use *sort_edges* to restore the original order.

Among our implementations of matching algorithms the algorithm by Alt, Blum, Mehlhorn, and Paul seems to be the best, see Section 7.6.7 for an experimental comparison of all implementations. We therefore use it as our default implementation.

⟨*_mcb_matching*⟩+≡

```
list<edge> MAX_CARD_BIPARTITE_MATCHING(graph& G, node_array<bool>& NC)
{ list<node> A,B;
  node v; edge e;
  if ( !Is_Bipartite(G,A,B) )
    error_handler(1,"MAX_CARD_BIPARTITE_MATCHING: G is not bipartite");
  edge_array<int> edge_number(G); int i = 0;
  forall_nodes(v,G)
```

[18] We would not make this requirement anymore if we could start from scratch.

```
    forall_adj_edges(e,v) edge_number[e] = i++;
  list<edge> edges_out_of_B;
  forall(v,B)
  { list<edge> outedges = G.adj_edges(v);
    edges_out_of_B.conc(outedges);
  }
  forall(e,edges_out_of_B) G.rev_edge(e);
  list<edge> result = MAX_CARD_BIPARTITE_MATCHING_ABMP(G,A,B,NC);
  forall(e,edges_out_of_B) G.rev_edge(e);
  G.sort_edges(edge_number);
#ifndef LEDA_CHECKING_OFF
CHECK_MCB(G,result,NC);
#endif
  return result;
}
```

7.6.4 *The Ford and Fulkerson Algorithm*

In this section $G = (V, E)$ is a bipartite graph with $V = A \dot{\cup} B$. All edges have one endpoint in A and one endpoint in B and all edges are directed from A to B. Our goal is to compute a matching of maximum cardinality. We are allowed to reorder adjacency lists and to reorient edges but we must at the end again orient all edges from A to B.

We will give several implementations of the Ford and Fulkerson algorithm [FF63] already derived in Section 7.6.2.

M = some matching;
forall nodes v in A
{ **if** there is an augmenting path p with respect to M starting in v
  { augment M by p; }
}

The implementations differ:

- in the strategy used to search for augmenting paths (we will study depth-first and breadth-first search),
- in the choice of the initial matching (we will either use the empty matching or the matching produced by the so-called greedy heuristic),
- in the data structures used.

All implementations have a worst case running time of $O(nm)$. They have different best case behaviors and different average case behaviors and they behave drastically differently in practice.

A First Implementation: We implement the algorithm above and call the resulting procedure MAX_CARD_BIPARTITE_MATCHING_FFB; FFB stands for basic version of the Ford and Fulkerson algorithm. It starts by declaring all nodes as free and then iterates over all nodes in A. For each node v in A it tries to find an augmenting path starting in v by calling *find_aug_path_by_dfs*$(G, f, \mathit{free}, \mathit{reached})$ for the edges f out of v. A call *find_aug_path_by_dfs*$(G, f, \ldots)$ returns true if there is an augmenting path starting with f and returns false otherwise. In the former case it also augments the current matching by the path (by reversing all its edges) and labels the endpoint in B of the path as non-free. In either case it labels all visited nodes (by setting $\mathit{reached}[w]$ to true for each visited node w). If an augmenting path starting with a particular edge f is found, v is made non-free and the next node in A is considered.

When all nodes in A have been considered the result list is prepared, all edges are directed from A to B (as this is required by our interface convention), and a node cover is computed.

⟨*_FFB_matching*⟩≡

```
  ⟨FFB: dfs⟩
  list<edge> MAX_CARD_BIPARTITE_MATCHING_FFB(graph& G,
                                const list<node>& A, const list<node>& B,
                                node_array<bool>& NC)
  { node v; edge e;
    node_array<bool> free(G,true);
    // check that all edges are directed from A to B
    forall(v,B) assert(G.outdeg(v) == 0);
    forall(v,A)
    { edge f;
      node_array<bool> reached(G,false);
      forall_adj_edges(f,v)
      { if (find_aug_path_by_dfs(G,f,free,reached))
        { free[v] = false;
          break;
        }
      }
    }
    ⟨MCB: prepare result and node cover and restore orientations⟩
  }
```

We give the details of *find_aug_path_by_dfs*$(G, f, \mathit{free}, \mathit{reached})$. It is a variant of depth-first search; later in the section we will also consider breadth-first search. In a general call, f is some edge and the recursion stack contains a path p starting at a free node in A and ending in f. In the procedure we distinguish cases according to whether the target node of f is free or not.

If the target node w of f is free, we have found an augmenting path. We label w as non-free and then reverse all edges in p. This can be done by unwinding the recursion stack and reversing all edges contained in it. More precisely, we reverse f and return true. The enclosing call receives true and knows that an augmenting path has been found. It reverses its argument and returns true. In this way all edges on the path are reversed.

If w is not free, we try to extend the path. Let $e = (w, z)$ be any edge out of w. If z was already reached then there is no need to explore e as we know already that no free node in B can be reached from z. If z was not reached yet we make a recursive call for e.

⟨*FFB: dfs*⟩≡

```
static bool find_aug_path_by_dfs(graph& G, edge f,
                          node_array<bool>& free, node_array<bool>& reached)
{ node w = G.target(f);
  reached[w] = true;
  if (free[w])
  { free[w] = false;
    G.rev_edge(f);
    return true;
  }
  edge e;
  forall_adj_edges(e,w)
  { node z = G.target(e);
    if ( reached[z] ) continue;
    if ( find_aug_path_by_dfs(G,e,free,reached) )
    { G.rev_edge(f);
      return true;
    }
  }
  return false;
}
```

We complete the description of our first matching algorithm by discussing how to produce the matching, the node cover, and how to orient all edges from A to B. The matching M consists of all edges that are directed from B to A. Their directions need to be reversed.

How can we find a node cover NC? We claim that the following rule determines a node cover. For each matched edge we select the endpoint in B, if this endpoint can be reached from a free node in A, and the endpoint in A otherwise, see Figure 7.21.

Clearly, each matching edge is incident to a node in NC. We now consider a non-matching edge $e = (v, w)$ with $v \in A$ and $w \in B$. If v is free then w must be matched (by optimality of M), and w was selected according to the rule above. If v is matched and was not selected then there must be a matching edge $f = (v, w')$ with w' selected. This means that w' can be reached from a free node in A. Extend this path by f and e to see that w is selected according to the rule above.

⟨*MCB: prepare result and node cover and restore orientations*⟩≡

```
list<edge> result;
forall(v,B)
  forall_adj_edges(e,v) result.append(e);
forall_nodes(v,G) NC[v] = false;
node_array<bool> reachable(G,false);
```

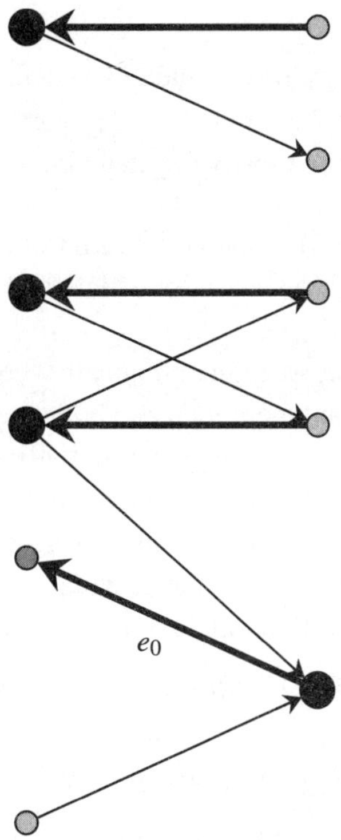

Figure 7.21 The edges of a matching M are shown in bold. The nodes on the left belong to A and the nodes on the right belong to B. Matching edges are directed from A to B and non-matching edges are directed from B to A. The edge e_0 is the only matching edge whose endpoint in B is reachable from a free node in A. The node cover is shown as large solid circles.

```
forall(v,A)
  if (free[v]) DFS(G,v,reachable);
forall(e,result)
  if ( reachable[G.source(e)] )
    NC[G.source(e)] = true;
  else
    NC[G.target(e)] = true;
forall(e,result) G.rev_edge(e);
return result;
```

What is the time complexity of our implementation? The worst case complexity is $O(nm)$ since we search at most n times for an augmenting path and since each search takes time $O(m)$ in the worst case. On many graphs the running time is smaller. However, the running time of the implementation above is never better than $\Omega(n^2)$. *This is due to very poor algorithmics* which lets each search for an augmenting path take time $\Omega(n)$. The culprit is the innocent looking statement

```
node_array<bool> reached(G,false);
```

which consumes $\Theta(n)$ time and is executed in each of the n phases. We will next describe two improvements. None of them improves the worst case running time, but both of them improve the running time dramatically for many inputs.

Improving the Best Case: We show how to improve the best case from $\Omega(n^2)$ to $O(m)$. We will see that the optimization has a dramatic effect on the observed running time of our implementation.

Consider the first search for an augmenting path when the current matching is still empty. At this point any edge is an augmenting path and hence the first call of *find_aug_path_by_dfs* returns with success immediately. However, in the implementation above the search will take time $\Omega(n)$ since the initialization of the *node_array<bool> reached* takes linear time. We aim for a design where the cost for reinitializing *reached* is proportional to the number of nodes that were actually reached in the previous search and not proportional to the total number of nodes. We call this the principle of

paying only for what we actually touched
and not
for what we could have conceivably touched.

We describe three ways to realize the principle.

The first method uses a *stack<node> reached_stack* in addition to the boolean array *reached*. Whenever *reached*$[w]$ is set to true for a node w we also push w onto *reached_stack* and after a successful augmentation we use *reached_stack* to reset *reached* to false for all nodes on the stack. In this way reinitialization takes time proportional to the number of elements reached. We obtain the following code. In *find_aug_path_by_dfs* we write

```
reached[w] = true; reached_stack.push(w);
```

and in the body of MAX_CARD_BIPARTITE_MATCHING_FFB we write

```
node_array<bool> reached(G,false);
stack<node> reached_stack;
forall(v,A)
{ edge f;
  forall_adj_edges(f,v)
  { if (find_aug_path_by_dfs(G,f,free,reached))
    { free[v] = false;
      while ( !reached_stack.empty() )
        reached[reached_stack.pop()] = false;
      break;
    }
  }
}
```

The second method uses the data type *node_slist*. This data type offers the functions *member*, *push*, *pop*, and *empty* and hence combines the functionality of a boolean array with a stack. We leave it to the reader to rewrite the algorithm so that a *node_slist* is used instead of *reached* and *reached_stack*.

The third method uses a counter *number_of_augmentations* and a *node_array*<*int*> *mark* instead of *reached*. The counter is increased whenever an augmentation occurs[19] and the mark *number_of_augmentations* is assigned to all nodes reached in the current search for an augmenting path. The test whether a node w has already been reached in the current search amounts to *mark*[w] == *number_of_augmentations*. We obtain the following code. In this code we have also made provisions for our second improvement in form of the program chunk ⟨*MCB: greedy heuristic*⟩.

⟨*_FF_DFS_matching*⟩≡

```
⟨FF: dfs⟩
list<edge> MAX_CARD_BIPARTITE_MATCHING_FF_DFS(graph& G,
                              const list<node>& A, const list<node>& B,
                              node_array<bool>& NC)
{ node v; edge e;
  node_array<bool> free(G,true);
  node_array<int> mark(G,-1);
  // check that all edges are directed from A to B
  forall(v,B) assert(G.outdeg(v) == 0);
  ⟨MCB: greedy heuristic⟩
  number_of_augmentations = 0;
  forall(v,A)
  { if ( !free[v] ) continue;
    edge f;
    forall_adj_edges(f,v)
    { if (find_aug_path_by_dfs(G,f,free,mark))
      { free[v] = false;
        number_of_augmentations++ ;
        break;
      }
    }
  }
  ⟨MCB: prepare result and node cover and restore orientations⟩
}
```

where

⟨*FF: dfs*⟩≡

```
static int number_of_augmentations;
static bool find_aug_path_by_dfs(graph& G, edge f,
             node_array<bool>& free, node_array<int>& mark)
{ node w = G.target(f);
  mark[w] = number_of_augmentations;
  if (free[w])
  { free[w] = false;
```

[19] There are 2^{32} numbers of type *int* and hence this counter will never overflow.

```
      G.rev_edge(f);
      return true;
    }
    edge e;
    forall_adj_edges(e,w)
    { node z = G.target(e);
      if ( mark[z] == number_of_augmentations ) continue;
      if ( find_aug_path_by_dfs(G,e,free,mark))
      { G.rev_edge(f);
        return true;
      }
    }
    return false;
  }
```

The third method has an interesting side effect (which we did not intend). Suppose that we searched for an augmenting path from a and did not succeed. Then all nodes reached by this search are marked (and stay marked) and hence the search from the next free node in A will not explore them. In this way the worst case time between successive augmentations is $O(m)$.

Table 7.3 compares the running times of the implementations FFB and FF in columns FFB- and FF- on random bipartite graphs; the other columns will be explained in the next section. Observe that FF is much faster than FFB. We conclude that the principle of

paying only for what we actually touched
and not
for what we could have conceivably touched

is worth being observed.

The Greedy Heuristic: We come to our second improvement. In our considerations at the beginning of the section we started the matching algorithm with the line

```
M = some matching;
```

So far, we have chosen the empty matching as out initial matching. We will now do something more clever and use the so-called *greedy heuristic* to find an initial matching. The greedy heuristic considers all edges in turn and adds an edge to the current matching if both of its endpoints are free.

⟨*MCB: greedy heuristic*⟩≡

```
forall_edges(e,G)
{ node v = G.source(e);
  node w = G.target(e);
  if ( free[v] && free[w] )
  { free[v] = free[w] = false;
    G.rev_edge(e);
  }
}
```

n	m	FFB-	FFB+	FF-	FF+	Check
1000	2000	1.17	0.32	0.04	0.03	0
1000	4000	1.26	0.3	0.11	0.08	0.01
1000	8000	1.2	0.18	0.08	0.1	0.01
2000	4000	4.57	1.22	0.09	0.07	0
2000	8000	5.04	1.2	0.27	0.25	0.01
2000	16000	4.67	0.57	0.21	0.25	0.01
4000	8000	18.32	4.51	0.29	0.18	0.009998
4000	16000	20.57	4.82	0.97	0.51	0.02
4000	32000	18.47	2.09	0.64	0.7	0.04
8000	16000	72.05	18.1	0.67	0.46	0.04001
8000	32000	82	19.82	2.79	1.47	0.04999
8000	64000	74.05	7.63	1.78	1.54	0.07999

Table 7.3 The running times of four versions of the basic bipartite matching algorithm. FFB and FF refer to the two programs above, a minus sign indicates that no heuristic was used to find an initial matching and a plus sign indicates that the greedy heuristic was used. The last column shows the time required to check the results. The programs were run on random bipartite graphs with n nodes on each side and m edges (generated by *random_bigraph*(G, n, n, m, A, B)). FFB and FF use depth-first search to find augmenting paths. You may perform your own experiments by calling FF_matching_time in the demo directory.

The greedy heuristic is frequently highly effective. We support this statement by analysis and also by experimental evidence.

For the analysis we consider random graphs where $|A| = |B| = n$ and each node in A has d incident edges for some integer d. The edges go to random destinations, e.g., for each edge the endpoint in B is chosen uniformly at random from the nodes in B.

Let us consider the case $d = 1$ first. We consider the nodes in A one by one. When the node v is considered and its incident edge is $e = (v, w)$ we add e to the matching if w is free and we discard e if w is already matched. This shows that every node in B which has degree at least one will be matched by the greedy heuristic. The probability that a node w in B has degree zero is $(1 - 1/n)^n \approx e^{-1} \approx 0.37$ since the probability that the edge starting in any particular node in A does not end in w is $(n - 1)/n = 1 - 1/n$ and hence the probability that none of the n edges starting in a node in A ends in w is $(1 - 1/n)^n$. Thus about $(1 - e^{-1})n \approx 0.63n$ nodes will be matched by the greedy heuristic in the case $d = 1$. Of course, even more nodes will be matched on average for larger d. We give a plausibility argument of what to expect; the remainder of this paragraph is not rigorous.

Consider $d = 2$. About $e^{-1}n$ nodes in A will not be matched by only considering the first edge incident to any node. For these nodes the second incident edge will be considered and hence a total number of about $n + n/e$ edges will be considered. The probability that a node in B stays unmatched reduces to $(1 - 1/n)^{n+n/e} \approx e^{-(1+1/e)} \approx 0.25$.

We turn to experiments. Table 7.4 shows the effect and the cost of the greedy heuristic. We used the program below. The effect of the heuristic is as predicted by our analysis, i.e, for $m = n$ about 63% of the nodes are matched by the heuristic and for $m = 2n$ about 75% of the nodes are matched by the heuristic. The running time of the heuristic is insignificant. Even for the graphs with $m = 10n$ the running time of the heuristic is less than 10 times the time required to initialize the node array *free* and the time to check that all edges are directed from A to B.

⟨*mcb: effect of heuristic*⟩≡

```
double MCB_EFFECT_OF_HEURISTIC(graph& G,
                               const list<node>& A, const list<node>& B)
{ node v; edge e;
  node_array<bool> free(G,true);
  forall(v,B) assert(G.outdeg(v) == 0);
  if (use_heuristic == 0) return 0;
  ⟨MCB: greedy heuristic⟩
  int n = 0;
  forall(v,A) if (!free[v]) n++;
  return double(n)/A.size();
}
```

Table 7.3 shows the running time of four variants of our basic algorithm. The table indicates that both refinements have a tremendous impact on running time at least for random graphs. The greedy heuristic finds a large initial matching and hence saves many searches for augmenting paths and the refined implementation of the set of reached nodes keeps the cost of searching for augmenting paths low. Observe that the running time of both versions of FFB is quadratic in n. FFB+ (that is, FFB with greedy heuristic) has a smaller constant in the n^2 term in the running time since the expensive search for augmenting paths is only started from those nodes in A that are left free by the greedy heuristic. Also FFB+ runs faster for denser graphs since the matching found by the greedy heuristic is larger for denser graphs. FF is always much better than FFB and the time to check the output of our algorithms is negligible compared to the running times of the algorithms.

We summarize the findings of this section:

- The use of a heuristic to find a good initial solution can speed up graph algorithms tremendously. *We recommend exploring the use of a heuristic always.* The value of a heuristic is usually the highest for the least sophisticated algorithm.
- If graph exploration, e.g., a depth-first or a breadth-first search or a shortest-path computation, is used as a subroutine in a graph algorithm, the initialization of the data

n	m	No heuristic		Greedy heuristic	
		%	time	%	time
10000	10000	0	0.02	0.632	0.07
10000	20000	0	0.03	0.764	0.08
10000	30000	0	0.02	0.823	0.1
10000	40000	0	0.02	0.858	0.11
10000	50000	0	0.03	0.881	0.11
10000	60000	0	0.03	0.9	0.12
10000	70000	0	0.02	0.912	0.13
10000	80000	0	0.03	0.927	0.14
10000	90000	0	0.02	0.931	0.14
10000	100000	0	0.03	0.937	0.14

Table 7.4 Percentage of nodes matched by the greedy heuristic and cost of the greedy heuristic. The experiments were performed on random bigraphs with n nodes on each side and m edges (generated by *random_bigraph*(G, n, n, m, A, B)). You can perform your own experiments by calling mcb_effect_of_heuristic in the demo directory.

structures should be performed outside the subroutine. Only those parts of the data structure which are actually touched inside the subroutine should be reinitialized.

Breadth-First versus Depth-First Search: In the previous section we used depth-first search for finding augmenting paths. In this section we will investigate the use of breadth-first search. We will see that breadth-first search is more effective than depth-first search in finding augmenting paths.

Before we give the code we briefly argue that this should be the case. Assume that a is a free node in A, that the shortest augmenting path starting in a consists of k edges, and that the outdegree of all nodes in A is bounded by d. When breadth-first search from a is used in a search for an augmenting path then only nodes in distance at most $k+1$ from a are visited in the search. The number of such nodes is bounded by $d^{(k+1)/2}$. Observe that we have fan-out only at the nodes in A since nodes in B have at most outgoing edge. Actually, the stronger bound $d(d-1)^{(k+1)/2-1}$ holds since each of the nodes in A reachable from a must have one matching edge incident to it and hence there are only $d-1$ outgoing edges left. For example for $d=3$ and $k=9$ the number of nodes visited is bounded by $3 \cdot 2^4 = 48$.

How will depth-first search do? Well, it might explore a large fraction of the graph in the worst case. Even, if there is an augmenting path of length one, it might explore the entire graph.

We turn to the implementation of breadth-first search. Let a be any free node in A. We start a breadth-first search from a. We maintain a queue Q that contains all nodes in A reached by the search from which we have not yet explored the outgoing edges. Initially, Q contains only a. A node (in A or B) has been reached by the search iff *mark*$[v]$ == *number_of_augmentations* and for a reached node v, *pred*$[v]$ contains the edge through which v was reached. When the procedure finds an augmenting path it augments the path and returns true, otherwise it returns false.

The procedure starts by putting a into the queue and marking a. As long as the queue is not empty, the first node is removed from Q. Call the node v; v is a node in A. We explore all edges out of v. Let $e = (v, w)$ be any such edge. If w has been reached before, we do nothing. Otherwise we set *pred*$[w]$ to e and mark w. If w is free, we augment by the path from a to w and return *true*. The path can be found by tracing edges as given by *pred*. If w is not free, let $f = (w, x)$ be the matching edge incident to w; note that f is the only edge out of w. We set *pred*$[x]$ to f, mark x, and append x to Q.

⟨*FF: bfs*⟩≡

```
#include <LEDA/queue.h>
static bool find_aug_path_by_bfs(graph& G, node a,
                           node_array<bool>& free, node_array<edge>& pred,
                           node_array<int>&  mark)
{ queue<node> Q;
  Q.append(a); mark[a] = number_of_augmentations;
  edge e;
  while ( !Q.empty() )
  { node v = Q.pop();  // v is a node in A
    forall_adj_edges(e,v)
    { node w = G.target(e); // w is a node in B
      if (mark[w] == number_of_augmentations) continue;
      // w has not been reached before in this search
      pred[w] = e; mark[w] = number_of_augmentations;
      if (free[w])
      { // augment path from a to w
        free[w] = free[a] = false;
        while ( w != a)
        { e = pred[w];
          w = G.source(e);
          G.rev_edge(e);
        }
        return true;
      }
      // w is not free
      edge f = G.first_adj_edge(w);
      node x = G.target(f);
      pred[x] = f; mark[x] = number_of_augmentations;
      Q.append(x);
    }
```

```
    }
    return false;
  }
```

The matching algorithm is as we already know it. We use either breadth-first or depth-first search for finding augmenting paths. The choice is made by the variable *use_bfs*. In both methods we declare all nodes unreached (by increasing *number_of_augmentations*) whenever an augmenting path has been found.

⟨*_FF_matching*⟩≡

```
  ⟨FF: dfs⟩
  ⟨FF: bfs⟩
  list<edge> MAX_CARD_BIPARTITE_MATCHING_FF(graph& G,
                              const list<node>& A, const list<node>& B,
                              node_array<bool>& NC,
                              bool use_heuristic, bool use_bfs)
  { node v; edge e;
    node_array<bool> free(G,true);
    node_array<int>  mark(G,-1);
    node_array<edge> pred(G);
    number_of_augmentations = 0;
    // check that all edges are directed from A to B
    forall(v,B) assert(G.outdeg(v) == 0);
    if (use_heuristic) ⟨MCB: greedy heuristic⟩
    forall(v,A)
    { if ( !free[v] ) continue;
      if (use_bfs)
      { if (find_aug_path_by_bfs(G,v,free,pred,mark) )
          number_of_augmentations++ ;
      }
      else
      { edge f;
        forall_adj_edges(f,v)
        { if (find_aug_path_by_dfs(G,f,free,mark))
          { free[v] = false;
            number_of_augmentations++ ;
            break;
          }
        }
      }
    }
    ⟨MCB: prepare result and node cover and restore orientations⟩
  }
```

Table 7.5 shows the running time of the procedure above on random bipartite graphs. The table shows that breadth-first search is almost always superior to depth-first search (as we already argued above). It also shows that breadth-first search is not helped at all by the greedy heuristic. We explain this observation. The greedy heuristic considers augmenting

n	m	k	dfs-	dfs+	bfs-	bfs+
10000	15000	1	0.26	0.26	0.28	0.27
10000	15000	10	0.25	0.24	0.26	0.25
10000	15000	100	0.24	0.23	0.24	0.25
10000	15000	1000	0.24	0.23	0.24	0.25
10000	15000	10000	0.23	0.23	0.25	0.24
10000	25000	1	8.46	3.56	2.89	2.91
10000	25000	10	5.44	3.11	2.34	2.33
10000	25000	100	5.34	3.11	2.54	2.53
10000	25000	1000	2.04	2.19	1.92	1.92
10000	25000	10000	0.31	0.29	0.29	0.28
10000	35000	1	5.38	2.28	2.51	2.52
10000	35000	10	7.62	2.55	2.75	2.76
10000	35000	100	22.78	2.24	2.37	2.37
10000	35000	1000	17.91	2.21	2.09	2.09
10000	35000	10000	2.15	1.12	0.92	0.93

Table 7.5 Depth-first versus breadth-first search. The table shows the running time of MAX_CARD_BIPARTITE_MATCHING_FF. Either no heuristic (indicated by a minus sign) or the greedy heuristic (indicated by a plus sign) is used to find an initial matching. To complete the matching, a search for an augmenting path is started from each free node in A that was not matched by the heuristic. Either breadth-first or depth-first search is used to find an augmenting path. The programs were run on random bipartite group graphs with n nodes on each side and m edges (generated by *random_bigraph*(G, n, n, m, A, B, k)). The nodes on either side are divided into k groups and the nodes in the i-th group are connected to nodes in groups $i-1$ and $i+1$ on the other side. The generator is described in detail in Section 7.6.7. You may perform your own experiments by calling mcb_dfs_vs_bfs in the demo directory.

paths of length one. It finds an augmenting path of length one by inspecting all the edges incident to a node. Breadth-first search does exactly the same when an augmenting path of length one exists.

7.6.5 *The Algorithm of Hopcroft and Karp*

In this and the next section we give algorithms whose worst case running time is $O(\sqrt{n}m)$.

The first such algorithm is due to Hopcroft and Karp [HK73]. They suggested organizing the execution into phases, restricting augmentation to shortest augmenting paths, and augmenting a maximal number of node disjoint augmenting paths in each phase. Observe that Lemma 15 guarantees the existence of many short augmenting paths when the current matching is still far from optimality.

The overall structure of the program is the same as for our previous algorithms. The differences are that we maintain some additional data structures, in particular a list of the free nodes in A, and that the search for augmenting paths is organized differently.

```
⟨_HK_matching⟩≡
  ⟨HK: bfs⟩
  ⟨HK: dfs⟩
  list<edge> MAX_CARD_BIPARTITE_MATCHING_HK(graph& G,
                              const list<node>& A, const list<node>& B,
                              node_array<bool>& NC, bool use_heuristic)
  { node v;
    edge e;
    node_array<bool> free(G,true);
    //check that all edges are directed from A to B
    forall(v,B) assert(G.outdeg(v) == 0);
    if (use_heuristic) { ⟨MCB: greedy heuristic⟩ }
    node_list free_in_A;
    forall(v,A) if (free[v]) free_in_A.append(v);
    ⟨HK: data structures⟩
    while ( ⟨there is an augmenting path⟩ )
    { ⟨find a maximal set and augment⟩ }
    ⟨MCB: prepare result and node cover and restore orientations⟩
  }
```

We now give the details of how the Hopcroft and Karp algorithm searches for augmenting paths.

The length (= number of edges) of the shortest augmenting path can be found by breadth-first search. The search starts from all free nodes in A. We give a variant of breadth-first search which does a bit more. It constructs a so-called *layered network*. In a layered network the nodes of a graph are partitioned into *layers* according to their distance with respect to the starting layer, i.e., a node v belongs to layer k if there is a path from the starting layer to v consisting of k edges and there is no path with fewer edges. For any edge in a layered network the distance of the target node is at most one more than the distance of the source node. Only edges that connect different layers can be contained in shortest augmenting paths and hence we mark them *useful* in the program below; the mark is an integer *phase_number* in which we count the number of phases executed[20]. The construction of the layered network starts by putting all free nodes in A into the zeroth layer, then proceeds by standard breadth-first search, and stops as soon as the first layer is completed that contains free nodes in B. We achieve the latter goal by stopping to put nodes into the queue as soon as the first free node in B has been removed from the queue.

[20] Observe that we are reusing the marking technique introduced in section 7.6.4. Incrementing *phase_counter* will unmark all edges.

The program returns *true* if there is an augmenting path and returns *false* otherwise.

⟨*HK: data structures*⟩≡

```
edge_array<int> useful(G,0);
node_array<int> dist(G);
node_array<int> reached(G,0);
phase_number = 1;
```

and

⟨*HK: bfs*⟩≡

```
#include <LEDA/b_queue.h>
#include <LEDA/node_list.h>
static int phase_number;
static bool bfs(graph& G, const node_list& free_in_A,
                const node_array<bool>& free, edge_array<int>& useful,
                node_array<int>& dist, node_array<int>& reached)
{
  list<node> Q;
  node v,w;
  edge e;
  forall(v,free_in_A)
  { Q.append(v);
    dist[v] = 0;  reached[v] = phase_number;
  }
  bool augmenting_path_found = false;
  while (!Q.empty())
  { v = Q.pop();
    int dv = dist[v];
    forall_adj_edges(e,v)
    { w = target(e);
      if (reached[w] != phase_number )
      { dist[w] = dv + 1; reached[w] = phase_number;
        if (free[w]) augmenting_path_found = true;
        if (!augmenting_path_found) Q.append(w);
      }
      if (dist[w] == dv + 1) useful[e] = phase_number;
    }
  }
  return augmenting_path_found;
}
```

With this procedure we can refine the test for the existence of an augmenting path in the main loop.

⟨*there is an augmenting path*⟩≡

```
bfs(G,free_in_A,free,useful,dist,reached)
```

The layered graph contains all augmenting paths of shortest length. We determine a maximal set P of augmenting paths. Distinct paths in P will be node disjoint and P is maximal in the sense that no augmenting path can be added to P without violating the disjointness property. We find P by a variant of depth-first search. The procedure *find_aug_path*$(G, f,$ *free*, *pred*, *useful*$)$ attempts to find a path in the layered network starting with the edge f, ending in a free vertex in B, and being node-disjoint from all previously constructed paths. In the main loop we will call this procedure for all edges out of free nodes in A. The call returns the last edge on the path if it succeeds and returns *nil* otherwise. It also records, for each node, the first edge through which the node was reached in a *node_array*<*edge*> *pred*.

The details of *find_aug_path*$(G, f, \ldots)$ are simple. Let w be the endpoint of f. We set *pred*$[w]$ to f and then distinguish cases. If w is a free node (it is necessarily in B then), we return f. If w is not a free node, we scan through all edges $e = (w, z)$ out of w. If e does not belong to the layered network or we have already tried to construct a path out of z, we ignore e. Otherwise, we recurse. The recursive call either returns *nil* or a proper edge. In the latter case we know that a new augmenting path has been found and forward the edge to the enclosing call.

⟨*HK: dfs*⟩≡

```
static edge find_aug_path(graph& G, edge f, const node_array<bool>& free,
                         node_array<edge>& pred, const edge_array<int>& useful)
{ node w = G.target(f);
  pred[w] = f;
  if (free[w]) return f;
  edge e;
  forall_adj_edges(e,w)
  { node z = G.target(e);
    if ( pred[z] != nil || useful[e] != phase_number ) continue;
    edge g  = find_aug_path(G,e,free,pred,useful);
    if ( g ) return g;
  }
  return nil;
}
```

In the main loop we call *find_aug_path* for all edges out of free nodes in A that belong to the layered network and where the target node of the edge has not been reached by a previous search and collect the (terminal edges of the) paths found in a list *EL*. We then augment all paths. Let e be an arbitrary edge in *EL*. We trace the path ending in e by means of the *pred*-array and for each path reverse all edges on the path. We complete the phase by incrementing *phase_number*.

⟨*find a maximal set and augment*⟩≡

```
node_array<edge> pred(G,nil);
list<edge> EL;

forall(v,free_in_A)
{ forall_adj_edges(e,v)
    if (pred[G.target(e)] == nil && useful[e] == phase_number)
```

```
    { edge f = find_aug_path(G,e,free,pred,useful);
      if ( f )  { EL.append(f); break; }
    }
}
while (!EL.empty())
{ edge e = EL.pop();
  free[G.target(e)] = false;
  node z;
  while (e)
  { G.rev_edge(e);
    z = G.target(e);
    e = pred[z];
  }
  free[z] = false;
  free_in_A.del(z);
}
// prepare for next phase
phase_number++;
```

We close our discussion of the Hopcroft–Karp matching algorithm with a word on running time. Each phase of the algorithm takes time $O(m)$ for the breadth-first and depth-first search and the augmentation and hence the total running time is $O(Dm)$ where D is the number of phases. It can be shown (see for example [HK73] or [AMO93, section 8.2] or [Meh84c, IV.9.2]) that the number of phases is $O(\sqrt{n})$. On many graphs the number of phases is much smaller. In particular, Motwani [Mot94] has shown that the number of phases is $O(\log n)$ for random graphs.

7.6.6 *The Algorithm of Alt, Blum, Mehlhorn, and Paul*

We discuss a variant of the Hopcroft–Karp algorithm due to Alt, Blum, Mehlhorn, and Paul [ABMP91]. It uses ideas first propagated for flow algorithms [AO89, GT88] to integrate the breadth-first and depth-first search used in the Hopcroft–Karp algorithm. The resulting algorithm is usually faster.

As above, we direct all edges in the current matching from B to A and all other edges from A to B. In this directed graph every path is an alternating path. For each node $v \in V$ we maintain a distance label *layer*[v]. Nodes in B will occupy even layers, and all free nodes in B will be in layer zero. Nodes in A will occupy odd layers, and all free nodes in A will be in two adjacent layers L and $L+2$, for some L. Observe that this layering is "opposite" to the layering used in the Hopcroft and Karp algorithm. Now free nodes in B are in the bottom layer (= layer zero) and free nodes in A are in the two topmost layers (= layers L and $L+2$). Initially, we put all nodes in B into layer zero, all nodes in A into layer one, direct all edges from A to B, and set L to one.

⟨*ABMP: initialization*⟩≡

```
node_array<bool> free(G,true);
node_array<int> layer(G);
if (use_heuristic) {⟨MCB: greedy heuristic⟩}
```

```
list<node> free_in_A;
forall(v,B) layer[v] = 0;
forall(v,A)
{ layer[v] = 1;
  if (free[v]) free_in_A.append(v);
}
int L = 1;
```

In *free_in_A* we collect all free nodes in A. We maintain the invariant that the free nodes in level L precede the free nodes in level $L + 2$. In this way L is always the layer of the first node in *free_in_A*.

We maintain the "layered graph invariant" that no edge reaches downwards by two or more layers, i.e.,

$$\text{for all edges } e = (v, w)\text{: } \mathit{layer}[v] \le \mathit{layer}[w] + 1.$$

It follows that $\mathit{layer}[v]$ is a lower bound on the length of an alternating path starting in v and ending in a free node in B. Call an edge $e = (v, w)$ *eligible*, if $\mathit{layer}[v] = \mathit{layer}[w] + 1$, and let $ce(v)$ be a function which returns an eligible edge starting in v, if there is one, and *nil* otherwise. We call *ce* the current edge function. Its implementation will be discussed at the end of the section.

We search for augmenting paths as follows: starting from a free node v in layer L we construct a path p of eligible edges. Let w be the last node of p. There are three cases to distinguish:

Case 1 (breakthrough): w is a free node in layer zero:
Then p is an augmenting path with respect to the current matching. We augment the current matching by reversing all edges of p and terminate the search.

Case 2 (advance): w is not a free node in layer zero and $ce(w)$ exists:
We extend p by adding $ce(w)$.

Case 3 (retreat): w is not a free node in layer zero and $ce(w) = \mathit{nil}$:
We increase $\mathit{layer}[w]$ by two and remove the last edge from p. If there is no last edge in p, i.e., w is equal to the free node v from which we started the search for an augmenting path, we terminate the search and add w to the end of *free_in_A*. Observe that this maintains the invariant that the nodes on layer L precede the nodes on layer $L + 2$ in *free_in_A*.

The following program chunk realizes this strategy. The edges of the path are stored in a stack p of edges and w is the last node of the path. In the case of a breakthrough v and w are declared matched and all edges of p are reversed. In the case of an advance we push the current edge of w onto p and set w to the target node of the edge. In the case of a retreat we increase the layer of w by two and pop the last edge from p and set w to the source node of the edge popped. If there is no edge to be popped we terminate the search and add w to the rear end of *free_in_A*.

⟨*search for an augmenting path from v*⟩≡

```
node w = v;
while (true)
{ if ( free[w] && layer[w] == 0 )
  { // breakthrough
    free[w] = free[v] = false;
    while ( !p.empty() )
    {  e = p.pop();
       ⟨breakthrough: current edge function⟩
       G.rev_edge(e);
    }
    break;
  }
  else
  { if ( (e = ce(w,G,layer,cur_edge)) )
    { // advance
      p.push(e);
      w = G.target(e);
    }
  else
    { // retreat
      layer[w] += 2;
      ⟨relabel: current edge function⟩
      if (p.empty())
      { free_in_A.append(w);
        break;
      }
      w = G.source(p.pop());
    }
  }
}
```

After a breakthrough or a retreat, which leaves us with an empty path, we start the next search for an augmenting path. If there are no more free nodes in layer L, we increase L by two and repeat. In the program below this increase of L is implicit; L is simply the layer of the first node in *free_in_A*. In this way we proceed until L exceeds *Lmax* where *Lmax* is a parameter of the algorithm or until the number of free nodes is smaller than δL where δ is a parameter (which we set rather arbitrarily to 50 in our implementation). The parameter *Lmax* can either be set by the user or is set to $\gamma\sqrt{n}$ where γ is a parameter (which we set rather arbitrarily to 0.1 in our implementation). Once L exceeds *Lmax* or the number of free nodes in A has fallen below δL we determine the remaining augmenting paths by breadth-first search as in the Ford and Fulkerson algorithm.

⟨*_ABMP_matching*⟩≡

```
static int number_of_augmentations;
⟨FF: bfs⟩  // for the basic algorithm
edge ce(const node v, const graph& G,
        const node_array<int>& layer, node_array<edge>& cur_edge)
{ ⟨implementation of current edge function⟩ }
```

```
list<edge> MAX_CARD_BIPARTITE_MATCHING_ABMP(graph& G,
                                   const list<node>& A, const list<node>& B,
                                   node_array<bool>& NC,
                                   bool use_heuristic, int Lmax)
{ node v; edge e;
  //check that all edges are directed from A to B
  forall(v,B) assert(G.outdeg(v) == 0);
  ⟨ABMP: initialization⟩
  node_array<edge> cur_edge(G,nil); // current edge iterator
  if (Lmax == -1) Lmax = (int)(0.1*sqrt(G.number_of_nodes()));
  b_stack<edge> p(G.number_of_nodes());
  while ( L <= Lmax && free_in_A.size() > 50 * L)
  { node v = free_in_A.pop();
    L = layer[v];
    ⟨search for an augmenting path from v⟩
  }
  ⟨complete by basic algorithm⟩
  ⟨MCB: prepare result and node cover and restore orientations⟩
}
```

where

⟨*complete by basic algorithm*⟩≡

```
node_array<int> mark(G,-1);
node_array<edge> pred(G);
number_of_augmentations = 0;
forall(v,free_in_A)
{ if ( find_aug_path_by_bfs(G,v,free,pred,mark) )
     number_of_augmentations++;
}
```

We establish correctness.

Lemma 17 *At all times during the execution of the algorithm, the following invariants hold:*

(I1) *For all edges* (v, w)*: layer*$[w] \geq$ *layer*$[v] - 1$.

(I2) *layer*$[v]$ *is even iff* $v \in B$.

(I3) *Let* $p = [e_0, e_1, \ldots, e_{l-1}]$ *with* $e_i = (v_i, v_{i+1})$*. Then* p *is a path in the current graph with layer*$[v_i] = L - i$ *for all* i, $0 \leq i < l$, *and* v_0 *is a free node in* A.

(I4) *All free nodes* $v \in A$ *are in layers* L *or* $L + 2$.

(I5) *The set* M *of edges that are directed from* B *to* A *forms a matching in* G*; furthermore free*$[v]$ *is true iff* v *is free with respect to* M.

Proof We use induction on the number of executions of the loop. All invariants hold initially. For the induction step we address the invariants in turn.

Only relabeling a node or reversing the direction of an edge may invalidate (I1). When a node v is relabeled there are no eligible edges out of v and hence $layer[w] \geq layer[v]$ for all $(v, w) \in E$. Since nodes in A live on odd layers and nodes in B live on even layers we even have $layer[w] > layer[v]$ for all $(v, w) \in E$. Hence increasing $layer[v]$ by two preserves (I1) for all edges $(v, w) \in E$. For edges $(w, v) \in E$ the invariant also stays true. Reversing the edges of the path p in the case of a breakthrough maintains (I1) as well, since all edges in p are eligible. Altogether, we have shown that (I1) is maintained.

Since layer labels are always increased by two, (I2) remains true.

The path p always starts at a free node in A in layer L and is only extended by eligible edges.

When a node is relabeled, it must be on the path p. Thus no free node in layer $L + 2$ can be relabeled by (I3). When L is increased by two, there is no free node v in layer L. Thus, (I4) is preserved.

In the case of a breakthrough, p is an alternating path from a free node $w \in A$ to a free node $v \in B$ by (I3) and the induction hypothesis, i.e., an augmenting path with respect to the current matching. Thus (I5) is preserved in the case of a breakthrough. □

The correctness of our algorithm is now established. Next we show that it is a derivative of the Hopcroft–Karp algorithm.

Lemma 18 *The algorithm always increases the matching along a shortest augmenting path.*

Proof Any augmenting path p found has length L. (I4) and (I5) imply that all free nodes in A are in layers L or $L + 2$, and those of B are in layer zero. Now the claim follows from (I1). □

Lemma 19 *Let M^* be a matching of maximum cardinality in G and M the matching computed by our algorithm when ⟨complete by basic algorithm⟩ is reached. Then $|M^*| - |M| \leq \max(\gamma Lmax, n/Lmax)$. Furthermore, ⟨complete by basic algorithm⟩ takes time $O(\max(\gamma Lmax, n/Lmax) \cdot m)$.*

Proof When ⟨*complete by basic algorithm*⟩ is reached then either $L > Lmax$ and there is no augmenting path with respect to the current matching M of length less than $Lmax$ or the number of free nodes in A is smaller than γL which in turn is smaller than $\gamma Lmax$. In the latter case we have established the claimed bound on $|M^*| - |M|$. In the former case we observe that $M^* \oplus M$ must contain $|M^*| - |M|$ node-disjoint augmenting paths with respect to M. The total length of these paths is at most n and each path has length at least $Lmax$. Thus $(|M^*| - |M|) \cdot Lmax \leq n$.

In ⟨*complete by basic algorithm*⟩ we need time $O(m)$ for each node in A which is still free. By the previous paragraph there are at most $\max(\gamma Lmax, n/Lmax)$ such nodes when the chunk is reached. □

The previous lemma suggests our choice of $Lmax$. In order to balance the contribution of the two choices we should set *Lmax* to $\Theta(\sqrt{n})$. Unfortunately, the theoretical analysis is not strong enough to suggest the "correct" factor of proportionality.

Lemma 20 *The total number of increases of layer labels and the total of number of calls to the eligible edge function ce is* $O(n \cdot Lmax)$.

Proof (I4) implies that the maximum layer of a node during an execution of the algorithm is $Lmax + 2$. Thus any node is relabeled at most $(Lmax + 2)/2$ times.

Each time the function *ce* returns an eligible edge (v, w), we extend the current path p by this edge. Either it still belongs to the path when p becomes augmenting for the next time, or $layer[w]$ is increased by two when (v, w) is deleted from p. Thus the number of calls to the function *ce* is bounded by the total number of increases of layer labels plus the total length of all augmenting paths. Since the length of an augmenting path is at most *Lmax*, because of (I4), and since there are at most n of them, the bound follows from the bound for the number of relabels. □

Lemma 20 implies that the total time spent outside ⟨*complete by basic algorithm*⟩ is $O(n \cdot Lmax)$ plus the time spent in calls to the current edge function. We now show how to implement the current edge function efficiently. We maintain for each node v an edge $cur\_edge[v]$ out of v such that all edges preceding $cur\_edge[v]$ in v's adjacency list are not eligible; when $cur\_edge[v]$ is *nil* all edges in v's adjacency list may be eligible. Recall that an edge (v, w) is eligible if the layer of w is one less than the layer of v and that no edge goes down more than one layer. Thus relabeling w cannot make (v, w) eligible and reversing an edge in an augmentation cannot make the edge eligible (because all edges in the augmenting path go from lower layers to higher layers after the augmentation). Only relabeling v can make an edge out of v eligible. With these observations it is easy to maintain the invariant that all edges preceding $cur\_edge[v]$ in v's adjacency list are not eligible:

When w is relabeled we set $cur\_edge[w]$ to nil.

When we search for a current edge we start searching at the current value of $cur\_edge[v]$ (at the first edge out of v if the current value is *nil*) until an eligible edge is found.

When an edge $e = (v, w)$ is reversed and e is the current value of $cur\_edge[v]$ we advance $cur\_edge[v]$ to the successor edge of v.

⟨*relabel: current edge function*⟩≡

```
cur_edge[w] = nil;
```

⟨*implementation of current edge function*⟩≡

```
edge e = cur_edge[v];
if ( e == nil ) e = G.first_adj_edge(v);
while (e && layer[G.target(e)] != layer[v] - 1) e = G.adj_succ(e);
cur_edge[v] = e;
return e;
```

⟨*breakthrough: current edge function*⟩≡

```
if (e == cur_edge[G.source(e)])
  cur_edge[G.source(e)] = G.adj_succ(e);
```

In this way the time spent in calls $ce(v)$ between relabelings of v is O(number of calls + $outdeg(v)$). Since each node is relabeled at most *Lmax* times and since the total number of calls to *ce* is $O(n \cdot Lmax)$ we conclude that the total time spent in calls to the current edge function is $O(m \cdot Lmax)$.

We summarize in:

Theorem 4 *A maximum cardinality matching in a bipartite graph with n nodes and m edges can be computed in time* $O(\sqrt{n}m)$.

Proof This follows from the discussion above and the choice $Lmax = \Theta(\sqrt{n})$. □

7.6.7 *An Experimental Comparison*

We compare the algorithms *FF*, *HK*, and *ABMP* experimentally on bipartite graphs of the form shown in Figure 7.22. We call these graphs *bipartite group graphs*. They were suggested by [CGM+97].

The following program generates bipartite group graphs with *na* nodes in A and *nb* nodes in B. We divide both sides into $k + 1$ groups numbered 0 to k. For all i, $0 \le i \le k - 1$, the i-th group on side X contains nodes $i \cdot Kx$ to $(i + 1) \cdot Kx - 1$ where $Kx = \lfloor nx/k \rfloor$. The final group contains nodes $k \cdot Kx$ to $n - 1$; it is empty if k divides nx.

We generate the edges in two phases. In the first phase we generate $d = \lfloor m/na \rfloor$ edges for each node in groups 0 to $k - 1$ of A. For a node in the i-group the destination of these edges are random nodes in groups $i - 1 \bmod k$ and $i + 1 \bmod k$ of B. In the second phase we add $m - d \cdot k \cdot Ka$ random edges.

⟨*random_bigraph.c*⟩≡

```
void random_bigraph(graph& G, int na, int nb, int m,
                    list<node>& A, list<node>& B, int k)
{ G.clear();
  if ( na < 0 || nb < 0 || m < 0 )
    error_handler(1,"random_bigraph: one of na, nb, or m < 0");
  node* AV = new node[na];
  node* BV = new node[nb];
  A.clear();
  B.clear();
  int a, b;
  for(a = 0; a < na; a++)  A.append(AV[a] = G.new_node());
  for(b = 0; b < nb; b++)  B.append(BV[b] = G.new_node());
  if ( na == 0 || nb == 0 || m == 0 ) return;
  if ( k < 1) error_handler(1,"random_bigraph: k < 1");
  int  d = m/na;
  if (k > na) k = na; if (k > nb) k = nb;
```

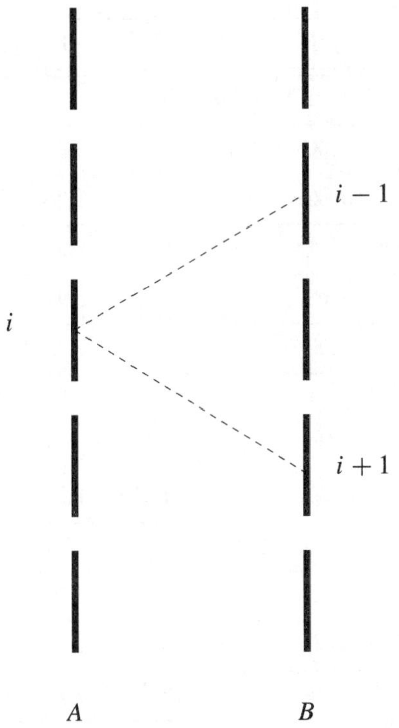

Figure 7.22 A bipartite graph with n nodes on each side. On each side the nodes are divided into k groups of size n/k each (this assumes that k divides n). Each node in A has degree $d = m/n$ and the edges out of a node in group i of A go to random nodes in groups $i+1$ and $i-1$ of B.

```
  int  Ka = na/k;      // group size in A
  int  Kb = nb/k;      // group size in B

  node v;
  int i;
  a = 0;

  forall(v,A)
  { int l = a/Ka;    // group of v
    if ( l == k) break;
    int base1 = (l == 0 ? (k-1)*Kb : (l-1)*Kb);
    int base2 = (l == k-1 ? 0 : (l+1)*Kb);
    for(i = 0; i < d; i++)
    { b = ( rand_int(0,1) == 0? base1 : base2 );
      G.new_edge(v,BV[b + rand_int(0,Kb-1)]);
    }
    a++;
  }
  int r = m - a*d;

  while (r--) G.new_edge(AV[rand_int(0,na-1)], BV[rand_int(0,nb-1)]);

  delete[] AV;
  delete[] BV;
}
```

n	m	k	FF-	FF+	HK-	HK+	AB-	AB+	Check
4	8	1	4.99	4.38	5.63	5.24	3.46	3.4	0.28
4	8	100	3.45	2.47	3.83	3.54	2.45	2.45	0.24
4	8	10000	1.11	1.04	3.76	3.51	2.16	2.16	0.22
4	12	1	155.7	50.02	8.37	7.95	4.91	4.95	0.36
4	12	100	69.07	44.09	5.94	5.78	3.19	3.1	0.26
4	12	10000	1.36	1.28	7.79	7.21	2.34	2.33	0.2599
4	16	1	42.75	21.34	9.71	9.16	4.95	5.33	0.43
4	16	100	48.75	41.59	6.99	6.57	3.02	3.37	0.29
4	16	10000	1.56	1.43	12.5	12.15	2.17	2.2	0.27
8	16	1	11.98	11.34	11.79	11.16	8.96	8.95	0.63
8	16	100	8.15	6.76	8.79	8.33	6.28	6.13	0.45
8	16	10000	2.33	2.15	7.83	7.29	5.42	5.44	0.46
8	24	1	611.6	188.6	19.49	18.56	12.28	12.35	0.77
8	24	100	349.8	221.4	13.14	12.69	8.33	8.36	0.54
8	24	10000	5.38	4.67	15.47	14.53	6.25	6.29	0.51
8	32	1	153.3	60.37	20.89	19.6	15.26	15.34	0.9099
8	32	100	247.1	208.2	13.9	13.22	9.73	9.76	0.6001
8	32	10000	13.58	12.46	26.38	25.96	6.75	6.71	0.5601

Table 7.6 The running times of the bipartite matching algorithms *FF*, *HK*, and *ABMP* on random bipartite group graphs with $n \cdot 10^4$ nodes on each side, $m \cdot 10^4$ edges and k groups (generated by *random_bigraph*(G, n, n, m, A, B, k)). The plus sign indicates the use of the greedy heuristic and the minus sign indicates that the algorithm started with the empty matching. The last column shows the time required to check the results. *FF* uses breadth-first search. You may perform your own experiments by calling mcb_matching_time in the demo directory.

Table 7.6 shows the outcome of our experiments. *FF* does very badly for some of the parameters and very well for others. It is always helped by the heuristic and frequently helped considerably. It shows the highest fluctuations of running time. *HK* and *ABMP* are more stable and *ABMP* is the fastest for most settings of the parameters. *HK* is always helped by the heuristic. For *ABMP* the effect of the heuristic is very small. If it is noticeable at all, it is negative. We have therefore chosen ABMP with the heuristic turned off as our default implementation. The time required for checking the result is negligible in all cases.

Exercises for 7.6

1 We described three methods to implement the principle of only paying for what is actually touched but gave the details of only two of them. Explore the third alternative. Rewrite MAX_BIPARTITE_CARD_MATCHING_FFB such that it uses a *node_slist* instead of *reached* and *reached_stack*.

2 In our implementations of matching algorithms we explicitly reverse the direction of matching edges by *rev_edge*. Explore the possibility of making the reversal only implicitly. Use a *node_array<edge> matching_edge* such that *matching_edge*[v] is *nil* if v is free and is the matching edge incident to v otherwise.

3 Rewrite the ABMP-implementation such that it uses depth-first search instead of breadth-first search in ⟨*complete by basic algorithm*⟩. Compare the running times.

4 Develop a strategy for choosing the parameter *Lmax* in the ABMP-algorithm (the authors have no good solution to this exercise).

5 Construct graphs where our maximum cardinality bipartite matching algorithms assume their worst case running time. Please inform the authors about your solution (as they can only partially solve this exercise).

7.7 Maximum Cardinality Matchings in General Graphs

A *matching* M in a graph G is a subset of the edges no two of which share an endpoint, see Figure 7.23. The cardinality $|M|$ of a matching M is the number of edges in M.

A node v is called *matched* with respect to a matching M if there is an edge in M incident to v and it is called *free* or *unmatched* otherwise. An edge e is called matching if $e \in M$. A matching is called *perfect* if all nodes of G are matched and is called *maximum* if it has maximum cardinality among all matchings.

The structure of this section is as follows. In Section 7.7.1 we discuss the functionality of our matching algorithms, in Section 7.7.2 we derive the so-called blossom shrinking algorithm for maximum matchings, and in Section 7.7.3 we give an implementation of it.

7.7.1 *Functionality*

The function

```
list<edge> MAX_CARD_MATCHING(const graph& G, int heur = 0)
```

returns a maximum matching in G. The underlying algorithm is the so-called blossom shrinking algorithm of Edmonds [Edm65b, Edm65a]. The worst case running time of the algorithm is $O(nm\alpha(m,n))$ ([Gab76]), the actual running time is usually much better. Table 7.7 contains some experimental data.

With *heur* $= 1$, the greedy heuristic is used to construct an initial matching which is then extended to a maximum matching by the blossom shrinking algorithm. As Table 7.7 shows, the influence of the greedy heuristic on the running time is small. It sometimes helps, it

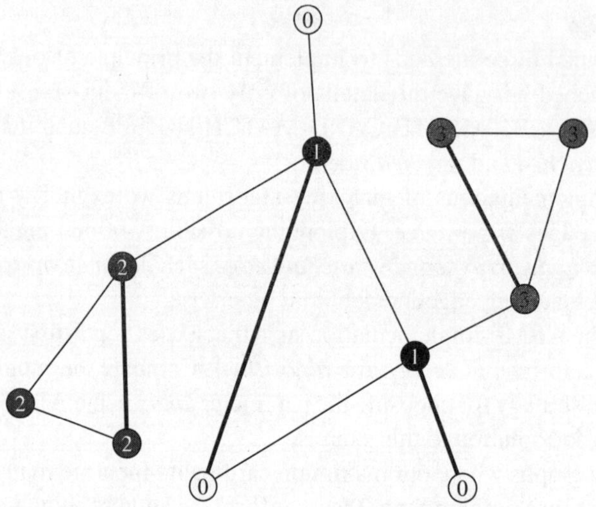

Figure 7.23 A maximum matching and a proof of optimality: The edges of the matching are shown in bold. The node labels prove the optimality of the matching. Observe that every edge is either incident to a node labeled 1 or connects two nodes that are labeled 2 or connects two nodes that are labeled 3. There are two nodes labeled 1, three nodes labeled 2, and three nodes labeled 3. Thus no matching can have more than $2 + \lfloor 3/2 \rfloor + \lfloor 3/2 \rfloor = 4$ edges. The matching shown has four edges and is hence optimal. You may generate similar figures with the xlman-demo gw_mc_matching.

sometimes harms, and it never causes a dramatic change. The cost of checking optimality is negligible in all cases.

In the remainder of this section we discuss the check of optimality. A labeling l of the nodes of G with non-negative integers is said to *cover* G (or to be a cover for G) if every edge of G (which is not a self-loop) is either incident to a node labeled 1 or connects two nodes labeled with the same i, for some $i \geq 2$. The *capacity* of l is defined as

$$cap(l) = n_1 + \sum_{i \geq 2} \lfloor n_i/2 \rfloor,$$

where n_i is the number of nodes labeled i. Observe that there may be nodes that are labeled zero. The capacity of a covering[21] is an upper bound on the cardinality of any matching.

Lemma 21 *If l covers G and M is any matching then $|M| \leq cap(l)$.*

Proof Since l covers every edge of G and hence every edge in M, each edge in M is either incident to a node labeled one or connects two nodes labeled i for some $i \geq 2$. There can be at most n_1 edges of the former kind and at most $\lfloor n_i/2 \rfloor$ edges of the second kind for any $i, i \geq 2$. Thus $|M| \leq cap(l)$. □

[21] In bipartite graphs only the labels zero and one are needed. The nodes labeled one form a node cover in the sense of Section 7.6.1.

n	m	MCM	MCM+	Check
10000	10000	0.287	0.223	0.024
20000	20000	0.905	0.717	0.074
40000	40000	2.178	1.758	0.184
80000	80000	4.857	3.934	0.413
10000	15000	1.049	1.03	0.027
20000	30000	3.799	3.862	0.102
40000	60000	11.45	11.9	0.262
80000	120000	30.51	33.57	0.583
10000	20000	1.247	1.304	0.04199
20000	40000	4.876	5.357	0.136
40000	80000	14.2	15.3	0.343
80000	160000	38.42	43.81	0.789
10000	25000	1.322	1.347	0.05099
20000	50000	4.761	4.782	0.169
40000	100000	13.95	14.22	0.422
80000	200000	35.2	37.3	0.959

Table 7.7 Running times of the general matching algorithm: The table shows the running time of the maximum cardinality matching algorithm without (MCM) and with the greedy heuristic (MCM+) and the time to check the result for random graphs with n nodes and m edges (generated by *random_graph*(G, n, m)). In all cases the time for checking the result is negligible compared to the time for computing the maximum matching. In each of the four blocks we used $n = 2^i \cdot 10^4$ for $i = 0, 1, 2, 3$ and a fixed relationship between n and m ($m/n = 1, 3/2, 2, 5/2$). The time to compute the maximum matching seems approximately to triple if n and m are doubled. Each entry is the average of ten runs. Except on the very sparse instances ($m \approx n$) it does not pay to use the greedy heuristic.

We will see in the next section that there is always a covering whose capacity is equal to the size of the maximum matching. The function

```
list<edge> MAX_CARD_MATCHING(const graph& G, node_array<int>& OSC,
                             int heur = 0)
```

returns a maximum matching M and a labeling *OSC* (*OSC* stands for odd set cover, a name to be explained in the next section) with:

- *OSC* covers G and

- $|M| = cap(OSC)$.

Thus *OSC* proves the optimality of *M*. Figure 7.23 shows an example. The additional running time for computing the proof of optimality is negligible.

The function

```
void CHECK_MAX_CARD_MATCHING(const graph& G, const list<edge>& M,
                                           const node_array<int>& OSC)
```

checks whether *OSC* is a node labeling that covers *G* and whose capacity is equal to the cardinality of *M*. The function aborts if this is not the case. It runs in linear time.

The implementation of the checker is trivial. We determine for each i the number n_i of nodes with label i and then compute $S = n_1 + \sum_{i \ge 2} \lfloor n_i/2 \rfloor$. We assert that S is equal to the size of the matching.

We also check whether all edges are covered by the node labeling. Every edge must either be incident to a node labeled one or connect two nodes labeled i for some $i \ge 2$.

⟨*MCM: checker*⟩≡

```
static bool False(string s)
{ cerr << "CHECK_MAX_CARD_MATCHING: " << s << "\n";
  return false;
}
bool CHECK_MAX_CARD_MATCHING(const graph& G, const list<edge>& M,
                                           const node_array<int>& OSC)
{ int n = Max(2,G.number_of_nodes());
  int K = 1;
  array<int> count(n);
  int i;
  for (i = 0; i < n; i++) count[i] = 0;
  node v; edge e;
  forall_nodes(v,G)
  { if ( OSC[v] < 0 || OSC[v] >= n )
      return False("negative label or label larger than n - 1");
    count[OSC[v]]++;
    if (OSC[v] > K) K = OSC[v];
  }
  int S = count[1];
  for (i = 2; i <= K; i++) S += count[i]/2;
  if ( S != M.length() )
    return False("OSC does not prove optimality");
  forall_edges(e,G)
  { node v = G.source(e); node w = G.target(e);
    if ( v == w || OSC[v] == 1 || OSC[w] == 1 ||
             ( OSC[v] == OSC[w] && OSC[v] >= 2) ) continue;
      return False("OSC is not a cover");
  }
  return true;
}
```

7.7.2 *The Blossom Shrinking Algorithm*

We derive the *blossom shrinking* algorithm of Edmonds [Edm65b, Edm65a] for maximum cardinality matching in non-bipartite graphs. In its original form the running time of the algorithm is $O(n^4)$. Gabow [Gab76] and Lawler [Law76] improved the running time to $O(n^3)$ and Gabow [Gab76] showed how to use the partition data structure of Section 5.5 to obtain a running time of $O(nm\alpha(m, n))$. Tarjan [Tar83b] gave a very readable presentation of Edmond's algorithm and Gabow's improvement. Our presentation and our implementation is based on [Law76] and [Tar83b].

The algorithm follows the general paradigm for matching algorithms: repeated augmentation by augmenting paths until a maximum matching is obtained. We assume familiarity with the paradigm, which can, for example, be obtained by reading Section 7.6.2. The natural way to search for an augmenting path starting in a node v is to grow a so-called *alternating tree* rooted at v.

The root of an alternating tree is a free node, the nodes on odd levels are reached by odd length alternating paths (and hence their incoming tree edge is a non-matching edge) and the nodes on even levels are reached by even length alternating paths (and hence their incoming tree edge is a matching edge). The root is even. All leaves in an alternating tree are even and odd nodes have exactly one child (namely their mate). Figure 7.24 shows an alternating tree. A node on an even level is called an *even* node and a node on an odd level is called an *odd* node. In the implementation an even node is labeled EVEN, an odd node is labeled ODD, and every node belonging to no alternating tree carries the label UNLABELED. This suggests calling a node *labeled* if it belongs to some alternating tree and calling it *unlabeled* otherwise.

We start the algorithm by making every free node the root of a trivial alternating tree (consisting only of the free node itself) and by labeling all free nodes even. We will maintain the following invariants:

- For each free node there is an alternating tree rooted at the free node.
- All nodes belonging to one of the alternating trees are labeled EVEN or ODD. Nodes on even levels are labeled EVEN and nodes on odd levels are labeled ODD.
- All nodes belonging to no alternating tree are unlabeled (= labeled UNLABELED).
- All unlabeled nodes are matched and if a node is unlabeled then its mate is also unlabeled.

An alternating tree is extended by exploring an edge $\{v, w\}$ incident to an even node v. It is a matter of implementation strategy which alternating tree is extended and which edge is chosen to extend it. There are four cases to be distinguished: w may be unlabeled, w may be odd, w may be even and in a different tree, and w may be even and in the same tree. The first three cases occur also in the bipartite case.

Case 1, w is unlabeled: We make w the child of v and the mate of w the child of w, see Figure 7.25. In this way, w becomes an odd node, its mate becomes an even node, and

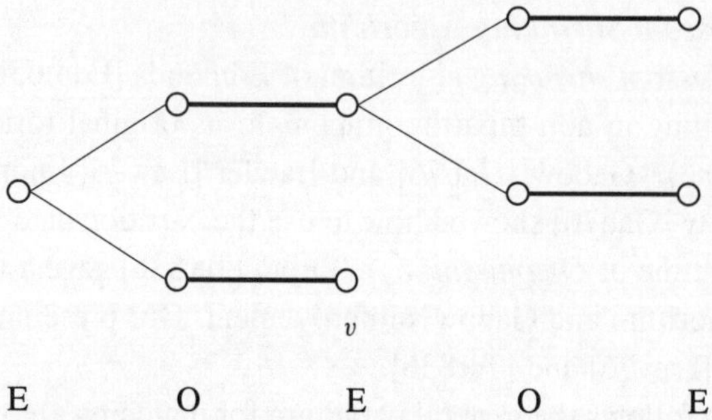

Figure 7.24 An alternating tree: It is rooted at a free node, nodes on odd levels (= odd nodes) are reached by odd length alternating paths, and nodes on even levels (= even nodes) are reached by even length alternating paths.

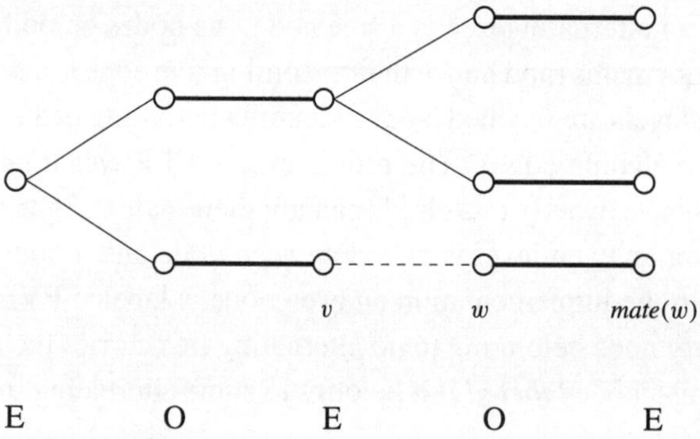

Figure 7.25 Growing an alternating tree: Exploration of the edge (v, w) turns w and its mate into labeled nodes, w becomes an odd node, and its mate becomes an even node.

both nodes become labeled. Observe that the growth action maintains the invariant that a matched node and its mate are either both labeled or both unlabeled.

Case 2, w is an odd node: We have discovered another odd length alternating path to w and do nothing.

Case 3, w is an even node in a different tree: We have discovered an augmenting path consisting of the edge $\{v, w\}$ and the tree paths from v and w to their respective roots, see Figure 7.26. We augment the matching by the augmenting path and unlabel all nodes in both trees. This makes all nodes in both trees matched (recall, that the root of an alternating tree is the only node in the tree that is unmatched) and destroys both trees. Observe that the remaining alternating trees, i.e., the ones whose roots are still free, are not affected by the augmentation. They are still augmenting trees with respect to the increased matching.

The three cases above also occur for bipartite graphs. The fourth and last case is new.

Case 4, w is an even node in the same tree as v: We have discovered a so-called *blossom*, see Figure 7.27. Let b be the lowest common ancestor of v and w, i.e., v and w

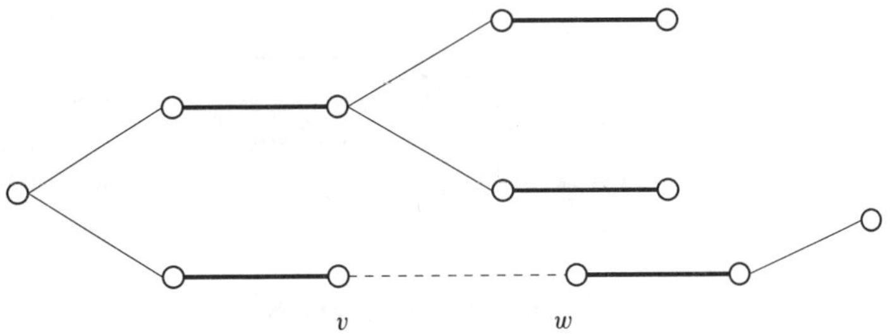

Figure 7.26 Discovery of an augmenting path: v and w are even nodes in distinct trees. The edge $\{v, w\}$ and the tree paths from v and w to their respective roots form an augmenting path.

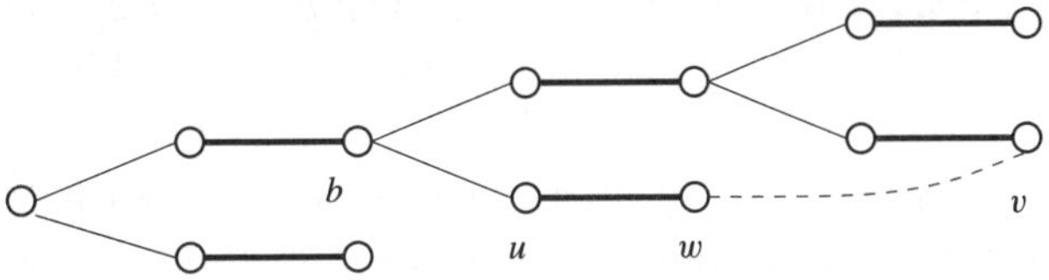

Figure 7.27 Discovery of a blossom: v and w are even nodes in the same tree. The node b is their lowest common ancestor. The blossom consists of the edge $\{v, w\}$ and the tree paths from b to v and w, respectively. The *stem* of the blossom consists of the tree path to b. The node b is the base of the blossom. The blossom consists of seven edges, three of which are matching. The even length alternating path to u follows the tree path to v, uses the edge $\{v, w\}$ and then proceeds down the tree to u.

are both descendants of b and there is no proper descendant of b with the same property. Since only even nodes can have more than one child, b is an even node. The blossom consists of the edge $\{v, w\}$ and the tree paths from b to v and w, respectively. The *stem* of the blossom consists of the tree path to b and b is called the *base* of the blossom. The stem is an even length alternating path ending in a matching edge; if the stem has length zero then b is free. The blossom is an odd length cycle of length $2k + 1$ containing k matching edges for some k, $k \geq 1$. All nodes in the blossom (except for the base) are reachable by an even and odd length alternating path from the root of the tree. For an even node u the even length path is simply the tree path to u and for an odd node u, say lying on the tree path from b to w, the even length path is the tree path to v followed by the edge $\{v, w\}$, followed by the path down the tree from w to u. For the odd length paths, the situation is reversed.

The action to take is to *shrink the blossom*. To shrink a blossom means to collapse all nodes of the blossom into the base of the blossom. This removes all edges from the graph which connect two nodes in the blossom and replaces any edge $\{u, z\}$ where u belongs to the blossom and z does not belong to the blossom by the edge $\{b, z\}$, see Figure 7.28. The node b is free after the shrinking iff it was free before the shrinking.

Lemma 22 *Let G' be obtained from G by shrinking a blossom with base b. If G' contains an augmenting path then so does G.*

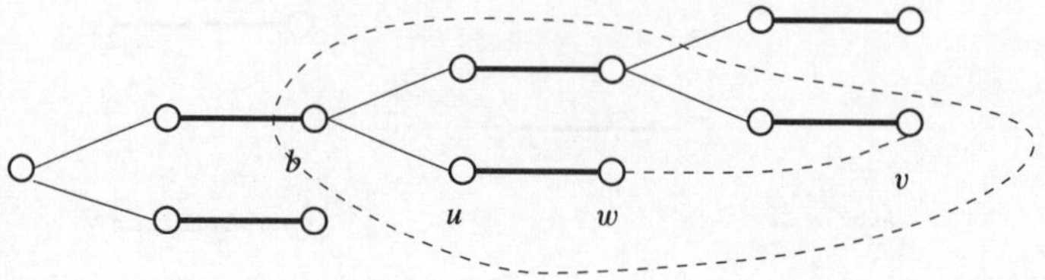

Figure 7.28 Shrinking a blossom: All nodes of the blossom are collapsed into the base of the blossom. After the shrinking, b stands for all the nodes enclosed by the dashed line.

Proof Suppose G' contains an augmenting path p. If p avoids b then p is an augmenting path in G and we are done. So let us assume that b lies on p. We break p at b into two pieces p_1 and p_2 and assume w.l.o.g that p_2 uses a non-matching edge e incident to b (in G'). The path p_1 is either empty (if b is free) or uses the matching edge incident to b. The edge $e = \{b, z\}$ in G' is induced by an edge $\{u, z\}$ in G where u is some node of the blossom. An augmenting path in G is obtained by first using p_1 then using the even length alternating path from b to u in the blossom, and then using p_2 (with its first edge replaced by $\{u, z\}$). □

We can now summarize the blossom shrinking algorithm. We grow alternating trees from the free nodes. Whenever a blossom is encountered it is shrunk. Whenever an augmenting path is discovered (this will in general happen after several shrinkings occurred), Lemma 22 is used to lift the augmenting path to the original graph. The matching is augmented by the augmenting path, the two trees involved are destroyed, all nodes in both trees are unlabeled, and the search for augmenting paths continues. The algorithm terminates when no alternating tree can be extended anymore. At this point the matching is maximum. Of course, this requires proof.

In order to show correctness we need the concept of an *odd-set cover*. It refines the notion of a covering introduced in Section 7.7.1.

For a subset N of an odd number of vertices of G we define the set of edges covered by N and the capacity of N as follows. If $|N| = 1$ then N covers all edges incident to the node in N and the capacity of N is equal to one. If $|N| = 2k + 1$ for some $k \geq 1$ then N covers all edges which have both endpoints in N and the capacity of N is k.

An *odd-set cover*[22] OSC of G is a family $\{N_1, \ldots, N_r\}$ of odd cardinality subsets of V such that each edge of G is covered by at least one of the sets in OSC. The capacity $c(OSC)$ of OSC is the sum of the capacities of the sets in OSC.

Lemma 23 *Let OSC be an odd-set cover in a graph G. Then the cardinality of any matching in G is at most $c(OSC)$.*

Proof Let M be any matching and let e be any edge in M. Then e must be covered by some

[22] An odd-set cover gives rise to an integer labeling of the nodes as follows: nodes that are contained in no set of the cover are labeled zero, nodes that are contained in a singleton set are labeled one, and nodes that are contained in an odd set of cardinality larger than one are labeled i for some $i > 1$. Distinct i's are used for distinct sets.

set in OSC. Moreover, the number of edges in M covered by any particular set in OSC is at most the capacity of the set. □

We are now ready for the correctness proof of the blossom shrinking algorithm. We will show that if the blossom shrinking algorithm does not find an augmenting path with respect to a matching M then there is an odd-set cover whose capacity is equal to the size of M, thus proving the optimality of M.

Let $G^{(0)} = G$ be our graph and let M be a matching in G. Suppose that the blossom shrinking algorithm does not discover an augmenting path. The blossom shrinking algorithm constructs a sequence $G^{(0)}, G^{(1)}, G^{(2)}, \ldots, G^{(h)}$ of graphs where for all i, $0 < i \le h$, $G^{(i)}$ is obtained from $G^{(i-1)}$ by shrinking a blossom. Each node v of every $G^{(i)}$ stands for a set of nodes of G. In $G^{(0)}$ every node represents itself, and a node v in $G^{(i)}$ either stands for the same set as in $G^{(i-1)}$ or, if v is equal to the base node of the shrunken blossom, stands for all nodes represented by the nodes of $G^{(i-1)}$ collapsed into it.

Lemma 24 *For every i and every node v of $G^{(i)}$:*

- *v stands for an odd set of nodes in G,*
- *if v is odd or unlabeled then v stands for the singleton set consisting of v itself,*
- *if v stands for a set B of $2k + 1$ nodes in G for some $k \ge 1$ then the number of edges in M connecting nodes in B is equal to k.*

Proof The claim is certainly true for i equal to zero. When a blossom is shrunk an odd number of nodes is collapsed into a single node. By induction hypothesis each collapsed node represents an odd number of nodes of G. The sum of an odd number of odd numbers is odd.

The result of a shrinking operation is an even node. Thus odd and unlabeled nodes represent only themselves.

Consider a shrinking operation that collapses $2r + 1$ nodes into one. Out of these nodes, $r + 1$ were even before the shrinking (namely the base v and every even node on the two tree paths belonging to the blossom) and r were odd. Every odd node represents a single node of G and every even node stands for an odd set of nodes of G. Suppose that the i-th odd node represents a set B_i of $2k_i + 1$ nodes in G.

After the shrinking operation v stands for the r odd nodes and the union of the B_i's. Thus B consists of

$$r + \sum_{1 \le i \le r+1} (2k_i + 1) = 2(r + \sum_{1 \le i \le r+1} k_i) + 1$$

nodes and hence $k = r + \sum_{1 \le i \le r+1} k_i$. The number of edges in M running between nodes of B_i is k_i, and the number of edges of M belonging to the blossom is r. We conclude that k edges of M connect nodes in B. □

Consider now the graph $G^{(h)}$. In $G^{(h)}$ we have an alternating tree rooted at each free node and the tree growing process has come to a halt. Thus there cannot be an edge connecting two even nodes (because this would imply the existence of either an augmenting path or a blossom) and there cannot be an edge connecting an even node to an unlabeled node (as this would allow us to grow one of the alternating trees). Thus every edge either connects two nodes contained in the same blossom, or is incident to an odd node, or connects two unlabeled nodes. Every unlabeled node is matched to an unlabeled node (since a matched node and its mate are either both unlabeled or both matched) and hence the number of unlabeled nodes is even. We construct an odd-set cover OSC whose capacity is equal to M. OSC consists of:

- all odd nodes (interpreted as singleton sets),
- for each even node that stands for a set of cardinality at least three: the set represented by the node,
- no further set if there is no unlabeled node, a singleton set consisting of an arbitrary unlabeled node if there are exactly two unlabeled nodes, and a singleton set consisting of an arbitrary unlabeled node and a set consisting of the remaining unlabeled nodes if there are more than two unlabeled nodes.

Lemma 25 *The capacity of the odd-set cover OSC is equal to the cardinality of M.*

Proof The number of edges in M that still exist in $G^{(h)}$, i.e., have not been shrunken into a blossom in the course of the algorithm, is equal to the number of odd nodes plus half of the number of unlabeled nodes. For each even node v of $G^{(h)}$, representing a set B of $2r + 1$ nodes of G, the number of edges in M connecting nodes in B is equal to r by Lemma 24. This concludes the proof. □

Theorem 5 *The blossom shrinking algorithm is correct.*

Proof The algorithm terminates when it does not find an augmenting path. When this happens, there is, by Lemma 25, an odd-set cover whose capacity is equal to the size of M. Thus M is optimal. □

7.7.3 *The Implementation*

The goal of this section is to implement the blossom shrinking algorithm. Our implementation refines the implementation described in [Tar83b] and is similar to the implementation given in [KP98]. The refinement does not change the worst case running time, but improves the best case running time from $\Omega(n^2)$ to $O(m)$. The observed behavior on random graphs with $m = O(n)$ seems to be much better than $O(n^2)$, see Table 7.7.

The overall structure of our implementation is given below. In the main loop we iterate over all nodes of G. Let $v_1, \ldots, v_n$ be an arbitrary ordering of the nodes of G. When $v = v_i$ is considered, every free node v_j with $j \geq i$ is the root of a trivial alternating tree, and the collection of alternating trees rooted at free nodes v_j with $j < i$ is *stable*. A collection $\mathcal{T}$

of alternating trees is stable if every edge $\{u, w\}$ incident to an even node u in $\mathcal{T}$ connects u to an odd node w in $\mathcal{T}$. In other words, every edge $\{u, w\}$ connecting a node u in $\mathcal{T}$ to a node outside $\mathcal{T}$ has u odd, and every edge connecting two nodes contained in $\mathcal{T}$ has at least one odd endpoint. It follows from our tree growing rules that the trees in $\mathcal{T}$ will not change in the future.

When $v = v_i$ is considered and v is already matched we do nothing. If v is still unmatched we grow the alternating tree T with root v until either an augmenting path is found or the growth comes to an end. We use a *node_list* Q to store all even nodes in T which have unexplored incident edges. We organize Q as a queue and hence grow the tree in breadth-first manner.

The growth process comes to an end when Q becomes empty. We claim that $\mathcal{T} \cup \{T\}$ is stable when Q becomes empty. Consider any edge $\{u, w\}$ with u an even node in T. Then w is odd, since otherwise the growth of T would not have come to an end. Moreover, w belongs to a tree in $\mathcal{T} \cup \{T\}$, since trees outside $\mathcal{T} \cup \{T\}$ are rooted at free nodes v_j, $j > i$, and consist only of a root and roots are even. Thus T can be added to our stable collection of alternating trees (this requires no action in the implementation) and the next free node can be considered.

When an augmenting path is found by exploring an edge $\{u, w\}$ with u an even node in T and w an even node in a tree different from T, w must be a free node v_j with $j > i$. Observe, that w cannot belong to T (since u and w are in distinct trees) and that w cannot belong to a tree in $\mathcal{T}$ (since $\mathcal{T}$ is stable). Thus w must belong to a tree rooted at some v_j, $j > i$, and hence must be equal to some v_j, $j > i$ (since the trees rooted at these nodes are trivial). When the matching is augmented by the augmenting path from v to w, all nodes in $T \cup w$ become matched and unlabeled. In order to be able to unlabel all nodes in $T \cup w$ in time proportional to the size of T we collect all nodes in T in a list of nodes (which we call T). We also set the variable *breakthrough* to *true* whenever an augmenting path is found in order to guarantee that we proceed to the next node in the main loop.

```
⟨_mc_matching⟩≡
  enum LABEL {ODD, EVEN, UNLABELED};
  ⟨MCM: helpers⟩
  list<edge> MAX_CARD_MATCHING(const graph& G,
                               node_array<int>& OSC, int heur)
  {
    ⟨MCM: data structures⟩
    ⟨MCM: heuristics⟩
    node v; edge e;
    forall_nodes(v,G)
    { if ( mate[v] != nil ) continue;
      node_list Q; Q.append(v);
      list<node> T; T.append(v);
      bool breakthrough = false;
      while (!breakthrough && !Q.empty()) // grow tree rooted at v
      {
```

```
      node v = Q.pop();
      ⟨explore edges out of the even node v⟩
    }
  }
  list<edge> M;
  ⟨MCM: compute M⟩
  ⟨general checking: compute OSC⟩
  return M;
}
```

The Main Data Structures: We next discuss the main data structures used in the program. We use a *node_array<node> mate* to keep track of the current matching and we use a *node_partition base* to keep track of the blossoms.

⟨*MCM: data structures*⟩≡

```
node_array<node> mate(G,nil);
node_partition base(G);     // now base(v) = v for all nodes v
```

If two nodes v and w are matched then $mate[v] = w$ and $mate[w] = v$ and if a node v is free then $mate[v] = nil$. At the beginning, all nodes are free.

The node partition (see Section 6.8) *base* establishes the relationship between the current graph G' and the original graph G; recall that the current graph is obtained from the original graph by a sequence of shrinkings of blossoms, that a node partition partitions the nodes of a graph into disjoint sets called blocks, and that for a node v, $base(v)$ is the canonical representative of the block containing v. The relationship between G and G' is as follows:

- For any node v of G: if $base(v) = v$ then v is a node of G' and if $base(v) \neq v$ then v was collapsed into $base(v)$. Thus $\{base(v)\ ;\ v \in V\}$ is the set of nodes of G'.
- An edge $\{v, w\}$ represents the edge $\{base(v), base(w)\}$ of G'.

Every node is labeled as either EVEN, ODD, or UNLABELED. A node is labeled UNLABELED if it does not belong to any alternating tree and it is labeled EVEN or ODD otherwise. A node is labeled when it is added to an alternating tree. It retains its label when it is collapsed into another node. At the beginning all nodes are free and hence the root of an alternating tree. Thus all nodes are EVEN at the beginning. For an odd node v we use $pred[v]$ to store its parent node in the alternating tree. The pred value is set when a node is added to an alternating tree; it is not changed when the node is collapsed into another node.

⟨*MCM: data structures*⟩+≡

```
node_array<int>  label(G,EVEN);
node_array<node> pred(G,nil);
```

Figure 7.29 shows an example.

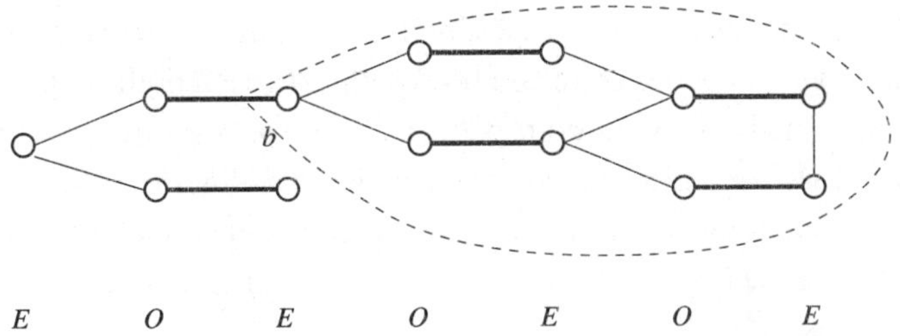

Figure 7.29 Snapshot of the data structure: The node labels are indicated by the labels "E" and "O". All nodes enclosed by the dashed line form a blossom and hence a block of the partition *base*. The canonical element of this block is b.

Exploring an Edge: Having defined most of the data structures we can give the details of exploring edges. Assume that v is an even node and let $e = \{v, w\}$ be an edge incident to v. Recall that e stands for the edge $\{base(v), base(w)\}$ in the current graph.

We do nothing if e is a self-loop or if $base(w)$ is ODD. If $base(w)$ is UNLABELED (this is equivalent to w being unlabeled) we grow the alternating tree containing v and if $base(w)$ is EVEN we have either discovered an augmenting path or a blossom.

⟨*explore edges out of the even node v*⟩≡

```
forall_inout_edges(e,v)
{ node w = G.opposite(v,e);
  if ( base(v) == base(w) || label[base(w)] == ODD )
     continue;   // do nothing
  if ( label[w] == UNLABELED )
    { ⟨grow tree⟩ }
  else   // base(w) is EVEN
    { ⟨augment or shrink blossom⟩ }
}
```

Growing the Tree: Let us first give the details of growing a tree. We label w as odd, make v the parent of w, label the mate of w as even, add the mate of w to Q, and add w and the mate of w to T.

⟨*grow tree*⟩≡

```
label[w] = ODD;             T.append(w);
pred[w] = v;
label[mate[w]] = EVEN;      T.append(mate[w]);
Q.append(mate[w]);
```

Discovery of a Blossom or an Augmenting Path: The node $base(w)$ is even. We have either found an augmenting path or a blossom. We have found an augmenting path if $base(v)$ and $base(w)$ belong to distinct trees and we have discovered a blossom if they belong to the same tree. We distinguish the two cases by tracing both tree paths in lock-step fashion until we either encounter a node that lies on both paths or reach both roots[23].

We discover a node lying on both paths as follows. We keep a counter *strue* which we increment in every execution of ⟨*augment or shrink blossom*⟩. Since there are at most n augmentations and at most n shrinkings between two augmentations the maximal value of the counter is bounded by n^2. It would therefore be unsafe to use type *int* for the counter, but type *double* is safe.

We use the counter as follows. As we trace the two tree paths we set *path1*[*hv*] to *strue* for all even nodes *hv* on the first path and *path2*[*hw*] to *strue* for all even nodes *hw* on the second path. The two paths meet iff *path1*[*hw*] or *path2*[*hv*] is equal to *strue* for some even *hw* on the second path or some even *hv* on the first path. The first node for which this is true is the base of the blossom. Recall that the base of a blossom is always even.

The cost of tracing the paths is proportional to the size of the blossom found, if a blossom is discovered, and is proportional to the length of the augmenting path found otherwise. Also observe that we define the arrays *path1* and *path2* outside the loop that searches for augmenting paths. Thus the cost for their initialization arises only once.

⟨*MCM: data structures*⟩+≡

```
double strue = 0;
node_array<double>  path1(G,0);
node_array<double>  path2(G,0);
```

⟨*augment or shrink blossom*⟩≡

```
node hv = base(v);
node hw = base(w);

strue++;
path1[hv] = path2[hw] = strue;

while ((path1[hw] != strue && path2[hv] != strue) &&
       (mate[hv] != nil || mate[hw] != nil) )
{ if (mate[hv] != nil)
  { hv = base(pred[mate[hv]]);
    path1[hv] = strue;
  }
  if (mate[hw] != nil)
  { hw = base(pred[mate[hw]]);
    path2[hw] = strue;
  }
}

if (path1[hw] == strue || path2[hv] == strue)
  { ⟨shrink blossom⟩ }
else
  { ⟨augment path⟩ }
```

[23] An alternative strategy is as follows: we have found an augmenting path if w is the root of a tree outside $\mathcal{T} \cup \{T\}$. We could, for each node, keep a bit to record this fact. The alternative simplifies the distinction between blossom shrinking and augmentations. However, it does not simplify the code overall, as all the information gathered in the program chunk ⟨*augment or shrink blossom*⟩ is needed in later steps of the algorithm.

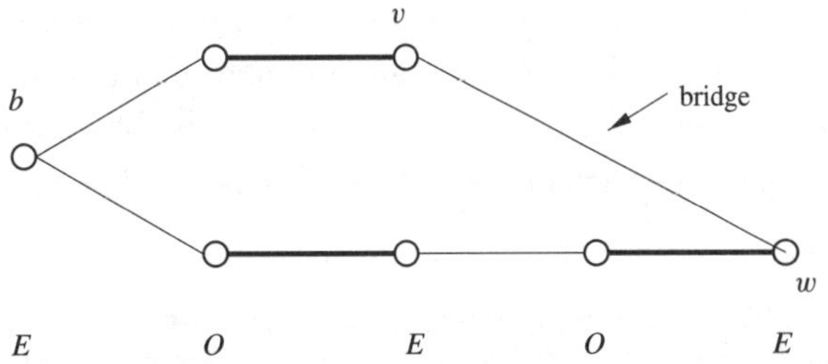

Figure 7.30 The bridge of a blossom: The edge $\{v, w\}$ closes a blossom with base b. For the odd nodes on the tree path from b to v we set *source_bridge* to v and *target_bridge* to w and for the odd nodes on the tree path from b to w we set *source_bridge* to w and *target_bridge* to v.

Shrinking a Blossom: Let us see how to shrink a blossom. The base b of the blossom[24] is either *hv* or *hw*. It is *hw* if *hw* also lies on the first path and it is *hv* otherwise. We shrink the blossom by shrinking the two paths that form the blossom.

The call *shrink_path*$(b, v, w, \ldots)$ collapses the path from v to b into b and the call *shrink_path*$(b, w, v, \ldots)$ collapses the path from w to b into b. Both calls also have the other end of the edge that closes the blossom as an argument.

⟨*shrink blossom*⟩≡

```
node b = (path1[hw] == strue) ? hw : hv;     // Base
shrink_path(b,v,w,base,mate,pred,source_bridge,target_bridge,Q);
shrink_path(b,w,v,base,mate,pred,source_bridge,target_bridge,Q);
```

Before we can give the details of the procedure *shrink_path* we need to introduce two more node labels. When an edge $\{v, w\}$ closes a blossom, all odd nodes in the blossom also get an even length alternating path to the root of their alternating tree. This path goes through the edge that closes the blossom. We call this edge the *bridge* of the blossom. The odd nodes on the tree path from v to b use the bridge in the direction from v to w and the odd nodes on the tree path from w to b use the bridge in the direction from w to v. We use the node arrays *source_bridge* and *target_bridge* to record for each odd node shrunken into a blossom the source node and the target node of its bridge (now viewed as a directed edge).

⟨*MCM: data structures*⟩+≡

```
node_array<node> source_bridge(G,nil);
node_array<node> target_bridge(G,nil);
```

The details of collapsing the tree path from v to b into b are now simple. For each node x on the path we perform *union_blocks*(x, b) to union the blocks containing x and b, for each odd node we set *source_bridge* to v and *target_bridge* to w, and we add all odd nodes to Q (because the edges out of the odd nodes now emanate from the even node b), see Figure 7.30.

24 With the alternative case distinction between blossom shrinking and augmentation we would have to compute *hv* and *hw* at this point.

There is one subtle point. After a union operation the canonical element of the newly formed block is unspecified (it may be any element of the resulting block). It is important, however, that b stays the canonical element of the block containing it. We therefore explicitly make b the canonical element by *base.make_rep*(b).

⟨*MCM: helpers*⟩≡

```
static void shrink_path(node b, node v, node w,
          node_partition& base, node_array<node>& mate,
          node_array<node>& pred, node_array<node>& source_bridge,
          node_array<node>& target_bridge, node_list& Q)
{ node x = base(v);
  while (x != b)
  {
    base.union_blocks(x,b);
    x = mate[x];
    base.union_blocks(x,b);
    base.make_rep(b);
    Q.append(x);
    source_bridge[x] = v;  target_bridge[x] = w;
    x = base(pred[x]);
  }
}
```

Augmentation: We treat the discovery of an augmenting path. The nodes v and w belong to distinct alternating trees with roots *hv* and *hw*, respectively. In fact, w is a root itself. The augmenting path consists of the edge $\{w, v\}$ plus the even length alternating path from v to its root *hv*.

For a node v let $p(v)$ be the even length alternating path from v to its root (if it exists). The path $p(v)$ can be defined inductively as follows:

If v is a root then $p(v)$ is the trivial path consisting solely of v.

If v is EVEN, $p(v)$ goes through the mate of v to the predecessor of the mate and then follows $p(pred[mate[v]])$.

If v is ODD, $p(v)$ consists of the alternating path from v to *source_bridge*[v] concatenated with $p(target\_bridge[v])$.

Lemma 26 *The above characterization of* $p(v)$ *is correct.*

Proof The claim is certainly true when v is a root. So assume otherwise and consider the time when $p(v)$ is discovered in the course of the algorithm. For an even node this is the time when v is labeled EVEN and for an odd node this is the case when it becomes part of a blossom. In either case the characterization is correct. □

How can we find the alternating path from v to *source_bridge*[v] when v is odd? The problem is that the *pred*-pointers are directed towards the roots of alternating trees and hence

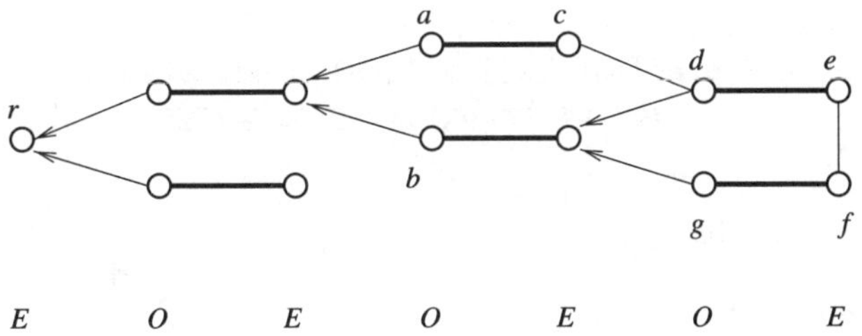

Figure 7.31 Tracing augmenting paths: The node labels are indicated by the labels "E" and "O". The predecessor pointer of the odd nodes are shown. When the bridge $\{e, f\}$ was explored we set *source_bridge*$[d]$ to e, *target_bridge*$[d]$ to f, *source_bridge*$[g]$ to f, and *target_bridge*$[g]$ to e, and when the bridge $\{c, d\}$ was explored we set *source_bridge*$[a]$ to c, *target_bridge*$[a]$ to d, *source_bridge*$[b]$ to d, and *target_bridge*$[b]$ to c.
The even length alternating path from b to its root r consists of the reversal of the path from d = *source_bridge*$[b]$ to b followed by the even length alternating path from c = *target_bridge*$[b]$ to r. The former path consists of the reversal of the alternating path from e = *source_bridge*$[d]$ to d followed by the alternating path from f = *target_bridge*$[d]$ to b.

there is no direct way to walk from v to *source_bridge*$[v]$. We walk from *source_bridge*$[v]$ to v instead and then take the reversal of the resulting path. The path from *source_bridge*$[v]$ to v is the prefix of $p($*source_bridge*$[v])$ ending in v, see Figure 7.31.

We cast this reasoning into a program by defining a procedure *find_path*$(P, x, y, \ldots)$ that takes two nodes x and y, such that y lies on $p(x)$ and such that the prefix of $p(x)$ ending in y has even length (the program would be slightly less elegant without the second assumption), and appends the prefix of $p(x)$ ending in y to the list P. *Find_path* distinguishes three cases:

If x is equal to y then the path consists of the single node x.

If $x \neq y$ and x is EVEN the path consists of x, *mate*$[x]$, followed by the path from *pred*$[$*mate*$[x]]$ to y.

If $x \neq y$ and x is ODD, let *P1* and *P2* be the paths from *target_bridge*$[x]$ to y and from *source_bridge*$[x]$ to *mate*$[x]$, respectively. Then path consists of x followed by the reversal of *P2* followed by *P1*.

⟨*MCM: helpers*⟩+≡

```
static void find_path(list<node>& P, node x, node y,
                      node_array<int>&  label, node_array<node>& pred,
                      node_array<node>& mate,
                      node_array<node>& source_bridge,
                      node_array<node>& target_bridge)
{ if ( x == y )
  {
    P.append(x);
    return;
  }
  if ( label[x] == EVEN )
  {
    P.append(x);
```

```
      P.append(mate[x]);
      find_path(P,pred[mate[x]],y,label,pred,mate,
                      source_bridge,target_bridge);
      return;
    }
    else // x is ODD
    {
      P.append(x);
      list<node> P2;
      find_path(P2,source_bridge[x],mate[x],label,pred,mate,
                                  source_bridge,target_bridge);
      P2.reverse_items();
      P.conc(P2);
      find_path(P,target_bridge[x],y,label,pred,mate,
                          source_bridge,target_bridge);
      return;
    }
  }
```

Given *find_path*, it is trivial to construct the augmenting path. We construct the path from v to hv in P and append w to the front of the path. We augment the current matching by the path by walking along the path and changing *mate* accordingly.

It remains to prepare for the next search for an augmenting path. All nodes in $T \cup \{w\}$ are now matched. We unlabel all nodes in $T \cup \{w\}$ and split the blocks of *base* containing nodes of T. No action is required for the other alternating trees.

Finally, we set *breakthrough* to *true* and break from the forall-inout-edges loop. Setting *breakthrough* to *true* makes sure that we also leave the grow tree loop. The next action will therefore be to grow an alternating tree from the next free node.

⟨*augment path*⟩≡

```
list<node> P;
find_path(P,v,hv,label,pred,mate,source_bridge,target_bridge);
P.push(w);
while(! P.empty())
{ node a = P.pop();
  node b = P.pop();
  mate[a] = b;
  mate[b] = a;
}
T.append(w);
forall(v,T) label[v] = UNLABELED;
base.split(T);
breakthrough = true;
break;
```

Computing the Node Labeling *OSC*: We compute the node labeling *OSC* as described in the paragraph preceding Lemma 25. We initialize $OSC[v]$ to -1 for all nodes v. This will allow us to recognize nodes without a proper *OSC*-label later. We then determine the number of unlabeled nodes (= nodes labeled *UNLABELED* and select an arbitrary unlabeled node. If there are unlabeled nodes, the selected unlabeled node is labeled one and all other unlabeled nodes are either labeled zero (if there are exactly two unlabeled nodes) or two (if there are more than two unlabeled nodes). We then set K to the smallest unused label larger than one.

Next we determine the number of sets of cardinality at least three and assign distinct labels to their representatives. We do so by iterating over all nodes. Every node v with $base(v) \neq v$ indicates a set of cardinality at least three. If its base is still unlabeled, we label it.

Finally, we label all other nodes. Nodes belonging to a set of cardinality at least two inherit the label of the base, and nodes that belong to sets of cardinality one (they satisfy `base(v) == v && OSC[base(v)] == -1`) are labeled one iff they are ODD and are labeled zero if they are EVEN.

⟨*general checking: compute OSC*⟩≡

```
forall_nodes(v,G) OSC[v] = -1;
int number_of_unlabeled = 0;
node arb_u_node;
forall_nodes(v,G)
 if ( label[v] == UNLABELED )
 { number_of_unlabeled++;
   arb_u_node = v;
 }
if ( number_of_unlabeled > 0 )
{ OSC[arb_u_node] = 1;
  int L = ( number_of_unlabeled == 2 ? 0 : 2 );
  forall_nodes(v,G)
    if ( label[v] == UNLABELED && v != arb_u_node ) OSC[v] = L;
}
int K = ( number_of_unlabeled <= 2 ? 2 : 3);
forall_nodes(v,G)
 if ( base(v) != v && OSC[base(v)] == -1 ) OSC[base(v)] = K++;
forall_nodes(v,G)
{ if ( base(v) == v && OSC[v] == -1 )
  { if ( label[v] == EVEN ) OSC[v] = 0;
    if ( label[v] == ODD  ) OSC[v] = 1;
  }
  if ( base(v) != v ) OSC[v] = OSC[base(v)];
}
```

Computing the List of Matching Edges: The list M of matching edges is readily constructed. We iterate over all edges. Whenever an edge is encountered whose endpoints are matched with each other, the edge is added to the matching. We also "unmate" the endpoints in order to avoid adding parallel edges to M.

⟨*MCM: compute M*⟩≡

```
forall_edges(e,G)
{ node v = source(e);
  node w = target(e);
  if ( v != w  &&  mate[v] == w )
  { M.append(e);
    mate[v] = v;
    mate[w] = w;
  }
}
```

Heuristics: If $heur = 1$, the greedy heuristic is used to compute an initial matching. We iterate over all edges. If both endpoints of an edge are unmatched, we match the endpoints and declare both endpoints unlabeled. Recall that matched nodes that do not belong to an alternating tree are UNLABELED.

⟨*MCM: heuristics*⟩≡

```
switch (heur) {
case 0: break;
case 1: { edge e;
          forall_edges(e,G)
          { node v = G.source(e); node w = G.target(e);
            if ( v != w && mate[v] == nil && mate[w] == nil)
            { mate[v] = w; label[v] = UNLABELED;
              mate[w] = v; label[w] = UNLABELED;
            }
          }
          break;
        }
}
```

Summary: We summarize and complete the running time analysis. The algorithm computes a maximum matching in phases. In each phase an alternating tree T from a free node is grown to find an augmenting path. If the search for an augmenting path is successful, the matching is increased and all nodes in the alternating tree are unlabeled, and if the search is unsuccessful, the tree will stay around and will never be looked at again.

The running time of a phase is $O((n_T + m_T)\alpha(n_T, m_T))$, where n_T is the number of nodes included into T, m_T is the number of edges having at least one endpoint in T, and $\alpha(n, m_T)$ is the cost of m_T operations on a node partition of n nodes. This can be seen as follows. In a phase zero or more blossoms are shrunken. The search for a blossom (if successful) has cost proportional to the size of the blossom, and shrinking a blossom of size $2k + 1$ removes $2k$ nodes from the graph. Therefore the total size of all blossoms shrunk in a phase is $O(n_T)$. In each phase each edge is explored at most twice (once from each endpoint). Each exploration of an edge and each removal of a node involves a constant number of operations on the node partition *base*. We conclude that the total cost of a phase is $O((n_T + m_T)\alpha(n, m_T)) = O((n+m)\alpha(n, m)) = O(m\alpha(n, m))$, since $n_T \le n \le m$ and $m_T \le m$.

There are at most n phases and hence the total running time is $O(nm\alpha(n, m))$ in the worst case. One may hope that n_T is significantly smaller than n and m_T is significantly smaller than m for many phases. The running times reported in Section 7.7.1 show that the hope is justified in the case of random graphs. There are no analytical results concerning the average case behavior of general matching algorithms.

In an earlier implementation of the blossom shrinking algorithm we did not collect the nodes of the alternating tree grown into a set T. Rather, we iterated over all nodes at the beginning of a phase and labeled all free nodes EVEN and all matched nodes UNLABELED. With this implementation the running time is $\Omega(n^2)$. The implementation discussed in this section is significantly faster. It is superior for two reasons. Firstly, the cost of a phase is proportional to the size of the alternating tree grown in the phase and hence may be sublinear, and secondly, an alternating tree that does not lead to a breakthrough is not destroyed, but kept till the end of the execution.

Exercises for 7.7

1 Compare the running time of the general matching algorithm and the bipartite matching algorithm on bipartite graphs.

2 Exhibit a family of graphs where the running time of our matching algorithm is $\Omega(nm)$. Write a program to generate such graphs and provide it as an LEP.

7.8 Maximum Weight Bipartite Matching and the Assignment Problem

Throughout this section $G = (A \dot\cup B, E)$ denotes a bipartite graph and $c : E \mapsto \mathbb{R}$ denotes a *cost function* on the edges of G. We also say *weight* instead of cost. A *matching* M is a subset of E such that no two edges in M share an endpoint. The *cost of a matching* M is the sum of the cost of its edges, i.e.,

$$c(M) = \sum_{e \in M} c(e).$$

A node v is called *matched* with respect to a matching M if there is an edge in M incident to v and it is called *free* or *unmatched* otherwise. An edge e is called *matching* if $e \in M$. For a matched node v the unique node w connected to v by a matching edge is called the *mate* of v. A matching is called *perfect* or an *assignment* if all nodes of G are matched.

A matching is called:

- a *maximum weight matching* if its cost is at least as large as the cost of any other matching,
- a *maximum weight assignment* if it is a heaviest perfect matching,

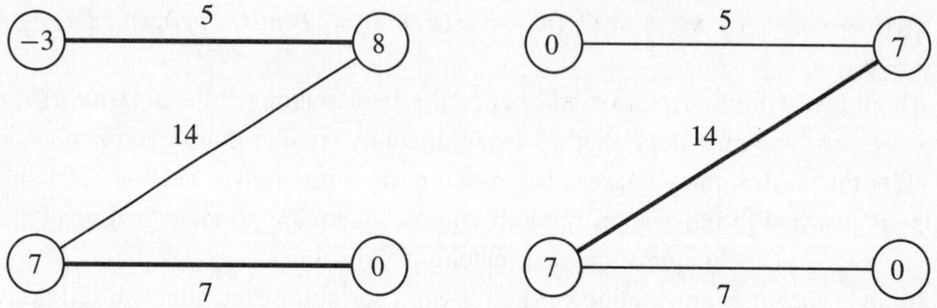

Figure 7.32 Maximum weight assignment and maximum weight matching. The matching on the left is a maximum weight perfect matching and the matching on the right is a maximum weight matching; the edges in the matchings are shown in bold in both cases. A potential function that proves the optimality of the matching is also given in both cases. The potential of each node and the cost of each edge is shown. For every edge the cost of the edge is bounded by the sum of the potentials of its endpoints. In an assignment every node is incident to exactly one edge of the assignment and hence the total cost of the assignment is bounded by the total potential. In the graph on the left the two quantities are equal and hence the assignment is optimal. In the graph on the right the potential function has the additional property that all potentials are non-negative and that all free nodes have potential zero. This implies (see Lemma 27) that the cost of any matching is bounded by the total potential. The two quantities are equal in the graph on the right and hence the matching is a maximum weight matching. The xlman-demo gw_mwb_matching allows the reader to experiment with weighted matchings in bipartite graphs.

- a *minimum weight assignment* if it is a lightest perfect matching,
- a *maximum weight maximum cardinality matching* if it is a heaviest matching among the matchings of maximum cardinality.

Figure 7.32 shows a a maximum weight assignment and a maximum weight matching. Clearly, a maximum or minimum weight assignment exists if and only if G contains a perfect matching.

In the next section we give the functionality of our algorithms and derive checkers of optimality. Sections 7.8.2 and 7.8.3 discuss an algorithm for maximum weight matchings and its implementation. In Sections 7.8.4 and 7.8.6 we modify our algorithms to compute assignments and maximum weight matchings of maximum cardinality. Finally, in Section 7.8.5 we show how to reduce the shortest path problem to the assignment problem.

7.8.1 *Functionality*

All functions in this section are function templates that work for an arbitrary number type *NT*. We use the convention that names of function templates for graph algorithms end with *_T*. In order to use the templates one must include <*LEDA/templates/mwb_matching.t*>. LEDA also contains pre-compiled instantiations for the number types *int* and *double*. The function names for the instantiated versions are *without* the suffix *_T*. In order to use the instantiated versions one must include <*LEDA/graph_alg.h*>. Section 7.1 discusses the relationship between templates and instantiated versions in more detail.

The function

```
list<edge> MAX_WEIGHT_BIPARTITE_MATCHING_T(graph& G,
                          const edge_array<NT>& c, node_array<NT>& pot)
```

returns a matching of maximal cost; the graph G is required to be bipartite. The worst case running time of the algorithm is $O(n \cdot (m + n \log n))$, the average case running time is much better. The function computes a proof of optimality in the form of the potential function *pot*. We discuss potential functions later in the section.

If a bipartition $V = A \dot{\cup} B$ is known and all edges are directed from A to B, the function

```
list<edge> MAX_WEIGHT_BIPARTITE_MATCHING_T(graph& G,
                          const list<node>& A, const list<node>& B,
                          const edge_array<NT>& c, node_array<NT>& pot)
```

can be used. If A and B have different sizes then it is advisable that A is the smaller set; in general, this leads to smaller running time.

The functions

```
list<edge> MAX_WEIGHT_ASSIGNMENT_T(graph& G,
                          const edge_array<NT>& c, node_array<NT>& pot);
list<edge> MIN_WEIGHT_ASSIGNMENT_T(graph& G,
                          const edge_array<NT>& c, node_array<NT>& pot);
```

return a maximum and minimum weight assignment, respectively. Both functions require that G is bipartite. If G does not contain a perfect matching the empty set of edges is returned.

All functions above are also available in the form where A and B are given as additional arguments and also without the argument *pot*.

The function

```
list<edge> MWMCB_MATCHING_T(graph& G,
                          const list<node>& A, const list<node>& B,
                          const edge_array<NT>& c, node_array<NT>& pot);
```

returns a maximum weight matching among the matchings of maximum cardinality. The potential function *pot* proves the optimality of the matching, see Section 7.8.6.

Potential Functions: We have mentioned the concept of a *potential function* several times already. It is time to define it. A function $\pi : V \mapsto \mathbb{R}$ is called a potential function. For an edge $e = (v, w)$ we call

$$\overline{c}(e) = \pi(v) + \pi(w) - c(e)$$

the *reduced cost* of e with respect to π. An edge is called *tight* iff its reduced cost is zero and the tight subgraph consists of all tight edges. For a subset U of the nodes we use $\pi(U)$ to denote $\sum_{v \in U} \pi(v)$. The following four properties of potential functions will play a role:

(1) Non-negativity of reduced costs, $\overline{c}(e) \geq 0$ for all $e \in E$.
(2) Tightness of matched edges, $\overline{c}(e) = 0$ for $e \in M$.

(3) Non-negativity of node potentials, $\pi(v) \geq 0$ for all $v \in V$.
(4) Tightness of free nodes, $\pi(v) = 0$ for all v that are free with respect to M.

The importance of potential functions stems from the following lemma.

Lemma 27 *Let M be any matching, let π be any potential function, and let F be the set of nodes that are free with respect to M.*

If all reduced costs are non-negative then $c(M) \leq \pi(V) - \pi(F)$. If, in addition, M is an assignment or all node potentials are non-negative then $c(M) \leq \pi(V)$.

If all reduced costs are non-negative and all matched edges have reduced cost zero then $c(M) = \pi(V) - \pi(F)$. If, in addition, M is an assignment or all free nodes have potential zero then $c(M) = \pi(V)$.

Proof If all reduced costs are non-negative then $c(e) \leq \pi(v) + \pi(w)$ for every edge $e = (v, w)$. Thus

$$\begin{aligned} c(M) &= \sum_{e \in M} c(e) \\ &\leq \sum_{e=(v,w) \in M} \pi(v) + \pi(w) \\ &= \sum_{v \in V;\, v \text{ is matched}} \pi(v) \;=\; \pi(V) - \pi(F), \end{aligned}$$

where the next to last equality follows from the fact that M is a matching and hence every matched node contributes exactly once to the sum on the second line and no free node contributes, and the last equality follows from the fact that the matched nodes are precisely the nodes that are not free. This establishes the first claim. For the third claim we observe that the inequality above becomes an equality if all matching edges have reduced cost zero.

The second and fourth claim follow from the first and third claim, respectively, and the additional observation that $\pi(F) \geq 0$ if node potentials are non-negative and that $\pi(F) = 0$ if M is an assignment or if the potential of all free nodes is zero. □

We call a potential function *feasible* if it satisfies (1), *non-negative* if it satisfies (3), and *tight* if it satisfies (1), (2), and (4). A tight non-negative potential function proves the optimality of a maximum weight matching and a tight potential function proves the optimality of a maximum weight assignment. Our algorithms return proofs of optimality in the form of tight potential functions.

The optimality conditions (1) to (4) are the basis for checkers of optimality. The function CHECK_MWBM_T takes a cost function c, a list of edges M, and a potential function *pot*, and checks that M is a matching and that the properties (1) to (4) above are satisfied.

⟨*mwb_matching.t*⟩+≡

```
bool False(const string s)
{ cerr << "CHECK_MWBM_T: " << s << "\n" << flush; return false;}
template <class NT>
```

```
bool CHECK_MWBM_T(const graph& G, const edge_array<NT>& c,
                  const list<edge>& M, const node_array<NT>& pot)
{ node v; edge e;
  // M is a matching
  node_array<int> deg_in_M(G,0);
  forall(e,M)
  { deg_in_M[G.source(e)]++;
    deg_in_M[G.target(e)]++;
  }
  forall_nodes(v,G)
    if ( deg_in_M[v] > 1) return False("M is not a matching");
  // node potentials are non-negative
  forall_nodes(v,G)
    if ( pot[v] < 0) return False("negative node potential");;
  // edges have non-negative reduced cost
  forall_edges(e,G)
  { node v = G.source(e); node w = G.target(e);
    if ( c[e] > pot[v] + pot[w])
      return False("negative reduced cost");
  }
  // edges in M have reduced cost equal to zero
  forall(e,M)
  { node v = G.source(e); node w = G.target(e);
    if ( c[e] != pot[v] + pot[w] )
     return False("non-tight matching edge");
  }
  // free nodes have potential equal to zero
  forall_nodes(v,G)
    if ( deg_in_M[v] == 0 && pot[v] != 0 )
      return False("free node with non-zero potential");
  return true;
}
```

The analogous functions

```
bool CHECK_MIN_WEIGHT_ASSIGNMENT_T(G,c,M,pot);
bool CHECK_MAX_WEIGHT_ASSIGNMENT_T(G,c,M,pot);
```

check minimum and maximum weight assignments, respectively. We do not give their implementations here. It is a good exercise to provide the implementations.

Potential Functions and Linear Programming Duality: We relate Lemma 27 to linear programming duality. Readers unfamiliar with linear programming may skip this material, although there is no harm in reading it anyway.

The maximum matching problem can be formulated as an integer program. We associate a variable $x(e)$ with every edge e, constrain it to the values 0 and 1, and consider the integer program

$$\max \sum_{e \in E} c(e)x(e)$$

subject to

$$\begin{array}{rcll} \sum_{e; e \text{ is incident to } v} x(e) & \le & 1 & \text{for all } v \in V \\ x(e) & \in & \{0,1\} & \text{for all } e \in E. \end{array}$$

Let M be the set of edges e with $x(e) = 1$. The first constraint states that for each node v at most one of the incident edges belongs to M, i.e., it guarantees that M is a matching. The objective function states that we are looking for a matching of maximal weight. It was shown by Edmonds [Edm65b, Edm65a] that the integrality constraints $x(e) \in \{0, 1\}$ may be replaced by the linear constraints $x(e) \ge 0$ without changing the problem[25]. Assume that the integrality constraint $x(e) \in \{0, 1\}$ has been replaced by $x(e) \ge 0$. We now consider the dual linear program. The dual has one variable for each node and one constraint for each edge. We use $\pi(v)$ for the variable corresponding to node v and obtain

$$\min \sum_{v \in V} \pi(v)$$

subject to

$$\begin{array}{rcll} c(e) & \le & \pi(v) + \pi(w) & \text{for all } e = (v, w) \in E \\ \pi(v) & \ge & 0 & \text{for all } v \in V. \end{array}$$

Linear programming duality states that the objective value of any feasible solution of the primal problem (= a matching) is no larger than the objective value of any feasible solution of the dual problem (= a potential function satisfying (1) and (3)) and that the value of the optimal solutions are equal. Complementary slackness implies in addition that the reduced cost of an edge in the matching must be zero and that the node potential of a free node must be zero. In fact, the proof of Lemma 27 is simply an adaption of the standard proofs of weak linear programming duality and complementary slackness to matchings.

[25] We sketch a proof of this fact. We first observe that the non-negativity constraints $x(e) \ge 0$ together with the matching constraints $\sum_{e; e \text{ is incident to } v} x(e) \le 1$ guarantee $0 \le x(e) \le 1$. It therefore suffices to prove that the linear program has an optimal integral solution. The optimal solution to the linear program is given by a basic feasible solution, i.e., by the solution to a system $Bx = 1$ where B is a square submatrix of the constraint matrix and 1 is a vector of ones. Thus $x = B^{-1}1$. It therefore suffices to prove that B^{-1} is integral. By Cramer's rule, each entry of B^{-1} is the quotient of the determinant of a submatrix of B and the determinant of B. It therefore suffices to prove that the determinant of B is in $\{-1, 0, +1\}$. We prove more generally that the determinant of any square submatrix of the constraint matrix has determinant -1, 0, or $+1$, i.e., that the constraint matrix is a so-called *totally unimodular* matrix. Let B be any square submatrix. We need to compute the determinant of B. Each entry of B is either zero or one, each column of B corresponds to an edge of G, each row of B corresponds to a node of G, and each column contains at most two ones, one for each endpoint. As long as B contains a row or column with at most one one, we expand the determinant along this row or column. Each such reduction step reduces the dimension by one and yields a factor -1, 0, or $+1$. When no further reduction step applies, we have either reduced the dimension to zero and are done or reached a matrix B in which every row and column contains at least two ones. We will show that B is singular. Since a column contains at most two ones, we conclude that every column contains exactly two ones. Since B is square and since every row contains at least two ones we conclude that every row contains exactly two ones. In other words in the graph defined by B every node has degree two and thus the graph consists of a set of cycles. Each cycle has even length since G is bipartite (this is where we use the fact that G is bipartite). Let $v_1, v_2, \ldots, v_{2k}$ be any one of the cycles and consider the following linear combination formed by the rows corresponding to these nodes. Rows corresponding to nodes with odd index are multiplied by $+1$ and rows corresponding to nodes with even index are multiplied by -1. This linear combination yields the zero vector since, in each column corresponding to an edge of the cycle, one contribution is $+1$ and the other contribution is -1; this argument relies on the fact that the cycle has even length. Altogether we have now shown that the determinant of B is either -1, 0, or $+1$.

Arithmetic Demand: Special care should be taken when using the template functions with a number type *NT* that can incur rounding error, e.g., the type *double*. Section 7.2 contains a general discussion of this issue. The template functions are only guaranteed to perform correctly if all arithmetic performed is without rounding error. This is the case if all numerical values in the input are integers (albeit stored as a number of type *NT*) and if none of the intermediate results exceeds the maximal integer representable by the number type ($2^{53} - 1$ in the case of *doubles*). All intermediate results are sums and differences of input values, in particular, the algorithms do not use divisions and multiplications.

The algorithms have the following arithmetic demands. Let C be the maximal absolute value of any edge cost. If all weights are integral then all intermediate values are bounded by $3C$ in the case of maximum weight matchings and by $4nC$ in the case of the other matching algorithms. We will prove these bounds when we discuss the algorithms. For the sequel let $f = 3$ in the case of the maximum weight matchings and let $f = 4n$ in the other cases.

The pre-instantiations for number type *int* issue a warning if C is larger than $MAXINT/f$.

The pre-instantiations for number type *double* compute the optimal matching for a modified weight function $c1$, where for every edge e

$$c1[e] = sign(c[e])\lfloor |c[e]| \cdot S \rfloor / S$$

and S is the largest power of two such that

$$S < 2^{53}/(f \cdot C).$$

The weight of the optimal matching for the modified weight function and the weight of the optimal matching for the original weight function differ by at most $n \cdot f \cdot C \cdot 2^{-52}$.

The weight modification can also be performed explicitly and we advise you to do so. The functions

```
bool MWBM_SCALE_WEIGHTS(const graph& G, edge_array<double>& c);
bool MWA_SCALE_WEIGHTS( const graph& G, edge_array<double>& c);
```

replace $c[e]$ by $c1[e]$ for every edge e, where $c1[e]$ was defined above and $f = 3$ for the first function and $f = 4n$ for the second function. The first scaling function is appropriate for the maximum weight matching algorithm and the second function is appropriate for all other matching algorithms. The functions return *false* if the scaling changed some weight, and return *true* otherwise.

7.8.2 *Maximum Weight Bipartite Matching: An Algorithm*

We describe an algorithm for maximum weight bipartite matching. The algorithm works iteratively. It starts with the empty matching and the graph spanned by B and the empty subset of A and then adds the nodes in A one by one. After each addition of a node it computes a new maximum weight matching and a new tight non-negative potential function.

Let $a_1, \ldots, a_n$ be an enumeration of the elements in A, let $A_i = \{a_1, \ldots, a_i\}$, let G_i be the subgraph spanned by $V_i = A_i \dot\cup B$, let M_i be a maximum weight matching in G_i, and let $\pi_i : V_i \mapsto I\!R_{\geq 0}$ be a non-negative potential function that is tight with respect to M_i. Our algorithm will construct M_i and π_i for $i = 0, 1, \ldots, n$. We assume that all matching edges are directed from B to A and all non-matching edges are directed from A to B.

M_0 and π_0 are trivial; M_0 is the empty matching and π_0 assigns zero to all nodes in B. Let us also construct M_1 and π_1. Let e be the heaviest edge incident to a_1. If e does not exist or has negative weight then M_1 is empty and π_1 assigns zero to all nodes in V_1. If e has non-negative weight then M_1 consists of e and π_1 assigns $c(e)$ to a_1 and zero to all nodes in B.

Assume now that we know M_{i-1} and π_{i-1} for some i, $i \geq 1$. We show how to construct M_i and π_i. An alternative interpretation of the construction will be given at the end of the section.

We start the construction of M_i and π_i by extending π_{i-1} to a feasible non-negative potential function $\overline{\pi}_i$ for V_i; this can be done by setting $\overline{\pi}_i(a_i)$ to any value that makes the reduced cost of all edges incident to a_i non-negative. Let $M = M_{i-1}$ and $\pi = \overline{\pi}_i$ and observe that M and π satisfy the optimality conditions (1), (2), and (3), and that $a = a_i$ is the only free node which violates (4). We now modify π (maintaining (1), (2), and (3), and (4) for all free nodes different from a) until there is an alternating path of tight edges from a either to a node a' in A having potential zero ($a = a'$ is possible) or to a free node in B. We set $M_i = M \oplus p$ and $\pi_i = \pi$. This re-establishes all four optimality conditions.

The potential function π is modified in phases. In each phase (except the last) we decrease $\pi(V_i)$ and we leave $\pi(V_i)$ unchanged in the last phase.

We now describe a phase. In each phase we determine the set R of nodes that are reachable from $a = a_i$ by tight edges and then distinguish three cases.

R contains a node in A of potential zero: Let v be a node in $A \cap R$ with $\pi(v) = 0$ and let p be a path of tight edges from a to v. We augment M by p, see Figure 7.33, and observe that π is tight with respect to $M \oplus p$. It is conceivable that $v = a$ and p is a path of length zero.

R contains a free node in B: Let w be a free node in $B \cap R$ and let p be a path of tight edges from a to w. We augment M by p, see Figure 7.34, and observe that π is tight with respect to $M \oplus p$.

Neither of the above: We define a value $\delta = \min(\alpha, \beta)$. Let α be the minimal value $\pi(v)$ for any node $v \in R \cap A$ and let β be the minimal value $\overline{c}(e)$ of any edge e leaving R. Then $\alpha > 0$ since R contains no node in A with potential zero and $\beta > 0$ since only non-tight edges can leave R, see Figure 7.35. We decrease the potential of all nodes in $R \cap A$ by δ, we increase the potential of all nodes in $R \cap B$ by δ, and recompute R. We continue in this fashion until one of the first two cases occurs.

The correctness of the method follows from the following lemma.

Lemma 28 *In the first two cases, π is tight with respect to $M \oplus p$. In the third case, the update of π preserves feasibility and non-negativity. The total potential decreases by δ. Moreover, all edges in M stay tight and a_i is the only free node whose potential can be*

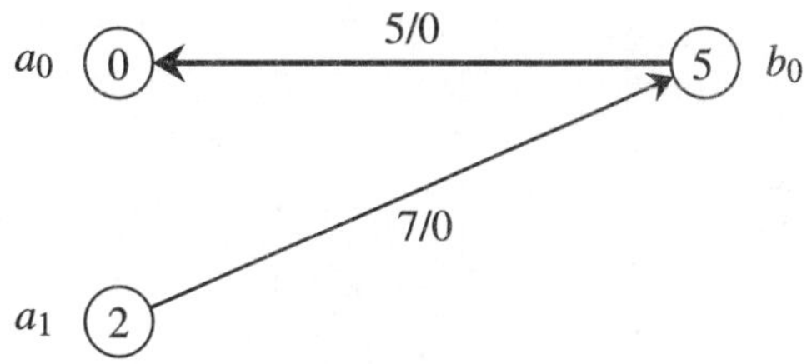

Figure 7.33 The edges of a matching M are shown in bold. The potential of each node is shown inside the node and the cost c and the reduced cost $\overline{c}$ of each edge is shown as $c(e)/\overline{c}(e)$. The path a_1, b_0, a_0 consists of tight edges and can be used for augmentation. The resulting matching has the same cardinality as the current matching.

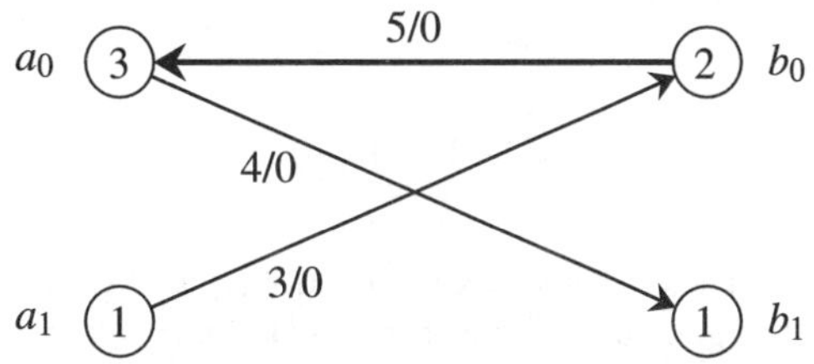

Figure 7.34 The edges of a matching M are shown in bold. The potential of each node is shown inside the node and the cost c and the reduced cost $\overline{c}$ of each edge is shown as $c(e)/\overline{c}(e)$. The path a_1, b_0, a_0, b_1 consists of tight edges and can be used for augmentation. The resulting matching has cardinality one larger than the current matching.

positive after the potential update. After the update there is either a node in $R \cap A$ whose potential is zero (if $\delta = \alpha$) or R grows (if $\delta = \beta$).

Proof In cases 1 and 2 we augment along a path of tight edges. Hence any edge in $M \oplus p$ is tight. Also, in case 1 we expose a node in A that has potential zero. Thus π is tight with respect to $M \oplus p$.

We turn to the third case. We start with a feasible potential function in which all edges in M are tight and in which all free nodes except for a_i have potential zero; a_i may or may not have potential zero. The set R contains one more node in A than in B since every node in R except a is matched and since for every matched edge either both endpoints or no endpoint is in R. Thus a potential update decreases the total potential by δ.

Let e be any edge. We show that the reduced cost of e stays non-negative. The reduced cost of e decreases only if one endpoint lies in $R \cap A$ and the other endpoint lies in $B \setminus R$. Then e is non-matching (since matching edges always have both or no endpoint in R) and hence the reduced cost of e before the potential update is at least β.

All edges in M stay tight since for any edge in M either both endpoints belong to R or neither endpoint does.

No free node in B can get a positive potential since there is no free node in $R \cap B$;

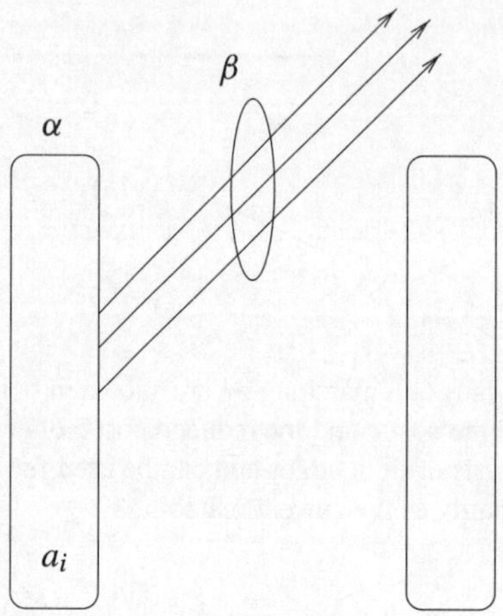

Figure 7.35 R is the set of nodes reachable from a_i by tight edges. The sets $A \cap R$ and $B \cap R$ are indicated as large ovals. The number of nodes in $A \cap R$ is one larger than the number of nodes in $B \cap R$, since each node in $A \cap R \setminus a_i$ is matched to a node in $B \cap R$ and vice versa. β is the minimum reduced cost of any edge leaving R (any such edge has its source node in A since all edges out of B are in M and hence tight) and α is the minimum potential of any node in $A \cap R$, and $\delta = \min(\alpha, \beta)$. We reduce the potential of all nodes in $A \cap R$ by δ and increase the potential of all nodes in $B \cap R$ by δ.

otherwise we would be in case 2. The potential of nodes in A does not increase and hence a_i can stay the only node with positive potential. □

At this point we have arrived at a first version of our algorithm.

```
M = the empty matching;
pot(b) = 0 for all b in B;

forall a ∈ A
{ set pot(a) to some value that makes the reduced cost of all edges incident to a
  non-negative;

  while (true)
  { determine the set R of nodes reachable from a by tight edges.

    if R contains a node in A with potential zero or a free node in B
    { augment by a path of tight edges from a to this node;
      break;
    }
    compute α, β, and δ and adjust the potentials;
  }
}
```

We leave it to the reader to implement the basic version of the algorithm.

Let us take a closer look at the inner loop of this algorithm. It grows a set R until R contains either a free node in B or a node in A with potential zero. Let R_k be the set R in the k-th iteration of the loop for $k = 1, 2, \ldots, K + 1$, let δ_k be the value of δ determined in the k-th iteration, let $\Delta_k = \delta_1 + \ldots + \delta_k$, and let $\Delta = \Delta_K$ be the sum of all δ's. Then the total change of potential of the nodes in $R_k \setminus R_{k-1}$ is $\delta_k + \delta_{k+1} + \ldots = \Delta - (\delta_1 + \ldots + \delta_{k-1})$.

Also $\delta_k = \beta_k < \alpha_k$ for $k < K$ since $\delta_i = \alpha_i$ implies that case 1 occurs in the next iteration and hence $i = K$. Finally, $\delta_K = \alpha_K$ implies that $R_{K+1} = R_K$. We relate the growth of R to a shortest-path computation with source a_i.

Lemma 29 *Let w be any node and let $\mu(w)$ be the shortest-path distance of w from a_i with respect to the reduced costs defined by $\overline{\pi}_i$. Then w is added to R after a total potential change of $\mu(w)$.*

Proof Let w be any node and consider a shortest path p from a_i to w. Let $e_1, e_2, \ldots$ be the edges on p that are not tight initially in the order in which they occur on p. The source node of e_1 belongs to R_1. The reduced cost of e_1 is decreased by δ_1 in the first phase, by δ_2 in the second phase, and becomes zero at the end of some phase, say the $(l - 1)$-th, i.e., the original reduced cost of e_1 was equal to $\delta_1 + \ldots + \delta_{l-1}$. In phase l the source node of e_2 belongs to R and the next potential updates reduce the cost of e_2 to zero. In this way w is added to R after a total potential change of $\mu(w)$. □

Lemma 30 *Let $\pi = \overline{\pi}_i$ and for any node w let $\mu(w)$ be the shortest-path distance of w from a_i with respect to the reduced costs defined by π. Let minA $= \min\{\mu(a) + \pi(a)\ ;\ a \in A\}$ and let minB $= \min\{\mu(b)\ ;\ b \in B$ and b is free$\}$. Then $\Delta = \min(minA, minB)$ and the total potential change for any node v is equal to $\max(0, \Delta - \mu(v))$.*

If $\Delta = minA$, let z be the node that defines minA and if $\Delta = minB$, let z be the node that defines minB (if minA = minB, define z by either half-sentence). In either case let p be a path of length $\mu(z)$ from a_i to z. Then all edges of p are tight after the change of potential.

Proof Consider an arbitrary node $a \in A$. It is added to R when the total potential change is equal to $\mu(a)$. Subsequent potential changes decrease $\pi(a)$ and hence the total potential change cannot be more than $\mu(a) + \pi(a)$ (since node potentials always stay non-negative).

Consider a free node $b \in B$. It is added to R when the total potential change is equal to $\mu(b)$. Thus the total potential change cannot be more than $\mu(b)$.

We stop changing the potentials once a node in $A \cap R$ reaches potential zero or a free node in B is added to R. Thus the total potential change Δ is equal to $\min(minA, minB)$.

A node v participates in potential changes after it has been added to R. Thus the total change of potential of v is equal to $\max(0, \Delta - \mu(v))$.

Let p be as defined in the statement of the lemma and let $e = (v, w)$ be any edge of p. Then $\mu(v) + \overline{c}(e) = \mu(w)$ since p is a shortest path. Also $\mu(v), \mu(w) \le \Delta$ since p is a shortest path to the node that defines Δ. We show that e is tight after the potential change.

If e is matching and hence $v \in B$ and $w \in A$ and $\overline{c}(e) = 0$, we have $\mu(v) = \mu(w)$. Thus $\pi(v)$ is increased by $\Delta - \mu(v)$ and $\pi(w)$ is decreased by the same amount. Thus e stays tight.

If e is non-matching and hence $v \in A$ and $w \in B$, we have $\mu(v) + \overline{c}(e) = \mu(w)$. Thus $\pi(w)$ is increased by $\Delta - \mu(w) = \Delta - \mu(v) - \overline{c}(e)$ and $\pi(v)$ is decreased by $\Delta - \mu(v)$. The reduced cost of e is therefore reduced by $\overline{c}(e)$ and hence e becomes tight. □

Lemmas 29 and 30 allow us to refine our basic algorithm.

M = the empty matching;
$pot(b) = 0$ for all b in B;

forall $a \in A$
{ set $pot(a)$ to some value that makes the reduced cost of all edges incident to a non-negative;

for any node v let $dist(v)$ be the shortest-path distance of v from a and let

$minA = \min\{dist(v) + pot(v) \; ; \; v \in A\}$;
best_node_in_A = a node in A that defines *minA*;

$minB = \min\{dist(v) \; ; \; v \in B$ and free$\}$;
best_node_in_B = a node in B that defines *minB*;
$Delta = \min(minA, minB)$;

forall $v \in A$: $pot(v) = pot(v) - \max(0, Delta - dist(v))$;
forall $v \in B$: $pot(v) = pot(v) + \max(0, Delta - dist(v))$;

augment by the alternating path of tight edges from a to *best_node_in_A*, if $Delta = minA$, and from a to *best_node_in_B*, otherwise;
}

The description above suggests that it is necessary to compute $dist[v]$ for all nodes v in each execution of the inner loop. This is not true. It is only necessary to compute *Delta* and the node defining it and to compute $dist[v]$ for all nodes v with $dist[v] < Delta$. Given this information all potentials can be updated correctly and the augmentation can be made.

How can we compute *Delta* without computing $dist[v]$ for all nodes v? We exploit the fact that Dijkstra's algorithm computes dist-values in increasing order. Let $v_0, v_1, \ldots$ with $v_0 = a$ be the order in which the nodes are reached by the shortest-path computation. Then $dist[v_0] \le dist[v_1] \le \ldots$. We observe:

(1) If

$$\min\{dist[v_i] + pot[v_i] \; ; \; i < k \text{ and } v_i \in A\} \le dist[v_k]$$

then some v_i with $i < k$ and $v_i \in A$ defines *minA*. This follows from the fact that all node potentials are non-negative.

(2) *minB* is the dist-value of the first free node in B that is reached by the shortest-path computation. If no v_j with $j < k$ is a free node in B then $dist[v_k] \le minB$.

(3) If

$$\min\{dist[v_i] + pot[v_i]\,;\, i < k \text{ and } v_i \in A\} \le dist[v_k]$$

and no v_j with $j < k$ is a free node in B then $Delta = minA$. This follows from (1) and (2).

(4) If

$$\min\{dist[v_i] + pot[v_i]\,;\, i < k \text{ and } v_i \in A\} \ge dist[v_k]$$

and v_k is a free node in B then $Delta = minB$. This follows from (1) and (2).

(5) Let k be minimal such that either (3) or (4) holds. Then $Delta \le dist[v_j]$ for all $j > k$ and the potentials of all nodes v_j with $j > k$ are not changed.

We will use items (3) and (4) as the stopping criteria for the shortest-path computation in our implementation. Item (5) implies that only nodes that are reached by the shortest-path computation can be affected by the potential change.

7.8.3 *Maximum Weight Bipartite Matching: An Implementation*

After all this preparatory work we are ready for the implementation.

We start by declaring the data structures required by the algorithm, then use one of three heuristics to initialize the potential function and the matching, then call *augment*$(a, \ldots)$ for each node in A that is left unmatched by the heuristic, and finally restore the graph and prepare the list of edges comprising the matching.

The data structures used by the algorithm are two boolean arrays to keep track of the free nodes and the nodes in A and the data structures needed for the shortest-path computations (arrays *pred* and *dist*, and a node priority queue *PQ*).

We describe three heuristics. The simplest heuristic (called naive in the program below) sets the potential of all nodes in B equal to zero, the potential of all nodes in A equal to the maximal cost of all edges, and sets the matching to the empty matching. The other heuristics are described later in the section.

⟨*mwb_matching.t*⟩+≡

```
  ⟨mwb_matching: helpers⟩
  static int which_heuristic = 2;
  template <class NT>
  list<edge> MAX_WEIGHT_BIPARTITE_MATCHING_T(graph& G,
                            const list<node>& A, const list<node>& B,
                            const edge_array<NT>& c, node_array<NT>& pot)
  { node a,b,v; edge e;
    list<edge> result;
    forall_nodes(v,G) pot[v] = 0;
    if (G.number_of_edges() == 0 ) return result;
    // check that all edges are directed from A to B
    forall(b,B) assert(G.outdeg(b) == 0);
    node_array<bool> free(G,true);
```

```
  node_array<edge> pred(G,nil);
  node_array<NT>   dist(G,0);
  node_pq<NT>      PQ(G);
  switch (which_heuristic)
  { case 0: { // naive heuristic
              NT C = 0;
              forall_edges(e,G) if (c[e] > C) C = c[e];
              forall(a,A)       pot[a] = C;
              break;
            }
    case 1: { // simple heuristic
              ⟨simple heuristic⟩
              break;
            }
    default: { // refined heuristic
              mwbm_heuristic( G, A, c, pot, free);
              break;
            }
  }
  forall(a,A)
    if (free[a]) augment(G,a,c,pot,free,pred,dist,PQ);
  forall(b,B)
    { forall_out_edges(e,b) result.append(e); }
  forall(e,result) G.rev_edge(e);
  return result;
}
```

We give the details of *augment*$(G, a, \ldots)$. It is a variant of Dijkstra's algorithm.

```
⟨mwb_matching: helpers⟩≡
  ⟨procedure augment_path_to⟩
  template <class NT>
  inline void augment(graph& G, node a, const edge_array<NT>& c,
                      node_array<NT>& pot, node_array<bool>& free,
                      node_array<edge>& pred, node_array<NT>& dist,
                      node_pq<NT>& PQ)
  { ⟨augment: initialization⟩
    while ( true )
    { ⟨select from PQ the node b with minimal distance db⟩
      ⟨distinguish three cases⟩
    }
   ⟨augment: potential update and reinitialization⟩
  }
```

We compute shortest paths starting in a. The priority queue PQ contains nodes in B (we will explain shortly why nodes in A are not put into the queue) together with their tentative distance from a, *minA* contains the minimum value of $\{\mu(v) + \pi(v)\ ;\ v \in A\}$ that we have seen so far, and *best_node_in_A* contains a node realizing *minA*. We use an array *dist* to record distances and an array *pred* to record predecessor edges in the shortest-path tree; this is as in Section 7.5.

Initially, the distance of a is zero, *minA* is equal to the potential of a, *best_node_in_A* is equal to a, and *PQ* contains all neighbors of a (recall that we store only nodes in B in the priority queue).

We do not define *PQ* within *augment* nor do we initialize *pred* within *augment*. This is absolutely vital for efficiency. We assume that *PQ* is empty and *pred*$[v]$ = *nil* for all v when *augment* is called. Within *augment* we collect, in stacks *RA* and *RB*, all nodes v (in A and B, respectively) that are added to *PQ* or for which *pred*$[v]$ is set. At the end of *augment* we use these stacks to reset *PQ* and *pred*. In this way augmentations can have sublinear running time.

⟨*augment: initialization*⟩≡

```
dist[a] = 0;

node best_node_in_A = a;
NT   minA           = pot[a];
NT   Delta;

stack<node> RA;  RA.push(a);
stack<node> RB;

node a1 = a; edge e;
⟨relax all edges out of a1⟩
```

where

⟨*relax all edges out of a1*⟩≡

```
forall_adj_edges(e,a1)
{ node b = G.target(e);
  NT db = dist[a1] + (pot[a1] + pot[b] - c[e]);
  if ( pred[b] == nil )
  { dist[b] = db; pred[b] = e; RB.push(b);
    PQ.insert(b,db);
  }
  else
  if ( db < dist[b] )
  { dist[b] = db; pred[b] = e;
    PQ.decrease_p(b,db);
  }
}
```

For each edge $e = (a1, b)$ we compute *db* as *dist*$[a1]$ plus the reduced cost of e. If b is reached for the first time, we add it to *PQ* and to *RB*, and if w has been reached before but *db* is smaller than the current distance value of b, we update the distance value accordingly. We will reuse the program chunk above below and hence have formulated it for an arbitrary node $a1$ in A. In the main loop we remove the node with smallest distance from *PQ*. Let b be this node and let *db* be its distance; b is a node in B.

⟨*select from PQ the node b with minimal distance db*⟩≡

```
node b;
NT db;
if (PQ.empty()) b = nil;
else { b = PQ.del_min(); db = dist[b]; }
```

We distinguish three cases according to the discussion at the end of Section 7.8.2.

If *b* does not exist, i.e., *PQ* is empty, or $db \geq minA$, we augment by a path to node *best_node_in_A*. *Delta* is equal to *minA*.

If *b* exists, $db < minA$, and *b* is free, we augment by a path to *b*. *Delta* is equal to *db*.

If *b* exists, $db < minA$, and *b* is matched, we continue the shortest-path computation.

⟨*distinguish three cases*⟩≡

```
if ( b == nil || db >= minA )
{ Delta = minA;
  ⟨augmentation by path to best node in A⟩
}
else
{ if ( free[b] )
  { Delta = db;
    ⟨augmentation by path to b⟩
  }
  else
  { ⟨continue shortest-path computation⟩ }
}
```

Augmentation to the best node in *A* is done by *augment_path_to*(*best_node_in_A*, ...), which simply reverses the direction of all edges on the path from *a* to *best_node_in_A*. The path is given by the *pred*-array. We also declare *a* matched and *best_node_in_A* unmatched. It is important that we do the latter actions in this order, since *a* may be the best node in *A*, in which case we do not want to change the current matching.

⟨*augmentation by path to best node in A*⟩≡

```
augment_path_to(G,best_node_in_A,pred);
free[a] = false; free[best_node_in_A] = true; // order is important
break;
```

where

⟨*procedure augment_path_to*⟩≡

```
inline void augment_path_to(graph& G, node v,
                            const node_array<edge>& pred)
{ edge e = pred[v];
  while (e)
  { G.rev_edge(e);
    e = pred[G.target(e)]; // not source (!!!)
  }
}
```

Augmentation by a path to b is equally simple. We augment and declare a and b matched.

⟨*augmentation by path to b*⟩≡

```
augment_path_to(G,b,pred);
free[a] = free[b] = false;
break;
```

We come to the case where the shortest-path computation is to be continued. Then b is matched. Let e be the matching edge incident to b and consider the mate $a1$ of b. The mate has the same distance value as b and its predecessor edge is e.

If $db + pot[a1]$ is smaller than $minA$ we update $minA$ and $best\_node\_in\_A$.

We also relax the edges out of $a1$. This may put more nodes in B into PQ. Observe that only nodes in B are put into PQ.

⟨*continue shortest-path computation*⟩≡

```
e = G.first_adj_edge(b);
node a1 = G.target(e);
pred[a1] = e; RA.push(a1);
dist[a1] = db;
if (db + pot[a1] < minA)
{ best_node_in_A = a1;
  minA = db + pot[a1];
}
⟨relax all edges out of a1⟩
```

This completes the description of the main loop.

We break from the main loop as soon as an augmenting path has been found. At this point $RA \cup RB$ contains all nodes that have been reached in the shortest-path computation and $Delta$ contains the value required for the potential updates. For each node v in $RA \cup RB$ we reset $pred[v]$ to nil, remove v from the priority queue (only nodes in B can be in the queue), and update its potential. The potential change is $\max(0, Delta - dist[v])$. It is a decrease for the nodes in A and an increase for the nodes in B. For the nodes outside $RA \cup RB$ the potential does not change (by item (5) of the discussion at the end of Section 7.8.2).

⟨*augment: potential update and reinitialization*⟩≡

```
while (!RA.empty() )
{ node a = RA.pop();
  pred[a] = nil;
  NT pot_change = Delta - dist[a];
  if (pot_change <= 0 ) continue;
  pot[a] = pot[a] - pot_change;
}
while (!RB.empty() )
{ node b = RB.pop();
  pred[b] = nil;
  if (PQ.member(b)) PQ.del(b);
  NT pot_change = Delta - dist[b];
```

```
  if (pot_change <= 0 ) continue;
  pot[b] = pot[b] + pot_change;
}
```

We come to the heuristics.

The simple heuristic sets *pot*[*a*] to the largest non-negative cost of any edge incident to a for every $a \in A$. This will make the heaviest edge incident to a tight (since the potential of all nodes in B is initially zero). The edge is added to the matching iff its endpoint in B is free.

⟨*simple heuristic*⟩≡

```
forall(a,A)
{ edge e_max = nil; NT C_max = 0;
  forall_adj_edges(e,a)
    if (c[e] >  C_max) { e_max = e; C_max = c[e]; }
  pot[a] = C_max;
  if ( e_max != nil && free[b = G.target(e_max)] )
  { G.rev_edge(e_max);
    free[a] = free[b] = false;
  }
}
```

The refined heuristic augments along paths of length one and length three. When it is called, the potential of all nodes in B is zero. It considers the nodes in A in turn. For each node $a \in A$ it determines the two incident edges with largest non-negative reduced cost. Call them *eb* and *e2*, respectively, and their reduced costs *max* and *max2*, respectively. If *e2* does not exist, then $max2 = 0$, and if *eb* does not exist, then $max = 0$.

We then distinguish cases. If *eb* does not exist, we set *pot*[*a*] to zero. If *eb* exists, let b be the target of *eb*. If b is free, we add *eb* to the matching, record *e2* as the second best edge of a, and set *pot*[*a*] to *max2* and *pot*[*b*] to *max* - *max2*. This makes *eb* tight, and it makes *e2* tight if it leads to a free node in B. Finally, if b is not free we set *pot*[*a*] to *max* and consider the second best edge, say e, incident to the mate of b. If e exists and the target of e is free, we use the path of length three for augmentation.

⟨*mwb_matching: helpers*⟩+≡

```
template <class NT>
void mwbm_heuristic(graph& G, const list<node>& A,
                    const edge_array<NT>& c, node_array<NT>& pot,
                    node_array<bool>& free)
{
  node a, b;  edge e, e2, eb;
  node_array<edge> sec_edge(G,nil);
  forall( a, A )
  { NT max2 = 0; NT max = 0; eb = e2 = nil;
    // compute edges with largest and second largest slack
    forall_adj_edges( e, a )
    { NT we = c[e] - pot[target(e)];
```

```
        if ( we >= max2 )
        { if( we >= max )
          { max2 = max;   e2 = eb;
            max = we;     eb = e;
          }
          else
          { max2 = we;    e2 = e;
          }
        }
      }
      if( eb )
      { b = target(eb);
        if( free[b] )
        { // match eb and change pot[] to make slack of e2 zero
          sec_edge[a] = e2;
          pot[a] = max2;
          pot[b] = max-max2;
          G.rev_edge(eb);
          free[a] = free[b] = false;
        }
        else
        { // try to augment matching along
          // path of length 3 given by sec_edge[]
          pot[a] = max;
          e2 = G.first_adj_edge(b);
          e = sec_edge[target(e2)];
          if( e && G.outdeg(target(e)) == 0 )
          { free[a] = free[G.target(e)] = false;
            G.rev_edge(e); G.rev_edge(e2); G.rev_edge(eb);
          }
        }
      }
      else pot[a] = 0;
    }
  }
```

The worst case running time of our matching algorithm is n times the worst case running time of the shortest-path computation. The worst case running time of the shortest-path computation depends on the implementation of the priority queue. Priority queues are discussed in Section 5.4. With either the Fibonacci heap or the pairing heap implementation we obtain a worst case running time of $O(n(m + n \log n))$ and with the redistributive heap implementation we obtain a worst case running time of $O(n(m + n \log C))$ where C is the largest edge weight (edge weights are assumed to be integral for the latter time bound). The implementation given has worst case running time $O(n(m + n \log n))$. The average case running time seems to be much better as Table 7.8 shows.

Arithmetic Demand: How large are the numbers that are handled by the program above? Let us assume that all edge weights are integers whose absolute value is bounded by C.

We observe first that all node potentials are non-negative integers less than or equal to C.

This is clear for the nodes in A since their potential is initialized to a value less than or equal to C and is only decreased afterwards. For the nodes in B it follows from the observation that the potential of any matched node is at most C (since the reduced cost of a matched edge is zero) and that the potential of free nodes in B is zero.

The fact that node potentials are bounded by C implies that the reduced cost of any edge is bounded by $2C$. Thus the largest number handled in any of the shortest-path computations is at most $2\min(|A|, |B|) \cdot C$. This bound holds since matched edges have reduced cost zero and hence no simple path can contain more than $\min(|A|, |B|)$ edges of non-zero reduced cost.

We will next establish a much better bound. The quantity *minA* is always bounded by C, since it is initialized to the potential of a node in A and is only decreased afterwards. The shortest-path computation stops as soon as a distance value larger than *minA* is selected from the queue. Thus only distance values less than *minA* (and hence less than C) can lead to the insertion of additional distance values into the queue. We conclude that the maximal value ever put into the queue is bounded by C plus the maximal reduced cost of any edge and is hence bounded by $3C$. We summarize.

Lemma 31 *If all edge weights are integers whose absolute value is bounded by C then the largest number handled by the maximum weight bipartite matching algorithm is bounded by $3C$.*

Experimental Data: Table 7.8 contains some running times. We used random bipartite graphs with n nodes on each side and m edges, and three different kinds of edge weights:

- Uniform edge weights, i.e., all edge weights equal to one.
- Random edge weights in [1..1000].
- Large random edge weights in [10000..10005].

In all cases we also solved the corresponding unweighted matching problem.

The instances with random edge weights are by far the simplest, followed by the instances with large random edge weights, followed by the uniform instances. We expected that random edge weights from a large range lead to simple problems because heavy edges are much more favorable than light edges. We were surprised to find that the uniform problems are the hardest and have no explanation for it.

The density of the problem has a big influence on running time. For very sparse problems ($m = 2n$) the weighted matching algorithm is faster than the unweighted matching algorithm. This is due to the use of the potential function.

Consider the graph shown in Figure 7.36. It consists of a connected graph H which has a perfect matching and additional nodes $a_1, a_2, \ldots, a_k$. Each a_1 is connected to a node on the B-side of H. In the figure, all a_i are connected to the same node in B, but this is not essential. Assume that the perfect matching in H has already been constructed and that the nodes $a_1, a_2, \ldots, a_k$ are considered in turn. In the unweighted matching algorithm

C	n	m	No	Simple	Refined	Check	Unweighted
U	20000	40000	0.995	0.997	0.994	0.186	2.633
U	20000	60000	61.1	60.41	58.43	0.213	3.679
U	20000	80000	116.2	114.2	109.9	0.239	6.248
U	40000	80000	2.139	2.153	2.144	0.39	6.791
U	40000	120000	212.2	210.3	204.3	0.4539	9.61
U	40000	160000	410	402.8	387.8	0.5081	9.217
R	20000	40000	0.84	0.849	0.8467	0.1836	2.73
R	20000	60000	1.399	1.401	1.391	0.2189	3.811
R	20000	80000	2.635	2.509	2.578	0.2402	6.32
R	40000	80000	1.812	1.82	1.817	0.3922	7.056
R	40000	120000	3.001	2.941	2.973	0.4621	9.855
R	40000	160000	5.667	5.364	5.512	0.5168	9.532
L	20000	40000	1.293	1.31	1.307	0.1838	2.811
L	20000	60000	20.84	20.89	20.65	0.2305	3.922
L	20000	80000	41.6	40.69	41.05	0.2529	6.726
L	40000	80000	2.815	2.816	2.816	0.4213	7.222
L	40000	120000	57.06	56.9	54.67	0.4834	9.98
L	40000	160000	116.5	113.9	103.1	0.5283	9.595

Table 7.8 The running times of three versions of the weighted bipartite matching algorithm. The first three columns contain the running times of the algorithm above with the three different heuristics, the fourth column shows the time to verify the result and the last column shows the time required to solve the unweighted problem (by *MAX_CARD_BIPARTITE*). The graphs were generated by $random\_bigraph(G, n, n, m, A, B)$ and three kinds of edge weights were used: uniform edge weights (denoted U), i.e., all edge weights were set to one, random edge weights (denoted R) in [1 .. 1000] and random edge weights (denoted L) in [10000 .. 10005]. Each number is the average of ten runs. The function mwb_matching_time in the demo directory allows readers to perform their own experiments.

every search for an augmenting path will explore H in its entirety. Not so in the weighted matching algorithm. After the search from a_1, a_1 will have potential equal to zero (since it is free) and hence the node in B connected to it will have potential equal to one. Since H is assumed to be connected, every node in $H \cap B$ will have potential equal to one. Consider next a search for an augmenting path starting at a_i, $i \geq 2$. The node a_i is given potential one (since one is the largest cost of an edge incident to a_i), and hence all edges out of a_i have

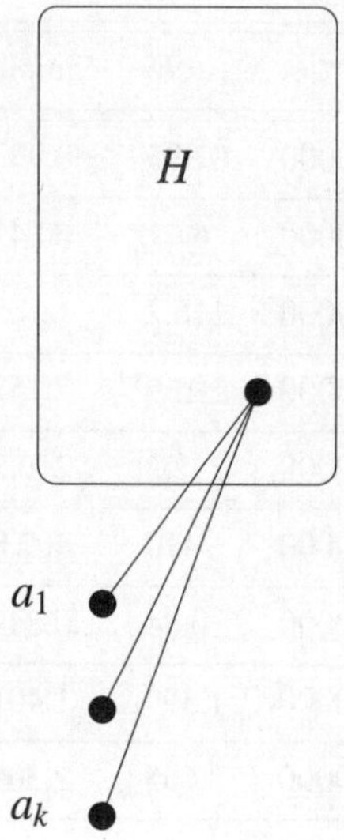

Figure 7.36 H has a perfect matching and each a_i, $1 \le i \le k$ is connected to some node on the B-side of H. After the search for an augmenting path from a_1 all nodes in B will have potential one. The searches from a_i, $i > 2$, take constant time.

reduced cost one. When the first neighbor of a_i is removed from the queue (with distance value one), the condition `dv >= minA` holds and hence the search for an augmenting path terminates. In this way, the fact that a_i cannot be matched is detected in time $O(1)$. We conclude that node potentials help tremendously in the example of Figure 7.36. Of course, this example is very special and hence we need to generalize the argument.

Our algorithm considers the nodes in A in turn. Let $A_{i-1} = \{a_1, \ldots, a_{i-1}\}$. After having considered the nodes in A_{i-1}, it has computed a maximal matching M_{i-1} in the subgraph G_{i-1} spanned by $A_{i-1} \cup B$ and a potential function π_{i-1} which proves the optimality of M_{i-1}. Observe now that a node in B which can be reached from a free node in A must have potential one (since free nodes have potential zero and hence their neighbors have potential one, and hence the neighbors of the neighbors have potential zero, ...).

Consider now the search for an augmenting path from a_i. We claim that it will not enter the subgraph H of G_{i-1} consisting of all nodes that can be reached from a free node in A_{i-1}. This is most easily seen for what we called the basic version of the algorithm in Section 7.8.2. We observe first that the node a_i is given potential one (since one is the largest cost of an edge incident to a_i) and hence an edge (a_i, b) will have reduced cost equal to zero or equal to one depending on whether the potential of b is zero or one. The edges connecting a_i to nodes in H will have reduced cost equal to one. The search will first explore all nodes that can be reached by tight edges. If a free node in B is reached, the matching will be increased. If no free node in B can be reached, a potential change will be made. The change reduces the potential of a_i to zero and hence no further search will be performed. We conclude that H is never entered.

For random edge weights the weighted matching algorithm is faster than the unweighted matching algorithm on the corresponding unweighted problem.

Alternative Interpretation: We close this section with an alternative interpretation of our algorithm. The alternative interpretation may be skipped.

We consider only the construction of M_i from M_{i-1} and π_{i-1}. For any alternating path starting in $a = a_i$ let $d(p)$ be the total cost of the edges in p that belong to M minus the total cost of the edges that do not belong to M, i.e.,

$$d(p) = \sum_{e \in p \cap M} c(e) - \sum_{p \setminus M} c(e).$$

Consider the matching $M \oplus p$ obtained by augmenting M by p. It has cost $c(M) - d(p)$ and hence $M \oplus p$ is "better" than M iff $d(p)$ is negative. This observation suggests the following definition. We call a path p improving with respect to M if $d(p)$ is negative. The observation also suggests the following algorithm for finding an improving path.

We orient all matching edges from B to A and all non-matching edges from A to B. We assign weight $c(e)$ to any matching edge and assign weight $-c(e)$ to any non-matching edge and search for a path of negative cost starting in a. If there is no such path then M is also a maximum cost matching in G_i. If there is such a path then let p be the most negative such path, i.e., the one with the most negative $d(p)$, and obtain M_i by augmenting M by p. A simple way to find p is to solve a single-source shortest-path problem with source a.

The previous paragraph leaves many questions unanswered. Why is M also a maximum cost matching in G_i if no path of negative cost exists, why is $M \oplus p$ a maximum cost matching in G_i if p is a most negative path, and why can there be no negative cycles?

In answering these questions the potential function $\pi = \pi_{i-1}$ comes handy. Recall that the first action in the construction of M_i is to extend π to a potential function on $A_i \cup B$ by setting $\pi(a)$ to any value that makes the reduced cost of every edge out of a non-negative. Consider any alternating path p with respect to M starting in a. Let $p = [e_1, \ldots, e_k]$ with $e_j = (v_{j-1}, v_j)$. Then v_0 is equal to a, $v_0, v_2, \ldots$ are nodes in A, $v_1, v_3, \ldots$ are nodes in B, $e_1, e_3, \ldots$ are edges not in M and $e_2, e_4, \ldots$ are edges in M, and if k is odd, then v_k is a free node in B. We have

$$d(p) = \sum_{j;j \text{ even}} c(e_j) - \sum_{j;j \text{ odd}} c(e_j).$$

Since π is tight with respect to M, we have $c(e_j) = \pi(v_{j-1}) + \pi(v_j)$ for all even j. Thus

$$\begin{aligned} d(p) &= \sum_{j;j \text{ even}} (\pi(v_{j-1}) + \pi(v_j)) - \sum_{j;j \text{ odd}} c(e_j) \\ &= -\pi(a) + \sum_{j;j \text{ odd}} (\pi(v_{j-1}) + \pi(v_j) - c(e_j)) + (-1)^k \pi(v_k) \\ &= -\pi(a) + \sum_{j;j \text{ odd}} \overline{c}(e_j) + (-1)^k \pi(v_k) \;=\; -\pi(a) + \sum_j \overline{c}(e_j) + (-1)^k \pi(v_k) \end{aligned}$$

$$= \quad -\pi(a) + \sum_j \overline{c}(e_j) + \pi(v_k).$$

This derivation deserves explanation. The first equality amounts to rearranging the sum. For example, if $k = 4$ then

$$-c(e_1) + (\pi(v_1) + \pi(v_2)) - c(e_3) + (\pi(v_3) + \pi(v_4)) =$$
$$-\pi(a) + (\pi(v_0) - c(e_1) + \pi(v_1)) + (\pi(v_2) - c(e_3) + \pi(v_3)) + \pi(v_4)$$

and if $k = 3$ then

$$-c(e_1) + (\pi(v_1) + \pi(v_2)) - c(e_3) =$$
$$-\pi(a) + (\pi(v_0) - c(e_1) + \pi(v_1)) + (\pi(v_2) - c(e_3) + \pi(v_3)) - \pi(v_3).$$

The second equality follows from $\overline{c}(e) = \pi(v) + \pi(w) - c(e)$ for any edge $e = (v, w)$, the third equality follows from the fact that $\overline{c}(e) = 0$ for any $e \in M$, and the last equality follows from the fact that $\pi(v_k) = 0$ if k is odd (since in this case v_k is a free node in B).

The derivation above is extremely powerful. It tells us that $d(p)$ is equal to the cost of p with respect to the reduced costs $\overline{c}$ plus the potential of the target node of p minus the potential of the source node of p. The source node of p is equal to a and hence the latter contribution is independent of p. In other words, searching for a path that minimizes $d(p)$ amounts to searching for a path that minimizes $\overline{c}(p) + \pi(v_k)$. For fixed v_k this amounts to searching for the path p from a to v_k that minimizes $\overline{c}(p)$. This problem is easily solved by Dijkstra's algorithm. For any node $v \in V_i$ let $\mu(v)$ be the minimum cost of a path from a to v with respect to the cost function $\overline{c}$. The iterative step from $M = M_{i-1}$ to M_i is then performed as follows:

Compute $\mu(v)$ for all v by Dijkstra's algorithm.

Let v be the node that minimizes $d = -\pi(a) + \mu(v) + \pi(v)$ and let p be a path from a to v that realizes $\mu(v)$.

If $d < 0$, augment M by p.

This completes our alternative derivation of the algorithm.

The first algorithm for the assignment problem was given by Kuhn [Kuh55]. In the early 60's, Jewell [Jew58], Iri [Iri60] and Busacker and Gowen [BG61] observed that the assignment problem can be solved by a sequence of shortest-path computations in general graphs. In the early 70's Tomizawa [Tom71] and Edmonds and Karp [EK72] showed that the use of node-potentials restricts the shortest-path computations to non-negative edge costs. Recent surveys of algorithms for the assignment problem can be found in an article by Galil [Gal86] and the book by Ahuja, Magnanti, and Orlin [AMO93]. In his master's thesis Markus Paul [Pau89] extended the algorithms to the maximum weight matching problem; he also implemented the algorithm for LEDA. His implementation always searched for augmenting paths from all nodes in A. Uli Finkler [Fin97] observed, in his PhD-thesis, that substantial improvements (not asymptotically but on average) can be obtained by considering the nodes in A one by one. The implementation given here follows his suggestion.

7.8.4 *The Assignment Problem*

The assignment problem asks for a perfect matching of maximum or minimum weight. A simple modification of the algorithm of the preceding section solves the maximum weight assignment problem.

We only need to change the way we search for augmenting paths. We insist that every augmentation increases the size of the matching and hence we continue our search for an augmenting path until a free node in B is found. When no free node in B is ever found, we return false to indicate that the graph has no perfect matching.

We obtain:

⟨*procedure augment for max weight assignment*⟩≡

```
#include <LEDA/stack.h>

template <class NT>
bool max_weight_assignment_augment(graph& G,
                    node a, const edge_array<NT>& c,
                    node_array<NT>& pot, node_array<bool>& free,
                    node_array<edge>& pred, node_array<NT>& dist,
                    node_pq<NT>& PQ)
{ ⟨augment: initialization⟩
  while ( true )
  { node b; NT db;
    if (PQ.empty()) { return false; }
    else { b = PQ.del_min(); db = dist[b]; }

    if ( free[b] )
    { Delta = db;
      ⟨augmentation by path to b⟩
    }
    else
    {⟨continue shortest-path computation⟩ }
  }
 ⟨augment: potential update and reinitialization⟩
 return true;
}
```

The minimum weight assignment problem is easily reduced to the maximum weight assignment problem. We only have to change the sign of all weights.

⟨*mwb_matching.t*⟩+≡

```
template <class NT>
list<edge> MIN_WEIGHT_ASSIGNMENT_T(graph& G,
                    const list<node>& A, const list<node>& B,
                    const edge_array<NT>& c, node_array<NT>& pot)
{ edge_array<NT> w(G);
  edge e;
  forall_edges(e,G) w[e] = - c[e];

  list<edge> M = MAX_WEIGHT_ASSIGNMENT_T(G,A,B,w,pot);
  node v;
```

```
    forall_nodes(v,G) pot[v] = -pot[v];
    return M;
  }
```

The worst case running time of the maximum and minimum weight assignment algorithms is the same as for the maximum weight bipartite matching algorithm, namely $O(n(m + n \log n))$.

Arithmetic Demand: How large are the numbers that are handled by the assignment algorithms? We assume that all edge weights are integers whose absolute value is bounded by C. Let $k = |A| = |B|$.

We will first derive a bound on the node potentials. Let v be any node and consider a change[26] of $\pi(v)$. After a change of $\pi(v)$ there is an undirected path p of tight edges from a node $b \in B$ that was just matched to v. Let $p = [b = v_0, v_1, \ldots, v_s = v]$, where $s \le 2k$. We claim that $\pi(v_i) \in [-iC \mathinner{..} iC]$ for all i after the potential update. This is true for $i = 0$, since b was just matched and hence has potential equal to zero. For $i > 0$ the claim follows from the fact that the edge $\{v_i, v_{i-1}\}$ has reduced cost equal to zero and cost in $[-C \mathinner{..} C]$. We conclude that $\pi(a) \in [-(2k-1)C \mathinner{..} (2k-1)C]$ for $a \in A$ and $\pi(b) \in [-(2k-2)C \mathinner{..} (2k-2)C]$ for $b \in B$ after a potential change. These bounds also hold before the first change of $\pi(v)$ since the potential of nodes in B is initialized to zero and since the potential of nodes in A is initialized such that there is a tight edge incident to the node.

The reduced cost of any edge is therefore bounded by $C + (2k-1)C + (2k-2)C \le (4k-2)C$.

When we search for an augmenting path from a free node $a \in A$ we start a shortest-path computation from a. The computation stops when the first free node in B is encountered. Let p be an augmenting path from a to a free node in B. The maximal number handled in the shortest-path calculation is the cost of p (with respect to the reduced cost function) plus the maximal reduced cost of any edge. The cost of p is the difference between the old and the new potential of a and is therefore bounded by $4kC$. We conclude that the absolute value of all integers handled by the algorithm is bounded by $8kC$.

We summarize.

Lemma 32 *If all edge weights are integers whose absolute value is bounded by C then the absolute value of all numbers handled by the maximum and minimum weight assignment algorithm is bounded by $8kC = 4nC$, where $k = |A| = |B|$ and $n = 2k$.*

7.8.5 *Shortest Paths via Assignment*

Our algorithms for the maximum weight matching problem and the assignment problem use an algorithm for the shortest-path problem (for non-negative edge weights) as a subroutine. We show in this section that any algorithm for the assignment problem can be used to solve

[26] We will derive a bound on the initial value of $\pi(v)$ later in the section.

n	m	D	A	BF	A	BF	A
5000	50000	0.76	2.51	2.51	185.9	0.67	2.01

Table 7.9 A comparison of the running time of the shortest path via assignment algorithm (denoted A) with the shortest-path algorithms of Section 7.5. Columns three and four contain a comparison with Dijkstra's algorithm (D) and columns five and six and seven and eight contain a comparison with the Bellman–Ford algorithm (BF). We used random graphs with non-negative edge weights for the first comparison, random graphs with arbitrary edge weights but no negative cycle for the second comparison, and graphs generated by *BF_GEN* for the third comparison. The program shortest_path_via_assignment_time in the demo directory allows readers to perform their own experiments.

the shortest-path problem with arbitrary edge weights. This will give us an alternative to the algorithms in Section 7.5. The alternative is of considerable theoretical interest and has led to the asymptotically most efficient shortest-path algorithm for arbitrary edge costs, see [AMO93, sections 12.4 and 12.7]. We wrote this section to find out whether it also leads to efficient programs. At least in our implementation it does not, see Table 7.9.

Let $G = (V, E)$ be a directed graph. We construct a bipartite network $G' = (V' \dot{\cup} V'', E')$; see Figure 7.37 for an illustration. G' contains two copies of each node of G, one in V' and one in V''. For each node $v \in V$ we use v' to denote the copy in V' and v'' to denote the copy in V''. For each edge (v, w) there is an edge $\{v', w''\}$ of the same cost in E'. In addition, for each node $v \in V$ we have an edge $\{v', v''\}$ of cost zero in E'. Clearly, the set $\big\{\{v', v''\}\ ;\ v \in V\big\}$ is an assignment of cost zero. It is a minimum cost matching iff G has no negative cycle.

Lemma 33 *G' contains a perfect matching of negative cost iff G contains a negative cycle.*

Proof Let $C = [e_0, e_1, \ldots, e_{k-1}]$ with $e_i = (v_i, v_{i+1})$ and $v_k = v_0$ be a simple cycle of negative cost in G. We construct a perfect matching of the same cost in G'. It consists of the edges $\{v_i', v_{i+1}'\}$ for i, $0 \le i < k$, and the edges $\{v', v''\}$ for all nodes v that do not lie on C.

For the reverse direction consider any perfect matching M of negative cost in G'. We show that M corresponds to a set of cycles in G and that one of these cycles has negative cost. Consider any edge $\{v_0', v_1''\} \in M$ with $v_0 \ne v_1$; there must be at least one such edge since M has negative cost. The node v_1' must also be matched. Let v_2'' be its mate. Continuing in this fashion we construct a sequence of edges $\{v_0', v_1''\}, \{v_1', v_2''\}, \ldots, \{v_{k-1}', v_k''\}$ in M. We stop as soon as we encounter a node v_k'' such that v_k' appeared previously in the sequence. We must have $v_k = v_0$ since $v_k = v_j$ for some j, $j > 0$, implies that two matching edges are incident to v_k''. We conclude that $[v_0, v_1, \ldots, v_k]$ is a simple cycle in G. Thus M induces a set of simple cycles in G and the total cost of these cycles is equal to the cost of M. Hence, one of the cycles must have negative cost. □

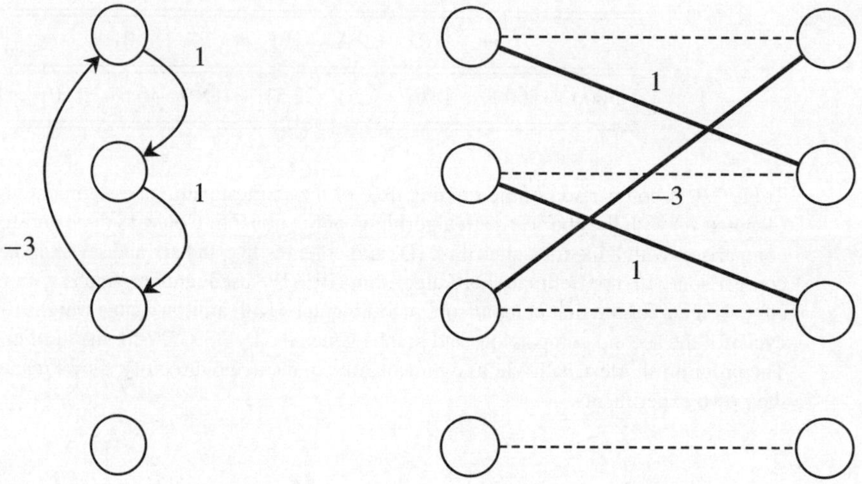

Figure 7.37 A directed graph and the derived bipartite graph. All dashed edges in the graph on the right have cost zero. The dashed edges define a perfect matching of cost zero. The solid edges together with the lowest dashed edge define a perfect matching of negative cost. It corresponds to the negative cycle in the graph on the left.

Assume now that G contains no negative cycle, let M be a minimum weight assignment in G' and let π' be a potential function that proves the optimality of M. We show that π' can be used to transform the cost function c into a non-negative cost function. M has cost zero[27] and hence $\sum_{v \in V} \pi'(v') + \pi'(v'') = 0$. Also $\pi'(v') + \pi'(v'') \leq 0$ for all $v \in V$ and hence

$$\pi'(v') = -\pi'(v'') \text{ for all } v \in V.$$

We define a potential function π on V by

$$\pi(v) = \pi'(v'') \text{ for all } v \in V.$$

Consider any edge $e = (v, w)$ in G and let $\overline{c}(e) = \pi(v) + c(e) - \pi(w)$ be its reduced cost. We have:

$$\begin{aligned} \overline{c}(e) &= \pi(v) + c(e) - \pi(w) = \pi'(v'') + c(e) - \pi'(w'') \\ &= -\pi'(v') + c(e) - \pi'(w'') \geq 0, \end{aligned}$$

where the inequality follows from the fact that $c(e) \geq \pi'(v') + \pi'(w'')$ for all edges $e = \{v, w\}$.

We conclude that $\overline{c}$ is a non-negative cost function on G. The shortest-path problem with respect to $\overline{c}$ can be solved by Dijkstra's algorithm. Also, if $\mu(v)$ and $\overline{\mu}(v)$ are the shortest-path distances from s to v with respect to c and $\overline{c}$, respectively, then

$$\mu(v) = -\pi(s) + \overline{\mu}(v) + \pi(v),$$

27 It is possible that one of the edges (v', v'') is not contained in M. How?

see Section 7.5.10.

The discussion above leads to the following program.

⟨*shortest_path_via_assignment.c*⟩≡

```
template <class NT>
bool shortest_path_via_assignment(const graph& G, node s,
                                  const edge_array<NT>& c,
                                  node_array<NT>& dist,
                                  node_array<edge>& pred)
{ node v,w; edge e;
  GRAPH<NT,NT> G1;
  list<node> A,B;
  node_array<node> left_copy(G), right_copy(G);
  forall_nodes(v,G)
  { A.append(left_copy[v]  = G1.new_node());
    B.append(right_copy[v] = G1.new_node());
    G1.new_edge(left_copy[v],right_copy[v],0);
  }
  forall_edges(e,G)
  { v = G.source(e); w = G.target(e);
    G1.new_edge(left_copy[v],right_copy[w],c[e]);
  }
  list<edge> M =
     MIN_WEIGHT_ASSIGNMENT_T(G1,A,B,G1.edge_data(),G1.node_data());
  NT sum = 0;
  forall_nodes(v,G1) sum += G1[v];
  if (sum < 0) return false;
  node_array<NT> pot(G);
  forall_nodes(v,G) pot[v] = G1[right_copy[v]];
  edge_array<NT> red_cost(G);
  forall_edges(e,G)
    red_cost[e] = pot[G.source(e)] + c[e] - pot[G.target(e)];
  DIJKSTRA_T(G,s,red_cost,dist,pred);
  forall_nodes(v,G) dist[v] += pot[v] - pot[s];
  return true;
}
```

7.8.6 *Maximum Weighted Matchings of Maximum Cardinality*

We show how to compute a matching of maximum weight among the matchings of maximum cardinality[28]. Let L be a real number and consider the weight function c_L defined by adding L to the weight of every edge, i.e.,

$$c_L(e) = c(e) + L \text{ for every } e \in E.$$

It is intuitively clear that larger values of L favor matchings of larger cardinality. We make this precise.

[28] For graphs that have a perfect matching this is the same as looking for a maximal weight perfect matching.

We observe first that $c_L(M) = c(M) + L|M|$ for any matching M. Thus, for two matchings M and N of the same cardinality the relative weight of the matchings does not change. Let C be the largest absolute value of any edge weight and let $k = \min(|A|, |B|)$. Then $|c(M)| \le kC$ for any matching M (since a matching consists of at most k edges) and hence $|c(N) - c(M)| \le 2kC$ for any two matchings M and N. We conclude that $|M| < |N|$ implies $c_L(M) < c_L(N)$ for $L > 2kC$. Thus in order to find a maximum weight matching of maximum cardinality we only have to find a maximum weight matching with respect to the cost function c_L where $L = 2kC + 1$.

⟨*mwb_matching.t*⟩+≡

```
template <class NT>
list<edge> MWMCB_MATCHING_T(graph& G,
                            const list<node>& A, const list<node>& B,
                            const edge_array<NT>& c, node_array<NT>& pot)
{ NT C = 0;
  edge e;
  forall_edges(e,G)
  { if (c[e] > C) C = c[e];
    if (-c[e] > C) C = -c[e];
  }
  int k = Max(A.size(),B.size());
  C = 1 + 2*k*C;
  edge_array<NT> c_L(G);
  forall_edges(e,G) c_L[e] = c[e] + C;
  list<edge> M = MAX_WEIGHT_BIPARTITE_MATCHING_T(G,A,B,c_L,pot);
#ifndef LEDA_CHECKING_OFF
  if ( !CHECK_MWBM_T(G,c_L,M,pot) )
    error_handler(0,"check in MWMCB_MATCHING_T failed");
#endif
  return M;
}
```

Be aware that the computed potential function proves optimality with respect to the cost function c_L, where $L = 1 + 2kC$. The function has an arithmetic demand similar to the programs for the assignment problem. Recall that the maximum weight matching algorithm deals with numbers up to $3D$ when all edges costs are bounded by D in absolute value. We have $D = C + 1 + 2kC$ and hence the numbers handled by the algorithm may be as large as $3 + (6k + 3)C$. Since $C \ge 1$ and $k \ge 1$ we have $3 + (6k + 3)C \le 4nC$.

Exercises for 7.8

1 Write a checker for the maximum weight assignment problem.

2 Write a checker for the maximum weight assignment problem that takes only a matching M as input. Hint: Direct all edges in the matching from B to A, give each edge in the matching cost $c(e)$ and each edge outside the matching cost $-c(e)$. Show that the matching is optimal iff the resulting graph has no negative cycle.

3 Formulate Lemma 27 for the minimum weight assignment problem and write a checker for it.

4 Implement the basic version of the weighted bipartite matching algorithm.
5 Extend the function *shortest_path_via_assignment* such that it can also deal with graphs with negative cycles.
6 Show that the following strategy computes a maximum weight matching among the matchings of maximum cardinality: when searching for augmenting path from $a = a_i$ choose the shortest path to a free node in B (if there is one) and choose the path to the best node in A otherwise.
7 Write a program that computes a minimum weight matching among the matchings of maximal cardinality.
8 Write a program that computes a maximum weight matching of cardinality k, where k is a parameter of the algorithm. You may assume that the graph is connected.

7.9 Weighted Matchings in General Graphs

A *matching* M in a graph G is a subset of the edges no two of which share an endpoint, see Figure 7.38. The cardinality $|M|$ of a matching M is the number of edges in M. If w is a weight function on the edges of G then the weight $w(M)$ of a matching M is the sum of the weights of its edges, i.e.,

$$w(M) = \sum_{e \in M} w(e).$$

A node v is called *matched* with respect to a matching M if there is an edge in M incident to v and it is called *free* or *unmatched* otherwise. An edge e is called matching if $e \in M$. A matching is called a *maximum weight matching* if its weight is at least as large as the weight of any other matching. Figure 7.38 shows an example.

The function

```
list<edge> MAX_WEIGHT_MATCHING(const graph& G, const edge_array<int>& w)
```

returns a maximum weight matching in G with respect to the weight function w. Observe that the algorithm is only available for integer weights. The underlying algorithm is the so-called blossom shrinking algorithm of Edmonds[Edm65b, Edm65a]. Its worst case running time is $O(n^3)$ ([Law76]). The implementation is due to Thomas Ziegler [Zie95]. There are algorithms with better performance, both theorically [GMG86, Gal86] and practically [AC93]. At present the function cannot be asked to return a proof of optimality.

7.10 Maximum Flow

Let $G = (V, E)$ be a directed graph, let s and t be distinct vertices in G and let $cap : E \longrightarrow \mathbb{R}_{\geq 0}$ be a non-negative function on the edges of G. For an edge e, we call $cap(e)$

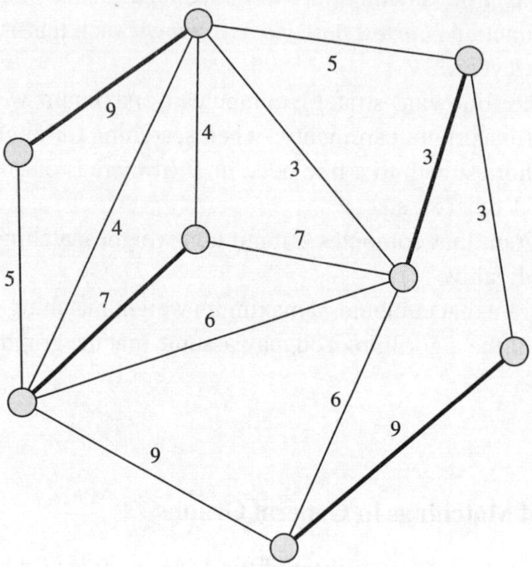

Figure 7.38 A maximum weight matching: The edges of the matching are shown in bold and the edge weights are indicated. We used the xlman-demo gw_mw_matching to generate this figure.

the *capacity* of e. An (s, t)-*flow* or simply *flow* is a function $f : E \longrightarrow \mathbb{R}_{\geq 0}$ satisfying the capacity constraints and the flow conservation constraints:

(1) $\quad 0 \leq f(e) \leq cap(e) \quad$ for every edge $e \in E$

(2) $\quad \sum_{e;source(e)=v} f(e) = \sum_{e;target(e)=v} f(e) \quad$ for every node $v \in V \backslash \{s, t\}$

The capacity constraints state that the flow across any edge is bounded by the capacity of the edge, and the flow conservation constraints state that for every node v different from s and t, the total flow out of the node is equal to the total flow into the node.

We call s and t the source and the sink of the flow problem, respectively, and we use V^+ to denote $V \backslash \{s, t\}$. For a node v, we call

$$excess(v) = \sum_{e;target(e)=v} f(e) - \sum_{e;source(e)=v} f(e)$$

the *excess* of v. Flow conservation states that all nodes except for s and t have zero excess.

The *value* of a flow f, denoted $|f|$, is the excess of the sink, i.e.,

$$|f| = excess(t).$$

A flow is called *maximum*, if its value is at least as large as the value of any other flow. Figure 7.39 shows an example.

In Section 7.10.1 we define the functionality of max flow algorithms and derive a checker,

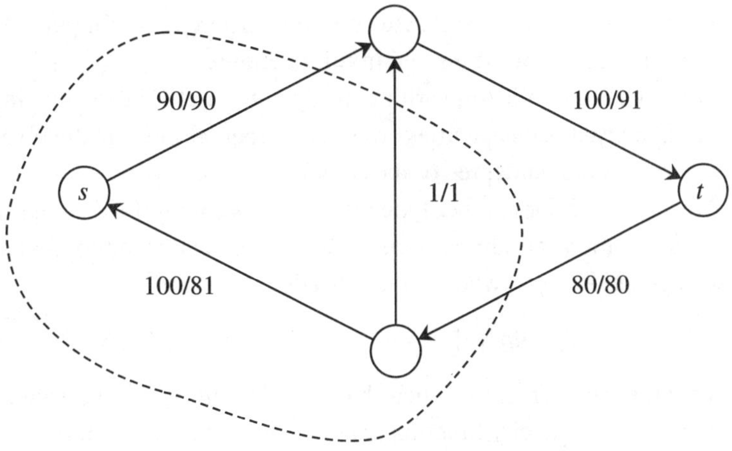

Figure 7.39 A maximum (s, t)-flow: For every edge e its capacity $cap(e)$ and the flow $f(e)$ across it are shown as $cap(e)/f(e)$. The value of the flow is equal 171. A saturated cut is indicated by the dashed line. It proves the maximality of the flow. The xlman-demo gw_max_flow visualizes maximum flows.

in Section 7.10.2 we discuss the generic preflow push algorithm, in Section 7.10.3 we give a first implementation of the preflow push algorithm, in Section 7.10.4 we describe several heuristic improvements, and in Section 7.10.5 we discuss the arithmetic demand of the algorithm and the danger of using the network flow algorithm with a number type that may incur rounding error.

7.10.1 *Functionality*

The function

```
NT MAX_FLOW_T(const graph& G, node s, node t
              const edge_array<NT>& cap, edge_array<NT>& f)
```

computes a maximum flow f in the network (G, s, t, cap) and returns the value of the flow. The function can be used with an arbitrary number type *NT*. There are pre-instantiated versions for the number types *int* and *double*. The function name of the pre-instantiated versions is MAX_FLOW, i.e., without the suffix _T. In order to use the pre-instantiated versions one must include <*LEDA/maxflow.h*>, and in order to use the template version, one must include <*LEDA/templates/maxflow.t*>.

Special care should be taken when using the template function with a number type *NT* that can incur rounding error, e.g., the type *double*. Section 7.2 contains a general discussion of this issue and Section 7.10.5 gives an example of what can go wrong in the computation of a maximum flow. The template function is only guaranteed to perform correctly if all arithmetic performed is without rounding error. This is the case if all numerical values in the input are integers (albeit stored as a number of type *NT*) and if none of the intermediate results exceeds the maximal integer representable by the number type ($2^{53} - 1$ in the case of *doubles*). All intermediate results are sums and differences of input values, in particular, the algorithms do not use divisions and multiplications.

The algorithm has the following arithmetic demand. Let C be the maximal absolute value of any edge capacity. If all capacities are integral then all intermediate values are bounded by $d \cdot C$, where d is the outdegree of the source.

The pre-instantiation for number type *int* issues a warning if C is larger than $MAXINT/d$.

The pre-instantiation for number type *double* computes the optimal matching for a modified capacity function *cap1*, where for every edge e

$$cap1[e] = sign(cap[e])\lfloor |cap[e]| \cdot S\rfloor / S$$

and S is the largest power of two such that $S < 2^{53}/(d \cdot C)$. The value of the maximum flow for the modified capacity function and the value of the maximum flow for the original capacity function differ by at most $m \cdot d \cdot C \cdot 2^{-52}$.

The weight modification can also be performed explicitly and we advise you to do so. The function

```
bool MAX_FLOW_SCALE_CAPS(const graph& G, node s, edge_array<double>& cap)
```

replaces $cap[e]$ by $cap1[e]$ for every edge e, where $cap1[e]$ is as defined above. The function returns *false* if the scaling changed some weight, and returns *true* otherwise.

In the remainder of this section we discuss a check of optimality and derive the famous max-flow-min-cut theorem of Ford and Fulkerson [FF63]. We need a technical lemma that generalizes the notion of excess to a set of nodes.

Lemma 34 *Let $S \subseteq V$ and let $T = V \setminus S$. Then*

$$\sum_{u \in S} excess(u) = \sum_{e \in E \cap (T \times S)} f(e) - \sum_{e \in E \cap (S \times T)} f(e).$$

Proof We have

$$\sum_{u \in S} excess(u) = \sum_{u \in S} \left(\sum_{e;\, target(e)=u} f(e) - \sum_{e;\, source(e)=u} f(e) \right),$$

by definition of excess. We now observe that each edge $e \in E \cap (T \times S)$ contributes $f(e)$ to this sum, each edge $e \in E \cap (S \times T)$ contributes $-f(e)$ to this sum, and each edge $e \in E \cap (S \times S)$ contributes $f(e) - f(e)$ to this sum. □

We draw a quick consequence. An application with $S = V$ and hence $T = \emptyset$ yields

$$excess(s) + excess(t) = 0,$$

i.e., $excess(s) = -|f|$. This agrees with the intuition that the flow arriving at t must originate at s.

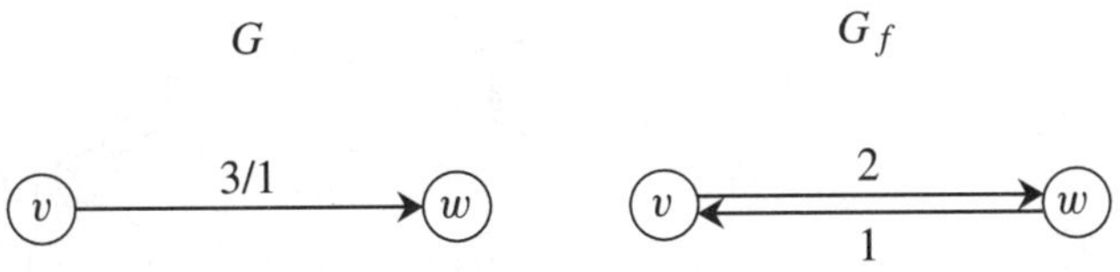

Figure 7.40 The residual network G_f: The left part shows an edge $e = (v, w)$ with capacity 3 and flow 1. It gives rise to two edges in the residual network shown on the right. The edge (v, w) has residual capacity 2 and the edge (w, v) has residual capacity 1.

An (s, t)-*cut* or simply *cut* is a set S of nodes with $s \in S$ and $t \notin S$. The *capacity* of a cut is the total capacity of the edges leaving the cut, i.e.,

$$cap(S) = \sum_{e \in E \cap (S \times T)} cap(e).$$

A cut S is called *saturated* if $f(e) = cap(e)$ for all $e \in E \cap (S \times T)$ and $f(e) = 0$ for all $e \in E \cap (T \times S)$.

The next lemma relates flows and cuts: the capacity of any (s, t)-cut is an upper bound for the value of any (s, t)-flow. Conversely, the value of any (s, t)-flow is a lower bound for the capacity of any (s, t)-cut.

Lemma 35 *Let f be any (s, t)-flow and let S be any (s, t)-cut. Then*

$$|f| \le cap(S).$$

If S is saturated then $|f| = cap(S)$.

Proof We have

$$\begin{aligned} |f| &= -excess(s) = -\sum_{u \in S} excess(u) \\ &= \sum_{e \in E \cap (S \times T)} f(e) - \sum_{e \in E \cap (T \times S)} f(e) \le \sum_{e \in E \cap (S \times T)} cap(e) \\ &= cap(S). \end{aligned}$$

For a saturated cut, the inequality is an equality. □

A saturated cut proves the maximality of f. A saturated cut is easily extracted from a maximum flow by means of the so-called residual network.

The *residual network* G_f with respect to a flow f has the same node set as G. Every edge of G_f is induced by an edge of G and has a so-called *residual capacity*. Let e be an arbitrary edge of G. If $f(e) < cap(e)$ then e is also an edge of G_f. Its residual capacity is $r(e) = cap(e) - f(e)$. If $f(e) > 0$ then e^{rev} is an edge of G_f. Its residual capacity is $r(e^{rev}) = f(e)$. Figure 7.40 shows an example.

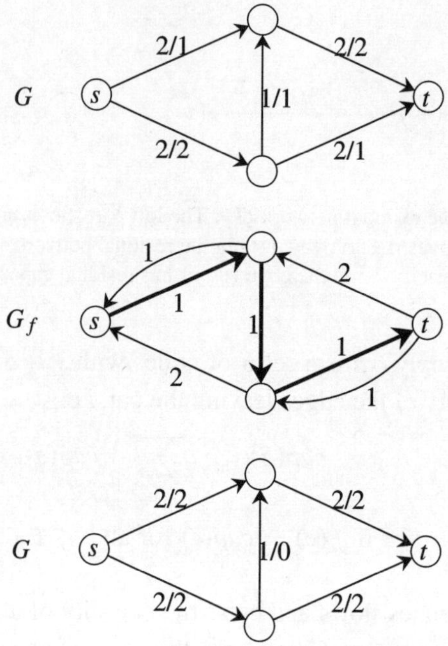

Figure 7.41 A path in the residual network and the resulting change of flow: A graph and an (s, t)-flow is shown at the top. The corresponding residual network is shown in the middle. A path p from s to t in the residual network is shown in bold. The flow obtained from augmentation by p is shown at the bottom.

Theorem 6 *Let f be an (s, t)-flow, let G_f be the residual network with respect to f, and let S be the set of nodes that are reachable from s in G_f.*

a) *If $t \in S$ then f is not maximum.*

b) *If $t \notin S$ then S is a saturated cut and f is maximum.*

Proof a) Let p be any simple path from s to t in G_f and let δ be the minimum residual capacity of any edge of p. Then $\delta > 0$. We construct a flow f' of value $|f| + \delta$. Let (see Figure 7.41)

$$f'(e) = \begin{cases} f(e) + \delta & \text{if } e \text{ is in } p \\ f(e) - \delta & \text{if } e^{rev} \text{ is in } p \\ f(e) & \text{if neither } e \text{ nor } e^{rev} \text{ belongs to } p. \end{cases}$$

Then f' is a flow and $|f'| = |f| + \delta$.

b) There is no edge (v, w) in G_f with $v \in S$ and $w \in T$. Hence, $f(e) = cap(e)$ for any e with $e \in E \cap (S \times T)$ and $f(e) = 0$ for any e with $e \in E \cap (T \times S)$, i.e., the cut S is saturated. Thus f is maximal. □

The function

```
bool CHECK_MAX_FLOW_T(const graph& G, node s, node t
                      const edge_array<NT>& cap, const edge_array<NT>& f)
```

checks whether f is a maximum (s,t)-flow. It returns *false* if this is not the case. The implementation is easy.

We check the capacity condition for each edge and compute the excess of all nodes. All nodes but s and t must have excess equal to zero. We then use breadth-first search to compute the set of nodes reachable from s in the residual graph; t must not be reachable.

⟨*max_flow_check*⟩≡

```
bool False_MF(string s)
{ cerr <<"\n\nCHECK_MAX_FLOW: " << s << "\n";
  return false;
}

template <class NT>
bool CHECK_MAX_FLOW_T(const graph& G, node s, node t,
                      const edge_array<NT>& cap, const edge_array<NT>& f)
{ node v; edge e;
  forall_edges(e,G)
    if ( f[e] < 0 && f[e] > cap[e] )
      return False_MF("illegal flow value");

  node_array<NT> excess(G,0);
  forall_edges(e,G)
  { node v = G.source(e); node w = G.target(e);
    excess[v] -= f[e]; excess[w] += f[e];
  }
  forall_nodes(v,G)
  { if ( v == s || v == t || excess[v] == 0 ) continue;
    return False_MF("node with non-zero excess");
  }

  node_array<bool> reached(G,false);
  queue<node> Q;

  Q.append(s); reached[s] = true;
  while ( !Q.empty() )
  { node v = Q.pop();
    forall_out_edges(e,v)
    { node w = G.target(e);
      if ( f[e] < cap[e] && !reached[w] )
      { reached[w] = true; Q.append(w); }
    }
    forall_in_edges(e,v)
    { node w = G.source(e);
      if ( f[e] > 0 && !reached[w] )
      { reached[w] = true; Q.append(w); }
    }
  }
  if ( reached[t] ) return False_MF("t is reachable in G_f");
  return true;
}
```

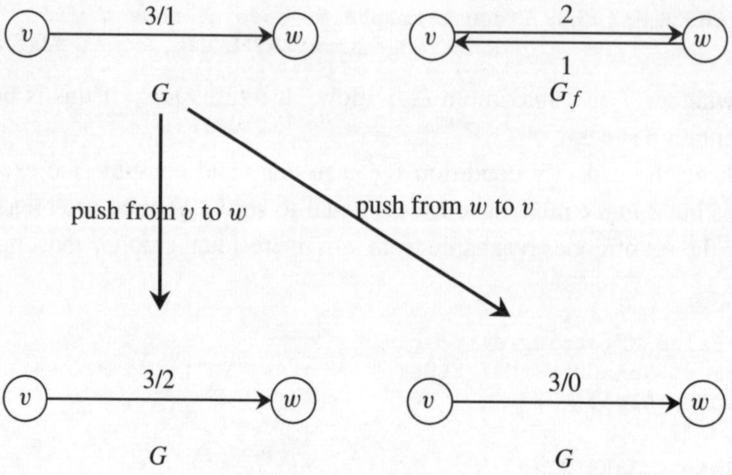

Figure 7.42 A push: The top left shows an edge $e = (v, w)$ in G with capacity three and flow one. This gives rise to two edges in the residual network shown on the right. A push of one unit of flow across e increases the flow across e by one and a push across e^{rev} decreases the flow across e by one.

7.10.2 *Algorithms*

The maximum flow problem is a widely studied problem and numerous algorithms have been proposed for it [FF63, EK72, Din70, Kar74, AO89, Gol85, GT88, CH95, CHM96, GR97].

Our implementations are based on the preflow-push method of Goldberg and Tarjan [GT88]. It manipulates a preflow that gradually evolves into a flow. Detailed computational studies of the preflow-push method can be found in [CG97, AKMO97] and in Section 7.10.4.

A *preflow* f is a function $f : E \longrightarrow I\!R_{\geq 0}$ with

$$\begin{array}{lll} (1) & 0 \leq f(e) \leq cap(e) & \text{for every edge } e \in E \text{ and} \\ (2) & excess(v) \geq 0 & \text{for every node } v \in V^+ \end{array}$$

i.e., the flow conservation constraint is replaced by the weaker constraint that no node in V^+ has negative excess. We call a node $v \in V^+$ *active* if its excess is positive. The residual network G_f with respect to a preflow f is defined as in the case of a flow.

The basic operation to manipulate a preflow is a *push*. Let v be an active node, let $e = (v, w)$ be a residual edge out of v, and let $\delta \leq \min(excess(v), r(e))$. A push of δ across e changes f as follows: it increases $f(e)$ by δ if e is an edge of G, and it decreases $f(e^{rev})$ by δ if e is the reversal of an edge of G, see Figure 7.42.

A push of δ across e increases $excess(w)$ by δ and decreases $excess(v)$ by δ. A push is called *saturating* if $\delta = r(e)$ and is called *non-saturating* otherwise. A saturating push across e removes e from the residual network and either kind of push adds e^{rev} to the residual network (if it is not already there).

The question is now which pushes to perform? Goldberg and Tarjan suggested to put the nodes of G (and hence G_f) onto layers with t on the bottom-most layer and to perform only pushes with transport excess to a lower layer. We use $d(v)$ to denote the (number of the) layer containing v. We call an edge $e = (v, w) \in G_f$ *eligible* if $d(w) < d(v)$.

Let us summarize: a push across an edge $e = (v, w) \in G_f$ can be performed if v is active and e is eligible. It moves $\delta \leq \min(excess(v), r(e))$ units of flow from v to w. If e is also an edge of G then $f(e)$ is increased by δ, and if e is the reversal of an edge of G then $f(e)$ is decreased by δ.

What are we going to do when v is active but there is no eligible edge out of v? In this situation v is *relabeled* by increasing $d(v)$ by one.

We are now ready for the generic preflow-push algorithm.

```
/* initialization */
set f(e) = cap(e) for all edges with source(e) = s;
set f(e) = 0 for all other edges;
set d(s) = n and d(v) = 0 for all other nodes;

/* main loop */
while there is an active node
{ let v be any active node;
  if there is an eligible edge e = (v, w) in G_f
  { push δ across e for some δ ≤ min(excess(v), r(e)); }
  else
  { relabel v; }
}
```

We will show that the algorithm terminates with a maximum flow (if it terminates). Call an edge $e = (v, w) \in G_f$ *steep* if $d(w) < d(v) - 1$, i.e., if it reaches down by two or more levels.

Lemma 36 *The algorithm maintains a preflow and does not generate steep edges. The nodes s and t stay on levels 0 and n, respectively.*

Proof The algorithm clearly maintains a preflow.

After the initialization, each edge in G_f either connects two nodes on level zero or connects a node on level zero to a node on level n. Thus, there are no steep edges (there are not even any eligible edges). A relabeling of a node v does not create a steep edge since a node is only relabeled if there are no eligible edges out of it. A push across an edge $e = (v, w) \in G_f$ may add the edge (w, v) to G_f. However, this edge is not even eligible.

Only active nodes are relabeled and only nodes different from s and t can be active. Thus, s and t stay on layers n and 0, respectively. □

The preceding lemma has an interesting interpretation. Since there are no steep edges, any path from v to t must have length (= number of edges) at least $d(v)$ and any path from

v to s must have length at least $d(v) - n$. Thus, $d(v)$ is a lower bound on the distance from v to t and $d(v) - n$ is a lower bound on the distance from v to s.

The next lemma shows that active nodes can always reach s in the residual network (since they must be able to send their excess back to s). It has the important consequence that d-labels are bounded by $2n - 1$.

Lemma 37 *If v is active then there is a path from v to s in G_f. No distance label ever reaches $2n$.*

Proof Let S be the set of nodes that are reachable from v in G_f and let $T = V \setminus S$. Then

$$\sum_{u \in S} excess(u) = \sum_{e \in E \cap (T \times S)} f(e) - \sum_{e \in E \cap (S \times T)} f(e),$$

by Lemma 34. Please convince yourself that this lemma holds for preflows and not only for flows.

There is no edge $(v, w) \in G_f$ with $v \in S$ and $w \notin S$. Thus, $f(e) = 0$ for every $e \in E \cap (T \times S)$. We conclude $\sum_{u \in S} excess(v) \leq 0$.

Since s is the only node whose excess may be negative and since $excess(v) > 0$ we must have $s \in S$.

Assume that a node v is moved to level $2n$. Since only active nodes are relabeled this implies the existence of a path (and hence simple path) in G_f from a node on level $2n$ to s (which is on level n). Such a path must contain a steep edge, a contradiction to Lemma 36. □

Theorem 7 *When the algorithm terminates, it terminates with a maximum flow.*

Proof When the algorithm terminates, there are no active nodes and hence the algorithm terminates with a flow. Call it f.

In G_f there can be no path from s to t since any such path must contain a steep edge (since s is on level n, t is on level 0). Thus, f is a maximum flow by Theorem 6. □

There is no guarantee that the generic preflow-push algorithm terminates, as it may choose to perform arbitrarily small pushes. However, it is fairly easy to bound the number of relabels and the number of saturating pushes.

Lemma 38 *There are at most $2n^2$ relabels and at most nm saturating pushes.*

Proof No distance label ever reaches $2n$ by Lemma 37 and hence each node is relabeled at most $2n$ times. The total number of relabels is therefore at most $2n^2$.

A saturating push across an edge $e = (v, w) \in G_f$ removes e from G_f. We claim that v has to be relabeled at least twice before the next push across e and hence there can be at most n saturating pushes across any edge. To see the claim, observe that only a push across e^{rev} can again add e to G_f. Since pushes occur only across eligible edges, w must be relabeled at least twice after the saturating push across e and before the next push across e^{rev}. Similarly, it takes two relabels of v before e becomes eligible again. □

It is more difficult to bound the number of non-saturating pushes. It depends heavily on which active node is selected for pushing, which edge is selected for pushing, and how much flow is pushed across the selected edge. In fact, without further assumptions, the number of non-saturating pushes is unbounded since we may choose to send only miniscule portions of flow. We make two assumptions for the remainder of the section:

Maximality: Every push moves the maximal possible amount, i.e., when flow is pushed across an eligible edge $e = (v, w)$ out of an active node v, the amount pushed is

$$\delta = \min(excess(v), r(e)).$$

This rule guarantees that every non-saturating push makes the source of the push inactive.

Persistence: When an active node v is selected, pushes out of v are performed until either v becomes inactive (because of a non-saturating push out of v) or until there are no eligible edges out of v anymore. In the latter case v is relabeled.

We study three rules for the selection of active nodes.

Arbitrary: An arbitrary active node is selected. Goldberg and Tarjan have shown that the number of non-saturating pushes is $O(n^2m)$ when the Arbitrary-rule is used. We will give their proof below.

FIFO: The active nodes are kept in a queue and the first node in the queue is always selected. When a node is relabeled or activated the node is added to the rear of the queue. The number of non-saturating pushes is $O(n^3)$ when the FIFO-rule is used. This bound is due to Goldberg.

Highest-Level: An active node on the highest level, i.e., with maximal dist-value, is selected. Observe that when a maximal level active node is relabeled it will be the unique maximal active node after the relabel. Thus, this rule guarantees that, when a node is selected, pushes out of the node will be performed until the node becomes inactive. The number of non-saturating pushes is $O(n^2\sqrt{m})$ when the highest-level-rule is used. This bound is due to Cheriyan and Maheshwari [CM89]. The proof given below is due to Cheriyan and Mehlhorn [CM99].

Lemma 39 *When the Arbitrary-rule is used, the number of non-saturating pushes is $O(n^2m)$.*

Proof The proof makes use of a potential function argument. Consider the potential function

$$\Phi = \sum_{v;\, v \text{ is active}} d(v).$$

We will show:

(1) $\Phi \geq 0$ always, and $\Phi = 0$ initially.

(2) A non-saturating push decreases Φ by at least one.
(3) A relabeling increases Φ by one.
(4) A saturating push increases Φ by at most $2n$.

Suppose that we have shown (1) to (4). By (3) and (4) and Lemma 38 the total increase of Φ is at most $n^2 + nm2n = n^2(1+2m)$. By (1), the total decrease can be no larger than this. Thus, the number of non-saturating pushes can be at most $n^2(1+2m)$ by (3).

It remains to show (1) to (4). (1) is obvious. For (2) we observe that a non-saturating push deactivates a node. It may or may not activate a node at the level below. In either case, Φ decreases by at least one. For (3) we observe that a relabeling of v increases $d(v)$ by one, and for (4) we observe that a saturating push may activate a node and that all distance labels are bounded by $2n$. □

We turn to the FIFO-rule. Recall that it keeps the active nodes in a queue and always selects the head of the queue. Relabeled and activated nodes are added to the rear of the queue.

It is convenient to split the execution into phases. The first phase starts at the beginning of the execution and a phase ends when all nodes that were active at the beginning of the phase have been selected from the queue. In this way each node is selected at most once in each phase and hence the number of non-saturating pushes is at most n times the number of phases.

Lemma 40 *When the FIFO-rule is used, the number of non-saturating pushes is $O(n^3)$.*

Proof By the discussion preceding the lemma it suffices to show that the number of phases is $O(n^2)$.

We use a potential function argument. Consider

$$\Phi = \max\{d(v)\ ;\ v \text{ is active }\}.$$

We show:

(1) $\Phi \geq 0$ always, and $\Phi = 0$ initially.
(2) A phase containing no relabel operation decreases Φ by at least one.
(3) A phase containing a relabel operation increases Φ by at most one.

Suppose that we have shown (1) to (3). By (3) and Lemma 38, the total increase is bounded by $2n^2$. By (1), the total decrease can be no larger. Thus the number of phases containing no relabel operation is bounded by $2n^2$ by (3). The total number of phases is therefore bounded by $4n^2$.

It remains to show (1) to (3). (1) is obvious. For (2) we observe that if a phase contains no relabel operation then all nodes selected in the phase get rid of their excess and push it to a lower layer. Thus, Φ decreases by at least one (it can decrease by more than one if an active node on level $n+1$ pushes its excess back to s). For (3), we observe that pushes move excess to a lower layer and that a relabeling of a node moves the node to one higher level. □

We turn to the highest-level selection rule. Recall that it always selects an active node with maximal distance label.

Lemma 41 *When the Highest-Level-rule is used, the number of non-saturating pushes is* $O(n^2\sqrt{m})$.

Proof We use a potential function argument. Let $K = \sqrt{m}$; this choice of K will become clear at the end of the proof. For a node v, let

$$d'(v) = |\{w;\, d(w) \le d(v)\}|/K$$

and consider

$$\Phi = \sum_{v;\, v \text{ is active}} d'(v).$$

We split the execution into phases. We define a phase to consist of all pushes between two consecutive changes of

$$d^* = \max\{d(v)\ ;\ v \text{ is active }\}$$

and call a phase *expensive* if it contains more than K non-saturating pushes, and *cheap* otherwise.

We show:

(1) The number of phases is at most $4n^2$.
(2) The number of non-saturating pushes in cheap phases is at most $4n^2K$.
(3) $\Phi \ge 0$ always, and $\Phi \le n^2/K$ initially.
(4) A relabeling or a saturating push increases Φ by at most n/K.
(5) A non-saturating push does not increase Φ.
(6) An expensive phase containing $Q \ge K$ non-saturating pushes decreases Φ by at least Q.

Suppose that we have shown (1) to (6). (4) and (5) imply that the total increase of Φ is at most $(2n^2 + mn)n/K$ and hence the total decrease can be at most this number plus n^2/K by (3). The number of non-saturating pushes in expensive phases is therefore bounded by $(2n^3+n^2+mn^2)/K$. Together with (2) we conclude that the total number of non-saturating pushes is at most

$$(2n^3 + n^2 + mn^2)/K + 4n^2K.$$

Observing that $n = O(m)$ and that the choice $K = \sqrt{m}$ balances the contributions from expensive and cheap phases, we obtain a bound of $O(n^2\sqrt{m})$.

It remains to prove (1) to (6). For (1) we observe that $d^* = 0$ initially, $d^* \ge 0$ always, and that only a relabel can increase d^*. Thus, d^* is increased at most $2n^2$ times, decreased no more than this, and hence changed at most $4n^2$ times. (2) follows immediately from (1) and the definition of a cheap phase. (3) is obvious. (4) follows from the observation that $d'(v) \le n/K$ for all v and at all times. For (5) observe that a non-saturating push across an edge (v, u) deactivates v, activates u (if it is not already active), and that $d'(u) \le d'(v)$.

For (6) consider an expensive phase containing $Q \geq K$ non-saturating pushes. By definition of a phase, d^* is constant during a phase, and hence all Q non-saturating pushes must be out of nodes at level d^*. The phase is finished either because level d^* becomes empty or because a node is moved from level d^* to level $d^* + 1$. In either case, we conclude that level d^* contains $Q \geq K$ nodes at all times during the phase. Thus, each non-saturating push in the phase decreases Φ by at least one (since $d'(u) \leq d'(v) - 1$ for a push from v to u). □

7.10.3 *A First Implementation*

We describe a first implementation of the generic preflow-push algorithm. The implementation is straightforward. We initialize a preflow, refine the flow into a flow, check that the computed flow is maximal, and return the value of the flow.

We want to execute the program with different rules for selection from the set of active nodes and therefore give the function two template parameters: the number type *NT* and the implementation of the set U of active nodes.

We want to count the number of pushes, the number of relabels, and the number of inspections of edges and therefore introduce appropriate parameters.

⟨*max_flow_basic*⟩≡

```
template<class NT, class SET>
NT MAX_FLOW_BASIC_T(const graph& G, node s, node t,
                    const edge_array<NT>& cap, edge_array<NT>& flow,
                    SET& U,
                    int& num_pushes, int& num_edge_inspections,
                    int& num_relabels)
{ if (s == t) error_handler(1,"MAXFLOW: source == sink");
  ⟨MF_BASIC: initialization⟩
  ⟨MF_BASIC: main loop⟩
#ifndef LEDA_CHECKING_OFF
  assert(CHECK_MAX_FLOW_T(G,s,t,cap,flow));
#endif
  return excess[t];
}
```

Initialization and Data Structures: We use the following data structures and variables: for each edge e we store the flow across e in *flow*$[e]$ and for each node v we store the level of v and the excess of v in *dist*$[v]$ and *excess*$[v]$, respectively. We store the active nodes in U.

We initialize the flow and the excess to zero, we put all nodes except for s on level zero, we put s on level n, we saturate all edges out of s, and initialize U with all nodes of positive excess. Thus

⟨*MF_BASIC: initialization*⟩≡
 ⟨*initialize flow and excess and saturate edges out of s*⟩
 ⟨*MF_BASIC: initialize dist and U*⟩
 ⟨*MF_BASIC: initialize counters*⟩

where

⟨*initialize flow and excess and saturate edges out of s*⟩≡

```
flow.init(G,0);
if (G.outdeg(s) == 0) return 0;
int n = G.number_of_nodes(); int max_level = 2*n - 1;
int m = G.number_of_edges();
node_array<NT>  excess(G,0);
// saturate all edges leaving s
edge e;
forall_out_edges(e,s)
{ NT c = cap[e];
  if (c == 0) continue;
  node v = target(e);
  flow[e] = c;
  excess[s] -= c;
  excess[v] += c;
 }
```

⟨*MF_BASIC: initialize dist and U*⟩≡

```
node_array<int> dist(G,0); dist[s] = n;
node v;
forall_nodes(v,G)
 if ( excess[v] > 0 ) U.insert(v,dist[v]);
```

⟨*MF_BASIC: initialize counters*⟩≡

```
num_relabels = num_pushes = num_edge_inspections = 0;
```

Implementations of the Set of Active Nodes: The implementation of U must support the following operations:

node U.del( ); delete a node from U and return it (return *nil* if U is empty).

U.insert(*node* v, *int* d); insert a node v with dist-value d. This version is to be used in the initialization phase and when a node is reinserted into the set of active nodes after a relabel.

U.insert0(*node* v, *int* d); insert a node v with dist-value d. This version is to be used when a node gets activated by a push into it.

bool U.empty( ); return true if U is empty.

U.clear( ); remove all elements from U.

Construction and Destruction.

We give three implementations:

The *FIFO implementation* keeps the nodes in U in a queue. Insertions add to the end of the queue, and deletions remove from the front of the queue.

⟨*FIFO implementation of SET*⟩≡

```
#include <LEDA/list.h>
class fifo_set{
  list<node> L;
public:
  fifo_set(){}
  node del() { if (!L.empty()) return L.pop(); else return nil; }
  void insert(node v, int d)  { L.append(v); }
  void insert0(node v, int d) { L.append(v); }
  bool empty() { return L.empty(); }
  void clear() { L.clear(); }
  ~fifo_set(){}
};
```

The *MFIFO (modified FIFO) implementation* keeps the nodes in U in a linear list and always selects the first node from the list. Nodes that are reinserted after a relabel operation are added to the front of the linear list, and nodes that get activated by a push into them are added to the rear of the list. In this way the same node is selected until all excess is removed from the node. The MFIFO implementation guarantees an $O(n^3)$ bound on the number of non-saturating pushes, see the exercises.

⟨*MFIFO implementation of SET*⟩≡

```
#include <LEDA/list.h>
class mfifo_set{
  list<node> L;
public:
  mfifo_set(){}
  node del() { if ( !L.empty() ) return L.pop(); else return nil; }
  void insert(node v, int d) { L.push(v); }
  void insert0(node v, int d){ L.append(v); }
  bool empty() { return L.empty(); }
  void clear() { L.clear(); }
  ~mfifo_set(){}
};
```

The *highest-level implementation* of U maintains an array A of linear lists with index range $[0 .. max\_level]$, where *max_level* is an argument of the constructor. The list $A[d]$ contains all nodes v that were inserted by $insert(v, d)$ or $insert0(v, d)$. The implementation maintains a variable *max* such that $A[d]$ is empty for $d > max$. In *insert0* we exploit the fact that it always inserts below the maximal level.

⟨*Highest level implementation of SET*⟩≡

```
#include <LEDA/list.h>
#include <LEDA/array.h>
class hl_set{
  int max, max_lev;
  array<list<node> > A;
public:
  hl_set(int max_level):A(max_level+1)
  { max = -1; max_lev = max_level;}
  node del()
  { while (max >= 0 && A[max].empty()) max--;
    if (max >= 0) return A[max].pop(); else return nil;
  }
  void insert(node v, int d)
  { A[d].push(v);
    if (d > max) max = d;
  }
  void insert0(node v, int d) { A[d].append(v); }
  bool empty()
  { while (max >= 0 && A[max].empty()) max--;
    return ( max < 0 );
  }
  ~hl_set(){}
  void clear()
  { for (int i = 0; i <= max_lev; i++) A[i].clear();
    max = -1;
  }
};
```

The Main Loop: In the main loop we select a node v from U. We call v the *current* node. If v does not exist, we break from the main loop, and if v is equal to t, we continue to the next iteration of the main loop. So assume otherwise. We try to push the excess of v to its neighbors in the residual graph. We inspect first the residual edges that correspond to edges out of v in G and then the residual edges that correspond to edges into v in G.

If v remains active after saturating all residual edges out of it, we relabel v and reinsert it into U.

⟨*MF_BASIC: main loop*⟩≡

```
for(;;)
{
  node v = U.del();
  if (v == nil) break;
  if (v == t) continue;
```

```
NT  ev = excess[v]; // excess of v
int dv = dist[v];   // level of v
edge e;
⟨MF_BASIC: push across edges out of v⟩
if ( ev > 0 )
{ ⟨MF_BASIC: push across edges into v⟩ }
excess[v] = ev;
if (ev > 0)
{ dist[v]++;
  num_relabels++;
  U.insert(v,dist[v]);
}
}
```

Pushing Excess Out of a Node: Let v be a node with positive excess. We want to push flow out of v along eligible edges. An edge $e \in G_f$ is either also an edge of G (and then $flow[e] < cap[e]$) or the reversal of an edge of G (and then $flow[e^{rev}] > 0$). We therefore iterate over all edges out of v and all edges into v.

For each edge e out of v we push $\max(excess[v], cap[e] - flow[e])$. If a push decreases the excess of v to zero we break from the loop.

⟨*MF_BASIC: push across edges out of v*⟩≡

```
for (e = G.first_adj_edge(v); e; e = G.adj_succ(e))
{ num_edge_inspections++;
  NT& fe = flow[e];
  NT  rc = cap[e] - fe;
  if (rc == 0) continue;
  node w = target(e);
  int dw = dist[w];
  if ( dw < dv ) // equivalent to ( dw == dv - 1 )
  { num_pushes++;
    NT& ew = excess[w];
    if (ew == 0) U.insert0(w,dw);
    if (ev <= rc)
    { ew += ev; fe += ev;
      ev = 0;   // stop: excess[v] exhausted
      break;
    }
    else
    { ew += rc; fe += rc;
      ev -= rc;
    }
  }
}
```

The code for the edges into v is symmetric.

⟨*MF_BASIC: push across edges into v*⟩≡

```
for (e = G.first_in_edge(v); e; e = G.in_succ(e))
{ num_edge_inspections++;
  NT& fe = flow[e];
  if (fe == 0) continue;
  node w = source(e);
  int dw = dist[w];
  if ( dw < dv ) // equivalent to ( dw == dv - 1 )
  { num_pushes++;
    NT& ew = excess[w];
    if (ew == 0) U.insert0(w,dw);
    if (ev <= fe)
    { fe -= ev; ew += ev;
      ev = 0;    // stop: excess[v] exhausted
      break;
    }
    else
    { ew += fe; ev -= fe;
      fe = 0;
    }
  }
}
```

Our first implementation is now complete. Let us see how it performs. We investigate the worst case complexity first and then give experimental data.

Worst Case Running Time: The running time of our implementation, not counting the time spent in the implementation of U, is proportional to the number of edge inspections. We bound the number of edge inspections first and then turn to the time spent in the implementation of U.

Consider an arbitrary iteration of the main loop and let v be the node selected in the iteration. In the iteration we inspect all edges incident to v, and either perform a push across an edge incident to v or relabel v. Thus the number of inspections of an edge e is bounded by the number of relabels of the endpoints of e plus the number of pushes out of the endpoints of e. No node is relabeled more than $2n$ times and hence the total number of edge inspections due to relabels is $O(nm)$. If P denotes the total number of pushes then the number of edge inspections due to pushes is $O(deg^* \cdot P)$, where deg^* is the maximal degree of any node. The number of pushes is $O(n^3)$ with the FIFO or MFIFO implementation for the set of active nodes and is $O(n^2\sqrt{m})$ with the highest-level implementation.

We turn to the time spent in maintaining the set of active nodes. For the FIFO and MFIFO implementation each operation on U takes constant time, and for the highest-level implementation each operation on U takes constant time plus the number of decreases of *max*. The number of decreases of *max* is bounded by the total increase of *max* and *max* is only increased by relabel operations. A relabel increases *max* by one. We conclude the total change of *max* is bounded by $O(n^2)$ by Lemma 38. The time spent in maintaining the set of active nodes is therefore $O(n^2)$ plus the number of operations on U. The number of operations on U is certainly bounded by the number of edge inspections.

We summarize in:

Theorem 8 *The worst case running time of our implementation is $O(n^3 \cdot deg^*)$ with the FIFO- or MFIFO-rule and is $O(n^2\sqrt{m} \cdot deg^*)$ with the highest-level-rule, where deg^* is the maximum degree of any node.*

The deg^*-factor in the running time is easily removed by means of the so-called *current edge data structure*. We used it already in Section 7.6. We found that the improvement is theoretical and does not show positively in the observed running times for all graphs where the average degree is bounded by 20. We therefore did not include the current edge data structure in our implementations.

We maintain for each node v a current out-edge $cur\_out\_edge[v]$ and a current in-edge $cur\_in\_edge[v]$ with the property that:

- no edge preceding $cur\_out\_edge[v]$ in the list of edges out of v is eligible and
- no edge preceding $cur\_in\_edge[v]$ in the list of edges into v is eligible.

When we push excess out of v we start searching for eligible edges at $cur\_out\_edge[v]$ and $cur\_in\_edge[v]$, respectively. When we relabel v we reset $cur\_out\_edge[v]$ and $cur\_in\_edge[v]$ to the first edge out of v and into v, respectively.

The implementation is correct since the only way a non-eligible edge $e = (v, w)$ can become eligible is through a relabeling of v.

The current edge implementation has the property that for any node v and between consecutive relabels of v the time spent in searching for eligible edges incident to v is proportional to the degree of v plus the number of pushes performed. The total time spent in searching for eligible edges is therefore bounded by $O(nm)$ plus the number of pushes.

Theorem 9 *The worst case running time of our implementation with the current edge data structure is $O(n^3)$ with the FIFO- or MFIFO-rule and is $O(n^2\sqrt{m})$ with the highest-level-rule.*

Four Generators: We describe four generators for max flow problems.

The first generator produces a graph with n nodes and $2n + m$ edges. It first produces a random graph with n nodes and m edges and makes s and t the first and the last node of G, respectively. It then adds edges (s, v) and (v, t) for all nodes v. The capacities are random numbers between 2 and 11 for all edges leaving s and between 1 and 10 for all other edges.

⟨*_max_flow_gen.c*⟩+≡

```
void max_flow_gen_rand(GRAPH<int,int>& G, node& s, node& t, int n, int m)
{ G.clear();
  random_graph(G,n,m);
  s = G.first_node(); t = G.last_node();
```

```
  node v; edge e;
  forall_nodes(v,G) { G.new_edge(s,v); G.new_edge(v,t); }
  forall_edges(e,G)
    G[e] = ( G.source(e) != s ? rand_int(1,10) : rand_int(2,11) );
}
```

The next two generators are due to Cherkassky and Goldberg [CG97]. For each integer k, $k \geq 1$, they generate the networks shown in Figure 7.43.

⟨*_max_flow_gen.c*⟩+≡

```
void max_flow_gen_CG1(GRAPH<int,int>& G, node& s, node& t, int n)
{ G.clear();
  if (n < 1)
    error_handler(1,"max_flow_gen_CG1: n must be at least one");
  array<node> V(2*n);
  int i;
  for(i = 0; i < 2*n; i++) V[i] = G.new_node();
  s = V[0]; t = V[2*n - 1];
  node v = V[n];
  for (i = 0; i < n; i++)
  { G.new_edge(V[i],V[i + 1], n - i);
    G.new_edge(V[i],v, 1);
  }
  G.new_edge(V[n - 1],V[2*n - 1], 1);
  G.new_edge(V[n - 1],V[n], 1);
  for (i = n; i <= 2*n - 2 ; i++ ) G.new_edge(V[i],V[i + 1],n);
}
void max_flow_gen_CG2(GRAPH<int,int>& G, node& s, node& t, int n)
{ G.clear();
  if (n < 1)
    error_handler(1,"max_flow_gen_CG2: n must be at least one");
  array<node> V(2*n);
  int i;
  for(i = 0; i < 2*n; i++) V[i] = G.new_node();
  s = V[0]; t = V[2*n-1];
  for (i = 0; i < n; i++ ) G.new_edge(V[i],V[2*n - 1 - i], 1);
  for (i = 0; i <= n - 1; i++ ) G.new_edge(V[i],V[i + 1], 2*n);
  for (i = n; i <= 2*n - 2; i++ ) G.new_edge(V[i],V[i + 1], n);
}
```

Observe the order in which we generate the edges out of node i: the edge from i to $2n-1-i$ precedes the edge to node $i+1$.

The fourth generator was suggested by Ahuja, Magnanti, and Orlin [AMO93]. The generated network is also shown in Figure 7.43.

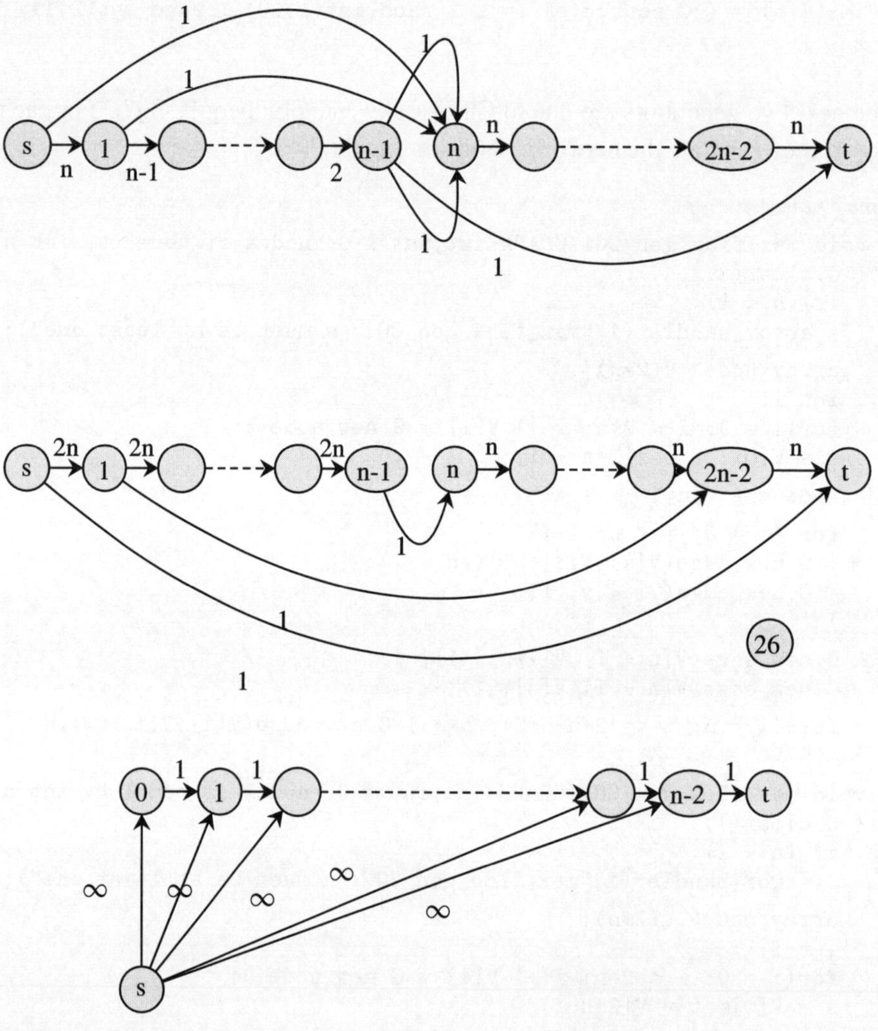

Figure 7.43 The generators *max_flow_gen_CG1*, *max_flow_gen_CG2*, and *max_flow_gen_AOM* generate the graphs shown. All three generators take the parameter *n* as an input.

⟨*_max_flow_gen.c*⟩+≡

```
void max_flow_gen_AMO(GRAPH<int,int>& G, node& s, node& t, int n)
{ G.clear();
  if (n < 1)
    error_handler(1,"max_flow_gen_AMO: n must be at least one");
  array<node> V(n);
  s = G.new_node();
  int i;
```

```
  for(i = 0; i < n; i++) V[i] = G.new_node();
  t = G.last_node();
  for (i = n - 2; i >= 0; i-- )
  { G.new_edge(s,V[i], 10000);
    G.new_edge(V[i],V[i + 1], 1);
  }
}
```

Running Times: Table 7.10 shows the behavior of our first implementation of the preflow-push method with three different selection rules and for four different kinds of graphs. For each of the four generators above we ran the cases $n = 500$ and $n = 1000$. For the random graph generator we used $m = 3n$. The number of pushes, the number of edge inspections, the number of relabels, and the running time quadruples or more than quadruples when n is doubled.

In the next section we will describe several optimizations which will lead to a dramatic improvement of observed running time. None of them improves the worst case behavior, however.

7.10.4 *Optimizations*

What is the best case running time of our implementation? The running time is $\Omega(n^2)$ if $\Omega(n)$ nodes need to be lifted above level n. This is usually the case. The best case behavior of the other parts of the algorithm is $O(m)$ and hence the cost of relabeling dominates the best case running time. In this section we will describe several heuristics that frequently reduce the time spent in relabeling nodes and as a side-effect reduce the time spent in all other operations. The heuristics will turn the preflow-push algorithm into a highly effective algorithm for solving flow problems.

Consider the example shown in Figure 7.44. We have nodes 0 to $n-1$, $s = 0$, $t = n-1$, and edges $(i, i+1)$ for all i, $0 \le i < n-1$. All edges have capacity two, except for edge $(n-2, n-1)$ which has capacity one.

Let us see what the preflow-push method does. In the initialization phase we saturate the edge $(0, 1)$, put s on level n, and all other nodes on level 0. Node 1 has positive excess. We lift node 1 to level 1 and push its excess to node 2. We lift node 2 to level 1 and push its excess to node 3. Continuing in this way the excess is pushed to node $n-2$. Only one unit can be forwarded to t and one unit remains on node $n-2$. At this point the value of the maximum flow has been determined. There is one unit of flow into t and this is the maximum possible. However, the algorithm does not know this fact yet and it will take the algorithm a long time to discover it. We lift node $n-2$ to level 2 and push the unit back to node $n-3$. Continuing in this way we lift nodes $n-2, n-3, \ldots, 2$ to level 2 and push the excess back to node 1. Then we lift node 1 to level 2 and then level 3, and .... Continuing in this way, we will invest $\Omega(n^2)$ relabels (and pushes) until nodes 1 to $n-1$ end up at level $n+1$. At this point we can push the excess back to s and the algorithm terminates.

Generator	Rule	Pushes	Inspections	Relabels	Time
rand	FIFO	1.764e+05	2.467e+06	2.34e+05	1.42
		6.831e+05	9.833e+06	9.28e+05	5.88
	HL	1.775e+05	2.672e+06	2.34e+05	1.47
		7.442e+05	1.073e+07	9.28e+05	6.04
	MFIFO	2.262e+05	2.566e+06	2.34e+05	1.28
		8.524e+05	1.018e+07	9.28e+05	5.25
CG1	FIFO	1.761e+05	9.63e+05	2.281e+05	0.81
		6.835e+05	4.121e+06	8.92e+05	3.94
	HL	1.875e+05	6.009e+06	1.885e+05	2.75
		7.5e+05	4.486e+07	7.52e+05	20.47
	MFIFO	1.682e+05	8.629e+05	2.207e+05	0.68
		6.713e+05	3.608e+06	8.801e+05	3.08
CG2	FIFO	2.864e+06	1.367e+07	2.751e+06	12.12
		1.149e+07	5.479e+07	1.1e+07	50.97
	HL	1.695e+06	1.226e+07	2.752e+06	11.33
		6.764e+06	4.902e+07	1.1e+07	43.17
	MFIFO	2.864e+06	1.367e+07	2.751e+06	11.02
		1.149e+07	5.479e+07	1.1e+07	45.14
AMO	FIFO	500	4.498e+06	1.5e+06	3.27
		1000	1.8e+07	6e+06	13.13
	HL	500	4.498e+06	1.5e+06	3.79
		1000	1.8e+07	6e+06	15.25
	MFIFO	500	4.498e+06	1.5e+06	2.74
		1000	1.8e+07	6e+06	11.13

Table 7.10 The basic implementation of the preflow-push algorithm. We show its behavior for four different kinds of graphs and three different selection rules. For each generator we ran the cases $n = 500$ and $n = 1000$. For the random graph generator we used $m = 3n$. The program max_flow_basic_time in the demo directory allows readers to make their own experiments.

We describe five optimizations. The first optimization is based on the observation that nodes on layer n and above can be treated more simply than nodes below level n. The

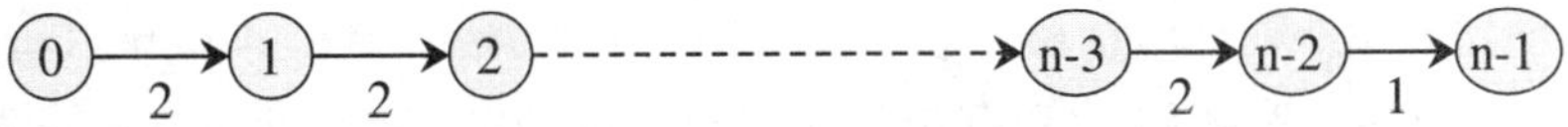

Figure 7.44 A network with nodes $0, \ldots, n-1$ and edges $(i, i+1)$ for all $i, 0 \le i < n-1$. All edges have capacity two except for edge $(n-2, n-1)$ which has capacity one.

second and third optimizations increase distance labels more aggressively, the fourth optimization splits the execution into two phases (where a maximum preflow is computed in the first phase and the remaining excess is pushed back to s in the second phase), and the fifth optimization recognizes nodes that have no chance of forwarding their flow to t. The combined effect of the five heuristics is to reduce the running time dramatically for many instances of the max flow problem, see Table 7.16 on page 485.

Large Distance Labels: We call a node v *high* if $d(v) \ge n$ and *low* otherwise and show that high nodes can be treated simpler than low nodes.

What distinguishes high nodes from low nodes? There can never be a path of residual edges from a high node to t as any such path would necessarily contain a steep edge. All excess of active high nodes must therefore flow back to s. The situation is different for active low nodes. Some of their excess can be pushed to t and some of their excess must flow back to s.

How can we exploit the difference? All excess of active high nodes must flow back to s. The excess reaches the active high nodes through edges $e \in E$ with $f(e) > 0$. This suggests that it can be sent back through such edges.

We therefore define

$$E_f^* = \left\{e^{rev} \; ; e \in E \text{ and } f(e) > 0\right\}$$

and use only edges in E_f^* when pushing out of high active nodes. We relabel a high active node when there are no eligible edges in E_f^* out of it.

```
/* initialization */
set f(e) = cap(e) for all edges with source(e) = s;
set f(e) = 0 for all other edges;
set d(s) = n and d(v) = 0 for all other nodes;

/* main loop */
while there is an active node
{ let v be any active node;
  if d(v) < n and there is an eligible edge e = (v, w) ∈ E_f or
     d(v) ≥ n and there is an eligible edge e = (v, w) ∈ E_f^*
  { push δ across e for δ = min(excess(v), r(e)); }
  else
  { relabel v; }
}
```

We need to show that the modified algorithm is correct. We adapt the correctness proof of the basic preflow-push algorithm. The modified algorithm may create steep edges. We show that no steep edge can end below level $n-1$ and that every steep edge belongs to $E_f \setminus E_f^*$; this modifies Lemma 36.

Lemma 42 *Any residual edge* $e = (v, w)$ *that becomes steep in the modified algorithm satisfies* $e \in E_f \setminus E_f^*$ *and* $d(w) \geq n - 1$.

Proof A steep edge $e = (v, w)$ can only be created by a relabeling of v. A node v is only relabeled when there is no eligible edge $(v, w) \in E_f^*$. Thus only edges in $E_f \setminus E_f^*$ can become steep.

A node v with $d(v) < n$ is only relabeled when there is no eligible edge out of it. Thus a relabeling of v that creates a steep edge $e = (v, w)$ can only occur when $d(v) \geq n$. The edge e was not steep before the relabeling of v and hence $d(w) \geq n - 1$. □

We next show that every active node can reach s in G_f^*; this modifies Lemma 37. The proof carries over almost literally.

Lemma 43 *If* v *is active then there is a path from* v *to* s *in* G_f^*. *No distance label ever reaches* $2n$.

Proof Let S be the set of nodes that are reachable from v in G_f^* and let $T = V \setminus S$. Then

$$\sum_{u \in S} excess(u) = \sum_{e \in E \cap (T \times S)} f(e) - \sum_{e \in E \cap (S \times T)} f(e),$$

by Lemma 34.

There is no edge $(v, w) \in G_f^*$ with $v \in S$ and $w \notin S$. Thus, $f(e) = 0$ for every $e \in E \cap (T \times S)$. We conclude $\sum_{u \in S} excess(v) \leq 0$.

Since s is the only node whose excess may be negative and since $excess(v) > 0$ we must have $s \in S$.

Assume that a node u is moved to level $2n$. Since only active nodes are relabeled this implies the existence of a path (and hence simple path) in G_f^* from a node on level $2n$ to s (which is on level n). Such a path must contain a steep edge, a contradiction to Lemma 42. □

Theorem 10 *When the modified algorithm terminates it terminates with a maximum flow. All bounds on the number of relabels and the number of pushes shown for the basic algorithm hold also true for the modified algorithm.*

Proof When the algorithm terminates there are no active nodes and hence the algorithm terminates with a flow. Call it f.

Assume that there is a path p in G_f from s to t. Write $p = p_1 \odot p_2$ where p_1 ends in a node with level at least n and p_2 contains no node with level n or more. Then p_2 starts with a node on level $n-1$ and contains no steep edges. Both claims follow from Lemma 42. However, p_2 contains at most $n-1$ nodes (since it cannot contain s) and hence must contain a steep edge.

Thus there is no path from s to t in G_f and hence f is optimal by Theorem 6. □

The changes in the program are minor. We push across the edges out of v only when v lives on a layer less than n.

⟨*MF_LH: main loop*⟩≡

```
for(;;)
{
  node v = U.del();
  if (v == nil) break;
  if (v == t) continue;
  NT  ev = excess[v]; // excess of v
  int dv = dist[v];   // level of v
  edge e;
  if ( dist[v] < n )
  { ⟨MF_BASIC: push across edges out of v⟩ }
  if ( ev > 0 )
  { ⟨MF_BASIC: push across edges into v⟩ }
  excess[v] = ev;
  if (ev > 0)
  { dist[v]++;
    num_relabels++;
    U.insert(v,dist[v]);
  }
}
```

The procedure MAX_FLOW_LH_T results from MAX_FLOW_BASIC_T by replacing the main loop. Table 7.11 shows the effect of distinguishing between low and high nodes. The effect is small and significant savings are only observed for the CG2-generator.

The Local Relabeling Heuristic: The *local relabeling heuristic* applies whenever a node is relabeled. It increases the dist-value of v to

$$1 + \min\left\{d(w)\ ;\ (v, w) \in G_f\right\}.$$

Observe that v is active whenever it is relabeled and that an active node has at least one outgoing edge in G_f. The expression above is therefore well defined. When v is relabeled, none of the outgoing edges is eligible and hence $d(w) \geq d(v)$ for all $(v, w) \in G_f$. Thus, the local relabeling heuristic increases $d(v)$ by at least one. It may increase it by more than one.

The correctness of the heuristic follows from the following alternative description: when a node is relabeled, continue to relabel it until there is an eligible edge out of it.

The local relabeling heuristic is easily incorporated into our implementation. We maintain a variable *dmin*, which we initialize to MAXINT before we scan the edges incident

Generator	Rule	Pushes	Inspections	Relabels	Time
rand	FIFO	1.728e+05	2.426e+06	2.27e+05	1.51
		1.726e+05	2.422e+06	2.269e+05	1.54
	HL	1.811e+05	2.654e+06	2.27e+05	1.6
		1.81e+05	2.649e+06	2.269e+05	1.64
	MFIFO	2.164e+05	2.513e+06	2.27e+05	1.36
		2.16e+05	2.508e+06	2.268e+05	1.4
CG1	FIFO	1.761e+05	9.63e+05	2.281e+05	0.85
		1.761e+05	9.63e+05	2.281e+05	0.9
	HL	1.875e+05	6.009e+06	1.885e+05	2.83
		1.875e+05	6.009e+06	1.885e+05	2.88
	MFIFO	1.682e+05	8.629e+05	2.207e+05	0.73
		1.682e+05	8.629e+05	2.207e+05	0.89
CG2	FIFO	2.864e+06	1.367e+07	2.751e+06	12.82
		2.54e+06	1.221e+07	2.544e+06	11.98
	HL	1.695e+06	1.226e+07	2.752e+06	11.31
		1.57e+06	1.12e+07	2.627e+06	11.24
	MFIFO	2.864e+06	1.367e+07	2.751e+06	11.6
		2.54e+06	1.221e+07	2.544e+06	10.87

Table 7.11 Effect of low-high distinction. We show the behavior for three different kinds of graphs and three different selection rules. For each generator we ran the case $n = 500$. For the random graph generator we used $m = 3n$. For each case we give the running time of MAX_FLOW_BASIC_T (first line) and of MAX_FLOW_LH_T (second line). Use the program max_flow_lh_time in the demo directory to perform your own experiments.

to the current active node v. Let $e = (v, w)$ be a residual edge. If e is eligible, i.e., $d(w) < d(v)$, we push across e, and if e is not eligible, i.e, $d(w) \geq d(v)$, we set *dmin* to $\min(dmin, d(w))$. If v is still active after scanning all residual edges incident to it, we can set $d(v)$ to $1 + dmin$.

We obtain

⟨*push across edges out of v*⟩≡

```
for (e = G.first_adj_edge(v); e; e = G.adj_succ(e))
{ num_edge_inspections++;
  NT& fe = flow[e];
```

```
    NT  rc = cap[e] - fe;
    if (rc == 0) continue;
    node w = target(e);
    int dw = dist[w];
    if ( dw < dv ) // equivalent to ( dw == dv - 1 )
    { num_pushes++;
      NT& ew = excess[w];
      if (ew == 0) U.insert0(w,dw);
      if (ev <= rc)
      { ew += ev; fe += ev;
        ev = 0;   // stop: excess[v] exhausted
        break;
      }
      else
      { ew += rc; fe += rc;
        ev -= rc;
      }
    }
    else { if ( dw < dmin ) dmin = dw; }
  }
```

The code for the edges into v is symmetric.

⟨*push across edges into v*⟩≡

```
  for (e = G.first_in_edge(v); e; e = G.in_succ(e))
  { num_edge_inspections++;
    NT& fe = flow[e];
    if (fe == 0) continue;
    node w = source(e);
    int dw = dist[w];
    if ( dw < dv ) // equivalent to ( dw == dv - 1 )
    { num_pushes++;
      NT& ew = excess[w];
      if (ew == 0) U.insert0(w,dw);
      if (ev <= fe)
      { fe -= ev; ew += ev;
        ev = 0;   // stop: excess[v] exhausted
        break;
      }
      else
      { ew += fe; ev -= fe;
        fe = 0;
      }
    }
    else { if ( dw < dmin ) dmin = dw; }
  }
```

The main loop turns into

⟨*MF_LRH: main loop*⟩≡
```
for(;;)
{
  node v = U.del();
  if (v == nil) break;
  if (v == t) continue;
  NT  ev   = excess[v]; // excess of v
  int dv   = dist[v];   // level of v
  int dmin = MAXINT;     // for local relabeling heuristic
  edge e;
  if (dv < n)
  { ⟨push across edges out of v⟩ }
  if ( ev > 0 )
  { ⟨push across edges into v⟩ }
  excess[v] = ev;
  if (ev > 0)
  { dist[v] = 1 + dmin;
    num_relabels++;
    U.insert(v,dist[v]);
  }
}
```

The procedure MAX_FLOW_LRH_T results from MAX_FLOW_BASIC_T by replacing the main loop. Table 7.12 shows the combined effect of the local relabeling heuristic and the low-high distinction.

The Global Relabeling Heuristic: The *global relabeling heuristic* updates the dist-values of all nodes. It sets

$$d(v) = \begin{cases} \mu(v,t) & \text{if there is a path from } v \text{ to } t \text{ in } G_f \\ n + \mu^*(v,s) & \text{if there is a path from } v \text{ to } s \text{ in } G_f^* \text{ but no} \\ & \text{path from } v \text{ to } t \text{ in } G_f \\ 2n-1 & \text{otherwise} \end{cases}$$

Here $\mu(v,t)$ and $\mu^*(v,s)$ denote the lengths (= number of edges) of the shortest paths from v to t in G_f and from v to s in G_f^*, respectively. The reader should convince himself that the global relabeling heuristic does not generate any steep edges.

The global relabeling heuristic can be implemented by breadth-first search and requires time $O(m)$. It should therefore not be applied too frequently. We will apply it every $h \cdot m$ edge inspections for some suitable constant h. In this way $\Omega(m)$ time is spent between applications of the global relabel heuristic and hence the worst case running time is increased by at most a constant factor. The best case can improve significantly.

In our example from the beginning of the section, the global relabeling heuristic is highly effective. Assume that it is applied after the edge $(n-2, n-1)$ is saturated. It will put node i on level $n+i$ for all i, $1 \le i \le n-2$, and the excess on node $n-2$ will flow back to s in a series of n pushes. In this way the running time decreases from $\Omega(n^2)$ to $O(n)$.

Generator	Rule	Pushes	Inspections	Relabels	Time
rand	FIFO	1.878e+05	2.554e+06	2.349e+05	1.51
		1.945e+05	1.949e+06	1.498e+05	1.25
	HL	1.915e+05	2.768e+06	2.349e+05	1.6
		1.915e+05	2.04e+06	1.36e+05	1.27
	MFIFO	2.332e+05	2.644e+06	2.348e+05	1.39
		2.332e+05	1.986e+06	1.457e+05	1.17
CG1	FIFO	1.761e+05	9.63e+05	2.281e+05	0.85
		2.234e+05	7.007e+05	1.403e+05	0.68
	HL	1.875e+05	6.009e+06	1.885e+05	2.8
		1.875e+05	5.726e+06	9.438e+04	2.67
	MFIFO	1.682e+05	8.629e+05	2.207e+05	0.71
		1.682e+05	5.482e+05	1.16e+05	0.52
CG2	FIFO	2.54e+06	1.221e+07	2.544e+06	11.35
		2.216e+06	9.529e+06	1.82e+06	9.19
	HL	1.57e+06	1.12e+07	2.627e+06	10.35
		1.57e+06	7.51e+06	1.377e+06	7.41
	MFIFO	2.54e+06	1.221e+07	2.544e+06	10.35
		2.54e+06	9.996e+06	1.796e+06	8.99

Table 7.12 Effect of low-high distinction and local relabeling heuristic. We show the behavior for three different kinds of graphs and three different selection rules. For each generator we ran the case $n = 500$. For the random graph generator we used $m = 3n$. For each case we give the running time of MAX_FLOW_LH_T (first line) and of MAX_FLOW_LRH_T (second line). The local relabeling heuristic results in a considerable saving in all cases. Use max_flow_lrh_time in the demo directory to perform your own experiments.

We turn to the implementation.

We define two functions *compute_dist_t* and *compute_dist_s* that compute the distance to t and s, respectively. Both functions need access to the residual graph and hence have parameters G, *flow*, and *cap*. We also provide them with the node t and the node s, respectively. The functions store the computed distances in *dist*. It is assumed that $dist[v] \geq n$ for all nodes v prior to a call of *compute_dist_t* and that $dist[v] = 2 * n - 1$ for all nodes v that cannot reach t in G_f prior to a call of *compute_dist_s*; the latter function also assumes that nodes that can reach t in G_f have a distance value less than n.

The calls insert all active nodes with their new distance labels into U. It is assumed that U is empty prior to a call of *compute_dist_t* and that U contains all active nodes that can reach t in G_f prior to a call of *compute_dist_s*.

The functions are realized by breadth-first search and hence need a queue Q. We provide it as a parameter. It is assumed that the queue is empty prior to a call of both functions. Both functions leave Q empty when they terminate.

The function *compute_dist_t* also computes for each d, $0 \le d < n$, the number of nodes v with $dist[v] = d$ and stores the number in $count[d]$; this count will be needed in the so-called gap heuristic to be described later.

The details of both functions are fairly simple. In *compute_dist_t* we perform a "backward" breadth-first search starting at t. Whenever a new node w is reached, say from node v, we set $dist[w]$ to $1 + dist[v]$, we insert w into U if it is active, we increase $count[dist[w]]$, and we add w to the rear of Q. Since we are computing distances to t and s, respectively, all edges are considered in their reverse direction.

⟨*max_flow_dist_st*⟩+≡

```
template<class NT, class SET>
void compute_dist_t(const graph& G, node t, const edge_array<NT>& flow,
                    const edge_array<NT>& cap,
                    const node_array<NT>& excess, node_array<int>& dist,
                    SET& U, b_queue<node>& Q, array<int>& count)
{
  int n = G.number_of_nodes();
  Q.append(t);
  dist[t] = 0;
  count.init(0);
  count[0] = 1;
  while ( !Q.empty() )
  { node v = Q.pop();
    int  d = dist[v] + 1;
    edge e;
    for(e = G.first_adj_edge(v); e; e = G.adj_succ(e))
    { if ( flow[e] == 0 ) continue;
      node u = target(e);
      int& du = dist[u];
      if ( du >= n )
      { du = d;
        Q.append(u); count[d]++;
        if ( excess[u] > 0 ) U.insert(u,d);
       }
    }
    for(e = G.first_in_edge(v); e; e = G.in_succ(e))
    { if ( cap[e] == flow[e] ) continue;
      node u = source(e);
      int& du = dist[u];
      if ( du >= n )
      { du = d;
        Q.append(u); count[d]++;
```

```
          if (excess[u] > 0) U.insert(u,d);
        }
      }
    }
  }
```

The "backward" breadth-first search from s is simpler because it only needs to consider edges in G_f^*.

⟨*max_flow_dist_st*⟩+≡

```
template<class NT, class SET>
void compute_dist_s(const graph& G, node s, const edge_array<NT>& flow,
                    const node_array<NT>& excess, node_array<int>& dist,
                    SET& U, b_queue<node>& Q)
{
  int n = G.number_of_nodes();
  int max_level = 2*n - 1;
  Q.append(s);
  dist[s] = n;
  while ( !Q.empty() )
  { node v = Q.pop();
    int  d = dist[v] + 1;
    edge e;
    for(e = G.first_adj_edge(v); e; e = G.adj_succ(e))
    { if ( flow[e] == 0 ) continue;
      node u = target(e);
      int& du = dist[u];
      if ( du == max_level )
      { du  = d;
        if (excess[u] > 0) U.insert(u,d);
        Q.append(u);
      }
    }
  }
}
```

Before we describe the required changes to the initialization phase and the main loop we describe one further optimization.

Two-Phase Approach: We partition the execution into two phases. The first phase ends when there is no active node at a level below n anymore. At this point of the execution the algorithm has determined a maximum preflow, i.e., a preflow which maximizes *excess*[t]. This follows from the observation that there can be no path in G_f from an active node to t at the end of phase one.

In the first phase we push only out of nodes with level below n and in the second phase we push only out of nodes with level at least n. Phase two ends when there are no active nodes anymore.

For the first phase we initialize *dist*[v] with the distance from v to t (if v can reach t in

G_f where f is the flow obtained by saturating all edges out of s) and we initialize $dist[v]$ with n otherwise.

⟨*MF_GRH: initialize dist and U for first phase*⟩≡

```
node_array<int> dist(G);
dist.init(G,n);
compute_dist_t(G,t,flow,cap,excess,dist,U,Q,count);
```

The other initializations are as before:

⟨*MF_GRH: initialization*⟩≡

```
⟨initialize flow and excess and saturate edges out of s⟩
⟨MF_GRH: additional data structures⟩
⟨MF_GRH: initialize dist and U for first phase⟩
⟨MF_GRH: initialize counters⟩
```

⟨*MF_GRH: initialize counters*⟩≡

```
num_relabels = num_pushes = num_edge_inspections = 0;
num_global_relabels  = 0;
```

We need some additional data structures: the global distance calculations need a queue and we need to know which phase we are in. We also need to introduce the array *count*: $count[d]$ is to contain the number of nodes at level d for $0 \le d < n$. It will be required by the gap heuristic to be explained below.

⟨*MF_GRH: additional data structures*⟩≡

```
b_queue<node> Q(n);
int phase_number = 1;
array<int> count(n);
```

The main loop has the same structure as before.

⟨*MF_GRH: main loop*⟩≡

```
for(;;)
{
  ⟨MF_GRH: extract v from queue⟩
  NT  ev   = excess[v]; // excess of v
  int dv   = dist[v];   // level of v
  int dmin = MAXINT;
  edge e;
  if ( dist[v] < n )
  { ⟨push across edges out of v⟩ }
  if ( ev > 0 )
  { ⟨push across edges into v⟩ }
  excess[v] = ev;
  if (ev > 0)
  { ⟨MF_GRH: update distance label(s)⟩ }
}
```

We still need to describe how nodes are selected from the queue and how distance labels are updated.

Let v be the node selected from the set U of active nodes. If v does not exist and we are in the second phase, we break from the main loop. If v does not exist and we are in the first phase, we start the second phase. If v is equal to t, we ignore v. In all other cases, we proceed and attempt to push out of v.

How do we start the second phase? We need to initialize the distance labels and also the set of active nodes for the second phase. We first compute the set of nodes that can still reach t (none of them is active) and collect its complement in a set S. None of the nodes in S can reach t. We then compute the distance labels for all nodes in S by computing their distances to s in G_f^*.

⟨*MF_GRH: extract v from queue*⟩≡

```
node v = U.del();
if (v == nil)
{
  if ( phase_number == 2 ) break; // done
  dist.init(G,n);
  compute_dist_t(G,t,flow,cap,excess,dist,U,Q,count);
  node u;
  forall_nodes(u,G)
  { if (dist[u] == n)
    { S.append(u);
      dist[u] = max_level;
    }
  }
  phase_number = 2;
  compute_dist_s(G,s,flow,excess,dist,U,Q);
  continue;
}
if (v == t) continue;
```

The set S needs to be declared.

⟨*MF_GRH: additional data structures*⟩+≡

```
list<node> S;
```

It remains to describe how we update distance labels. We mentioned already that the global relabeling heuristic has a cost of $\Theta(m)$ and that we want to apply it every $h \cdot m$ edge inspections for some constant h.

We therefore introduce two integer variables *limit_heur* and *heuristic*, initialize *heuristic* to $h \cdot m$, increment *limit_heur* by *heuristic* whenever the global relabel heuristic is applied, and apply the global relabel heuristic whenever the number of edge inspections exceeds *limit_heur*. Thus

⟨*MF_GRH: update distance label(s)*⟩≡

```
if (num_edge_inspections <= limit_heur)
  { ⟨MF_GRH: update the distance label of v⟩ }
else
  { limit_heur += heuristic;
    num_global_relabels++;
    ⟨MF_GRH: global relabel⟩
  }
```

and

⟨*MF_GRH: additional data structures*⟩+≡

```
int heuristic  = (int) (h*m);
int limit_heur = heuristic;
```

In order to update the distance label of v we increment *dmin* and then distinguish cases. If we are in phase one and *dmin* is at least n, we set *dist*$[v]$ to n and do not insert v into the set of active nodes (since v cannot reach t in G_f anymore). In all other cases, we set *dist*$[v]$ to *dmin* and insert v into U.

⟨*MF_GRH: update the distance label of v*⟩≡

```
dmin++; num_relabels++;
if ( phase_number == 1 && dmin >= n) dist[v] = n;
else { dist[v] = dmin;
       U.insert(v,dmin);
     }
```

A global relabel operation clears U and then distinguishes cases. In phase two the distance to s is recomputed for all nodes in S; recall that the nodes in $V \setminus S$ can reach t in G_f and hence are irrelevant for phase two.

In phase one we compute the distance from v to t in G_f for all nodes v. For nodes that cannot reach t we set the distance label to n. If no active node can reach t, phase one ends. We set S to all nodes that cannot reach t and then proceed as described above for phase two.

⟨*MF_GRH: global relabel*⟩≡

```
U.clear();
if (phase_number == 1)
{ dist.init(G,n);
  compute_dist_t(G,t,flow,cap,excess,dist,U,Q,count);
  if ( U.empty() )
  { node u;
    forall_nodes(u,G)
    { if (dist[u] == n)
      { S.append(u);
        dist[u] = max_level;
      }
    }
    phase_number = 2;
```

```
      compute_dist_s(G,s,flow,excess,dist,U,Q);
    }
  }
  else
  { node u;
    forall(u,S) dist[u] = max_level;
    compute_dist_s(G,s,flow,excess,dist,U,Q);
  }
```

The function MAX_FLOW_GRH_T incorporates the distinction between low and high nodes, the local and the global relabel heuristic, and the distinction between phases one and two.

⟨*max_flow_GRH*⟩≡

```
template<class NT, class SET>
NT MAX_FLOW_GRH_T(const graph& G, node s, node t,
                  const edge_array<NT>& cap, edge_array<NT>& flow,
                  SET& U, int& num_pushes, int& num_edge_inspections,
                  int& num_relabels, int& num_global_relabels, float h)
{ if (s == t) error_handler(1,"MAXFLOW: source == sink");
  ⟨MF_GRH: initialization⟩
  ⟨MF_GRH: main loop⟩
#ifndef LEDA_CHECKING_OFF
  assert(CHECK_MAX_FLOW_T(G,s,t,cap,flow));
#endif
  return excess[t];
}
```

Table 7.13 shows that the combined effect of the global relabel heuristic and the two-phase approach is dramatic. The running times decrease considerably for all generators and for all three selection rules.

The Gap Heuristic: We come to our last optimization.

Consider a relabeling of a node v in phase one and let *dv* be the layer of v before the relabeling. If the layer *dv* becomes empty by the relabeling of v, then v cannot reach t anymore in G_f after the relabeling, since any edge crossing the now empty layer would be steep.

If v cannot reach t in G_f then no node reachable from v in G_f can reach t. We may therefore move v and all nodes reachable from v to layer n whenever the old layer of v becomes empty by the relabeling of v. This is called the *gap heuristic*.

We realize the heuristic as follows. For each d, $0 \leq d < n$ we keep a count of the number of nodes in layer d. For this purpose we use the array *count* introduced in the previous section.

The array *count* is recomputed in *compute_dist_t* and is updated whenever a node is relabeled. When a node v is moved from a layer *dv* to a layer *dmin*, we decrement *count*[*dv*] and increment *count*[*dmin*] (if *dv* or *dmin* is smaller than n).

When *count*[*dv*] is decremented to zero we move v and all nodes reachable from v in G_f

Gen	Rule	Pushes	Inspections	Relabels	GR	Time
rand	FIFO	7.377e+05	7.354e+06	5.794e+05	—	4.8
		6978	5.181e+04	4119	2	0.06
	HL	7.254e+05	7.749e+06	5.32e+05	—	4.82
		5.412e+04	5.264e+05	4.2e+04	21	0.43
	MFIFO	8.907e+05	7.498e+06	5.631e+05	—	4.5
		8048	5.171e+04	3918	2	0.06
CG1	FIFO	8.908e+05	2.789e+06	5.581e+05	—	2.87
		5.02e+05	5.05e+05	994	6	0.91
	HL	7.5e+05	4.373e+07	3.763e+05	—	20.92
		5.015e+05	5.045e+05	988	12	1.22
	MFIFO	6.713e+05	2.352e+06	4.619e+05	—	2.3
		5.02e+05	5.05e+05	994	6	0.91
CG2	FIFO	8.851e+06	3.807e+07	7.277e+06	—	37.29
		9.793e+05	9.939e+05	4710	9	1.76
	HL	6.265e+06	3.002e+07	5.504e+06	—	29.81
		1.928e+04	5.53e+04	6518	1	0.17
	MFIFO	1.019e+07	4.012e+07	7.16e+06	—	36.53
		5.033e+05	5.085e+05	1992	9	0.98

Table 7.13 Effect of low-high distinction, the local relabeling heuristic, the global relabeling heuristic, and the two-phase approach. We show the behavior for three different kinds of graphs and three different selection rules. For each generator we ran the case $n = 1000$. For the random graph generator we used $m = 3n$. For each case we give the running time of MAX_FLOW_LRH_T (first line) and of MAX_FLOW_GRH_T (second line). The savings are dramatic in all cases. The column GR shows the number of times the global relabeling heuristic was applied. The parameter h of MAX_FLOW_GRH_T was set to 5. Use max_flow_grh_time in the demo directory to perform your own experiments.

to layer n. We find these nodes by a breadth-first search starting in v. We reuse the queue Q, which we introduced for the distance calculations, for the breadth-first search.

⟨*MF_GAP: update the distance label of v*⟩≡

```
num_relabels++;
if (phase_number == 1)
{ if ( --count[dv] == 0 || dmin >= n - 1)
  { // v cannot reach t anymore
```

```
      ⟨move all vertices reachable from v to level n⟩
    }
    else
    { dist[v] = ++dmin; count[dmin]++;
      U.insert(v,dmin);
    }
  }
  else // phase_number == 2
  { dist[v] = ++dmin;
    U.insert(v,dmin);
  }
```

Let us see the details of the breadth-first search. The layer *dmin* is the highest layer containing a node reachable from v. If this layer is less than n, we start the breadth-first search from v. We visit all nodes that are reachable from v in G_f and that live on a layer less than n. We move all such nodes to layer n. We count the number of nodes moved by the gap heuristic in *num_gaps*.

⟨*move all vertices reachable from v to level n*⟩≡

```
  dist[v] = n;
  if ( dmin < n )
  { Q.append(v);
    node w,z;
    while ( !Q.empty() )
    { edge e;
      w = Q.pop(); num_gaps++;
      forall_out_edges(e,w)
      { if ( flow[e] < cap[e] && dist[z = G.target(e)] < n)
        { Q.append(z);
          count[dist[z]]--; dist[z] = n;
        }
      }
      forall_in_edges(e,w)
      { if ( flow[e] > 0 && dist[z = G.source(e)] < n)
        { Q.append(z);
          count[dist[z]]--; dist[z] = n;
        }
      }
    }
  }
```

The main loop has the same structure as before and only one change is required. When the gap heuristic moves a node to layer n it does not remove it from the set of active nodes (which it should because the node should stay inactive till the beginning of phase two). We remedy the situation as follows. Whenever a node on level n is removed from the set of active nodes in phase one we ignore the node and continue to the next iteration.

⟨*MF_GAP: main loop*⟩≡

```
for(;;)
{
  ⟨MF_GRH: extract v from queue⟩
  if (dist[v] == n && phase_number == 1) continue;
  NT  ev   = excess[v]; // excess of v
  int dv   = dist[v];   // level of v
  int dmin = MAXINT;
  edge e;
  if ( dist[v] < n ) { ⟨push across edges out of v⟩ }
  if ( ev > 0 ) { ⟨push across edges into v⟩ }
  excess[v] = ev;
  if (ev > 0) { ⟨MF_GAP: update distance label(s)⟩ }
}
```

⟨*MF_GAP: update distance label(s)*⟩≡

```
if (num_edge_inspections <= limit_heur)
  { ⟨MF_GAP: update the distance label of v⟩ }
else
  { limit_heur += heuristic;
    num_global_relabels++;
    ⟨MF_GRH: global relabel⟩
  }
```

Finally, we give the function MAX_FLOW_GAP_T a further parameter *num_gaps*, in which we count the number of nodes that are moved by the gap heuristic.

⟨*max_flow_GAP*⟩≡

```
template<class NT, class SET>
NT MAX_FLOW_GAP_T(const graph& G, node s, node t,
                  const edge_array<NT>& cap, edge_array<NT>& flow,
                  SET& U, int& num_pushes, int& num_edge_inspections,
                  int& num_relabels, int& num_global_relabels,
                  int& num_gaps, float h)
{ if (s == t) error_handler(1,"MAXFLOW: source == sink");
  ⟨MF_GRH: initialization⟩
  num_gaps = 0;
  ⟨MF_GAP: main loop⟩
#ifndef LEDA_CHECKING_OFF
  assert(CHECK_MAX_FLOW_T(G,s,t,cap,flow));
#endif
  return excess[t];
}
```

Table 7.14 shows the combined effect of all heuristics.

Gen	Rule	Pushes	Inspections	Relabels	GR	Gaps	Time
rand	FIFO	1.394e+04	1.036e+05	8154	2	—	0.18
		1.39e+04	1.036e+05	8142	2	2	0.17
	HL	1.911e+05	1.929e+06	1.444e+05	38	—	1.59
		2.536e+04	1.959e+05	1.258e+04	3	934	0.27
	MFIFO	1.589e+04	1.033e+05	7674	2	—	0.15
		1.589e+04	1.033e+05	7672	2	11	0.15
CG1	FIFO	2.002e+06	2.008e+06	1988	12	—	4.49
		2.002e+06	2.008e+06	1988	12	0	4.05
	HL	2.003e+06	2.009e+06	1975	25	—	5.41
		2.003e+06	2.009e+06	1975	25	0	5.67
	MFIFO	2.004e+06	2.01e+06	1988	12	—	3.64
		2.004e+06	2.01e+06	1988	12	0	4.08
CG2	FIFO	3.951e+06	3.971e+06	6846	18	—	8.85
		3.982e+06	3.992e+06	3983	18	2015	7.88
	HL	3.852e+04	1.106e+05	1.302e+04	1	—	0.36
		1.599e+04	4.396e+04	4002	0	3995	0.28
	MFIFO	2.079e+06	2.098e+06	6684	18	—	3.93
		2.001e+06	2.012e+06	3983	18	2017	4.27

Table 7.14 Effect of low-high distinction, the local relabeling heuristic, the global relabeling heuristic, the two-phase approach, and the gap heuristic. We show the behavior for three different kinds of graphs and three different selection rules. For each generator we ran the case $n = 2000$. For the random graph generator we used $m = 3n$. For each case we give the running time of MAX_FLOW_GRH_T (first line) and of MAX_FLOW_GAP_T (second line). The effect of the gap heuristic is small. The column GR shows the number of global relabels and the column Gaps shows the number of nodes moved by the gap heuristic. Use max_flow_gap_time in the demo directory to perform your own experiments.

Choice of H: How often should the heuristics be applied? Table 7.15 shows the behavior for different values of h. The choice of h does not have a big influence on running time. We have chosen $h = 5$ as the default value of h.

Summary and Implementation History: Table 7.16 summarizes our experiments. It shows the running times of our different implementations for four different kinds of graphs, three selection rules, and two different graph sizes ($n = 1000$ and $n = 2000$). The heuristics

Gen	Rule	h	Pushes	Inspections	Relabels	GR	Gaps	Time
rand	FF	0.5	9988	6.362e+04	4850	3	9	0.14
		2.5	1.18e+04	8.356e+04	6562	2	4	0.14
		4.5	1.6e+04	1.236e+05	9753	2	7	0.19
		6.5	1.989e+04	1.636e+05	1.287e+04	2	9	0.22
	HL	0.5	1.425e+04	8.442e+04	5506	16	1333	0.36
		2.5	1.967e+04	1.403e+05	9113	5	280	0.26
		4.5	2.563e+04	1.998e+05	1.28e+04	4	811	0.28
		6.5	2.347e+04	1.812e+05	1.18e+04	2	1279	0.25
	MF	0.5	1.112e+04	5.376e+04	3592	10	17	0.2
		2.5	1.328e+04	7.814e+04	5729	3	0	0.15
		4.5	1.476e+04	9.33e+04	6992	2	0	0.15
		6.5	1.956e+04	1.333e+05	9943	2	0	0.18
CG1	FF	0.5	1.992e+06	1.998e+06	1970	30	0	4.23
		2.5	1.996e+06	2.002e+06	1985	15	0	4.11
		4.5	2e+06	2.006e+06	1990	10	0	4.06
		6.5	2.004e+06	2.01e+06	1993	7	0	4.05
	HL	0.5	2.003e+06	2.009e+06	1750	250	0	8.6
		2.5	2.003e+06	2.009e+06	1950	50	0	5.67
		4.5	2.003e+06	2.009e+06	1973	27	0	5.33
		6.5	2.003e+06	2.009e+06	1981	19	0	5.21
	MF	0.5	2.004e+06	2.01e+06	1874	126	0	5.08
		2.5	2.004e+06	2.01e+06	1975	25	0	4.19
		4.5	2.004e+06	2.01e+06	1986	14	0	4.11
		6.5	2.004e+06	2.01e+06	1991	9	0	4.06

Table 7.15 Effect of the choice of h. We show the behavior for two different kinds of graphs and three different selection rules. For each generator we ran the case $n = 2000$. For the random graph generator we used $m = 3n$. For each case we give the running time of MAX_FLOW_GAP_T for different values of h. FF stands for FIFO and MF stands for MFIFO.

lead to dramatic savings in all cases, the global relabeling heuristic being the main source

Gen	Rule	BASIC	HL	LRH	GRH	GAP	LEDA
rand	FF	5.84	6.02	4.75	0.07	0.07	—
		33.32	33.88	26.63	0.16	0.17	—
	HL	6.12	6.3	4.97	0.41	0.11	0.07
		27.03	27.61	22.22	1.14	0.22	0.16
	MF	5.36	5.51	4.57	0.06	0.07	—
		26.35	27.16	23.65	0.19	0.16	—
CG1	FF	3.46	3.62	2.87	0.9	1.01	—
		15.44	16.08	12.63	3.64	4.07	—
	HL	20.43	20.61	20.51	1.19	1.33	0.8
		192.8	191.5	193.7	4.87	5.34	3.28
	MF	3.01	3.16	2.3	0.89	1.01	—
		12.22	12.91	9.52	3.65	4.12	—
CG2	FF	50.06	47.12	37.58	1.76	1.96	—
		239	222.4	177.1	7.18	8	—
	HL	42.95	41.5	30.1	0.17	0.14	0.08002
		173.9	167.9	120.5	0.3599	0.28	0.1802
	MF	45.34	42.73	37.6	0.94	1.07	—
		198.2	186.8	165.7	4.11	4.55	—
AMO	FF	12.61	13.25	1.17	0.06	0.06	—
		55.74	58.31	5.01	0.1399	0.1301	—
	HL	15.14	15.8	1.49	0.13	0.13	0.07001
		62.15	65.3	6.99	0.26	0.26	0.1399
	MF	10.97	11.65	0.04999	0.06	0.06	—
		46.74	49.48	0.1099	0.1301	0.1399	—

Table 7.16 The effect of the different heuristics. We show the behavior for four different kinds of graphs and three selection rules. For each generator we ran the cases $n = 1000$ and $n = 2000$. The last column stands for the default implementation in LEDA. It uses one further optimiziation which we have not explained in the text.

of improvement. You may use the program max_flow_summary_time in the demo directory to perform your own experiments.

Gen	Rule	GRH			GAP			LEDA		
rand	FF	0.16	0.41	1.16	0.15	0.42	1.05	—	—	—
	HL	1.47	4.67	18.81	0.23	0.57	1.38	0.16	0.45	1.09
	MF	0.17	0.36	1.06	0.14	0.37	0.92	—	—	—
CG1	FF	3.6	16.06	69.3	3.62	16.97	71.29	—	—	—
	HL	4.27	20.4	77.5	4.6	20.54	80.99	2.64	12.13	48.52
	MF	3.55	15.97	68.45	3.66	16.5	70.23	—	—	—
CG2	FF	6.8	29.12	125.3	7.04	29.5	127.6	—	—	—
	HL	0.33	0.65	1.36	0.26	0.52	1.05	0.15	0.3	0.63
	MF	3.86	15.96	68.42	3.9	16.14	70.07	—	—	—
AMO	FF	0.12	0.22	0.48	0.11	0.24	0.49	—	—	—
	HL	0.25	0.48	0.99	0.24	0.48	0.99	0.12	0.24	0.52
	MF	0.11	0.24	0.5	0.11	0.24	0.48	—	—	—

Table 7.17 The asymptotic behavior of our implementations. We show the behavior for four different kinds of graphs and three selection rules. For each generator we ran the cases $n = 5000 \cdot 2^i$ for $i = 0$, 1, and 2. For the random graph generator we used $m = 3n$. FF stands for FIFO and MF stands for MFIFO. You may use the program max_flow_large_time in the demo directory to perform your own experiments. The program max_flow_time in the demo directory times the default implementation.

The FIFO and MFIFO selection rule are superiour to the HL-rule on three of our four generators, although never by a large margin. However, on the generator CG2 both rules do very badly compared to the HL-rule. Figure 7.17 shows this even more clearly. For generators rand and AMO the running time seems to grow linearly (or maybe slightly more) for all three selection rules, for generator CG1 the running time seems to grow quadratically for all three selection rules, and for generator CG2 the running time seems to grow quadratically for the FIFO and the MFIFO-rule and seems to grow linearly for the HL-rule.

We have chosen the HL-rule as the default selection rule for our max flow algorithm. This is also what other researchers recommend [CG97, AKMO97].

The worst case running time of our max flow algorithm is $O(mdeg \cdot n^2\sqrt{m})$, where $mdeg$ is the maximal degree of any node. This can be improved to $O(n^2\sqrt{m})$ with the current edge data structure. Theoretically more efficient algorithms are known. Goldberg and Tarjan [GT88] have shown that the so-called dynamic tree data structure can be used to improve the running time of the preflow-push method to $O(nm \log n)$. In [CH95, CHM96] this was further improved to $O(nm+n^2 \log n)$. The dynamic tree data structure is available in LEDA. Monika Humble [Hum96] has implemented the preflow-push algorithm with the dynamic tree data structure. The observed running time was not impressive. Recently, Goldberg and

Rao [GR97] improved the running time to $O(\min(n^{2/3}, m^{1/2}m \log(n^2/m) log U)$, where U is the largest capacity of any edge (the capacities must be integral for their algorithm). It remains to be seen whether the improved bound also leads to better observed running times. A first experimental evaluation can be found in [HST98].

The first implementation of the preflow-push algorithm for LEDA was done by Cheriyan and Näher in 1989. It used the FIFO selection rule, the distinction between low and high nodes, and the local and global relabeling heuristic. Stefan Näher refined the implementation over the years and added the highest-level selection rule. For the book we added the two-phase approach, the gap heuristic, and the possibility of choosing the selection rule.

7.10.5 *Network Flow and Floating Point Arithmetic*

The preflow-push algorithm computes the maximum flow iteratively (and so do all other maximum flow algorithms). It starts with a preflow which it gradually transforms into a flow. The flow across any single edge is changed by pushes across the edge. These pushes may be in forward and backward direction, i.e., the flow across an edge is changed by additions and subtractions: the final flow across an edge is a sum of flow portions and these flow portions may be positive and negative.

What happens when the algorithm is executed with an arithmetic which may incur rounding error, e.g., floating point arithmetic? Then there may be cancellation in forming this sum. As a consequence the correctness of the algorithm is no longer guaranteed. The algorithm may not terminate or compute a function f which is not a flow (because it violates one of the constraints) or is a flow but not a maximal flow. Figure 7.45 shows an example of the disastrous effect that rounding error may have.

The preflow-push algorithm uses only additions and subtractions to manipulate flow and determines the flow to be sent across an edge as the maximum of the available excess and the residual capacity of the edge. This implies that all flow values are integral when the capacities are integral. Also the maximum excess of any node is bounded by D, where D is the sum of the capacities of the edges out of s.

If the number type *double* is used and all edge capacities are integral, there will be no overflow as long as $D < 2^{53}$. If the number type *double* is used and the edge capacities are not integral, we replace the edge capacities by

$$cap1[e] = sign(cap[e]) \lfloor |cap[e]| \cdot S \rfloor / S,$$

where S is the largest power of two such that $S < 2^{53}/D$, and apply the results of Section 7.2. They guarantee that there is no rounding error in the computation of the maximum flow with respect to *cap1* and that the value of the maximum flows with respect to *cap* and *cap1*, respectively, differ by at most $m \cdot D \cdot 2^{-52}$. The bound follows from the fact that the value of the maximum flow is equal to the capacity of a minimum cut, that the capacity of a minimum cut is the sum of at most m edge capacities and that the choice of S guarantees that for each edge the difference between the orginal capacity and the modified capacity is at most $D \cdot 2^{-52}$.

The paragraph above bounds the absolute error in the value of the flow resulting from

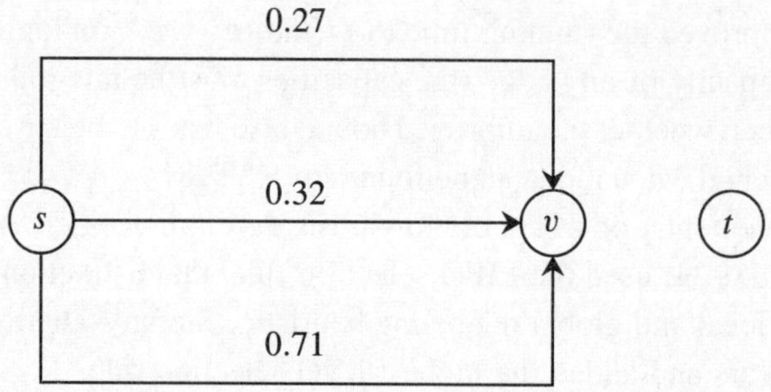

Figure 7.45 The effect of rounding error on the preflow-push algorithm: The capacities of the edges are as shown. The preflow-push algorithm starts by saturating all edges out of s. This will create an excess of $0.27 + 0.32 + 0.71 = 1.3$ in v. In the course of the execution, the algorithm will determine that none of this excess can be forwarded to t and hence the excess will be shipped back to s by sending 0.27, 0.32, and 0.71, respectively, across the three edges (v, s). The final excess in v is $1.3 - 0.27 - 0.32 - 0.71 = 0$.
Assume now that all calculations are carried out in a *floating point system with a mantissa of two decimal places and rounding by cut-off*. Then the excess in v after saturating all edges out of s will still be 1.3 as there is no cancellation in the summation. However, when the flow is pushed back to s the first subtraction $1.3 \ominus 0.29$ yields 1.1 as the last digit of 0.29 is dropped when the two summands are aligned for the subtraction; here $\ominus$ denotes floating point subtraction. The effect of this is that v ends up with an excess of 0.09, but no outgoing edge across which to push flow. This may put the algorithm into an infinite loop.

scaling. It does not bound the relative error. Observe that the quotient between D and the maximum flow may be arbitrarily large. Althaus and Mehlhorn [AM98] have shown that a slightly more elaborate scaling scheme can be used to bound the relative error. The idea is as follows. One modifies the edge capacities as described above and computes a maximum flow f with respect to them. Then

$$|val(f_{opt}) - val(f)| \le m \cdot D \cdot 2^{-52},$$

where f_{opt} is a maximum flow with respect to the original edge capacities. One now distinguishes cases. If $m \cdot D \cdot 2^{-52} \ll val(f)$, the relative error in the value of the flow is small. Otherwise, let $B = val(f) + m \cdot D \cdot 2^{-52}$ and observe that $val(f_{opt}) \le B$ and hence any capacity which is larger than B may be decreased to B without changing the maximum flow. Next they recompute D and S and repeat. After a smaller number of iterations the relative error will be small.

Exercises for 7.10

1 Let $G = (V, E)$ be a directed graph, let $cap : E \longrightarrow \mathbb{R}_{\ge 0}$ be a non-negative capacity function, and let $d : V \longrightarrow \mathbb{R}$ be a function with $\sum_{v \in V} d(v) = 0$. A node v with $d(v) > 0$ is called a *supply node*, a node v with $d(v) < 0$ is called a *demand node*, and d is called a demand function. A flow f is a function $f : E \longrightarrow \mathbb{R}_{\ge 0}$ satisfying the capacity constraints and the supply-demand constraints $excess(v) = d(v)$ for all $v \in V$. Design an algorithm that decides whether a flow exists and, if so, computes a flow. Hint:

Add two vertices s and t, an edge (s, v) with capacity $d(v)$ for every supply node, an edge (v, t) with capacity $-d(v)$ for every demand node, and compute a maximum (s, t)-flow.

2 The problem is as above but a lower bound $lb(e)$ on the flow across any edge e is also specified, i.e., for each edge two values $lb(e)$ and $ub(e)$ with $0 \le lb(e) \le ub(e)$are specified and the flow across any edge must lie between the lower and the upper bound. Hint: For any edge $e = (v, w)$ introduce two additional vertices a_e and b_e, replace e by the edges (v, a_e), (a_e, b_e), and (b_e, w), give a_e demand $-lb(e)$, give b_e supply $lb(e)$, and give (a_e, b_e) capacity $ub(e) - lb(e)$. Solve the problem above.

3 Show that the number of non-saturating pushes is $O(n^3)$ when the MFIFO-rule is used. Hint: Reuse the proof for the FIFO-rule.

4 Study alternative implementations of the highest-level-rule: *Insert*(v, d) and *insert0*(v, d) may add v to the front or the rear of the d-th list.

5 Incorporate the current edge data structure into our implementations.

6 Experiment with the global relabel heuristic but without the two-phase approach.

7.11 Minimum Cost Flows

The minimum cost maximum flow problem generalizes the maximum flow problem of the preceding section.

Let $G = (V, E)$ be a directed graph. For each edge $e \in E$ let *lcap*(e) and *ucap*(e) be lower and upper bounds for the flow across e (we assume $0 \le lcap(e) \le ucap(e)$) and let *cost*$(e)$ be the cost of shipping one unit of flow across e, and for each node v let *supply*(v) be the supply or demand at node v. We talk about a supply if $supply(v) > 0$ and we talk about a demand if $supply(v) < 0$. We assume that the supplies and demands balance, i.e.,

$$\sum_{v \in V} supply(v) = 0.$$

A flow f is a function on the edges satisfying the capacity constraints and the mass balance conditions, i.e.,

$$lcap(e) \le f(e) \le ucap(e)$$

for every edge e and

$$supply(v) \quad = \quad \sum_{e;\ source(e)=v} f(e) \quad - \quad \sum_{e;\ target(e)=v} f(e)$$

for every node v.

For every edge e, *cost*(e) is the cost of sending one unit of flow across the edge. The total cost of a flow f is therefore given by

$$cost(f) = \sum_{e \in E} f(e) \cdot cost(e).$$

A *minimum cost flow* is a flow of minimum cost. The function

```
bool MIN_COST_FLOW(graph& G, const edge_array<int>& lcap,
                             const edge_array<int>& ucap,
                             const edge_array<int>& cost,
                             const node_array<int>& supply,
                                   edge_array<int>& flow)
```

returns *true* if a flow exists and returns *false* otherwise. If a flow exists, it returns a minimum cost flow in *flow*. Observe that capacities and costs must be integers. The algorithm is based on capacity scaling and successive shortest-path computation (cf. [EK72] and [AMO93]) and has running time $O(m \log U(m + n \log n))$, where n is the number of nodes of G, m is the number of edges of G, and U is the largest absolute value of any capacity.

There is also a variant of this function where the lower bound on all flows is assumed to be zero.

```
bool MIN_COST_FLOW(graph& G, const edge_array<int>& cap,
                             const edge_array<int>& cost,
                             const node_array<int>& supply,
                                   edge_array<int>& flow);
```

The function

```
int MIN_COST_MAX_FLOW(graph& G, node s, node t,
                      const edge_array<int>& cap,
                      const edge_array<int>& cost,
                      edge_array<int>& flow)
```

computes a minimum cost maximal flow, i.e., it computes a maximal flow from s and t and among these flows a flow of minimum cost. The value of the flow is returned.

The xlman-demo gw_min_cost_flow illustrates minimum cost flows.

Exercises for 7.11

1 Consider an edge $e = (u, v)$ with $c = lcap(e) > 0$. Change the problem as follows: decrease $lcap(e)$ and $ucap(e)$ by c, decrease $supply(u)$ by c, and increase $supply(v)$ by c. Show that a solution to the modified problem yields a solution of the original problem.

2 Allow negative lower bounds. Describe a transformation that gets rid of negative lower bounds.

3 Assume that $lcap(e) = 0$ for all e. Introduce auxiliary nodes s and t and edges (s, v) with capacity $c = supply(v)$ for all nodes v with $supply(v) > 0$ and edges (u, t) with capacity $c = -supply(u)$ for all nodes u with $supply(u) < 0$. Show that there is a flow satisfying the capacity constraints and the bass balance constraints in the original network iff there is a flow from s to t in the modified network that saturates all edges out of s (and hence all edges into t). Based on this insight derive a necessary and sufficient condition for the existence of a flow satisfying the capacity constraints and the mass balance constraints.

4 Let f be a flow satisfying the capacity constraints and the mass balance constraints and let G_f be the residual network with respect to f. If $e = (v, w)$ is an edge in G with $f(e) < ucap(e)$ then there is an edge (v, w) in G_f with capacity $ucap(e) - f(e)$ and cost $cost(e)$ and if $e = (v, w)$ is an edge in G with $f(e) > lcap(e)$ then there is an edge

(w, v) in G_f with capacity $f(e) - lcap(e)$ and cost $-cost(e)$. Show that f is a minimum cost flow iff there is no negative cycle in G_f.

5 Derive a checker for minimum cost flows based on the preceding items.

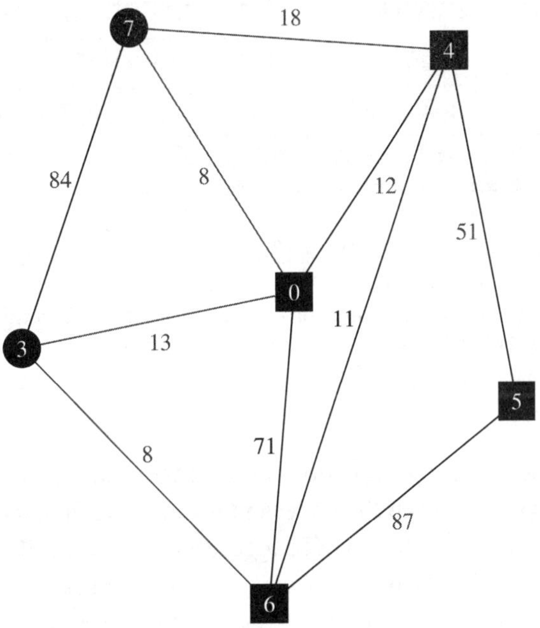

Figure 7.46 A minimum cut C in a graph. The nodes in C are shown as circles and the nodes outside C are shown as squares. The value of the cut is 47. You may generate your own figures with the xlman-demo gw_min_cut.

7.12 Minimum Cuts in Undirected Graphs

Let $G = (V, E)$ be an undirected graph (self-loops and parallel edges are allowed) and let $w : E \to \mathbb{R}_{\geq 0}$ be a *non-negative* weight function on the edges of G. A cut C of G is any subset of V with $\emptyset \neq C \neq V$. The weight of a cut is the total weight of the edges crossing the cut, i.e.,

$$w(C) = \sum_{e \in E; |e \cap C| = 1} w(e).$$

A *minimum cut* is a cut of minimum weight. Figure 7.46 shows an example. The function

```
int MIN_CUT(const graph& G,const edge_array<int>& weight,
            list<node>& C, bool use_heuristic = true)
```

	5000		10000		15000		20000	
	NOH	WH	NOH	WH	NOH	WH	NOH	WH
1000	9.22	3.52	17.11	17.11	27.86	29.36	38.88	39.46
2000	29.58	1.26	54.32	2.76	82.14	33.77	117.6	98.68
3000	62.51	3.71	107.2	3.64	145.6	8.76	191.1	85.17
4000	91.66	5.51	157	4.84	205.7	4.98	279.5	8.99
5000	144.2	15.62	213.5	11.8	273.8	11.7	378.6	18.22

Table 7.18 Running times of the minimum cut algorithms. We used random graphs with n nodes and m edges and random edge weights. The rows are indexed by n and the columns are indexed by m. For each combination of n and m we ran the algorithm without (NOH) and with the heuristic (WH). The use of the heuristic is the default.

takes a graph G and a *weight* function on the edges and computes a minimum cut. The value of the cut is returned and the nodes in the cut are assigned to C. The running time of the algorithm is $O(nm + n^2 \log n)$. The algorithm is due to [NI92, SW97]. The algorithm can be asked to use a heuristic. In some cases the heuristic improves the running time dramatically; it never seems to harm, see Table 7.18. There is also a version of the function where the cut C is the return value of the function.

```
list<node> MIN_CUT(const graph& G, const edge_array<int>& weight)
```

The function

```
int CUT_VALUE(const graph& G,const edge_array<int>& weight,
              const list<node>& C)
```

returns the value of the cut C.

We use a particularly simple and nevertheless efficient min-cut algorithm due to Nagamochi and Ibaraki [NI92] and later refined by Stoer and Wagner [SW97]. The algorithm runs in time $O(nm + n^2 \log n)$. Alternative minimum cut algorithms can be found in [PR90, HO92, KS96]. The papers [CGK$^+$97, JRT97] contain experimental comparisons of minimum cut algorithms.

We need the notion of an s-t cut. For a pair $\{s, t\}$ of distinct vertices of G a cut C is called an s-t cut if C contains exactly one of s and t.

The algorithm works in phases. In each phase it determines a pair of vertices s and t and a minimum s-t cut C. If there is a minimum cut of G separating s and t then C is a minimum cut of G. If not then any minimum cut of G has s and t on the same side and therefore the graph obtained from G by *combining* s and t has the same minimum cut as G. So a phase determines vertices s and t and a minimum s-t cut C and then combines s and t

into one node. After $n-1$ phases the graph is shrunk to a single node and one of the phases must have determined a minimum cut of G.

⟨*min_cut*⟩≡

```
⟨combine s and t⟩
int MIN_CUT(const graph& G0, const edge_array<int>& weight,
            list<node>& C, bool use_heuristic)
{ node v; edge e;
  forall_edges(e,G0)
    if ( weight[e] < 0 )
      error_handler(1,"MIN_CUT: no negative weights");
  ⟨initialization⟩
  while ( G.number_of_nodes() >= 2 ) { ⟨a phase⟩ }
  return best_value;
}
```

We call our input graph *G0* and our current Graph G. Every node of G represents a set of nodes of *G0*. This set is stored in a linear list pointed to by $G[v]$ and hence we use the type *GRAPH<list<node>*, int>* for G. Every edge $e = \{v, w\}$ of G represents a set of edges of *G0*, namely $\{\{x, y\}\ ; x \in G[v] \text{ and } y \in G[w]\}$. The total weight of these edges is stored in $G[e]$.

It is easy to initialize G. We simply make G a copy of *G0* (except for self-loops) and initialize $G[v]$ to the appropriate singleton set for every vertex v of G.

⟨*initialization*⟩≡

```
typedef list<node>* nodelist_ptr;
GRAPH<nodelist_ptr, int> G;
G.make_undirected();
node_array<node> partner(G0);
forall_nodes(v,G0)
{ partner[v] = G.new_node(new list<node>);
  G[partner[v]]->append(v);
}
forall_edges(e, G0)
if ( source(e) != target(e) )
  G.new_edge(partner[source(e)], partner[target(e)],weight[e]);
```

We also fix a particular node a of G and introduce variables to store the currently best cut.

⟨*initialization*⟩+≡

```
node a = G.first_node();
int best_value = MAXINT;
int cut_weight = MAXINT;
```

We now come to the heart of the matter, a phase. A phase initializes a set A to the singleton set $\{a\}$ and then successively merges all other nodes of G into A. In each stage the node

$v \notin A$ which maximizes

$$w(v, A) = \sum_{e;\ e=\{v,y\} \text{ for some } y \in A} w(e)$$

is merged into A. Let s and t be the last two vertices added to A in a phase. The cut C computed by the phase is the cut consisting of node t only; in the graph *G0* this corresponds to the cut $G[t]$.

Lemma 44 *Let s and t be the last two nodes merged into A during a phase. Then $\{t\}$ is a minimum s-t cut.*

Proof Let C' be any s-t cut. We show that $w(C') \geq w(\{t\})$. Let $v_1, \ldots, v_n$ be the order in which the nodes are added to A. Then $v_1 = a$, $v_{n-1} = s$, and $v_n = t$.

Call a vertex $v = v_i$ critical if $i \geq 2$ and v_i and v_{i-1} belong to different sides of C'. Note that t is critical. Let k be the number of critical nodes and let $i_1, i_2, \ldots, i_k$ be the indices of the critical nodes. Then $i_k = n$. For integer i use A_i to denote the set $\{v_1, \ldots, v_i\}$. Then

$$w(\{t\}) = w(v_{i_k}, A_{i_k - 1})$$

and

$$w(C') \geq \sum_{j=1}^{k} w(v_{i_j}, A_{i_j - 1} \setminus A_{i_{j-1} - 1}),$$

since any edge counted on the right side is also counted on the left and edge costs are non-negative. We now show for all integers l, $1 \leq l \leq k$, that

$$w(v_{i_l}, A_{i_l - 1}) \leq \sum_{j=1}^{l} w(v_{i_j}, A_{i_j - 1} \setminus A_{i_{j-1} - 1}).$$

For $l = 1$ we have equality. So assume $l \geq 2$. We have

$$\begin{aligned}
w(v_{i_l}, A_{i_l - 1}) &= w(v_{i_l}, A_{i_{l-1} - 1}) + w(v_{i_l}, A_{i_l - 1} \setminus A_{i_{l-1} - 1}) \\
&\leq w(v_{i_{l-1}}, A_{i_{l-1} - 1}) + w(v_{i_l}, A_{i_l - 1} \setminus A_{i_{l-1} - 1}) \\
&\leq \sum_{j=1}^{l-1} w(v_{i_j}, A_{i_j - 1} \setminus A_{i_{j-1} - 1}) + w(v_{i_l}, A_{i_l - 1} \setminus A_{i_{l-1} - 1}) \\
&\leq \sum_{j=1}^{l} w(v_{i_j}, A_{i_j - 1} \setminus A_{i_{j-1} - 1}).
\end{aligned}$$

Here the first inequality follows from the fact that $v_{i_{l-1}}$ is added to $A_{i_{l-1} - 1}$ and not v_{i_l} and the second inequality uses the induction hypothesis. □

⟨*a phase*⟩≡

⟨*determine s and t and the value of the cut V-t,t*⟩;

```
bool new_best_cut = false;
if ( cut_weight < best_value )
```

```
{ C = *(G[t]);
  best_value = cut_weight;
  new_best_cut = true;
}
combine_s_and_t(G,s,t);
⟨heuristic⟩
```

How can we determine the order in which the vertices are merged into A? This can be done in a manner akin to Prim's minimum spanning tree algorithm. We keep the vertices v, $v \notin A$, in a priority queue ordered according to $w(v, A)$. In each stage we select the node, say u, with maximal $w(u, A)$ and add it to A. This increases $w(v, A)$ by $w(\{v, u\})$ for any vertex $v \notin A$ and $v \neq u$. Since LEDA priority queues select minimal values we store $-w(v, A)$ in the queue. The node added last to A is the vertex t. The value *cut_weight* is $w(t, A_t)$.

⟨*determine s and t and the value of the cut V-t,t*⟩≡

```
node t = a;
node s;
node_array<bool> in_PQ(G,false);
node_pq<int> PQ(G);
forall_nodes(v,G)
if (v != a)
{ PQ.insert(v,0);
  in_PQ[v] = true;
}
forall_adj_edges(e,a)
  PQ.decrease_inf(G.opposite(a,e),PQ.prio(G.opposite(a,e)) - G[e]);
while (!PQ.empty())
{ s = t;
  cut_weight =  -PQ.prio(PQ.find_min());
  t = PQ.del_min();
  in_PQ[t] = false;
  forall_adj_edges(e,t)
  { if (in_PQ[v = G.opposite(t,e)])
        PQ.decrease_p(v,PQ.prio(v) - G[e]);
  }
}
```

It remains to combine s and t. We do so by deleting t from G and moving all edges incident to t to s. More precisely, we need to do three things:

- Add $G[t]$ to $G[s]$ ($G[s] \rightarrow conc(*(G[t]))$).
- Increase $G[\{s, v\}]$ by $G[\{t, v\}]$ for all vertices v with $\{t, v\} \in E$ and $v \neq s$.
- Delete t and all its incident edges from G (*G.del_node*(t)).

The second step raises two difficulties: the edge $\{s, v\}$ might not exist and there is no simple way to go from the edge $\{t, v\}$ to the edge $\{s, v\}$. We overcome these problems by first recording the edge $\{s, v\}$ in *s_edge*$[v]$ for every neighbor v of s. We then go through the neighbors v of t: if v is connected to s then we simply increase $G[\{s, v\}]$ by $G[\{t, v\}]$, if v is not connected to s and different from s then we add a new edge $\{s, v\}$ with weight $G[\{t, v\}]$.

We formulate the piece of code to combine s and t as a procedure because we want to reuse it in the heuristic.

⟨*combine s and t*⟩≡

```
static void combine_s_and_t(GRAPH<list<node>*,int>& G, node s, node t)
{ G[s]->conc(*(G[t]));
  node_array<edge> s_edge(G,nil);
  edge e;
  forall_adj_edges(e,s)  s_edge[G.opposite(s,e)] =  e;
  forall_adj_edges(e,t)
  { node v = G.opposite(t,e);
    if ( v == s) continue;
    if (s_edge[v] == nil) G.new_edge(s,v,G[e]);
    else G[s_edge[v]] += G[e];
  }
  G.del_node(t);
}
```

This completes the description of the algorithm. The running time of our algorithm is clearly at most n times the running time of a phase. A phase takes time $O(m + n \log n)$ to merge all nodes into the set A (the argument is the same as for Prim's algorithm) and time $O(n)$ to record the cut computed and to merge s and t. The total running time is therefore $O(nm + n^2 \log n)$.

We next discuss a heuristic improvement. Clearly, any edge whose weight is at least *best_value* cannot cross a minimum cut whose value is smaller than *best_value*. We therefore might as well shrink any such edge.

Which edges might have weight at least as large as *best_value*? If *best_value* decreased in the current phase, then all edges of G are candidates, and if *best_value* stayed unchanged in the current phase, then all edges incident to s are candidates, because their weight may have increased .

⟨*heuristic*⟩≡

```
if ( use_heuristic )
{ bool one_more_round = true;
  while ( one_more_round )
  { one_more_round = false;
    forall_adj_edges(e,s)
    { node t = G.opposite(s,e);
      if ( G[e] >= best_value )
      { combine_s_and_t(G,s,t); one_more_round = true; break; }
    }
  }
  if ( new_best_cut )
```

```
  { bool one_more_round = true;
    while ( one_more_round )
    { one_more_round = false;
      forall_edges(e,G)
      { node s = G.source(e);
        node t = G.target(e);
        if ( G[e] >= best_value )
        { combine_s_and_t(G,s,t); one_more_round = true; break; }
      }
    }
  }
}
```

Table 7.18 shows that the heuristic can lead to dramatic improvements in running time. We will now argue that is does increase the asymptotic running time. If the phase did not decrease *best_value*, the running time of the heuristic is $O((1+k)n)$, where k is the number of edges shrunken by the heuristic. If the phase decreased *best_value*, the running time of the heuristic is $O((1+k)m)$, where k is the number of edges shrunken by the heuristic. In either case the asymptotic running time of our procedure is not increased, since a phase has cost $\Omega(m + n\log n)$.

We considered an alternative implementation of the heuristic. We kept the edges of G in a priority queue according to negative weight and at the end of each phase selected all edges from the queue which had weight at least as large as *best_value*. The alternative implementation was slower than the simple implementation described above.

8

Embedded Graphs

Drawings of graphs are ubiquitous. In this chapter we introduce important mathematical concepts related to embedded graphs and we discuss algorithms that draw and embed graphs and that deal with embedded graphs. We provide only a minimum of the required mathematics and refer the reader to [Whi73] for a detailed treatment.

We start with the definition of what it means to draw a graph and an example of a drawing algorithm. We discuss bidirected graphs and maps, our technical vehicle for dealing with embedded graphs, in Section 8.2 and the concepts of embedding and planar embedding in Section 8.3. In this section we also introduce functions that test the planarity of a graph, that construct a plane embedding of a planar graph, and that exhibit a Kuratowski subgraph in a non-planar graph. Their implementation is discussed in Section 8.7. Sections 8.4 and 8.5 introduce order-preserving embeddings, plane maps, face cycles, and the genus of maps. In Section 8.6 and 8.12 we relate combinatorics and geometry. In particular, we prove that a map is plane if and only if its genus is zero, we derive an upper bound on the number of edges of any planar graph and we show how to construct the map induced by geometric positions assigned to the nodes of a graph. In Section 8.8 we show how to modify maps, in Section 8.9 we discuss the generation of random plane maps, and in Section 8.13 we introduce functions that five-color a planar graph and choose a large independent set in a planar graph. Section 8.10 introduces face items as a means of dealing with faces in the same way as with nodes and edges. In Section 8.11 we discuss our design choice of representing maps by directed graphs instead of undirected graphs.

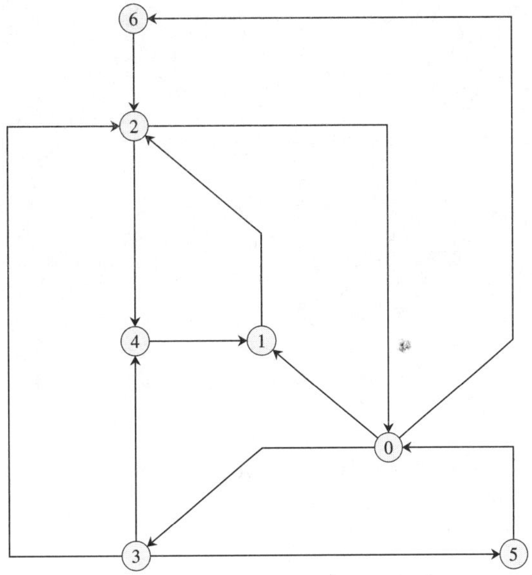

Figure 8.1 A drawing produced by one of the graph drawing algorithms in AGD [JMN].

8.1 Drawings

We have already seen many drawings of graphs in this book. We have never defined what we mean by a drawing, embedding, and planar embedding.

Let G be a graph and let S be a surface, e.g., the plane or the sphere or the torus. We will be almost exclusively concerned with the plane in this book. However, the concepts also apply to more complex surfaces.

A *drawing* I of G in S assigns a point $I(v) \in S$ to every node v of G and a Jordan curve[1] $I(e)$ to every edge $e = (v, w)$ such that:

(1) distinct points are assigned to distinct nodes, i.e., $I(v) \neq I(w)$ for $v \neq w$,
(2) the curve assigned to any edge connects the endpoints of the edge, i.e., if $e = (v, w)$ then $I(e)(0) = I(v)$ and $I(e)(1) = I(w)$.

A drawing in the plane is called a straight line drawing if every edge is drawn as a straight line segment. Figure 8.2 shows some drawings.

An algorithm, that takes a graph and produces a drawing for it, is called a *graph drawing algorithm*[2]. LEDA provides some graph drawing algorithms; see the section on graph drawing in the manual and *try the button layout in a GraphWin for a demonstration.* Many more graph drawing algorithms are available in the systems AGD [JMN] and GDToolkit [Bat].

[1] A Jordan curve c is a curve without self-intersections, i.e., a continuous mapping $c : [0, 1] \longrightarrow S$ with $c(x) \neq c(y)$ for $0 \leq x < y < 1$.

[2] Graph drawing is an active area of research, see [BETT94, EM98, DETT98] for surveys.

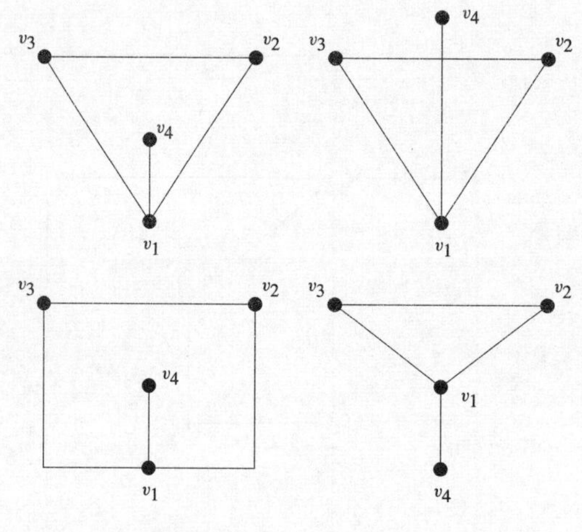

Figure 8.2 Some drawings of the same graph. All drawings except for the right upper drawing are embeddings.

Both systems are based on LEDA. Figure 8.1 shows a drawing produced by an algorithm in AGD.

The functions

```
void SPRING_EMBEDDING(const graph& G,
        node_array<double>& xpos, node_array<double>& ypos,
        double xleft, double xright, double ybottom, double ytop,
        int iterations = 250);
void SPRING_EMBEDDING(const graph& G, const list<node>& fixed,
        node_array<double>& xpos, node_array<double>& ypos,
        double xleft, double xright, double ybottom, double ytop,
        int iterations = 250);
```

compute straight line drawings of a graph G using a so-called *spring embedder*[3]. A spring embedder works iteratively. It models the nodes of a graph as points in the plane that repulse each other, and it models each edge as a spring between the endpoints of the edge. In each iteration the force acting on any node is computed as the sum of repulsive forces (from all other nodes) and attractive forces (from incident edges), and the node is moved accordingly. The number of iterations is determined by the parameter *iterations*.

The x- and y-coordinates of the positions assigned to the nodes of G are returned in *xpos* and *ypos*, respectively, and the points are constrained to lie in the rectangle defined by *xleft*, *xright*, *ybottom*, and *ytop*. The second version of the function keeps the positions of the nodes in *fixed* fixed.

Drawings in which edges do not cross are particularly nice. We call such drawings embeddings. Out of the four drawings shown in Figure 8.1 three are embeddings. Embeddings are the topic of Section 8.3. The graphs in Figure 8.2 are undirected. For the purposes of

[3] The name spring drawer would be more appropriate, as spring embedders do not produce embeddings, but drawings. However, the name spring embedder is in general use.

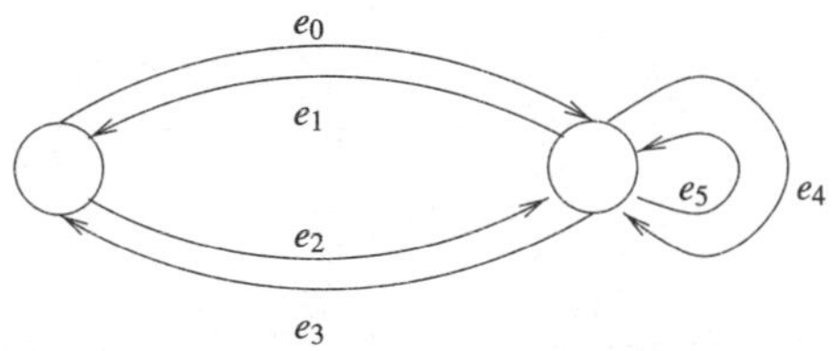

Figure 8.3 A bidirected graph: We have $reversal(e_{2i}) = e_{2i+1}$ and $reversal(e_{2i+1}) = e_{2i}$ for all i with $0 \leq i \leq 2$. Requirement (2) excludes the possibility that $reversal(e_0) = e_1$, and $reversal(e_3) = e_0$, and requirement (3) excludes the possibility that $reversal(e_4) = e_4$ and $reversal(e_5) = e_5$.

this chapter it is convenient to distinguish between the two orientations of an edge. This leads to the concepts of bidirected graphs and maps, which we treat in the next section.

Exercise for 8.1

1 Implement a spring embedder.

8.2 Bidirected Graphs and Maps

A directed graph $G = (V, E)$ is called *bidirected* if there is a bijective function *reversal* : $E \to E$ such that for every edge $e = (v, w)$ with $e^R = reversal(e)$:

(1) $e^R = (w, v)$, i.e., $source(e) = target(e^R)$ and $target(e) = source(e^R)$,
(2) $reversal(e^R) = e$, and
(3) $e \neq e^R$.

Property (1) ensures that reversal deserves its name, and properties (2) and (3) ensure that reversal behaves properly in the presence of parallel edges and self-loops. Figure 8.3 shows an example of a bidirected graph and also illustrates properties (2) and (3). A bidirected graph has an even number of edges.

The function

```
bool G.is_bidirected();
```

returns *true* if G is bidirected and returns *false* otherwise. The function

```
void G.make_bidirected(<list<edge>& R);
```

adds a minimum number of edges to G so as to make G bidirected. The added edges are returned in R.

Every edge e of any graph G has a reversal information associated with it. It is accessed through

```
G.reversal(e)
```

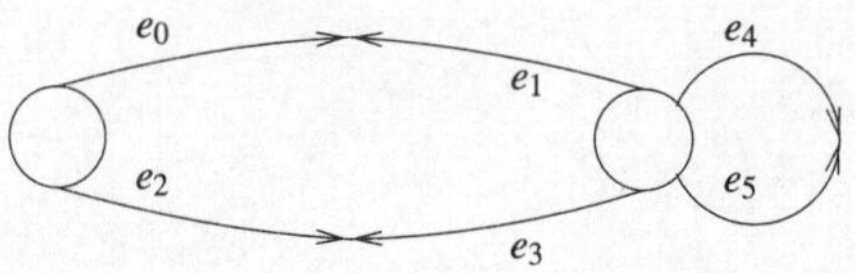

Figure 8.4 A map: Every pair of edges $\{e, e^R\}$ with *reversal*$(e) = e^R$ and *reversal*$(e^R) = e$ is drawn as two half-edges. For each half-edge the name of the half-edge is shown on the left side of the half-edge.

and has type *edge*. The reversal information of an edge is either undefined (= *nil*) or is an edge e^R satisfying (1) to (3). The operation

```
G.set_reversal(e,f)
```

sets the reversal information of e to f and the reversal information of f to e. The function checks whether the created reversal information is legal and aborts if it is not. If the reversal information of e was defined prior to the operation, the reversal information of e^R is set to *nil* by the operation. The same holds true for f.

A *map* is a graph in which the reversal information of every edge is defined. A map is always a bidirected graph and every bidirected graph can be turned into a map by setting the reversal information appropriately. The function

```
bool G.is_map()
```

returns *true* if G is a map and the functions

```
bool G.make_map()
void G.make_map(list<edge>& R)
```

turn G into a map by setting the reversal information of every edge. The first function requires that G is bidirected (if G is not bidirected, the function returns *false* and sets the reversal information of a maximal number of edges), the second function adds a minimum number of edges to G so as to make G bidirected and then turns G into a map. Both functions preserve reversal information, i.e., if *reversal*(e) is defined before the call, then *reversal*(e) is not changed by either call.

We call a pair of edges $\{e, e^R\}$ with *reversal*$(e) = e^R$ (and hence *reversal*$(e^R) = e$) a *uedge* (undirected edge) and say that e and e^R form the uedge. The uedge comprising e and e^R is denoted $\{e, e^R\}$ or $\{v, w\}$, where v and w are the two endpoints of e. The latter notation is ambiguous in the presence of parallel edges. We depict maps as shown in Figure 8.4. For every uedge $\{e, e^R\}$ we draw "two half-edges that meet" and label them e and e^R, respectively.

We have no iteration statement that iterates over the uedges of a graph. However, it is easy to obtain the effect of iterating over uedges.

```
forall_edges(e,G)
{ if ( index(e) > index(G.reversal(e)) ) continue;
  <body of loop>
}
```

Observe that the body of the loop is executed for exactly one edge in each uedge, namely the one with smaller index.

We describe the implementations of some of the functions introduced above. We also introduce a function that checks whether the reversal information of all edges is properly defined. This section may be skipped on first reading.

We start with a function *check_reversal_inf* that checks whether the reversal information of every edge is either nil or satisfies (1) to (3) and raises an error if this is not the case[4]. The function is non-trivial to write because it cannot assume that the reversal information of an edge has a meaningful value, i.e., the function has to cope with the possibility that *G.reversal(e)* is non-nil and not an edge of *G* for some *e*.

We proceed as follows. We introduce a map *is_edge_of_G* from edges to bool that we initialize to *false*. We then set *is_edge_of_G*[*e*] to *true* for all edges *e* of *G*. Next, we iterate again over all edges *e* of *G* and make sure that *reversal(e)* is either *nil* or an edge of *G*. In a third step we make sure that (1) to (3) holds for all edges *e* whose reversal information is not *nil*.

⟨*check_reversal_inf.c*⟩+≡

```
bool check_reversal_inf(const graph& G)
{ map<edge,bool> is_edge_of_G(false);
  edge e;
  forall_edges(e,G) is_edge_of_G[e] = true;
  forall_edges(e,G)
  { edge r = G.reversal(e);
    if ( r == nil || !is_edge_of_G[r]) return false;
  }
  forall_edges(e,G)
  { edge r = G.reversal(e);
    if (r == e || G.reversal(r) != e ||
       G.source(e) != G.target(r) || G.target(e) != G.source(r) )
    return false;
  }
  return true;
}
```

It is instructive to investigate what can go wrong when only the third *forall_edges* loop is executed. It would then be possible that *r* is different from *nil* but not an edge of *G*. The access to the reversal, target, or source of *r* could then result in a segmentation fault. The

[4] We use the function *check_reversal_inf* for testing purposes. Of course, all functions of the LEDA system are designed to preserve the invariant that the reversal of every function is either nil or an edge of *G* satisfying (1) to (3) and hence, if none of the implementers of LEDA had ever made a mistake, the function would have never raised an error.

program above guards against this possibility by ensuring first that the reversal of any edge e of G is either *nil* or an edge of G.

We next show the implementation of the function *make_map*. Its implementation is derived from the function *Is_Bidirected* given in Section 6.12.

A call of *G.make_map*( ) sets the reversal information of a maximal number of edges. We proceed as follows: let $v_1, v_2, \ldots, v_n$ be an arbitrary order on the nodes of G, e.g., the ordering given by the internal numbering of the nodes[5]. We make two lists *EST* and *ETS* of all edges whose reversal information is undefined. *EST* starts with all edges out of v_1, followed by all edges out of $v_2, \ldots$. For each i, the edges out of v_i are in increasing order of their target node. *ETS* starts with all edges into v_1, followed by all edges into $v_2, \ldots$. For each i, the edges into v_i are in increasing order of the source node. We also want the self-loops incident to any v_i to appear in reverse order in the two lists.

The lists *EST* and *ETS* are easy to generate. We collect all edges whose reversal information is undefined in a list *EST* and use bucket sort to rearrange *EST* in increasing lexicographic order. We use the index of the source node of an edge as the primary key and the index of the target node as the secondary key. For *ETS* we interchange the roles of the primary and the secondary key, and we initialize *ETS* to the reversal of *EST*. The effect of initializing *ETS* with the reversal of *EST* instead of with *ETS* is that the self-loops incident to any v_i appear in reverse order in the two lists; this follows from the fact that bucket sort is stable.

Having rearranged both lists we establish the reversal information. *EST* starts with all edges out of v_1 sorted in order of increasing target and *ETS* starts with all edges into v_1 sorted in order of increasing source. Both lists start with all self-loops incident to v_1.

We scan over both lists and check whether the first edge on *EST*, call it e, can be paired with the first edge on *ETS*, call it r. We can pair e and r if none of them was paired previously and if $source(e) = target(r)$, $target(e) = source(r)$, and $e \neq r$. If e and r can be paired, we pair them by setting their reversal information appropriately. The function succeeds if all edges can be paired.

So assume that e and r cannot be paired. We show that at least one of e and r will never find a partner.

Assume first that $source(e) \neq target(r)$. If $source(e) < target(r)$ then *ETS* contains no further edge which ends in $source(e)$. Thus e cannot be paired. Similarly, if $source(e) > target(r)$ then *EST* contains no further edge that starts in $target(r)$. Thus r cannot be paired.

Assume next that $source(e) = target(r)$ and $target(e) \neq source(r)$. If $target(e)$ is less than $source(r)$ then *ETS* contains no further edge that starts in $source(e)$ and ends in $target(e)$ and hence e cannot be paired. If $target(e)$ is greater than $source(r)$ then *EST* contains no further edge that ends in $target(r)$ and starts in $source(r)$ and hence r cannot be paired.

Assume finally that $source(e) = target(r)$ and $target(e) = source(r)$ and $e = r$, i.e., e is a self-loop. Since *EST* and *ETS* contain the self-loops incident to any node in reverse order

[5] The internal number of a node v is given by $index(v)$.

this can only happen if there is an odd number of self-loops incident to *source*(*e*) and if *e* is the middle element of the block of self-loops incident to *source*(*e*). In this situation it is OK if *e* stays unpaired and all other self-loops incident to *source*(*e*) are paired.

⟨*make_map.c*⟩≡

```
static int map_edge_ord1(const edge& e) { return index(source(e)); }
static int map_edge_ord2(const edge& e) { return index(target(e)); }

bool graph::make_map()
{
  int n = max_node_index();
  int count = 0;

  edge e,r;

  list<edge> EST;
  forall_edges(e,(*this)) if (e->rev == nil) EST.append(e);

  int number_of_undefined_reversals = EST.length();

  list<edge> ETS = EST; ETS.reverse();

  EST.bucket_sort(0,n,&map_edge_ord2); // secondary key
  EST.bucket_sort(0,n,&map_edge_ord1); // primary key

  ETS.bucket_sort(0,n,&map_edge_ord1); // secondary key
  ETS.bucket_sort(0,n,&map_edge_ord2); // primary key

  // merge EST and ETS to find corresponding edges

  while (! EST.empty() && ! ETS.empty())
  { e = EST.head();
    r = ETS.head();

    if ( e->rev != nil ) { EST.pop(); continue; }
    if ( r->rev != nil ) { ETS.pop(); continue; }

    if ( target(r) == source(e) )
    { if ( source(r) == target(e) )
      { ETS.pop(); EST.pop();
        if ( e != r )
        { e->rev = r; r->rev = e;
          count += 2;
        }
        continue;
      }
      else // target(r) == source(e) && source(r) != target(e)
      { if (index(source(r)) < index(target(e)))
          ETS.pop();  // r cannot be matched
        else
          EST.pop();  // e cannot be matched
      }
    }
    else // target(r) != source(e)
    { if (index(target(r)) < index(source(e)))
        ETS.pop();  // r cannot be matched
      else
        EST.pop();  // e cannot be matched
    }
```

```
  }
  return count == number_of_undefined_reversals;
}
```

Given the function above, it is trivial to extend a graph G to a map. A call *G.make_map*( ) determines the reversal information of a maximal number of edges. For any edge whose reversal information is still undefined, we add the reversed edge to G and set the reversal information accordingly.

⟨*make_map.c*⟩+≡

```
void graph::make_map(list<edge>& R)
{ if (make_map()) return;
  list<edge> el = all_edges();
  edge e;
  forall(e,el)
  { if (e->rev == nil)
    { edge r = new_edge(target(e),source(e));
      e->rev = r;
      r->rev = e;
      R.append(r);
    }
  }
}
```

Exercises for 8.2

1 Does the function *check_reversal_inf* work if the map *is_edge_of_G* is replaced by an edge array?

2 Does the function *check_reversal_inf* work if the last two *forall_edges* loops are combined into one?

8.3 Embeddings

Embeddings are special drawings, namely drawings where no edge is drawn across a node, where the images of distinct edges do not cross, and where the two edges comprising a uedge are embedded the same. Formally, we define as follows:

A drawing I of a graph G into a surface S is called an *embedding* if the images of edges contain no images of points in their relative interiors[6], if the images of edges belonging to distinct uedges are disjoint except for endpoints[7], and if the curves assigned to edges belonging to the same uedge are reversals of each other[8].

Figure 8.1 shows three embeddings of a map M_0 into the plane; M_0 has nodes v_1, v_2, v_3,

[6] $I(e)(x) \neq I(v)$ for any edge e, node v, and real x with $0 < x < 1$

[7] $I(e)(x) \neq I(e')(y)$ for edges e and e' with $e \neq e'$ and $e' \neq reversal(e)$ and all x and y with $0 < x, y < 1$

[8] $I(e^R)(x) = I(e)(1-x)$ for all edges e, $e^R = reversal(e)$, and all x, $0 \leq x \leq 1$

and v_4 and uedges $\{v_1, v_2\}$, $\{v_1, v_3\}$, $\{v_1, v_4\}$, and $\{v_2, v_3\}$, and will be used as the running example in this chapter. An embedding into the plane is called a *planar embedding*, and a planar embedding in which every edge is mapped to a straight line segment is called a *straight line embedding*. A graph G is called *planar* if it has a planar embedding.

The function

```
bool Is_Planar(const graph& G)
```

tests whether the graph $G = (V, E)$ has a planar embedding. It returns *true* if G is planar and *false* otherwise. The running time is $O(n + m)$.

The functions

```
bool    PLANAR(graph& G, bool embed = false);
bool HT_PLANAR(graph& G, bool embed = false);
bool BL_PLANAR(graph& G, bool embed = false);
```

also test whether the graph G is planar. When *embed* is *true*, G is a map, and G is planar (the functions rise an error when *embed* is *true* and G is not a map), the functions in addition reorder the adjacency lists of G such that G becomes a plane map. The notion of plane map is explained in Section 8.4. All of this takes time $O(n + m)$.

There are two implementations of the planarity test and planar embedding algorithm: HT_PLANAR realizes the planarity testing algorithm of Hopcroft and Tarjan, see [HT74] or [Meh84c, IV.10], and the embedding algorithm of Mehlhorn and Mutzel, see [MM95]. BL_PLANAR realizes the planarity testing algorithm of Lempel, Even, and Cederbaum, and Booth and Lueker, see [LEC67, Eve79, BL76], and the embedding algorithm of Nishizeki and Chiba, see [NC88]. The implementation of HT_PLANAR is documented in [MMN94] and the implementation of BL_PLANAR is discussed in Section 8.7. BL_PLANAR is the faster of our implementations and hence PLANAR is synonymous to BL_PLANAR.

The functions

```
bool    PLANAR(graph& G, list<edge>& el, bool embed = false);
bool HT_PLANAR(graph& G, list<edge>& el, bool embed = false);
bool BL_PLANAR(graph& G, list<edge>& el, bool embed = false);
```

behave like the functions above when G is planar. If G is non-planar, the functions also return a proof of non-planarity in the form of the edges *el* of a Kuratowski subgraph. The identification of Kuratowski subgraphs takes linear time $O(n + m)$ in BL_PLANAR and PLANAR, and takes quadratic time $O(n^2)$ in HT_PLANAR. We explain the notion of *Kuratowski subgraph*.

Figure 8.5 shows two non-planar graphs, the complete graph K_5 on five nodes and the complete bipartite graph $K_{3,3}$ with three nodes on each side. The non-planarity of both graphs will be shown in Lemma 47 in Section 8.6. It is a famous theorem of Kuratowski, see [Kur30, Whi73], that every non-planar graph G contains a subdivision[9] of either K_5 or $K_{3,3}$, i.e., there is a set *el* of edges in G forming a subdivision of either K_5 or $K_{3,3}$. Figure 8.6 shows a Kuratowski subgraph of a non-planar graph.

[9] Let K be an arbitrary graph. A subdivision of K is obtained from K by subdividing edges. To subdivide an edge means to split the edge into two by placing a new vertex on the edge.

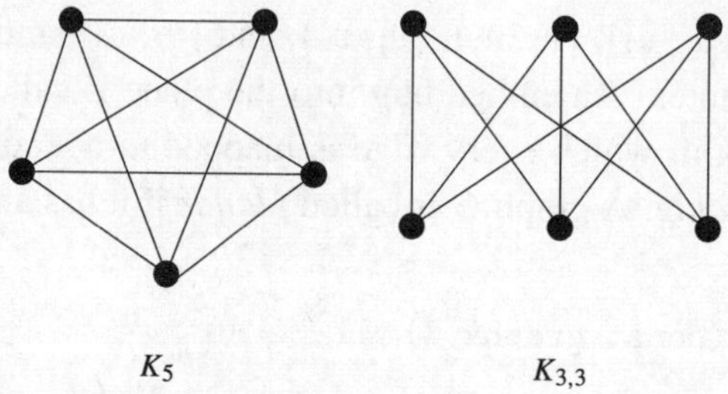

Figure 8.5 The Kuratowski graphs K_5 and $K_{3,3}$.

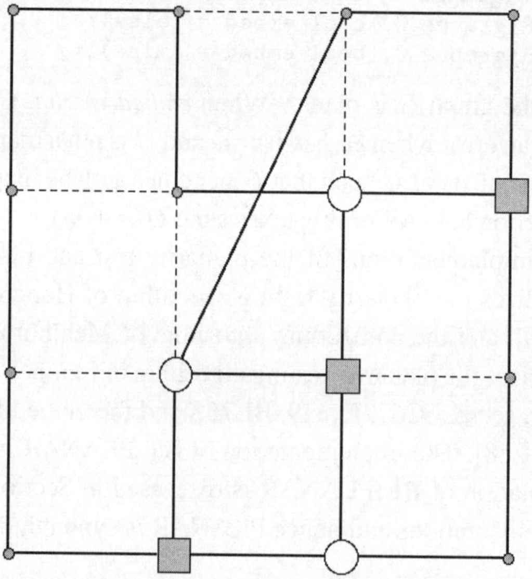

Figure 8.6 A non-planar graph and the Kuratowski subgraph proving non-planarity. The edges of the Kuratowski subgraph are shown in bold. This figure was generated with the xlman-demo gw_plan_demo.

There is also a function that gives more information about the Kuratowski subgraph than just the list of its edges.

```
int KURATOWSKI(graph& G, list<node>& V, list<edge>& E,
               node_array<int>& deg);
```

returns zero if G is planar and returns one otherwise. If G is non-planar, it computes a Kuratowski subdivision K of G as follows: V is the list of all nodes and subdivision points of K. For all $v \in V$ which are subdivision points, the degree $deg[v]$ is equal to 2. If K is a K_5, then $deg[v]$ is equal to 4 for all nodes $v \in V$ that are not subdivision points. If K is a $K_{3,3}$, then $deg[v]$ is equal to -3 $(+3)$ for the nodes v on the left (right) side of the $K_{3,3}$.

If G is a plane map, the function

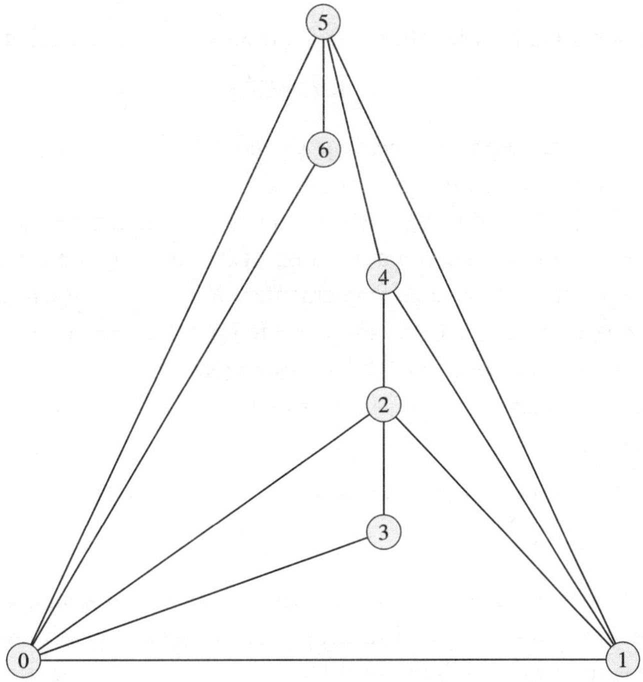

Figure 8.7 A straight line drawing produced by STRAIGHT_LINE_EMBEDDING. This figure was generated with the xlman-demo gw_plan_demo.

```
int STRAIGHT_LINE_EMBEDDING(graph& G, node_array<int>& xcoord,
                                       node_array<int>& ycoord);
```

computes for each node v of G a point (*xcoord*[v], *ycoord*[v]) with integer coordinates in the range $[0 \mathinner{..} 2(n-1)]$ such that the straight line embedding defined by these node positions is an order preserving embedding of G. The algorithm [Fár48, dFPP88] has running time $O(n^2)$. G must not have parallel edges and it must not have self-loops (since the existence of parallel edges or self-loops excludes the existence of a straight line embedding). Figure 8.7 shows a straight line drawing produced by this algorithm.

The function *Is_Planar* played an important role in the development of LEDA. We added the function to the system in 1991. The function had been implemented as part of a master's thesis and had been tested on a small number of examples (we did not have a large collection of planar graphs available to us). The master's thesis described the implementation; the actual program was not part of the thesis.

In 1993 we were sent a planar graph which, however, our program declared non-planar. When we started to revise the program we learned two things. First, we learned that writing a function

```
bool Is_Planar(const graph& G)
```

means asking for trouble. A function that answers a complex question like

Is G planar?

should not just return "YES" or "NO"; *it should justify its answer in a way that is easily checked by the caller of the function.*

Second, we learned that documentation and implementation had to be tied together more closely by the use of literate programming. Literate programming, first advocated by D.E. Knuth, suggests to embed an implementation into a document that describes the algorithm. All programs in this book are presented in a literate programming style. We first used CWEB [KL93] and later switched to noweb [Ram94].

In the case of planarity testing, the learning process led to reports [MMN94, MM95, HMN96] and to function

```
bool PLANAR(graph& G, list<edge>& el, bool embed)
```

which justifies its answers:

- When G is non-planar the function returns a proof of non-planarity in the form of the set *el* of edges of a Kuratowski subgraph. The caller can easily check that the edges in *el* form a Kuratowski subdivision of G.

- When G is planar, *embed* is set to *true*, and G is a map, the function reorders the adjacency lists of G such that G becomes a plane map. A caller of PLANAR has two ways to check whether the returned map is plane. He can either produce a planar drawing of G with the help of STRAIGHT_LINE_EMBEDDING and visually inspect the result, or he can compute the genus of G. The genus of maps will be discussed in Section 8.6 and it will be shown there that a map is plane iff its genus is zero. The genus of a map can be computed by a simple program.

The fact that PLANAR justifies its answers and that the answers are easily checked can be used to test the function on any input. Observe that testing is usually restricted to inputs where the answer is known by other means. The following test program exploits the fact that PLANAR can be tested on any input.

We choose integers n and m such that a random map with n nodes and m uedges has a fair chance of being planar and a fair chance of being non-planar, generate random maps with n nodes and about m edges, test them for planarity, and check the answer.

⟨*planar_test.c*⟩+≡

```
main(){
int n = read_int("n = ");  int m = read_int("m = ");
graph G;
list<edge> el;
int P = 0; int K = 0;
while (P + K < 1000)
```

```
{ random_graph(G,n,m);
  list<edge> R;
  G.make_map(R);
  if ( PLANAR(G,el,true) )
    { assert(Genus(G) == 0); P++; }
  else
    { assert(CHECK_KURATOWSKI(G,el)); K++; }
}
cout << "\n\nnumber of plane graphs = " << P;
cout << "\n\nnumber of non-plane graphs = " << K; newline;
}
```

In a run with $n = 50$ and $m = 55$, the program above found 308 planar graphs and 692 non-planar graphs.

The function PLANAR was the first function in LEDA that justified its answers. By now, many functions do. We have seen many examples already in the preceding chapters and we will see more in the chapters to come. A general discussion of the role of program checking in LEDA can be found in Section 2.14.

Exercises for 8.3

1 Let G be a non-planar graph. Show that the following strategy identifies the edges of a Kuratowski subgraph. Iterate over all edges e of G. If $G \setminus e$ is non-planar, remove e from G, and if $G \setminus e$ is planar leave G unchanged. The edges remaining in G form a Kuratowski subgraph.

2 Write a function

```
bool CHECK_KURATOWSKI(const graph& G, const list<edge>& el)
```

that returns *true* if the edges in *el* form a Kuratowski subdivision of G.

8.4 Order-Preserving Embeddings of Maps and Plane Maps

We define the notion of an order preserving embedding of a map.

For a vertex v, we use $A(v)$ to denote the set of edges with source v. The set $A(v)$ is stored as a cyclic list. For an edge e,

```
G.cyclic_adj_succ(e);
G.cyclic_adj_pred(e);
```

return the successor and predecessor of e, respectively, in the cyclic list $A(source(e))$.

We will, from now on, assume that the adjacency lists of the map M_0, our running example, are ordered as follows:

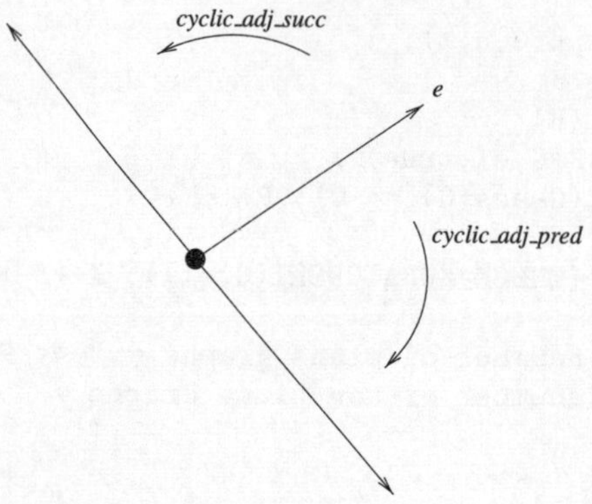

Figure 8.8 Order-preserving embeddings: The cyclic order of the edges in $A(v)$ agrees with the counter-clockwise ordering of the edges around v in the drawing.

$$\begin{aligned} v_1 &: \quad e_1 = (v_1, v_2), e_2 = (v_1, v_4), e_3 = (v_1, v_3) \\ v_2 &: \quad e_4 = (v_2, v_3), e_1^R = (v_2, v_1) \\ v_3 &: \quad e_3^R = (v_3, v_1), e_4^R = (v_3, v_2) \\ v_4 &: \quad e_2^R = (v_4, v_1). \end{aligned}$$

Consider a drawing of a map M into the plane (more generally, into any orientable surface) and let v be any node of M. The drawing defines a cyclic ordering on the edges $A(v)$ emanating from v, namely the counter-clockwise ordering[10] of the curves $I(e)$, $e \in A(v)$, around $I(v)$. A drawing is called *order-preserving* or *order-compatible* if for every node v the counter-clockwise ordering of the curves $I(e)$, $e \in A(v)$, around $I(v)$ agrees with the cyclic ordering of the edges in $A(v)$, see Figure 8.8. In Figure 8.9 one of the embeddings of M_0 is order-preserving and one is not. In all further drawings of maps in this chapter we will use order-preserving drawings.

A map is called *plane* if it has an order-preserving planar embedding. The function

```
bool Is_Plane_Map(const graph& G)
```

returns *true* if G is a plane map and returns *false* otherwise. We will see its implementation in Section 8.6.

8.5 The Face Cycles and the Genus of a Map

We define a partition of the edges of a map into cycles, the so-called *face cycles*. We introduce face cycles as purely combinatorial objects and will interpret them geometrically in the next section. Based on the concept of face cycles we will define the *genus* of a map.

[10] A precise definition is as follows: for a positive real ϵ consider the first intersections of the curves $I(e)$, $e \in A(v)$, with the circle of radius ϵ around $I(v)$. For small enough ϵ this ordering does not depend on ϵ.

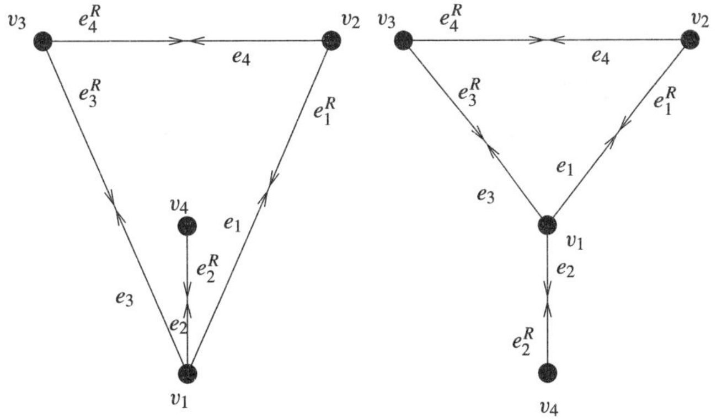

Figure 8.9 Two planar embeddings of the map M_0: In the embedding on the left the counter-clockwise ordering of the edges in $A(v_1)$ is e_1, e_2, e_3 and in the embedding on the right the ordering is e_1, e_3, e_2. The embedding on the left is order-preserving.

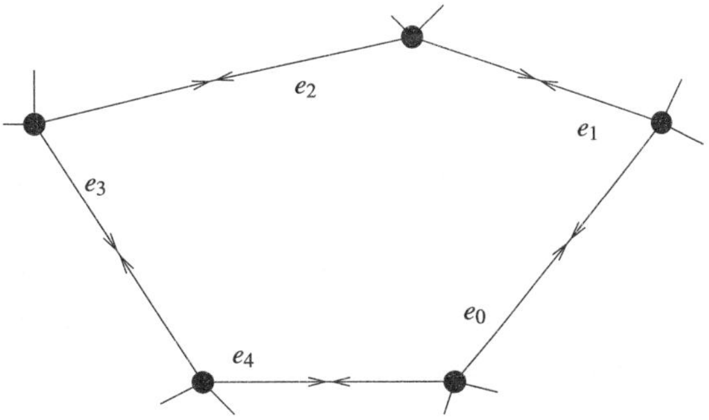

Figure 8.10 Face cycle successors and predecessors: We have $e_{i+1} = \mathit{face\_cycle\_succ}(e_i)$ for all $i, 0 \le i < 5$. Indices are mod 5. The drawing convention for maps is used.

For an edge e of a map M we define the face cycle successor and face cycle predecessor of e by:

```
face_cycle_succ(e) = cyclic_adj_pred(reversal(e))
face_cycle_pred(e) = reversal(cyclic_adj_succ(e)).
```

Figure 8.10 illustrates these definitions. The next lemma justifies the use of the names *succ* and *pred* and also shows that the function *face_cycle_succ* decomposes the edges of a map into cycles.

Lemma 45 *Let M be a map and let e be an edge of M. Then*

(a) $\mathit{face\_cycle\_pred}(\mathit{face\_cycle\_succ}(e)) = e$

(b) $\mathit{face\_cycle\_succ}(\mathit{face\_cycle\_pred}(e)) = e$

(c) *Let $e_0 = e$ and set $e_{i+1} = \mathit{face\_cycle\_succ}(e_i)$ for $i \ge 0$. Then there is a k such that $e_{k+1} = e_0$ and $e_i \ne e_j$ for all i and j with $0 \le i < j \le k$.*

Proof (a) and (b) We have

$$\begin{aligned} & face\_cycle\_pred(face\_cycle\_succ(e)) \\ &= reversal(cyclic\_adj\_succ(cyclic\_adj\_pred(reversal(e)))) \\ &= reversal(reversal(e)) \\ &= e \end{aligned}$$

and

$$\begin{aligned} & face\_cycle\_succ(face\_cycle\_pred(e)) \\ &= cyclic\_adj\_pred(reversal(reversal(cyclic\_adj\_succ(e)))) \\ &= cyclic\_adj\_pred(cyclic\_adj\_succ(e)) \\ &= e \end{aligned}$$

(c) Let k be minimal such that $e_{k+1} = e_i$ for some $i \le k$. Assume $i > 0$. From $e_{k+1} = face\_cycle\_succ(e_k)$ and $e_i = face\_cycle\_succ(e_{i-1})$ and part (a) we conclude $e_k = e_{i-1}$, a contradiction to the definition of k. Thus $i = 0$. □

For an edge e of a map M we define the *face cycle* containing e as the cycle $[e_0, e_1, \ldots, e_k]$ where $e_0 = e$, $e_{i+1} = face\_cycle\_succ(e_i)$ for $i \ge 0$, $e_{k+1} = e$, and $e_j \ne e_i$ for $0 \le i < j \le k$. Part (c) of the lemma above guarantees that this is a good definition. Every edge of M belongs to exactly one face cycle and the face cycles partition the edges of M.

We illustrate the concept of face cycle on our running example, the map M_0. The face cycle containing the edge $e_1 = (v_1, v_2)$ is

$$[e_1, e_4, e_3^R, e_2, e_2^R],$$

and the face cycle containing the edge $e_1^R = (v_2, v_1)$ is

$$[e_1^R, e_3, e_4^R].$$

Let us verify that this is indeed the case. We have

$$face\_cycle\_succ(e_1^R) = cyclic\_adj\_pred(reversal(e_1^R)) = cyclic\_adj\_pred(e_1) = e_3,$$

$$face\_cycle\_succ(e_3) = cyclic\_adj\_pred(reversal(e_3)) = cyclic\_adj\_pred(e_3^R) = e_4^R,$$

and

$$face\_cycle\_succ(e_4^R) = cyclic\_adj\_pred(reversal(e_4^R)) = cyclic\_adj\_pred(e_4) = e_1^R.$$

We want to stress that the concept of face cycles is purely combinatorial. It is made without any reference to a drawing of a map. A geometric interpretation is given in the next section.

We close this section with the definition of the *genus* of a map. Let M be a map with m edges, c connected components, n nodes, nz isolated nodes, and fc face cycles. Then

$$genus(M) = (m/2 + 2c - n - nz - fc)/2.$$

The genus of a map is always a non-negative integer, as we will show in the next section, and characterizes the surfaces into which a map can be embedded. For the map M_0 we have $m = 8$, $c = 1$, $n = 4$, $nz = 0$, and $f = 2$, and hence $genus(M_0) = 0$. We will see in the next section that this implies that M_0 is a plane map.

The following program computes the genus of a map. We determine the number of nodes and edges and the number of isolated nodes in the obvious way, and we call *COMPONENTS* to determine the number of connected components. We determine the number of face cycles by tracing them one by one. We iterate over all edges e of G. If the face cycle of e has not been traced yet, we trace it and mark all edges on the cycle as considered.

⟨*genus.c*⟩≡

```
int Genus(const graph& G)
{ if ( !Is_Map(G) ) error_handler(1,"Genus only applies to maps");
  int n = G.number_of_nodes();
  if ( n == 0 ) return 0;
  int nz = 0;
  node v;
  forall_nodes(v,G) if ( outdeg(v) == 0 )  nz++;
  int m = G.number_of_edges();
  node_array<int> cnum(G);
  int c = COMPONENTS(G,cnum);

  edge_array<bool> considered(G,false);
  int fc = 0;
  edge e;
  forall_edges(e,G)
  { if ( !considered[e] )
    { // trace the face to the left of e
      edge e1 = e;
      do { considered[e1] = true;
           e1 = G.face_cycle_succ(e1);
         }
      while (e1 != e);
      fc++;
    }
  }
  return (m/2 - n - nz - fc + 2*c)/2;
}
```

8.6 Faces, Face Cycles, and the Genus of Plane Maps

The purpose of this section is to relate combinatorics and geometry. We will define the faces of an embedding and relate it to the face cycles of a map. We will prove that a map is plane if and only if its genus is zero. We will also show that K_5 and $K_{3,3}$ are non-planar graphs.

Consider a map M and an embedding I of M into an orientable surface S. The removal of the embedding from S leaves us with a family of open connected subsets of S, called the *faces of the embedding*. In an embedding into the plane exactly one of the faces is unbounded and all other faces are bounded. The unbounded face is also called the *outer face*. We associate a set of edges with each face F, the boundary of F. An edge e belongs to the boundary of F if the "left side" of $I(e)$ is contained in F, formally, if for every point p in the relative interior of the embedding $I(e)$ of e and every sufficiently small disk centered at p, the part of the disk lying to the left of $I(e)$ is contained in F.

Consider the embeddings of M_0 shown in Figure 8.9. In the embedding on the left, the boundary of the unbounded face consists of the edges e_1^R, e_3, and e_4^R, and the boundary of the bounded face consists of the edges e_1, e_4, e_3^R, e_2, and e_2^R. In the embedding on the right, the boundary of the unbounded face consists of the edges e_1^R, e_2, e_2^R, e_3, and e_4^R, and the boundary of the bounded face consists of the edges e_1, e_4, and e_3^R. In the embedding on the left the face boundaries correspond to the face cycles of M_0.

The boundary of a face consists of one or more cycles[11], which we call *boundary cycles*. In the case of an order-preserving embedding boundary cycles and face cycles are the same.

Lemma 46 *Let I be an order-preserving embedding of a map M. The boundary cycles of the faces of I are in one-to-one correspondence to the face cycles of M.*

Proof Let $e = (v, w)$ be any edge of M and consider the boundary cycle C containing $I(e)$. Let $g = (w, z)$ be the edge such that $I(g)$ follows $I(e)$ in C. Then $I(g)$ follows $I(\mathit{reversal}(e))$ in the clockwise ordering of the embedded edges around $I(v)$. Since I is an order-preserving embedding we have $g = \mathit{face\_cycle\_pred}(e)$. Thus, boundary cycles and face cycles are the same. □

The next theorem shows that the genus of a map gives a combinatorial condition whether a map is plane. It is more generally true, see [Whi73], that the genus of a map M characterizes the oriented surfaces into which M can be embedded in an order-preserving way. The following theorem is due to Euler [Eul53] and Poincaré [Poi93].

Theorem 11 *Let M be any map. Then $\mathit{genus}(M) \geq 0$. Moreover, M is a plane map iff $\mathit{genus}(M) = 0$.*

Proof We observe first that it suffices to prove the claims for a connected map M. Let M_1, ..., M_c be the connected components of M. Then[12] $m = \sum m_i$, $n = \sum n_i$, $nz = \sum nz_i$, $fc = \sum fc_i$, and $c = \sum c_i$ and hence

$$\mathit{genus}(M) = \sum \mathit{genus}(M_i).$$

Let us assume for the moment that the claims hold for connected maps, i.e., we have $\mathit{genus}(M_i) \geq 0$ and M_i is plane iff $\mathit{genus}(M_i) = 0$ for all i. We conclude $\mathit{genus}(M) \geq 0$. If M is plane then all M_i's are plane. Thus, $\mathit{genus}(M_i) = 0$ for all i and hence $\mathit{genus}(M) = 0$. Conversely, $\mathit{genus}(M) = 0$ implies $\mathit{genus}(M_i) = 0$ for all i (since $\mathit{genus}(M_i) > 0$ for some

[11] In a connected graph the boundary of each face consists of exactly one cycle.

[12] We use m_i to denote the number of edges in M_i and analogously for n_i, nz_i, fc_i, and c_i.

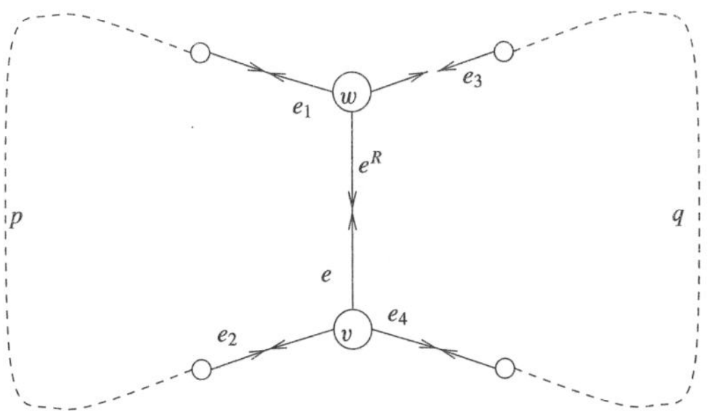

Figure 8.11 The edges e and e^R belong to distinct face cycles $e \circ p$ and $e^R \circ q$. Removal of e and e^R leaves us with a connected graph since p and q provide alternative connections between v and w. Let $e_1 = \mathit{face\_cycle\_succ}(e)$, $e_2 = \mathit{face\_cycle\_pred}(e)$, $e_3 = \mathit{face\_cycle\_pred}(e^R)$, and $e_4 = \mathit{face\_cycle\_succ}(e^R)$. Removal of e and e^R makes e_1 the face cycle successor of e_3, and e_4 the face cycle successor of e_2. No other successor relationship is affected. We conclude that the removal of e and e^R generates the face cycle $p \circ q$ and affects no other face cycles. Thus, $fc' = fc - 1$.

i would imply $\mathit{genus}(M_j) < 0$ for some j). Thus, M_i is plane for all i and hence M is plane.

For connected maps we use induction on the number of edges. If $m = 0$ then $n = nz = 1$ and $fc = 0$. Thus, M is plane and $\mathit{genus}(M) = 0$. We turn to the induction step.

Assume first that M contains a uedge $\{e, e^R\}$ such that e and e^R belong to different face cycles. Removal of e and e^R generates a map M' with $m' = m-2$, $n' = n$, $c' = c = 1$, $nz' = nz = 0$, and $fc' = fc - 1$, see Figure 8.11. Thus, $\mathit{genus}(M) = \mathit{genus}(M')$. By induction hypothesis, $\mathit{genus}(M') \geq 0$ and M' is plane iff $\mathit{genus}(M') = 0$. From $\mathit{genus}(M') \geq 0$ we conclude $\mathit{genus}(M) \geq 0$. We next show that M is plane iff $\mathit{genus}(M) = 0$. If M is plane then M' is plane (since an order-preserving embedding of M' is obtained from an order-preserving embedding of M by removing the images of e and e^R). Thus $\mathit{genus}(M') = 0$ by induction hypothesis and hence $\mathit{genus}(M) = 0$. Conversely, if $\mathit{genus}(M) = 0$ then $\mathit{genus}(M') = 0$ and hence there is an order-preserving embedding I' of M', by induction hypothesis. By Lemma 46 there is a face F in the embedding I' with boundary cycle $p \circ q$. We embed e and e^R into F and obtain an order-preserving embedding I of M.

Assume next that for every uedge $\{e, e^R\}$ of M, e and e^R belong to the same face cycle. Consider any node v and let $A(v) = (e_0, e_1, \ldots, e_{k-1})$ be the cyclic list of edges out of v. Then

$$e_i = \mathit{face\_cycle\_succ}(e_{i+1}^R)$$

for all i, $0 \leq i < k$, by the definition of face cycles, see Figure 8.12. Since e_i and e_i^R belong to the same face cycle by assumption, all edges incident to v belong to the same face cycle and, since M is connected, all edges of M belong to the same face cycle. Thus, $fc = 1$. Since M is connected, the number of uedges is at least $n - 1$. Thus, $m \geq 2(n - 1)$, $c = 1$, $nz = 0$, and hence $\mathit{genus}(M) \geq 0$. We next show that M is plane iff $\mathit{genus}(M) = 0$. If M is plane consider an order-preserving embedding I of M. The face cycles of M are

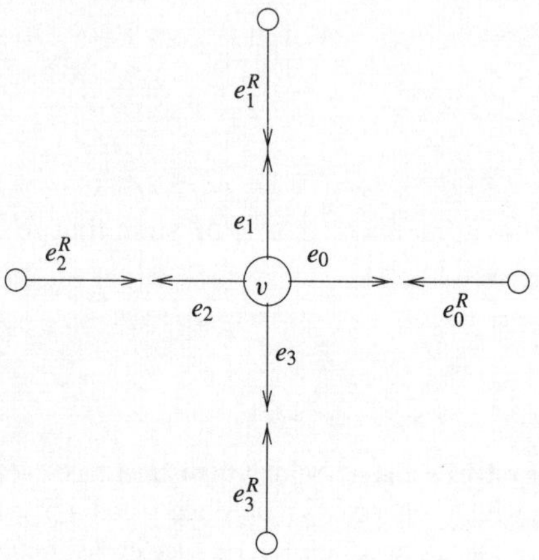

Figure 8.12 A node v with $A(v) = (e_0, e_1, e_2, e_3)$. There is a face cycle containing e_{i+1}^R and e_i for all i, $0 \le i < 4$. Indices are modulo 4.

in one-to-one correspondence to the faces of the embedding. Since there is only one face cycle, there is only one face, and hence M cannot contain a cycle. Thus, $m = 2(n-1)$ and hence $genus(M) = 0$. Conversely, if $genus(M) = 0$ then $(m/2 + 2 - n - 1) = 0$ and hence $m = 2(n-1)$. The number of uedges is therefore equal to $n-1$ and hence the uedges form a tree. For a tree there is clearly an order-preserving embedding. □

The theorem above implies that the test of whether a graph G is a plane map is trivial to implement. We only have to test whether G is a map and whether the genus of G is zero.

```
bool Is_Plane_Map(const graph& G) { return Is_Map(G) && Genus(G) == 0; }
```

We draw some more consequences of Theorem 11. It implies an upper bound on the number of edges in a planar graph (without self-loops and parallel edges) and it implies that the Kuratowski graphs K_5 and $K_{3,3}$ are non-planar.

Lemma 47

(a) Let M be a connected plane map in which every face cycle consists of at least d edges, where $d \ge 3$. Then

$$m/2 \le \frac{d}{d-2}(n-2),$$

i.e., M has at most $(d/(d-2)) \cdot (n-2)$ uedges.

(b) Let M be a connected planar map without self-loops and without parallel edges. Then M has at most $3n-6$ uedges, if $n > 3$, and a node of degree at most five.

(c) Let M be a connected bipartite planar map without self-loops and without parallel edges. Then M has at most $2n-4$ uedges, if $n \ge 4$.

(d) The complete graph K_5 on five nodes is not planar.

(e) The complete bipartite graph $K_{3,3}$ with three nodes on each side is not planar.

Proof (a) If every face cycle consists of at least d edges then $m \geq d \cdot fc$. Thus,

$$0 = genus(M) = m/2 + 2 - n - fc \geq m/2 + 2 - n - m/d$$

and hence $(m/2) \cdot (1 - 2/d) \leq n - 2$ or $m/2 \leq (d/(d-2)) \cdot (n-2)$.

(b) and (c) Reorder the adjacency lists of M such that M becomes a plane map. If M has no self-loops and no parallel edges, every face cycle of M consists of at least three edges. If, in addition, M is bipartite, every face cycle of M consists of at least four edges. The bounds on the number of edges now follow from part (a). If every node would have degree six or more, the total number of edges would be at least $6n/2 = 3n$.

(d) A planar graph with five nodes and no self-loops and no parallel edges has at most nine uedges by part (b). The graph K_5 has $5 \cdot 4/2 = 10$ uedges.

e) A planar bipartite graph with six nodes and no self-loops and no parallel edges has at most eight uedges by part (c). The graph $K_{3,3}$ has $3 \cdot 3 = 9$ uedges. □

Exercise for 8.6

1 It is obvious from the definition of $genus(M)$ that $2 \cdot genus(M)$ is an integer. The purpose of this exercise is to show that $genus(M)$ is an integer. In the proof of Theorem 11 we have constructed for every connected map M a connected map M' such that $genus(M) = genus(M')$ and such that M' has a single face cycle. Let M'' be obtained from M' by removing an edge e and its reversal e^R. Determine the number of edges, nodes, face cycles, and connected components of M'' and conclude that $genus(M') - genus(M'')$ is an integer. Use this observation and induction to show that the genus of every map is an integer.

8.7 Planarity Testing, Planar Embeddings, and Kuratowski Subgraphs

This section is joint work with D. Ambras, R. Hesse, Christoph Hundack, and E. Kalliwoda.

We give the details of the planarity test, the planar embedding algorithm, and the algorithm for finding Kuratowski subgraphs. For each algorithm we will first derive the required theory and then give an implementation. All implementations run in linear time and are collected in the file

⟨*_bl_planar.c*⟩≡

```
#include <LEDA/graph_alg.h>
#include <LEDA/pq_tree.h>
#include <LEDA/array.h>
#include <assert.h>
⟨auxiliary functions⟩
⟨planarity test⟩
⟨planar embedding of biconnected maps⟩
```

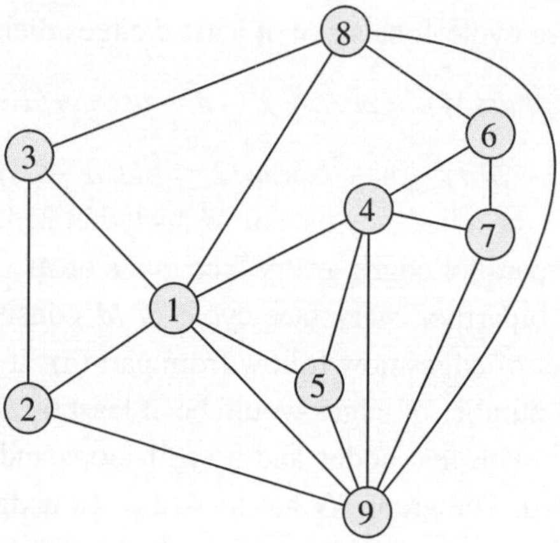

Figure 8.13 A biconnected st-numbered graph G. Node s is labeled 1 and node t is labeled 9.

⟨*planar embedding of arbitrary maps*⟩
⟨*Kuratowski graphs in biconnected maps*⟩
⟨*Kuratowski graphs in arbitrary graphs*⟩

8.7.1 *The Lempel–Even–Cederbaum Planarity Test*

We discuss the planarity testing algorithm invented by Lempel, Even, and Cederbaum [LEC67, Eve79]. We assume that $G = (V, E)$ is a biconnected graph[13], that $e_0 = (s, t)$ is an arbitrary edge of G, and that the nodes of G are st-numbered, i.e., s is numbered 1, t is numbered n, and every node distinct from s and t has a lower and a higher numbered neighbor.

We will first discuss the required theory and then describe an implementation based on PQ-trees.

The Theory: We identify nodes with their st-number, i.e., $V = \{1, \ldots, n\}$. Figure 8.13 shows an example of an st-numbered biconnected graph. We will use it as our running example.

Let $V_k = \{1, \ldots, k\}$ and let $G_k = (V_k, E_k)$ be the graph induced by V_k, i.e., E_k consists of all edges of G whose endpoints are both in V_k. We extend G_k to a graph B_k. For each edge (v, w) of G with $v \leq k$ and $w \geq k + 1$ there is a node and an edge in B_k. They are called virtual nodes and virtual edges, respectively. We label every virtual node with its counterpart in G. Figure 8.14 shows the graph B_7 for our running example.

If G is planar, B_k has a plane embedding which resembles a bush: node v, $1 \leq v \leq k$, is drawn at height v, all virtual nodes are put on a horizontal line at height $k + 1$, and all edges are drawn as y-monotone curves[14]. We call such an embedding a *bush form* for B_k and we

[13] The rather trivial extension to arbitrary graphs will be given at the end of the section.
[14] A curve is y-monotone if any horizontal line intersects the curve at most once.

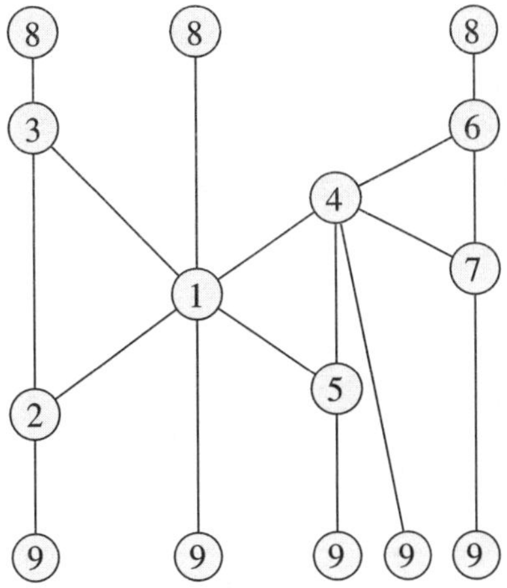

Figure 8.14 The graph B_7 for the graph G of Figure 8.13. There are three virtual nodes labeled 8, one for each edge connecting node 8 to a node labeled 7 or less in G, and there are five virtual nodes labeled 9, one for each edge connecting node 9 to a node labeled 7 or less in G. The nodes 4, 6, and 7 comprise a biconnected component which we denote H_0 for later reference.

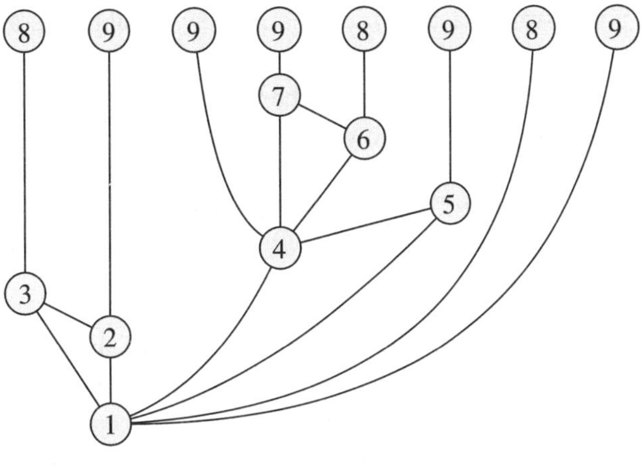

Figure 8.15 A bush form for the graph B_7 of Figure 8.14.

call the horizontal line at height $k + 1$ the horizon. The existence of bush forms will follow from the discussion to come. Figures 8.15 and 8.16 shows two bush forms for the graph of Figure 8.14.

The *leaf word* of a bush form is a sequence in $\{N, E\}^*$, where E represents a virtual node labeled $k + 1$, N represents a virtual node labeled $k + 2$ or larger, and the virtual nodes are listed in their left-to-right order on the horizon. The bush form in Figure 8.15 has leaf word $ENNNENEN$ and the bush form in Figure 8.16 has leaf word $NEEENNNN$. A bush form for B_k is called *extendible* if all virtual nodes labeled $k + 1$ are consecutive on the horizon, i.e., if its leaf word is in $N^*E^*N^*$. An extendible bush form $\hat{B}_k$ is readily extended

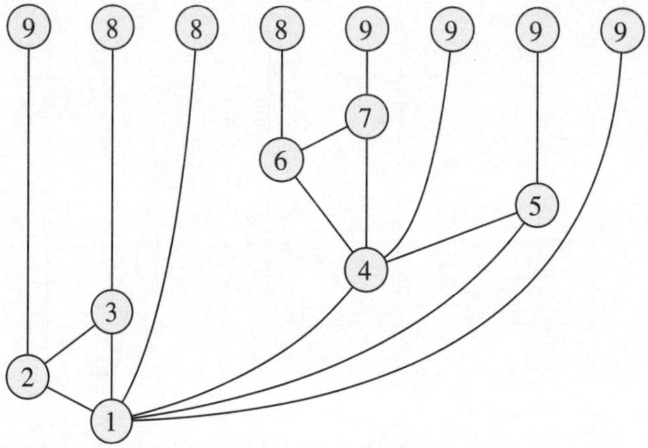

Figure 8.16 An extendible bush form for B_7.

to a bush form $\hat{B}_{k+1}$ for B_{k+1}. We move all nodes v, $v > k+1$, to height $k+2$, we merge all virtual nodes labeled $k+1$ into a single node (since they are consecutive on the horizon, merging does not destroy planarity), and add a new virtual edge and node for each edge $(k+1, w)$ with $w > k+1$.

The question is now how to decide whether B_k has an extendible bush form, and how to find an extendible bush form. We show:

Theorem 12 *B_{k+1} has a bush form iff B_k has a bush form and no obstructions. Moreover, if B_k has no obstructions then any bush form $\hat{B}_k$ of B_k can be transformed into an extendible bush form of B_k by a sequence of permutations and flippings.*

We still need to define several of the terms used in the theorem above. An obstruction is either an obstructing articulation point or an obstructing biconnected component. In the definition of either kind of obstruction we need the concepts of clean, mixed, or full subgraph. A subgraph of B_k is called *clean*, *mixed*, or *full* if none, some but not all, or all of its virtual nodes are labeled $k+1$.

An articulation point v of B_k is *obstructing* if there are three or more components of $B_k \setminus v$ that are mixed.

Consider the graph B_7 of Figure 8.14. Node 4 is an articulation point and $B_7 \setminus 4$ has three components: Two of them are mixed and one is full. Node 4 is non-obstructing. Please convince yourself that none of the articulation points is obstructing.

We come to biconnected components of B_k. A node y of a biconnected component H is called an *attachment* node of H if it is also the endpoint of an edge outside H. Attachment nodes are articulation points of B_k and hence are embedded on the boundary of the outside face in every bush form of B_k. In the graph B_7 the biconnected component H_0 has attachment nodes 4, 6, and 7.

Let $y_0, y_1, \ldots, y_{p-1}$ be the attachment nodes of a biconnected component H of B_k. We

Figure 8.17 y_h and y_{h+1} are adjacent attachment nodes on the boundary cycle of H in $\hat{B}_k$, but are separated by y_i and y_j in the boundary cycle of H in $\hat{B}'_k$.

use y_0 for the lowest numbered attachment node; y_0 is also the lowest numbered node in H. Any bush form $\hat{B}_k$ of B_k induces an embedding of H (simply remove all nodes outside H and their incident edges). In this embedding of H the boundary of the outside face of H is a simple cycle, which we call the *boundary cycle*[15] of H in $\hat{B}_k$. A counter-clockwise traversal of the boundary cycle yields a cyclic order on the attachment nodes, which we call the cyclic order induced by the bush form. Consider Figures 8.15 and 8.16. The cyclic order of the attachment nodes 4, 6, and 7 is 4, 6, 7 in the first figure and is 4, 7, 6 in the second figure.

Lemma 48 *Let $y_0, y_1, \ldots, y_{p-1}$ be the attachment nodes of a biconnected component H of B_k in the cyclic order induced by some bush form $\hat{B}_k$ of B_k. Then any other bush form of B_k induces either the same cyclic order or its reversal.*

Proof Assume otherwise, i.e., there is a bush form $\hat{B}'_k$ such that the attachment nodes appear in a different cyclic order in $\hat{B}'_k$. Then there must be indices h, i, and j such that y_h and y_{h+1} (indices are mod p) are separated by y_i and y_j in the boundary cycle of H in $\hat{B}'_k$, see Figure 8.17. The embedding $\hat{B}'_k$ implies that any pair of paths connecting y_h to y_{h+1} and y_i to y_j, respectively, must cross. On the other hand, the embedding $\hat{B}_k$ implies the existence of non-crossing paths. □

Let $y_0, y_1, \ldots, y_{p-1}$ be the attachments of H in one of their cyclic orders[16]. The *component of B_k opposite to H at y_i* is the subgraph of B_k spanned by all nodes that are reachable from y_i without using an edge of H. We denote it by C_i. Each C_i is either clean, mixed, or full. We define the signature of H as the word

$$s_0 s_1 \ldots s_{p-1} \in \{\text{clean,mixed,full}\}^*$$

where s_i describes the status of C_i. In the graph B_7, the component opposite to H_0 at 6 is full, the component opposite to H_0 at 7 is clean, and the component opposite to H_0 at 4 is

[15] A node of H which is not an attachment node of H may lie on the boundary cycle of H in some bush forms and may not lie on the boundary cycle in others. Attachment nodes belong to the boundary cycle in every bush form.

[16] There are two by the preceding lemma. For the definition in this paragraph it does not matter which one is chosen.

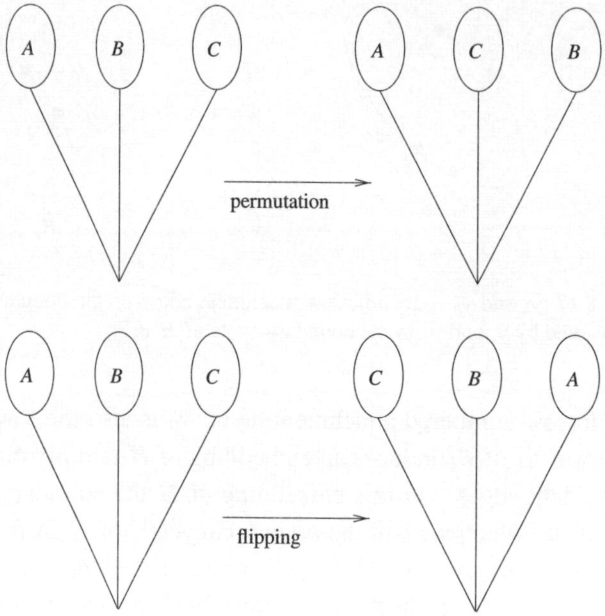

Figure 8.18 Permuting and flipping.

mixed. The signature of H_0 is "mixed clean full" for the ordering 4, 7, 6 and "mixed full clean" for the ordering 4, 6, 7.

A biconnected component H is *non-obstructing* iff a cyclic shift of its signature is in

$$\text{clean}^* \ \text{mixed}_0^1 \ \text{full}^* \ \text{mixed}_0^1 \ \text{clean}^*,$$

where mixed_0^1 denotes zero or one occurrence of mixed, and is obstructing otherwise.

We come to permutations and flippings. Permutations apply to articulation points of B_k. Let v be an articulation point of B_k. Then, if $v > 1$, exactly one component of B_k with respect to v contains nodes lower than v, and if $v = 1$, no component does[17]. We call the component containing lower numbered nodes the *root component* of v and all other components *non-root components* of v.

In the graph B_7 of Figure 8.14 the root component of node 4 contains nodes 5, 1, 2, 3, two copies of 8, and three copies of 9.

Consider now any bush form $\hat{B}_k$ of B_k. A *sub-bush* of $\hat{B}_k$ with lowest numbered node v is the restriction of $\hat{B}_k$ to the union of some non-root components with respect to v. In particular, each non-root component of v corresponds to a sub-bush of $\hat{B}_k$. A *permutation operation* permutes the sub-bushes corresponding to the non-root components with respect to an articulation point v and a *flipping operation* flips over a sub-bush, see Figure 8.18.

We are now ready for the if-direction of Theorem 12.

[17] Observe that any node u with $u < v$ can reach 1 without passing through v by the virtue of st-numberings.

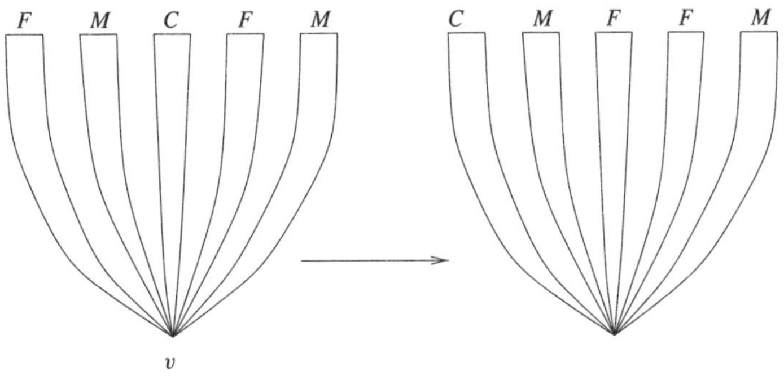

Figure 8.19 Permuting the sub-bushes of $\hat{B}$ with respect to v. C, M, and F stand for clean, mixed, and full sub-bushes, respectively.

Lemma 49 *If B_k has a bush form and no obstructions then any bush form $\hat{B}_k$ can be transformed into an extendible bush form by a sequence of permutations and flippings.*

Proof We want to use induction over sub-bushes and therefore prove a slightly stronger claim. We call a sub-bush *incomplete* if there is a virtual node labeled $k+1$ outside the sub-bush and we call a sub-bush *strongly extendible* if its leaf word is in N^*E^* or E^*N^*. We show that every sub-bush can be transformed into an extendible sub-bush, i.e., a sub-bush whose leaf word is in $N^*E^*N^*$, and that every incomplete sub-bush can be transformed into a strongly extendible sub-bush.

Let $\hat{B}$ be any sub-bush. If $\hat{B}$ has only one virtual node, the claims are obvious. So, assume otherwise and let v be the lowest numbered node in $\hat{B}$. We distinguish cases according to whether v is an articulation point of $\hat{B}$ or not.

If v is an articulation point of $\hat{B}$ then at most two of the components of $\hat{B}$ with respect to v are mixed. We can therefore permute the components such that all full and all clean components are consecutive and such that the two mixed components bracket the full components, see Figure 8.19. We apply the induction hypothesis to the sub-bushes and therefore may assume that the sub-bushes are extendible or even strongly extendible (for incomplete sub-bushes). We complete the induction step with two observations. First, the mixed sub-bushes are incomplete except if there is at most one mixed sub-bush and this sub-bush contains all virtual nodes labeled $k+1$. Second, if $\hat{B}$ is incomplete then there is at most one mixed sub-bush since the root component of B_k with respect to v is mixed. Thus, $\hat{B}$ can be transformed into an extendible bush form and into a strongly extendible bush form if $\hat{B}$ is incomplete. The transformation consists of transformations of the sub-bushes, permuting the sub-bushes, and maybe flipping one of the mixed sub-bushes.

If v is not an articulation point of $\hat{B}$, let H be the biconnected component of $\hat{B}$ containing v. Let $y_0, y_1, \ldots, y_{p-1}$ with $v = y_0$ be the attachment points of H in B_k in one of their two cyclic orders. We have a sub-bush $\hat{B}_i$ of $\hat{B}$ for the component C_i of B_k opposite to y_i for all i, $1 \le i \le p-1$. Since H is non-obstructing and since C_0 is either clean or mixed (it cannot be full since it contains the edge (s, t)), we have

$$s_1 \ldots s_{p-1} \in \text{clean}^* \ \text{mixed}_0^1 \ \text{full}^* \ \text{mixed}_0^1 \ \text{clean}^*$$

if C_0 is clean and we have

$$s_1 \ldots s_{p-1} \in \text{clean}^* \ \text{mixed}_0^1 \ \text{full}^* \cup \text{full}^* \ \text{mixed}_0^1 \ \text{clean}^*$$

if C_0 is mixed. In either case we conclude that $\hat{B}$ can be transformed into an extendible bush form and into a strongly extendible bush form if $\hat{B}$ is incomplete and hence C_0 is mixed. The transformation consists of transformations of sub-bushes followed (maybe) by a flipping of the two mixed sub-bushes. □

Figure 8.20 illustrates Lemma 49. It shows a sequence of transformations that transform the bush form of Figure 8.15 into the extendible bush form of Figure 8.16.

We summarize. The Lempel–Even–Cederbaum planarity test constructs a sequence $\hat{B}_0$, $\hat{B}_1$, $\hat{B}_2$, …, $\hat{B}_n$ of bush forms. In iteration $k + 1$ the bush form $\hat{B}_k$ is first transformed into an extendible bush form $\hat{B}'_k$ and then extended to a bush form $\hat{B}_{k+1}$. The transformation to an extendible bush form uses permutations and flippings and is possible if $\hat{B}_k$ contains no obstructions.

The running time of the Lempel–Even–Cederbaum test is $O(n^2)$ in its original form. Booth and Lueker improved the running time to $O(n + m)$ by the introduction of the *PQ-tree* data structure, which we will discuss in the next section. In Section 8.7.3 we will show that the existence of an obstruction implies the existence of a Kuratowski graph in G.

The PQ-Tree Data Structure: Booth and Lueker [BL76] introduced the PQ-data structure to keep track of the sequence of bush forms arising in the Lempel–Even–Cederbaum planarity test. PQ-trees have wider applications than planarity testing but we will not discuss them here.

PQ-trees have the following interface.

```
pq_tree T(m);
```

declares a PQ-tree T which can represent bush forms in which every edge is labeled with an integer in $[1 .. m]$. After the declaration T represents the empty bush form with no nodes and no edges. We use S to denote the set of virtual edges in the current bush form. S is empty initially.

The operation

```
bool T.replace(list<int>& L, list<int>& U, list<int>& I)
```

adds a node to the current bush form. The node is incident to the virtual edges L in the current bush form and introduces new virtual edges U. We must have $L \subseteq S$, U is a set of integers (= edges) that have never been in S before, and $L = \emptyset$ iff $S = \emptyset$; the latter requirement corresponds to the fact that only node 1 is incident to no edge from below. The new set of virtual edges becomes $(S \setminus L) \cup U$.

The function returns *true* if the current bush form is extendible, i.e., can be transformed to

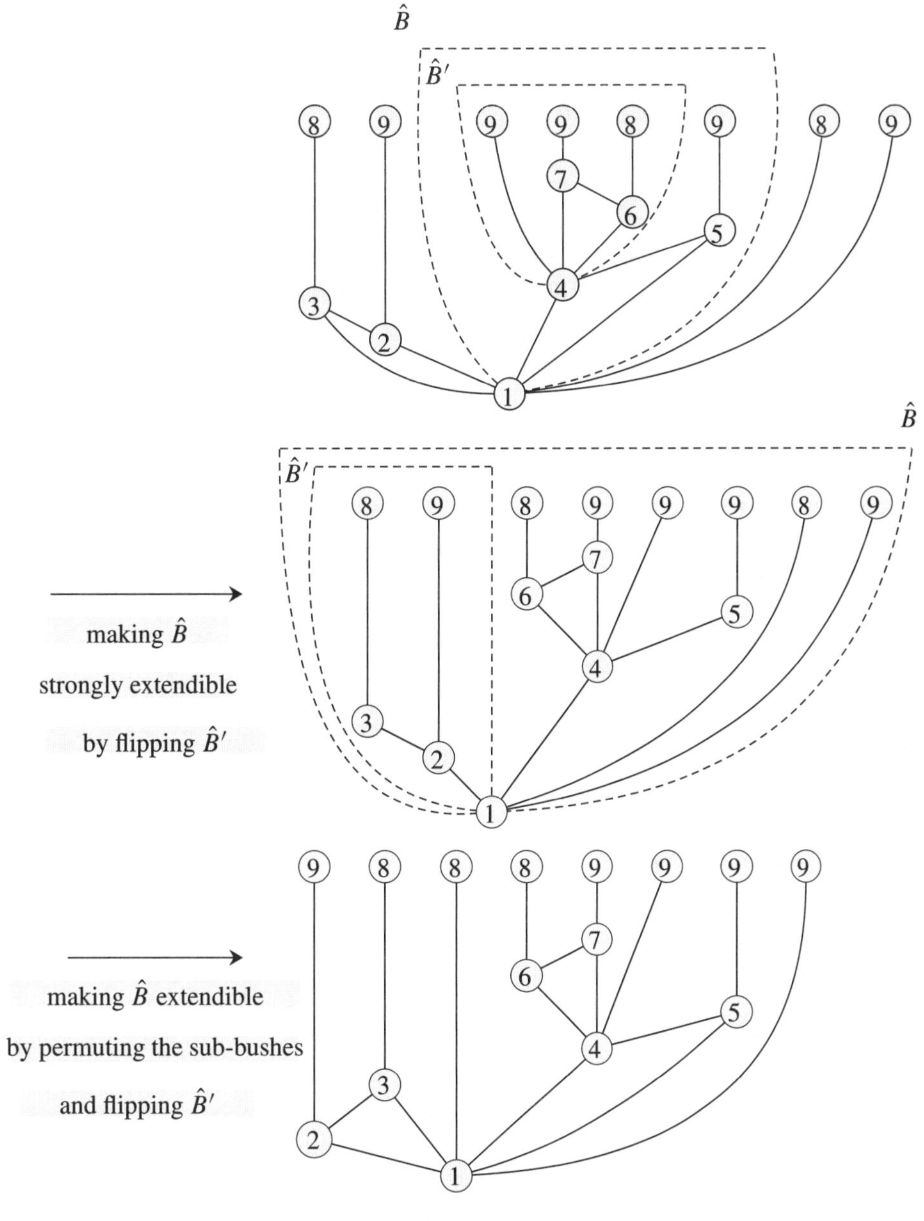

Figure 8.20 Transforming the bush form of Figure 8.15 into an extendible bush form.

a bush form in which all edges in L are contiguous on the horizon. The function returns *false* otherwise. Once a call of *replace* has returned *false*, the PQ-tree becomes non-functional and no further operations can be applied to it.

The last argument I is irrelevant for the planarity test and is only required for the construction of a planar embedding. We will discuss it in the next section.

The amortized running time of *replace* is proportional to the length of L plus the length of U and the running time of the declaration $T(m)$ is $O(m)$.

We are now ready for the planarity test. The function PLANTEST expects a biconnected graph G, an st-numbering *st_num* of its nodes, and a list *st_list* containing the nodes of G in increasing order of st-number, and returns *true* iff G is a planar graph.

If G has less than five nodes then G is planar. So assume that G has at least five nodes. We declare a PQ-tree $T(m)$, where m is one larger than the maximal index of any edge[18]. We use T to maintain the bush forms $\hat{B}_k$ for $k = 0, 1, 2, \ldots$.

We iterate over the nodes in increasing order of st-number. For each v, we collect the edges that connect v to lower numbered nodes in L, and we collect the edges that connect v to higher numbered nodes in U. Self-loops are ignored as they do not affect planarity. We update the bush form by

```
T.replace(L,U,I),
```

where I is a dummy argument. If the call is not successful, we break from the loop and return *false*, if the call is successful, we proceed to the next node. If all nodes can be added to the bush form we return *true*.

⟨*planarity test*⟩≡

```
static bool PLANTEST(graph& G, node_array<int>& st_num,
                                list<node>& st_list)
{
  int n = G.number_of_nodes();
  int m = G.max_edge_index() + 1;
  if (n < 5)  return true;
  pq_tree  T(m);
  int stv = 1;
  node v;
  forall(v,st_list)
  {
    list<int> L, U, I;
    edge e;
    forall_inout_edges(e,v)
    { node w = G.opposite(v,e);
      int stw = st_num[w];
      if (stw < stv) L.push(index(e)+1);
      if (stw > stv) U.push(index(e)+1);
    }
    if ( !T.replace(L,U,I) ) break;
    stv++;
  }
  return stv == n+1;
}
```

18 The data type graph numbers edges with non-negative integers. The number of an edge is called its index. Since PQ-trees expect positive numbers, we identify any edge with its index plus one.

The program above performs the planarity test in time $O(n+m)$. This follows from the fact that the declaration of T requires time $O(m)$ and that the total cost of all *replace* operations is $O(n+m)$ and that an st-numbering can be computed in linear time (see Section 7.4).

The program above is short and elegant. It performs a complex task, namely, to test whether a graph is planar, in linear time and a few lines of code. Of course, all the complexity is hidden in the implementation of PQ-trees.

Can you trust the program above? "*Yes, you can trust it*", but "*it would be unwise to do so*". We have not explained the inner workings of PQ-trees, their implementation is complex (almost 2000 lines), and most seriously there is no way to check the answer of the program above. It just says "yes" or "no". In the sections to come we will extend the program above to a program that can be checked. We show how to compute planar embeddings of planar graphs and Kuratowski subgraphs of non-planar graphs.

8.7.2 *Planar Embeddings*

Chiba et al [CNAO85, NC88] have shown how to extend the planarity test of Lempel, Even, and Cederbaum to an embedding algorithm. We review their algorithm and give the implementation of functions

```
static bool PLAN_EMBED(graph& G, node_array<int>& st_num,
                                         list<node>& st_list);
bool BL_PLANAR(graph& G, bool embed);
```

The first function takes a biconnected map G, an st-numbering *st_num* of G, and the list of nodes of G in increasing order of st-number, and tests whether G is planar. If G is planar, it reorders the adjacency lists of G such that G becomes a plane map.

The second function applies to any map G. It returns *true* if G is planar and it returns *false* otherwise. If G is planar and *embed* is *true*, G is turned into a plane map. If *embed* is *true* and G is not a map, the function aborts. If *embed* is *false*, the function applies to any graph G.

Biconnected st-numbered Maps: We discuss the function PLAN_EMBED. The planarity testing algorithm constructs a sequence of bush forms $\hat{B}_0, \hat{B}_1, \hat{B}_2, \ldots, \hat{B}_n$. The construction is implicit in the sense that the bush forms are hidden in the internal structure of the PQ-tree. We want $\hat{B}_n$. The construction of $\hat{B}_{k+1}$ from $\hat{B}_k$ consists of two steps: first, $\hat{B}_k$ is transformed into an extendible bush form $\hat{B}'_k$ and then node $k+1$ is added to obtain $\hat{B}_{k+1}$.

For a node v let $L(v)$ be the set of edges (v, w) with $w < v$, and for any integer k with $k \geq v$ let $L_k(v)$ be the counter-clockwise order of the edges in $L(v)$ in the bush form $\hat{B}_k$. The embedding algorithm is based on the following observations:

- The cyclic order of the adjacency lists $A(v)$, $v \in V$, can be constructed from the lists $L_n(v)$, $v \in V$.
- The sequence $L_k(k)$ is readily extracted from the PQ-tree data structure.
- The sequence $L_{k+1}(v)$ is equal to $L_k(v)$ or $L_k^{rev}(v)$ for $k \geq v$.

We provide more details on the last item and postpone the discussion of the other two items.

Bush forms are transformed by permutations and flippings. Permutations have no effect on the order of the lists $L(v)$ for any v. They have a dramatic effect on the order of the lists $U(v)$, where $U(v)$ is the set of edges (v, w) with $v < w$. For this reason we do not keep track of the order of the $U(v)$'s during the construction process but determine their orders in a second phase (this is the subject of the first item). A flipping of a sub-bush with lowest numbered vertex w reverses the order of $L(v)$ for all v in the sub-bush with $v \neq w$ and does not affect the order of $L(v)$ for any other v. We conclude that $L_{k+1}(v)$ is equal to either $L_k(v)$ or $L_k^{rev}(v)$ for any v with $v \leq k$. We say that node v is flipped in iteration $k+1$ if $L_{k+1}(v) = L_k^{rev}(v)$. If v is not flipped in iteration $k+1$ then $L_{k+1}(v) = L_k(v)$.

We conclude that $L_n(v)$ is equal to $L_v(v)$ if v is flipped an even number of times and is equal to $L_v^{rev}(v)$ if v is flipped an odd number of times. We next show how to determine efficiently how often nodes are flipped. We could maintain a counter for each node and increment it whenever the node is flipped. Since a linear number of nodes may be flipped in each iteration, this would result in a quadratic algorithm. We are aiming for linear running time and hence need a more compact way to maintain the counters.

In the graph B_{k+1} there is a unique biconnected component H_{k+1} having $k+1$ as its highest numbered node. We call H_{k+1} the biconnected component formed in iteration $k+1$.

Lemma 50 *All edges in* $L(k+1)$ *are contained in* H_{k+1} *and any biconnected component* H *of* B_k *is either contained in* H_{k+1} *or edge-disjoint from* H_{k+1}*, see Figure 8.21.*

Proof Consider any two lower neighbors u and v of $k+1$. They are connected by a path of length two through $k+1$ and they are connected by a path which avoids $k+1$, the second half-sentence being a consequence of st-numbering. Thus, all edges in $L(k+1)$ belong to H_{k+1} and the first part of the lemma is shown.

Any two edges belonging to the same biconnected component of B_k belong to the same biconnected component of B_{k+1}. This proves the second part of the lemma. □

For a biconnected component H of B_k let $V^+(H)$ denote the set of nodes of H except for the lowest numbered node of H. A flipping operation changes either the order of $L(v)$ for all nodes $v \in V^+(H)$ or for no node $v \in V^+(H)$. This follows from the fact that a biconnected component is either contained in a sub-bush or disjoint from it. We say that a biconnected component H is flipped in iteration $k+1$ if all nodes in $V^+(H)$ are flipped in iteration $k+1$.

Lemma 51 *There is a transformation of* $\hat{B}_k$ *to an extendible bush form in which only biconnected components* H *of* B_k *are flipped that become part of* H_{k+1}.

Proof Let $\hat{B}'_k$ be the extendible bush form produced by the strategy of Lemma 49 and assume that some biconnected component H that does not become part of H_{k+1} is flipped

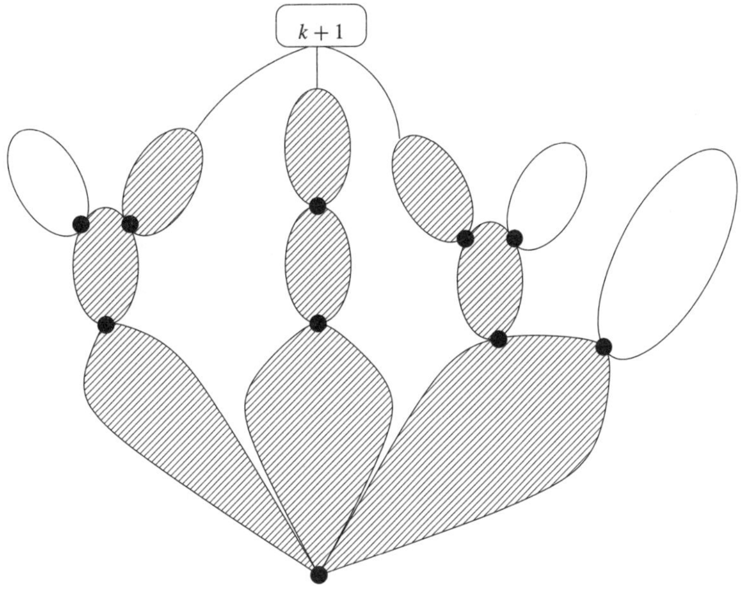

Figure 8.21 The biconnected components of B_k are indicated as ovals and articulation points are indicated as solid circles. The hatched biconnected components become part of H_{k+1}.

by the transformation from $\hat{B}_k$ to $\hat{B}'_k$. Let $y = y(\hat{B}'_k)$ be the lowest numbered node that is part of a biconnected component H that is flipped by the transformation to $\hat{B}'_k$ and does not become part of H_{k+1}. Consider the bush form $\hat{B}''_k$ obtained by flipping the smallest sub-bush $\hat{B}$ that contains H. $\hat{B}''_k$ is extendible since no leaf labeled $k+1$ is contained in $\hat{B}$. Moreover, either no biconnected component that does not become part of H_{k+1} is flipped in $\hat{B}''_k$ or $y(\hat{B}''_k) > y(\hat{B}'_k)$.

We conclude that $\hat{B}_k$ can be transformed into an extendible bush form in which only biconnected components are flipped that become part of H_{k+1}. □

We can now explain the third argument of function *replace* of class *pq_tree*. It consists of three parts, which in iteration $k+1$ are as follows (see Figure 8.22):

- An integer l specifying the number of components of B_k that are merged into H_{k+1}.
- A sequence $j_0, j_1, \ldots, j_{l-1}$ of integers, where $H_{|j_0|}, \ldots, H_{|j_{l-1}|}$ are the biconnected components of B_k that are merged into H_{k+1}, and j_i is positive if H_{j_i} is not flipped in iteration $k+1$ and is negative otherwise.
- The edges[19] in $L(k+1)$ in their counter-clockwise order around $k+1$ in $\hat{B}_{k+1}$.

We denote the third argument of *replace* by I because it contains the instructions of how to obtain $\hat{B}_{k+1}$ from $\hat{B}_k$.

[19] More precisely, the sequence of numbers identifying the edges.

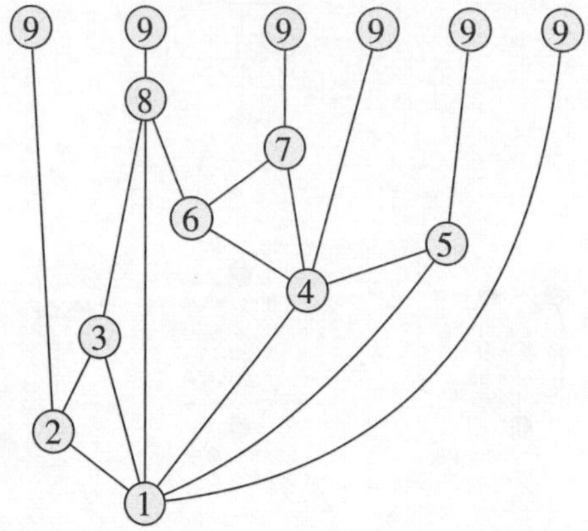

Figure 8.22 The bush form $\hat{B}_8$ obtained from adding node 8 to the bush form of Figure 8.16. The biconnected component H_8 consists of the biconnected components H_3, H_5, and H_7 and the edges in $L(8)$. The counter-clockwise order of the edges in $L(8)$ is (8, 3), (8, 1), (8, 6). The biconnected components H_3 and H_7 are flipped when going from the bush form $\hat{B}_7$ of Figure 8.15 to $\hat{B}_8$. Thus $I = 3, -3, 5, -7, (8, 3), (8, 1), (8, 6)$, where the first 3 indicates that three components are merged into H_8, the sequence $-3, 5, -7$ indicates that the merged components are H_3, H_5, and H_7 and that H_3 and H_7 are flipped, and where (8, 3), (8, 1), (8, 6) form $L(8)$.

We are now ready for the implementation of PLAN_EMBED. It consists of three phases. In the first phase, we run the planarity test of the preceding section with three changes:

- We are now dealing with a map and therefore store only one direction of each edge in the PQ-tree. In phase one we are dealing with lists $L(v)$ and hence we store the direction from larger to smaller nodes. We construct the lists $L(v)$ and $U(v)$ by iterating over all edges out of v: edges to lower numbered nodes are put into $L(v)$ and the reversals of edges to higher numbered nodes are put into $U(v)$. We put edge reversals into $U(v)$ in order to guarantee that for each uedge the direction going from higher to smaller st-number is put into the PQ-tree. Self-loops are ignored in phase one.
- We define an array *EDGE* that stores for each integer in $[1..m]$ the edge corresponding to it.
- In iteration k we store the output I of PQ-tree operation *replace* in $I[k]$.

Here comes phase one.

⟨*PLAN_EMBED: phase 1*⟩≡

```
int n = G.number_of_nodes();
if ( G.number_of_edges() == 0 ) return true;
int m = G.max_edge_index() + 1;
// interface for pq_tree
pq_tree  T(m);
list<int>* I = new list<int>[n+1];
```

```
edge* EDGE  = new edge[m+1];  // EDGE[i+1] = edge with index i
edge  e;
forall_edges(e,G) EDGE[index(e)+1] = e;
// planarity test
int stv = 1;
node v;
forall(v,st_list)
{
  list<int> L, U;
  edge e;
  forall_adj_edges(e,v)
  { int stw = st_num[target(e)];
    if (stw < stv) L.push(index(e) + 1);
    if (stw > stv) U.push(index(G.reversal(e)) + 1);
  }
  if ( !T.replace(L,U,I[stv]) ) break;
  stv++;
}
```

At the end of phase one, we either have $stv < n+1$ and then G is non-planar, or $stv = n+1$ and then G is planar and $I[k]$ contains the instruction list of the k-th iteration for all k, $1 \le k \le n$. Thus:

⟨*planar embedding of biconnected maps*⟩≡

```
static int PLAN_EMBED_K(graph& G, node_array<int>& st_num,
                                   list<node>& st_list)
{ ⟨PLAN_EMBED: phase 1⟩
  if (stv == n+1) { ⟨PLAN_EMBED: phase 2⟩ }
  delete[] EDGE;
  delete[] I;
  return stv - 1;
}
static bool PLAN_EMBED(graph& G, node_array<int>& st_num,
                                   list<node>& st_list)
{ return PLAN_EMBED_K(G,st_num,st_list) == G.number_of_nodes(); }
```

The first version of the function is needed for the search for Kuratowski subgraphs in the next section. It returns the largest integer k such that B_k has a bush form.

We come to the second phase. The purpose of the second phase is to determine for each node the order of $L(v)$ in $\hat{B}_n$. This is either $L_v(v)$ or $L_v^{rev}(v)$ depending on whether v is flipped an even or an odd number of times.

Node n is not flipped at all. Consider now a node $j < n$ and assume that H_j is merged into H_k in iteration k. Then j is not flipped in iterations $j+1$ to $k-1$, is flipped in iteration k if $I[k]$ contains $-j$ in its second part and is not flipped in iteration k if $I[k]$ contains $+j$ in its second part, and is flipped in iterations later than k iff node k is flipped. Thus it is easy to compute the number of times any node v is flipped by iterating over all nodes in downward order of st-number.

It actually suffices to compute the parity of the number of times a node is flipped; the parity is $+1$ if the number is even and is -1 otherwise. Assume that we process node k and let j be such that H_j is merged into H_k in iteration k. Then the parity of j is equal to the sign of the occurrence of j in $I[k]$ times the parity of k. In the piece of code below, node k tells node j, if the parity of j is odd, by putting the indicator ODD as the first element of $I[j]$.

The order of $L_n(v)$ is equal to the third part of $I(v)$, if v is flipped an even number of times, and is equal to the reversal of the third part of $I(v)$ otherwise.

⟨*PLAN_EMBED: phase 2*⟩≡

```
node_array<list<edge> > L_n(G);
const int EVEN = +1; const int ODD = -1;
int stv = n;
forall_rev(v,st_list)
{
  if (stv == 1) break;    // for v = t down to s+1
  list<int>* I_v = &I[stv];
  int d  = 1;
  int l  = I_v->pop();
  if ( l == ODD )
  { d = -1;
    l = I_v->pop();
  }
  // l = number of components merged into H_v
  int i;
  for( i = 0; i < l; i++)
  { int j = d * I_v->pop();
    if (j < 0) I[-j].push(ODD);  // tell j that it is odd
  }
  if (d > 0)
    forall(i,*I_v) L_n[v].append(EDGE[i]);
  else
    forall(i,*I_v) L_n[v].push(EDGE[i]);
  stv--;
}
```

⟨*PLAN_EMBED: phase 3*⟩

We come to the third and last phase of PLAN_EMBED. We know $L_n(v)$ for every node v and want to compute the counter-clockwise order of the edges in $U(v)$, where $U(v)$ is the set of edges connecting v to higher numbered nodes. Self-loops will be treated as an add-on. We compute the ordering of the edges in $U(v)$ by so-called *leftmost depth-first search*.

Consider a depth-first search starting in t that uses only edges in $L(v)$ and that considers the edges in $L(v)$ in their counter-clockwise order. Such a depth-first search is called a

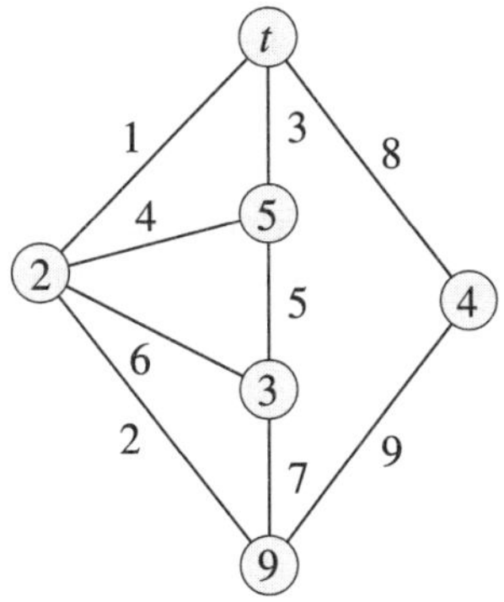

Figure 8.23 A leftmost depth-first search starting in t. For every node v the edges going to lower numbered neighbors are explored in left-to-right order. The edge labels indicate the order in which the edges are explored.

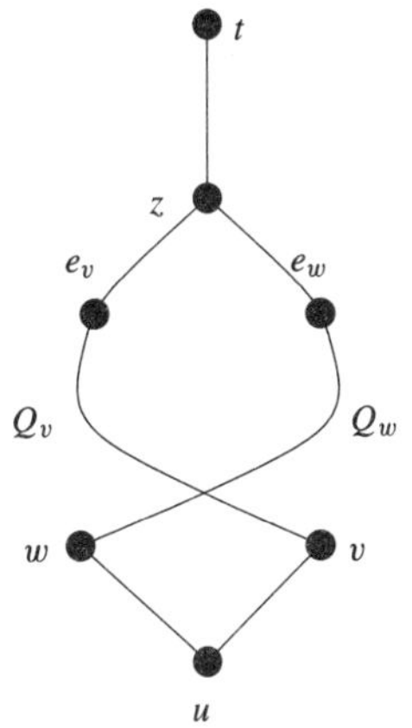

Figure 8.24 The edge (u, v) is after (u, w) in the clockwise order of edges in $U(u)$ but (v, u) is explored before (w, u).

leftmost depth-first search, as the edges in $L(v)$ are explored in left-to-right order (if drawn downwards from v) for any v and, more generally, the graph $\hat{B}_n$ is explored in a left-to-right fashion. This implies that for any node v, the edges in $U(v)$ are explored in left-to-right fashion, i.e., clockwise order, see Figure 8.23.

Lemma 52 *A leftmost depth-first search explores the edges in $U(u)$ in clockwise order for any node u.*

Proof Assume otherwise. Let u be the highest numbered node such that $U(v)$ is ordered incorrectly, say edge (u, v) is after edge (u, w) in the clockwise order of edges in $U(u)$, but (v, u) is explored before (w, u). Consider the paths P_v and P_w from t to u, which follow the tree paths to v and w in the depth-first search tree, respectively, and then take the edge (v, u)

or (w, u), respectively, see Figure 8.24. Let z be the node furthest from t and different from u that is common to both path. Let Q_v and Q_w be the induced paths from z to u passing through v and w, respectively, and let e_v and e_w be the first edges on these paths. Then e_v precedes e_w in the counter-clockwise order of the edges in $L(z)$.

The paths Q_v and Q_w are y-monotone, Q_v is left of Q_w "near" z, and Q_v is right of Q_w "near" u, and hence the two paths must cross. By definition of z they do not cross in a node and hence $\hat{B}_n$ is not a bush form of B_n. □

The following function LMDFS realizes leftmost depth-first search and builds a list *embed_list* containing all edges in $\cup_u U(u)$ in the order in which they are explored; the edge which is explored first comes last in the list, and the edge which is explored last comes first (since edges are pushed on the list and not appended). In other words, for each node u the edges in $U(u)$ occur in counter-clockwise order in *embed_list*. The edges do not necessarily occur consecutively.

LMDFS reuses the array *st_num* to record whether a node has been visited. leftmost depth-first search

⟨*auxiliary functions*⟩≡

```
static void LMDFS(graph& G, node v, const node_array<list<edge> >& L_n,
                  node_array<int>& st_num, list<edge>& embed_list)
{
  if (st_num[v] < 0) return;
  st_num[v] = -1;
  edge e;
  forall(e,L_n[v])
  { embed_list.push(G.reversal(e));
    LMDFS(G,target(e),L_n,st_num,embed_list);
  }
}
```

We use LMDFS in a function *embedding* that reorders the adjacency lists. We first build a list *embed_list* containing for each node v the set of edges in $A(v)$ in counter-clockwise order but not necessarily consecutively, and then use the sorting function *G.sort_edges*(*embed_list*) to rearrange the adjacency lists accordingly.

We build *embed_list* in three steps. In the first step we copy the lists *L_n*$[v]$ to *embed_list*, in the second step we call LMDFS to add the edges in $\cup_v U(v)$ in their counter-clockwise order, and in the third step we deal with all self-loops. The self-loops can be added in any order, we only have to make sure that the two directions of a self-loop are placed next to each other. In this way there will be no crossings between self-loops.

⟨*auxiliary functions*⟩+≡

```
static void embedding(graph& G, node t, node_array<int>& st_num,
                      node_array<list<edge> >& L_n)
{
  list<edge> embed_list;
  node v; edge e;
```

```
    forall_nodes(v,G)
      forall(e,L_n[v]) embed_list.append(e);
    LMDFS(G,t,L_n,st_num,embed_list);
    // append self-loops at the end of the list
    edge_map<bool> treated(G,false);
    forall_nodes(v,G)
    { edge e;
      forall_adj_edges(e,v)
        if (target(e) == v && !treated[e])
        { embed_list.append(e); embed_list.append(G.reversal(e));
          treated[e] = treated[G.reversal(e)] = true;
        }
     }
    G.sort_edges(embed_list);
  }
```

After all this preparatory work phase three reduces to a call of embedding.

⟨*PLAN_EMBED: phase 3*⟩≡

```
  node t = st_list.tail();
  embedding(G,t,st_num,L_n);
```

The running time of PLAN_EMBED is $O(n + m)$. We have already argued that phase one takes linear time. Phase two touches every edge once and hence takes also linear time. Phase three consists of a depth-first search followed by extracting the adjacency lists from *embed_list* and hence takes linear time.

Arbitrary Maps: We give the implementation of *BL_PLANAR*(G, *embed*). Recall that G must be a map if *embed* is *true*. The implementation is fairly simple.

We extend G to a biconnected graph (if *embed* is *false*) and to a biconnected map (if *embed* is *true*), compute an st-numbering of G, call the planarity test for biconnected graphs and maps, respectively, and remove the added edges. The function *Make_Biconnected* is discussed in the exercises of Section 7.4. It makes a graph biconnected by adding edges. It does so without destroying planarity.

⟨*planar embedding of arbitrary maps*⟩≡

```
  bool BL_PLANAR(graph& G, bool embed)
  { if (G.number_of_edges() <= 0)  return true;
    // prepare graph
    list<edge> el;
    if (embed)
    { if ( !G.make_map() )
        error_handler(1,"BL_PLANAR: can only embed maps.");
      Make_Biconnected(G,el);
      edge e;
      forall(e,el)
      { edge x = G.new_edge(target(e),source(e));
```

```
        el.push(x);
        G.set_reversal(e,x);
      }
    }
    else
      Make_Biconnected(G,el);

    node_array<int> st_num(G);
    list<node> st_list;
    ST_NUMBERING(G,st_num,st_list);
    bool plan;
    if (embed)
      plan = PLAN_EMBED(G,st_num,st_list);
    else
      plan = PLANTEST(G,st_num,st_list);

    // restore graph
    edge e; forall(e,el) G.del_edge(e);
    return plan;
}
```

8.7.3 *Kuratowski Subgraphs*

We describe functions to extract Kuratowski subgraphs. We first give a simple algorithm with quadratic running time, then a linear time algorithm for biconnected graphs, and finally a linear time algorithm for arbitrary graphs.

We start with a simple algorithm that computes Kuratowski subgraphs in quadratic time $O((n+m)m)$. We iterate over all edges e of G. We hide e and check the planarity of $G \setminus e$. If $G \setminus e$ is non-planar, we leave e hidden, and if $G \setminus e$ is planar, we add e to the set of edges of the Kuratowski subgraph and restore it. At the end we restore all edges. The running time of this algorithm is m times the running time of the planarity test. The running time can be improved to $O(n^2)$ by observing that it suffices to consider $3n+7$ uedges of G, since a planar graph with n nodes can have at most $3n + 6$ edges according to Lemma 47. We leave it to the exercises to implement this improvement.

⟨*auxiliary functions*⟩+≡

```
static void KURATOWSKI_SIMPLE(graph& G, list<edge>& K)
{ K.clear();
  if ( BL_PLANAR(G,false) )
    error_handler(1,"KURATOWSKI_SIMPLE: G is planar");
  list<edge> L = G.all_edges();
  edge e;
  forall(e,L)
  { G.hide_edge(e);
    if (BL_PLANAR(G,false))
    { G.restore_edge(e);
      K.append(e);
    }
```

```
  }
  G.restore_all_edges();
}
```

We turn to the linear time algorithm of Karabeg and Hundack, Mehlhorn, and Näher [Kar90, HMN96] to find Kuratowski subgraphs. We assume that G a biconnected non-planar map without self-loops and parallel edges.

When the planarity test algorithm is run on G there will be a minimal k such that B_k has a bush form but B_{k+1} does not, because B_k contains an obstruction. Then $k+1 < n$ since $\hat{B}_{n-1}$ can always be extended. We show

Lemma 53 *If B_k has a bush form and contains an obstruction then G contains a Kuratowski subgraph.*

An obstruction is either an obstructing articulation point or an obstructing biconnected component. We deal with obstructing articulation points first and then with obstructing biconnected components. For both cases we need some simple facts about trees. For a tree T and a subset S of its nodes we use $T(S)$ to denote the smallest subtree of T connecting all nodes in S. If $|S| \le 3$ then $T(S)$ contains a node r, called the *join* of S in T, such that the paths from r to the nodes in S are pairwise edge-disjoint ($r \in S$ is allowed). If $|S| = 3$, the join is unique.

Lemma 54 *Let v be an articulation point of B_k and let T be a depth-first search tree of B_k rooted at v. If w and z are distinct virtual nodes in some connected component C of B_k with respect to v then the join of $\{v, w, z\}$ in T is distinct from v, w, and z.*

Proof Let u be the first node reached in a depth-first search of C starting in v. Since C is a component with respect to v, $C \backslash v$ is connected. This implies that all nodes in $C \backslash v$ are descendants of u in T. □

In the sequel we use T_t to denote a tree on nodes $\{k+1, \ldots, n\}$ rooted at $t(= n)$ and where each node v, $v < n$, has an incoming edge from a higher numbered node. Such a tree exists since G is st-numbered.

We also use T_s to denote a depth-first search tree of B_k. T_s is rooted at s except if explicitly specified otherwise.

An Obstructing Articulation Point: Let v be an obstructing articulation point, i.e., at least three of the components with respect to v are mixed. Let C_i, $0 \le i \le 2$, be a mixed component with respect to v, let w_i be a leaf[20] labeled $k+1$ in C_i and let z_i be a large[21] leaf in C_i. Let T_s be a depth-first search tree of B_k rooted at v.

Let T_i be the subgraph of T_s spanned by v, w_i, and z_i, and let x_i be the join of T_i. Consider the subgraph K of G consisting of:

[20] We will use leaf and virtual node as synonyms in this section.

[21] A large leaf is a leaf that is labeled $k+2$ or larger.

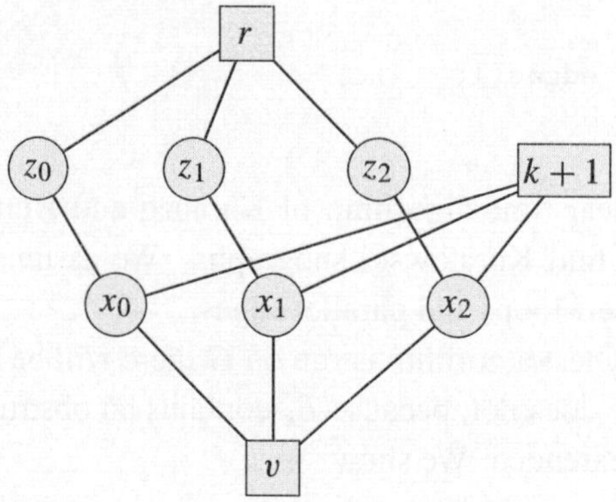

Figure 8.25 A $K_{3,3}$ with sides $\{x_0, x_1, x_2\}$ and $\{v, k+1, r\}$.

- T_0, T_1, T_2, and the tree $T_t(z_0, z_1, z_2)$.

Let r be the join of z_0, z_1, and z_2 in T_t. Then $r \neq k+1$ and hence K is a subdivision of $K_{3,3}$ with sides $\{x_0, x_1, x_2\}$ and $\{k+1, v, r\}$, see Figure 8.25.

An Obstructing Biconnected Component: Let H be a biconnected component with attachment nodes $y_0, y_1, \ldots, y_{p-1}$. We assume that y_0 is the lowest numbered attachment node and that $y_0, y_1, \ldots, y_{p-1}$ appear in this order on the boundary cycle of H in $\hat{B}_k$, where $\hat{B}_k$ is a bush form of B_k. Let C_i be the part of B_k opposite to H at y_i and let $s(C_i) \in$ {clean, mixed, full} be the status of C_i. We have

$$s(C_0)s(C_1)\ldots s(C_{p-1}) \notin \text{clean}^* \,\text{mixed}_0^1\, \text{full}^* \,\text{mixed}_0^1\, \text{clean}^*,$$

since H is obstructing.

Lemma 55 *One of the cases below arises:*

(1) There are indices a, b, c, and d such that y_a, y_b, y_c, and y_d occur in this order on the boundary cycle of H, and C_a and C_c are non-clean and C_b and C_d are non-full.
(2) There are indices a, b, and c such that y_a, y_b, and y_c occur in this order on the boundary cycle of H, and C_a, C_b, and C_c are mixed.

In either case, 0 is among the selected indices.

Proof Observe first, that C_0 is either clean or mixed, but never full (since there is a leaf labeled n in C_0 and $k+1 < n$). If

$$s(C_1)\ldots s(C_{p-1}) \notin \text{clean}^* \,\text{mixed}_0^1\, \text{full}^* \,\text{mixed}_0^1\, \text{clean}^*,$$

then there are a, b, c with $1 \leq a < b < c \leq p-1$ and C_a and C_c are non-clean and C_b is non-full. Since C_0 is non-full we are in case (1) with $d = 0$. So assume that

$$s(C_1)\ldots s(C_{p-1}) \in \text{clean}^* \,\text{mixed}_0^1\, \text{full}^* \,\text{mixed}_0^1\, \text{clean}^*.$$

Then C_0 is non-clean (and hence mixed) and $p - 1 \geq 2$ since H is non-obstructing otherwise.

If case (1) does not arise with $a = 0$ then there are no b, c, and d with $1 \leq b < c < d \leq p - 1$ with C_b and C_d non-full and C_c non-clean, i.e., any C_c between two non-full C_b and C_d is clean. Thus, either $p - 1 = 2$ or

$$s(C_1) \ldots s(C_{p-1}) \in \text{ clean}^* \text{ mixed}_0^1 \text{ full}^* \cup \text{ full}^* \text{ mixed}_0^1 \text{ clean}^*.$$

In the latter situation H is non-obstructing, and hence this case is excluded. In the former situation C_1 and C_2 must be mixed since H is non-obstructing otherwise. Thus, (2) arises. □

We next exhibit Kuratowski subgraphs for cases (1) and (2).

Assume first that there are indices a, b, c, and d such that y_a, y_b, y_c, and y_d occur in this order on the boundary cycle of H, C_a and C_c are non-clean and C_b and C_d are non-full. We call this an *obstructing cycle with four alternating attachments*. Consider the subgraph K of G consisting of:

- the boundary cycle of H,
- a path from y_a to a copy of $k + 1$ in C_a,
- a path from y_c to a copy of $k + 1$ in C_c,
- a path from y_b to a large leaf z_b in C_b,
- a path from y_d to a large leaf z_d in C_d,
- the tree $T_t(\{k + 1, z_b, z_d\})$.

Let r be a join of $k + 1$, z_b, and z_d in T_t; we may assume that $r \neq k + 1$ (observe that $z_b \neq k + 1$ and $z_d \neq k + 1$). K is a subdivision of $K_{3,3}$ with sides $\{y_b, y_d, k + 1\}$ and $\{y_a, y_d, r\}$, see Figure 8.26.

Assume next that there are indices a, b, and c such that y_a, y_b, and y_c occur in this order on the boundary cycle of H and C_a, C_b, and C_c are mixed. We call this a *cycle with three mixed attachments*. Consider the subgraph K of G consisting of:

- the boundary cycle of H,
- trees $T_s(\{y_i, w_i, z_i\})$ where $i \in \{a, b, c\}$, w_i is a leaf labeled $k + 1$ in C_i, and z_i is a large leaf in C_i,
- tree $T_t(\{k + 1, z_1, z_2, z_3\})$.

Let y'_i be the join of y_i, z_i, and w_i. Then y'_i is distinct from z_i and w_i but may be equal to y_i. Figure 8.27 illustrates the situation.

We can obtain a K_5 from K by contracting the paths connecting y_i with y'_i for $i \in \{a, b, c\}$ and by contracting the edges in $T_t(\{z_a, z_b, z_c\})$. We can now appeal to the fact that if a graph K can be contracted to a subdivision of $K_{3,3}$ or K_5 then it contains a subdivision

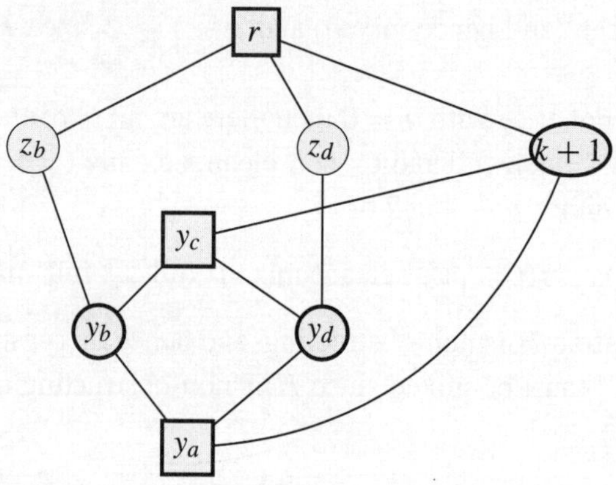

Figure 8.26 An obstructing cycle with four alternating attachments gives rise to a $K_{3,3}$ with sides $\{y_a, y_c, r\}$ and $\{y_b, y_d, k+1\}$.

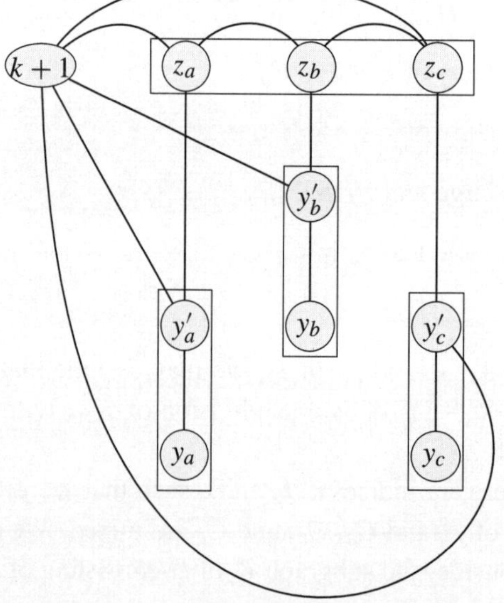

Figure 8.27 An obstructing cycle with three mixed attachments yields a K_5 after contraction of the paths from y_i to y'_i for $i \in \{a, b, c\}$ and contraction of the edges in tree $T_t(\{z_a, z_b, z_c\})$.

of $K_{3,3}$ or K_5 before the contraction, see [NC88, Lemma 1.2] and the exercises. We will exploit this fact in our implementation.

For completeness we also exhibit the Kuratowski subgraphs directly. We distinguish three cases.

If $y_i = y'_i$ for all $i \in \{a, b, c\}$ and $T_t(\{k+1, z_a, z_b, z_c\})$ contains a node of degree four then K is a subdivision of K_5.

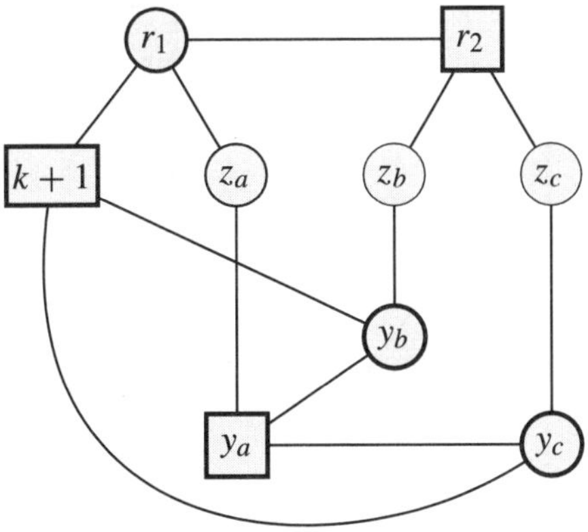

Figure 8.28 An obstructing cycle with three mixed attachments yields a $K_{3,3}$ if $y_i = y_i'$ for $i \in \{a, b, c\}$ and $T_t(\{k+1, z_a, z_b, z_c\})$ contains no node of degree four. In the figure, $k+1$ is paired with z_a.

If $y_i = y_i'$ for all $i \in \{a, b, c\}$ and $T_t(\{k+1, z_a, z_b, z_c\})$ contains no node of degree four then $T_t(\{k+1, z_a, z_b, z_c\})$ contains two nodes of degree three, say r_1 and r_2. The removal of the path joining r_1 and r_2 pairs $k+1$ with some z_i. We remove from K the path from y_i to the copy of $k+1$ in C_i and the part of the boundary cycle of H joining the other two y's and obtain a subdivision of $K_{3,3}$, see Figure 8.28, with sides $\{y_a, k+1, r_2\}$ and $\{y_b, y_c, r_1\}$.

If $y_i \neq y_i'$ for some $i \in \{a, b, c\}$, say $y_a \neq y_a'$, let r be the join of z_a, z_b, z_c in $T_t(\{z_a, z_b, z_c\})$. We obtain a subdivision of $K_{3,3}$ with sides $\{y_a, k+1, r\}$ and $\{y_b', y_c', y_a'\}$ from K by deleting the part of the boundary cycle of H that connects y_b and y_c, and by replacing $T_t(\{k+1, z_a, z_b, z_c\})$ by $T_t(\{z_a, z_b, z_c\})$, see Figure 8.29.

This completes the proof of Lemma 53.

We turn to a linear time implementation. The following function assumes that G is a biconnected non-planar map without self-loops and parallel edges. It computes the set of edges of a Kuratowski subgraph of G in K.

⟨*Kuratowski graphs in biconnected maps*⟩≡
```
static void Kuratowski(graph& G, list<edge>& K)
{ node v; edge e;
  string current_case;  // for debugging purposes
  ⟨compute st-numbering⟩
  int k = PLAN_EMBED_K(G,st_num,st_list);
  if ( k == G.number_of_nodes() )
    error_handler(1,"Kuratowski: G must be non-planar");
  ⟨compute bush form B for B_k⟩
  ⟨obstructing articulation point⟩
  ⟨obstructing biconnected component⟩
}
```

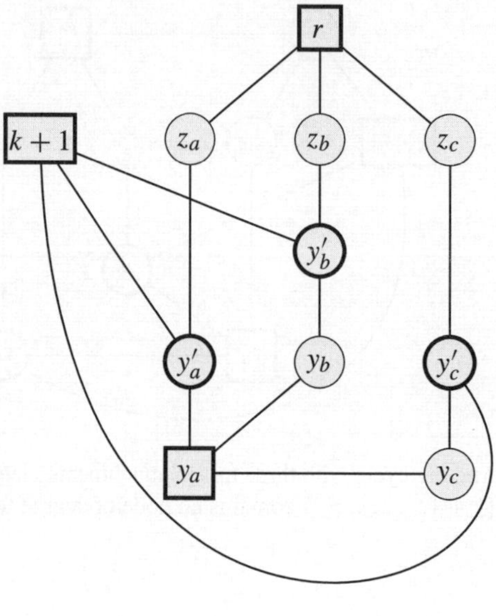

Figure 8.29 An obstructing cycle with three mixed attachments yields a $K_{3,3}$ if $y_a \neq y'_a$.

We start by computing an st-numbering of G. Next we call PLAN_EMBED_K to find k such that B_k has a bush form but B_{k+1} has not. We compute a bush form B for B_k and then search for an obstruction in B. This will be the most difficult part of the implementation. Having found an obstruction we extract a Kuratowski subgraph as shown in the proof of Lemma 53.

Compute st-Numbering: We compute an st-numbering and the nodes s and t.

⟨*compute st-numbering*⟩≡

```
node_array<int> st_num(G);
list<node> st_list;
ST_NUMBERING(G,st_num,st_list);
node s = st_list.head();
node t = st_list.tail();
```

The Bush Form B for B_k: We construct a bush form B for B_k. We declare B of type *GRAPH<node, edge>* and let every node and edge of B know its original in G. We add a node *top_B* to B and connect it to every virtual node (by a uedge). In this way B becomes a biconnected map.

We st-number the nodes of B by first numbering the non-virtual nodes, then the virtual nodes, and finally the node *top_B*. We store the st-numbering in *st_numB*, the ordered list of nodes in *st_listB*. Finally, *sB* is the node in B that corresponds to s and *tB* is a virtual node in B that is connected to *sB* by an edge. *tB* is a large leaf in the root component of every articulation point and in the part of B opposite to y_0 for any biconnected component H with lowest attachment node y_0.

Having constructed the st-numbering we call PLAN_EMBED to compute a planar embedding of B. We restore the st-numbers as they are destroyed by the planar embedding program, and we delete the auxiliary node *top_B* from B and *st_listB*.

⟨*compute bush form B for B_k*⟩≡

```
GRAPH<node,edge> B;
list<node>       st_listB;
node_array<node> v_in_B(G,nil);
forall(v,st_list)
{ if ( st_num[v] > k ) break;
  node vB = v_in_B[v] = B.new_node(v);
  st_listB.append(vB);
}
node top_B = B.new_node();
forall_nodes(v,G)
{ if (st_num[v] > k) continue;
  forall_adj_edges(e,v)
  { node w = G.target(e);
    if ( st_num[w] < st_num[v] )  continue;
    edge r = G.reversal(e);
    node wB;
    if ( st_num[w] > k )
    { wB = B.new_node(w);
      st_listB.append(wB);
      B.set_reversal(B.new_edge(wB,top_B),B.new_edge(top_B,wB));
    }
    else
      wB =  v_in_B[w];
    edge e1 = B.new_edge(v_in_B[v],wB,e);
    edge r1 = B.new_edge(wB,v_in_B[v],r);
    B.set_reversal(e1,r1);
  }
}
node sB = v_in_B[s];  node tB;
forall_adj_edges(e,sB)
  if ( B[B.target(e)] == t) tB = B.target(e);
B.set_reversal(B.new_edge(sB,top_B),B.new_edge(top_B,sB));
st_listB.append(top_B);
node_array<int> st_numB(B);
int stn = 1;
forall(v,st_listB) st_numB[v] = stn++;
PLAN_EMBED(B,st_numB,st_listB); // destroys st-numbers
stn = 1;
forall(v,st_listB) st_numB[v] = stn++;
B.del_node(top_B); st_listB.Pop();  // remove top_B
```

Obstructing Articulation Points: We search for an obstructing articulation point and, if successful, extract a Kuratowski subgraph.

⟨*obstructing articulation point*⟩≡

```
array<node> z(3);
array<node> spec(3);
```

A successful search for an obstructing articulation point will store the obstructing articulation point in v, and for i, $0 \le i < 3$, will store a large leaf in the i-th mixed component with respect to v in $z[i]$ and a leaf labeled $k + 1$ in $spec[i]$.

The search (successful or not) will also compute some auxiliary information for internal use and for later use in the search for obstructing biconnected components.

We define an enum that we use to distinguish between leafs labeled $k+1$ and large leafs, and we define two functions so that node arrays can be used as type parameters.

⟨*auxiliary functions*⟩+≡

```
enum { K_PLUS_1 = 0, OTHERS = 1};
ostream& operator<<(ostream& o, const node_array<node>&) { return o; }
istream& operator>>(istream& i, node_array<node>&)        { return i; }
```

We give the declarations of the auxiliary informations and explain them below.

⟨*obstructing articulation point*⟩+≡

```
list<node> dfs_list;
node_array<edge> tree_edge(B,nil);
node_array<int> dfs_num(B,-1);
int dfs_count = 0;
DFS(B,sB,dfs_list,dfs_num,dfs_count,tree_edge);
edge_array<int> comp_num(B);
int num_comps = BICONNECTED_COMPONENTS(B,comp_num);
node_array<edge> up_tree_edge(G,nil);
array<node_array<node> > leaf(2);
leaf[K_PLUS_1] = leaf[OTHERS] = node_array<node>(B,nil);
array<node_array<node> > leaf_in_upper_part(2);
leaf_in_upper_part[K_PLUS_1] =
        leaf_in_upper_part[OTHERS] = node_array<node>(B,nil);
node_array<int>  num_mixed_non_root_comps(B,0);
node_array<node> spec_leaf_in_root_comp(B,nil);
array<node_array<node> > child(1,2);  // want indices one and two
child[1] = child[2] = node_array<node>(B,nil);
```

The auxiliary information is as follows: let T_s be a depth-first search tree of B rooted at s. *tree_edge*$[v]$ is the tree edge into v in T_s for $v \neq s$ and is *nil* for $v = s$, *dfs_num*$[v]$ is the dfs-number of v, and *dfs_list* is the list of nodes of B in increasing order of dfs-number. All quantities just mentioned are computed by a call of the auxiliary function DFS, see below. *num_comps* is the number of biconnected components, and *comp_num*$[e]$ is the number of the biconnected component containing e for any edge e of B. Both values are computed by calling the biconnected components function. We call *comp_num*[e] the component number of e.

up_tree_edge[v] is for any node v of G with *st_num*[v] $> k$ and $v \neq t$ an edge from a higher numbered node. It is *nil* for all other nodes of G. The up-tree edges define a tree T_t rooted at t on the nodes labeled $k + 1$ and larger.

leaf[*K_PLUS_1*][v] is a leaf labeled $k + 1$ in the subtree of T_s rooted at v (*nil* if no such leaf exists).

leaf[*OTHERS*][v] is a large leaf in the subtree of T_s rooted at v (*nil* if no such leaf exists).

The next four pieces of information are only defined for articulation points. The *upper part with respect to an articulation point* is the union of the non-root components with respect to the articulation point.

leaf_in_upper_part[*K_PLUS_1*][v] is a leaf labeled $k + 1$ in the upper part of v (*nil* if there is no such leaf).

leaf_in_upper_part[*OTHERS*][v] is a large leaf in the upper part of v (*nil* if there is no such leaf).

child[1][v] is a child of v in T_s that lies in a mixed non-root component with respect to v (*nil* if there is no such child).

child[2][v] is a child of v in T_s that lies in a second mixed non-root component with respect to v (*nil* if there is no such child).

We next discuss how the auxiliary information is computed. The quantities related to depth-first search are computed by a variant of depth-first search.

⟨*auxiliary functions*⟩+≡

```
void DFS(const graph& G, node v,
         list<node>& dfs_list, node_array<int>& dfs_num,
         int& dfs_count, node_array<edge>& tree_edge)
{ dfs_list.append(v);
  dfs_num[v] = dfs_count++;
  edge e;
  forall_adj_edges(e,v)
  { node w = G.target(e);
    if ( dfs_num[w] == -1 )
    { tree_edge[w] = e;
      DFS(G,w,dfs_list,dfs_num,dfs_count,tree_edge);
    }
  }
}
```

The up-tree is easily computed. We simply select for each node labeled larger than k an edge going to a node with higher st-number and then put the reversal of the edge into the tree.

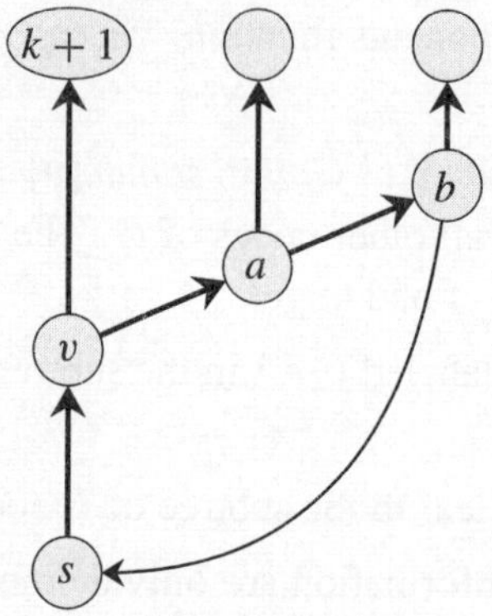

Figure 8.30 The root component of v consists of the nodes s, v, a, and b. Tree edges are drawn in bold. The tree edge (v, a) belongs to the same biconnected component as the tree edge into v, but the tree edge $(v, k+1)$ does not. The tree edge $(v, k+1)$ belongs to a non-root component with respect to v.

⟨*obstructing articulation point*⟩+≡

```
forall_nodes(v,G)
{ if (st_num[v] <= k ) continue;
  edge e;
  forall_adj_edges(e,v)
  { node w = G.target(e);
    if ( st_num[w] > st_num[v] )
    { up_tree_edge[v] = G.reversal(e); break; }
  }
}
```

All other auxiliary information is computed by scans over T_s. We start with some simple observations, see Figure 8.30. We have, for any node v, the following:

- The tree edge into v belongs to the root component with respect to v.
- A tree edge out of v belongs to the root component with respect to v iff it belongs to the same biconnected component as the tree edge into v iff it has the same component number as the tree edge into v.
- A tree edge out of v belongs to a non-root component with respect to v iff its component number is different from the component number of the tree edge into v or if v is equal to (the copy of) s in B.
- The non-root components with respect to v are in one-to-one correspondence to the tree edges out of v.

The node labels *leaf*[*K_PLUS_1*] and *leaf*[*OTHERS*] are computed by a leaf to root scan of T_s.

⟨*obstructing articulation point*⟩+≡
```
forall_nodes(v,B)
{ if (st_numB[v] <= k) continue;
  if ( st_num[B[v]] == k + 1 )
    leaf[K_PLUS_1][v] = v;
  else
    leaf[OTHERS][v] = v;
}
forall_rev(v,dfs_list)  // down the tree
{ if (v == sB) continue;
  node pv = B.source(tree_edge[v]);
  assign(leaf[K_PLUS_1][pv],leaf[K_PLUS_1][v]);
  assign(leaf[OTHERS][pv],  leaf[OTHERS][v]);
}
```

where we used the following conditional assignment function *assign* to propagate information.

⟨*auxiliary functions*⟩+≡
```
void assign(node& x, const node& y) { if ( x == nil) x = y; }
```

We next compute for each articulation point v the number of mixed non-root components with respect to v and *leaf_in_upper_part*[][v].

A node v identifies a non-root component of its parent *pv* if either *pv* is equal to *sB* and *sB* has more than one child or if the tree edges into v and *pv* belong to different biconnected components. Actually, *sB* always has at least two children, one is a copy of t and the other contains a copy of $k+1$ in its subtree. Note that $k+1 \neq t$ since the planarity test cannot fail when node t is to be added.

The non-root component of *pv* identified by v is mixed if it contains a leaf labeled $k+1$ as well as a large leaf.

We are propagating information from the leaves to the root and hence know the number of mixed non-root components of v when v is reached. If a node v has three mixed non-root components we extract a Kuratowski subgraph.

⟨*obstructing articulation point*⟩+≡
```
forall_rev(v,dfs_list)   // down the tree
{ if (num_mixed_non_root_comps[v] >= 3)
  { ⟨v has three mixed non-root components⟩ }
  if ( v == sB) continue;
  node pv = B.source(tree_edge[v]);
  if ( pv == sB || comp_num[tree_edge[v]] != comp_num[tree_edge[pv]] )
  { if ( leaf[K_PLUS_1][v] && leaf[OTHERS][v] )
      num_mixed_non_root_comps[pv]++;
    assign(leaf_in_upper_part[K_PLUS_1][pv],leaf[K_PLUS_1][v]);
    assign(leaf_in_upper_part[OTHERS][pv],leaf[OTHERS][v]);
  }
}
```

Assume that v has three mixed non-root components. We iterate over all children of v and search for three children that define mixed non-root components. Whenever such a child is found we copy its two leaves to $y[i]$ and $spec[i]$ for $i = 0, 1$, and 2.

⟨*v has three mixed non-root components*⟩≡

```
current_case = "three mixed non-root components";
int i = 0;
forall_adj_edges(e,v)
{ node w = B.target(e);
  if ( w == sB || v != B.source(tree_edge[w]) ) continue;
  if ( leaf[K_PLUS_1][w] && leaf[OTHERS][w] )
  { z[i] = leaf[OTHERS][w]; spec[i] = leaf[K_PLUS_1][w];
    i++;
    if ( i == 3) break;
  }
}
```

⟨*obstructing articulation point: extract Kuratowski graph*⟩

The actual extraction of the Kuratowski subgraph will be discussed below.

If no articulation point has three mixed non-root components, we need to check whether there is an articulation point with two mixed non-root components and a mixed root component. It is slightly tricky to determine whether root components are mixed. We observe first that node s and hence node t is contained in any root component. Thus there is always a large leaf in the root component. In fact, it is the node tB.

We want to compute for each node v a leaf labeled $k + 1$ in its root component (if any). Consider any path p in T_s from v to a leaf labeled $k + 1$. The leaf belongs to the root component of v iff the target of the first edge of p belongs to the root component of v. This is the case if the first edge of p is the tree edge into v or is a tree edge out of v which belongs to the same biconnected component as the tree edge into v. We compute *spec_leaf_in_root_comp* by considering the two kinds of paths separately.

For the second kind of path we propagate information down the tree. We pass information about a leaf along a tree edge (v, w) if this edge belongs to the root component of v, i.e., if it has the same component number as the tree edge into v.

⟨*obstructing articulation point*⟩+≡

```
forall_rev(v,dfs_list)  // down the tree
{ if (v == sB) continue;
  node pv = B.source(tree_edge[v]);
  if ( pv != sB && comp_num[tree_edge[v]] == comp_num[tree_edge[pv]] )
    assign(spec_leaf_in_root_comp[pv],leaf[K_PLUS_1][v]);
}
```

For the first kind of path we compute for every node v, *spec_leaf_via_tree_edge*$[v]$, a leaf labeled $k + 1$ in the root component of v that is reachable through the tree edge into v (*nil* if there is no such leaf). A leaf labeled $k + 1$ in the root component is then either a leaf that was already computed above or the leaf that can be reached via the tree edge into v.

spec_leaf_via_tree_edge is computed from the root towards the leaves of T_s. Let v be any node and consider the time when we process v. Let c be any child of v. A leaf in the root component of c that is reachable through the tree edge into c is either reachable through the tree edge into v or through a sibling of c.

If v has a leaf labeled $k+1$ that is reachable through the tree edge into v we simply pass this leaf to all children of v.

So assume that v has no leaf labeled $k+1$ that is reachable through the tree edge into v. We try to determine two children c_1 and c_2 of v that have a leaf labeled $k+1$ in their subtree. If there is none, then no child of v can reach a leaf labeled $k+1$ through one of its siblings, if there is exactly one child, then all siblings of this child can reach a leaf labeled $k+1$ through it, and if there are two children, then all children of v can reach a leaf labeled $k+1$ through a sibling.

When a node v is encountered that has two mixed non-root components and a mixed root component we have found an obstructing articulation point and proceed to extract a Kuratowski subgraph.

⟨*obstructing articulation point*⟩+≡

```
node_array<node> spec_leaf_via_tree_edge(B,nil);
forall(v,dfs_list)  // up the tree
{ assign(spec_leaf_in_root_comp[v],spec_leaf_via_tree_edge[v]);
  if ( num_mixed_non_root_comps[v] == 2 && spec_leaf_in_root_comp[v] )
  { ⟨v has two mixed non-root and a mixed root component⟩ }
  if ( spec_leaf_via_tree_edge[v] != nil )
  { forall_adj_edges(e,v)
    { node c = B.target(e);
      if ( c == sB || v != B.source(tree_edge[c]) ) continue;
      spec_leaf_via_tree_edge[c] = spec_leaf_via_tree_edge[v];
    }
  }
  else
  { forall_adj_edges(e,v)
    { node c = B.target(e);
      if ( c == sB || v != B.source(tree_edge[c]) ) continue;
      if ( leaf[K_PLUS_1][c] )
      { if ( child[1][v] == nil )
          child[1][v] = c;
        else
          child[2][v] = c;
      }
    }
    if ( child[1][v] )
    { forall_adj_edges(e,v)
      { node c = B.target(e);
        if ( c == sB || v != B.source(tree_edge[c]) ) continue;
        if ( c != child[1][v] )
          spec_leaf_via_tree_edge[c] = leaf[K_PLUS_1][child[1][v]];
        else
          if ( child[2][v] )
```

```
          spec_leaf_via_tree_edge[c] = leaf[K_PLUS_1][child[2][v]];
        }
      }
    }
  }
```

Assume that v has two mixed non-root and a mixed root component. A leaf labeled $k+1$ in the root component of v is given by *spec_leaf_in_root_comp*$[v]$ and a large leaf is given by *tB*. For the other components we find the leaf labeled $k+1$ and the large leaf as in the case of three mixed non-root components.

⟨*v has two mixed non-root and a mixed root component*⟩≡

```
current_case = "two mixed non-root and a mixed root component";
z[0] = tB;
spec[0] = spec_leaf_in_root_comp[v];
int i = 1;
forall_adj_edges(e,v)
{ node w = B.target(e);
  if ( w == sB || v != B.source(tree_edge[w]) ) continue;
  if ( v != sB && comp_num[e] == comp_num[tree_edge[v]] ) continue;
  if ( leaf[K_PLUS_1][w] && leaf[OTHERS][w] )
  { z[i] = leaf[OTHERS][w]; spec[i] = leaf[K_PLUS_1][w];
    i++;
    if ( i == 3) break;
  }
}
⟨obstructing articulation point: extract Kuratowski graph⟩
```

Obstructing Articulation Point: Extraction of Kuratowski Graph: The node v is an obstructing articulation point. For every i, $0 \leq i < 3$, we have a large leaf in the i-th component in $z[i]$ and a leaf labeled $k+1$ in *spec*$[i]$.

We reroot the depth-first search tree at v and then extract the Kuratowski subgraph as described in the proof of Lemma 53.

⟨*obstructing articulation point: extract Kuratowski graph*⟩≡

```
// reroot the DFS-tree at v
dfs_list.clear();
dfs_num.init(B,-1);
tree_edge.init(B,nil);
int dfs_count = 0;
DFS(B,v,dfs_list,dfs_num,dfs_count,tree_edge);
list<edge> join_edges;
for (i = 0; i < 3; i++)
{ join(z[i],spec[i],v,tree_edge,B,join_edges);
  translate_to_G(join_edges,B); K.conc(join_edges);
}
join(B[z[0]],B[z[1]],B[z[2]],up_tree_edge,G,join_edges);
```

```
K.conc(join_edges);
check_before_return(G,K,st_num,leaf,tree_edge,dfs_num,k,
                    B,st_numB,sB,current_case);
return;
```

The function *check_before_return* calls *CHECK_KURATOWSKI*(G, K) to check whether K is a Kuratowski subgraph. If not, it opens two *GraphWins* and displays the edges in K in one of them and the bush form B in the other. We do not give details here. This visual debugging aid proved very valuable during the development phase of the algorithm.

The Join Function: Let T be a tree and let a, b, and c be the three nodes to be joined in T. For each node v the tree edge into v is stored in *tree_edge*$[v]$.

We trace the paths to the root from all three nodes and count, for each node of T, the number of paths containing it. Let r be the highest node which is reachable from all three nodes. The subtree joining the three nodes is the union of the paths from the three nodes to r. This union is not necessarily a disjoint union. We want to output each edge in the subtree only once and therefore mark nodes as we trace the paths. When a node is marked, its tree edge is added to the set L of edges comprising the subtree. The function returns r.

⟨*auxiliary functions*⟩+≡

```
node join(node a, node b, node c, const node_array<edge>& tree_edge,
          graph& B, list<edge>& L)
{ L.clear();
  node_array<int> num_desc(B,0);
  array<node> A(3); A[0] = a; A[1] = b; A[2] = c;
  int i;
  for (i = 0; i < 3; i++)
  { node v = A[i];
    num_desc[v]++;
    while ( tree_edge[v] != nil )
    { v = B.source(tree_edge[v]);
      num_desc[v]++;
    }
  }
  node r;
  for (i = 0; i < 3; i++)
  { node v = A[i];
    while (num_desc[v] < 3)
    { L.append(tree_edge[v]);
      num_desc[v] = 3;
      v = B.source(tree_edge[v]);
    }
    if ( i == 0 ) r = v;
  }
  return r;
}
void translate_to_G(list<edge>& L, const GRAPH<node,edge>& B)
```

```
{ list_item it;
  forall_items(it,L) L[it] = B[L[it]];
}
```

The function *translate* takes a list L of edges of B and replaces each edge by its counterpart in G.

Obstructing Biconnected Component: We come to obstructing biconnected components. We describe the search for an obstructing biconnected component and the extraction of a Kuratowski subgraph once an obstructing component has been found.

We exploit the fact that B is a plane map in our search for obstructing biconnected components. Consider any node v and the cyclic list $A(v)$ of edges out of v. If v is not an articulation point then all edges in $A(v)$ belong to the same biconnected component. If v is an articulation point then $A(v)$ decomposes into blocks, one for each biconnected component containing v. This follows from the fact that the boundary cycles of all biconnected component are part of the boundary of the outer face in every bush form.

Blocks that consist of at least two edges indicate the boundary cycle of a biconnected component. We find such blocks as follows. We iterate over all edges f out of v. If the cyclic predecessor of f in $A(v)$ belongs to a different biconnected component and the cyclic successor belongs to the same biconnected component, then f belongs to the boundary cycle of a non-trivial biconnected component, i.e., a biconnected component which is not just a single uedge. We maintain an edge array *treated_component* to record which biconnected components have already been treated.

If the component having f in its boundary cycle has not been treated yet, we determine its boundary cycle in *cycle_edges* and then determine whether one of the cases (1) or (2) of Lemma 55 applies.

In our search for biconnected components we iterate over the nodes of T_s from the root to the leaf. This has the advantage that we hit every biconnected component at its lowest node.

Let H be a biconnected component with attachment cycle $[y_0, y_1, \ldots, y_k]$, where y_0 is the lowest numbered node in the biconnected component. We need to know whether the component of B opposite to H at y_0 is mixed, i.e., contains a leaf labeled $k + 1$. We compute such a leaf in *spec_leaf_in_opposite_part*. For all i different from zero, the part of B opposite to H at y_i is simply the upper part of B with respect to y_i. We have collected information about upper parts already.

If the search for an obstructing biconnected component is unsuccessful, we give debugging information. After all, there must be either an obstructing articulation point or an obstructing biconnected component.

⟨*obstructing biconnected component*⟩≡

```
array<bool> treated_component(num_comps);
edge f;
forall(v,dfs_list)        // upwards
```

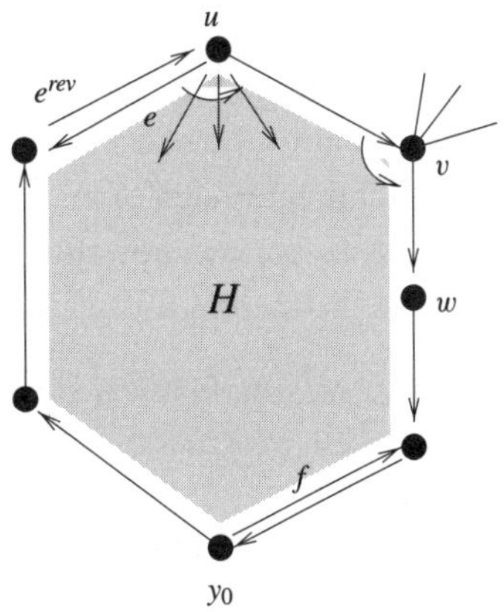

Figure 8.31 Scanning the boundary of a biconnected component H. We scan the boundary in clockwise direction. At each node, the reversal of a boundary edge is turned clockwise (i.e., through H) until the next boundary edge is reached. Two stopping criteria apply to the turning process: we stop if the cyclic adjacency successor does not belong to H or if all edges incident to the boundary node have been considered.
The edge e^{rev} is a boundary edge into u. We turn its reversal e clockwise until the next boundary edge is reached. At node v the first stopping criterion applies and at a node which has no incident edge outside H the second stopping criterion applies.

```
{ forall_adj_edges(f,v)
  { edge e1 = B.cyclic_adj_succ(f);
    edge e_pred = B.cyclic_adj_pred(f);
    if ( comp_num[e1] != comp_num[f] ||
         comp_num[f] == comp_num[e_pred] ) continue;
    if ( treated_component[comp_num[f]] ) continue;
    list<edge> cycle_edges;
    treated_component[comp_num[f]] = true;
    ⟨determine boundary cycle of component with lowest node y_0 = v⟩
    node spec_leaf_in_opposite_part = nil;
    ⟨compute leaf labeled k+1 in part opposite to y_0⟩
    ⟨obstructing cycle with four alternating attachments⟩
    if ( spec_leaf_in_opposite_part )
    { ⟨obstructing cycle with three mixed attachments⟩ }
  }
}
⟨unreachable point: give debugging information⟩
```

The boundary cycle of a biconnected component H is easily traced. We start with an edge f that emanates from v, the lowest node in the component, and that lies on the boundary cycle of the component. The unbounded face is to the right of f, see Figure 8.31. We will trace the boundary cycle in clockwise direction, i.e., keeping the unbounded face to our left, and store it in *cycle_edges*.

Assume that e is an edge such that its reversal belongs to the boundary cycle. Initially, e

is equal to f. We show how to find the successor edge of e^{rev} in the boundary cycle. Let e_1 be the cyclic adjacency successor of e. We advance e_1 until the successor of e_1 belongs to a different biconnected component or the successor of e_1 is equal to e. The former case happens for nodes v that are attachment nodes of H and the latter case happens for nodes that lie on the boundary cycle of H but are not attachment nodes of H. Edge e_1 is the successor of e^{rev} on the cycle. We proceed in this way until the cycle is completely traced.

⟨*determine boundary cycle of component with lowest node y_0 = v*⟩≡

```
edge e0 = f;
node y0 = v;
edge e = f; // e1 was set to B.cyclic_adj_succ(f) above
do { while ( comp_num[B.cyclic_adj_succ(e1)] == comp_num[e] &&
             B.cyclic_adj_succ(e1) != e )
       { e1 = B.cyclic_adj_succ(e1); }
       cycle_edges.append(e1);
       e = B.reversal(e1);
       e1 = B.cyclic_adj_succ(e);
} while ( e != e0 );
```

We next show how to compute a leaf labeled $k+1$ in the part of B opposite to H at y_0 in constant time. Constant time is needed since y_0 can be the lowest numbered node of many biconnected components.

The part of B opposite to H at y_0 consists of the root component of y_0 and all non-root components with respect to y_0 that do not contain H. We have computed above two children of y_0 (if they exist) that define mixed non-root components. A leaf labeled $k+1$ can be found in either the root component or in one of the mixed children that does not contain H. A non-root component does not contain H if the tree edge into the child does not belong to H.

⟨*compute leaf labeled k+1 in part opposite to y_0*⟩≡

```
spec_leaf_in_opposite_part = spec_leaf_in_root_comp[v];
for (int i = 1; i <= 2; i++)
{ node c = child[i][v];
  if ( spec_leaf_in_opposite_part == nil
       && c &&  comp_num[tree_edge[c]] != comp_num[e0] )
     spec_leaf_in_opposite_part = leaf[K_PLUS_1][c];
}
```

Obstructing Cycle with Four Alternating Attachments: We search for a cycle with four alternating attachments. By Lemma 55 there are two ways such a cycle may occur: The component opposite to y_0 contributes either a large leaf or a leaf labeled $k+1$. We therefore perform two searches. In the first search we set *y0_type* to OTHERS and let C_0 contribute a large leaf and in the second search we set *y0_type* to K_PLUS_1 and let C_0 contribute a leaf labeled $k+1$. The second search is only performed when *spec_leaf_in_opposite_part* is defined.

For an attachment y_i different from y_0 the part opposite to H at y_i is equal to the upper part of B with respect to y_i.

We store the four attachments in $y[0]$ to $y[3]$ and we store the selected leaf in the i-th component in $z[i]$.

⟨*obstructing cycle with four alternating attachments*⟩≡

```
list<int> kinds;
kinds.append(OTHERS); kinds.append(K_PLUS_1);
int y0_type;
forall(y0_type, kinds)
{ array<node> y(4);
  y[0] = y0; y[1] = y[2] = y[3] = nil;
  array<node> z(4);
  if (y0_type == OTHERS)
  { z[0] = tB;
    current_case = "cycle with 4 attachments; y_0 connects to t";
  }
  else
  { z[0] = spec_leaf_in_opposite_part;
    current_case = "cycle with 4 attachments; y_0 connects to k + 1";
    if ( !spec_leaf_in_opposite_part ) break;
  }
  list_item it0 = cycle_edges.first();
  list_item it = cycle_edges.cyclic_succ(it0);
  int i = 1;
  while (it != it0)
  { node v = B.source(cycle_edges[it]);
    int kind = (i == 2 ? y0_type : 1 - y0_type);
    if ( leaf_in_upper_part[kind][v] )
    { y[i] = v;
      z[i] = leaf_in_upper_part[kind][v];
      i++;
    }
    if ( i == 4 )
    { ⟨build the Kuratowski graph⟩
      return;
    }
    it = cycle_edges.cyclic_succ(it);
  }
}
```

Assume that we have found an obstructing cycle with four alternating attachments. We have the four attachments in $y[0]$ to $y[3]$ and the selected leaf in the i-th component in $z[i]$. Also *y0_type* tells us the type of the component C_0.

In the upper tree we need to take the subtree spanned by the two large leaves and node $k+1$.

⟨*build the Kuratowski graph*⟩≡
```
translate_to_G(cycle_edges,B); K.conc(cycle_edges);
list<edge> join_edges;
int i;
for (i = 0; i < 4; i++)
{ join(y[i],z[i],z[i],tree_edge,B,join_edges);
  translate_to_G(join_edges,B); K.conc(join_edges);
}
// subtree of T_t spanned by k+1 and two large leaves.
if (y0_type == OTHERS) i = 0; else i = 3;
join(B[z[i]],B[z[1]],B[z[2]],up_tree_edge,G,join_edges);
K.conc(join_edges);
check_before_return(G,K,st_num,leaf,tree_edge,dfs_num,k,
                    B,st_numB,sB,current_case);
```

Obstructing Biconnected Component with Three Mixed Opposing Parts: For case (2) we need that the component opposite to y_0 is mixed and that there are y_a, y_b such that C_a and C_b are mixed.

⟨*obstructing cycle with three mixed attachments*⟩≡
```
array<node> y(3);
array<node> spec_leaf_opposing(3);
array<node> other_leaf_opposing(3);
y[0] = y0;
spec_leaf_opposing[0] = spec_leaf_in_opposite_part;
other_leaf_opposing[0] = tB;
int i = 1;
list_item it0 = cycle_edges.first();
list_item it = cycle_edges.cyclic_succ(it0);
while (it != it0)
{ node v = B.source(cycle_edges[it]);
  if ( leaf_in_upper_part[OTHERS][v] && leaf_in_upper_part[K_PLUS_1][v])
  { y[i] = v;
    spec_leaf_opposing[i] = leaf_in_upper_part[K_PLUS_1][v];
    other_leaf_opposing[i] = leaf_in_upper_part[OTHERS][v];
    i++;
  }
  if ( i == 3 )
  { ⟨obstructing cycle with three mixed attachments: extract Kuratowski⟩
    return;
  }
  it = cycle_edges.cyclic_succ(it);
}
```

It remains to extract the Kuratowski subgraph. We proceed as described in the proof of Lemma 53. We collect all edges shown in Figure 8.27 in K. K is not a Kuratowski graph yet, but is guaranteed to contain one.

⟨*obstructing cycle with three mixed attachments: extract Kuratowski*⟩≡

```
current_case = "obstructing cycle with three mixed attachments";
translate_to_G(cycle_edges,B); K.conc(cycle_edges);
list<edge> join_edges;
for(int j = 0; j <= 2; j++)
{ join(spec_leaf_opposing[j], other_leaf_opposing[j], y[j],
       tree_edge,B,join_edges);
  translate_to_G(join_edges,B); K.conc(join_edges);
}
node r = join(B[other_leaf_opposing[1]], B[other_leaf_opposing[2]],
              B[spec_leaf_opposing[0]], up_tree_edge,G,join_edges);
K.conc(join_edges);
join(r,r,t,up_tree_edge,G,join_edges);
K.conc(join_edges);
{ ⟨thin out K⟩ }
check_before_return(G,K,st_num,leaf,tree_edge,dfs_num,k,
                    B,st_numB,sB,current_case);
```

Thinning Out: K is now an appropriate set of edges in G. It might still be too big. We want to thin it out so that only a $K_{3,3}$ or a K_5 remains. This is easy to do. We construct an auxiliary graph *AG*, which has a node for each node of G that has degree three or more in K and which has an edge for each path in K connecting two such nodes and having only intermediate nodes of degree two. We associate with every edge of *AG* the path in G represented by it.

AG is a small graph; in fact, it has at most twelve nodes. We call the quadratic version of the Kuratowski algorithm to find a Kuratowski subgraph of *AG* and then translate is back to G.

⟨*thin out K*⟩≡

```
node v; edge e;
edge_array<bool> in_K(G,false);
node_array<int> deg_in_K(G,0);
forall(e,K)
{ in_K[e] = true;
  deg_in_K[G.source(e)]++; deg_in_K[G.target(e)]++;
}
GRAPH<node,list<edge> > AG;
node_array<node> link(G,nil);
forall_nodes(v,G)
  if ( deg_in_K[v] > 2 ) link[v] = AG.new_node(v);
forall_nodes(v,G)
{ if ( !link[v] ) continue;
  edge e;
  forall_inout_edges(e,v)
  { if ( in_K[e] )
```

```
      { // trace path starting with e
        list<edge> path;
        edge f = e; node w = v;
        while (true)
        { in_K[f] = false; path.append(f);
          w = G.opposite(w,f);
          if ( link[w] ) break;
          // observe that w has degree two and hence ...
          forall_inout_edges(f,w)
            if ( in_K[f] ) break;
        }
        edge e_new = AG.new_edge(link[v],link[w]);
        AG[e_new].conc(path);  // O(1) assignment
      }
    }
}
list<edge> el;
KURATOWSKI_SIMPLE(AG,el);
K.clear();
forall(e,el) K.conc(AG[e]);
```

There is a small optimization in the program above which we want to mention. Instead of

```
edge e_new = AG.new_edge(link[v],link[w]);
AG[e_new].conc(path); // O(1) assignment
```

we could have written more elegantly

```
AG.new_edge(link[v],link[w],path);
```

The second version calls the copy constructor to construct a copy of *path* as the edge information of the new edge of *AG*, the first version concatenates *path* to the edge information of the new edge (which is initialized to the default value of lists, i.e., the empty list, by the new edge operation). Concatenation is a constant time operation. Concatenation empties *path* and this is all right. We have now completed the implementation of the linear time Kuratowski graph finder for biconnected graphs.

Arbitrary Graphs: We extend the algorithm to arbitrary graphs G. We first call the embedding algorithm to find out if G is planar. If it is, we are done.

So assume that G is non-planar. Then one of the biconnected components of G is non-planar. The idea is to search for a non-planar biconnected component of G and to call the algorithm of the preceding section for the biconnected component.

We give more details. A call *BICONNECTED_COMPONENTS*(G, *comp_num*) returns the number *num_c* of biconnected components of G and computes for each edge of G the index of the biconnected component containing e.

We iterate over all edges of G and construct for every c, $0 \le c < num\_c$, the set $E[c]$ of edges in the component and the set $V[c]$ of nodes of the component. We determine the set $V[c]$ as the set of endpoints of edges in $E[c]$ and hence this set may contain duplicates.

When the edge and node sets of all biconnected components are determined, we iterate over all components. For each c, $0 \le c < num\_c$, we construct a copy of the component in H. The nodes and edges of H know their counterparts in G. Since $V[c]$ may contain duplicates, we maintain a node array *link*, in which we store for each node v in G, whether a copy of v has already been constructed in H. We reset *link* when the construction of H is completed. In this way the extraction of a biconnected component has cost proportional to the size of the component.

When the extraction of a component is completed, we test it for planarity. We break from the loop once a non-planar biconnected component is found.

If G is biconnected we take a short cut and make H a copy of G.

The identification of Kuratowski graphs is simplified if H is a map without self-loops and parallel edges. We therefore remove self-loops (or do not put them into H in the first place) and parallel edges, and we turn H into a map by adding edges. Every added edge is made to point to the same edge in G as its reversal. We then call *Kuratowski* to find a Kuratowski subgraph K of H. We turn K into a Kuratowski subgraph of G by replacing every edge by its counterpart in G.

⟨*Kuratowski graphs in arbitrary graphs*⟩≡

```
bool BL_PLANAR(graph& G, list<edge>& K, bool embed)
{
  if (BL_PLANAR(G, embed)) return true;
  edge_array<int> comp_num(G);
  int num_c = BICONNECTED_COMPONENTS(G,comp_num);
  GRAPH<node,edge> H;
  edge e;
  if ( num_c == 1 )
  { CopyGraph(H,G);
    Delete_Loops(H);
  }
  else
  { node_array<node> link(G,nil);
    array<list<edge> > E(num_c);
    array<list<node> > V(num_c);
    forall_edges(e,G)
    { node v = source(e);  node w = target(e);
      if (v == w) continue;
      int c = comp_num[e];  E[c].append(e);
      V[c].append(v); V[c].append(w);
    }
    int c; node v;
    for(c = 0; c < num_c; c++)
    { H.clear();
      forall(v,V[c]) if ( link[v] == nil ) link[v] = H.new_node(v);
      forall(e,E[c])
      { node v = source(e); node w = target(e);
        H.new_edge(link[v],link[w],e);
      }
```

```
      forall(v,V[c]) link[v] = nil;
      if (!BL_PLANAR(H,false)) break;
    }
  }
  K.clear();
  // H is a biconnected non-planar graph; we turn it into map
  Make_Simple(H);
  list<edge> R;
  H.make_map(R);
  forall(e,R) H[e] = H[H.reverse(e)];
  // auxiliary edges inherit original edge from their reversal
  Kuratowski(H,K);
  list_item it;
  forall_items(it,K) K[it] = H[K[it]];
  return false;
}
```

8.7.4 *Running Times*

Table 8.1 shows the running times of the functions discussed in this section. We used five kinds of graphs:

- Random planar maps with n nodes and $m = 2n$ uedges (P).
- Random planar maps with n nodes and $m = 2n$ uedges plus a $K_{3,3}$ on six randomly chosen nodes (P + $K_{3,3}$).
- Random planar maps with n nodes and $m = 2n$ uedges plus a K_5 on five randomly chosen nodes (P + K_5).
- Maximal planar maps with n nodes (MP).
- Maximal planar maps on n nodes plus one additional edge between two random nodes that are not connected in G (MP + e).

We constructed the graphs using the generators discussed in Section 8.9 and then permuted the adjacency lists, so as to hide the graph structure.

We ran the following algorithms:

- *BL_PLANAR*(G), the Booth–Lueker planarity test (T) that gives a yes-no answer, but does not justify its answer.
- *BL_PLANAR*(G, K, *true*), the Booth–Lueker planarity test that justifies its answers (T + J). If G is planar, it turns G into a planar map, and if G is non-planar, it exhibits a Kuratowski subgraph of G.

Graph	Gen	BL_PLANAR		Check	HT_PLANAR	
		T	T + J		T	T + J
P	0.76	1.59	1.82	0.23	2.6	4.18
	1.72	3.27	3.71	0.47	5.41	8.87
	3.47	6.67	7.43	0.95	11.38	19.22
P + $K_{3,3}$	0.97	1.1	5.66	0.17	2.54	–
	1.74	2.4	12.65	0.34	5.16	–
	3.56	5.47	20.01	0.69	11.02	–
P + K_5	1	0.98	5.72	0.16	2.61	–
	1.75	1.81	12.91	0.34	5.35	–
	3.58	3.26	22.06	0.67	10.86	–
MP	0.87	2.28	2.41	0.33	3.88	6.24
	1.5	4.59	4.84	0.66	7.81	12.98
	3.05	9.23	9.66	1.34	16.06	26.84
MP + e	0.87	1.26	5.47	0.23	1.05	–
	1.49	2.19	9.61	0.49	2.1	–
	3.06	5.87	23.81	0.96	4.28	–

Table 8.1 The running times of functions related to planarity: The column labeled Gen contains the time needed to generate the input graph. All other columns are as described in the text. We used $n = 2^i \cdot 5000$ for $i = 0, 1,$ and 2. This table was generated with the program planarity_time in the demo directory.

- The check whether the algorithm in the previous item worked correctly, i.e., the check *Genus*$(G) == 0$, if G is planar, and *CHECK_KURATOWSKI*(G, K), if G is non-planar.
- *HT_PLANAR*(G), the Hopcroft–Tarjan planarity test (T) that gives a yes-no answer, but does not justify its answer.
- *HT_PLANAR*(G, K, true), the Hopcroft–Tarjan planarity test that justifies its answers (T + J). This algorithm was only run when the previous item declared G planar. The extraction of the Kuratowski subgraph would have taken hours, since there is no efficient Kuratowski finder implemented for the Hopcroft–Tarjan planarity test.

Exercises for 8.7

1 Show that the number of distinct permutations in which the virtual leaves of B_k can appear on the horizon is

$$2^C \cdot P,$$

where C is the number of biconnected components of B_k with three or more attachments and $P = \prod p_v!$ where the product is over all articulation points of B_k and p_v is the number of non-root components of B_k with respect to v.

2 Improve the running time of the simple search for Kuratowski subgraphs to $O(n^2)$. Make sure that your algorithm works in the presence of parallel edges and self-loops.

3 Let G be a graph, let $e = (a, b)$ be an edge of G, and let G' be obtained from G by contraction of e. Show that if G' contains a Kuratowski subgraph then G does.

4 We have shown in Lemma 53 that the existence of an obstruction in B_k guarantees the existence of the Kuratowski subgraph of G. Show that it guarantees that B_{k+1} has no bush form.

8.8 Manipulating Maps and Constructing Triangulated Maps

In the chapter on graphs we saw functions that allow us to add new nodes and edges to a graph G. In particular,

```
edge G.new_edge(node v, node w)
```

adds a new edge (v, w) to G and returns it. The edge is appended to *out_edges*(v) and to either *in_edges*(w) (if G is directed) or *out_edges*(w) (if G is undirected).

In this chapter the cyclic ordering of the adjacency lists plays a crucial role and hence we need much finer control over the positions where edges are inserted into adjacency lists. The following function gives full control:

```
edge G.new_edge(edge e1, edge e2,
                int d1 = LEDA::after, int d2 = LEDA::after)
```

adds a new edge $x = (v, w)$ to G, where $v = source(e1)$ and $w = target(e2)$, and returns the new edge. The new edge is inserted before or after edge *e1* into *out_edges*(v) as directed by *d1*. If G is directed, it is also inserted before or after edge *e2* into *in_edges*(w) as directed by *d2*. If G is undirected, it is also inserted before or after edge *e2* into *out_edges*(w) as directed by *d2*. The constants *LEDA::after* and *LEDA::before* are predefined constants.

If control about the position of insertion is needed at only one endpoint of the edge (or if the new edge is the first edge incident to a node) the functions

```
edge G.new_edge(edge e, node w, int dir = LEDA::after)
edge G.new_edge(node v, edge e, int dir = LEDA::after)
```

should be used. The former function adds a new edge $x = (source(e), w)$ to G. x is inserted before or after edge e into *out_edges*$(source(e))$ as directed by *dir* and appended to *in_edges*(w) (if G is directed) or *out_edges*(w) (if G is undirected). The operation returns the new edge x. If G is undirected we must have *source*(e) $\neq w$. The latter function is symmetric to the former.

Related to the *new_edge* function is the *move_edge* function. The call

```
G.move_edge(edge e, node v, node w)
```

requires that e is an edge of G. It makes v the source of e and w the target of e. For all versions of the *new_edge* function mentioned above, there is a corresponding version of the *move_edge* function, which takes the edge to be moved as an additional argument. The effect of *move_edge*(e, v, w) is similar, but distinct to the combined effect of *del_edge*(e) followed by *new_edge*(v, w). The effect is similar as e ceases to make the connection between its old source and target and as there is now an edge from v to w. The effect is distinct, as *move_edge* moves an already existing edge (which may for example have associated entries in edge arrays) and *new_edge* creates a new edge.

For maps it is frequently convenient to add an edge and its reversal in a single operation.

```
edge M.new_map_edge(edge e1, edge e2)
```

inserts a new edge e = (*source*(*e1*), *source*(*e2*)) after *e1* into the adjacency list of *source*(*e1*) and the reversal to e after *e2* into the adjacency list of *source*(*e2*).
The following function splits a uedge in a map M.

```
edge M.split_map_edge(edge e)
```

splits edge $e = (v, w)$ and its reversal $r = (w, v)$ into edges (v, u), (u, w), (w, u), and (u, v), where u is a new node. It returns the edge (u, w).

We give an application of the functions above. We show how to *triangulate a map*. Let M be a map. The task is to add edges to M such that:

- the genus is not increased, in particular, a plane map stays plane, and
- every face cycle of the resulting map consists of at most three edges.

Both items are easy to achieve. As long as M is not connected we take any two nodes v and w in distinct components and join them by a uedge. This increases the number of edges by two, decreases the number of components by one, and either decreases the number of isolated nodes by two and increases the number of face cycles by one, or decreases the number of isolated nodes by one and leaves the number of face cycles unchanged, or leaves the number of isolated nodes unchanged and decreases the number of face cycles by one. In either case the genus is unchanged.

So assume that M is connected. As long as there is a face cycle consisting of four or more edges, we consider any such face cycle C and two nodes v and w on C that are not neighbors on C, say

$$C = [\ldots, e_2, v, e_4, \ldots, e_3, w, e_1, \ldots].$$

We split C by adding edges (v, w) and (w, v). The edge (v, w) is added after e_4 to the list of out-edges of v and the edge (w, v) is added after e_1 to the list of out-edges of w; this is the reverse of the operation illustrated in Figure 8.11. Adding the two edges increases the number of face cycles by one; thus the genus is not changed.

We use the triangulation routine as a subroutine in our straight line drawing routine for planar graphs. The straight line drawing routine assumes that its input is a triangulated graph without parallel edges. We therefore have to make sure that the triangulation routine does not introduce parallel edges. Unfortunately, when face cycles are split independently, parallel edges may be introduced. We want to avoid this.

- If the genus of M is zero then no new edge is parallel to another edge of the graph (new or old).

Christian Uhrig and Torben Hagerup suggested a triangulation algorithm that achieves all three items above. Their algorithm runs in linear time $O(n+m)$. The algorithm steps through the nodes of M. For each node v, it triangulates all faces incident on v. For each node v, this consists of the following:

First, the neighbors of v are marked. During the processing of v, a node will be marked exactly if it is a neighbor of v.

Then the faces incident on v are processed in any order. A face with boundary $[v = x_1, x_2, \ldots, x_n]$ is triangulated as follows: if $n \leq 3$, nothing is done. Otherwise,

(1) if x_3 is not marked, a uedge $\{x_1, x_3\}$ is added, x_3 is marked, and the same strategy is applied to the face with boundary $[x_1, x_3, x_4, \ldots, x_n]$.

(2) if x_3 is marked, a uedge $\{x_2, x_4\}$ is added, and the same strategy is applied to the face with boundary $[x_1, x_2, x_4, x_5, \ldots, x_n]$.

When all faces incident to v are triangulated, all neighbors of v are unmarked.

The algorithm just described clearly triangulates all face cycles. We need to show that it does not introduce parallel edges.

During the processing of a node v, the marks on neighbors of v clearly prevent the addition of a parallel edge with endpoint v. After the processing of v, such an edge is not added because all faces incident on v have been triangulated. This takes care of the edges added in (1).

Whenever a uedge $\{x_2, x_4\}$ is added in step (2), the presence of a uedge $\{x_1, x_3\}$ implies that x_2 and x_4 are incident on exactly one common face, namely the face currently being processed, see Figure 8.32. Hence another edge $\{x_2, x_4\}$ will never be added.

The linear running time can be seen as follows. The time to process a node v is proportional to the degree of v plus the number of edges added during the processing of v. The total running time is therefore proportional to $O(n+m')$ where m' is the number of edges in the final graph. The number of uedges in the final graph is at most $3n$ by Lemma 47.

The following program implements the algorithm. We first add edges to make the graph connected, then make sure that all reversal informations are properly set, and finally add edges to triangulate the graph.

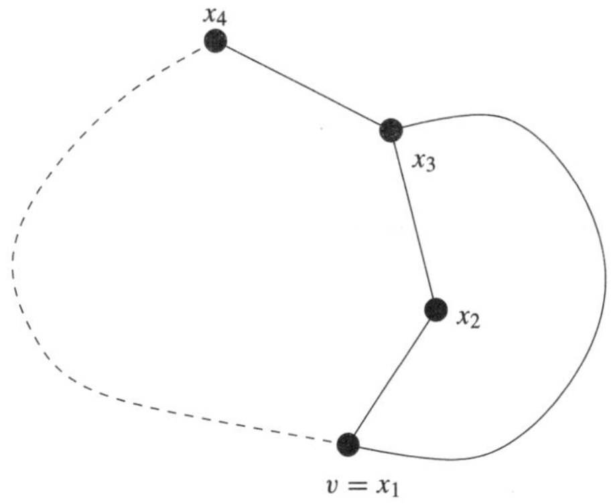

Figure 8.32 x_1, x_2, x_3, and x_4 are consecutive nodes on a face and the uedge $\{x_1, x_3\}$ exists. Then $\{x_2, x_4\}$ cannot exist.

⟨*triangulate.c*⟩≡

```
list<edge> graph::triangulate_map()
{ node v;
  edge x, e, e1, e2, e3;
  list<edge> L;
  ⟨add edges to make the graph connected⟩
  if ( !make_map() )
  error_handler(1,"TRIANGULATE_PLANAR_MAP: graph is not a map.");
  node_array<int>  marked(*this,0);
  forall_nodes(v,*this)
  { list<edge> El = adj_edges(v);
    // mark all neighbors of v
    forall(e1,El) marked[target(e1)] = 1;
    ⟨process faces incident to v⟩
    //unmark all neighbors of v
    node w;
    forall_adj_nodes(w,v) marked[w] = 0;
  } // end of stepping through nodes
 return L;
}
```

The two sub-steps are both fairly easy to implement. For the first sub-step we call COMPONENTs to determine the number of connected components and to label each node with its component number. If there is more than one component, we create an array *still_disconnected* with index set $[0..c-1]$, where c is the number of connected components. For each component except the one which contains s, the first node of G, we state that the component still needs to be connected with the component containing s. We then iterate over all nodes. Whenever we encounter a node v whose component still needs to be connected with s, we add the uedge $\{v, s\}$, and record that the component of v is now connected with the component of s.

⟨*add edges to make the graph connected*⟩≡

```
node_array<int>  comp(*this);
int c = COMPONENTS(*this, comp);
if ( c > 1 )
{ node s = first_node();
  array<bool> still_disconnected(c);
  for (int i = 0; i < c; i++)
    still_disconnected[i] = ( i == comp[s] ? false : true);
  forall_nodes(v,(*this))
  { if ( still_disconnected[comp[v]] )
    { set_reversal(e1 = new_edge(s,v), e2 = new_edge(v,s));
      L.append(e1); L.append(e2);
      still_disconnected[comp[v]] = false;
    }
  }
}
```

The faces incident to a node v are processed as described above. We store three consecutive edges of the face in *e1*, *e2*, and *e3*, respectively. If either of the three edges ends in v, the face cycle has length at most three and we are done.

So assume otherwise and let w be the endpoint of *e2*.

If w is not marked, we mark w and add the uedge $\{v, w\}$ inside the current face, i.e., we add the edge (w, v) after *e3* to $A(w)$ and we add the edge (v, w) after *e1* to $A(v)$. Also (v, w) becomes the new *e1*, *e2* becomes *e3*, and *e3* becomes the face cycle successor of *e2*.

If w is marked, we add the uedge $\{source(e2), target(e3)\}$ inside the current face, i.e., after edge *e2* at *source*(*e2*) and after the face cycle successor of *e3* at *target*(*e3*).

⟨*process faces incident to v*⟩≡

```
forall(e,El)
{
  e1 = e;
  e2 = face_cycle_succ(e1);
  e3 = face_cycle_succ(e2);
  if (target(e1) == v || target(e2) == v || target(e3) == v) continue;
  while (target(e3) != v)
  { node w = target(e2);
    if ( !marked[w] )
    { // we mark w and add the uedge {v,w}
      marked[w] = 1;
      L.append(x  = new_edge(e3,v));
      L.append(e1 = new_edge(e1,w));
      set_reversal(x,e1);
      e2 = e3;
      e3 = face_cycle_succ(e2);
    }
    else
```

```
    { //add the uedge {source(e2),target(e3)}
      e3 = face_cycle_succ(e3);
      L.append(x  = new_edge(e3,source(e2)));
      L.append(e2 = new_edge(e2,source(e3)));
      set_reversal(x,e2);
    }
  }//end of while
} //end of stepping through incident faces
```

8.9 Generating Plane Maps and Graphs

We discuss the generation of random plane maps and random plane graphs. We describe two methods to generate plane maps, a combinatorial method and a geometric method. We warn the reader that neither method generates plane maps according to the uniform distribution.

Combinatorial Constructions: The function

```
void maximal_planar_map(graph& G, int n);
```

generates a plane map with n nodes and $3n-6$ uedges, no self-loops and no parallel edges. The number of edges is the maximal possible, see Lemma 47, and, if $n \geq 3$, every face cycle is a triangle.

We give the implementation. If $n = 0$ we return the empty graph, if $n = 1$ we return the graph consisting of a single isolated node, and if $n = 2$ we return the graph consisting of two nodes and a single uedge. So let $n > 2$ and assume, that we have already constructed a maximal planar map with $n-1$ nodes. We select one of the existing edges, say e, at random and put a new node v into the face to the left of e.

Let $[e_1, e_2, e_3]$ be the face cycle containing e (when the third node is inserted the face cycle has length 2 instead of 3). For each i we add the edge $(source(e_i), v)$ to $A(source(e_i))$ after e_i and we append the edge $(v, source(e_i))$ to $A(v)$.

⟨*generate_planar_map.c*⟩≡

```
void maximal_planar_map(graph& G, int n)
{
  G.clear();
  if (n <= 0 ) return;
  node a = G.new_node();
  n--;
  if (n == 0) return;
  node b = G.new_node();
  n--;
  edge* E = new edge[n == 0? 2 : 6*n];
  E[0] = G.new_edge(a,b); E[1] = G.new_edge(b,a);
```

```
  G.set_reversal(E[0],E[1]);
  int m = 2;
  while (n--)
  { edge e = E[rand_int(0,m-1)];
    node v = G.new_node();
    while (target(e) != v)
    { edge x = G.new_edge(v,source(e));
      edge y = G.new_edge(e,v,LEDA::after);
      E[m++] = x; E[m++] = y;
      G.set_reversal(x,y);
      e = G.face_cycle_succ(e);
    }
  }
  delete[] E;
}
```

The function

```
void random_planar_map(graph& G, int n, int m);
```

generates a plane map with n nodes and $\min(m, 3n-6)$ uedges. It first generates a maximal plane map and then deletes a random set of uedges until the desired number of edges is obtained.

The functions

```
void maximal_planar_graph(graph& G, int n);
void random_planar_graph( graph& G, int n, int m);
```

first construct a plane map with the same parameters and then keep only one of the edges comprising each uedge.

Geometric Constructions: Geometry is a rich source of planar graphs. A simple way to generate a planar map is to choose n random points in the plane and to triangulate the resulting point set. We will see how to triangulate a point set in Section 10.3. Alternatives are to compute the Delaunay triangulation of a set of random points, see Section 10.4, or to choose a random set of segments and to compute the arrangement of the segments, see Section 10.7.

The functions

```
void triangulation_map(graph& G, int n);
void triangulation_map(graph& G, node_array<double>& xcoord,
                       node_array<double>& ycoord, int n);
void triangulation_map(graph& G, list<node>& outer_face,
                       node_array<double>& xcoord,
                       node_array<double>& ycoord,
                       int n);
```

choose n random points in the unit square and set G to some triangulation. G will be a plane map. The first function only returns the triangulation, the second function also returns the point coordinates, and the third function also returns the list of vertices lying on the convex hull (in clockwise order).

The function

```
void random_planar_map(graph& G, node_array<double>& xcoord,
                       node_array<double>& ycoord, int n, int m);
```

first constructs a triangulated planar map and then deletes all but m edges.

All functions above are also available with *map* replaced by *graph* in the function name. The modified functions keep only one edge of each uedge.

8.10 Faces as Objects

The face cycles of maps played an important role in the preceding sections. It is therefore only natural to introduce them as a type of their own. For succinctness, we use the type name *face*.

8.10.1 *Concepts*

The operation

```
M.compute_faces()
```

computes the set of face cycles of the map M; the function aborts if M is not a map. After this operation and till the next modification of M by a *new_node*, *new_edge*, *del_node*, or *del_edge* operation, the face cycles of M are available in much the same way as the edges and nodes of M are available.

For example,

```
int         M.number_of_faces();
list<face> M.all_faces();
```

return the number of faces and the list of all faces of M, respectively. If f is a face, the predecessor and successor face of f in the list of all faces is returned by *M.succ_face*(f) and *M.pred_face*(f), respectively, and the first and last face in the list of all faces is returned by *M.first_face*( ) and *M.last_face*( ), respectively. The four functions just mentioned return *nil* if the requested object does not exist. The iteration statement

```
forall_faces(f,M)
```

iterates over all face cycles of M.

The function

```
face M.face_of(edge e)
```

returns the face cycle of M which contains the edge e and the functions

```
list<edge> M.adj_edges(face f)
edge       M.first_face_edge(face f)
int        M.size(face f)
```

return the list of all edges in the face cycle f, the first edge in this cycle, and the number of edges in the face cycle, respectively. The iteration statement

```
forall_face_edges(e,f)
```

iterates over all edges e in the face cycle f.

For a node v, the function

```
list<face> M.adj_faces(node v)
```

returns the list of faces incident to v. More precisely, if $A(v) = [e_0, e_1, \ldots, e_{k-1}]$ is the list of edges out of v then the list $[\mathit{face\_of}(e_0), \ldots, \mathit{face\_of}(e_{k-1})]$ is returned.

Similarly, for a face f, the function

```
list<node> M.adj_nodes(face f)
```

returns the list of all nodes of M incident to f. More precisely, if $f = [e_0, e_1, \ldots, e_{k-1}]$, the list $[\mathit{source}(e_0), \ldots, \mathit{source}(e_{k-1})]$ is returned.

There is a small number of update operations which do not destroy the list of faces of a map. The operation

```
edge M.split_face(edge e1, edge e2)
```

inserts the edge $e = (\mathit{source}(e_1), \mathit{source}(e_2))$ and its reversal into M and returns e. The edges e_1 and e_2 must belong to the same face. This face cycle is split into two by the operation by inserting e after e_1 into the list of edges out of $\mathit{source}(e_1)$ and by inserting e^R after e_2 into the list of edges out of $\mathit{source}(e_2)$. The operation

```
face M.join_faces(edge e)
```

deletes the edge e and its reversal from M and updates the list of faces accordingly. Let f and g be the face cycles containing e and e^R, respectively. Assume first that $f \neq g$. If both f and g consist of a single edge[22] then the number of face cycles goes down by two and *nil* is returned. If at least one of f or g consists of more than one edge, then f and g are joined into a single face and this face is returned. When we coined the name for the operations we assumed that the latter case would be the "normal" use of the operation. Assume next that $f = g$. If f consists of exactly two edges, namely e and e^R then the number of face cycles goes down by one and *nil* is returned. If f consists of at least three edges and either e or e^R is the face cycle successor of the other then the number of face cycles is unchanged and f is returned. Finally, if neither e nor e^R is the face cycle successor of the other, then the number of faces goes up by one and one of the new faces is returned.

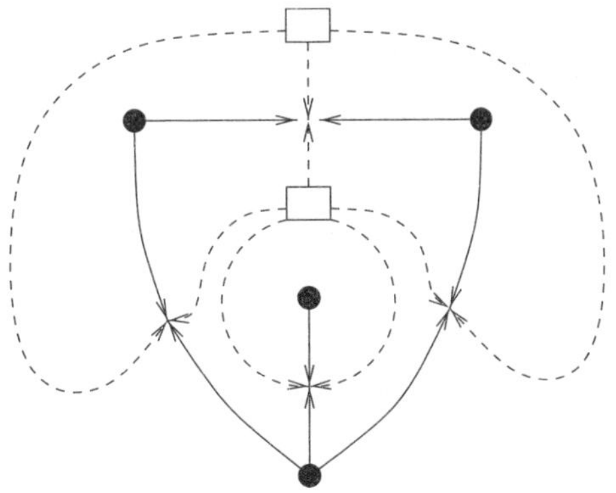

Figure 8.33 The dual of our map M_0. The dual has two nodes (shown as squares) and four uedges (drawn dashed).

8.10.2 *The Dual of a Map*

The (combinatorial) dual of a map M is another map D, see Figure 8.33:

- D has one node for each face cycle of M. More precisely, the nodes of D and the face cycles of M are in one-to-one-correspondence. We use $d(f)$ to denote the node of D corresponding to the face cycle f of M.
- D has one edge for each edge of M. Let e be any edge of M, let f be the face cycle containing e, and let g be the face cycle containing e^R. Then D contains the edge $d(e) = (d(f), d(g))$.
- Let $f = [e_0, e_1, \ldots, e_{k-1}]$ be a face cycle of M. Then the cyclic adjacency list of the node $d(f)$ of D is equal to $[d(e_0), d(e_1), \ldots, d(e_{k-1})]$.

The following program computes the dual D of a map M. We first compute the face cycles of M. We then put a node into D for each face cycle of M and record the correspondence in a *face_array<node>*. We then iterate over all face cycles of M and for each face cycle over the edges comprising the face cycle. For each edge we constructs its dual and record the correspondence. Observe that the edges incident to any dual node are constructed in the order in which they are supposed to appear in the adjacency list of the dual node. Finally, we establish the reversal information of all dual edges.

⟨*dual.c*⟩≡

```
void graph::dual_map(graph& D) const
{ D.clear();
  graph& M = *((graph*)this); // cast away the const
  M.compute_faces();
  face f; edge e;
  face_array<node> dual(M);
  forall_faces(f,M) dual[f] = D.new_node();
```

[22] This case occurs, for example, in a graph with one node and one uedge.

```
    edge_array<edge> dual_edge(M);
    forall_faces(f,M)
    { node df = dual[f];
      forall_face_edges(e,f)
      { face g = M.face_of(M.reversal(e));
        dual_edge[e] = D.new_edge(df,dual[g]);
      }
    }
    forall_edges(e,M)
      D.set_reversal(dual_edge[e],dual_edge[M.reversal(e)]);
  }
```

8.10.3 *Faces of Planar Maps*

There are two functions that deal with faces of planar maps. The function

```
void M.make_planar_map()
```

assumes that M is a bidirected graph. It first calls *M.make_map*( ) to turn M into a map. It then calls *PLANAR*(M, *true*) to turn M into a plane map. It finally calls *M.compute_faces*( ) to compute the face cycles of M.

The function

```
list<edge> M.triangulate_planar_map()
```

calls *M.triangulate_map*( ) followed by *M.compute_faces*( ) and returns the list of edges added to M by the former call.

Exercise for 8.10

1 Is the dual of the dual of a map M isomorphic to M? Give a counterexample. Under which conditions does the claim hold? State and prove a lemma.

8.11 Embedded Graphs as Undirected Graphs

The reader may wonder about the use of directed graphs in this chapter. After all, in maps we always combine a pair of directed edges into a uedge. We chose bidirected graphs to represent maps mainly for two reasons.

Although maps are basically undirected graphs, the two orientations of an undirected edge play a major role in the functions operating on maps. In particular, the face cycle successor of an edge and the reversal of an edge are "directed concepts" and hence would require additional arguments if maps were realized by undirected graphs. For example, one could distinguish the two orientations of an undirected edge by specifying a node to indicate the source node of the oriented edge. This would, however, not work for self-loops.

The second reason is that maps are frequently constructed incrementally and that the two orientations of an edge are constructed at different moments of time. We saw one example already in the program *dual_map* that constructs the dual of a map. Such constructions are difficult to implement with a representation that can only represent maps. The problem is that we arrive at a map at the end of the construction process but have no map during the construction process.

Our choice of directed graphs to represent maps wastes space, since the two edges comprising a uedge are stored in two lists at each endpoint of the uedge. One list for each endpoint would suffice for most functions presented in this chapter.

8.12 Order from Geometry

The following problem arises frequently. A graph is constructed by drawing it in a *GraphWin* and the combinatorial structure of the graph is supposed to reflect the drawing, i.e., for every node v the cyclic order of $A(v)$ is supposed to agree with the counter-clockwise order of the edges out of v in the drawing.

Let us be more precise. For every edge e let $d(e)$ be a vector (not necessarily, non-zero) in the plane. We define an order on two-dimensional vectors. For a non-zero vector d let $\alpha(d)$ be the angle between the positive x-axis and d, i.e., the angle by which the positive x-axis has to be turned in counter-clockwise direction until it aligns with d. A vector d_1 *precedes* a vector d_2 if $\alpha(d_1) < \alpha(d_2)$ and a vector d_1 is *equivalent* to a vector d_2 if $\alpha(d_1) = \alpha(d_2)$. The zero vector precedes all other vectors. The implementation of this order on vectors is discussed in Chapter 9 on geometry kernels.

The functions

```
bool SORT_EDGES(graph &G,
                const edge_array<NT>& dx, const edge_array<NT>& dy)
bool SORT_EDGES(graph &G,
                const node_array<NT>& x, const node_array<NT>& y)
```

reorder all adjacency lists in non-decreasing order of the vectors $d(e)$, $e \in E$. For the first function, the vector associated with an edge e is $(dx[e], dy[e])$, and for the second function, the vector associated with an edge $e = (v, w)$ is $(x[w] - x[v], y[w] - y[v])$.

The functions return *true* if G is a plane map after the reordering. When will this be the case? Assume that G is a map and that the vectors $d(e)$ come from a planar drawing of G, i.e., $d(e)$ is a vector tangent to the image of e as it leaves its source. If G has no self-loops and no parallel edges[23] then G will be a plane map after the call of *SORT_EDGES*. In fact, it will be a plane map for which the given drawing is an order-preserving embedding.

We next give an application of the function SORT_EDGES to the task described in the

[23] Observe that sorting edges by angle leaves the relative order of self-loops and the relative order of parallel edges undefined.

introductory paragraph. The goal is to deduce a plane map from a straight line drawing of the map. Assume that *gw* is a GraphWin with an associated graph *G*, i.e., defined by

⟨*gw_sort_edges_demo*⟩≡

```
graph G;
⟨gw_sort_edges_demo: auxiliary functions⟩
int main()
{ GraphWin gw(G,"Plane Map from Geometry");
  gw.set_init_graph_handler(init_handler);
  gw.set_new_edge_handler(new_edge_handler);
  gw.set_del_edge_handler(del_edge_handler);
  gw.set_new_node_handler(new_node_handler);
  gw.set_del_node_handler(del_node_handler);
  gw.set_directed(true);
  gw.display();
  gw.edit();
  return 0;
}
```

We define an auxiliary function *sort* that queries for each node *v* of *G* its position in *gw* and then calls SORT_EDGES. We call *sort* whenever an edge is added to the graph (and hence the new edge handler is called) or if a new graph is read in by *gw* (and hence the init handler is called). When an edge is added, we also add the reversal to make sure that we deal with a map.

The effect of the call of *sort* is to rearrange the adjacency lists according to the counter-clockwise order in which the edges incident to any node appear in the drawing. We print the graph at the end of sort in order to allow a visual comparison between the drawing and the representation of the graph. The graph will be a plane map as long as the drawing is a planar embedding.

⟨*gw_sort_edges_demo: auxiliary functions*⟩≡

```
void sort(GraphWin& gw)
{
  node_array<double> x(G), y(G);
  node v;
  forall_nodes(v,G)
  { point p = gw.get_position(v);
    x[v] = p.xcoord(); y[v] = p.ycoord();
  }
  SORT_EDGES(G,x,y);
  cout << "\n\nThe adjacency lists are:\n";
  G.print();
}
void init_handler(GraphWin& gw)
{ list<edge> L;
  G.make_map(L);
```

```
  sort(gw);
}
void new_edge_handler(GraphWin& gw, edge e)
{ G.set_reversal(e,gw.new_edge(G.target(e),G.source(e)));
  sort(gw);
}
bool del_edge_handler(GraphWin& gw, edge e)
{ gw.del_edge(G.reversal(e)); return true; }
void new_node_handler(GraphWin& gw,node)    {}
void del_node_handler(GraphWin& gw)         {}
```

We will see more functions that relate geometry and graphs in Chapter 10 on geometric algorithms.

Exercises for 8.12

1 Extend the gw_drawing_demo.c such that it can also cope with edges that contain bends.

2 Write a function that checks whether the geometric positions assigned to the nodes of a map define a straight line embedding of the map. Hint: Read Section 10.7.2 on line segment intersection before working on this exercise.

8.13 Miscellaneous Functions on Planar Graphs

There are many problems that are simpler for planar graphs than for arbitrary graphs. We collect two in this section.

8.13.1 *Five Coloring*

Every planar graph can be four-colored, i.e., the nodes of the graph can be labeled with the integers 1 to 4 such that any edge connects two nodes of distinct color. We have not implemented a four coloring algorithm but only a five coloring algorithm.

The function

```
void FIVE_COLOR(graph& G, node_array<int>& C);
```

attempts to color the nodes of G using five colors, more precisely, it computes for every node v a color $C[v] \in \{1, \ldots, 5\}$, such that $C[source(e)] \neq C[target(e)]$ for every edge e. The function runs in linear time and is guaranteed to succeed if G is planar and contains no self-loops and no parallel edges[24].

We sketch how the algorithms works. In a planar graph there is always a node with at most five neighbors (Lemma 47). Let v be a node with at most five neighbors. If v has less than five neighbors, we recursively five-color the graph $G \setminus v$ and then use a color for v which is not used by any of its neighbors. If v has degree 5, we have to work slightly

[24] Self-loops are clearly an obstruction to colorability. Parallel edges are no "real" problem; it is just that our algorithm is not able to handle them.

harder. We observe that there must be two neighbors of G which are not connected by an edge (otherwise the neighbors of v would form a complete graph on five nodes; this is, however, impossible in a planar graph by Lemma 47). Let w and z be two neighbors of v that are not connected by an edge. We remove v and merge w and z into a single node. This can be done without destroying planarity as Figure 8.34 shows. When merging w and z we also delete any parallel edges which may result from the merging process. We five-color the resulting graph G' recursively. In order to obtain a coloring of G we unmerge w and z, give w and z the color of the node that represented them both in G', and give v a color which is not used on its neighbors.

To obtain linear running time is slightly tricky and we leave it for the exercises.

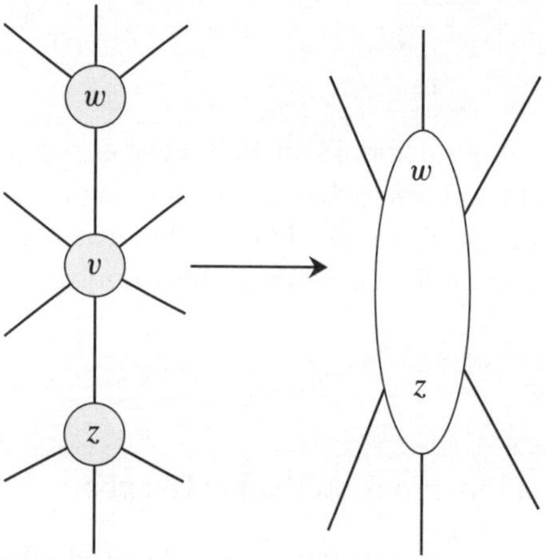

Figure 8.34 Merging the neighbors w and z of v.

8.13.2 *Independent Sets of Small Degree*

An independent set in a graph G is a set I of nodes no two of which are connected by an edge. A five coloring of a graph yields an independent set of size at least $n/5$, since at least one of the colors is used on at least $n/5$ of the nodes and since all edges have their endpoints in different color classes. Sometimes, it is desirable to have an independent set all of whose nodes have small degree.

The function

```
void INDEPENDENT_SET(const graph& G, list<node>& I)
```

computes an independent set I all of whose nodes have degree at most 9. If G is planar and has no parallel edges, it is guaranteed that $|I| \geq n/6$. The algorithm is due to David Kirkpatrick and Jack Snoeyink [KS97] and is extremely simple and elegant.

The algorithm starts by removing all nodes that have degree 10 or more. It then repeatedly chooses a node v of smallest degree, adds v to I, and removes v and its neighbors from G.

We describe an implementation. We start by making an isomorphic copy H of G; H is of type *GRAPH<node, edge>*, and each node v of H stores in $H[v]$ the node of G to which it corresponds. We saw the implementation of *CopyGraph* in Section 6.1. We will work on H.

We delete all self-loops from H and turn H into a map. Recall that turning a graph into a map pairs a maximum number of edges and adds reversals for the unpaired edges. After turning H into a map, each edge is part of a uedge.

We then determine all nodes of degree at least 10 and delete all such nodes.

Next we collect all nodes of H of degree i, $0 \le i \le 9$ in a linear list *LD*$[i]$. In the course of the algorithm the lists *LD*$[i]$ may contain nodes that were already deleted from H. We need to be able to identify those nodes and therefore maintain an array *node_of_H*.

The construction of the independent set can now begin. As long as H is not empty, we select a node v from the lowest indexed non-empty list. We continue the selection process until we select a node that belongs to the current H. We add $H[v]$ to I (recall that $H[v]$ is the node in G that corresponds to v), and we delete v and its neighbors from H; we do not remove them from the lists *LD* though (this could be done by maintaining an array *pos_in_LD* that stores for each node v the item in *LD* that contains v). We collect all neighbors of v in a list *affected_nodes* and add them to the lists *LD* according to their new degrees.

⟨*_independent_set*⟩≡

```
void INDEPENDENT_SET(const graph& G, list<node>& I)
{ I.clear();
  GRAPH<node,edge> H;
  CopyGraph(H,G);
  node v; edge e;
  list<edge> E = H.all_edges();
  forall(e,E) { if (H.source(e) == H.target(e) ) H.del_edge(e); }
  H.make_map(E); // E is a dummy argument
  list<node> HD; // high degree nodes
  forall_nodes(v,H) if (H.degree(v) >= 10) HD.append(v);
  forall(v,HD) H.del_node(v);
  array<list<node> > LD(10);
  forall_nodes(v,H) LD[H.degree(v)].append(v);
  node_array<bool> node_of_H(H,true);
  while (H.number_of_nodes() > 0)
  { int i = 0;
    while (i < 10)
    { if ( LD[i].empty() ) { i++; continue; }
      v = LD[i].pop();
      if ( node_of_H[v] ) break;
    }
    I.append(H[v]);
    list<node> affected_nodes;
    forall_inout_edges(e,v)
    { node w = H.opposite(v,e);
      edge f;
      forall_inout_edges(f,w)
```

```
        affected_nodes.append(H.opposite(w,f));
      H.del_node(w); node_of_H[w] = false;
    }
    H.del_node(v); node_of_H[v] = false;
    forall(v,affected_nodes)
      if ( node_of_H[v] ) LD[H.degree(v)].append(v);
  }
}
```

Exercises for 8.13

1 Extend the function *FIVE_COLORING* so that it can handle parallel edges.

2 Implement the function FIVE_COLORING. Try to achieve linear running time.

3 Modify the implementation of INDEPENDENT_SET such that the lists *LD* contain only nodes of *H* and every node at most once.

4 A separator in a graph G is a set S of nodes of G such that removal of S decomposes G into two or more subgraphs none of which has more than $2n/3$ nodes. Planar graphs have separators of size $O(\sqrt{n})$ and there are linear time algorithms to compute them, see [LT77] or [Meh84c, IV.10]. Implement the planar separator theorem and provide it as a LEP.

9

The Geometry Kernels

A geometry kernel offers basic geometric objects, such as points, lines, segments, rays, planes, circles, ..., and geometric primitives operating on these objects, e.g, the computation of the area of the triangle defined by three points and the computation of the intersection of two lines.

LEDA offers geometric kernels for plane geometry, for three-dimensional geometry, and for geometry in higher dimensional space. We discuss the kernels for two- and three-dimensional geometry in the first eight sections. The kernel for higher dimensional geometry will be discussed in Section 9.9.

The two- and three-dimensional kernels come in two kinds: the rational kernel and the floating point kernel. Write one of

```
#include <LEDA/rat_kernel.h>
#include <LEDA/float_kernel.h>

#include <LEDA/d3_rat_kernel.h>
#include <LEDA/d3_float_kernel.h>
```

to select a kernel. The kernels for two-dimensional geometry provide points, lines, segments, rays, vectors, circles, polygons, generalized polygons, and affine transformations. We use the type names *point*, *line*, *segment*, *ray*, *vector*, *circle*, *polygon*, *gen_polygon*, and *transform* for the corresponding classes of the floating point kernel and the names *rat_point*, *rat_line*, *rat_segment*, *rat_ray*, *rat_vector*, *rat_circle*, *rat_polygon*, *rat_gen_polygon*, and *rat_transform* for the corresponding classes of the rational kernel. If the distinction between rational and floating point kernel is immaterial, we use capital letters: POINT, LINE, SEGMENT, ... The three-dimensional kernels provide lines and planes.

The header files above simply collect the header files of all relevant classes into one. For example,

⟨*rat_kernel.h*⟩≡
```
#include <LEDA/rational.h>
#include <LEDA/rat_point.h>
#include <LEDA/random_rat_point.h>
#include <LEDA/rat_segment.h>
#include <LEDA/rat_ray.h>
#include <LEDA/rat_line.h>
#include <LEDA/rat_circle.h>
#include <LEDA/rat_vector.h>
#include <LEDA/rat_polygon.h>
#include <LEDA/rat_gen_polygon.h>
```

It is important to understand the difference between the rational and the floating point kernel.

In the rational kernel the Cartesian coordinates of points are rational numbers (in the sense of mathematics) and the geometric primitives are exact, i.e., always give the correct result.

In the floating point kernel the Cartesian coordinates of points are double precision floating point numbers and the geometric primitives are approximate, i.e, they usually give the correct result but there is no guarantee. The use of the floating point kernel is therefore not without risk.

Why do we have the floating point kernel at all? There are several reasons: (1) the outside world, e.g., the graphics systems used to visualize the results of geometric computations, wants floating point numbers, (2) we started with the floating point kernel, and (3) floating point computation is faster than computation with rational numbers. The last sentence requires further explanation. First, floating point computation is unreliable and hence the cost of efficiency is a reliability problem. The dangers of floating point arithmetic in geometric computations are discussed in Section 9.6. Second, the overhead of exact computation is surprisingly small due to our extensive use of so-called floating point filters. Our experiments show that the cost of exact arithmetic is never more than a factor of three in running time and usually much smaller. The efficient realization of exact geometric computation and floating point filters are discussed in Section 9.7.

In our own work we do program development exclusively with the rational kernel. Only when a program is stable, we might consider switching to the floating point kernel. We switch only if the use of the rational kernel does not give the desired performance. A switch to the floating point kernel should always be accompanied by a careful analysis of its limits, see Section 9.8.

This chapter is organized as follows: the first two sections deal with geometric objects and geometric predicates, respectively. Every user of LEDA geometry should read them. The next three sections treat special topics: affine transformations, generators for geometric objects, and writing kernel independent code. They may be skipped on first reading. We then have three sections on arithmetic. We first discuss the danger of using floating point arithmetic as an implementation of mathematics' real numbers, then describe the efficient implementation of exact geometric predicates in the rational kernel, and finally comment on the safe use of the floating point kernel. The last three sections give a glimpse at the higher-dimensional kernel, briefly review the history of geometry in LEDA, and discuss the relation between LEDA and CGAL.

9.1 Basics

We discuss points, segments, lines, rays, vectors, and circles.

Cartesian and Homogeneous Coordinates: We assume that the ambient space is equipped with the standard Cartesian coordinate system and specify points by their Cartesian coordinates. For a point p in the plane the functions

```
p.xcoord();
p.ycoord();
```

return the x- and y-coordinate of p, respectively. Of course, the z-coordinate of a point in space is returned by *p.zcoord( )*. The Cartesian coordinates of a *point* are of type *double* and the Cartesian coordinates of a *rat_point* are of type *rational*. We use RAT_TYPE as the generic name, i.e., RAT_TYPE stands for *double* when the floating point kernel is used and stands for *rational* when the rational kernel is used.

Points are stored by their Cartesian coordinates. For *rat_points* it is more efficient to store them by their homogeneous coordinates, i.e., to use the same denominator for the x- and the y-coordinate. The homogeneous coordinates of a point in the plane are a triple (x, y, w) with $w \neq 0$; here w is called the homogenizing coordinate. The Cartesian coordinates of a point with homogeneous coordinates (x, y, w) are $(x/w, y/w)$. Observe that the homogeneous coordinates of a point are not unique. Two triples that are multiples of each other specify the same point. The homogeneous coordinates of a point p in the plane are returned by

```
p.X();
p.Y();
p.W();
```

respectively. The homogeneous coordinates of a *rat_point* are of type *integer*. Do *points* also have homogeneous coordinates? Yes, for compatibility with *rat_points* they do. The homogenizing coordinate of a *point* is the constant 1.0 and the X- and Y-coordinate is simply the corresponding Cartesian coordinate. Thus the homogeneous coordinates of a *point* are of type *double*. We use INT_TYPE to denote the type of the homogeneous coordinates[1], i.e., INT_TYPE stands for *integer* when the rational kernel is used, and stands for *double* when the floating point kernel is used.

We said above that homogeneous coordinates are not unique. We guarantee, however, that all accesses to the homogeneous coordinates of a point return the same value. We do

[1] We chose RAT_TYPE and INT_TYPE as the names for the types of the Cartesian and the homogeneous coordinates because we prefer the rational kernel.

not guarantee, however, that these values are the homogeneous coordinates specified in the constructor for the point. The constructor may simplify the representation by cancelling out common factors. Moreover, we always store a positive value for the homogenizing coordinate.

In mathematical context we also use x_p and y_p for the Cartesian coordinates of a point p and X_p, Y_p, and W_p for the homogeneous coordinates.

Construction: Points are constructed by either specifying their Cartesian or their homogeneous coordinates. Thus

```
point      p(0.2,0.8);
point      q(1,4,5);
rat_point r(1,4,5);
rat_point s(rational(1,5),rational(4,5));
```

are four different ways of defining a point with coordinates $(1/5, 4/5)$. In the first constructor we have defined a *point* by specifying its Cartesian coordinates, in the second constructor we have specified a *point* by giving a triple of doubles (the Cartesian coordinates are obtained by performing the floating point divisions 1/5 and 4/5), in the third constructor we have specified a *rat_point* by a triple of *integers*, and in the fourth constructor we have specified a *rat_point* by a pair of rational numbers.

The generic form of the constructor is

```
POINT p(RAT_TYPE x, RAT_TYPE y)
```

for the construction from Cartesian coordinates, and

```
POINT p(INT_TYPE X, INT_TYPE Y, INT_TYPE W = 1)
```

for the construction from homogeneous coordinates. The default constructor

```
POINT p;
```

constructs the origin. It is bad programming style to exploit this fact. We recommend writing

```
POINT p(0,0);
```

to construct the origin.

We turn to segments, lines, and rays. A segment is constructed by specifying its two endpoints. Thus

```
segment     s(point p, point q);
rat_segment s(rat_point p, rat_point q);
```

define a *segment* and a *rat_segment*, respectively. The second point may also be specified by a vector which defines the relative position of the second point with respect to the first point. The generic forms are

```
SEGMENT s(POINT p, POINT q);
SEGMENT s(POINT p, VECTOR v);      // same as s(p,p+v)
```

The defining points of a segment can be accessed by

```
s.source();
s.target();
```

Lines and rays are also defined by two points or by a point and a vector.

```
LINE l(POINT p, POINT q);
LINE l(POINT p, VECTOR v);  // same as l(p,p+v)
RAY r(POINT p, POINT q);
RAY r(POINT p, VECTOR v);  // same as r(p,p+v)
```

Of course, the two defining points must not be equal and the vector must not be the zero-vector.

The default constructors

```
SEGMENT s;
LINE    l;
RAY     r;
```

introduce variables of the appropriate type. They are initialized to some object of the type (the manual even specifies which), but it is bad programming style to rely on this fact.

Vectors can be specified by either their Cartesian or their homogeneous coordinates.

```
vector     v(double x, double y);
rat_vector v(rational x, rational x);
rat_vector v(integer X, integer Y, integer W = 1);
```

Observe that the analogy between *vectors* and *rat_vectors* is not complete. There is no way to define a two-dimensional *vector* by a triple of doubles. The reason is that *vectors* and *rat_vectors* exist for arbitrary dimensions and that

```
vector     v(double x, double y, double z);
```

constructs a three-dimensional vector. The default constructor defines the zero vector.

Circles can be constructed in many ways. We describe two:

```
CIRCLE C(POINT a, POINT b, POINT c);
CIRCLE C(POINT a, POINT b);
```

define a circle passing through points a, b, and c, and a circle with center a and passing through b respectively. If $a = b$ in the second constructor, the circle has radius zero.

Some triples of points are unsuitable for defining a circle. A triple is *admissible* if $|\{p_1, p_2, p_3\}| \neq 2$. Assume now that p_1, p_2, p_3 are admissible. If $|\{p_1, p_2, p_3\}| = 1$, they define the circle with center p_1 and radius zero. If p_1, p_2, and p_3 are collinear, C is a straight line passing through them and the center of C is undefined. If p_1, p_2, and p_3 are not collinear, C is the circle passing through them.

Affine transformations are discussed in Section 9.3 and polygons and generalized polygons are discussed in Section 10.8.

Points and Vectors: Points and vectors are related but clearly distinct geometric objects. In order to work out the relationship between points and vectors it is useful to identify a point with an arrow extending from the origin to the point. In this view a point is an arrow attached to the origin. A vector is an arrow which is allowed to float freely in space[2].

Points and vectors can be combined by arithmetical operations: for two points p and q the difference $p - q$ is a vector[3] and for a point p and a vector v, $p + v$ is a point.

For two vectors v and w their sum $v + w$ and their difference $v - w$ are also vectors. However, it does not make sense to add two points. The unary operator $-$ reverses a vector.

The coordinates of a vector v are accessed by

```
RAT_TYPE v.coord(int i);  // i-th Cartesian coordinate
RAT_TYPE v[int i];        // i-th Cartesian coordinate
INT_TYPE v.hcoord(int i); // i-th homogeneous coordinate
```

For a vector v in d-space the Cartesian coordinates are indexed from 0 to $d - 1$ and the homogeneous coordinates are indexed from 0 to d. The homogenizing coordinate has index d. The homogenizing coordinate of a *vector* is the constant 1. In two-dimensional space the Cartesian and homogeneous coordinates can also be accessed by *xcoord*( ), *ycoord*( ), X( ), Y( ), and W( ), respectively.

Vectors may be stretched or shrunk. If v is a vector and r has INT_TYPE or RAT_TYPE then

```
r * v;
v / r;
```

compute the vectors whose Cartesian coordinates are multiplied by r and divided by r, respectively.

If v and w are vectors then

```
v * w
```

returns the scalar product of v and w. This is the component-wise product of the Cartesian coordinates and has RAT_TYPE.

The scalar product of a vector with itself yields the squared length of the vector. Instead of writing $v * v$ one can also write

```
v.sqr_length();
```

Handle Types, Identity and Equality: All geometric types are so-called handle types or independent item types, see Sections 2.2 and 2.2.2, i.e., an object of any geometric type is a (smart) pointer to a representation object. For example, a *rat_point* is a pointer to a *rat_point_rep* and a *segment* is a pointer to a *segment_rep*. The objects of the representation class contain the defining information about the geometric object and possibly additional information for internal use.

[2] More precisely, a vector is an equivalence class of arrows where two arrows are equivalent if one can be moved into the other by a translation of space.

[3] More precisely, it is the equivalence class of arrows containing the arrow extending from p to q.

We give more details for *rat_points*. The classes *rat_point* and *rat_point_rep* are derived from *handle_base* and *handle_rep*, respectively. The class *handle_base* contains a data member *PTR*, which is a pointer to a *handle_rep*. In *rat_point* we have a private member function *ptr* which casts this pointer to a pointer to a *rat_point_rep*. The class *handle_rep* is discussed in Section 13.7. A *rat_point_rep* contains the homogeneous coordinates of a point (three *integers*), floating point approximations of the homogeneous coordinates (three *doubles*) and the id-number of the point. The floating point approximations of the homogeneous coordinates are used in the floating point filter and will be discussed in Section 9.7. The id-number is used as the hash key in maps and hashing arrays. Any two *point_reps* have distinct id-numbers.

```
class rat_point_rep  : public handle_rep {
  integer x,  y,  w;
  double  xd, yd, wd;
  unsigned long id;
};
class rat_point  : public handle_base {
  rat_point_rep* ptr() const { return (rat_point_rep*)PTR; }
};
```

We distinguish between identical and equal objects. Two points p and q are *identical* (function *identical*(p, q)) if they point to the same *point_rep*, and two points p and q are *equal* (binary operator ==) if they agree as geometric objects, i.e., have the same Cartesian coordinates.

The assignment statement and the copy constructor preserve identity, i.e., are realized by pointer assignment.

```
POINT p(0,0);
POINT q(0,0);
POINT r = p;
identical(p,q); // evaluates to false
p == q;         // evaluates to true
identical(p,r); // evaluates to true
p == r;         // evaluates to true
```

Linear Orders: There are several linear orders defined on points.

- *cmp_x* compares points by their x-coordinate.
- *cmp_y* compares points by their y-coordinate.
- *cmp_xy* compares points by their x-coordinates. Points with equal x-coordinate are compared by their y-coordinate.
- *cmp_yx* compares points by their y-coordinates. Points with equal y-coordinate are compared by their x-coordinate.
- *cmp* is the same as *cmp_xy*. It is the default order for points.

Associating Information with Geometric Objects: Points, lines, segments, rays, and circles have id-numbers and hence *maps* and *h_arrays* can be defined for them. Observe that *maps* and *h_arrays* associate information with representation objects, i.e, only identical objects share their information. For example,

```
map<POINT,int> color;
POINT p(0,0); color[p] = 0;
POINT q(0,0); color[q] = 1;
POINT r = p;
cout << color[p] << color[q] << color[r];  // outputs 010
```

For points we can also use dictionaries and dictionary arrays to associate information (for the other geometric types this requires the definition of a compare function). In dictionaries and dictionary arrays equal objects share their information. For example,

```
d_array<POINT,int> color;
POINT p(0,0); color[p] = 0;
POINT q(0,0); color[q] = 1;
POINT r = p;
cout << color[p] << color[q] << color[r];  // outputs 111
```

Observe that p and q are equal and hence the assignment to $color[q]$ also changes the color of p.

Dictionary arrays are useful for removing multiple occurrences of equal objects. For example, if L is a list of points, then

```
d_array<POINT,bool> first_occurrence(true);
list_item it;
forall_items(it,L)
{ if ( !first_occurrence[ L[it] ] )
     L.del_item(it);
  else
     first_occurrence[ L[it] ] = false;
}
```

removes all but the first occurrence of every point from L. What will this program do when a *map* is used instead of a *d_array*?

Converting between the Rational and the Floating Point Kernel: Floating point objects can be converted to rational objects and rational objects can be converted to floating point objects. We illustrate conversion for points.

If p is a *point* or *rat_point* then

```
point p.to_point();
```

returns a *point*. If p is a point the call is equivalent to the call of the copy constructor, and if p is a *rat_point*, the Cartesian coordinates of the point returned are floating point approximations of the Cartesian coordinates of p.

The conversion from rational objects to floating point objects needs to be used whenever an object is to be displayed in a window. For example, if W is a *window* and p is a POINT, then

```
W << p.to_point();
```

draws p in W. The output statement above could be written even more elegantly as $W \ll p$ if the class *rat_point* provided a conversion operator to *point*. We opted for the less elegant code since the use of conversion operators can lead to unexpected side effects.

Both point classes have a constructor

```
POINT(const point& p, int prec = 0);
```

If POINT is *rat_point* and *prec* is positive the constructor is equivalent to

```
rat_point(integer(p.xcoord() * P), integer(p.ycoord() * P), P),
```

where $P = 2^{prec}$, i.e., the Cartesian coordinates of p are approximated as rational numbers with denominator P. If *prec* is non-positive, the value of *prec* is chosen such that there is no loss of precision in the conversion.

When POINT is *point* and *prec* is positive, the point constructed has Cartesian coordinates $(\lfloor P * x \rfloor / P, \lfloor P * x \rfloor / P)$, where $p = (x, y)$ and $P = 2^{prec}$. If *prec* is non-positive, the new point has coordinates x and y.

Immutability: All geometric objects are *immutable*. There are no operations that change a geometric object, there are only operations to generate new geometric objects from already existing ones. For example, the operation

```
p.translate(1,1);
```

returns a point which is obtained from p by translating it by the vector $(1, 1)$; it does not change the coordinates of the point p. Of course, the translated point may be assigned to p:

```
p = p.translate(1,1);
```

Input and Output: Geometric objects can be written on files and read from files. For example, if p is a POINT then

```
cout << p;
cin  >> p;
```

writes p on standard output, and reads p from standard input, respectively. The input operators $\gg$ are designed such that output written by $\ll$ can be read by $\gg$.

Graphical input and output is very important for geometric objects. The *window* class knows how to draw geometric objects and supports the construction of geometric objects by mouse input. The simplest way to draw a geometric object is to use the operator $\ll$, for example,

```
W << p.to_point();    // W << p can be used if p is a point
W << s.to_segment();  // W << s can be used if s is a segment
W << r.to_ray();      // W << r can be used if r is a ray
W << l.to_line();     // W << l can be used if l is a line
W << C.to_circle();   // W << C can be used if C is a circle
W << P.to_polygon();  // W << P can be used if P is a polygon
```

If more control is needed, e.g, concerning the color or whether a circle should be drawn as a disk, the *draw* functions need to be used. For example,

```
W.draw_segment(s,red);            // draws s in red
W.draw_disk(C,blue);              // draws a blue filled circle
W.draw_filled_polygon(P,green); // draws a filled green polygon
```

Observe that s, C, and P must be floating point objects. Rational objects must be converted to floating point objects first. For example,

```
W.draw_filled_polygon(P.to_polygon(),green);
```

has to be used to draw a filled *rat_polygon.* Observe that the call will also work for *polygons.*

Why did we not overload the *draw*-functions such that they also work for rational objects? The reason is that this would have required to include the header files of the rational kernel into the header file of the window class. The header file of *window* is very large already and we wanted to avoid a further increase in size.

We come to mouse input. The operator ≫ can be used to read a point, segment, line, ray, circle, or polygon. For example,

```
W >> p;  // p is a point
W >> s;  // s is a segment
```

read a point and a segment, respectively. The reading operations are blocking and wait for mouse clicks. A point is constructed by a single click of the left mouse button, and a segment, line, ray, and circle is constructed by two clicks of the left mouse button.

What happens when a mouse button different from the left mouse button is clicked? Windows have an internal state in the same way as C++ input streams do. The state indicates whether there is more input to read or not. The state is initially true and is set to false by a click of the right mouse button (this is similar to ending stream input by the "eof" character). If an input statement is used in the test of a conditional, an object of type *window* is automatically converted to a boolean whose value is the internal state. For example,

```
list<point> L;
point p;
while ( W >> p ) L.append(p);
```

reads a sequence of points from W. Every click of the left mouse button inputs a point and a click of the right mouse button terminates the sequence. The three lines above are essentially the implementation of the input operator for polygons.

In window.h the input operator ≫ is only defined for the floating point objects. If you want to use them for rational objects you must include the header file rat_window.h. For example,

```
#include <LEDA/rat_window.h>

rat_point p;
while (W >> p)  W << p.to_point();
```

reads a sequence of *rat_points* and echos them in W.

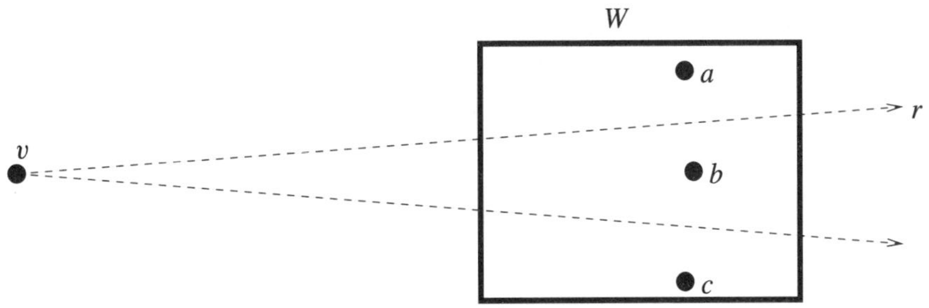

Figure 9.1 The Voronoi vertex v is the center of the circle passing through the points a, b, and c. The three points lie in the window W (indicated as a solid frame) but v lies far outside W. It is a bad strategy to draw the ray r as a ray starting in v and having direction orthogonal to the direction from a to b. A slight error in the computation of the coordinates of v due to round-off may change the appearance of r in W dramatically.

Input and Output: A Warning: As already mentioned, the *window* class offers functions to draw lines, rays, and segments, and many other geometric objects. For example,

```
W.draw_segment(point p, point q);
W.draw_ray(point p, point q);
```

will draw the segment with endpoints p and q and the ray with start point p passing through q, respectively. These functions have the desired effect if the points p and q lie in a rectangle whose side lengths are about 1000 times the side lengths of W. If one of the points lies further away from W, the use of these functions is ill-advised.

Consider the following situation. We are given three points a, b, and c in a window W and want to display their Voronoi diagram. Voronoi diagrams are discussed in Section 10.5. Except when the points lie on a common line, the Voronoi diagram will consist of a single vertex v from which three rays emanate. The Voronoi vertex is the center of the circle passing through the three points. When the three points lie almost on a line, v will lie far outside W, see Figure 9.1. Each ray is part of the perpendicular bisector of two sites. It is natural to draw the ray which is part of the perpendicular bisector of a and b by the following piece of code:

```
POINT v = CIRCLE(a,b,c).center();
VECTOR vec = b - a;
POINT ray_point = v + vec.rotate90();
W.draw_ray(v.to_point(),ray_point.to_point());
```

The drawing produced by this program will be a disappointment, if a, b, and c lie sufficiently close to a common line, since the conversion of v and *ray_point* to points of the floating point kernel (note that this conversion cannot be avoided since the windows class knows only floating point objects) will incur rounding error. Moving either v or *ray_point* slightly has a dramatic effect on the appearance of r in W.

We recommend using a different strategy to draw rays and segments whose defining points may lie far outside W. In this situation the underlying line l is frequently known by other means. In our example, l is the perpendicular bisector of the points a and b.

```
LINE l = p_bisector(a,b);
```

The defining elements of l lie in W and are hence known with high precision. The window class offers functions

```
W.draw_segment(point p, point q, line l, color c);
W.draw_ray(point p, point q, line l, color);
```

that draw the part of the line l between p and q, respectively, the part of l on the ray with source p and second point q. Of course, p and q must lie on l or at least close to it. We give the implementation of the second function.

If p is contained in W we simply draw the ray with source p and second point q. If p lies outside the window we clip the line l on W and call the resulting segment s. The segment s has the property that its source precedes its target in the lexicographic order of points; equality is possible. We draw s either if p is smaller than the source of s and q is larger than p, or if p is larger than the target of s and q is smaller than p, or if p lies lexicographically between the source and the target of s. The latter case cannot happen mathematically, but it can happen numerically, if p lies close to either the source or the target of s but not exactly on l.

```
void window::draw_ray(point p, point q, line l, color col)
{
  if ( contains(p) ) { draw_ray(p,q,col); return; }
  segment s;
  point llc(xmin(),ymin()); // left lower corner
  point rrc(xmax(),ymax()); // right upper corner
  if ( !l.clip(llc,rrc,s) ) return;
  if ( compare(p,s.source()) < 0 && compare(p,q) < 0 ||
       compare(s.target(),p) < 0 && compare(q,p) < 0 ||
       compare(s.source(),p) <= 0 && compare(p,s.target()) <= 0 )
    draw_segment(s,col);
}
```

We will see an application of the refined drawing functions in Section 10.10.

Exercises for 9.1

1. Write a program that allows to input points in a graphics window and colors the points randomly red and blue.
2. Write a program that allows to input points in a graphics window and always highlights a pair of points with smallest distance. For two points p and q, *p.sqr_dist(q)* computes the squared distance between p and q.
3. Write a program that removes duplicates from a list of segments.

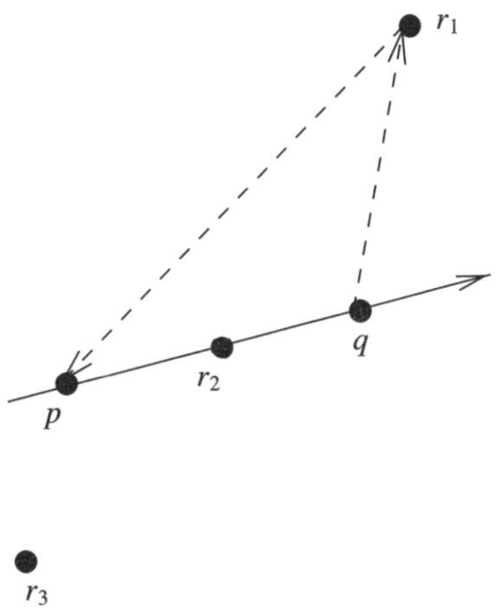

Figure 9.2 *orientation*$(p, q, r_1) = 1$, *orientation*$(p, q, r_2) = 0$, and *orientation*$(p, q, r_3) = -1$. The triangle $\Delta(p, q, r_1)$ is shown dashed.

9.2 Geometric Primitives

We discuss some of the geometric primitives available in LEDA, in particular, the orientation function and its variants, lengths and distances, angles, and intersections.

9.2.1 *The Orientation Function in the Plane*

The *orientation function* is probably the most useful geometric primitive. Let p, q, and r be three points in the plane. The tuple (p, q, r) is said to have *positive orientation* if p and q are distinct and r lies to the left of the oriented line passing through p and q and oriented from p to q, the tuple is said to have *negative orientation* if r lies to the right of the line, and the tuple is said to have *orientation zero* if the three points are collinear, see Figure 9.2. An alternative way to define positive orientation is to say that p, q, and r form a counter-clockwise oriented triangle. The function

```
int orientation(POINT p, POINT q, POINT r)
```

computes the orientation of the triple (p, q, r). It returns $+1$ in the case of positive orientation, -1 in the case of negative orientation, and 0 in the case of zero orientation. There are also predicates that test for special cases.

```
bool leftturn(p,q,r);  // same as orientation(p,q,r) >  0
bool rightturn(p,q,r); // same as orientation(p,q,r) <  0
bool collinear(p,q,r); // same as orientation(p,q,r) == 0
```

We next derive a determinant formula for the orientation function. For points p, q, and r we use $\Delta(p, q, r)$ to denote the triangle with vertices p, q, and r. We define the *signed area* of the triangle $\Delta(p, q, r)$ as its area times the orientation of the triple (p, q, r).

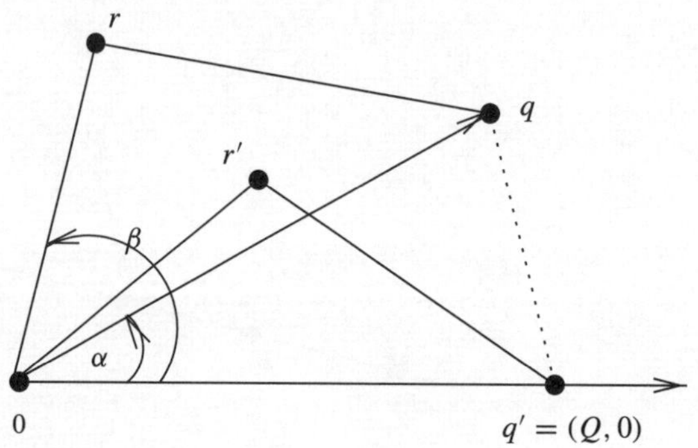

Figure 9.3 Proof of Lemma 56.

Lemma 56 *Let p, q, and r be points in the plane.*
(a) The signed area of the triangle $\Delta(p, q, r)$ *is given by*

$$\frac{1}{2}\begin{vmatrix} 1 & 1 & 1 \\ x_p & x_q & x_r \\ y_p & y_q & y_r \end{vmatrix}$$

(b) The orientation of (p, q, r) *is equal to the sign of the determinant above.*

Proof Part(b) follows immediately from part (a) and the definition of signed area. So we only need to show part (a). We do so in two steps. We first verify the formula for the case that p is the origin and then extend it to arbitrary p. So let us assume that p is equal to the origin. We need to show that the signed area A of $\Delta(p, q, r)$ is equal to $(x_q y_r - x_r y_q)/2$.

Let α be the angle between the positive x-axis and the ray Oq and let Q be the length of the segment Oq, cf. Figure 9.3. Then $\cos\alpha = x_q/Q$ and $\sin\alpha = y_q/Q$. Rotating the triangle $\Delta(O, q, r)$ by $-\alpha$ degrees about the origin yields a triangle $\Delta(O, q', r')$ with $q' = (Q, 0)$ and the same signed area. Thus, $A = Q \cdot y_{r'}/2$.

Next observe that $y'_r = R\sin(\beta - \alpha)$, where R is the length of the segment Or and β is the angle between the positive x-axis and the ray Or. Since $\sin(\beta - \alpha) = \sin\beta\cos\alpha - \cos\beta\sin\alpha$ and $R\cos\beta = x_r$ and $R\sin\beta = y_r$ we conclude that

$$\begin{aligned} A &= Q \cdot y_{r'}/2 = Q \cdot R \cdot \sin(\beta - \alpha)/2 \\ &= (Q\cos\alpha \cdot R\sin\beta - Q\sin\alpha \cdot R\cos\beta)/2 = (x_q y_r - x_r y_q)/2. \end{aligned}$$

This verifies the formula in the case where p is the origin.

Assume next that p is different from the origin. Translating p into the origin yields the triangle $\Delta(O, q', r')$ with $q' = q - p$ and $r' = r - p$[4]. On the other hand subtracting the

[4] Strictly speaking, we would have to write $q' = 0 + (q - p)$ and similarly for r'.

first column from the other two columns of the determinant yields

$$\begin{vmatrix} 1 & 1 & 1 \\ x_p & x_q & x_r \\ y_p & y_q & y_r \end{vmatrix} = \begin{vmatrix} 1 & 0 & 0 \\ x_p & x_q - x_p & x_r - x_p \\ y_q & y_q - y_p & y_r - y_p \end{vmatrix} = \begin{vmatrix} x_{q'} & x_{r'} \\ y_{q'} & y_{r'} \end{vmatrix}$$

which by the above is twice the area of the translated triangle. □

Part (b) of the lemma above is the implementation of the orientation function.

9.2.2 *The Orientation Function in Higher-Dimensional Space*

We define the orientation function for an arbitrary dimensional space and derive a determinant formula for it. Less mathematically inclined readers may skip the proofs of the lemmas to follow.

Let $(p_0, p_1, \ldots, p_d)$ be a $d+1$-tuple of points in d-dimensional space. Their orientation is zero if the points lie in a common hyperplane. If they do not, their orientation is either positive or negative as determined by the following rules:

- Let o be the origin and let e_i for i, $0 \leq i < d$, be the endpoint of the i-th coordinate vector of d-dimensional space. The tuple $(o, e_0, \ldots, e_{d-1})$ has positive orientation.
- Two tuples $(p_0, p_1, \ldots, p_d)$ and $(q_0, q_1, \ldots, q_d)$ have the same orientation if the affine map that maps p_i into q_i for i, $0 \leq i \leq d$, has positive determinant.

Lemma 57 *Let $(p_0, p_1, \ldots, p_d)$ be a $d+1$-tuple of points in d-dimensional space. Then*

$$\textit{orientation}(p_0, p_1, \ldots, p_d) = \textit{sign}\det\begin{pmatrix} 1 & \cdots & 1 \\ p_0 & \cdots & p_d \end{pmatrix},$$

where the i-th column of the determinant consists of a 1 followed by the vector of Cartesian coordinates of p_i for all i, $0 \leq i \leq d$.

Proof Observe first that the points $p_0, \ldots, p_d$ have orientation zero iff they lie in a common hyperplane which is true iff the homogeneous linear system

$$\sum_{0 \leq i \leq d} \lambda_i = 0$$

$$\sum_{0 \leq i \leq d} \lambda_i p_{i,l} = 0\ ,\ \ 0 \leq l \leq d-1$$

in variables $\lambda_0, \lambda_1, \ldots, \lambda_d$ has a non-trivial solution. The determinant above is the determinant of this system. We conclude that $\textit{orientation}(p_0, \ldots, p_d) = 0$ iff the sign of the determinant above is zero.

Assume next that $\textit{orientation}(p_0, p_1, \ldots, p_d) \neq 0$. The affine transformation that maps $(o, e_0, \ldots, e_{d-1})$ into $(p_0, p_1, \ldots, p_d)$ is given by $x \mapsto p_0 + P \cdot x$ where P has columns $p_1 - p_0, p_2 - p_0, \ldots, p_d - p_0$. Thus

$$\det P \ = \ \det\begin{pmatrix} p_1 - p_0 & p_2 - p_0 & \cdots & p_d - p_0 \end{pmatrix}.$$

Adding an additional first row and first column to this determinant with the first entry in the new row equal to one and all other entries in the new row equal to zero does not change the value of the determinant (develop the determinant according to the new row). Therefore

$$\begin{aligned}\det P &= \det\begin{pmatrix} p_1 - p_0 & p_2 - p_0 & \cdots & p_d - p_0 \end{pmatrix} \\ &= \det\begin{pmatrix} 1 & 0 & \cdots & 0 \\ p_0 & p_1 - p_0 & \cdots & p_d - p_0 \end{pmatrix} = \det\begin{pmatrix} 1 & 1 & \cdots & 1 \\ p_0 & p_1 & \cdots & p_d \end{pmatrix},\end{aligned}$$

where the last equality follows from adding the first column to all other columns. We conclude that $(p_0, p_1, \ldots, p_d)$ has the same orientation as $(o, e_0, \ldots, e_{d-1})$ if and only if the determinant above is positive. □

The lemma above generalizes Lemma 56. Observe that both lemmas give the same formula for points in the plane.

We have already given an intuitive definition of orientation in the plane: three points (p_0, p_1, p_2) in the plane have orientation zero if they are collinear, have positive orientation if they form a counter-clockwise oriented triangle, and have negative orientation if they form a clockwise oriented triangle.

In three-dimensional space there is also an intuitive definition. Four points (p_0, p_1, p_2, p_3) in three-dimensional space have orientation zero if they are coplanar, have positive orientation if they form a right-handed system, and have negative orientation if they form a left-handed system. We need to explain the terms right- and left-handed system. Imagine that you place the base of your thumb at point p_0 and let the thumb (index finger, middle finger) point to p_1, p_2, and p_3, respectively. Only one of your hands will work and this determines the handedness of the system. For four three-dimensional points p, q, r, and s

```
int orientation(p,q,r,s);
```

computes their orientation.

An alternative definition of orientation in three-dimensional space is to say that the four-tuple (p_0, p_1, p_2, p_3) has positive orientation if p_3 sees (p_0, p_1, p_2) in counter-clockwise orientation. The last sentence connects orientation in three-dimensional space with orientation in two-dimensional space. The next lemma generalizes this connection to higher dimensions.

Lemma 58 *Let $(p'_0, p'_1, \ldots, p'_{d-1})$ be a d-tuple of points in $(d-1)$-dimensional space with positive orientation and let $(p_0, p_1, \ldots, p_d)$ be a $d+1$-tuple of points in d-dimensional space such that p_i projects into p'_i for i, $1 \le i < d$, i.e., the Cartesian coordinate vector of p'_i is the Cartesian coordinate vector of p_i with the last entry removed. Let h be the hyperplane spanned by $p_0, \ldots, p_{d-1}$. Then $(p_0, p_1, \ldots, p_d)$ has positive orientation if p_d lies above h, has orientation zero if p_d lies on h, and has negative orientation if p_d lies below h.*

Proof Let q be the projection of p_d into h. Then $p_d = q + c \cdot e_{d-1}$ where e_{d-1} is the

$(d-1)$-th coordinate vector and c is positive if p_d lies above h, is zero if p_d lies on h, and is negative if p_d lies below h. Moreover there are $\lambda_0, \lambda_1, \ldots, \lambda_{d-1}$ such that

$$\sum_{0 \le i \le d-1} \lambda_i = 1,$$

and

$$\sum_{0 \le i \le d-1} \lambda_i p_i = q.$$

Thus

$$\begin{aligned}
\det\begin{pmatrix} 1 & 1 & \cdots & 1 \\ p_0 & p_1 & \cdots & p_d \end{pmatrix} &= \det\begin{pmatrix} 1 & 1 & \cdots & 1 & 1 \\ p_0 & p_1 & \cdots & p_{d-1} & q + c \cdot e_{d-1} \end{pmatrix} \\
&= \det\begin{pmatrix} 1 & 1 & \cdots & 1 & 0 \\ p_0 & p_1 & \cdots & p_{d-1} & c \cdot e_{d-1} \end{pmatrix} \\
&= c \cdot \det\begin{pmatrix} 1 & 1 & \cdots & 1 \\ p'_0 & p'_1 & \cdots & p'_{d-1} \end{pmatrix},
\end{aligned}$$

where the second equality follows from subtracting the λ_i-th multiple of the i-th column from the last column for i, $0 \le i < d$, and the last equality follows by expanding the determinant according to the last column. Observe that the last column has only one non-zero entry and that this entry is in the last row. □

In the plane we connected the orientation of a triple (p, q, r) to the signed area of the triangle defined by the points. A similar connection holds in higher-dimensional space. The signed area of the simplex with vertices $p_0, p_1, \ldots, p_d$ is equal to $\frac{1}{d!}$ times the determinant defined by the points.

9.2.3 *Sidedness*

Many geometric objects, such as lines and circles in the plane, planes and spheres in three-dimensional space, and more generally hyperplanes and hyperspheres in d-dimensional space, partition ambient space into two parts. We designate one of the parts as positive and one as negative. The function

```
int O.side_of(x);
```

where O is a geometric object and x is a point in ambient space returns a positive number (zero, a negative number, respectively) if x lies in the positive part (lies on O, lies in the negative part, respectively). Examples are

```
int l.side_of(x);      // l is a line
int C.side_of(x);      // C is a circle
int P.side_of(x);      // P is a polygon
```

What is the positive subspace with respect to a line or circle or hyperplane? We use the orientation function for points to formulate general rules:

- For a hyperplane h in d-space defined by points $p_0, p_1, \ldots, p_{d-1}$ (in this order) the positive subspace consists of all points p_d such that $(p_0, p_1, \ldots, p_d)$ has positive orientation. Thus $line(p, q).side\_of(x)$ is the same as $orientation(p, q, x)$, if p and q are distinct.

- For a hypersphere S in d-space defined by points $p_0, p_1, \ldots, p_d$ (in this order) the positive subspace consists of the interior of the sphere if $(p_0, p_1, \ldots, p_d)$ is positively oriented and consists of the exterior of the sphere otherwise. The same rule applies to simplices.

In two-dimensional space the following alternative rule is also worth remembering. Two points defining a line and three points defining a circle impose a sense of direction on the line or circle respectively (from the first point to the second point in the case of a line, and from the first point through the second point to the third point in the case of a circle). *The positive subspace is the region to the left of the object.*

Let p, q, and r be points in the plane. We may want to inquire about the position of a point x with respect to $circle(p, q, r)$. We could write $circle(p, q, r).side\_of(x)$. Since this test incurs overhead for the construction of a circle we also have an alternative syntactic format that avoids this overhead and also gives an answer in the case where the p, q, and r do not define a circle.

```
int side_of_circle(p,q,r,x);
```

returns $+1$ if x is to the left of the oriented circle through p, q, and r, returns -1 if x is to the right of the oriented circle through p, q, and r, and returns 0 if either $|\{p, q, r\}| \leq 2$ or x lies on the oriented circle through p, q, and r. We give some more explanations.

Three points p, q, and r that are not collinear define a unique circle passing through them. We give this circle an orientation by insisting that p, q, and r occur in this order on the circle. Consider now a fourth point x. It is either left of, on, or right of the oriented circle through p, q, and r. Note that left corresponds to inside if the circle is counter-clockwise oriented and to outside otherwise, see Figure 9.4. The case that the points p, q, and r are collinear deserves special attention. If the three points are not pairwise distinct then the *which_side* function returns zero. If they are pairwise distinct then we orient the line passing through them such that the order of the points along the line is a circular permutation of (p, q, r), i.e., either (p, q, r) or (q, r, p) or (r, p, q), and use again $+1$ for the left side and -1 for the right side of the line.

Circles, spheres, triangles, simplices, simple polygons, and many other geometric objects partition ambient space into a bounded and an unbounded region. Since there is no standard convention in mathematics that connects boundedness and unboundedness with positive and negative respectively, we have an enumeration type for the outcome of the *region_of* function.

```
enum region_kind { BOUNDED_REGION, ON_REGION, UNBOUNDED_REGION };
region_kind O.region_of(x);              // the generic form
region_kind C.region_of(x);              // C is a circle
```

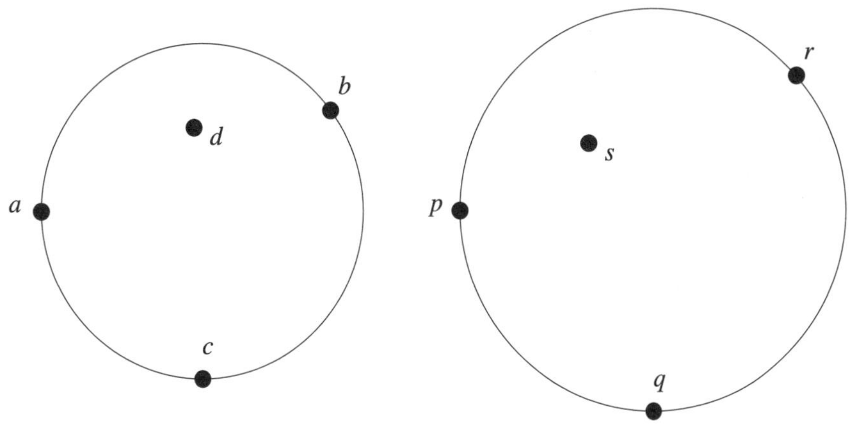

Figure 9.4 The sides of a circle: d lies on the negative side of the circle defined by points a, b, and c, and s lies on the positive side of the circle defined by points p, q, and r.

Frequently, one only wants to test for one of the outcomes. We have appropriate predicates.

```
bool O.inside(x);        // O.region_of(x) == bounded_region
bool O.on_boundary(x); // O.region_of(x) == on_region
bool O.outside(x);       // O.region_of(x) == unbounded_region
```

9.2.4 *Length and Distance*

If p and q are POINTs and l is a LINE,

```
RAT_TYPE p.sqr_dist(q);
RAT_TYPE l.sqr_dist(q);
```

compute the square of the distance between q and p or l, respectively.

In the rational kernel there are no functions to compute distances, in the floating point kernel there are, but think twice before using them. Why?

The distance between two points p and q is equal to $((x_p - x_q)^2 + (y_p - y_q)^2)^{1/2}$ and is hence, in general, not a rational number. The squared distance is a rational number and hence the rational kernel provides only functions to compute squared distances. The floating point kernel uses the *sqrt* function from the standard math-library to compute distances.

We find that the computation of distances is rarely needed. Consider the following problem. Let p and q be points. We want to define the circle centered at p whose radius is ρ times the distance between p and q. This is best written as

```
CIRCLE C(p, p + rho * (q - p));
```

Observe that $q - p$ is the vector from p to q and hence $rho * (q - p)$ is a vector whose length is ρ times the distance between p and q.

The distances between p and q and r, respectively, can be compared by

```
int p.cmp_dist(q,r); // same as cmp(p.sqr_dist(q),p.sqr_dist(r));
```

This is more efficient than computing the two squared distances and comparing them.

9.2.5 *Angles*

There is no type angle in either the rational or the floating point kernel. There are, however, a number of functions related to angles. In particular, two vectors v_1 and v_2 can be compared by the angle which they form with the positive x-axis. For a vector v let $\alpha(v)$ be the angle by which the positive x-axis has to be turned counter-clockwise until it aligns with v. The zero vector defines the angle zero.

```
int compare_by_angle(VECTOR v1, VECTOR v2);
```

returns $cmp(\alpha(v1), \alpha(v2))$.

We describe the implementation. If one of the vectors is the zero vector the comparison is easily made. If both vectors are zero, they are equal, and if only one is zero, it is the smaller. So assume that both vectors are non-zero. We say that a non-zero vector (x, y) belongs to the upper half-plane if either $y > 0$ or $y = 0$ and $x > 0$, and we say that it belongs to the lower half-plane otherwise. Let *upper1* and *upper2* be the half-planes to which our vectors belong (the value is $+1$ for a vector in the upper half-plane and -1 for a vector in the lower half-plane). If the two vectors belong to distinct half-planes, the vector in the upper half-plane is smaller and hence we may return the sign of *upper2* − *upper1*. If the two vectors lie in the same half-plane, v_1 precedes v_2 iff the triangle $(O, O+v_1, O+v_2)$ is counter-clockwise oriented iff the orientation of $(O, O + v_1, O + v_2)$ is positive iff its signed area is positive. The signed area is the length of the cross-product of v_1 and v_2, i.e., $x_1y_2 - x_2y_1$. We may therefore return $-sign(x_1y_2 - x_2y_1)$.

Rational vectors are stored by their homogeneous coordinates. Since the ordering of angles does not depend on the length of vectors and since the homogenizing coordinate is guaranteed to be non-negative, we may ignore it.

⟨*_angle_order.c*⟩+≡

```
int compare_by_angle(const rat_vector& v1, const rat_vector& v2)
{ const integer& x1 = v1.hcoord(0);
  const integer& y1 = v1.hcoord(1);
  const integer& x2 = v2.hcoord(0);
  const integer& y2 = v2.hcoord(1);
  if ( x1 == 0 && y1 == 0 ) return ( x2 == 0 && y2 == 0 ? 0 : -1);
  if ( x2 == 0 && y2 == 0 ) return 1;
  // both vectors are non-zero
  int sy1 = sign(y1); int sy2 = sign(y2);
  int upper1 = ( sy1 != 0 ? sy1 : sign(x1) );
  int upper2 = ( sy2 != 0 ? sy2 : sign(x2) );
  if ( upper1 == upper2 ) return sign(x2*y1 - x1*y2);
  return sign(upper2 - upper1);
}
```

9.2.6 *Intersections*

There are functions to compute the intersections between lines, rays, and segments. For example, if l is a LINE and s is a SEGMENT then

```
bool l.intersection(s, p);
```

returns *true* if l and s have a single point in common and returns *false* otherwise. In the latter case, the unique point of intersection is assigned to p.

Exercises for 9.2

1 Write a function *circum_center* that takes three points p, q, and r and returns the center of the circle passing through p, q, and r. The three points are assumed to be non-collinear.

2 Use the left-turn predicate to write a function that tests whether four points p, q, r, and s in the plane form a convex quadrilateral.

3 Modify the test from the previous exercise such that it decides whether the four points form a counter-clockwise oriented convex quadrangle.

4 Let p, q, r, and s be four points in three space not lying in a plane. Position your left or right hand such that p coincides with the base of your thumb, and q, r, and s coincide with the tips of your thumb, index finger, and middle finger, respectively. Convince yourself that only one of the two hands will work and relate the choice of hand to the orientation of the four points.

9.3 Affine Transformations

An affine transformation T of the plane is specified by a 3×3 matrix T with $T_{2,0} = T_{2,1} = 0$ and $T_{2,2} \neq 0$. It maps the point p with homogeneous coordinate vector (p_x, p_y, p_w) to the point $T \cdot p$. Transformations are called *transform* in the floating point kernel and are called *rat_transform* in the rational kernel. We use TRANSFORM as the generic name.

```
TRANSFORM T;
TRANSFORM T1(M);
```

declares T as the identity transform and declares *T1* as the transform with transformation matrix M. M must be a 3×3 *matrix* in the floating point kernel and a 3×3 *integer_matrix* in the rational kernel. Functional notation is used to apply an affine transformation to a geometric object. For example,

```
p = T(q);     // p and q are points
P = T(Q);     // P and Q are polygons
v = T(w);     // v and w are vectors
C = T(D);     // C and D are circles; T must be rigid
```

The norm of an affine transformation T is defined as

$$|T| = (T_{0,0}T_{1,1} - T_{0,1}T_{1,0})/T_{2,2}^2.$$

A transformation is called *rigid* iff its norm has absolute value one.

```
RAT_TYPE T.norm();
```

returns the norm of T.

If T and $T1$ are transformations then

```
T(T1);
```

is the transformation obtained by first applying $T1$ and then T.

Translations, rotations, and reflections are special cases of affine transformations.

A matrix of the form

$$\begin{pmatrix} w & 0 & x \\ 0 & w & y \\ 0 & 0 & w \end{pmatrix}$$

realizes a translation by the vector $(x/w, y/w)$ and a matrix of the form

$$\begin{pmatrix} a & -b & 0 \\ b & a & 0 \\ 0 & 0 & w \end{pmatrix}$$

where $a^2 + b^2 = w^2$ realizes a rotation by the angle α about the origin, where $\cos\alpha = a/w$ and $\sin\alpha = b/w$. Rotations are in counter-clockwise direction.

It is inconvenient to specify transformations by their transformation matrix. We have several functions that construct transformations. Observe that these functions are not constructors but functions that return transformations. For example

```
TRANSFORM T = translation(const INT_TYPE& dx, const INT_TYPE& dy,
                          const INT_TYPE& dw);
TRANSFORM T = translation(const RAT_TYPE& dx, const RAT_TYPE& dy);
```

construct translations by the vector $(dx/dw, dy/dw)$ and the vector (dx, dy), respectively.

```
TRANSFORM T = reflection(const POINT& q, const POINT& r);
TRANSFORM T = reflection(const POINT& q);
```

construct the reflection across the straight line passing through q and r and the reflection across the point q, respectively.

```
TRANSFORM T = rotation90(const POINT & q);
TRANSFORM T = rotation(const POINT& q, double alpha, double eps);
```

construct rotations about the point q. In the first case the rotation is by $\pi/4$ and in the second case the rotation is approximately by α. ϵ is a tolerance parameter.

We show the implementations of the last two functions. Rotation by $\pi/4$ is achieved by the rotation matrix

$$\begin{pmatrix} 0 & -1 & 0 \\ 1 & 0 & 0 \\ 0 & 0 & 1 \end{pmatrix}$$

and rotation about an arbitrary point q is achieved by first translating by the vector $O - q$, rotating about the origin, and finally translating back by the vector $q - 0$.

⟨*rotation*⟩≡

```
static TRANSFORM rotation90_origin(const POINT& q)
{
  INT_MATRIX M(3,3);
  for (int i = 0; i < 3; i++)
    for (int j = 0; j < 3; j++)
      M(i,j) = 0 ;
  M(0,1) = -1;    M(1,0) = +1;
  M(2,2) = 1;
  return TRANSFORM(M);
}
TRANSFORM rotation90(const POINT& q)
{
  TRANSFORM R = rotation90_origin(q);
  TRANSFORM T0 = translation(-q.X(),-q.Y(),q.W());
  TRANSFORM T1 = translation( q.X(), q.Y(),q.W());
  TRANSFORM T = T1(R(T0));
  T.simplify();
  return T;
}
```

Observe that we have given the function *rotation90_origin* an artificial argument of type *POINT* so that we can use the same code for both kernels. In the piece of code above, we declared *rotation90_origin* static, as it is an auxiliary function that should not be visible outside the file _transform.c.

We come to the rotation by an arbitrary angle α. We only show how to construct the transformation matrix for the rotation about the origin. We construct a point p on the unit circle and in direction α (this is a member function of CIRCLE) and then use the coordinates of p as the sine and cosine of α.

⟨*rotation*⟩+≡

```
static TRANSFORM rotation_origin(const POINT& q,
                                 double alpha, double eps)
{ POINT origin(0,0);
  POINT X(1,0);
  CIRCLE C(origin,X); // unit circle centered at origin
  POINT p = C.point_on_circle(alpha,eps);
  INT_MATRIX M(3,3);
  M(0,2) = M(1,2) = M(2,0) = M(2,1) = 0;
  M(0,0) = M(1,1) = p.X() ;
  M(0,1) = -p.Y();    M(1,0) = p.Y();
  M(2,2) = p.W();
  return TRANSFORM(M);
}
```

It remains to explain the function *point_on_circle*. In the floating point kernel we use the sine and cosine function from the math-library to construct p; *eps* plays no role in this construction. In the rational kernel we use the method described in [CDR92] to find integers a, b, and w and an angle α'such that

$$\begin{aligned} a^2 + b^2 &= w^2 \\ \cos\alpha' &= a/w \\ \sin\alpha' &= b/w \\ |\alpha' - \alpha| &\leq \epsilon. \end{aligned}$$

General affine transformations are a fairly recent addition to our geometry kernels. In earlier versions we had only functions for special affine transformations. They were member functions of the geometric classes. For example,

```
p.translate(RAT_TYPE dx,RAT_TYPE dy);
```

returns the point $p + v$ where $v = (dx, dy)$.

Transformations are a good tool to generate difficult inputs for geometric algorithms. In Section 10.8.4 we perform the following experiment. We first construct a regular n-gon P, $n = 20000$, with its vertices on the unit circle. We then construct $Q = T(P)$ where T is a rotation by $2\pi/(nm)$ and m is a large integer, e.g., $m = 10^9$. We finally compute the union of P and Q.

Exercises for 9.3

1 Implement the function that composes two transformations.
2 Implement the function that applies a transformation to a point.
3 Implement the function that applies a transformation to a vector. This is different from the solution to the previous exercise.
4 Implement the function that constructs the transformation matrix for reflection at a point.
5 Implement the function that constructs the transformation matrix for reflection at a line.

9.4 Generators for Geometric Objects

There is a frequent need to generate geometric objects, random or otherwise. We describe generators for random points in the plane and generators for polygons. There are also generators for random points in space.

Generators for Random Points: We have generators for random points in squares, in discs, near circles, and on circles. For each generator there is a version that generates a single point and a version that generates a list of points.

```
random_point_in_square(POINT& p, int maxc);
random_points_in_square(int n, int maxc, list<POINT>& L);
```

generate a random point with integer coordinates in the range $[-maxc \,..\, +maxc]$ and a list of n such points, respectively.

```
random_point_in_unit_square(POINT& p, int D = (1<<30) - 1 );
random_points_in_unit_square(int n, int D, list<POINT>& L);
random_points_in_unit_square(int n, list<POINT>& L);
```

generate a point in the unit square, i.e., a point whose coordinates are of the form i/D for a random integer i, $0 \leq i \leq D$, n such points, and n such points with the default value of D, respectively.

For the remaining generators we only give the form that generates a single point.

```
random_point_in_disc(POINT& p, int R);
random_point_in_unit_disc(POINT& p, int D = (1<<30) - 1);
```

generate a random point with integer coordinates in the disc with radius R and a random point with coordinates of the form i/D for integer i in the unit disc, respectively.

```
random_point_near_circle(POINT& p, int R);
random_point_near_unit_circle(POINT& p, int D = (1<<30) - 1);
```

generate a random point with integer coordinates near the circle with radius R and a random point with coordinates of the form i/D for integer i near the unit circle, respectively.

The latter function is implemented as follows. We generate a random double x in the unit interval, set $\phi = 2\pi x$, and construct the point $(\lfloor D\cos\phi\rfloor, \lfloor D\sin\phi\rfloor, D)$.

```
void random_point_near_unit_circle(POINT& p, int D)
{ double a;
  Rand_Source >> a;
  double phi = 2*a*LEDA_PI;
  int x = int(D*cos(phi));
  int y = int(D*sin(phi));
  p = POINT(x,y,D);
}
```

With the rational kernel we can also generate points that lie *exactly* on a circle.

```
random_point_on_circle(POINT& p, int R, int C = 1000000);
random_point_on_unit_circle(POINT& p, int C = 1000000);
```

constructs a point on the circle with radius R and on the unit circle, respectively. This assumes that the rational kernel is used. In both cases the point is chosen at random from a set of at least C candidates. With the floating point kernel the function is equivalent to the *near_circle* and the *near_unit_circle* function with $D = 1.0/C$, respectively.

The implementation of *random_point_on_unit_circle* with the rational kernel is as follows:

```
void random_point_on_unit_circle(rat_point& p, int C)
{ rat_point origin(0,0);
  rat_circle Circ(origin,origin + rat_vector::unit(1));
  double a; Rand_Source >> a;
```

```
    double eps = 1.0/(2*C);
    p = Circ.point_on_circle(2*LEDA_PI*a,eps);
  }
```

where the function *point_on_circle* is as described at the end of Section 9.3.

The last two generators are much slower than all other generators when the rational kernel is used. We have therefore generated files of 50000 random points (with $C = 10^6$). They are available as:

LEDAROOT/data/geo/rat_points_unit_circle_random_50000.ex
LEDAROOT/data/geo/points_unit_circle_random_50000.ex

Generating Polygons: We have two generators for polygons.

```
POLYGON P = reg_n_gon(int n, CIRCLE C, double epsilon);
POLYGON P = n_gon(    int n, CIRCLE C, double epsilon);
```

The first generator generates a nearly regular n-gon. The i-th point is generated by the call $C.point\_on\_circle(2\pi i/n, epsilon)$. With the rational kernel the vertices of the n-gon are guaranteed to lie on the circle, with the floating point kernel they are only guaranteed to lie near C.

The second generator generates a (nearly) regular n-gon whose vertices lie near the circle C. For the floating point kernel the function is equivalent to the function above. For the rational kernel the function first generates an n-gon with floating point arithmetic and then converts the resulting *polygon* to a *rat_polygon*.

9.5 Writing Kernel Independent Code

We use the C++ precompilation mechanism to write code that is independent of the kernel. Recall that the kernels are designed such that all functions that are available in a rational kernel are also available in the corresponding floating point kernel.

The only difference between the rational kernel and the floating point kernel is the interpretation of the generic names POINT, SEGMENT, LINE, ... In order to give the generic names the interpretation required in a particular kernel one of the files must be included:

```
#include <LEDA/rat_kernel_names.h>
#include <LEDA/float_kernel_names.h>
#include <LEDA/d3_rat_kernel_names.h>
#include <LEDA/d3_kernel_names.h>
```

Every one of these files consists of a sequence of define-statements which define the generic names for the corresponding kernel. For example,

```
// part of rat_kernel_names.h

#define KERNEL       RAT_KERNEL
#define INT_TYPE     integer
#define RAT_TYPE     rational
```

```
#define VECTOR          rat_vector
#define POINT           rat_point
#define SEGMENT         rat_segment
#define TRANSFORM       rat_transform
```

We also have files that undefine all names used in a kernel. They are:

```
#include <LEDA/kernel_names_undef.h>
#include <LEDA/d3_kernel_names_undef.h>
```

Suppose now that we want to write a program that is supposed to work for both two-dimensional kernels. We write a generic version of the program using only the generic names and then derive the two specialized versions from it. For example,

⟨*FOO.c*⟩≡

```
main(){
window W; W.display();
POINT p;
while ( W >> p) W << p.to_point();
}
```

⟨*rat_foo_test.c*⟩≡

```
#include <LEDA/rat_point.h>
#include <LEDA/window.h>
#include <LEDA/rat_window.h> // lets W >> p work for rat_points
#include <LEDA/rat_kernel_names.h>
⟨FOO.c⟩
#include <LEDA/kernel_names_undef.h>
```

⟨*foo_test.c*⟩≡

```
#include <LEDA/point.h>
#include <LEDA/window.h>
#include <LEDA/float_kernel_names.h>
⟨FOO.c⟩
#include <LEDA/kernel_names_undef.h>
```

The header file window.h is included in both specializations and it is hence tempting to write

⟨*BAD_FOO.c*⟩≡

```
#include <LEDA/window.h>
main(){
window W; W.display();
POINT p;
while ( W >> p) W << p.to_point();
}
```

This will lead to a disaster. Never include a file in a piece of code that is subject to renaming, except if you are absolutely sure that the renaming mechanism is not used in the included file. Window.h includes the entire floating point kernel which in turn includes files like transform.h. The latter file uses the renaming mechanism.

Why did we undefine all names at the end of foo_test.c and rat_foo_test.c? We found that it helps to guard against the error pointed out in the preceding paragraph. If foo_test.c is included in a file that uses the renaming mechanism the compiler will generate a message that certain names are undefined. For example

```
#include <LEDA/rat_kernel_names.h>
#include "rat_foo_test.c"
POINT p;  // POINT is undefined here
```

We use the renaming mechanism just described for all source files in src/plane_alg and for some source files in src/plane. We also use the mechanism for the header files for polygons, generalized polygons, transformations, point sets, and generation of random points. In these cases the generic header files are stored in incl/LEDA/generic.

Sometimes, a small part of the code is specific to a particular kernel. We use conditional compilation in this situation. For example,

```
// an error was just discovered
#if ( KERNEL == FLOAT_KERNEL )
cerr << "Please move to the rational kernel.";
#else
cerr << "Please report this error.";
#endif
```

The conversion functions between floating point objects and rational objects form a more substantial example. In the case of POLYGONs we have:

```
// part of POLYGON.h
POLYGON(const POLYGON& P) : handle_base(P) {} // copy constructor
#if ( KERNEL == RAT_KERNEL )
rat_polygon(const polygon& Q, int prec = 0);
#endif
#if ( KERNEL == FLOAT_KERNEL )
polygon(const polygon& Q, int prec);
#endif
polygon     to_polygon() const;
```

The first declaration defines the copy constructor for both instantiations and the last declaration defines the conversion function to *polygons* for both instantiations. The middle declaration is conditional. In class *rat_polygon* we also have the constructors

```
rat_polygon(const polygon&, int);
rat_polygon(const polygon&);
```

and in class *polygon* we also have the constructor

```
polygon(const polygon&, int prec);
```

It is important that *prec* is not an optional argument in the latter case as this would clash with the copy constructor.

We summarize: the pre-compilation mechanism of C++ allows us to write kernel independent code. Files that use the renaming mechanism must never be included in a piece of code that is subject to renaming.

9.6 The Dangers of Floating Point Arithmetic

We give two examples for the dangers of floating point arithmetic in geometric computation. Both examples show that floating point geometric objects can exhibit bizarre behavior that deviates widely from the behavior predicted by mathematics. We will see more examples in the chapter on geometry algorithms.

9.6.1 *Convex Hulls*

The first example was suggested by Stefan Schirra. Consider the following piece of code. We define a segment s and construct a set L of points consisting of the endpoints of L and the intersections between s and some number of random lines.

⟨*float_hull_test*⟩≡
```
point p0(-LEDA_PI, -LEDA_PI);
point p1(+LEDA_PI, +LEDA_PI);
segment s(p0,p1);
list<point> L; L.append(p0); L.append(p1);
for (int i = 0; i < 10000; i++)
{ double ax, ay;
  rand_int >> ax; rand_int >> ay; point p(ax*LEDA_PI, ay*LEDA_PI);
  rand_int >> ax; rand_int >> ay; point q(ax*LEDA_PI, ay*LEDA_PI);
  line l(p,q); point r;
  if ( l.intersection(s,r) ) L.append(r);
}
list<point> CH = CONVEX_HULL(L);
```

We then compute the convex hull of L, see Section 10.1. Since all points in L lie on s, the convex hull should have exactly two vertices. Figure 9.5 shows the output of a sample run of the program. The convex hull has more than two vertices, contrary to what mathematics tells us. The explanation is simple. When the intersection between s and a line l is computed with the floating point kernel, the point of intersection does not necessarily lie on s but only near s.

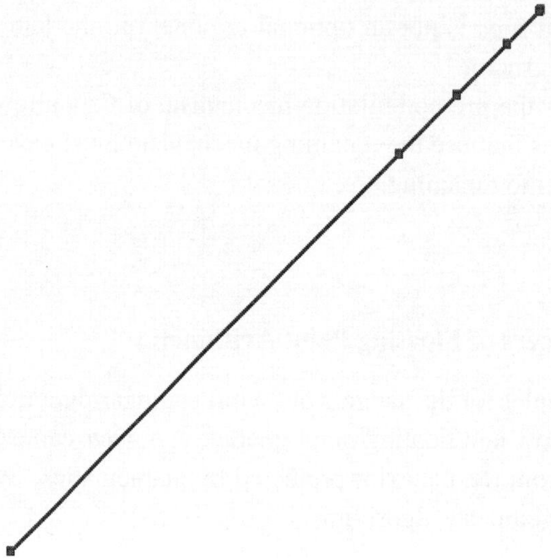

Figure 9.5 The convex hull of points contained in a common line segment computed with the floating point kernel. The hull has five vertices although there should be only two.

9.6.2 *Braided Lines (Verzopfte Geraden)*

The second example was suggested by Lyle Ramshaw who also coined the name braided lines (verzopfte Geraden in German) for it. Consider the lines

$$l_1 : y = 9833 \cdot x/9454 \quad \text{and} \quad l_2 : y = 9366 \cdot x/9005.$$

Both lines pass through the origin and the slope of l_1 is slightly larger than the slope of l_2. At $x = 9454 \cdot 9005$ we have $y_1 = 9833 \cdot 9005 = 9366 \cdot 9454 + 1 = y_2 + 1$.

The following program runs through multiples of 0.001 between 0 and 1 and computes the corresponding y-values y_1 and y_2. It compares the two y-values and, if the outcome of the comparison is different than in the previous iteration, prints x together with the current outcome.

⟨*braided_lines_test.c*⟩≡

```
#include <stream.h>
main(){
cout.precision(12);
float delta = 0.001;
int last_comp = -1;
float a = 9833, b = 9454, c = 9366, d = 9005;
for (float x = 0; x < 0.1; x = x + delta)
{ float y1 = a*x/b;   // l1 is steeper
  float y2 = c*x/d;
  int comp = (y1 < y2? -1 : (y1 == y2? 0 : +1));
```

```
    if (comp != last_comp)
    { cout <<"\n" << x << ": ";
      if (comp == -1) cout << "l1 is below l2";
      if (comp ==  0) cout << "l1 intersects l2";
      if (comp == +1) cout << "l1 is above l2";
    }
    last_comp = comp;
  }
  cout <<"\n\n";
  }
```

Clearly, we should expect the program to print

```
0.000: l1 intersects l2
0.001: l1 is above l2
```

Well, the first few lines of the actual output are[5] :

```
0: l1 intersects l2
0.00300000002608: l1 is above l2
0.00400000018999: l1 intersects l2
0.0050000003539: l1 is above l2
0.00800000037998: l1 intersects l2
0.00900000054389: l1 is below l2
0.0100000007078: l1 is above l2
0.0110000008717: l1 intersects l2
0.0120000010356: l1 is above l2
0.0130000011995: l1 intersects l2
0.0140000013635: l1 is above l2
0.0150000015274: l1 is below l2
0.01600000076: l1 intersects l2
0.0180000010878: l1 is below l2
0.0190000012517: l1 intersects l2
```

We conclude that the lines intersect many times, contrary to what mathematics teaches us.

What went wrong? The type *float* consists of only a finite number of values and hence a line is really a step function as shown in Figure 9.6. The width of the steps of our two lines l_1 and l_2 are distinct and hence the lines intersect.

9.6.3 *Overcoming the Dangers of Floating Point Arithmetic*

The examples above show that the implementation of geometric algorithms may be a difficult task. How can we overcome the difficulties?

The first approach sticks with inexact arithmetic but uses it more carefully. The papers [Mil88, Mil89a, Mil89b, FM91, LM90, GSS93, GSS89] develop algorithms for line

[5] This output is produced on the first author's workstation. If the program is run on the same author's notebook, it produces the correct result. The explanation for this behavior is that on the notebook double precision arithmetic is used to implement floats. According to the C++ standard floats must not offer more precision than doubles; they are not required to provide less.

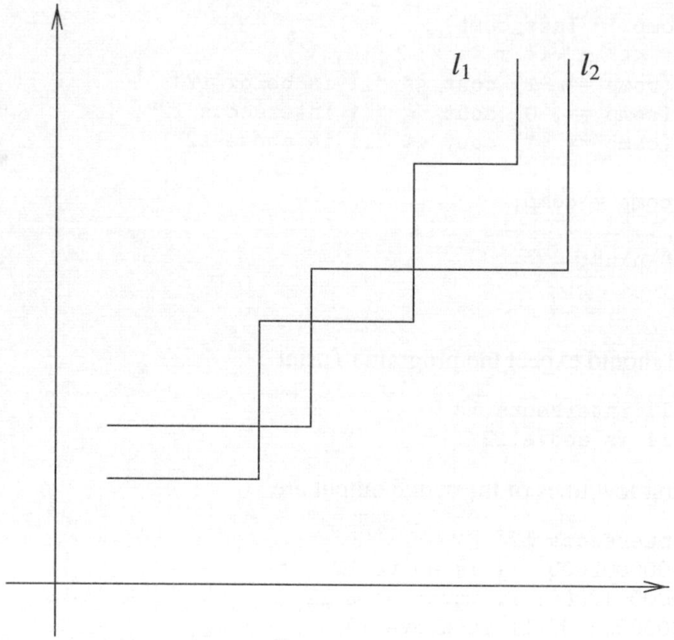

Figure 9.6 Lines as step functions and their multiple intersections.

arrangements, intersections, convex hulls, and Voronoi diagrams based on imprecise primitives. We suggest that the reader has a look at at least one of these papers in order to appreciate the ingenuity needed to overcome the shortcomings of floating point arithmetic. We were afraid of the required ingenuity and therefore did not adopt this approach for LEDA.

The alternative approach is to switch to exact arithmetic. This approach was pioneered by Karasick, Lieber, and Nackman [KLN91]. They discussed the computation of Delaunay diagrams by exact rational arithmetic. The use of exact arithmetic overcomes the correctness problems associated with floating point arithmetic, however, at the cost of a much increased running time. Fortune and van Wyk [FvW96] showed that the use of floating point filters can give exact geometric computation at low cost. We adapted their ideas[6] to the LEDA system [MN94b, MN94a]. Floating point filters are the topic of the next section.

Exercises for 9.6

1 Give a version of the intertwined lines for *double* arithmetic.

2 Play with the voronoi demo (in xlman) and try to find examples where it works incorrectly when run with the floating point kernel. Try to explain what goes wrong.

[6] The conference version of their paper appeared in 1993.

9.7 Floating Point Filters

Floating point filters apply to the evaluation of geometric predicates as used in the conditionals of geometric programs. For example,

```
switch ( orientation(a,b,c) )
{ case -1: // negative orientation
  case  0: // collinear points
  case +1: // positive orientation
}
```

Evaluating a geometric predicate is tantamount to determining the sign of an arithmetic expression. For example, the test above is equivalent to

```
switch (sign((ax*bw-bx*aw)*(ay*cw-cy*aw)-(ay*bw-by*aw)*(ax*cw-cx*aw)))
{ case -1: //
  case  0: //
  case +1: //
}
```

where *ax*, *ay*, *aw* denote the homogeneous coordinates of point *a* and similarly for the points *b* and *c*. The homogeneous coordinates of a *rat_point* are *integers* and hence evaluating the conditional involves ten multiplications and four additions of *integers*. Unfortunately, *integer* arithmetic is considerably more expensive than floating point arithmetic and hence we might expect to pay a tremendous price for exact computation.

The observation that paves the way for floating point filters is that we only want to know the sign of the arithmetic expression but not its value. It is frequently possible to determine the sign of an expression with floating point arithmetic although it is impossible to determine its value with floating point arithmetic.

In order to compute the sign of an expression[7] E, a floating point filter computes an approximation $\tilde{E}$ of E using floating point arithmetic and also a bound B on the maximal difference between $\tilde{E}$ and the (unknown) exact value E, i.e.,

$$|E - \tilde{E}| \leq B,$$

or ,

$$\tilde{E} - B \leq E \leq \tilde{E} + B.$$

Thus:

- if $\tilde{E} > B$ then $E > 0$,
- if $\tilde{E} < -B$ then $E < 0$,
- if neither of the above, $B < 1$ and E and $\tilde{E}$ are integral then $E = 0$.

For the third item observe that if neither of the first two cases applies then $|\tilde{E}| \leq B$. If $\tilde{E}$ is integral and $B < 1$ this implies $\tilde{E} = 0$. If E is integral this implies further that $E = 0$.

In order to derive a specific floating point filter one has to:

[7] We use E in the usual double meaning: it denotes an expression and also the value of the expression.

E	$\tilde{E}$	mes_E	ind_E
a, integer	$fl(a)$	$\lvert fl(a)\rvert$	1
a, float integer	$fl(a)$	$\lvert fl(a)\rvert$	0
$A + B$	$\tilde{A} \oplus \tilde{B}$	$mes_A \oplus mes_B$	$1 + \max(ind_A, ind_B) \cdot \delta$
$A - B$	$\tilde{A} \ominus \tilde{B}$	$mes_A \oplus mes_B$	$1 + \max(ind_A, ind_B) \cdot \delta$
$A \cdot B$	$\tilde{A} \odot \tilde{B}$	$mes_A \odot mes_B$	$1 + ind_A \cdot \delta + ind_B \cdot \delta^2$

Table 9.1 The recursive definition of mes_E and ind_E. The first column contains the case distinction according to the syntactic structure of E, the second column contains the rule for computing $\tilde{E}$ and the third and fourth columns contain the rules for computing mes_E and ind_E; $\oplus$ and $\odot$ denote the floating point implementations of addition and multiplication. We use the abbreviations $\delta = 1 + 2^{-53}$ and $fl(a) = a.to\_double(\,)$. For the entry in the last row and last column one may assume $ind_B \le ind_A$.

- specify how the approximation $\tilde{E}$ is computed,
- specify how the bound B is computed, and
- prove that $|E - \tilde{E}| \le B$ holds.

In the next section we will describe a variant of the floating point filter used in the rational kernel. In later sections we comment on other filters, we discuss an expression compiler for the automatic generation of floating point filters, and we give theoretical and experimental evidence for the efficacy and efficiency of floating point filters.

9.7.1 *A Floating Point Filter*

We discuss a variant of the filter used in the rational kernel. The filter described here is slightly stronger that the one described in [MN94b, MN94a]. In the current kernel you will find a mixture of both filters. The filter works for expressions with integer operands and operations addition, subtraction, and multiplication. An extension to expressions with real operands and the additional operations division and square root was later devised in [Bur96, Fun97, BFS98].

The approximation $\tilde{E}$ is simply the value obtained by evaluating E with double precision floating point arithmetic.

The bound B is computed according to the rules given in Table 9.1. This table contains the recursive definitions of the index ind_E and the measure mes_E of an expression E; B is defined as

$$B = 2^{-53} \cdot ind_E \cdot mes_E.$$

Before we prove that $\tilde{E}$ and B have the property required for a floating point filter, we apply the filter to the orientation predicate. We obtain:

```
// convert arguments to double
double axd = ax.to_double(), ayd = ay.to_double();
// and similarly for the other coordinates
// evaluate E with floating point arithmetic
double E_tilde = (axd*bwd - bxd*awd) * (ayd*cwd - cyd*awd) -
                 (ayd*bwd - byd*awd) * (axd*cwd - cxd*awd);
// compute mes by replacing all arguments by their absolute
// values and by replacing - by + in E.
double axd = fabs(axd), ayd = fabs(ayd);
// and similarly for the other coordinates
double mes = (axd*bwd + bxd*awd) * (ayd*cwd + cyd*awd) +
                 (ayd*bwd + byd*awd) * (axd*cwd + cxd*awd);
double ind = 11.0;  // see below
double B   =  ind * mes * eps;  // eps = 2^{-53}.
if ( E_tilde >  B ) return  1;
if ( E_tilde < -B ) return -1;
if ( B < 1)         return  0;
// resort to integer arithmetic
return sign((ax*bw-bx*aw)*(ay*cw-cy*aw)-(ay*bw-by*aw)*(ax*cw-cx*aw));
```

Some comments on this program are in order.

(1) How did we compute the index? We have:
The index of an integer a is $s_1 = 1$;
The index of an expression of the form $a \cdot a$ is $s_2 = 1 + s_1(\delta + \delta^2) \approx 3$.
The index of an expression of the form $a \cdot a + a \cdot a$ is $s_3 = 1 + s_2\delta \approx 4$.
The index of an expression of the form $(a \cdot a + a \cdot a) \cdot (a \cdot a + a \cdot a)$ is $s_4 = 1 + s_3(\delta + \delta^2) \approx 9$.
The index of the orientation predicate is $s_5 = 1 + s_4\delta \approx 10$.
s_5 is slightly larger than 10 and certainly less than 11. We may therefore use 11 as the index of the expression predicate. This overestimate of ind_E will also cover any rounding error in the computation of B. Note that we defined B as $2^{-53} \cdot ind_E \cdot mes_E$ but compute $2^{-53} \odot ind_E \odot mes_E$, where $\odot$ denotes floating point multiplication.

(2) The computation of $\tilde{E}$ starts with the conversion of the homogeneous coordinates of a, b, and c from *integer* to *double*. In the rational kernel we make this conversion when the points are constructed. In this way the conversion is made only once for each *rat_point* and not every time a predicate is evaluated for a *rat_point*.

(3) The computation of mes_E involves the same number of arithmetic operations as the computation of $\tilde{E}$. The computation of B requires, in addition, to take the absolute values of the arguments and to multiply ind_E, mes_E, and 2^{-53}. The number of operations to compute B is therefore at least the number of operations to compute $\tilde{E}$. The actual time required to compute $\tilde{E}$ and B is usually less than twice the time to compute $\tilde{E}$ alone (see Section 9.7.4 for some measurements), since modern micro-processors have highly effective floating point units with multiple pipelined arithmetic units and since the cost of arithmetic is small once the data is in the processing unit.

(4) Our expressions have integer operands and operations $+$, $-$, and $\cdot$. Hence E and $\tilde{E}$ are integral.

We will next prove that Table 9.1 indeed defines a valid bound B. We need to review some properties of the IEEE floating point standard [Gol90, Gol91, IEE87].

A floating point number consists of a sign s, a mantissa m, and an exponent e. In double format s has one bit, m consists of fifty-two bits $m_1, \ldots, m_{52}$, and e consists of the remaining eleven bits of a double word. The number represented by the triple (s, m, e) is defined as follows:

- e is interpreted as an integer in $[0\,..\,2^{11} - 1] = [0\,..\,2047]$.
- If $m_1 = \ldots = m_{52} = 0$ and $e = 0$ then the number is $+0$ or -0 depending on s.
- If $1 \le e \le 2046$ then the number is $s \cdot (1 + \sum_{1 \le i \le 52} m_i 2^{-i}) \cdot 2^{e-1023}$.
- If some m_i is non-zero and $e = 0$ then the number is $s \cdot \sum_{1 \le i \le 52} m_i 2^{-i} 2^{-1023}$. This is a so-called denormalized number.
- If all m_i are zero and $e = 2047$ then the number is $+\infty$ or $-\infty$ depending on s.
- In all other cases the triple represents NaN (= not a number).

The largest positive double (except for ∞) is `MAXDOUBLE` $= (2 - 2^{-52}) \cdot 2^{1023}$ and the smallest positive double is `MINDOUBLE` $= 2^{-52} \cdot 2^{-1023}$.

In this section we are interested in *floating point integers*, i.e., integers that can be represented as floating point numbers. The set of floating point integers consists of:

- the number zero,
- all integers of the form $s \cdot (1 + \sum_{1 \le i \le 52} m_i 2^{-i}) \cdot 2^e$ with $0 \le e \le 1023$ (we must have $m_i = 0$ for $i > e$),
- the numbers $+\infty$ and $-\infty$.

We call an integer *representable* if $|a| \le 2 \cdot 2^{1023}$. For a representable integer a, let $fl(a)$ be a floating point number nearest to a. For a non-representable integer let $fl(a) = \pm\infty$ depending on the sign of a.

Floating point arithmetic incurs rounding error. It is therefore important to distinguish between the mathematical operations addition, subtraction, multiplication and their floating point implementations. We use $+$, $-$, and $\cdot$ for the exact operations and $\oplus$, $\ominus$, and $\odot$ for their floating point implementations.

We need the following facts:

(a) If a is an integer then

$$|a - fl(a)| \le 2^{-53} \cdot |fl(a)|,$$

where $eps = 2^{-53}$ is called the *machine precision*. If a is a non-representable integer (including $\pm\infty$) then $fl(a) = \infty$ and the claim is true. So, assume that a is representable. The floating point approximation of a is obtained by "rounding" in the 53-rd bit. More precisely, if $|a| < 2^{53}$ then $fl(a) = a$ and if $|a| \geq 2^{53}$ and a has the binary representation

$$a = s \cdot \sum_{0 \leq i \leq L} m_i \cdot 2^{L-i}$$

with $m_0 = 1$ and $L \geq 53$, then

$$fl(a) = s \cdot (\sum_{0 \leq i \leq 52} m_i \cdot 2^{L-i} + \delta \cdot 2^{L-52}),$$

where $\delta \in \{0, 1\}$ is chosen such that the better approximation of a is obtained. Clearly, $|a - fl(a)| \leq 2^{L-52}/2$ and $|fl(a)| \geq 2^L$. Thus, $|a - fl(a)| \leq 2^{-53} \cdot |fl(a)|$.

We want to remark that the assumption that a is integer is crucial for claim (a). If $|a| \leq$ *MinDouble*$/2$, the best floating point approximation of a is zero. Thus, there is no bound on the error $|a - fl(a)|$ in terms of $fl(a)$. Life is easier for integers.

(b) If a is an integer then $fl(a)$ is a floating point integer.

(c) If f_1 and f_2 are floating point integers, op $\in \{+, -, \cdot\}$, $f = f_1$ op f_2, and $\widetilde{\text{op}}$ is the floating point implementation of op, then

$$f_1 \widetilde{\text{op}} f_2 = fl(f),$$

i.e., the floating point operation returns a floating point integer closest to f. There is no need to argue here. It is an "axiom" of the IEEE standard that every arithmetic operation is implemented with the least possible error.

(d) Under the same hypothesis as in the preceding item:

$$|f_1 \widetilde{\text{op}} f_2 - f_1 \text{ op } f_2| \leq 2^{-53} |f_1 \widetilde{op} f_2|.$$

Let $\tilde{f} = f_1 \widetilde{\text{op}} f_2$ and $f = f_1$ op f_2. Then $\tilde{f} = fl(f)$ by (c) and hence $|\tilde{f} - f| \leq 2^{-53} |\tilde{f}|$ by part (a).

(e) If f is an *integer* then *a.to_double*( ) returns $fl(a)$. That is the way we implemented the function *to_double*.

(f) Floating point arithmetic is monotone, i.e., if $a_1 \leq a_2$ and $b_1 \leq b_2$ then $a_1 \oplus a_2 \leq b_1 \oplus b_2$ and if $0 \leq a_1 \leq a_2$ and $0 \leq b_1 \leq b_2$ then $a_1 \odot a_2 \leq b_1 \odot b_2$.

(g) Multiplication by a power of two incurs no rounding error, i.e., if a is a power of two and b is a floating point integer such that $2a$ and $a \cdot b$ are representable, then $a \oplus a = 2 \cdot a$ and $a \odot b = a \cdot b$.

Theorem 13 *If mes_E and ind_E are computed according to Table 9.1 then $|\tilde{E}| \leq mes_E$ and*

$$\begin{aligned} |\tilde{E} - E| &\leq 2^{-53} \cdot ind_E \cdot mes_E \\ &\leq 2^{-53} \odot ind_E \odot mes_E \odot (1 + 2^{-52}). \end{aligned}$$

Proof We use induction on the structure of the expression E. The claim $|\tilde{E}| \le mes_E$ follows immediately from the monotonicity of floating point arithmetic. For the other claims we have to work slightly harder. We first prove

$$|\tilde{E} - E| \le 2^{-53} \cdot ind_E \cdot mes_E.$$

Assume first that E is an integer a. Then

$$|a - fl(a)| \le 2^{-53} \cdot |fl(a)|$$

by item (a) and the claim is certainly true. If a is a floating point integer then $fl(a) = a$ and hence the index can be set to zero for floating point integers.

We come to the induction step. Let A and B be the two subexpressions of E and let $\tilde{A}$ and $\tilde{B}$ be their floating point values. Then

$$\begin{aligned} |\tilde{A}| &\le mes_A \\ |\tilde{A} - A| &\le 2^{-53} \cdot ind_A \cdot mes_A \\ |\tilde{B}| &\le mes_B \\ |\tilde{B} - B| &\le 2^{-53} \cdot ind_B \cdot mes_B \end{aligned}$$

by induction hypothesis.

We now make a case distinction according to the operation combining A and B.

Assume $E = A + B$. Then

$$|\tilde{E} - E| = |\tilde{A} \oplus \tilde{B} - (A + B)| \le |\tilde{A} \oplus \tilde{B} - (\tilde{A} + \tilde{B})| + |\tilde{A} - A| + |\tilde{B} - B|.$$

Item (d) with $f_1 = \tilde{A}$ and $f_2 = \tilde{B}$ implies that the first term is bounded by $2^{-53}|\tilde{A} \oplus \tilde{B}|$ and monotonicity of floating point arithmetic implies that

$$|\tilde{A} \oplus \tilde{B}| \le mes_A \oplus mes_B = mes_E.$$

For the other two terms we use the induction hypothesis to conclude

$$\begin{aligned} |\tilde{A} - A| + |\tilde{B} - B| &\le 2^{-53} \cdot (ind_A \cdot mes_A + ind_B \cdot mes_B) \\ &\le 2^{-53} \cdot \max(ind_A, ind_B) \cdot (mes_A + mes_B) \\ &\le 2^{-53} \cdot \max(ind_A, ind_B) \cdot (1 + 2^{-53}) \cdot (mes_A \oplus mes_B) \\ &= 2^{-53} \cdot \max(ind_A, ind_B) \cdot (1 + 2^{-53}) \cdot mes_E. \end{aligned}$$

Putting the two bounds together completes the induction step for the case of an addition. The argument for subtractions is completely analogous.

We turn to multiplications, $E = A \cdot B$. We have

$$|\tilde{E} - E| = |\tilde{A} \odot \tilde{B} - A \cdot B| \le |\tilde{A} \odot \tilde{B} - \tilde{A} \cdot \tilde{B}| + |\tilde{A} \cdot \tilde{B} - A \cdot \tilde{B}| + |A \cdot \tilde{B} - A \cdot B|.$$

Item (d) with $f_1 = \tilde{A}$ and $f_2 = \tilde{B}$ implies that the first term is bounded by $2^{-53}|\tilde{A} \odot \tilde{B}|$ and monotonicity of floating point arithmetic implies that

$$|\tilde{A} \odot \tilde{B}| \le mes_A \odot mes_B = mes_E.$$

For the second term we use the induction hypothesis to conclude

$$\begin{aligned}
|\tilde{A}\cdot\tilde{B} - A\cdot\tilde{B}| &= |\tilde{A} - A|\cdot|\tilde{B}| \\
&\le 2^{-53}\cdot ind_A\cdot mes_A\cdot mes_B \\
&\le 2^{-53}\cdot ind_A\cdot(1+2^{-53})\cdot(mes_A \odot mes_B) \\
&= 2^{-53}\cdot ind_A\cdot(1+2^{-53})\cdot mes_E,
\end{aligned}$$

and for the third term we conclude analogously

$$\begin{aligned}
|A\cdot\tilde{B} - A\cdot B| &= |A|\cdot|\tilde{B} - B| \\
&\le (1+2^{-53})\cdot|\tilde{A}|\cdot 2^{-53}\cdot ind_B\cdot mes_B \\
&\le (1+2^{-53})\cdot mes_A\cdot 2^{-53}\cdot ind_B\cdot mes_B \\
&\le 2^{-53}\cdot ind_B\cdot(1+2^{-53})^2(mes_A \odot mes_B) \\
&= 2^{-53}\cdot ind_B\cdot(1+2^{-53})^2\cdot mes_E.
\end{aligned}$$

Putting the three bounds together completes the induction step for the case of a multiplication.

It remains to prove the inequality

$$2^{-53}\cdot ind_E\cdot mes_E \le 2^{-53}\odot ind_E \odot mes_E \odot (1+2^{-52}).$$

It follows from

$$ind_E\cdot mes_E \le (ind_E \odot mes_E)\cdot(1+2^{-53}) \le ind_E \odot mes_E \odot (1+2^{-52})$$

and the fact that the multiplication by 2^{-53} incurs no rounding error. □

9.7.2 *Alternative Filters*

We discuss the filter originally (and still mostly) used in the kernel, static and dynamic filters, special methods for determinants, and specialized arithmetics.

The Filter Used Originally in the Kernel: In our original filter we computed ind_E and mes_E according to Table 9.2. In this table we also define a quantity P_E. P_E is a power of two with $|E| \le P_E$, $|\tilde{E}| \le P_E$, and $P_E \le mes_E$. The bound $B(E)$ is defined as

$$B = 2^{-53} \odot ind_E \odot mes_E.$$

In order to see that this bound is correct one proves that

$$|E - \tilde{E}| \le 2^{-53}\cdot ind_E\cdot P_E \quad\text{and}\quad P_E \le mes_E$$

and observes that

$$2^{-53}\cdot ind_E\cdot P_E = 2^{-53} \odot ind_E \odot P_E \le 2^{-53} \odot ind_E \odot mes_E,$$

since 2^{-53} and P_E are powers of two and since floating point arithmetic is monotonic.

The inequality

$$|E - \tilde{E}| \leq 2^{-53} \cdot ind_E \cdot P_E$$

is again shown by induction on the structure of E. The base case is obvious. The induction steps are as follows.

In the case of an addition we have

$$\begin{aligned} |E - \tilde{E}| &= |\tilde{A} \oplus \tilde{B} - (A + B)| = |\tilde{A} \oplus \tilde{B} - (\tilde{A} + \tilde{B})| + |\tilde{A} - A| + |\tilde{B} - B| \\ &\leq 2^{-53}(|\tilde{A} \oplus \tilde{B}| + ind_A P_A + ind_B P_B) \\ &\leq 2^{-53}(P_A \oplus P_B + (ind_A + ind_B)\max(P_A, P_B)) \\ &\leq 2^{-53}(1 + (ind_A + ind_B)/2) \cdot 2 \cdot \max(P_A, P_B)), \end{aligned}$$

where the last inequality follows from

$$\begin{aligned} P_A \oplus P_B &\leq \max(P_A, P_B) \oplus \max(P_A, P_B) \\ &= \max(P_A, P_B) + \max(P_A, P_B) = 2 \cdot \max(P_A, P_B). \end{aligned}$$

In the case of multiplication we have

$$\begin{aligned} |E - \tilde{E}| &= |\tilde{A} \odot \tilde{B} - \tilde{A} \cdot \tilde{B}| + |\tilde{A}| \cdot |\tilde{B} - B| + |B| \cdot |\tilde{A} - A| \\ &\leq 2^{-53}(|\tilde{A} \odot \tilde{B}| + |\tilde{A}||\tilde{B} - B| + |B||\tilde{A} - A|) \\ &\leq 2^{-53}(P_A \odot P_B + P_A \cdot ind_B \cdot P_B + P_B \cdot ind_A \cdot P_A) \\ &\leq 2^{-53}(1 + ind_A + ind_B) \cdot P_A \cdot P_B. \end{aligned}$$

The inequality $P_E \leq mes_E$ is also shown by induction on the structure of E. We leave the induction step to the reader. For the basis of the induction we observe that $2^{\log |a|} \leq 2 \cdot fl(a) = mes_a$ for an integer a.

This concludes the proof that Table 9.2 defines a filter.

For the orientation predicate Table 9.2 gives an index of 5 and a measure of $8 \cdot M$, where M is the measure according to Table 9.1. Thus $B = 40 \cdot M$. Table 9.1 gives $B = 11 \cdot M$, which is significantly better.

Static Filters: Fortune and van Wyk [FvW96] invented the idea of a floating point filter. They proposed a static filter in which B is precomputed completely. Assume that it is known a priori that $|a| \leq 2^L$ for all integer arguments of an expression E. Then $mes_a \leq 2^L$ for all arguments a and we may *precompute* mes_E by replacing mes_a by 2^L for all arguments a. This yields $B = 2^{-53} \cdot 11 \cdot 2^{4L+3}$ with Table 9.1. The filter of Fortune and van Wyk is called *static* because B is precomputed entirely. In contrast, the filter used in the rational kernel precomputes ind_E but computes mes_E on the fly. Such a filter may be called *semi-dynamic*.

Static filters are faster than semi-dynamic filters, but they are less precise and they are less convenient to use. For example, they cannot be used at all in an on-line algorithm, where no a priori bound on the size of the arguments is known. We decided against static filters because of their less convenient use.

E	$\tilde{E}$	P_E	mes_E	ind_E
a, integer	$fl(a)$	$2^{\lceil \log \lvert a \rvert \rceil}$	$2\lvert fl(a) \rvert$	1
a, float integer	$fl(a)$	$2^{\lceil \log \lvert a \rvert \rceil}$	$2\lvert fl(a) \rvert$	0
$A + B$	$\tilde{A} \oplus \tilde{B}$	$2\max(P_A, P_B)$	$2(mes_A \oplus mes_B)$	$1 + (ind_A + ind_B)/2$
$A - B$	$\tilde{A} \ominus \tilde{B}$	$2\max(P_A, P_B)$	$2(mes_A \oplus mes_B)$	$1 + (ind_A + ind_B)/2$
$A \cdot B$	$\tilde{A} \odot \tilde{B}$	$P_A P_B$	$mes_A \odot mes_B$	$1 + ind_A + ind_B$

Table 9.2 The recursive definition of mes_E and ind_E in the original filter. P_E is a power of two with $|E| \le P_E$, $|\tilde{E}| \le P_E$, and $P_E \le mes_E$; it is only needed for the correctness proof of the filter. We set $2^{\lceil \log 0 \rceil} = 0$.

Dynamic Filters: Consider the expression

$$E = (a + b) - a$$

when a and b are float integers and $a \gg b$. The semi-dynamic filter of Section 9.7.1 assumes that the error in the subtraction may be as large as

$$2^{-53} mes_E \approx 2^{-53}(2a + b).$$

However, the actual error is approximately

$$2^{-53} \cdot \tilde{E} \approx 2^{-53} \cdot b,$$

which is much smaller.

Dynamic filters attempt to exploit this differency by estimating the round-off error more carefully. They use the formulae

$$\begin{aligned} |\tilde{A} \oplus \tilde{B} - (A + B)| &\le |\tilde{A} \oplus \tilde{B} - (\tilde{A} + \tilde{B})| + |\tilde{A} - A| + |\tilde{B} - B| \\ &\le 2^{-53}|\tilde{A} \oplus \tilde{B}| + |\tilde{A} - A| + |\tilde{B} - B| \end{aligned}$$

and

$$\begin{aligned} |\tilde{A} \odot \tilde{B} - A \cdot B| &= |\tilde{A} \odot \tilde{B} - \tilde{A} \cdot \tilde{B} + \tilde{A} \cdot \tilde{B} - A \cdot \tilde{B} + A \cdot \tilde{B} - A \cdot B| \\ &\le 2^{-53}|\tilde{A} \odot \tilde{B}| + |\tilde{A} - A| \cdot |\tilde{B}| + |A||\tilde{B} - B| \end{aligned}$$

to recursively compute a bound on the error. More precisely, in the case of an addition the error err_E for the expression E is computed as

$$err_e = (2^{-53} \odot |\tilde{E}| \oplus err_A \oplus err_B) \odot (1 + 2^{-51}),$$

where the multiplication by $1 + 2^{-51}$ accounts for the error in the computation of the error bound. We leave it to the reader to derive the corresponding formula for multiplication.

Dynamic filters are more costly but also more precise than semi-dynamic filters. Observe that the computation of err_E in the case of an addition requires two additions and two multiplications. The computation of mes_E requires only one addition. We concluded from our experiments in [MN94b] that the additional cost is not warranted for the rational kernel.

We do use dynamic filters in the number type *real*, see Section 4.4, since the cost of exact computation is very high for *reals* and hence a higher computation time for the filter is justified.

Determinants: Many geometric predicates, e.g., the orientation and the insphere predicates, are naturally formulated as the sign of a determinant. The efficient computation of the signs of determinants has therefore received special attention [Cla92, ABDP97, BEPP97]. None of the methods is available in LEDA.

Specialized Arithmetics: Consider again the orientation predicate

```
sign((ax*bw-bx*aw)*(ay*cw-cy*aw) - (ay*bw-by*aw)*(ax*cw-cx*aw) )
```

and assume that it is known that the absolute value of all arguments is less than 2^L. The arguments are assumed to be integer. It is then easy to compute an a priori bound on the maximal number of binary digits required for any of the intermediate results. We have:
The integer a requires L bits;
An expression of the form $a \cdot a$ requires $2L$ bits.
An expression of the form $a \cdot a + a \cdot a$ requires $2L + 1$ bits.
An expression of the form $(a \cdot a + a \cdot a) \cdot (a \cdot a + a \cdot a)$ requires $4L + 2$ bits.
The orientation predicate requires at most $4L + 3$ bits.
Given this knowledge one could try to optimize the arithmetic, i.e., instead of using a general purpose package for the computation with arbitrary precision integers (such as the class *integer*) one could design integer arithmetic optimized for a particular bit length. This avenue is taken in [FvW96, She97].

9.7.3 *Expression Compilers*

The incorporation of the floating point filter into the rational kernels was a tedious task; it was done to a large extent by Ulrike Bartuschka. For each predicate she had to derive manually the formulae for ind_E and mes_E. For example, the code for the orientation test contains the following comment:

```
-------------------------------------------------------------------------
 ERROR BOUNDS
-------------------------------------------------------------------------
 mes(E) = 2*(mes(aybw-byaw)*mes(axcw-cxaw) + mes(axbw-bxaw)*mes(aycw-cyaw))
        = 2*(4*(fabs(aybw)+fabs(byaw)) * (fabs(axcw)+fabs(cxaw)) +
             4*(fabs(axbw)+fabs(bxaw)) * (fabs(aycw)+fabs(cyaw)))
        = 8*((fabs(aybw)+fabs(byaw)) * (fabs(axcw)+fabs(cxaw)) +
             (fabs(axbw)+fabs(bxaw)) * (fabs(aycw)+fabs(cyaw)))

 ind(E) = ((ind(aybw-byaw) + ind(axcw-cxaw) +0.5) +
           (ind(axbw-bxaw) + ind(aycw-cyaw) +0.5) + 1 ) / 2
```

```
        = (4.5 + 4.5 + 1) / 2  =  5

 eps(E) = ind(E) * mes(E) * eps0
        = 40 * ((fabs(aybw)+fabs(byaw))*(fabs(axcw)-fabs(cxaw)) +
                (fabs(axbw)-fabs(bxaw))*(fabs(aycw)-fabs(cyaw))) * eps0;
----------------------------------------------------------------------------
```

Already Fortune and Wyk [FvW96] observed that the generation of the filters can be automated. Stefan Funke [Fun97, BFS98] adopted the idea for LEDA and generalized it to a larger class of expressions and number types. His expression compiler generates floating point filters automatically from suitably decorated expressions. For example, in order to generate a filter for the orientation predicate one writes

```
int orientation(const rat_point& a, const rat_point& b,
                                    const rat_point& c)
{ int res_sign;
BEGIN_PREDICATE
{
DECLARE_ATTRIBUTES integer_type FOR a.X() a.Y() a.W() b.X()
                               b.Y() b.W() c.X() c.Y() c.W();
  integer AX=a.X(); integer AY=a.Y(); integer AW=a.W();
  integer BX=b.X(); integer BY=b.Y(); integer BW=b.W();
  integer CX=c.X(); integer CY=c.Y(); integer CW=c.W();
  integer D= (AX*BW-BX*AW) * (AY*CW-CY*AW) -
           (AY*BW-BY*AW) * (AX*CW-CX*AW);
  res_sign=sign(D);
}
END_PREDICATE
  return res_sign;
}
```

The expression compiler produces a (very lengthy) program of the following form.

```
int orientation(const rat_point& a, const rat_point& b,
                                    const rat_point& c)
{ int res_sign;
{
  /* a floating point evaluation of the predicate which assigns
     one of -1, 0, +1, NO_IDEA to res_sign   */
  if (res_sign == NO_IDEA)
  { /* exact evaluation of predicate with result in res_sign */
  }
}
 return res_sign;
}
```

The expression compiler is available as an LEP.

9.7.4 *Efficacy and Efficiency of Filters*

We discuss the efficacy and the efficiency of floating point filters. Efficacy refers to the percentage of tests, for which the filter is able to deduce the sign of the test, and efficiency refers to the cost of the evaluation of the filter and the relationship of this cost to the cost of a computation with integers.

A floating point filter for an expression E computes an approximation $\tilde{E}$ of E and a bound B for the maximal difference between the approximation and the exact value. The following lemma is trivial but useful.

Lemma 59 *If E and $\tilde{E}$ are integral and $B < 1$ then $sign(\tilde{E}) = sign(E)$.*

Under what conditions can we claim that $B < 1$ without actually computing it? Consider the orientation predicate for points with integer homogeneous coordinates $(x, y, 1)$ with $|x|, |y| \leq 2^L$. We assume that L is small enough such that the coordinates are floating point integers. The orientation predicate for points a, b, and c is given by the expression

```
E = (AX - BX) * (AY - CY) - (AY - BY) * (AX - CX)
```

and hence $B \leq 8 \cdot 2^{-53} \cdot 2^{2L+3}$ according to Theorem 13; the index of the expression is 7 when computed with $\delta = 1$. We rounded up to 8 to account for the fact that $\delta = 1 + 2^{-53}$.

We have $8 \cdot 2^{-53} \cdot 2^{2L+3} < 1$ iff $3 - 53 + 2L + 3 < 0$ iff $L < 47/2$. We conclude that double precision floating point arithmetic is guaranteed to give the correct result if the x- and y-coordinates are at most 2^{23}.

What happens if L is larger? The floating point computation is able to deduce the sign of E if $|\tilde{E}| > B$. Since E is twice the signed area (see Lemma 56) of the triangle with vertices (a, b, c), the floating point computation is able to deduce the correct sign for any triple of points which span a triangle whose area is at least $8 \cdot 2^{-53} \cdot 2^{2L+3}/2$. Devillers and Preparata [DP98] have shown that for a random triple of points and for L going to infinity, the probability that the area of the spanned triangle is at least $8 \cdot 2^{-53} \cdot 2^{2L+3}/2$ goes to one. Thus for large L and for triples of random points, the floating point computation will almost always be able to deduce the sign of E and exact computation will be rarely needed.

Observe that the result cited in the previous paragraph depends crucially on the fact that the points are chosen randomly. In an actual computation orientation tests will not be performed for random triples of points even if the input consists of random points. It is therefore not clear what the result says about actual computations.

The class *rat_point* has a static member function *print_statistics* which gives information about the efficacy of its floating point filter. The call

```
rat_point::print_statistics();
```

prints a statistic of the following form:

```
compare:            167 / 44330    (0.38 %)
orientation:         71 / 48975    (0.14 %)
side of circle:    3194 / 22317    (14.31 %)
```

The statistic states for each of the functions *compare*, *orientation*, and *side_of_circle* how many times it was evaluated and how many times the filter failed and an exact computation was necessary. In this particular execution, 22317 side of circle tests were performed out of which 3194 required exact computation. This amounts to 14.31 percent.

Table 9.3 shows the results of a more substantial experiment. The table was generated by the program below. We first generate a list *L0* of *n* random points either on the unit circle or in the unit square. We then construct a list *L1* of points whose homogeneous coordinates are *d* bit binary numbers for different values of *d* by truncating the Cartesian coordinates to *d* bits; for $d = 60$ no truncation takes place (this is indicated by the infinity-sign in Table 9.3. We construct the Delaunay diagram for the points in *L1*.

⟨*produce efficacy of filter table*⟩≡

```
int n = 10000;
list<rat_point> L0;
for (int k = 0; k < 2; k++)
{ if ( k == 0 ) random_points_on_unit_circle(n,L0);
  else          random_points_in_unit_square(n,L0);
  for (int d = 8; d <= 60; d += d < 12 ? 2 : 10)
  { list<rat_point> L1;
    rat_point p;
    I.write_table("\n");
    if ( d <= 50 )
    { double D = ldexp(1,d);
      forall(p,L0) L1.append(rat_point(integer(p.xcoordD()*D),
                                       integer(p.ycoordD()*D),1));
      I.write_table("",d);
    }
    else
    { L1 = L0;
      I.write_table("$ \\infty $");
    }
    ⟨reset counters to zero⟩
    GRAPH<rat_point,int>  DT;
    DELAUNAY_TRIANG(L1,DT);
    ⟨write a line of the table⟩
  }
  I.write_table(" \\hline");
}
```

For each experiment we generate one line in Table 9.3. The class *rat_point* has static data members that keep a count of the number of compare, orientation, and side of circle tests performed and also of the number of tests where the filter fails. Before each experiment we set the counters to zero. After each experiment we print a line of the table.

⟨*reset counters to zero*⟩≡

```
rat_point::cmp_count = 0;
rat_point::exact_cmp_count = 0;
rat_point::orient_count = 0;
rat_point::exact_orient_count = 0;
rat_point::soc_count = 0;
rat_point::exact_soc_count = 0;
```

		Compare			Orientation			Side of circle		
d	N	number	exact	%	number	exact	%	number	exact	%
8	1883	157814	0	0.00	19909	0	0.00	7242	0	0.00
10	5298	187379	0	0.00	58263	0	0.00	20736	5743	27.70
12	8383	216679	0	0.00	89307	0	0.00	35931	24693	68.72
22	9999	230556	0	0.00	98899	0	0.00	46410	42454	91.48
32	9999	231656	0	0.00	90664	137	0.15	40003	39797	99.49
42	9999	231665	0	0.00	91205	152	0.17	40083	40083	100.00
∞	9999	231665	125	0.05	44279	87	0.20	13082	13082	100.00
8	9267	230060	0	0.00	130431	0	0.00	64176	0	0.00
10	9953	236690	0	0.00	147814	0	0.00	77409	136	0.18
12	9996	236661	0	0.00	149233	0	0.00	78693	105	0.13
22	10000	235727	0	0.00	149057	0	0.00	78695	113	0.14
32	10000	235729	0	0.00	149059	0	0.00	78695	115	0.15
42	10000	235729	0	0.00	149059	0	0.00	78695	115	0.15
∞	10000	235729	574	0.24	149059	0	0.00	78695	115	0.15

Table 9.3 Efficacy of floating point filter: The top part contains the results for random points on the unit circle and the lower part contains the results for random points in the unit square. In each case we generated 10000 points. The first column shows the precision (= number of binary places) used for the homogeneous coordinates of the points, the second column contains the number of distinct points in the input. The other columns contain the number of tests, the number of exact tests, and the percentage of exact tests performed for the compare, the orientation, and the side of circle primitive.

Table 9.3 confirms the theoretical considerations from the beginning of the section. For each test there is a value of d below which the floating point computation is able to decide all tests. For the orientation test this value of d is somewhere between 22 and 32 (we argued above that the value is $47/2$) and for the side of circle test the value is somewhere between 8 and 10 (we ask the reader in the exercises to compute the exact value). Also, the percentage of the tests, where the filter fails, is essentially an increasing function of d.

The compare, orientation, and side of circle functions seem to be tests of increasing difficulty. This is easily explained. The compare function decides the sign of a linear function of the Cartesian coordinates of two points, the orientation function decides the sign of a quadratic function of the Cartesian coordinates of three points, and the side of circle function decides the sign of a polynomial of degree four in the Cartesian coordinates of four points. The larger the degree of the polynomial of the test, the larger the arithmetic demand of the test.

Among the two sets of inputs, the random points on the unit circle are much more difficult than the random points in the unit square, in particular, for the side of circle test. Again this is easily explained.

For the side of circle test, four almost co-circular points or four exactly co-circular points are the most difficult input, and for sufficiently large d the situation that $|\tilde{E}| \leq B$ and $B > 1$ arises frequently. Points on (or near) the unit circle cause no particular difficulty for the compare and the orientation function. Points on (or near) a segment would prove to be difficult for the orientation test.

For random points in the unit square the filter is highly effective for all three tests; the filter fails only for a very small percentage of the tests.

We turn to the question of how much a filter saves with respect to running time. Table 9.4 was produced by the following program.

⟨*produce efficiency of filter table*⟩≡

```
forall(p,L1) Lf.append(p.to_point());

GRAPH<rat_point,int>  DT;
GRAPH<rat_point,int>  DT_no_filter;
GRAPH<     point,int>  DT_FK;

float T = used_time();
DELAUNAY_TRIANG(Lf,DT_FK);
I.write_table(" & ", used_time(T));
```

⟨*efficiency table: check correctness of float computation*⟩

```
used_time(T);  // to set the timer
DELAUNAY_TRIANG(L1,DT);
I.write_table(" & ", used_time(T));

rat_point::use_filter = 0;
DELAUNAY_TRIANG(L1,DT_no_filter);
I.write_table(" & ", used_time(T));
rat_point::use_filter = 1;
```

We generated the same list *L1* of *rat_points* as above. We then converted each *rat_point* to a *point* to obtain a list *Lf* of *points*. Finally, we computed the Delaunay triangulation in three different ways: first with the floating point kernel, then with the rational kernel, and finally with the rational kernel without its floating point filter. The class *rat_point* has a static variable *use_filter* which controls the use of the floating point filter.

Table 9.4 has to be interpreted with care. Let us first inspect the individual columns.

The running time with the floating point kernel does not increase with the precision of the input. Observe, that for $d < 22$ and points on the unit circle, the input contains a significant fraction of multiple points (see the second column of Table 9.3) and hence the first three lines really refer to simpler problem instances. For $d \geq 22$ and points on the unit circle and for $d \geq 10$ and points in the unit square the input contains almost no multiple points and the running times are independent of the precision. The computation with the floating point

d	Float kernel	Rational kernel	RK without filter
8	0.73	1.12	4.35
10	1.3	2.43	7.8
12	1.85	5.09	11.18
22	2.17	7.93	14.4
32	2.02	7.79	13.29
42	2.01	8.32	15.46
∞	2*	5.09	9.19
8	2.58	3.59	16.33
10	2.8	3.98	18.36
12	2.83	4.04	18.63
22	2.82	4.02	20.51
32	2.86	3.96	20.77
42	2.83	4.01	26.02
∞	2.83	3.99	33.2

Table 9.4 Efficiency of the floating point filter: The top part contains the results for random points on the unit circle and the lower part contains the results for random points in the unit square. The first column shows the precision (= number of binary places) used for the Cartesian coordinates of the points. The other columns show the running time with the floating point filter, with the rational kernel with the floating point filter, and with the rational kernel without its floating point filter. A star in the second column indicates that the computation with the floating point kernel produced an incorrect result. geometry kernels!running time

kernel is not guaranteed to give the correct result. In fact, it produced an incorrect result in one of the experiments (indicated by a *). We come back to this point below.

The running time with the rational kernel and no filter increases sharply as a function of the precision. This is due to the fact that larger precision means larger integers and hence larger computation time for the integer arithmetic. We see one exception in the table. For points on the unit circle the computation on the exact points is faster than the computation with the rounded points. The explanation can be found in Table 9.3. The number of tests performed is much smaller for exact inputs than for rounded inputs. Observe, that for points that lie exactly on a circle any triangulation is Delaunay.

The running time for the rational kernel (with the filter) increases only slightly for the second set of inputs and increases more pronouncedly for the points on the unit circle. This is to be expected because the filter fails more often for the points on the unit circle.

Let us next compare columns.

The comparison between the last two columns shows the efficiency gained by the floating point filter. The gains are impressive, in particular, for the easier set of inputs. For random points in the unit square, the computation without the filter is between five and almost ten times slower. For random points on a unit circle the gain is less impressive, but still substantial. The running time without the filter is between two and five times higher than with the filter.

The comparison between the second and the third column shows what we might gain by further improving our filter technology. For our easier set of inputs the computation with the rational kernel is about 50% slower than the computation with the floating point kernel. This increase in running time stems from the computation of the error bound B in the filter. For our harder set of inputs the difference between the rational kernel and the floating point kernel is more pronounced. This is to be expected since the rational kernel resorts to exact computation more frequently for the harder inputs. The floating point kernel produced the incorrect result in one of the experiments.

We used the following piece of code to check the correctness of the computation with the floating point kernel. We make a copy *DT_FK1* of the graph computed with the floating point kernel, in which every *point* is converted to a *rat_point*. This conversion is without loss of precision. We then check whether the copy is a Delaunay triangulation; the check is discussed in Section 10.4.3. The check is executed with the rational kernel and is therefore exact.

⟨*efficiency table: check correctness of float computation*⟩≡

```
GRAPH<rat_point,int> DT_FK1;
node v; edge e;
node_array<node> copy_of(DT_FK);
forall_nodes(v,DT_FK) copy_of[v] = DT_FK1.new_node(rat_point(DT_FK[v]));
forall_nodes(v,DT_FK)
  forall_adj_edges(e,v)
    DT_FK1.new_edge(copy_of[v],copy_of[DT_FK.target(e)],DT_FK[e]);
DT_FK1.make_map();
if ( !Is_Delaunay_Triangulation(DT_FK1,NEAREST) ) I.write_table("$^*$");
```

We were very surprised when we first saw Table 9.4. We expected that the floating point computation would fail more often, not only when the full 52 bits are used to represent Cartesian coordinates of points. After all, the rational kernel resorts to integer arithmetic most of the time already for much smaller coordinate length and the difficult set of inputs.

We generated Table 9.5 to gain more insight[8]. It gives more detailed information for d ranging from 43 to 52. For our difficult inputs the floating point computation fails when d

[8] While writing this section, our work was very much guided by experiments. We had a theory of what floating point filters can do. Based on this theory we had certain expectations about the behavior of filters. We made experiments to confirm our intuition. In some cases the experiments contradicted our intuition and we had to revise the theory.

d	43	44	45	46	47	48	49	50	51	52
diff	C	C	C	F	F	F	F	F	F	F
easy	C	C	C	C	C	C	C	C	C	C

Table 9.5 Correctness of floating point computation: A detailed view for d ranging from 43 to 52. The second row corresponds to points on the unit circle and the last row corresponds to points in the unit square. A "C" indicates that the computation produced the correct result and a "F" indicates that a incorrect result was produced.

is 46 or larger and for our easy inputs it never fails. For $d < 45$ and both sets of inputs it produces the correct result. Our theoretical considerations give a guarantee only for $d < 10$.

In the remainder of this section we try to explain this discrepancy. We find the explanation interesting[9] but do not know at present whether it has any consequences for the design of floating point filters.

Let $D = 2^d$ and consider four points a, b, c, and d on the unit circle[10]. We use points a', b', c', and d' with integer Cartesian coordinates $\lfloor a_x D \rfloor$, $\lfloor a_y D \rfloor$, … . The side of circle function is the sign of the determinant

$$\begin{vmatrix} 1 & 1 & 1 & 1 \\ a_x & b_x & c_x & d_x \\ a_y & b_y & c_y & d_y \\ a_x^2 + a_y^2 & b_x^2 + b_y^2 & c_x^2 + c_y^2 & d_x^2 + d_y^2 \end{vmatrix}$$

as will be shown in Section 10.9. The value of this determinant is a homogeneous fourth degree polynomial $p(a_x, a_y, \ldots)$. We need to determine the sign of $p(a'_x, a'_y, \ldots)$. Let us relate $p(a_x, a_y, \ldots)$ and $p(a'_x, a'_y, \ldots)$.

We have

$$a'_x = \lfloor a_x D \rfloor = a_x D + \delta_{a_x},$$

where $-1 < \delta_{a_x} \le 0$, and analogous equalities hold for the other coordinates. Thus

$$\begin{aligned} p(a'_x, a'_y, \ldots) &= p(a_x D + \delta_{a_x}, a_y D + \delta_{a_y}, \ldots) \\ &= p(a_x D, a_y D, \ldots) + q_3(a_x D, \delta_{a_x}, a_y D, \delta_{a_y}, \ldots) \\ &\quad + q_2(a_x D, \delta_{a_x}, a_y D, \delta_{a_y}, \ldots) + q_1(a_x D, \delta_{a_x}, a_y D, \delta_{a_y}, \ldots) \\ &\quad + q_0(a_x D, \delta_{a_x}, a_y D, \delta_{a_y}, \ldots), \end{aligned}$$

where q_i has degree i in the $a_x D$, $a_y D$, … and degree $4 - i$ in the δ_{a_x}, δ_{a_y}, … . Since the four points a, b, c, and d are co-circular, we have

$$p(a_x D, a_y D, \ldots) = D^4 p(a_x, a_y, \ldots) = 0.$$

9 We all know from our physics classes that the important experiments are the ones that require a new explanation.

10 In the final round of proof-reading we noticed that we use d with two meanings. In the sequel d is a point, except in the final sentence of the section.

Up to this point our argumentation was rigorous. From now on we give only plausibility arguments. Since the values $a_x D$ may be as large as D and since the values δ_{a_x} are smaller than one, the sign of $p(a'_x, a'_y, \ldots)$ is likely to be determined by the sign of q_3. Since q_3 is a third degree polynomial in the $a_x D$ we might expect its value to be about $f \cdot D^3$ for some constant f. The constant f is smaller than one but not much smaller. Expansion of the side of circle determinant shows that the coefficient of δ_{a_x} in q_3 is equal to

$$\begin{vmatrix} 1 & 1 & 1 \\ b_y D & c_y D & d_y D \\ (b_x^2 + b_y^2) \cdot D^2 & (c_x^2 + c_y^2) \cdot D^2 & (d_x^2 + d_y^2) \cdot D^2 \end{vmatrix} = D^3(c_y - a_y - b_y),$$

where we used the fact that $p_x^2 + p_y^2 = 1$ for a point p on the unit circle. We conclude that f has the same order as the y-coordinate of a random point on the unit circle and hence $f \approx 1/2$.

We evaluate $p(a'_x, a'_y, \ldots)$ with floating point arithmetic. By Theorem 13, the maximal error in the computation of p is $g \cdot D^4 \cdot 2^{-53}$ for some constant g; the actual error will be less. The argument in the proof of Lemma 60 shows that $g \leq 2^8$. Thus we might expect that the floating point evaluation of $p(a'_x, a'_y, \ldots)$ gives the correct sign as long as $g \cdot D^4 \cdot 2^{-53} < f \cdot D^3$ or $d < 53 - \log g + \log f \approx 53 - 8 - 1 = 44$. This agrees quite well with Table 9.5.

9.7.5 *Conclusion*

We discussed the floating point filter in the rational kernel. We have seen that floating point filters give an exact implementation of geometric primitives at a reasonable cost.

Exercises for 9.7

1 The side of circle predicate determines for a four tuple (a, b, c, d) of points, whether d lies to the left, on, or to the right of the circle defined by the first three points. Derive a formula for the side of circle predicate for points given by Cartesian coordinates and for points given by homogeneous coordinates.

2 (Continuation) Derive a filter for both versions of the side of circle predicate according to Tables 9.1 and 9.2. Compare your results with the implementation of the side of circle predicate for *rat_points*.

3 Dynamic Filter: Derive a formula to compute err_E from $\tilde{E}$, err_A, and err_B for $E = A \cdot B$.

4 In ⟨*produce efficacy of filter table*⟩ we generated points by truncating the Cartesian coordinates to D bits, i.e., we generated *rat_points* by

```
rat_point(integer(p.xcoordD()*D),integer(p.ycoordD()*D),1).
```

What will change if we generate the points by

```
rat_point(integer(p.xcoordD()*D),integer(p.ycoordD()*D),D).
```

instead? Predict and then experiment.

5 Produce tables similar to Tables 9.3 and 9.4 for points that lie on a segment. Predict the outcome of the experiment before making it.

9.8 Safe Use of the Floating Point Kernel

The discussion of floating point filters in the previous section paves the way for a safe use of the floating point kernel. The following statement is trivial but nevertheless important.

It is safe to use the floating point kernel if it is guaranteed to give the correct result.

Lemma 59 gives a sufficient condition for the correctness of a floating point computation. If all arguments of an expression are integers, if the expression is a polynomial, i.e., uses only operations addition, subtraction, and multiplication, and if $B < 1$ then the evaluation with floating point arithmetic gives the correct sign of the expression. We have seen in Section 9.7.4 that the condition $B < 1$ is guaranteed if the arguments of the expression are sufficiently small; of course, the meaning of sufficiently small depends on the test. The following lemma gives information.

Lemma 60 *Assume that all points have integer Cartesian coordinates whose absolute value is less than 2^L. Then the floating point kernel correctly evaluates the compare function if $L \le 50$, correctly evaluates the orientation function if $L \le 24$, and correctly evaluates the side of circle function if $L \le 11$.*

Proof We give the proof for the side of circle function. Let a, b, c and d be points. We use ax and ay to denote the Cartesian coordinates of a and similarly for the other points.

The side of circle function is the sign of the determinant

$$\begin{vmatrix} 1 & 1 & 1 & 1 \\ ax & bx & cx & dx \\ ay & by & cy & dy \\ ax^2+ay^2 & bx^2+by^2 & cx^2+cy^2 & dx^2+dy^2 \end{vmatrix}$$

as will be shown in Section 10.9.

If a is equal to the origin the determinant above reduces to a 3×3 determinant. If a is not equal to the origin, we may shift a into the origin without changing the side of circle function. Shifting a into the origin replaces any point p by the point $O + (p - a)$.

This leads to the following program to compute the side of circle function. In this program we indicate the bit length of all intermediate results as comments.

```
int side_of_circle(const point& a, const point& b, const point& c,
                                                    const point& d)
{ // comments indicate bit lengths of values if coordinates have
  // at most L bits.
  double ax = a.xcoord();       // L bits
  double ay = a.ycoord();
  double bx = b.xcoord() - ax;  // L + 1 bits
  double by = b.ycoord() - ay;
  double bw = bx*bx + by*by;    // 2L + 3 bits
  double cx = c.xcoord() - ax;  // L + 1 bits
  double cy = c.ycoord() - ay;
  double cw = cx*cx + cy*cy;    // 2L + 3 bits
```

```
    double D1 = cy*bw - by*cw;  // 2L + 3 + L + 1 + 1 = 3L + 5 bits
    double D2 = bx*cw - cx*bw;  // 3L + 5 bits
    double D3 = by*cx - bx*cy;  // 2L + 3

    double dx = d.xcoord() - ax;  // L + 1 bits
    double dy = d.ycoord() - ay;

    double D  = D1*dx  + D2*dy + D3*(dx*dx + dy*dy);
                                // 3L + 5 + L + 1 + 2 = 4L + 8 bits

    if (D != 0)
       return (D > 0) ? 1 : -1;
    else
       return 0;
}
```

The comments show that the maximal number of bits required for the determinant D is $4L+8$. Thus D can be represented provided that $4L+8 \leq 53$; observe that the mantissa of a double precision floating point number consists of 53 bits m_0, m_1, ..., m_{52}, of which the bit m_0 is not stored, since it is always 1 (except if the number is zero or underflow occurred). □

The computation of, for example, Delaunay diagrams uses only the compare, orientation, and side of circle functions applied to input points and hence is safe as long as all input points have integer Cartesian coordinates whose absolute value is less than $2^{11} = 2048$.

If the coordinates of the inputs come from a larger range, it is frequently possible to round the input coordinates to a smaller precision without affecting the meaning of the computation, for example, if the coordinates come from a physical measurement whose precision is limited.

The following function *truncate* is useful in this situation. It takes a list *L0* of points and an integer *prec* and returns a list L of points. If all points in *L0* are equal to the origin, L is equal to *L0*. So assume otherwise and let M be the smallest power of two larger than the absolute value of all coordinates of all points in *L0*, and let $P = 2^{prec}$. For each point $p = (x, y)$ the point $(\lfloor (x/M) \cdot P \rfloor \cdot (M/P), \lfloor (y/M) \cdot P \rfloor \cdot (M/P))$ is added to L. Observe that x/M (and similarly y/M) is less than 1 and hence $(x/M) \cdot P$ is less than 2^{prec}. The multiplication by M/P (which is a power of two) moves the binary point for all points in the same way. Thus the theorem above applies to the modified points (with $L = prec$).

The implementation is simple. We first determine the maximum absolute value of any coordinate. If it is zero we are done. Otherwise, we set M to the smallest power of two larger than any absolute value. This is easily done using the functions *frexp* and *ldexp* from the math-library. Recall that *frexp*(M, $*exp$) assigns to *exp* the exponent of the smallest power of two larger than M and that *ldexp*$(1, k)$ returns 2^k.

⟨_truncate.c⟩+≡

```
list<point> truncate(const list<point>& L0, int prec)
{ double M = 0;
  point p;
  forall(p,L0)
    M = leda_max(M,leda_max(fabs(p.xcoord()),fabs(p.ycoord())));
```

```
  if ( M == 0 ) return L0;
  int exp;
  frexp(M,&exp);        // 2^(exp - 1) <= max < 2^exp
  M = ldexp(1,exp);  // round max to next power of two
  double C =      ldexp(1,prec - exp);  // P/M
  double C_inv = ldexp(1,exp - prec); // M/P
  list<point> L;
  forall(p,L0) L.append(point(floor(p.xcoord() * C)*C_inv,
                              floor(p.ycoord() * C)*C_inv));
  return L;
}
```

There is also a version of truncate which operates on a list of *rat_points*. It simply converts every *rat_point p* to a point by calling *p.to_point*( ), then applies the function above to the resulting list of points, and finally converts every *point q* in the resulting list to a *rat_point* by calling the constructor *rat_point*(*q*).

9.9 A Glimpse at the Higher-Dimensional Kernel

The higher-dimensional kernel provides points, vectors, directions, hyperplanes, segments, lines, affine transformations, and operations connecting these types in d-dimensional Euclidean space for arbitrary finite d. Points have rational coordinates, hyperplanes have rational coefficients, and analogous statements hold for the other types. All geometric primitives are exact, since they are implemented using rational arithmetic. The computational basis for the kernel is provided by the classes integer, integer vector, and integer matrix discussed in Chapter 4. We refer the reader to [MMN+98] for details. The higher-dimensional kernel is available as an LEP and was developed as part of the CGAL project.

9.10 History

The geometric part of LEDA evolved slowly and not without pain. We started with plane geometry in 1991. We introduced classes point, line, and segment and some algorithms operating on them, e.g., line segment intersection, Voronoi diagram construction, and convex hull construction. The programs provided in 1991 were not robust; on some inputs they failed by either delivering a wrong result or by crashing. The non-robustness of our original implementations was mainly due to three reasons:

- The programs were only designed to handle so-called non-degenerate inputs, e.g., the line segment intersection program assumed that no two input segments overlapped and the convex hull program assumed that the first three points were not collinear.

- Floating point arithmetic was used as the underlying arithmetic. We have seen in Section 9.6 that floating point arithmetic can lead to bizarre behavior of geometric objects.

- We had no checkers for geometric objects and hence were limited in our ability to test our algorithms.

Based on the bad experiences made by us and many others, we and others laid the theoretical foundations for correct and efficient implementations of geometric algorithms [FvW96, For96, CDR92, Yap93, Cla92, MN94b, BMS94b, BMS94a, BFS98, BFMS97, MNS+96, DLPT97, BR96, YD95, Sch, BEPP97].

Starting in 1994 we reimplemented the geometric classes and algorithms and simultaneously extended them considerably. We introduced the rational kernel with its built-in floating point filter, we redesigned all geometric algorithms and freed them from the assumption of non-degenerate inputs, and we added many new algorithms and checkers.

9.11 LEDA and CGAL

In 1997 the geometry effort of LEDA became part of project CGAL (= Constructing a Geometry Algorithms Library), a research project carried out by ETH Zürich, Freie Universität Berlin, INRIA Sophia Antipolis, Martin-Luther Universität Halle-Wittenberg, Max-Planck-Institut für Informatik and Universität des Saarlandes, RISC Linz, Tel-Aviv University, and Universiteit Utrecht, and funded by the European Union. The project was coordinated by Mark Overmars from Utrecht and ran for twenty-four months. The successor project is called GALIA and will be coordinated by the Max-Planck-Institut.

One of the goals of the projects is to build a comprehensive library for computational geometry called CGAL (Computational Geometry Algorithms Library). CGAL [CGA] goes much beyond LEDA geometry. Its distinctive features are:

- A geometry kernel [FGK+96] that can be instantiated with any number type. In LEDA we only have a floating point kernel and a rational kernel. It would be a non-trivial task to build a kernel based on the number type *real*. In CGAL this is easily possible.

- Geometric algorithms that are decoupled from the geometry kernel and can be used with any geometry kernel. Observe that LEDA's geometric algorithms are tied to the LEDA kernels and also to LEDA's graphs and data structures. CGAL achieves the new flexibility by the use of so-called *generic programming*. In this paradigm the kernel and the data structures are specified as template arguments of any geometric algorithm. The algorithm can then be instantiated with different kernels and data structures.

- A large variety of geometric data structures and algorithms which will go beyond what is offered by LEDA.

- An open architecture that makes it easy to import modules from other libraries.

The development of CGAL will not make LEDA geometry obsolete. The systems can be used side by side and both systems offer functionality which the other system does not have.

10

Geometry Algorithms

We discuss convex hulls, triangulations, the verification of geometric structures, Delaunay triangulations and Delaunay diagrams, Voronoi diagrams, applications of Delaunay and Voronoi diagrams, geometric dictionaries, line segment intersection, polygons, and close with a glimpse at higher-dimensional computation geometry. For each problem we introduce the required mathematics and derive algorithms and their implementations. The books [Meh84d, Ede87, PS85, Mul94, Kle97, BY98, dBKOS97] provide a wider view of computational geometry.

The chapter uses results of all preceding chapters and is, in this sense, the culmination point of the book, e.g., we use lists and arrays from the basic data types, integers and rationals from the number types, dictionaries, maps and sorted sequences from the advanced data types, graphs and graph algorithms, embedded graphs, and the geometry kernels.

Computational geometry is a very rich area and LEDA certainly does not provide everything there is to it. Other good sources of geometric software are CGAL [CGA] and the LEDA extension packages [LEP].

10.1 Convex Hulls

The convex hull problem in the plane is one of the simplest geometric problems and hence a good starting point for our exploration of geometry algorithms. It will allow us to address five important themes in a simple setting:

- The *sweep paradigm*: In this paradigm the input points are first sorted according to the lexicographic order and then the desired geometric structure is constructed incrementally during a single sweep over the points. We will derive and implement a

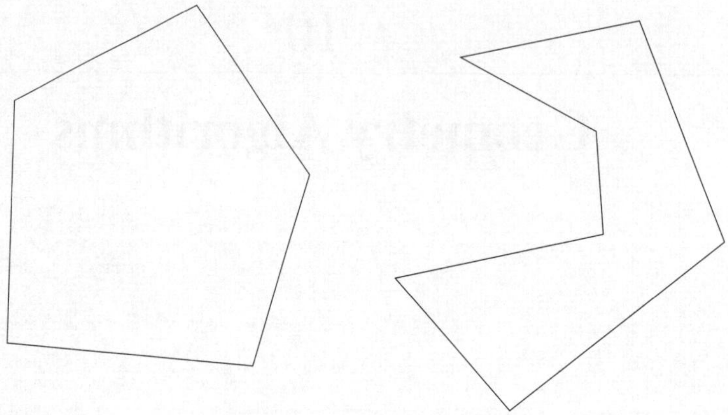

Figure 10.1 A convex and a non-convex set.

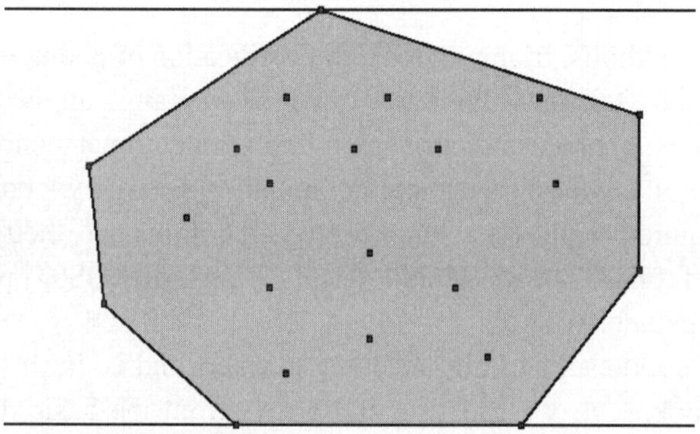

Figure 10.2 A point set, its convex hull, and its width. The figure was generated with the xlman-demo voronoi_demo. The width of point sets is discussed in Section 10.1.3.

sweep algorithm for convex hulls. We will see more applications of the sweep paradigm in later sections.

- The *(randomized) incremental construction paradigm*: In this paradigm the input points are considered one by one in either arbitrary or random order and the desired geometric structure is constructed incrementally. We will derive and implement an incremental algorithm for convex hulls.

- The careful handling of *degeneracies*: The literature on computational geometry frequently makes the so-called *general position assumption* which states that only inputs are considered for which none of the geometric predicates required by the algorithm (recall that the evaluation of a geometric predicate calls for the evaluation of the sign of an expression) ever evaluates to zero. For example, the incremental

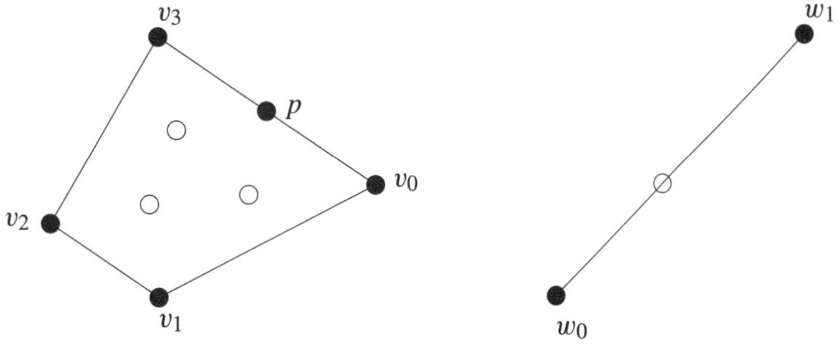

Figure 10.3 Two point sets and their convex hulls. The hulls are represented as cyclic lists of points, namely v_0, v_1, v_2, v_3 for the example on the left and w_0, w_1 for the example on the right.

algorithm for convex hulls uses the orientation predicate and hence the general position assumption excludes all inputs containing three collinear points. Of course, we do not want to exclude any inputs and hence cannot make the general position assumption. Dropping the general position assumption typically requires a more careful formulation of the algorithms. The sweep as well as the incremental algorithm for convex hulls will work for all inputs. In fact, all algorithms in this chapter do.

- *Verification of geometric structures*: Geometric programs require checking. Although the convex hull problem is one of the simplest geometric problems, the programs derived in this section will be non-trivial. We will see how to partially check the output of convex hull programs in Section 10.3.
- The importance of *exact geometric primitives*: In the preceding chapter we introduced the rational geometry kernel; in this section we will profit from it.

A set C is called *convex* if for any two points p and q in C the entire line segment pq is contained in C, see Figure 10.1. The *convex hull* conv S of a set S of points is the smallest (with respect to set inclusion) convex set containing S. A point $p \in S$ is called an *extreme point* of S if there is a closed halfspace containing S such that p is the only point in S that lies in the boundary of the halfspace. A point $p \in S$ is called a *weak extreme point* of S if there is a closed halfspace containing S such that p lies in the boundary of the halfspace. Clearly, an extreme point is also a weak extreme point, but there may be weak extreme points that are not extreme points. The point p in Figure 10.3 is an example.

From now on we restrict our discussion to the plane. If S contains no three collinear points then every weak extreme point is also extreme, i.e., under the general position assumption there is no need to distinguish between weak extreme points and extreme points. We define the convex hull problem as the problem of computing the extreme points of a finite set of points as a cyclically ordered list of point, see Figure 10.3. The cyclic order is the clockwise order in which the extreme points appear on the hull.

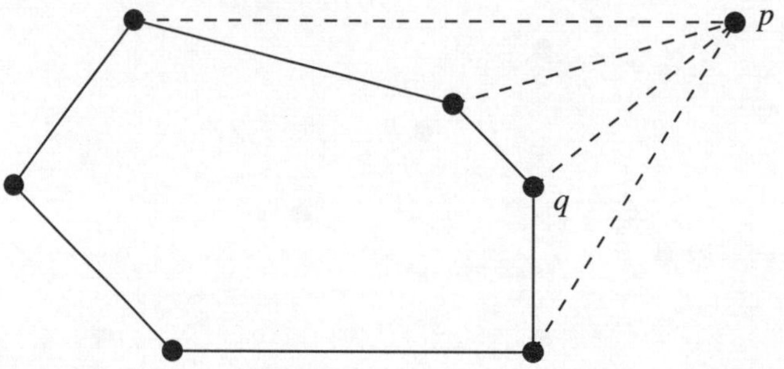

Figure 10.4 Adding point p. We determine the two tangents from p by a clockwise and counter-clockwise walk along the current hull starting at the most recently added point q.

The function

```
list<POINT> CONVEX_HULL(const list<POINT>& L);
```

computes the convex hull of the points in L and returns its list of vertices. The cyclic order of the vertices in the result corresponds to the clockwise order of the vertices on the hull. The algorithm uses randomized incremental construction and its expected running time is $O(n \log n)$.

10.1.1 *The Sweep Algorithm*

The sweep algorithm for convex hulls consists of the following three steps:

- The input points are sorted in increasing lexicographic order.
- The convex hull is initialized with the two lexicographically smallest points in L.
- The remaining points are considered in increasing lexicographic order and the convex hull is updated for each point. Assume that p is the next point to be considered and that we have already constructed the convex hull of the preceding points. The new hull can be obtained from the old hull by constructing the two tangents from p. The construction of the tangents is simple since p is guaranteed to see the point q added just before p. We only have to walk from q in clockwise and counter-clockwise direction along the hull in order to determine the other endpoints of the tangents, see Figure 10.4.

We now turn this strategy into a program. We assume that the set S is given as a list *L0* of points. We allow multiple occurrences of points. We follow the general outline above and proceed in three steps. We first make a local copy L of *L0* and sort L. Next we initialize the list of hull vertices with the first two points (in the sorted version) of L, and finally, we add all other points of L. We call the resulting program CONVEX_HULL_S since it uses the sweep paradigm to compute convex hulls.

⟨*convex_hull.c*⟩≡

```
list<POINT>  CONVEX_HULL_S(const list<POINT>& L0)
{ list<POINT> CH;
  list<POINT> L = L0;
  L.sort();
  ⟨initialize hull with two points⟩
  ⟨add all other points⟩
  return CH;
}
```

We prepare for the sweep by sorting the points according to the lexicographic order. A point p precedes a point q in the *lexicographic order* if either its x-coordinate is smaller or the two x-coordinates are equal and its y-coordinate is smaller. The default ordering on points is the lexicographic ordering and hence *L.sort*( ) rearranges L in the desired way.

We can now start building the hull. We begin with the first two points in L and make them the vertices of the first hull. As said above we represent the hull as a linear list *CH* that contains the hull vertices in *clockwise* order. The list is to be interpreted as a cyclic list. We maintain an item *last_vertex* into the list; it contains the point added last.

⟨*initialize hull with two points*⟩≡

```
if ( L.empty() ) return CH;
POINT last_p;
CH.append(last_p = L.pop());
// remove duplicates of first point
while ( !L.empty() && last_p == L.head() ) L.pop();
if ( L.empty() ) return CH;
list_item last_vertex = CH.append(last_p = L.pop());
```

We process the remaining points. If the next point p is equal to the last point added we do nothing. If the current hull consists of only two vertices and the new point p is collinear with these vertices we replace the second vertex by p. Otherwise, we determine two items *up_item* and *down_item* in *CH* which correspond to the other endpoints of the two tangents starting at p. To determine *up_item* we scan the hull in counter-clockwise direction starting at *last_vertex*. If the point stored at the predecessor of *up_item*, the point stored at *up_item*, and p do not form a right turn we move *up_item* to its predecessor vertex. We determine *down_item* by the symmetric procedure.

After having determined *up_item* and *down_item* we update the hull. We delete all items strictly between *up_item* and *down_item* and insert p instead of them. Note that *up_item* and *down_item* are guaranteed to be different since p sees at least one of the edges incident to the most recently added vertex.

⟨*add all other points*⟩≡

```
POINT p;
forall(p,L)
{ if ( p == last_p ) continue; // duplicate point
```

```
    last_p = p;
    if (CH.length() == 2 && collinear(CH.head(),CH.tail(),p))
      { CH[last_vertex] = p; continue; }
    // the interesting case
    // compute up_item
    list_item up_item = last_vertex;
    while (!right_turn(CH[CH.cyclic_pred(up_item)], CH[up_item], p))
    { up_item = CH.cyclic_pred(up_item); }
    // compute down_item
    list_item down_item = last_vertex;
    while (!left_turn(CH[CH.cyclic_succ(down_item)], CH[down_item], p))
    { down_item = CH.cyclic_succ(down_item); }
    // update hull
    while (down_item != CH.cyclic_succ(up_item))
    { CH.del_item(CH.cyclic_succ(up_item)); }
    last_vertex = CH.insert(p,up_item,after);
  }
```

The running time of the convex hull program is $O(n \log n)$. It takes time $O(n \log n)$ to sort the points lexicographically. After that everything is linear as the following amortization argument shows. Adding a point to the hull takes constant time plus time proportional to the number of points removed from the hull. Since any point can disappear from the hull at most once, the total time to add all points is linear. The running time of the algorithm is never better than $n \log n$ since it takes $\Theta(n \log n)$ time to sort the points. The sweep algorithm for convex hulls is due to Andrew ([And79]); it refines an earlier algorithm of Graham ([Gra72]).

The convex hull program makes use of the primitives provided by the geometry kernels. The rational kernel guarantees that all geometric primitives behave according to their mathematical specification and hence binding the program with the rational kernel will yield a correct executable. The program may behave incorrectly if bound with the floating point kernel. Consider the following example.

We compute the convex hull of the set $\{(-M+1, -M), (0, 0), (M, M+1), (0, -2)\}$ for $M = 2^m$ and increasing values of m. All four points are extreme and hence the following program will print "everything went fine", when executed with the rational kernel.

⟨*convex hull and kernel*⟩≡

```
for (int m = 20; m < 50; m++)
{
  double M = ldexp(1.0,m);
  INT_TYPE IM(M);
  POINT p(-IM + 1, -IM) , q(0, 0), r(IM, IM + 1), s(0, -2);
  list<POINT> L;
  L.append(p); L.append(q); L.append(r); L.append(s);
  list<POINT> CH = CONVEX_HULL_S(L);
  if ( CH.length() != 4 )
```

```
    { cout << "\n\nlength = " << CH.length() << " for  m = " << m;
      return 0;
    }
  }
  cout << "\n\neverything went fine";
```

However, when executed with the floating point kernel the program will print

```
  length = 3 for  m = 27,
```

since the floating point kernel believes that the triple (p, q, r) is collinear for $m \geq 27$.

10.1.2 *Incremental Construction*

We will next describe an alternative algorithm to compute convex hulls. The algorithm is based on the paradigm of *(randomized) incremental construction.* The algorithm has a worst case running time of $O(n^2)$, an average running time of $O(n \log n)$, and a best case running time of $O(n)$.

The algorithm starts by searching for three non-collinear points a, b, and c. If there are none, then all points are collinear and the vertices of the hull are simply the lexicographically smallest and largest point.

⟨*convex_hull.c*⟩+≡

```
  ⟨ch_edge_class⟩
  list<POINT>  CONVEX_HULL_IC(const list<POINT>& L)
  {
    if (L.length() < 2) return L;
    list<POINT> CH;
    POINT a = L.head(), b = L.tail();
    POINT c, p;
    if ( a == b ) { forall(p,L) if (p != a) { b = p; break; } }
    if ( a == b ) { // all points are equal
                    CH.append(a);
                    return CH;
                  }
    int orient;
    forall(c,L) if ( (orient = orientation(a,b,c)) != 0 ) break;
    if ( orient == 0 )
    { // all points are collinear
      forall(p,L) { if ( compare(p,a) < 0 ) a = p;
                    if ( compare(p,b) > 0 ) b = p;
                  }
      CH.append(a); CH.append(b);
      return CH;
    }
    // a, b, and c are not collinear
    if ( orient < 0 ) leda_swap(b,c);
    ⟨full-dimensional case: initialization⟩
```

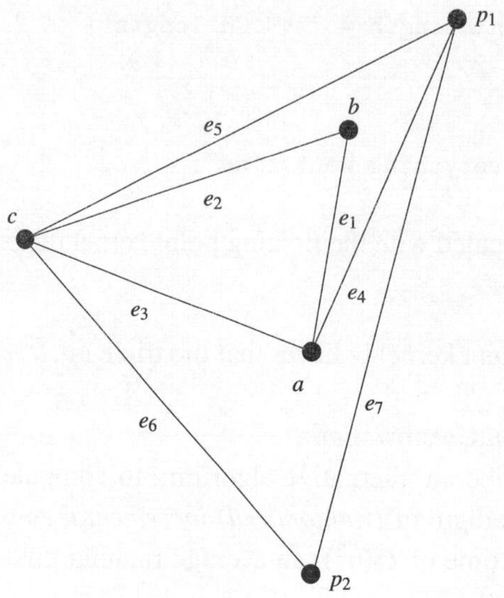

Figure 10.5 The initial convex hull consists of the points a, b, and c. When point p_1 is added the edges e_1 and e_2 are deleted from the hull and the edges e_4 and e_5 are added, and when p_2 is added to the hull the edges e_3 and e_4 are deleted from the hull and the edges e_6 and e_7 are added. The boundary of the current hull consists of edges e_7, e_5, and e_6 in counter-clockwise order. Every edge ever deleted from the hull points to the two edges that replaced it, e.g., e_3 and e_4 point to e_6 and e_7.

```
  forall(p,L) { ⟨full-dimensional case: insertion of p⟩ }
  ⟨full-dimensional case: prepare result and clean-up⟩
  return CH;
}
```

We come to the interesting case that not all points in L are collinear. We have already determined three non-collinear points a, b, and c. Their orientation is positive, i.e., the three points form a counter-clockwise oriented triangle.

The algorithm maintains the current hull as a cyclically linked list of edges and also keeps all edges that ever belonged to a hull. Every edge that is not on the current hull anymore points to the two edges that replaced it. More precisely, assume that S is the set of points already seen and that p is a point outside the current hull $CH(S)$. There is a chain C of edges of the boundary of $CH(S)$ that do not belong to the boundary of $CH(S \cup p)$. The chain is replaced by the two tangents from p to the previous hull. All edges in C are made to point to the two new edges, see Figure 10.5.

We use a class *ch_edge* to represent convex hulls. Every edge stores its two endpoints, three links *succ*, *pred*, and *link* to other edges, and a boolean flag *outside*. We use *link* to collect all edges into a linear list in the order of their creation; every edge points to the edge created just before it and *last_edge* points to the edge created last. The only purpose of this linear list is to help in the destruction of edges.

The boolean flag *outside* indicates whether an edge belongs to the current hull or not. All edges in the current hull form a cyclic doubly linked list with *succ* pointing to the clockwise successor and *pred* pointing to the clockwise predecessor. All edges that do not belong to the current hull anymore use their *succ* and *pred* fields to point to the two replacement edges.

⟨*ch_edge_class*⟩≡

```
class ch_edge;
static ch_edge* last_edge = nil;
class ch_edge {
public:
  POINT    source, target;
  ch_edge* succ, pred, link;
  bool     outside;
  ch_edge(const POINT& a, const POINT& b) : source(a), target(b)
  { outside = true;
    link = last_edge;
    last_edge = this;
  }
 ~ch_edge() {}
};
```

In order to initialize the data structure we create the edges (a, b), (b, c) and (c, a), store them in an array T, and turn them into a doubly-linked cyclic list. We initialize *last_edge* to *nil* before doing any of this, such that the list of all edges has the correct anchor.

⟨*full-dimensional case: initialization*⟩≡

```
last_edge = nil;
ch_edge* T[3];
T[0] = new ch_edge(a,b);
T[1] = new ch_edge(b,c);
T[2] = new ch_edge(c,a);
int i;
for(i = 0; i < 2; i++)  T[i]->succ = T[i+1];
T[2]->succ = T[0];
for(i = 1; i < 3; i++)  T[i]->pred = T[i-1];
T[0]->pred = T[2];
```

We are now ready to deal with the insertion of a point p. We proceed in two steps. We first determine whether p is outside the current hull and then update the hull (if p is outside).

In order to find out whether p lies outside the current hull, we walk through the history of hulls. We first find out whether p can see one of the edges of the initial triangle: p lies outside the initial triangle if there is an edge e of the initial triangle such that p lies to the right of the edge.

More generally, p is outside one of the intermediate hulls $CH(S)$ if there is an edge e on

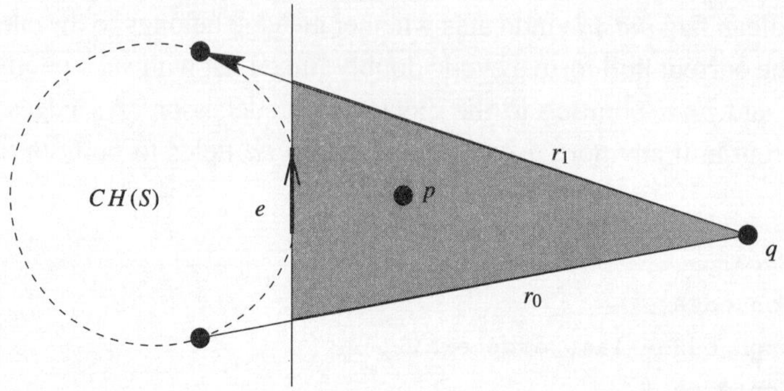

Figure 10.6 e is a (counter-clockwise) edge of the current hull and p lies to the right of it; e is replaced by r_0 and r_1 when the point q is added. If p lies neither to the right of r_0 nor to the right of r_1 then p lies in the shaded region and hence in $CH(S \cup q)$.

its boundary such that p lies to the right of the edge. If e is an edge on the boundary of the current hull then p lies outside the current hull. If e is not an edge on the boundary of the current hull, let r_0 and r_1 be the two edges that replaced e when $CH(S)$ was enlarged to $CH(S \cup q)$. p is outside $CH(S \cup q)$ if it lies to the right of either r_0 or r_1, see Figure 10.6.

⟨*full-dimensional case: insertion of p*⟩≡

```
int i = 0;
while (i < 3 && !right_turn(T[i]->source,T[i]->target,p) ) i++;
if (i == 3) { // p inside initial triangle
            continue;
          }
ch_edge* e = T[i];
while (! e->outside)
{ ch_edge* r0 = e->pred;
  if ( right_turn(r0->source,r0->target,p) ) e = r0;
  else { ch_edge* r1 = e->succ;
         if ( right_turn(r1->source,r1->target,p) ) e = r1;
         else { e = nil; break; }
       }
}
if (e == nil) continue;  // p inside current hull
```

⟨*insertion of p: p is outside current hull*⟩

Assume now that p lies outside the current hull and to the right of the counter-clockwise hull edge e. We determine all edges visible from p by walking along the hull in both directions. This is exactly as in the previous algorithm. Let *low* be the first predecessor of e that is not visible and let *high* be the first successor that is not visible.

We then add the new tangents between *low* and *high* and mark all edges that were deleted from the hull as inside and make the two new tangents the replacement edges of all deleted edges.

⟨*insertion of p: p is outside current hull*⟩≡

```
// compute "upper" tangent (p,high->source)

ch_edge* high = e->succ;
while (orientation(high->source,high->target,p) <= 0) high = high->succ;

// compute "lower" tangent (p,low->target)

ch_edge* low = e->pred;
while (orientation(low->source,low->target,p) <= 0) low = low->pred;

e = low->succ;  // e = successor of low edge

// add new tangents between low and high

ch_edge* e_l = new ch_edge(low->target,p);
ch_edge* e_h = new ch_edge(p,high->source);

e_h->succ = high;
e_l->pred = low;
high->pred = e_l->succ = e_h;
low->succ  = e_h->pred = e_l;

// mark edges between low and high as "inside"
// and define refinements

while (e != high)
{ ch_edge* q = e->succ;
  e->pred = e_l;
  e->succ = e_h;
  e->outside = false;
  e = q;
}
```

Having computed the hull we prepare the output and delete all edges. We prepare the output by running around the hull once and we clean up by deleting all edges.

⟨*full-dimensional case: prepare result and clean-up*⟩≡

```
ch_edge* l_edge = last_edge;

CH.append(l_edge->source);
for(ch_edge* e = l_edge->succ; e != l_edge; e = e->succ)
  CH.append(e->source);

// clean up

while (l_edge)
{ ch_edge* e = l_edge;
  l_edge = l_edge->link;
  delete e;
}
```

What is the running time of the incremental construction of convex hulls?

The worst case running time is $O(n^2)$ since the time to insert a point is $O(n)$. The time to insert a point is $O(n)$ since there are at most $2(k+1)$ edges after the insertion of k points and since every edge is looked at at most once in the insertion process.

The best case running time is $O(n)$. An example for the best case is when the points a, b, and c span the hull.

The average case running time is $O(n \log n)$ as we will show next. What are we averaging over? We consider a fixed but arbitrary set S of n points and average over the $n!$ possible insertion orders. The following theorem is a special case of the by now famous *probabilistic analysis of incremental constructions* started by Clarkson and Shor [CS89]. The books [Mul94, BY98, MR95, dBKOS97] contain detailed presentations of the method. The reader may skip the proof of Theorem 14. Why do we include a proof at all given the fact that the method is already well treated in textbooks? We give a proof because the cited references prove the theorem only for points in general position. We want to do without the general position assumption in this book.

Theorem 14 *The average running time of the incremental construction method for convex hulls is* $O(n \log n)$.

Proof We assume for simplicity that the points in S are pairwise distinct. The theorem is true without this assumption; however, the notation required in the proof is more clumsy.

The running time of the algorithm is linear iff all points in S are collinear. So let us assume that S contains three points that are not collinear. In this case we will first construct a triangle and then insert the remaining points. Let p be one of the remaining points. When p is inserted, we first determine the position of p with respect to the initial triangle (time $O(1)$), then search for a hull edge e visible by p, and finally update the hull. The time to update the hull is $O(1)$ plus some bounded amount of time for each edge that is removed from the hull. We conclude that the total time (= time summed over all insertions) spent outside the search for a visible hull edge is $O(n)$.

In the search for a visible hull edge we perform tests *rightturn*(x, y, p) where x and y are previously inserted points. We call a test *successful* if it returns true and observe that in each iteration of the while-loop at most two rightturn tests are performed and that in all iterations except the last at least one rightturn test is successful. It therefore suffices to bound the number of successful rightturn tests.

For an ordered pair (x, y) of distinct points in S we use $K_{x,y}$ to denote the set of points z in S such that *rightturn*(x, y, z) is true plus[1] the set of points on the line through (x, y) but not between x and y, see Figure 10.7. We use k_{xy} to denote the cardinality of $K_{x,y}$, F_k to denote the set of pairs (x, y) with $k_{xy} = k$, $F_{\le k}$ to denote the set of pairs (x, y) with $k_{xy} \le k$, and f_k and $f_{\le k}$ to denote the cardinalities of F_k and $F_{\le k}$, respectively. We have

Lemma 61 *The average number A of successful rightturn tests is bounded by* $\sum_{k \ge 1} 2 f_{\le k}/k^2$.

Proof Consider a pair (x, y) with $k_{xy} = k$. If some point in $K_{x,y}$ is inserted before both x and y are inserted then (x, y) is never constructed as a hull edge and hence no rightturn

[1] The set to be defined next is empty if S is in general position. The probabilistic analysis of incremental constructions usually assumes general position. We do not want to assume it here and hence have to modify the proof somewhat.

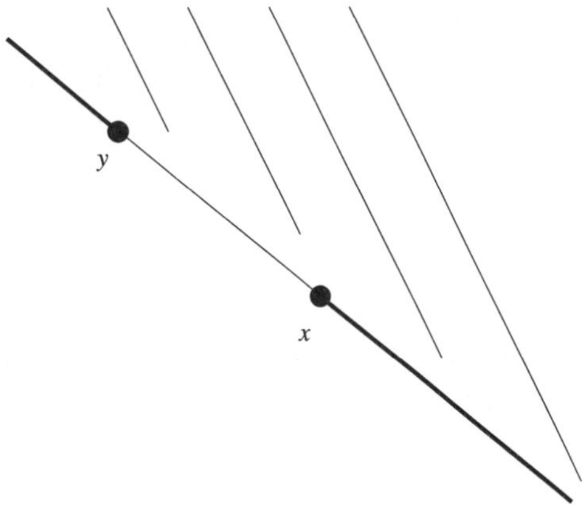

Figure 10.7 $K_{x,y}$ consists of all points in the shaded region plus the two solid rays.

tests $(x, y, -)$ are performed. However, if x and y are inserted before all points in $K_{x,y}$ then up to k successful rightturn tests (x, y, z) are performed.
The probability that x and y are inserted before all points in $K_{x,y}$ is

$$2!k!/(k+2)!$$

since there are $(k + 2)!$ permutations of $k + 2$ points out of which $2!k!$ have x and y as their first two elements. Thus the expected number of successful rightturn tests (x, y, z) is bounded by

$$2!k!/(k+2)! \cdot k = 2 \cdot k/(k+1)(k+2) < 2/(k+1).$$

The argument above applies to any pair (x, y) and hence the average number of successful rightturn tests is bounded by

$$\sum_{k \geq 1} 2 f_k/(k+1).$$

We next write $f_k = f_{\leq k} - f_{\leq k-1}$ and obtain

$$\begin{aligned} A &\leq \sum_{k \geq 1} 2(f_{\leq k} - f_{\leq k-1})/(k+1) = \sum_{k \geq 1} 2 f_{\leq k}(1/(k+1) - 1/(k+2)) \\ &= \sum_{k \geq 1} 2 f_{\leq k}/((k+1)(k+2)). \end{aligned}$$

□

It remains to bound $f_{\leq k}$. We use random sampling to derive a bound.

Lemma 62 *$f_{\leq k} \leq 2e^2 n \cdot k$ for all k, $1 \leq k \leq n$.*

Proof There are only n^2 pairs of points of S and hence we always have $f_{\leq k} \leq n^2$. Thus, the claim is certainly true for $n \leq 10$ or $k \geq n/4$.
So assume that $n \geq 10$ and $k \leq n/4$ and let R be a random subset of S of size r. We will

fix r later. Clearly, the convex hull of R consists of at most r edges. On the other hand, if for some $(x, y) \in F_{\leq k}$, x and y are in R but none of the points in $K_{x,y}$ is in R, then (x, y) will be an edge of the convex hull of R. The probability of this event is

$$\frac{\binom{n-i-2}{r-2}}{\binom{n}{r}} \geq \frac{\binom{n-k-2}{r-2}}{\binom{n}{r}},$$

where $i = k_{x,y}$. Observe that the event occurs if x and y are chosen and the remaining $r-2$ points in R are chosen from $S \setminus \{x, y\} \setminus K_{x,y}$. The expected number of edges of the convex hull of R is therefore at least

$$f_{\leq k} \cdot \frac{\binom{n-k-2}{r-2}}{\binom{n}{r}}.$$

Since the number of edges is at most r we have

$$f_{\leq k} \cdot \binom{n-k-2}{r-2} / \binom{n}{r} \leq r$$

or

$$f_{\leq k} \leq r \cdot \binom{n}{r} / \binom{n-k-2}{r-2} = r \cdot \frac{n(n-1)}{r(r-1)} \cdot \frac{[n-2]_{r-2}}{[n-k-2]_{r-2}},$$

where $[n]_i = n(n-1)\cdots(n-i+1)$. Next observe that

$$\begin{aligned}\frac{[n-2]_{r-2}}{[n-k-2]_{r-2}} &\leq \frac{[n]_r}{[n-k]_r} = \prod_{i=0}^{r-1} \frac{n-i}{n-k-i} = \prod_{i=0}^{r-1} \left(1 + \frac{k}{n-k-i}\right) \\ &= \exp\left(\sum_{i=0}^{r-1} \ln(1 + k/(n-k-i))\right) \leq \exp\left(rk/(n-k-r)\right),\end{aligned}$$

where the last inequality follows from $\ln(1+x) \leq x$ for $x \geq 0$ and the fact that $k/(n-k-i) \leq k/(n-k-r)$ for $0 \leq i \leq r-1$. Setting $r = n/(2k)$ and using the fact that $n-k-r \geq n/4$ for $k \leq n/4$ and $n \geq 10$, we obtain

$$f_{\leq k} \leq e^2 n^2 / r = 2e^2 nk.$$

□

Putting our two lemmas together completes the proof of Theorem 14

$$A \leq 4e^2 \sum_{k \geq 1} nk/k^2 = O(n \log n).$$

□

There are two important situations when the assumptions of the theorem above are satisfied:

- When the points in S are generated according to a probability distribution for points in the plane.
- When the points are randomly permuted before the incremental construction process is started. We then speak about a *randomized incremental construction.*

CONVEX_HULL_RIC realizes the randomized incremental construction of convex hulls.

⟨*convex_hull.c*⟩+≡
```
list<POINT> CONVEX_HULL_RIC(const list<POINT>& L)
{ list<POINT> L1 = L;
  L1.permute();
  return CONVEX_HULL_IC(L1);
}
list<POINT> CONVEX_HULL(const list<POINT>& L)
{ return CONVEX_HULL_RIC(L); }
```

It is important to understand the difference between _IC and _RIC. The former is a *deterministic* procedure whose average running time is $O(n \log n)$ if the assumptions of Theorem 14 are satisfied. The latter is a randomized algorithm whose expected running time for any input is $O(n \log n)$. Table 10.1 shows the difference. We generated a list L of n random points for each of three distributions: random points in the unit square, random points in the unit disk, and random points close to the boundary of the unit circle. We also generated a second input set *LS* by sorting L lexicographically. On the random inputs _IC does slightly better than _RIC because the latter does something that is unnecessary for random inputs: it randomly permutes an input that is already random. However, for the sorted inputs the situation is completely different. _RIC behaves about the same as for random inputs. However, _IC behaves much worse. For the points on the circle the behavior seems to be quadratic and for the points in the square and the disk the behavior seems to be n^δ for some $\delta > 1$. For this reason RIC is to be preferred over IC.

We next compare the sweep line algorithm with the randomized incremental construction algorithm. Table 10.2 shows the results. Observe that we use much larger inputs sizes for this table. The randomized incremental algorithm is faster than the sweep algorithm for inputs with only few hull vertices and is somewhat slower for points on the unit circle. Observe that the proof of Theorem 14 implies that the running time of randomized incremental construction is $o(n \log n)$ if a random subset of the input points has a small convex hull.

There are many more convex hull algorithms than sweep and (randomized) incremental construction. Schirra [Sch98] discusses implementations.

10.1.3 *The Width of a Point Set*

The *width* of a point set L is the minimal width of a stripe containing all points in L. A stripe is the region of the plane between two parallel lines. Minimum width stripes are illustrated in the xlman-demo voronoi-demo, see Figure 10.2. The function

```
RAT_TYPE WIDTH(const list<POINT>& L, LINE& l1, LINE& l2)
```

assumes that L is non-empty and returns the square[2] of the minimum width stripe containing L and the boundaries of the stripe.

We show how to compute the minimum width stripe by the so-called *rotating caliber method.* We start with a partial characterization of the minimum width stripe.

[2] We return the square of the width instead of the width because this choice avoids the use of square roots.

				IC		RIC	
K	n	Gen	V	Random	Sorted	Random	Sorted
S	4000	0.29	18	0.09	0.27	0.11	0.13
S	8000	0.64	23	0.16	0.76	0.22	0.21
S	16000	1.34	29	0.33	2.53	0.42	0.41
D	4000	0.27	59	0.1	0.45	0.11	0.1
D	8000	0.59	66	0.17	1.26	0.23	0.2
D	16000	1.25	87	0.43	3.48	0.5	0.41
C	4000	9.32	4000	0.32	15.57	0.34	0.37
C	8000	18.87	7995	0.7	65.93	0.75	0.71
C	16000	37.62	1.599e+04	1.47	253.4	1.53	1.57

Table 10.1 A comparison of incremental and randomized incremental construction: We generated n points according to one of three distributions, either points with random integer coordinates in $[-R\,..\,R]$, or random points with integer coordinates in the disc with radius R centered at the origin, or random points with integer coordinates that lie approximately on the circle with radius R centered at the origin. We used $R = 16000$. The columns show from left to right the kind of the point set (S for points in a square, D for points in the disc, and C for points on a circle), the number n of points, the time to generate the n points, the number of vertices of the hull, the running time of the incremental algorithm (_IC), and the running time of the randomized incremental algorithm (_RIC). For both algorithms the first column gives the time for random inputs and the second column gives the time for lexicographically sorted inputs. Observe the bad behavior of _IC on sorted inputs. Also observe that the time to compute the hull is usually smaller than the time to generate the points.

Lemma 63 *Let S be a minimum width stripe containing L. Then one of the boundaries contains an edge of the convex hull of L and the other boundary contains at least one vertex of the convex hull of L.*

Proof Clearly, both boundaries of S must contain at least one vertex of the convex hull of S. Assume that neither boundary contains an edge of the convex hull and let p and q be the two vertices of the convex hull of L that are contained in the boundary of S. Since the boundary of L contains no edge of the convex hull we can rotate both lines around p and q, respectively. Let α be the acute angle between the segment pq and the boundary of S incident to p, see Figure 10.8. Then

$$width(S) = |pq| \cdot \sin\alpha$$

and hence the width decreases when α is decreased. □

				Sweep		RIC	
K	n	Gen	V	Random	Sorted	Random	Sorted
S	20000	1.72	25	1.68	1.54	0.55	0.55
S	40000	3.77	29	3.6	3.26	1.26	1.43
S	80000	7.92	31	7.72	6.98	2.06	2.07
D	20000	1.62	106	1.75	1.59	0.55	0.56
D	40000	3.49	109	3.76	3.33	1.17	1.25
D	80000	7.32	152	8	7.02	2.42	2.58
C	20000	47	1.999e+04	1.82	1.67	2.13	2.12
C	40000	94.68	3.994e+04	3.96	3.57	4.46	4.41
C	80000	188.8	7.979e+04	8.6	7.78	10.31	10.04

Table 10.2 The running times of the sweep algorithm and the randomized incremental construction algorithm for convex hulls. The meaning of the columns is the same as for Table 10.1.

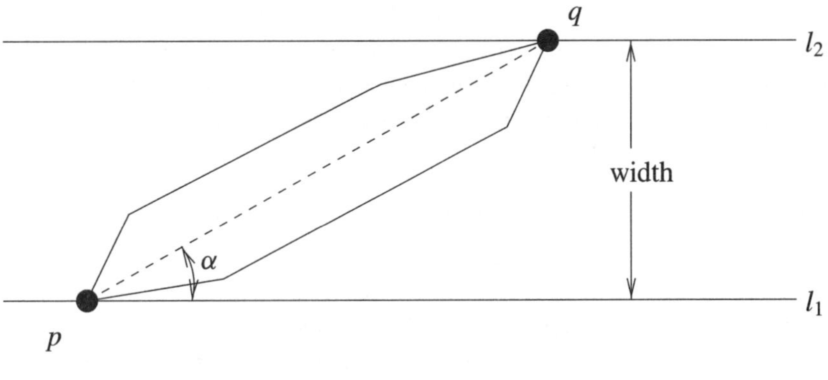

Figure 10.8 The stripe S with boundaries l_1 and l_2 contains all points of L, but neither boundary contains an edge of the convex hull of L. Rotating its boundaries decreases the width of the stripe.

We conclude from the lemma above that the minimum width stripe is defined by an edge of the convex hull and the vertex of maximum distance from the line supporting this edge. The next lemma constrains the part of the convex hull where this vertex of maximal distance may lie.

Lemma 64 *Let $v_0, v_1, \ldots, v_{k-1}$ be the vertices of the convex hull of L, let $l = l(v_{k-1}, v_0)$ be the line passing through v_{k-1} and v_0, and let v_m be the vertex of maximal distance from l.*

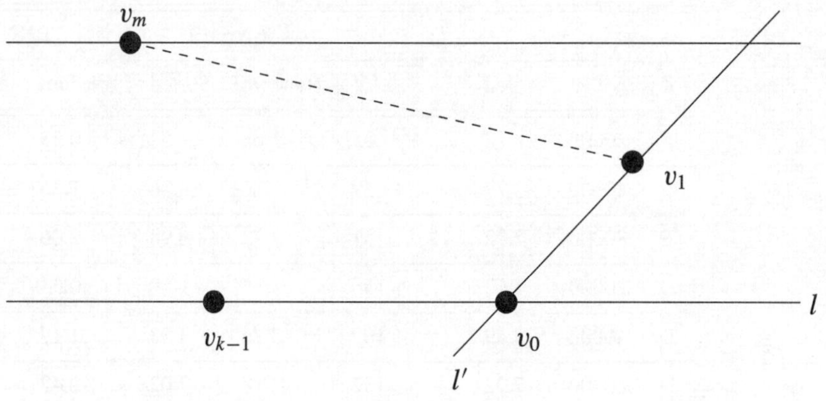

Figure 10.9 Illustration of the proof of Lemma 64.

Let $l' = l(v_0, v_1)$, let $v_{m'}$ be the vertex of maximal distance from l'. Then $m \le m' \le k-1$. Also m' is minimal such that $v_{m'+1}$ has smaller distance to l' than $v_{m'}$.

Proof Consider Figure 10.9. All vertices v_i with $1 \le i \le m$ are contained in the triangle with corners v_1, v_m, and the intersection between l' and the line parallel to l through v_m. Any point in this triangle has smaller distance to l' than v_m. Thus $m \le m' \le k-1$.

For the second claim consider the distance between l' and v_i as a function of i and as i ranges from 1 to $k-1$. It follows from convexity that this function is first strictly increasing then reaches its maximum for either one or two vertices and is then again strictly decreasing. □

It is easy to derive an algorithm from the preceding lemma. We determine for each hull edge pq the vertex m of maximal distance from the line $l(p,q)$. We initialize p and q to the first two hull vertices and find m by a search over all vertices. We then scan once around the convex hull of L in order to check all other edges.

We maintain the square of the width of the currently best stripe in *min_sqr_width* and the boundaries of the stripe in *l1* and *l2*.

⟨*width.c*⟩≡
```
RAT_TYPE WIDTH(const list<POINT>& L, LINE& l1, LINE& l2)
{
  if ( L.empty() )
    error_handler(1,"WIDTH applies only to non-empty sets");
  list<POINT> CH = CONVEX_HULL(L);
  if ( CH.length() == 1 )
  { l1 = l2 = LINE(L.head(), VECTOR(INT_TYPE(1),INT_TYPE(1)));
    return 0;
  }
  if ( CH.length() == 2 )
```

```
    { l1 = l2 = LINE(CH.head(), CH.tail()); return 0; }
    list_item p_it = CH.first();
    list_item q_it = CH.cyclic_succ(p_it);
    list_item m_it = q_it;
    list_item it;
    LINE l(CH[p_it],CH[q_it]);
    RAT_TYPE min_sqr_width = 0; RAT_TYPE sqr_dist;
    // find vertex with maximal distance from l
    forall_items(it,CH)
    { if ( (sqr_dist = l.sqr_dist(CH[it])) > min_sqr_width )
      { min_sqr_width = sqr_dist;
        m_it = it;
      }
    }
    l1 = l; l2 = LINE(CH[m_it], CH[q_it] - CH[p_it]);
    ⟨rotate caliber around CH⟩
    return min_sqr_width;
  }
```

Let r be the successor vertex of q. We want to determine the vertex m' with maximal distance from $l' = l(q, r)$. The last sentence of the lemma above implies that m' is the closest successor of m (inclusive) such that the successor of m' has smaller distance to l' than m'.

⟨*rotate caliber around CH*⟩≡

```
    do  // move caliber to next edge
    {
      list_item r_it = CH.cyclic_succ(q_it);
      LINE l(CH[q_it],CH[r_it]);
      RAT_TYPE cur_sqr_dist = l.sqr_dist(CH[m_it]);
      list_item new_m_it = m_it;
      it = CH.cyclic_succ(m_it);
      while ( (sqr_dist = l.sqr_dist(CH[it])) >= cur_sqr_dist )
      { new_m_it = it; it = CH.cyclic_succ(it);
        cur_sqr_dist = sqr_dist;
      }
      if ( cur_sqr_dist < min_sqr_width )
      { min_sqr_width = cur_sqr_dist;
        l1 = l; l2 = LINE(CH[new_m_it], CH[r_it] - CH[q_it]);
      }
      p_it = q_it; q_it = r_it; m_it = new_m_it;
    } while ( p_it != CH.first() );
```

The running time of the width computation is the time to compute the convex hull plus an amount of time that is linear in the number of vertices of the convex hull. It takes linear time to compute the vertex of maximal distance from the first hull edge and it takes linear time to compute the vertex of maximal distance for all other edges. The latter follows from the observation that both the edge and the vertex of maximal distance "travel around the convex hull once".

Exercises for 10.1

1 Design an example where the running time of CONVEX_HULL_IC is quadratic.
2 Design an example where the running time of CONVEX_HULL_IC is linear.
3 Redo the proof of Theorem 14 under the assumption that the expected number of hull edges in the convex hull of r random points is $r^{1-\delta}$ for some $\delta > 0$.
4 Modify either convex hull algorithm such that it returns all points that lie on the boundary of the convex hull.
5 Let P and Q be two disjoint convex polygons given by their cyclic list of vertices. Write a program that computes the common tangents of P and Q.
6 Use the solution of the previous exercise to compute the convex hull by divide-and-conquer. Sort the points lexicographically and split them into two halves. Compute the hull of both halves recursively. Merge the two hulls by constructing the common tangents.
7 Generate n random points in the unit square and compute their convex hull. Do so for different values of n and derive a conjecture concerning the expected number of extreme points in a set of n random points. Try to prove your conjecture or do a literature search to find out what is known about the problem. Do the same for random points in the unit disk.

10.2 Triangulations

A *triangulation* of S is a partition of the convex hull of S into triangles. This assumes that not all points of S are collinear. Each triangle in the partition has three points of S as its vertices and any two triangles in the partition are either disjoint, or share a vertex, or an edge and two vertices. The union of all triangles is the convex hull of S, see Figure 10.10 for two examples. What is a triangulation of conv S if all points of S are collinear? It is simply a partition of conv S into line segments[3], see Figures 10.10 and 10.11.

Triangulations are a versatile data structure. We will use them for point location queries, nearest neighbor queries, and range queries in Section 10.6 and describe their use in *interpolation* now. Assume that we are given the values of some function f at some finite set S of points and want to interpolate f for all points in the convex hull of S. Triangulations offer an elegant way to approach this problem. We compute a triangulation T of S and lift it to three-dimensional space. More precisely, for every triangle (p, q, r) of T we define

[3] More generally, if S has affine dimension d then a triangulation of S is a partition of conv S into d-dimensional simplices. A d-dimensional *simplex* is the convex hull of $d + 1$ affinely independent points. Thus, triangles are two-dimensional simplices and line segments are one-dimensional simplices and hence a triangulation of a one-dimensional set S is a partition of its convex hull into line segments, a triangulation of a two-dimensional set is a partition of its convex hull into triangles, and a triangulation of a three-dimensional set is the partition of its convex hull into tetrahedra.

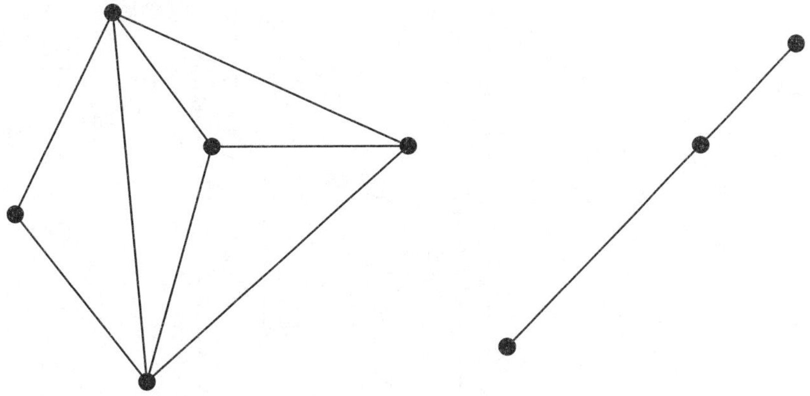

Figure 10.10 A triangulation of a two-dimensional and of a one-dimensional point set.

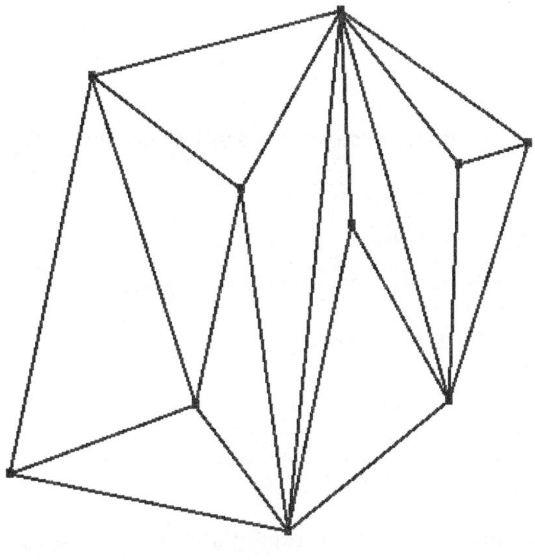

Figure 10.11 A triangulation computed by the function TRIANGULATE_POINTS discussed in this section.

a triangle $((p, f(p)), (q, f(q)), (r, f(r)))$ in three-space, see Figure 10.12. In this way we obtain a surface in three-space. In order to determine the interpolating value at a point $x \in \text{conv } S$ we determine the height of the interpolating surface above x and return it. This requires us to find the triangle of T containing x (a point location query) and to determine the height at x by linear interpolation from the height at the vertices of the triangle containing x. Assume that x lies in the triangle with vertices p, q, and r. We write x as a convex

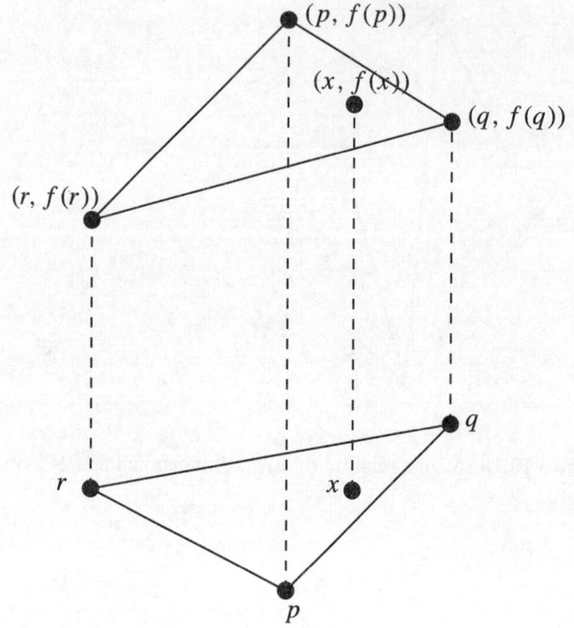

Figure 10.12 A triangle in the plane and its lifting to three-space.

combination of p, q, and r, i.e.,

$$x = c_p p + c_q q + c_r r,$$

where $c_p + c_q + c_r = 1$ and compute $f(x)$ as

$$f(x) = c_p f(p) + c_q f(q) + c_r f(r).$$

The coefficients c_p, c_q, and c_r are called the *barycentric coordinates* of x with respect to the triangle (p, q, r).

We next discuss how to represent triangulations. We represent triangulations as straight line embedded plane maps; embedded graphs are the subject of Chapter 8 and we recommend that you read the first four sections of that chapter before proceeding. Let T be a triangulation of a set S of points. We use a graph G of type *GRAPH<POINT*, *int*> to represent T; G has the following properties, see Figure 10.14:

- The nodes of G are in one-to-one correspondence to the points in S. For a node v of G the point in S corresponding to it is stored as $G[v]$.
- G is a directed graph whose edges will be called *darts*. We use the word dart instead of edge in order to distinguish the edges of the representing graph from the edges of the represented geometric object. The darts of G come in pairs. For every dart $e = (v, w)$ of G the reversed dart $e^R = (w, v)$ is also a dart of G. Moreover, the member function *reversal* maps each dart to its reversal, i.e., *G.reversal(e)* $= e^R$ and

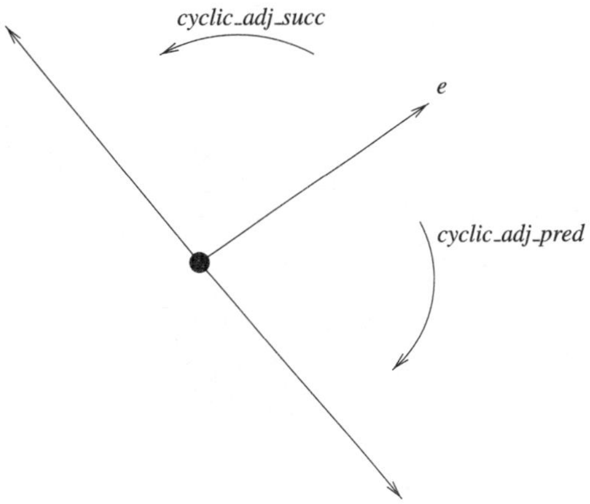

Figure 10.13 The relationship between the cyclic ordering of the adjacency list $A(v)$ of a node v and the counter-clockwise ordering of the edges incident to $G[v]$.

G.reversal$(e^R) = e$. We call a pair consisting of a dart and its reversal a uedge (= undirected edge). The uedges of G correspond to the edges of T and a dart (v, w) of G corresponds to the oriented edge $(G[v], G[w])$ of T.

- For each node v of G the list $A(v)$ of edges out of v is ordered cyclically. For an edge e with source v the functions

```
G.cyclic_adj_succ(e);
G.cyclic_adj_pred(e);
```

  return the cyclic successor and the cyclic predecessor of e in $A(v)$. The cyclic ordering of the edges in $A(v)$ agrees with the counter-clockwise ordering of the edges incident to $G[v]$ in the triangulation, i.e., *G.cyclic_adj_succ*(e) is the next dart out of v in counter-clockwise direction and *G.cyclic_adj_pred*(e) is the next dart out of v in clockwise direction, see Figure 10.13.

- The preceding items guarantee that the faces of the triangulation correspond to the face cycles of G. For each counter-clockwise triangle $(G[u], G[v], G[w])$ of the triangulation the edges (u, v), (v, w), (w, u) form a face cycle of G. There is also a face cycle corresponding to the unbounded face of T. As a face cycle is traversed the face lies to the left of the face cycle. The functions

```
G.face_cycle_succ(e);
G.face_cycle_pred(e);
```

  support the convenient traversal of the face cycles of a map. They give the successor and predecessor of e in the face cycle containing e, respectively. The face cycle successor is the cyclic adjacency predecessor of the reversal of e, see Figure 8.10.

- Each dart has an integer label (available as $G[e]$) that gives information about the dart. The labels come from the enumeration type

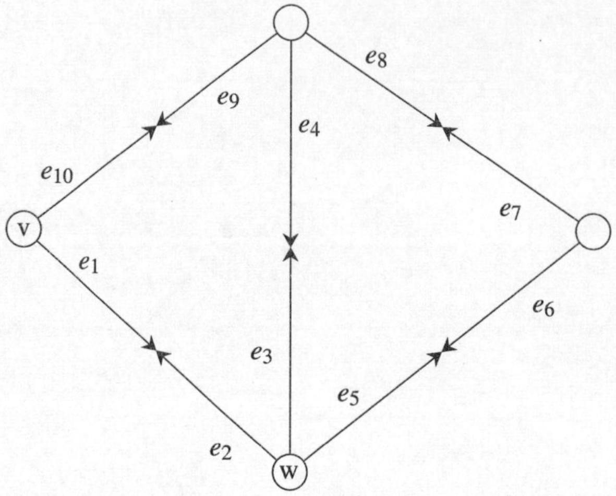

Figure 10.14 A graph G representing a triangulation. For each edge of the triangulation there are two darts in G, e.g., the edge $G[v]G[w]$ is represented by the darts $e_1 = (v, w)$ and $e_2 = (w, v)$. We have $G.reversal(e_1) = e_2$ and $G.reversal(e_2) = e_1$. For each dart its name is shown near the source of the dart and to the left of the dart. The list $A(w)$ of edges out of w is a cyclic shift (it is not specified which) of (e_5, e_3, e_2). The two triangles correspond to the face cycles (e_1, e_3, e_9) and (e_5, e_7, e_4). The unbounded face corresponds to the face cycle (e_6, e_2, e_{10}, e_8).

```
enum delaunay_edge_info{ DIAGRAM_EDGE = 0, DIAGRAM_DART = 0,
                         NON_DIAGRAM_EDGE = 1, NON_DIAGRAM_DART = 1,
                         HULL_EDGE = 2, HULL_DART = 2
                       };
```

defined in <*LEDA/geo_global_enums.h*>. We discuss them in Section 10.4.

A dart is called a *hull dart* if the unbounded face of G lies to its left. If *hull_dart* is any hull dart, the following lines of code traverse all hull darts.

```
edge e = hull_dart;
do { e = G.face_cycle_succ(e); } while (e != hull_dart);
```

We next extend the hull program of the preceding section to a triangulation program. This algorithm was first described in [Meh84b]. Again, we start by sorting the points lexicographically. Then we set up the triangulation of the first two points and finally add point by point to the triangulation.

⟨*triangulation.c*⟩≡

```
inline int left_bend(const POINT& p, const GRAPH<POINT,int>& G,
                     const edge& e)
{ return (orientation(p,G[source(e)],G[target(e)]) > 0); }
edge TRIANGULATE_POINTS(const list<POINT>& L0, GRAPH<POINT,int>& G)
{
  G.clear();
```

```
  if (L0.empty()) return nil;
  list<POINT> L = L0;
  L.sort();
  if ( L.empty() ) return nil;
  // initialize G with a single edge starting at the first point
  POINT last_p = L.pop();              // last visited point
  node  last_v = G.new_node(last_p); // last inserted node
  while (!L.empty() && last_p == L.head()) L.pop();
  if (!L.empty())
  { last_p = L.pop();
    node v = G.new_node(last_p);
    edge x = G.new_edge(last_v,v,0);
    edge y = G.new_edge(v,last_v,0);
    G.set_reversal(x,y);
    last_v = v;
  }
  ⟨triangulate points: scan remaining points⟩
}
```

In order to facilitate the addition of points we maintain the dart *e_last*; it is the hull dart that leaves the most recently added vertex. Let p be the point to be added and let *e_up* and *e_low* be hull darts such that exactly the hull vertices between the target of *e_up* and the source of *e_low* are visible from p, see Figure 10.15. All edges e between *e_up* and *e_low* are such that p, the source of e, and the target of e form a left turn, but *e_up* and *e_low* do not have this property. Moreover, *e_up* is a proper face cycle predecessor of *e_last*, and *e_low* is a face cycle successor of *e_last*. Thus it is easy to determine *e_up* and *e_low*. For example, the former is the first proper face cycle predecessor e of *e_last* such that p, the source of e, and target of e do not form a left turn.

Having determined *e_up* we walk to *e_low* and extend the triangulation by adding edges between v, where v is a new node corresponding to point p, and the hull vertices visible from p. We must be careful to add the new edges in a way that reflects the triangulation. We iterate over the hull darts between *e_up* inclusive and *e_low* exclusive, starting at *e_up* and walking towards *e_low*. Consider any such e and let *e_succ* be its face cycle successor. We add the dart $(source(e\_succ), v)$ after *e_succ* to $A(source(e\_succ))$ and we append the dart $(v, source(e\_succ))$ to $A(v)$. Observe that this way of adding darts builds $A(v)$ in counter-clockwise order and adds the dart $(source(e\_succ), v)$ at the proper position to $A(source(e\_succ))$.

The update step just described works correctly even if the new point is collinear with all preceding points. In this situation only a line segment is added to the triangulation.

⟨*triangulate points: scan remaining points*⟩≡

```
POINT p;
forall(p,L)
{ if (p == last_p) continue;
  edge e =  G.last_adj_edge(last_v);
```

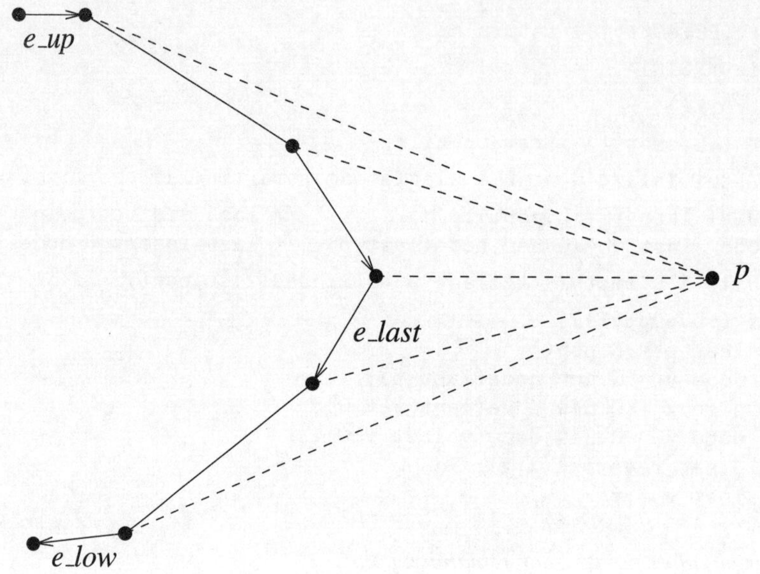

Figure 10.15 Edges *e_last*, *e_up*, and *e_low*.

```
    last_v = G.new_node(p);
    last_p = p;
    // walk up to upper tangent
    do e = G.face_cycle_pred(e); while (left_bend(p,G,e));
    // now e = e_up
    // walk down to lower tangent and triangulate
    do { edge succ_e = G.face_cycle_succ(e);
         edge x = G.new_edge(succ_e,last_v,after,0);
         edge y = G.new_edge(last_v,source(succ_e),0);
         G.set_reversal(x,y);
         e = succ_e;
       } while (left_bend(p,G,e));
  }
```
⟨*mark edges of convex hull as HULL_DARTS*⟩

In the pieces of code above we labeled all new edges with zero. We now relabel all hull darts as such. The last edge added to the triangulation is a hull dart and all other hull darts are reached by tracing the face cycle containing it. The labeling of the hull darts will prove useful in the section on Delaunay diagrams.

We return a hull dart.

⟨*mark edges of convex hull as HULL_DARTS*⟩≡

```
edge hull_dart = G.last_edge();
if (hull_dart)
{ edge e = hull_dart;
  do { G[e] = HULL_DART;
```

K	n	Gen	V	Hull	Hull check	Triang	Triang check
S	20000	0.44	25	1.71	0	3.1	23.7
S	40000	0.92	29	3.65	0	6.43	47.35
S	80000	1.84	35	7.52	0	13.04	94.31
D	20000	0.41	91	1.9	0	3.13	24.72
D	40000	0.73	123	3.6	0	6.26	47.29
D	80000	1.47	147	7.72	0	13.15	94.36
C	20000	47.3	19992	1.69	0.17	2.62	21.19
C	40000	95.59	39958	3.59	0.42	5.47	42.01
C	80000	190.9	79756	8.08	1.32	11.65	86.39

Table 10.3 The running times of the sweep algorithms for convex hulls and triangulations. We generated unsorted lists of n points according to the same distributions as in Table 10.1. The meaning of the first four columns is as in Table 10.1. The column "Hull" shows the time to compute the convex hull, the column "Hull check" shows the time to verify that any three consecutive vertices of *CH* form a right turn, the column "Triang" shows the time to compute the triangulation, and the column "Triang check" shows the time to run *Is_Triangulation*(G).

```
      e = G.face_cycle_succ(e);
    } while (e != hull_dart);
}
return hull_dart;
```

Table 10.3 compares the running times of the sweep algorithms for convex hulls and triangulations. We generated n random points in a square, a disc, and on a circle, respectively, The triangulation algorithm takes about twice as long as the convex hull program. The table also shows the time for partially checking the output of either program. For the convex hull program we checked that any three consecutive vertices form a right turn and in the case of triangulations we called the checker *Is_Triangulation*(G), which will be discussed in the next section.

Both checks are only partial. In the case of triangulations we do not check that exactly the input points appear as vertices of the triangulation. This omission could be corrected by the use of a dictionary. In the case of the convex hull program we do not verify that all input points lie inside the produced convex chain. This is an omission which is not easily corrected; the obvious approach takes quadratic time.

Exercises for 10.2

1 Write a program that verifies that the nodes of a *GRAPH<POINT, int>* agree with the points in a *list<POINT>*. Add this to the check of the triangulation program.

2 Extend the randomized incremental construction of convex hulls to an incremental construction of triangulations.

10.3 Verification of Geometric Structures, Basics

We have by now seen programs to compute convex hulls, minimum width stripes, and triangulations. The programs are non-trivial and we will see more complex programs in later sections. Although we wrote the documentation and the correctness proofs in parallel to the development of these programs, we nevertheless made mistakes, some minor, like testing for positive orientation instead of non-negative orientation, and some major, like assuming that every set of points contains three non-collinear points. *Visual debugging*, i.e., displaying the output of a geometric computation, was an indispensable aid in getting the programs correct, but visual debugging has its limits. Visual debugging is most useful in the plane; already displaying a partition of three-space is next to impossible. Also, the representation underlying a geometric object may be incorrect, although the object itself "looks correct".

One of our key experiences was the development of a program to compute convex hulls in arbitrary dimensions. It took some time to get the programs working for points in the plane, but after some time it produced convex hulls which "looked right". We moved to three-space and a few hours later the convex hulls in three-space looked right. We got adventurous and tried an example in seven-dimensional space. The program ran to completion and claimed that it had computed the convex hull. Given our past experience we had every reason to believe the contrary. At that time we had no way to check the result of the convex hull computation. We teamed up with some colleagues and wrote [MNS+96]. In this paper we discuss how to verify convex hulls, triangulations, Delaunay diagrams, and Voronoi diagrams. Alternative checkers are discussed in [DLPT97].

In this section and in Sections 10.4.3, 10.4.6, and 10.5.3 we derive procedures to verify properties of *geometric graphs*. A geometric graph is a straight line embedded map. Every node is mapped to a point in the plane and every dart is mapped to the line segment connecting its endpoints. We start with procedures to check that the edges around vertices are cyclically ordered, that face cycles define convex polygons, and that a graph defines a convex subdivision or a triangulation. In later sections we will extend these functions to check Delaunay triangulations, Delaunay diagrams, and Voronoi diagrams.

We use *geo_graph* as a template parameter for geometric graphs. Any instantiation *geo_graph_inst* of *geo_graph* must provide a function

```
VECTOR edge_vector(const geo_graph_inst& G, const edge& e)
```

that returns a vector from the source to the target of e. We will use two instantiations of *geo_graph* in this chapter: *GRAPH<POINT, int>* for triangulations, Delaunay triangulations, and Delaunay diagrams, and *GRAPH<CIRCLE, POINT>* for Voronoi diagrams. In the first case, the position of a node v is given by the point $G[v]$ and hence the edge vector function can be realized as

⟨*GRAPH⟨POINT,int⟩: edge vector function*⟩≡

```
static VECTOR edge_vector(const GRAPH<POINT,int>& G, const edge& e)
{ return G[G.target(e)] - G[G.source(e)]; }
```

In the second case, the position of a node v is given by the center of the circle $G[v]$. We will define the corresponding edge vector function in the section on Voronoi diagrams.

All functions that check properties of geometric graphs are collected in the file

⟨*geo_check.t*⟩≡
  ⟨*comparing edges by angle*⟩
  ⟨*cyclically ordered lists*⟩
  ⟨*verifying the order of adjacency lists and the convexity of faces*⟩

in directory LEDA/templates. This file must be included to use any of these functions.

10.3.1 *Monotone and Cyclically Monotone Sequences*

Let $x = (x_1, x_2, \ldots, x_n)$ be a sequence of elements from some ordered set; x is called *non-decreasing* if $x_i \le x_{i+1}$ for all i, $1 \le i < n$, and x is called *increasing* if $x_i < x_{i+1}$ for all i, $1 \le i < n$, x is called *cyclically non-decreasing* iff some cyclic shift of x is non-decreasing, and x is called *cyclically increasing* iff some cyclic shift of x is increasing. The notions non-increasing, decreasing, cyclically non-increasing, and cyclically decreasing are defined analogously.

The functions *Is_C_Nondecreasing* and *Is_C_Increasing* check whether a sequence is cyclically non-decreasing or increasing. They take a list L of elements of some type T and a compare object *cmp* for type T.

The implementation is simple. We iterate over the elements of L and compare every element with its cyclic successor. We count how often the successor is smaller (smaller or equal for the second function). If the count reaches two, the sequence violates the property.

⟨*cyclically ordered lists*⟩≡

```
template <class T>
bool Is_C_Nondecreasing(const list<T>& L, const leda_cmp_base<T>& cmp)
{ list_item it;
  int number_of_less = 0;
  forall_items(it,L)
    if ( cmp(L[L.cyclic_succ(it)],L[it]) < 0 ) number_of_less++;
  return (number_of_less < 2);
}
template <class T>
bool Is_C_Increasing(const list<T>& L, const leda_cmp_base<T>& cmp)
{ list_item it;
```

```
    int number_of_lesseq = 0;
    forall_items(it,L)
      if ( cmp(L[L.cyclic_succ(it)],L[it]) <= 0 ) number_of_lesseq++;
    return (number_of_lesseq < 2);
  }
```

The functions *Is_C_Nonincreasing* and *Is_C_decreasing* are defined analogously. We leave their implementation to the reader.

10.3.2 *Comparing Edges by Angle*

For a non-zero two-dimensional vector v let $\alpha(v)$ be the angle between the positive x-axis and v, i.e., the angle by which the positive x-axis has to be turned counter-clockwise until it aligns with v. The geo kernels provide functions

```
int compare_by_angle(const VECTOR& v1,const VECTOR& v2)
```

that compare vectors by angle, i.e., the functions return -1 if *v1* precedes *v2*, 0 if *v1* and *v2* define the same angle, and $+1$ if *v1* succeeds *v2*. The zero vector precedes all non-zero vectors in the ordering by angle.

In a geometric graph G the function *edge_vector*(G, e) returns the vector from the source to the target of edge e. The compare object *cmp_edges_by_angle* compares the edges of any *geo_graph* G according to the vectors defined by the edges of G. It is derived from *leda_cmp_base<edge>*, has a constructor that takes a geometric graph G and stores a reference to it in the object, and a function operator that takes two edges e and f and compares them according to the vectors defined by them.

⟨*comparing edges by angle*⟩≡

```
template <class geo_graph>
class cmp_edges_by_angle: public leda_cmp_base<edge> {
  const geo_graph& G;
public:
  cmp_edges_by_angle(const geo_graph& g): G(g){}
  int operator()(const edge& e, const edge& f) const
  { return compare_by_angle(edge_vector(G,e), edge_vector(G,f)); }
};
```

10.3.3 *Counter-Clockwise Ordered Adjacency Lists*

The function

```
bool Is_CCW_Ordered(const geo_graph& G)
```

returns true if for all nodes v the neighbors of v are in increasing counter-clockwise order around v, and the function

```
bool Is_CCW_Ordered_Plane_Map(const geo_graph& G)
```

returns true if, in addition, G is a plane map. The function

```
void SORT_EDGES(geo_graph& G)
```

reorders the adjacency lists such that for every node v of G the edges in $A(v)$ are in non-decreasing order by angle.

All three functions are very easy to implement. For the first function, we define a compare object *cmp* to compare the darts of G by angle, and then check whether the darts out of every node v are cyclically increasing. The second function calls the first and checks whether G is a plane map, and the third function sorts the set of darts and then rearranges the adjacency lists.

⟨*verifying the order of adjacency lists and the convexity of faces*⟩≡

```
template <class geo_graph>
bool Is_CCW_Ordered(const geo_graph& G)
{ node v;
  cmp_edges_by_angle<geo_graph> cmp(G);
  forall_nodes(v,G)
    if ( !Is_C_Increasing(G.out_edges(v),cmp) ) return false;
  return true;
}
template <class geo_graph>
bool Is_CCW_Ordered_Plane_Map(const geo_graph& G)
{ return Is_Plane_Map(G) && Is_CCW_Ordered(G); }
template <class geo_graph>
bool Is_CCW_Weakly_Ordered(const geo_graph& G)
{ node v;
  cmp_edges_by_angle<geo_graph> cmp(G);
  forall_nodes(v,G)
    if ( !Is_C_Nondecreasing(G.out_edges(v),cmp) ) return false;
  return true;
}
template <class geo_graph>
bool Is_CCW_Weakly_Ordered_Plane_Map(const geo_graph& G)
{ return Is_Plane_Map(G) && Is_CCW_Weakly_Ordered(G); }
template <class geo_graph>
void SORT_EDGES(geo_graph& G)
{
  cmp_edges_by_angle<geo_graph> cmp(G);
  list<edge> L = G.all_edges();
  L.sort(cmp);
  G.sort_edges(L);
}
```

10.3.4 *Convex Faces*

We define functions that check for convexity of faces. Consider any face cycle f of a geometric graph G; f defines a closed polygonal chain C in the plane. We want to know

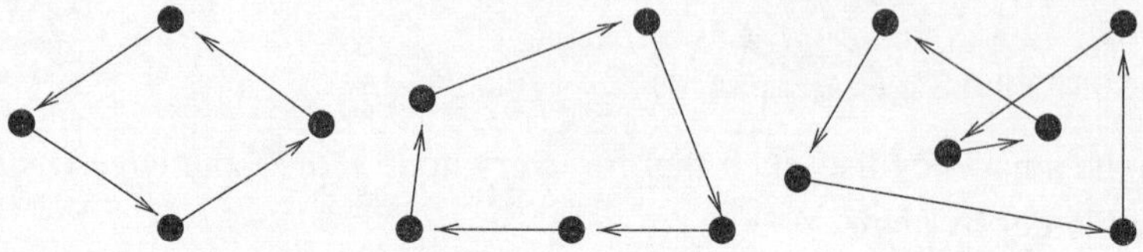

Figure 10.16 A strictly convex counter-clockwise polygonal chain, a weakly convex clockwise polygonal chain, and a chain which is not simple.

whether the polygonal chain is the boundary of a convex region. More precisely, we call C a *weakly convex counter-clockwise polygonal chain* if C is simple, i.e., does not intersect itself, and the region to the left of C is convex. We call C a *strictly convex counter-clockwise polygonal chain* or simply *convex counter-clockwise polygonal chain* if, in addition, any two consecutive edges of C do not have the same direction, see Figure 10.16. For clockwise chains the region to the right of C must be convex.

In a convex subdivision, e.g., a triangulation, the face cycles of all bounded faces form convex counter-clockwise polygonal chains, and the face cycle of the unbounded face forms a weakly convex clockwise polygonal chain.

Let $p_0, p_1, \ldots, p_{k-1}$ be the points associated with the nodes of C.

Lemma 65 *C is a counter-clockwise weakly convex polygonal chain iff the sequence $s = (p_1 - p_0, p_2 - p_1, \ldots, p_0 - p_{k-1})$ is cyclically non-decreasing.*

Proof If C is a counter-clockwise weakly convex polygonal chain then s is clearly cyclically non-decreasing.

Assume next that s is cyclically non-decreasing. Then no pair of consecutive vectors forms a right turn and the angles between all pairs of consecutive vectors sum to 2π. We conclude that C is simple, i.e, does not intersect itself, and that the region to the left of C is convex. □

The functions

```
bool Is_CCW_Convex_Face_Cycle(const geo_graph& G,const edge e)
bool Is_CCW_Weakly_Convex_Face_Cycle(const geo_graph& G, const edge e)
bool Is_CW_Convex_Face_Cycle(const geo_graph& G, const edge e)
bool Is_CW_Weakly_Convex_Face_Cycle(const geo_graph& G, const edge e)
```

return true if the face cycle of G containing e has the stated property, i.e., if the face cycle forms a cyclically increasing, non-decreasing, decreasing, or non-increasing, respectively, sequence of edges according to the compare-by-angles ordering.

We give the implementation of the first function. We collect the edges of the face cycle in a list L, define a compare object *cmp* that compares edges of G, and then check whether L is cyclically increasing.

⟨*verifying the order of adjacency lists and the convexity of faces*⟩+≡

```
template <class geo_graph>
bool Is_CCW_Convex_Face_Cycle(const geo_graph& G, const edge& e)
{
  list<edge> L;
  edge e1 = e;
  do { L.append(e1);
       e1 = G.face_cycle_succ(e1);
  } while ( e1 != e );
  cmp_edges_by_angle<geo_graph> cmp(G);
  return Is_C_Increasing(L,cmp);
}
```

10.3.5 *Convex Subdivisions*

A geometric graph G is a *convex planar subdivision*, if G is a plane map and if the positions assigned to the nodes of G define a straight line embedding of G in which all bounded faces are strictly convex and in which the unbounded face is weakly convex.

The function

```
bool Is_Convex_Subdivision(const GRAPH<POINT,int>& G)
```

returns true if G is a convex planar subdivision, and the function

```
bool Is_Triangulation(const geo_graph& G)
```

returns true if G is a convex planar subdivision in which every bounded face is a simplex. More precisely, if all nodes of G lie on a common line, then every face cycle of a bounded face is simply a pair of anti-parallel edges, and if the nodes of G do not lie on a common line, then every bounded face of G is a triangle.

Both functions are implemented in terms of the function

```
bool Is_Convex_Subdivision(const GRAPH<POINT,int>& G,
                           bool& is_triangulated)
```

that returns true if G is a convex subdivision and sets *is_triangulated* to true if, in addition, G is a triangulation.

We discuss the theory behind the latter function and then give its implementation. If G is a convex subdivision, then the following conditions are certainly satisfied:

- G is a connected plane map.
- All nodes of G have counter-clockwise ordered adjacency lists.
- If all vertices lie on a common line, i.e., the underlying point set has affine dimension less than 2, then G is a path which reflects the ordering of its vertices on the line.
- If the underlying point set has affine dimension 2, then each face is either a bounded counter-clockwise oriented convex polygon or a clockwise oriented weakly convex polygon. There is only one face of the latter kind.

Lemma 66 *If G satisfies the four conditions above, then G is a convex planar subdivision.*

Proof Assume first that all vertices of G lie on a line l and let $v_1, v_2, \ldots, v_n$ be the ordering of the vertices on l. Then the points assigned to adjacent vertices must be distinct, v_1 and v_n must have degree one, and v_i must have neighbors v_{i-1} and v_{i+1} for $1 < i < n$. The number of edges of G is $2n - 2$ where n is the number of nodes of G.

Assume next that not all vertices of G lie on a common line. Let R be the region that is enclosed by the unique face cycle f which is a weakly convex clockwise polygon. We claim that all vertices that are not part of f lie in the interior of R. Assume otherwise. Then there must be a vertex v that is not part of f and a direction d such that v is a maximal vertex of G in direction d (note that we said "a maximal vertex" and not "the maximal vertex"). Since v is maximal there must be a pair of edges incident to v which span an angle of at least π and hence v must be part of a weakly convex chain. Thus v belongs to f, a contradiction.

Every face cycle of G different from f defines a counter-clockwise oriented convex polygonal region in the plane. We need to show that these regions form a partition of R. Consider a point p moving in the plane such that it avoids vertices of G. Whenever p crosses a directed edge e it will enter another region (namely, the one to the left of *reversal*(e)) except when *reversal*(e) belongs to f. This shows that all points in the interior of R are covered by the same number of regions. Also, since all vertices on the boundary of R are part of f, exactly one bounded region is incident to each edge of f. Altogether we have shown that the regions defined by the face cycles different from f partition R. The number of edges of G must be at least $2n$ since every node must have degree at least two. □

We turn to the implementation. We first check whether G is a connected plane map in which all adjacency lists are counter-clockwise ordered. Then we compare m and n. If $m = 2n - 2$ we must be in the situation that all vertices of G are collinear and if $m > 2n - 2$ we must be in the situation that the underlying point set has affine dimension 2.

⟨*subdivision_check.c*⟩+≡

```
static bool False(const string& s)
{ cerr << "Is_Convex_Subdivision: " << s; return false; }
bool Is_Convex_Subdivision(const GRAPH<POINT,int>& G,
                           bool& is_triangulated)
{
  is_triangulated = true;
  if ( !Is_Connected(G) ) return False("G is not connected");
  if ( !Is_CCW_Ordered_Plane_Map(G) )
    return False("G is not a CCW-ordered plane map");
  int n = G.number_of_nodes();
  int m = G.number_of_edges();
  cmp_edges_by_angle<GRAPH<POINT,int> > cmp(G);
  if ( m == 2*n - 2) { ⟨ICS: collinear points⟩ }
  ⟨ICS: affine dimension is two⟩
}
```

If $m = 2n - 2$, the fact that G is a connected bidirected graph guarantees that G is a tree. It therefore suffices to check that there is no vertex of degree three and that for every vertex of degree two the two incident edges point in opposite directions.

⟨*ICS: collinear points*⟩≡
```
node v;
if ( n <= 1 ) return true;
forall_nodes(v,G)
{ if ( G.outdeg(v) > 2 ) return False("G is a tree but not a chain");
  if (G.outdeg(v) == 1) continue;
  edge e1 = G.first_adj_edge(v), e2 = G.last_adj_edge(v);
  node w = G.target(e1);
  node u = G.target(e2);
  if ( G[v] == G[w] || G[v] == G[u] )
    return False("nodes at equal positions");
  if ( cmp(e1,G.reversal(e2)) != 0 )
     return False("direction not opposite");
}
return true;
```

It remains to deal with the situation that the affine dimension of the underlying point set is 2. We trace all face cycles of G. One face cycle must be a weakly convex clockwise oriented polygon and all other face cycles must be strongly convex counter-clockwise polygons. We make the distinction by considering three consecutive nodes of a face cycle and determining their orientation. If the orientation is positive, the face cycle must be a strongly convex counter-clockwise polygon, and if the orientation is non-positive, the face cycle must be the boundary of the unbounded face.

If the number of edges of the face cycle is three, the orientation test itself guarantees strong convexity and there is no need to trace the face cycle to check convexity.

⟨*ICS: affine dimension is two*⟩≡
```
edge e;
edge_array<bool> considered(G,false);
bool already_seen_unbounded_face = false;
forall_edges(e,G)
{ if ( !considered[e] )
  { // check the face to the left of e
    POINT a = G[source(e)];
    POINT b = G[target(e)];
    POINT c = G[target(G.face_cycle_succ(e))];
    int orient = orientation(a,b,c);
    int n = 0;
    edge e0 = e;
    do { considered[e] = true;
         e = G.face_cycle_succ(e);
         n++;
    } while ( e != e0);
    if ( orient > 0 )
```

```
    { if ( n > 3 )
      { is_triangulated = false;
        if ( !Is_CCW_Convex_Face_Cycle(G,e) )
          return False("non-convex bounded face");
      }
    }
    else
    { if ( already_seen_unbounded_face )
        return False("two faces qualify for unbounded face");
      already_seen_unbounded_face = true;
      if ( !Is_CW_Weakly_Convex_Face_Cycle(G,e) )
        return False("unbounded face is not weakly convex");
    }
  }
}
return true;
```

Exercises for 10.3

1 Improve the implementation of *Is_CWW_Ordered* and the functions checking convexity of faces. In our implementation we first construct a list of edges and then check this list for cyclic monotonicity. Avoid the construction of the list.

2 Improve the theory underlying *Is_Convex_Subdivision*. Is it necessary to check whether the edges in $A(v)$ are CCW-ordered or does this property follow from the condition that all bounded faces are counter-clockwise strongly convex polygonal chains?

3 Extend the function *Is_Convex_Subdivision* such that it works for *geo_graph* and not only for *GRAPH<POINT, int>*.

10.4 Delaunay Triangulations and Diagrams

A point set may in general be triangulated in many different ways. Depending on the application one triangulation is preferable over another. A triangulation that is useful in many contexts is the so-called *Delaunay triangulation*. A triangulation of a point set S is called *Delaunay* if the interior of the circumcircle of any triangle in the triangulation contains no point of S. Figure 10.17 shows a Delaunay triangulation. The voronoi_demo and the point_set_demo in xlman illustrate Delaunay diagrams.

In this section we will first show the existence of Delaunay triangulations. The existence proof is constructive and yields a simple algorithms for the construction of Delaunay triangulation, the so-calling *flipping algorithm*. We give an implementation of the algorithm based on the so-called *incircle test*, a powerful geometric primitive. The Delaunay triangulation of a point set is in general not unique (if the point set contains co-circular points); it has, however, a substructure which is unique, the so-called *Delaunay diagram*. We characterize Delaunay diagrams and give some applications of Delaunay diagrams and triangulations.

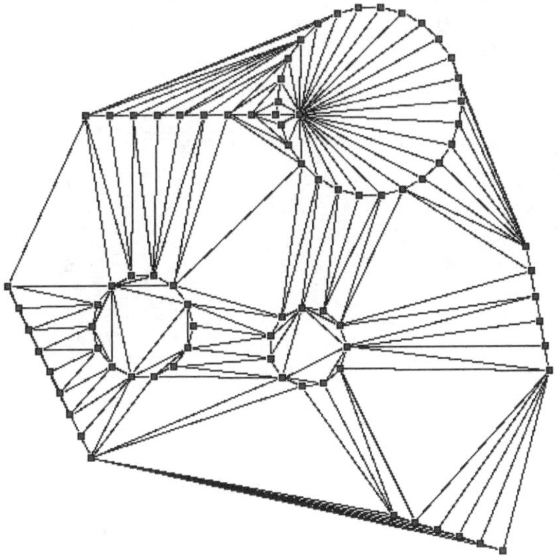

Figure 10.17 A Delaunay triangulation. The figure was produced with the voronoi_demo in xlman.

10.4.1 *Delaunay Triangulations and the Flipping Algorithm*

Our immediate goal is to prove that Delaunay triangulations exist. Consider the simplest situation first, four points p, q, r, and s forming the corners of a convex quadrilateral. There are two triangulations corresponding to the chords pr and qs, respectively, see Figure 10.18. We show that at least one of the two triangulations is Delaunay. Assume that the triangulation corresponding to the chord pr is not Delaunay, say because s is contained in the circumcircle of triangle $\Delta(p, q, r)$. Then q is also contained in the circumcircle of triangle $\Delta(p, r, s)$. We can obtain the circumcircle of triangle $\Delta(p, q, s)$ from the circumcircle of $\Delta(p, q, r)$ by reducing the size of the circle while simultaneously insisting that it passes through p and q. This shows that r is outside the circumcircle of triangle $\Delta(p, q, s)$ and that the radius of the circumcircle of $\Delta(p, q, s)$ is smaller than the radius of the circumcircle of $\Delta(p, q, r)$. The symmetric argument shows that p is outside the circumcircle of triangle $\Delta(q, r, s)$ and that the radii of both circles in the Delaunay triangulation are smaller than the radii of the circles in the other triangulation.

Let us next turn to point sets of larger cardinality. We show that any triangulation which is not Delaunay contains two adjacent triangles, i.e., triangles sharing an edge, that form a convex quadrilateral and such that the circumcircles of both triangles contain the third vertex of the other triangle. Clearly, a triangulation which is not Delaunay contains a triangle, say $\Delta(p, q, r)$ whose circumcircle is non-empty. Assume w.l.o.g. that there is a point s contained in the region R formed by the chord pq and the circular arc connecting p and q and not containing r, see Figure 10.19. Consider the other triangle incident to edge pq. If

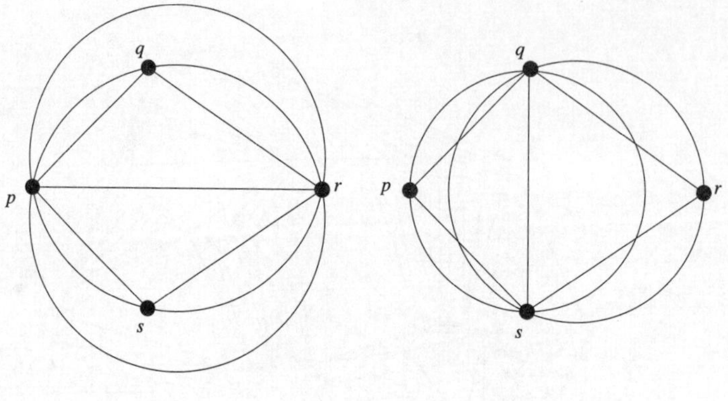

Figure 10.18 The two triangulations of a convex quadrilateral.

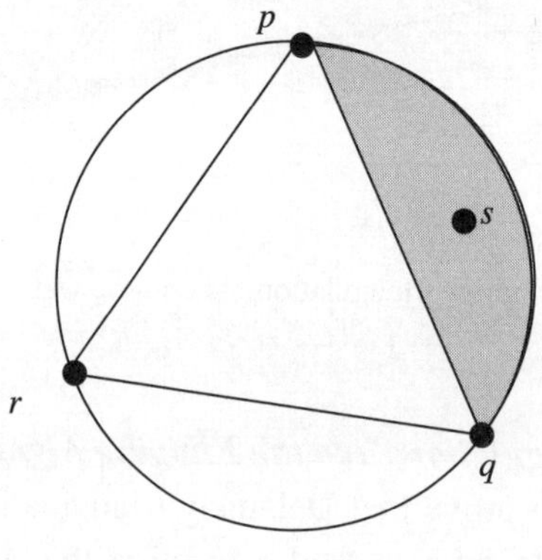

Figure 10.19 A triangle $\Delta(p, q, r)$ with non-empty circumcircle. Region R is shown shaded.

the third vertex of this triangle is also contained in R, we have identified the desired pair of triangles. If the third vertex, say t, is outside R then s is also contained in the circumcircle of triangle $\Delta(p, q, t)$ and s is closer to $\Delta(p, q, t)$ than to $\Delta(p, q, r)$. Here, the distance of a point to a triangle is the distance to the closest point of the triangle. We repeat the argument with triangle $\Delta(p, q, t)$ and point s. After a finite number of steps we must arrive at the first case.

We have now shown that any triangulation that is not Delaunay contains a convex quadrilateral formed by two adjacent triangles such that the triangulation of this quadrilateral is not Delaunay. The deletion of the common edge of both triangles and the insertion of the other diagonal of the quadrilateral is called a *diagonal-flip* or simply *flip*. A flip makes the triangulation locally Delaunay and also decreases the sum of the radii of the circumcircles of all triangles. We have thus arrived at the so-called flipping algorithm for Delaunay triangulations:

T = some triangulation;
while (T is not Delaunay)
{ find a pair of adjacent triangles that form a convex quadrilateral and whose triangulation is not Delaunay;

flip the diagonal of the quadrilateral;
}

The algorithm terminates since every flip reduces the sum of the radii of all circumcircles and hence no triangulation can repeat. The maximal number of flips performed by the flipping algorithm is $\Theta(n^2)$. We ask you in the exercises to construct a worst case point set. The upper bound follows from the fact that once a segment pq is flipped away it will never be reintroduced into the triangulation. The flipping algorithm is due to Lawson ([Law72]).

For points in convex position[4] there is also a so-called *furthest site Delaunay triangulation*. In a furthest site Delaunay triangulation of a set S the circumcircle of any triangle has no point of S in its exterior. The flipping algorithm can also be used to construct furthest site Delaunay triangulation. We start with an arbitrary triangulation of a set of points in convex position and flip as long as the triangulation is not furthest site Delaunay. Of course, this time we flip the diagonal of a convex quadrilateral if the third vertex of the other triangle is outside the circumcircle.

When it is necessary to emphasize the difference between ordinary Delaunay triangulations and furthest site Delaunay triangulations we call the former nearest site Delaunay triangulations. Some algorithms work for nearest and furthest site Delaunay triangulations. In these algorithms we use the enumeration type

```
enum delaunay_voronoi_kind { NEAREST, FURTHEST };
```

defined in LEDA/geo_global_enums.h to distinguish between the two kinds of triangulations.

As in the preceding section we use the type *GRAPH<POINT, int>* to represent triangulations. For every node v of G the associated point is given by $G[v]$. For every edge e of G, $G[e]$ is an integer in the enumeration type *delaunay_edge_info*. In the Delaunay triangulation all hull darts are labeled HULL_DART, and every other dart is labeled either DIAGRAM_DART or NON_DIAGRAM_DART. A non-hull dart is labeled DIAGRAM_DART if the circumcircles of the triangles incident to it are distinct and is labeled NON_DIAGRAM_DART otherwise. The reversals of hull darts are labeled DIAGRAM_DART.

The functions

```
void DELAUNAY_TRIANG(  const list<POINT>& L, GRAPH<POINT,int>& G);
void F_DELAUNAY_TRIANG(const list<POINT>& L, GRAPH<POINT,int>& G);
```

compute the nearest site and the furthest site Delaunay triangulation of a list L of points.

[4] A set S of points is in convex position if every point in S is a vertex of the convex hull of S.

10.4.2 *The Flipping Algorithm*

We turn the flipping algorithm into a program[5]. The flipping algorithm works for nearest and furthest site Delaunay triangulations.

We assume that we start with a triangulation G in which all hull darts are labeled with the label HULL_DART and in which all other darts have a label different from HULL_DART. The algorithm terminates with a Delaunay triangulation and returns the number of flips performed. For furthest site triangulations we assume further that the vertices of G are in convex position.

The algorithm maintains a set S of darts which may potentially violate the Delaunay property. Initially, S consists of one dart in each uedge of G. The algorithm terminates when S is empty. As long as S is non-empty, an arbitrary dart e of S is chosen. If it violates the Delaunay property, a flip is performed.

We define the integer f to be $+1$ if we are aiming for a nearest site diagram and to be -1 if we are aiming for a furthest site diagram. It will be used in the test for the Delaunay property.

⟨*flip_delaunay.c*⟩≡

```
int DELAUNAY_FLIPPING(GRAPH<POINT,int>& G, delaunay_voronoi_kind kind)
{
  if (G.number_of_nodes() <= 3) return 0;
  int f = ( kind == NEAREST ? +1 : -1);
  list<edge> S;
  edge e;
  forall_edges(e,G) if ( index(e) < index(G.reversal(e)) ) S.append(e);
  int flip_count = 0;
  while ( !S.empty() )
  { edge e = S.pop();
    edge r = G.reversal(e);
    ⟨check e for the Delaunay property and flip if necessary⟩
  }
  return flip_count;
}
```

Let e be a dart of the current triangulation. If e is a hull dart or the reversal of a hull dart, then no action is required as hull darts belong to every Delaunay triangulation. If e is not a hull dart, define edges r, e_1, and e_3, and points a, b, c, and d as in Figure 10.20; r is the reversal of e, e_1 is the face cycle successor of r, e_3 is the face cycle successor of e, a and b are source and target of e_1, and c and d are source and target of e_3. The quadrilateral (a, b, c, d) is convex if and only if the interior angles at vertices a and c are less than 180°, i.e., if (d, a, b) and (b, c, d) are left turns.

5 The program delaunay_flip_anim in LEDAROOT/book/Geo animates the algorithm.

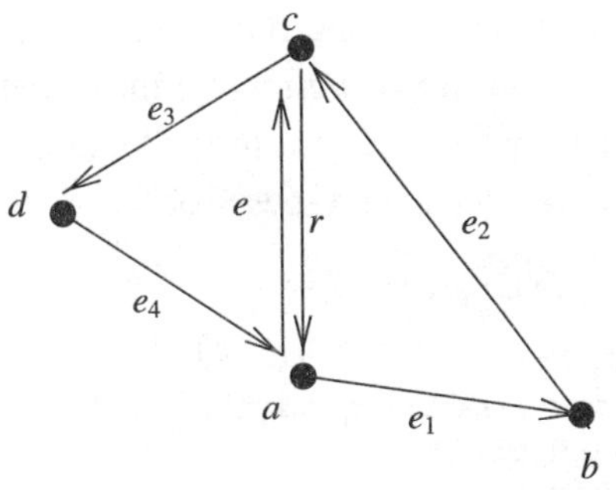

Figure 10.20 The edges e, r, e_1, e_2, e_3, and e_4, and the points a, b, c, and d.

⟨*check e for the Delaunay property and flip if necessary*⟩≡

```
if (G[e] == HULL_DART || G[r] == HULL_DART) continue;
G[e] = DIAGRAM_DART;
G[r] = DIAGRAM_DART;
// e1,e2,e3,e4: edges of quadrilateral with diagonal e
edge e1 = G.face_cycle_succ(r);
edge e3 = G.face_cycle_succ(e);
// flip test
POINT a = G[source(e1)];
POINT b = G[target(e1)];
POINT c = G[source(e3)];
POINT d = G[target(e3)];
if ( left_turn(d,a,b) && left_turn(b,c,d) )
{ // the quadrilateral is convex
  ⟨check circle property and flip if necessary⟩
}
```

Assume now that the quadrilateral (a, b, c, d) is convex. The triangulation is locally Delaunay if d does not lie inside the circle defined by (a, b, c), and can be improved by a flip if d lies inside the circle. For the furthest site triangulation the situation is reversed. The test

```
side_of_circle(a,b,c,d)
```

returns

$+1$ if d is left of the oriented circle through a, b, and c,
0 if $|\{a, b, c\}| \leq 2$ or d lies on the oriented circle through a, b, and c,
-1 if d is right of the oriented circle through a, b, and c.

Let $soc = f \cdot side\_of\_circle(a, b, c, d)$. If *soc* is zero, the four points are co-circular, and no flip is required. However, e and r have to be relabeled with NON_DIAGRAM_DART. If *soc* is positive, d lies inside the circumcircle of the triangle (a, b, c) (outside for furthest site triangulations) and a flip is required. Let e_2 and e_4 be the other two edges of the quadrilateral (a, b, c, d). We move e and r to the other diagonal of the quadrilateral. More precisely, we insert e after e_2 into $A(source(e_2))$[6] and make $source(e_4)$ the target of e, and we insert r after e_4 into $A(source(e_4))$ and make $source(e_2)$ the target of r. We also add all four sides of the quadrilateral to S to make sure that their Delaunay property is rechecked. Observe that flipping e may affect the "Delaunay-ness" of the sides of the quadrilateral.

⟨*check circle property and flip if necessary*⟩≡

```
int soc = f * side_of_circle(a,b,c,d);
if (soc == 0) // co-circular quadrilateral(a,b,c,d)
{ G[e] = NON_DIAGRAM_DART;
  G[r] = NON_DIAGRAM_DART;
}
if (soc > 0) // flip
{ edge e2 = G.face_cycle_succ(e1);
  edge e4 = G.face_cycle_succ(e3);
  S.push(e1);
  S.push(e2);
  S.push(e3);
  S.push(e4);
  // flip diagonal
  G.move_edge(e,e2,source(e4));
  G.move_edge(r,e4,source(e2));
  flip_count++;
}
```

In order to construct the Delaunay triangulation for a set of points we first triangulate the set of points and then call the flipping algorithm to turn the triangulation into a Delaunay triangulation.

In the case of the furthest site Delaunay triangulation we first extract the vertices of the convex hull, then construct a triangulation of them, and finally use the flipping algorithm to obtain a furthest site Delaunay triangulation.

⟨*flip_delaunay.c*⟩+≡

```
int DELAUNAY_FLIP(const list<POINT>& L, GRAPH<POINT,int>& G)
{ TRIANGULATE_POINTS(L,G);
  if (G.number_of_edges() == 0) return 0;
  return DELAUNAY_FLIPPING(G,NEAREST);
}

int F_DELAUNAY_FLIP(const list<POINT>& L, GRAPH<POINT,int>& G)
{
  list<POINT> H = CONVEX_HULL(L);
  TRIANGULATE_POINTS(H,G);
  if (G.number_of_edges() == 0) return 0;
  return DELAUNAY_FLIPPING(G,FURTHEST);
}
```

[6] Recall that for a node v, $A(v)$ is the counter-clockwise ordered cyclic list of darts out of v.

10.4.3 *Verifying Delaunay Triangulations*

The function

```
bool Is_Delaunay_Triangulation(const GRAPH<POINT,int>& G,
                               delaunay_voronoi_kind kind);
```

checks whether G is a Delaunay triangulation of the points associated with its nodes. The flag *kind* allows us to choose between nearest and furthest site diagrams.

Let S be the set of points associated with the nodes of G. G is a Delaunay triangulation of S, if G is a triangulation and every triangle of G has the Delaunay property.

Thus the implementation is simple. First we check whether G is a triangulation. If the affine dimension of S is less than 2 this suffices; the affine dimension is less than 2 if $m = 2n - 2$. Otherwise, we walk over all edges. If an edge separates two triangles that form a convex quadrilateral we check the Delaunay property.

⟨*delaunay_check.c*⟩+≡

```
static bool False(const string& s)
{ cerr << "Is_Delaunay_Triangulation: " << s; return false; }
bool Is_Delaunay_Triangulation(const GRAPH<POINT,int>& G,
                               delaunay_voronoi_kind kind)
{ if ( !Is_Triangulation(G) ) return False("G is no triangulation");
  if (G.number_of_edges() == 2*G.number_of_nodes() - 2) return true;
  ⟨check Delaunay property⟩
  return true;
}
```

where

⟨*check Delaunay property*⟩≡

```
edge e;
edge_array<bool> considered(G,false);
forall_edges(e,G)
{ if (!considered[e])
  { // check the faces incident to e and reversal(e)
    considered[e] = considered[G.reversal(e)] = true;
    POINT a = G[source(e)];
    POINT b = G[target(G.cyclic_adj_pred(e))];
    POINT c = G[target(e)];
    POINT d = G[target(G.face_cycle_succ(e))];
    if (left_turn(a,b,c) && left_turn(b,c,d) &&
           left_turn(c,d,a) && left_turn(d,a,b) )
    { // the faces to the left and right of e are bounded
      int s = side_of_circle(a,b,c,d);
      /* +1 for inside, -1 for outside */
      if ( (kind == NEAREST && s > 0) || (kind == FURTHEST && s < 0) )
        return False("violated Delaunay property");
    }
  }
}
```

K	n	Flipping	Guibas–Stolfi	Dwyer	Check
S	20000	26.4	17.36	8.57	25.63
S	40000	56.89	37.45	17.44	51.66
S	80000	122.1	79.61	36.35	102.7
D	20000	26.13	17.22	8.71	25.53
D	40000	56.28	37.1	17.62	51.09
D	80000	120.8	78.49	36.92	102.7
C	20000	14.66	10.6	11.09	27.72
C	40000	29.74	21.73	22.89	55.87
C	80000	60.74	44.55	45.29	111

Table 10.4 The running times of Delaunay triangulation algorithms. The first column designates the kind of input (S for random points in a square, D for random points in a disk, C for random points near a circle), and the other columns show the number of points, the running time of the flipping algorithm, the running time of the algorithm of Guibas and Stolfi, the running time of the algorithm of Dwyer, and the time to verify the correctness of the result, respectively.

10.4.4 *Other Algorithms for Delaunay Triangulations*

The flipping approach yields a simple but not the most efficient Delaunay triangulation algorithm. There are $O(n \log n)$ algorithms based on sweeping [For87], on divide-and-conquer [GS85, Dwy87], and on randomized incremental construction [BT93]. The paper [SD97] compares many Delaunay algorithms.

In LEDA the divide-and-conquer algorithms of Guibas and Stolfi and of Dwyer are available. Table 10.4 shows an experimental comparison of the flipping algorithm with the two divide-and-conquer algorithms. The algorithm of Dwyer is consistently the best and therefore we use it as our default implementation. For the furthest site diagram we only have the flipping algorithm.

⟨*delaunay.c*⟩≡

```
void DELAUNAY_TRIANG(const list<POINT>& L, GRAPH<POINT,int>& G)
{ DELAUNAY_DWYER(L,G); }

void F_DELAUNAY_TRIANG(const list<POINT>& L, GRAPH<POINT,int>& G)
{ F_DELAUNAY_FLIP(L,G); }
```

10.4.5 *Delaunay Diagrams*

The Delaunay triangulation of a set S is in general not unique, e.g., if S consists of the corners of a square, or more generally of four co-circular points, then both triangulations of S

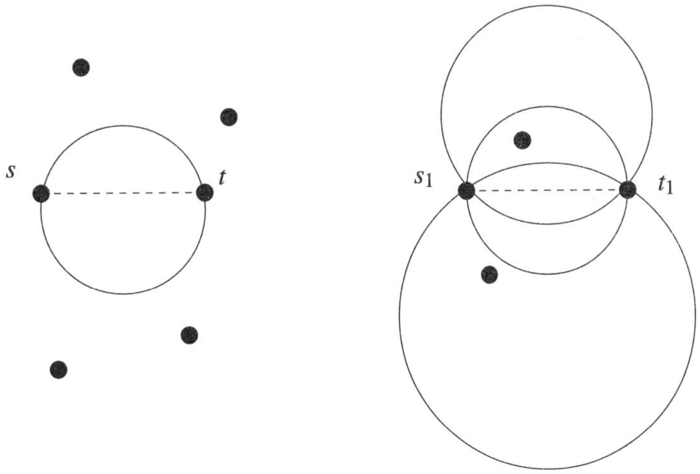

Figure 10.21 st is essential but s_1t_1 is not.

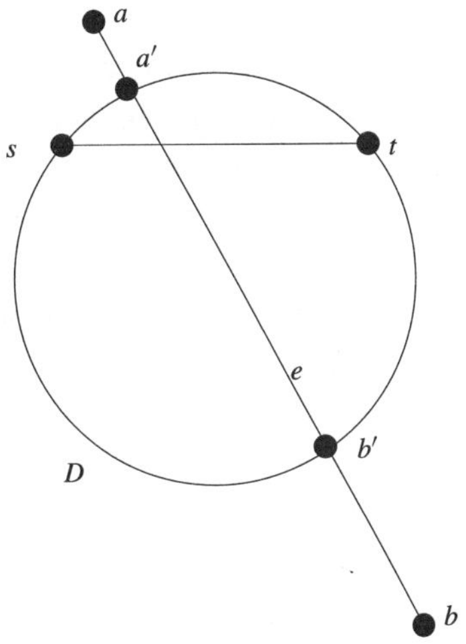

Figure 10.22 An essential segment st with its disk D and an edge $e = (a, b)$ of a Delaunay triangulation intersecting st.

are Delaunay. We now characterize the segments that belong to all Delaunay triangulations. Let s and t be two distinct points in S. A segment st is called *essential* if there is a closed disk D with $S \cap D = \{s, t\}$. In other words, there is a circle passing through s and t such that s and t are the only points of S contained in the closure of the circle, see Figure 10.21. We have

Lemma 67 *Let S be a finite set of points in the plane and let s and t be distinct points in S. The segment st is essential if and only if it belongs to every Delaunay triangulation of S.*

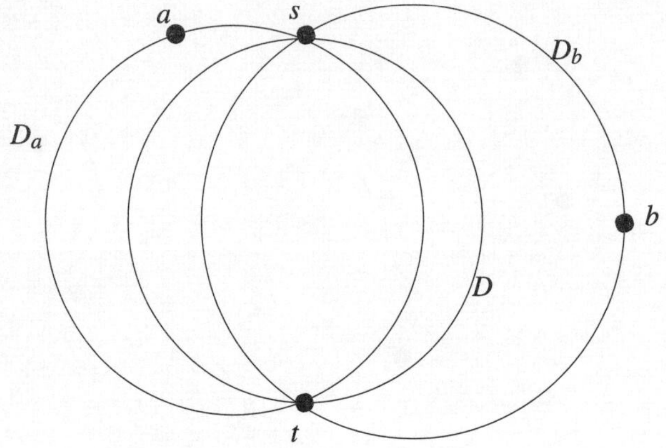

Figure 10.23 The discs D_a, D_b, and D.

Proof We first show that essential segments belong to all Delaunay triangulations. Assume otherwise, say st is essential but does not belong to some Delaunay triangulation T. Then st cannot be an edge of the convex hull of S because any such edge belongs to every triangulation. The open segment st is therefore contained in the interior of conv S. Imagine traveling along the segment st from s to t. In the vicinity of s the segment st runs inside some triangle of T and in the vicinity of t it runs inside some other triangle of T. We conclude that the segment st must intersect an edge $e = (a, b)$ of T. Since st is essential there is a closed disk D with $S \cap D = \{s, t\}$. Let a' and b' be the intersections of the boundary of D with edge e, see Figure 10.22. The four points a', s, b', and t form the corners of a convex quadrilateral and are co-circular. This implies that any closed disk containing the segment $a'b'$ must also contain either s or t. Consider next any of the triangles of T incident to e. The circumcircle of this triangle contains the segment $a'b'$ in its interior and hence also contains either s or t in its interior. The triangle is therefore not Delaunay, a contradiction. This proves that essential edges are part of every Delaunay triangulation.

To show the converse consider a non-essential segment st. We will construct a Delaunay triangulation that does not contain st. Let T be any Delaunay triangulation of S. If st is not an edge of T we are done. Otherwise, consider the two triangles Δ and Δ' incident to st in T; it is easy to see that st is not a hull edge and hence the two triangles exist. Let a and b be the third vertices of Δ and Δ', respectively. If the four points s, a, t, b are co-circular then we may replace st by ab and stay Delaunay. So, assume that the four points are not co-circular. Then b is outside the closed disk D_a having s, a, and t on its boundary and a is outside the closed disk D_b having s, b, and t on its boundary, see Figure 10.23. Consider the closed disk D having s and t on its boundary and having its center at the midpoint of the centers of D_a and D_b; all of D (except for s and t) is contained in the interior of $D_a \cup D_b$. Thus, $D \cap S \subseteq \{s, t\}$ and st is essential, a contradiction. □

We can now define the *Delaunay diagram* of a set S of points. It consists of all essential segments defined by the points in S and is denoted $DD(S)$. The Delaunay diagram is a subgraph of every Delaunay triangulation. The Delaunay diagram is a planar graph whose bounded faces are convex polygons all of whose vertices are co-circular. If no four points of S are co-circular then all bounded faces are triangles and the Delaunay diagram is a triangulation.

It is trivial to construct the Delaunay diagram from a Delaunay triangulation. We only have to delete all edges that are labeled NON_DIAGRAM_DART.

⟨*delaunay.c*⟩+≡

```
void DELAUNAY_DIAGRAM(const list<POINT>& L, GRAPH<POINT,int>& DD)
{
  DELAUNAY_TRIANG(L,DD);
  list<edge> el;
  edge e;
  forall_edges(e,DD) if ( DD[e] == NON_DIAGRAM_DART) el.append(e);
  forall(e,el) DD.del_edge(e);
}
```

For furthest site diagrams the construction is completely analogous and therefore not shown.

10.4.6 *Verifying Delaunay Diagrams*

We show how to verify Delaunay diagrams. The function

```
bool Is_Delaunay_Diagram(const GRAPH<POINT,int>& G,
                         delaunay_voronoi_kind kind);
```

checks whether G is a Delaunay diagram of the points associated with its nodes. The flag *kind* allows us to choose between nearest and furthest site diagrams. Let S be the set of points associated with the nodes of G.

It is clearly necessary that G is a convex subdivision in which the vertices of every bounded face (= a face whose face cycle is a convex counter-clockwise polygon) are co-circular. Assume this is the case. Then G is a Delaunay diagram if an arbitrary triangulation of G is a Delaunay triangulation. It therefore suffices to check the Delaunay property of all edges of G as in ⟨*check Delaunay property*⟩.

⟨*delaunay_check.c*⟩+≡

```
static bool False_IDD(const string& s)
{ cerr << "Is_Delaunay_Diagram: " << s; return false; }
bool Is_Delaunay_Diagram(const GRAPH<POINT,int>& G,
                         delaunay_voronoi_kind kind)
{
  if ( !Is_Convex_Subdivision(G) )
    return False_IDD("G is no convex subdivision");
  edge e;
  edge_array<bool> considered(G,false);
  forall_edges(e,G)
```

```
  { if (!considered[e])
    { // check the face to the left of e
      POINT a = G[source(e)];
      POINT c = G[target(e)];
      POINT d = G[target(G.face_cycle_succ(e))];
      if ( left_turn(a,c,d) )
      { // face is bounded
        CIRCLE C(a,c,d);
        edge e0 = e;
        do { considered[e] = true;
             if ( !C.contains(G[source(e)]) )
             return False_IDD("face with non-co-circular vertices");
             e = G.face_cycle_succ(e);
        } while ( e != e0 );
      }
      else
      { // face is unbounded
        edge e0 = e;
        do { considered[e] = true;
             e = G.face_cycle_succ(e);
        } while ( e != e0 );
      }
    }
  }
  { ⟨check Delaunay property⟩ }
  return true;
}
```

10.4.7 *Applications*

Delaunay triangulations have several useful properties. We mention three:

- For a triangulation T let $\mu(T)$ be the smallest interior angle of any triangle in T. Delaunay triangulations maximize $\mu(T)$.
- Delaunay triangulations tend to produce "rounder" triangles than other triangulations, see Figure 10.24, a property desirable for numerical applications of triangulations. For example, the interpolation scheme presented at the beginning of Section 10.2 is numerically more stable if the triangulation contains no "skinny" triangles.
- The Euclidean minimum spanning tree of a set S is a tree of minimum cost connecting all points in S, where the cost of an edge is its Euclidean length. The Euclidean minimum spanning tree is a subgraph of the Delaunay diagram.

The function

```
void MIN_SPANNING_TREE(const list<POINT>& L, GRAPH<POINT,int>& T)
```

computes the Euclidean minimum spanning tree for the points in L. It first constructs the Delaunay diagram T for L, then runs the minimum spanning tree algorithm on T, and finally deletes all edges from T that do not belong to the minimum spanning tree.

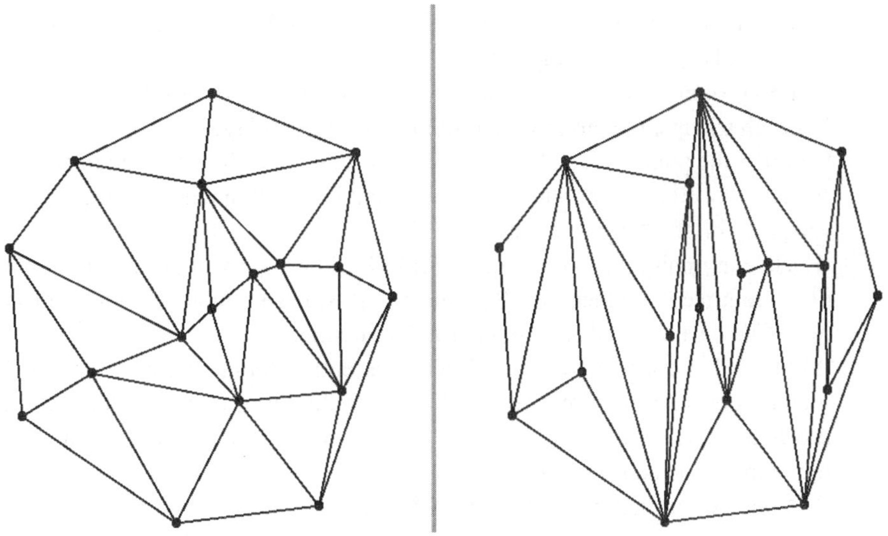

Figure 10.24 A Delaunay triangulation and a triangulation produced by sweeping. The Delaunay triangulation is shown on the left. The triangles in the Delaunay triangulation are "rounder" than in the triangulation by sweeping. The figure was generated with the triangulation_demo (see LEDAROOT/demo/book/Geo).

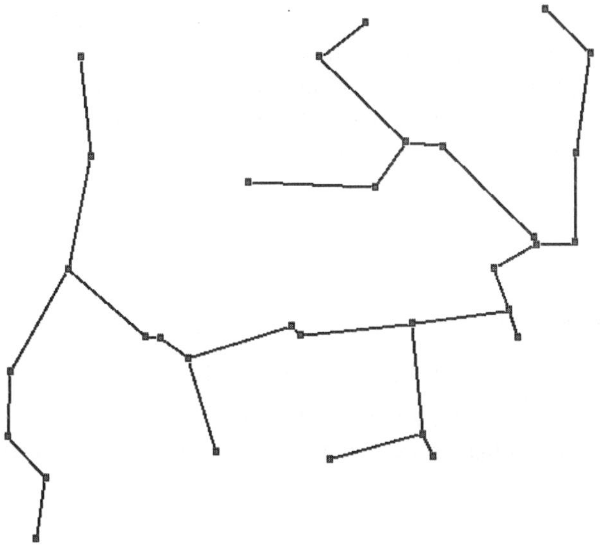

Figure 10.25 A point set and its Euclidean minimum spanning tree. The figure was generated with the voronoi_demo in xlman.

Exercises for 10.4

1 Show that the flipping algorithm constructs a furthest site Delaunay triangulation for a set of points in convex position.

2 Extend the functions for checking Delaunay triangulations and Delaunay diagrams such that they also check the edge labels.

3 Write a program that takes a Delaunay triangulation and draws it into a graphics *window*. For each triangle the circumcircle should also be displayed.

4 Consider the points (i, i^2), $0 \le i < n$. Show that the Delaunay triangulation of this point set has a fan-like shape. Show that the flipping algorithm may perform $\Omega(n^2)$ flips when starting with the "opposite fan".

5 (Euclidean minimum spanning tree (EMST)) For a set S of points in the plane a tree T of minimum cost connecting all points in S is called a Euclidean minimum spanning tree of S. The cost of an edge is defined as its Euclidean length.
(a) Show that every edge of an EMST is essential. (Hint: For an edge e with endpoints a and b consider the circle centered at the midpoint of e and passing through a and b. Assume that it contains a point $c \in S \setminus \{a, b\}$. Show that a better tree can be obtained by removing e and adding either (a, c) or (c, b).)
(b) Conclude from part (a) that an EMST is a subgraph of the Delaunay diagram. Write a program to compute an EMST. Make use of programs for Delaunay diagrams and minimum spanning trees. Try to work with the squared length of edges instead of their length.

6 For a triangulation T let $\alpha(T)$ be the sorted tuple of all interior angles of all triangles in T. Consider Figure 10.18 and let T_1 and T_2 be the two triangulations shown with T_2 being Delaunay. Show that $\alpha(T_1) \le \alpha(T_2)$ where the ordering on tuples is the lexicographic one. Consider next any triangulation T of a set S that is not Delaunay and let T' be obtained from T by a diagonal flip. Show that $\alpha(T) \le \alpha(T')$. Conclude that Delaunay triangulations maximize the smallest interior angle.

7 Improve the implementation of the flipping algorithm by ensuring that, for any pair of darts in a uedge, at most one is in S. Observe that we ensure this property only at the time of initialization. Does the running time improve?

10.5 Voronoi Diagrams

We discuss Voronoi diagrams. We define them and discuss their representation by graphs. We relate them to Delaunay triangulations and show how to obtain Voronoi diagrams from Delaunay triangulations. Finally, we discuss applications and the verification of Voronoi diagrams.

10.5.1 *Definition and Representation*

A structure closely related to the Delaunay diagram is the so-called *Voronoi diagram*. Let S be a set of points in the plane. We will refer to the elements of S as *sites*. For any point p of the plane let $close(p)$ be the set of sites that realize the closest distance between p and the sites in S, i.e., $s \in close(p)$ if $dist(s, p) \le dist(t, p)$ for all $t \in S$. In other words, there is a circle with center p passing through all points in $close(p)$ and having no points of S in its interior, see Figure 10.26. For most points p of the plane $close(p)$ consists of only a

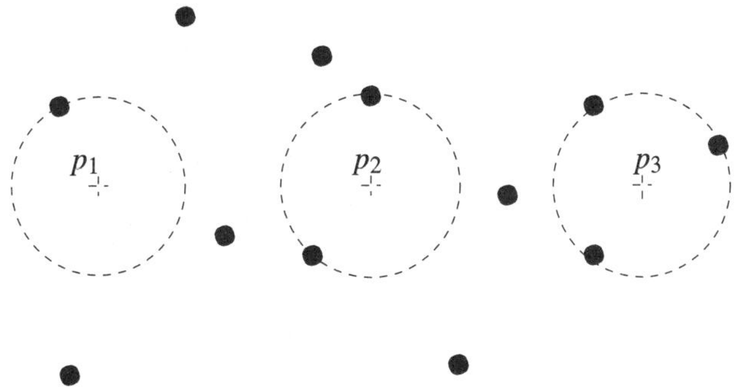

Figure 10.26 Sites are shown as dots. The point p_i has i sites in $close(p_i)$.

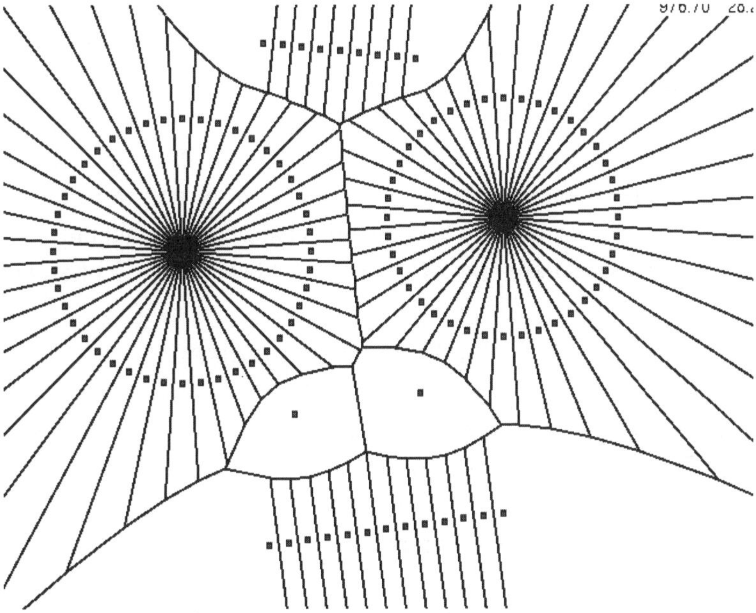

Figure 10.27 A Voronoi diagram. The figure was generated with the voronoi_demo in xlman.

single site. For some points p, $close(p)$ contains two or more sites. These points form the so-called Voronoi diagram $VD(S)$ of S.

$$VD(S) = \{p \in \mathbb{R}^2; |close(p)| \geq 2\}.$$

Figure 10.27 shows a Voronoi diagram. The Voronoi diagram is a graph-like structure. Its vertices are all points p with $|close(p)| \geq 3$, its edges are maximal connected sets of points p with $|close(p)| = 2$, and its faces are maximal connected sets of points p with $|close(p)| = 1$.

We derive some more properties of edges and faces. Consider any edge e of the Voronoi diagram, and let s and t be the two sites of S such that $close(p) = \{s, t\}$ for all points p

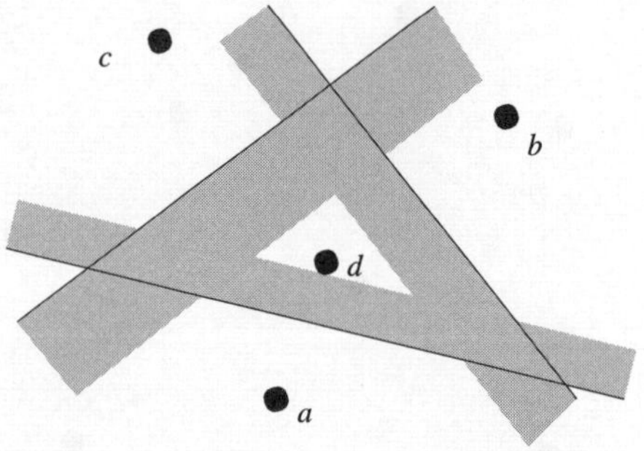

Figure 10.28 The Voronoi region of d is the intersection of three open halfspaces $VR(d, a)$, $VR(d, b)$, and $VR(d, c)$.

of e. Any such p lies on the perpendicular bisector of s and t and hence e is a straight line segment contained in the perpendicular bisector of s and t.

Consider next any face f of the Voronoi diagram and let s be the site of S such that $close(p) = \{s\}$ for all points p of f. Then $dist(s, p) < dist(t, p)$ for all $t \in S \setminus \{s\}$ and hence f is contained in the open halfplane bounded by the perpendicular bisector of s and t and containing s. We use $VR(s, t)$ to denote this halfplane, see Figure 10.28, and call it the halfplane where s *dominates over* t. We have just shown that $f \subseteq VR(s, t)$ for all $t \in S \setminus \{s\}$ and hence

$$f \subseteq VR(s) := \bigcap_{t \in S \setminus \{s\}} VR(s, t).$$

We even have equality since $p \in VR(s)$ implies $p \in VR(s, t)$ for all $t \in S \setminus \{s\}$ which in turn implies that p is closer to s than to any other site in S. We call $VR(s)$ the *Voronoi region* of site s. It is the intersection of open halfspaces and hence an open convex polygonal region.

How are we going to represent Voronoi diagrams? We use plane maps of type

```
GRAPH<CIRCLE,POINT>.
```

We defined the Voronoi diagram $VD(S)$ as a set of points. We turn it into a graph G by placing a "vertex at infinity" on every unbounded edge of $VD(S)$[7] and by deleting the portion of the edge that goes beyond the vertex at infinity, see Figure 10.29. A node v of G has either degree one or degree three or more. We call v a node at infinity in the former case and a proper node in the latter case.

The node and edge labels give information about the positions of the node of G in the plane and about the Voronoi regions:

- Every dart is labeled with the site whose region lies to its left.
- Every proper node v is labeled by a circle $CIRCLE(a, b, c)$, where a, b, and c are

[7] If all sites are collinear and hence $VD(S)$ consists of a set of parallel lines, we put two vertices at infinity on every line.

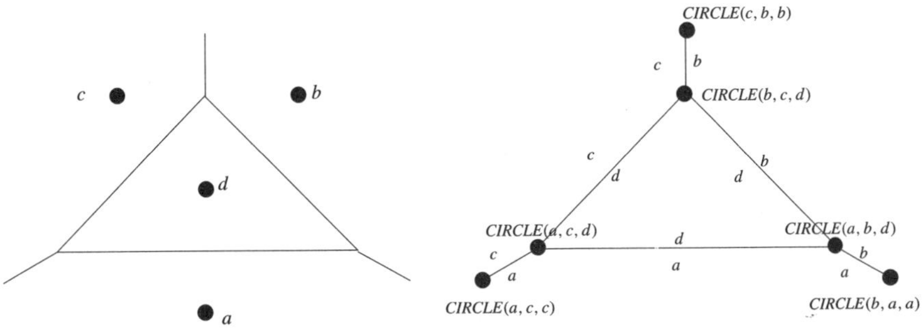

VD(S) G

Figure 10.29 A Voronoi diagram for a set of four sites and its graph representation.

distinct sites whose regions are incident to v. The center of this circle is the position of v in the plane.

- Every node v at infinity lies on the perpendicular bisector of two sites a and b. We label v by $CIRCLE(a, x, b)$, where x is an arbitrary point collinear to a and b (e.g., a) and v lies to the left of the oriented segment from a to b.

The function

```
void VORONOI(const list<POINT>& L, GRAPH<CIRCLE,POINT>& VD);
```

computes the Voronoi diagram of the sites in L in time $O(n \log n)$.

There is also a so-called *furthest site Voronoi diagram*, see Figure 10.30 for an example. Its definition is the same as for (nearest site) Voronoi diagrams except for replacing closest by furthest. For any point p let *furthest*(p) be the set of sites that realize the furthest distance between p and the sites in S, i.e., $s \in \mathit{furthest}(p)$ if $\mathit{dist}(s, p) \geq \mathit{dist}(t, p)$ for all $t \in S$. In other words, there is a circle with center p passing through all points in *furthest*(p) and having no points of S in its exterior. For most points p of the plane *furthest*(p) consists of only a single site. For some points p, *furthest*(p) contains two or more sites. These points form the so-called furthest site Voronoi diagram $FVD(S)$ of S.

$$FVD(S) = \{p \in R^2;\ |\mathit{furthest}(p)| \geq 2\}.$$

The furthest site Voronoi region of a site s is given by

$$FVR(s) := \bigcap_{t \in S \setminus \{s\}} FVR(t, s).$$

Only vertices of the convex hull have non-empty regions in the furthest site Voronoi digram. The rules for the graph representation of furthest site diagrams are the same as for nearest site diagrams.

The function

```
void F_VORONOI(const list<POINT>& L, GRAPH<CIRCLE,POINT>& FVD);
```

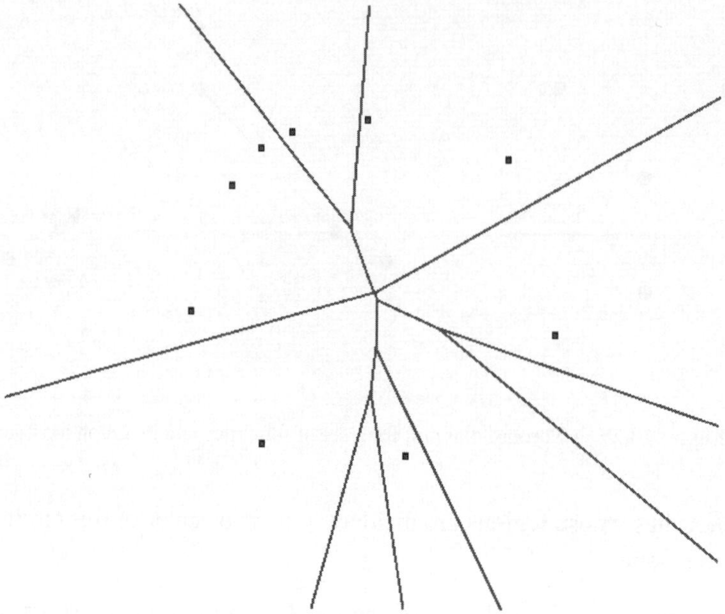

Figure 10.30 A furthest site Voronoi diagram. The figure was generated with the voronoi_demo in xlman.

computes the furthest site Voronoi diagram of the points in L.

We recommend that the readers exercise the Voronoi demo in xlman before proceeding.

10.5.2 *The Duality between Voronoi and Delaunay Diagrams*

Voronoi diagrams and Delaunay diagrams are closely related structures. In fact, each one of them is easily obtained from the other. Let S be a set of sites and let $VD(s)$ and $DD(S)$ be its Voronoi and Delaunay diagram, respectively. We show how to obtain $VD(S)$ from $DD(S)$.

(1) For every bounded face f of $DD(S)$ there is a vertex $c(f)$ of $VD(S)$ located at the center of the circumcircle of f.

(2) Consider an edge st of $DD(S)$ and let f_1 and f_2 be the faces incident to the two sides of the edge.

(a) If f_1 and f_2 are both bounded, then the edge $c(f_1)c(f_2)$ belongs to $VD(S)$.

(b) If f_1 is unbounded and f_2 is bounded, then a ray with source $c(f_2)$ and contained in the perpendicular bisector of s and t belongs to $VD(S)$. It extends into the halfplane containing the unbounded face.

(c) If f_1 and f_2 are unbounded[8] and hence $f_1 = f_2$, then the entire perpendicular bisector of s and t belongs to $VD(S)$.

[8] Case (c) arises only if all sites in S are collinear. Then cases (a) and (b) do not arise.

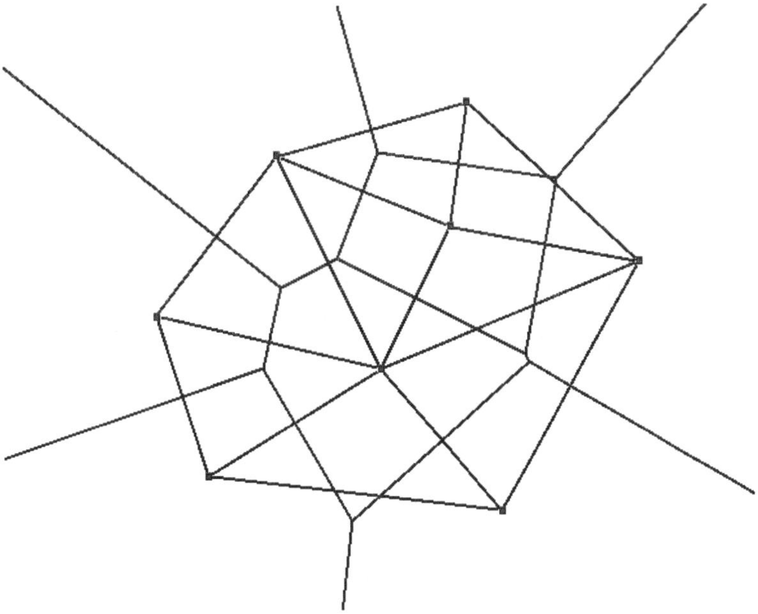

Figure 10.31 A Voronoi diagram and a Delaunay diagram for the same set of sites. This figure was generated with the voronoi_demo in xlman.

(3) That's all.

Figure 10.31 shows a Delaunay and a Voronoi diagram for the same set of sites. Use the Voronoi demo to construct your own examples. The rules above are called a *duality* relation because they map faces (= 2-dimensional objects) into vertices (= 0-dimensional objects), edges into edges, and vertices into faces. The latter map is implicit. There is a corresponding set of rules that construct the Delaunay diagram from the Voronoi diagram. We leave them to the exercises.

Theorem 15 *The rules above construct the Voronoi diagram from the Delaunay diagram.*

Proof We proceed in two steps. We first show that everything that is constructed by the rules does indeed belong to the Voronoi diagram and in a second step we show that the complete Voronoi diagram is obtained.

Consider any bounded face f of $DD(S)$. The vertices of f are co-circular and hence the circumcenter $c(f)$ is a point with $|close(p)| \geq 3$, i.e., a vertex of $VD(S)$.

Consider next any edge st of $DD(S)$. View it as oriented from s to t and let f_1 and f_2 be the faces to its left and right, respectively. Assume first that f_1 and f_2 are both bounded. The centers $c(f_1)$ and $c(f_2)$ of the circumcircles of f_1 and f_2 both lie on the perpendicular bisector of s and t and any point between $c(f_1)$ and $c(f_2)$ is the center of a disk D with $D \cap S = \{s, t\}$, see Figure 10.32. Thus, $c(f_1)c(f_2)$ is an edge of $VD(S)$.

Assume next that f_1 is unbounded and f_2 is bounded, i.e, st is a clockwise convex hull

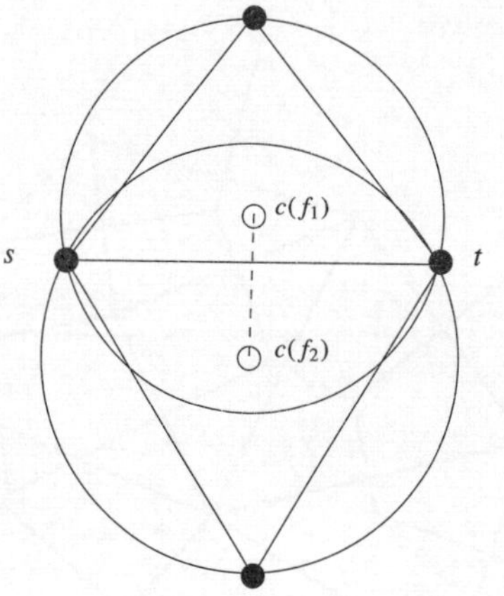

Figure 10.32 An edge $e = (s, t)$ of $DD(S)$, the two incident faces f_1 and f_2 and the circumcircles of f_1 and f_2. Each point on the open line segment $c(f_1)c(f_2)$ is the center of an empty circle passing through s and t.

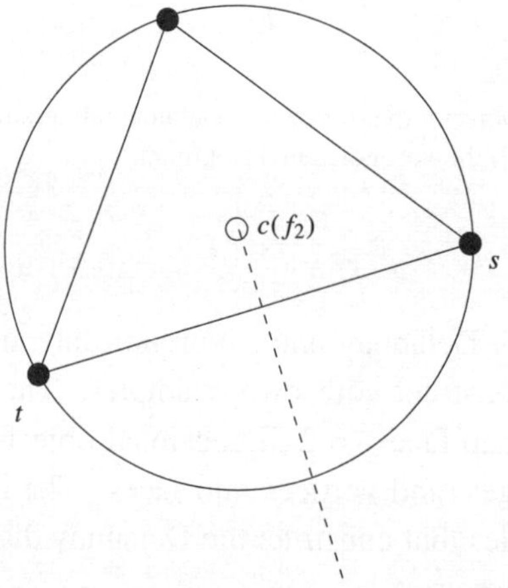

Figure 10.33 st is a clockwise convex hull edge and the face f_2 to its right is bounded.

edge, see Figure 10.33. Then the same argument shows that the ray starting in $c(f_2)$, contained in the perpendicular bisector of s and t, and extending into the left halfplane with respect to st belongs to $VD(S)$.

Finally, if f_1 and f_2 are both unbounded then the entire perpendicular bisector of s and t is an edge of $VD(S)$.

We have now shown that the rules above construct only features of the Voronoi diagram. We show next that the entire Voronoi diagram is constructed. Consider any edge e of $VD(S)$, say separating the regions $VR(s)$ and $VR(t)$. Then $close(p) = \{s, t\}$ for every point $p \in e$, i.e., every $p \in e$ witnesses that the segment st is essential and hence is an edge of $DD(S)$. Imagine a disk centered at p and having s and t in its boundary as p moves along e. When

p moves into an endpoint of e (e may have 0, 1, or 2 endpoints), $close(p)$ grows to at least three points, namely the vertices of a face of $DD(S)$ incident to st. Thus, applying the appropriate rule 2a, 2b, or 2c to st yields e. Moreover, applying rule 1 to the bounded faces incident to st produces the endpoints of e (if any). We have now shown that all edges of $VD(S)$ are constructed and since every vertex of $VD(S)$ is incident to at least one (actually three) edge we have also shown that all vertices are constructed. □

We next give the program that constructs a Voronoi diagram from a Delaunay diagram. The Voronoi diagram is empty if the number of sites is less than two. So assume that there are at least two sites. We first determine a hull edge, then create all nodes of the Voronoi diagram and finally all darts of the Voronoi diagram. We use an edge array *vnode* in order to associate with each dart e of DD the node of VD that lies in the face to the left of e.

⟨*voronoi.c*⟩≡
```
void DELAUNAY_TO_VORONOI(const GRAPH<POINT,int>& DD,
                         GRAPH<CIRCLE,POINT>& VD)
{
  VD.clear();
  if (DD.number_of_nodes() < 2) return;
  // determine a hull dart
  edge e;
  forall_edges(e,DD) if (DD[e] == HULL_DART) break;
  edge hull_dart = e;
  edge_array<node> vnode(DD,nil);
  ⟨DD to VD: create Voronoi nodes⟩
  ⟨DD to VD: create Voronoi darts⟩
}
```

We create the Voronoi nodes in two phases. We first create the nodes at infinity and then the proper nodes.

There is one node at infinity for each hull dart. If e is a hull dart and a and b are the sites associated with the source and target of e, respectively, then the label of the node at infinity is *CIRCLE*(a, x, b), where x is any point collinear with a and b. We use the midpoint of a and b for x.

If e is not a hull dart then there is a proper node v associated with the face cycle of e. We label v with *CIRCLE*(a, b, c), where a, b, and c are any three vertices of the face cycle, and associate v with every dart of the face cycle.

⟨*DD to VD: create Voronoi nodes*⟩≡
```
// create Voronoi nodes for outer face
POINT a = DD[source(e)];
do { POINT b = DD[target(e)];
     vnode[e] =  VD.new_node(CIRCLE(a,center(a,b),b));
     e = DD.face_cycle_succ(e);
     a = b;
   } while ( e != hull_dart );
```

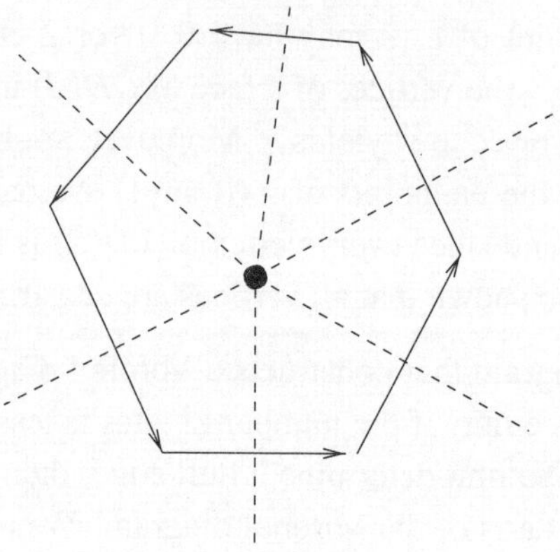

Figure 10.34 Tracing a face cycle in forward direction generates the darts incident to the node dual to the face in counter-clockwise order.

```
// and for all other faces
forall_edges(e,DD)
{ if (vnode[e]) continue;
  edge  x = DD.face_cycle_succ(e);
  POINT a = DD[source(e)];
  POINT b = DD[target(e)];
  POINT c = DD[target(x)];
  node  v = VD.new_node(CIRCLE(a,b,c));
  vnode[e] = v;
  do { vnode[x] = v;
       x = DD.face_cycle_succ(x);
     } while( x != e );
}
```

We come to the construction of the Voronoi darts. Let *e* be a dart of *DD*, let *r* be its reversal, and let *p* be the point associated with the target of *e*. The dart dual to *e* starts at the node associated with *e*, ends at the node associated with *r*, and is labeled by *p*.

We want to construct the darts incident to any node of *VD* in their proper counter-clockwise order. For the nodes at infinity this is no problem since they have degree one. We therefore construct the Voronoi darts in two phases. We first construct the Voronoi darts out of the nodes at infinity and then the Voronoi darts out of the proper nodes. Finally, we link the two darts in each. For each dart *e* of *DD* we record the dart dual to it in the edge array *vedge*.

Consider a proper node *v*. It corresponds to a bounded face of *DD* and has one incident dart for each dart of the face cycle. We construct the darts in their proper counter-clockwise order if we trace the face cycle in forward direction, see Figure 10.34.

⟨*DD to VD: create Voronoi darts*⟩≡

```
edge_array<edge> vedge(DD,nil);
// construct Voronoi darts out of nodes at infinity
e = hull_dart;
do { edge r = DD.reversal(e);
     POINT p = DD[target(e)];
     vedge[e] = VD.new_edge(vnode[e],vnode[r],p);
     e = DD.cyclic_adj_pred(r); // same as DD.face_cycle_succ(e)
   } while ( e != hull_dart );
// and out of all other nodes.
forall_edges(e,DD)
{ node v = vnode[e];
  if (VD.outdeg(v) > 0) continue;
  edge x = e;
  do { edge  r = DD.reversal(x);
       POINT p = DD[target(x)];
       vedge[x] = VD.new_edge(v,vnode[r],p);
       x = DD.cyclic_adj_pred(r);
     } while ( x != e);
}
// assign reversal edges
forall_edges(e,DD)
{ edge r = DD.reversal(e);
  VD.set_reversal(vedge[e],vedge[r]);
}
```

This completes the construction of Voronoi diagrams from Delaunay diagrams. The construction runs in linear time.

In order to construct the Voronoi diagram for a set L of points we first construct the Delaunay diagram and then the Voronoi diagram from the Delaunay diagram.

⟨*voronoi.c*⟩+≡

```
void VORONOI(const list<POINT>& L, GRAPH<CIRCLE,POINT>& VD)
{ GRAPH<POINT,int> DD;
  DELAUNAY_DIAGRAM(L,DD);
  DELAUNAY_TO_VORONOI(DD,VD);
}
```

The construction of furthest site Voronoi diagrams from furthest site Delaunay triangulations is completely analogous. We leave it to the exercises.

10.5.3 *Verifying Voronoi Diagrams*

Let G be a graph of type *GRAPH<CIRCLE, POINT>*. We want to verify that G is the Voronoi diagram of the sites associated with its nodes. The procedure to be described is fairly complicated and we wished we had a simpler one. The procedure is probably the least elegant piece of code contained in this book. We considered to drop this section, but decided against it for two reasons. We had invested a lot of time in it, and more importantly, the check discovered several mistakes.

G must satisfy the following conditions:

- G is a CCW-ordered plane map.
- The site information associated with edges is consistent, i.e., if e and e' are consecutive edges on some face cycle then both edges have the same associated site.
- The sites associated with e and *reversal*(e) are distinct.
- Call a vertex whose associated circle is non-degenerate non-trivial and call it trivial otherwise. Every non-trivial vertex has degree at least three and every trivial vertex has degree one.
- For each non-trivial vertex each of the three points defining the associated circle is associated with one of the incident edges and the sites associated with all incident edges lie on the associated circle.
- Each trivial vertex has an associated circle of the form *CIRCLE*$(a, \_, c)$, where a and c are distinct. Let e be the unique outgoing edge. In a nearest site diagram the site associated with the face to the left of e is c and the site associated with the face to the right of e is a and in a furthest site diagram the roles of a and c are interchanged.
- For every edge $e = (v, w)$ such that v and w are non-trivial, the centers of the circles associated with v and w are distinct. Let p and q be these centers and let a be the site associated with e. In a nearest site diagram a lies to the left of the segment pq and in a furthest site diagram a lies to the right of the segment pq.
- Each face is a convex polygonal region and the regions associated with the different sites partition the plane.

In the implementation we first check the first six conditions and then distinguish cases according to whether G is connected or not. For the first item we want to use the function *Is_CCW_Ordered_Plane_Map* and therefore we need to define the *edge_vector* function for circle-points. Let e be an edge and let C and D be the circles associated with the source and the target of e, respectively. If both circles are non-degenerate the edge vector is simply the vector from the center of C to the center of D. So assume that one of the circles is degenerate. If D is degenerate then D = *CIRCLE*$(a, \_, c)$ and D represents a point at infinity on the perpendicular bisector of a and c and to the right of the line segment ac. Let m be the midpoint of a and c and let a_1 be the point obtained by rotating a by $90°$ in a clockwise direction about m. We may return the vector $m - a_1$. The case that C is degenerate is symmetric.

⟨*voronoi_check: edge vector function*⟩≡

```
static VECTOR edge_vector(const GRAPH<CIRCLE,POINT>& G, const edge& e)
{ const CIRCLE& C = G[G.source(e)];
  const CIRCLE& D = G[G.target(e)];
  if ( D.is_degenerate() ) { POINT a = D.point1();
                             POINT c = D.point3();
                             POINT m = midpoint(a,c);
                             return m - a.rotate90(m);
                           }
  if ( C.is_degenerate() ) { POINT a = C.point1();
                             POINT c = C.point3();
                             POINT m = midpoint(a,c);
                             return a.rotate90(m) - m;
                           }
  // both circles are non-degenerate
  return D.center() - C.center();
}
```

and

⟨*voronoi_check.c*⟩+≡

⟨*voronoi_check: edge vector function*⟩

```
static bool False_IVD(const string& s)
{ cerr << "Is_Voronoi_Diagram: " << s; return false; }
bool Is_Voronoi_Diagram(const GRAPH<CIRCLE,POINT>& G,
                        delaunay_voronoi_kind kind)
{ if ( G.number_of_nodes() == 0 ) return true;
  node v,w; edge e;
  if ( !Is_CCW_Ordered_Plane_Map(G) )
     return False_IVD("G is not CCW-ordered plane map");
  forall_edges(e,G)
  { if ( G.outdeg(target(e)) != 1 )
    { // e does not end at a vertex at infinity
      if ( G[e] != G[G.face_cycle_succ(e)] )
        return False_IVD("inconsistent site labels");
    }
    if ( G[e] == G[G.reversal(e)] )
        return False_IVD("same site on both sides");
  }
  forall_nodes(v,G)
  { CIRCLE C = G[v];
    if ( C.is_degenerate() )
    { // vertex at infinity
      if ( G.outdeg(v) != 1 )
        return  False_IVD("degree of vertex at inf");
      edge e = G.first_adj_edge(v); edge r = G.reversal(e);
      POINT a = C.point1(); POINT c = C.point3();
      if ( (kind == NEAREST) && (c != G[e] || a != G[r]) ||
           (kind == FURTHEST) && (a!= G[e] || c != G[r]) )
        return False_IVD("vertex at inf: wrong edge labels");
```

```
      }
      else
      { // finite vertex
        if ( G.outdeg(v) < 3 )
          return False_IVD("degree of proper vertex");
        forall_adj_edges(e,v)
        { if ( !C.contains(G[e]) )
            return False_IVD("label of proper vertex");
        }
        for (int i = 1; i <= 3; i++)
        {
          POINT a = ( i == 1 ? C.point1() :
                      (i == 2 ? C.point2() : C.point3() ) );
          bool found_a = false;
          forall_adj_edges(e,v) if ( a == G[e] ) found_a = true;
          if ( !found_a ) return False_IVD("wrong cycle");
        }
        forall_adj_edges(e,v)
        { w = G.target(e);
          if ( G.outdeg(w) == 1 ) continue;
          if ( C.center() == G[w].center() )
            return False_IVD("zero length edge");
          int orient = orientation(C.center(),G[w].center(),G[e]);
          if ( kind == NEAREST && orient <= 0 ||
               kind == FURTHEST && orient >= 0 )
            return False_IVD("orientation");
        }
      }
    }
    if ( Is_Connected(G) )
      { ⟨G is connected⟩ }
    else
      { ⟨G is not connected⟩ }
    return true;
  }
```

When G has passed all tests above we can construct a geometric object from it as follows. We assign a position $pos(v)$ to each non-trivial vertex v and a segment, ray, or line $geo(e)$ to each edge e. For a non-trivial vertex v let $pos(v)$ be the center of the circle associated with v. For an edge $e = (v, w)$ let a and c be the sites separated by e, i.e., one of a and c is associated with e and the other with $reversal(e)$. If v is non-trivial then a and c lie on the circle associated with v and hence $pos(v)$ lies on the perpendicular bisector of a and b. Define $geo(e)$ as follows. First assume that v and w are both non-trivial. Then $geo(e)$ is the segment directed from $pos(v)$ to $pos(w)$. Note that this segment has non-zero length and is part of the perpendicular bisector of a and c. Next assume that exactly one of v and w is non-trivial. Assume w.l.o.g. that the triple of points associated with the trivial vertex is of the form $(a, \_, c)$. If w is trivial then $geo(e)$ is the ray starting at $pos(v)$, running along the perpendicular bisector of a and c, and extending to infinity to the right of the segment

ac. If w is trivial then $geo(e)$ is the ray ending in $pos(v)$, running along the perpendicular bisector of a and c, and coming from infinity to the right of the segment ac. Finally, assume that v and w are trivial and assume w.l.o.g. that the triple of points associated with v is of the form $(a, \_, c)$. Then $geo(e)$ is the bisector of a and c oriented such that a lies to its left.

Now we distinguish cases according to whether G is connected or not.

G is connected: Define a face chain as a minimal sequence $e_0, e_1, \ldots, e_k$ of edges such that e_{i+1} is the face cycle successor of e_i for all i, $0 \le i < k$, and either $target(e_k) = source(e_0)$ or $source(e_0)$ and $target(e_0)$ have degree one. We call face chains of the former kind closed and face chains of the latter kind open. All face chains are strictly convex counter-clockwise oriented. Moreover, the rays going to infinity wind around the origin once and open face chains cover only a half-circle. There is no need to check the second half-sentence as it is implied by the first half-sentence.

Below, we first search for a vertex of degree one and then check the open face chains one by one. Simultaneously we build the list of all rays; note that they will wind clockwise around the origin. Having checked all open face chains we turn to the closed face chains.

⟨*G is connected*⟩≡

```
cmp_edges_by_angle<GRAPH<CIRCLE,POINT> > cmp(G);
node v;
forall_nodes(v,G) if ( G.outdeg(v) == 1 ) break;
edge_array<bool> considered(G,false);
list<edge> rays;
edge e = G.first_adj_edge(v);
do { rays.push(e);
     list<edge> D;
     do { considered[e] = true;
          D.append(e);
          e = G.face_cycle_succ(e);
        } while ( G.outdeg(source(e)) != 1);
     if ( !Is_C_Increasing(D,cmp)  ) return False_IVD(": wrong order");
} while ( G.source(e) != v);
if ( !Is_C_Nondecreasing(rays,cmp) )
  return False_IVD("wrong order, rays");
forall_edges(e,G)
{ if ( !considered[e] )
  { edge e0 = e;
    do { considered[e] = true;
         if ( G.outdeg(target(e)) == 1 )
           return False_IVD("unexpected vertex of degree one");
         e = G.face_cycle_succ(e);
       } while ( e != e0);
    if ( !Is_CCW_Convex_Face_Cycle(G,e) )
      return False_IVD("wrong order");
  }
}
```

We claim that we are done at this point. Let us see why. Consider any face chain f. All edges on the boundary of f have the same associated site, say a, the circles associated with all non-trivial vertices of f pass through a, for each edge e of f, *geo*(e) is part of the perpendicular bisector of a and the site associated with the other side of e, and a lies to the left of *geo*(e) if *kind* is *NEAREST* and to the right of it if *kind* is *FURTHEST*. Define

$$reg(f) = \bigcap_{e;\ e \text{ is an edge of } f} H(a, site\_of\_reversal(e)),$$

where $b = site\_of\_reversal(e)$ is the site associated with the reversal of e and $H(a, b)$ is the halfplane defined by a and b and containing a if *kind* is *NEAREST* and not containing a otherwise. Then $reg(f)$ is a convex polygonal region which contains the Voronoi region of site a (since in the definition of a Voronoi region the intersection is over all sites different from a). We still need to show that the regions partition the plane. Consider a point moving in the plane and avoiding vertices of regions. Such a point is always covered by the same number of regions. Moreover, when the point travels along a cycle at infinity it is always covered by exactly one region since the rays of the diagram wind around the origin once. Altogether we have shown that the regions partition the plane.

G is not connected: If G is not connected it can only be the Voronoi diagram of a set of collinear sites. As such it must have the following additional properties:

- All nodes have out-degree one.
- All sites are collinear.
- No site is associated with three edges of G.
- The number of distinct sites is equal to $m/2 + 1$.

We show that these conditions suffice. Clearly, the geometric interpretation of G is a set of parallel line segments. Consider the placement of the sites on their common underlying line. For each site s which is associated with two edges, it is guaranteed that the two adjacent sites (= sites for which there is an edge having s on one of its sides) lie on opposite sides of s; this follows from the fact that we have already checked that each edge incident to a trivial node separates the sites it is supposed to separate. We conclude that the conditions above suffice.

⟨*G is not connected*⟩≡

```
forall_nodes(v,G)
  if ( G.outdeg(v) > 1 ) return False_IVD("degree larger than 1");
d_array<POINT,int> count(0);
int n_dual = 0;
edge e = G.first_edge();
LINE l(G[e],G[G.reversal(e)]);
forall_edges(e,G)
{ if ( !l.contains(G[e]) )
```

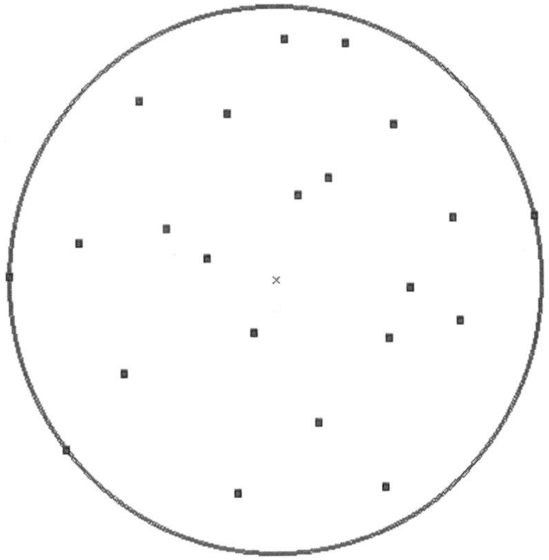

Figure 10.35 The smallest circle enclosing a set of points. The figure was generated with the voronoi_demo in xlman.

```
    return False_IVD("non-collinear sites");
  int& pc = count[G[e]];
  if (pc == 0) n_dual++;
  pc++;
  if (pc == 3) return False_IVD(": site mentioned thrice");
}
if ( n_dual != (G.number_of_edges()/2 + 1) )
  return False_IVD(": two many sites");
```

10.5.4 *Applications of Voronoi Diagrams*

We discuss some applications of Voronoi diagrams. All of them are illustrated in the voronoi-demo of xlman.

Extremal Circles: The *smallest enclosing circle* for a set L of points is the circle with the smallest radius containing all points in L, see Figure 10.35. The smallest enclosing circle is the best approximation of L by a circle. It is easy to see that such a circle has at least two points in L on its boundary and hence its center lies on the furthest site Voronoi diagram of L.

We conclude that the center of the minimum enclosing circle is either a vertex of the furthest site diagram (and then has three points in L on its boundary) or lies on an edge of the furthest site diagram (and then is the circle of minimum radius passing through the two sites defining the edge). In this way each edge and vertex of the furthest site Voronoi

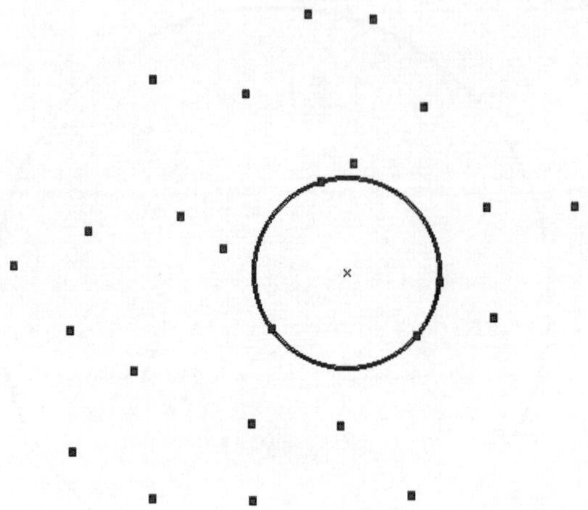

Figure 10.36 The largest empty circle for a set of points. The figure was generated with the voronoi_demo in xlman.

diagram defines a candidate circle. The minimum enclosing circle is the smallest of these circles.

The function

```
CIRCLE SMALLEST_ENCLOSING_CIRCLE(const list<POINT>& L);
```

computes a smallest enclosing circle according to the strategy just described.

The *largest empty circle* for a set L of points is the circle with the largest radius whose interior is void of points in L and whose center lies inside the convex hull of L, see Figure 10.36. We know of no good motivation for considering largest empty circles. It is easy to see that such a circle has at least two points in L on its boundary and hence its center lies on the nearest site Voronoi diagram of L.

We conclude that the center of the largest empty circle is either a vertex of the nearest site diagram (and then has three points in L on its boundary) or lies on an edge of the nearest site diagram (and then is the circle of maximum radius passing through the two sites defining the edge and having its center inside the convex hull). In this way each edge and vertex of the nearest site Voronoi diagram defines a candidate circle. The largest empty circle is the largest of these circles.

The function

```
CIRCLE LARGEST_EMPTY_CIRCLE(const list<POINT>& L);
```

computes a largest empty circle according to the strategy just described.

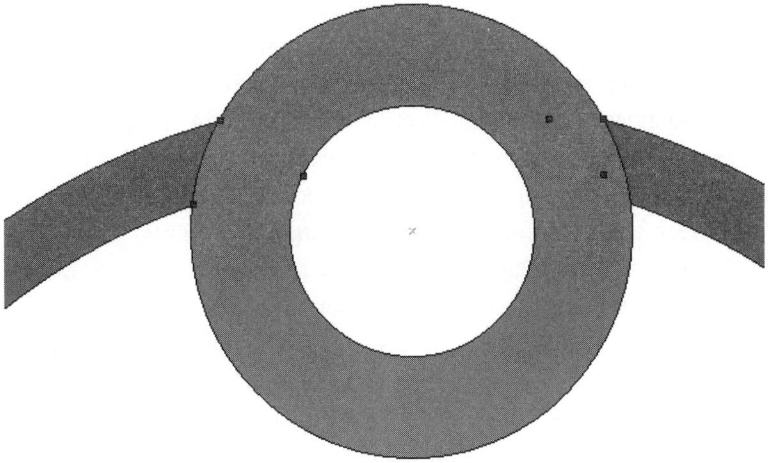

Figure 10.37 The minimum width and the minimum area annulus for a set of points. The figure was generated with the voronoi_demo in xlman.

The application of Voronoi diagrams to find enclosing and empty circles is due to Shamos and Hoey ([SH75]).

Minimum Width and Minimum Area Annuli: An *annulus* A is the region between two concentric circles. When the common center of the circles is a point at infinity, an annulus degenerates to a stripe between parallel lines. Annuli are closed sets. An annulus covers a set L of points if all points in L are contained in the annulus. The *width* of an annulus is the difference between the radius of the outer circle and the radius of the inner circle of the annulus (in the case of a stripe the width is the distance between the two boundaries of the stripe). The *area* of an annulus is the area of the region between the outer and the inner circle (it is infinite in the case of a stripe of non-zero width and is zero in the case of a stripe of width zero). We are interested in computing minimum width and minimum area annuli covering a given set L of points, see Figure 10.37 for an example. Minimum width and minimum area annuli are used to estimate the "roundness" of a set of points.

It can be shown that there is always a minimum annulus covering a given set L of points that is either:

- the minimum width stripe covering the points, or
- a pair of concentric circles whose center is either a vertex of the nearest site Voronoi diagram, or a vertex of the furthest site diagram, or an intersection between an edge of

the nearest site diagram and an edge of the furthest site diagram. This observation was made in [SH75].

The idea for the proof is as follows. Consider an annulus covering the points in L. Clearly, if one of the boundaries does not contain a point in L then the annulus can be improved. So both boundaries must contain at least one point in L. If the two boundaries together contain a total of four points of L then the center of the annulus is either a vertex of one the diagrams (if one boundary contains three points and the other contains one) or an intersection between edges (if both boundaries contain two points). So assume that the boundaries together contain less than four points, say there are two points p and q on one of the boundaries and one point r on the other boundary. Then the center c lies on the perpendicular bisector of p and q. Let d be a vector in the direction of the perpendicular bisector and consider the annulus $A(\epsilon)$ with center $c + \epsilon \cdot d$ and having p, q and r on its boundaries. For small enough ϵ, $A(\epsilon)$ covers L. Consider the optimization criterion as a function of ϵ and conclude that the center can be moved either in the direction $+d$ or the direction $-d$ without increasing the objective value. Move until a further point lies on one of the boundaries. For example, if the objective value is the area, the area of $A(\epsilon)$ is proportional to

$$dist(p, c + \epsilon \cdot d)^2 - dist(r, c + \epsilon \cdot d)^2 = (p - c)^2 - (r - c)^2 + 2\epsilon(p - r) \cdot d,$$

i.e., is a linear function of ϵ. If $(p-r) \cdot d \neq 0$ then the annulus can be improved by moving the center, and if $(p - r) \cdot d = 0$ then the center can be moved in either direction without increasing the area of the annulus.

The two items above suggest a strategy to compute minimum width and minimum area annuli. One simply checks all the candidates listed. This results in quadratic algorithms. The functions

```
bool MIN_AREA_ANNULUS(const list<POINT>& L, POINT& center,
                      POINT& ipoint, POINT& opoint, LINE&  l1);
bool MIN_WIDTH_ANNULUS(const list<POINT>& L, POINT& center,
                       POINT& ipoint, POINT& opoint,
                       LINE& l1, LINE& l2);
```

compute minimum area and minimum width annuli covering the points in L, respectively. The functions return *true*, if the optimal annulus is the region between two circles, and return *false* if the optimal annulus is a stripe. In the former case the center of the annulus and a point on the inner and the outer circle are returned in *center*, *ipoint* and *opoint*, respectively. In the latter case the boundaries of the stripe are returned in *l1* and *l2*. In the case of the a minimum area annulus a stripe can only be optimal if it has width zero. Hence only one line is returned in the former function.

Both functions have quadratic running time and hence should be used only for small input size. There are much faster algorithms: the minimum area annulus can be computed in linear time by linear programming ([Sei91]) and the minimum width annulus can be computed in time $O(n^{8/5+\epsilon})$ by parametric search ([AST94]).

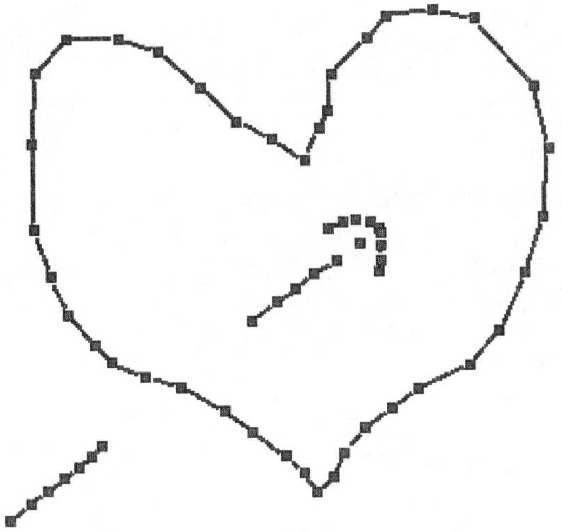

Figure 10.38 A set of points in the plane and the curve reconstructed by *CRUST*. The figure was generated by the Voronoi_demo in xlman.

Curve Reconstruction: The reconstruction of a curve from a set of sample points is an important problem in computer vision. We describe a reconstruction algorithm due to Amenta, Bern, and Eppstein [ABE98]. Figure 10.38 shows a point set and the curves reconstructed by their algorithm.

The precise problem formulation is as follows. Let F be a smooth curve in the plane and let $S \subset F$ be a finite set of sample points from F. A *polygonal reconstruction* of F is a graph that connects every pair of samples adjacent along F, and no others.

The algorithm *CRUST* to be described takes a list S of points and returns a graph G. The graph G is guaranteed to be a polygonal reconstruction of F if F is sufficiently densely sampled by S. We refer the reader to [ABE98] to the definition of sufficient dense sampling density.

The algorithm proceeds in three steps:

- It first constructs the Voronoi diagram *VD* of the points in S.
- It then constructs a set $L = S \cup V$, where V consists of all proper vertices of *VD*.
- Finally, it constructs the Delaunay triangulation *DT* of L and makes G the graph of all edges of *DT* that connect points in L.

The algorithm is very simple to implement[9].

[9] In 1997 the authors attended a conference, where Nina Amenta presented the algorithm. We were supposed to give a presentation of LEDA later in the day. We started the presentation with a demo of algorithm *CRUST*.

⟨*crust.c*⟩+≡

```
void CRUST(const list<POINT>& S, GRAPH<POINT,int>& G)
{
  list<POINT> L = S;
  map<POINT,bool> voronoi_vertex(false);
  GRAPH<CIRCLE,POINT> VD;
  VORONOI(L,VD);
  // add Voronoi vertices and mark them
  node v;
  forall_nodes(v,VD)
  { if (VD.outdeg(v) < 2) continue;
    POINT p = VD[v].center();
    voronoi_vertex[p] = true;
    L.append(p);
  }
  DELAUNAY_TRIANG(L,G);
  forall_nodes(v,G)
    if (voronoi_vertex[G[v]]) G.del_node(v);
}
```

The program above owes much of its elegance to the fact that we use graphs to represent Delaunay diagrams and hence have the full power of the graph data type available to us. Observe that after having constructed the Delaunay triangulation of L in G, we treat G as an "ordinary graph". We simply delete all auxiliary nodes from it, a step that does not make sense on the level of Delaunay triangulations.

10.5.5 *Voronoi Diagrams of Line Segments*

The Voronoi diagram of a set of point sites under the Euclidean metric is just one instance in a wide class of Voronoi diagrams. Other diagrams are obtained by choosing a different metric and/or a different class of sites.

Figure 10.39 shows a Voronoi diagram of line segments. In such a diagram the sites are points and open line segments; the endpoints of every line segment must belong to the point sites. The edges of a Voronoi diagram of line segments are part of angular bisectors between line segments, of parabola, and of lines perpendicular to segments at their endpoints.

Michael Seel [See97] has written a package to compute Voronoi diagrams of line segments. It is available as a LEDA extension package.

The Voronoi diagram of line segments has played an important role in the development of the number types in LEDA, see Section 4.4. Our first program for Voronoi diagrams of line segments used floating point arithmetic in a naive way and worked only for a small number of examples. The main difficulty was a correct implementation of the incircle test. Observe that the coordinates of Voronoi vertices are non-rational algebraic numbers and hence the incircle test requires to compute the sign of certain algebraic numbers. This computation is very error-prone when executed with floating point arithmetic.

In [Bur96, BMS94a, BFMS97] we laid the theoretical basis for an efficient and correct

Figure 10.39 A Voronoi diagram of line segments. The figure was generated with Michael Seel's extension package for Voronoi diagrams of line segments.

sign test of simple algebraic numbers which is used in [BMS96] to implement the number type *real*. Michael Seel uses this number type in his implementation.

Exercises for 10.5

1 Construct a set S where the Voronoi diagram contains no vertices and S has at least three points. What is the Delaunay diagram of S?

2 Give the rules for obtaining the Delaunay diagram from the Voronoi diagram for the same set of sites.

3 Write a program that constructs the Delaunay diagram of a set S given its Voronoi diagram.

4 Write a program to compute the largest empty circle.

5 Write a program to compute the smallest enclosing circle.

10.6 Point Sets and Dynamic Delaunay Triangulations

The class *POINT_SET*[10] maintains a set of points in the plane under insertions and deletions. It offers dictionary operations, nearest neighbor queries, point location queries, and circular, triangular and rectangular range queries. A point set is maintained as a Delaunay triangulation of its elements and hence the class may equally well be called dynamic Delaunay triangulation[11]. The class is derived from *GRAPH<POINT, int>* and hence all graph algorithms and all operations for graphs are available for point sets[12].

In this section we will first give an impression of the functionality and then give part of the implementation. The full implementation can be found in [MN98a]. We close the section with some experimental data. POINT_SETS are illustrated by the point_set_demo in xlman, see Figure 10.40.

10.6.1 *Functionality*

The constructors

```
point_set T;                    // set of points
point_set T(list<point> L);
rat_point_set RT;               // set of rat_points
rat_point_set RT(list<rat_point> L);
```

create a point set for the empty set and the set of points in L, respectively. We mentioned already that *POINT_SET* is derived from *GRAPH<POINT, int>*. Every instance of class *POINT_SET* is an embedded planar map. The position of a vertex v is given by $T.pos(v)$ and also by $T[v]$ and we use

$$S = \{T.pos(v) \mid v \in T\}$$

to denote the underlying point set. Each edge is labeled by an element in the enumeration type *delaunay_edge_info* defined in Section 10.2. If the list L in the constructor contains multiple occurrences of equal points, only the last occurrence of each point is retained and the others are discarded.

The function

```
int T.dim()
```

returns the affine dimension of the point set, i.e., -1 if S is empty, 0 if S consists of only one point, 1 if S consists of at least two points and all points in S are collinear, and 2 otherwise.

The functions *lookup*, *insert* and *del* give point sets the functionality of a *dictionary for points*.

```
node T.lookup(POINT p)
```

[10] The instantiations are *point_set* for *points* and *rat_point_set* for *rat_points*.

[11] In an earlier version of LEDA we called the class *delaunay_triang*. We found, however, that the typical use of the class emphasizes the query operations and hence we now find the name point set more appropriate.

[12] Only *const* graph operations and graph algorithms should be used as others may destroy the additional invariants imposed by POINT_SET.

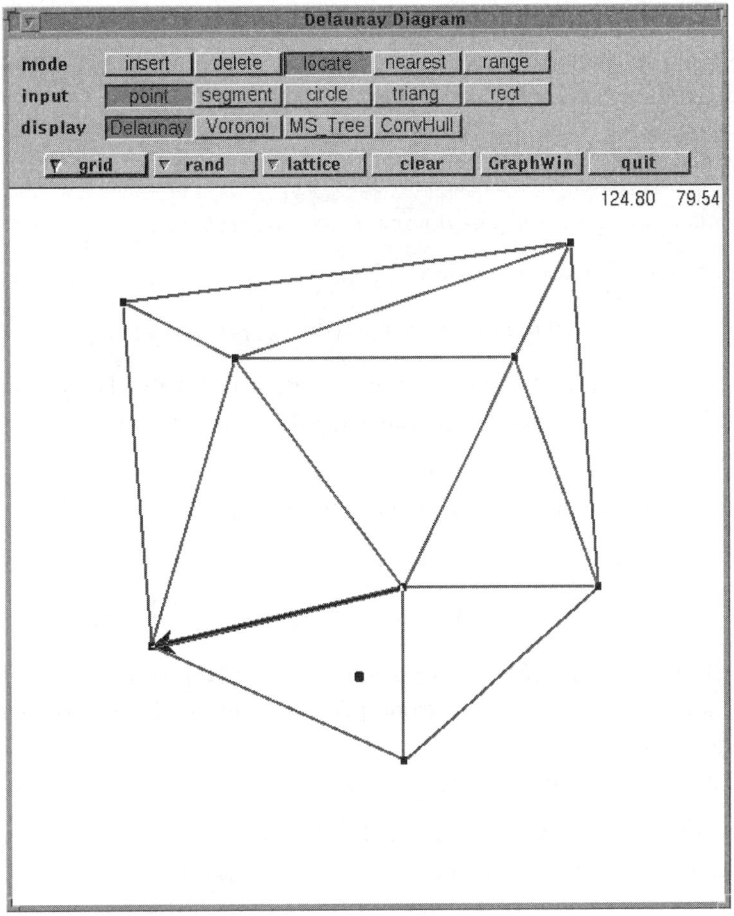

Figure 10.40 A screenshot of the point_set_demo in xlman. A locate query for the highlighted point was performed. The edge returned by the query is highlighted.

returns a node v of T with $T.pos(v) = p$, if there is such a node, and returns *nil* otherwise.

```
node  T.insert(POINT p)
```

inserts p into T and returns the corresponding node. More precisely, if there is already a node v in T positioned at p (i.e., $pos(v)$ is equal to p) then $pos(v)$ is changed to p (i.e., $pos(v)$ is made identical to p) and if there is no such node then a new node v with $pos(v) = p$ is added to T. In either case, v is returned.

```
void T.del(node v)
```

removes node v, i.e., makes T a point set for $S \setminus \{pos(v)\}$.

We come to *point location* and *nearest neighbor* queries. The function

```
edge T.locate(POINT p)
```

performs point location. It returns a dart e (*nil* if T has no edge) such that p lies in the closure of the face to the left of e, see Figure 10.40.

The functions

```
node         T.nearest_neighbor(POINT p);
list<node> T.k_nearest_neighbors(POINT p, int k);
```

return a node v of T that is closest to p, i.e.,

$$dist(p, pos(v)) = \min\{dist(p, pos(u))\ ;\ u \in T\}$$

and the list of the $\min(k, |S|)$ closest points to p, respectively. The points in the result list are ordered by distance from p. One can also ask for the nearest neighbor(s) of a node.

```
node         T.nearest_neighbor(node w);
list<node> T.k_nearest_neighbors(node w, int k);
```

return a node v of T that is closest to $T[w]$, i.e.,

$$dist(p, pos(v)) = \min\{dist(p, pos(u))\ ;\ u \in T \setminus w\}$$

and the list of the $\min(k, |S| - 1)$ closest points to $T[w]$, respectively. The points in the result list are ordered by distance from $T[w]$. Figure 10.41 illustrates nearest neighbor queries and the deletion of nodes.

The next three functions concern *range queries*.

```
list<node> T.range_search(const CIRCLE& C);
list<node> T.range_search(node v,const POINT& b);
list<node> T.range_search(const POINT& a,const POINT& b,const POINT& c);
list<node> T.range_search(const POINT& a,const POINT& b);
```

return the list of points contained in the closure of disk C, in the closure of the disk centered at $T[v]$ and having b in its boundary, in the closure of the triangle (a, b, c), and in the closure of the rectangle with diagonal (a, b), respectively. Figure 10.42 illustrates circular range queries.

```
list<edge> T.minimum_spanning_tree()
```

returns a list of edges of T that comprise a minimum spanning tree of S and

```
void    T.compute_voronoi(GRAPH<CIRCLE,POINT>& V)
```

computes the Voronoi diagram V for the sites in S. Each node of V is labeled with its defining circle and each edge is labeled with the site lying in the face to its left.

The class POINT_SET also provides functions that support the drawing of Delaunay triangulations, Delaunay diagrams, and Voronoi diagrams. For example,

```
void T.draw_nodes(void (*draw_node)(const POINT&))
```

calls *draw_node*($pos(v)$) for every node v of T.

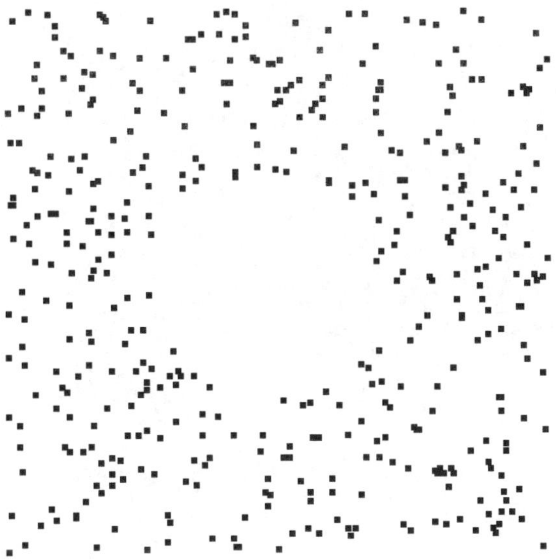

Figure 10.41 Illustration of nearest neighbor searching plus deletion. We generated a point set of 500 random point and then performed the following operation about thirty times: Locate the nearest neighbor of a point in the center of the screen and delete it. The resulting point set is displayed.

10.6.2 *Implementation*

We start with an overview and explain how point sets are represented.

⟨*POINT_SET.h*⟩+≡

```
class __exportC POINT_SET : public  GRAPH<POINT,int>
{
private:
   edge cur_dart;
   edge hull_dart;
   bool check; // functions are checked if true
   // for marking nodes in search procedures
   int cur_mark;
   node_map<int> mark;
   ⟨handler functions for animation⟩
   ⟨functions to mark nodes⟩
   ⟨auxiliary functions⟩
public:
   ⟨public member functions⟩
   ⟨public member functions for checking⟩
};
⟨inline functions⟩
```

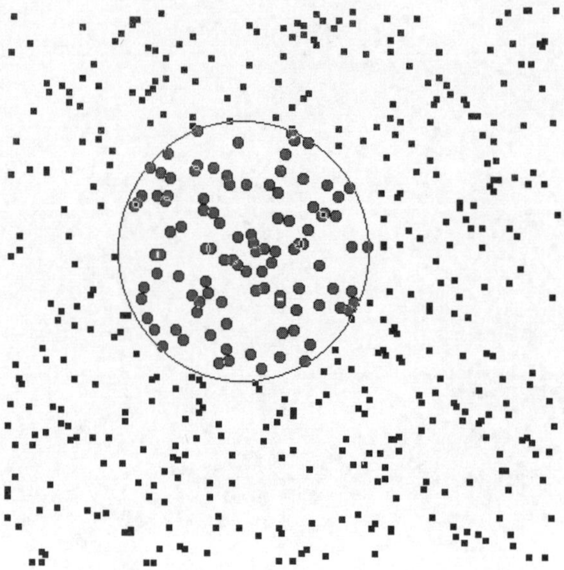

Figure 10.42 We generated a point set of 500 random points and then performed a circular range query. The points returned by the query are highlighted.

We store a POINT_SET as a planar map *GRAPH<POINT, int> T* plus two edges *cur_dart* and *hull_dart*. For each node v of T we store its position in the plane in $T[v]$ and for each edge e we store its type in $T[e]$. The edge type is an element of the global enumeration type *delaunay_edge_info* defined in geo_global_enums.

```
enum delaunay_edge_info { DIAGRAM_EDGE = 0,  DIAGRAM_DART = 0,
                          NON_DIAGRAM_EDGE = 1, NON_DIAGRAM_DART = 1,
                          HULL_EDGE = 2, HULL_DART = 2
                        }
```

The darts of T are labeled as defined in Section 10.4 on static Delaunay diagrams. Hull darts are labeled *HULL_DART* and non-hull darts are labeled either *DIAGRAM_DART* or *NON_DIAGRAM_DART*. The former label is used for non-hull darts that belong to the Delaunay diagram.

In *hull_dart* we always store a dart of the convex hull and in *cur_dart* we store an arbitrary dart of the triangulation. We use *cur_dart* as the starting point for searches.

Many member functions of *POINT_SET* come with a checker. The boolean *check* controls whether checking is done or not.

Most query operations require graph searches. We use a *node_map<int> mark* and an integer *cur_mark* to mark visited nodes in these searches. More precisely, a node v is marked if $mark[v] == cur\_mark$ and in order to unmark all nodes we increase *cur_mark* by one. We start with *cur_mark* equal to zero and all node marks equal to -1 and hence this solution is safe as long as *cur_mark* does not wrap around by overflow. Overflow occurs after MAXINT search operations. Assuming that a query takes at least 100 instructions one can do at most 10^6 (about 2^{20}) queries per second. Thus the solution would work for at least 2^{12} seconds or about an hour. We conclude that we should guard against this error, in particular, since it will be very difficult to locate once it occurs. The solution is simple. Whenever *cur_mark* reaches MAXINT we reinitialize.

⟨*functions to mark nodes*⟩≡

```
void init_node_marks() { mark.init(*this,-1);
                         cur_mark = 0;
                       }
void mark_node(node v) const { ((node_map<int>&)mark)[v] = cur_mark; }
void unmark_node(node v) const
{ ((node_map<int>&)mark)[v] = cur_mark - 1; }
bool is_marked(node v)  const { return mark[v] == cur_mark; }
void unmark_all_nodes() const
{ ((int&)cur_mark)++;
  if ( cur_mark == MAXINT)
  ((POINT_SET*)this) -> init_node_marks(); //cast away constness
}
```

Checking: We have two general routines for purposes of checking:

- *save_state(POINT p)* saves the current state of the data structure and the point *p* (which is typically the argument of a query operation) to a file, and
- *check_state(string loc)* checks the state of the data structure and prints diagnostic information to *cerr* if an error is found.

Checking is controlled by the boolean flag *check*, i.e., if *check* is *true*, *save_state* and *check_state* perform as described, and if *check* is *false*, they do nothing and *check_state* returns *true*.

A typical function *F* of class POINT_SET has a body of the following form.

```
if ( check ) save_state(POINT p);
/* proper body of F */
if ( check && !check_state("POINT_SET::F") )
{  cerr << additional information ; }
```

Assume now that check is set to *true* and that some function *F* contains an error. The error will be caught by *check_state*. Since the state before the execution of *F* was saved, the error is reproducible. We added this feature to POINT_SET because an earlier version of POINT_SET contained errors which arose very infrequently. For example, at one point we ran a test program for more than an hour before it failed.

Auxiliary Functions: The function *mark_edge* is used to assign a *delaunay_edge_info* to an edge. The call to *mark_edge_handler* is for the purposes of animation which we do not discuss here. Readers interested in the animation of the point set class should read [MN98a].

⟨*auxiliary functions*⟩≡

```
void mark_edge(edge e, delaunay_edge_info k)
{ assign(e,k);
  if (mark_edge_handler) mark_edge_handler(e);
}
```

The Constructors: The constructors allow us to construct a point set for either the empty set of points or for a set S of points. In the latter case the Delaunay triangulation algorithm of Section 10.4 is used, i.e., an arbitrary triangulation is constructed by plane sweep and then Delaunay flips are performed to obtain a Delaunay triangulation. The work horse for the second step is a member function *make_delaunay*(E) that takes a list of edges (it is required that all edges not in E have the Delaunay property) and turns the current triangulation into a Delaunay triangulation.

Locate: The function

```
edge T.locate(POINT p)
```

is the basis for all query functions. It returns an edge e of T (*nil* if T has no edge) with the following properties:

- If there is an edge of T containing p, such an edge is returned. If p lies on the boundary of the convex hull then a hull dart is returned (and not the reversal of a hull dart).

- If p lies in the interior of a face f of T (if p lies outside the convex hull of S, f is the unbounded face) then a dart on the boundary of f is returned. This dart has p to its left, except if all points in S are collinear and p lies on the line passing through the points in S. In this case, *target*(e) is the point in S closest to p.

The implementation of *locate* is non-trivial. We therefore define a function *check_locate* that checks the output of *locate*.

⟨*auxiliary functions*⟩+≡

```
void check_locate(edge answer,const POINT& p) const;
```

The implementation of *check_locate* is left to the reader; it can be found in [MN98a]. We turn to the implementation of *locate*. We distinguish cases according to the dimension of the triangulation.

⟨*POINT_SET.c*⟩+≡

```
edge POINT_SET::locate(POINT p) const
{
  if (number_of_edges() == 0) return nil;
```

```
  if (dim() == 1) { ⟨locate: one-dimensional case⟩ }
  ⟨locate: two-dimensional case⟩
}
```

If the dimension is less than one we return nil.

Let us assume next that the affine dimension of S is one. If p does not lie in the affine hull of S, i.e., p does not lie on the line supporting *hull_dart*, we return either *hull_dart* or its reversal. If p lies on the line supporting *hull_dart* we determine the answer by a walk in the triangulation. triangulations!walk through a triangulation

We initialize e to either *hull_dart* or its reversal such that p lies in the halfspace orthogonal[13] to e. We walk in the direction of e. Let $e1$ be the face cycle successor of e. As long as $e1$ points into the same direction as e, i.e., is not the reversal of e, and contains p in the halfspace orthogonal to it, we advance e to $e1$.

The walk ends when $e1$ is either the reversal of e or does not contain p in the halfspace orthogonal to it. In the former case p lies on e or $target(e)$ is the point in S closest to p and in the latter case p lies on e. In either case we may therefore return e.

⟨*locate: one-dimensional case*⟩≡

```
edge e = hull_dart;
int orient = orientation(e,p);
if (orient != 0) { if (orient < 0) e = reversal(e);
                   if (check) check_locate(e,p);
                   return e;
                 }
// p is collinear with the points in S. We walk
if ( !IN_HALFSPACE(e,p) ) e = reversal(e);
// in the direction of e. We know IN_HALFSPACE(e,p)
edge e1 = face_cycle_succ(e);
while ( e1 != reversal(e) && IN_HALFSPACE(e1,p) )
{ e = e1;
  e1 = face_cycle_succ(e);
}
if (check) check_locate(e ,p);
return e;
```

We come to the two-dimensional case. Assume w.l.o.g that *cur_dart* is not a hull dart (otherwise, replace *cur_dart* by its reversal).

If p is equal to the source of *cur_dart*, we are done and return the reversal of *cur_dart*; recall that we want to return a hull dart if p lies on the boundary of the convex hull.

So assume that p is distinct from the source of *cur_dart*. The face cycle containing *cur_dart* is a triangle since *cur_dart* is not a hull dart and hence p either does not lie on the line supporting *cur_dart* or the line supporting *face_cycle_pred*(*cur_dart*). Let e be the

[13] The halfspace orthogonal to e has normal vector e, has $source(e)$ in its boundary, and contains the target of e. We need this definition only for this paragraph.

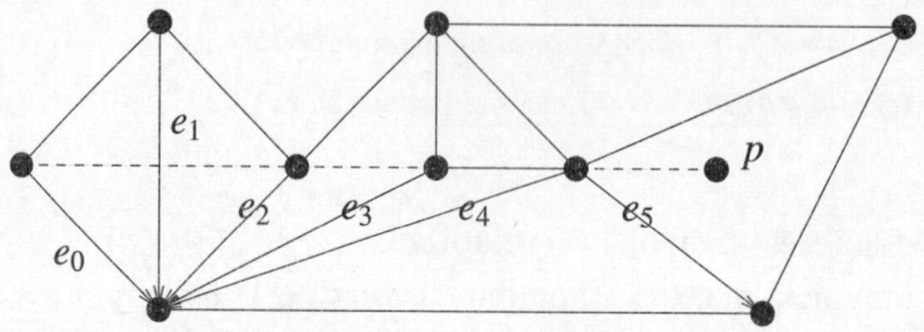

Figure 10.43 In order to locate p we walk along the segment s from $source(e_0)$ to p; s intersects the half-closures of the darts $e_0, e_1, \ldots, e_5$; $e_0, \ldots, e_5$ are directed downwards.

appropriate dart and assume that p lies in the positive halfspace[14] of e (replace e by its reversal otherwise).

We walk along the ray s starting in the source of e and ending in p, see Figure 10.43. We will maintain the following invariant during the walk:

- p lies in the positive subspace with respect to e.
- s intersects the half-closure of e, where the half-closure of e consists of the interior of e plus its source. However, the target of the dart does not belong to the half-closure.

⟨*locate: two-dimensional case*⟩≡

```
edge e = is_hull_dart(cur_dart) ? reversal(cur_dart) : cur_dart;
if (p == pos_source(e) ) return reversal(e);
int orient = orientation(e,p);
if (orient == 0) { e = face_cycle_pred(e);
                   orient = orientation(e,p);
                 }
if (orient < 0) e = reversal(e);
SEGMENT s(pos_source(e),p);
while ( true )
{
  if (is_hull_dart(e)) break;
  ⟨locate: determine the next edge e or break from the loop⟩
}
if (check) check_locate(e ,p);
((edge&)cur_dart) = e;
return e;
```

The while-loop performs the walk. We distinguish cases according to whether e is a hull dart or not. If e is a hull dart, we stop and return e.

Otherwise, let e, e_1, e_2 be the face cycle of the triangle F to the left of e. We need to find out whether the walk ends in F or whether we are leaving the triangle through e_1 or through

[14] The positive halfspace with respect to e is the halfspace to the left of the oriented line supporting e.

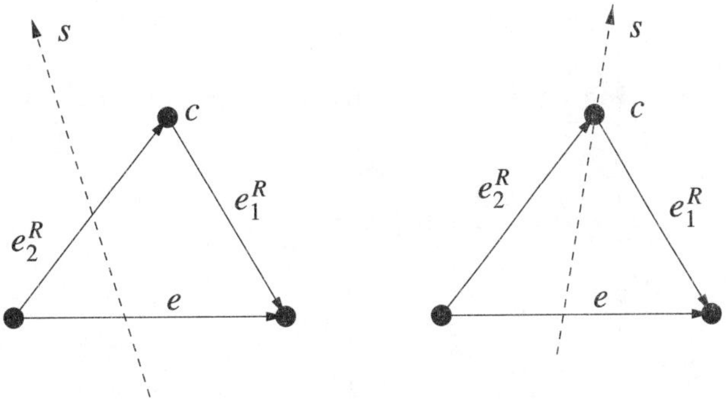

Figure 10.44 A step of the walk through the triangulation: In the left part of the figure, c lies to the right of s and in the right part it does not.

e_2. Let c be the common endpoint of e_1 and e_2. We distinguish cases according to whether c lies to the right of s or not.

Assume first that c lies to the right of s, i.e., s intersects the half-closure of the reversal e_2^R of e_2, see Figure 10.44. If p lies to the left of e_2^R, we replace e by e_2^R and continue. If p lies on e_2^R, we return e_2^R, and if p lies to the right of e_2^R and hence in the interior of F, we return e.

Assume next that c does not lie to the right of s, i.e., s intersects the half-closure of the reversal e_1^R of e_1, see Figure 10.44. If p lies to the left of e_1^R, we replace e by e_1^R and continue. If p lies on e_1^R, we return e_1^R, and if p lies to the right of e_1^R and hence in the interior of F or on e_2 (the latter case can only occur when s passes through the source of e and p lies on e_2), we return e in the former case and e_2^R in the latter.

⟨*locate: determine the next edge e or break from the loop*⟩≡

```
edge e1 = face_cycle_succ(e);
edge e2 = face_cycle_pred(e);
int d = ::orientation(s,pos_target(e1));
edge e_next = reversal( (d < 0) ? e2 : e1 );
int orient = orientation(e_next,p);
if ( orient > 0 )  { e = e_next; continue; }
if ( orient == 0 ) { e = e_next; break; }
if ( d == 0 && orient < 0 && orientation(e2,p) == 0 ) e = reversal(e2);
break;
```

This completes the description of *locate*. We still need to argue termination. We clearly make progress when the new dart e intersects s closer to p than the old dart e. It may, however, be the case that the intersections are the same. In this situation the new dart e forms a smaller angle with s than the old one.

Having *locate*, we can easily implement the *lookup* operation.

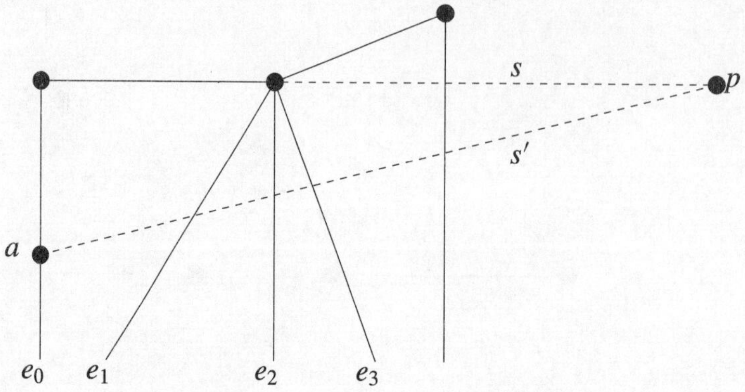

Figure 10.45 The node a lies in the interior of dart e_0 but infinitesimally close to the source node of e_0. The darts e_0, e_1, ... have p on their left and are directed downwards. The ray s' intersects only the interior of darts.

⟨*POINT_SET.c*⟩+≡

```
node POINT_SET::lookup(POINT p) const
{ if (number_of_nodes() == 1) { node v = first_node();
                                return (pos(v) == p) ? v : nil;
                              }
  edge e = locate(p);
  if (pos(source(e)) == p) return source(e);
  if (pos(target(e)) == p) return target(e);
  return nil;
}
```

It took us a long time to come up with the short and elegant inner loop for *locate* given above. Earlier attempts were longer and less elegant (and some were plain wrong). Why did we have such difficulties and how did we finally arrive at the program given above? The difficulties stemmed from degeneracies; we had difficulties handling the case that the ray s passes through some node of the triangulation or even runs on top of an edge of the triangulation. Under the additional assumption that there are no degeneracies, i.e., that s enters and leaves triangles through relative interiors of edges, it was easy to write a correct program. We had difficulties extending the solution to the case where s enters and/or leaves through a vertex. Our original solution was clumsy because we used the weaker invariant that s intersects the closure of e (and not only the half-closure as we stated above). This resulted in a lengthy case distinction.

The key to the simpler program was a thought experiment using *perturbation*. Recall that we locate p by a walk through the triangulation starting at the source node of some dart e_0. The idea of perturbation is to simulate the walk along a perturbed ray s' that starts in a node a that lies in the interior of e_0 but infinitesimally close to the source of e_0, see Figure 10.45. The perturbed ray will only pass through the interior of darts (except maybe at p); it may pass infinitesimally close to the source of a dart but not infinitesimally close to the target. We concluded that source nodes of darts play a different role than target nodes of darts and came up with the concept of the half-closure of a dart. Once we had the concept of a half-closure, we arrived at a correct program within an hour.

We close this section with a remark about the efficiency of *locate*. Clearly, the running time of locate is proportional to the number of darts of the Delaunay triangulation crossed by the segment s. Bose and Devroye [BD95] have shown that the expected number of edges of a Delaunay triangulation of random points crossed by a line segment of length l is $O(l\sqrt{\gamma})$, where γ is the point density.

Insert: The function

```
node T.insert(POINT p);
```

inserts the point p into T and returns the corresponding node. More precisely, if there is already a node v in T positioned at p (i.e., $pos(v)$ is equal to p) then $pos(v)$ is changed to p (i.e., $pos(v)$ is made identical to p) and if there is no such node then a new node v with $pos(v) = p$ is added to T. In either case, v is returned.

We first define our return statement

⟨*insert::check and return v*⟩≡

```
if ( check && !check_state("POINT_SET::insert") )
{ cerr << "The point inserted was " << p;
  exit(1);
}
return v;
```

and then give an overview. We first deal with the case that T has at most one node. If T has more than one node, we locate p in the triangulation. Let e be the edge returned by *locate*(p). If p is equal to an endpoint of e, we replace the endpoint by p and return.

Otherwise, we determine whether p lies on e and then distinguish cases according to the dimension of the triangulation after the insertion. The dimension is one if the current dimension is one and p lies in the affine subspace of S.

⟨*POINT_SET.c*⟩+≡

```
node POINT_SET::insert(POINT p)
{ if ( check ) save_state(p);
  node v;
  ⟨T has zero or one node⟩
  edge e = locate(p);
  if (p == pos_source(e)) { assign(source(e),p); return source(e); }
  if (p == pos_target(e)) { assign(target(e),p); return target(e); }
  bool p_on_e = seg(e).contains(p);
  if ( dim() == 1 && orientation(e,p) == 0 )
  { ⟨dimension is one after the insertion⟩ }
  ⟨dimension is two after the insertion⟩
}
```

Assume first that T has at most one node. If T has no node, we create a node, label it with p and return it, if T has one node, we either relabel this node with p or we create a new node with label p and connect it to the old node.

⟨*T has zero or one node*⟩≡

```
if (number_of_nodes() == 0)
{ v = new_node(p); ⟨insert::check and return v⟩ }
if (number_of_nodes() == 1)
{ node w = first_node();
  if (p == pos(w))
  { assign(w,p);
    v = w;
    ⟨insert::check and return v⟩
  }
  else
  { v = new_node(p);
    edge x = new_edge(v,w); edge y = new_edge(w,v);
    mark_edge(x,HULL_DART); mark_edge(y,HULL_DART);
    set_reversal(x,y);
    hull_dart = cur_dart = x;
    ⟨insert::check and return v⟩
  }
}
```

If *dim* is one and p lies in the affine hull of S there are two cases. If p is on e then we split e into two edges and if p does not lie on e we simply add new edges between p and *target*(e).

⟨*dimension is one after the insertion*⟩≡

```
v = new_node(p);
edge x = new_edge(v,target(e)); edge y = new_edge(target(e),v);
mark_edge(x,HULL_DART);          mark_edge(y,HULL_DART);
set_reversal(x,y);
if (p_on_e)
{ x = new_edge(v,source(e));
  y = new_edge(source(e),v);
  mark_edge(x,HULL_DART);
  mark_edge(y,HULL_DART);
  set_reversal(x,y);
  hull_dart = cur_dart = x;
  del_edge(reversal(e));
  del_edge(e);
}
⟨insert::check and return v⟩
```

In the remaining case the hull is guaranteed to be two-dimensional after the insertion. We now have to triangulate the face that contains p. p lies in the interior of the convex hull iff e is not a hull dart.

If p lies in a bounded face (= triangle), we connect it to all (three) nodes of the face.

One of the three new triangles could have height zero. We made sure that *make_delaunay* handles this case correctly.

If p lies in the outer face or on its boundary, we first determine the set of hull darts visible from p by walking in both directions along the hull starting in e. We call the two extreme darts reached by these walks *e1* and *e2*. We then add an edge for each visible vertex, i.e. for all vertices from *target(e1)* to *source(e2)*.

There is one subtle point. It is important how ties are broken when p lies on a hull dart. Only one triangle should be added to the triangulation and not three (the latter would be the case if we break the tie in favor of the triangle incident to the hull dart). In order to guarantee that ties are broken correctly, we have *locate* return a hull dart if p does not lie in the interior of the triangulation.

In the implementation we retriangulate the outer face and bounded faces in a uniform way; we add new edges for all nodes from *target(e1)* to *source(e2)* for two darts *e1* and *e2*. In the case of a bounded face we choose *e1* = *e2* = *e* and in the case of the outer face we set *e1* and *e2* to the extreme (tangent) darts as described above.

⟨*dimension is two after the insertion*⟩≡

```
v  = new_node(p);
edge e1 = e;
edge e2 = e;
list<edge> E;
bool outer_face = is_hull_dart(e);

if (outer_face)
{ //  move e1/e2 to compute upper/lower tangents
  do e1 = face_cycle_pred(e1); while (orientation(e1,p) > 0);
  do e2 = face_cycle_succ(e2); while (orientation(e2,p) > 0);
}
// insert edges between v and target(e1) ... source(e2)
e = e1;
do { e = face_cycle_succ(e);
     edge x = new_edge(e,v);
     edge y = new_edge(v,source(e));
     set_reversal(x,y);
     mark_edge(e,DIAGRAM_DART);
     E.append(e);
     E.append(x);
   } while (e != e2);

if (outer_face)
{ // mark last visited and new edges as hull edges
  mark_edge(face_cycle_succ(e1),HULL_DART);
  mark_edge(face_cycle_pred(e2),HULL_DART);
  mark_edge(e2,HULL_DART);
  hull_dart = e2;
}

make_delaunay(E); // restores Delaunay property
```

⟨*insert::check and return v*⟩

Deletion: The functions

```
void T.del(node v)
void T.del(POINT p)
```

remove the node v and the point p, respectively, i.e., make T a Delaunay triangulation for $S \setminus \{pos(v)\}$ and $S \setminus p$, respectively.

The strategy to remove a node is simple. Removal of a node from the interior of a two-dimensional triangulation (of course, the program also has to handle the removal of a node from a triangulation that is not two-dimensional or of a node which lies on the boundary of the convex hull) creates a cavity in the triangulation. The cavity is retriangulated in an arbitrary way and then *make_delaunay*(E) is called to restore the Delaunay property, where E is the set of new edges and the set of edges on the boundary of the cavity.

After this general outline we define our return statement and give an overview of the deletion procedure.

⟨*del: check and return*⟩≡

```
if ( check && !check_state("POINT_SET::del(node v)") )
{ cerr << "deleted the node with position " << pos(v);
  exit(1);
}
return;
```

⟨*POINT_SET.c*⟩+≡

```
void POINT_SET::del(node v)
{
  if (v == nil) error_handler(1,"POINT_SET::del: nil argument.");
  if (number_of_nodes() == 0)
    error_handler(1,"POINT_SET::del: graph is empty.");
  if (check) save_state(inf(v));
  if ( dim() < 2 )
  {
    if ( outdeg(v) == 2)
    { node s = target(first_adj_edge(v));
      node t = target(last_adj_edge(v));
      edge x = new_edge(s,t);  edge y = new_edge(t,s);
      mark_edge(x,HULL_DART);  mark_edge(y,HULL_DART);
      set_reversal(x,y);
    }
    del_node(v);
    cur_dart = hull_dart = first_edge();
    ⟨del: check and return⟩
  }
  ⟨removal of v from a two-dimensional triangulation⟩
  ⟨del: check and return⟩
}
```

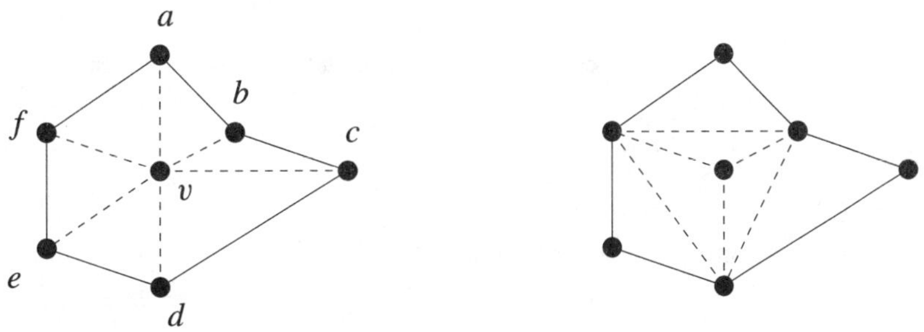

Figure 10.46 The right part of the figure shows the effect of flipping the edges (v, a), (v, c) and (v, e).

If the dimension of the triangulation is less than two, the removal of v is trivial. If the dimension is zero or the dimension is one and v is an extreme node of the triangulation (i.e., the outdegree of v is one), we simply remove v. If v has outdegree two, we connect the two neighbors of v by a new edge and then delete v. Of course, *cur_dart* or *hull_dart* could have been incident to v and hence have to be given new values.

We come to the interesting case, the removal of v from a two-dimensional triangulation. We first discuss the case that v lies in the interior of the triangulation. We will later see that the same strategy also handles the case where v lies on the boundary of the convex hull.

Removal of v creates a face P that is, in general, not a triangle. It is only a triangle if the degree of v is three. We need to retriangulate this face. A natural approach would be to remove v and to retriangulate after the removal of v. However, this approach does not exploit the fact that P is a so-called *star-shaped polygon* with respect to v, i.e., that v can see all vertices of P. We will exploit this fact as follows in the retriangulation process. We will show below that there is always an edge e incident to v such that the two triangles incident to v form a convex quadrilateral. We "flip e away from v" by replacing it by the other diagonal of the triangle. In this way the degree of v is decreased by one. We continue until the degree of v is three. At this point, v is removed and the created face is a triangle, see Figure 10.46.

We now give the details. We need a slightly more general definition of star-shapedness than was alluded to in the text above. The more general definition is needed to cope with the case that three or more points of S lie on a common line.

We call a polygon P *star-shaped* with respect to a point v if either:

- v lies in the interior of P and for every vertex p of P the open line segment vp is contained in the interior of P, or

- v lies in the relative interior of an edge e of P and for every vertex p of P that is not an endpoint of e the open line segment vp is contained in the interior of P.

Lemma 68 *Let P be a polygon which has at least four vertices and is star-shaped with respect to some point v. Then there are three consecutive vertices p, q, r of P such that*

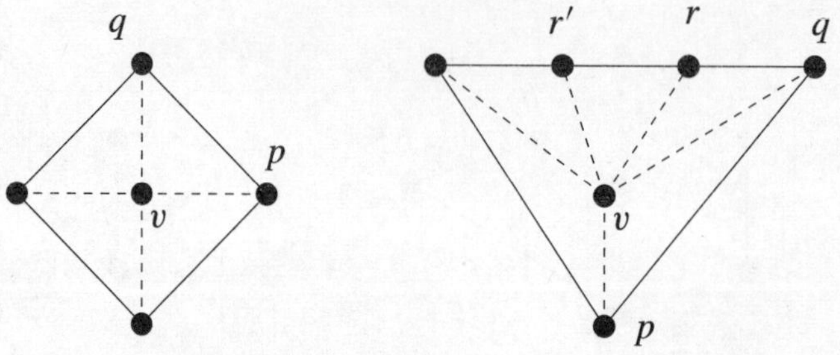

Figure 10.47 (p, q, r, v) forms a convex quadrilateral. In the situation on the left v will lie on an edge of P' after the flip of edge (v, q) and in the situation on the right it will still lie in the interior of P'. The quadruple (q, r, r', v) does not qualify for a flip.

(v, p, q, r) form a convex quadrilateral. In this quadrilateral the angle at v maybe equal to π. The angle at v can be equal to π only if v lies in the interior of P, see Figure 10.47.

Let P' be the polygon obtained from P by replacing the edges pq and qr by the edge pr. Then P' is star-shaped with respect to v.

Proof Consider any triangulation T of P. T consists of at least two triangles. Since the dual of a triangulation is a tree and every tree has at least two leaves, there must be at least two triangles in T whose edges consist of two consecutive edges of P plus the chord connecting the source of the first edge with the target of the second edge and hence there must be at least one such triangle which, in addition, does not contain v in its interior. Consider one such triangle, say t, and let $e_1 = (p, q)$ and $e_2 = (q, r)$ be the edges of P that are contained in its boundary. Since (p, q, r) is a triangle of T the angle at q is less than π.

Since v is not contained in the interior of t, (v, p, q, r) forms a convex quadrilateral. In this quadrilateral the angles at p and r must be less than π since P is star-shaped with respect to v. Also by the star-shapedness, the angle at v can be equal to π only if v lies in the interior of P.

P' is clearly star-shaped with respect to v. □

Call an edge incident to v *flipable* if the two triangles incident to it form a convex quadrilateral. As long as there is a flipable edge incident to v flip it. The lemma above guarantees that the process does not terminate before v has degree three.

How can we find flipable edges quickly? We scan through the edges incident to v. Let e be the current edge. If e is not flipable, we advance e to the cyclic successor of e, and if e is flipable, we flip it and set e to the cyclic predecessor of e.

When do we terminate? We terminate when v has degree three. Since we want to use the same procedure also for nodes on the hull we develop a more general termination condition. We terminate when the degree of v reaches *min_deg*, where *min_deg* is three for nodes in the interior and is two for hull nodes. We also keep a counter *count* which is a lower bound on the number of edges out of v that are certainly not flipable. We increment *count* whenever a non-flipable edge is found, we decrement *count* by two whenever a flip is performed, as this may make the two neighbors of the flipped edge flipable, and we terminate if *count* reaches $outdeg(v)$.

Why is this correct? Call an edge *certified non-flipable* if it has been tested for flipping and its two neighbors have not changed since. In the procedure just outlined the edges that are certified non-flipable are consecutive in the cyclic adjacency list of v and *count* is a lower bound on their number. This shows correctness.

The running time of retriangulation is linear in the initial degree of v. This follows from the fact that the total decrement of *count* is bounded by twice the initial degree of v and hence the total increase of *count* is bounded by thrice the initial degree of v.

We obtain the following code.

⟨*removal of v from a two-dimensional triangulation*⟩≡

```
list<edge> E;
int min_deg = 3;
edge e;
forall_adj_edges(e,v)
{ E.append(face_cycle_succ(e));
  if (is_hull_dart(e)) min_deg = 2;
}
int count = 0;
e = first_adj_edge(v);
while ( outdeg(v) > min_deg && count < outdeg(v) )
{ edge e_pred = cyclic_adj_pred(e);
  edge e_succ = cyclic_adj_succ(e);
  POINT a = pos_target(e_pred); POINT c = pos_target(e_succ);
  if ( !right_turn(a,c,pos(v)) && right_turn(a,c,pos_target(e)) )
  { // e is flipable
    edge r = reversal(e);
    move_edge(e,reversal(e_succ),target(e_pred));
    move_edge(r,reversal(e_pred),target(e_succ),LEDA::before);
    mark_edge(e,DIAGRAM_DART);
    mark_edge(r,DIAGRAM_DART);
    E.append(e);
    e = e_pred;
    count = count - 2;
    if ( count < 0 ) count = 0;
  }
  else
  { e = e_succ;
    count++;
  }
}
if ( min_deg == 2 )
{ ⟨adjust marks of new hull darts and their reversals⟩ }
```

```
cur_dart = E.head();
del_node(v);
make_delaunay(E);
```

We give some more explanations. The while-loop performs the retriangulation. During the retriangulation we build up a list E of edges whose Delaunay property needs to be checked: E consists of all edges in the boundary of the cavity created by the removal of v and all edges created during retriangulation.

After retriangulation we remove v and add call *make_delaunay*(E) to restore the Delaunay property.

We also have to take care of *cur_dart*. It may have been incident to v. We set it to an arbitrary edge in the boundary of the cavity created by the removal of v.

This completes the discussion of the case when a node in the interior of the triangulation is removed. We will next argue that the same retriangulation strategy works when v is a node in the boundary of the triangulation.

Again we flip edges away from v until no further edges are flipable. When this is the case, the neighbors of v form a chain that is concave as seen from v and hence removal of v leaves us with a triangulation of the remaining nodes. Removal of v also turns some darts into hull darts. Their labels have to be changed to *HULL_DART* and the edges of their reversal have to be changed to *DIAGRAM_DART*. There is a small exception to the latter rule, namely when a reversal is a hull dart itself. This will be the case when the removal of v reduces the dimension of the triangulation from two to one.

⟨*adjust marks of new hull darts and their reversals*⟩≡

```
edge e,x;
forall_adj_edges(e,v)
{ x = face_cycle_succ(e);
  mark_edge(x,HULL_DART);
  if ( !is_hull_dart(reversal(x)) ) mark_edge(reversal(x),DIAGRAM_DART);
}
hull_dart = x;
```

Nearest Neighbor Searching: The functions

```
node          T.nearest_neighbor(POINT p);
list<node> T.k_nearest_neighbors(POINT p, int k);
```

return a node v closest to p, i.e., $dist(p, pos(v)) = \min\{dist(p, pos(u))\ ;\ u \in T\ \}$, and the list of the $\min(k, |S|)$ closest points to p, respectively. The points in the result list are ordered by distance from p. One can also ask for the nearest neighbor(s) of a node.

```
node          T.nearest_neighbor(node w);
list<node> T.k_nearest_neighbors(node w, int k);
```

return a node v different from w that is closest to $T[w]$ and the list of the $\min(k, |S| - 1)$ closest points to $T[w]$, respectively.

The following observation paves the way for a simple algorithm for both problems and is also the basis of the range query algorithms to be discussed in the next section.

Lemma 69 *Let s and t be two nodes of a Delaunay triangulation T and let d be their distance. Then there is a path from s to t in T such that all intermediate nodes have distance less than d from s.*

Proof We use induction on d. Let D be the disk with radius d centered at s. If st is an edge of T, we are done. Otherwise let a and b be the two neighbors of t such that the segment st runs between the edges ta and tb of T. The points a, b, and t form a triangle of T. If one of a and b has distance less than d from s, we can apply the induction hypothesis and are done. So assume otherwise, i.e., neither a nor b lies in the interior of D. The segments st and ab intersect (since s cannot lie in the interior of the triangle with corners a, b, and t) and hence (s, a, t, b) is a convex quadrilateral. The disk D proves that the segment ab does not belong to the Delaunay triangulation of $\{a, b, s, t\}$ and hence cannot be an edge of T. □

The lemma suggests a simple strategy to find the k-nearest neighbors of $p = T[v]$. If the number of points in T is no more than k, we simply return all nodes in T. So assume otherwise. We start a graph search starting in v. We keep all reached nodes in a priority queue according to their (squared) distance from v and always continue the exploration from a node with smallest distance. The lemma above guarantees that this strategy explores the nodes of T in order of increasing distance from v.

⟨*POINT_SET.c*⟩+≡

```
#include <LEDA/p_queue.h>
list<node> POINT_SET::nearest_neighbors(node v, int k) const
{ list<node> result;
  int n = number_of_nodes();
  if ( k <= 1 ) return result;
  if ( n + 1 <= k )
  { node w;
    forall_nodes(w,*this) if ( w != v ) result.append(w);
    return result;
  }
  POINT p = pos(v);
  unmark_all_nodes();
  p_queue<RAT_TYPE,node> PQ;
  PQ.insert(0,v); mark(v);
  while ( k > 0 )
  { pq_item it = PQ.find_min();
    node w = PQ.inf(it); PQ.del_item(it);
    if ( w != v ) { result.append(w); k--; }
    node z;
    forall_adj_nodes(z,w)
```

```
    { if ( !is_marked(z) ) { PQ.insert(p.sqr_dist(pos(z)),z);
                             mark(z);
                           }
    }
  }
  return result;
}
```

We come to the case where we want to search for the nearest neighbors of a point p. We simply insert p into T and then use the procedure above.

A small complication arises from the fact that p may lie on a node of T. We test for this case by performing a lookup for p. If p does not lie on a node of v, we insert it. Of course, it has to removed again after calling the procedure above and p has to be removed from the list of answers.

⟨*POINT_SET.c*⟩+≡

```
list<node> POINT_SET::nearest_neighbors(POINT p, int k)
{ list<node> result;
  int n = number_of_nodes();
  if ( k <= 0 ) return result;
  if ( n <= k ) return all_nodes();
  // insert p and search neighbors graph starting at p
  node v = lookup(p);
  bool old_node = true;
  if ( v == nil ) { v = ((POINT_SET*)this)->insert(p);
                    old_node = false;
                  }
  else k--;
  result = nearest_neighbors(v,k);
  if ( old_node )
    result.push_front(v);
  else
    ((POINT_SET*)this)->del(v);
  return result;
}
```

The nearest neighbor of a node v in a Delaunay diagram is a node adjacent to v. Thus one only has to find the minimum (squared) distance between v and its neighboring nodes.

⟨*POINT_SET.c*⟩+≡

```
node POINT_SET::nearest_neighbor(node v) const
{
  if (number_of_nodes() <= 1) return nil;
  POINT p = pos(v);
  edge e = first_adj_edge(v);
  node min_v = target(e);
  while ((e = adj_succ(e)) != nil)
```

n	I	NN	NNA
50000	128.2	2.32	18.08

Table 10.5 We constructed a point set of n random points in the unit square and performed a nearest neighbor query for each node in the triangulation. NN shows the time for the function *nearest_neighbor* and NNA shows the time with alternative implementation of the inner loop. Column I shows the time for the n insertions. The table was made with the rational kernel.

```
    { node u = target(e);
      if ( p.cmp_dist(pos(u),pos(min_v)) < 0 ) min_v = u;
    }
    return min_v;
}
```

An alternative way to write the inner loop is:

⟨*alternative inner loop*⟩≡

```
node min_v = target(e);
RAT_TYPE min_d = p.sqr_dist(pos(min_v));
while ((e = adj_succ(e)) != nil)
{ node u = target(e);
  RAT_TYPE d_u = p.sqr_dist(pos(u));
  if ( d_u < min_d ) { min_v = u;
                       min_d = d_u;
                     }
}
```

This is much slower, see Table 10.5. Why is the alternative so much slower; aren't the two programs doing exactly the same thing? Both programs compute the squared distance from v to all its neighbors and find the minimum.

The difference is that the alternative version computes all squared distances *exactly* as rational numbers[15] and finds the minimum of these rational numbers. The original version asks the kernel to compare distances. The kernel first computes floating point approximations to the squared distances and uses them in the comparisons. If the floating point approximation suffices to decide the comparison, the exact squared distance is never computed and a lot of work is saved.

Range Searches: We have functions for circular, triangular, and rectangular range searches.

In order to perform a circular range query with center v we perform a DFS starting at v. The search is restricted to the nodes that lie in the circular range. Correctness follows from Lemma 69.

[15] We assume for this paragraph that the rational kernel is used.

⟨*POINT_SET.c*⟩+≡

```
void POINT_SET::dfs(node s, const POINT& pv,
                    const POINT& p, list<node>& L) const
{ L.append(s);
  mark_node(s);
  node u;
  forall_adj_nodes(u,s)
      if (!is_marked(u) && pv.cmp_dist(pos(u),p) <= 0 ) dfs(u,pv,p,L);
}

list<node> POINT_SET::range_search(node v,const POINT& p) const
{
  list<node> L;
  POINT pv = pos(v);
  unmark_all_nodes();
  dfs(v,pv,p,L);
  return L;
}
```

The other two kind of queries can be reduced to circular queries by first performing a range query with the circumcircle of the triangle or rectangle and then filtering the returned list of points with the triangle or rectangle, respectively. We leave the implementation of the other queries to the reader.

Experimental Data: Table 10.6 contains running times. The table shows that nearest neighbor queries for nodes are very efficient in comparison to nearest neighbor queries for points. This comes from the fact that the latter involve a lookup, an insertion, a deletion, as well as a nearest neighbor query for a node. For queries that ask for the ten nearest neighbors the difference is not as pronounced. This stems from the fact that k-nearest neighbor queries involve rational arithmetic.

Exercises for 10.6

1 Implement circular range queries.

2 Implement triangular and rectangular range queries. You may use circular range queries.

3 Animate the Delaunay class such that the actions performed after the insertion of a point are visualized.

4 The *nearest_neighbors* algorithm uses a *p_queue*<*RAT_TYPE*, *node*>. The code becomes slightly simpler if a *node_pq*<*RAT_TYPE*> is used. Why is it better to use a *p_queue* instead of a *node_pq*? Time both programs and explain.

5 Develop a version of the k-nearest neighbor search that cuts down on the use of rational arithmetic.

6 Our implementation of *nearest_neighbor*(*POINT p*) modifies the Delaunay triangulation by an insertion and a deletion. It is not guaranteed that the original Delaunay triangulation is restored. Can you modify the implementation such that it becomes a const-operation? Try to determine the set L of all edges of the current triangulation whose

K	n	I	L	NNP	NNV	NNP(10)	NNV(10)	D
S	1000	1.14	0.66	2.15	0.05	10.22	7.07	1.16
	2000	2.79	1.83	4.92	0.09	21.25	14.06	2.77
	4000	6.83	5.36	11.68	0.2	44.4	28.29	6.65
D	1000	1.15	0.68	2.22	0.03999	10.27	7.03	1.18
	2000	2.78	1.89	4.99	0.11	21.21	14.04	2.75
	4000	6.76	5.23	11.53	0.2	44.25	28.25	6.65
C	1000	0.82	0.41	0.99	0.03	5.43	4.65	2.84
	2000	1.75	0.9	2.08	0.06	11.09	9.31	8.2
	4000	3.78	2.03	4.42	0.13	22.35	18.48	29.09

Table 10.6 The performance of point sets. As in the other tables of this chapter we used three kinds of inputs: random points in the unit square, random points in the unit disk, and random points on the unit circle. We generated two sets L and LQ of n points, built a point set T by inserting the points in L (I), performed n lookups for the points in LQ (L), performed nearest neighbor queries for the points in LQ (NNP), performed nearest neighbor queries for the nodes of T (NNV), computed the ten nearest neighbor queries for the points in LQ (NNP(10)), computed the ten nearest neighbor queries for nodes of T (NNV(10)), and finally deleted all points.

Delaunay property is destroyed by p. The nearest neighbor of p must be a vertex of the triangle containing p or an endpoint of an edge in L.

10.7 Line Segment Intersection

The line segment intersection problem asks to compute the set of intersections of a set S of line segments in the plane. It is one of the basic geometric problems and has numerous applications, e.g., in computer aided design, geographic information systems, and cartography. We will see an application to boolean operations on polygons in Section 10.8. Many different algorithms have been designed for the problem and several of them are available in LEDA. The line segment intersection problem comes in many different flavors as different applications have different output requirements. One may be interested in the number of intersections, or one may want to trigger an action for every pair of intersecting segments, or one may want to compute the graph induced by the segments, or one may want to compute the trapezoidal decomposition induced by the set of segments. In LEDA we provide functions for several output conventions which we survey in Section 10.7.1. We also give

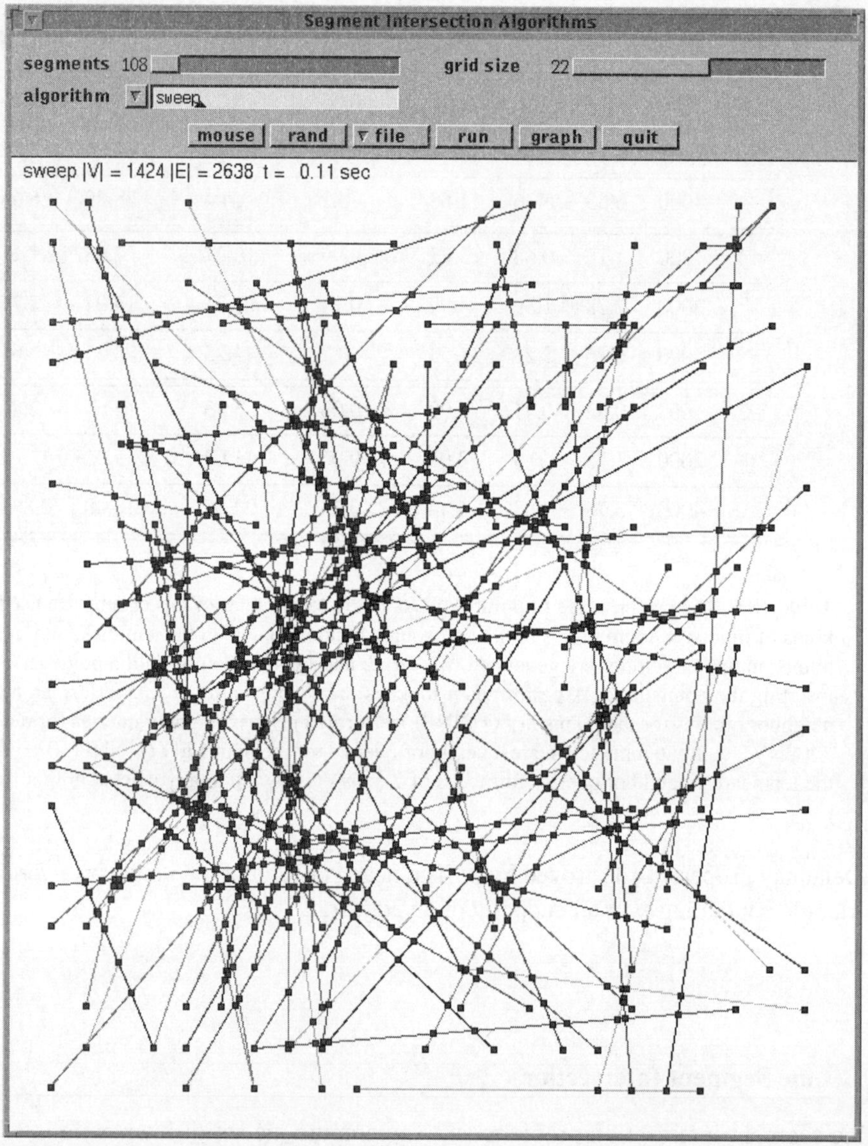

Figure 10.48 A screen shot of the intersect_segments demo in xlman. The sweep line algorithm was used to compute the graph induced by a set of 203 random segments. The induced graph has 1424 nodes and 2638 edges.

some experimental data in this section. In the remaining sections we discuss the sweep line algorithm for segment intersection.

The algorithms discussed in this section are illustrated by the intersect_segments demo in xlman. Figure 10.48 shows a screen shot.

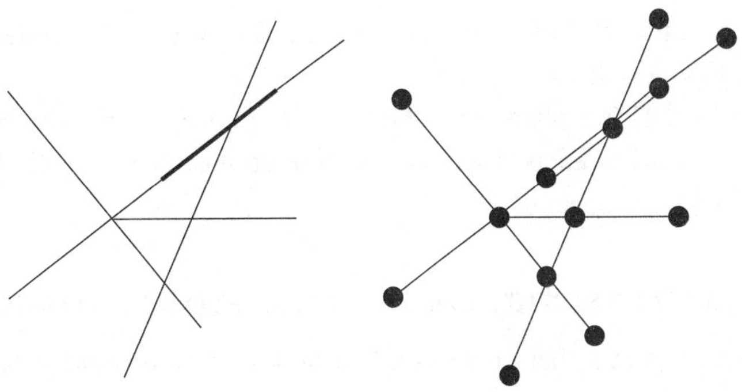

Figure 10.49 A set S of segments and the induced planar graph.

10.7.1 *Functionality*

We first introduce some terminology. Two segments s_1 and s_2 *intersect* if they have at least one point in common and *overlap* if they have more than one point in common. Two segments s_1 and s_2 are said to have a *proper intersection* if they share exactly one point and this point lies in the relative interior of both segments. A segment of length zero is called a trivial segment.

The undirected graph $U(S)$ induced by S is defined as follows. The vertices of $U(S)$ are all endpoints of segments and all proper intersection points between segments in S. The edges of U are the maximal relatively open and connected subsets of segments in S that contain no vertex of $U(S)$. Figure 10.49 shows an example. Note that the graph $U(S)$ contains parallel edges if S contains segments that overlap. We use n to denote the number of segments in S, s to denote the number of nodes of U, m to denote the number of edges of U, and k to denote the number of pairs of intersecting segments. If S contains no overlapping segments, $m = O(n + s)$. If S contains overlapping segments, m may be as large as $n(n + s)$ since an input segment may be divided into $n + s$ pieces by the endpoints and intersection points. The number of nodes of U is at most $n + k \leq n + n(n - 1)/2$. If many segments have a common intersection, k may be much larger than s. For example, if all n segments pass through a common point then $s = n + 1$ and $k = n(n - 1)/2$.

The function

```
void SEGMENT_INTERSECTION(const list&lt;SEGMENT&gt;& S,
                          GRAPH<POINT,SEGMENT>& G, bool embed = false)
```

computes a directed graph $G(s)$ representing $U(S)$. The algorithm makes no assumption about the segments in S. They may be overlapping, they may have multiple intersections, they may share endpoints, they may have length zero, ...

G and U have the same set of nodes; each node of G is labeled by its position in the plane.

The edges of G correspond to the edges of U. If *embed* is false, there is exactly one dart in G for each edge in U; the dart is labeled by the segment in S containing it and inherits

its direction from the segment containing it, i.e, if $e = (v, w)$ is a dart of G then $G[e]$ is directed from $G[v]$ to $G[w]$.

If *embed* is true, G is a plane map. For each edge of U there are two darts in G and the two darts are reversal of each other. For each node v of G the cyclic list of darts out of v are counter-clockwise ordered.

The function

```
void SEGMENT_INTERSECTION(const list&lt;SEGMENT&gt;& S, list<POINT>& P)
```

returns the list of points that correspond to nodes of G of degree two or more and the function

```
SEGMENT_INTERSECTION(const list&lt;SEGMENT&gt;& S,
                     void (*report)(const SEGMENT&, const SEGMENT&) )
```

calls *report*(s_1, s_2) for every pair (s_1, s_2) of intersecting segments. Observe that the points in P are a subset of the points for which *report* is called. For example, if S consists of two identical trivial segments, then $G(S)$ consists of a single node and no edge and hence P will be empty. On the other hand, *report* will be called for this pair of segments.

For all functions above several implementations are available. The implementations are based on the algorithms of Bentley and Ottmann ([BO79]), Mulmuley ([Mul90]), and Balaban ([Bal95]). For the reporting version of segment intersection we also have the trivial implementation which simply checks every pair of segments in S for an intersection.

```
void MULMULEY_SEGMENTS(const list&lt;SEGMENT&gt;& S, GRAPH<POINT,SEGMENT>& G,
                       bool embed = false);
void SWEEP_SEGMENTS(const list&lt;SEGMENT&gt;& S, GRAPH<POINT,SEGMENT>& G,
                    bool embed = false, bool use_optimization = true);
void SWEEP_SEGMENTS(const list&lt;SEGMENT&gt;& S , list<POINT>& P);
void BALABAN_SEGMENTS(const list&lt;SEGMENT&gt;& S,
                      void (*report)(const SEGMENT&, const SEGMENT&));
void TRIVIAL_SEGMENTS(const list&lt;SEGMENT&gt;& S,
                      void (*report)(const SEGMENT&, const SEGMENT&));
```

The asymptotic running time of the Bentley–Ottmann algorithm is $O(m+(n+s)\log n)$, the asymptotic running time of the Mulmuley algorithm is $O(m+s+n\log n)$. Both algorithms can be used for all functions above. If *embed* is true the running time of the Bentley–Ottmann algorithm increases by $O(m\log m)$, since an additional sorting step is required. The asymptotic running time of the Balaban algorithm is $O(n\log^2 n + k)$. It can only be used for the functions that report intersections. The asymptotic running time of the trivial implementation is $O(n^2)$.

Table 10.7 compares the running time of our various implementations. In the examples, Balaban's algorithm is always better than the trivial algorithm. Mulmuley's algorithm is better than the Bentley–Ottmann algorithm when the number of intersections is large. It also incurs a smaller additional cost for turning $G(S)$ into a planar map (as it always computes an undirected planar map). When the number of intersections is small, the Bentley–Ottmann algorithm and Mulmuley's algorithm behave about the same.

n	d	V	E	S	S + E	M	M + E	T	B
2000	22	4007	2014	1.14	1.3	1.74	1.76	15.13	1.94
2000	23	4026	2052	1.18	1.37	2.25	2.29	14.87	2.07
2000	24	4136	2272	1.25	1.42	2.91	2.91	15.26	2.17
2000	25	4428	2856	1.39	1.63	3.44	3.47	15.06	2.33
2000	26	5857	5714	1.81	2.37	4.44	4.5	15.31	2.48
2000	27	10954	15908	3.03	5.02	5.93	6.02	15.41	2.74
2000	28	29683	53366	7.57	16.43	9.71	10.02	16.01	3.22
2000	29	91789	177578	22.84	58.31	20.04	20.94	16.62	5.38
2000	30	267045	528090	70.24	193.7	48.96	51.95	18.42	11.54

Table 10.7 The running time of the functions related to segment intersections. S stands for the sweep line algorithm of Bentley and Ottmann ([BO79]), M and B stand for the algorithms of Mulmuley and Balaban ([Mul90, Bal95]), and T stands for the trivial algorithm that checks every pair of segments for an intersection. The "+ E" indicates that the graph $G(S)$ is returned as a planar map. The first three columns contain the number of input segments, the number of nodes of G, and the number of edges of G, respectively.
We chose n segments. For each segment we chose random k bit integer for the Cartesian coordinates of the first endpoint and obtained the second endpoint from the first by adding a vector with random d bit integer coordinates. We used $k = 30$ and different values of d. The number of intersections is an increasing function of d.

Let us interpret the experimental findings in terms of asymptotic running time. When the number of intersections is very large, the $O(k \log n)$ term[16] in the time bound of the sweep algorithm dominates the $O(k)$ term in the time bound of the other algorithms. The trivial algorithm has a running time $\Theta(n^2 + k \cdot T_{report})$, where T_{report} is the cost of calling the function *report*. In our tests, *report* increases a counter and hence does minimal work. Thus the constant factor in the big-O expression is small. This explains why the running time of the trivial algorithm depends very little on the number of intersections and why the trivial algorithm is competitive when the number of intersections is large. When the number of intersections is small the Bentley–Ottmann algorithm and Mulmuley's algorithm have running time $O(n \log n)$ and Balaban's algorithm has running time $O(n \log^2 n)$. We should therefore expect that the former two algorithms are superior when the number of intersections is small. This is confirmed by the experiments.

10.7.2 *The Sweep Line Algorithm*

We discuss the Bentley–Ottmann sweep line algorithm for line segment intersection and give an implementation of the function

[16] In our examples, there are hardly any intersections of three or more segments and hence $s \approx k$. Observe that if all nodes are endpoints or proper intersections of exactly two segments then $E = n + 2(V - 2n)$, as $U(S)$ contains $2n$ nodes of degree one and $(V - 2n)$ nodes of degree four. In our examples we have $E \approx n + 2(V - 2n)$. We will therefore replace s by k in the discussion to follow.

SWEEP_SEGMENTS(const list<SEGMENT>& S, GRAPH<POINT,SEGMENT>& G,
bool embed, bool use_optimization)

that takes a list S of segments and computes the graph G induced by it. For each vertex v of G it also computes its position in the plane, and for each edge e of G it computes the segment containing it.

If *embed* $=$ *true*, the algorithm turns G into a planar map, i.e., G is made bidirected and the adjacency lists are sorted according to the geometric embedding in clockwise order.

If *use_optimization* $=$ *true*, an optimization described below is used.

The algorithm runs in time $O((n+s)\log(n+m)+m)$, where n is the number of segments, s is the number of nodes of G, and m is the number of edges of G. If S contains no overlapping segments then $m = O(n+s)$. If *embed* is *true*, the running time is increased by an additive factor of $O(m \log m)$. Note that $s \leq 3(n+k)$ and that k can be as large as s^2.

We want to stress that the implementation makes no assumptions about the input, in particular, segments may have length zero, may be vertical or may overlap, several segments may intersect in the same point, endpoints of segments may lie in the interior of other segments, ...

We achieve this generality by reformulating the plane sweep algorithm so that it can handle all geometric situations. The reformulation makes the description of the algorithm shorter and it also makes the algorithm faster, since k is replaced by s in the time bound[17]. The only previous algorithm that could handle all degeneracies is due to Myers [Mye85]. Its expected running time for random segments is $O(n \log n + k)$ and its worst case running time is $O((n+k)\log n)$.

In the sweep line paradigm a vertical line is moved from left to right across the plane and the output (here the graph $G(S)$) is constructed incrementally as it evolves behind the sweep line. One maintains two data structures to keep the construction going: the so-called *Y-structure* contains the intersection of the sweep line with the scene (here the set S of line segments) and the so-called *X-structure* contains the events where the sweep has to be stopped in order to add to the output or to update the X- or Y-structure. In the line segment intersection problem an event occurs when the sweep line hits an endpoint of some segment or an intersection point. When an event occurs, some nodes and edges are added to the graph $G(S)$, the Y-structure is updated, and maybe some more events are generated. When the input is in general position (no three lines intersecting in a common point, no endpoint lying on a segment, no two endpoints or intersections having the same x-coordinate, no vertical lines, no overlapping segments, ...) then at most one event can occur for each position of the sweep line and there are three clearly distinguishable types of events (left endpoint, right endpoint, intersection) with easily describable associated actions, cf. [Meh84d, VII.8]. We want to place no restrictions on the input and therefore need to proceed slightly differently. We now describe the required changes.

We define the sweep line by a point $p\_sweep = (x\_sweep, y\_sweep)$. Let ϵ be a positive

[17] Bentley and Ottmann formulated their algorithm for line segments in general position and stated a time bound of $O((n+k)\log n)$.

infinitesimal (readers not familiar with infinitesimals may think of ϵ as an arbitrarily small positive real). Consider the directed line L consisting of a vertical upward ray ending in point $(x\_sweep+\epsilon^2, y\_sweep+\epsilon)$ followed by a horizontal segment ending in $(x\_sweep-\epsilon^2, y\_sweep+\epsilon)$ followed by a vertical upward ray. We call L the *sweep line*. Note that[18] no endpoint of any segment lies on L, that no two segments of S intersect L in the same point except if the segments overlap, and that no non-vertical segment of S intersects the horizontal part of L. All three properties follow from the fact that ϵ is arbitrarily small but positive. Figure 10.50 illustrates the definition of L and the main data structures used in the algorithm: the Y-structure, the X-structure, and the graph G.

The Y-structure contains all segments intersecting the sweep line L ordered as their intersections with L appear on the directed line L. Overlapping segments are ordered by their *ID-numbers*. Every segment has an associated ID-number; distinct segments are guaranteed to have distinct IDs.

The X-structure contains all endpoints that are to the right of the sweep line and also some intersection points between segments in the Y-structure. More precisely, for each pair of segments adjacent in the Y-structure their intersection point is contained in the X-structure (if it exists and is to the right of the sweep line). The X-structure may contain other intersection points. The graph G contains the part of $G(S)$ that is to the left of the sweep line.

Initially, the Y-structure and the graph G are empty and the X-structure contains all endpoints of all input segments. The events in the X-structure are then processed in left to right order. Events with the same x-coordinate are processed in bottom to top order.

Assume that we need to process an event at point p and that the X-structure and Y-structure reflect the situation for $p\_sweep = (p.x, p.y - 2\epsilon)$. Note that this is true initially, i.e., before the first event is removed from the X-structure. We now show how to establish the invariants for $p\_sweep = p$. We proceed in seven steps.

1. We add a node v at position p to our graph G.
2. We determine all segments in the Y-structure containing the point p. These segments form a possibly empty subsequence of the Y-structure.
3. For each segment in the subsequence we add an edge to the graph G.
4. We delete all segments ending in p from the Y-structure.
5. We update the order of the subsequence in the Y-structure. This amounts to moving the sweep line across the point p.
6. We insert all segments starting in p into the Y-structure.
7. We generate events for the segments in the Y-structure that become adjacent by the actions above and insert them into the X-structure.

This completes the description of how to process the event p. The invariants now hold for $p\_sweep = p$ and hence also for $p\_sweep = (p'.x, p'.y - 2\epsilon)$ where p' is the new first element of the X-structure.

[18] We defined the sweep line in this seemingly complicated way in order to be able to write this "Note that". The note will allow us to define a linear order on the segments intersecting the sweep line.

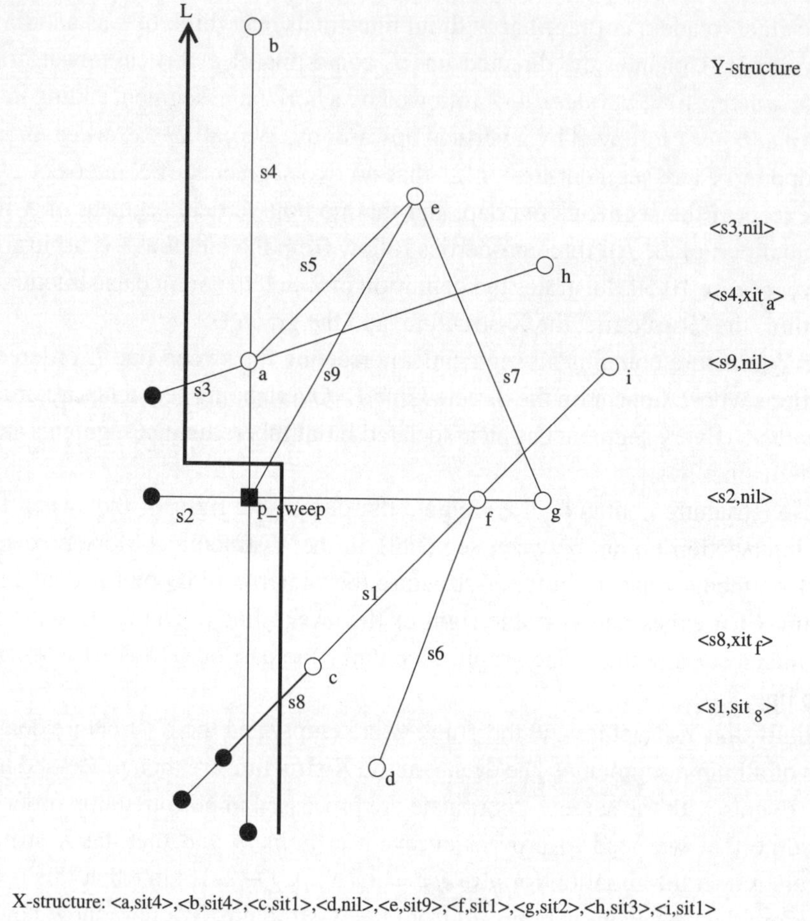

Figure 10.50 A scene of nine segments. The segments s_1 and s_8 overlap. The sweep line is shown in bold. The part of $G(S)$ to the left of the sweep line is already constructed. Its nodes are shown filled. The sweep line intersects the segments s_1, s_8, s_2, s_9, s_4, and s_3 and in this order. The Y-structure contains one item for each one of them. The X-structure contains points a, b, c, d, e, f, g, h, and i and in this order.
The information associated with the items in the X- and Y-structure will be explained in the next section.

10.7.3 *The Implementation of the Sweep Line Algorithm*

This section is joint work with Ulrike Bartuschka.

The implementation follows the algorithm closely. It makes use of several data types discussed in earlier chapters. The main "ingredients" are the basic geometric objects and primitives, sorted sequences for the X- and Y-structure, priority queues for storing events, and graphs for representing the output.

To make this section self-contained we briefly review the data types used.

Points and Segments: The types *rat_point* and *rat_segment* realize points and segments in the plane with rational coordinates and are part of the rational kernel. A *rat_point* is specified by its homogeneous coordinates of type *integer* – the type of arbitrary precision integers. If p is a *rat_point* then $p.X(\,)$, $p.Y(\,)$, and $p.W(\,)$ return its homogeneous coordinates and *p.xcoord*( ) and *p.ycoord*( ) return its Cartesian coordinates. If x, y, and w are of type *integer* with $w \neq 0$ then *rat_point*(x, y) and *rat_point*(x, y, w) create the *rat_point* with homogeneous coordinates $(x, y, 1)$ and (x, y, w), respectively. Two points are equal (*operator==*) if they agree in their Cartesian coordinates. A *rat_segment* is specified by its two endpoints; so if p and q are *rat_points* then *rat_segment*(p, q) is the directed segment with source p and target q. If s is a *rat_segment* then *s.source*( ) and *s.target*( ) return the source and target of s, respectively.

There are also points (class *point*) with coordinates of type *double*. The corresponding segment class is called *segment*. The classes *point* and *segment* have the same interface as *rat_point* and *rat_segment*. However, the internal representation is different: instead of storing the homogeneous coordinates as *integers*, the Cartesian coordinates are stored as *doubles*.

The sweep program can be executed with either the rational or the floating point geometry kernel. Be aware, however, that the instantiation with the floating point kernel is not fully reliable, see Section 10.7.2. In the sequel we use POINT to denote the point class and SEGMENT to denote the segment class.

POINTS and SEGMENTS come with a large number of geometric primitives. In the sweep program the following primitives are used:

- *int compare(POINT p, POINT q)*
  compares points by their lexicographic order; p precedes q if either
  p.xcoord( ) < *q.xcoord*( ) or
  p.xcoord( ) = *q.xcoord*( ) and *p.ycoord*( ) < *q.ycoord*( ). The function returns -1 if p precedes q, returns 0 if p and q are equal, and returns $+1$ otherwise. The lexicographic order of points is the default order on points.

- *int orientation(POINT p, POINT q, POINT r)*
  computes the orientation of points p, q, and r in the plane, i.e., 0 if the points are collinear, -1 if they define a clockwise oriented triangle, and $+1$ if they define a counter-clockwise oriented triangle.

- *int orientation(SEGMENT s, POINT p)*
  computes *orientation*(*s.source*( ), *s.target*( ), p).

- *int cmp_slopes(SEGMENT s1, SEGMENT s2)*
  compares the slopes of *s1* and *s2*. If one of the segments is degenerate, i.e., has length zero, the result is zero. Otherwise, the result is the sign of *slope*(*s1*) − *slope*(*s2*).

- *bool intersection_of_lines(SEGMENT s1, SEGMENT s2, POINT& p)*
  returns *false* if segments *s1* and *s2* are parallel or one of them is degenerate. Otherwise, it computes the point of intersection of the two straight lines supporting the segments, assigns it to the third parameter q, and returns *true*.

Our program maintains its own set of segments which we call *internal segments* or simply segments and store in the list *internal*; input segments are called input segments or original segments when the need for distinction arises. Internal segments are directed from left to right; vertical segments are directed upwards. There is one internal segment for every non-trivial input segment. The *map<SEGMENT, SEGMENT> original* stores for each internal segment the corresponding original segment.

⟨*local declarations*⟩≡

```
list&lt;SEGMENT&gt;          internal;
map&lt;SEGMENT,SEGMENT&gt; original;
```

Sorted Sequences: The type *sortseq<K, I>* realizes sorted sequences of pairs in $K \times I$, see Section 5.6; K is called the key type and I is called the information type of the sequence. The key type must be linearly ordered, i.e., the function *int compare(const K&, const K&)* must be defined for the type K and the relation $<$ on K defined by $k_1 < k_2$ iff $compare(k_1, k_2) < 0$ must be a linear order on K. An object of type *sortseq<K, I>* is a sequence of items (type *seq_item*) each containing a pair in $K \times I$. We use *<k, i>* to denote an item containing the pair (k, i) and call k the key and i the information of the item. The keys in a sorted sequence $\langle k_1, i_1\rangle, \langle k_2, i_2\rangle, \ldots, \langle k_m, i_m\rangle$ form an increasing sequence, i.e., $k_l < k_{l+1}$ for $1 \le l < m$.

Let S be a sorted sequence of type *sortseq<K, I>* and let k and i be of type K and I, respectively. The operation *S.lookup(k)* returns the item $it = \langle k, .\rangle$ in S with key k if there is such an item and returns *nil* otherwise. If *S.lookup(k) == nil* then *S.insert(k, i)* adds a new item $\langle k, i\rangle$ to S and returns this item. If *S.lookup(k) == it* then *S.insert(k, i)* changes the information in the item *it* to i. If $it = \langle k, i\rangle$ is an item of S then *S.key(it)* and *S.inf(it)* return k and i, respectively, and *S.succ(it)* and *S.pred(it)* return the successor and predecessor item of *it*, respectively; the latter operations return *nil* if these items do not exist. The operation *S.min( )* returns the first item of S, *S.empty( )* returns *true* if S is empty and *false* otherwise. Finally, if *it1* and *it2* are items of S with *it1* before *it2* then *S.reverse_items(it1, it2)* reverses the subsequence of S starting at item *it1* and ending at item *it2*.

In our implementation the X-structure has type *sortseq<POINT, seq_item>* and the Y-structure has type *sortseq<SEGMENT, seq_item>*. The Y-structure has one item for each segment intersecting the sweep line. The information field in the Y-structure is used for cross-links with the X-structure and for linking overlapping segments.

The X-structure is ordered according to the default order of points and the Y-structure is ordered according to the intersections of the segments with the directed sweep line L. The position of the sweep line is determined by *p_sweep* and the comparison object *cmp* realizes the order in the Y-structure. The class *sweep_cmp* will be defined below.

⟨*local declarations*⟩+≡

```
POINT p_sweep;
sweep_cmp cmp(p_sweep);
sortseq<POINT,seq_item>      X_structure;
sortseq&lt;SEGMENT, seq_item&gt;  Y_structure(cmp);
```

In the example of Figure 10.50 the sweep line intersects the segments s_1, s_8, s_2, s_9, s_4, and s_3. The Y-structure therefore consists of six items, one each for segments s_1, s_8, s_2, s_9, s_4, and s_3.

The X-structure contains an item for each endpoint of an input segment that is to the right of the sweep line and an item for each intersection point between segments that are adjacent in the Y-structure and that intersect to the right of the sweep line. It may also contain intersection points between segments that are not adjacent in the Y-structure.[19] The points in the X-structure are ordered according to the lexicographic ordering of their Cartesian coordinates. As mentioned above this is the default order on points.

In the example of Figure 10.50 the X-structure contains items for the endpoints b, c, d, e, g, h, i and for intersections a and f. Here, a and f are the intersections between segments s_4 and s_3, and s_1 and s_2, respectively.

The informations associated with the items of both structures serve as cross-links between the two structures: the information associated with an item in the X-structure is either *nil* or an item in the Y-structure; the information associated with an item in the Y-structure is either *nil* or an item of either structure. The precise definition follows: consider first an item $\langle s, it\rangle$ in the Y-structure and let s' be the segment associated with the successor item it' in the Y-structure. If s and s' overlap then $it = it'$. If s and s' do not overlap and $s \cap s'$ exists and lies to the right of the sweep line then it is the item in the X-structure with key $s \cap s'$. In all other cases we have $it = nil$.

Consider next an item $\langle p, sit\rangle$ in the X-structure. If $sit \neq nil$ then sit is an item in the Y-structure and the segment associated with it contains p. Moreover, if there is a pair of adjacent segments in the Y-structure that intersect in p then $sit \neq nil$. We may have $sit \neq nil$ even if there is no pair of adjacent segments intersecting in p.

In our example, the Y-structure contains the items $\langle s_1, sit_8\rangle$, $\langle s_8, xit_f\rangle$, $\langle s_2, nil\rangle$, $\langle s_9, nil\rangle$, $\langle s_4, xit_a\rangle$, and $\langle s_3, nil\rangle$ where sit_8 is the item of the Y-structure with associated segment s_8 and xit_a and xit_f are the items of the X-structure with associated points a and f, respectively. Let's turn to the items of the X-structure next. All items except $\langle d, nil\rangle$ point back to the Y-structure. If sit_i denotes the item $\langle s_i, \ldots\rangle$, $i \in \{1, 2, 9, 4, 3\}$, of the Y-structure then the items of the X-structure are $\langle a, sit_4\rangle$, $\langle b, sit_4\rangle$, $\langle c, sit_1\rangle$, $\langle d, nil\rangle$, $\langle e, sit_9\rangle$, $\langle f, sit_1\rangle$, $\langle g, sit_2\rangle$, $\langle h, sit_3\rangle$, and $\langle i, sit_1\rangle$.

The Order on the Y-structure: The segments in the Y-structure are ordered according to their intersection with the sweep line. Overlapping segments are ordered according to their

[19] Our X-structure may contain intersection points between segments that are no longer adjacent in the Y-structure. These events could be removed from the X-structure. Removing these events would guarantee an X-structure of linear size, however, at the cost of complicating the code. Since the size of the X-structure is always bounded by the size of the output graph we do not remove these events.

ID-number. All segments in the Y-structure are non-trivial and the position of the sweep line is determined by *p_sweep*.

The Y-structure is realized as a sorted sequence. In a sorted sequence comparisons between keys are only made during insertions and lookups and then one of the keys involved in the comparison is an argument of the operation. We conclude that compare is only called for segments s_1 and s_2 where one of the segments has its source point equal to *p_sweep*. Also, at least one of the segments is non-trivial and if one of the segments is trivial it has both endpoints equal to *p_sweep*. Let us assume first that both segments are non-trivial.

Assume s_i has its source point equal to *p_sweep*. If *p_sweep* does not lie on s_{1-i}, i.e., $orientation(s_{1-i}, p\_sweep) \neq 0$, then the orientation test is also the outcome of compare.

If both segments contain *p_sweep* we compare the slopes of s_1 and s_2 (*orientation*(*s2*, *s1.target*( ))). Only overlapping segments are equal after this comparison. They are ordered according to their ID-numbers. Since only internal segments are stored in the Y-structure and since internal segments are pairwise non-identical, any two internal segments have different ID-numbers.

The compare class *sweep_cmp* is derived from *leda_cmp_base*, see Section 2.10. It has a private data member *p_sweep* whose value will always be equal to the position of the sweep line; in the constructor the data member is initialized to the initial position of the sweep line and *set_position* is used to inform the compare object about any advance of the sweep line.

⟨*geometric primitives*⟩≡

```
class sweep_cmp : public leda_cmp_base&lt;SEGMENT&gt;
{
  POINT p_sweep;
public:
  sweep_cmp(const POINT& p) : p_sweep(p) {}
  void set_position(const POINT& p) { p_sweep = p; }
  int operator()(const SEGMENT& s1, const SEGMENT& s2) const
  { // Precondition:
    // p_sweep is identical to the left endpoint of either s1 or s2.
    if (identical(s1,s2)) return 0;
    int s = 0;
    if ( identical(p_sweep,s1.source()) )  s = orientation(s2,p_sweep);
    else
     if ( identical(p_sweep,s2.source()) ) s = orientation(s1,p_sweep);
     else error_handler(1,"compare error in sweep");
    if (s || s1.is_trivial() || s2.is_trivial()) return s;
    s = orientation(s2,s1.target());
    // overlapping segments will be ordered by their ID_numbers :
    return s ? s : (ID_Number(s1) - ID_Number(s2));
  }
};
```

We still need to explain the purpose of the tests *is_trivial*. We will also have to locate trivial segments in the Y-structure. These segments will have both endpoints equal to *p_sweep*. We want the search to be successful iff the Y-structure contains a segment passing through *p_sweep*. In the order defined above, the trivial segment (*p_sweep*, *p_sweep*) is larger than all segments intersecting the sweep line before *p_sweep*, is equal to all segments passing through *p_sweep*, and is larger than all segments intersecting the sweep line after *p_sweep*. We conclude that a search for the trivial segment will return a segment passing through *p_sweep* if there is one.

It is important to observe that the compare function for segments changes as the sweep progresses. What does it mean then that the keys of the items in a sorted sequence form an increasing sequence? The requirement is that whenever a lookup or insert operation is applied to a sorted sequence, the sequence must be sorted with respect to the current compare function. The other operations may be applied even if the sequence is not sorted.

The Graph G: The graph G has type *GRAPH<POINT, SEGMENT>*, i.e., it is a directed graph where a *POINT*, respectively *SEGMENT*, is associated with each node, respectively edge, of the graph. The graph G is the part of $G(S)$ that is left of the sweep line. The point associated with a vertex defines its position in the plane and the segment associated with an edge is an input segment containing the edge. We use two operations to extend the graph G. If p is a *POINT* then *G.new_node*(p) adds a new node to G, associates p with the node, and returns the new node. If v and w are nodes of G and s is a *SEGMENT* then *G.new_edge*(v, w, s) adds the edge (v, w) to G, associates s with the edge, and returns the new edge. In order to facilitate the addition of edges we maintain a *map<SEGMENT, node> last_node*: it gives for each segment in the Y-structure the rightmost vertex lying on the segment.

$\langle$*local declarations*$\rangle{+}{\equiv}$

```
map&lt;SEGMENT,node&gt;              last_node(nil);
```

The Priority Queue: We use a priority queue *seg_queue* to drive the insertion of segments into the Y-structure. The queue contains all internal segments that are ahead of the sweep line ordered according to their left endpoint. In particular, the first segment in *seg_queue* is always the segment that is encountered next by the sweep line. *Seg_queue* has type *p_queue<POINT, SEGMENT>*.

The data type *p_queue*$<P, I>$ realizes priority queues with priority type P and information type I. P must be linearly ordered. Priority queues are an item-based data type. Every item (of type *pq_item*) stores a pair (p, i) from $P \times I$, p is called the priority and i is called the information of the item. The usual operations on priority queues (*insert*, *delete_min*, *find_min*) are available.

$\langle$*local declarations*$\rangle{+}{\equiv}$

```
p_queue<POINT,SEGMENT>       seg_queue;
```

We are now ready for the program. It has the following structure:

⟨*sweep_segments.c*⟩+≡
  ⟨*geometric primitives*⟩
  ⟨*embedding*⟩
```
void SWEEP_SEGMENTS(const list&lt;SEGMENT&gt;& S, GRAPH&lt;POINT, SEGMENT&gt;& G,
                              bool embed, bool use_optimization)
```
  { ⟨*local declarations*⟩
    ⟨*initialization*⟩
    ⟨*sweep*⟩
    ⟨*post processing*⟩
  }

Initialization: We describe the initialization of the data structures. We clear the graph G, we compute a coordinate *Infinity* that is larger than the absolute value of the coordinates of all endpoints and that plays the role of ∞ in our program, we insert the endpoints of all input segments into the X-structure, and we create for each non-trivial input segment an internal segment with the same endpoints, insert this segment into *seg_queue* and link the input segment to it (through map *original*), we create two sentinel segments at $-\infty$ and $+\infty$, respectively, and insert them into the Y-structure, we put the sweep line at its initial position by setting *p_sweep* to $(-\infty, -\infty)$, and we add a stopper point with coordinates $(+\infty, +\infty)$ to *seg_queue*. The sentinels avoid special cases and thus simplify the code. Finally, we introduce a variable *next_seg* that always contains the first segment in *seg_queue*.

⟨*initialization*⟩≡
```
G.clear();
COORD Infinity = 1;
SEGMENT s;
forall(s,S)
{
  COORD x1 = s.xcoord1(), y1 = s.ycoord1();
  COORD x2 = s.xcoord2(), y2 = s.ycoord2();
  if (x1 < 0) x1 = -x1;
  if (y1 < 0) y1 = -y1;
  if (x2 < 0) x2 = -x2;
  if (y2 < 0) y2 = -y2;
  while (x1 >= Infinity || y1 >= Infinity ||
         x2 >= Infinity || y2 >= Infinity )   Infinity *= 2;
  seq_item it1 = X_structure.insert(s.source(), seq_item(nil));
  seq_item it2 = X_structure.insert(s.target(), seq_item(nil));
  if (it1 == it2) continue;  // ignore zero-length segments
  POINT p = X_structure.key(it1);
  POINT q = X_structure.key(it2);
  SEGMENT s1 = ( compare(p,q) < 0 ? SEGMENT(p,q) : SEGMENT(q,p) );
  original[s1] = s;
  internal.append(s1);
  seg_queue.insert(s1.source(),s1);
}
SEGMENT lower_sentinel(-Infinity,-Infinity,Infinity,-Infinity);
```

```
SEGMENT upper_sentinel(-Infinity, Infinity,Infinity, Infinity);
p_sweep = lower_sentinel.source();
cmp.set_position(p_sweep);
Y_structure.insert(upper_sentinel,seq_item(nil));
Y_structure.insert(lower_sentinel,seq_item(nil));
POINT pstop(Infinity,Infinity);
seg_queue.insert(pstop,SEGMENT(pstop,pstop));
SEGMENT next_seg = seg_queue.inf(seg_queue.find_min());
```

There is one subtle point in the code above. An insert operation into a sorted sequence with a key that is already present in the sorted sequence returns the item containing the key; it does not add a new item to the sequence and its does not change the key of the item returned. We exploit this feature of sorted sequences to ensure that internal segments share endpoints. Assume for concreteness that s_1 and s_2 are two input segments with a common source point and assume that s_1 is processed first. When the source point of s_2 is inserted into the X-structure, the item containing the source point of s_1 will be returned and hence the internal segments corresponding to s_1 and s_2 have the same (not just equal) source point[20].

Processing Events: We now come to the heart of procedure sweep: processing events. Let $event = \langle p, sit\rangle$ be the first event in the X-structure and assume inductively that our data structure is correct for $p\_sweep = (p.x, p.y - 2\epsilon)$. Our goal is to change *p_sweep* to p, i.e., to move the sweep line across point p. As long as the X-structure is not empty we perform the following actions.

We first extract the next event point *p_sweep* from the X-structure by assigning the minimal key in the X-structure to *p_sweep*, adjusting the compare function for segments to the new position of the sweep line, and adding a vertex v with position *p_sweep* to the output graph G. Then, we handle all segments passing through or ending at *p_sweep*. Finally, we insert all segments starting at *p_sweep* into the Y-structure, check for possible intersections between pairs of segments now adjacent in the Y-structure, and update the X-structure. Finally, we delete the event from the X-structure.

⟨*sweep*⟩≡

```
while ( !X_structure.empty() )
{ seq_item event = X_structure.min();
  p_sweep = X_structure.key(event);
  cmp.set_position(p_sweep);
  node v = G.new_node(p_sweep);
  ⟨handle passing and ending segments⟩
  ⟨insert starting segments⟩
  ⟨compute new intersections and update X-structure⟩
  X_structure.del_item(event);
}
```

20 A point is realized as a pointer to a representation class. Two points are equal if they have the same Cartesian coordinates and two points are identical if they share the representation. Testing two points for identity is faster than testing them for equality.

Handling Passing and Ending Segments: We first determine the segments passing through or ending in *p_sweep* and then handle them by reversing their order in the Y-structure.

⟨*handle passing and ending segments*⟩≡
```
seq_item sit = X_structure.inf(event);
if (sit == nil) sit = Y_structure.lookup(SEGMENT(p_sweep,p_sweep));
seq_item sit_succ      =  nil;
seq_item sit_pred      =  nil;
seq_item sit_pred_succ =  nil;
seq_item sit_first =      nil;
if (sit != nil)
{ ⟨determine passing and ending segments⟩
  ⟨reverse order of passing segments⟩
}
```

We first determine whether there is any segment passing through or ending in *p_sweep*. Recall that the current event is ⟨*p_sweep*, *sit*⟩.

If $sit \neq nil$, the segment associated with *sit* contains *p_sweep*. If $sit = nil$, there is no pair of adjacent non-overlapping segments in the Y-structure intersecting in *p_sweep*. However, there may be a bundle of overlapping segments in the Y-structure that contain *p_sweep*. We can decide whether there is such a bundle and determine some segment in the bundle by locating the point *p_sweep* in the Y-structure[21]. We defined the comparison function for segments such that a search for the trivial segment (*p_sweep*, *p_sweep*) in the Y-structure is successful iff the Y-structure contains a segment containing *p_sweep*.

If there is no segment in the Y-structure containing *p_sweep*, there is nothing to do. Assume otherwise. Then *sit* points to one such segment. We determine all such segments. The corresponding items form a subsequence of the Y-structure, see Figure 10.51. We compute the first (*sit_first*) and last (*sit_last*) item of this bundle of items and also the predecessor (*sit_pred*) and successor (*sit_succ*) item of the bundle. We also store in *sit_pred_succ* the successor of *sit_pred* before the insertion, i.e, *sit_first*.

The items in the bundle are easily recognized by their informations. The information of every item in the bundle except for the last is either equal to the current event item *event* or equal to the successor item in the Y-structure (in the case of a segment overlapping with its successor). The information of the last item in the bundle is either *nil* or an item in the X-structure different from *event* (such an item stands for an intersection with *sit_succ*).

We determine the items in the bundle as follows. Starting at *sit* we first walk up until *sit_succ* is reached. Then we walk down to *sit_pred*. During the downward walk we also start to update the data structures. For every segment *s* in the bundle we do the following:

- We add an edge to *G* connecting *last_node*[*s*] and *v* and label it with *s*. The new edge gets its direction from the original segment containing it, if *embed* is *false*, and is directed from *v* to *last_node*[*s*], if *embed* is *true*.

21 The Y-structure contains segments and hence only segments can be located in it. In order to locate the point *p_sweep* in the Y-structure, we locate the zero-length segment (*p_sweep*, *p_sweep*) instead.

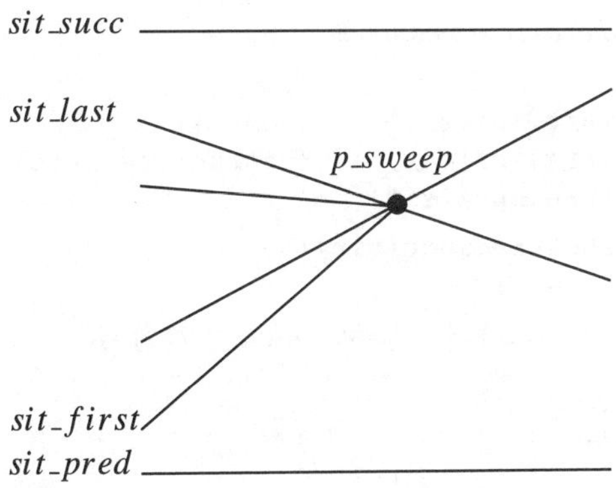

Figure 10.51 The items *sit_pred*, *sit_first*, *sit_last*, and *sit_succ*.

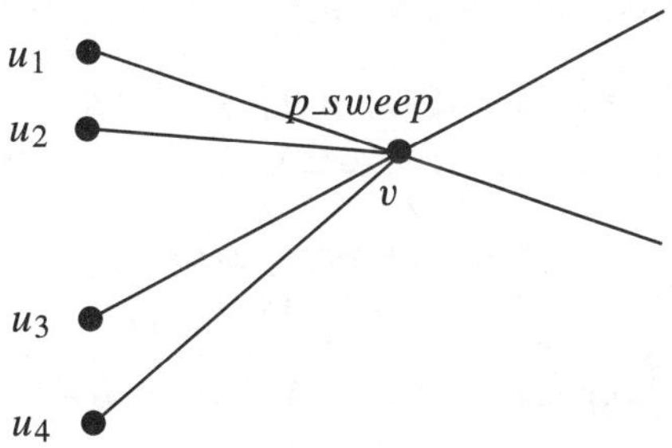

Figure 10.52 The edges out of v are constructed in the order (v, u_1), (v, u_2), (v, u_3), (v, u_4).

- If s ends at *p_sweep* then we delete it from the Y-structure. If the predecessor segment overlaps with s, we copy the information about the successor segment of s (if any) to the predecessor and set a flag that the downward walk is not finished yet.
- If s continues through *p_sweep* then we change the intersection information associated with it to *nil* and set *last_node* to v.

We explain why we direct the edge constructed for a segment s from v to *last_node*[s] if *embed* is *true*. Since *new_edge* appends the edge constructed to the list of outgoing edges of v and since we construct edges during the downward walk the edges out of v will be constructed in their proper counter-clockwise order, see Figure 10.52. We will exploit this fact when we construct the planar embedding of G in the post-processing step.

The identification of the subsequence of segments incident to *p_sweep* takes constant time per element of the sequence. Moreover, the constant is small since the test of whether p is incident to a segment involves no geometric computation but only identity tests between items. The code is particularly simple due to our sentinel segments: *sit* can never be the first or last item of the Y-structure.

⟨*determine passing and ending segments*⟩≡

```
// walk up
while ( Y_structure.inf(sit) == event ||
        Y_structure.inf(sit) == Y_structure.succ(sit) )
  sit = Y_structure.succ(sit);
sit_succ = Y_structure.succ(sit);
seq_item sit_last = sit;
if ( use_optimization ) { ⟨optimization, part 1⟩ }
// walk down
bool overlapping;
do
{ overlapping = false;
  s = Y_structure.key(sit);
  if ( !embed && s.source() == original[s].source() )
    G.new_edge(last_node[s], v, s);
  else
    G.new_edge(v, last_node[s], s );
  if ( identical(p_sweep,s.target()) )   // ending segment
  {
     seq_item it = Y_structure.pred(sit);
     if ( Y_structure.inf(it) == sit )
     { overlapping = true;
       Y_structure.change_inf(it, Y_structure.inf(sit));
     }
     Y_structure.del_item(sit);
     sit = it;
   }
   else  // passing segment
   {
     if ( Y_structure.inf(sit) != Y_structure.succ(sit) )
       Y_structure.change_inf(sit, seq_item(nil));
     last_node[s] = v;
     sit = Y_structure.pred(sit);
   }
} while ( Y_structure.inf(sit) == event || overlapping ||
          Y_structure.inf(sit) == Y_structure.succ(sit) );
sit_pred = sit;
sit_first = Y_structure.succ(sit_pred);
sit_pred_succ = sit_first;
```

All segments in the bundle starting with *sit_first* and ending in *sit_last* pass through node v and moving the sweep line through *p_sweep* changes the order of these segments in the Y-structure. More precisely, if s and s' are two segments passing through *p_sweep* then moving the sweep line through *p_sweep* reverses their order iff s and s' do not overlap.

If the bundle is non-empty, we update its order as follows: first we reverse all subsequences of overlapping segments and then we reverse the entire bundle, see Figure 10.53.

The bundle of segments passing through *p_sweep* is empty iff *sit_first* is equal to *sit_succ*.

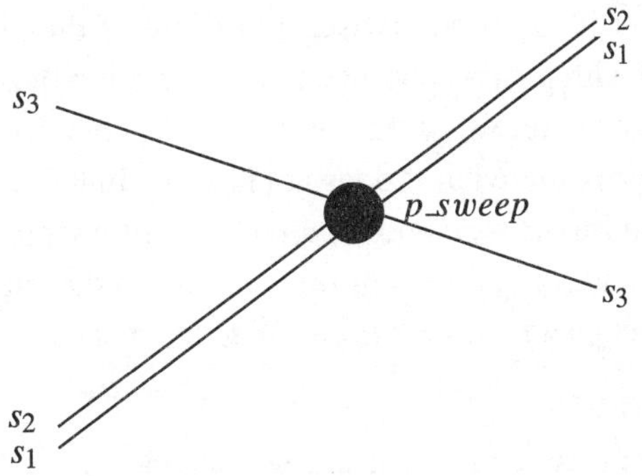

Figure 10.53 Three segments passing through *p_sweep*, two of them overlapping. The order of the segments is reversed, but the order within the sub-bundle of overlapping segments is retained.

⟨*reverse order of passing segments*⟩≡

```
sit = sit_first;
// reverse subsequences of overlapping segments  (if existing)
while ( sit != sit_succ )
{ seq_item sub_first = sit;
  seq_item sub_last  = sub_first;
  while (Y_structure.inf(sub_last) == Y_structure.succ(sub_last))
    sub_last = Y_structure.succ(sub_last);
  if ( sub_last != sub_first )
    Y_structure.reverse_items(sub_first, sub_last);
  sit = Y_structure.succ(sub_first);
}
// reverse the entire bundle
if ( sit_first != sit_succ )
  Y_structure.reverse_items(Y_structure.succ(sit_pred),
                            Y_structure.pred(sit_succ));
```

Insertion of Starting Segments: The last step in handling the event point *p_sweep* is to insert all segments starting at *p_sweep* into the Y-structure and to test the new pairs of adjacent items (*sit_pred*, ...) and (..., *sit_succ*) for possible intersections. If there were no segments passing through or ending in *p_sweep* then the items *sit_succ* and *sit_pred* still have the value *nil* and we have to compute them now.

We use the priority queue *seg_queue* to find the segments to be inserted. As long as the first segment in *seg_queue* starts at *p_sweep*, i.e., *next_seg.source*( ) is identical[22] to *p_sweep*, we remove it from the queue and locate it in the Y-structure. Let *s_it* be the item returned by *locate* and let *p_sit* be its predecessor.

We insert *next_seg* after *s_it* into the Y-structure; this will add an item *sit* to the Y-structure. We set the information of *sit* to *s_sit* if the new segment overlaps with the segment associated

22 Recall that we ensured that endpoints of internal segments that are equal are identical.

with *s_sit* and we set it to *nil* otherwise. Similarly, if the new segment overlaps with the segment associated with *p_sit* we change the information of *p_sit* to *sit*.

We associate the new item *sit* with the right endpoint of *next_seg* in the X-structure; note that the point is already there but it does not have its link to the Y-structure yet. We also set *last_node*[*s*] to *v*, and if *sit_succ* and *sit_pred* are still undefined, i.e, there was no segment passing through or ending in *p_sweep*, we set them to the successor and predecessor of the new item, respectively, and we set *sit_pred_succ* to *sit_succ*.

⟨*insert starting segments*⟩≡

```
while ( identical(p_sweep,next_seg.source()) )
{ seq_item s_sit = Y_structure.locate(next_seg);
  seq_item p_sit = Y_structure.pred(s_sit);
  s = Y_structure.key(s_sit);
  if ( orientation(s, next_seg.start()) == 0 &&
       orientation(s, next_seg.end()) == 0 )
     sit = Y_structure.insert_at(s_sit, next_seg, s_sit);
  else
     sit = Y_structure.insert_at(s_sit, next_seg, seq_item(nil));
  s = Y_structure.key(p_sit);
  if ( orientation(s, next_seg.start()) == 0 &&
       orientation(s, next_seg.end()) == 0 )
    Y_structure.change_inf(p_sit, sit);
  X_structure.insert(next_seg.end(), sit);
  last_node[next_seg] = v;
  if ( sit_succ == nil )
  { sit_succ = s_sit;
    sit_pred = p_sit;
    sit_pred_succ = sit_succ;
  }
  // delete minimum and assign new minimum to next_seg
  seg_queue.del_min();
  next_seg = seg_queue.inf(seg_queue.find_min());
}
```

Computing New Intersections: If *sit_pred* still has the value *nil*, there were no ending, passing or starting segments and hence *p_sweep* is an isolated point and we are done. Isolated points result from segments of length zero.

So assume that *sit_pred* exists. We have to update its information field (which still has the value from before the event). We set it to *nil* if there is no intersection between *sit_pred* and its successor. If the intersection exists, we insert it into the X-structure and set the information field of *sit_pred* to it. If there are segments leaving *p_sweep*, i.e, *sit_pred* is not the predecessor of *sit_succ*, we also check for an intersection between *sit_succ* and its predecessor.

⟨*compute new intersections and update X-structure*⟩≡
```
if ( sit_pred != nil )
{ if ( !use_optimization )
  { Y_structure.change_inf(sit_pred,seq_item(nil));
    compute_intersection(X_structure, Y_structure, sit_pred);
    sit = Y_structure.pred(sit_succ);
    if ( sit != sit_pred )
      compute_intersection(X_structure, Y_structure, sit);
  }
  else
  { ⟨optimization, part 2⟩ }
}
```

The function *compute_intersection* takes an item *sit0* of the Y-structure and determines whether the segment associated with *sit0* intersects the segment associated with its successor item *sit1* to the right of the sweep line. If so, it updates the X- and the Y-structure. Let s_0 and s_1 be the segments associated with *sit0* and *sit1*, respectively, and let ℓ_0 and ℓ_1 be the supporting lines of s_0 and s_1, respectively.

We know that s_0 intersects the sweep line L before s_1. Thus s_0 and s_1 intersect right of the sweep line if the right endpoint of s_1 lies below or on ℓ_0 (*orientation(s0, s1.target( ))* ≥ 0) and the right endpoint of s_0 lies above or on ℓ_1 (*orientation(s1, s0.target( ))* ≤ 0).

If the segments intersect, we compute the point of intersection, call it q, by a call of *s0.intersection_of_lines*(*s1*, q), insert a new pair (q, *sit0*) into the X-structure and associate this pair with *sit0* in the Y-structure.

⟨*geometric primitives*⟩+≡
```
static void compute_intersection(sortseq<POINT,seq_item>& X_structure,
                         sortseq&lt;SEGMENT,seq_item&gt;& Y_structure, seq_item sit0)
{ seq_item sit1 = Y_structure.succ(sit0);
  SEGMENT  s0   = Y_structure.key(sit0);
  SEGMENT  s1   = Y_structure.key(sit1);
  if ( orientation(s0,s1.target()) <= 0 &&
       orientation(s1,s0.target()) >= 0 )
  { POINT q;
    s0.intersection_of_lines(s1,q);
    Y_structure.change_inf(sit0, X_structure.insert(q,sit0));
  }
}
```

Post Processing: We associate with each edge of G an input segment containing it. This is easily done as each edge has an internal segment associated with it. Thus we only have to replace $G[e]$ by *original*$[G[e]]$.

The graph G constructed during the sweep is planar but is not in the form of a planar map yet. In particular, the order of the adjacency lists depends on the insertion order.

When *embed* is *true*, we turn G into a planar map.

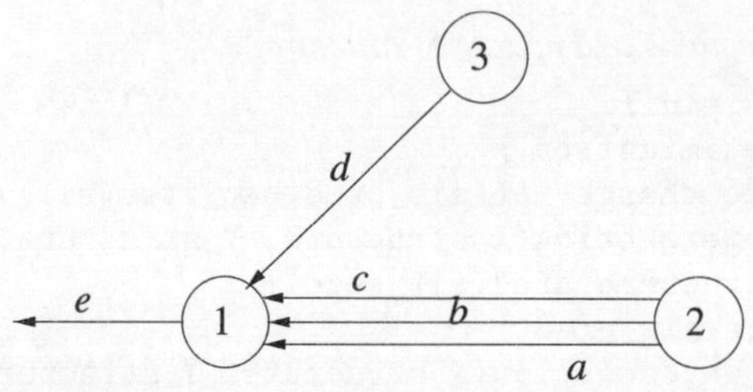

Figure 10.54 Before the call of embedding there is only one edge leaving node 1, namely, the edge *e*. There are three parallel edges (2, 1); their counter-clockwise order around node 2 is in decreasing order of ID-number. We need to add the reversals of the edges *a*, *b*, *c*, and *d* to the list of edges out of 1. Sorting the edges by increasing slope and edges of equal slope by ID-number gives the desired order.

⟨*post processing*⟩≡
```
if (embed) construct_embedding(G);
edge e;
forall_edges(e, G) G[e] = original[G[e]];
```

When *embed* is *true* all edges of *G* are directed from right to left (vertical edges are directed downwards). Moreover, the edges out of any node are already in their proper counter-clockwise order.

In order to turn *G* into a planar map we need to add the reversal of every edge and to insert the new edges at their proper position into the adjacency lists.

Edge reversals are directed from left to right (the reversal of a vertical edge is directed upwards). The proper order of edge reversals is therefore by slope. Reversals of parallel edges should be ordered by ID-number. Consider Figure 10.54.

Let *R* be a copy (!!!) of the set of all edges of *G*. We use *R* instead of *E* to indicate that *R* represents the set of edge reversals. We sort the edges in *R* according to slope and then add for each edge *e* in *R* the edge (*target*(*e*), *source*(*e*)) to *G*. Since new edges are appended to the lists of outgoing edges, this will result in properly ordered adjacency lists.

⟨*embedding*⟩≡
```
class sweep_cmp_edges : public leda_cmp_base<edge>
{
  const GRAPH<POINT,SEGMENT>& G;
public:
  sweep_cmp_edges(const GRAPH<POINT,SEGMENT>& g): G(g) {}
  int operator()(const edge& e1, const edge& e2) const
  { SEGMENT s1 = G[e1];
    SEGMENT s2 = G[e2];
    int c = cmp_slopes(s1,s2);
    if (c == 0) c = compare(ID_Number(s1),ID_Number(s2));
    return c;
  }
};
```

```
static void construct_embedding(GRAPH<POINT,SEGMENT>& G)
{
  list<edge> R = G.all_edges();
  sweep_cmp_edges cmp(G);
  R.sort(cmp);
  edge e;
  forall(e,R)
  { edge r = G.new_edge(target(e),source(e),G[e]);
    G.set_reversal(e,r);
  }
}
```

In the post-processing step we first compute the embedding and then replace internal segments by input segments. It would be incorrect to change the order of two steps: first, the ordering of the Y-structure is an ordering on internal segments and we must use the same ordering in the embedding step. Second, the input may contain multiple occurrences of the same segment and the ordering by ID-number does not break ties between identical segments.

An Optimization: The running time of SWEEP_SEGMENTS is $O((n+s)\log(n+m)+m)$ where n is the number of segments, s is the number of nodes of G and m is the number of edges of G. If there are no overlapping segments then $m = O(n+s)$ since G is planar. In the presence of overlapping segments, m may be as large as $n(n+s)$. The time bound can be seen as follows. There are $O(n+k)$ lookups, insertions, and deletions in the X- and Y-structure, each for a cost of $O(\log(n+m))$. Observe that $n+m$ is an upper bound on the number of items in the Y-structure and that $n+s$ is an upper bound on the number of items in the X-structure. Since $s \le n^2$ we have $\log(n+s+m) = O(\log(n+m))$. The total number of items handled by the *reverse_items* operations on the Y-structure is $O(m)$. Since the cost of *reverse_items* is proportional to the number of items reversed, the total cost for all *reverse_items* operations is $O(m)$. The number of operations on G is $O(n+k+m)$, each for a cost of $O(1)$.

Experiments show that a significant fraction of the running time is spent in the geometric primitives *sweep_cmp* and *compute_intersection*, in particular, if the rational kernel is used (which we recommend). The rational kernel has a built-in floating point filter, i.e., all geometric tests are first performed in floating point arithmetic, the rounding error is estimated, and only if the error estimation indicates that the result of the floating point computation may be wrong, the computation is repeated with exact arithmetic. The floating point filter is discussed in detail in Section 9.7.

The function *compute_intersection* performs orientation tests and computes an intersection point. The floating point filter applies to the orientation tests but does not apply to the computation of intersection points since constructions of new points are always performed with exact arithmetic.

The function *compute_intersection* is called whenever two segments become adjacent in

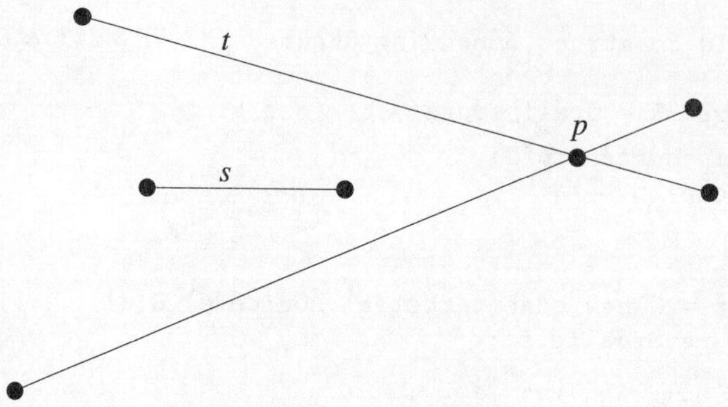

Figure 10.55 The intersection p is first discovered when t is inserted into the Y-structure and is rediscovered when s is removed from the Y-structure.

the Y-structure. Segments may become adjacent in the Y-structure more than once, see Figure 10.55. We show how to avoid the recomputation of intersections.

We maintain a dictionary *inter_dic* which maps pairs of segments to items in the X-structure. The appropriate data type is a two-dimensional map.

⟨*local declarations*⟩+≡

```
map2&lt;SEGMENT,SEGMENT,seq_item&gt;   inter_dic(nil);
```

Whenever a pair of segments that is adjacent in the Y-structure becomes non-adjacent we store their intersection in the dictionary and whenever a pair of segments becomes adjacent we consult the dictionary to find out whether their intersection was already computed.

When processing an event two intersections may get lost. Consider the sequence of items corresponding to segments passing through or ending in *p_sweep*. Let *sit_last* be the last item in this sequence and let *sit_pred* and *sit_succ* be the items before and after the sequence, respectively; *sit_last* does not exist if there are no segments passing through or ending in *p_sweep*.

Sweeping through *p_sweep* reverses the subsequence starting with *sit_first* and ending with *sit_last* and hence two intersections can get lost, the intersection stored in *sit_last* and the intersection stored in *sit_pred*. The intersection stored in *sit_last* is with the segment associated with *sit_succ* and the intersection stored in *sit_pred* is with the segment associated with the successor of *sit_pred*. This is the item *sit_pred_succ*.

⟨*optimization, part 1*⟩≡

```
seq_item xit = Y_structure.inf(sit_last);
if (xit) { SEGMENT s1 = Y_structure.key(sit_last);
           SEGMENT s2 = Y_structure.key(sit_succ);
           inter_dic(s1,s2) = xit;
         }
```

⟨*optimization, part 2*⟩≡

```
seq_item xit = Y_structure.inf(sit_pred);
```

```
if ( xit )
{ SEGMENT s1 = Y_structure.key(sit_pred);
  SEGMENT s2 = Y_structure.key(sit_pred_succ);  // sit_first
  inter_dic(s1,s2) = xit;
  Y_structure.change_inf(sit_pred, seq_item(nil));
}
compute_intersection(X_structure, Y_structure,inter_dic,sit_pred);
sit = Y_structure.pred(sit_succ);
if ( sit != sit_pred )
  compute_intersection(X_structure, Y_structure,inter_dic,sit);
```

We also need to change the function *compute_intersection*. Before computing an intersection point we check whether the two segments already have an intersection event in the X-structure by a lookup in *inter_map*. If the lookup fails we compute the intersection and add it to the X-structure.

⟨*geometric primitives*⟩+≡

```
static void compute_intersection(sortseq<POINT,seq_item>& X_structure,
                  sortseq&lt;SEGMENT,seq_item&gt;& Y_structure,
                  const map2&lt;SEGMENT,SEGMENT,seq_item&gt;& inter_dic,
                  seq_item sit0)
{ seq_item sit1 = Y_structure.succ(sit0);
  SEGMENT  s0   = Y_structure.key(sit0);
  SEGMENT  s1   = Y_structure.key(sit1);

  if ( orientation(s0,s1.target()) <= 0 &&
       orientation(s1,s0.target()) >= 0 )
  {
    seq_item it = inter_dic(s0,s1);
    if ( it == nil)
    { POINT q;
      s0.intersection_of_lines(s1,q);
      it = X_structure.insert(q,sit0);
    }
    Y_structure.change_inf(sit0, it);
  }
}
```

10.7.4 *Experimental Evaluation of the Sweep Line Algorithm*

We report about tests for *three kinds of test data*, namely random, difficult, and highly degenerate inputs, *three different implementations of points and segments*, namely the floating point kernel (FK), the rational kernel (RK) and the rational kernel with turned-off floating point filter (FK^-), and *with and without the optimization*. We describe the test data, list running times, and comment on the results.

Random Inputs: The random data set consists of n segments whose endpoints have random k bit coordinates. Table 10.8 gives the number of nodes and edges of the output graph and the running time for $n = 200$ and different values of k. The experiments indicate that the optimization described above and the floating point filter are effective. The optimization

k	V	E	RK⁻	RK⁻O	RK	RKO	FK	FKO
10	4813	9028	2.27	2	1.2	1.09	0.73	0.67
20	4742	8884	2.63	2.19	1.31	1.1	0.7	0.67
30	5467	10334	3.07	2.57	1.52	1.26	0.8	0.77
40	5478	10356	3.78	3.13	1.69	1.38	0.81	0.77
50	5168	9736	3.66	3.13	1.62	1.3	0.76	0.73
60	5558	10516	4.36	3.59	1.81	1.43	0.82	0.79
70	5909	11218	5.2	4.23	2.13	1.6	0.86	0.83
80	5174	9748	4.75	3.78	1.86	1.43	0.78	0.74
90	4808	9016	4.86	3.82	1.77	1.34	0.71	0.68
100	5080	9560	5.92	4.5	2.12	1.54	0.75	0.73

Table 10.8 200 random segments, coordinates are random k-bit integers. An "O" indicates the use of the optimization.

is more effective for the rational kernels because the computation of intersections is more costly in exact arithmetic. Floating point arithmetic is faster than exact arithmetic but the difference is never more than a factor of two in running time. We have to admit though that the difference can be made arbitrarily larger by choosing larger values of k.

Difficult Inputs: Let $size = 2^k$ and let $y = 2size/(n-1)$. The random data set consists of n segments where the i-th segment has endpoints $(size + rx1, size + i \cdot y + ry1)$ and $(3 \cdot size + rx2, 3 \cdot size - i \cdot y + ry2)$ and $rx1, rx2, ry1, ry2$ are random integers in $[-s, s]$ for some small integer s. For $s = 0$ all segments in the difficult data set pass through the point $(2 \cdot size, 2 \cdot size)$, and for small but non-zero values of s they intersect in the neighborhood of this point. Table 10.9 gives the results for the difficult data set with $s = 10$, $k = 10, 20, \ldots$, 100, and $n = 200$. The floating point filter and the optimization are again quite effective. The floating point implementation produced incorrect results for all values of k; the floating point implementation does, however, work correctly for smaller values of n and/or larger values of s.

Highly Degenerate Inputs: The highly degenerate test set consists of n segments with random coordinates in a small grid with side length s. For example, for $n = 100$ and $s = 10$ one should expect a large number of degeneracies. We used this test set to support our claim that the algorithm handles all degeneracies. We do not report running times for the highly degenerate inputs.

The readers may perform their own experiments by running either the sweep_time program in the demo directory or the sweep_segments_demo in xlman.

We were surprised by two outcomes of our experiments.

First, we expected the implementation using the rational kernel to be much slower than the floating point computation and not just by a factor of two. We achieve the small factor by the use of the floating point filter, by the optimization which avoids the costly recomputation

k	V	E	RK$^-$	RK$^-$O	RK	RKO	FK	FKO
10	20134	39669	9.84	8.29	5.07	4.46	error	error
20	20298	39997	11.75	9.71	5.65	4.64	error	error
30	20296	39994	12.33	10.5	6.04	4.88	error	error
40	20298	39997	14.79	11.71	6.5	5.13	error	error
50	20300	40000	16.12	12.5	6.7	5.22	error	error
60	20298	39997	16.32	12.95	6.91	5.45	error	error
70	20300	40000	18.77	14.84	7.51	5.69	error	error
80	20300	40000	19.82	15.91	7.62	5.72	error	error
90	20298	39997	21.27	16.25	7.68	5.71	error	error
100	20296	39994	24.61	18.39	8.58	6.24	error	error

Table 10.9 The difficult example with 200 segments. An "O" indicates the use of the optimization and error indicates that the computation with the floating point kernel gave the incorrect result.

of intersections, and by the observation that many equality tests for points can be replaced by tests for identity of points.

Second, we expected the floating point implementation to have difficulties with the difficult example. However, we were surprised by the fact that it never crashed. It always produced an output, albeit an incorrect one. We try to explain this phenomenon by arguing that the program does not crash as long as the sentinels are handled correctly, i.e, the segments *lower_sentinel* and *upper_sentinel* have all segments between them and all intersection points precede *pstop*. We do not care what the geometric tests do with segments that are not sentinels. If sentinels are handled correctly, every lookup in the Y-structure will return an item different from the first item in the Y-structure[23]. Also the walks performed in the Y-structure will determine a subsequence that does not include the sentinel items. For this reason none of the operations on the Y-structure will fail; i.e., it will never happen that we ask for the successor of the last or the predecessor of the first item. Also since *pstop* is handled correctly, we will never attempt to extract *next_seg* from an empty *seg_queue*.

Exercises for 10.7

1 Let G_0 and G_1 be graphs of type *GRAPH<POINT, SEGMENT>*. Write a function that checks whether the graphs are isomorphic, i.e., whether there are bijections $i_V : V_0 \rightarrow V_1$ and $i_E : E_0 \rightarrow E_1$ such that $G_0[v] = G_1[i_V(v)]$ for all nodes of G_0 and such that $i_E(e) = (i_V(v), i_V(w))$ and $G_0[e] = G_1[i_E(e)]$ for all edges $e = (v, w)$ of G_0.

2 Use the solution to the previous exercise to write a function that runs two implementations of SEGMENT_INTERSECTION and then checks the computed graphs for isomorphism.

[23] This sentence requires knowledge of the implementation of sorted sequences. The implementation is such that if the comparisons with the first and the last element of the sorted sequence are correct and the outcome of any other comparison is arbitrary, lookup will not return the first element.

3 Write a trivial implementation of *SEGMENT_INTERSECTION*$(G, report)$ that simply checks every pair of segments for an intersection.

4 Extend the sweep line algorithm or any of the other algorithms such that it computes the trapezoidal decomposition induced by a set of segments.

10.8 Polygons

We define the types polygon and generalized polygon. A polygon is an open region of the plane whose boundary is a closed polygonal chain[24] and a generalized polygon is anything that can be obtained from polygons by regularized set operations. Both classes offer functions for point location, for intersection with lines and segments, and for moving objects around. Generalized polygons offer, in addition, the regularized set operations complement, union, intersection, difference, and symmetric difference.

This section is structured as follows: in Section 10.8.1 we discuss the functionality of polygons and generalized polygons, in Section 10.8.2 we give the essentials of the implementation of polygons, in Section 10.8.3 we give the mathematics underlying the representation of generalized polygons, and in Section 10.8.4 we give the highlights of the implementation of generalized polygons.

We advise you to exercise the polygon demo in xlman before reading this section, see Figure 10.56.

10.8.1 *Functionality*

A closed polygonal chain P is a cyclic sequence $(p_0, p_1, \ldots, p_{n-1})$ of points. The points are called the vertices of the chain and the number of vertices is called the size of the chain. The vertices of a closed polygonal chain are indexed modulo the size n of the chain, in particular, $p_n = p_0$. A closed polygonal chain induces a set $S(P)$ of segments, namely the set of segments $p_i p_{i+1}$, $0 \le i \le n-1$, connecting consecutive vertices. A closed polygonal chain is called *simple* if all nodes of the graph $G(S(P))$ defined by the segments in $S(P)$ have degree equal to two, i.e., if no two segments in $S(P)$ except for consecutive segments share a point. A closed polygonal chain P is called *weakly simple* if the segments in $S(P)$ are disjoint except for common endpoints[25] and if the chain does not cross itself. Figure 10.57 shows some examples.

A weakly simple polygonal chain splits the plane into an unbounded region and one or more bounded regions. For a simple polygonal chain there is just one bounded region. When a weakly simple polygonal chain P is traversed either the bounded region is consistently to the left of P or the unbounded region is consistently to the left of P; this follows from the fact that a weakly simple chain does not cross itself. We say that P is positively oriented in the former case and negatively oriented in the latter case. We call the region

[24] A precise definition is given below.

[25] It is allowed that segments that are not consecutive on P share an endpoint.

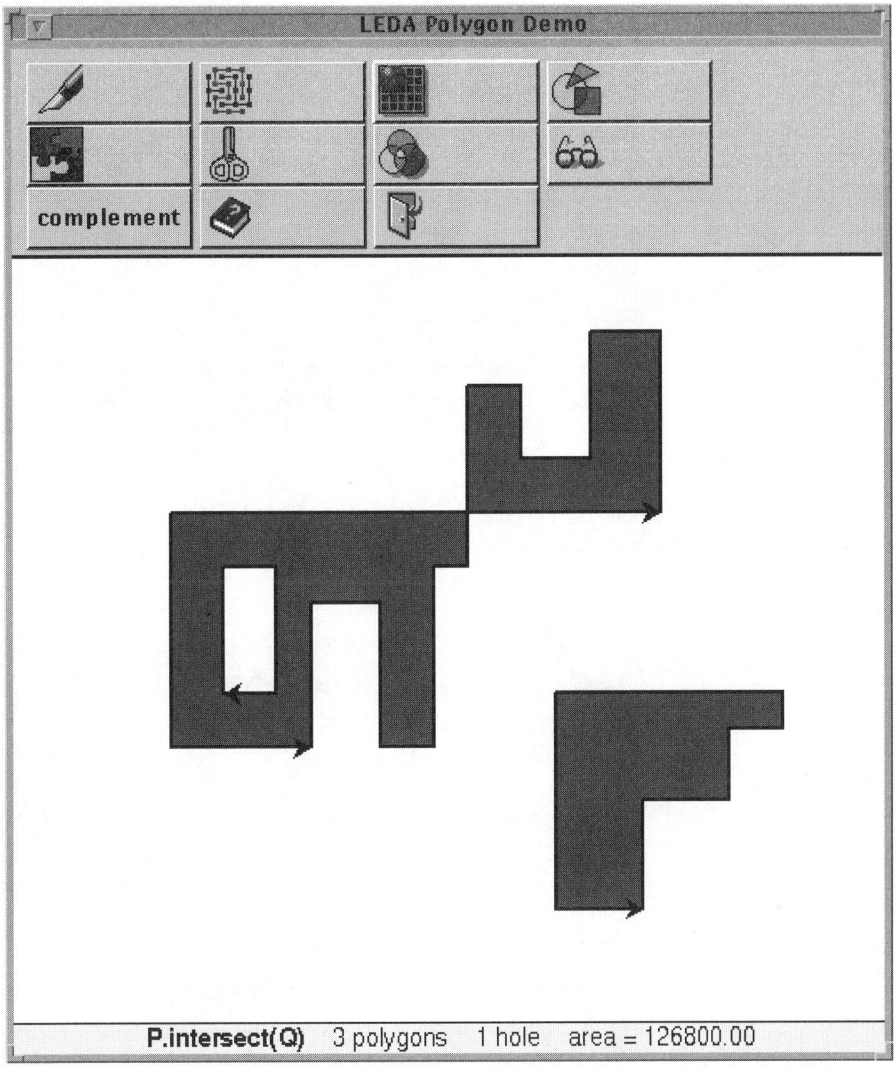

Figure 10.56 A screen shot of the polygon demo in xlman. The display shows a generalized polygon. The boundary cycles are indicated by arrows and the inside of the polygon is shaded. The various buttons allow the user to construct polygons by mouse input or by calling generators, to force vertices to a grid, to compute intersections, unions, differences, and symmetric differences, to perform point location queries, and to compute complements.

to the left of P the positive side of P. We overload notation and use P also to denote the positive side of P, see Figure 10.58. The positive side of P is an open set and P is its boundary.

Frequently, we do not want to distinguish between a polygonal chain and the polygonal region defined by it. We use the word *polygon* to cover both aspects.

We have two classes of polygons: *rat_polygons* have *rat_points* as their vertices and *polygons* have *points* as their vertices. Both classes offer essentially the same function-

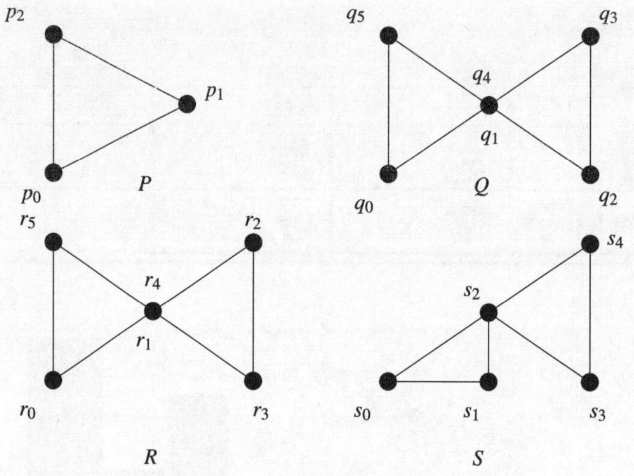

Figure 10.57 P is simple and Q is weakly simple but not simple. R is not weakly simple because it crosses itself at $r = r_1 = r_4$, and S is not weakly simple since s_2 lies in the interior of another segment.

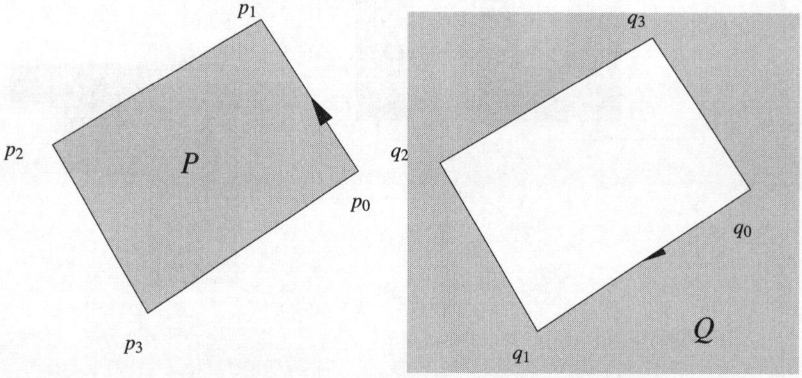

Figure 10.58 The bounded region is to the left of P; P is positively oriented. The unbounded region is to the left of Q, Q is negatively oriented.

ality, but, of course, only *rat_polygons* guarantee correct results. We use *rat_polygons* in this section.

The declarations

```
rat_polygon P1;
rat_polygon P2(const list<rat_point>& pl,
               CHECK_TYPE check = rat_polygon::SIMPLE,
               bool respect_orientation =
                            rat_polygon::RESPECT_ORIENTATION);
```

introduce polygons *P1* and *P2*; *P1* is initialized to the empty polygon and *P2* is initialized to the polygon with vertex sequence *pl*. The second argument takes one of the values

NO_CHECK, SIMPLE, WEAKLY_SIMPLE of a local enumeration type CHECK_TYPE. If *check* is SIMPLE, the polygon must be simple, and if *check* is WEAKLY_SIMPLE, the polygon must be weakly simple. The third argument takes one of the values RESPECT_ORIENTATION or DISREGARD_ORIENTATION. If *respect_orientation* is DISREGARD_ORIENTATION, the orientation of *pl* is chosen such that the bounded region with respect to *pl* lies to the left of *pl*. The meaning of this flag is undefined if *pl* is not weakly simple.

Simplicity and weak simplicity can also be checked by the functions

```
bool P.is_simple();
bool P.is_weakly_simple();
```

Assignment and copy constructor are available for polygons. The functions

```
list<rat_point>   P.vertices();
list<rat_segment> P.edges();
```

return the list of vertices and the list of segments of P, respectively. The second function is also available as *P.segments( )*.

Let l be a line and let s be a segment. The functions

```
list<rat_point> P.intersection(l);
list<rat_point> P.intersection(s);
```

return the crossings between the chain P and l or s, respectively. The function

```
rat_polygon P.complement()
```

returns the polygon whose list of vertices is the reversal of P's list. If P is weakly simple, the positive side of the complement is the negative side of P and vice versa.

The remaining functions for polygons assume that P is weakly simple. Their meaning is undefined if P is not weakly simple. Recall that a weakly simple polygon P splits the plane in an unbounded region and one or more bounded regions. Also recall that we designated the region(s) to the left of P as the positive side of P and use P also for the positive side of P.

Let p be a point. The function

```
int P.side_of(p);
```

returns the side of P to which p belongs, i.e., $+1$ if p belongs to the positive side, 0 if p lies on P, and -1 if p belongs to the negative side, see Figure 10.59. The function

```
region_kind P.region_of(p);
```

returns the region with respect to P to which p belongs, i.e., BOUNDED_REGION if p lies in the bounded region of P, ON_REGION if p lies on P, and UNBOUNDED_REGION if p lies in the unbounded region. One can also ask for the containment in a specific region by

```
bool P.inside(p);
bool P.on_boundary(p);
bool P.outside(p);
```

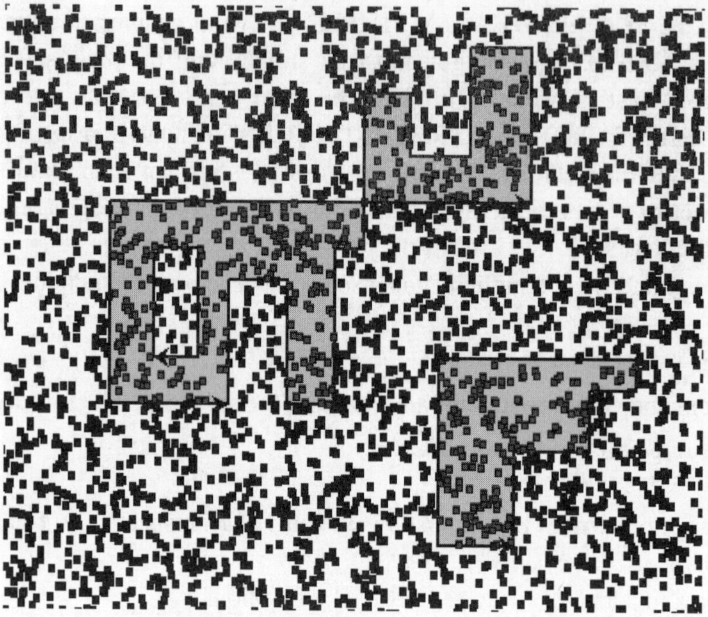

Figure 10.59 Side-of tests: We performed side-of tests with respect to the generalized polygon of Figure 10.56 for 5000 random points. The points on the different sides are shown at different grey level.

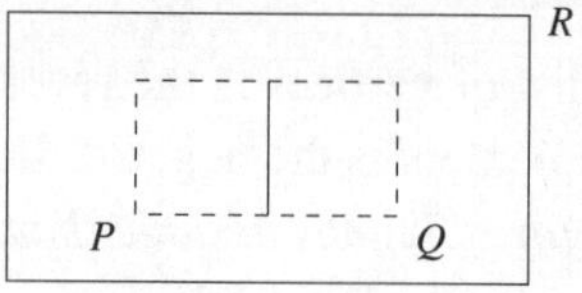

Figure 10.60 The intersection of P and Q is a line segment; $R \setminus (P \cap Q)$ is a rectangle minus a line segment.

The function

```
RAT_TYPE P.area();
```

returns the signed area of the bounded region of P. The sign of the area is positive if P is positively oriented and is negative if P is negatively oriented.

We come to generalized polygons. The class of polygons is not closed under boolean operations. In fact, very strange objects can be generated from polygons by boolean operations, see Figure 10.60. The class of generalized polygons encompasses all sets that can be constructed from polygons by the so-called *regularized set operations*, see [Req80, TR80, Hof89]. We refer the reader to [Nef78] for the general case.

In order to define the regularized set operations we need to review some elementary concepts of topology. For a set X we use int X, cl X, bd X, and cpl X to denote its *interior*, *closure*, *boundary*, and *complement*, respectively. An open set X is called *regular* if $X =$

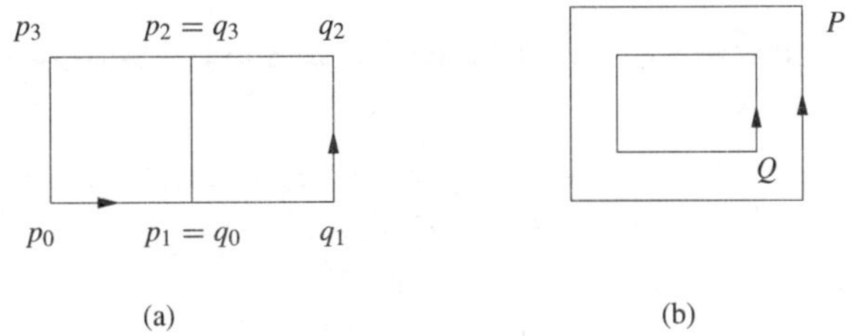

Figure 10.61 In (a), the polygons $P$and Q share an edge, $P \cap Q$ is a closed line segment, and $\text{reg}(P \cap Q)$ is the empty set. In (b), $P \setminus Q$ is the half-closed region between the cycles P and Q; the chain P does not belong to $P \setminus Q$ and the chain Q belongs to it. The regularized set difference $\text{reg}(P \setminus Q)$ is the open region with boundaries P and Q.

$\text{int}\,\text{cl}\,X$. The following sets are non-regular: the plane minus a single point or the plane minus a line. A set is called *polygonal* if its boundary consists of a finite number of points and open line segments. The regularization of a set X is defined as $\text{int}\,\text{cl}\,X$; we use $\text{reg}\,X$ as a shorthand for $\text{int}\,\text{cl}\,X$. We show that regularization generates regular sets and that the regularized set operations[26] applied to regular polygonal regions generate regular polygonal regions, see Figure 10.61.

Lemma 70

(a) *Let X be any set. Then* $\text{reg}\,X$ *is regular.*
(b) *Let X be any open set. X is regular iff X and* $\text{int}\,\text{cpl}\,X$ *have the same boundary.*
(c) *Let P be a weakly simple polygonal chain. Then the bounded region and the unbounded region with respect to P are regular polygonal sets.*
(d) *If P and Q are regular polygonal regions then so are* $\text{reg}\,\text{cpl}\,P$, $\text{reg}(P \cap Q)$, $\text{reg}(P \cup Q)$, $\text{reg}(P \setminus Q)$, *and* $\text{reg}(P \oplus Q)$.

Proof We start with part (a). Let X be any set and let $Y = \text{reg}\,X$. We need to show that Y is regular. We have $Y \subseteq \text{cl}\,Y$ and hence $Y \subseteq \text{int}\,\text{cl}\,Y$ since Y is open. We have $Y \subseteq \text{cl}\,X$ by definition of Y and hence $\text{cl}\,Y \subseteq \text{cl}\,\text{cl}\,X = \text{cl}\,X$. Thus $\text{int}\,\text{cl}\,Y \subseteq \text{int}\,\text{cl}\,X = Y$.

We turn to part (b). Assume first that X is regular, i.e., $X = \text{int}\,\text{cl}\,X$, and let x be any point in the boundary of X. Then $x \in \text{cl}\,X \setminus X$ since X is open. Assume that there is a neighborhood U of x such that $U \cap \text{int}\,\text{cpl}\,X = \emptyset$. Then $U \subseteq \text{cl}\,X$ and hence $x \in \text{int}\,\text{cl}\,X$, a contradiction to the regularity of X.

To prove the converse we observe that $X \subseteq \text{int}\,\text{cl}\,X$ since X is open. We need to show that the containment is not proper. Consider any point $x \in \text{bd}\,X$. By assumption every neighborhood U of x has $U \cap \text{int}\,\text{cpl}\,X \neq \emptyset$. Thus $x \notin \text{int}\,\text{cl}\,X$ and hence $\text{int}\,\text{cl}\,X \subseteq X$.

For part (c) we observe that the boundary of the bounded as well as the unbounded region with respect to P is equal to P and hence both regions are certainly polygonal. The regularity of both regions follows from part (b) and the fact that P is weakly simple.

[26] The regularized union of two sets X and Y is defined as $\text{reg}(X \cup Y)$; the definition of the other regularized set operations is analogous.

The results of the regularized set operations are certainly polygonal; regularity follows from part (a). □

The classes *rat_gen_polygon* and *gen_polygon* represent regular polygonal regions over the rational and the floating point kernel, respectively. In our examples we use *rat_gen_polygons*; *gen_polygon* stands for generalized polygon.

The constructors

```
rat_gen_polygon P;
rat_gen_polygon Q(rat_polygon R);
```

construct the empty generalized polygon and the generalized polygon corresponding to R, respectively. The second constructor requires that R is a weakly simple polygon. There are two special generalized polygons, the empty one and the full one. The *full polygon* is the entire plane.

The functions

```
bool P.is_empty();
bool P.is_full();
```

return true if P is the empty set or the entire plane, respectively.

If p is a point and P is a generalized polygon then

```
bool P.side_of(p)
```

returns $+1$ if $p \in P$, returns 0 if p lies on P, and returns -1 otherwise, see Figure 10.59.

The function

```
region_kind P.region_of(p);
```

returns the region with respect to P to which p belongs, i.e., BOUNDED_REGION if p lies in the bounded region of P, ON_REGION if p lies on P, and UNBOUNDED_REGION if p lies in the unbounded region. The bounded region of the empty polygon is empty and the bounded region of the full polygon is the entire plane.

The function

```
RAT_TYPE P.area();
```

returns the signed area of the bounded region of P. The sign of the area is positive if P is bounded and is negative if P is unbounded. This function cannot be applied to the full polygon.

For the following operations let P and Q be generalized polygons.

```
rat_gen_polygon P.complement()
```

returns the regularized complement of P and

```
gen_rat_polygon P.unite(Q);
gen_rat_polygon P.intersection(Q);
gen_rat_polygon P.diff(Q);
gen_rat_polygon P.sym_diff(Q);
```

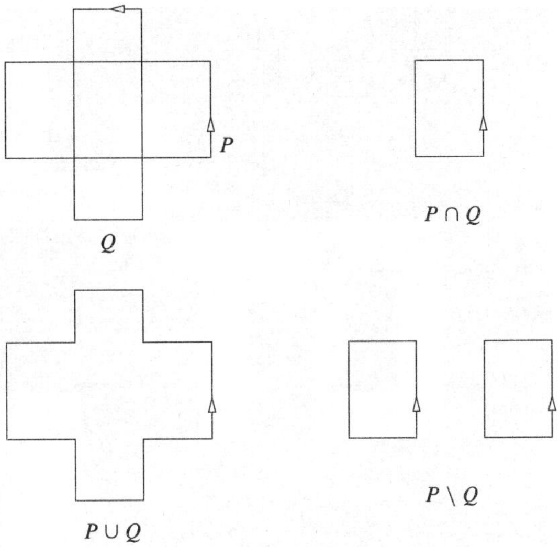

Figure 10.62 Two polygons P and Q and the results of the three boolean operations $\cap$, $\cup$, and $\setminus$.

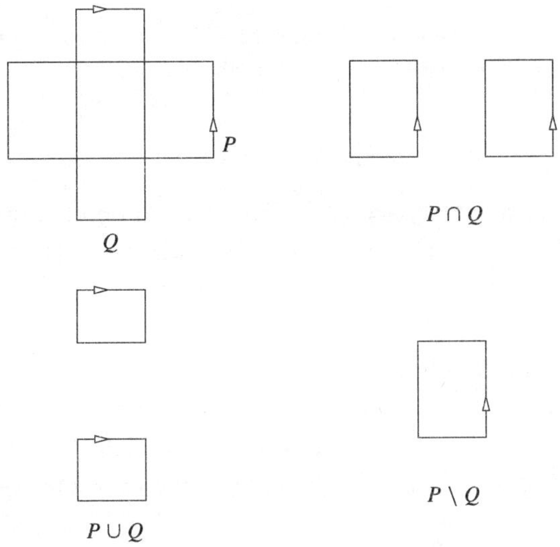

Figure 10.63 Two polygons P and Q and the results of the three boolean operations $\cap$, $\cup$, and $\setminus$. Observe that the positive side of Q is unbounded.

return reg($P \cup Q$), reg($P \cap Q$), reg($P \setminus Q$), and reg($P \oplus Q$), respectively. The word *union* is a reserved word of C++, hence the name *unite* for the union-operation. Figures 10.62 and 10.63 show some examples.

A generalized polygon can be represented by its boundary cycles as will be explained in Section 10.8.3. The function

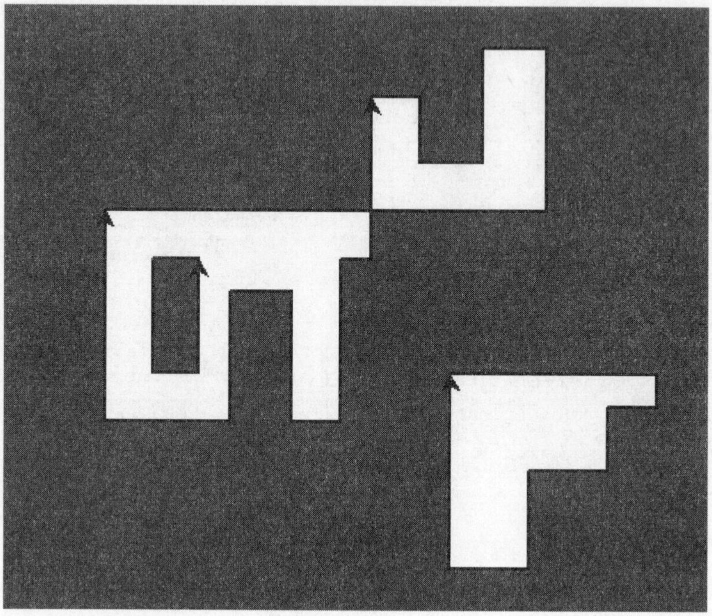

Figure 10.64 The complement of the generalized polygon of Figure 10.56. Observe that the orientation of all boundary cycles is reversed.

```
list<rat_polygon> P.polygons();
```

returns the list of boundary cycles of P. The list is ordered according to nesting, i.e., if a boundary cycle D is nested in a boundary cycle C, then C is before D in the list of boundary cycles.

10.8.2 *The Implementation of Polygons*

Polygons are a handle type, i.e., a polygon is realized as a pointer to a representation class (called *polygon_rep* and *rat_polygon_rep*, respectively) which contains the actual representation. The member function *ptr*( ) of class polygon returns the pointer to the representation object.

The representation consists of a list of points, a list of segments, four extreme points, and an integer which stores the orientation of the polygon. The orientation is positive if the bounded region is to the left of the polygon and is negative otherwise.

```
list<POINT>   pt_list;
list&lt;SEGMENT&gt; seg_list;
POINT xmin, ymin, xmax, ymax;
int orient;
```

Here, *pt_list* contains the list of points, *seg_list* contains the list of segments (the i-th segment in *seg_list* connects the i-th point in *pt_list* to the $i + 1$-th point in *pt_list*), and *xmin*, *ymin*, *xmax*, and *ymax* are vertices with minimal x-coordinate, minimal y-coordinate, maximal x-coordinate, and maximal y-coordinate, respectively.

We will next discuss some of the member functions of *polygon*.

The Signed Area of a Simple Polygon: Assume that *seg_list* is the list of boundary segments of a simple polygon P. We show how to compute the signed area $A(P)$ of the bounded face of P. The sign of the area is positive if the bounded face lies to the left of P and is negative otherwise.

Lemma 71 *Let P be a simple polygon and let n be the number of segments in the boundary of P. For $0 \le i < n$, let p_i be the source point of the i-th boundary segment. Let p be an arbitrary point in the plane and let $A_i = A(\Delta_i)$ be the signed area of the triangle $\Delta_i = (p, p_i, p_{i+1})$. Then*

$$A(P) = \sum_{0 \le i < n} A_i$$

is the signed area of A.

Proof We use induction on n and assume w.l.o.g. that the signed area is positive. Assume first that P is a triangle, see Figure 10.65. If p lies in the bounded face of P or on P, the bounded face of P is partitioned by the triangles Δ_0, Δ_1, and Δ_2, and hence $A(P) = A(\Delta_0) + A(\Delta_1) + A(\Delta_2)$. If p lies in the unbounded face of P, then p can see either one or two edges of P. If p can see one edge of P, say $p_0 p_1$, then

$$A(P) = |A(\Delta_1)| + |A(\Delta_2)| - |A(\Delta_0)| = A(\Delta_1) + A(\Delta_2) + A(\Delta_0),$$

where the second equality follows from the fact that Δ_1 and Δ_2 are positively oriented and Δ_0 is negatively oriented. If p can see two edges of P, say $p_0 p_1$ and $p_1 p_2$, then

$$A(P) = |A(\Delta_2)| - |A(\Delta_1)| - |A(\Delta_0)| = A(\Delta_2) + A(\Delta_1) + A(\Delta_0),$$

where the second equality follows from the fact that the orientation of Δ_2 is positive and the orientations of Δ_0 and Δ_1 are negative. This completes the base step of the induction.

Assume next that $n \ge 4$. Then there is an i such that the segment $p_i p_{i+2}$ is contained in the interior of P [27]. Let Q be the polygon obtained from P by replacing the segments $p_i p_{i+1}$ and $p_{i+1} p_{i+2}$ by the segment $p_i p_{i+2}$. Then

$$A(P) = A(Q) + A(\Delta)$$

where $\Delta = (p_i, p_{i+1}, p_{i+2})$. Applying the induction hypothesis to Q yields

$$A(Q) = \sum_{j=0}^{i-1} A(\Delta_j) + A(p, p_i, p_{i+2}) + \sum_{j=i+2}^{n-1} A(\Delta_j)$$

and applying the induction hypothesis to Δ yields

$$A(\Delta) = A(\Delta_i) + A(\Delta_{i+1}) + A(p_{i+2}, p_i, p) = A(\Delta_i) + A(\Delta_{i+1}) - A(p, p_i, p_{i+2}).$$

[27] Consider an arbitrary triangulation of P. The dual of the triangulation is a tree and hence there is at least one triangle in the triangulation which has two edges of P in its boundary. The two edges are $p_i p_{i+1}$ and $p_{i+1} p_{i+2}$ for some i.

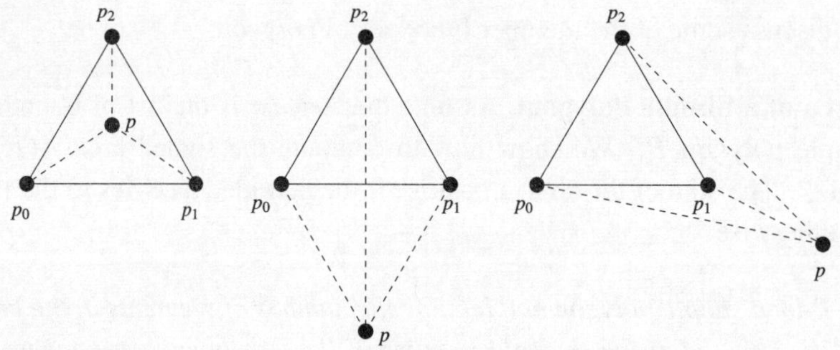

Figure 10.65 Let $\Delta_i = (p, p_i, p_{i+1})$ for $i = 0, 1, 2$, and let $P = (p_0, p_1, p_2)$. Then $A(P) = A(\Delta_0) + A(\Delta_1) + A(\Delta_2)$ in all three cases.

Adding the two equations completes the induction step. □

The implementation follows directly from the lemma above.

⟨*polygon: compute area*⟩≡

```
static RAT_TYPE compute_area(const list&lt;SEGMENT&gt;& seg_list)
{
  if (seg_list.length() < 3) return 0;
  list_item it = seg_list.get_item(1);
  POINT     p  = seg_list[it].source();
  it = seg_list.succ(it);
  RAT_TYPE A   = 0;
  while (it)
  { SEGMENT s = seg_list[it];
    A += ::area(p,s.source(),s.target());
    it = seg_list.succ(it);
  }
  return A;
}
```

The time to compute the signed area of a polygon is $O(n)$. The constant factor in the O-expression is fairly large, in particular, with the rational kernel. Observe that the areas of n triangles are computed and that an area computation of a triangle amounts to the evaluation of a 3×3 determinant.

Determining the Orientation: The simplest way to compute the orientation of a polygon P is to take the sign of the area. This takes linear time but is slow; see the remark at the end of the preceding section. A faster approach is as follows.

Let q be the lexicographically smallest vertex of P and let p and r be the predecessor and successor vertices of q on P. Then the orientation of P is equal to the orientation of the triple (p, q, r), see Figure 10.66. Observe that this statement is not true for an arbitrary vertex q; it is only true for a vertex that is extreme in some direction.

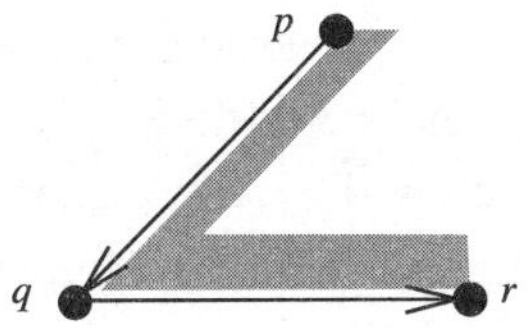

Figure 10.66 The triple (p, q, r) has positive orientation. If q is the lexicographically smallest vertex of the polygon, the region to the left of the polygonal chain is bounded. This conclusion cannot be drawn for an arbitrary vertex.

The implementation of *orientation* follows directly from the preceding paragraph.

⟨*polygon: compute orientation*⟩≡

```
static int compute_orientation(const list&lt;SEGMENT&gt;& seg_list)
{ list_item q_it = seg_list.first();
  POINT q = seg_list[q_it].source();
  list_item it;
  forall_items(it,seg_list)
    if ( compare(seg_list[it].source(),q) < 0 )
    { q_it = it;
      q = seg_list[q_it].source();
    }
  POINT p = seg_list[seg_list.cyclic_pred(q_it)].source();
  POINT r = seg_list[seg_list.cyclic_succ(q_it)].source();
  return ::orientation(p,q,r);
}
```

Point Containment: Let P be a weakly simple polygon. The function

```
region_kind P.region_of(const POINT& p) const
```

returns the region of P containing p. In order to decide containment we first use the extreme vertices for a quick test. If p lies to the left of *xmin* or to the right of *xmax* or below *ymin* or above *ymax*, we return UNBOUNDED_REGION. Next we check whether p lies on P. Assume this is not the case, i.e., p lies either in the bounded face or the unbounded face of P.

We use the following observation. Consider a vertical upward ray r_p starting in p and assume that r_p does not pass through any vertex of P. Then r_p intersects an odd number of segments of P iff p lies in the bounded region of P. The observation solves the problem iff r_p does not pass through any vertex of P.

We use *perturbation* to extend the solution to arbitrary points p. If p does not lie on P, the point q obtained from p by moving p by an infinitesimal amount to the right belongs to the same face with respect to P as p. Moreover, the vertical upward ray r_q starting at q does not pass through any vertex of P. In particular, r_q does not intersect any vertical edge of P.

Consider a segment s of P. If s is vertical, r_q does not intersect it. So assume that s is not vertical. Let a be the endpoint of s with the smaller x-coordinate and let b be the other endpoint of s. Then r_q intersects s if $x_a < x_q < x_b$ and q lies to the right of the oriented line ℓ through a and b. Here, we used x_z to denote the x-coordinate of a point z. Since $x_q = x_p + \epsilon$ for an infinitesimal ϵ, the first condition is equivalent to $x_a \le x_p < x_b$ and the second condition is equivalent to p being to the right of ℓ.

We obtain the following code.

⟨*polygon: region_of and side_of*⟩≡

```
region_kind POLYGON::region_of(const POINT& p) const
{
  // use extreme vertices for a quick test.
  int cx1 = POINT::cmp_xy(p,ptr()->xmin);
  int cx2 = POINT::cmp_xy(p,ptr()->xmax);
  int cy1 = POINT::cmp_yx(p,ptr()->ymin);
  int cy2 = POINT::cmp_yx(p,ptr()->ymax);
  if (cx1 < 0 || cx2 > 0 || cy1 < 0 || cy2 > 0) return UNBOUNDED_REGION;
  list&lt;SEGMENT&gt;& seglist = ptr()->seg_list;
  // check boundary segments
  list_item it;
  forall_items(it,seglist)
  { SEGMENT s = seglist[it];
    if (s.contains(p)) return ON_REGION;
  }
  // count intersections with vertical ray starting in p
  int count  = 0;
  forall_items(it,seglist)
  { SEGMENT s = seglist[it];
    POINT a = s.source();  POINT b = s.target();
    int orient = POINT::cmp_x(a,b);
    if ( orient == 0 ) continue;
    if ( orient > 0 ) { // a is right of b
                        leda_swap(a,b);
                      }
    if ( POINT::cmp_x(a,p) <= 0 && POINT::cmp_x(p,b) < 0
        && ::orientation(a,b,p) < 0 )
      count++;
  }
  return ( count % 2 == 0 ? UNBOUNDED_REGION : BOUNDED_REGION );
}
```

Given the function *region_of* it is easy to implement *side_of*. The positive side of P is equal to the bounded region if P is positively oriented and is equal to the unbounded region otherwise.

⟨*polygon: region_of and side_of*⟩+≡

```
int POLYGON::side_of(const POINT& p) const
{ region_kind k = region_of(p);
  switch (k) {
    case ON_REGION:          return 0;
    case BOUNDED_REGION:     return    ptr()->orient;
    case UNBOUNDED_REGION: return -(ptr()->orient);
    default:                 assert( 0 == 1); return 0;
  }
}
```

The Complement of a Polygon: The complement of a weakly simple polygon is easy to compute. We simply reverse the list of segments. The complement has the opposite orientation.

⟨*polygon: complement*⟩≡

```
POLYGON POLYGON::complement() const
{ list&lt;SEGMENT&gt; R;
  SEGMENT s;
  forall(s,ptr()->seg_list) R.push(SEGMENT(s.target(),s.source()));
  return POLYGON(R, - orientation());
}
```

10.8.3 *The Mathematics of Generalized Polygons*

The purpose of this section is to give the mathematical underpinning for the representation of regular polygonal sets. We show that a regular polygonal set can be represented by its list of boundary cycles.

If X is a regular polygonal set and p is an arbitrary point in the plane the intersection $U \cap X$ for U a sufficiently small neighborhood of p has one of the following three forms:

- If p is contained in (the interior of) X then $U \cap X \subseteq X$.
- If p is contained in the interior of the complement of X then $U \cap X = \emptyset$.
- If p is contained in the boundary of X then $U \cap X$ and $U \cap \operatorname{int} \operatorname{cpl} X$ are unions of "pieces of pie" as shown in Figure 10.67.

We call a set X trivial if either $X = \emptyset$ or $X = \mathbb{R}^2$. Let X be a non-trivial polygonal set. We call a collection $P_1, \ldots, P_k$ of weakly simple polygons a *representation* of X if:

- the set of segments in the boundary of X is the disjoint union of the set of segments of the P_i's, and
- the orientation of each P_i is such that X is locally to the left of P_i, and
- the P_i are pairwise non-crossing, i.e., there are no consecutive segments pq and qr on some P_i and xq and qy on some P_j with $i \neq j$ and the segments interleaving around q, see Figure 10.68.

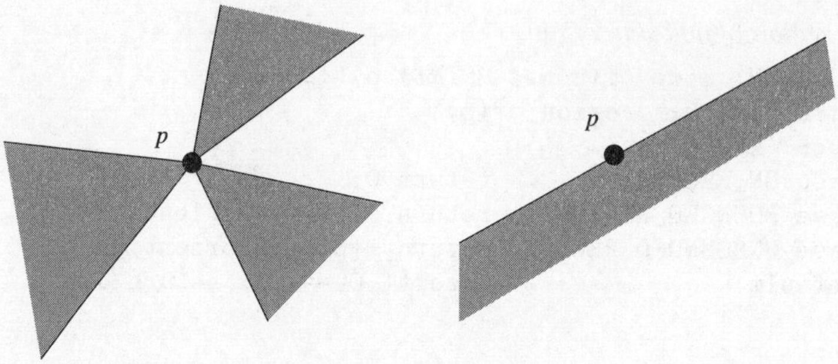

Figure 10.67 The shaded part of the plane belongs to the polygonal region X and p lies in the boundary of X. If p is a vertex of X and U is a sufficiently small neighborhood of p then $U \cap X$ and $U \cap \text{int cpl } X$ are unions of pieces of pie. If p lies in the relative interior of a boundary segment of X then X looks like an open half-plane in the vicinity of p.

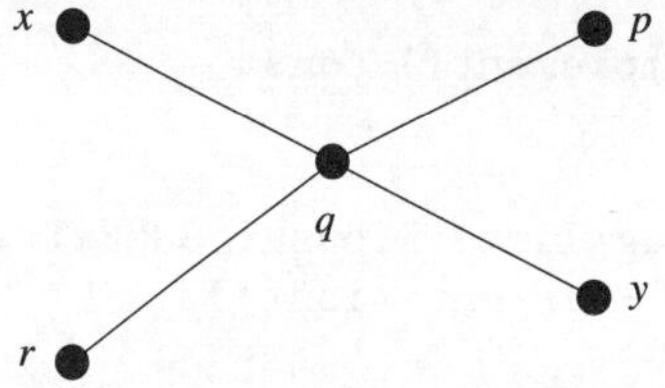

Figure 10.68 The chains $(\ldots, p, q, r, \ldots)$ and $(\ldots, x, q, y, \ldots)$ cross in q.

Figure 10.69 shows an example.

Lemma 72 *Every non-trivial polygonal set has a representation.*

Proof Consider a boundary segment s of X. Since X is regular, X lies on only one of the sides of s and hence s can be oriented such that X is locally to the left of s.

Consider next a point p as shown in Figure 10.67. Since X is the union of pieces of pie in the neighborhood of p we can join the boundary segments of X incident to p such that any two consecutive segments define one of the pieces of the pie. In this way no crossings are introduced. Also, since none of the pieces of the complement of X is degenerated to a line, every boundary segment incident to p is used only once.

The construction guarantees that the polygons formed are weakly simple and satisfy the two properties of a representation stated above. □

The representation of a polygonal set is not unique as Figure 10.69 shows. We still need to justify the choice of the name representation. In what sense does a representation of a polygonal set "represent" the set?

We start with the observation that the polygons in a representation form a so-called nested family. Let P_i and P_j be two polygons in a representation. Since P_i and P_j do not cross, we have either $\text{bR}\, P_i \cap \text{bR}\, P_j = \emptyset$ or $\text{bR}\, P_i \subset \text{bR}\, P_j$ or $\text{bR}\, P_j \subset \text{bR}\, P_i$, where $\text{bR}\, P$

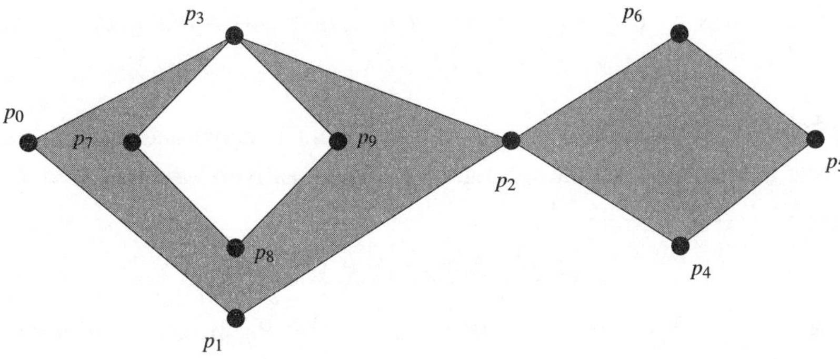

Figure 10.69 The open shaded region consists of two connected sets, one of which is simple. X can be represented by $(p_0, p_1, p_2, p_4, p_5, p_6, p_2, p_3, p_9, p_8, p_7, p_3)$ or by (p_0, p_1, p_2, p_3), (p_2, p_4, p_5, p_6), (p_3, p_9, p_8, p_7), or by $(p_0, p_1, p_2, p_4, p_5, p_6, p_2, p_3)$, (p_9, p_8, p_7, p_3).

denotes the bounded region with respect to a polygon P. We say that P_j is *nested* in P_i if $\text{bR}\, P_j \subset \text{bR}\, P_i$.

We can now define a forest F on the polygons in a representation. A polygon P_j is a child of a polygon P_i if P_j is nested in P_i and there is no P_k such that P_j is nested in P_k and P_k is nested in P_i. If P_j is a child of P_i in F, we say that P_j is directly nested in P_i. We have:

Lemma 73 *If P_j is a child of P_i in F then P_j and P_i have different orientations. All roots of F have the same orientation.*

Proof If P_i is positively oriented then $\text{bR}\, P_i$ belongs to X in the vicinity of P_i and to the left of P_i. Since P_j is directly nested in P_i and since it is part of the boundary of X, P_j must be negatively oriented. If P_i is negatively oriented then $\text{bR}\, P_i$ belongs to $\text{int}\,\text{cpl}\, X$ in the vicinity of P_i and to the left of P_i. Since P_j is directly nested in P_i and since it is part of the boundary of X, P_j must be positively oriented.

If X is bounded, all roots of F are positively oriented and if X is unbounded, all roots of P are negatively oriented. □

It is convenient to turn the forest F into a tree by adding an artificial root. The polygon associated with the root represents the "circle at infinity". The circle at infinity is positively oriented if X is unbounded and is negatively oriented if X is bounded. We use P_0 to denote the artificial polygon representing the circle at infinity. Every point of the plane is contained in the bounded region with respect to the circle at infinity.

We assume from now on that the polygons $P_0, P_1, \ldots, P_k$ in a representation are ordered such that no P_i is nested in a P_j for $i < j$. In other words, parents precede their children.

Lemma 74 *Let $P_0, P_1, \ldots, P_k$ be a representation of a polygonal set X and let p be a point in the plane that does not lie on any of the polygons in the representation. Let i be maximal* *such that* $p \in \text{bR}\, P_i$. *If* P_i *is positively oriented then* $p \in X$ *and if* P_i *is negatively oriented then* $p \notin X$.

Proof Observe first that i exists since every point is contained in the bounded region of the circle at infinity. Assume w.l.o.g. that P_i is positively oriented. Let P_{j_1} to P_{j_l} be the children of P_i in F. We have

$$\text{bR}\, P_i \setminus (\text{bR}\, P_{j_1} \cup \ldots \cup \text{bR}\, P_{j_l}) \subseteq X$$

and $i < j_1, \ldots, i < j_l$. Thus $p \notin (\text{bR}\, P_{j_1} \cup \ldots \cup \text{bR}\, P_{j_l})$ by the definition of i. This shows that $p \in X$. □

10.8.4 *The Implementation of Generalized Polygons*

Generalized polygons are a handle type, i.e., a generalized polygon is realized as a pointer to a representation class (called *gen_polygon_rep* and *rat_gen_polygon_rep*, respectively) which contains the actual representation. The member function *ptr*( ) returns the pointer to the representing object.

The representation consists of a flag k which indicates whether the polygon is trivial and a list *pol_list* of polygons. More precisely, we have a local enumeration type *kind* with elements EMPTY, FULL, and NON_TRIVIAL and k is equal to EMPTY or FULL iff the polygon is empty or full and is equal to NON_TRIVIAL, otherwise. If the polygon is trivial, *pol_list* is empty, and if the polygon is non-trivial, *pol_list* is the list of boundary cycles.

```
enum kind { EMPTY, FULL, NON_TRIVIAL };
kind k;
list<rat_polygon> pol_list;
```

We next discuss some member function of generalized polygons.

Checking a Representation: We define a function *check_representation* that applies to a list *pol_list* of polygons. It returns true if *pol_list* is a legal boundary representation, i.e., if:

- the segments of the polygons in *pol_list* meet only at endpoints, i.e, the planar map G defined by them has $2m$ edges, where m is the number of segments, and no parallel edges.
- there are no crossings between polygons,
- if D is directly nested in C then D and C have alternate orientations, and C is before D in the list of polygons, and
- all outermost polygons have the same orientation.

In the following program we check only the first two items. We know of no method to check the other items that is substantially different from our method to compute boundary representations. The latter method will be described in Section 10.8.4.

⟨*gen_polygon: check representation*⟩≡

```
static bool check_rep(const list<POLYGON>& pol_list)
{ GRAPH<POINT,SEGMENT> G;
  list&lt;SEGMENT&gt; seg_list;
  POLYGON P;
  forall(P,pol_list)
  { list&lt;SEGMENT&gt; SL = P.segments();
    seg_list.conc(SL);
  }
  SEGMENT_INTERSECTION(seg_list,G,true);
  if ( G.number_of_edges() != 2*seg_list.length() )
    return False("check_rep: wrong number of edges");
  // no parallel edges
  node v; edge e;
  forall_edges(e,G)
   if ( target(e) == target(G.cyclic_adj_succ(e)) )
     return False("check_rep: parallel edges");
  ⟨check_representation: check for crossings⟩
  return true;
}
bool GEN_POLYGON::check_representation() const
{ if ( trivial() ) return polygons().empty();
  return check_rep(polygons());
}
```

We describe how to check for crossings. Consider any node v of G. Each edge e out of v corresponds to a segment s of one of the polygons in *pol_list*. The polygons running through v introduce a pairing on the edges incident to v, where two edges are paired if they correspond to consecutive edges of one of the polygons. We number the pairs and replace each edge by the label of its pair. Then it must not happen that we have distinct labels a and b interlacing around v, i.e., the cyclic sequence of labels induced by the edges out of v must not contain a subsequence of the form $a, \ldots, b, \ldots, a, \ldots, b$. This is easily checked by means of a push down store S. We iterate over the edges e out of v. If the edge label of e agrees with the label on the top of S, we pop S, if it does not agree, we push the label of e. There is no crossing at v iff the push down store is empty at the end of the iteration.

⟨*check_representation: check edge labels*⟩≡

```
forall_nodes(v,G)
{ stack<int> S;
  forall_adj_edges(e,v)
  { if ( S.empty() || label[e] != S.top() )
      S.push(label[e]);
    else
```

```
    S.pop();
  }
  if ( !S.empty() ) return False("check_rep: crossing");
}
```

It remains to compute the edge labels. We do so in a two step process. We first construct a dictionary that stores for every segment s the edge $e(s)$ in G corresponding to it, i.e., having the same source and sink. We then iterate over all pairs (s, t) of consecutive segments and give $e(s)^{rev}$ and $e(t)$ the same label.

⟨*check_representation: check for crossings*⟩≡

```
map&lt;SEGMENT,edge&gt; segment_to_edge;
forall_edges(e,G)
{ SEGMENT s = G[e];
  node v = G.source(e);
  segment_to_edge[s] = ( s.source() == G[v] ? e : G.reversal(e) );
}
edge_array<int> label(G);
int count = 0;
forall(P,pol_list)
{ list_item it;
  const list&lt;SEGMENT&gt;& seg_list = P.segments();
  forall_items(it,seg_list)
  { edge e = segment_to_edge[seg_list[it]];
    e = G.reversal(e);
    edge f = segment_to_edge[seg_list[seg_list.cyclic_succ(it)]];
    label[e] = label[f] = count++;
  }
}
```
⟨*check_representation: check edge labels*⟩

Point Containment: The implementation of *side_of* follows directly from Lemma 74. If P is either empty or full, the answer is obvious. If P is non-trivial, we scan through the list of polygons in the representation. If p lies on one of the polygons, we return ON_REGION. Otherwise, we find the last P_i such that p lies in the bounded region of P_i; P_i might not exist, i.e., be equal to the fictitious polygon P_0. We return the orientation of P_i.

⟨*gen_polygon: side_of*⟩≡

```
int GEN_POLYGON::side_of(const POINT& p) const
{ if ( empty() ) return -1;
  if ( full() ) return +1;
  POLYGON P, P_i;
  bool P_i_exists = false;
  forall(P,polygons())
  { region_kind k = P.region_of(p);
    if ( k == ON_REGION ) return 0;
    if ( k == BOUNDED_REGION ) { P_i = P; P_i_exists = true; }
  }
```

```
    if ( P_i_exists ) return P_i.orientation();
    P = (ptr()->pol_list).front();
    return -P.orientation(); // = P0.orientation()
  }
```

Boolean Operations: We only discuss the binary boolean operations and leave the implementation of *complement* as an exercise. The implementations of all binary boolean operations follow a common principle. Let P_0 and P_1 be two generalized polygons and let R be the result of the boolean operation. We construct R in stages:

(1) We first deal with the case that either P_0 or P_1 is trivial. The remaining stages are not needed if this is the case.
(2) We construct the planar map G induced by P_0 and P_1.
(3) We classify the face cycles of G, i.e., compute for each face its status with respect to P_1 and P_2.
(4) Given the classification of the edges computed in the preceding stage, we mark all edges of G that are relevant for the result R of the boolean operation. An edge is relevant if the face to its left belongs to R.
(5) We simplify the graph G by deleting edges. We keep only those edges that separate a face belonging to R from a face belonging to the complement of R.
(6) We trace the face cycles of G and compute the representation of R.

Only the first and the fourth stage depend on the boolean operation. All other stages are generic and apply to all boolean operations. In the sequel we concentrate on the *intersection* routine.

We define constants `P0_face`, `non_P0_face`, `P1_face`, and `non_P1_face` which we use to label edges in stages two and three. The constants are chosen such that boolean operations are possible on them. After stages two and three every edge e of G will have a label describing the status of the face to its left with respect to P_0 and P_1.

The functions defined in ⟨*construct labeled map*⟩ realize stages two and three, the functions defined in ⟨*simplify graph*⟩ realize stage five, and the functions defined in ⟨*collect polygon*⟩ realize stage six. We will discuss them below.

Stage one is easy. If either argument is empty the intersection is empty, and if either argument is full the result is the other argument.

In stage four we label those edges as relevant which border a face of G which belongs to P_0 and P_1. These are precisely the edges whose label is equal to *P0_face* + *P1_face*.

⟨*gen_polygon: boolean operations*⟩+≡

```
static int P0_face     = 1;
static int not_P0_face = 2;
static int P1_face     = 4;
static int not_P1_face = 8;
```

⟨*construct labeled map*⟩

⟨*simplify graph*⟩

⟨*collect polygon*⟩

```
GEN_POLYGON GEN_POLYGON::intersection(const GEN_POLYGON& P1) const
{ // stage I
  if ( empty() || P1.empty() )
    return GEN_POLYGON(GEN_POLYGON_REP::EMPTY);
  if ( full() ) return P1;
  if ( P1.full() ) return *this;
  // stages II and III
  ⟨gen boolean operations: set up labeled map⟩
  // label relevant edges, stage IV
  edge_array<bool> relevant(G,false);
  int d = P0_face + P1_face;
  edge e;
  forall_edges(e,G) if (label[e] == d) relevant[e] = true;
  // stages V and VI
  ⟨gen boolean operations: extract result⟩
}
```

We come to stages two and three. We define the graph G, we introduce $P0$ as a synonym for the *this*-argument of the intersection, we define an edge array *label*, and call *construct_labeled_map*. It computes the planar map defined by the segments of P_0 and P_1 and labels all edges of this map.

⟨*gen boolean operations: set up labeled map*⟩≡

```
GRAPH<POINT,SEGMENT> G;
const GEN_POLYGON& P0 = *this;
edge_array<int> label;
construct_labeled_map(P0,P1,G,label);
```

The function *construct_labeled_map* realizes stages two and three. It first calls *construct_initial_map* for stage two and then uses *extend_labeling* for stage three. A call of *extend_labeling* with argument e labels the edges of the face cycle of G containing e.

⟨*construct labeled map*⟩≡

```
⟨construct initial map⟩
⟨extend labeling⟩
static void construct_labeled_map(const GEN_POLYGON& P0,
                                  const GEN_POLYGON& P1,
                                  GRAPH<POINT,SEGMENT>& G,
                                  edge_array<int>& label)
{ construct_initial_map(P0,P1,G,label);
  edge_array<bool> visited(G,false);
  edge e;
  forall_edges(e,G)
  { if (visited[e]) continue;
```

```
    extend_labeling(P0,P1,G,e,visited,label);
  }
}
```

Stage two is realized by *construct_initial_map*. It takes two generalized polygons P_0 and P_1 and computes the planar map G induced by their segments using the segment intersection algorithm of Section 10.7. It also computes a label for every dart of G. The label of a dart $e = (v, w)$ of P_0 is `P0_face` if P_0 is locally to the left of e and is `non_P0_face` otherwise. The analogous statement holds true for darts of P_1.

We proceed in several steps. In the first step we collect the segments of P_0 and P_1 into a list *seg_list* and label each segment with the gen_polygon to which it belongs. Note that a segment may belong to P_0 and P_1. We therefore use the labels 1, 2 and 3, where 3 indicates that a segment belongs to both polygons and label i, $1 \le i \le 2$, indicates that the segment belongs to P_{i-1}.

In a second step we compute the planar map induced by the segments in *seg_list*. In this planar map every node must have even degree. If the floating point kernel is used the map returned by SEGMENT_INTERSECTION may be non-plane or have a vertex of odd degree; if this is the case we recommend use of the rational kernel.

In the third step we compute the label of each dart. We discuss it below.

⟨*construct initial map*⟩+≡

```
static void construct_initial_map(const GEN_POLYGON& P0,
                                  const GEN_POLYGON& P1,
                                  GRAPH<POINT,SEGMENT>& G,
                                  edge_array<int>& label)
{
  list&lt;SEGMENT&gt; seg_list;
  map&lt;SEGMENT,int&gt; seg_label(0);
  const list&lt;SEGMENT&gt;& L0 = P0.edges();
  const list&lt;SEGMENT&gt;& L1 = P1.edges();
  SEGMENT s;
  forall(s,L0) { seg_label[s] = 1;
                 seg_list.append(s);
               }
  forall(s,L1) { seg_label[s] += 2;
                 seg_list.append(s);
               }
  SEGMENT_INTERSECTION(seg_list,G,true);
  node v;
#if ( KERNEL == FLOAT_KERNEL )
  if ( Genus(G) != 0 ) error_handler(1,mes + "Genus(G) != 0.");
  forall_nodes(v,G)
  { int deg = G.outdeg(v);
    if (deg % 2 != 0) error_handler(1,mes + "odd degree vertex.");
  }
```

```
#endif
  ⟨construct_initial_map: compute dart labels⟩
}
```

It remains to compute the dart labels.

Consider a dart e and its reversal. We assign a polygon to e as follows. If the segment $s = G[e]$ belongs to a unique polygon, e inherits the polygon from $G[e]$. Otherwise, either the cyclic adjacency predecessor or the cyclic adjacency successor of e must be parallel to e, i.e., have the same target as e. We arbitrarily assign e to P_0 in the former case and to P_1 in the latter case.

The polygon P_i is locally to the left of e if s and e point into the same direction, i.e., if the dot product of the underlying vectors is positive.

⟨*construct_initial_map: compute dart labels*⟩≡

```
label.init(G,0);
edge e0;
forall_edges(e0,G)
{ if ( label[e0] != 0 ) continue;
  edge e = e0; edge e_rev = G.reversal(e);
  POINT   a = G[source(e)];
  POINT   b = G[target(e)];
  SEGMENT s = G[e];
  if ( (b - a) * (s.target() - s.source()) <= 0 )
    leda_swap(e,e_rev);
  // now s and e point into the same direction
  switch ( seg_label[s] )
  { case 1: label[e] = P0_face;
            label[e_rev] = not_P0_face;
            break;
    case 2: label[e] = P1_face;
            label[e_rev] = not_P1_face;
            break;
    case 3: { edge f = G.cyclic_adj_pred(e);
              if ( target(f) != target(e) ) f = G.cyclic_adj_succ(e);
              label[e] = P0_face;
              label[e_rev] = not_P0_face;
              label[f] = P1_face;
              label[G.reversal(f)] = not_P1_face;
            }
  }
}
```

The function *extend_labeling* classifies the face F to the left of dart e. It scans the face cycle containing e, marks all darts of the cycle as visited, and computes the "or" of all dart labels on the cycle in d.

If all darts of the face cycle originate from either P_0 (d is less than four) or P_1 (d is divisible by four), we still have to classify the face cycle with respect to the other polygon and update d accordingly. This will be discussed below.

Finally, the label d is propagated to all darts of the cycle. If the label is contradictory, i.e., claims that the face is a P_i-face and a not-P_i-face, we raise an error.

⟨*extend labeling*⟩≡

```
static void extend_labeling(const GEN_POLYGON& P0,const GEN_POLYGON& P1,
                            const GRAPH<POINT,SEGMENT>& G, edge e,
                            edge_array<bool>& visited,
                            edge_array<int>& label)
{ int d = 0; int length = 0;
  edge x = e;
  do { visited[x] = true;   length++;
       //node v = source(x);
       //if (G.outdeg(v) == 2) v2 = v;
       d |= label[x];
       x = G.face_cycle_succ(x);
  } while (x != e);
  if ( d % 4 == 0 || d < 4 )
    { ⟨extend_labeling: face cycle has only darts from one polygon⟩ }
  x = e;
#if ( KERNEL == FLOAT_KERNEL )
  if ( d % 4 == P0_face + not_P0_face ||
      (d/4)*4 == P1_face + not_P1_face )
    error_handler(1,mes + "contradicting edge labels.");
#endif
  do { label[x] = d;
       x = G.face_cycle_succ(x);
  } while (x != e);
}
```

It remains to deal with the case that all darts of the face cycle F belong to the same *gen_polygon*, say P_i. Let v be the source of e. We distinguish two cases: either no dart out of v has a determined status with respect to P_{1-i} or this is not the case. In the former case v cannot lie on the boundary of P_{1-i} and hence v's side with respect to P_{i-1} determines the status of F with respect to P_{i-1}. In the latter case let f be the nearest adjacency predecessor of e such that the status of f with respect to P_{1-i} is already known. For all darts between e and f the status is still unknown and hence none of them can be contained in the boundary of P_{1-i}; f may be contained in the boundary of P_{1-i} or not (in the latter case, f belongs to a face cycle which was already considered and hence its status with respect to both polygons is known). In either case the status of F with respect to P_{1-i} is given by the status of f with respect to P_{1-i}, see Figure 10.70.

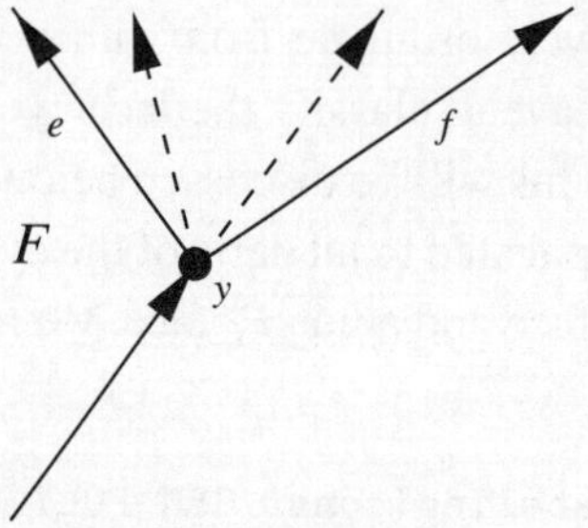

Figure 10.70 The dart f is the nearest adjacency predecessor of e whose status with respect to P_{1-i} is known. The edges between e and f do not belong to the boundary of P_{1-i} and hence F and the face to the left of f have the same status with respect to P_{1-i}.

⟨*extend_labeling: face cycle has only darts from one polygon*⟩≡

```
edge f;
for ( f = G.cyclic_adj_pred(e); f != e; f = G.cyclic_adj_pred(f) )
{ if ( d % 4 == 0 && label[f] % 4 != 0 || d < 4  && label[f] > 4 )
    break;
}
if ( f == e )
{ node v = source(e);
  if ( d % 4 == 0 )
      d |= ( P0.side_of(G[v]) == 1 ? P0_face : not_P0_face );
  if ( d < 4 )
      d |= ( P1.side_of(G[v]) == 1 ? P1_face : not_P1_face );
}
else
{ if ( d % 4 == 0 ) d |= ( label[f] % 4 );
  if ( d < 4 )      d |= ( ( label[f] / 4 ) * 4 );
}
```

We come to stage five. At this point all darts of G are labeled as relevant or non-relevant. A dart is labeled relevant if the face to its left belongs to the result R of the boolean operation.

We simplify the graph by removing darts. We proceed in two steps. In the first step we remove parallel darts that come from overlapping segments in the two arguments of the boolean operation, see Figure 10.71. This turns all face cycles of G into weakly simple polygons. In the second step we remove all edges from the graph that do not separate R from its complement.

The details of the first step are as follows. Let e and f be two parallel darts and assume that f is the cyclic adjacency successor of e. This implies that we have a face cycle (e, f^{rev}) of length two. This face cycle defines a polygon of area zero which we can remove. We remove the face cycle by removing its two constituent darts and making f and e^{rev} reversals of each other. There cannot be a set of three parallel darts and hence the target of f should be different from the target of its cyclic adjacency successor.

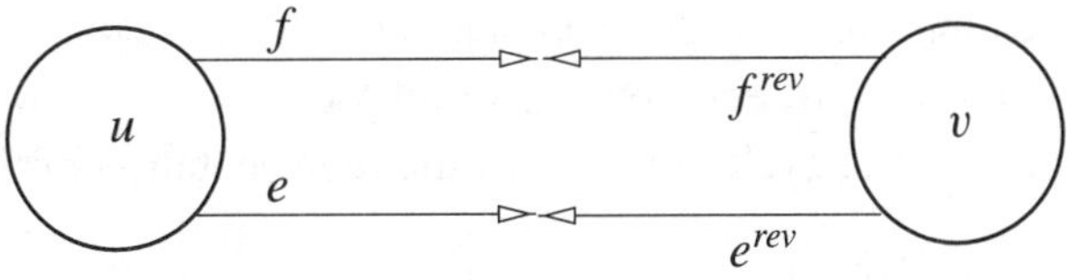

Figure 10.71 The darts e and f come from a segment of P_0 and P_1, respectively. The face cycle (e, f^{rev}) consists of only two darts. We remove e and f^{rev} and make f and e^{rev} reversals of each other.

The first simplification step leaves us with a planar map without parallel darts. This implies that all face cycles are weakly simple polygons. In the second step we merge adjacent faces that belong to the same side of the result polygon.

A dart e does not separate R from its complement if e and e^{rev} are either both relevant or both irrelevant. In the former case R exists on both sides of the edge and in the latter case the complement of R lives on both sides of the edge.

The second step may remove all edges from the graph. This will be the case if the result is either empty or full. We need to distinguish these cases. We have the former case if there are no relevant edges before simplification and we have the latter case if all edges are relevant before simplification. We return true in the latter case.

⟨*simplify graph*⟩≡

```
static bool simplify_graph(GRAPH<POINT,SEGMENT>& G,
                           edge_array<bool>& relevant)
{ edge e; node v;
  forall_nodes(v,G)
  { list<edge> E = G.out_edges(v);
    forall(e,E)
    { edge f = G.cyclic_adj_succ(e);
      if ( target(e) != target(f) ) continue;
      edge e_rev = G.reversal(e);
      G.del_edge(e); G.del_edge(G.reversal(f));
      G.set_reversal(e_rev,f);
    }
  }
  bool non_trivial_result = false;
  forall_nodes(v,G)
  { list<edge> E = G.out_edges(v);
    forall(e,E)
    { if ( relevant[e] || relevant[G.reversal(e)] )
        non_trivial_result = true;
      if ( relevant[e] == relevant[G.reversal(e)] )
      { G.del_edge(G.reversal(e)); G.del_edge(e); }
    }
  }
  return non_trivial_result;
}
```

After simplification every uedge of G separates R from its complement and hence belongs to the boundary representation. Also all face cycles are weakly simple polygons. We conclude that the face cycles of G form the representation of the result of the boolean operation.

The following function *collect_polygon* takes a dart e, marks all darts in the face cycle of e as visited, and collects the segments corresponding to the face cycle in a list *pol*.

⟨*collect polygon*⟩≡

```
static void collect_polygon(const GRAPH<POINT,SEGMENT>& G, edge e,
                            edge_array<bool>& visited,
                            list&lt;SEGMENT&gt;& pol)
{ pol.clear();
  edge x = e;
  do { visited[x] = true;
       node v = source(x);
       node w = target(x);
       POINT   a = G[v];
       POINT   b = G[w];
       pol.append(SEGMENT(a,b));
       x = G.face_cycle_succ(x);
  } while (x != e);
}
```

The function above is the main ingredient for the last stage. We first simplify G. If this trivializes G, i.e., removes all edges from it, we either return the full gen_polygon or the empty gen_polygon; the return value of *simplify_graph* tells us which.

⟨*gen boolean operations: extract result*⟩≡

```
bool non_trivial_result = simplify_graph(G,relevant);
if (G.number_of_edges() == 0 )
{ if ( non_trivial_result )
    return GEN_POLYGON(GEN_POLYGON_REP::FULL);
  else
    return GEN_POLYGON(GEN_POLYGON_REP::EMPTY);
}
edge_array<bool> visited(G,false);
list<POLYGON> result;
⟨gen boolean operations: form boundary cycles⟩
return GEN_POLYGON(result,GEN_POLYGON::NO_CHECK);
```

So assume that G is non-trivial. We cycle over all darts of G and collect all face cycles consisting of relevant darts.

⟨*gen boolean operations: form boundary cycles, first try*⟩≡

```
forall_edges(e,G)
{ if ( visited[e] || !relevant[e] ) continue;
  list&lt;SEGMENT&gt; pol;
```

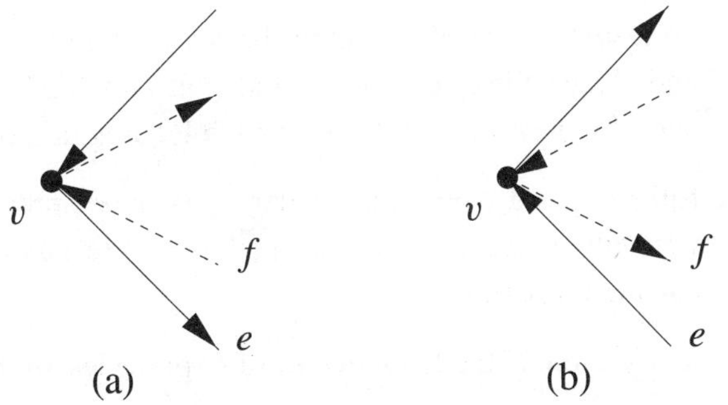

Figure 10.72 The dashed boundary cycle is nested in the solid cycle and both cycles have v as their leading node. In situation (a) the leading dart of the solid cycle is e and the leading dart of the dashed cycle is f^{rev}. In situation (b) the leading dart of the solid cycle is e^{rev} and the leading dart of the dashed cycle is f. In either case the leading dart of the solid cycle has smaller slope.

```
    collect_polygon(G,e,visited,pol);
    POLYGON P(pol);
    result.append(P);
}
```

The code above generates the boundary cycles in no particular order. We want an order that reflects nesting, i.e., no polygon should be nested in a polygon following it.

There are several ways to achieve a proper ordering. Our first solution took time $O(n + k \log k)$ and, moreover, was burdened with a fairly large constant factor. We exploited the fact that if D is nested in C then D has smaller unsigned area than C. We generated the polygons in an arbitrary order and then sorted the polygons in decreasing order of their unsigned area.

We describe an alternative approach. We show that one can rearrange the darts of G such that the code above generates the polygons in the proper order. Our approach is based on the following definition and observation. Define the leading node and dart of a boundary cycle as follows:

- The leading node $v(C)$ of a boundary cycle C is the lexicographically smallest node of the boundary cycle.
- The leading dart $e(C)$ of a boundary cycle is the shallowest (= smallest slope) dart of C starting in $v(C)$ if C is positively oriented, and is the reversal of the shallowest dart in C ending in $v(C)$ if C is negatively oriented.

Lemma 75 *If D is nested in C then either:*

- *$v(C)$ is lexicographically smaller than $v(D)$ or*
- *$v(C)$ is equal to $v(D)$ and $e(C)$ has smaller slope than $e(D)$.*

Proof Clearly the leading node of C cannot be lexicographically larger than the leading node of D. If C and D have the same leading node, the situation is as shown in Figure 10.72 and the leading dart of C has smaller slope than the leading dart of D. □

Consider the following order on darts. A dart $e = (v, w)$ precedes a dart $f = (x, y)$ if either v lexicographically precedes x or v is equal to x and e has smaller slope than f. This order has the following properties:

- For any boundary cycle C the leading dart of C precedes all darts of C.
- If D is nested in C then the leading dart of C precedes the leading dart of D.

The following compare class realizes the dart ordering; the base class *leda_cmp_base* is discussed in Section 2.10.

⟨*collect polygon*⟩+≡

```
template <class POINT, class SEGMENT>
class cmp_for_cycle_tracing : public leda_cmp_base<edge> {
const GRAPH<POINT,SEGMENT>& G;
public:
  cmp_for_cycle_tracing(const GRAPH<POINT,SEGMENT>& g): G(g) {}
  int operator()(const edge& e1, const edge& e2) const
  { node v = G.source(e1);
    node w = G.source(e2);
    if ( v != w ) return compare(G[v],G[w]);
    SEGMENT s1 = G[e1];
    SEGMENT s2 = G[e2];
    return cmp_slopes(s1,s2);
  }
};
```

It is now easy to generate the boundary cycles in the appropriate order. We sort the darts of G according to the ordering above and then iterate over all darts of G. Whenever we encounter a uedge that is not contained in a boundary cycle yet, we collect the boundary cycle. The uedge is a pair $\{e, e^{rev}\}$ and either e or its reversal is relevant (but not both). If e is relevant, the cycle to be traced is positively oriented, and if e^{rev} is relevant, the cycle to be traced is negatively oriented, see Figure 10.72. Thus there is no need to compute the orientations of the boundary cycles; our method of generating boundary cycles in an ordered fashion yields the orientations as a by-product.

We obtain:

⟨*gen boolean operations: form boundary cycles*⟩≡

```
cmp_for_cycle_tracing<POINT,SEGMENT> cmp(G);
list<edge> E = G.all_edges();
E.sort(cmp);
edge e0;
forall(e0,E)
```

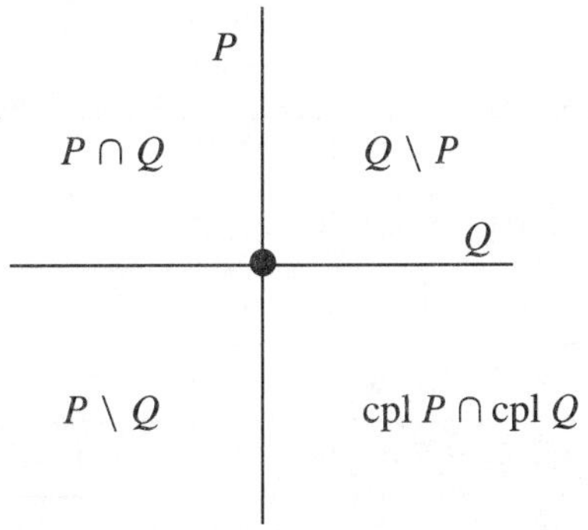

Figure 10.73 The vertex v is an intersection between the boundaries of P and Q. There are four faces incident to v and at least one but not all of them belong to the result of the boolean operation.

```
{ edge e = e0;
  if ( visited[e] || visited[G.reversal(e)]) continue;
  int orient;
  if ( relevant[e] )
    { orient = +1; }
  else
    { e = G.reversal(e); orient = -1; }
  list&lt;SEGMENT&gt; pol;
  collect_polygon(G,e,visited,pol);
  POLYGON P(pol,orient);
  result.append(P);
}
```

We conclude our treatment of boolean operations on polygons with a discussion of their asymptotic running time. Consider a boolean operation with input polygons P and Q and result polygon R. Let n be the total number of vertices of P, Q, and R, and let G be the graph induced by the two input polygons. Any vertex of G is either a vertex of one of the input polygons or is an intersection between the boundaries of the input polygons. In the latter case it will be a vertex of the result polygon, as Figure 10.73 shows. We conclude that G has at most n vertices and hence can be computed in time $O(n \log n)$. The time required to sort the edges before tracing the boundary cycles is also $O(n \log n)$. Let f be the number of face cycles of G which have darts from only one of the polygons; f can be as large as $O(n)$. For each such face cycle we spend time $O(n)$ to classify it with respect to the other polygon for a total time of $O(fn)$ (this time bound could be reduced to $O(f \log n)$ by using a more refined data structure for point location). All other steps take time $O(n)$. We conclude that the total time to compute boolean operations is $O(n + n \log n + fn)$.

A Demo Program: We give a small demo program. We construct an n-gon P with vertices near the unit circle. We also construct an affine transformation T that rotates the plane by

n	m	P	T	Q	$P \cap Q$	$\|P \cap Q\|$
5000	6.175e+06	1.35	0	0.36	12.92	20000
5000	2.47e+07	1.33	0	0.37	13.06	20000
5000	9.88e+07	1.35	0.01	0.39	13.44	20000
5000	3.952e+08	1.35	0	0.35	13.71	20000
5000	1.581e+09	1.35	0	0.36	–	–
20000	2.47e+07	5.65	0	1.47	56.13	80000
20000	9.88e+07	5.71	0	1.61	–	–

Table 10.10 Execution times with floating point kernel: The first two columns show n and m, respectively, the next four columns show the time to construct P, T, $Q = T(P)$, and $P \cap Q$, respectively, and the last column shows the number of vertices of $P \cap Q$. A dash in the next to last column indicates that the program produced an error message and recommended use of rat_polygons.

an angle $\alpha = 2\pi/(2nm) \pm eps$ about the origin, where $eps = 1/(10nm)$. Let $Q = T(P)$ be the result of turning P by angle α and let R be the union of P and Q.

$\langle$*n_gon_time*$\rangle \equiv$

```
double eps = 1/(10.0*n*m);
POLYGON P = N_GON(n,C,eps);
GEN_POLYGON PG(P,GEN_POLYGON::NO_CHECK);
report_time("time to generate P = ");
TRANSFORM T = rotation(ORIGIN, LEDA_PI/(n * m), eps);
report_time("time to generate the transformation T = ");
POLYGON Q = T(P);
GEN_POLYGON QG(Q,GEN_POLYGON::NO_CHECK);
report_time("time to compute T(P) = ");
GEN_POLYGON R = PG.unite(QG);
report_time("time to compute P union T(P) = ");
```

Tables 10.10 and 10.11 show the execution times for the floating point and the rational kernel and different values of n and m. Observe that we ran extreme examples. We took 5000-gons and 20000-gons and rotated them by angles $2\pi/(2 * n * m)$, where m ranges between 10^6 and 10^9. This amounts to rotations by angles between 10^{-8} and 10^{-10} degrees.

The floating point kernel did not always obtain a result. In the two cases where it did not obtain a result, it discovered that there is a problem. For $n = 5000$ and $n = 1.581 \cdot 10^9$ it reported that the map computed by SEGMENT_INTERSECTION is not planar and for $n = 20000$ and $m = 9.88 \cdot 10^7$ it reported that there is a node of odd degree in the map.

n	m	P	T	Q	$P \cap Q$	$\|P \cap Q\|$
5000	6.175e+06	1.69	0	1.4	30.8	20000
5000	2.47e+07	1.73	0	1.41	31.45	20000
5000	9.88e+07	1.74	0.01	1.4	33.93	20000
5000	3.952e+08	1.77	0	1.41	34.01	20000
5000	1.581e+09	1.78	0.009995	1.41	34.7	20000
20000	2.47e+07	7.25	0	5.66	140.9	80000
20000	9.88e+07	7.37	0	5.69	141.6	80000
20000	3.952e+08	7.45	0	5.66	143.2	80000
20000	1.581e+09	7.52	0.01001	5.58	145.1	80000
20000	6.323e+09	7.53	0	5.6	149.2	80000

Table 10.11 Execution times with rational kernel: The meaning of the columns is the same as for Table 10.10.

It is instructive to study the output of the program when the test for the planarity of G is not made. The graph G constructed by SEGMENT_INTERSECTION had 19994 nodes (and so 6 nodes are missing) and 59952 edges, 10012 nodes had degree two (12 too many) and 9982 nodes had degree four (18 too few). The genus of G was one. G had face cycles of length two and three and *only one* face cycle of length larger than three (there should be two). *All* edges of the graph were declared relevant and hence removed by *simplify_graph*. The full polygon was returned. It took several hours of detective work to discover this explanation for the behavior of the floating point implementation. The detective work was considerably helped by the fact that the execution with the rational kernel produced the correct result and hence we *knew* that the error must be in the floating point arithmetic.

It would be fantastic if the floating point implementation would always degrade gracefully, i.e., either compute the correct result or tell that the problem is too difficult for a floating point computation. We are not making this claim.

Although the floating point implementation did not always obtain the correct result it can handle surprisingly difficult cases.

The rational kernel always worked correctly, as it is supposed to do. There is about a factor three overhead for the use of the rational kernel.

Exercises for 10.8

1 Implement the function *complement* for generalized polygons.

2 Implement the function *unite* for generalized polygons. Start from the implementation of *intersection* and describe the required modifications.

10.9 A Glimpse at Higher-Dimensional Geometric Algorithms

We give an overview of the extension package for higher-dimensional computational geometry, exhibit a relationship between convex hulls and Delaunay triangulations, and use it to derive the formula for the side-of-sphere test. For a detailed treatment of higher-dimensional geometry we refer the reader to [Ede87].

10.9.1 *The Extension Package for Higher-Dimensional Geometry*

The extension package [MMN+98] features a higher-dimensional kernel, simplicial complexes, convex hulls and Delaunay diagrams.

The *higher-dimensional kernel* offers points, lines, segments, rays, vectors, hyperplanes, spheres, affine transformations, and geometric operations and predicates in d-dimensional Euclidian space for arbitrary dimension d. Examples for geometric predicates are the orientation test, the side-of-sphere test, the test of whether a point is contained in a simplex, and the computation of the affine rank of a set of points. Examples for geometric constructions are the construction of a hyperplane from a set of points, or the computation of the intersection of a line and a hyperplane.

The extension package offers three geometric data structures: regular simplicial complexes, convex hulls and Delaunay diagrams.

A *simplicial complex* is a collection of simplices in which the intersection of any two simplices in the collection is a face of both[28]. A simplicial complex is *regular* iff all maximal simplices of the collection[29] have the same dimension and if its maximal simplices are connected under the neighboring relation[30]. The data type *regl_complex* realizes regular simplicial complexes. It supports navigation in the complex (go to the i-th neighbor) and update operations on the complex (add a new simplex and make it the neighbor of some existing simplices). Regular simplicial complexes generalize triangulations to arbitrary dimension.

Convex hulls are represented as regular simplicial complexes, namely by a complex arising from a triangulation of the hull. Figure 10.11 shows an example in two-dimensional space.

The convex hull complex is built by a natural generalization of the incremental hull algorithm of Section 10.1.2. Whenever a point p is added to a convex hull, a simplex with peak p is added to the convex hull for every facet of the hull visible from p.

The data type *convex_hull* supports navigation through the underlying triangulation, navigation over the boundary of the hull, visibility queries (find all facets visible from a point p), point location queries (does a point p lie in the interior, on the boundary, or in the exterior of the hull) and insertion of new points.

Delaunay triangulations are also represented as simplicial complexes. The data type

[28] The empty set is a face of any simplex.

[29] A simplex is maximal if it is not contained in any other simplex.

[30] Two simplices of dimension k are neighbors if they share a face of dimension $K - 1$.

delaunay extends the functionality of the type *point_set* of Section 10.6 to higher dimensions. It supports navigation in the complex, insertion of new points, point location queries (return the simplex containing a query point p), nearest neighbor queries (return the point closest to a query point p), and range searches with spheres and simplices (return all points contained in a query sphere or query simplex, respectively).

10.9.2 *Delaunay Diagrams and Convex Hulls*

The implementation of Delaunay diagrams in higher-dimensional space is based on a powerful relationship between Delaunay diagrams, Voronoi diagrams, and convex hulls in one higher dimension.

Let d be a positive integer. We use $x_0, x_1, \ldots, x_{d-1}$, and z for the Cartesian coordinates of a $d+1$-dimensional space. Our Delaunay triangulations live in the d-dimensional subspace with coordinates $x_0, x_1, \ldots, x_{d-1}$ and the corresponding convex hulls will live in the $d+1$-dimensional space with coordinates $x_0, x_1, \ldots, x_{d-1}$, and z. We call the former space the *base space*.

The *paraboloid of revolution* P is defined by

$$z = x_0^2 + x_1^2 + \ldots + x_{d-1}^2.$$

It is obtained by rotating the two-dimensional parabola $z = x_0^2$ about the z-axis. The key for the entire section is the following observation.

Lemma 76 *The intersection between P and any hyperplane h that is not parallel to the z-axis is a curve C whose projection into the base space is a sphere and any sphere in the base space can be obtained in that way.*

Proof Since h is not parallel to the z-axis it is defined by an equation

$$z = a_0x_0 + a_1x_1 + \ldots + a_{d-1}x_{d-1} + a_d.$$

Any point $(x_0, x_1, \ldots, x_{d-1}, z)$ in the intersection between P and h satisfies

$$x_0^2 + x_1^2 + \ldots + x_{d-1}^2 = z = a_0x_0 + a_1x_1 + \ldots + a_{d-1}x_{d-1} + a_d$$

and hence

$$(x_0 - a_0/2)^2 + \ldots + (x_{d-1} - a_{d-1}/2)^2 = a_d + (a_0^2 + \ldots + a_{d-1}^2)/4.$$

This is the equation of a sphere in base space with center c and radius r where

$$c = (a_0/2, \ldots a_{d-1}/2) \text{ and } r = \sqrt{a_d + (a_0^2 + \ldots + a_{d-1}^2)/4}.$$

Thus the projection of $P \cap h$ into base space is a sphere. Conversely, if we start with any sphere B with center c and radius r in base space and define coefficients $a_0, a_1, \ldots, a_d$ through $c = (a_0/2, \ldots a_{d-1}/2)$ and $r^2 = a_d + (a_0^2 + \ldots + a_{d-1}^2)/4$ then the hyperplane $z = a_0x_0 + a_1x_1 + \ldots + a_{d-1}x_{d-1} + a_d$ will intersect P in a curve projecting into B. □

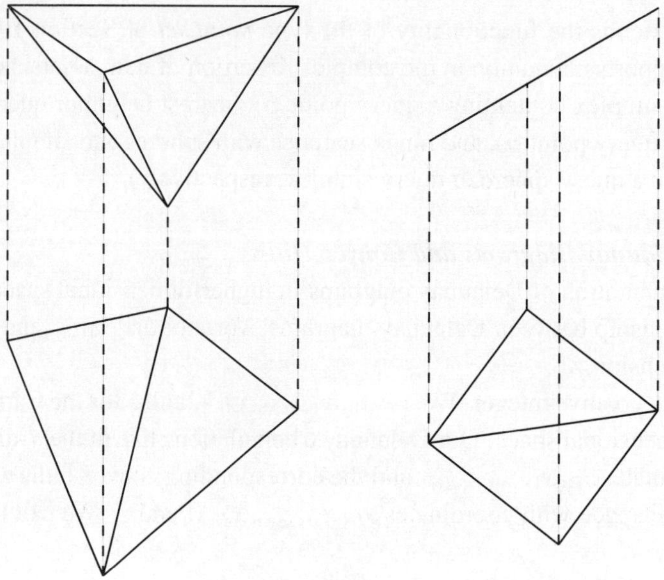

Figure 10.74 The connection between Delaunay diagrams in the plane and convex hulls in three-space. The lifting map is indicated by dashed lines. The four points on the left are not co-circular and hence the convex hull of the lifted points is a tetrahedron. The Delaunay diagram is the projection of the lower part of the tetrahedron.
The four points on the right are co-circular and hence the lifted points lie in a common plane. The convex hull of the lifted points is a rectangle contained in this plane. The Delaunay diagram is the projection of the rectangle and the projection of any triangulation of the rectangle is a Delaunay triangulation.

For a point $p = (x_0, x_1, \ldots, x_{d-1})$ in base space we call

$$lift(p) = (x_0, x_1, \ldots, x_{d-1}, x_0^2 + x_1^2 + \ldots + x_{d-1}^2)$$

its *lifting* onto P, i.e., the intersection of P with a vertical upward ray starting in p. We use the lifting map to establish a surprising connection between Delaunay diagrams and convex hulls.

Let S be any full-dimensional finite set of points in base space and let $p_0, p_1, \ldots, p_d$ be $d+1$ affinely independent points in S. The lifted points $lift(p_0), lift(p_1), \ldots, lift(p_d)$ define a hyperplane h. By the above, this hyperplane intersects P in a curve C whose projection into the base space is a sphere B. Of course, B passes through $p_0, p_1, \ldots, p_d$. In other words, B is the circumsphere of the simplex spanned by $p_0, p_1, \ldots, p_d$.

Next consider an arbitrary additional point p in base space. If p lies inside B then $lift(p)$ lies below h, if p lies on B then $lift(p)$ lies on h, and if p lies outside B then $lift(p)$ lies above h. We conclude that the interior of the circumsphere of $p_0, p_1, \ldots, p_d$ is void of points of S if and only if no point of

$$lift(S) = \{lift(p) \mid p \in S\}$$

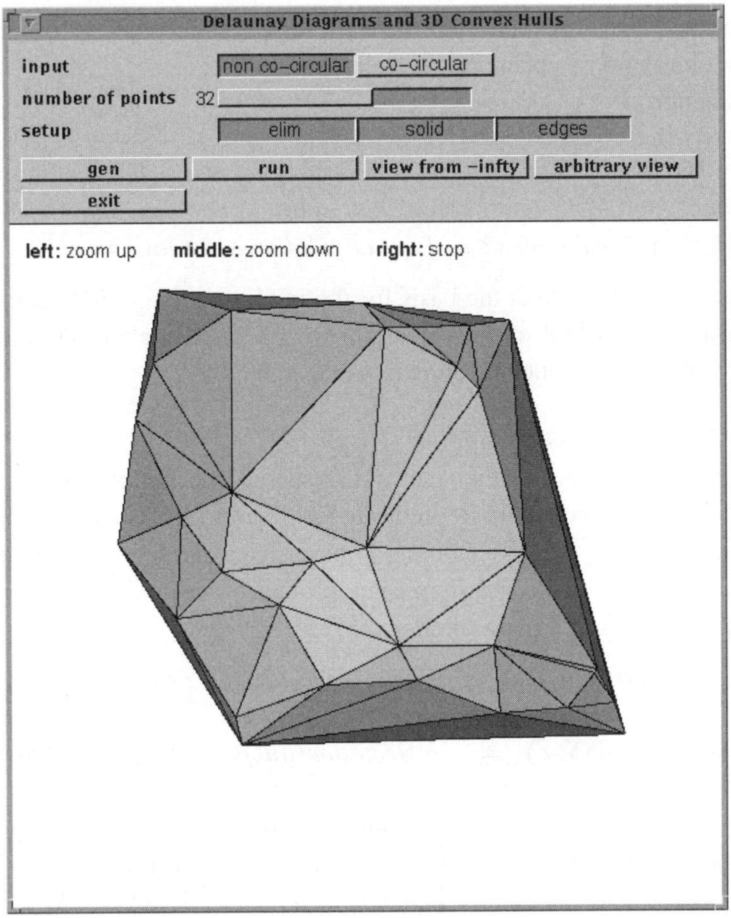

Figure 10.75 A screen shot of the delaunay_and_convex_hull_demo (in demo/book/Geo). The screen shot shows the lower convex hull of 32 random points in the unit square lifted to the paraboloid of revolution.

lies below h, or in other words, if h supports the lower convex hull of $lift(S)$. The *lower convex hull* of a point set consists of all points of the convex hull which are visible from $z = -\infty$.

Let us take a closer look at the lower convex hull. We need to distinguish cases according to whether the points in S are co-spherical or not, see Figures 10.74 and 10.75.

If the points in S are not co-spherical, the dimension of $list(S)$ is one higher than the dimension of S and hence $list(S)$ is full-dimensional. The convex hull of $lift(S)$ is a $d+1$-dimensional object. The lower convex hull consists of all facets with a downward normal.

If the points in S are co-spherical, the points in $lift(S)$ lie in a common hyperplane and the dimension of $lift(S)$ is the same as the dimension of S. The Delaunay diagram of S is identical to the convex hull of S and any triangulation of the convex hull is a Delaunay triangulation. The convex hull of *lift*(S) is a d-dimensional object; it is simply the lifting of the convex hull of S to a plane in $d+1$-dimensional space.

We summarize.

Theorem 16 *For any finite point set S in base space the Delaunay diagram $DD(S)$ is the vertical projection of the lower convex hull of lift(S) into base space*[31]. *A Delaunay triangulation is the vertical projection of a triangulation of the lower hull.*

The preceding theorem is the basis for the implementation of Delaunay diagrams. We maintain the convex hull of the lifted points. All queries about Delaunay diagrams are translated into queries about the corresponding hull.

10.9.3 *Sidedness and Orientation*

In this section we show how the results of the preceding section can be used to define the orientation, side-of, and region-of predicate for spheres.

Let $p_0, p_1, \ldots, p_d$ be $d+1$ points in base space and let p be an additional point in base space and let S be the sphere passing through $p_0, p_1, \ldots, p_d$. Define *orientation*(S), *side_of_sphere*(S, p), and *region_of_sphere*(S, p) by

$$\mathit{orientation}(S) = \mathit{orientation}(p_0, p_1, \ldots, p_d),$$

$$\mathit{side\_of\_sphere}(S, p) = -\mathit{orientation}(\mathit{lift}(p_0), \mathit{lift}(p_1), \ldots, \mathit{lift}(p_d), \mathit{lift}(p)),$$

$$\mathit{region\_of\_sphere}(S, p) = \begin{cases} \mathit{bounded\_region} & \text{if } o(S) \cdot o(S, p) > 0 \\ \mathit{on\_region} & \text{if } o(S) \cdot o(S, p) = 0 \\ \mathit{unbounded\_region} & \text{if } o(S) \cdot o(S, p) < 0 \end{cases}$$

where we used o as an abbreviation for *orientation* in the last formula to save space.

We will next show that *side_of_sphere*(S, p) and *region_of_sphere*(S, p) have their intended meaning.

Lemma 77 *Let $p_0, p_1, \ldots, p_d$ be $d+1$ affinely independent points in base space and let p be an additional point in base space. Then we have*

$$\mathit{side\_of\_sphere}(S, p) = \begin{cases} +1 & \text{if } p \text{ lies inside } S \\ 0 & \text{if } p \text{ lies on } S \\ -1 & \text{if } p \text{ lies outside } S \end{cases}$$

if orientation$(S) > 0$ and

$$\mathit{side\_of\_sphere}(S, p) = \begin{cases} +1 & \text{if } p \text{ lies outside } S \\ 0 & \text{if } p \text{ lies on } S \\ -1 & \text{if } p \text{ lies inside } S \end{cases}$$

[31] In the discussion above we assumed that S is full-dimensional. If S is contained in a lower dimensional subspace, we only need to restrict the discussion to this subspace. More precisely, assume that S is contained in a k-dimensional subspace. We may assume w.l.o.g that the first k coordinates span this subspace and can then use the argument above with d replaced by k.

if orientation$(S) < 0$. *Also*

$$region\_of\_sphere(S, p) = \begin{cases} bounded\_region & \text{if } p \text{ lies inside } S \\ on\_region & \text{if } p \text{ lies on } S \\ unbounded\_region & \text{if } p \text{ lies outside } S \end{cases}$$

Proof Observe first that the assumption that $p_0, p_1, \ldots, p_d$ are affinely independent implies that

$$orientation(S) = orientation(p_0, p_1, \ldots, p_d) \neq 0.$$

Furthermore, by symmetry, we may assume without loss of generality that the points p_0, $p_1, \ldots, p_d$ are positively oriented. Under the assumption that $p_0, p_1, \ldots, p_d$ are positively oriented the following three statements are equivalent:

(a) p is inside (on, outside) the sphere S.
(b) $lift(p)$ lies below (on, above) the hyperplane through points $lift(p_0), lift(p_1), \ldots, lift(p_d)$.
(c) $(lift(p_0), lift(p_1), \ldots, lift(p_d), lift(p))$ is negatively oriented.

We argued the equivalence of the first two items in the preceding section. The equivalence between the last two items follows from Lemma 58 in Section 9.2.2. This establishes the first claim. The second claim follows directly from the first. □

Exercises for 10.9

1 Let $p_0, p_1, \ldots, p_d$ be $d+1$ affinely dependent points ($orientation(p_0, p_1, \ldots, p_d) = 0$) in base space and let p be an additional point. Discuss the possible values of *side_of_sphere* and *region_of_sphere* for the $d+2$ tuple $(p_0, p_1, \ldots, p_d, p)$.

2 Assume that the base space is two-dimensional and that all points in S lie on the line $x_0 + x_1 = 1$. What does the convex hull of $lift(S)$ look like?

3 Assume that the base space is two-dimensional and that all points in S lie on a circle. What does the convex hull of $lift(S)$ look like?

4 Consider a circular range query with a square C in a set S. Translate the query by the lifting map. What is the result?

5 Show how to implement a nearest neighbor query by use of the lifting map.

10.10 A Complete Program: The Voronoi Demo

We discuss the voronoi_demo in xlman. The demo illustrates many of the geometric algorithms available in LEDA and we have already seen several screen shots. The demo is also a representative example for the design of geometric demos in LEDA and useful as a starting point for the development of further demos. We start with an overview, then give the details of the implementation, and end with a discussion of what can go wrong when the demo is run with the floating point kernel.

It is best to have the demo running while reading this section. Figure 10.76 shows yet another screen shot of the demo. The window consists of a panel part and a display part. The panel part in turn is structured in four parts. There is a list of eleven choice items which control which geometric structures are to be displayed; in the situation shown only the button for the Delaunay diagram is pressed and hence only the Delaunay diagram is shown. There is a list of three choice items which control how mouse clicks in the display part of the window are to be interpreted. In the situation shown every click of the left mouse button adds a point. The other two buttons allow the user to input points and circles respectively. There is a choice item which allows the user to switch between the rational kernel and the floating point kernel, and there is a boolean item and a slider item that control whether the input points are rounded to a grid and how many grid lines there are. Finally, there are six buttons for opening sub-menus, for clearing the window, for asking for help, and for exiting the demo.

10.10.1 *Overview*

The Voronoi demo allows the user to construct a scene of points and to visualize several fundamental geometric data structures for it: the nearest and furthest site Delaunay diagram, the nearest and furthest site Voronoi diagram, the convex hull and the width, the minimum spanning tree, the minimum enclosing and the maximum empty circle, the minimum width and the minimum area annuli, and the crust of the point set.

The point set is constructed either by mouse input or by calling one of the generators (sub-menu points). For mouse input there is the choice between single points, points on a line segment, and points on a circle. The current set of points is maintained as a list *p_list* of *rat_points*. The list is initially empty and is cleared by the clear-button. Any newly constructed point is added to it. It is important to remember that adding a line segment or adding a circle adds points that lie *exactly* on a line or a circle.

The geometric structures to be displayed can be computed with the use of three different geometry kernels: the rational kernel with the built-in floating point filter (this is the default), the rational kernel without the built-in floating point filter, and the floating point kernel. This allows the user to compare the relative speeds of the kernels and also to check visually whether the floating point kernel worked correctly. *When the floating point kernel is used, the program may abort or produce incorrect results.*

The geometric structures are not computed directly for the points in *p_list* but for a derived set of points. The derived set of points is called *rp_list* for use with the rational kernel and is called *fp_list* for use with the floating point kernel. The following procedure adds a point to *rp_list* and *fp_list*.

⟨*manipulate p_list, rp_list, and fp_list*⟩≡

```
void move_point(const rat_point& p)
{ point fp = p.to_point();
  if ( !round_to_grid )
  { fp_list.append(fp); rp_list.append(p); return; }
  double x = truncate(fp.xcoord(),truncation_prec);
  double y = truncate(fp.ycoord(),truncation_prec);
```

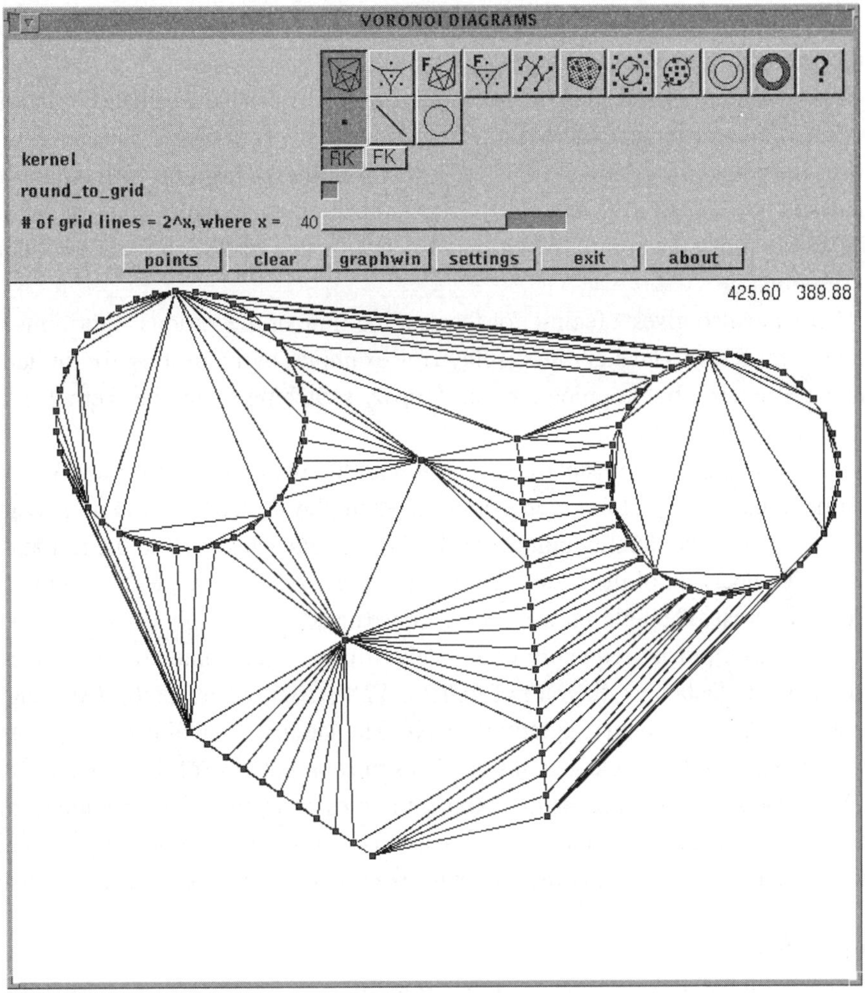

Figure 10.76 A screen shot of the Voronoi demo. A Delaunay triangulation is displayed.

```
    point tp(x,y);
    fp_list.append(tp);
    rp_list.append(rat_point(tp));
  }
```

The addition of a point is controlled by variables *round_to_grid* and *truncation_prec*. Let p be a *rat_point*. If *round_to_grid* is false, p is added to *rp_list* and $fp = p.to\_point(\,)$ is added to *fp_list*; the Cartesian coordinates of *fp* are the optimal approximations of the rational coordinates of p by *doubles*. Observe that when *round_to_grid* is false, the points p and *fp* are in general distinct. In particular, if *p_list* contains points on a circle or segment, the corresponding points in *fp_list* will lie close to the circle or segment but not exactly on it.

Such inputs will frequently overburden the floating point kernel, e.g., try to construct the crust of co-circular points.

When *round_to_grid* is true, the mantissae of the Cartesian coordinates of *fp* are truncated to *truncation_prec* binary places, i.e., all but the first *truncation_prec* bits are set to zero. This moves the points on a grid with $2^{truncation\_prec}$ grid lines. The point with the truncated coordinates is then added to *fp_list* and *rp_list*. Truncation with small values of *truncation_prec* will visibly move the points. When *round_to_grid* is true, *rp_list* and *fp_list* contain the same set of points.

The demo also gives a feeling for the running time of the various algorithms. Whenever the user requests to change the display (for example, by requesting for an additional geometric structure, by dropping a request, or by switching to another kernel) *all* requested structures are recomputed.

The demo can make mistakes when run with the floating point kernel. When using the floating point kernel, set *round_to_grid* to true and play with *truncation_prec* to get a feeling for the limits of the floating point kernel. You can always switch to the rational kernel for a visual comparison of the result. We want to point out one frequently occurring mistake. When the crust of points on a circle is constructed and a high value of *truncation_prec* is used, the output is frequently completely wrong. This comes from the fact that crust constructs the Delaunay diagram of $fp\_list \cup VD(fp\_list)$, where $VD(fp\_list)$ denotes the set of vertices of the Voronoi diagram of *fp_list*. The latter set contains many points crowding near the center of the circle and this confuses the computation of the Delaunay diagram.

When the scene contains many points on circles or segments, the running time with the rational kernel may go up sharply. The reason is that these inputs are very difficult, because our generators guarantee that the points lie exactly on a circle or line, respectively.

10.10.2 *Implementation*

We start with the global structure of the program.

We use a global variable *p_list* to store the current set of points, a list *fp_list* to store the corresponding list of float points, a pointer *Wp* to the display window, and integers *display* and *input* that govern which geometric structure to display and which kind of geometric object is selected for input. The variable *kernel* controls which kernel is used and the variable *use_filter* controls whether the filter is used in the rational kernel (it can be changed in the settings menu). We have already explained the role of *use_to_grid* and *truncation_prec*.

In the main program we first set up the display window *W* and then go into an infinite loop. At the beginning of the loop we wait for a mouse button to be pressed. The mouse button is either pressed on one of the seven buttons in the lower row of the panel section (cases zero to six) or in the display part of the window (case MOUSE_BUTTON(1)); the buttons in the top row of the display part are handled elsewhere as will be explained below.

In case of the event MOUSE_BUTTON(1) we put back the event, so that the mouse click can be processed again, and call *get_input*(*W*, *input*) to further process the mouse click.

At the end of the inner loop we draw the window as governed by the variable *display*.

⟨*voronoi_demo.c*⟩≡

```
#include <LEDA/plane_alg.h>
#include <LEDA/vector.h>
#include <LEDA/rat_vector.h>
#include <LEDA/window.h>
#include <LEDA/graphwin.h>
#include <LEDA/bitmaps/button32.h>
#include <math.h>
#include <LEDA/rat_window.h>
⟨definition of bit maps⟩
⟨definition of display mask⟩
static list<rat_point>  p_list, rp_list;
static list<point> fp_list;
static window* Wp;
static int display = 0;
static int input = 0;
enum { RK = 0, FK = 1};
static int kernel = RK;
static bool use_filter = true;
static int truncation_prec = 40;
static bool round_to_grid = true;
⟨further global variables⟩
⟨manipulate p_list, rp_list, and fp_list⟩
#include <LEDA/rat_kernel_names.h>
⟨displaying geometric structures⟩
⟨graph edit for graphwin⟩
#include <LEDA/kernel_names_undef.h>
#include <LEDA/float_kernel_names.h>
⟨displaying geometric structures⟩
⟨graph edit for graphwin⟩
#include <LEDA/kernel_names_undef.h>
⟨global drawing functions⟩
⟨action functions⟩
⟨point generators⟩
⟨adding a geometric object⟩
int main()
{
  window W(630,720,"VORONOI DIAGRAMS");
  Wp = &W;
  ⟨set up window⟩
for(;;)
{
 int but = W.read_mouse();
  rat_point::use_filter = use_filter;
  if (but == 0) break;
```

```
switch (but) {
case MOUSE_BUTTON(1): put_back_event();
                      get_input(W,input);
                      break;
case 1: { ⟨generate points menu⟩; break; }
case 2: { ⟨settings menu⟩; break; }
case 3:  clear_all(); break;
case 4:  // start GraphWin
         if ( kernel == FK )
            graph_edit(display,fp_list);
         else
            graph_edit(display,rp_list);
         break;
case 5:  // help
         help_win.open(W); break;
 }
 draw(display);
}
rat_point::print_statistics();
return 0;
}
```

The drawing functions are needed for both kernels and hence are included twice. We comment below why we did not use templates.

We give more details.

Setting up the Window: We start by defining a *help_string* and the panel *help_win* that pops up when the "about"-button is pressed. We then define the panel section of W. It consists of three sets of *choice_items*, a boolean item, a slider item, and a set of six buttons. We come back to them below.

Having defined the panel part we open the display, state that window coordinates for the x-coordinate are between 0 and 1000 and that they start at 0 for the y-coordinate (the upper bound for the y-coordinate depends on the actual geometry of W, state that nodes are drawn with width two, and that coordinates are to be shown.

⟨*set up window*⟩≡

```
string help_string;
help_string += "This program demonstrates some of the algorithms ";
help_string += "for two dimensional geometry of points based on ";
help_string += "Delaunay triangulations and Voronoi Diagrams.";

panel help_win;
help_win.text_item("\\bf Voronoi Demo");
help_win.text_item("");
help_win.text_item("K. Mehlhorn and S. Naeher (1997)");
help_win.text_item("");
help_win.text_item("see LEDAROOT/demo/documentation/voronoi_demo.ps");
```

```
help_win.text_item(help_string);
help_win.button("ok");
W.set_bitmap_colors(black,blue);
W.choice_mult_item("",display,11,32,32,display_bits,draw);
W.choice_item("",input,3,32,32,input_bits);
list<string> kernel_choices;
kernel_choices.append("RK"); kernel_choices.append("FK");
W.choice_item("kernel",kernel,kernel_choices,change_kernel);
W.bool_item("round_to_grid",round_to_grid,change_round_to_grid);
W.int_item("# of grid lines = 2^x, where x =",truncation_prec,
           2,52,change_truncation_prec);
W.button("points",   1, "Opens a point generator panel.");
W.button("clear",    3, "Clears point set and window.");
W.button("graphwin", 4, "Loads graph into GraphWin.");
W.button("settings", 2, "Opens an option setting dialog.");
W.button("exit",     0, "Exits the program.");
W.button("about",    5, "Displays information about this program.");
W.display();
W.init(0,1000,0);
W.set_redraw(redraw);
W.set_node_width(2);
W.set_show_coordinates(true);
```

We need to say a few more words about the panel part of the window. The first choice item controls the variable *display* and consists of eleven items. Whenever the i-th button is pressed the i-th bit of *display* is flipped and the function call *draw*(*display*) is made. Each item is drawn as a 32x32 pixel map taken from the collection of pixel maps defined in LEDA/bitmaps/button32.h. The pixel maps selected are defined by the array *display_bits*. The pixel maps are shown black when the corresponding button is released and are shown in blue when the button is pressed.

The second choice item controls the variable *input*. The effect of pressing one of the buttons in this collection of buttons is to set *input* to the number of the button.

The third choice item controls the use of the filter, the boolean item controls whether the input is rounded to a grid, and the slider item controls the number of grid lines.

The other buttons are added by the seven *button* statements. Each button is given a name, a number, and a help string that is displayed when the mouse rests over the button for an extended period of time.

⟨*definition of bit maps*⟩≡

```
static char* input_bits [] = { point_bits, line_bits, circle_bits };
static char* display_bits [] = { triang_bits, voro_bits, f_triang_bits,
  f_voro_bits, tree_bits, hull_bits, empty_circle_bits,
  encl_circle_bits, w_annulus_bits, a_annulus_bits, help_bits };
```

Action Functions: Some of the items in the menu part of the window have action functions associated with them. Recall that action functions are called with the new value of the variable associated with the item (the value of the variable itself is only changed after return from the action function such that new and old values of the variable are available during the action). All action functions follow the same scheme. They set the corresponding variable to the new value (since we want the new value during the execution of the action), clear the window and redraw the sites, recompute *rp_list* and *fp_list*, and recompute the display. The function *draw* will be discussed below.

⟨*action functions*⟩≡
```
void change_truncation_prec(int new_prec)
{ truncation_prec = new_prec;
  Wp->clear();
  draw_sites(p_list);
  recompute_rp_and_fp_list();
  draw(display);
}
void change_round_to_grid(int new_mode)
{ round_to_grid = new_mode;
  Wp->clear();
  draw_sites(p_list);
  recompute_rp_and_fp_list();
  draw(display);
}
void change_kernel(int new_kernel)
{ kernel = new_kernel;
  Wp->clear();
  draw_sites(p_list);
  draw(display);
}
```

The function *recompute_rp_and_fp_list* clears both lists and then moves all points from *p_list*. The function *add_point* will be called whenever a new point is added to *p_list* and *clear_all* clears the window and all lists.

⟨*manipulate p_list, rp_list, and fp_list*⟩+≡
```
void add_point(const rat_point& p)
{ p_list.append(p);
  move_point(p);
}
void recompute_rp_and_fp_list()
{ fp_list.clear(); rp_list.clear();
  rat_point p;
  forall(p,p_list) move_point(p);
}
void clear_all()
{ Wp->clear();
  p_list.clear(); fp_list.clear(); rp_list.clear();
}
```

Global Drawing Functions: The function *draw_area*(*disp*, *x0*, *y0*, *x1*, *y1*, *L*) draws the part of *W* covered by the rectangle with lower corner (*x0*, *y0*) and upper corner (*x1*, *y1*). It is our master drawing function. The geometric structures shown are governed by *disp* and *L* is either *p_list* or *fp_list*. If *L* is *p_list* the drawing functions use the rational kernel and if *L* is *fp_list* the drawing functions use the floating point kernel.

⟨*global drawing functions*⟩≡

```
template <class POINT>
void draw_area(int disp, double x0, double y0, double x1, double y1,
               const list<POINT>& L)
{
  if (L.empty()) return;

  Wp->start_buffering();
  Wp->clear();

  if (disp & MWA_MASK)   draw_min_width_annulus(L);
  if (disp & MAA_MASK)   draw_min_area_annulus(L);
  if (disp & HULL_MASK)  draw_convex_hull(L);
  if (disp & DT_MASK)    draw_delaunay(L);
  if (disp & VD_MASK)    draw_voronoi(L);
  if (disp & FDT_MASK)   draw_f_delaunay(L);
  if (disp & FVD_MASK)   draw_f_voronoi(L);
  if (disp & LEC_MASK)   draw_max_empty_circle(L);
  if (disp & SEC_MASK)   draw_min_encl_circle(L);
  if (disp & MST_MASK)   draw_min_span_tree(L);
  if (disp & CRUST_MASK) draw_crust(L);

  draw_sites(L);

  Wp->flush_buffer(x0,y0,x1,y1);
  Wp->stop_buffering();
}
```

If our current set of sites is empty, *draw_area* has nothing to do. Otherwise we clear the window, draw the selected geometric structures (the constants MWA_MASK, MAA_MASK, ... are defined in an enumeration type and denote 2^0, 2^1, 2^2, ...), and draw the sites. The appearance of the window is better if the sites are displayed after the selected geometric structures. We want the new drawing to appear in a single blow and therefore put the window in buffering mode before constructing the drawings of the selected geometric structures.

Once all drawings are constructed we flush the buffer and stop the buffering mode.

⟨*definition of display mask*⟩≡

```
enum display_mask {
  DT_MASK    =    1,  VD_MASK    =    2,  FDT_MASK   =    4,
  FVD_MASK   =    8,  MST_MASK   =   16,  HULL_MASK  =   32,
  LEC_MASK   =   64,  SEC_MASK   =  128,  MWA_MASK   =  256,
  MAA_MASK   =  512,  CRUST_MASK = 1024
};
```

The master drawing function is used by the functions *draw_area*, *draw* and *redraw*.

Draw_area (now without the *list<POINT>*-argument) makes the distinction between the use of the rational kernel and the floating point kernel.

Draw is called whenever one of the choice items changing *display* is called and at the end of each iteration of the main loop and redraw is called whenever the geometry of the window is changed. Accordingly, we redraw either only the display part of the window (in *draw*) or the entire window (in *redraw*).

⟨*global drawing functions*⟩+≡

```
void draw_area(int disp, double x0, double y0, double x1, double y1)
{
  if ( kernel == FK ) draw_area(disp,x0,y0,x1,y1,fp_list);
  else                draw_area(disp,x0,y0,x1,y1,rp_list);
}
void draw(int disp)
{ draw_area(disp,Wp->xmin(),Wp->ymin(),Wp->xmax(),Wp->ymax()); }
void redraw(window* wp, double x0, double y0, double x1, double y1)
{ draw_area(display,x0,y0,x1,y1); }
```

Displaying Specific Geometric Structures: For each of our geometric structures we have a function that displays it. We discuss only a representative sample of the functions.

We draw each site as a filled node of color *site_color*, where *site_color* is a global variable defined in ⟨*further global variables*⟩. This code is not shown. The default value of *site_color* is red; the color can be changed in the settings menu.

⟨*displaying geometric structures*⟩≡

```
void draw_sites(const list<POINT>& L)
{ POINT p;
  forall(p,L) Wp->draw_filled_node(p.to_point(),site_color);
}
```

Most of our geometric structures are graphs. We have to deal with two kinds of graphs. Voronoi diagrams have type *GRAPH<CIRCLE, POINT>* and Delaunay diagrams have type *GRAPH<POINT, int>*. We define a drawing function for each kind of graph. Recall that we use bidirected graphs to represent Delaunay diagrams and Voronoi diagrams. We therefore have to draw uedges and not edges.

In order to draw a *GRAPH<POINT, int>* we simply draw each uedge as the segment defined by the endpoints of the edge.

⟨*displaying geometric structures*⟩+≡

```
void draw_graph_edges(const GRAPH<POINT,int>& T, color col)
{ edge_array<bool> drawn(T,false);
  edge e;
  forall_edges(e,T)
    if (!drawn[e])
    { drawn[e] = true;
      edge r = T.reversal(e);
```

```
        if (r) drawn[r] = true;
        POINT p = T[source(e)];
        POINT q = T[target(e)];
        Wp->draw_edge(p.to_point(),q.to_point(),col);
      }
  }
```

Voronoi diagrams are a bit harder to draw. The positions of the nodes are determined by the circles associated with them. A proper node, i.e., a node of degree at least three, is positioned at the center of the circle associated with it. A node of degree one is positioned at the circle at infinity. If its circle is *CIRCLE*(a, _, b) then the node lies on the perpendicular bisector of a and b, and to the left of the oriented segment from a to b. Each edge is labeled by the site owning the region to the left of the edge. An edge e is part of the perpendicular bisector of sites a and b, where $a = G[e]$ and $b = G[G.reversal(e)]$.

After these preliminaries it is clear how to draw a Voronoi edge (v, w). An edge connecting two improper nodes is drawn as the perpendicular bisector of the points a and b, an edge connecting a proper node and an improper node is drawn as a ray starting at the proper node, running along the perpendicular bisector of points a and b and extending towards the position of the improper node at the circle at infinity, and an edge connecting two proper nodes is drawn as a segment connecting the nodes. We obtain the following code.

⟨*draw_voro_edges*⟩≡

```
void draw_voro_edges(const GRAPH<CIRCLE,POINT>& VD, color col)
{
  edge_array<bool> drawn(VD,false);
  edge e;
  forall_edges(e,VD)
  { if (drawn[e]) continue;
    drawn[VD.reversal(e)] = drawn[e] = true;
    node v = source(e);
    node w = target(e);
    POINT a = VD[e];
    POINT b = VD[VD.reversal(e)];
    VECTOR vec = (b - a).rotate90();
    line l = p_bisector(a,b).to_line();
    if (VD.outdeg(v) == 1 && VD.outdeg(w) == 1){ Wp->draw_line(l,col); }
    else
      if (VD.outdeg(w) == 1)
      { POINT cv = VD[v].center();
        VECTOR vec = VD[w].point3() - VD[w].point1();
        POINT  rp  = cv + vec.rotate90();
        Wp->draw_ray(cv.to_point(),rp.to_point(),col);
      }
      else
        if (VD.outdeg(v) == 1)
        { POINT cw = VD[w].center();
          VECTOR vec = VD[v].point3() - VD[v].point1();
          POINT  rp  = cw + vec.rotate90();
```

```
        Wp->draw_ray(cw.to_point(),rp.to_point(),col);
      }
      else
      { POINT cv = VD[v].center();
        POINT cw = VD[w].center();
        Wp->draw_segment(cv.to_point(),cw.to_point(),col);
      }
  }
}
```

The procedure above has serious numerical differences. Consider the following example. Assume that we compute the Voronoi diagram of three points that lie almost on a common line. The Voronoi diagram consists of one vertex and three rays. The vertex has very large coordinates and even if its coordinates are computed exactly (as they will be with the rational kernel) the conversion to point in *draw_ray* will suffer some loss of accuracy. We are now drawing a ray from a distant point. It is unlikely that this ray intersects the window in the desired form.

The window class offers drawing functions that are appropriate for this situation as discussed in Section 9.1. The modified drawing functions have an additional argument l of type *line*, which is supposed to be the line underlying the segment s or ray r to be drawn. In our case l is the bisector of a and b and hence determined with high precision. The additional argument is used as follows.

If the source of r lies in W or the two endpoints of s lie in W, l is ignored. Otherwise, the intersection t between l and the window is determined and the part of t which also belongs to r or s is drawn.

⟨*displaying geometric structures*⟩+≡

```
// template <class POINT, class CIRCLE, class VECTOR, class LINE>
void draw_voro_edges(const GRAPH<CIRCLE,POINT>& VD, color col)
{
  edge_array<bool> drawn(VD,false);
  edge e;
  forall_edges(e,VD)
  { if (drawn[e]) continue;
    drawn[VD.reversal(e)] = drawn[e] = true;
    node v = source(e);
    node w = target(e);
    POINT a = VD[e];
    POINT b = VD[VD.reversal(e)];
    line l = p_bisector(a,b).to_line();
    if (VD.outdeg(v) == 1 && VD.outdeg(w) == 1){ Wp->draw_line(l,col); }
    else
      if (VD.outdeg(w) == 1)
      { POINT cv = VD[v].center();
        VECTOR vec = VD[w].point3() - VD[w].point1();
        POINT  rp  = cv + vec.rotate90();
        Wp->draw_ray(cv.to_point(),rp.to_point(),l,col);
```

```
      }
      else
        if (VD.outdeg(v) == 1)
        { POINT cw = VD[w].center();
          VECTOR vec = VD[v].point3() - VD[v].point1();
          POINT  rp  = cw + vec.rotate90();
          Wp->draw_ray(cw.to_point(),rp.to_point(),1,col);
        }
        else
        { POINT cv = VD[v].center();
          POINT cw = VD[w].center();
          Wp->draw_segment(cv.to_point(),cw.to_point(),1,col);
        }
  }
}
```

The function above uses points, lines, circles, and vectors and hence would require four template arguments. Moreover, we would have to add artificial arguments of type LINE and VECTOR such that the appropriate type inference can be made by the compiler. We decided to use our primitive renaming mechanism instead. An alternative would be to introduce a class *rat_kernel*

```
class rat_kernel{
  typedef rat_point   POINT;
  typedef rat_segment SEGMENT;
  // and so on
}
```

and a similar class *float_kernel*, to use a single template argument called *kernel*, and to use qualified type names such as *kernel::POINT* and *kernel::SEGMENT* in *draw_voro_edges*. This design is used extensively in CGAL [CGA].

We come to the drawing functions for the individual geometric structures. Nearest and furthest sites Delaunay diagrams, crusts, and minimum spanning trees are drawn by first computing the structure and then calling *draw_graph_edges*. For example,

⟨*displaying geometric structures*⟩+≡

```
void draw_delaunay(const list<POINT>& L)
{ GRAPH<POINT,int>  DT;
  DELAUNAY_TRIANG(L,DT);
  draw_graph_edges(DT,triang_color);
}
```

Nearest and furthest site Voronoi diagrams are drawn by computing the structure and calling *draw_voro_edges*. For example,

⟨*displaying geometric structures*⟩+≡

```
void draw_voronoi(const list<POINT>& L)
{ GRAPH<CIRCLE,POINT> VD;
  VORONOI(L,VD);
  draw_voro_edges(VD,voro_color);
}
```

In order to display the convex hull and the width of our set of points we compute the convex hull (a list of POINTs), convert the list to a list of *points*, and draw the list of points as a filled polygon of *hull_color* and as a black polygonal line. We also compute the minimum width slab containing our set of points and display the two lines bounding the slab.

⟨*displaying geometric structures*⟩+≡

```
void draw_convex_hull(const list<POINT>& L)
{ list<POINT> CH = CONVEX_HULL(L);
  list<point> pol;
  POINT p;
  forall(p,CH) pol.append(p.to_point());
  Wp->draw_filled_polygon(pol,hull_color);
  Wp->draw_polygon(pol,black);
  // width
  LINE l1,l2;
  WIDTH(L,l1,l2);
  Wp->draw_line(l1.to_line(),blue);
  Wp->draw_line(l2.to_line(),blue);
}
```

In order to draw a minimum width annulus we either draw the two circles or the two parallel lines defining the annulus. In the first case we want the annulus to be shown in orange. We therefore draw the larger disk in orange first and then the smaller disk in white. This leaves the annulus in orange.

⟨*displaying geometric structures*⟩+≡

```
void draw_min_width_annulus(const list<POINT>& L)
{ POINT a,b,c; LINE l1,l2;
  if ( MIN_WIDTH_ANNULUS(L,a,b,c,l1,l2) )
  { // proper annulus
    circle c1(a.to_point(),b.to_point());
    circle c2(a.to_point(),c.to_point());
    Wp->draw_disc(c2,orange);
    Wp->draw_disc(c1,white);
    Wp->draw_circle(c1,black);
    Wp->draw_circle(c2,black);
    Wp->draw_point(a.to_point(),orange);
  }
  else
  { // strip
    Wp->draw_line(l1.to_line(),black);
```

```
      Wp->draw_line(l2.to_line(),black);
    }
  }
```

Adding a Geometric Object: We come to the mouse input of points, lines, and circles. The function *get_input*(*W*, *input*) reads either a point, or a segment, or a circle and then calls the appropriate insertion function.

```
⟨adding a geometric object⟩≡
  ⟨adding a point, segment or circle⟩
  void get_input(window& W, int inp)
  { rat_point p; rat_segment s; rat_circle c;
    switch (inp) {
      case 0: if (W >> p) insert_point(p);   break;
      case 1: if (W >> s) insert_segment(s); break;
      case 2: if (W >> c) insert_circle(c);  break;
    }
  }
```

```
⟨adding a point, segment or circle⟩≡
  void insert_point(rat_point p)
  { Wp->draw_filled_node(p.to_point(),site_color);
    add_point(p);
  }
```

Addition of a point does the obvious. In order to add points on a segment we generate n points on the segment, where n is determined by the ratio between the length of the segment and the global variable *point_dist*.

In order to add a circle we generate n uniformly spaced points on the circle, where n is determined by the ratio between the circumference of the circle and the global variable *point_dist*.

```
⟨adding a point, segment or circle⟩+≡
  void insert_segment(rat_segment s)
  {
    double l = s.to_segment().length();
    int n = Wp->real_to_pix(l)/point_dist + 1;
    list<rat_point> L;
    points_on_segment(s,n,L);
    rat_point p;
    forall(p,L)
    { add_point(p);
      Wp->draw_filled_node(p.to_point(),site_color);
    }
  }
  void insert_circle(rat_circle C)
```

```
{
  double L = 2 * C.to_circle().radius() * LEDA_PI;
  int n = Wp->real_to_pix(L)/point_dist + 1;
  double d = (2*LEDA_PI)/n;
  double eps = 0.001;
  double a = 0;
  for(int i = 0; i < n; i++)
  { rat_point q = C.point_on_circle(a,eps);
    add_point(q);
    Wp->draw_filled_node(q.to_point(),site_color);
    a += d;
  }
}
```

Point Generators: The point generator menu allows the user to select between three generators. A generator for random points in a square, a generator for regularly spaced points, and a generator for random points near a circle. The third generator produces inputs which are useful to illustrate the computation of annuli.

⟨*generate points menu*⟩≡

```
panel P;
P.text_item("\\bf Generate input points");
P.text_item("");
P.choice_item("",k_gen,"random","lattice","near circle");
P.int_item("",n_gen,0,500);
P.button("create",0);
P.button("cancel",1);
if (P.open(W) == 0)
{ switch (k_gen) {
    case 0: random_square(n_gen); break;
    case 1: lattice_points(n_gen); break;
    case 2: near_circle(n_gen,point_dist); break;
    }
}
```

We only show the *near_circle* generator. It generates points in an annulus with inner radius *rmin* and outer radius *rmax*; *rmax* is chosen such that the annulus fits nicely on the screen and *rmin* is chosen as 90% of *rmax*.

For each point to be generated we generate a random point on a circle of radius *r* where *r* is randomly chosen between *rmin* and *rmax*.

⟨*point generators*⟩+≡

```
void near_circle(int n, int point_dist)
{
  double x0 = Wp->xmin(), y0 = Wp->ymin();
  double x1 = Wp->xmax(), y1 = Wp->ymax();
  point cent((x0+x1)/2,(y0+y1)/2);
  int rmax = int(0.35 * (x1-x0));
```

```
    int rmin = int(0.9 * rmax);
    clear_all();
    for(int i=0; i < n; i++)
    { //circle C(cent,rand_int(rmin,rmax));
      circle C(cent,(double)rand_int(rmin,rmax));
      double a;
      rand_int >> a;
      point q = C.point_on_circle(2*a*LEDA_PI);
      int x = (int)q.xcoord();
      int y = (int)q.ycoord();
      add_point(rat_point(x,y,1));
      Wp->draw_filled_node(x,y,site_color);
    }
  }
```

Calling GraphWin: The function *graph_edit* visualizes the graphs underlying our geometric structures. We do not discuss it here.

Settings: The settings menu allows the user to set some of the global variables. It is self-explanatory.

⟨*settings menu*⟩≡
```
  panel SP("SETTINGS");
  SP.bool_item("use filter in rat kernel", use_filter);
  SP.bool_item("draw lines with width 2",thick_lines);
  SP.int_item("grid", grid_width,0,50,10);
  SP.int_item("pix dist", point_dist,1,64);
  SP.color_item("sites ", site_color);
  SP.color_item("voro  ", voro_color);
  SP.color_item("triang", triang_color);
  SP.color_item("hull",   hull_color);
  SP.color_item("tree",   tree_color);
  SP.button("continue");
  SP.open(W);
  W.set_grid_mode(grid_width);
  W.clear();
  W.set_line_width( thick_lines ? 2 : 1);
  draw_sites(p_list);
  recompute_rp_and_fp_list();
  draw(display);
```

10.10.3 *Floating Point Errors*

What can go wrong when the demo is executed with the floating point kernel?

When a segment or circle is added a certain number of points on the segment or circle are added to *p_list*. The rational kernel guarantees that these points lie exactly on the segment or circle, respectively. When the *rat_points* are converted to *points*, they will lie only almost on the circle or segment.

Consider now a scene that consists of points on two segments. The Delaunay triangulation will contain extremely flat triangles. This can cause the computation of the Delaunay diagram and the Voronoi diagram to fail.

Crust is also a good source of error. It computes the Delaunay diagram of the points is *fp_list* plus the vertices of the Voronoi diagram of *p_list*. When *fp_list* contains points that lie almost on a circle there will be many Voronoi vertices near the center of the circle and the Delaunay diagram computation will get confused. This can lead to strange crusts.

11

Windows and Panels

The data type *window* is the base type for all visualization and animation support in the LEDA system. It provides an interface for the graphical input and output of basic geometric objects for both the *X11* system on Unix platforms and for Microsoft *Windows* systems.

An instance W of type *window* is a rectangular window on the display screen. The width w and height h of W are measured in pixels and can be defined in the constructor. The default constructor initializes the width and height of W to default values depending on the system and screen resolution of the display. The position on the display is given by the pixel coordinates of the upper left corner of W. It can be specified in the *display* operation.

A window consists of two rectangular regions, a *panel section* in the upper part and a *drawing section* in the rest of the window. Either section may be empty. The panel section contains *panel items* such as sliders, choice fields, string items, and buttons. They have to be created by the operations described in Section 11.14 before the window is displayed for the first time. Figure 11.1 shows a typical LEDA window. If a window has no drawing section we call it a *panel*. Figure 11.2 shows the LEDA panel used for the *xlman* manual reader.

The drawing section can be used to draw geometric objects such as points, lines, segments, arrows, circles, polygons, graphs, ...and to input any of these objects using the mouse input device.

In this chapter we discuss LEDA windows and show how to use them in demo and visualization programs.

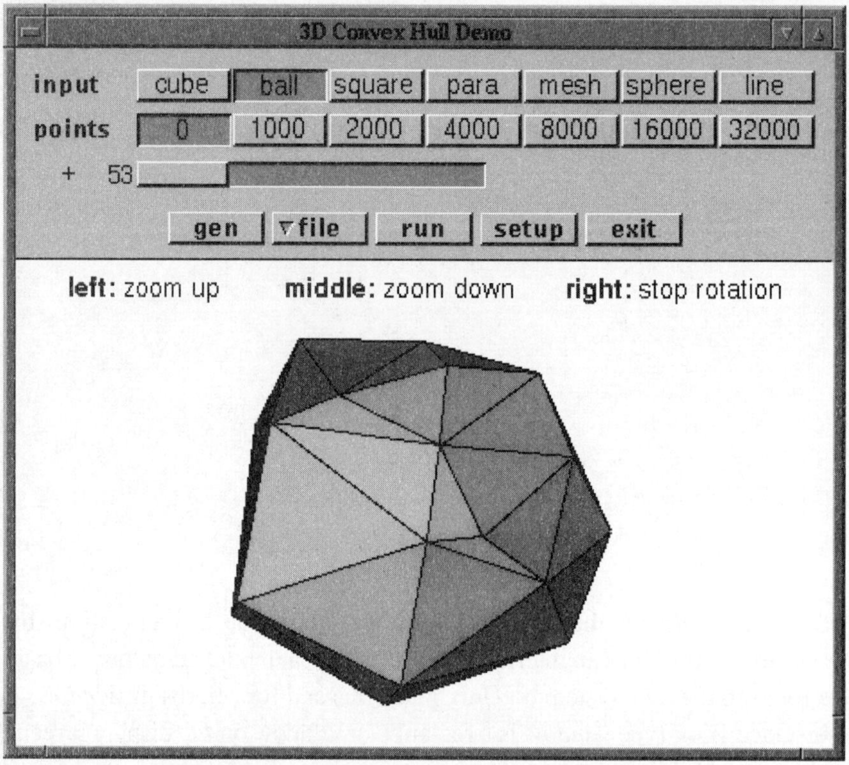

Figure 11.1 A typical LEDA window.

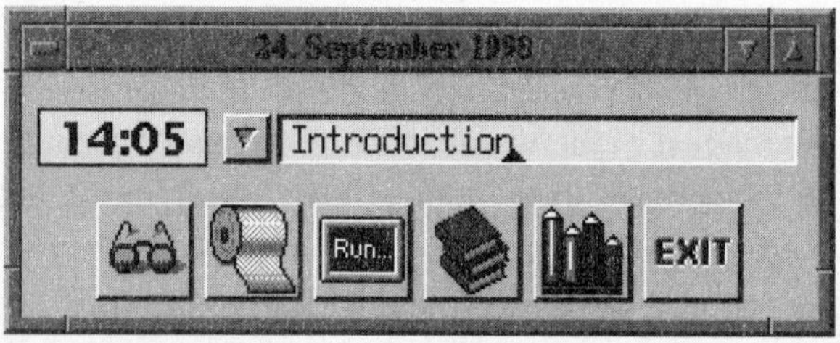

Figure 11.2 A typical LEDA panel: xlman.

11.1 Pixel and User Coordinates

The underlying graphics systems (X11 or Windows) maps windows to rectangular regions of the display screen using a pixel based coordinate system. In this *pixel coordinate system*, the upper left corner of the window rectangle has coordinates (0, 0), x-coordinates increase from left to right, and y-coordinate increase from top to bottom. This is illustrated in Figure 11.3.

All drawing and input operations in the drawing section use the *user coordinate system* whose y-axis is oriented in the usual mathematical way, i.e., from bottom to top. The

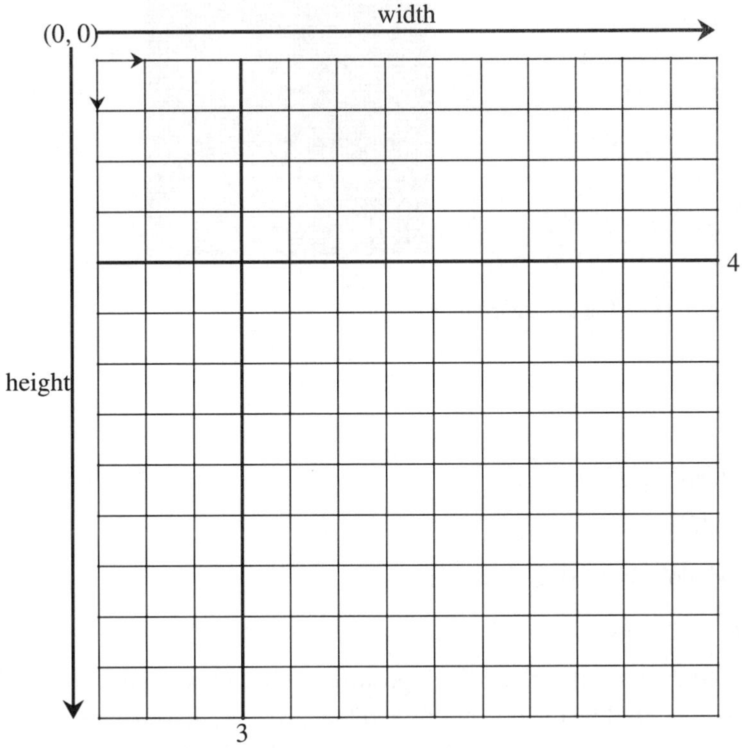

Figure 11.3 The pixel coordinate system: the pixel with coordinates (3, 4).

user coordinate system is defined by three numbers of type *double*: *xmin*, the minimal x-coordinate, *xmax*, the maximal x-coordinate, and *ymin*, the minimal y-coordinate. The two parameters *xmin* and *xmax* define the scaling factor

$$scaling = w/(xmax - xmin),$$

where w is the width of the window in pixels. The maximal y-coordinate *ymax* of the drawing section is equal to $ymin + h \cdot scaling$, where h is the height of the drawing section in pixels. The user coordinates (x, y) correspond to the pixel

$$(scaling \cdot (x - xmin), scaling \cdot (y - ymin)).$$

The window type provides operations for translating user coordinates into window coordinates and vice versa.

11.2 Creation, Opening, and Closing of a Window

We describe how to create, open, and close a window.

Figure 11.4 The LEDA default icon.

```
window W;
```

creates a window of default size.

```
window W(int w, int h);
```

creates a window W of size $w \times h$ pixels.

```
void W.display();
```

opens W and displays it at the default position on the screen. Note that W.display() has to be called before all drawing operations and that all operations adding panel items to W (cf. Section 11.14) have to be called before the first call of W.display().

```
void W.display(int x, int y);
```

opens W and displays it with its left upper corner at position (x, y) in pixel coordinates. The three special constants *window*::*min*, *window*::*center*, *window*::*max* can be used for positioning W at the minimal or maximal x- or y-coordinate or centering it horizontally or vertically on the screen.

```
void W.display(window W0, int x=window::center, int y=window::center);
```

opens W and displays it at position (x, y) above window $W0$ which must be displayed already.

```
void W.iconify();
```

closes W and displays it as a small icon. If no user-defined icon is specified (see the *icon pixrect* parameter) the LEDA default icon, as shown in Figure 11.4, is used.

```
void W.close();
```

closes W and removes it from the display.

11.3 Colors

The data type *color* represents all colors available in drawing operations.

Each color value corresponds to a triple of integers (r, g, b) with $0 \leq r, g, b \leq 255$, the so-called *rgb-value* of the color. The number of available colors is restricted and depends on the underlying hardware. A color can be created from rgb-values,

```
color col(int r, int g, int b);
```

from a color name in a system data base (X11 only)

```
color col(string color_name);
```

or from one of the integer color constants defined in *<LEDA/impl/x_window.h>*

```
color col(int color_const);
```

where *color_const* is one of the constants from the enumeration

```
enum { black, white, red, green, blue, yellow, violet,
       orange, cyan, brown, pink, green2, blue2,
       grey1, grey2, grey3, ivory, invisible }
```

A drawing operation with the special color *invisible* has no effect on the display.

The definition of a color may fail due to one of the following reasons:

- There is a system dependent limitation on the total number of different colors any application may use and the construction exceeds this limit.
- One of the specified (r, g, b)-values is illegal, i.e., not in the range $[0, \ldots, 255]$.
- The color name is not present in the systems color data base or the system does not support this method of specifying colors.

If the definition of a color fails, we say that the constructed color is *bad*; it is called *good* otherwise. The operation

```
bool col.is_good()
```

tests whether a color is good or bad.

It is also possible to retrieve the (r, g, b)-values of a color by

```
void col.get_rgb(int& r, int& g, int& b);
```

The following program tries to construct all 256 possible grey colors and reports how many of them are available.

⟨*greyscales.c*⟩≡

```
#include <LEDA/window.h>
#include <LEDA/array.h>

main()
{
  array<color> grey(256);
  int n = 0;
```

```
  for(int i = 0; i < 256; i++)
  { color c(i,i,i);
    if (c.is_good()) grey[n++] = c;
   }
  cout << n << " different greys available." << endl;
  return 0;
}
```

Exercises for 11.3

1 How man different versions of "red" are available on your system? Write a program to find out.

2 Write a program that displays a rainbow.

11.4 Window Parameters

Every window has a list of parameters which control its appearance and the way drawing operations are performed on the window. In this section we will first survey the available window parameters and then show how to read and to change them.

The Available Parameters: We list the parameters together with their type, default value, and a short description of their meaning.

background color: A parameter of type *color* (default value *white*) defining the default background color (e.g., used by *W.clear*( ) to erase the drawing area).

background pixrect: A parameter of type *char∗* (default value: *NULL*) defining a pixrect (see Section 11.8) that is used to tile the background of the window. If it is different from *NULL* the background color parameter is ignored.

foreground color: A parameter of type *color* (default value: *black*) defining the default color to be used in all drawing operations. All drawing operations have an optional color argument that can be used to override the default foreground color temporarily.

mouse cursor: A parameter of type *int* (default value: -1) defining the shape of the mouse cursor. Its value must be either the default value or one of the values listed in *<LEDA/X11/cursorfont.h>*.

text font: A parameter of type string (default value: system dependent) defining the name of the font to be used in text drawing operations. Possible values are strings of the form: *T<num>*, *F<num>*, *I<num>*, and *B<num>*. Here *T* stands for (normal) text, *F* for fixed size, *I* for italic, and *B* for bold, and *num* gives the font size in points. These special names are used by the window class to provide a platform independent way of specifying fonts. For example, "*B14*" specifies a "usual" 14pt bold font of the underlying operating system. Note, however, that, in general, a font specified in this way will look different for different platforms. On Unix systems fonts can also be specified by an X11 font name as for instance `-adobe-helvetica-medium-r-*-*-14-*-*-*-*-*-*-*`.

window coordinates (xmin, xmax, ymin): Parameters of type *double* (default values: (0, 100, 0)) defining the user coordinate space of the window, i.e., *xmin* is the minimal x-coordinate, *xmax* the maximal x-coordinate, and *ymin* the minimal y-coordinate of the drawing area. The maximal y-coordinate *ymax* depends on the shape and size of the drawing area.

grid width: A parameter of type *int* (default value: 0) defining the width of the grid used in the drawing area. A grid width of 0 indicates that no grid is to be used.

grid style: A parameter of type *grid_style* (default value: *point_grid*) defining how a grid is represented in the window. Possible values are *invisible_grid*, *point_grid*, and *line_grid*.

frame label: A parameter of type *string* (default value: LEDA header) defining the frame label of the window that is used by the graphics system or window manager.

icon label: A parameter of type *string* (default value: empty) defining the icon label of the window.

icon pixrect: A parameter of type *char∗* (default value: *NULL*) defining a pixrect (see Section 11.8) that is used as the icon of the window. If it has value *NULL* the default icon is used.

show coordinates: A parameter of type *bool* (default value: *false*) determining whether the current coordinates of the mouse pointer are displayed in the upper right corner of the window.

line width: A parameter of type *int* (default value: 1) defining the width of all kinds of lines (segments, arrows, edges, circles, polygons) in pixels.

line style: A parameter of type *line_style* (default value: *solid*) defining the style of all kinds of lines. Possible styles are *solid*, *dashed*, *dotted*, and *dashed_dotted*.

node width: A parameter of type *int* (default value: 10) defining the diameter of nodes created by the *draw_node* and *draw_filled_node* operations.

text mode: A parameter of type *text_mode* (default value: *transparent*) defining how text is inserted into the window. Possible values are *transparent* and *opaque*.

drawing mode: A parameter of type *drawing_mode* (default value: *src_mode* defining the logical operation that is used for setting pixels in all drawing operations. Possible values are *src_mode* and *xor_mode*. In *src_mode* pixels are set to the respective color value, in *xor_mode* the value is bitwise added to the current pixel value.

clip region: A parameter defining the clipping region of the window, i.e., the region of the window to which drawing operations are applied (default value: the entire drawing area). In the current implementation clip regions are restricted to rectangles (defined by *set_clip_rectangle*) and ellipses (defined by *set_clip_ellipse*).

redraw function: A parameter of type *void* (*∗func*)(*window∗*) (default value: NULL).

Its value is a pointer to a function that is called with a pointer to the corresponding window, whenever a redrawing of the window is necessary, e.g., if the shape of the window is changed or previously hidden parts of the window become visible.

client data: A parameter of type *void∗* (default value: NULL). Its value is an arbitrary pointer value that can be set or read by client applications. In most cases it is used to associate user-defined data with a window for use in *redraw* or other call-back functions.

buttons per line: A parameter of type *int* (default value: ∞) defining the maximal number of buttons in one line of the panel section.

Reading and Changing Parameters: Most parameters may be retrieved or changed by *get* and *set* functions. We use *param* to denote any of the window parameters and *param_t* to denote its type.

```
param_t W.get_param()
```

returns the current value of parameter *param*, and

```
param_t W.set_param(param_t val)
```

sets the value of parameter *param* of type *param_t* to the new value *val* and returns the former value of the parameter.

Here are some simple examples:

```
line_style = W.get_line_style();
int lw = W.get_line_width();

W.set_cursor(XC_dotbox);
W.set_bg_pixrect(leda_pixmap);
W.set_grid_dist(10);
W.set_grid_style(line_grid);
W.set_line_width(1)
W.set_bg_color(ivory)
W.set_color(blue)
W.set_redraw(redraw_func);
```

The fact that the *set*-operation returns the old value of the parameter is very convenient when a parameter is to be changed only temporarily. For instance, in order to change the mouse cursor to a "watch symbol" during the execution of a time consuming operation, one writes:

```
int old_cursor = W.set_cursor(XC_watch);
// some time consuming computation
W.set_cursor(old_cursor);
```

There are a few operations for changing parameters that do not follow the scheme described above, e.g., the *init* operation for changing the user coordinate system that is explained in the next section.

11.5 Window Coordinates and Scaling

We discuss the connection between coordinates and pixels. We use w and h for the width and the height of the drawing section in pixels. Both values are determined by the appearance of a window on the screen. The coordinate system underlying the drawing area is defined by the *init* operation.

```
void W.init(double x0, double x1, double y0, int grid_dist=0);
```

defines the coordinate system underlying the drawing area of W by setting *xmin* to x_0, *xmax* to x_1, and *ymin* to y_0. It also defines implicitly a scaling factor *scaling* and the maximal y-coordinate *ymax* of the drawing area.

$$\mathit{scaling} = w/(\mathit{xmax} - \mathit{xmin}) \quad \text{and} \quad \mathit{ymax} = \mathit{ymin} + h \cdot \mathit{scaling}.$$

If, in addition, a *grid_dist* argument is supplied, it is used to initialize the grid distance of the window. The following function give information about the window coordinates and the scaling factor:

```
double W.xmin()
```

returns *xmin*, the minimal x-coordinate of the drawing area of W, i.e., the coordinate of the left window border in user space. The analogous functions *W.xmax*( ), *W.ymin*( ), and *W.ymax*( ) are also available.

```
double W.scale()
```

returns the scaling factor of the drawing area of W, i.e. the number of pixels of a unit length line segment in user space.

```
double W.pix_to_real(int p)
```

translates pixel distances into user space distances, more precisely, returns the length of a p pixel horizontal or vertical line segment in the user coordinate system.

```
double W.real_to_pix(double d)
```

translates user space distances into pixel distances, more precisely, returns the number of pixels contained in a horizontal or vertical line segment of length d.

11.6 The Input and Output Operators $\ll$ and $\gg$

For the input and output of basic two-dimensional geometric objects of the floating point kernel (*point*, *segment*, *ray*, *line*, *circle*, *polygon*) the $\ll$ and $\gg$ operators can be used. In analogy to C++ input streams, windows have an internal state indicating whether there was more input to read or not. The state is true initially and is turned to false if an input sequence is terminated by clicking the right mouse button (similar to ending stream input by the *eof*-character). In conditional statements, objects of type *window* are automatically converted to boolean by returning this internal state. Thus, window-objects can be used in conditional statements in the same way as C++ input streams. For example, to read a sequence of points terminated by a right button click, use

```
while (W >> p) { .... }
```

The following program uses the $\gg$ operator to read points defined by mouse clicks and draws each point using the $\ll$ operator until input is terminated by clicking the right mouse button.

⟨*draw_points.c*⟩≡

```
#include <LEDA/window.h>
main()
{
  window W(400,400);
  W.display(window::center,window::center);
  point p;
  while (W >> p) W << p;
  W.screenshot("draw_points.ps");
}
```

Graphical input and output for LEDA windows can be extended to user-defined types by overloading the $\ll$ and $\gg$ operators. This is in analogy to C++ stream input and output. For example, *<LEDA/rat_window.h>* contains input and output operators for the objects of the rational kernel.

```
window& operator<<(window& W, const rat_point& p)
{ return W << p.to_point(); }
window& operator>>(window& W, rat_point& p)
{ point q;
  W >> q;
  p = rat_point(q);
  return W;
 }
```

Exercises for 11.6

1 Modify the program *draw_points.c* such that segments (circles, line, or polygons) are echoed. The modified program is supposed to work for only one of the mentioned objects.

2 Write operators $\ll$ and $\gg$ for *rat_polygons*.

11.7 Drawing Operations

The *W* $\ll$ *object* output operators apply to the basic objects of the floating point kernel. The windows class also provides a large number of additional drawing operations that give more flexibility. In this book we can only give a few examples. For the complete list of operations we refer the reader to the LEDA User Manual.

There are two kinds of drawing operations

```
void W.draw_object(coords, color col=window::fg_color);
void W.draw_object(object, color col=window::fg_color);
```

For the first variant, a geometric object is given by its coordinates in the user coordinate system of the window, and for the second variant, the object is given as an object of the floating point kernel. For example,

```
W.draw_circle(double x, double y, double r, color col);
```

draws a circle with center (x, y) and radius r,

```
W.draw_polygon(list<point> P, color col);
```

draws a polygon with vertex sequence P,

```
W.draw_circle(circle C, color col);
```

draws the circle C, and

```
W.draw_polygon(polygon P, color col);
```

draws the polygon P.

The allowed objects are points, pixels, segments, lines rays, ellipses, circles and disks, triangles (unfilled and filled), polygons (unfilled and filled), rectangles and boxes, arcs, Bezier curves, splines, arrows, text, nodes, and edges. The window data type can draw many more types of objects than are available in the geometry kernel. For these types only the first variant exists that takes an explicit coordinate representation as input.

The optional color argument at the end of the parameter list can be used to specify a color that is to be used as foreground color by the operation. If it is omitted the current value of the foreground color parameter (cf. Section 11.4) is used.

The clear operation erases the window by painting it with the background color or tiling it using the background pixrect (if defined).

```
void W.clear();
void W.clear(double x0, double y0, double x1, double y1);
```

The second variant only clears rectangle (x_0, y_0, x_1, y_1).

Exercises for 11.7

1 Write a program that draws a red circle, a green line segment, and a blue filled polygon.

2 Write a program that draws a filled box for each available shading of grey.

11.8 **Pixrects and Bitmaps**

Pixrects and bitmaps are rectangular regions of pixels and bits, respectively.

11.8.1 *Pixrects*

Pixrects (often called pixmaps) are rectangles of pixels of a certain width and height. Each pixel has a color value from the possible set of colors available in the underlying graphics system. In this way pixrects represent rectangular pictures.

There are operations to copy a pixrect into a rectangle of the drawing area of a displayed window of the appropriate size and to construct a pixrect from a rectangle of the drawing area. Pixrects can also be constructed from external representations of pictures stored in *xpm* files or *xpm* data strings. xpm data strings are of type *char* $**$, i.e., they are represented by arrays of C++_ strings. An xpm file contains the (C++) definition of an xpm data string, see Figure 11.5 for an example. For the exact definition of the xpm format we refer the reader to one of the *X11* handbooks or manuals [Nye93]. LEDA provides a small collection of icon pictures stored in xpm files in the *<LEDA/pixmaps/button32>* directory. A typical X11 system provides tools for the construction and manipulation of xpm files.

In the current implementation of LEDA pixrects and bitmaps are not realized by real data types but by pointers (of type *char*$*$). In particular, there is no constructor and destructor, i.e., the user must explicitly create and destroy pixrects or bitmaps by calling *create* and *destroy* operations.

Constructing and Destroying Pixrects: We discuss functions for constructing and destroying pixrects.

```
char* W.create_pixrect(double x0, double y0, double x1, double y1)
```

constructs a pixrect of all pixels contained in the rectangle (x_0, y_0, x_1, y_1) of the drawing area of W and returns it.

```
char* W.get_window_pixrect()
```

constructs a pixrect of all pixels in the drawing area of W and returns it.

```
char* W.create_pixrect(char** xpm)
```

constructs a pixrect from the xpm pixmap data string xpm.

```
char* W.create_pixrect(string xpm_file)
```

constructs a pixrect from the xpm pixmap data in file *xpm_file*.

```
void W.del_pixrect(char* prect)
```

destroys pixrect *prect*.

Drawing Pixrects: We discuss the functions for drawing picrects.

```
void W.put_pixrect(double x, double y, char* prect)
void W.put_pixrect(point p, char* prect)
```

copies the pixels of pixrect *prect* into a rectangle of the drawing area of W which is placed with its left lower corner at the specified position of the drawing area.

```
<<xpm_example_file.h>>=

/* XPM */
static char *example_xpm[] = {
/* width height ncolors chars_per_pixel */
"32 32 6 1",
/* colors */
"` c #000000",
"a c #F5DEB3",
"b c #E6E6FA",
"c c #DBDBDB",
"d c #CC9933",
"e c #FFFFCC",
/* pixels */
"cccccccccccccccccccccccccccccccc",
"cccccccccccccccccccccccccccccccc",
"cccccccccccc````cccccccccccccccc",
"ccccccccc```bebe``c```cccccccccc",
"ccccccc``bebebebeb`beb`c`ccccccc",
"cccccc`ebebebebebebebeb`b`cccccc",
"cccccc`bebebebebeb`bebebeb`ccccc",
"ccccccc`bebe``b```bebebebe`ccccc",
"cccccccc```b```dd``beb`bebe`cccc",
"ccccccccc```ddddddd````ebebe`ccc",
"ccccc``````dddddddddddd`ebeb`ccc",
"cccc``aaa``d`ddd`ddd`dd```be`ccc",
"ccc``aaaa``dd`ddd`ddd`d```eb`ccc",
"ccc`aaa````dd`ddd`ddd`d``eb`cccc",
"ccc`aa`cc``dd`ddd`ddd`d``be`cccc",
"ccc`aa`cc``dd`ddd`ddd`d``eb`cccc",
"ccc`aa`cc``dd`ddd`ddd`d``b`ccccc",
"ccc`aa`cc``dd`ddd`ddd`d``e`ccccc",
"ccc`aa`cc``dd`ddd`ddd`d``b`ccccc",
"ccc`aa`cc``dd`ddd`ddd`d``e`ccccc",
"ccc`aa`cc``dd`ddd`ddd`d``b`ccccc",
"ccc`aa`cc``dd`ddd`ddd`d``e`ccccc",
"ccc`aaa````dd`ddd`ddd`d``b`ccccc",
"ccc``aaaa``dd`ddd`ddd`d``e`ccccc",
"cccc``aaa``dd`ddd`ddd`d```cccccc",
"ccccc``````dd`ddd`ddd`d``ccccccc",
"ccccccccc``d`dd``dd``dd``ccccccc",
"ccccccccc``dddddddddddd``ccccccc",
"ccccccccc````````````````ccccccc",
"cccccccccc``````````````cccccccc",
"cccccccccccccccccccccccccccccccc",
"cccccccccccccccccccccccccccccccc"
};
```

Figure 11.5 A pixrect stored in *xpm* format.

```
void W.center_pixrect(double x, double y, char* prect)
void W.center_pixrect(point p, char* prect)
```

copies the pixels of pixrect *prect* into a rectangle of the drawing area of *W* that is placed with its center at the specified position of the drawing area.

In the following example we construct a pixrect representing the LEDA icon and put it (with its lower left corner) at positions defined by mouse clicks. Figure 11.6 shows a screenshot.

Figure 11.6 A screenshot of the put_pixrect program.

⟨*put_pixrect.c*⟩≡

```
#include <LEDA/window.h>
#include <LEDA/pixmaps/leda_icon.xpm>
main()
{
  window W(400,400);
  W.display();
  char* pr = W.create_pixrect(leda_icon);
  point p;
  while (W >> p) W.put_pixrect(p,pr);
  W.del_pixrect(pr);
  W.screenshot("put_pixrect.ps");
  return 0;
}
```

11.8.2 *Bitmaps*

Bitmaps are pixrects containing pixels of only two possible colors: black and white. The name indicates that each pixels in a bitmap can be represented by a single bit and that is exactly the way bitmaps are usually represented: by a triple (w, h, s), where w and h give the width and height of the bitmap and s is a string of bits (of type *char∗*). A file that contains the (C++) definition of such a string is called a bitmap file. Usually the suffix *xbm* (x bit map) is used for such a file. LEDA provides a small collection of bitmap pictures stored in *xbm* files in the <*LEDA/bitmaps/button32.h*> directory. As for pixmaps there are many programs for constructing and manipulating *xbm* files.

Bitmap Operations:

```
char* W.create_bitmap(int w, int h, char* xbm)
```

creates a bitmap of width w and height h from the bits in the xbm string xbm. The length of xbm must be at least $w \cdot h$ bits, i.e., $\lceil (w \cdot h)/8 \rceil$ characters.

```
void W.put_bitmap(double x, double y, char* bmap, color c)
void W.put_bitmap(point p, char* bmap, color c)
```

places the bitmap *bmap* with its left lower corner at the specified position of the drawing area and draws with color *c* all pixels in the drawing area that correspond to a pixel of *bmap* with value one.

```
void W.del_bitmap(char* bmap)
```

destroys bitmap *bmap*.

The following program is very similar to the last example program but uses a bitmap instead of a pixrect. First, we construct a bitmap representing the LEDA icon and put it (with its lower left corner) at positions defined by mouse clicks.

⟨*bitmap.c*⟩≡

```
#include <LEDA/window.h>
#include <LEDA/bitmaps/leda_icon.xbm>

main()
{
  window W(400,400);
  W.set_bg_color(yellow);
  W.display();

  // construct bitmap from the bitmap data in
  // <LEDA/bitmaps/leda_icon.xbm>

  char* bm = W.create_bitmap(leda_icon_width, leda_icon_height,
                             (char*)leda_icon_bits);

  // copy copies of bm into the window

  point p;
  while (W >> p) W.put_bitmap(p.xcoord(),p.ycoord(),bm,blue);

  W.del_bitmap(bm);

  W.screenshot("bitmap.ps");
  return 0;
}
```

Exercises for 11.8

1 Write a program that converts a bitmap into a pixrect.

2 Construct a pixrect containing your picture.

3 What is shown in the pixrect of Figure 11.5

11.9 Clip Regions

Sometimes it is necessary to limit the effect of a drawing operation to some restricted area, a so-called *clipping region* of the window. The following operations allow us to define clipping regions.

```
void W.set_clip_rectangle(double x0, double y0, double x1, double y1);
```

sets the clipping region to rectangle (x_0, y_0, x_1, y_1).

```
void W.set_clip_ellipse(double x0, double y0, double r1, double r2);
```

sets the clipping region to the ellipse with center (x_0, y_0), horizontal radius r_1 and vertical radius r_2.

```
void W.reset_clipping();
```

resets the clipping region to the entire drawing area of the window.

We give an example for the usefulness of clipping. We show how to fill a circle with a pixrect picture. In this situation, we have to restrict the effect of a *put_pixrect* operation to the interior of this circle. This can be done by defining a corresponding clip-ellipse. Here is the program and the resulting picture (Figure 11.7).

⟨*clip_pixrect.c*⟩≡

```
#include <LEDA/window.h>
#include <LEDA/pixmaps/leda_icon.xpm>

void draw_pix_circle(window& W, const circle& C, char* prect)
{
  point  p = C.center();
  double x = p.xcoord();
  double y = p.ycoord();

  double r = C.radius();

  W.draw_disc(C,black);

  W.set_clip_ellipse(x,y,r,r);
  W.center_pixrect(x,y,prect);
  W.reset_clipping();
}

main()
{
   window W(400,400, "Clipping a Pixmap");
   W.display();

   // create a pixrect using LEDA's xpm icon

   char* leda_pix = W.create_pixrect(leda_icon);

   circle c;
   while (W >> c) draw_pix_circle(W,c,leda_pix);

   W.del_pixrect(leda_pix);

   W.screenshot("clip_pixrect.ps");
   return 0;
}
```

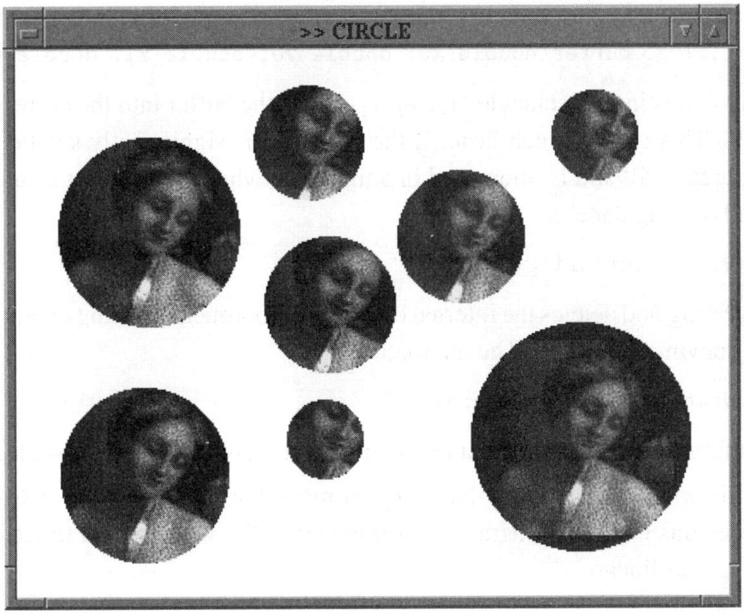

Figure 11.7 A screenshot demonstrating the effect of clip regions.

11.10 Buffering

The default behavior of all drawing operations discussed in the preceding sections is to draw immediately into the drawing area of the displayed window. There are, however, situations where this behavior is not desired, and where it is very useful to construct an entire drawing in a memory buffer before copying it (or parts of it) into the drawing area.

Buffering allows us to draw complex objects, which require several primitive drawing operations, in a single blow. One draws the complex object into a buffer and then copies the buffer to the drawing area. In this way, the illusion is created that the entire object is drawn by a single drawing operation. The ability to draw complex objects in a single operation is frequently needed in *animations*, where one wants to display a sequence of snapshots of a scene that changes over time. Another application of buffering is to create a pixrect copy of a drawing without displaying it in the drawing area. At the end of this section we will give example programs for both applications.

These are the most important buffering operations:

```
void W.start_buffering()
```

starts buffering of window W, i.e, all subsequent drawing operations have no effect in the drawing area of the displayed window, but draw into an internal buffer with the same size and coordinates as the drawing area of W.

```
void W.flush_buffer()
```

copies the contents of the internal buffer into W.

```
void W.flush_buffer(double x0, double y0, double x1, double y1)
```

copies all pixels in the rectangle (x_0, y_0, x_1, y_1) of the buffer into the corresponding rectangle of W. This can be much faster if the rectangle is significantly smaller than the entire drawing area of W and is often used in animations when the drawing changes only locally in a small rectangular area.

```
void W.stop_buffering()
```

stops buffering and deletes the internal buffer; all subsequent drawing operations again draw into the drawing area of W. The alternative

```
void W.stop_buffering(char*& pr)
```

stops buffering and converts the internal buffer into a picrect that is assigned to *pr*.

The following program uses buffering to move the LEDA pixrect ball that was drawn by the previous example program smoothly across the window and to let it bounce at the window border lines.

⟨*buffering1.c*⟩≡

```
#include <LEDA/window.h>
#include <LEDA/pixmaps/leda_icon.xpm>
void move_ball(window& W, circle& ball, double& dx, double& dy,
                                                    char* prect)
{
  ball = ball.translate(dx,dy);
  point  c = ball.center();
  double r = ball.radius();
  if (c.xcoord()-r < W.xmin() || c.xcoord()+r > W.xmax()) dx = -dx;
  if (c.ycoord()-r < W.ymin() || c.ycoord()+r > W.ymax()) dy = -dy;
  W.clear();
  W.set_clip_ellipse(c.xcoord(),c.ycoord(),r,r);
  W.center_pixrect(c.xcoord(),c.ycoord(),prect);
  W.reset_clipping();
  W.draw_circle(ball,black);
}
main()
{
  window W(300,300, "Bouncing Leda");
  W.set_bg_color(grey1);
  W.display(window::center,window::center);
  circle ball(50,50,16);
  double dx = W.pix_to_real(2);
  double dy = W.pix_to_real(1);
  char* leda = W.create_pixrect(leda_icon);
  W.start_buffering();
  for(;;)
  { move_ball(W,ball,dx,dy,leda);
```

```
      W.flush_buffer();
     }
    W.stop_buffering();
    W.del_pixrect(leda);
    W.screenshot("buffering1.ps");
    return 0;
  }
```

We next show how to use buffering to construct a pixrect copy of a drawing. The following program uses an auxiliary window *W1* in buffering mode to create a pixrect picture that is used as an icon for the primary window *W*.

⟨*buffering2.c*⟩≡

```
  #include <LEDA/window.h>
  main()
  {
    window W1(100,100);
    W1.set_bg_color(grey3);
    W1.init(-1,+1,-1);
    W1.start_buffering();
    W1.draw_disc(0,0,0.8,blue);  W1.draw_circle(0,0,0.8,black);
    W1.draw_disc(0,0,0.6,yellow);W1.draw_circle(0,0,0.6,black);
    W1.draw_disc(0,0,0.4,green); W1.draw_circle(0,0,0.4,black);
    W1.draw_disc(0,0,0.2,red);   W1.draw_circle(0,0,0.2,black);
    char* pr;
    W1.stop_buffering(pr);
    window W(400,400);
    W.set_icon_pixrect(pr);
    W.display(window::center,window::center);
    point p;
    while (W >> p) W.put_pixrect(p,pr);
    W.del_pixrect(pr);
    W.screenshot("buffering2.ps");
    return 0;
  }
```

Exercises for 11.10

1 Draw ten random line segments, once without buffering and once with buffering.

2 Extend the "Bouncing LEDA" program, such that the ball is compressed when it hits the boundary of the window.

11.11 **Mouse Input**

The main input operation for reading positions, mouse clicks, and buttons from a window *W* is the operation *W.read_mouse*( ). This operation is blocking, i.e., waits for a button to be pressed which is either a "real" button on the mouse device or a button in the panel section of *W*. In both cases, the number of the selected button is returned. Mouse buttons have predefined numbers *MOUSE_BUTTON*(1) for the left button, *MOUSE_BUTTON*(2) for the middle button, and *MOUSE_BUTTON*(3) for the right button. The numbers of the panel buttons can be defined by the user. If the selected button has an associated action function or sub-window, this function/window is executed/opened (cf. Section 11.14 for details).

There is also a non-blocking input operation *W.get_mouse*( ), it returns the constant *NO_BUTTON* if no button was pressed since the last call of *get_mouse* or *read_mouse*, and there are even more general input operations for reading window events. Both will be discussed at the end of this section.

Read Mouse: The function

```
int W.read_mouse();
```

waits for a mouse button to be pressed inside the drawing area or for a panel button of the panel section to be selected. In both cases, the number n of the button is returned. The number is one of the predefined constants *MOUSE_BUTTON*(i) with $i \in \{1, 2, 3\}$ for mouse buttons and a user defined value (defined when adding the button with *W.button*( )) for panel buttons. If the button has an associated action function, this function is called with parameter n. If the button has an associated window M, M is opened and *M.read_mouse*( ) is returned.

The functions

```
int W.read_mouse(double& x, double& y)
int W.read_mouse(point& p)
```

wait for a button to be pressed. If the button is pressed inside the drawing area, the position of the mouse cursor (in user space) is assigned to (x, y) or p, respectively. If a panel button is selected, no assignment takes place. In either case the operation returns the number of the pressed button.

The following program shows a trivial but frequent application of *read_mouse*. We exploit the fact that *read_mouse* is blocking to stop the program at the statement *W.read_mouse*( ). The user may then leisurely view the scene drawn. Any click of a mouse button resumes execution (and terminates the program).

⟨*read_mouse1.c*⟩≡

```
#include <LEDA/window.h>

main()
{
  window W;
  W.init(-1,+1,-1);
  W.display();

  W.draw_disc(0,0,0.5,red);

  W.read_mouse();
```

```
  W.screenshot("read_mouse1.ps");
  return 0;
}
```

The next program prints the different return values of *read_mouse* for clicks on mouse and panel buttons.

⟨*read_mouse2.c*⟩≡

```
#include <LEDA/window.h>
main()
{
  window W;
  W.button("button 0"); W.button("button 1");
  W.button("button 2"); W.button("button 3");
  int exit_but = W.button("exit");
  W.display();
  for(;;)
  { int but =  W.read_mouse();
    if (but == exit_but) break;
    switch (but) {
     case MOUSE_BUTTON(1): cout << "left button click" << endl;   break;
     case MOUSE_BUTTON(2): cout << "middle button click" << endl; break;
     case MOUSE_BUTTON(3): cout << "right button click" << endl;  break;
     default:  cout << string("panel button: %d",but) << endl;    break;
    }
  }
  W.screenshot("read_mouse2.ps");
  return 0;
}
```

Get Mouse: The functions

```
int W.get_mouse()
int W.get_mouse(double& x, double& y)
int W.get_mouse(point& p)
```

are non-blocking variants of *read_mouse*, i.e., they do not wait for a mouse click, but check whether there is an unprocessed click in the input queue of the window. If a click is available, it will be processed in the same way as by the corresponding *read_mouse* operation. If there is no click, the special button value *NO_BUTTON* is returned.

The following program draws random points. It uses *get_mouse* at the beginning of every execution of the main loop to check whether a mouse button has been clicked or not. If the right button has been clicked the loop is terminated, if the left button has been clicked the drawing area is erased.

⟨*get_mouse.c*⟩≡

```
#include <LEDA/window.h>

random_source& operator>>(random_source& ran, point& p)
{ int x,y;
  ran >> x >> y;
  p = point(x,y);
  return ran;
}

main()
{
  window W(400,400);
  W.display(window::center,window::center);
  W.message("left button: clear     right button: stop");
  random_source ran(0,100);
  int but;
  while ( (but = W.get_mouse()) != MOUSE_BUTTON(3) )
  {
    if (but == MOUSE_BUTTON(1)) W.clear();
    point p;
    ran >> p;
    W.draw_point(p,blue);
  }
  W.screenshot("get_mouse.ps");
  return 0;
}
```

Exercises for 11.11

1 The following lines of code wait for a mouse click.

```
int but;
do but = W.get_mouse(); while (but == NO_BUTTON);
```

What is the difference to $but = W.read\_mouse()$?

2 Write a program that implements the input operator $\ll$ for polygons.

11.12 Events

In window systems like the *X11* or *Windows* system, the communication between input devices such as the mouse or the keyboard and application programs is realized by so-called *events*. For example, if the mouse pointer is moved across a window, the system generates motion events that can be handled by an application program to keep track of the current position of the mouse pointer, or, if a mouse button is clicked, an event is generated that carries the information which button was pressed at what position of the mouse pointer, or, if a key is pressed, a keyboard event is triggered that tells application programs which key was pressed and what window had the input focus, i.e., should receive this character input.

Events are buffered in an *event queue* such that applications can access them in a similar way as character input of a C++ input stream. It is possible to read and remove the next event from this queue, to test whether the queue is empty, and to push events back into the queue.

LEDA supports only a restricted set of events. Each event is represented by a five-tuple with the fields type, window, value, position, and time stamp.

The *type* of an event defines the kind of input reported by this event, e.g., a click on a mouse button or pressing a key on the keyboard. Event types are specified by integers from the enumeration

```
enum {button_press_event, button_release_event, key_press_event,
      key_release_event, motion_event, configure_event, no_event}
```

The *window* of an event specifies the window to which the event refers. This is usually the window under the mouse cursor.

The *value* of an event is an integer whose interpretation depends on the type of the event, e.g., the number of a mouse button for a button press event. See below for a description of the possible values for each event type.

The *position* of an event gives the position of the mouse pointer in the user coordinate system of the window at the time the event occurred.

The *time stamp* of an event is the time of a global system clock at which the event occurred. It is measured in milliseconds.

The following event types are recognized by LEDA and can be handled in application programs:

button_press_event indicates that a mouse button has been pressed. The value of the event is the number of the pressed button. The mouse buttons are numbered *MOUSE_BUTTON*(1), *MOUSE_BUTTON*(2), and *MOUSE_BUTTON*(3).

button_release_event indicates that a mouse button has been released. The value of the event is the number of the released button.

key_press_event indicates that a keyboard key has been pressed down. The value of the event is the character associated with the key or in the case of a special key (such as a cursor or function key) a special key code.

key_release_event indicates that a keyboard key has been released, value as above.

motion_event indicates that the mouse pointer has been moved inside the drawing area. The value of this event is unspecified.

configure_event indicates that the window size has changed.

Blocking Event Input: Similar to the *read_mouse* input operation, there is a *read_event* operation that removes the first event of the system's event queue. This operation is blocking, i.e., if the event queue is empty, the program waits until a new event occurs.

```
int W.read_event(int& val, double& x, double& y, unsigned long& t)
```

waits for an event with window W (discarding all events with a different window field) and returns its type, assigns the value of the event to *val*, its position to (x, y), and the time stamp of the event to t.

```
int  W.read_event(int& val, double& x, double& y,
                            unsigned long& t, int timeout)
```

is similar, but waits (if no event for W is available) for at most *timeout* milliseconds; if no event occurs during this period of time, the special event *no_event* is returned.

The next program implements a click and drag input routine for the definition of rectangles. In its main loop the program waits for a mouse click and stores the corresponding position in a variable p by calling *W.read_mouse*(p). If the right button was clicked, the program terminates. Otherwise, we take p as the first endpoint of the diagonal of the rectangle to be defined, wait until the mouse button is released, say at some position q, and take q as the other endpoint of the diagonal of the rectangle. Waiting for the release of the button is implemented by the inner loop

```
while (W.read_event(val,x,y) != button_release_event) { ... }
```

This loop handles all events of window W and terminates as soon as a *button_release* event occurs. For every event processed the value of the event is assigned to *val* and the position is assigned to (x, y), in particular for motion events, the pair (x, y) keeps track of the position of the mouse pointer in the drawing area of W. In the body of the inner loop we draw the (intermediate) rectangle with diagonal from p to (x, y) as a yellow box with a black border on top of the current drawing. The current drawing is kept as a pixrect *win_buf* and is constructed by a call to *W.get_window_pixrect*( ) before the execution of the inner loop. This allows us to restore the picture without the intermediate rectangles by copying the pixels of *win_buf* into the drawing area (*W.put_pixrect*(*win_buf*)). Of course, *win_buf* has to be destroyed after the inner loop has terminated.

In addition, we use buffering as discussed in Section 11.10, to prevent any flickering effects. Figure 11.8 shows a screenshot.

⟨*event.c*⟩≡

```
#include <LEDA/window.h>
#include <math.h>
int main()
{
  window W(450,500,"Event Demo");
  W.display();
  W.start_buffering();
  for(;;)
  {
    // read the first corner p of the rectangle
    // terminate if the right button was clicked
    point p;
```

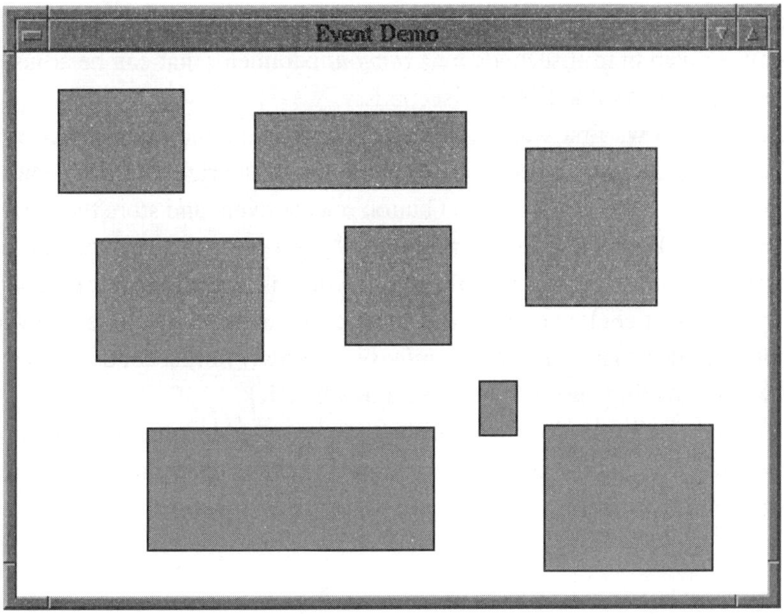

Figure 11.8 A screenshot of the Event Demo.

```
      if (W.read_mouse(p) == MOUSE_BUTTON(3)) break;
      // draw rectangle from p to current position
      // while button down
      int  val;
      double x,y;
      char* win_buf = W.get_window_pixrect();
      while (W.read_event(val,x,y) != button_release_event)
      { point q(x,y);
        W.put_pixrect(win_buf);
        W.draw_box(p,q,yellow);
        W.draw_rectangle(p,q,black);
        W.flush_buffer();
       }
      W.del_pixrect(win_buf);
    }
    W.stop_buffering();
    W.screenshot("event.ps");
    return 0;
}
```

The next example program uses the timeout-variant of *read_event* to implement a function that recognizes *double clicks*. But what is a double click?

A double click is a sequence of three button events, a button press event followed by button release event followed by a second button press event, with the property that the time interval between the two button press events is shorter than a given time limit. Usually, the time limit is given in milliseconds by a *timeout* parameter that can be adjusted by the user. In our example we fix it at 500 milliseconds.

In the program we first wait for a button press event and store the corresponding time stamp in a variable *t_press*. If the pressed button was the right button the program is terminated, otherwise, we wait for the next button release event and store the corresponding time stamp in a variable *t_release*. Now *t_release* − *t_press* gives the time that has passed between the pressing and releasing of the button. If this time is larger than our timeout parameter we know that the next click cannot complete a double click. Otherwise, we wait for the next click but no longer than *timeout* − (*t_release* − *t_press*) milliseconds. If and only if a click occurs within this time interval, we have a double click.

The program indicates double clicks by drawing a red ball and simple clicks by drawing a yellow ball. The middle button can be used to erase the window. Figure 11.9 shows a screenshot of the program.

⟨*dblclick.c*⟩≡

```
#include <LEDA/window.h>

int main()
{
  unsigned long timeout = 500;

  window W(400,400,"Double Click Demo");
  W.set_grid_dist(6);
  W.set_grid_style(line_grid);
  W.display(window::center,window::center);

  for(;;)
  {
    int b;
    double x0,y0,x,y;
    unsigned long t, t_press, t_release;

    while (W.read_event(b,x0,y0,t_press) != button_press_event);
    // a button was pressed at (x0,y0) at time t_press

    // the middle button erases the window
    if (b == MOUSE_BUTTON(2) ) { W.clear(); continue; }

    // the right button terminates the program
    if (b == MOUSE_BUTTON(3) ) break;

    while (W.read_event(b,x,y,t_release) != button_release_event);
    // the button was released at time t_release

    color col = yellow;

    // If the button was held down no longer than timeout msecs
    // we wait for the remaining msecs for a second press, if the
    // the button is pressed again within this period of time we
    // have a double click and we change the color to red.

    if (t_release - t_press < timeout)
    { unsigned long timeout2 = timeout - (t_release - t_press);
      if (W.read_event(b,x,y,t,timeout2) == button_press_event)
        col = red;
```

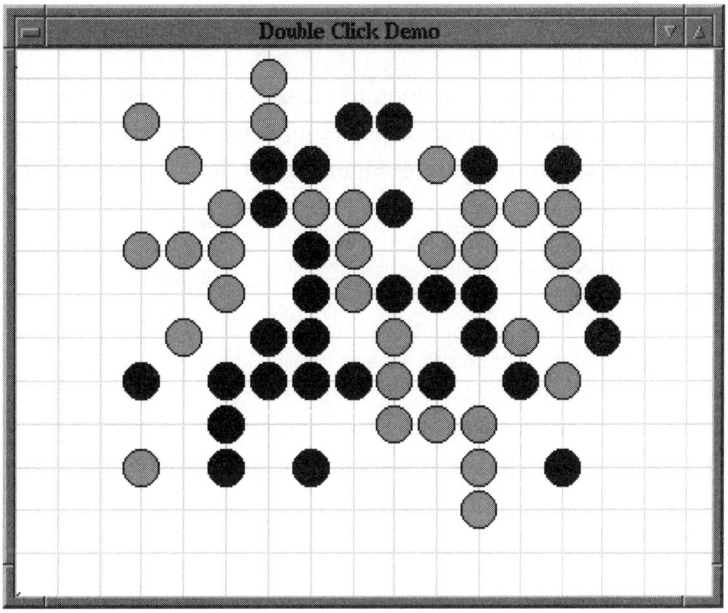

Figure 11.9 A screenshot of the double click program.

```
      }
      W.draw_disc(x0,y0,2.5,col);
      W.draw_circle(x0,y0,2.5,black);
    }
    W.screenshot("dblclick.ps");
    return 0;
  }
```

Putting Back Events: The function[1]

```
void put_back_event();
```

puts the event handled last back to the system's event queue, such that it will be processed again by the next *read_event* or *read_mouse* or basic input operation.

The function is very useful in programs that have to handle different types of input objects using the basic input operators. We give an example. We partition the drawing area of a window into four quadrants and want to draw points in the first, segments in the second, circles in the third, and polygons in the fourth quadrant. The kind of object to be drawn is defined by the position of the first mouse click. The main loop of the program waits for a mouse click and performs, depending on the quadrant that contains the position of this click, the corresponding input and output operation. The difficulty is that already the first

[1] Observe that this function is a global function and not a member function of class *window*.

click that we use to distinguish between the different input objects is part of the definition of the object.

We use the *put_back_event*( ) function to push the first mouse click back into the event queue and to make it available as the first event for the following basic input operator. The details are given in the following code. Figure 11.10 shows a screenshot.

⟨*putback.c*⟩≡

```
#include <LEDA/window.h>
int main()
{
  window W(400,400, "Putback Event Demo");
  W.init(-100,+100,-100);
  W.display(window::center,window::center);
  // partition the drawing area in four quadrants
  W.draw_hline(0);
  W.draw_vline(0);
  for(;;)
  {
    double x,y;
    // wait for first click
    int but = W.read_mouse(x,y);
    // middle button erases the window
    if (but == MOUSE_BUTTON(2))
    { W.clear();
      W.draw_hline(0);
      W.draw_vline(0);
      continue;
     }
    // right button terminates the program
    if (but == MOUSE_BUTTON(3)) break;
    // now we put the mouse click back to the event queue
    put_back_event();
    // and distinguish cases according to its position
    if (x < 0)
      if (y > 0)
        { point p;
          if (W >> p) W.draw_point(p,red);
        }
      else
        { segment s;
          if (W >> s) W.draw_segment(s,green);
        }
    else
      if (y > 0)
        { polygon pol;
          if (W >> pol) W.draw_polygon(pol,blue);
        }
      else
```

Figure 11.10 A screenshot of the putback program.

```
        { circle c;
          if (W >> c) W.draw_circle(c,orange);
        }
  }
  W.screenshot("putback.ps");
  return 0;
}
```

Non-Blocking Event Input: Similar to the non-blocking versions of the *read_mouse* operation, there are non-blocking variants of the *read_event* operation.

```
int  W.get_event(int& val, double& x, double& y)
```

looks for an event for W. More precisely, if there is an event for window W in the event queue, a *W.read_event* operation is performed, otherwise the integer constant *no_event* is returned.

There is also a more general non-member variant that allows us to read events of arbitrary windows.

```
int read_event(window*& wp, int& val, double& x, double& y)
```

waits for an event. When an event occurs, it returns its type, assigns a pointer to the corresponding window to *wp*, the value to *val*, and the position to (x, y).

This version of *read_event* can be used to write programs that can handle events for several windows simultaneously. The following program opens two windows *W1* and *W2*. The main loop reads all events, determines for each event in which of the two windows it occurred, and puts the event back to the systems event queue. If the event occurred in W_1, it reads and draws a point in W_1, if the event occurred in W_2, it reads and draws a segment in *W2* using the basic input and output operators discussed in Section 11.6.

⟨*two_windows.c*⟩≡

```
#include <LEDA/window.h>

main()
{
   window W1(500,500,"Window 1: points");
   W1.display(window::min,window::min);

   window W2(500,500,"Window 2: segments");
   W2.display(window::max,window::min);

   for(;;)
   { window* wp;
     double x,y;
     int val;

     if (read_event(wp,val,x,y) != button_press_event) continue;

     if (val == MOUSE_BUTTON(3)) break;

     put_back_event();

     if (wp == &W1) { point p;   W1 >> p; W1 << p; }
     if (wp == &W2) { segment s; W2 >> s; W2 << s; }
    }
  return 0;
}
```

Exercises for 11.12

1 Write a "click and drag" program for drawing circles.

2 Write a program that displays text written on the keyboard of your computer in a LEDA window.

3 Implement a simple graph editor that can be used to draw the nodes and edges of a graph. Your program should allow you to move a node by clicking on it and dragging it with the mouse to a new position.

11.13 Timers

Each LEDA window has a *timer* clock that can be used to execute periodically a user-defined function. The function and the time interval between two consecutive calls of the function are specified in the start operation

```
void W.start_timer(int msec,void (*func)(window*));
```

A call of this operation starts the timer of *W* and makes it call the function *func* with a pointer to *W* as the actual parameter (*func*(&*W*)) every *msec* milliseconds.

Figure 11.11 A screenshot of the dclock program.

```
void W.stop_timer();
```

stops the timer.

We show the usefulness of timers by writing a simple digital clock demo program. Figure 11.11 shows a screenshot of the clock.

⟨*dclock.c*⟩≡

```
#include <LEDA/window.h>
#include <time.h>
void display_time(window* wp)
{
  window& W = *wp;
  // get the current time
  time_t clock;
  time(&clock);
  tm* T = localtime(&clock);
  // and display it (centered in W)
  double  x = (W.xmax() - W.xmin())/2;
  double  y = (W.ymax() - W.ymin())/2;
  W.clear();
  W.draw_ctext(x,y,string("%2d:%02d:%02d",
                         T->tm_hour,T->tm_min,T->tm_sec));
}
int main()
{
  window W(150,50, "dclock");
  W.set_bg_color(grey1);
  W.set_font("T32");
  W.set_redraw(display_time);
  W.display(window::center,window::center);
  W.start_timer(1000,display_time);
  W.read_mouse();
  W.screenshot("dclock.ps");
  return 0;
}
```

Exercises for 11.13

1 Implement an analog clock.

2 Write a program that draws randomly colored balls are random times.

11.14 The Panel Section of a Window

The panel section of a window is used for displaying text messages and for updating the values of variables. It consists of a list of panel items and a list of panel buttons. We discuss panel items and panel buttons in separate subsections.

11.14.1 *Panel Items*

A panel item consists of a string label and an associated variable of a certain type. The value of this variable is visualized by the appearance of the item in the window (e.g. by the position of a slider) and can be manipulated through the item (e.g. by dragging the slider with the mouse) during a *read_mouse* or *get_mouse* operation.

There are five types of items. Figure 11.12 shows the representation of the items in a panel. It also shows some menu buttons at the bottom of the panel. The program that generates this panel can be found in *LEDAROOT/demo/win/panel_demo.c*.

Text items have only an associated string, but no variable. The string is formatted and displayed in the panel section of the window.

Simple items have an associated variable of type *int*, *double*, and *string*. The item displays the value of the variable as a string. The value can be updated in a small sub-window by typing text and using the cursor keys. For string items there exists a variant called *string menu item* that in addition displays a menu from which strings can be selected.

Choice items have an associated variable of type *int* whose possible values are from an interval of integers $[0..k]$. With every value i of this range there is a choice string s_i associated. These strings are arranged in a horizontal array of buttons and the current value of the variable is displayed by drawing the corresponding button as pressed down and drawing all other buttons as non-pressed (if the value of the variable is out of the range $[0..k]$ no button is pressed). The value of the variable is set to i by pressing the button with label s_i. Pressing a button will release the previously pressed button. It is tempting to confuse the semantics of the string s_i with the integer i. LEDA will not hinder you to use the string "seven" for the third button. Pressing the button with name "seven" will assign 3 to the variable assigned with the button.

For *multiple choice items* the state (pressed or unpressed) of the button with label s_i indicates the value of the i-th bit in the binary representation of the integer value of the associated variable. Multiple choice buttons allow several buttons to be pressed at the same time. For example, the value of the variable associated with the item named "multiple choice" in Figure 11.12 is $1 \cdot 2^0 + 0 \cdot 2^1 + 1 \cdot 2^2 + 1 \cdot 2^3 + 0 \cdot 2^4 = 13$.

In both cases there exist variants that use bitmaps $b_0, \ldots, b_k$ instead of strings to label the choice buttons. Furthermore, there are special choice items for choosing colors (*color_item*) and line styles (*line_style_item*).

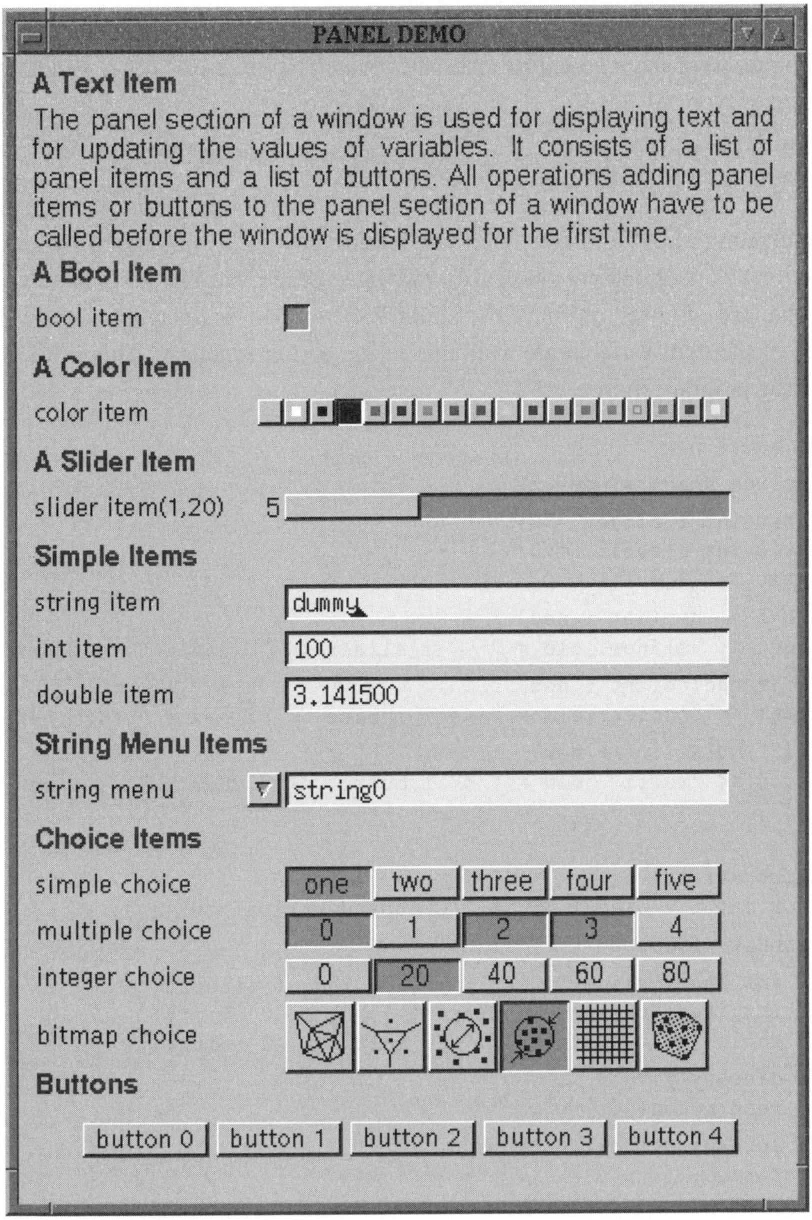

Figure 11.12 Panel items and buttons.

Slider items have associated variables of type *int* with values from an interval [*low* .. *high*]. The current value is shown by the position of a slider in a horizontal box. It can be changed by moving the slider with the mouse.

Boolean items are used for variables of type *bool*. They consist of a single small button whose state (pressed or unpressed) represents the two possible values (*true* or *false*).

We discuss the operations for adding panel items to a panel in Section 11.14.4. It is possible to associate a so-called *call-back* or *action* function with a panel item. This is a function of type

```
void (*action)(T x)
```

where T is the type of the variable of the item. The action function is called after each item manipulation (e.g. dragging a slider or pressing down a choice button) with the *new* value of the item as its argument. However, the value of the variable associated with the item is only changed *after* the return of the action function. In this way, both the old and the new values of the item variable are available in the action function. This is very useful as the following program shows.

⟨*callback.c*⟩≡

```
#include <LEDA/window.h>
static int i_slider = 0;
static int i_choice = 0;
static int i_multi = 0;
void f_slider(int i_new)
{ cout << "slider: old = " << i_slider << ", new = " << i_new << endl; }
void f_choice(int i_new)
{ cout << "choice: old = " << i_choice << ", new = " << i_new << endl; }
void f_multi(int i_new)
{ cout << "multi:  old = " << i_multi  << ", new = " << i_new << endl; }
main()
{
  list<string> L;
  for(int i = 0; i < 8; i++) L.append(string("%d",i));
  window W(300,300);
  W.int_item("slider", i_slider, 0, 100, f_slider);
  W.int_item("choice", i_choice, 1, 8, 1, f_choice);
  W.choice_mult_item("multi", i_multi, L, f_multi);
  W.display();
  W.read_mouse();
  W.screenshot("callback.ps");
  return 0;
}
```

In the main program we define three panel items, each with an associated action function. In each case the action function prints the old value and the new value of the variable. The slider item has a range $[0..100]$, the choice item has eight buttons with associated values 1 to 8 (the smallest value is one, values are increased by one, and the largest value is no larger than eight), and the multiple choice item has eight buttons labeled with strings "0", "1", ..., "7". The button with label i represents the i-th bit of variable *i_multi*.

An action function associated with a panel item of a window W may obtain a pointer to W by calling the static member function *window::get_call_window*( ).

The program below implements a simple color definition panel. It uses three slider items for adjusting the (r, g, b)-values of the color. With each slider a call-back function is associated that paints the window background with the current color. A screenshot is shown in Figure 11.13.

⟨*defcolor.c*⟩≡

```
#include <LEDA/window.h>
static int r,g,b;
void slider_red(int x){window::get_call_window()->clear(color(r,g,b));}
void slider_green(int x){window::get_call_window()->clear(color(r,g,b));}
void slider_blue(int x)window::get_call_window()->clear(color(r,g,b));}
int main()
{
  window W(320,300,"define color");
  color col = green2;
  col.get_rgb(r,g,b);
  W.int_item("red  ",r,0,255,slider_red);
  W.int_item("green",g,0,255,slider_green);
  W.int_item("blue ",b,0,255,slider_blue);
  W.set_bg_color(col);
  W.display(window::center, window::center);
  W.read_mouse();
  W.screenshot("defcolor.ps");
  return 0;
}
```

The values of item variables may also be changed in the program. This has *no* effect on the display until the panel is redrawn for the next time. The *redraw_panel* operation redraws the panel area.

We use a simple progress indicator as an example. It uses a slider item to visualize the increasing value of a counter. Figure 11.14 shows a screenshot.

⟨*progress.c*⟩≡

```
#include <LEDA/window.h>
main()
{
  int count = 0;
  window W(400,100);
  W.set_item_width(300);
  W.int_item("progress",count,0,1000);
  W.display(window::center, window::center);
  for(;;)
  { count = 0;
    while (count < 1000)
    { W.redraw_panel();
      W.flush();
```

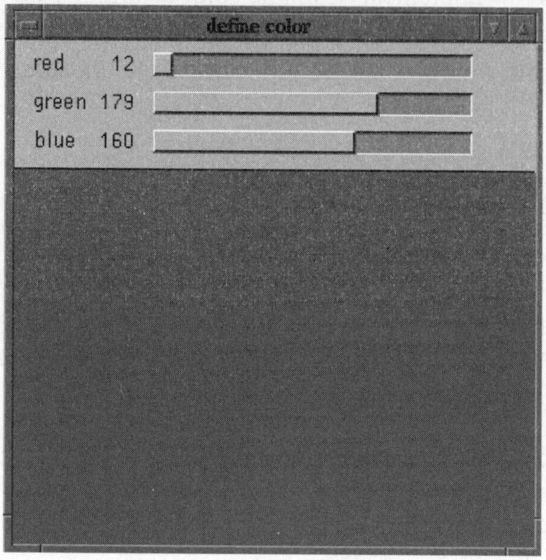

Figure 11.13 A screenshot of the defcolor program.

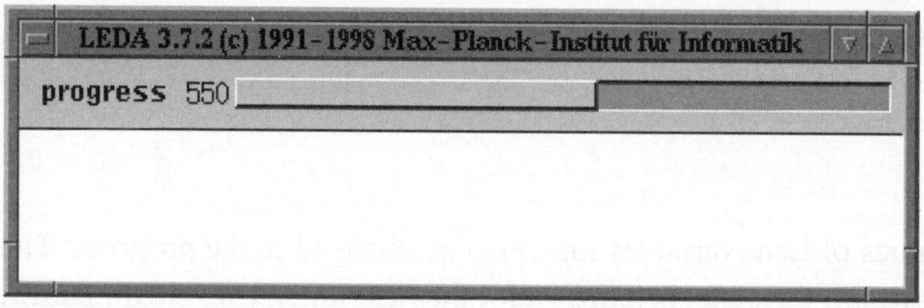

Figure 11.14 A screenshot of the progress program.

```
        leda_wait(0.05);
        count++;
      }
     if (W.read_mouse() == MOUSE_BUTTON(3)) break;
    }
   W.screenshot("progress.ps");
   return 0;
 }
```

11.14.2 *Panel Buttons*

Panel buttons are special panel items. They can be pressed by clicking a mouse button when the mouse pointer is positioned inside their area. Pressing a panel button during a *read_mouse* or *get_mouse* call has the same effect as pressing a mouse button in the drawing area: the operation terminates and the number of the pressed button is returned.

Each panel button has a label or a pixrect image (displayed on the button) and an associated number. The number of a button is either defined by the user or is the rank of the button in the list of all buttons. If a button is pressed (i.e. selected by a mouse click) during a *read_mouse* operation its number is returned. Buttons can have *action functions* of type

```
void (*action)(int but)
```

Whenever a button with an associated action function is pressed this function is called with the number of the button as its actual parameter.

Instead of an action function, a button may have an attached sub-window, in which case we call it a *menu button* (since in most cases such a sub-window is used to realize a menu). Whenever a menu button is pressed the attached sub-window (or menu) M will open and the result of *M.read_mouse*( ) will be returned by the currently active *read_mouse* operation. Of course, M again can have menu buttons, ...

11.14.3 *Panels and Menus*

The data types *panel* and *menu* are two special types representing windows that have no drawing area. Panels (windows of type *panel*) support all panel operations of the general *window* type described in the following section. In addition, they have a special *P.open*( ) operation that displays a panel P, executes *P.read_mouse*( ), closes P, and returns the result of the *read_mouse* operation. There are variants of the *open* operation allowing us to pass parameters for the (initial) positioning of the panel (see the *display* operations for windows for an explanation).

```
int P.open(int xpos=window::center, int ypos=window::center);
int P.open(window& W, int xpos=window::center, int ypos=window::center);
```

Menus (windows of type *menu*) are special panels that only consist of a vertical array of buttons. They support only one kind of panel operation, the addition of buttons, and can be used as sub-windows attached to (menu) buttons only.

11.14.4 *Adding Panel Items*

The operations in this section add panel items or buttons to the panel section of W. Note that they have to be called before the window is displayed the first time.

All operations return a pointer to the corresponding panel item (type *panel_item*)

The generic interface of an operation for adding a panel item (of kind *XXX_item*) for a variable x of type T is as follows:

```
panel_item W.XXX_item(string label, T x&, void (*action)(T) );
```

The last parameter is optional. We give some examples. In all examples we use ... to indicate the optional action function argument.

Simple Items: The following functions add simple items with name *s* and associated variable *x*.

```
panel_item W.bool_item(string s, bool& x, ...);
panel_item W.double_item(string s, double& x, ...);
panel_item W.int_item(string s, int& x, ...);
panel_item W.string_item(string s, string& x, ...);
panel_item W.color_item(string s, color& x, ...);
```

String Menu Items: The functions

```
panel_item W.string_item(string s, string& x, list<string> L, ...);

panel_item W.string_item(string s, string& x, list<string> L,int h,...);
```

add string menu items with name *s*, associated variable *x*, and a menu list *L* of candidate values for *x*. The first version displays the strings of *L* in a rectangular table of appropriate size. The second version uses a scroll box of height *h* with a vertical slider that can be used to scroll through the list.

Choice Items: The functions

```
panel_item W.int_item(string s, int& x, int l, int h, int step);
panel_item W.choice_item(string s, int& x, const list<string>& L, ...);
panel_item W.choice_item(string s, int& x, int n, int w, int h,
                                                    char** bm, ...);
panel_item W.choice_mult_item(string s, int& x,
                                      const list<string>& L, ...);
panel_item W.choice_mult_item(string s, int& x, int n, int w,
                                           int h, char** bm, ...);
```

define choice and multi-choice items with name *s* and associated variable *x*. The first variant defines a choice item with buttons l, $l + step$, ..., the second variant defines a choice item whose buttons are labeled by the strings in *L*, the third variant defines a choice item with *n* buttons each of which is labeled by a bitmap of width *w* and height *h* (*bm* is the array that contains the bitmaps). The fourth and fifth variant are analogous to the second and third variant, but define multi-choice items instead of choice items.

Slider Items: The function

```
panel_item W.int_item(string s, int& x, int l, int h);
```

adds a slider item with name *s*, associated variable *s*, and range $[l .. h]$.

11.14.5 *Adding Buttons*

The following operations add buttons to the panel section of a window. Note that buttons are always positioned at the bottom of the panel area. There are three basic kinds of buttons: buttons with string labels, buttons with bitmaps, and buttons with pixrects.

String Buttons:

```
int W.button(string label, int n);
```

adds a new button to W with label s and number n.

```
int W.button(string label);
```

adds a new button to W with label s and number equal to its rank in the list of all buttons.

```
int W.button(string s, int n, void *(F)(int));
```

adds a button with label s, number n, and action function F to W. Function F is called with actual parameter n whenever the button is pressed.

```
int W.button(string s, void (*F)(int));
```

adds a button with label s, number equal to its rank, and action function F to W. Function F is called with the value of the button as argument whenever the button is pressed.

```
int W.button(string s, int n, window& M);
```

adds a button with label s, number n, and attached sub-window (menu) M to W. Window M is opened whenever the button is pressed.

```
int W.button(string s, window& M);
```

adds a button with label s and attached sub-window M to W. The number returned by *read_mouse* is the number of the button selected in sub-window M.

Bitmap Buttons: Bitmap buttons are labeled with bitmaps instead of string labels. Each bitmap button has an associated bitmap (w, h, bm) that is specified in the operation for adding the button (see below). There exist the same variants (with and without a user-defined number, with action function or with sub-window) as for string buttons.

```
int W.button(int w, int h, char* bm, string s, int n);
int W.button(int w, int h, char* bm, string s);
int W.button(int w, int h, char* bm, string s, int n, void (*F)(int));
int W.button(int w, int h, char* bm, string s, void (*F)(int));
int W.button(int w, int h, char* bm, string s, int n, window& M);
int W.button(int w, int h, char* bm, string s, window& M);
```

The following program creates the panel shown in Figure 11.15.

⟨*bm_buttons.c*⟩≡

```
#include <LEDA/window.h>
#include <LEDA/bitmaps/button32.h>

int main()
{
  panel P("Bitmap Buttons");
  P.buttons_per_line(8);
  P.set_button_space(3);

  for(int i=0; i < num_button32; i++)
    P.button(32,32,bits_button32[i],string(name_button32[i]));
```

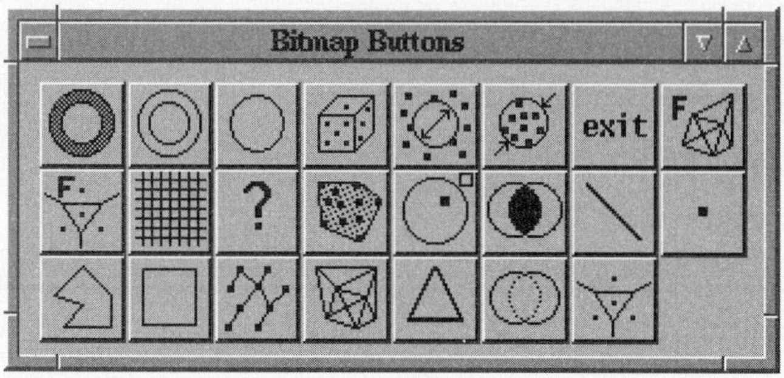

Figure 11.15 A screenshot of the bm_buttons program.

```
  int button = P.open();
  P.screenshot("bm_buttons.ps");
  return 0;
}
```

Pixrect Buttons: Pixrect buttons are labeled with pixrects instead of string labels. Each button has two pixrects, the first one (*pr1*) is used for unpressed buttons and the second (*pr2*) is used for pressed-down buttons. Again we have the same variants as for string buttons.

```
int  button(char* pr1, char* pr2, string s, int n);
int  button(char* pr1, char* pr2, string s);
int  button(char* pr1, char* pr2, string s, int n, void (*F)(int));
int  button(char* pr1, char* pr2, string s, void (*F)(int));
int  button(char* pr1, char* pr2, string s, int n, window& M);
int  button(char* pr1, char* pr2, string s, window& M);
```

The following program creates the panel shown in Figure 11.16. For simplicity, we have used the same pixrect for unpressed and pressed buttons.

⟨*pm_buttons.c*⟩≡

```
#include <LEDA/window.h>
#include <LEDA/pixmaps/button32.h>
int main()
{
  panel P("Pixrect Buttons");
  P.buttons_per_line(10);
  P.set_button_space(3);
  for(int i = 0; i < num_button32; i++)
  { char* pr = P.create_pixrect(xpm_button32[i]);
    P.button(pr,pr,name_button32[i],i);
  }
  int button = P.open();
```

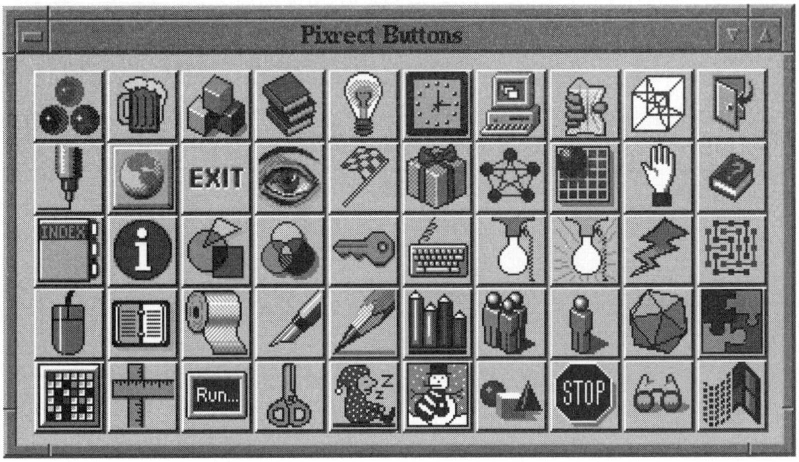

Figure 11.16 A screenshot of the pm_buttons program.

```
  P.screenshot("pm_buttons.ps");
  return 0;
}
```

Creating a Menu Bar: There are two styles for menu buttons, i.e., buttons with an attached sub-window. In the default style menu buttons are displayed as buttons with an additional menu-sign. In the second style the menu buttons are arranged into a menu bar at the top of the panel section. Figure 11.17 shows both styles. The call

```
void W.make_menu_bar()
```

selects the menu button style.

The following program and the screenshots in Figure 11.17 demonstrate both alternatives. With the command line argument “menu bar”, the menu bar version is chosen.

⟨*menu_bar.c*⟩≡

```
#include <LEDA/window.h>
int main(int argc, char** argv)
{
  menu M;
  M.button("button 1"); M.button("button 2"); M.button("button 3");
  M.button("button 4"); M.button("button 5");
  window W(400,300,"Menu Demo");
  W.button("File",M); W.button("Edit",M); W.button("Help",M);
  W.button("exit");
  if (argc > 1 && string(argv[1]) == "menu_bar") W.make_menu_bar();
  W.display();
  W.read_mouse();
```

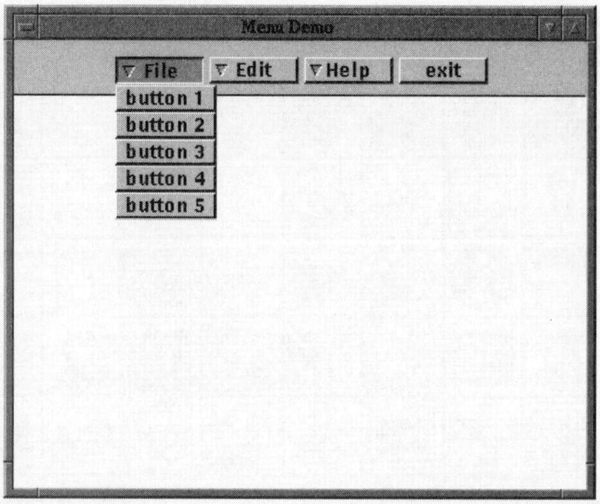

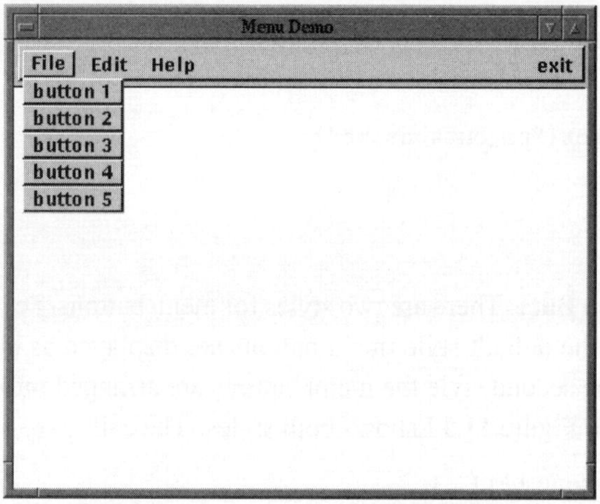

Figure 11.17 Menu buttons: The upper screenshot shows the default style and the lower screenshot shows the menu bar style.

```
    W.screenshot("menu_bar.ps");
    return 0;
}
```

Exercises for 11.14

1. Implement a simple desk calculator with a graphical input.
2. Implement quicksort and use a panel to monitor the values of all variables.
3. Implement a simple file viewer program with a menu bar containing a "File" menu with operations for loading and saving text, and an "Option" menu for defining global parameters such as the font and color of the text.

11.15 Displaying Three-Dimensional Objects: d3_window

The data type *d3_window* uses a LEDA window to visualize and animate three-dimensional drawings of graphs. If the graph to be shown is a planar map (as in the following application) the faces are drawn in different grey scales.

The following program uses a *d3_window* to visualize the convex hull of a set of three-dimensional points. Figure 11.1 at the beginning of this chapter shows a screenshot of the program *LEDAROOT/demo/geo/d3hull_demo.c* which expands on the program below.

The convex hull algorithm

```
CONVEX_HULL(const list<d3_rat_point>& L, GRAPH<d3_rat_point,int>& H)
```

takes a list L of three-dimensional points and constructs the surface graph H of their convex hull. H is a planar map that is embedded into three-dimensional space.

To visualize this graph we create a d3-window *d3win*, whose constructor takes a window W (that has to be displayed before), the graph H, and a node array *pos* of vectors that gives for every node v of H the position $H[v]$ of v in space as a three-dimensional vector.

Finally, we call *d3win.read_mouse*( ) that does something very similar to the *read_mouse* operation for (two-dimensional) windows. It waits for a mouse click and returns the number of the mouse button pressed. While waiting for a click, the graph H is shown in a two-dimensional projection and is, depending on the current position of the mouse pointer, rotated in space. If H is a planar map (as it is in this case), the d3-window, in addition, computes its faces and paints them in different grey scales.

There are many parameters for controlling the appearance of the graph, e.g., whether faces should be painted as described above, for the center and speed of rotation, for changing colors of nodes and edges, ... For details, we refer the reader to the user manual.

⟨*d3_hull.c*⟩≡

```
#include <LEDA/d3_hull.h>
#include <LEDA/d3_window.h>
main()
{
  // construct a random set of points L
  list<d3_rat_point> L;
  random_d3_rat_points_in_ball(50,75,L);
  // construct the convex hull H of L
  GRAPH<d3_rat_point,int> H;
  CONVEX_HULL(L,H);
  // open a window W
  window W(400,400,"d3 hull demo");
  W.init(-100,+100,-100);
  W.display(window::center,window::center);
  // extract the node positions into an array of vectors
  node_array<rat_vector> pos(H);
  node v;
  forall_nodes(v,H) pos[v] = H[v].to_vector();
  // and display H in a d3_window for window W
```

```
    d3_window d3win(W,H,pos);
    d3win.read_mouse();
    W.screenshot("d3_hull.ps");
    return 0;
}
```

Exercise for 11.15

1 Extend the 3d convex hull program by adding a panel section to the window that allows you to choose between different types of input points and to specify the size of the input point set. Your window should look like the window of Figure 11.1.

12

GraphWin

The *GraphWin* data type combines the *graph* and the *window* data type. An object of type *GraphWin* (short: a GraphWin) is a window, a graph, and a drawing of the graph, all at once. The graph and its drawing can be modified either by mouse operations or by running a graph algorithm on the graph. The GraphWin data type can be used to:

- construct and display graphs,
- visualize graphs and the results of graph algorithms,
- write interactive demos for graph algorithms,
- animate graph algorithms.

All demos and animations of graph algorithms in LEDA are based on GraphWin, many of the drawings in this book have been made with GraphWin, and many of the geometry demos in LEDA have a GraphWin button that allows us to view the graph structure underlying a geometric object.

In this chapter we discuss GraphWins and teach the reader the use of GraphWin. We give an overview and discuss the interactive interface of GraphWins. Next we discuss the node and edge attributes and the global parameters that control how graphs are displayed. In the remaining section we discuss the programming interface of GraphWins and show how to write demos using GraphWins. You will see that it is surprisingly simple to write nice demos of graph algorithms.

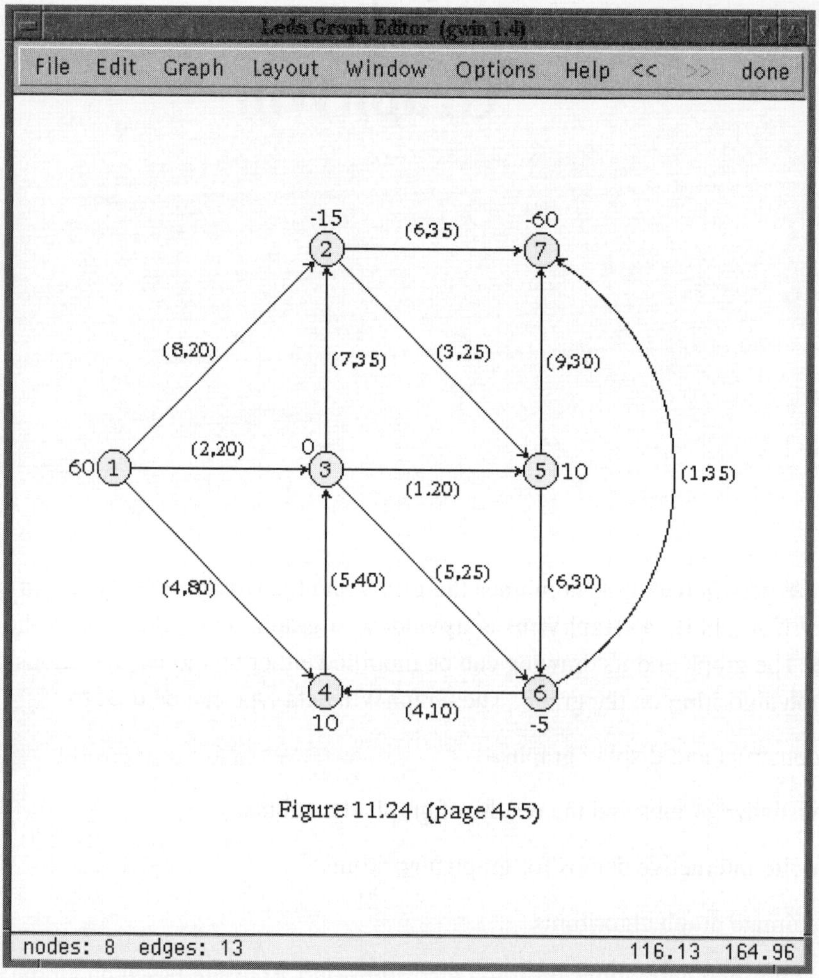

Figure 12.1 A GraphWin. The display part of the window shows a graph *G* and the panel part of the window features the default menu of a GraphWin. We discuss the default menu in Section 12.1. *G* can be edited interactively, e.g., nodes and edges can be added, deleted, and moved around. It is also possible to run graph algorithms on *G* and to display their result or to animate their execution.

12.1 Overview

Figure 12.1 shows a GraphWin. We advise that you open a GraphWin before reading on, e.g., by starting the program gwin in xlman. A window as shown in Figure 12.1 will pop up, but with an empty display region. Press the Help button to learn about the interactive use of GraphWin and then construct a graph.

Most of the interaction is with the *left mouse button*. A *single click on the background*

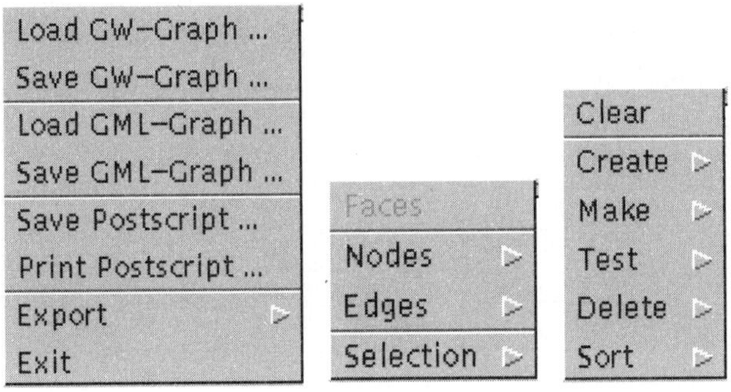

Figure 12.2 GraphWin: File, Edit, and Graph menus.

creates a new node. A *single click on a node* selects the node as the source of a new edge. The next click defines the target of the edge which is either an existing node or a new node (if clicked on the window background). Before defining the target node, bends may be introduced using the middle button. The creation of the new edge can be canceled by clicking the right button.

Nodes can be moved by dragging. Select the node with the left mouse button, hold the button down, and drag the object by moving the cursor. Simultaneously pressing a SHIFT key will move the connected component containing the node. The entire graph can be moved by selecting the background. Of course, when a node is moved all edges incident to it will move with it.

A node is *resized* by clicking on its boundary and dragging the border line of the node.

A *double click* on a node or edge opens a dialog box for setting or changing its attributes. We will discuss the geometric and visual attributes of nodes and edges in Section 12.2.

For the functionality of the middle and the right mouse button we refer the reader to the help menu of GraphWin. Please construct and edit a graph before reading on.

Let us next have a look at the default menu of a GraphWin . We have menu buttons "File", "Edit", "Graph", "Layout","Window", "Options", "Help", "undo (≪)", "redo (≫)", and "done". The first six buttons give access to sub-menus as shown in Figures 12.2 and 12.3. We next briefly discuss all buttons and the associated menus.

File: A menu that offers file I/O operations for graphs in either of two formats, allows one to export drawings of graphs, and contains the *exit* button (see Section 12.3 for the effect of the *exit* button).

Edit: A menu with panels for setting the (default) attributes of nodes and edges.

Graph: A menu that offers graph generators, modifiers, and checkers. The generators allow us to construct random, planar, complete, bipartite, grid graphs, ... The modifiers change the current graph (e.g., by removing or adding edges) to make it connected, biconnected, bidirected, ... The checkers can be used to check graph properties, like con-

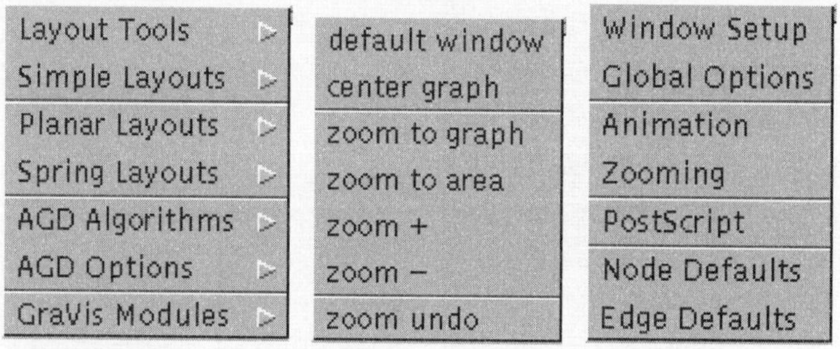

Figure 12.3 GraphWin: Layout, Window and Option menus.

nectedness, biconnectedness, and planarity. Figure 12.4 shows the output of the planarity test for a graph that is non-planar. Many of the checkers can be asked for a proof by clicking the *proof* button. In the case of the planarity test this will either generate a planar drawing or highlight a Kuratowski subgraph as shown in Figure 12.5.

Layout: A menu that gives access to tools for simple layout manipulations (e.g., removing all edge bends or fitting the graph into a box or window) and a collection of graph drawing algorithms. If the graph drawing systems AGD [JMN] or GraVis [Lau98] are installed, their layout algorithms are included into the menu as shown in Figure 12.3.

Window: A menu with (zoom) operations for changing the user space of the drawing window, e.g., the *zoom graph* button adjusts the window coordinates to the bounding box of the current graph.

Options:A menu with various sub-panels for editing the various window and editor parameters.

undo ($\ll$):A button to undo the last update operation.

redo ($\gg$): A button to undo the undo.

done: The done button, see Section 12.3.

The drawing of a graph in a GraphWin is controlled by node and edge attributes and by global parameters. We discuss attributes and parameters in the next section.

Exercises for 12.1

1 Call a GraphWin, construct a graph, and test whether it is biconnected.

2 Construct a graph and then change all node shapes from circular to rectangular.

3 Construct the dependency graph for the chapters of this book as shown in the preface. Apply some of the layout algorithms to the graph.

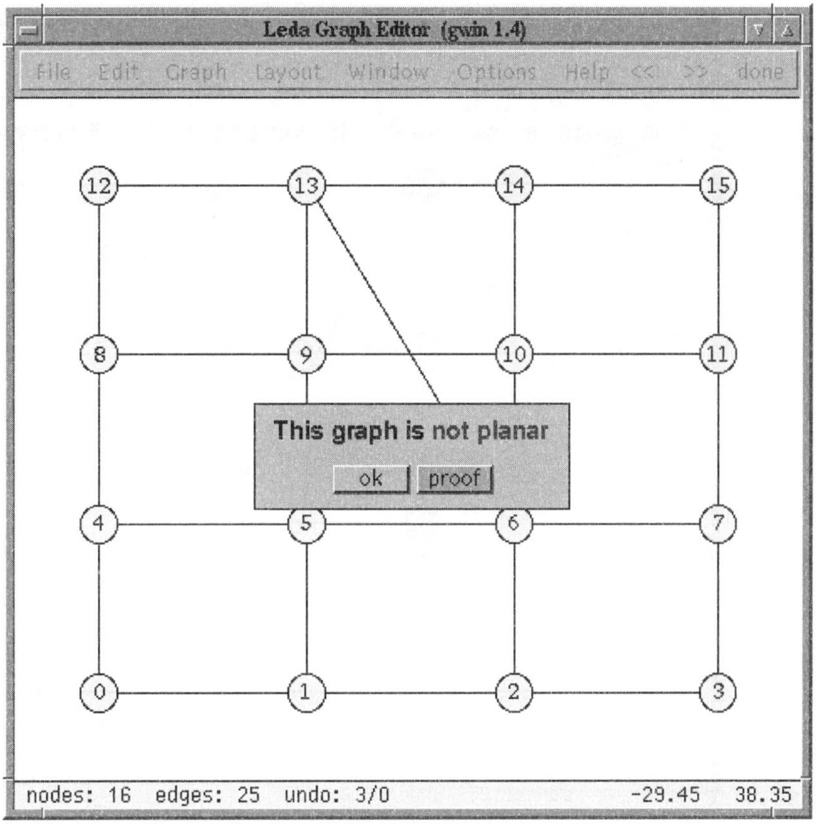

Figure 12.4 An outcome of a planarity test.

12.2 Attributes and Parameters

In this section we discuss global parameters and node and edge attributes. The node and edge attributes control how nodes and edges are drawn and the global parameters control the general behavior of a GraphWin. Attributes and parameters can be changed either by setup panels (as shown in Figures 12.6 and 12.7) or by operations of the programming interface as discussed in Section 12.3.

Node Attributes: The node attributes are:

position:An attribute of type *point* (default value: (0, 0)) defining the position of the center of the node in the user coordinate system of the window.

shape:An attribute of type *gw_node_shape* (default value: *circle_node*) defining the shape of the node. Possible values are *circle_node*, *ellipse_node*, *square_node*, and *rectangle_node*. The size of a node is determined by its width and its height. Width and height are measured

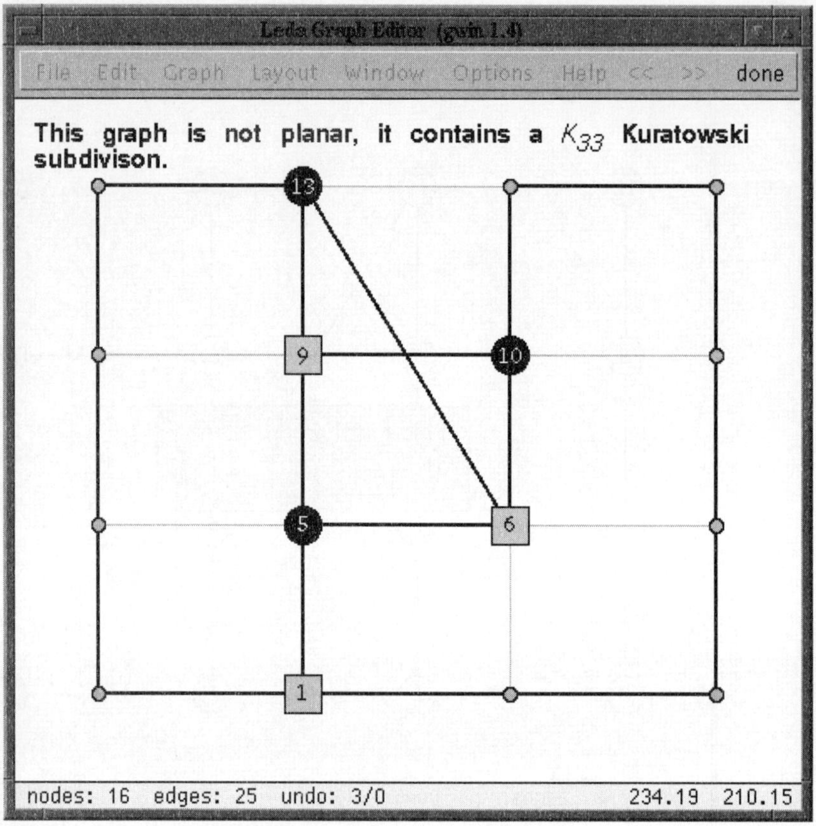

Figure 12.5 The effect of clicking the proof button in Figure 12.4.

in pixels. The horizontal and vertical dimension can also be measured in user space; we use *radius1* and *radius2* for the dimensions in user space.

width:An attribute of type *int* (default value: 20) defining the width of the node in pixels. The horizontal dimension of a node is also available as an attribute with name *radius1* that gives the horizontal dimension of the node in user space. Any change of one of these two attributes also changes the other, maintaining the relation $radius1 = W.pix\_to\_real(width)/2$.

height:An attribute of type *int* (default value: 20) defining the height of the node in pixels. As for the *width* attribute the vertical dimension of a node can be accessed or changed through a *radius2* attribute giving the vertical dimension of the node in user space.

color:An attribute of type *color* (default value: *ivory*) defining the color used to fill the interior of the node.

pixmap:An attribute of type *char*∗ (default value: *NULL*) defining a pixrect used to fill the interior of the node.

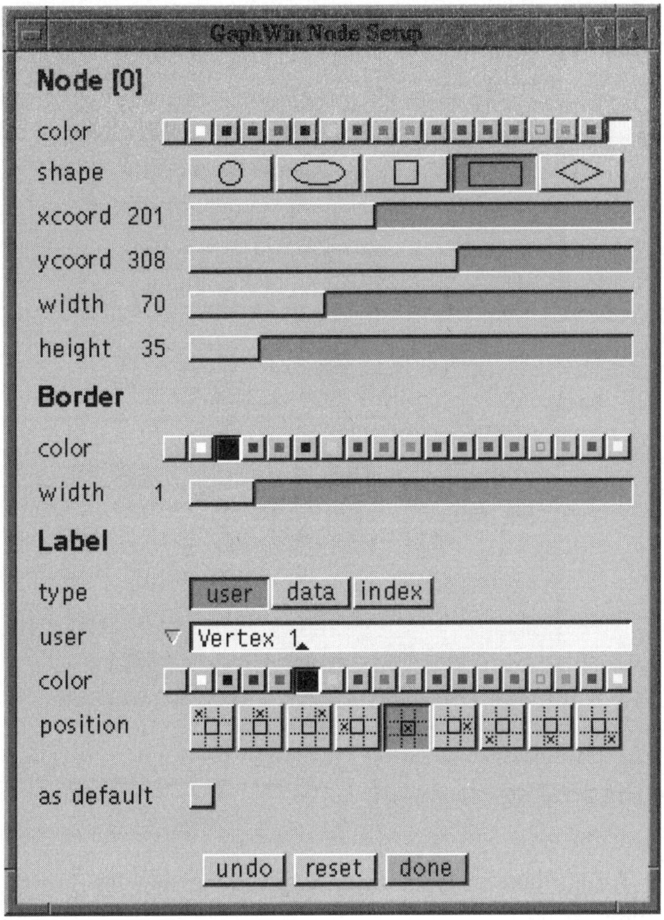

Figure 12.6 The node setup panel.

border color:An attribute of type *color* (default value: black) defining the color used to draw the boundary line of the node.

border width: An attribute of type *int* (default value: 1) defining the line width in pixels used to draw the border line of the node. We also have a user space variant of this attribute called *border thickness*: *border_width* and *border_thickness* are related through the equation $border\_thickness = W.pix\_to\_real(border\_width)$.

label type:An attribute of type *gw_label_type* (default value: *index_label*) specifying which label of a node is displayed. Possible values are *no_label*, *user_label*, *data_label*, and *index_label*. Every node of a GraphWin has three labels associated with it: an index label generated automatically from the internal numbering of the nodes, a user label (of type *string*), and a data label that is used to represent the node data of parameterized graphs.

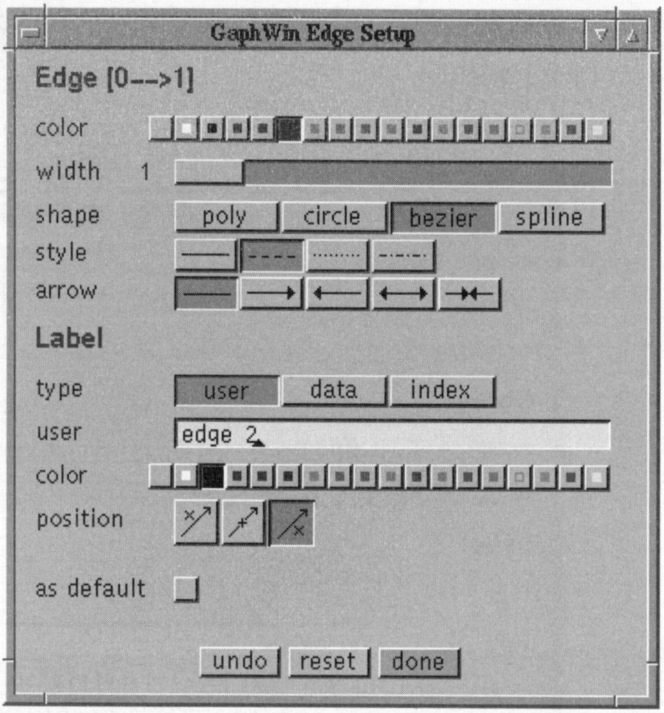

Figure 12.7 The edge setup panel.

user label: An attribute of type *string* defining the user label of the node. The default value is the empty string.

label position: An attribute of type *gw_position* (default value: *central_pos*) defining the position of the label. Possible values are *central_pos*, *northwest_pos*, *north_pos*, *northeast_pos*, *east_pos*, *southeast_pos*, *south_pos*, *southwest_pos*, and *west_pos*. Each value defines one of the eight neighboring cells in a rectangular grid of appropriate dimension or the node position itself as the position of the label.

label color: An attribute of type *color* (default value: *black*) defining the color used to draw the label of the node.

Edge Attributes: Edges have the following attributes:

shape:An attribute of type *gw_edge_shape* (default value: *poly_edge*) defining the shape of the edge. Possible values are *poly_edge* (polygonal edges), *circle_edge* (circular arcs), *bezier_edge* (Bezier curves), *spline_edge* (spline curves).

bends:An attribute of type *list<point>* (default value: empty list) defining the sequence of bends of the edge. The interpretation of the bends depends on the shape of the edge. For *poly_edge* this list defines the sequence of bends of the poly-line. For *circle_edge* only the first point p of the sequence is used. Together with the two terminal node positions it defines a circular arc starting at the source position, passing through p and ending in the target position of the edge. For *bezier_edge* and *spline_edge* edges the list gives the sequence of control points that define the corresponding Bezier or spline curve.

direction:An attribute of type *gw_edge_dir* defining whether the edge is drawn as a directed or an undirected edge. Possible values are *undirected_edge* (the edge is drawn undirected), *directed_edge* (the edge is drawn directed from source to target), *redirected_edge* (the edge is drawn directed from target to source), and *bidirected_edge* (the edge is drawn bidirected).

width:An attribute of type *int* (default value: 1) defining the width of the edge in pixels. The width of an edge can also be specified by an attribute called *thickness* that gives the line width of the edge in user coordinates; *thickness* and *width* are related through $thickness = W.pix\_to\_real(width)$.

color:An attribute of type *color* (default value: *black*) defining the color of the edge.

style:An attribute of type *gw_edge_style* (default value: *solid*) defining the line style of the edge. Possible values are *solid*, *dashed*, *dotted*, and *dashed_dotted*.

label type:An attribute of type *gw_label_type* (default value: *no_label*) defining the type of the label of the edge. Possible values are *no_label*, *user_label*, *data_label*, and *index_label* (see the corresponding attribute for nodes for an explanation).

user label: An attribute of type *string* defining the user label of the edge. The default value is the empty string.

label position: An attribute of type *gw_position* (default value: *west_pos*) defining the position of the label. Possible values are *central_pos* (the label is placed centered on the edge), *east_pos* (the label is placed to the right of the edge), and *west_pos* (the edge is placed to the left of the edge).

label color: An attribute of type *color* (default value: *black*) defining the color of the edge label.

slider positions:Every edge has three *sliders* associated with it. They are only visible if the corresponding handler (see Section 12.5.1) is defined. For each slider the *slider position* is an attribute of type *double* (default value: 0) defining the relative position of the slider on the (directed) edge. The value of slider position lies between zero and one. Edge sliders can be used to adjust the value of an edge label interactively.

Global Parameters: A *GraphWin* has the *window* parameters background color, background pixmap, grid style, and grid distance, and the following additional parameters.

flush: A parameter of type *bool* (default value: *true*) that controls whether changes of node and edge attributes are shown directly or not. If *flush* is false, changes are invisible up to the next call of the *redraw* operation. In this way, it is possible to hide all intermediate steps of a sequence of operations and to show only the end result.

animation steps: A parameter of type *int* (default value: 16) that defines the number of intermediate drawings used in the animation of layout changes and zoom operations. Setting animation steps to 0 disables all animations.

zoom objects: A parameter of type *bool* (default value: true). If this flag is true, the size of nodes and edges is adjusted automatically during zoom operations. If the flag is false, the pixel width and height of all objects is preserved during zoom operations.

show status:A parameter of type *bool* (default value: true). If this flag is true, some selected parameters, e.g., the number of nodes and edges and the current position of the mouse cursor in user coordinates, is shown in a status line at the bottom of the display region.

12.3 The Programming Interface

So far we have concentrated on the interactive interface of GraphWins, as most LEDA users will become acquainted with GraphWins through their interactive use. We now turn to the programming interface. You must read this section if you want to write programs that use a GraphWin.

The *GraphWin* data type offers a large variety of operations. We discuss the most important one in the remainder of this chapter and refer the reader to the manual for the complete list of operations.

12.3.1 *Creating and Opening a Graph Window*

A GraphWin has an associated graph and an associated window. Either one of them may or may not be specified in the constructor.

```
GraphWin gw;
```

creates a graph window *gw* that uses its own (private) graph G and window W. G is initialized with the empty graph. Three optional arguments may be passed to initialize W: a *label* of type *string*, the initial *width*, and the initial *height* both of type *int*.

```
GraphWin gw(graph& G);
```

creates a graph window *gw* and associates the graph G with it. G may also be a parameterized graph of some type *GRAPH<vtype, etype>*. In this case, every node v has an additional *data label* attribute that contains a string representation of the *vtype* value $G[v]$ associated with v. This representation is constructed using the stream output operator (*operator* $\ll$ (*ostream&*, *const vtype&*)). In the same way, every edge e has a data label representing $G[e]$. In Section 12.6 we give a program that uses *GraphWin* to display a graph of type *GRAPH<point, int>* representing a Delaunay triangulation.

```
GraphWin gw(window& W);
GraphWin gw(graph& G, window& W);
```

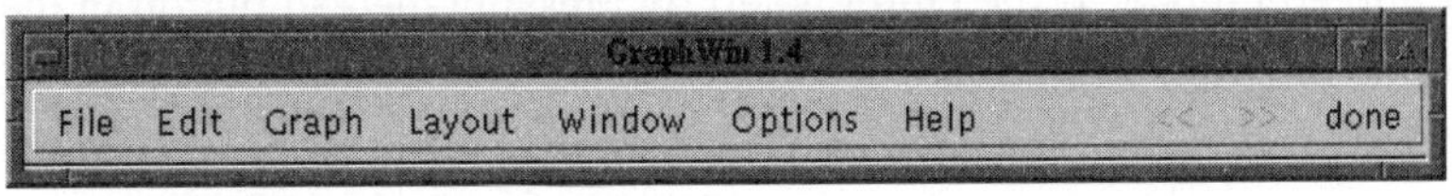

Figure 12.8 A GraphWin panel for editing a graph in a different window.

do not create their own window but use the supplied window *W* for displaying the graph. In this case, the display operation opens a small panel window (see Figure 12.8) containing only the standard menu.

References to the graph and window of a *GraphWin gw* can be retrieved by

```
window& W = gw.get_window();
```

and

```
graph&  G = gw.get_graph();
```

respectively.

A graph window is opened and displayed by calling one of the two following *display* operations:

```
gw.display()
```

opens *gw* and displays it at the default position of data type *window* and

```
gw.display(x,y)
```

opens *gw* and displays it with its left upper corner at the position with pixel coordinates (x, y). As for windows the special coordinate *window::center* can be used to center the graph window in either coordinate on the screen.

The interactive interface is started by the *edit* operation.

```
bool gw.edit();
```

puts *gw* into *edit mode* (also called *interactive mode*). The buttons of *gw* are now enabled; in particular the graph associated with *gw* may now be changed interactively. The edit session is terminated when either the *done* button is pressed or *exit* is selected from the file menu. The edit operation returns *true* in the first case and *false* in the second case.

We are now ready for the first example program. We declare and display a graph window *gw*, and then start an edit loop (*while* (*gw.edit*( ))) that lets the user construct or modify the graph *G* associated with *gw*. If *edit* is terminated by pressing the *done* button, the graph is tested for planarity, the outcome of the planarity test is written to standard output, and *gw* is again put into edit mode. If the editor is left by pressing the *exit* button in the *file* menu, the loop and the program terminate.

⟨*gw.c*⟩≡

```
#include <LEDA/graphwin.h>
#include <LEDA/graph_alg.h>
main()
{
  GraphWin gw("Leda Graph Editor");
  graph& G = gw.get_graph();
  gw.display(window::center,window::center);
  while ( gw.edit() )
  { if (PLANAR(G))
      cout << "This graph is planar." << endl;
    else
      cout << "This graph is non-planar." << endl;
   }
  return 0;
}
```

The structure of the program above is generic for many simple interactive demos of graph algorithms. The program runs in a loop. In each iteration the graph is edited and the graph algorithm is run. We call this scheme the *edit-and-run paradigm* for interactive demos. We will see a more elaborate use of the paradigm in Section 12.4.

12.3.2 *Graph Operations*

A GraphWin has an associated graph. There are two methods to update this graph through the programming interface.

The first method uses the update operations offered by GraphWin. For example,

```
node gw.new_node(const point& p);
```

creates a new node v with default attributes. The position of v is set to p.

```
void gw.del_node(node v);
```

removes v from the graph,

```
edge gw.new_edge(node v, node w);
```

creates a new edge $e = (v, w)$ with default attributes,

```
void gw.del_edge(edge e);
```

removes e from the graph, and

```
void gw.clear_graph();
```

makes the graph empty.

The second method reuses the update operations for graphs. We obtain a reference to the graph associated with *gw* by calling *gw.get_graph*( ) and then apply graph update operations to it.

```
graph& G = gw.get_graph();
// some update operations on G
G.new_node();
G.del_edge(e);
gw.update_graph();                    // CRUCIAL
```

Observe the *gw.update_graph*( ) statement at the end of the sequence. This statement informs *gw* about the fact that its graph was modified and allows it to update its internal data structures. Without the statement the graph *G* and the internal data structures of *gw* will go out of sync and disasters may occur.

We illustrate the use of *update_graph* operation by giving an implementation of the *new_node* operation of GraphWin; the actual implementation is different and more efficient.

```
node gw_new_node(GraphWin& gw, const point& p)
{ graph& G = gw.get_graph();
  node v = G.new_node();
  gw.update_graph();
  gw.set_position(v,p);
  return v;
}
```

12.3.3 *Attribute and Parameter Operations*

Attributes of nodes and edges and global parameters are manipulated by *get* and *set* operations. In the case of attributes we distinguish between the individual attributes of existing nodes and edges and the *default attributes* which are used to initialize the attributes of new nodes and edges.

Individual Attributes of Nodes and Edges: The attributes of existing nodes and edges can be retrieved or changed by the following operations. We use *object* for either *node* or *edge* and *attrib* (of type *attrib_type*) for an arbitrary attribute.

```
attrib_type  gw.get_attrib(object x);
```

returns the current value of attribute *attrib* of object x.

```
attrib_type  gw.set_attrib(object x, attrib_type a);
```

sets the attribute *attrib* of object x to a and returns the previous value of the attribute.

```
void  gw.set_attrib(list<object>& L, attrib_type a);
```

sets attribute *attrib* for all objects in L to a.

```
void gw.reset_attributes();
```

resets the attributes of all objects to their default values.

The current attributes of all nodes and edges may be saved and restored later to the saved values by the following functions.

```
void gw.save_node_attributes();
void gw.save_edge_attributes();
void gw.restore_node_attributes();
void gw.restore_edge_attributes();
```

These functions are very useful if the appearance of the graph has to be changed temporarily, e.g., to highlight a substructure of the graph.

We give an example. We replace all nodes of elliptic shape by yellow rectangular nodes and all blue edges by black dashed edges, wait five seconds, and then restore all attributes to their original values.

```
graph& G = gw.get_graph();
void gw.save_node_attributes();
void gw.save_edge_attributes();

node v;
forall_nodes(v,G) {
 if (gw.get_shape(v) == ellipse_node)
 { gw.set_shape(v,rectangle_node);
   gw.set_color(v,yellow);
  }
}
edge e
forall_edge(e,G) {
 if (gw.get_color(e) == blue)
 { gw.set_style(e,dashed);
   gw.set_color(e,black);
  }
}
gw.redraw();
leda_wait(5);
void gw.restore_node_attributes();
void gw.restore_edge_attributes();
```

Default Attribute Values: Every attribute has a default value which is used to initialize the attributes of new objects. The default attribute values can be changed by the following operations. Note that changing a default attribute also affects all existing objects, unless the optional boolean flag *apply* in the corresponding *set_node_attrib* operation is set to *false*.

```
attrib_type  get_node_attrib();
attrib_type  set_node_attrib(attrib_type x, bool apply=true);
```

reads or sets the default value of node attribute *attrib*. If *apply* is true, the *attrib* attribute of all existing nodes is changed in the same way.

```
attrib_type  get_edge_attrib();
attrib_type  set_edge_attrib(attrib_type x, bool apply=true);
```

reads or sets the default value of edge attribute *attrib*. If *apply* is true, the *attrib* attribute of all existing edges is changed in the same way.

The current default values of all attributes can be saved to a file and later reloaded by the following operations.

```
void gw.save_defaults(string fname);
void gw.read_defaults(string fname);
```

We close with an example. We declare a GraphWin *gw*, change the default values of some attributes, open the window associated with *gw*, and put *gw* into edit mode.

⟨*gw_attributes.c*⟩≡

```
#include <LEDA/graphwin.h>
main()
{
  GraphWin gw;
  // default attributes of nodes
  gw.set_node_shape(rectangle_node);
  gw.set_node_color(yellow);
  // default attributes of edges
  gw.set_edge_width(2);
  gw.set_edge_color(blue);
  gw.set_edge_direction(undirected_edge);
  gw.display();
  gw.edit();
}
```

Almost every program using a GraphWin starts with a small preamble that changes default attributes to settings that are appropriate for the application.

Global Parameters: Global parameters can be retrieved or changed by a collection of *get*- and *set*-operations. We use *param_type* for the type and *param* for the value of the corresponding parameter.

There is a *get* and *set* operation for each global parameter *param*.

```
param_type  gw.get_param();
param_type  gw.set_param(param_type x);
```

The set operation returns the previous value of the corresponding parameter.

In the following example we set the *flush* parameter to *false* before changing the individual attributes of some nodes. Then we redraw the graph to display the changes and reset the *flush* parameter to its previous value;

```
gw.set_animation_steps(12);
bool fl = gw.set_flush(false);
forall(v,L) {
 gw.set_color(v,blue);
```

```
    gw.set_shape(v,rhombus_node);
  }
  gw.redraw();
  gw.set_flush(fl);
```

12.3.4 *I/O Operations*

GraphWin supports two file formats for the permanent storage of graphs and their attributes, the (native) gw-format and the GML-format [Him97] of Himsolt. It can also generate a Postscript representation of the current drawing that can easily be included into LaTeX documents. Many of the figures of this book have been produced in this way. The operations in this section are available in the file-menu.

The read operations

```
int  gw.read_gw(istream& istr);
int  gw.read_gw(string fname);
int  gw.read_gml(istream& istr);
int  gw.read_gml(string fname);
```

clear the current graph and read a new graph and its attributes from the input stream *istr* or file *fname*, respectively. The operations return 0 on success and a special error code if something goes wrong (see the manual for details). The write operations

```
int gw.save_gw(ostream& ostr);
int gw.save_gw(string fname);
int gw.save_gml(ostream& ostr);
int gw.save_gml(string fname);
```

write the current graph and its layout to output stream *ostr* or to file *fname*, respectively. The operations return 0 on success and a non-zero error code if something goes wrong.

Postscript representations of drawings are generated by

```
bool gw.save_ps(ostream& ostr);
bool gw.save_ps(string fname);
```

which write the current drawing as a Postscript file to output stream *ostr* or to file *fname*, respectively.

12.3.5 *Layout Operations*

We discuss operations for manipulating the *layout* of the graph associated with a GraphWin, i.e., the positions of the nodes and the sequence of bends of the edges. The operations are, for example, used to realize the functions in the layout-menu.

The arguments of the layout operations specify new node positions and/or new sequences of bends. The layout operation moves the nodes and changes the drawings of edges accordingly. The animation of the layout operations (and also of the zooming operations) is controlled by the *animation_steps* parameter. *GraphWin* animates changes in the layout by linear interpolation. It shows a sequence of *animation_steps* intermediate layouts, where each node and edge moves a fraction of $1/animation\_steps$ of its total movement in each step. If *animation_steps* is set to zero, the layout change is performed instantaneously.

In most layout operations the new node position can be specified either as *points* or as pairs of *doubles*. We list both versions for the first layout function and only one for the others. The operations

```
void gw.set_position(const node_array<double>& xpos,
                     const node_array<double>& ypos);
void gw.set_position(const node_array<point>& pos);
```

move every node v of *gw* from its old position to position ($xpos[v]$, $ypos[v]$) or $pos[v]$, respectively, and leave the bends of all edges unchanged,

```
void gw.set_layout(const node_array<point>& pos,
                   const edge_array<list<point> >& bends);
```

moves every node v to position $pos[v]$ and sets the bend sequence of every edge e to $bends[e]$,

```
void gw.set_layout(const node_array<point>& pos);
```

moves every node v of the graph to position $pos[v]$ and removes all edge bends from the layout,

```
void gw.remove_bends();
```

removes all bends from the layout and leaves the node positions unchanged,

```
void gw.place_into_box(double x0, double y0, double x1, double y1);
```

moves the graph into the rectangular box $(x0, y0, x1, y1)$ by scaling and translating the layout, and

```
void gw.place_into_win();
```

moves the graph into the drawing window by scaling and translating.

Layout coordinate computations: Consider the following situation. We have a graph window *gw* and its associated graph G. We have computed a new layout for G, but the new layout does not conform to the coordinate space of *gw*. We want to adjust the layout data before applying it. Section 12.4 gives an application.

The operations in this section are very helpful in this situation. They apply the transformations *place_into_box* and *place_into_win* to the layout data supplied separately in node and edge arrays.

```
void gw.adjust_coords_to_box(node_array<double>& xpos,
                             node_array<double>& ypos,
                             edge_array<list<double> >& xbends,
                             edge_array<list<double> >& ybends,
                             double x0, double y0, double x1, double y1);
```

transforms the layout given by *xpos*, *ypos*, *xbends*, and *ybends* in the same way as a call *place_into_box*(*x0*, *y0*, *x1*, *y1*) would do. However, the actual layout of the current graph is not changed by this operation.

```
void gw.adjust_coords_to_box(node_array<double>& xpos,
                             node_array<double>& ypos,
                             double x0, double y0, double x1, double y1);
```

transforms the layout given by *xpos*, *ypos* as *gw.place_into_box*(*x0*, *y0*, *x1*, *y1*) would do. It ignores any edge bends. The actual layout of the current graph is not changed by this operation.

```
void gw.adjust_coords_to_win(node_array<double>& xpos,
                             node_array<double>& ypos,
                             edge_array<list<double> >& xbends,
                             edge_array<list<double> >& ybends);
```

calls *adjust_coords_to_box*(*xpos*, *ypos*, *xbends*, *ybends*, *wx0*, *wy0*, *wx1*, *wy1*) with the current window rectangle (*wx0*, *wy0*, *wx1*, *wy1*). Finally,

```
void gw.adjust_coords_to_win(node_array<double>& xpos,
                             node_array<double>& ypos);
```

calls *adjust_coords_to_box*(*xpos*, *ypos*, *wx0*, *wy0*, *wx1*, *wy1*), where as in the preceding operation (*wx0*, *wy0*, *wx1*, *wy1*) is the current window rectangle.

12.3.6 *Zoom Operations*

Zoom operations change the coordinate system of the window but do not change the layout of the graph. A zoom operation is a combination of a stretch or shrink transformation (changing the scaling factor of the window) with a translation of the window in user space. The *animation step* parameter specifies the number of intermediate window positions to be shown in the animation of the zoom operation; if the parameter is zero the zoom is performed instantaneously.

```
void gw.zoom(double f)
```

zooms the window by the factor f; this multiplies the scaling factor by f and leaves the coordinates of the center of the window unchanged.

```
void gw.zoom_area(double x0, double y0, double x1, double y1)
```

zooms the window to rectangle (*x0*, *y0*, *x1*, *y1*). More precisely, if the aspect ratio of the zoom rectangle $r = (y1 - y0)/(x1 - x0)$ is equal to the aspect ratio wr of the current window, the window coordinates are set to (*x0*, *y0*, *x1*, *y1*). Otherwise, if r is smaller than wr the new window coordinates are $(x0, y0, x1, y')$ with $y' = y0 + wr * (x1 - x0)$ and if r is greater than wr the new coordinates are $(x0, x', y0, y1)$ with $x' = x0 + (y1 - y0)/wr$.

```
void gw.center_graph()
```

performs a zoom operation that does not change the scaling of the window and moves the center of the bounding box of the current graph layout to the center of the window.

```
void gw.zoom_graph();
```

calls *gw.zoom_area*(*x0*, *y0*, *x1*, *y1*) such that $x0$, $x1$, and $y0$ are the left, right and lower coordinates of the bounding box of the current layout of the graph.

12.3.7 *Miscellaneous Operations*

We close our discussion of the programming interface with a list of small, but useful functions.

```
void gw.message(string msg);
```

displays *msg* at the top of the window. If *msg* is the empty string, the previous message is deleted.

```
bool gw.wait(const msg);
```

displays *msg* and waits until the done-button is pressed or exit is selected from the file menu. The result of the operation is *true* in the first case and *false* in the second case.

```
int gw.open_panel(panel& P)
```

displays panel P centered on *gw* and returns the result of *P.open*( ). During the execution of *P.open*( ) all menus of *gw* are disabled.

```
node gw.ask_node();
```

asks the user to select a node by clicking with the left mouse button on it. The selected node is returned; *nil* is returned if the click does not hit a node.

```
edge gw.ask_edge();
```

asks the user to select an edge by clicking with the left mouse button on it. The selected edge is returned; *nil* is returned if the click does not hit an edge.

```
void gw.get_bounding_box(double& x0, double& y0, double& x1, double& y1);
```

computes the coordinates (*x0*, *y0*, *x1*, *y1*) of a minimal area rectangular bounding box containing the current layout of the graph.

12.4 Edit and Run: A Simple Recipe for Interactive Demos

We implement a simple demo that illustrates planarity testing based on the edit-and-run paradigm for interactive demos of graph algorithms. The demo illustrates many of the functions discussed in the preceding sections.

We define a GraphWin *gw* with frame label "Planarity Test Demo" and open it. We then enter the edit-loop. After each edit operation, we run the graph algorithm on the graph G associated with *gw* and display the result.

⟨*gw_plandemo.c*⟩≡

```
#include <LEDA/graphwin.h>
#include <LEDA/graph_alg.h>
⟨plandemo: highlight⟩
int main()
{
  GraphWin gw("Planarity Test Demo");
  gw.display(window::center,window::center);
  while (gw.edit())
  {
    graph& G = gw.get_graph();
    ⟨run graph algorithm and display result⟩
  }
  return 0;
}
```

So far the program is generic (except for the frame label). We now come to the part specific to the planarity demo.

We test G for planarity. If G is planar and has at least three nodes (otherwise the current drawing is already without crossings), we compute a straight line embedding and display it. The computation of the straight line embedding returns the coordinates of a straight line embedding in some coordinate system. We adjust the coordinates to the coordinate space of *gw* by calling *adjust_coords_to_win*. Finally, we display the straight line embedding by calling *gw.set_layout*(...).

If the graph is non-planar, we compute a Kuratowski subdivision $K = (V_k, E_k)$ and display it by calling the *high_light* function. We wait until the user clicks done and then restore the old drawing. The function KURATOWSKI computes the set of nodes and edges of the subdivision and for each node of G the degree of the node in the subdivision. For all $v \in V$ the degree $deg[v]$ is equal to 2 for subdivision points, 4 for all other nodes if K is a K_5, and -3 ($+3$) for the nodes of the left (right) side if K is a $K_{3,3}$.

⟨*run graph algorithm and display result*⟩≡

```
if (PLANAR(G))
{ if (G.number_of_nodes() < 3) continue;
  node_array<double> xcoord(G);
  node_array<double> ycoord(G);
  STRAIGHT_LINE_EMBEDDING(G,xcoord,ycoord);
  gw.adjust_coords_to_win(xcoord,ycoord);      // !!!
  gw.set_layout(xcoord,ycoord);
}
else
{ list<node> V_k;
  list<edge> E_k;
```

```
    node_array<int> kind(G);
    KURATOWSKI(G,V_k,E_k,kind);
    gw.save_all_attributes();
    highlight(gw,V_k,E_k,kind);
    gw.wait("This Graph is not planar. I show you a\
             Kuratowski Subdivision (click done).");
    gw.restore_all_attributes();
  }
```

We still have to define the function *highlight* that highlights the Kuratowski subgraph. We set *flush* to false at the beginning of *highlight* and call *redraw* and restore the old value of *flush* at the end. This ensures that all changes made by *highlight* will become effective at the same time.

We highlight the Kuratowski subgraph by drawing its edges with width two and black (all other edges are drawn grey and with width one) and by using color and shape codes to highlight its nodes. Figure 12.9 shows an example.

⟨*plandemo: highlight*⟩≡

```
void highlight(GraphWin& gw, list<node> V, list<edge> E,
                                 node_array<int>& kind)
{
  const graph& G = gw.get_graph();
  bool flush0 = gw.set_flush(false);
  node v;
  forall_nodes(v,G) {
    switch (kind[v]) {
      case  0: gw.set_color(v,grey1);
               gw.set_border_color(v,grey1);
               gw.set_label_color(v,grey2);
               break;
      case  2: gw.set_color(v,grey1);
               gw.set_label_type(v,no_label);
               gw.set_width(v,8);
               gw.set_height(v,8);
               break;
      case  3:
      case  4: gw.set_shape(v,rectangle_node);
               gw.set_color(v,red);
               break;
      case -3: gw.set_shape(v,rectangle_node);
               gw.set_color(v,blue2);
               break;
    }
  }
  edge e;
  forall_edges(e,G) gw.set_color(e,grey1);
  forall(e,E)
  { gw.set_color(e,black);
```

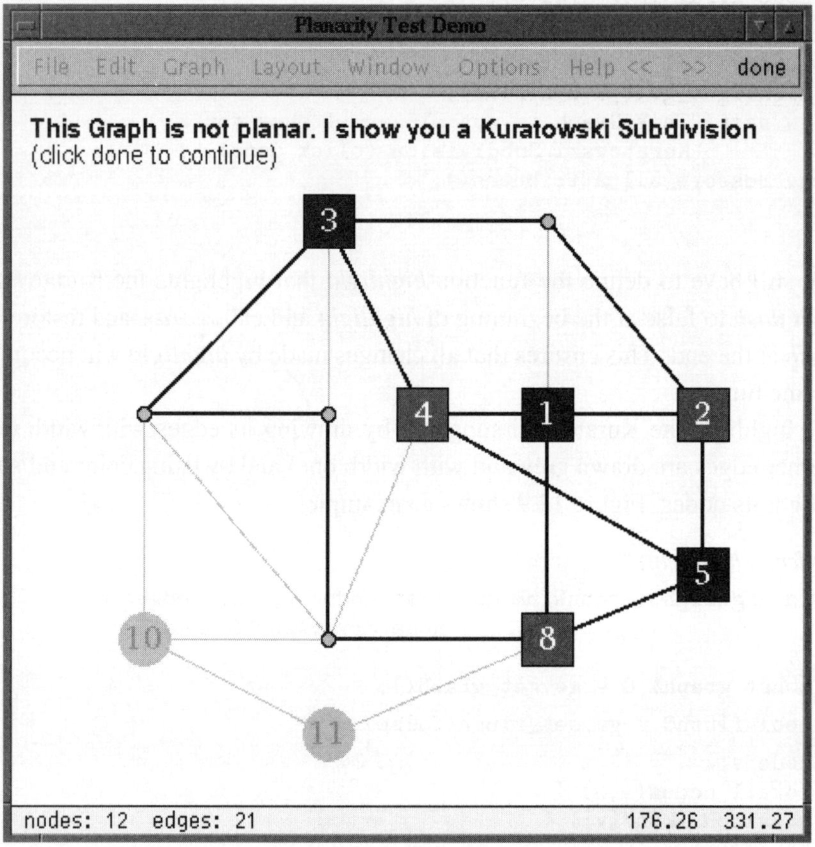

Figure 12.9 The planarity test demo: Highlighting a Kuratowski subdivision.

```
      gw.set_width(e,2);
    }
  gw.redraw();
  gw.set_flush(flush0);
}
```

Exercises for 12.4

1 Write a program that animates quicksort. Have a graph with one node for each input and no edges. Change the layout of the graph as the sort progresses.

2 Write a program that animates heapsort.

3 Write a program that always shows a DFS-structure of the currently edited graph by drawing the different edge types (tree, backward, forward, cross) in different colors or styles.

12.5 Customizing the Interactive Interface

We describe three ways for customizing the interactive interface:

- Call-back functions,
- Extended and/or additional menus, and
- Redefined edit actions.

Each method will allow us to write nicer demos.

12.5.1 *Call-Back Functions*

Call-back or handler functions can be used to associate arbitrary functionality with the edit operations of *GraphWin*. Two handlers can be defined for every operation. The first one, the so-called *pre-handler*, is called immediately before the corresponding edit operation. The second one, the so-called *post-handler*, is called at the end of the operation. For move operations of nodes and sliders, there is a third handler, the so-called *move-handler* which is called for all intermediate positions.

The pre-handlers have a boolean return value which tells *GraphWin* whether the corresponding edit operation is to be executed or not. This provides a simple way of disallowing edit operations under certain conditions. In general, pre- and post-handler also have different parameter lists.

The null-handler (*NULL*) can be used to remove a pre- or post-handler from an edit operation.

We give a list of the most important handlers and the corresponding *set* operations. There are two versions of each *set_handler*, one each for defining the pre- and post-handler. The functions have the same name and differ in the type of the function pointer argument: functions for setting pre-handlers take an argument of type *bool* $(*func)$(*GraphWin&*, ...) and functions for setting post-handlers take an argument of type *void* $(*func)$(*GraphWin&*, ...).

```
void gw.set_new_node_handler(bool (*f)(GraphWin&,point));
```

sets the pre-handler of the new-node operation to f, i.e., $f(gw, p)$ is called before a node is created at position p.

```
void gw.set_new_node_handler(void (*f)(GraphWin&,node));
```

sets the post-handler of the new-node operation to f, i.e., $f(gw, v)$ is called after a new node v has been created.

```
void gw.set_new_edge_handler(bool (*f)(GraphWin&,node,node));
```

sets the pre-handler of the new-edge operation to f, i.e., $f(gw, v, w)$ is called before a new edge (v, w) is created.

```
void gw.set_new_edge_handler(void (*f)(GraphWin&,edge));
```

sets the post-handler of the new-edge operation to f, i.e., $f(gw, e)$ is called after a new edge e has been created.

```
void gw.set_del_node_handler(bool (*f)(GraphWin&,node));
```

sets the pre-handler of the del-node operation to f, i.e., $f(gw, v)$ is called each time before a node v is deleted.

```
void gw.set_del_node_handler(void (*f)(GraphWin&));
```

sets the post-handler of the del-node operation to f, i.e., $f(gw)$ is called each time a node has been deleted.

```
void gw.set_del_edge_handler(bool (*f)(GraphWin&,edge));
```

sets the pre-handler of the del-edge operation to f, i.e., $f(gw, e)$ is called each time before an edge e is deleted.

```
void gw.set_del_edge_handler(void (*f)(GraphWin&));
```

sets the post-handler of the del-edge operation to f, i.e., $f(gw)$ is called each time an edge has been deleted.

```
void gw.set_init_graph_handler(bool (*f)(GraphWin&));
```

sets the pre-handler of the init-graph operation to f, i.e., $f(gw)$ is called every time before any global update of the graph, e.g., in a clear, generate, or load operation.

```
gw.set_init_graph_handler(void (*f)(GraphWin&));
```

sets the post-handler of the init-graph operation to f, i.e., f is called after each global update of the graph.

Node moving and edge slider moving operations may have three different handlers. The first is called before the moving starts, the second is called for every intermediate position, and the third one is called at the final position of the node after the moving has been finished. The handlers are set by:

```
gw.set_start_move_node_handler(bool (*f)(GraphWin&,node));
gw.set_move_node_handler(bool (*f)(GraphWin&,node,point));
gw.set_end_move_node_handler(void (*f)(GraphWin&,node));
gw.set_start_edge_slider_handler(
                    void (*f)(GraphWin& gw,edge,double),int i);
gw.set_edge_slider_handler(
                    void (*f)(GraphWin& gw,edge,double),int i);
gw.set_end_edge_slider_handler(
                    void (*f)(GraphWin& gw,edge,double),int i);
```

Recall that each edge has three sliders associated with it. The integer argument i in the last three functions selects the slider, $0 \le i \le 2$.

12.5.2 *A Recipe for On-line Demos of Graph Algorithms*

The edit-and-run paradigm for demos of graph algorithms requires an explicit user action, namely a click on the done-button, to start the graph algorithm to be demonstrated. Callback or handler functions allow us to write on-line demos which show the result of a graph algorithm while the graph is edited and not only after editing.

We give the generic structure of a demo that calls a graph algorithm after every addition or deletion of a node or edge and after the initialization of the graph (for example, by reading it from a file). We define a function *run_and_display* that runs the graph algorithm on the graph associated with *gw* and updates the display. We then define post-handlers for the *new_node*, *new_edge*, *del_node*, *del_edge*, and *init_graph* operations; each handler simply calls *run_and_display(gw)*. In the main program we tell *GraphWin* which handlers to use by calling the corresponding *set_handler* functions, display the window, and call *gw.edit( )*. That's all.

⟨*gw_handler.c*⟩≡

```
#include <LEDA/graph_alg.h>
#include <LEDA/graphwin.h>

void run_and_display(GraphWin& gw)
{ ⟨run algorithm and update display⟩ }

void new_node_handler(GraphWin& gw, node) { run_and_display(gw); }
void new_edge_handler(GraphWin& gw, edge) { run_and_display(gw); }
void del_edge_handler(GraphWin& gw)       { run_and_display(gw); }
void del_node_handler(GraphWin& gw)       { run_and_display(gw); }
void init_graph_handler(GraphWin& gw)     { run_and_display(gw); }

int main()
{
  GraphWin gw;

  gw.set_init_graph_handler(init_graph_handler);
  gw.set_new_edge_handler(new_edge_handler);
  gw.set_del_edge_handler(del_edge_handler);
  gw.set_new_node_handler(new_node_handler);
  gw.set_del_node_handler(del_node_handler);

  gw.display();
  gw.edit();

  return 0;
}
```

We will next derive a specific demo from this framework by instantiating the *run_and_display* function. We illustrate the strongly connected components of the graph associated with *gw*; all nodes belonging to the same component should be colored the same and nodes in different components should be colored differently.

The "work horse" of our demo is a function *void run_and_display(GraphWin&)* that uses the graph algorithm *STRONG_COMPONENTS* to compute a numbering *comp_num* of the nodes of the current graph, such that all nodes of a strongly connected component receive the same number. Each node is painted with the number of its component.

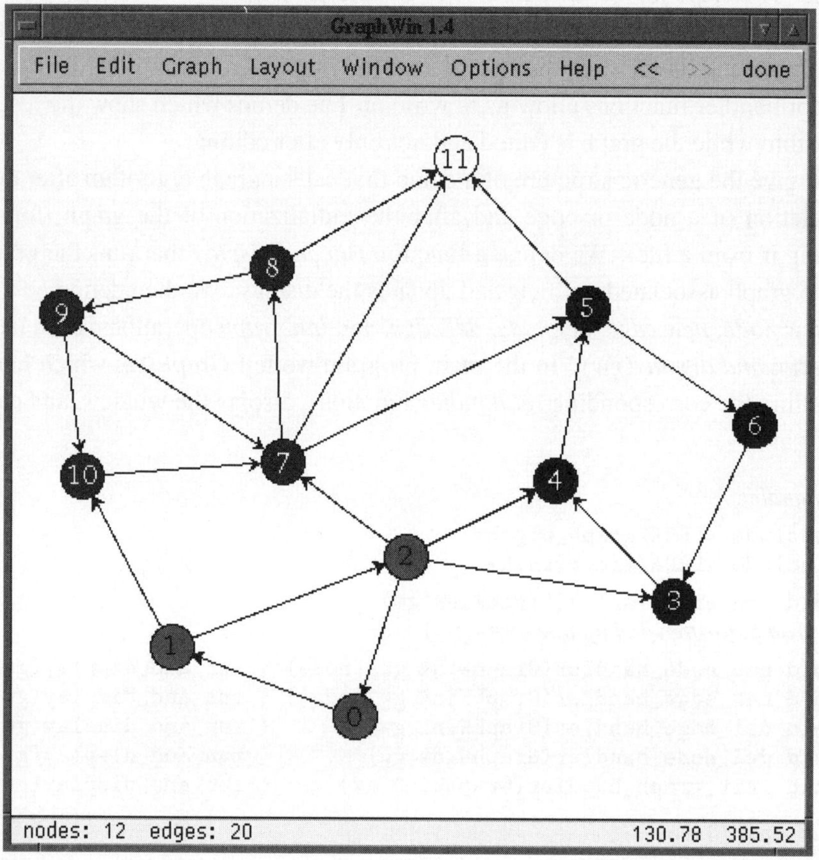

Figure 12.10 An screen shot of an on-line demo for the strongly connected components of a graph.

⟨*run algorithm and update display*⟩≡

```
graph& G = gw.get_graph();
node_array<int> comp_num(G);
STRONG_COMPONENTS(G,comp_num);
node v;
forall_nodes(v,G) gw.set_color(v,color(comp_num[v]));
```

Figure 12.10 shows a screen shot of the program after a few editing operations.

12.5.3 *Defining and Changing Menus*

The menus of *GraphWin* are not fixed. New sub-menus and buttons can be added to the main window and any sub-menu, in this way extending the set of functions and algorithms that can be applied to the current graph by a mouse click. Furthermore, the set of default menus in the main window's menu bar can be changed by removing standard menus. All operations for changing menus have to be called before the window is displayed for the first time.

Changing the Standard Main Menu: The default menus in *GraphWin*'s menu bar are determined by a bit mask that is the bitwise-or of an arbitrary subset of the predefined constants *M_FILE*, *M_EDIT*, *M_GRAPH*, *M_LAYOUT*, *M_WINDOW*, *M_OPTIONS*, *M_HELP*, and *M_DONE*. Each of these constants represents the corresponding standard menu discussed in Section 12.1. The value *M_COMPLETE* is defined as the bitwise-or of all constants above, i.e., it specifies a menu bar containing all standard menus. The operation

```
long gw.set_default_menu(long mask);
```

defines the set of standard menus, where *mask* is the bitwise-or of an arbitrary subset of the predefined constants listed above. The operation

```
void gw.del_menu(long mask);
```

removes all menus corresponding to 1-bits in *mask* from the menu bar.

Adding New Menus: New sub-menus can be added to an existing menu (or the main menu bar) by calling the *add_menu* operation. Each menu is represented by an integer (*menu_id*) from an internal numbering of all menus. The main menu bar has *menu_id* zero.

```
int  gw.add_menu(GraphWin& gw, string label, int menu_id = 0),
```

creates a sub-menu in menu with id *menu_id*. The corresponding button is labeled with *label*. The operation returns the menu id of the new menu. The menu id of a standard menu can be obtained by calling *get_menu*(*string*) with the name of the menu, e.g.,

```
get_menu("Help");
```

returns the menu id of the help menu.

Adding Simple Functions: We call functions of type *void func*(*GraphWin& gw*) *simple*. The *add_simple_call* operation of *GraphWin* can be used to add (buttons for starting) simple functions to an existing menu or the main menu bar.

```
void  gw.add_simple_call(void (*func)(GraphWin&),
                          string label, int menu_id = 0);
```

adds a new button with label *label* to the menu with menu id *menu_id*. Whenever this button is pressed during edit mode *func*(*gw*) is called.

We give an example. Assume we want to add a button to the main menu that runs a DFS algorithm of type

```
void dfs(graph& G, node s, node_array<bool>& reached)
```

on the current graph. We write a simple function *void* (*run_dfs*)(*GraphWin&*) that tells *GraphWin* how to call *dfs* and how to display its result.

```
void run_dfs(GraphWin& gw)
{
  // provide arguments
  graph& G = gw.get_graph();
  node s = gw.ask_node();
  node_array<bool> reached(G,false);
  // call function
  dfs(G,s,reached);
  // display result
  node v;
  forall_nodes(v,G) if (reached[v]) gw.set_color(v,red);
}
```

and add the function to the main menu by calling

```
gw.add_simple_call(run_dfs,"dfs");
```

The string argument "dfs" will be used as the label of the new menu button. We may also want to extend the help menu. We define a simple function *about_dfs* that opens a panel and displays a help string

```
void about_dfs(GraphWin& gw)
{ window& W = gw.get_window();

  panel P;
  P.set_panel_bg_color(win_p->mono() ? white : ivory);
  P.text_item("The dfs-button runs dfs on the current graph.");
  P.button("OK");
  W.disable_panel();
  P.open(W);
  W.enable_panel();
}
```

and add it to the help menu.

```
int h_menu = gw.get_menu("Help");
gw.add_simple_call(about_dfs, "About DFS",h_menu);
```

Adding GraphWin Member Functions: Not every operation of the programming interface of *GraphWin* is available in the interactive interface. However, there is an easy way of adding operations of type *void GraphWin::func( )*, i.e., member functions without parameters and without a result. The operation

```
gw.add_member(void (*GraphWin::func)(), string label, int menu_id = 0);
```

adds a new button with label *label* to the menu with menu id *menu_id*. Whenever this button is pressed during edit mode *gw.func( )* is called.

As an example, we add a "redraw" button, that calls the *gw.redraw( )* operation, to the main panel.

```
gw.add_member_call(&GraphWin::redraw,"redraw");
```

Adding Families of Functions: Sometimes, one wants to add an entire group of functions, all with the same interface, to a menu. In this case it would be tedious to write a wrapper for each of these functions. It is more convenient to write only a single *caller* function that can deal with all functions of the group. The caller takes a reference to a *GraphWin* and a pointer to the function to be called as arguments. More precisely, if the function to be called is of type *function_t*, the caller has type *void* $(*caller)$(*GraphWin&*, *function_t*).

The *gw_add_call* function template adds a function together with its caller to a menu. This operation should better be realized by a member function template. However, only a few compilers currently support this feature of C++.

```
template <class function_t>
void  gw_add_call(GraphWin& gw, function_t func,
                  void (*caller)(GraphWin&, function_t),
                  string label, int menu_id=0);
```

adds a new button with label *label* to the menu with menu id *menu_id*. Whenever this button is pressed in edit mode, the function *caller* is called with arguments *gw* and *func*.

We use a family of graph drawing functions as an example. Assume we have a library of graph drawing algorithms (e.g., the AGD library [JMN]) and want to build a *graph_draw* menu which makes all functions in the library available on a mouse click. We assume that all graph drawing algorithms take a graph G and compute for every node v of G a position ($xcoord[v]$, $ycoord[v]$).

```
void draw_alg1(const graph& G, node_array<double> xcoord,
                               node_array<double> ycoord);
void draw_alg2(const graph& G, node_array<double> xcoord,
                               node_array<double> ycoord);
...
```

A generic caller function for this type of graph algorithm is as follows:

```
typedef void (*draw_alg)(graph&, node_array<double>&,
                                 node_array<double>&);
void call_draw_alg(GraphWin& gw, draw_alg draw)
{
  // provide arguments
  graph& G = gw.get_graph();
  node_array<double> xcoord(G);
  node_array<double> ycoord(G);
  // call function
  draw(G,xcoord,ycoord);
  // display result
  gw.adjust_coords_to_win(xcoord,ycoord);
  gw.set_layout(xcoord,ycoord);
  if (!gw.get_flush()) gw.redraw();
}
```

The new menu is now easily created.

```
int draw_menu = gw.add_menu("graph drawing");
gw_add_call(gw,draw_alg1,call_draw_alg,"draw_alg1",draw_menu)
gw_add_call(gw,draw_alg2,call_draw_alg,"draw_alg2",draw_menu)
...
```

A Complete Example: We give a complete example that illustrates the possibilities to extend and modify menus. We will write a demo that illustrates dfs, spanning trees, connected components, and strongly connected components.

For dfs and spanning trees we use simple functions.

⟨*simple functions*⟩≡

```
void dfs_num(GraphWin& gw)
{ graph& G = gw.get_graph();
  node_array<int> dfsnum(G);
  node_array<int> compnum(G);
  DFS_NUM(G,dfsnum,compnum);
  node v;
  forall_nodes(v,G) gw.set_label(v,string("%d|%d",dfsnum[v],compnum[v]));
  if (gw.get_flush() == false) gw.redraw();
}
void span_tree(GraphWin& gw)
{ graph& G = gw.get_graph();
  list<edge> L = SPANNING_TREE(G);
  gw.set_color(L,red);
  gw.set_width(L,2);
  if (gw.get_flush() == false) gw.redraw();
}
```

The LEDA functions to compute components of a graph all have the same interface. They take a graph and compute a node array of *ints*, and return an int. Any such function can be added to a GraphWin using the caller

⟨*components caller*⟩≡

```
// a caller for component algorithms
void call_comp(GraphWin& gw,
               int (*comp)(const graph& G, node_array<int>& compnum) )
{ graph& G = gw.get_graph();
  node_array<int> compnum(G);
  comp(G,compnum);
  node v;
  forall_nodes(v,G)
  { int i = compnum[v];
    gw.set_label(v,string("%d",i));
    gw.set_color(v,(color)(i%16));
   }
  if (gw.get_flush() == false) gw.redraw();
}
```

In the main program we define a GraphWin, delete some of the standard menus (just to illustrate how it is done), add our simple calls, add a reset button, and finally create a sub-menu for the components functions.

⟨*gw_menu.c*⟩≡

```
#include <LEDA/graphwin.h>
#include <LEDA/graph_alg.h>
#include <LEDA/graph_misc.h>
⟨components caller⟩
⟨simple functions⟩
int main()
{
  GraphWin gw;
  // we delete some of the standard menus
  gw.set_default_menu(M_COMPLETE & ~M_LAYOUT & ~M_HELP);
  // add two simple function calls
  gw.add_simple_call(dfs_num,   "dfsnum");
  gw.add_simple_call(span_tree, "spanning");
  // a member call
  gw.add_member_call(&GraphWin::reset,"reset");
  // and a menu with three non-simple functions using
  // a common call function
  int menu1 = gw.add_menu("components");
  gw_add_call(gw,COMPONENTS,       call_comp,"simply connected",  menu1);
  gw_add_call(gw,COMPONENTS1,      call_comp,"simply connected1", menu1);
  gw_add_call(gw,STRONG_COMPONENTS,call_comp,"strongly connected",menu1);
  gw.display();
  gw.edit();
  return 0;
}
```

Figure 12.11 shows a screen shot of this demo.

12.5.4 *Defining Edit Actions*

Mouse operations in the display region of a *GraphWin* generate events. An event is characterized by its event bit mask *event_mask* (which is the or of elementary masks to be defined below) and the current position *mouse_position* of the mouse pointer. Event masks have associated *edit actions*. All edit actions are functions of type

```
void action(GraphWin& gw, const point& pos);
```

When an event occurs, the associated action function is called with the *GraphWin* object and the current mouse pointer position *mouse_position* as arguments. The object (node or edge) under the current position can be queried by the *get_edit_node* or *get_edit_edge* operation.

Event masks are the bitwise-or of some of the following predefined constants:

A_LEFT, A_MIDDLE, A_RIGHT: If one of these bits is set, the corresponding mouse button (left, middle, or right) has been clicked.

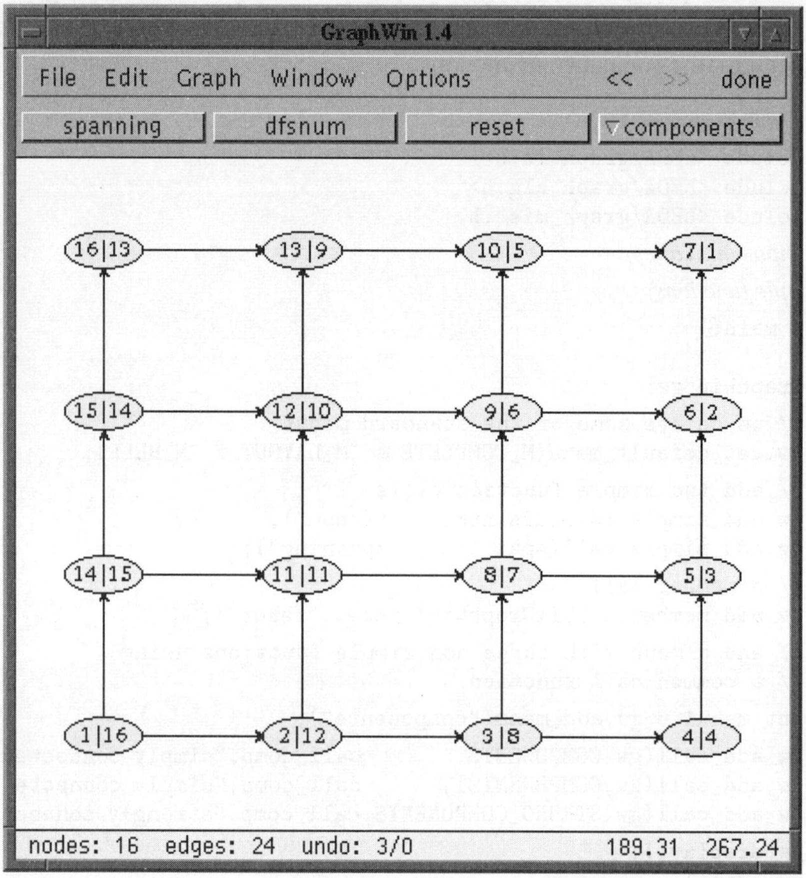

Figure 12.11 Extending the menu: Computing a DFS-numbering.

A_DRAG: This bit indicates that the mouse is moved with one or more buttons (specified by the bits discussed above) held down.

A_DOUBLE: This bit indicates a double click , i.e., the event that a mouse button has been clicked twice.

A_SHIFT, A_CTRL, A_ALT: If one of these bits is set, the corresponding keyboard control key (Shift,Ctrl,Alt) is pressed.

A_NODE: If this bit is set, the mouse pointer is located over a node and the node can be queried by the *gw.get_edit_node*( ) operation.

A_EDGE: If this bit is set, the mouse pointer is located over an edge and the edge can be queried by the *gw.get_edit_edge*( ) operation.

A_SLIDER: If this bit is set, the mouse pointer is located over a slider of an edge. The corresponding edge can be queried as above and the number of the slider (0,1, or 2) can be obtained by calling the *gw.get_edit_slider*( ) operation.

An event mask is defined by a combination of these bits, for instance

```
( A_LEFT | A_NODE | A_DOUBLE )
```

describes a double click of the left mouse button on a node.

Setting Edit Actions: The following operations can be used to change the action functions associated with events.

```
gw_action gw.set_action(long mask,void (*func)(GraphWin&, const point&));
```

sets the action on condition *mask* to *func* and return the previous action of this condition. After this call *func* is called with the *GraphWin* object and the current edit position as arguments whenever the condition defined by *mask* becomes true.

```
void gw.reset_actions();
```

resets all actions to their default values and

```
void gw.clear_actions();
```

sets all actions to *NULL*.

The following piece of code shows part of the initialization of the default edit actions.

```
// left button (create,move,scroll,zoom)
set_action( A_LEFT                                , gw_new_node);
set_action( A_LEFT |                   A_NODE     , gw_new_edge);
set_action( A_LEFT |            A_DRAG |A_NODE    , gw_move_node);
set_action( A_LEFT |            A_DRAG |A_EDGE    , gw_move_edge);
set_action( A_LEFT |            A_DRAG            , gw_scroll_graph);
set_action( A_LEFT |            A_DRAG |A_SLIDER, gw_move_edge_slider);

set_action( A_LEFT |A_SHIFT   |A_DRAG |A_NODE    , gw_move_component);
set_action( A_LEFT |A_DOUBLE |         A_NODE    , gw_setup_node);
set_action( A_LEFT |A_DOUBLE |         A_EDGE    , gw_setup_edge);
```

An Example Program: The following program redefines some of the default actions, for example, when the left mouse button is clicked over a node with the control key pressed, the node color will be increased by one.

⟨*gw_action.c*⟩≡

```
#include<LEDA/graphwin.h>
void change_node_color(GraphWin& gw, const point&)
{ node v  = gw.get_edit_node();
  int col = (gw.get_color(v) + 1) % 16;
  gw.set_color(v,color(col));
 }
void change_edge_color(GraphWin& gw, const point&)
```

```
{ edge e  = gw.get_edit_edge();
  int col = (gw.get_color(e) + 1) % 16;
  gw.set_color(e,color(col));
 }

void center_node(GraphWin& gw, const point& p)
{ node v  = gw.get_edit_node();
  gw.set_position(v,p);
 }

void delete_node(GraphWin& gw, const point&)
{ node v  = gw.get_edit_node();
  gw.del_node(v);
 }

void zoom_up(GraphWin& gw, const point&)   { gw.zoom(1.5); }
void zoom_down(GraphWin& gw, const point&) { gw.zoom(0.5); }
main()
{
  GraphWin gw;
  gw.set_action(A_LEFT | A_NODE | A_CTRL,  change_node_color);
  gw.set_action(A_LEFT | A_EDGE | A_CTRL,  change_edge_color);
  gw.set_action(A_LEFT | A_NODE | A_SHIFT, center_node);
  gw.set_action(A_RIGHT| A_NODE, delete_node);
  gw.set_action(A_LEFT | A_CTRL, zoom_up);
  gw.set_action(A_RIGHT| A_CTRL, zoom_down);
  gw.display(window::center,window::center);
  gw.edit();
}
```

12.6 Visualizing Geometric Structures

Many geometric data structures of LEDA are implemented by labeled graphs, e.g., Delaunay diagrams are represented by graphs of type *GRAPH<point, int>* and Voronoi diagrams are represented as graphs of type *GRAPH<CIRCLE, POINT>*. Many geometry demos have a GraphWin-button for viewing the underlying graph structures.

We sketch how this button is realized. In the demo below we compute the Delaunay triangulation DT of a set L of twenty-five points on a regular grid. We then declare a GraphWin *gw* for DT, tell *gw* that we want each node v to be drawn at position $DT[v]$, as a circle of radius eight pixels, and without label, and that we want each edge to be drawn with a color indicating its label. Start the demo and the graph shown in Figure 12.12 will appear.

⟨*gw_delaunay.c*⟩≡

```
#include <LEDA/plane_alg.h>
#include <LEDA/graphwin.h>
main()
```

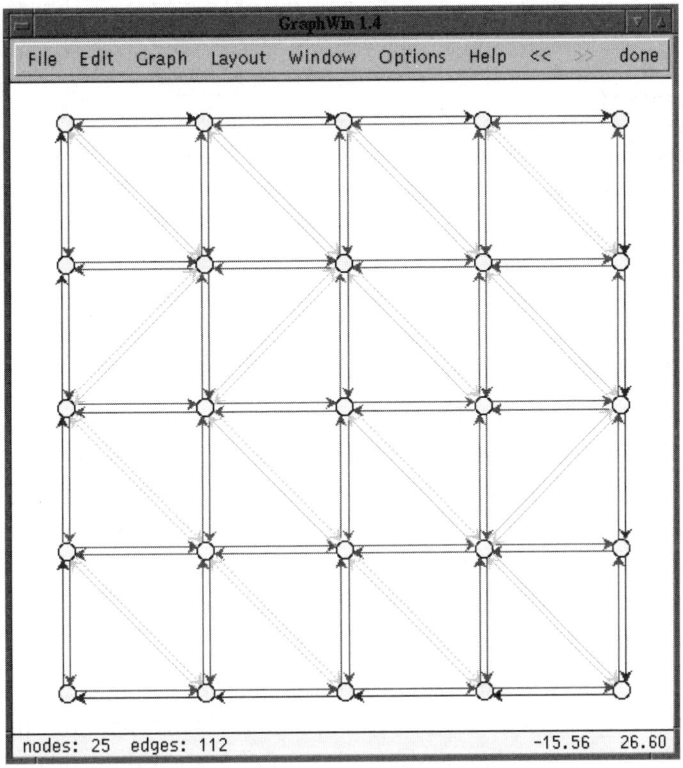

Figure 12.12 GraphWin displaying a Delaunay triangulation.

```
{
  GRAPH<rat_point,int> DT;
  list<rat_point> L;
  lattice_points(25,100,L);
  DELAUNAY_TRIANG(L,DT);
  GraphWin gw(DT);
  node v;
  forall_nodes(v,DT)
  { rat_point p = DT[v];
    gw.set_position(v,p.to_point());
    gw.set_label_type(v,no_label);
    gw.set_width(v,8);
    gw.set_height(v,8);
  }
  edge e;
  forall_edges(e,DT)
  { switch (DT[e]) {
     case DIAGRAM_EDGE:     gw.set_color(e,green2); break;
     case NON_DIAGRAM_EDGE: gw.set_color(e,yellow); break;
```

```
      case HULL_EDGE:          gw.set_color(e,red);    break;
    }
  }
  gw.display();
  gw.zoom_graph();
  gw.edit();
}
```

12.7 A Recipe for On-line Demos of Network Algorithms

Networks are graphs whose edges (and sometimes nodes) are labeled with numbers, e.g., capacities or costs. On-line demos of network algorithms should allow the user to edit the underlying graph as well as the edge capacities. We have already seen how to react on-line to update operations. In this section we will show how to implement capacity changes by edge sliders. All demos of network algorithms follow the paradigm presented in this section. We use the min cost flow algorithm as our example. All other demos are simpler. Figure 12.13 shows a screenshot.

The global structure of our demo is as follows. We define edge maps *cap* and *cost* in order to make edge capacities and edge costs globally available for the handler functions. We then define a function that runs the min cost flow algorithm and displays the result and we define handlers for edge events and handlers for slider events.

In the main program we generate the grid graph G shown in Figure 12.13 and associate the edge maps *cap* and *cost* with it. We define a GraphWin *gw* for G and set its header to "Min Cost Max Flow". We disable edge bends since sliders can be used for straight line edges only. We set the node and edge attributes to the colors hinted at in the figure, and we adjust the size of the layout such that it uses about 90% of the window. Finally, we open the window and put it into edit mode.

⟨*gw_mcmflow.c*⟩≡

```
#include <LEDA/graphwin.h>
#include <LEDA/graph_alg.h>
static edge_map<int> cap;
static edge_map<int> cost;
⟨run min cost flow and display result⟩
⟨edge handlers⟩
⟨capacity and cost sliders⟩
int main()
{
  // construct a (grid) graph
  graph G;
  node_array<double> xcoord;
  node_array<double> ycoord;
```

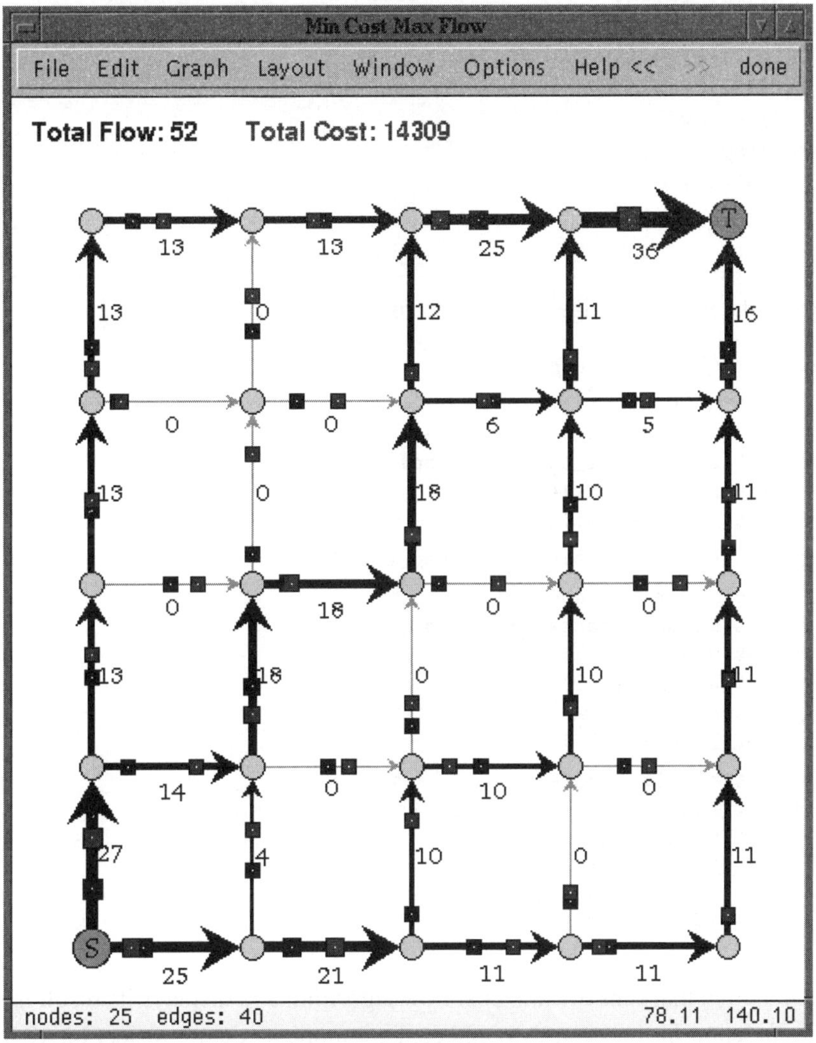

Figure 12.13 Animation of a min-cost-flow algorithm.

```
grid_graph(G,xcoord,ycoord,5);
// initialize cap and cost maps
cap.init(G);
cost.init(G);
GraphWin gw(G,"Min Cost Max Flow");
// disable edge bends
gw.set_action(A_LEFT | A_DRAG | A_EDGE , NULL);
```

⟨*set handlers*⟩

⟨*set attributes of nodes and edges*⟩

```
  //adjust layout
  gw.adjust_coords_to_win(xcoord,ycoord);
  gw.set_layout(xcoord,ycoord);
  gw.zoom(0.9);
  // open gw
  gw.display();
  gw.edit();
  return 0;
}
```

Setting the node and edge attributes is routine.

⟨*set attributes of nodes and edges*⟩≡

```
  gw.set_node_color(yellow);
  gw.set_node_shape(circle_node);
  gw.set_node_label_type(no_label);
  gw.set_node_width(14);
  gw.set_node_height(14);
  gw.set_edge_direction(directed_edge);
  node s = G.first_node();
  gw.set_shape(s,rectangle_node);
  gw.set_width(s,22);
  gw.set_height(s,22);
  gw.set_color(s,cyan);
  gw.set_label(s,"S");
  node t = G.last_node();
  gw.set_shape(t,rectangle_node);
  gw.set_width(t,22);
  gw.set_height(t,22);
  gw.set_color(t,cyan);
  gw.set_label(t,"T");
```

The function that runs the min cost flow algorithm and displays its result is similar to the display function in the strongly connected components demo of Section 12.4, but slightly more complex because we are aiming for a more elaborated visualization.

We obtain the graph G from *gw*, we set s and t to the first and last node, respectively, and compute the flow using the global edge maps *cap* and *cost*. We compute the flow value and the cost of the flow and we set the width of every edge proportional to the flow through the edge. Edges with flow zero are faded to grey. We reset flush, write a message containing flow value and cost, and redraw.

⟨*run min cost flow and display result*⟩≡

```
void run_mcm_flow(GraphWin& gw)
{ bool flush = gw.set_flush(false);
  graph& G = gw.get_graph();
  node   s = G.first_node();
  node   t = G.last_node();
  gw.message("\\bf Computing MinCostMaxFlow");
```

```
  edge_array<int> flow(G);
  int F = MIN_COST_MAX_FLOW(G,s,t,cap,cost,flow);
  int C = 0;
  // sum up total cost and indicate flow[e] by the width of e
  edge e;
  forall_edges(e,G)
  { C += flow[e]*cost[e];
    gw.set_label_color(e,black);
    gw.set_label(e,string("%d",flow[e]));
    gw.set_width(e,1+int((flow[e]+4)/5.0));
    if (flow[e] == 0)
      gw.set_color(e,grey2); // 0-flow edges are faded to grey
    else
      gw.set_color(e,black);
   }
  gw.set_flush(flush);
  gw.message(string("\\bf Flow: %d  \\bf Cost: %d",F,C));
  gw.redraw();
}
```

We come to the edge handlers. We first define an auxiliary function *init_edge* that sets the capacity and the cost of an edge to random values and sets the slider values for the zeroth and the first slider of the edge accordingly. The *init_handler* initializes all edges, computes a min cost flow and displays it. The new edge handler initializes the edge, computes a min cost flow and displays it.

The init handler and the node and edge handlers of *gw* are set in the obvious way.

⟨*edge handlers*⟩≡

```
void init_edge(GraphWin& gw, edge e)
{ // init capacity and cost to a random value
  cap[e] = rand_int(10,50);
  cost[e] = rand_int(10,75);
  // set sliders accordingly
  gw.set_slider_value(e,cap[e]/100.0,0);  // slider zero
  gw.set_slider_value(e,cost[e]/100.0,1); // slider one
}
void init_handler(GraphWin& gw)
{ edge e;
  forall_edges(e,gw.get_graph()) init_edge(gw,e);
  run_mcm_flow(gw);
}
void new_edge_handler(GraphWin& gw, edge e)
{ init_edge(gw,e);
  run_mcm_flow(gw);
}
```

⟨*set handlers*⟩≡

```
gw.set_init_graph_handler(init_handler);
```

```
gw.set_del_edge_handler(run_mcm_flow);
gw.set_del_node_handler(run_mcm_flow);
gw.set_new_edge_handler(new_edge_handler);
```

We come to the sliders. The cap slider handlers handle the change of capacities. We use the zeroth edge slider for the capacities. When an edge slider is picked up we display an appropriate message. As long as the slider is moved we display the new capacity. When the edge slider is released we recompute the flow and display it.

⟨*capacity and cost sliders*⟩≡

```
// capacity sliders
void start_cap_slider_handler(GraphWin& gw, edge, double)
{ gw.message("\\bf\\blue Change Edge Capacity"); }
void cap_slider_handler(GraphWin& gw,edge e, double f)
{ cap[e] = int(100*f);
  gw.set_label_color(e,blue);
  gw.set_label(e,string("cap = %d",cap[e]));
}
void end_cap_slider_handler(GraphWin& gw, edge, double)
{ run_mcm_flow(gw); }
```

⟨*set handlers*⟩+≡

```
gw.set_start_edge_slider_handler(start_cap_slider_handler,0);
gw.set_edge_slider_handler(cap_slider_handler,0);
gw.set_end_edge_slider_handler(end_cap_slider_handler,0);
gw.set_edge_slider_color(blue,0);
```

Cost sliders are treated completely analogously.

⟨*capacity and cost sliders*⟩+≡

```
// cost sliders
void start_cost_slider_handler(GraphWin& gw, edge, double)
{ gw.message("\\bf\\red Change Edge Cost"); }
void cost_slider_handler(GraphWin& gw, edge e, double f)
{ cost[e] = int(100*f);
  gw.set_label_color(e,red);
  gw.set_label(e,string("cost = %d",cost[e]));
}
void end_cost_slider_handler(GraphWin& gw, edge, double)
{ run_mcm_flow(gw); }
```

⟨*set handlers*⟩+≡

```
  gw.set_start_edge_slider_handler(start_cost_slider_handler,1);
  gw.set_edge_slider_handler(cost_slider_handler,1);
  gw.set_end_edge_slider_handler(end_cost_slider_handler,1);
  gw.set_edge_slider_color(red,1);
```

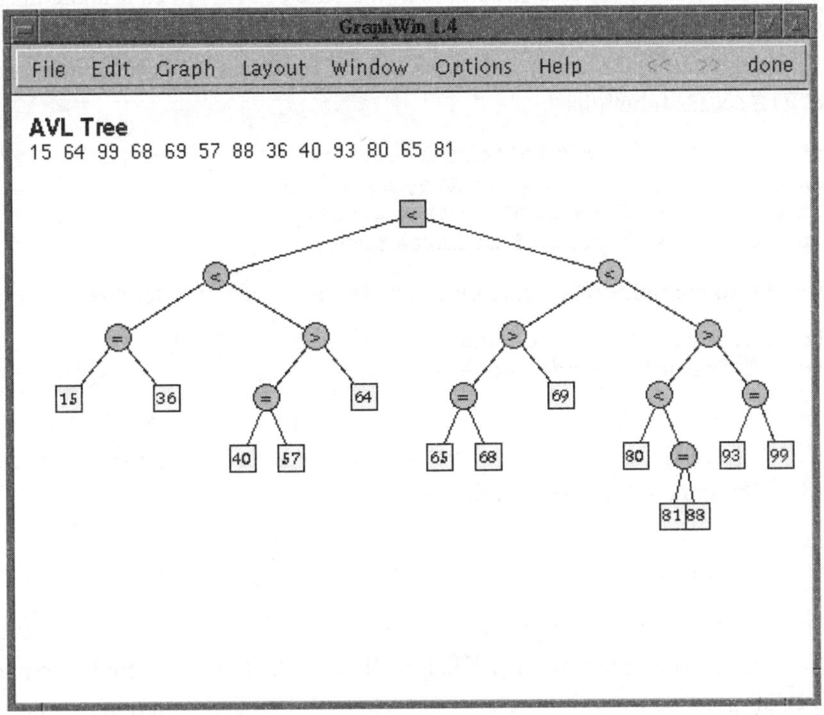

Figure 12.14 Visualization of an AVL tree.

Exercises for 12.7

1 Add menus to the main window for running and displaying the result of the different shortest-path and network flow algorithms of LEDA. Use edge sliders for the input of edge cost and capacities.

2 Design and implement an animation of the vertex addition planarity test algorithm discussed in Chapter 8.

3 Write an animation program of the generic preflow-push algorithm for computing a maximum flow in a network.

12.8 A Binary Tree Animation

We close this chapter with a demo which animates several implementations of balanced binary trees, namely AVL-trees, BB[α]-trees, and red-black trees.

All balanced binary tree implementations use a common base, the classes *bin_tree* and *bin_tree_node*. A *bin_tree* is a collection of *bin_tree_nodes*. Each *bin_tree_node* stores pointers to its parent and its children, and a balance of type *int*. The interpretation of the balance of a node depends on the tree structure. In the case of AVL-trees it is the height difference between the left and right subtree, in the case of BB[α]-trees it is the number of nodes in the subtree rooted at the node, and in the case of red-black trees it encodes the color of the node. The access functions

```
int             T.get_bal(bin_tree_node*)
bin_tree_node*  T.parent(bin_tree_node*)
bin_tree_node*  T.l_child(bin_tree_node*)
bin_tree_node*  T.r_child(bin_tree_node*)
```

give access to the fields of a node. One can also ask whether a node is a root or a leaf

```
bool  T.is_root(bin_tree_node*)
bool  T.is_leaf(bin_tree_node*)
```

and one can inquire about the type and name of a tree. The name of a tree is one of "AVL Tree", "BB[alpha] Tree", ..., and the type of a tree is an integer from an enumeration type encoding the same information as the name.

```
int   T.tree_type()
char* T.tree_name()
```

A pointer to a *bin_tree_node* is a *bin_tree_item*.

The overall structure of the demo is as follows. We define the control parameters n, the number of insertions, *input*, the choice between random and sorted insertions, and *kind*, the type of tree to be used, we define a panel that allows us to set the control parameters, and we define three *bin_trees* and initialize them to an empty AVL-tree, BB[α]-tree, and red-black tree, respectively. We then enter a loop.

In each iteration of the loop we open the panel and ask the reader to set the control parameters. We then define an object T of class *anim_bin_tree* for the GraphWin *gw* and the tree selected by *kind*. The class *anim_bin_tree* will be discussed below and does the bulk of the work. We perform n insertions on T with either random inputs or increasing inputs. Finally, we display the message "Press done to continue" and put *gw* into edit mode such that the user can reply.

⟨*gw_bintree.c*⟩≡

```
#include <LEDA/graphwin.h>
#include <LEDA/impl/bin_tree.h>
#include <LEDA/impl/avl_tree.h>
#include <LEDA/impl/bb_tree.h>
#include <LEDA/impl/rb_tree.h>
#include <LEDA/impl/rs_tree.h>
#include <LEDA/map.h>
⟨class anim_bin_trees⟩
int main()
{
  GraphWin gw(500,400);
  gw.set_node_width(18);
  gw.set_node_height(18);
```

```
    gw.set_node_label_type(no_label);
    gw.set_node_label_font(roman_font,10);
    gw.set_edge_direction(undirected_edge);
    gw.set_show_status(false);

    gw.display(window::center,window::center);

    int n = 16;
    int input = 0;
    int kind  = 0;

    // define a panel P to control n, input, and kind

    panel P;
    P.text_item("\\bf\\blue Binary Tree Animation");
    P.text_item("");
    P.choice_item("tree type",kind, "avl-tree","bb-tree","rb-tree");
    P.choice_item("input data",input,"random", "1 2 3 ...");
    P.int_item("# inserts",n,0,64);
    P.button("ok",0);
    P.button("quit",1);

    bin_tree* tree[3];
    tree[0] = new avl_tree;
    tree[1] = new bb_tree;
    tree[2] = new rb_tree;

    while ( gw.open_panel(P) == 0)
    {
      anim_bin_tree T(gw,tree[kind]);

      switch (input) {

      case 0: { // random
                for(int i=0;i<n;i++) T.insert(rand_int(0,99));
                break;
              }

      case 1: { // increasing
                for(int i=0;i<n;i++) T.insert(i);
                break;
              }
     }

     gw.message("Press done to continue.");
     gw.edit();
    }

    delete[] tree;

    return 0;
  }
```

It remains to explain the class *bin_tree_anim*. An object of this class consists of a reference *T* to a *bin_tree* and a reference *gw* to a GraphWin, a *GRAPH<point, int> G*, and a map *NODE* from tree items to graph nodes; *T* and *gw* are set in the constructor to references of our GraphWin and the selected tree, respectively.

The idea is that *G* represents a drawing of *T* and that *NODE* makes the translation from tree nodes to graph nodes. In the constructor we make *G* the graph of *gw* and set *flush* to false, and in the destructor we reset T to the empty tree. The other functions will be discussed below.

⟨*class anim_bin_trees*⟩≡

```
class anim_bin_tree {
  GraphWin& gw;
  bin_tree& T;
  GRAPH<point,int> G;
  map<bin_tree_item,node>  NODE;
  ⟨functions to compute a drawing of T⟩
public:
anim_bin_tree(GraphWin& gwin, bin_tree* tptr) : gw(gwin), T(*tptr)
{ gw.message(string("\\bf\\blue %s",T.tree_name()));
  //G.clear();
  gw.set_flush(false);
  gw.set_graph(G);
}
~anim_bin_tree() { T.clear(); }
⟨anim_bin_tree:: insert⟩
};
```

We next explain the function *scan_tree* that computes the layout and sets the visual parameters of the nodes by calling *set_node_params* for each item r of T. Setting the node parameters is easy. We draw leaves and the root as rectangles and all other nodes as ellipses. For non-leaves we display the balance of the node in an appropriate form: in the case of AVL-trees we use the labels $<$, $=$, and $>$, in the case of BB[α]-trees we display the balance, and in the case of red-black trees we display the balance as a color.

⟨*functions to compute a drawing of T*⟩≡

```
void set_node_params(bin_tree_item r)
{
  node v = NODE[r];
  if ( T.is_leaf(r) )
  { gw.set_color(v,ivory);
    gw.set_label(v,string("%d",T.key(r)));
    gw.set_shape(v,rectangle_node);
    return;
   }
  if ( T.is_root(r) )
    gw.set_shape(v,rectangle_node);
  else
    gw.set_shape(v,ellipse_node);
  gw.set_color(v,grey1);
  int bal = T.get_bal(r);
  switch ( T.tree_type() ) {
    case  LEDA_AVL_TREE:
```

```
            switch (bal) {
              case  0: gw.set_label(v,"="); break;
              case -1: gw.set_label(v,">"); break;
              case  1: gw.set_label(v,"<"); break;
             }
            break;
    case  LEDA_BB_TREE:
            gw.set_label(v,string("%d",bal));
            break;
    case  LEDA_RB_TREE:
            gw.set_label_type(v,no_label);
            gw.set_color(v,(bal == 0) ? red : grey3);
            break;
   }
}
```

The function *scan_tree* computes the layout for the subtree rooted at r and also adds the edges in the subtree to G. The subtree is placed in the rectangle with left boundary *x0*, right boundary *x1*, upper boundary y, and vertical displacement *dy* between parents and their children. Such a layout is easily computed. We set the x-coordinate of r to the midpoint of *x0* and *x1* and the y-coordinate to the upper boundary and then place the left subtree in the left half of the rectangle and the right subtree in the right half of the rectangle. In both halves we lower the upper boundary by *dy*.

⟨*functions to compute a drawing of T*⟩+≡

```
node scan_tree(bin_tree_item r,double x0, double x1, double y, double dy)
{
  set_node_params(r);
  node   v = NODE[r];
  double x = (x0 + x1)/2;
  G[v] = point(x,y);
  bin_tree_item left  = T.l_child(r);
  bin_tree_item right = T.r_child(r);
  if (left)  G.new_edge(v,scan_tree(left,x0,x,y-dy,dy));
  if (right) G.new_edge(v,scan_tree(right,x,x1,y-dy,dy));
  return v;
}
```

We finally explain the insertion procedure. We lookup x; our trees store generic pointers of type *void∗* as explained in Chapter 13. We therefore need to convert x to a generic pointer. If x is already in the tree, we do nothing. Otherwise, we insert the pair $(x, 0)$ into T and store the tree item returned in p. If p is the root of T, i.e., the current insertion was the first insertion into T, we add a node to *gw* (and hence G), place it at the origin, and associate it with p. If p is not the root of T and hence the current insertion is not the first, the insertion added two nodes to the tree as shown in Figure 12.15. The node p is a leaf of T and p and $r = T.get\_last\_node()$ are the new nodes of T.

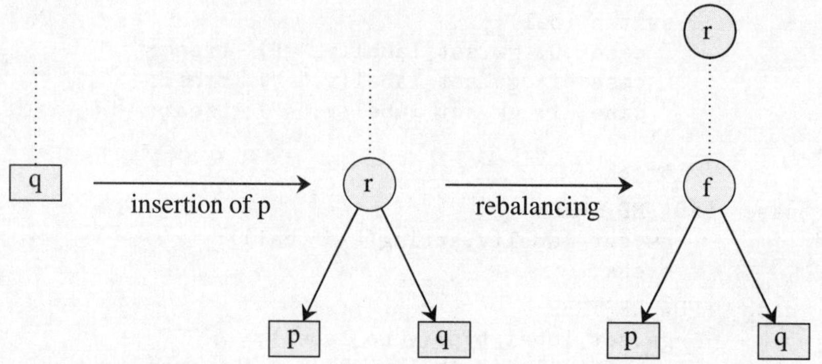

Figure 12.15 Insertion of a new key adds a new leaf *p* and a new node *r*. The search for the key of *p* in the old tree ended in *q* and the key of *p* is either smaller or larger than the key of *q*. In the former case, *p* will be the left child or *r* and in the latter case it will be the right child. After the addition of the new leaf the tree is rebalanced and *r* might move to a different position in the tree. A call of *T.get_last_node*( ) after the insertion returns *r*. We set the initial positions of *p* and *r* to the position of *q* before the insertion.

We add two new nodes to *gw*, one corresponding to *p* and the other one corresponding to *r*. We place both nodes on top of *q*. We next compute the drawing area and update the drawing. We compute the drawing area as follows. We leave four pixels unused on either side and we divide the *y*-extension of the window into ten (since our trees will never grow deeper than eight) strips. We leave the two top-most strips unused.

⟨*anim_bin_tree:: insert*⟩≡

```
void insert(int x)
{
  if (T.lookup(GenPtr(x))) return;
  bin_tree_item p = T.insert(GenPtr(x),0);
  if ( T.is_root(p) )
     NODE[p] = gw.new_node(point(0,0));
  else
  { bin_tree_item f = T.parent(p);
    bin_tree_item q = T.l_child(f);
    if (p == q) q = T.r_child(f);
    point pos = gw.get_position(NODE[q]);
    bin_tree_item r = T.get_last_node();
    NODE[p] = gw.new_node(pos);
    NODE[r] = gw.new_node(pos);
   }
  node v = NODE[p];
  // compute drawing area
  double dx = gw.get_window().pix_to_real(4);
  double x0 = gw.get_xmin() + dx;
  double x1 = gw.get_xmax() - dx;
  double y0 = gw.get_ymin();
  double y1 = gw.get_ymax();
```

```
    double dy = (y1-y0)/10;
    ⟨update drawing⟩
  }
```

It remains to explain how we update the drawing. We first remove all edges from G and then call *scan_tree* for the root of T and the entire drawing area. This builds T in G and computes a new layout in the node data of G. We then inform *gw* that G has changed and set the color of the new node to green. We set flush to true so that changes go into effect and change the node positions to the node data of G by the call *gw.set_position*(*G.node_data*( )). Because layout changes are animated this will make the tree move slowly into its new shape. You may change the speed in the options menu. When the tree is in its new form we reset the color of v and set flush back to false.

⟨*update drawing*⟩≡

```
    G.del_all_edges();
    scan_tree(T.root(),x0,x1,y1-2*dy,dy);
    gw.update_graph();
    color col = gw.set_color(v,green2);
    gw.set_flush(true);
    gw.set_position(G.node_data());
    gw.set_color(v,col);
    gw.set_flush(false);
```

Exercise for 12.8

1 Extend the binary tree animation of this chapter to allow deletions of keys by clicking on the corresponding leaves.

13

On the Implementation of LEDA

This chapter deals with the implementation of LEDA. It gives the details of the implementation of parameterized data types, implementation parameters, handle types, the memory management, and iteration macros. We close the chapter with a comprehensive example that illustrates all concepts discussed.

13.1 Parameterized Data Types

The definition of parameterized data types of LEDA has been discussed in Chapter 2. In the next sections we describe how they are implemented. We first describe the C++ template approach to parameterized data types using a simple list data type. Then we use the same example to explain the basic idea of the LEDA solution for implementing parameterized data types and discuss the reasons for choosing this solution. Finally, we extend the basic solution and apply it to more advanced data types and develop optimizations for the case where the actual type parameters are small (fit into one memory word) or are basic built-in types.

13.2 A Simple List Data Type

We start this section by giving a very simple implementation for a data type *list* of singly linked lists of integers. It offers about the same set of operations as the LEDA *stack* data type. There is a *push* operation that inserts a given integer at the front of the list and a *pop* operation that removes the first element from the list and returns it. Operation *head* returns the first element without changing the list, and finally, operation *size* returns the number of elements of the list. Of course, we also have to provide a constructor, destructor, copy constructor and an assignment operator in order to make *list* a fully equipped C++ data type.

Note that we use this simple type only as a first example for introducing some aspects of the LEDA mechanism for implementing parameterized data types. Of course, LEDA contains much more powerful and useful list types, see Section 3.2.

As usual, the declaration (or specification) of our list class is contained in a header file called *_list.h* and the implementations of its operations are contained in a separate source code file *_list.c*. We let the file names start with _, because we want to use the file names without the underscore later in the section.

The header file *_list.h* might look as follows:

⟨*_list.h*⟩≡

```
class list {
  struct list_elem
  { // a local structure for representing the elements of the list
    int entry;
    list_elem* succ;
    list_elem(const int& x, list_elem* s) : entry(x), succ(s) {}
    friend class list;
  };
  list_elem* hd;  // head of list
  int        sz;  // size of list
public:
  void push(int);
  void pop(int&);
  int  head() const;
  int  size() const;
  list();
 ~list();
  list(const list&);
};
```

The corresponding source code file *_list.c* is as follows:

⟨*_list.c*⟩≡

```
#include "_list.h"
#define NULL 0
int list::head() const { return hd->entry; }
void list::push(int x)
{ hd = new list_elem(x,hd);
  sz++;
}
void list::pop(int& y)
{ y = hd->entry;
  list_elem* p = hd;
  hd = p->succ;
  delete p;
  sz--;
```

```
}
list::list()
{ // construct an empty list
  hd = NULL;
  sz = 0;
}
list::list(const list& L)
{ // construct a copy of L
  hd = NULL;
  sz = L.sz;
  if (sz > 0)
  { hd = new list_elem(L.hd->entry,0); // first element
    list_elem* q = hd;
    // subsequent elements
    for (list_elem* p = L.hd->succ; p != NULL; p = p->succ)
    { q->succ = new list_elem(p->entry,NULL);
      q = q->succ;
    }
  }
}
list::~list()
{ // destroy the list
  while (hd)
  { list_elem* p = hd->succ;
    delete hd;
    hd = p;
  }
}
```

13.3 The Template Approach

Most data types in LEDA are parameterized. LEDA does not only offer lists of integers but lists of an arbitrary element type E. In this section we discuss the C++ standard approach to parameterized data types. We explain the approach and discuss why we have not taken it in LEDA. The solution which we adopted in LEDA is described in the next section.

C++ supports parameterized classes by means of its *template feature*. How can one obtain lists of *char* from our implementation of lists of *int*? It seems to be very simple. Replace in files list.h and list.c all occurrences of *int* by *char*. Well, that's not quite true. Actually, we should replace only those occurrences of *int* that refer to the element type of the list. So the declarations of variable *sz* and the return type of *size*( ) stay unchanged. Since it is completely mechanical to derive list of characters from lists of integers we might as well ask the compiler to do it. All we have to do is to mark those occurrences of *int* that are to be replaced. The template feature of C++ is an elegant way to automate this transformation. The following simple textual transformation changes the definition of our list class into the definition of a parameterized list class:

- Replace in list.h all occurrences of *int* that refer to the element type of our lists by a new class name, say E.
- Prefix the definition of class *list* in the file list.h and the definition of each member function in the file list.c by *template <class E>*. This informs the compiler that E is the name of a type parameter and not the name of a concrete type.
- Replace in list.c all occurrences of *list* that refer to the name of the list class by *t_list<E>*. This replacement is not really necessary. We make it so that we can later contrast classes *list* and *t_list*.

For concreteness, we include excerpts from the modified files t_list.h and t_list.c.

⟨*t_list.h*⟩≡

```
template <class E>
class t_list {
  struct list_elem
  { E entry;
    list_elem* succ;
    list_elem(const E& x, list_elem* s) : entry(x), succ(s) {}
  };
  list_elem* hd;  // head of list
  int        sz;  // size of list
public:
  void push(const E&);
  void pop(E&);
  const E& head() const;
  int  size() const;
  t_list();
  t_list(const t_list<E>&);
 ~t_list();
};
```

and file t_list.c is as follows:

⟨*t_list.c*⟩≡

```
#include "t_list.h"
#define NULL 0
template <class E> t_list<E>::t_list()
{ hd = NULL;
  sz = 0;
}
template <class E> const E& t_list<E>::head() const
{ return hd->entry; }
template <class E> void t_list<E>::push(const E& x)
{ hd = new list_elem(x,hd);
  sz++;
}
```

```
template <class E> void t_list<E>::pop(E& y)
{ y = hd->entry;
  list_elem* p = hd;
  hd = p->succ;
  delete p;
  sz--;
}
```

In an application program we can now write

```
t_list<char> L1;
t_list&lt;segment&gt; L2;
```

to define a list *L1* of *char* and a list *L2* of line segments, respectively. When the compiler encounters these definitions it constructs two versions of files list.c and list.h by substituting *E* by *char* and by *segment*, respectively, which it can then process in the standard way. Let us summarize:

- The template feature is powerful and elegant. The implementer of a data type simply prefixes his code by *template<class E>* and otherwise writes his code as usual, and the user of a parameterized data type only needs to specify the actual type parameter in angular brackets.

- The template feature duplicates code. This increases code length and compilation time. It has to duplicate code because the layout of the elements of a list in memory (type *list_elem*) depends on the size of the objects of type *E* and hence the code generating new list elements depends on the size of the objects of the actual type parameter.

- Separate compilation is impossible. Since the code to be generated depends on the actual type parameter one cannot precompile t_list.c to obtain an object file t_list.o. Rather both files t_list.h and t_list.c have to be included in an application and have to be compiled with the application. For an application, that uses many parameterized data types from the library, this leads to a large source and therefore large compilation times. Moreover, it forces the library designer to make his .c-files public.

- When we started this project, most C++ compilers did not support templates and, even today, many do not support them fully. Some compilers use repositories of precompiled object code to avoid multiple instantiations of the same template code. However, there is no standard way for solving this problem.

We found in particular the drawback of large compilation times unacceptable and therefore decided against the strategy of implementing parameterized data types directly by the template feature of C++.

The LEDA solution uses templates in a very restricted form. It allows separate compilation, it allows us to keep the .c-files private, and it does not over-strain existing C++ compilers. We discuss it in the next section.

Let us summarize. The template feature is an elegant method to realize parameterized data types (from a user's as well as an implementor's point of view). However, it also has a certain weakness. It duplicates code, it does not allow us to precompile the data types, and it is only partially supported by compilers.

13.4 The LEDA Solution

In LEDA every parameterized data type is realized by a pair of classes: a *class for the abstract data type* and a *class for the data structure*, e.g., we have a dictionary class (= the data type class) and a binary tree class (= the data structure class). Only the data type classes use the template mechanism. All data type classes are specified in the header file directory *LEDAROOT/incl/LEDA* and only their header files are to be included in application programs. All data structures are precompiled into the object code libraries (*libL, libG, libP, libW, ...)* and are linked to application programs by the C++ linker. Instead of abstract data type class we will also say data type class or data type template or abstract class and instead of data structure class we will also say implementation class or concrete class.

Precompilation of a data structure is only possible if its implementation does not depend on the actual type parameters of the corresponding parameterized data type. In particular:

- the layout of the data structure in memory must not depend on the size of the objects stored in it. We achieve this (in a first step) by always storing pointers to objects instead of the objects themselves in our data structures. Observe that the space requirement of a pointer is independent of the type of the object pointed to. In a second step (cf. Section 13.5.1) we show how to avoid this level of indirection in the case of small types (types whose size in memory is at most the size of a pointer).

- all functions used in the implementation whose meaning depends on the actual type parameters use the dynamic binding mechanism of C++, i.e., are realized as virtual functions. A prime example is the comparison function in comparison based data structures. The comparison function is defined as a virtual member function of the implementation class, usually called *cmp_key*. In the definition of the abstract data type template we bind *cmp_key* to a function *compare* that defines the linear order on the actual type parameter.

The remainder of this section is structured as follows. We first give the basic idea for parameterized data types in LEDA. Then we discuss the use of virtual functions and dynamic binding for the implementation of assignment, copy constructor, default constructor, and destruction. In the sections to follow we describe an improvement for so-called one-word or small types, and show how implementation parameters are realized. Finally, we give the full implementation of priority queues by Fibonacci heaps and illustrate all features in one comprehensive example.

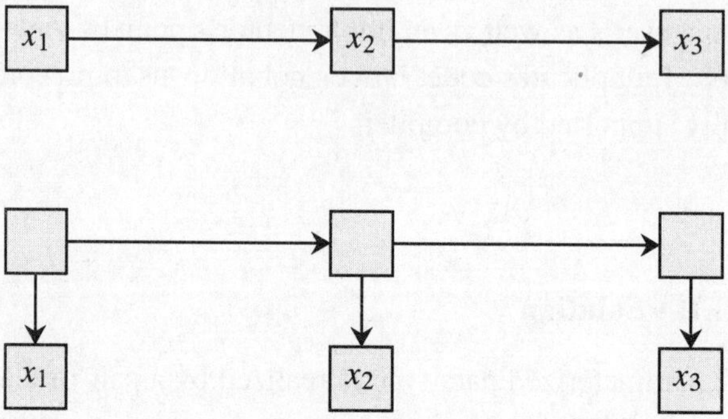

Figure 13.1 A *t_list* and a *list*: The top part shows a *t_list*<*E*> with three elements x_1, x_2, x_3. The bottom part shows the corresponding *list* data structure in the LEDA approach.

13.4.1 *The Basic Idea*

We introduce the basic idea for realizing parameterized data types in LEDA, the idea will be refined in later sections:

- The data fields in the containers of all data structures are of type *void*∗, the generic pointer type of C++. They contain pointers to objects of the actual type parameters. Consider a data structure whose containers have a slot for storing objects of a type T, e.g., the type *t_list*<*T*>.

- In the LEDA approach the objects of type T are not stored directly in the containers of the data structure but on the heap. The data slots of the containers have type *void*∗, the generic pointer type of C++, and contain pointers to the objects on the heap. More precisely, if a container has a slot of type T in the template solution and t is the object stored in it (at a particular time) then the corresponding container in the LEDA solution will have a field of type *void*∗ and this field will contain a pointer to t. See Figure 13.1 for an illustration.

- The abstract data type class uses the template mechanism and is derived from the implementation class.

- Type casting is used to bridge the gap between the untyped world of the implementation class (all data is *void*∗) and the typed world of the abstract class.

We use our singly linked list data type as a first example to illustrate our approach. We saw an implementation of lists, called *t_list*, using the template approach in the preceding section.

Our goal is to realize the parameterized data type *list*<*T*> by a concrete data structure *list_impl* that stores pointers of type *void*∗. The definition of *list_impl* is straightforward. It is essentially a list of type *t_list*<*void* ∗ >

⟨*list_impl.h*⟩≡

```
class list_impl {
  struct list_impl_elem
  { void* entry;
    list_impl_elem* succ;
    list_impl_elem(void* x,list_impl_elem* s):entry(x),succ(s) {}
    friend class list_impl;
  };
  list_impl_elem* hd;
  int sz;
 protected:
  list_impl();
 ~list_impl();
  void* head() const;
  void* pop();
  void  push(void* x);
  void  clear();
  int   size() const;
};
```

and

⟨*list_impl.c*⟩≡

```
#include "list_impl.h"
list_impl::list_impl() : hd(0), sz(0) {}
list_impl::~list_impl() { clear(); }
void* list_impl::head() const { return  hd->entry; }
void list_impl::push(void* x)
{ hd = new list_impl_elem(x,hd);
  sz++;
}
void* list_impl::pop()
{ void* x =  hd->entry;
  list_impl_elem* p = hd;
  hd = p->succ;
  delete p;
  sz--;
  return x;
}
void list_impl::clear() { while (hd) pop(); }
int  list_impl::size() const { return sz; }
```

We declared the member functions of *list_impl* protected so that they can only be used in derived classes. We can now easily derive the data type template *list*<*T*> for arbitrary types *T* from *list_impl*. We make *list_impl* a private base class of *list*<*T*> and implement the member functions of *list*<*T*> in terms of the member functions of the implementation class.

Making the implementation class a private base class makes it invisible to the users of the *list*<*T*> class. This guarantees type safety as we argue at the end of the section.

A member function of *list*<*T*> with an argument of type *T* first copies the argument into the dynamic memory (also called heap), then casts a pointer to the copy to *void*∗, and finally passes the pointer to the corresponding function of the implementation class.

All member functions of *list*<*T*> that return a result of type *T* call the corresponding function of the implementation class (which returns a result of type *void*∗), cast the pointer to *T*∗, and return the dereferenced pointer.

We next give the details.

```
template<class T>
class list : private list_impl {
public:
  list() : list_impl() {}
```

The constructor of *list*<*T*> constructs an empty *list_impl*.

```
  void push(const T& x) { list_impl::push((void*) new T(x)); }
```

L.push(*x*) makes a copy of *x* in dynamic memory (by calling the copy constructor of *T* in the context of the *new* operator) and passes a pointer to this copy (after casting it to *void*∗) to *list_impl*::*push*. The conversion from *T*∗ to *void*∗ is a built-in conversion of C++ and hence we may equivalently write

```
  void push(const T& x) { list_impl::push(new T(x)); }
```

Let us relate *list*<*T*>::*push*(*x*) to *t_list*<*T*>::*push*(*x*). The latter operation stores a copy of *x* directly in the entry-field of a new list element and the former makes a copy of *x* on the heap and stores a pointer to the copy in entry-field.

```
  const T&  head() const { return *(T*)list_impl::head(); }
```

L.head( ) casts the *void*∗ result of *list_impl*::*head*( ) to a *T*∗ pointer, dereferences the result, and returns the object obtained as a const-reference. It thus returns the element of the list that was pushed last.

```
  T pop()
  { T* p = (T*)list_impl::pop();
    T x = *p;
    delete p;
    return x;
  }
```

L.pop(*x*) casts and dereferences the *void*∗ result of *list_impl*::*pop*, assigns it to a local variable *x*, deletes the copy (made by *list*<*T*>::*push*), and returns *x*. Observe that the assignment to *x* makes a copy, and it is therefore OK to delete the copy made by *push*. It is also necessary to delete it, as we would have a memory leak otherwise.

```
  int size() const { return list_impl::size(); }
  void clear()
  { while (size() > 0) delete (T*)list_impl::pop();
```

```
      list_impl::clear();
    }
   ~list() { clear(); }
  };
```

The implementations of *clear* and of the destructor are subtle. *Clear* first empties the list and then calls *list_impl::clear*. The latter call is unnecessary as popping all elements from the list already has the effect of clearing the list. We make the call for reasons of uniformity (all *clear* functions of abstract classes in LEDA first destroy all objects contained in the data structure and then call the *clear* function of the implementation). It is, however, absolutely vital to destroy the objects stored in the list before calling *list_impl::clear*. An implementation

```
void clear()        { list_impl::clear(); }
```

has a memory leak as it leaves the elements contained in the list as orphans on the heap.

The destructor first calls *clear* and then the destructor of the base class (the latter call being automatically inserted by the compiler). The base class destructor *~list_impl* deletes all list elements. Observe that it does not suffice to call this destructor as this will leave all entries contained in the elements of the list on the heap.

If our list implementation class would support iteration in the LEDA *forall* style an alternative implementation of the clear function would be

```
void clear()
{ void* p;
  forall(p,*this) delete (T*)p;
  list_impl::clear();
}
```

Let us assess our construction:

- The construction is non-trivial. Please read it several times to make sure that you understand it and try to mimic the approach for other data types (see the exercises). The construction is certainly more complicated than the pure template approach presented in the preceding section.

- The data type *list<T>* simulates the data type *t_list<T>*. Suppose that we perform the same sequence of operations on a *list<T> S* and a *t_list<T> TS*. Assume that $x_0, \ldots,$ x_{t-1} are the entries of *TS* after performing the sequence. Then *S* also has *t* elements and the corresponding entries contain pointers to copies of x_0 to x_{t-1} in dynamic memory, see Figure 13.1.

- All operations of *list<T>* are implemented by very simple inline member functions. Except for *pop*, *clear*, and *~list( )* they do not produce any additional code. We will show in the next section how the code for *pop* and *clear* can also be moved into the data structure by the use of virtual functions. This will make the definition of the abstract class cleaner.

- The implementation class can be precompiled; see below.
- The above implementation of lists is incomplete. In particular, the definitions of the copy constructor and of the assignment operator are missing. We will discuss them in Section 13.4.3.

The abstract data type *list<T>* can be used in the usual way.

```
list<string> L;
L.push("fun");
L.push("is");
L.push("LEDA");
while (L.size() > 0) cout << L.pop() << endl;
```

Separate Compilation: We defined two classes, the implementation class *list_impl* and the abstract data type class *list<T>*, in three files: the file *list_impl.h* contains the skeleton of the class definition of *list_impl*, namely the definition of the private data of the class and the declarations of the member functions, the file *list_impl.c* contains the implementation of all member functions of *list_impl* and the file *list.h* contains the definition of the abstract data type and its implementation in terms of the implementation class. We have shown all three files above. It is still worthwhile to repeat their global structure.

⟨*list_impl.h* ⟩≡

```
class list_impl {
   ⟨definition of private data⟩
protected:
   ⟨declaration of member functions⟩
};
```

The file *list_impl.c* contains the implementations of all member functions. It must include *list_impl.h*

⟨*list_impl.c* ⟩≡

```
#include <list_impl.h>
⟨implementation of all member functions⟩
```

The abstract class template *list<T>* is defined in file *list.h*. It is derived from class *list_impl* and all member functions of the abstract class are realized by calling the corresponding member function of the implementation class as described above. The calls also do the appropriate type conversions from type T to $void*$ and vice versa. Since class *list_impl* is only used to implement its derived classes *list<T>* it is qualified as a private base class of the list template. Of course, we have to include *list_impl.h* before using *list_impl* as a base class.

⟨*list.h*⟩≡

```
#include "list_impl.h"

template<class T>
class list : private list_impl {
public:
  void push(const T& x)  { list_impl::push(new T(x)); }
  const T&  head() const { return *(T*)list_impl::head(); }
  void pop(T& x)
  { T* p = (T*)list_impl::pop();
    x = *p;
    delete p;
  }
  int size() const       { return list_impl::size(); }
  void clear()
  { while (size() > 0) delete (T*)list_impl::pop();
    list_impl::clear();
  }
  list() : list_impl() {}
 ~list() { clear(); }
};
```

The file *list_impl.c* can be compiled into the object code file *list_impl.o*. An application program, say *list_prog.c*, using lists needs to include *list.h* and can be compiled separately into file *list_prog.o*. Finally, *list_prog.o* and *list_impl.o* can be linked to an executable program.

In the LEDA system the header files of implementation classes are collected in the directory *LEDAROOT/incl/LEDA/impl* and the header files of abstract classes are collected in *LEDAROOT/incl/LEDA*. All .c-files are contained in the various subdirectories of *LEDAROOT/src*.

Type Safety: We next comment on the type safety of the construction described above. The implementation class *list_impl* is untyped in the sense that anything can be pushed onto a list of type *list_impl*, the class *list* is typed in the sense that only objects of type T can be pushed onto a list of type *list*<T>. In the definition of class *list* we make the transition from the safer (typed) world to a potentially unsafer (untyped) world. Since we declared all operations of *list_impl* protected and made *list_impl* a private base class of *list*, the untyped world is completely encapsulated inside class *list* and invisible to any application program. Only the implementation class *list_impl* works in the untyped world; we designed it carefully so as to avoid the dangers of the untyped world. We conclude that the construction is type safe.

Efficiency: The construction is also efficient. Note that no code needs to be generated for the type conversions; the casts simply tell the compiler how the entries of the list are to be interpreted. Also all member functions of the abstract class are trivial inline functions and their calls can be eliminated by optimizing compilers, i.e., there is, for example, no need first to call the abstract function *list::push* which in turn calls the concrete function *list_impl::push*. The compiler will directly call the concrete function.

Genericness: Finally, the construction is elegant, although not as elegant as the solution relying completely on templates. The definition of the implementation class is completely natural[1], it is essentially *t_list<void*>*. The definition of the abstract class in terms of the implementation class is somewhat inelegant because of the required type conversions. However, these type conversions follow a very simple rule. In-going values are converted to *void** and return-values are converted back to type T.

13.4.2 *Virtual Functions and Dynamic Binding*

In the example of the preceding section the implementation class *list_impl* required no knowledge about the actual type argument of the data type template *list*$<T>$. This is an exceptional situation; in most situations the implementation needs to have some knowledge about the actual type argument. We give two examples.

The first example is a print operation for our list type that prints all elements to the standard output. We want to realize this operation by a *print* member function in the implementation class *list_impl*. Of course, this function needs to know how to print an object of the actual type parameter. The second example is comparison-based implementations of dictionaries, e.g., binary search trees. Any comparison-based implementation of the parameterized data type *dictionary*$<K, I>$ (cf. Section 5.3) needs to know how to compare keys. In LEDA, the linear order on a key type K is defined by a global function *int compare*(*const* K&, *const* K&) (cf. Section 2.10) and hence the implementation class must be able to call this function.

In both examples we need a mechanism to transfer functionality of the actual type parameters from the abstract data type template to the implementation class. The appropriate C++ feature is dynamic binding and virtual functions. Detailed discussions of this concept can be found in [Str91, ES90]. The following should be clear even without prior knowledge of the concept.

In the first example, the class *list_impl* uses a virtual function *print_elem*(*void* $* \, p$) to print elements to standard output. This function is declared in the implementation class but its implementation is left undefined by labeling it as pure virtual. Syntactically, pure virtual functions are designated by the key word *virtual* and the assignment "=0" which replaces the body. The implementation class may use the virtual function in its other member functions, e.g., *list_impl* uses *print_elem* in a function *print* that prints the entire list.

```
class list_impl {
  ...
virtual void print_elem(void*) const = 0;
  ...
void print() const
```

[1] You may want to include a *typedef void** T; to make it look even more natural.

```
{ for(list_impl_elem* p = hd; p; p = p->succ) print_elem(p->entry); }
  ...
};
```

The implementation of *print_elem* is provided in the derived class *list*<*T*>. It converts its argument from *void*∗ to *T*∗ (observe that this conversion makes sense on the level of the data type template) and then hands the object pointed to to the output operator (≪) of type *T* (assuming that this operator is defined for *T*).

```
template<class T>
class list : private list_impl {
  ...
 void print_elem(void* p) const { cout << *(T*)p << endl; }
 void print() const { list_impl::print(); }
  ...
};
```

When a list is created, say through the declaration *list*<*char* ∗ > *L*; the definition of *print_elem* in terms of *operator* ≪ (*ostream*&, *char*∗) is associated with *L*. In a call *L.print*( ) which leads via *list_impl*::*print*( ) to a call of *list_impl*::*print_elem* the implementation of *print_elem* bound to *L* is used. In this way, information about the actual type parameter is transported into the implementation class.

We turn to our second example. All implementations of *dictionary*<*K*, *I*> use a virtual member function *int cmp_key*(*void*∗, *void*∗) for comparing keys. We discuss the implementation class *bin_tree*. As in the previous example, *cmp_key* is declared as pure virtual in the implementation class. In the derived class template *dictionary*<*K*, *I*> we define *cmp_key* in terms of the compare function of type *K*. We have

```
class bin_tree {
  ...
virtual int cmp_key(void*,void*) const = 0;
  ...
};
```

in the implementation class and

```
template<class K, class I>
class dictionary: private bin_tree {
  ...
int cmp_key(void* x, void* y) const { return compare(*(K*)x,*(K*)y); }
  ...
};
```

in the data type template (note the conversion from *void*∗ to *K*∗ in the implementation of *cmp_key*).

The construction associates the appropriate compare function with every dictionary, e.g., *compare*(*const int*&, *const int*&) with *dictionary*<*int*, *int*>. Furthermore, the compare function is available in the implementation class *bin_tree* and can be called by its member functions (e.g. lookup).

In the remainder of this section and in the next section we give more details of the *bin_tree* class. This will allow us to discuss further aspects of the LEDA approach to parameterized data types.

The nodes of a *bin_tree* are realized by a class *bin_tree_node*. Each node contains a key and an information, both of type *void*∗, and additional data members for building the actual tree. For unbalanced trees the pointers to the two children suffice. For balanced trees additional information needs to be maintained. All implementations of balanced trees in LEDA are derived from the *bin_tree* class.

In the remainder of this chapter we will use the type name *GenPtr* for the generic pointer type *void*∗.

```
typedef void* GenPtr;
class bin_tree_node {
  GenPtr key;
  GenPtr inf;
  bin_tree_node* left_child;
  bin_tree_node* right_child;
  // allow bin_tree to access all members
  friend class bin_tree;
};
```

The class *bin_tree* contains some private data, such as a pointer to the root of the tree. The member functions realizing the usual dictionary operations are declared protected to make them accessible for derived classes (e.g., *dictionary*$<K, I>$) and the *cmp_key* function is declared a private pure virtual function. Finally, we define the item type (cf. Section 2.2.2) for class *bin_tree* (*bin_tree*::*item*) to be equal to type *bin_tree_node*∗.

```
class bin_tree {
private:
   bin_tree_node*  root;
   int cmp_key(GenPtr,GenPtr) const = 0;
protected:
   typedef bin_tree_node* item;
   item  insert(GenPtr,GenPtr);
   item  lookup(GenPtr) const;
   void  del_item(item);
   GenPtr key(item p) const { return p->key; }
   GenPtr inf(item p) const { return p->inf; }
   bin_tree();
  ~bin_tree();
};
```

The virtual *cmp_key* function is used to compare keys, e.g., in the *lookup* member function that returns a pointer to the node storing a given key *k* or *nil* if *k* is not present in the tree.

```
bin_tree_node* bin_tree::lookup(GenPtr k) const
{ bin_tree_node* p = root;
  while (p)
```

```
  { int c = cmp_key(k,p->key);
    if (c == 0) break;
    p = (c > 0) ? p->right_child : p->left_child;
  }
  return p;
}
```

In the definition of the data type template *dictionary*<K, I> we define *cmp_key* in terms of the compare function for type K. The dictionary operations are realized by calling the corresponding member functions of *bin_tree*. As in the list example, we also need to perform the necessary type conversions. The item type of dictionaries (*dic_item*) is defined to be equal to the item type of the implementation class *bin_tree*::*item*.

```
typedef bin_tree::item dic_item;

template<class K, class I>
class dictionary : private bin_tree {
  int cmp_key(GenPtr x, GenPtr y) const
  { return compare(*(K*)x,*(K*)y); }
public:
  const K& key(dic_item it) const { return *(K*)bin_tree::key(it); }
  const I& inf(dic_item it) const { return *(I*)bin_tree::inf(it); }
  dic_item insert(const K& k const I& i)
  { return bin_tree::insert(new K(k), new I(i)); }
  dic_item bin_tree::lookup(const K& k) const
  { return bin_tree:lookup(&k); }
};
```

Observe that *bin_tree*::*lookup* expects a *GenPtr* and hence we pass the address of k to it.

The code for classes *bin_tree* and *dictionary*<K, I> is distributed over the files *bin_tree.h*, *bin_tree.c*, and *dictionary.h* as described in the previous section: classes *bin_tree_node* and *bin_tree* are defined in *LEDA*/*impl*/*bin_tree.h*, the implementation of *bin_tree* is contained in *LEDAROOT*/*src*/*dict*/*bin_tree.c*, and *dictionary*<K, I> is defined in *LEDA*/*dictionary.h*.

The above implementation of dictionaries has a weakness (which we will overcome in the next section). Consider the insert operation. According to the specification of dictionaries (see Section 5.3) a call *D.insert*(k, i) adds a new item $\langle k, i\rangle$ to D when there is no item with key k in D yet and otherwise replaces the information of the item with key k by i. However, in the implementation given above *dictionary*<K, I>::*insert*(k, i) makes a copy of k and then passes a pointer to this copy to *bin_tree*::*insert*. If k is already in the tree *bin_tree*::*insert* must destroy the copy again (otherwise, there would be a memory leak). It would be better to generate the copy of k only when needed.

In the next section we show how to shift the responsibility for copying and deleting data objects to the implementation class by means of virtual functions. We will also show how to implement the missing copy constructor, assignment operator, and destructor.

13.4.3 *Copy Constructor, Assignment, and Destruction*

Copying, assignment, and destruction are fundamental operations of every data type. In C++ they are implemented by copy constructors, assignment operators, and destructors. Let us see how they are realized in LEDA. As an example, consider the assignment operation $D1 = D2$ for the data type *dictionary*$\langle K, I\rangle$. A first approach would be to implement this operation on the level of abstract types, i.e., in the data type template *dictionary*$\langle K, I\rangle$. We could simply first clear *D1* by a call of *D1.clear*( ) and then insert the key/information pairs for all items *it* of *D2* by calling *D1.insert*(*D2.key*(*it*), *D2.inf*(*it*)) for every one of them. This solution is inflexible and inefficient; the assignment would take time $O(n \log n)$ instead of time $O(n)$.

A second approach is to realize the operation on the level of the implementation class *bin_tree*. This requires that *bin_tree* knows how to copy a key and an information. In the destructor it also needs to know how to destroy them. There are also many other reasons why the implementation class should have these abilities, as we will see. In LEDA, we use virtual functions and dynamic binding to provide this knowledge.

In the dictionary example, we have the following virtual member functions in addition to *cmp_key*:

void copy_key(*GenPtr*& x) and *void copy_inf*(*GenPtr*& x) that make a copy of the object pointed to by x and assign a pointer to this copy to x,

void clear_key(*GenPtr* x) and *void clear_inf*(*GenPtr* x) that destroy the object pointed to by x, and finally

void assign_key(*GenPtr* x, *GenPtr* y) and *void assign_inf*(*GenPtr* x, *GenPtr* y) that assign the object pointed to by y to the object pointed to by x.

We exemplify the use of the virtual copy and clear function in two recursive member functions *copy_subtree* and *clear_subtree* of *bin_tree* that perform the actual copy and clear operations for binary trees. The copy constructor, the assignment operator, the destructor, and the *clear* function of class *bin_tree* are then realized in terms of *copy_subtree* and *clear_subtree*. The use of *assign_inf* will be demonstrated later in the realization of the operation *change_inf*.

In the header file *bin_tree.h* we extend class *bin_tree* as follows.

```
class bin_tree {
private:
  ...
  virtual void copy_key(GenPtr&) const = 0;
  virtual void clear_key(GenPtr) const = 0;
  virtual void assign_key(GenPtr x, GenPtr y) const =0;

  virtual void copy_inf(GenPtr&) const = 0;
  virtual void clear_inf(GenPtr) const = 0;
  virtual void assign_inf(GenPtr x, GenPtr y) const =0;

  void clear_subtree(bin_tree_node* p);
  // deletes subtree rooted at p

  bin_tree_node* copy_subtree(bin_tree_node* p);
  // copies subtree rooted at p, returns copy of p
```

```
protected:
  void clear();
  bin_tree(const bin_tree& T);
  bin_tree& operator=(const bin_tree& T);
 ~bin_tree() { clear(); }
};
```

In the data type template *dictionary*<K, I> we realize the virtual copy, assign, and clear functions by type casting, dereferencing, and calling the new, assignment, or delete operators of the corresponding parameter types K and I. Copy constructor, assignment operator, and destructor of type *dictionary* are implemented by calling the corresponding operations of the base class *bin_tree*.

```
template<class K, class I>
class dictionary: private bin_tree {
  ...
  void copy_key(GenPtr& x) const { x = new K(*(K*)x); }
  void copy_inf(GenPtr& x) const { x = new I(*(I*)x); }
  void clear_key(GenPtr x) const { delete (K*)x; }
  void clear_inf(GenPtr x) const { delete (I*)x; }
  void assign_key(GenPtr x, GenPtr y) const { *(K*)x = *(K*)y; }
  void assign_inf(GenPtr x, GenPtr y) const { *(I*)x = *(I*)y; }
  ...
public:
  ...
  dictionary(const dictionary<K,I>& D) : bin_tree(D) {}
  dictionary<K,I>& operator=(const dictionary<K,I>& D)
  { bin_tree::operator=(D); return *this; }
 ~dictionary() { bin_tree::clear(); }
};
```

The functions *bin_tree::copy_subtree*, *bin_tree::clear_subtree*, *bin_tree::clear*, the copy constructor, the destructor, and the assignment operator are implemented in bin_tree.c.

```
bin_tree_node*  bin_tree::copy_subtree(bin_tree_node* p) {
  if (p == nil) return nil;
  bin_tree_node* q = new bin_tree_node;
  q->l_child = copy_subtree(p->l_child);
  q->r_child = copy_subtree(p->r_child);
  q->key = p->key;
  q->inf = p->inf;
  copy_key(q->key);
  copy_inf(q->inf);
  return q;
}

void  bin_tree::clear_subtree(bin_tree_node* p) {
  if (p == nil) return;
  clear_subtree(p->l_child);
  clear_subtree(p->r_child);
  clear_key(p->key);
  clear_inf(p->inf);
```

```
  delete p;
}
void bin_tree::clear() {
  clear_subtree(root);
  root = nil;
}
bin_tree& bin_tree::operator=(const bin_tree& T) {
  if (this != &T)
  { clear();
    root = copy_subtree(T.root);
  }
  return *this;
}
```

The implementation of the copy constructor is subtle. It is tempting to write (as in *operator* =)

```
bin_tree::bin_tree(const bin_tree& T) { root = copy_subtree(T.root); }
```

This will not work. The correct implementation is

```
bin_tree::bin_tree(const bin_tree& T) { root = T.copy_subtree(T.root); }
```

What is the difference? In the first case we call *copy_subtree* for the object under construction, and in the second case we call *copy_subtree* for the existing tree T. The body of *copy_subtree* seems to make no reference to either T or the object under construction. But note that all member functions of a class have an implicit argument, namely the instance to which they are applied. In particular, the functions *copy_key* and *copy_inf* are either T's versions of these functions or the new object's versions. The point is that these versions are different.

Object T belongs to class *dictionary*<K, I> and hence knows the correct interpretation of *copy_key* and *copy_inf*. The object under construction does not know them yet. It knows them only when the construction is completed. As long as it is under construction the functions *copy_key* and *copy_inf* are as defined in class *bin_tree* and not as defined in the derived class *dictionary*<K, I>. In other words, when an object of type *dictionary*<K, I> is constructed we first construct a *bin_tree* and then turn the *bin_tree* into a *dictionary*<K, I>. The definitions of the virtual functions are overwritten when the *bin_tree* is turned into a *dictionary*<K, I>.

What will happen when the wrong definition of the *copy_subtree* function is used, i.e., when the copy constructor of *bin_tree* is defined as

```
bin_tree::bin_tree(const bin_tree& T) { root = copy_subtree(T.root); }
```

In this situation, the original definition of *copy_key* is used. According to the specification of C++ the effect of calling a virtual function directly or indirectly for the object being constructed is undefined. The compilers that we use interpret a pure virtual function as a function with an empty body and hence the program above will compile but no copies will be made. One may guard against the inadvertent call of a pure virtual function by using a virtual function whose call rises an error instead, e.g., one may define

```
virtual void copy_key(GenPtr&) { assert(false); return 0; }
```

Destructors give rise to the same problem as constructors. In a destructor of a base class virtual member functions also have the meaning defined in the base class and not the meaning given in a derived class. What does this mean for the destructor of class *dictionary*<K, I>? It first calls *bin_tree*::*clear* and then the destructor of the base class *bin_tree* (the latter call is generated by the compiler). The destructor of *bin_tree* again calls *bin_tree*::*clear*. So why do we need the first call at all? We need it because the second call uses the "wrong" definitions of the virtual functions *clear_key* and *clear_inf*. When *bin_tree*::*clear* is called for the second time the object to be destroyed does not know anymore that it was a *dictionary*<K, I>. The second call of the *clear* is actually unnecessary. We put it for reasons of uniformity; it incurs only very small additional cost.

Since *bin_tree* now knows how to copy and destroy the objects of type K and I, respectively, we can write correct implementations of the operations *del_item* and *insert* on the level of the implementation class, i.e., use precompiled versions of these functions, too.

```
void bin_tree::del_item(bin_tree_node* p) {
  // remove p from the tree
     ...
  clear_key(p->key);
  clear_inf(p->inf);
  delete p;
}

bin_tree_node* bin_tree::insert(GenPtr k, GenPtr i) {
  bin_tree_node* p = lookup(k);
  if (p != nil) { // k already present
    change_inf(p,i);
    return p;
  }
  copy_key(k);
  copy_inf(i);
  p = new bin_tree_node();
  p->key = k;
  p->inf = i;
  // insert p into tree
  ...
  return p;
}
```

By using the virtual *assign_inf* function we can realize the *change_inf* operation on the level of the implementation class, too.

```
void bin_tree::change_inf(bin_tree_node* p, GenPtr i) {
  assign_inf(p->inf,i);
}
```

With this modification the corresponding operations in the *dictionary*<K, I> template do not need to copy or destroy a key or an information anymore. They just pass the addresses of their arguments of type K and I to the member functions of class *bin_tree*.

```
template<class K, class I>
class dictionary: private bin_tree {
  ...
public:
  ...
  void del_item(dic_item it) { bin_tree::del_item(it); }
  void change_inf(dic_item it, const I& i)
  { bin_tree::change_inf(it,&i); }
  dic_item insert(const K& k, const I& i) { bin_tree::insert(&k,&i); }
```

13.4.4 *Arrays and Default Construction*

Some parameterized data types require that the actual element type has a default constructor, i.e., a constructor taking no arguments, that initializes the object under construction to some default value of the data type. The LEDA data types *array* and *map* are examples for such types.

The declaration

```
array<string> A(1,100);
```

creates an array of 100 variables of type *string* and initializes every variable with the empty string (using the default constructor of type *string*).

The declaration

```
map<int,vector>  M;
```

creates a map with index type *int* and element type *vector*, i.e., a mapping from the set of all integers of type *int* to the set of variables of type *vector*. All variables are initialized with the vector of dimension zero (the default value of type *vector*).

Note that a default constructor does not necessarily need to initialize the object under construction to a unique default value. There are data types that have no natural default value (for example, a line segment) and there are others where initialization to a default value is not done for efficiency reasons. In these cases, the default constructor simply constructs some arbitrary object of the data type. Examples for such types are the built-in types of C++. The declaration

```
int x;
```

declares x as a variable of type *int* initialized to some unspecified integer, and the declaration

```
array<int> A(1,100);
```

creates an array of 100 variables of type *int* each holding some arbitrary integer.

As for copying, assignment, and destruction, LEDA implements default initialization of parameterized data types in the corresponding implementation class by virtual functions and dynamic binding. We use the array data type as an example.

The parameterized data type *array<T>* is derived from the implementation class *gen_array* of arrays for generic pointers. The class *gen_array* provides two operations *init_all_entries* and *clear_all_entries* which can be called to initialize or to destroy all entries of the array, respectively. They use the virtual member functions *void init_entry(GenPtr&)* and *void clear_entry(GenPtr)* to do the actual work, i.e., they use the first function to initialize an array entry and the second function to destroy one.

```
class gen_array {
  GenPtr* first;
  GenPtr* last;
  ...
  virtual void init_entry(GenPtr& x)  = 0;
  virtual void clear_entry(GenPtr x)  = 0;
  ...
protected:
  ...
  void init_all_entries()
  { for(GenPtr* p = first; p <= last; p++) init_entry(*p); }
  void clear_all_entries()
  { for(GenPtr* p = first; p <= last; p++) clear_entry(*p); }
};
```

In the data type class *array<T>* we define *init_entry* and *clear_entry* by calling the new and delete operator of type T, respectively. The constructor of *array<T>* uses *init_all_entries* to initialize all elements of the array and the destructor uses *clear_all_entries* to destroy all objects stored in the array.

```
template <class T>
class array : private gen_array {
  void init_entry(GenPtr& x) { x = new T; }
  void clear_entry(GenPtr x) { delete (T*)x; }
public:
  ...
  array(int l, int h) : gen_array(l,h) { init_all_entries(); }
 ~array() { clear_all_entries(); }
};
```

We give one more example of default construction, the *new_node* and *new_edge* operations of parameterized graphs *GRAPH<vtype, etype>*. There are two variants of these operations: the first one takes an argument that is used to initialize the information associated with the new object (node or edge).

```
node G.new_node(const vtype&)
edge G.new_edge(node, node, const etype&)
```

The second one does not take such an argument. Here the information associated with the object is initialized by the default constructor of the corresponding type (*vtype* or *etype*).

```
node G.new_node()
edge G.new_edge(node v, node w)
```

The following piece of code constructs a graph with two nodes v and w connected by an edge $e = (v, w)$. The nodes are labeled with the default value of type *string*, i.e., the empty string, and edge e is labeled with a vector of dimension zero, the default value of type *vector*.

```
GRAPH<string,vector> G;
node v = G.new_node();
node w = G.new_node();
edge e = G.new_edge(v,w);
```

Default initialization for nodes and edges is also used by LEDA's various graph generators. If G is a parameterized graph of type *GRAPH<vtype, etype>*, a call *random_graph*(G, n, m) constructs a random graph with n nodes and m edges where each node information is initialized by the default constructor of type *vtype* and each edge information is initialized by the default constructor of type *etype*.

13.4.5 *Some Useful Function Templates*

In *<LEDA/param_types.h>* we define five function templates that are useful to define the virtual functions required in the LEDA approach.

```
template <class T>
inline T& leda_access(const T*, const GenPtr& p) { return *(T*)p; }
```

returns a reference to the object of type T pointed to by p. The first argument of this function template is a dummy pointer argument of type $T*$ that is used for selecting the correct instantiation. For instance, to access an object of type T through a generic pointer p we write *leda_access*$((T*)0, p)$. As an abbreviation LEDA provides the macro.

```
#define LEDA_ACCESS(T,p)        leda_access((T*)0,p)
```

The function template

```
template <class T>
inline GenPtr leda_create(const T*) { return new T; }
```

returns a generic pointer to an object of type T initialized with the default value of type T. Again, there is a dummy pointer argument of type $T*$.

The function template

```
template<class T>
inline GenPtr leda_copy(const T& x) { return new T(x); }
```

returns a generic pointer to an object of type T initialized with a copy of x.

The function template

```
template <class T>
inline void leda_clear(T& x) { T* p = &x; delete p; }
```

destroys the object stored at x and the function template

```
template <class T>
inline GenPtr leda_cast(const T& x) { return (GenPtr)&x; }
```

returns the address of x casted to a generic pointer.

Given these function pointers it is easy to define the virtual function required in the LEDA approach in a generic way for every type parameter T.

```
void create_T(GenPtr& p)  { p = leda_create((T*)0); }
void copy_T  (GenPtr& p)  { p = leda_copy(LEDA_ACCESS(T,p)); }
void clear_T (GenPtr  p)  { leda_clear(LEDA_ACCESS(T,p)); }
void assign_T(GenPtr& p, GenPtr q)
                          { LEDA_ACCESS(T,p) = LEDA_ACCESS(T,q); }
```

We return to the dictionary and array data type templates to demonstrate the use of the above defined function templates and macros. We have

```
class dictionary : public bin_tree {
  int  cmp(GenPtr x, GenPtr y) const
            { return compare(LEDA_ACCESS(K,x), LEDA_ACCESS(K,x,y); }
  void clear_key(GenPtr& x) const { leda_clear(LEDA_ACCESS(K,x)); }
  void clear_inf(GenPtr& x) const { leda_clear(LEDA_ACCESS(I,x)); }
  void copy_key(GenPtr& x)  const { x = leda_copy(LEDA_ACCESS(K,x)); }
  void copy_inf(GenPtr& x)  const { x = leda_copy(LEDA_ACCESS(I,x)); }
  void assign_inf(GenPtr& x, GenPtr y) const
                          { LEDA_ACCESS(I,x) = LEDA_ACCESS(I,y); }
public:
  ...
  K key(dic_item it) const
  { return LEDA_ACCESS(K,bin_tree::key(it)); }
  I inf(dic_item it) const
  { return LEDA_ACCESS(I,bin_tree::inf(it)); }
  dic_item insert(const K& k, const I& i)
  { return bin_tree::insert(leda_cast(k),leda_cast(i)); }
  dic_item lookup(const K& k) const
  { return bin_tree::lookup(leda_cast(k)); }
  void change_inf(dic_item it, const I& i)
  { bin_tree::change_inf(it,leda_cast(i)); }
  ...
};
```

and

```
template <class T>
class array : private gen_array {
  void init_entry(GenPtr& x) { x = leda_create((T*)0); }
  void clear_entry(GenPtr x) { leda_clear(LEDA_ACCESS(T,x)); }
  ...
};
```

13.4.6 *Further Uses of Virtual Functions*

There are many other situations where LEDA uses virtual functions for transferring functionality of actual type arguments from the data type class to the implementation class. Examples are:

- Printing and Reading
- Hashing
- Id-Numbers
- Type Information (see the next section)
- Rebalancing of binary trees

We touched upon printing and reading in Section 5.7.3, an example of the use of id-numbers can be found in Section 5.1.2, and we will see type information in Section 13.5.3.

Exercises for 13.4

1 Write a template implementation of the LEDA data type *queue*.

2 Is it correct to change the interface of *pop* to `const T& pop()`?

3 The implementation of *list*<*T*>::*clear* which simply calls *list_impl*::*clear* has a memory leak, as it leaves the entries contained in the elements of the list as orphans on the heap. Why does *t_list*<*T*>::*clear* not have a memory leak?

4 Define a class *dlist*<*T*> that implements doubly linked lists for elements of type T. Use the template approach and convert the solution to the LEDA approach.

5 Add an operation $pop(T\& \ x)$ to the list data type that returns the result of the pop operation in the reference parameter x.

6 In the text we established a relationship between corresponding states of *t_list*<*T*> and *list*<*T*>. Argue that the implementations of the various functions of the list data type leave this correspondence invariant.

7 Consider the following skeleton for the function *bin_tree*::*insert*.

```
bin_tree_node* insert(void* k, void* i)
{ bin_tree_node *p = root, *q = nil;  // q is always the parent of p
  int c;
  while (p)
  { c = cmp_key(k,p->key);
    if (c == 0)
    { // something is missing here
      return p;
    }
    q = p;
    p = (c > 0) ? p->right_child : p->left_child;
  }
  if ( c > 0 ) return q->right_child = new bin_tree_node(k,i);
  else return q->left_child) = new bin_tree_node(k,i);
}
```

Complete the code. Make sure that your implementation has no memory leak.

13.5 Optimizations

In this section we describe some optimizations that can be applied to special type arguments of parameterized data types.

13.5.1 *Small Types*

The LEDA solution for parameterized data types presented in the preceding sections uses one additional (generic) pointer field for every value or object that is stored in the data type. The method incurs overhead in space and time, in space for the additional pointer and in time for the additional indirection. We show how to avoid the overhead for types whose values are no larger than a pointer. In C++ the space requirement of a type is easily detemined: $sizeof(T)$ returns the size of the objects of type T in bytes. We call a type T *small* if $sizeof(T) \leq sizeof(GenPtr)$ and *large* otherwise. By definition, all pointer types are small. On 32 bit systems the built-in types *char, short, int, long, float* are small as well, and type *double* is big. On 64 bit systems even the type *double* is small. Note that class types can be small too, e.g., a class containing a single pointer data member. An example for small class types are the LEDA *handle types* that will be discussed in Section 13.7.

Values of any small type T can be stored directly in a data field of type *void∗* or *GenPtr* by using the *in-place new operator* of C++. If p is a pointer of type *void∗*

```
new(p) T(x);
```

calls the copy constructor of type T to construct a copy of x at the address in memory that p points to, in other words with $this = p$. Similarly,

```
new(p) T;
```

calls the default constructor of type T (if defined) to construct the default value of type T at the location that p points to.

We use the in-place new operator as follows. If y is a variable corresponding to a data field of some container and T is a small type then

```
new(&y) T(x);
new(&y) T;
```

constuct a copy of x and the default value of T directly in y.

Of course, small objects have to be destroyed too. For this purpose we will use the *explicit destructor call* of C++. If z is a variable of some type T,

```
z.~T()
```

calls the destructor of T for the object stored in z. Destructor calls for named objects are constructed automatically in C++ when the scope of the object ends, and therefore few C++ programmers ever need to make an explicit destructor call.

We have to. Observe that we construct objects of type T in variables of type *void∗* and therefore cannot rely on the compiler to generate the destructor call. We destroy an object of type T stored in a variable y of type *void∗* by casting the address of y to a pointer of type $T*$ and calling the destructor explicitly as in

```
((T*)&y)->~T();
```

To access the value of a small type T stored in a *void*$*$ data field y we take the address of y, cast it into a $T*$ pointer, and dereference this pointer.

```
*((T*)&y)
```

13.5.2 *Summary of LEDA Approach to Parameterized Data Types*

We summarize the LEDA approach to parameterized data types. We store values of arbitrary types T in data fields of type *void*$*$ (also called *GenPtr*). We distinguish between small and large types.

For objects of a large type T (*sizeof*$(T) >$ *sizeof*$(GenPtr)$) we make copies in the dynamic memory using the *new* operator and store pointers to the copies.

For objects of a small type T (*sizeof*$(T) \leq$ *sizeof*$(GenPtr)$) we avoid the overhead of an extra level of indirection by copying the value directly into the *void*$*$ data field using the "in-place" variant of the *new* operator.

We next give versions of *leda_copy*, *leda_create*, *leda_clear*, *leda_access*, and *leda_cast* that can handle small and large types. The functions are defined in LEDA/param_types.h.

GenPtr leda_copy(*const* T& x) makes a copy of x and returns it as a generic pointer of type *GenPtr*. If T is a small type, the copy of x is constructed directly in a *GenPtr* variable using the in-place new operator of T, and if T is a big type, the copy of x is constructed in the dynamic memory (using the default new operator) and a pointer to this copy is returned.

```
template<class T>
inline GenPtr leda_copy(const T& x)
{ GenPtr p;
  if (sizeof(T) <= sizeof(GenPtr)) new(&p) T(x);
  if (sizeof(T) >  sizeof(GenPtr)) p = new T(x);
  return p;
}
```

GenPtr leda_create(*const* $T*$) constructs the default value of type T by a call of either the in-place new or the normal new operator of T.

```
template <class T>
inline GenPtr leda_create(const T*)
{ GenPtr p;
  if (sizeof(T) <= sizeof(GenPtr)) new(&p) T;
  if (sizeof(T) >  sizeof(GenPtr)) p = new T;
  return p;
}
```

void leda_clear(T& x) destroys the object stored in x either by calling the destructor of T explicitly or by calling the *delete* operator on the address of x.

```
template <class T>
inline void leda_clear(T& x)
{ T* p = &x;
```

```
    if (sizeof(T) <= sizeof(GenPtr)) p->~T();
    if (sizeof(T) >  sizeof(GenPtr)) delete p;
  }
```

T& leda_access(const T∗, const GenPtr& p) returns a reference to the object of type *T* stored in *p* or pointed to by *p* respectively.

```
  template <class T>
  inline T& leda_access(const T*, const GenPtr& p)
  { if (sizeof(T) <= sizeof(GenPtr)) return *(T*)&p;
    if (sizeof(T) >  sizeof(GenPtr)) return *(T*)p;
  }
```

GenPtr leda_cast(const T& x) either returns the value of *x* or the address of *x* casted to a generic pointer.

```
  template <class T>
  inline GenPtr leda_cast(const T& x)
  { GenPtr p;
    if (sizeof(T) <= sizeof(GenPtr)) *(T*)&p = x;
    if (sizeof(T) >  sizeof(GenPtr)) p = (GenPtr)&x;
    return p;
  }
```

The functions above incur no overhead at run time. Note that all comparisons between the size of *T* and the size of a pointer can be evaluated at compile-time when instantiating the corresponding function template and therefore do not cause any overhead at run time.

13.5.3 *Optimizations for Built-in Types*

Our method of implementing parameterized data types stores the objects of the data type in *void∗* data fields and uses virtual member functions for passing type-specific functionality from the data type template to the implementation class.

In a previous section we already showed how to avoid the space overhead of an additional pointer for small types. However, there is also an overhead in time. Every type-dependent operation, such as comparing two keys in a dictionary, is realized by a virtual member function. Calling such a function, e.g., in the inner loop when searching down a tree, can be very expensive compared to the cost of applying a built-in comparison operator.

LEDA has a mechanism for telling the implementation class that an actual type parameter is one of the built-in types in order to avoid this overhead. For the identification of these types we use an enumeration. For every built-in type *xyz* this enumeration contains an element *XYZ_TYPE_ID*. There is also an *UNKNOWN_TYPE_ID* member used for indicating that the corresponding type is unknown, i.e., is not one of the built-in types.

```
  enum { UNKNOWN_TYPE_ID, CHAR_TYPE_ID, SHORT_TYPE_ID, INT_TYPE_ID,
         LONG_TYPE_ID, FLOAT_TYPE_ID, DOUBLE_TYPE_ID };
```

To compute the type identification for a given type we use a global function *leda_type_id*. Given a pointer to some type *T* this function returns the corresponding type identification, e.g., if *T* = *int*, it will return *INT_TYPE_ID*, if *T* is not one of the recognized types, the result is *UNKNOWN_TYPE_ID*. We first define a default function template returning the special value *UNKNOWN_TYPE_ID* and then define specializations for all built-in types.

```
template <class T>
inline int leda_type_id(const T*)       { return UNKNOWN_TYPE_ID; }
inline int leda_type_id(const char*)   { return CHAR_TYPE_ID; }
inline int leda_type_id(const int*)    { return INT_TYPE_ID; }
inline int leda_type_id(const long*)   { return LONG_TYPE_ID; }
inline int leda_type_id(const double*){ return DOUBLE_TYPE_ID; }
...
```

Now we can add a virtual function *key_type_id* to the dictionary implementation and define it in the corresponding data type template by calling the *leda_type_id* function with an appropriate pointer value.

```
class bin_tree {
  ...
  virtual int key_type_id() = 0;
  ...
};
template <class K, class I>
class dictionary {
  ...
  int key_type_id() { return leda_type_id((K*)0); }
  ...
};
```

In the implementation of the various dictionary operations (in *bin_tree.c*) we can now determine whether the actual key type is one of the basic types and choose between different optimizations. We use the *bin_tree::search* member function as an example. Let us assume we want to write a special version of this function for the built-in type *int* that does not call the expensive *cmp_key* function but compares keys directly. First we call *type_id*( ) to get the actual key type id and in the case of *INT_TYPE_ID* we use a special searching loop that compares keys using the *LEDA_ACCESS* macro and the built-in comparison operators for type *int*.

```
bin_tree_node* bin_tree::search(GenPtr x) const
{
  bin_tree_node* p = root;
  switch ( type_id() ) {
   case INT_TYPE_ID: {
        int x_int =  LEDA_ACCESS(int,x);
        while (p)
        { int p_int =  LEDA_ACCESS(int,p->k);
          if (x_int == p_int) break;
          p = (x_int < p_int) ? p->left_child : p->right_child;
        }
        break;
     }
   default: {
```

n	myint	int
1000000	6.74	0.68

Table 13.1 The effect of the optimization for built-in types. The time to sort an array of n random elements is shown. The table was generated with the program built_in_types_optimization in directory LEDAROOT/demo/book/Impl.

```
        while (p)
        { int c = cmp(x,p->k);
          if (c == 0) break;
          p = (c < 0) ? p->left_child : p->right_child;
        }
        break;
      }
  }
  return p;
}
```

The above piece of code is easily extended to other built-in types.

Table 13.1 shows the effect of the optimization. We defined a class *myint* that encapsulates an *int*

⟨*class myint*⟩≡

```
class myint {
  int x;
public:
  myint() {}
  myint(const int _x): x(_x) {}
  myint(const myint& p)  {  x = p.x; }
  friend void operator>>(istream& is, myint& p) { is >> p.x; };
  friend ostream& operator<<(ostream& os, myint& p)
         { os << p.x; return os; };
  friend int compare(const myint&,const myint&);
};
int compare(const myint& p,const myint& q)
{
  if (p.x == q.x) return 0;
  if (p.x < q.x) return -1; else return +1;
}
```

and then built two arrays of size n, one filled with random *int*s and the other one filled with the same *myint*s. We then sorted both arrays. Table 13.1 shows that the optimization leads to a considerable reduction in running time.

Exercise for 13.5

1 Extend the search procedure for binary trees such that it uses the optimization also for *doubles*.

13.6 Implementation Parameters

There are many implementations of dictionaries: binary trees, skiplists, hashing, sorted arrays, self-adjusting lists, ... Which implementation should be included in a library?

If one provides only one implementation, then this implementation should clearly be the "best" possible. This was the direction taken in the first versions of LEDA. In the case of the dictionary data type, we included red-black trees because they are asymptotically as efficient as any other implementation. But, of course, only asymptotically. Also, there are better implementations for special cases, e.g., for integer keys from a bounded universe. For other data types, e.g., range trees, there are implementations with vastly differing performance parameters (time-space tradeoff) and so there is not even an asymptotically best implementation. All of this implies that providing only one implementation for each data type is not satisfactory.

So, one has to provide many and allow for the possibility of adding more. What properties should a mechanism for choosing between different implementations have?

(1) There should be a simple syntax for choosing between different implementations. In LEDA, the declaration

```
_dictionary<K,I,rb_tree> D;
```

creates an empty dictionary with key type K and information type I and selects red-black trees as the implementation variant, *_dictionary*$<K, I,$ *impl*$>$ selects the implementation *impl*. The actual type parameter for *impl* has to be a dictionary implementation, i.e., must be a class that provides a certain set of operations and uses virtual functions for type dependent operations. This will be discussed below. The declaration

```
dictionary<K,I> D;
```

selects the default implementation (skiplists in the current version).

Remark: Because templates cannot be overloaded in C++ we have to use different names *dictionary* and *_dictionary*. The general rule is that the data type variant with implementation parameter starts with an underscore.

(2) Applications can be written that work with any implementation. For example, applications that use a dictionary are written as functions with an additional parameter of the abstract dictionary type. Then the function can be called with any implementation of the dictionary type. We illustrate this feature with the word-count example.

```
void WORD_COUNT(const list<string>& L, dictionary<string,int>& D)
{ string s;
  forall(s,L)
```

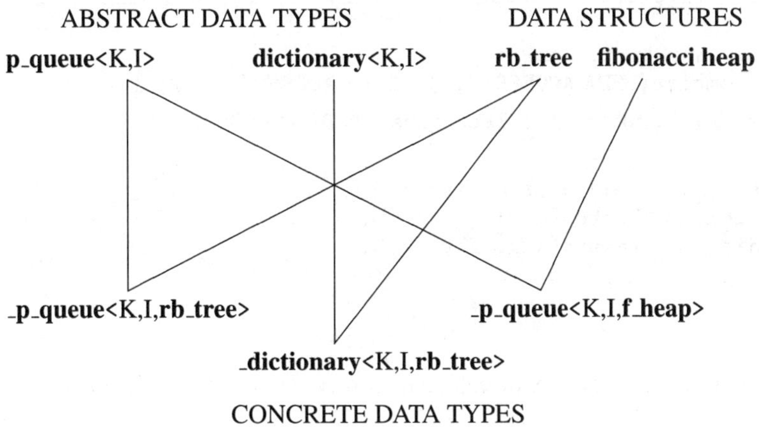

Figure 13.2 Multiple inheritance combines abstract data types and data structures to concrete data types.

```
  { dic_item it = D.lookup(s);
    if (it == nil)
       D.insert(s,1);
    else
       D.change_inf(it,D.inf(it)+1);
  }
  dic_item it;
  forall_items(it,D)
    cout << D.key(it) << " appeared " << D.inf(it) << " times.";
}
```

In the context of the declarations

```
dictionary<string, int>            SL_D;  // skiplists
_dictionary<string, int, rb_tree> RB_D;  // red-black trees
_dictionary<string, int, my_impl> MY_D;  // user implementation
```

the calls

```
WORD_COUNT(L,SL_D);
WORD_COUNT(L,RB_D);
WORD_COUNT(L,MY_D);
```

are now possible.

The realization of the implementation parameter mechanism makes use of multiple inheritance, cf. Figure 13.2. Every concrete data type, say dictionary with the rb_tree implementation, is derived from the abstract data type and the data structure used to implement it. In the abstract data type class, all functions are virtual, i.e., have unspecified implementations. In the data structure class the details of the implementation are given and the classes in the bottom line of Figure 13.2 are used to match the abstract functions with the concrete implementations.

```
template<class K,class I> class dictionary : private default_impl
{
  int cmp_key(GenPtr x, GenPtr y)
  { return compare(LEDA_ACCESS(K,x), LEDA_ACCESS(K,y)); }
  void clear_key(GenPtr x) { leda_clear(LEDA_ACCESS(K,x)); }
public:
  virtual K        key(dic_item it) = 0;
  virtual dic_item lookup(K y)      = 0;
  virtual dic_item insert(K x, I y) = 0;
  virtual void     del(K y)         = 0;
  ...
};
```

Dictionaries with implementation parameter can now be derived from the abstract dictionary class.

```
template<class K, class I,class IMPL>
class _dictionary : private IMPL, public dictionary<K,I>
{
  public:
  K        key(dic_item it) { return LEAD_ACCESS(K,IMPL::key(it)); }
  dic_item lookup(K y)      { return IMPL::lookup(leda_cast(y)); }
  dic_item insert(K x, I y)
           { return IMPL::insert(leda_cast(x),leda_cast(y)); }
  void     del(K y)         { IMPL::del(leda_cast(y)); }
  ...
};
```

Of course, an implementation class *IMPL* can be used as actual implementation parameter of a parameterized data type only if it provides all necessary operations and definitions and calls type-dependent functions through the appropriate virtual member functions. For item-based types, it must in addition define a local type *item* representing the items of the data type. In the case of dictionaries, any class *dic_impl* with the following definitions and declarations can be used as implementation class.

```
class dic_impl {
  // type dependent functions
  virtual int  cmp(GenPtr, GenPtr) const = 0;
  virtual int  type_id()           const = 0;
  virtual void clear_key(GenPtr&)  const = 0;
  virtual void clear_inf(GenPtr&)  const = 0;
  virtual void copy_key(GenPtr&)   const = 0;
  virtual void copy_inf(GenPtr&)   const = 0;
  virtual void assign_inf(GenPtr&, GenPtr) const = 0;
public:
  // definition of the item type
  typedef ... item;
  // construction, destruction, copying
  dic_impl();
```

```
  dic_impl(const dic_impl&);
 ~dic_impl();
  dic_impl& operator=(const dic_impl&);

  // dictionary operations
  GenPtr key(item)  const;
  GenPtr inf(item)   const;

  item insert(GenPtr,GenPtr);
  item lookup(GenPtr)  const;

  void change_inf(item,GenPtr);
  void del_item(item);
  void del(GenPtr);
  void clear();

  int size() const;

  // iteration
  item first_item() const;
  item next_item(item) const;
};
```

For most of its parameterized data types LEDA provides several implementation classes. Before using an implementation class *xyz* the corresponding header file *<LEDA/impl/xyz.h>* has to be included. The following dictionary implementations are currently available: AVL-Trees (*avl_tree*), (a,b)-Trees (*ab_tree*), BB[α]-Trees (*bb_tree*), Skiplists (*skiplist*), Red-Black-Trees (*rb_tree*), Randomized Search Trees (*rs_tree*), Dynamic Perfect Hashing (*dp_hashing*), and Hashing with Chaining (*ch_hashing*).

Section "Available Implementations" of the LEDA user manual gives the complete list of all available implementations.

Exercises for 13.6

1 Write an implementation class for dictionaries based on the *so_set* class of Section 3.2.

2 Write an implementation class for priority queues.

13.7 Independent Item Types (Handle Types)

All independent item types of LEDA (cf. Section 2.2.2) are implemented by so-called *handle types*. Basically, a handle type H is a pointer (or handle) to some representation class *H_rep* that contains all data members used for the representation of objects of type H. Assignment and copy operations translate to simple pointer assignments and the test for identity translates to the equality test for pointers. Thus assignment, copy operations, and identity functions are easily handled, but destruction of representation objects causes a problem.

A representation object has to be destroyed as soon as no handle is pointing to it anymore. To detect this situation we use a technique called *reference counting*. Every representation object has a reference counter *ref_count* that contains the number of handles which are still in scope and point to the object. The counters are updated in the copy constructor, assignment operator, and destructor of the corresponding handle class.

We use a two-dimensional point class *point* as an example. The representation class *point_rep* has three data members, a pair of floating-point coordinates (x, y) and a reference counter *ref_count*. A constructor initializing the coordinates to two given values and setting the reference counter to one is the only member function.

```
class point_rep {
  double x, y;
  int    ref_count;
  point_rep(double a, double b) :x(a),y(b),ref_count(1) {}
};
```

Now we could implement points by pointers to the representation class *point_rep*. However, just using the type *point_rep∗* for representing points, as in

```
typedef point_rep* point;
```

would not make reference counting work automatically when variables of type *point* are created, assigned to each other, or destroyed. Therefore *point* has to be implemented by a real C++ class with constructors, destructor, and assignment operator.

The only data member of class *point* is a pointer to the corresponding representation class *point_rep*.

```
class point {
  point_rep* ptr;
public:
  point(double,double);
  point(const point&);
  point& operator=(const point&);
 ~point();
  double xcoord()     const;
  double ycoord()     const;
  point  translate() const;
  friend bool identical(const point& x, const point& y);
};
```

The constructor of class *point* creates a new representation object (with *ref_count* equal to one) in the dynamic memory and assigns the pointer to *ptr*. The copy constructor copies the corresponding pointer and increases the reference counter of the representation object by one. The destructor decreases the corresponding reference counter by one and deletes the representation object if the new value of the counter is zero.

```
point::point(double x, double y) { ptr = new point_rep(x,y); }
point::point(const point& p)
{ ptr = p.ptr;
  ptr->count++;
}
point::~point() { if (--ptr->ref_count == 0) delete ptr; }
```

In an assignment operation $q = p$ we first increase the reference counter of the representation object pointed to by p and then decrease the counter of the representation object pointed to by q. If the counter of the representation object pointed to by q is zero afterwards then q was the only handle pointing to the representation object and we have to delete it. Note that in the case that p and q are identical the same reference counter is first increased and then decreased and hence is unchanged in the end.

```
point& point::operator=(const point& p)
{ p.ptr->count++;
  if (--ptr->count == 0) delete ptr;
  ptr = x.ptr;
  return *this;
}
```

Two handles are identical if they share a common representation object, i.e., the *identical* function reduces to pointer equality.

```
bool identical(const point& x, const point& y)
{ return x.ptr == y.ptr; }
```

The above defined member functions and operators are common to all handle types. We will show how to put them in a common base class for all handle types below.

In order to complete the definition of *points*, we still have to implement the individual operations specific to them. For example,

```
double point::xcoord() const { return ptr->x; }
double point::ycoord() const { return ptr->y; }
point  point::translate(double dx, double dy) const
{ return point(ptr->x+dx, ptr->y+dy); }
```

Classes handle_rep and handle_base: As mentioned above, there is a group of operations that is the same for all handle types (copy constructor, assignment, destructor, identity). LEDA encapsulates these operations in two classes *handle_rep* and *handle_base* (see *<LEDA/handle_types.h>*). Concrete handle types and their representation classes are derived from them. This will be demonstrated for the *point* type at the end of this section.

The *handle_rep* base class contains a reference counter of type *int* as its only data member, a constructor initializing the counter to 1, and a trivial destructor. Later we will derive representation classes of particular handle types (e.g., *point_rep*) from this base class adding type specific individual data members (e.g., x- and y-coordinates of type *double*).

```
class handle_rep  {
  int ref_count;
  handle_rep() : ref_count(1) {}
  virtual ~handle_rep() {}
  friend class handle_base;
};
```

The *handle_base* class has a data member *PTR* of type *handle_rep∗*, a copy constructor, an assignment operator, and a destructor. Furthermore, it defines a friend function *identical* that declares two *handle_base* objects identical if and only if their *PTR* fields point to the same representation object. Specific handle types (e.g., *point*) derived from *handle_base* use the *PTR* field for storing pointers to the corresponding representation objects (e.g., *point_rep*) derived from *handle_rep*.

```
class handle_base {
  handle_rep* PTR;
  handle_base(const handle_base& x)
  { PTR = x.PTR;
    PTR->ref_count++;
  }
  handle_base& operator=(const handle_base& x)
  { x.PTR->ref_count++;
    if (--PTR->ref_count == 0)  delete PTR;
    PTR = x.PTR;
    return *this;
  }
 ~handle_base() { if (--PTR->ref_count == 0)  delete PTR; }
  friend bool identical(const handle_base& x, const handle_base& y)
  { return x.PTR == y.PTR; }
};
```

This completes the definition of classes *handle_base* and *handle_rep*. We can now derive an independent item type *T* from *handle_base* and the corresponding representation class *T_rep* from *handle_rep*. We demonstrate the technique using the point example.

point_rep is derived from *handle_rep* adding two data members for the x- and y-coordinates and a constructor initializing these members.

```
class point_rep  : public handle_rep {
  double x, y;
  point_rep(double a, double b) x(a), y(b) { }
 ~point_rep() {}
};
```

We will next derive class *point* from *handle_base*. The class *point* uses the inherited *PTR* field for storing *pointer_rep∗* pointers. The constructor constructs a new object of type *point_rep* in the dynamic memory and stores a pointer to it in the *PTR* field, and copy constructor and assignment reduce to the corresponding function of the base class. In order to access the representation object we cast *PTR* to *point_rep∗*. This is safe since *PTR* always points to a *point_rep*. For convenience, we add an inline member function *ptr*( ) that performs this casting. Now we can write *ptr*( ) wherever we used *ptr* in the original *point* class at the beginning of this section. The full class definition is as follows:

```
class point  : public handle_base
{
  point_rep* ptr() const { return (point_rep*)PTR; }
public:
  point(double x=0, double y=0) { PTR = new point_rep(x,y); }
  point(const point& p) : handle_base(p) {}
 ~point() {}
  point& operator=(const point& p)
  { handle_base::operator=(p); return *this; }
  double xcoord()  const   { return ptr()->x; }
  double ycoord()  const   { return ptr()->y; }
  point  translate(double dx, double dy) const
  { return point(xcoord() + dx, ycoord() + dy); }
};
```

Note that all the "routine work" (copy construction, assignment, destruction) is done by the corresponding functions of the base class *handle_base*.

Exercises for 13.7

1. Explain why the destructor *handle_rep*::~*handle_rep*( ) is declared *virtual.*
2. How would the above code have to be changed if it were not *virtual*?
3. Implement a *string* handle type using the mechanism described above.
4. Add an array subscript operator *char& string*::*operator*[](*int i*) to your string class. What kind of problem is caused by this operator and how can you solve it?

13.8 Memory Management

Many LEDA data types are implemented by collections of small objects or nodes in the dynamic memory, e.g., lists consist of list elements, graphs consist of nodes and edges, and handle types are realized by pointers to small representation objects.

Most of these data types are dynamic and thus spend considerable time for the creation and destruction of these small objects by calling the *new* and *delete* operators.

Typically, the C++ default *new* operator is implemented by calling the *malloc* function of the *C* standard library

```
void* operator new(size_t bytes) { return malloc(bytes) }
```

and the default *delete* operator by calling the *free* library function

```
void operator delete(void* p) { free(p); }
```

Unfortunately, *malloc* and *free* are rather expensive system calls on most systems.

LEDA offers an efficient memory manager that is used for all node, edge and item types. The manager can easily be applied to a user defined class *T* by adding the macro call

"LEDA_MEMORY(T)" to the declaration of the class T. This redefines the new and delete operators for type T, such that they allocate and deallocate memory using LEDA's internal memory manager.

The basic idea in the implementation of the memory manager is to amortize the expensive system calls to *malloc* and *free* over a large sequence of requests (calls of *new* and *delete*) for small pieces of memory. For this purpose, LEDA uses *malloc* only for the allocation of large memory blocks of a fixed size (e.g., 4 kbytes). These blocks are sliced into chunks of the requested size and the chunks are maintained in a singly linked list. The strategy just outlined is efficient if the size of the chunks is small compared to the size of a block. Therefore the memory manager applies this strategy only to requests for memory pieces up to a certain size. Requests for larger pieces of memory (often called vectors) are directly mapped to *malloc* calls. The maximal size of memory chunks handled by the manager can be specified in the constructor. For the standard memory manager used in the *LEDA_MEMORY* macros this upper bound is set to 255 bytes.

The heads of all lists of free memory chunks are stored in a table *free_list*[256]. Whenever an application asks for a piece of memory of size $sz < 256$ the manager first checks whether the corresponding list *free_list*[*sz*] is empty. If the list is non-empty, the first element of the list is returned, and if the list is empty, it is filled by allocating a new block and slicing it as described above. Freeing a piece of memory of size $sz < 256$ in a call of the *delete* operator is realized by inserting it at the front of list *free_list*[*sz*].

Applications can call the global function *print_statistics* to get a summary of the current state of the standard memory manager. It prints for every chunk size that has been used in the program the number of free and still used memory chunks.

The following example illustrates the effect of the memory manager. We defined a class *pair* and a class *dumb_pair*. The definitions of the two classes are identical except that *dumb_pair* does not use the LEDA memory manager.

⟨*class pair*⟩≡
```
class pair {
  double x, y;
public:
  pair(double a=0, double b=0) : x(a), y(b) { }
  pair(const pair& p) : x(p.x), y(p.y) { }
  friend ostream& operator<<(ostream& ostr, const pair&) {return ostr;}
  friend istream& operator>>(istream& istr, pair&) { return istr; }
  LEDA_MEMORY(pair)  // not present in dumb_pair
};
```

We then built a list of n pairs or dumb pairs, respectively, and cleared them again. Table 13.2 shows the difference in running time. We also printed the memory statistics before and after the *clear* operation.

n	LEDA memory	C++ memory
1000000	0.94	2.77

Table 13.2 The effect of the memory manager. We built and destroyed a list of n pairs or dumb pairs, respectively. Pairs use the LEDA memory manager and dumb pairs do not. The table was generated with program memmgr_test.c in LEDAROOT/demo/book/Impl.

⟨*timing for dumb pair*⟩≡

```
list<dumb_pair> DL;
for (i = 0; i < n; i++ ) DL.append(dumb_pair());
print_statistics();
DL.clear();
print_statistics();
UT = used_time(T);
```

13.9 Iteration

For most of its item-based data types LEDA provides iteration macros . These macros can be used to iterate over the items or elements of lists, arrays, sets, dictionaries, and priority queues or over the nodes and edges of graphs. Iteration macros can be used similarly to the C++ *for*-statement. We give some examples.

For all item-based data types:

```
forall_items(it,D) { ... }
```

iterates over the items *it* of *D* and

```
forall_rev_items(it,D) { ... }
```

iterates over the items *it* of *D* in reverse order.

For sets, lists and arrays:

```
forall(x,D) { ... }
```

iterates over the elements x of D and

```
forall_rev(x,D) { ... }
```

iterates over the elements x of D in reverse order.

For graphs:

```
forall_nodes(v,G) { ... }
```

iterates over the nodes v of G,

```
STD_MEMORY_MGR (memory status)
+--------------------------------------------------+
|  size     used      free     blocks     bytes    |
+--------------------------------------------------+
|   12   1000001       388      1469   12004668    |
|   16   1000000       110      1961   16001760    |
|   20        29       379         1       8160    |
|   28         1       290         1       8148    |
|   40         2       201         1       8120    |
| > 255        -         -         1        300    |
+--------------------------------------------------+
|  time:  0.64 sec              space:27450.88 kb  |
+--------------------------------------------------+
```

```
STD_MEMORY_MGR (memory status)
+--------------------------------------------------+
|  size     used      free     blocks     bytes    |
+--------------------------------------------------+
|   12         1   1000388      1469   12004668    |
|   16         0   1000110      1961   16001760    |
|   20        29       379         1       8160    |
|   28         1       290         1       8148    |
|   40         2       201         1       8120    |
| > 255        -         -         1        300    |
+--------------------------------------------------+
|  time:  0.98 sec              space:27450.88 kb  |
+--------------------------------------------------+
```

Figure 13.3 Statistic of memory usage. We built a list of $n = 10^6$ pairs of doubles. A list of n pairs requires n list items of 12 bytes each and n pairs of 16 bytes each. The upper statistic shows the situation before the clear operations and the lower statistic shows the situation after the clear operations. The figure was generated with program memmgr_test.c in LEDAROOT/demo/book/Impl.

```
forall_edges(e,G) { ... }
```

iterates over the edges e of G,

```
forall_adj_edges(e,v) { ... }
```

iterates over all edges e adjacent to v, and

```
forall_adj_nodes(u,v) { ... }
```

iterates over all nodes e adjacent to v.

Inside the body of a forall loop insertions into or deletions from the collection iterated over are not allowed, with one exception, the current item or object of the iteration may be removed, as in

```
// remove self-loops
forall_edges(e,G) { if (G.source(e) == G.target(e)) G.del_edge(e); }
```

The *forall_item*(*it*, *S*) iteration macro can be applied to instances *S* of all item-based data types *T* that define *T*::*item* as the corresponding item type and that provide the following member functions:

```
T::item  S.first_item()
```

returns the first item of *S* and *nil* if *S* is empty

```
T::item  S.next_item(T::item it)
```

returns the successor of item *it* in *S* (*nil* if *it* = *S.last_item*( ) or *it* = *nil*).

The *forall_rev_items*(*it*, *S*) macro can be used if the following member functions are defined:

```
T::item  S.last_item()
```

returns the last item of *S* and *nil* if *S* is empty, and

```
T::item  S.pred_item(T::item it)
```

returns the predecessor of item *it* in *S* (*nil* if *it* = *S.first_item*( ) or *it* = *nil*).

The *forall*(x, *S*) and *forall_rev*(x, *S*) iteration macros in addition require that the operation *S.inf*(*T*::*item it*) is defined and returns the information associated with item *it*.

A first try of an implementation of the *forall_items* macro could be

```
#define forall_items(it,S)\
for(it = S.first_item(); it != nil; it = S.next_item(it))
```

However, with this implementation the current item of the iteration cannot be removed from *S*. To allow this operation we use a temporary variable p always containing the successor item of the current item *it*. Since our macro has to work for all item-based LEDA data types, the item type (e.g., *dic_item* for dictionaries) is not known explicitly, but is given implicitly by the type of the variable *it*. We therefore use a temporary iterator p of type *void*∗ and a function template *LoopAssign*(*item_type*& *it*, *void* ∗ p) to copy the contents of p to *it* before each execution of the for-loop body. The details are given by the following piece of code.

```
template <class T>
inline bool LoopAssign(T& it, void* p) { it = (T)p; }

#define forall_items(it,S)\
for( void* p = S.first_item(); \
    LoopAssign(it,p), p = S.next_item(it), it != nil; )

#define forall__rev_items(it,S)\
for( void* p = S.last_item();  \
    LoopAssign(it,p), p = S.pred_item(it), it != nil; )
```

With the above implementation of the *forall_items* loop the current item (but not its successor) may be deleted. There are many situations where this is desirable.

The following piece of code deletes all occurrences of a given number x from a list L of integers:

```
list_item it;
forall_items(it,L) if (L[it] == x) L.del_item(it);
```

The following piece of code removes self-loops from a graph G:

```
edge e;
forall_adj_edges(e,G) if (source(e) == target(e)) G.del_edge(e);
```

Exercises for 13.9

1 Design a forall macro allowing insertions at the end of the collection.

2 Implement an iteration macro for the binary tree class *bin_tree* traversing the nodes in in-order.

13.10 Priority Queues by Fibonacci Heaps (A Complete Example)

We give a comprehensive example that illustrates most of the concepts introduced in this chapter, the implementation of the priority queue data type *p_queue*<*P*, *I*> by Fibonacci heaps. The data type *p_queue*<*P*, *I*> was discussed in Section 5.4 and is defined in the header file <*LEDA/p_queue.h*>. We show the header file below, but without the manual comments that generate the manual page.

We call the implementation class *PRIO_IMPL*. There is one slight anomaly in the derivation of *p_queue*<*P*, *I*> from *PRIO_IMPL*: What is called *prio*rity in the data type template is called *key* in the implementation class, since in the first version of LEDA priorities were called keys and this still shows in the implementation class.

13.10.1 *The Data Type Template*

We start with the data type template.

⟨*p_queue.h*⟩≡

```
#define PRIO_IMPL f_heap
typedef PRIO_IMPL::item pq_item;
template<class P, class I>
class p_queue: private PRIO_IMPL
{
  int  key_type_id()          const { return leda_type_id((P*)0); }
  int  cmp(GenPtr x, GenPtr y) const
       { return compare(LEDA_ACCESS(P,x),LEDA_ACCESS(P,y)); }
  void clear_key(GenPtr& x)     const { leda_clear(LEDA_ACCESS(P,x)); }
  void clear_inf(GenPtr& x)     const { leda_clear(LEDA_ACCESS(I,x)); }
  void copy_key(GenPtr& x)      const { x = leda_copy(LEDA_ACCESS(P,x)); }
  void copy_inf(GenPtr& x)      const { x = leda_copy(LEDA_ACCESS(I,x)); }
public:
  p_queue()  {}
```

```
  p_queue(const p_queue<P,I>& Q):PRIO_IMPL(Q) {}
 ~p_queue()  { PRIO_IMPL::clear(); }

  p_queue<P,I>& operator=(const p_queue<P,I>& Q)
          { PRIO_IMPL::operator=(Q); return *this; }

  P       prio(pq_item it) const
          { return LEDA_CONST_ACCESS(P,PRIO_IMPL::key(it)); }
  I       inf(pq_item it)  const
          { return LEDA_CONST_ACCESS(I,PRIO_IMPL::inf(it)); }
  pq_item find_min()        const { return PRIO_IMPL::find_min(); }
  void    del_min()               { PRIO_IMPL::del_min(); }
  void    del_item(pq_item it)    { PRIO_IMPL::del_item(it); }

  pq_item insert(const P& x, const I& i)
          { return PRIO_IMPL::insert(leda_cast(x),leda_cast(i)); }

  void    change_inf(pq_item it, const I& i)
          { PRIO_IMPL::change_inf(it,leda_cast(i)); }

  void    decrease_p(pq_item it, const P& x)
          { PRIO_IMPL::decrease_key(it,leda_cast(x)); }

  int     size()  const { return PRIO_IMPL::size(); }
  bool    empty() const { return (size()==0) ? true : false; }
  void    clear()       { PRIO_IMPL::clear(); }

  pq_item first_item()           const { return PRIO_IMPL::first_item(); }
  pq_item next_item(pq_item it) const { return PRIO_IMPL::next_item(it); }
};
```

Every implementation class *PRIO_IMPL* for *p_queue*<*P*, *I*> has to provide the following operations and definitions.

```
class PRIO_IMPL
{
  virtual int  key_type_id()       const = 0;
  virtual int  cmp(GenPtr, GenPtr) const = 0;
  virtual void clear_key(GenPtr&)  const = 0;
  virtual void clear_inf(GenPtr&)  const = 0;
  virtual void copy_key(GenPtr&)   const = 0;
  virtual void copy_inf(GenPtr&)   const = 0;

public:

  typedef ... item;

protected:

  PRIO_IMPL();
  PRIO_IMPL(const PRIO_IMPL&);

  virtual ~PRIO_IMPL();

  PRIO_IMPL& operator=(const PRIO_IMPL&);

  item insert(GenPtr,GenPtr);
  item find_min() const;

  GenPtr key(item) const;
  GenPtr inf(item) const;

  void del_min();
  void del_item(item);
```

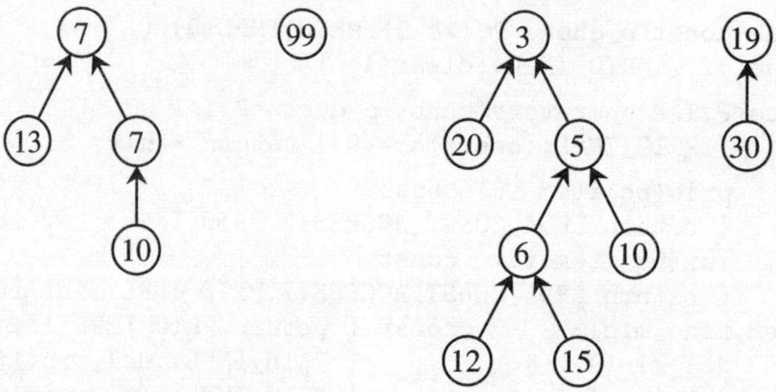

Figure 13.4 A heap-ordered forest.

```
    void decrease_key(item,GenPtr);
    void change_inf(item,GenPtr);
    void clear();
    int  size()  const;
    //iteration
    item first_item() const;
    item next_item(item) const;
};
```

13.10.2 *Fibonacci Heaps*

In the remainder of this section we give the Fibonacci heap realization of *PRIO_IMPL*.

Definition and Header File: Fibonacci heaps (class *f_heap*) are one of the best realizations of priority queues [FT87]. They represent priority queues as heap-ordered forests. The items of the priority queue are in one-to-one correspondence to the *nodes* of the forest; so it makes sense to talk about the key and the information of a node. A forest is *heap-ordered* if each tree in the forest is *heap-ordered*, and a tree is heap-ordered if the key of every non-root node is no less than the key of the parent of the node. In other words, the sequence of keys along any root to leaf path is non-decreasing. Figure 13.4 shows a heap-ordered forest.

In the storage representation of *f_heaps* every node contains a pointer to its parent (the parent pointer of a root is *nil*) and to one of its children. The child-pointer is *nil* if a node has no children. The children of each node and also the roots of the trees in a *f_heap* form a doubly-linked circular list (pointers *left* and *right*). In addition, every node contains the four fields *rank*, *marked*, *next*, and *pred*. The *rank* field of each node contains the number of children of the node and the *marked* field is a boolean flag whose purpose will be made clear below. The *next* and *pred* fields are used to keep all nodes of a Fibonacci heap in a doubly-linked linear list. This list is needed for the *forall_items*-iteration. An *f_heap*-item (type *F_heap*::*item*) is a pointer to a node. Figure 13.5 shows the storage representation of the heap-ordered forest of Figure 13.4.

The constructor of class *f_heap_node* creates a new node $\langle k, i\rangle$ and initializes some of the

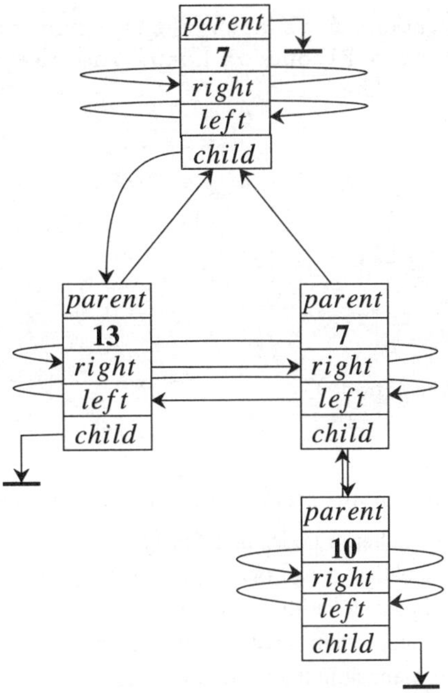

Figure 13.5 The storage representation of the heap-ordered forest of Figure 13.4. The *key*, *rank*, *marked*, *next*, and *pred* fields are not shown, informations are integers and nil-pointers are shown as pointing to "ground".

fields to their obvious values. It also adds the new item to the front of the list of all items of the heap. The LEDA memory management is used for *f_heap_node*s (cf. Section 13.8).

⟨*f_heap.h*⟩≡

```
#include <LEDA/basic.h>
class f_heap_node;
typedef f_heap_node* f_heap_item;
class f_heap_node {
  friend class f_heap;
  f_heap_item left;      // left and right siblings (circular list)
  f_heap_item right;
  f_heap_item parent;    // parent node
  f_heap_item child;     // a child
  f_heap_item next;      // list of all items
  f_heap_item pred;
  int  rank;             // number of children
  bool marked;           // mark bit
  GenPtr key;            // key
  GenPtr inf;            // information
f_heap_node(GenPtr k, GenPtr info, f_heap_item n)
{
```

```
    // the third argument n is always the first item in the list
    // of all items of a Fibonacci heap. The new item is added
    // at the front of the list
    key = k;
    inf = info;
    rank = 0;
    marked = false;
    parent = child = nil;
    next = n;
    if (n) n->pred = this;
  }
    LEDA_MEMORY(f_heap_node)
  };
```

The storage representation of an *f_heap* consists of five fields:

number_of_nodes	the number of nodes in the heap
power	the smallest power of two greater than or equal to *number_of_nodes*
logp	the binary logarithm of power
minptr	a pointer to a root with minimum key
node_list	first element in the list of all nodes

⟨*f_heap.h*⟩+≡

```
  class f_heap  {
    int number_of_nodes;
    int power;
    int logp;
    f_heap_item minptr;
    f_heap_item node_list;
    ⟨virtual functions related to keys and infs⟩
    ⟨auxiliary functions⟩
  public:
    typedef f_heap_item item;
  protected:
    // constructors, destructor, assignment
    f_heap();
    f_heap(const f_heap&);
    f_heap& operator=(const f_heap&);
    virtual ~f_heap();
    // priority queue operations
    f_heap_item insert(GenPtr, GenPtr);
    f_heap_item find_min()  const;
    void   del_min();
    void   decrease_key(f_heap_item,GenPtr);
    void   change_inf(f_heap_item,GenPtr);
    void   del_item(f_heap_item);
    void   clear();
    GenPtr key(f_heap_item) const;
    GenPtr inf(f_heap_item) const;
```

```
    int     size() const;
    bool    empty() const;

    // iteration
    f_heap_item first_item() const;
    f_heap_item next_item(f_heap_item) const;
};
```

We turn to the implementation of the member functions. The file _f_heap.c contains the implementations of all operations on *f_heaps*.

Construction: To create an empty *f_heap* set *number_of_nodes* to zero, *power* to one, *logp* to zero, and *minptr* and *node_list* to *nil*.

⟨*_f_heap.c*⟩≡

```
#include <LEDA/basic.h>
#include "f_heap.h"

f_heap::f_heap()
{ number_of_nodes = 0;
  power = 1;
  logp = 0;
  minptr = nil;
  node_list = nil;
}
```

Simple Operations on Heaps: We discuss create, findmin, size, empty, key, inf, and change_key. A *find_min* operation simply returns the item pointed to by *minptr*. The empty operation compares *number_of_nodes* to zero, and the *size* operation returns *number_of_nodes*. Both operations take constant time.

The *key* and *inf* operations apply to an item and return the appropriate component of the item.

The *change_inf* operations applies to an item x and an information *inf* and changes the information associated with x to a copy of *inf*. It also clears the memory used for the old information.

⟨*_f_heap.c*⟩+≡

```
f_heap_item f_heap::find_min()          const { return minptr; }
int         f_heap::size()              const { return number_of_nodes; }

bool        f_heap::empty()             const
                                         { return number_of_nodes == 0; }

GenPtr      f_heap::key(f_heap_item x) const { return x->key; }
GenPtr      f_heap::inf(f_heap_item x) const { return x->inf; }

void f_heap::change_inf(f_heap_item x, GenPtr i)
{ clear_inf(x->inf);
  copy_inf(i);
  x->inf = i;
}
```

We have used functions *clear_key* and *copy_key* without defining them. Both functions belong to the set of virtual functions of class *f_heap* which we need to make *f_heap* a parameterized data structure. We declare these functions as pure virtual and define them in the definition of the class *p_queue*<K, I> as discussed in Section 13.4.

The six virtual functions are: *cmp* compares two keys (of type P), *clear_key* and *clear_inf* deallocate a key and an information, respectively, *copy_key* and *copy_inf* return a copy of their argument, and *key_type_id*( ) determines whether its argument belongs to a built-in type as discussed in Section 13.5. It is used to bypass the calls to compare function for such types.

⟨*virtual functions related to keys and infs*⟩≡

```
virtual int     cmp(GenPtr,GenPtr)        const = 0;
virtual void    clear_key(GenPtr&)        const = 0;
virtual void    clear_inf(GenPtr&)        const = 0;
virtual GenPtr  copy_key(GenPtr&)         const = 0;
virtual GenPtr  copy_inf(GenPtr&)         const = 0;
virtual int     key_type_id()             const = 0;
```

Some Theory: The non-trivial operations are *insert*, *decrease_inf* and *del_min*. We discuss them in some detail now. The discussion will be on the level of heap-ordered forests. All implementation details will be given later.

An insert adds a new single node tree to the Fibonacci heap and, if necessary, adjusts the *minptr*. So a sequence of n inserts into an initially empty heap will simply create n single node trees. The cost of an insert is clearly $O(1)$.

A *del_min* operation removes the node indicated by *minptr*. This turns all children of the removed node into roots. We then scan the set of roots (old and new) to find the new minimum. To find the new minimum we need to inspect all roots (old and new), a potentially very costly process. We make the process even more expensive (by a constant factor) by doing some useful work on the side, namely combining trees of equal rank into larger trees. A simple method to combine trees of equal rank is as follows. Let *max_rank* be the maximal rank of any node. Maintain a set of buckets, initially empty and numbered from 0 to *max_rank*. Then step through the list of old and new roots. When a root of rank i is considered inspect the i-th bucket. If the i-th bucket is empty then put the root there. If the bucket is non-empty then combine the two trees into one (by making the root with the larger information a child of the other root). This empties the i-th bucket and creates a root of rank $i + 1$. Try to throw the new tree into the $i + 1$st bucket. If it is occupied, combine .... When all roots have been processed in this way, we have a collection of trees whose roots have pairwise distinct ranks. What is the running time of the *del_min* operation?

Let K denote the number of roots before the call of *del_min*. The cost of the operation is $O(K + max\_rank)$ (since the deleted node has at most *max_rank* children and hence there are at most $K + max\_rank$ roots to start with. Moreover, every combine reduces the number of roots by one). After the call there will be at most *max_rank* roots (since they all have

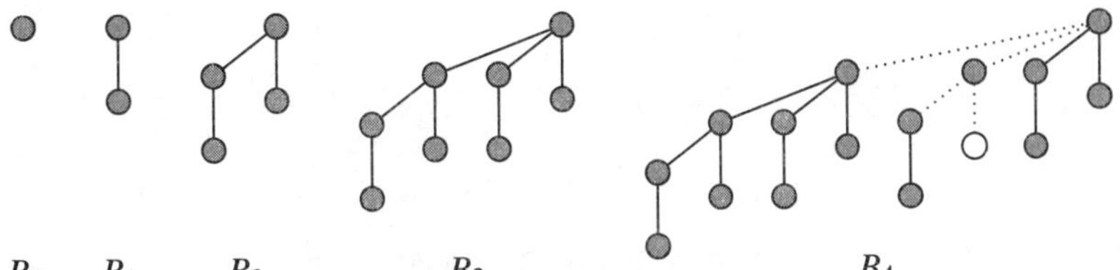

Figure 13.6 Binomial trees. Deletion of the high-lighted node and all high-lighted edges decomposes B_4 into binomial trees.

different ranks) and hence the number of roots decreases by at least $K - max\_rank$. Thus, if we use the potential function Φ_1 with

$$\Phi_1 = \text{number of roots}$$

then the amortized cost of a *delete_min* operation is $O(max\_rank)$. The amortized cost of an insert is $O(1)$; note that n inserts increase the potential Φ_1 by one. We will extend the potential by a second term Φ_2 below.

What can we say about the maximal rank of a node in a Fibonacci heap? Let us consider a very simple situation first. Suppose that we perform a sequence of inserts followed by a single *del_min*. In this situation, we start with a certain number of single node trees and all trees formed by combining are so-called *binomial trees* as shown in Figure 13.6. The binomial tree B_0 consists of a single node and the binomial tree B_{i+1} is obtained by joining two copies of the tree B_i. This implies that the root of the tree B_i has rank i and that the tree B_i contains exactly 2^i nodes. We conclude that the maximal rank in a binomial tree is logarithmic in the size of the tree. If we could guarantee in general that the maximal rank of any node is logarithmic in the size of the tree then the amortized cost of the *del_min* operation would be logarithmic.

We turn to the *decrease_key* operation next. It is given a node v and a new information *newkey* and decreases the information of v to *newkey*. Of course, *newkey* must not be larger than the old information associated with v. Decreasing the information associated with v will in general destroy the heap property. In order to maintain the heap property we delete the edge connecting v to its parent and turn v into a root. This has the side effect that for any ancestor w of v different from v's parent the size of w's subtree decreases by one but w's rank is unchanged. Thus, if we want to maintain the property that the maximal rank of any node is logarithmic in the size of the subtree rooted at the node, we need to do more than just cutting v's link to its parent.

An old solution suggested by Vuillemin [Vui78] is to keep all trees in the heap binomial. This can be done as follows: for any proper ancestor z of v delete the edge into z on the path from v to z, call it e, and all edges into z that were created later than e. In Figure 13.6 a node and a set of edges is high-lighted in the tree B_4. If all high-lighted edges are removed then B_4 decomposes into two copies of B_0 and one copy each of B_1, B_2, and B_3. It is not too hard to see that at most k edges are removed when a B_k is disassembled (since a B_k

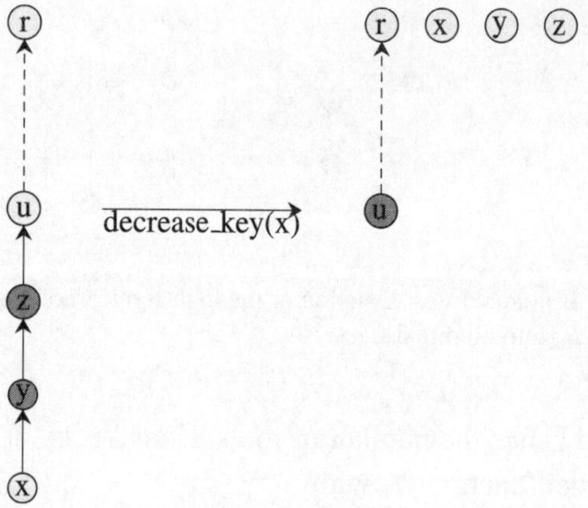

Figure 13.7 A decrease key on x is performed and y and z are marked but u is not; x, y, and z become roots, roots are unmarked, and u becomes marked. Marked nodes are shown shaded. A dashed edge stands for a path of edges.

decomposes into two B_j's and one each of $B_{j+1}, \ldots, B_{k-1}$ for some j, with $0 \le j \le k-1$) and hence this strategy gives a logarithmic time bound for the *decrease_key* operation.

In some graph algorithms the *decrease_key* operation is executed far more often than the other priority queue operations, e.g., Dijkstra's shortest-path algorithm (cf. Section 6.6) executes m *decrease_key*s and only n *insert*s and *del_min*s, where m and n are the number of edges and nodes of the graph, respectively. Since m might be as large as n^2 it is desirable to make the *decrease_key* operation cheaper than the other operations. Fredman and Tarjan showed how to decrease its cost to $O(1)$ without increasing the cost of the other operations. Their solution is surprisingly simple and we describe it next.

When a node x loses a child because *decrease_key* is applied to the child the node x is marked; this assumes that x has not already been marked. When a marked node x loses a child, we turn x into a root, remove the mark from x and attempt to mark x's parent. If x's parent is marked already then .... In other words, suppose that we apply *decrease_key* to a node v and that the k-nearest ancestors of v are marked, then turn v and the k-nearest ancestors of v into roots and mark the $k+1$st-nearest ancestor of v (if it is not a root). Also unmark all the nodes that were turned into roots, cf. Figure 13.7. Why is this a good strategy?

First, a *decrease_key* marks at most one node and unmarks some number k of nodes. No other operation marks a node and hence in an amortized sense k can be at most one (we cannot unmark more nodes than we mark). However, we also increase the number of roots by k which in turn increases the potential Φ_1 by k and therefore we have to argue more carefully. Let

$$\Phi_2 = 2 \cdot \text{number of marked nodes}$$

and let $\Phi = \Phi_1 + \Phi_2$. A *decrease_key* operation where the node v has k marked ancestors has actual cost $O(k+1)$ and decreases the potential by at least $2(k-1)-(k+1)=k-3$. Note that the number of marked nodes is decreased by at least $k-1$ (at least k nodes are unmarked and at most one node is marked) and that the number of roots is increased by $k+1$. The amortized cost of a *decrease_key* is therefore $O(1)$. *inserts* do not change Φ_2 and *del_mins* do not increase Φ_2 (it may decrease it because the marked children of the removed node become unmarked roots) and hence their amortized cost does not increase by the introduction of Φ_2.

How does the strategy affect the maximal rank. We show that it stays logarithmic. In order to do so we need some notation. Let $F_0=0$, $F_1=1$, and $F_i=F_{i-1}+F_{i-2}$ for $i\geq 2$ be the sequence of Fibonacci numbers. It is well-known that $F_{i+1}\geq(1+\sqrt{5}/2)^i\geq 1.618^i$ for all $i\geq 0$.

Lemma 78 *Let v be any node in a Fibonacci heap and let i be the rank of v. Then the subtree rooted at v contains at least F_{i+2} nodes. In a Fibonacci heap with n nodes all ranks are bounded by* $1.4404\log n$.

Proof Consider an arbitrary node v of rank i. Order the children of v by the time at which they were made children of v. Let w_j be the j-th child, $1\leq j\leq i$. When w_j was made child of v both nodes had the same rank. Also, since at least the nodes $w_1,\ldots,w_{j-1}$ were nodes of v at that time, the rank of v was at least $j-1$ at the time when w_j was made a child of v. The rank of w_j has decreased by at most 1 since then because otherwise w_j would be a root. Thus the current rank of w_j is at least $j-2$.

We can now set up a recurrence for the minimal number S_i of nodes in a tree whose root has rank i. Clearly $S_0=1$, $S_1=2$, and $S_i\geq 2+S_0+S_1+\ldots+S_{i-2}$. The last inequality follows from the fact that for $j\geq 2$, the number of nodes in the subtree with root w_j is at least S_{j-2}, and that we can also count the nodes v and w_1. The recurrence above (with = instead of $\geq$) generates the sequence 1, 2, 3, 5, 8,... which is identical to the Fibonacci sequence (minus its first two elements).

Let's verify this by induction. Let $T_0=1$, $T_1=2$, and $T_i=2+T_0+\ldots+T_{i-2}$ for $i\geq 2$. Then, for $i\geq 2$, $T_{i+1}-T_i=2+T_0+\ldots+T_{i-1}-2-T_0-\ldots-T_{i-2}=T_{i-1}$, i.e., $T_{i+1}=T_i+T_{i-1}$. This proves $T_i=F_{i+2}$.

For the second claim, we only have to observe that $F_{i+2}\leq n$ implies $i\cdot\log(1+\sqrt{5}/2)\leq\log n$ which in turn implies $i\leq 1.4404\log n$. □

This concludes our theoretical treatment of Fibonacci heaps. We have shown the following time bounds: an *insert* and a *decrease_key* take constant amortized time and a *del_min* takes logarithmic amortized time. The operations *size*, *empty*, and *findmin* take constant time.

We now return to the implementation.

Insertions: An *insert* operation takes a key k and an information i and creates a new heap-ordered tree consisting of a single node $\langle k,i\rangle$. In order to maintain the representation invariant it must also add the new node to the circular list of roots, increment *number_of_nodes*, and may be *power* and *logp*, and change *minptr* if *k* is smaller than the current minimum key in the queue.

⟨*f_heap.c*⟩+≡

```
f_heap_item f_heap::insert(GenPtr k, GenPtr i)
{
  k = copy_key(k);
  i = copy_inf(i);
  f_heap_item new_item = new f_heap_node(k,i,node_list);
  if ( number_of_nodes == 0 )
  { // insertion into empty queue
    minptr = new_item;
    // build trivial circular list
    new_item->right = new_item;
    new_item->left = new_item;
    // power and logp have already the correct value
  }
  else
  { // insertion into non-empty queue;
    // we first add to the list of roots
    new_item->left = minptr;
    new_item->right = minptr->right;
    minptr->right->left = new_item;
    minptr->right = new_item;
    if ( cmp(k,minptr->key) < 0 ) minptr = new_item; // new minimum
    if ( number_of_nodes >= power) // log number_of_nodes grows by one
    { power = power * 2;
      logp = logp + 1;
    }
  }
  number_of_nodes++;
  return new_item;
}
```

Delete_min: A *del_min* operation removes the item pointed to by *minptr*, i.e., an item of minimum *key*. This turns all children of the removed node into roots. We then scan the set of roots (old and new) to find the new minimum.

⟨*f_heap.c*⟩+≡

```
void f_heap::del_min()
{ // removes the item pointed to by minptr
  if ( minptr == nil )
    error_handler(1,"f_heap: deletion from empty heap");
  number_of_nodes--;
  if ( number_of_nodes==0 )
  { // removal of the only node
    // power and logp do not have to be changed.
```

```
      clear_key(minptr->key);
      clear_inf(minptr->inf);
      delete minptr;
      minptr = nil;
      node_list = nil;
      return;
    }
    /* removal from a queue with more than one item. */
    ⟨turn children of minptr into roots⟩;
    ⟨combine trees of equal rank and compute new minimum⟩;
    ⟨remove old minimum⟩;
}
```

We now discuss the removal of a node of minimum *key* from an *f_heap* with more than one item. Recall that *number_of_nodes* already has its new value. We first update *power* and *logp* (if necessary) and then turn all children of *minptr* into roots (by setting their parent pointer to nil and their mark bit to false and combining the list of children of *minptr* with the list of roots). We do not delete *minptr* yet. It is convenient to keep it as a sentinel.

The cost of turning the children of the *minptr* into roots is $O(maxrank)$;

Note that the body of the loop is executed for each child of the node *minptr* and that, in addition, to the children of *minptr* we access *minptr* and its right sibling.

⟨*turn children of minptr into roots*⟩≡

```
if ( 2 * number_of_nodes <= power )
{ power = power / 2;
  logp = logp - 1;
}
f_heap_item r1 = minptr->right;
f_heap_item r2 = minptr->child;
if ( r2 )
{ // minptr has children
  while ( r2->parent )
  { //  visit them all and make them roots
    r2->parent = nil;
    r2->marked = false;
    r2 = r2->right;
  }
  // combine the lists, i.e. cut r2's list between r2 and its left
  // neighbor and splice r2 to minptr and its left neighbor to r1
  r2->left->right = r1;
  r1->left = r2->left;
  minptr->right = r2;
  r2->left = minptr;
}
```

The task of the combining phase is to combine roots of equal rank into larger trees. The combining phase uses a procedure *link* which combines two trees of equal rank and returns the resulting tree.

⟨*f_heap.c*⟩+≡

```
f_heap_item f_heap::link(f_heap_item r1, f_heap_item r2)
{
  // r1 and r2 are roots of equal rank, both different from minptr;
  // the two trees are combined and the resulting tree is returned.
  f_heap_item h1;
  f_heap_item h2;
  if (cmp(r1->inf,r2->inf) <= 0)
   { // r2 becomes a child of r1
     h1 = r1;
     h2 = r2;
   }
   else
   { // r1 becomes a child of r2
     h1 = r2;
     h2 = r1;
   }
  // we now make h2 a child of h1. We first remove h2 from
  // the list of roots.
  h2->left->right = h2->right;
  h2->right->left = h2->left;
  /* we next add h2 into the circular list of children of h1 */
  if ( h1->child == nil )
  { // h1 has no children yet; so we make h2 its only child
    h1->child = h2;
    h2->left = h2;
    h2->right = h2;
  }
  else
  { // add h2 to the list of children of h1
    h2->left = h1->child;
    h2->right = h1->child->right;
    h1->child->right->left = h2;
    h1->child->right = h2;
  }
  h2->parent = h1;
  h1->rank++;
  return h1;
}
```

Let's not forget to add the declaration of link to the set of auxiliary functions of *class f_heap*.

⟨*auxiliary functions*⟩≡

```
f_heap_item link(f_heap_item, f_heap_item);
```

Next comes the code to combine trees of equal rank. The task is to step through the list of old and new roots, to combine roots of equal rank, and to determine the node of minimum key. We solve this task iteratively. We maintain an array *rank_array* of length *maxrank* of pointers to roots: *rank_array*[i] points to a root of rank i, if any and to *nil* otherwise.

Initially all entries point to *nil*. When a root of rank r is inspected and *rank_array*$[r]$ is *nil*, store r there. If it is non-empty, combine r with the array entry and replace r by the combined tree. The combined tree has rank one higher. We declare *rank_array* as an array of length $12 * sizeof(int)$. This is a save choice since the number of nodes in a heap is certainly bounded by $MAXINT = 2^{8*sizeof(int)}$. Hence $maxrank \leq 1.5 * \log(MAXINT) = 12 * sizeof(int)$.

There is a small subtlety in the following piece of code. We are running over the list of roots and simultaneously modifying it. This is potentially dangerous, but our strategy is safe. Imagine the list of roots drawn with the *minptr* at the far right. Then *current* points to the leftmost element initially. At a general step of the iteration *current* points at some arbitrary list element. All modifications of the list by calls of *link* take place strictly to the left of *current*. For this reason it is important to advance *current* at the beginning of the loop.

⟨*combine trees of equal rank and compute new minimum*⟩≡

```
f_heap_item rank_array[12*sizeof(int)];
for (int i = (int)1.5*logp; i >= 0; i--) rank_array[i] = nil;
f_heap_item new_min = minptr->right;
f_heap_item current = new_min;
while (current != minptr)
{ // old min is used as a sentinel
  r1 = current;
  int rank = r1->rank;
  // it's important to advance current already here
  current = current->right;
  while (r2 = rank_array[rank])
  { rank_array[rank] = nil;
    // link combines trees r1 and r2 into a tree of rank one higher
    r1 = link(r1,r2);
    rank++;
  }
  rank_array[rank] = r1;
  if ( cmp(r1->inf,new_min->inf) <= 0 ) new_min = r1;
}
```

We complete the operation by actually deleting the old minimum and setting *minptr* to its new value.

⟨*remove old minimum*⟩≡

```
minptr->left->right = minptr->right;
minptr->right->left = minptr->left;
clear_key(minptr->key);
clear_inf(minptr->inf);
r1 = minptr->pred;
r2 = minptr->next;
if (r2) r2->pred = r1;
```

```
  if (r1) r1->next = r2; else node_list = r2;
  delete minptr;
  minptr = new_min;
```

Decrease_key, Clear, and Del_item: *decrease_key* makes use of an auxiliary function *cut*(x) that turns a non-root node x into a root and returns its old parent.

⟨*auxiliary functions*⟩+≡

```
f_heap_item cut(f_heap_item);
```

⟨*_f_heap.c*⟩+≡

```
f_heap_item f_heap::cut(f_heap_item x)
{
  f_heap_item y = x->parent;
  if ( y->rank == 1 ) y->child = nil;  // only child
  else
  { /* y has more than one child. We first make sure that its childptr
       does not point to x and then delete x from the list of children */
    if ( y->child == x ) y->child = x->right;
    x->left->right = x->right;
    x->right->left = x->left;
  }
  y->rank--;
  x->parent = nil;
  x->marked = false;
  // add to circular list of roots
  x->left = minptr;
  x->right = minptr->right;
  minptr->right->left = x;
  minptr->right = x;
  return y;
}
```

Now we can give the implementation of *decrease_key*.

⟨*_f_heap.c*⟩+≡

```
void f_heap::decrease_key(f_heap_item v, GenPtr newkey)
{
  /* changes the key of f_heap_item v to newkey;
     newkey must be no larger than the old key;
     if newkey is no larger than the minimum key
     then v becomes the target of the minptr   */
  if (cmp(newkey,v->key) > 0)
    error_handler(1,"f_heap: key too large in decrease_key.");
  // change v's key
  clear_key(v->key);
  v->key = copy_key(newkey);
```

```
    if ( v->parent )
    { f_heap_item x = cut(v);            // make v a root
      while (x->marked) x = cut(x);     // a marked f_heap_node
                                         // is a non-root
      if (x->parent) x->marked = true; // mark x if it not a root
    }
    // update minptr (if necessary)
    if (cmp(newkey,minptr->key) <= 0) minptr = v;
  }
```

To clear a heap simply remove the minimum until the heap is empty. The cost of *clear* is bounded by n times the cost of *del_min*. We can also use *clear* as the destructor of class *f_heap*.

⟨*f_heap.c*⟩+≡

```
void   f_heap::clear()  { while (number_of_nodes > 0) del_min(); }
f_heap::~f_heap() { clear(); }
```

To remove an arbitrary item from a heap, we first decrease its *key* to the minimum key (this makes the item the target of the *minptr*) and then remove the minimum. The cost of removing an item is therefore bounded by $O(1)$ plus the cost of *decrease_key* plus the cost of *del_min*.

⟨*f_heap.c*⟩+≡

```
void f_heap::del_item(f_heap_item x)
{ decrease_key(x,minptr->key);       // the minptr now points to x
  del_min();
}
```

Assignment, Iteration, and Copy Constructor: Next comes the assignment operator. In order to execute $S = H$ we simply step through all the items of H and insert their key and information into S. We must guard against the trivial assignment $H = H$.

⟨*f_heap.c*⟩+≡

```
f_heap& f_heap::operator=(const f_heap& H)
{ if (this != &H)
  { clear();
    for (f_heap_item p = H.first_item(); p; p = H.next_item(p))
      insert(p->key,p->inf);
  }
  return *this;
}
```

The assignment operator makes use of the two functions *first_item* and *next_item*. They allow us to iterate over all items of a heap. We use these functions in the assignment operator, the copy constructor, and the *forall_items*-iteration. The last use forces us to make both functions public members of the class. However, we do not list them in the manual and so they are only semi-public. For this reason *next_item* does not check whether its argument is distinct from *nil*.

⟨*f_heap.c*⟩+≡

```
f_heap_item f_heap::first_item() const { return node_list; }
f_heap_node* f_heap::next_item(f_heap_node* p) const
                                        { return p ? p->next : 0; }
```

The last operation to implement is the copy constructor. It makes a copy of its argument *H*. The strategy is simple. For each item of *H* we create a single node tree with the same key and information.

There is a subtle point in the implementation. When a virtual function is applied to an object under construction then the default implementation of the function is used and not the overriding definition in the derived class. It is therefore important in the code below to call the virtual functions *copy_key*, *copy_inf* and *cmp* through the already existing object *H*; leaving out the prefix *H*. would select the default definitions (which do not do anything).

⟨*f_heap.c*⟩+≡

```
f_heap::f_heap(const f_heap& H)
{ number_of_nodes = H.size();
  minptr = nil;
  node_list = nil;
  f_heap_item first_node = nil;
  for(f_heap_item p = H.first_item(); p; p = H.next_item(p))
  { GenPtr k = H.copy_key(p->key);
    GenPtr i = H.copy_inf(p->inf);
    f_heap_item q = new f_heap_node(k,i,node_list);
    q->right = node_list->next;
    if (node_list->next) node_list->next->left = q;
    if (minptr == nil) { minptr = q; first_node = q; }
    else if ( H.cmp(k,minptr->key) < 0 ) minptr = q;
  }
  first_node->right = node_list;
  node_list->left = first_node;
}
```

14

Manual Pages and Documentation

This chapter is authored jointly with Evelyn Haak, Michael Seel, and Christian Uhrig.

Software requires documentation. In this chapter we explain:

- how to make LEDA-style manual pages,
- how to make a LEDA-style manual,
- and how to write documentations in the style of this book.

14.1 Lman and Fman

Lman and Fman are the LEDA tools for manual production and quick reference to manual pages. We will discuss Fman at the end of the section. The command

```
Lman T[.lw|.nw|.h] options
```

searches for a file with name T.lw, T.nw, T.h, or T (in this order) first in the current directory and then in the directory LEDAROOT/incl/LEDA and produces a LEDA-style manual page from it. Thus

```
Lman sortseq
Lman myproject.lw
```

produce the manual page of sorted sequences and of myproject, respectively.

The extraction of the manual page is guided by the so-called manual comments contained in the file-argument of Lman. A manual comment is any comment of the form

```
/*{\Mcommand ... arbitrary text ... }*/
```

```
/*{\Manpage {stack} {E} {Stacks} {S}}*/

template<class E> class _CLASSTYPE stack : private SLIST
{
/*{\Mdefinition
An instance |S| of the parameterized data type |\Mname| is a sequence of
elements of data type |E|, called the element type of |S|. Insertions or
deletions of elements take place only at one end of the sequence, called
the top of |S|. The size of |S| is the length of the sequence, a stack
of size zero is called the empty stack.}*/

  void copy_el(GenPtr& x)  const { x=Copy(ACCESS(E,x)); }
  void clear_el(GenPtr& x) const { Clear(ACCESS(E,x)); }
public:

/*{\Mcreation}*/

  stack() {}
/*{\Mcreate creates an instance |\Mvar| of type |\Mname| and initializes
it to the empty stack.}*/

  stack(const stack<E>& S) : SLIST(S) {}
 ~stack() { clear(); }
  stack<E>& operator=(const stack<E>& S)
    { return (stack<E>&)SLIST::operator=(S); }

/*{\Moperations 2.5 4}*/

E top()    const { return ACCESS(E,SLIST::head());}
/*{\Mop       returns the top element of |\Mvar|.\\
              \precond $S$ is not empty.}*/

void push(E x)  { SLIST::push(Copy(x)); }
/*{\Mop       adds $x$ as new top element to |\Mvar|.}*/

E pop()           { E x=top(); SLIST::pop(); return x; }
/*{\Mop       deletes and returns the top element of |\Mvar|.\\
              \precond $S$ is not empty.}*/

int  empty() { return SLIST::empty(); }
/*{\Mop       returns true if |\Mvar| is empty, false otherwise.}*/

}

\*{\Mimplementation
Stacks are implemented by singly linked linear lists.
All operations take time $O(1)$. }*/
```

Figure 14.1 A file decorated by manual comments. The file is part of the header file of the data type stack. Figure 14.2 shows the manual page produced by Lman.

where Mcommand is one of so-called manual commands. We discuss manual commands in Section 14.2.2. Every manual comment causes Lman to extract part of the manual. Figures 14.1 and 14.2 show a file augmented by manual comments and the manual page produced from it.

The layout of the manual page is fine-tuned by the options-argument of Lman. We will discuss the available options in Section 14.2.8. Options may also be put in a configuration file *Lman.cfg* in either the home directory or the working directory. Command line options

Stacks (stack)

1. Definition

An instance S of the parameterized data type *stack<E>* is a sequence of elements of data type E, called the element type of S. Insertions or deletions of elements take place only at one end of the sequence, called the top of S. The size of S is the length of the sequence, a stack of size zero is called the empty stack.

2. Creation

stack<E> S;	creates an instance S of type *stack<E>* and initializes it to the empty stack.

3. Operations

E	S.top()	returns the top element of S. *Precondition*: S is not empty.
void	S.push(E x)	adds x as new top element to S.
E	S.pop()	deletes and returns the top element of S. *Precondition*: S is not empty.
int	S.empty()	returns true if S is empty, false otherwise.

Figure 14.2 The manual page produced from the file in Figure 14.1.

take precedence over options in the working directory which in turn take precedence over options in the home directory.

Fman is our tool for quick reference to manual pages. The command

```
Fman T[.lw|.nw|.h] filter
```

searches for a file with name T.lw, T.nw, T.h, or T (in this order) first in the current directory and then in the directory LEDAROOT/incl/LEDA and extracts manual information from it. The information is displayed in ASCII-format. For example,

```
Fman sortseq insert
Fman sortseq creation
```

give information about operation insert of type sortseq and about the different ways of creating a sorted sequence, respectively.

```
Fman
```

gives information about Fman and the available filters.

Fman uses Perl [WS90] and Lman uses Perl, LaTeX [Lam86], and xdvi.

Please try out Lman and Fman before proceeding. If they do not work, the error is very likely to be one of the following (if not, you should refer to the LEDA installation guide):

- One of the required systems Perl, LaTeX, and xdvi is not installed.
- The environment variable LEDAROOT is not set to the root directory of the LEDA system.
- LEDAROOT/Manual/cmd is not part of your PATH.
- LEDAROOT/Manual/tex is not part of your TEXINPUTS.

14.2 Manual Pages

Figure 14.2 shows a typical LEDA manual page. It is produced from the file in Figure 14.1 by a call of the Lman utility. Observe that the file contains comments starting with `/*{\M...` and ending with `}*/`. They are called *manual comments*. They start with a so-called manual command, e.g., `Mdefinition` or `Mop` and control the extraction of the manual page from the header file. There are about twenty different manual commands. We will discuss them in turn in this section. Before doing so, we justify our decision to incorporate all manual information into the header files of the LEDA system.

In the early years of the LEDA project we kept the manual page of a data type separate from its implementation. The manual was contained in a tex-file and the implementation was contained in an h-file and a c-file. Updates of a data type usually required changes to all three files and this led to a consistency problem between the three files. The consistency between h-file and c-file is a minor issue since every compiler run checks it. However, we found it almost impossible to keep the manual pages consistent with the implementation. The inconsistencies between manual and implementation had two causes:

- Clerical errors: Frequently, things that were supposed to be identical were different, e.g., a type was spelled `sort_seq` in the manual and `sortseq` in the implementation, or the parameters of a function were permuted.
- Lack of discipline: We frequently forgot to make changes due to lack of time or other reasons. We were quite creative in this respect.

In 1994 we decided to end the separation between implementation and manual. We incorporated the manual into the h-files in the form of so-called manual comments and wrote a tool called *Lman* that extracts the tex-file for the manual page automatically from the h-file. Every manual comment produces part of the manual page, e.g., the manual comment starting with `\Mdefinition` produces the definition section of the manual page, and a comment starting with `\Mop` produces an entry for an operation of the data type. Such an entry consists of the return type, an invocation of the operation, and a definition of the semantics in the form of a text. Only the latter piece of information is explicitly contained in the Mop-comment, the other two pieces are generated automatically from the C++-text in the header file. Experience shows that our decision to incorporate manual pages into header files greatly alleviates the consistency problem:

- Clerical errors are reduced because things that should be identical are usually only typed once. For example, the fact that the C++-text in the manual is automatically generated from the C++-text in the header file guarantees the consistency between the two.
- Lack of discipline became a lesser issue since the fact that the header file of the implementation and the tex-file for the manual page are indeed the same file makes it a lot easier to be disciplined.

Lman produces manual pages in a two-step process. It first extracts a TEX-file from the header file and then applies LATEX. The first step is directed by the manual commands in the header file and the second step uses a specially developed set of TEX macros. We discuss the manual commands in Section 14.2.2 and the TEX macros in Section 14.2.5.

The first phase is realized by a Perl-program *lextract* that reads the file-argument and the options and produces a (temporary) TEX-file of the form:

```
\documentclass[a4paper,size pt]{article}
\usepackage{Lweb}
\begin{document}
  output of lextract
\end{document}
```

The program lextract is defined in the file *ext.nw* in LEDAROOT/Manual/noweb.

14.2.1 *The Structure of Manual Pages*

All manual pages of the LEDA system are organized in one of two ways depending on whether the page defines a data type or a collection of functions. Since manual pages are extracted from header files, the corresponding header files are organized accordingly. Examples of header files for data types are stack.h, sortseq.h, and list.h, and examples of header files for collections of functions are plane_alg.h, plane_graph_alg.h, and mc_matching.h.

All *header files for classes* follow the format shown in Figure 14.3. The *header files for collections of functions* have no particular structure.

14.2.2 *The Manual Commands*

We discuss the manual commands in the order in which they are typically used in the header file of a class.

The Manpage Command: A manual comment of the form

`/*{\Manpage {type} {parlist} {title} {varname}}*/`

produces the header line of the manual page for `type`. The argument `parlist` is the list of

```
  /*{\Manpage Comment }*/
class DT {
  /*{\Mdefinition comment }*/
  /*{\Mtypes comment }*/
  // type definitions
private:
  // private data and functions
public:
  /*{\Mcreation comment }*/
  // constructors and destructors and their manual entries
  /*{\Moperations comment }*/
  // operations and their manual entries
};
  // friends and their manual entries
  /*{\Mimplementation comment }*/
  /*{\Mexample comment }*/
```

Figure 14.3 The generic structure of a header file for a class. Any of the parts may be omitted.

type parameters of the type, `title` is the title of the manual page, and the optional argument `varname` is used in the manual page as the name of a canonical object of the type. The argument `parlist` is empty if the type has no type parameters. The following comments produce the header lines for character strings, linear lists, and sorted sequences, respectively.

```
/*{\Manpage {string}  {}    {Character Strings} {s}}*/
/*{\Manpage {list}    {E}   {Linear Lists}  {L}    }*/
/*{\Manpage {sortseq} {K,I} {Sorted Sequences} {S} }*/
```

The Manpage command produces the header line for the manual page and defines placeholders `\Mtype`, `\Mname`, and `\Mvar`. The first placeholder stands for `type`, the second placeholder stands for either `type` or `type<parlist>` depending on whether `parlist` is empty or not, and the third placeholder stands for `varname`. In the last example the placeholders `\Mtype`, `\Mname`, and `\Mvar` have values `sortseq`, `sortseq<K,I>`, and `S`, respectively.

The placeholders can be used instead of their values in later manual comments. This helps to maintain consistency. The placeholders are also used in the generation of the manual entries for the constructors and member functions, e.g., in Figure 14.2 all operations are applied to the canonical stack variable S.

What does lextract do when it encounters a Manpage-command? It records the values of all placeholders and outputs

```
\section*{title (type')}
```

where `type'` is obtained from `type` by quoting all occurrences of the underscore character (i.e., replacing _ by _). When LaTeX executes this line it will produce the header line of the manual page. If a manual page is to be included into a larger document, it is convenient to number the manual pages. The option `numbered=yes` causes the preprocessor to output

`\section{title (type')} \label{title}\label{type}`

The labels can be used to refer to the data type in other parts of an enclosing document.

The manual page of a class consists of sections *Definition*, *Types Creation*, *Operations*, *Implementation*, and *Example*; any of the sections may be omitted. Accordingly, we have the manual commands `\Mdefinition`, `\Mypes`, `\Mcreation`, `\Moperations`, `\Mimplementation`, and `\Mexample`.

The Mdefinition Command: A manual command of the form

`/*{\Mdefinition  body }*/`

produces the definition part of a manual page. For example,

```
template <class E>
class list {
/*{\Mdefinition
An instance [[\Mvar]] of class |\Mname| is a ...
}*/
```

produces

---

1. Definition

An instance L of class *list<E>* is a ...

The body of a definition comment (and of any of the other comments to come) is an arbitrary LaTeX text. As suggested by the literate programming tools CWEB [KL93] and noweb [Ram94] we added the possibility of quoting code. *Quoted code* is given special typographic treatment. There are two ways of quoting code:

- By enclosing it between verticals bars (| . . . |), or
- By enclosing it between double square brackets (`[[...]]`).

Quoted code is typeset according to the following rules: first all occurrences of the placeholders `\Mtype`, `\Mname`, and `\Mvar` are replaced by their values. We call this step *placeholder substitution*. In the example above this step yields[1]:

```
template <class E>
class list {
/*{\Mdefinition
An instance [[L]] of class |list<E>| is a ...
}*/
```

In a second step we apply what we call *C++ to LaTeX conversion* to quoted code. For code quoted by double square brackets this means using typewriter font for the quoted code and for code quoted by vertical bars this produces a math-like appearance, e.g., all identifiers

[1] We assume that the Mdefinition command is executed in the context of the Manpage comment for lists given above, i.e., L is the name of the canonical list and $list\langle E\rangle$ is the type of the list. We make the analogous assumption for all examples to follow.

are put into math-italics and <= is typeset as $\leq$. All code in this book is typeset using one of the two quoting mechanisms.

We give some examples of the quoting mechanisms. Be aware that putting an identifier between vertical bars is different from putting it between dollar signs except for identifiers consisting of a single character.

<table>
<tr><td><code>|diff|</code></td><td>produces</td><td>diff</td></tr>
<tr><td><code>$diff$</code></td><td>produces</td><td>$diff$</td></tr>
<tr><td><code>|x1|</code></td><td>produces</td><td>x1</td></tr>
<tr><td><code>$x1$</code></td><td>produces</td><td>$x1$</td></tr>
<tr><td><code>x</code></td><td>produces</td><td>x</td></tr>
<tr><td><code>|x|</code></td><td>produces</td><td>x</td></tr>
<tr><td><code>[[diff]]</code></td><td>produces</td><td><code>diff</code></td></tr>
</table>

Sometimes, one wants to produce vertical bars and/or double square brackets in the output. We provide TeX-macros to this effect. The macros `\Lvert`, `\DLK` and `\DRK` expand to |, `[[`, and `]]`, respectively. The TeX-macro `\Labs{...}` puts its argument between vertical bars, Lvert and Labs can only be used in math-mode.

We close this paragraph with a *warning*. The quoting mechanism by vertical bars is not perfect. In principle one can put any piece of text between vertical bars. The preprocessor attempts to understand the C++ structure of the text and generates output accordingly. Since the preprocessor has only limited knowledge of the syntax of C++, it succeeds only in simple cases:

<table>
<tr><td><code>|diff|</code></td><td>produces</td><td>diff</td></tr>
<tr><td><code>|diff + x1|</code></td><td>produces</td><td>$diff + x1$</td></tr>
<tr><td><code>|diff+x1|</code></td><td>produces</td><td>$diff + x1$</td></tr>
<tr><td><code>|list_item|</code></td><td>produces</td><td>list_item</td></tr>
<tr><td><code>|GRAPH<POINT,int>|</code></td><td>produces</td><td>GRAPH<POINT, int></td></tr>
<tr><td><code>|mark[v] <= cur_mark|</code></td><td>produces</td><td>$mark[v] \leq cur\_mark$</td></tr>
<tr><td><code>$|source|(e_0)$</code></td><td>produces</td><td>$source(e_0)$</td></tr>
</table>

The Mtypes and Mtypemember Commands: A manual command of the form

```
/*{\Mtypes w}*/
```

produces the header line of the type part of the manual. The argument `w` is optional. The argument w governs the layout of the entries for the local types of the data type. We will discuss it below. The manual entries for the local types are produced by Mtypemember commands. We give an example which is taken from the header file for the LEDA extension package for higher-dimensional geometry.

```
/*{\Mtypes 4}*/
typedef ch_Simplex<CHTRAITS,POINT,PLANE>* ch_simplex;
/*{\Mtypemember the item type for simplices of the complex.}*/

typedef ch_Simplex<CHTRAITS,POINT,PLANE>* ch_facet;
/*{\Mtypemember the item type for facets of the complex.}*/
```

```
typedef rc_Vertex<CHTRAITS,POINT>*        ch_vertex;
/*{\Mtypemember the item type for vertices of the complex.}*/
```

produces

---

2. Types

ch_simplex	the item type for simplices of the complex.
ch_facet	the item type for facets of the complex.
ch_vertex	the item type for vertices of the complex.

Each Mtypemember command produces a manual entry for a local type. Each manual entry is typeset on a line of its own and a two-column layout is followed. There is a column of width w containing the name of the local type and a column containing the text explaining the local type. The name of the type is extracted automatically from the type definition preceding the manual comment.

The Mcreation and Mcreate Commands: A manual command of the form

```
/*{\Mcreation name w}*/
```

produces the header line of the creation part of the manual. The arguments `name` and `w` are optional. If name is present, it is used as the value of the placeholder `\Mvar`. We recommend that you define `\Mvar` already in the Manpage command and keep the possibility to define it in the Mcreation command for reasons of backward compatibility. The argument w governs the layout of the entries for the constructors of the data type. We will discuss it below. The manual entries for the constructors are produced by Mcreate commands. We give an example.

```
/*{\Mcreation}*/

vector();
/*{\Mcreate creates an instance |\Mvar| of type |\Mname|;
|\Mvar| is initialized to the zero-dimensional vector.}*/

vector(int d);
 /*{\Mcreate creates an instance |\Mvar| of type |\Mname|;
|\Mvar| is initialized to the zero vector of dimension $d$.}*/
```

produces (assuming that Mvar stands for v and Mname stands for vector)

---

3. Creation

vector v;	creates an instance *v* of type *vector*; *v* is initialized to the zero-dimensional vector.
vector v(*int d*);	creates an instance *v* of type *vector*; *v* is initialized to the zero vector of dimension d.

Each Mcreate command produces a manual entry for a constructor. The manual entries are typeset in the form of a variable declaration for a variable Mvar of type Mname, i.e., for the default constructor the entry has the form

```
Mname Mvar;
```

and for a constructor taking arguments the entry has the form

```
Mname  Mvar(parameter list);
```

In the second case the parameter list is extracted automatically from the code unit preceding the manual comment. What is a code unit?

A *code unit* is a maximal sequence of consecutive non-blank lines not containing a comment. In other words, the line preceding a code unit is either empty or the end of a comment, the line following a code unit is either empty or the beginning of a comment, and all lines in a code unit are non-empty and do not belong to a comment. A code unit from which the preprocessor is supposed to extract a function declaration should contain exactly one such declaration. The general form for generating an entry for a constructor is therefore:

```
<empty line or end of a comment>
<code unit>
<zero or more empty lines>
/*{\Mcreate   body  }*/
```

The body of the Mcreate command contains the text that explains the constructor. Placeholder substitution and C++ to LaTeX conversion are applied to it. We give some more examples.

```
vector(double d, double e)
{ ... inline implementation of constructor ...}
/*{\Mcreate This is okay.}*/

vector(double d, double e, double f)
/*{\Mcreate This is also okay.}*/
{ ... inline implementation of constructor ...}

vector();
vector(int d);
/*{\Mcreate illegal, since code unit contains more
than one constructor.}*/

vector(double d)

{ ... inline implementation of constructor ...}
/*{\Mcreate illegal, since code unit preceding
the manual comment contains no constructor.}*/

vector(long d); /*{\Mcreate illegal, since manual comment
must start on a new line}*/
```

We still need to discuss the role of the optional argument w. The layout for the manual entry of a constructor follows either the two-column format shown in Figure 14.4 or the

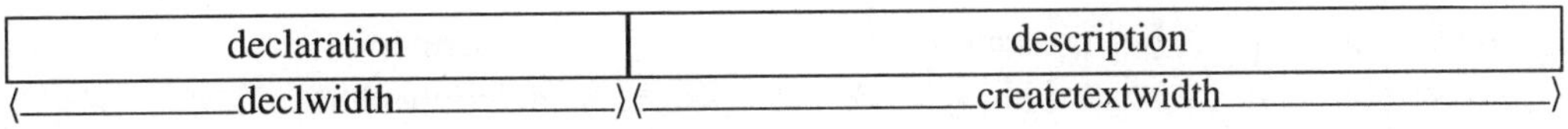

Figure 14.4 The two-column layout for constructors.

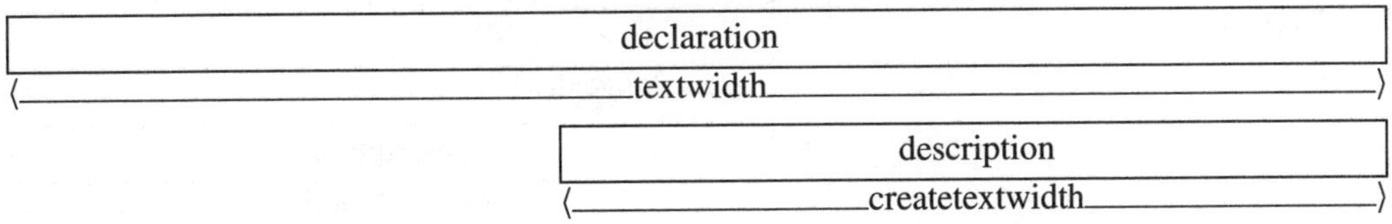

Figure 14.5 The two-row layout for constructors.

two-row format shown in Figure 14.5. The argument w defines the value of *declwidth*. The default value of declwidth is 40% of the textwidth. The value of *createtextwidth* is defined by

$$\text{createtextwidth} = \text{textwidth} - \text{declwidth}.$$

We use two-column layout if the declaration is short enough to fit into a box of width declwidth and use two-row layout otherwise. The argument w is either a pure number or a number followed by one of the TeX units of length (mm, cm, in, pt, or em). A missing unit is taken to be cm, i.e., 3.2 is equivalent to 3.2cm.

The Mdestruct Command: Mdestruct applies to the destructor of a class.

```
~vector();
/*{\Mdestruct The destructor ...}*/
```

produces

$\sim$*vector*()　　　　The destructor ...

It is customary in LEDA to produce *no* manual entry for the assignment operator, the copy constructor, and the destructor of a class because the semantics of these operations is defined in a uniform way for all LEDA types (see Section 2.3) and hence there is no need to define them again for each data type. In fact, it would be confusing. Think twice before you break this rule.

We now come to the section for the operations of a data type. It is started by a Moperations comment.

The Moperations Command: A comment of the form

```
/*{\Moperations a b }*/
```

generates the header line of the operations part. The length arguments a and b are optional. An entry in the operations part is displayed in either a three-column layout as shown in Figure 14.6 or a two-row layout as shown in Figure 14.7. The values of *typewidth* and

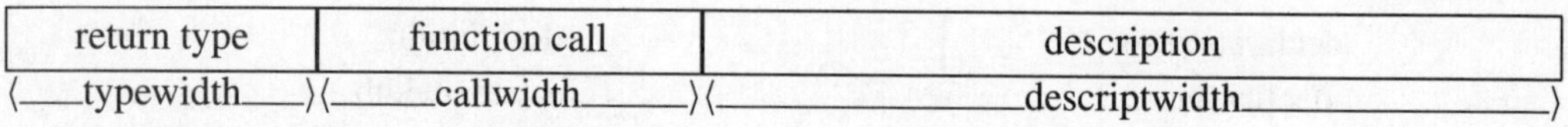

Figure 14.6 The three-column layout for the operations of a data type.

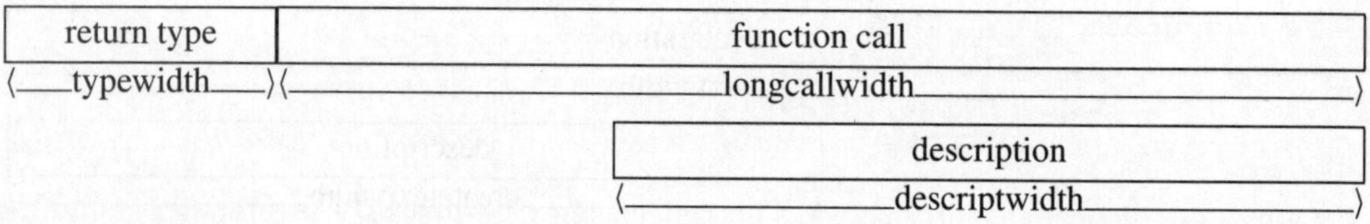

Figure 14.7 The two-row layout for the operations of a data type.

callwidth are set to a and b, respectively, and the value of *descriptwidth* is defined by the equation

$$\text{descriptwidth} = \text{textwidth} - \text{typewidth} - \text{callwidth}.$$

We choose the three-column layout if the function call fits into a box of width callwidth and the two-row layout otherwise[2]. If the return type does not fit into a box of width typewidth, we combine the return type and the function call into a single unit and attempt to put it into a box of width typewidth + callwidth. If the combined unit fits, we use a modified three-column layout, if it does not fit, we use a modified two-row layout.

An operation of a data type is either a member or a friend. In either case it can be a function or an operator. Operators may be binary or unary. We have a manual command for each case. The existence of distinct manual commands for the distinct cases is a historical relict. The current version of the extractor knows the syntax of C++ sufficiently well to be able to distinguish the cases without guidance by the manual command; this was not the case for an earlier version of the extractor. We find that the use of distinct manual commands increases readability.

The Mop Command: The Mop command applies to member functions of a data type. For example,

```
list_item  append(E x);
/*{\Mop appends a new item \Litem{x} to list |\Mvar| and
returns it
(equivalent to |\Mvar.insert(x,\Mvar.last(),after)|).}*/
```

generates (assuming that Mvar has value L)

list_item	L.append(E x)	appends a new item $\langle x\rangle$ to list L and returns it (equivalent to $L.insert(x,\ L.last(),\ after)$).

[2] In earlier versions of the preprocessor the choice between the two layout styles had to be done manually. We therefore had two versions of each manual command. The standard version selected three-column layout and the version with an appended character "l" selected two-row layout. You can still find manual commands Mopl and Mfuncl in many LEDA header files.

Note how the content of the first two columns is extracted from the code unit preceding the manual comment. Also note that we use member-function-call-syntax for the second column and that the function is applied to the canonical object of the type (which is the value of placeholder Mvar). We give some more examples.

```
list_item append(const E& x);
/*{\Mop        appends a new item \Litem{x} to
               list |\Mvar| and returns it\\
     (equivalent to |\Mvar.insert(x\,Mvar.last(),after)|).}*/
```

also produces the manual entry above. This reflects our view that a const-reference-parameter is equivalent to a value-parameter. The option `constref=yes` does not suppress const-ref pairs. The next function is long and hence is typeset in two-row layout.

```
list_item  insert(E x, list_item it, int direction = after);
/*{\Mop inserts a new item \Litem{x}  after or
        before item |it|.                   }*/
```

produces (assuming that Mvar has value L)

---

list_item L.insert(*E x*, *list_item it*, *int direction* $=$ *after*)

inserts a new item $\langle x \rangle$ after or before item *it*.

In either layout style it may happen that the return type does not fit into a box of width typewidth. In this case we combine return type and function call into a single unit for which we allot a box of width typewidth + callwidth. For example,

```
two_tuple<int,int> strange();
/*{\Mop a strange function. }*/
```

produces (assuming that Mvar has value L)

---

two_tuple<int, *int>* L.strange() a strange function.

The Mbinop Command: Mbinop applies to *binary operators* defined as member functions.

```
integer  operator+(const integer& y);
/*{\Mbinop   returns |\Mvar + y|. }*/
```

produces (assuming that Mvar has value x)

---

integer $x + y$ returns $x + y$.

There are two facts worth noting about this output. First, we use operator-call-syntax for the second column. Second, we suppress the type of the argument y. The rule is as follows. For an operator of class T the type of any value argument of type T is not shown. The option `partypes=yes` turns off this behavior.

The Munop Command: Munop applies to *unary operators* defined as member functions.

```
integer  operator++(){....}
 /*{\Munop returns the value of |\Mvar| and increments it.}*/
```

produces (assuming that Mvar has value x)

integer	$++x$	returns the value of x and increments it.

We put the operator applied to the canonical variable into the second column. Of course, unary operators are typeset as either prefix or postfix operators as prescribed by the syntax of C++.

The Marrop Command: Marrop applies to the *array access operator*.

```
E&  operator[](list_item it)  { ... return ... }
/*{\Marrop returns a reference to the
          entry |it| of |\Mvar|.}*/
```

produces (assuming that Mvar has value L)

E&	*L*[*list_item it*]	returns a reference to the entry *it* of *L*.

The Mfunop Command: Mfunop applies to the *function call operator*.

```
string operator()(int i, int j)  const { return sub(i,j); }
/*{\Mfunop    returns the substring of |\Mvar| ... }*/
```

produces (assuming that Mvar has value s)

string	s(*int i*, *int j*)	returns the substring of s ...

The Mstatic Command: Mstatic applies to *static member functions*. For example, the type bigfloat has a static member *round_mode* that determines the current rounding mode. A static member function *set_round_mode* is used to set the rounding mode.

```
static void set_round_mode(rounding_modes m =TO_NEAREST);
{round_mode = m;}
/*{\Mstatic sets |round_mode| to |m|.}*/
```

produces (assuming that Mname has value bigfloat)

void	*bigfloat*::set_round_mode(*rounding_modes m* = *TO_NEAREST*)	
		sets *round_mode* to *m*.

The Mfunc Command: Mfunc applies to *non-member functions* of a data type.

```
friend integer abs(const integer& x);
/*{\Mfunc returns the absolute value of |x|.}*/
```

produces

integer	abs(*integer x*)	returns the absolute value of *x*.

Note that the `friend` qualifier does not appear in the manual. After all, it has nothing to do with the semantics of the operation but is only an information for the compiler.

The Mbinopfunc Command: Mbinopfunc applies to *binary operators* that are non-member functions. You have probably got the rule by now. Commands ending with op apply to members and commands ending with func apply to non-members.

```
friend string  operator+(const string& x, const string& y);
/*{\Mbinopfunc  returns the concatenation of |x| and |y|.}*/

friend ostream& operator<<(ostream& O, const string& s);
/*{\Mbinopfunc  writes string |s| to output stream |O|. }*/
```

produces

string	$x + y$	returns the concatenation of *x* and *y*.
ostream&	*ostream&* $O \ll s$	writes string *s* to the output stream *O*.

The Munopfunc Command: Munopfunc applies to *unary operators* that are nonmember functions.

```
friend integer  operator-(const integer& x)
/*{\Munopfunc     unary minus ... }*/
```

produces

integer	$-x$	unary minus ...

The Mconversion Command: Mconversion applies to *user-defined conversion operators*. The following definition within class integer

```
operator rational()
/*{\Mconversion converts an |\Mtype| to a rational.}*/
```

produces (assuming that Mvar has value x)

rational	x	converts an *integer* to a rational.

Invisible Functions: Sometimes there is the need to generate a manual entry for a function or operator that does not exist. A typical situation is as follows. A type A is derived from a type B and inherits a function from B. We want the function to appear in the manual page of type A but we do not want the function to appear in the header file (because type A inherits it and including it in the header file would obscure the situation). The solution is to put the function inside a comment, e.g.,

```
/* inherited
void sort_edges() { graph::sort_edges(); }
*/
/*{\Mop the edges of $G$ are sorted increasingly according
        to their contents. }*/
```

The begin and the end of the comment must be on separate lines. The starting line may contain a text that explains the situation.

Code Units with More than One Function Definition: The restriction that a code unit contains only one function definition is sometimes unnatural. An example is two closely related functions for which one wants to produce only one manual entry.

```
friend bool operator==(const string& x, const char* y);
friend bool operator==(const string& x, const string& y);
/*{\Mbinopfunc     true iff $x$ and $y$ are equal.}*/
```

produces

bool	*string* x $==$ *string* y	true iff x and y are equal.

Another example is conditional definitions, e.g., the access function in the array data type which depends on the compiler flag LEDA_CHECKING_OFF.

```
#if defined(LEDA_CHECKING_OFF)
E&  operator[](int x) { return LEDA_ACCESS(E,v[x-Low]); }
#else
E&  operator[](int x) { return LEDA_ACCESS(E,entry(x)); }
#endif
/*{\Marrop       returns $A(x)$.\\
                 \precond $a\le x\le b$.  }*/
```

produces (assuming that Mvar has value A)

E&	A[*int* x]	returns $A(x)$. *Precondition*: $a \le x \le b$.

If a code unit contains more than one function definition our preprocessor attempts to extract the *last* definition. It outputs the extracted definition on standard output (except with option `warnings=no`) and asks for an acknowledgment (except with option `ack=no`).

The Mimplementation Command: A command of the form

```
/*{\Mimplementation body}*/
```

produces the header line of the implementation part and typesets body. For example,

```
/*{\Mimplementation The data type |\Mtype| is realized
by doubly linked linear lists. All operations take
constant time except
for the following operations: |search| and |rank|
take linear time $O(n)$, ...
}*/
```

produces (we assume here that the implementation part is preceded by four other manual parts, usually, definition, types, creation, and operations)

---

5. Implementation

The data type *list* is realized by doubly linked linear lists. All operations take constant time except for the following operations: *search* and *rank* take linear time $O(n)$, ...

The Mexample Command: The Mexample command is used to produce the header line of the example part and to include program code into the manual. The simplest way to include program code is to use the verbatim environment of LaTeX.

```
/*{\Mexample The following little example illustrates
the list data type.
\begin{verbatim}
#include <LEDA/list.h>
main()
{
  list<string> L;
  L.append("hello world");
}

\end{verbatim} }*/
```

produces

---

6. Example

The following little example illustrates the list data type.

```
#include <LEDA/list.h>
main()
{
  list<string> L;
  L.append("hello world");
}
```

The Mtext Command: The Mtext command can be used to add arbitrary text to the manual. For example,

```
/*{\Mtext
\headerline{Additional Operations for two-dimensional Points}
The following operations are only available for points
in two-dimensional space.
We will not mention this precondition in the sequel.
}*/
```

produces

Additional Operations for two-dimensional Points

The following operations are only available for points in two-dimensional space. We will not mention this precondition in the sequel.

Generally,

```
/*{\Mtext  body  }*/
```

adds body to the document. The body is subject to placeholder substitution and C++ to LaTeX conversion. The Mtext command can be used to include arbitrary LaTeX commands into the output of the preprocessor. We did this already for the header line command in the example above. Another frequent use of the Mtext command is to change the values of the parameters governing the layout. For example

```
/*{\Mtext
\settowidth{\typewidth}{|void|}
\addtolength{\typewidth}{\colsep}
\computewidths
}*/
```

sets the width of the first column to the width of *void* plus the value of colsep, where colsep is predefined as 1.5em. The command `\computewidths` causes the recomputation of the dependent variable descriptwidth.

The Moptions Command: The Moptions command allows us to include preprocessor options directly into the header file. For example, the header file for LEDA's window type contains

```
/*{\Moptions
usesubscripts=yes
}*/
```

and hence this section of the LEDA-manual is typeset with subscripts, see also Section 14.2.4.

The Msubst Command: The Msubst command allows us to define additional placeholders. For example,

```
/*{\Msubst
int_type integer
quot_type rational
}*/
```

introduces the placeholders `int_type` and `quot_type` with values `integer` and `rational`, respectively.

14.2.3 *Warnings and Acknowledgments*

The preprocessor issues warnings and error messages and asks the user to acknowledge them. With the option `ack=no` no acknowledgments are necessary and the option `warnings=no` suppresses the warnings. One can also suppress warnings for a single manual comment, e.g.,

```
/*{\Moptions nextwarning=no }*/
point head();
point start();
/*{\Mop returns the start point of |\Mvar|}*/
```

suppresses the warning that there is more than one function definition in the current code section. We recommend running Lman with `warnings=yes` and `ack=yes` and using the mechanism above to turn off warnings individually.

14.2.4 *Subscripts*

Sometimes program variables are numbered and it would be nice to typeset the numbers as subscripts. The option `usesubscripts=yes` does exactly this. Within the context of this option `|x0|` is typeset as x_0 and `|x11|` is typeset as x_{11}. Note that the subscript rule is applied only to identifiers consisting of a single character. Thus `|diff1|` is still typeset as *diff1*.

14.2.5 *TEX macros*

We defined a collection of TEX-commands that facilitate the production of manual pages; they are contained in `MANUAL.mac` in LEDAROOT/Manual/tex.

Many data types in LEDA are defined in terms of items. We have adopted the convention that items are enclosed in angular braces. The command `\Litem` produces items. It takes a single argument and encloses it in angular braces. The argument is typeset in math-mode, i.e., `\Litem{x}` produces $\langle x \rangle$, `\Litem{x,y}` produces $\langle x, y \rangle$, and `\Litem{diff}` produces $\langle diff \rangle$. The last example shows that identifiers of length more than one should be enclosed in vertical bars, e.g., `\Litem{|diff|}` produces ⟨*diff*⟩.

The word *Precondition* appears frequently in manual pages; `\precond` produces it. The macro `\CC` produces C++. The command `\headerline{arg}` produces a header line, i.e., it prints its argument in boldface and disallows pagebreaks after the header line. The commands `\DLK` and `\DRK` produce `[[` and `]]`, respectively.

Vertical bars require some care. Recall that vertical bars have a special meaning (they bracket C++ text) and therefore we need to make special provisions to produce vertical bars in LaTeX-text produced by our preprocessor. The command `\Lvert` expands to a vertical bar, i.e., the preprocessor leaves it alone and its TeX-definition is `\def\Lvert{|}`. A frequent use of vertical bars in mathematical text is to denote absolute values. The command `\Labs` produces absolute values, e.g., `$x + \Labs{|diff|} + z$` produces $x+|\mathit{diff}|+z$. The commands `\Lvert` and `\Labs` can only be used in math-mode, i.e., in order to produce a | within text you need to write `$\Lvert$`.

`MANUAL.mac` also defines the LaTeX environment *manual*. This environment sets parindent to zero, parskip to 14pt and increases baselineskip slightly above its standard value. The manual is typeset in this environment.

The file `MANUAL.pagesize` in LEDAROOT/Manual/tex defines textwidth, textheight, topmargin, evensidemargin, and oddsidemargin. Values which work well with European a4-size paper and US legal-size paper are predefined in this file.

14.2.6 *Applying Lman to Web-Files*

Followers of literate programming do not split their implementations into h-files and c-files but combine them into a single file. This causes no problem for Lman as it ignores all but the manual commands and the code units preceding them.

A problem may arise if the web-system in use allows the user to put formatting instructions into the code chunks as, for example, CWEB does. In this case the manual extractor must purge the code of formatting instructions. The standard version of ext knows how to remove CWEB's formatting instructions. In order to adapt the manual extractor to another web-system which allows formatting instructions in code chunks you need to edit the code chunk <*purge code unit . . .*> in ext.nw. We have used Lman successfully on CWEB, noweb, and Lweb-files.

14.2.7 *Redirecting Output*

Lman and Ldoc write the extracted manual page to the file `outfile`. In the case of Ldoc the default value of `outfile` is equal to `basename.man` where `basename.lw` is the input file to Ldoc. In the case of Lman the outfile is some internal file. You may redirect the output to a different file by assigning to `outfile` in an Moptions command, e.g., after

```
/*{\Moptions outfile=type.man }*/
```

the output will be written to file `type.man`. This feature is useful for at least two purposes.

The first use is to generate several manual pages from the same source. This can be achieved by always directing the output to the appropriate man-file. There is a small inconvenience: LaTeX expects manual pages to be enclosed in the manual environment. However, the required `\begin{manual}` and `\end{manual}` commands are generated automatically only for the default outfile. So one needs to write:

```
/*{\Moptions outfile=type.man }*/
/*{\Mtext \begin{manual} }*/
   now come the commands than generate the manual
/*{\Mtext \end{manual} }*/
```

The second use of redirecting output is to rearrange the material within a single manual page. It is conceivable that one wants to use a different order of presentation in the manual page and in the implementation. Assume that the manual consists of two parts and that we want to arrange the two parts in reverse order in the manual and in the documentation. Write:

```
\section{The Manual Page}

\begin{manual}
\input{part1.man}
\input{part2.man}
\end{manual}

\section{Code}

/*{Moptions outfile=part2.man }*/
        the stuff that goes into part 2

/*{Moptions outfile=part1.man }*/
        the stuff that goes into part 1
```

14.2.8 *The Lman Options*

The behavior of Lman can be fine-tuned by options. A call `Lman` without arguments gives a short survey of all available options. Options are specified in assignment syntax `variable=value`. There must be no blank on either side of the equality sign. In the list of options to follow we list the default value of each option first.

size={12, 11, 10}: Determines the font size.

constref={no, yes}: Determines how const-ref parameters are displayed. With the no-option a const-ref parameter `const T& x` is displayed as a value parameter `T x` and with the yes-option it is displayed in full.

partypes={no, yes}: Determines how parameters of unary and binary operators are displayed. Consider, for example, an operator + of a class number. With the no-option the operator `operator+(number x, number y)` is displayed as `x + y` and with the yes-option it is displayed as `number x + number y`.

numbered={no, yes}: Determines whether the header line of the manual page is numbered. You probably want it numbered when the manual page becomes part of a larger document.

title={yes, no}: If title is set to no, the manpage comment produces no output.

warnings={no, yes}: Determines whether Lman gives warnings. You probably want to use the no-option when you inspect LEDA manual pages and the yes-option when you design manual pages yourself.

ack={no, yes}: Determines whether Lman asks for acknowledgments of warnings.

usesubscripts={no, yes}: Determines whether variables consisting of a single character followed by a number are displayed as subscripted variables.

filter={all, signatures, definition, creation, operations, implementation, example, opname}: Determines which part of the manual page is shown. The all-option shows the complete manual page, the signature-option shows the signatures of all operations of the data type, the next five options show only the corresponding section of the manual page, and the opname-option shows only the operation with the same name.
outfile={string}: Determines whether the TEX-file generated is only written on a temporary file (the default option) or on the file with name string.

latexruns={1, 0, 2}: Determines the number of LATEX runs used to produce the manual page. LATEX needs to be run twice if the manual page contains cross references.

xdvi={yes, no}: Determines whether the manual page is displayed by xdvi. If latexruns is at least one and xdvi is no then the resulting dvi-file is copied into file T.dvi in the working directory.

Lman can be customized by putting options in a file Lman.cfg in either the home directory or the working directory. Command line options take precedence over options in the working directory which in turn take precedence over options in the home directory.

14.3 Making a Manual: The Mkman Command

Many manual pages combined into a single document make a manual. We explain a simple mechanism to produce LEDA-style manuals. Assume that we want to produce a document consisting of a title page, an introduction, and the manual pages of types A and B. Assume also that the manual information about types A and B is contained in files with extension `ext`[3] in a common directory `dir` and that the working directory contains a master TeX-file as shown in Figure 14.8 and also a file `Introduction.tex`. The command

```
Mkman dir ext
```

cycles through all files `f.ext` in `dir` and calls

```
lextract f.ext /extract/f.tex
```

for each one of them. This creates files `extract/A.tex` and `extract/B.tex` after which the master file may be processed with LATEX.

All header files of LEDA are contained in the directory LEDAROOT/incl/LEDA and the master file for manual production is called MANUAL.tex and is contained in the directory LEDAROOT/Manual/MANUAL. Thus an execution of

```
Mkman $LEDAROOT/incl/LEDA h
latex MANUAL.tex
```

in the latter directory produces the dvi-file of the LEDA manual. As LEDAROOT/incl/LEDA and h are the default values of the first and second argument of Mkman, respectively, the

[3] Typical extensions are h, nw, and lw.

```
\documentclass[12pt,a4paper]{book}
\usepackage{Lweb}
\begin{document}

\title {A Simple Manual}
\maketitle
\input{Introduction.tex}
\input{extract/A.tex}
\input{extract/B.tex}
\end{document}
```

Figure 14.8 A master tex-file for a simple manual.

```
#!/bin/csh -f
if ($1 == "") then
  set source = $LEDAROOT/incl/LEDA
  set ext = h
else
  set source = $1
  if ($2 == "") then
    set ext = h
  else
    set ext = $2
  endif
endif

\rm -r -f extract

mkdir extract

echo Extracting manual pages ...
echo " "

foreach f ($source/*.$ext)
   echo "extracting manual from $f"
   lextract $f  extract/`basename $f .$ext`.tex
end
```

Figure 14.9 The shell script Mkman for manual production.

first line may actually be abbreviated to Mkman. Figure 14.9 shows the shell script that realizes Mkman.

14.4 The Manual Directory in the LEDA System

The subdirectory Manual of the LEDA directory contains all files that are relevant for manual production, see Figure 14.10.

```
                LEDAROOT
                   |
                 Manual
   ----------------|-----------------
   |       |      |      |      |      |
  DVI  MANUAL  cmd  contrib  noweb  tex
```

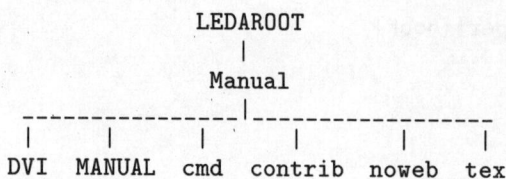

Figure 14.10 The subdirectory Manual of the LEDA directory.

- MANUAL contains the tex-sources for the LEDA-Manual.
- DVI contains the dvi-files obtained by applying Lman to all header files of the LEDA system. The dvi-files in DVI are accessed by the online manual viewer xlman.
- cmd contains the commands Lman, Mkman, … .
- contrib contains sources of contributions made by persons outside the LEDA group.
- noweb contains the noweb-sources for all programs used for manual production. In particular, the noweb-file `ext.nw` contains the Perl programs and shell scripts for lextract, Lman, Mkman, … .
- tex contains the TEX files required for manual production.

14.5 Literate Programming and Documentation

Many data types and algorithms of the LEDA system are documented in the literate programming system noweb [Ram94] and its LEDA-dialect Lweb. Ldoc and lweave are our tools to turn noweb- and Lweb-files into nice looking documents.

Literate programming advises to integrate specification, implementation, and documentation into a single file and to use tools (usually called *tangle* and *weave*) to extract program and to typeset documentation. Among the many literate programming systems we have used CWEB[KL93] and noweb[4]: our current favorite is noweb and its LEDA-dialect Lweb. We used Lweb to produce this book.

14.5.1 *Noweb and Lweb*

We start with a brief review of noweb, see also Section 2.7. Noweb provides commands *notangle* and *noweave* that can be applied to so-called noweb-files. A noweb-file foo.nw contains program source code interleaved with documentation. When notangle is given a noweb-file, it extracts the program and writes it to standard output, and when noweave is given a noweb-file it produces a LATEX source on standard output.

[4] noweb can be obtained by anonymous ftp from CTAN, the Comprehensive TeX Archive Network, in directory web/noweb.

```
      ______ noweave________  LaTeX source
     |
     |
  foo.nw
     |
     |______ notangle_______  program
```

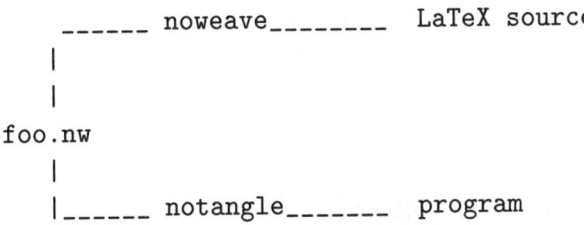

A noweb-file is a sequence of *chunks*. A chunk is either a *documentation chunk* or a *code chunk*. Documentation chunks begin with a line that starts with an at-sign (@) followed by a space or newline. Code chunks begin with

<<code chunk name >>=

on a line by itself. Chunks are terminated by the beginning of another chunk or by the end of the file. Several code chunks may have the same name. Notangle concatenates their definitions to produce a single chunk. Code chunks contain source code and references to other code chunks.

Notangle extracts code by expanding one code chunk. In the expansion process code chunk definitions behave like macro definitions, i.e., if the definition of chunk XXX contains references to other code chunks then these chunks are also expanded, and so on.

Noweave produces a LaTeX source from a noweb-file. To this end it copies the documentation chunks verbatim to standard output (except for quoted code, see below) and it typesets code chunks in typewriter font. Note that this implies that documentation chunks starting with an @-sign followed by a newline-character start a new paragraph in the sense of LaTeX and that documentation chunks containing non-white stuff on the same line as the @-sign do not. Code may be *quoted* within documentation chunks by placing double square brackets (`[[...]]`) around it. Noweave typesets quoted code in typewriter font.

This completes our review of noweb. *Lweb* is our local dialect of noweb which we developed for the production of this book and for the documentation of the LEDA system. Lweb-files have extension `.lw`. Figure 14.11 shows an Lweb-file and Figure 14.12 shows the result of applying lweave to it. The differences between Lweb and noweb are the following:

- Code can be quoted by either double square brackets (`[[...]]`) or vertical bars (| . . . |). Code quoted in double square brackets is set in typewriter font and code quoted in vertical bars is typeset in mathitalics font. This was already discussed in Section 14.2.2.
- Program examples can be included in documentation chunks by lines that start with `@c`. The text after the program example must start with an @-sign followed by a space-character or a newline-character.
- Empty lines in program chunks generate somewhat less vertical space than an empty line in a verbatim-like environment. This makes code chunks look better.
- Page breaks are forbidden between the first few and the last few lines of a code chunk.

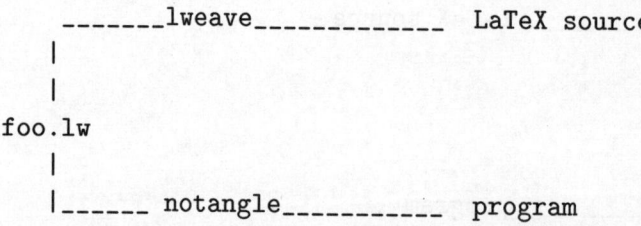

```
       _______lweave_____________  LaTeX source
      |
      |
   foo.lw
      |
      |______ notangle___________  program
```

Lweb-files have extension .lw. Notangle applies also to Lweb-files and noweave is replaced by lweave; lweave is realized as a pair of pre- and postprocessor to noweave. The preprocessor handles the code quoted by vertical bars and the program examples and the postprocessor takes care of empty lines in code chunks. The implementation of lweave is part of ext.nw in LEDAROOT/Manual/noweb.

14.5.2 *Documentation*

Many classes and programs of the LEDA-system are documented using Lweb and this book is also an Lweb document. We recommend having at least the following major sections in a documentation:

- A preamble consisting of the title page, the table of contents, and maybe an abstract and an introduction.
- A manual page as discussed in the previous section.
- A section containing the header file augmented by manual comments so as to allow manual extraction.
- A section containing the c-file.
- A section containing test, example, or demo programs.

Figure 14.13 shows a simple Lweb-file stack.lw having the recommended structure. More substantial examples can be found in the subdirectory Lweb of the LEDAROOT directory.

14.5.3 *Ldoc*

Ldoc combines the functionality of Lman and lweave. A call

```
Ldoc XXX[.lw] options
```

produces a file XXX.man in the working directory and a temporary file temp.lw. The former file contains the manual and is essentially the file produced by Lman (except for the preamble and postamble required by LaTeX). The Lman-options constref, partypes, warnings, ack, and usesubscripts apply. The file temp.lw is obtained by the deletion of all manual comments (except for Mpreamble comments) from the input file. The option `delman=no` suppresses the deletion. The temporary file temp.lw is then sent through lweave and the result is moved to XXX.tex in the working directory.

We introduced an additional manual comment for the use with Ldoc, the Mpreamble comment. As far as Lman is concerned it is equivalent to the Mtext command, i.e., its

```
\documentclass[a4paper]{article}
\usepackage{Lweb}
\begin{document}
\subsubsection{Jordan Sorting}

We proceed to describe an implementation. Its global
structure is given by:

@c
<<include statements>>;
<<typedefs and global variables>>;
<<class point>>;
<<class bracket>>;
<<procedure Jordan sort>>;

@ As outlined above, we construct three data structures
simultaneously: the sorted list of the numbers processed so
far, call it |L|, and the
upper and lower tree of brackets. Each item of the
list |L| contains its abscissa (a |float|) and pointers
to the brackets in the two trees containing it.

<<class point>>=

class point{
private:
float abscissa;
bracket* bracket_in_upper_tree;
bracket* bracket_in_lower_tree;

public:
<<member functions of class point>>
}

@ A node of either tree corresponds to a bracket.
A bracket needs to know its two endpoints
(as items in the list |L|), its sorted sequence
of sub-brackets (a |sortseq<bracket*,>| which we
abbreviate as |children_list|),
and its position among its siblings (a |seq_item|).
\end{document}
```

Figure 14.11 An Lweb file: It is part of the section on sorted sequences of this book.

body is included into the produced tex-file after placeholder substitution and C++ to LaTeX conversion. Ldoc produces two output files, namely XXX.man and temp.lw. The output of Mpreamble commands is put into the latter file instead of the former. A typical use of the Mpreamble command is the definition of a LaTeX-command whose body should be subjected to C++ to LaTeX-conversion. The following example is taken from LEDA's geo_rep class.

We proceed to describe an implementation. Its global structure is given by:

⟨*include statements*⟩ ;
⟨*typedefs and global variables*⟩ ;
⟨*class point*⟩ ;
⟨*class bracket*⟩ ;
⟨*procedure Jordan sort*⟩ ;

As outlined above, we construct three data structures simultaneously: the sorted list of the numbers processed so far, call it L, and the upper and lower tree of brackets. Each item of the list L contains its abscissa (a *float*) and pointers to the brackets in the two trees containing it.

⟨*class point*⟩≡

```
class point{
private:
float abscissa;
bracket* bracket_in_upper_tree;
bracket* bracket_in_lower_tree;
public:
⟨member functions of class point⟩
}
```

A node of either tree corresponds to a bracket. A bracket needs to know its two endpoints (as items in the list L), its sorted sequence of sub-brackets (a *sortseq<bracket*∗, > which we abbreviate as *children_list*), and its position among its siblings (a *seq_item*).

Figure 14.12 The result of applying lweave + latex to the file of Figure 14.11.

```
/*{\Mpreamble
\newcommand{\grsummary}
{The class |geo_rep| is used to represent points, hyperplanes,
directions, and vectors. The latter ...}
}*/
```

14.5.4 *The Implementation of Ldoc*

Ldoc is based on the commands *lextract*, *ldel*, and *weave*, where weave is noweave for noweb and lweave for Lweb.

```
foo.[lw|nw|w] - ldel - foo-del.[lw|nw|w] - weave - foo.tex
  |                                                        |
lextract                                        \input{foo.man}
  |                                                        |
foo.man ------------------------------------------------------
```

Ldoc first uses lextract to extract the manual and ldel to remove the manual comments, it then applies the appropriate weave command to the output of ldel, and it finally applies LaTeXand xdvi to the resulting file. All Lman options apply. In order to try out Ldoc copy

```
\documentclass[a4paper]{article}
\usepackage{Lweb}
\begin{document}
\title{Stack\\   |stack| }
\author{Kurt Mehlhorn}
\maketitle
\tableofcontents
\section{The Manual Page of Type Stack}
\input{stack.man}
@ \section{The Header File}
<<stack.h>>=  the file of Figure 1.2
@ \section{The Implementation}
<<stack.c>>= ...
@ \section{A Test Program}
<<stack-test.c>>= ...
@
\end{document}
```

Figure 14.13 The generic structure of a documentation.

sortseq.lw from LEDAROOT/Lweb to a directory where you have write-permission and then call Ldoc sortseq.

Bibliography

[ABDP97] F. Avnaim, J.-D. Boissonnat, O. Devillers, and F.P. Preparata. Evaluating signs of determinants with floating point arithmetic. *Algorithmica*, 17(2):111–132, 1997.

[ABE98] N. Amenta, M. Bern, and D. Eppstein. The crust and the β-skeleton: Combinatorial curve reconstruction. *Graphical Models and Image Processing*, pages 125–135, 1998.

[ABMP91] H. Alt, N. Blum, K. Mehlhorn, and M. Paul. Computing a maximum cardinality matching in a bipartite graph in time $O(n^{1.5}\sqrt{m/\log n})$. *Information Processing Letters*, 37(4):237–240, 1991.

[AC93] D. Applegate and W. Cook. Solving large-scale matching problems. In D. Johnson and C.C. McGeoch, editors, *Network Flows and Matchings*, volume 12 of *DIMACS Series in Discrete Mathematics and Theoretical Computer Science*, pages 557–576. American Mathematical Society, 1993.

[AHU74] A.V. Aho, J.E. Hopcroft, and J.D. Ullman. *The Design and Analysis of Computer Algorithms.* Addison-Wesley, 1974.

[AHU83] A.V. Aho, J.E. Hopcroft, and J.D. Ullman. *Data Structures and Algorithms*. Addison-Wesley, 1983.

[AKMO97] R.K. Ahuja, M. Kodialam, A.K. Mishra, and J.B. Orlin. Computational investigation of maximum flow algorithms. *European Journal on Operational Research*, 97:509–542, 1997.

[AL94] N.M. Amato and M.C. Loui. Checking linked data structures. In *Proceedings of the 24th Annual International Symposium on Fault-Tolerant Computing (FTCS'94)*, pages 164–173, 1994.

[AM98] E. Althaus and K. Mehlhorn. Maximum network flow with floating point arithmetic. *Information Processing Letters*, 66:109–113, 1998.

[AMO93] R.K. Ahuja, T.L. Magnanti, and J.B. Orlin. *Network Flows*. Prentice Hall, 1993.

[AMOT90] R.K. Ahuja, K. Mehlhorn, J.B. Orlin, and R.E. Tarjan. Faster algorithms for the shortest path problem. *Journal of the ACM*, 3(2):213–223, 1990.

[And79] A.M. Andrew. Another efficient algorithm for convex hulls in two dimensions. *Information Processing Letters*, 9:216–219, 1979.

[AO89] R.K. Ahuja and J.B. Orlin. A fast and simple algorithm for the maximum flow problem. *Operation Research*, 37:748–759, 1989.

[AS89] C. Aragon and R. Seidel. Randomized search trees. In *Proceedings of the 30th Annual Symposium on Foundations of Computer Science (FOCS'89)*, pages 540–545, 1989.

[ASE92] N. Alon, J.H. Spencer, and P. Erdös. *The Probabilistic Method.* John Wiley & Sons, 1992.

[AST94] P.K. Agarwal, M. Sharir, and S. Toledo. Applications of parametric searching in geometric optimization. *Journal of Algorithms*, 17:292–318, 1994.

[AVL62] G.M. Adel'son-Velskii and Y.M. Landis. An algorithm for the organization of information. *Soviet Math. Doklady*, 3:1259–1262, 1962.

[Bal95] I. Balaban. An optimal algorithm for finding segment intersections. In *Proceedings of the 11th Annual ACM Symposium on Computational Geometry (SCG'95)*, pages 211–219, 1995.

[Bat] G. Di Battista. GD-Toolkit. Check the item "friends" of the LEDA-web page for a pointer.

[BD95] P. Bose and L. Devroye. Intersections with random geometric objects. manuscript, 1995.

[Bel58] R.E. Bellman. On a routing problem. *Quart. Appl. Math.*, 16:87–90, 1958.

[BEPP97] H. Brönnimann, I. Emiris, V. Pan, and S. Pion. Computing exact geometric predicates using modular arithmetic with single precision. In *Proceedings of 13th Annual ACM Symposium on Computational Geometry (SCG'97)*, pages 174–182, 1997.

[BETT94] G. Di Battista, P. Eades, R. Tamassia, and I. Tollis. Algorithms for drawing graphs: An annotated bibliography. *Computational Geometry: Theory and Applications*, 4(5):235–282, 1994.

[BFMS97] C. Burnikel, R. Fleischer, K. Mehlhorn, and S. Schirra. A strong and easily computable separation bound for arithmetic expressions involving square roots. In *Proceedings of the 8th Annual ACM-SIAM Symposium on Discrete Algorithms (SODA'97)*, pages 702–709, 1997. `www.mpi-sb.mpg.de/~mehlhorn/ftp/sep-bound.ps`.

[BFMS99] C. Burnikel, R. Fleischer, K. Mehlhorn, and S. Schirra. Exact efficient computational geometry made easy. In *Proceedings of the 15th Annual Symposium on Computational Geometry (SCG'99)*, 1999. `www.mpi-sb.mpg.de/~mehlhorn/ftp/egcme.ps`.

[BFS98] C. Burnikel, S. Funke, and M. Seel. Exact arithmetic using cascaded computation. In *Proceedings of the 14th Annual Symposium on Computational Geometry (SCG'98)*, pages 175–183, 1998.

[BG61] R.G. Busacker and P.J. Gowen. A procedure for determining minimal-cost network flow patterns. Technical report, Operations Research Office, John Hopkins University, 1961.

[BK89] M. Blum and S. Kannan. Designing programs that check their work. In *Proceedings of the 21th Annual ACM Symposium on Theory of Computing (STOC'89)*, pages 86–97, 1989.

[BKM+95] C. Burnikel, J. Könemann, K. Mehlhorn, S. Näher, S. Schirra, and C. Uhrig. Exact geometric computation in LEDA. In *Proceedings of the 11th Annual Symposium on Computational Geometry (SCG'95)*, pages C18–C19, New York, NY, USA, June 1995. ACM Press.

[BL76] K.S. Booth and G.S. Lueker. Testing for the consecutive ones property, interval graphs, and graph planarity using *PQ*-tree algorithms. *Journal of Computer and System Sciences*, 13:335–379, 1976.

[BLR90] M. Blum, M. Luby, and R. Rubinfeld. Self-testing/correcting with applications to numerical problems. In *Proceedings of the 22nd Annual ACM Symposium on Theory of Computing (STOC'90)*, pages 73–83, 1990.

[BM80] N. Blum and K. Mehlhorn. On the average number of rebalancing operations in weight-balanced trees. *Theoretical Computer Science*, 11:303–320, 1980.

[BMS94a] C. Burnikel, K. Mehlhorn, and S. Schirra. How to compute the Voronoi diagram of line segments: Theoretical and experimenta l results. In *Proceedings of the 2nd Annual European Symposium on Algorithms - ESA'94*, volume 855 of *Lecture Notes in Computer Science*, pages 227–239. Springer, 1994.

[BMS94b] C. Burnikel, K. Mehlhorn, and S. Schirra. On degeneracy in geometric computations. In *Proceedings of the 5th Annual ACM-SIAM Symposium on Discrete Algorithms (SODA'94)*, pages 16–23, 1994.

[BMS96] C. Burnikel, K. Mehlhorn, and S. Schirra. The LEDA class real number. Technical Report MPI-I-96-1-001, Max-Planck-Institut für Informatik, January 1996.

[BO79] J.L. Bentley and T. Ottmann. Algorithms for reporting and counting geometric intersections. *IEEE Transaction on Computers C 28*, pages 643–647, 1979.

[BR96] R.P.K. Banerjee and J.R. Rossignac. Topologically exact evaluation of polyhedra defined in CSG with loose primitives. *Computer Graphics Forum*, 15(4):205–217, 1996. ISSN 0167-7055.

[BS94] J.D. Bright and G.F. Sullivan. Checking mergeable priority queues. In *Proceedings of the 24th Annual International Symposium on Fault-Tolerant Computing (FTCS'94)*, pages 144–153, Los Alamitos, CA, USA, June 1994. IEEE Computer Society Press.

[BS95] J.D. Bright and G.F. Sullivan. On-line error

monitoring for several data structures. In *Proceedings of the 25th Annual International Symposium on Fault-Tolerant Computing (FTCS'95)*, pages 392–401, Pasadena, California, 1995.

[BSM95] J.D. Bright, G.F. Sullivan, and G.M. Masson. Checking the integrity of trees. In *Proceedings of the 25th Annual International Symposium on Fault-Tolerant Computing (FTCS'95)*, pages 402–413, Pasadena, California, 1995.

[BSM97] J.D. Bright, G.F. Sullivan, and G.M. Masson. A formally verified sorting certifier. *IEEE Transactions on Computers*, 46(12):1304–1312, 1997.

[BT93] J.-D. Boissonnat and M. Teillaud. On the randomized construction of the Delaunay tree. *Theoretical Computer Science*, 112(2):339–354, 1993.

[Bur96] C. Burnikel. *Exact Computation of Voronoi Diagrams and Line Segment Intersections*. PhD thesis, Max-Planck-Institut für Informatik, 1996.

[BW96] M. Blum and H. Wasserman. Reflections on the pentium division bug. *IEEE Transaction on Computing*, 45(4):385–393, 1996.

[BY98] J.-D. Boissonnat and M. Yvinec. *Algorithmic Geometry*. Cambridge University Press, Cambridge, 1998.

[CDR92] J. Canny, B. Donald, and G. Ressler. A rational rotation method for robust geometric algorithms. In ACM-SIGACT ACM-SIGGRAPH, editor, *Proceedings of the 8th Annual ACM Symposium on Computational Geometry (SCG '92)*, pages 251–260, 1992.

[CFMP97] C. Cooper, A. Frieze, K. Mehlhorn, and V. Priebe. Average-case complexity of shortest-paths problems in the vertex-potential model. In José Rolim, editor, *Proceedings of the International Workshop on Randomization and Approximation Techniques in Computer Science (RANDOM'97)*, volume 1269 of *Lecture Notes in Computer Science*, pages 15–26. Springer, 1997.

[CG] B. Cherkassky and A. Goldberg. PRF, a Maxflow Code. `www.inter-trust.com/star/goldberg/index.html`.

[CG96] B.V. Cherkassky and A.V. Goldberg. Negative cycle detection algorithms. In *Proceedings of the 4th Annual European Symposium on Algorithms - ESA'96*, volume 1136 of *Lecture Notes in Computer Science*, pages 349–363, 1996.

[CG97] B.V. Cherkassky and A.V. Goldberg. On implementing the push-relabel method for the maximum flow problem. *Algorithmica*, 19(4):390–410, 1997.

[CGA] CGAL (Computational Geometry Algorithms Library). `www.cs.ruu.nl/CGAL`.

[CGK+97] C.S. Chekuri, A.V. Goldberg, D.R. Karger, M.S. Levine, and C. Stein. Experimental study of minimum cut algorithms. In *Proceedings of the 8th Annual ACM-SIAM Symposium on Discrete Algorithms (SODA'97)*, pages 324–333, New Orleans, Louisiana, 5–7 January 1997.

[CGM+97] B. Cherkassky, A. Goldberg, P. Martin, J. Setubal, and J. Stolfi. Augment or relabel? A computational study of bipartite matching and unit capacity maximum flow algorithms. Technical Report TR 97-127, NEC Research Institute, 1997.

[CGR94] B.V. Cherkassky, A.V. Goldberg, and T. Radzik. Shortest paths algorithms: Theory and experimental evaluation. In Daniel D. Sleator, editor, *Proceedings of the 5th Annual ACM-SIAM Symposium on Discrete Algorithms (SODA'94)*, pages 516–525. ACM Press, 1994.

[CGS97] B.V. Cherkassky, A.V. Goldberg, and C. Silverstein. Buckets, heaps, lists, and monotone priority queues. In *Proceedings of the 8th Annual ACM-SIAM Symposium on Discrete Algorithms (SODA'97)*, pages 83–92, 1997.

[CH95] J. Cheriyan and T. Hagerup. A randomized maximum-flow algorithm. *SIAM Journal of Computing*, 24(2):203–226, April 1995.

[CHM96] J. Cheriyan, T. Hagerup, and K. Mehlhorn. An $o(n^3)$-time maximum flow algorithm. *SIAM Journal of Computing*, 25(6):1144–1170, 1996.

[Cla92] K.L. Clarkson. Safe and effective determinant evaluation. In *Proceedings of the 31st Annual Symposium on Foundations of Computer Science (FOCS'92)*, pages 387–395, 1992.

[CLR90] T.H. Cormen, C.E. Leiserson, and R.L. Rivest. *Introduction to Algorithms*. MIT Press/McGraw-Hill Book Company, 1990.

[CM89] J. Cheriyan and A. Maheshwari. Analysis of preflow push algorithms for maximum network flow. *SIAM Journal of Computing*, 18:1057–1086, 1989.

[CM96] J. Cheriyan and K. Mehlhorn. Algorithms for dense graphs and networks on the random access computer. *Algorithmica*, 15(6):521–549, 1996.

[CM99] J. Cheriyan and K. Mehlhorn. An analysis of the highest-level selection rule in the preflow-push max-flow algorithm. *IPL*, 69:239–242, 1999. www.mpi-sb.mpg.de/~mehlhorn/ftp/maxflow.ps.

[CNAO85] N. Chiba, T. Nishizeki, S. Abe, and T. Ozawa. A linear algorithm for embedding planar graphs using PQ-trees. *Journal of Computer and System Sciences*, 30(1):54–76, 1985.

[CS89] K.L. Clarkson and P.W. Shor. Applications of random sampling in computational geometry, II. *Journal of Discrete and Computational Geometry*, 4:387–421, 1989.

[dBKOS97] M. de Berg, M. van Kreveld, M. Overmars, and O. Schwarzkopf. *Computational Geometry: Algorithms and Applications*. Springer, 1997.

[DETT98] G. Di Battista, P. Eades, R. Tamassia, and I.G. Tollis. *Graph Drawing: Algorithms for the Visualization of Graphs*. Prentice-Hall, 1998.

[dFPP88] H. de Fraysseix, J. Pach, and R. Pollack. Small sets supporting Fáry embeddings of planar graphs. In *Proceedings of the 20th Annual ACM Symposium on Theory of Computing (STOC'88)*, pages 426–433, 1988.

[DH91] P. Deuflhard and A. Hohmann. *Numerische Mathematik: Eine algorithmisch orientierte Einführung*. Walter de Gruyter, 1991.

[Dij59] E.W. Dijkstra. A note on two problems in connection with graphs. *Num. Math.*, 1:269–271, 1959.

[Din70] E.A. Dinic. Algorithm for solution of a problem of maximum flow in networks with power estimation. *Soviet Mathematics Doklady*, 11:1277–1280, 1970.

[DKM+94] M. Dietzfelbinger, A. Karlin, K. Mehlhorn, F. Meyer auf der Heide, H. Rohnert, and R. Tarjan. Dynamic perfect hashing: Upper and lower bounds. *SIAM Journal of Computing*, 23(4):738–761, 1994.

[DLPT97] O. Devillers, G. Liotta, F.P. Preparata, and R. Tamassia. Checking the convexity of polytopes and the planarity of subdivisions. Technical report, Center for Geometric Computing, Department of Computer Science, Brown Universi ty, 1997.

[DP98] O. Devillers and F. Preparata. A probabilistic analysis of the power of arithmetic filters. *Discrete & Computational Geometry*, 20:523–547, 1998.

[Dwy87] R.A. Dwyer. A faster divide-and-conquer algorithm for constructing Delaunay triangulations. *Algorithmica*, 2:137–151, 1987.

[Ede87] H. Edelsbrunner. *Algorithms in Combinatorial Geometry*. Springer, 1987.

[Edm65a] J. Edmonds. Maximum matching and a polyhedron with 0,1 - vertices. *Journal of Research of the National Bureau of Standards*, 69B:125–130, 1965.

[Edm65b] J. Edmonds. Paths, trees, and flowers. *Canadian Journal on Mathematics*, pages 449–467, 1965.

[EK72] J. Edmonds and R.M. Karp. Theoretical improvements in algorithmic efficiency for network flow problems. *Journal of the ACM*, 19:248–264, 1972.

[EM98] P. Eades and P. Mutzel. Graph drawing algorithms. In M.J. Athallah, editor, *Algorithms and Theory of Computation Handbook*. CRC Press, 1998.

[ES90] M.A. Ellis and B. Stroustrup. *The Annotated C++ Reference Manual*. Addison-Wesley, 1990.

[Eul53] L. Euler. Demonstratio nonulaarum insignium proprietatum, quibus solida hedris planis inclusa sunt praedita. *Novi Comm. Acad. Sci. Petropol.*, 4:140–160, 1752/53.

[Eve79] S. Even. *Graph Algorithms*. Pitman, 1979.

[Fár48] I. Fáry. On straight line representations of planar graphs. *Acta. Sci. Math. (Szeged)*, 11:229–233, 1948.

[Fel68] W. Feller. *An Introduction to Probability Theory and its Applications*. John Wiley & Sons, 1968.

[FF63] L.R. Ford and D.R. Fulkerson. *Flows in Networks*. Princeton University Press, Princeton, NJ, 1963.

[FGK+96] A. Fabri, G.-J. Giezeman, L. Kettner, S. Schirra, and S. Schönherr. The CGAL Kernel: A basis for geometric computation. In *Workshop on Applied Computational Geometry (WACG'96)*, volume 1148 of *Lecture Notes in Computer Science*, pages 191–202, 1996.

[Fin97] U. Finkler. *Design of Efficient and Correct Algorithms: Theoretical Results and Runtime Prediction of Programs in Practice*. PhD thesis, Fachbereich Informatik, Universität des Saarlandes, Saarbrücken, 1997.

[FKS84] M.L. Fredman, J. Komlos, and E. Szemeredi. Storing a sparse table with $o(1)$ worst case access time. *Journal of the ACM*, 31:538–544, 1984.

[FM91] S. Fortune and V.J. Milenkovic. Numerical stability of algorithms for line arrangements. In *Proceedings of the 7th Annual ACM Symposium on Computational Geometry (SCG'91)*, pages 334–341. ACM Press, 1991.

[FM97] U. Finkler and K. Mehlhorn. Runtime prediction of real programs on real machines. In *Proceedings of the 8th Annual ACM-SIAM Symposium on Discrete Algorithms (SODA'97)*, pages 380–389, 1997.

[For87] S.J. Fortune. A sweepline algorithm for Voronoi diagrams. *Algorithmica*, 2:153–174, 1987.

[For96] S. Fortune. Robustness issues in geometric algorithms. In *Proceedings of the 1st Workshop on Applied Computational Geometry: Towards Geometric Engineering (WACG'96)*, volume 1148 of *Lecture Notes in Computer Science*, pages 9–13, 1996.

[FPTV88] B.P. Flannery, W.H. Press, S.A. Teukolsky, and W.T. Vetterling. *Numerical Recipes in C: The Art of Scientific Computing*. Cambridge University Press, 1988.

[FT87] M.L. Fredman and R.E. Tarjan. Fibonacci heaps and their uses in improved network optimization algorithms. *Journal of the ACM*, 34:596–615, 1987.

[Fun97] S. Funke. Exact arithmetic using cascaded computation. Master's thesis, Fachbereich Informatik, Universität des Saarlandes, Saarbrücken, 1997.

[FvW96] S. Fortune and C. van Wyk. Static analysis yields efficient exact integer arithmetic for computational geometry. *ACM Transactions on Graphics*, 15:223–248, 1996.

[Gab76] H.N. Gabow. An efficient implementation of Edmond's algorithm for maximum matching on graphs. *Journal of the ACM*, 23:221–234, 1976.

[Gal86] Z. Galil. Efficient algorithms for finding maximum matching in graphs. *ACM Computing Surveys*, 18(1):23–37, 1986.

[GMG86] Z. Galil, S. Micali, and H.N. Gabow. An $O(EV \log V)$ algorithm for finding a maximal weighted matching in general graphs. *SIAM Journal of Computing*, 15:120–130, 1986.

[Gol85] A.V. Goldberg. A new max-flow algorithm. Technical Report MIT/LCS/TM-291, Lab. for Computer Science, MIT, Cambridge, Mass., 1985.

[Gol90] D. Goldberg. What every computer scientist should know about floating-point arithmetic. *ACM Computing Surveys*, 23(1):5–48, 1990.

[Gol91] D. Goldberg. Corrigendum: "What every computer scientist should know about floating-point arithmetic". *ACM Computing Surveys*, 23(3):413–413, 1991.

[GR97] A.V. Goldberg and S. Rao. Beyond the flow decomposition barrier. In *Proceedings of the 29th Annual ACM Symposium on the Theory of Computing (STOC'97)*, 1997.

[Gra72] R.L. Graham. An efficient algorithm for determining the convex hulls of a finite point set. *Information Processing Letters*, 1:132–133, 1972.

[GS78] L.J. Guibas and R. Sedgewick. A dichromatic framework for balanced trees. In *Proceedings of the 19th Annual Symposium on Foundations of Computer Science (FOCS'78)*, pages 8–21, 1978.

[GS85] L.J. Guibas and J. Stolfi. Primitives for the manipulation of general subdivisions and computation of Voronoi diagrams. *ACM Transactions on Graphics*, 4(2):74–123, 1985.

[GSS89] L.J. Guibas, D. Salesin, and J. Stolfi. Epsilon geometry: building robust algorithms from imprecise computations. In *Proceedings of the 5th Annual ACM-SIAM Symposium on Computational Geometry (SODA'89)*, pages 208–217. ACM Press, 1989.

[GSS93] L.J. Guibas, D. Salesin, and J. Stolfi. Constructing strongly convex approximate hulls with inaccurate primitives. *Algorithmica*, 9:534–560, 1993.

[GT85] H.N. Gabow and R.E. Tarjan. A linear-time algorithm for a special case of disjoint set union. *Journal of Computer and System Sciences*, 30(2):209–221, 1985.

[GT88] A.V. Goldberg and R.E. Tarjan. A new approach to the maximum-flow problem. *Journal of the ACM*, 35:921–940, 1988.

[Him97] M. Himsolt. The graphlet system. In *Proceedings of the Symposium on Graph Drawing (GD'96)*, volume 1190 of *Lecture Notes in Computer Science*, pages 233–240, 1997.

[HK73] J.E. Hopcroft and R.M. Karp. An $n^{5/2}$ algorithm for maximum matchings in bipartite graphs. *SIAM Journal of Computing*, 2(4):225–231, 1973.

[HM82] S. Huddlestone and K. Mehlhorn. A new data structure for representing sorted lists. *Acta Informatica*, 17:157–184, 1982.

[HMN96] C. Hundack, K. Mehlhorn, and S. Näher. A simple linear time algorithm for identifying Kuratowski subgraphs of non-planar graphs. Unpublished, 1996.

[HMRT85] K. Hoffmann, K. Mehlhorn,

P. Rosenstiehl, and R.E. Tarjan. Sorting Jordan sequences in linear time. *Proceedings of the 1st Annual ACM Symposium on Computational Geometry (SCG'85)*, pages 196–203, 1985.

[HO92] J. Hao and J.B. Orlin. A faster algorithm for finding the minimum cut in a graph. In *Proceedings on the 3rd Annual Symposium on Discrete Algorithms (SODA'92)*, volume 3, pages 165–174. ACM/SIAM, 1992.

[Hof89] C.M. Hoffmann. *Geometric and Solid Modeling : An Introduction. 2nd pr. 1993.* Series in computer graphics and geometric modeling. Morgan Kaufmann, 1989.

[HP90] J.L. Hennessy and D.A. Patterson. *Computer Architecture: A Quantitative Approach.* Morgan Kaufmann, 1990.

[HST98] T. Hagerup, P. Sanders, and J. Träff. An implementation of the binary blocking flow algorithm. In *Proceedings of the 2nd Workshop on Algorithm Engineering (WAE'98)*, pages 143–154. Max-Planck-Institut für Informatik, 1998.

[HT73] J.E. Hopcroft and R.E. Tarjan. Dividing a graph into triconnected components. *SIAM Journal of Computing*, 2(3):135–158, 1973.

[HT74] J.E. Hopcroft and R.E. Tarjan. Efficient planarity testing. *Journal of the ACM*, 21:549–568, 1974.

[Hum96] M. Humble. Implementierung von Flußalgorithmen mit dynamischen Bäumen. Master's thesis, Fachbereich Informatik, Universität des Saarlandes, Saarbrücken, 1996.

[IEE87] IEEE standard 754-1985 for binary floating-point arithmetic, 1987.

[Iri60] M. Iri. A new method for solving transportation-network problems. *Journal of the Operations Research Society of Japan*, 3:27–87, 1960.

[Jew58] W.S. Jewell. Optimal flow through networks. Technical report, Operations Research Center, MIT, 1958.

[JMN] M. Jünger, P. Mutzel, and S. Näher. The AGD graph drawing library. Search the WEB for AGD or one of the authors.

[JRT97] M. Jünger, G. Rinaldi, and S. Thienel. Practical performance of efficient minimum cut algorithms. Technical report, Institut für Informatik, Universität zu Köln, 1997.

[Kar74] A.V. Karzanov. Determining the maximum flow in a network by the method or preflows. *Soviet Mathematics Doklady*, 15:434–437, 1974.

[Kar90] A. Karabeg. Classification and detection of obstructions to planarity. *Linear and Multilinear Algebra*, 26:15–38, 1990.

[Kin90] J.H. Kingston. *Algorithms and Data Structures.* Addison-Wesley, 1990.

[KKT95] D.R. Karger, P.N. Klein, and R.E. Tarjan. A randomized linear-time algorithm for finding minimum spanning trees. *Journal of the ACM*, 42:321–329, 1995.

[KL93] D. Knuth and S. Levy. *The CWEB System of Structured Documentation, Version 3.0.* Addison-Wesley, 1993.

[Kle97] R. Klein. *Algorithmische Geometrie.* Addison-Wesley, 1997.

[KLN91] M. Karasick, D. Lieber, and L.R. Nackman. Efficient Delaunay triangulation using rational arithmetic. *ACM Transactions on Graphics*, 10(1):71–91, January 1991.

[Knu81] D.E. Knuth. *The Art of Computer Programming (Volume II): Seminumerical Algorithms.* Addison-Wesley, 1981.

[KO63] A. Karatsuba and Yu. Ofman. Multiplication of multidigit numbers on automata. *Soviet Physics—Doklady*, 7(7):595–596, January 1963.

[KP98] J.D. Kececioglu and J. Pecqueur. Computing maximum-cardinality matchings in sparse general graphs. In *Proceedings of the 2nd Workshop on Algorithm Engineering (WAE'98)*, pages 121–132. Max-Planck-Institut für Informatik, 1998.

[Kru56] J.B. Kruskal. On the shortest spanning subtree of a graph and the travelling salesman problem. In *Proceedings of the American Mathematical Society*, pages 48–50, 1956.

[KS96] D.R. Karger and C. Stein. A new approach to the minimum cut problem. *Journal of the ACM*, 43(4):601–640, 1996.

[KS97] D. Kirkpatrick and J. Snoeyink, 1997. personal communication.

[KSK76] G. Kemeny, L. Snell, and A.W. Knapp. *Denumerable Markov Chains.* Springer, 1976.

[Kuh55] H.W. Kuhn. The Hungarian method for the assignment problem. *Naval Research Logistics Quarterly*, 2:83–97, 1955.

[Kur30] C. Kuratowski. Sur le problème the courbes guaches en topologie. *Fundamenta Mathematicae*, 15:271–283, 1930.

[Lam86] L. Lamport. *LaTeX*. Addison-Wesley, 1986.

[Lau98] H. Lauer. *GraVis – Interactive Graph Visualization.* Herbert Utz Verlag – Wissenschaft, Tübingen, 1998.

[Law66] E.L. Lawler. Optimal cycles in doubly weighted directed linear graphs. In *Theory of*

Graphs: International Symposium, pages 209–213. Dunod, Paris, 1966.

[Law72] C.L. Lawson. Transforming triangulations. *Discrete Mathematics*, pages 365–372, 1972.

[Law76] E.L. Lawler. *Combinatorial Optimization: Networks and Matroids*. Holt, Rinehart, and Winston, 1976.

[LEC67] A. Lempel, S. Even, and I. Cederbaum. An algorithm for planarity testing of graphs. In P. Rosenstiehl, editor, *Theory of Graphs, International Symposium, Rome*, pages 215–232, 1967.

[LEP] LEDA Extension Packages. `www.mpi-sb.mpg.de/LEDA/friends/leps.html`.

[LL97] A. LaMarca and R.E. Ladner. The influence of caches on the performance of sorting. In *Proceedings of the 8th Annual ACM-SIAM Symposium on Discrete Algorithms (SODA'97)*, pages 370–379, 1997.

[LL98] S.B. Lippmann and J. Lajoie. *C++ Primer*. Addison-Wesley, 1998.

[LM90] Z. Li and V.J. Milenkovic. Constructing strongly convex hulls using exact or rounded arithmetic. In *Proceedings of the 6th Annual ACM Symposium on Computational Geometry (SCG'90)*, pages 235–243. ACM Press, 1990.

[LT77] R. Lipton and R.E. Tarjan. A separator theorem for planar graphs. In *Conference on Theoretical Computer Science, Waterloo*, pages 1–10, 1977.

[MCN91] J.-F. Mondou, T.G. Crainic, and S. Nguyen. Shortest path algorithms: A computational study with the C programming language. *Computers and Operations Research*, 18:767–786, 1991.

[Meg83] N. Megiddo. Applying parallel computation algorithms in the design of serial algorithms. *Journal of the ACM*, 30:852–865, 1983.

[Meh84a] K. Mehlhorn. *Data Structures and Algorithms 1: Sorting and Searching*. Springer, 1984.

[Meh84b] K. Mehlhorn. *Data Structures and Algorithms 1,2, and 3*. Springer, 1984.

[Meh84c] K. Mehlhorn. *Data Structures and Algorithms 2: Graph Algorithms and NP-Completeness*. Springer, 1984.

[Meh84d] K. Mehlhorn. *Data Structures and Algorithms 3: Multidimensional Searching and Computational Geometry*. Springer, 1984.

[Mil88] V.J. Milenkovic. *Verifiable Implementations of Geometric Algorithms Using Finite Precision Arithmetic*. PhD thesis, Carnegie Mellon University, 1988.

[Mil89a] V.J. Milenkovic. Calculating approximate curve arrangements using rounded arithmetic. In Kurt Mehlhorn, editor, *Proceedings of the 5th Annual ACM Symposium on Computational Geometry (SCG'89)*, pages 197–207. ACM Press, 1989.

[Mil89b] V.J. Milenkovic. Double precision geometry: A general technique for calculating line and segment intersections using rounded arithmetic. In *Proceedings of the 30th Annual Symposium on Foundations of Computer Science (FOCS'89)*, pages 500–505. IEEE, 1989.

[MM95] K. Mehlhorn and P. Mutzel. On the embedding phase of the Hopcroft and Tarjan planarity testing algorithm. *Algorithmica*, 16(2):233–242, 1995.

[MMN94] K. Mehlhorn, P. Mutzel, and S. Näher. An implementation of the Hopcroft and Tarjan planarity test and embedding algorithm, 1994. available at the first author's WEB-page.

[MMN+98] K. Mehlhorn, M. Müller, S. Näher, S. S. Schirra, M. Seel, C. Uhrig, and J. Ziegler. A computational basis for higher-dimensional computational geometry and its applications. *Computational Geometry: Theory and Applications*, 10:289–303, 1998. `www.mpi-sb.mpg.de/~seel`.

[MN89] K. Mehlhorn and S. Näher. LEDA: A library of efficient data types and algorithms. In Antoni Kreczmar and Grazyna Mirkowska, editors, *Proceedings of the 14th International Symposium on Mathematical Foundations of Computer Science (MFCS'89)*, volume 379 of *Lecture Notes in Computer Science*, pages 88–106. Springer, 1989.

[MN92] K. Mehlhorn and S. Näher. Algorithm design and software libraries: Recent developments in the LEDA project. In *Algorithms, Software, Architectures, Information Processing 92*, volume 1, pages 493–505. Elsevier Science Publishers B.V. North-Holland, 1992.

[MN94a] K. Mehlhorn and S. Näher. Implementation of a sweep line algorithm for the straight line segment intersection problem. Technical Report MPI-I-94-160, Max-Planck-Institut für Informatik, 1994.

[MN94b] K. Mehlhorn and S. Näher. The implementation of geometric algorithms. In *Proceedings of the 13th IFIP World Computer Congress*, volume 1, pages 223–231. Elsevier

Science B.V. North-Holland, Amsterdam, 1994.

[MN95] K. Mehlhorn and S. Näher. LEDA, a Platform for Combinatorial and Geometric Computing. *Communications of the ACM*, 38:96–102, 1995.

[MN98a] K. Mehlhorn and S. Näher. Dynamic Delaunay triangulations, 1998. www.mpi-sb.mpg.de/~mehlhorn.

[MN98b] K. Mehlhorn and S. Näher. From algorithms to working programs: On the use of program checking in LEDA. In *Proceedings of Mathematical Foundations of Computer Science (MFCS'98)*, volume 1450 of *Lecture Notes in Computer Science*, pages 84–93, 1998.

[MNS$^+$96] K. Mehlhorn, S. Näher, T. Schilz, S. Schirra, M. Seel, R. Seidel, and C. Uhrig. Checking Geometric Programs or Verification of Geometric Structures. In *Proceedings of the 12th Annual Symposium on Computational Geometry (SCG'96)*, pages 159–165, 1996.

[MNU97] K. Mehlhorn, S. Näher, and C. Uhrig. The LEDA platform for combinatorial and geometric computing. In *Proceedings of the 24th International Colloquium on Automata, Languages and Programming (ICALP'97)*, volume 1256 of *Lecture Notes in Computer Science*, pages 7–16, 1997.

[Moo61] E.F. Moore. U.S. Patent 2983904, 1961.

[Mot94] R. Motwani. Average-case analysis of algorithms for matching and related problems. *Journal of the ACM*, 41(6):1329–1356, 1994.

[MR95] R. Motwani and P. Raghavan. *Randomized Algorithms*. Cambridge University Press, 1995.

[MS96] D.R. Musser and A. Saini. *STL Tutorial and Reference Guide*. Addison-Wesley, Reading, 1996.

[Mul90] K. Mulmuley. A fast planar partition algorithm, I. *Journal of Symbolic Computation*, 10(3-4):253–280, 1990.

[Mul94] K. Mulmuley. *Computational Geometry*. Prentice Hall, 1994.

[Mur93] R.B. Murray. *C++ Strategies and Tactics*. Addison-Wesley, 1993.

[Mye85] E. Myers. An $O(E \log E + I)$ expected time algorithm for the planar segment intersection problem. *SIAM Journal of Computing*, 14(3):625–636, 1985.

[Näh93] S. Näher. LEDA: a library of efficient data types and algorithms. In Patrice Enjalbert, Alain Finkel, and Klaus W. Wagner, editors, *Proceedings of the 10th Annual Symposium on Theoretical Aspects of Computer Science (STACS'93)*, volume 665 of *Lecture Notes in Computer Science*, pages 710–723. Springer, 1993.

[NC88] T. Nishizeki and N. Chiba. *Planar Graphs: Theory and Algorithms*. Annals of Discrete Mathematics (32). North-Holland Mathematics Studies, 1988.

[Nef78] W. Nef. *Beitraege Zur Theorie Der Polyeder Mit Anwendungen in der Computergraphik (Contributions to the Theory of Polyhedra, with Applications in Computer Graphics)*. Verlag H. Lang & Cie. AG, Bern, 1978.

[NH93] J. Nievergelt and K.H. Hinrichs. *Algorithms and Data Structures*. Prentice Hall, 1993.

[NI92] H. Nagamochi and T. Ibaraki. Computing edge-connectivity in multigraphs and capacitated graphs. *SIAM Journal on Discrete Mathematics*, 5(1):54–66, 1992.

[NM90] S. Näher and K. Mehlhorn. LEDA: A library of efficient data types and algorithms. In Michael S. Paterson, editor, *Proceedings of the 17th International Colloquium on Automata, Languages, and Programming (ICALP'90)*, volume 443 of *Lecture Notes in Computer Science*, pages 1–5. Springer, 1990.

[Nos85] K. Noshita. A theorem on the expected complexity of Dijkstra's shortest path algorithm. *Journal of Algorithms*, 6(3):400–408, 1985.

[NR73] J. Nievergelt and E. Reingold. Binary search trees of bounded balance. *SIAM Journal of Computing*, 2:33–43, 1973.

[Nye93] Adrian Nye. *Xlib Programming Manual for Version 11*. O'Reilly & Associates, Inc., Sebastopol CA, 3rd edition, 1993.

[O'R94] J. O'Rourke. *Computational Geometry in C*. Cambridge University Press, 1994.

[OW96] T. Ottmann and P. Widmayer. *Algorithmen und Datenstrukturen*. Spektrum Akademischer Verlag, 1996.

[Pau89] M. Paul. Algorithmen für das Maximum Weight Matching Problem in bipartiten Graphen. Master's thesis, Fachbereich Informatik, Universität des Saarlandes, Saarbrücken, 1989.

[Poi93] H. Poincaré. Sur la généralisation d'un theorem d'Euler relativ aux polyédres. *Comptes Rend. Acad. Sci. Paris*, 117:144–145, 1893.

[PR90] M. Padberg and G. Rinaldi. An efficient algorithm for the minimum capacity cut problem. *Mathematical Programming*, 47:19–36, 1990.

[PS85] F.P. Preparata and M.I. Shamos. *Computational Geometry: An Introduction*.

Springer, 1985.

[Pug90a] W. Pugh. A skip list cookbook. Technical Report CS-TR-2286.1, Institute for Advanced Computer Studies, Department of Computer Science, University of Maryland, College Park, MD, June 1990.

[Pug90b] W. Pugh. Skip lists: A probabilistic alternative to balanced trees. *Communications of the ACM*, 33(6):668–676, 1990.

[Rab80] M.O. Rabin. A probabilistic algorithm for testing primality. *Journal of Number Theory*, 12, 1980.

[Ram94] N. Ramsey. Literate programming simplified. *IEEE Software*, 11:97–105, 1994.

[Req80] A.A.G. Requicha. Representations for rigid solids: Theory, methods and systems. *Computing Surveys*, 12(4):437–464, 1980.

[Sch] S. Schirra. Robustness and precision issues in geometric computation. to appear, preliminary version available as MPI report.

[Sch98] S. Schirra. Parameterized implementations of classical planar convex hull algorithms and extreme point computations. Technical Report MPI-I-98-1-003, Max-Planck-Institut für Informatik, 1998.

[SD97] P. Su and R.L.S. Drysdale. A comparison of seqential Delaunay triangulation algorithms. *Computational Geometry: Theory and Applications*, 7:361–385, 1997.

[Sed91] R. Sedgewick. *Algorithms*. Addison-Wesley, 1991.

[See97] M. Seel. The computation of abstract Voronoi diagrams. `www.mpi-sb.mpg.de/~seel`, 1997.

[Sei91] R. Seidel. Small-dimensional linear programming and convex hulls made easy. *Discrete and Computational Geometry*, 6:423–434, 1991.

[SH75] M.I. Shamos and D. Hoey. Closest-point problems. In *Proceedings of the 16th Annual Symposium on Foundations of Computer Science (FOCS'75)*, pages 151–165, 1975.

[Sha81] M. Sharir. A strong-connectivity algorithm and its applications in data flow analysis. *Computational Mathematics with Applications*, 7(1):67–72, 1981.

[She97] J.R. Shewchuk. Adaptive precision floating-point arithmetic and fast robust geometric predicates. *Discrete & Computational Geometry*, 18:305–363, 1997.

[Ski] S.S. Skiena. The Stony Brook algorithm repository. `www.cs.sunysb.edu/~algorith/index.html`.

[Ski98] S.S. Skiena. *The Algorithm Design Manual*. Springer, 1998.

[SM90] G.F. Sullivan and G.M. Masson. Using certification trails to achieve software fault tolerance. In Brian Randell, editor, *Proceedings of the 20th Annual International Symposium on Fault-Tolerant Computing (FTCS '90)*, pages 423–433. IEEE, 1990.

[SM91] G.F. Sullivan and G.M. Masson. Certification trails for data structures. In *Proceedings of the 21st Annual International Symposium on Fault-Tolerant Computing(FTCS'91)*, pages 240–247, 1991.

[SS77] R. Solovay and V. Strassen. A fast Monte-Carlo test for primality. *SIAM Journal of Computing*, 6(1):84–85, 1977.

[SSS97] S. Schirra, J. Schwerdt, and M. Smid. Computing the minimum diameter for moving points: An exact implementation using parametric search. In *Proceedings of the 13th Annual ACM Symposium on Computational Geometry (SCG'97)*, pages 466–468, 1997.

[Str91] B. Stroustrup. *The C++ Programming Language*. Addison-Wesley, 1991.

[SV87] J.T. Stasko and J.S. Vitter. Pairing heaps: Experiments and analysis. *Communications of the ACM*, 30:234–249, 1987.

[SW97] M. Stoer and F. Wagner. A simple min-cut algorithm. *Journal of the ACM*, 44(4):585–591, July 1997.

[SWM95] G.F. Sullivan, D.S. Wilson, and G.M. Masson. Certification of computational results. *IEEE Transactions on Computers*, 44(7):833–847, 1995.

[Tar72] R.E. Tarjan. Depth-first search and linear graph algorithms. *SIAM Journal of Computing*, 1:146–160, 1972.

[Tar75] R.E. Tarjan. Efficiency of a good but not linear set union algorithm. *Journal of the ACM*, 22:215–225, 1975.

[Tar81] R.E. Tarjan. Shortest paths. Technical report, AT&T Bell Laboratories, Murray Hill, New Jersey, 1981.

[Tar83a] R.E. Tarjan. Data structures and network algorithms. In *CBMS-NSF Regional Conference Series in Applied Mathematics*, volume 44, 1983.

[Tar83b] R.E. Tarjan. *Data Structures and Network Algorithms*. SIAM, 1983.

[Tom71] N. Tomizawa. On some techniques useful for solution of transportation network. *Networks*, 1:173–194, 1971.

[TR80] R.B. Tilove and A.A.G. Requicha. Closure

of boolean operations on geometric entities. *Computer-Aided Design*, 12:219–222, 1980.

[van88] C.J. van Wyk. *Data Structures and C Programs*. Addison-Wesley, 1988.

[Vui78] J. Vuillemin. A data structure for manipulating priority queues. *Communications of the ACM*, 21:309–314, 1978.

[Wal77] A.J. Walker. An efficient method for generating discrete random variables with general distributions. *ACM Transaction on Mathematical Software*, 3:253–256, 1977.

[WB97] H. Wasserman and M. Blum. Software reliability via run-time result-checking. *Journal of the ACM*, 44(6):826–849, 1997.

[Whi73] A.T. White. *Graphs, Groups, and Surfaces*. North Holland, 1973.

[Woo93] D. Wood. *Data Structures, Algorithms, and Performance*. Addison-Wesley, 1993.

[WS90] L. Wall and R.L. Schwartz. *Programming perl*. O'Reilly & Associates, 1990.

[Yap93] C.K. Yap. Towards exact geometric computation. In *Proceedings of the 5th Canadian Conference on Computational Geometry (CCCG'93)*, pages 405–419, 1993.

[YD95] C.K. Yap and T. Dube. The exact computation paradigm. In *Computing in Euclidean Geometry II*. World Scientific Press, 1995.

[Zie95] T. Ziegler. Max-Weighted-Matching auf allgemeinen Graphen. Master's thesis, Fachbereich Informatik, Universität des Saarlandes, Saarbrücken, 1995.

Index